FIFTH EDITION

ELECTRONIC COMMUNICATION

Robert L. Shrader

Former Chairman of Electronics,
Laney College
U.S. Merchant Marine Academy,
Kings Point, Ret.

McGRAW-HILL BOOK COMPANY

New York   Atlanta   Dallas   St. Louis
San Francisco   Auckland   Bogotá   Guatemala
Hamburg   Johannesburg   Lisbon   London
Madrid   Mexico   Montreal   New Delhi
Panama   Paris   San Juan   São Paulo
Singapore   Sydney   Tokyo   Toronto

Sponsoring Editor: George Z. Kuredjian
Editing Supervisor: Curt Berkowitz
Design and Art Supervisor: Patricia F. Lowy
Production Supervisor: Laurence Charnow

Text Designer: Levavi & Levavi
Cover Photographer: Ken Karp

Resonance 119
oscillators 201
RF AMPS 307
AM 365
Detectors 405
FM 435
ANTenna 451

Library of Congress Cataloging in Publication Data
Shrader, Robert L.
  Electronic communication.

  Includes index.
  1. Electronics.   2. Radio.   I. Title.
TK7815.S5 1985   621.38   84-964
ISBN 0-07-057151-1

ELECTRONIC COMMUNICATION, *Fifth Edition*

 234567890 DOCDOC 891098765

ISBN 0-07-057151-1

Contents

Preface

The fifth edition of *Electronic Communication* represents in one volume a complete textbook for courses in basic electricity, basic electronics, communication circuits and systems, and applications. A survey of instructors and industry helped to determine the direction of the fifth edition. It may be used for introductory electricity courses, introductory electronics courses, electronics communication courses, and other specialized communications courses. Some specialized courses may include two-way radiotelephone; AM, FM, and television broadcasting; maritime radiotelegraphy; microwaves; satellite communications; teleprinters; facsimile; navigational aids; and amateur radio.

Essentially, all modern electronic circuits use solid-state devices. However, high-power radio or audio-frequency amplifiers still use vacuum tubes. For this reason it is necessary to relate a minimal amount of vacuum-tube technology to the many similar discrete semiconductor circuits explained throughout the text.

In keeping with modern notations, the voltage symbols in formulas and elsewhere have been changed to V or v. In the past, E and e signified voltage and electromotive force in general.

The technical information contained in this book adequately prepares anyone to take any of the Federal Communications Commission (FCC) license examinations. In addition, engineer technicians working in any of the AM-FM-TV Broadcast Radio services or the Private Land Mobile Radio services may be required by employers to obtain industry backed certificates issued by FCC-endorsed non-governmental entities. They will particularly find the basic theory in this book an excellent foundation for the required fields of certification. The names and addresses of such certificating bodies should be available through the FCC Field Offices.

As an aid to the student, color has been used in this edition to highlight specific topics under discussion and to focus the student's attention to specific features in the illustrations. As a further aid, and to allow students to evaluate their progress, checkup quizzes on the subjects just discussed are inserted on every few pages. The quiz answers may be found on the next turn page. End-of-chapter questions test and reinforce the student's grasp of the chapter material, with answers to the even-numbered questions listed in the answers section at the back of the book. The answers are always suffixed with parenthetical reference numbers which specify the chapter and section where the discussion of that particular question may be found in the text.

Metric measurements have been used throughout this book except where prevailing usage makes the U.S. customary units of measurement (inch, foot, mile, etc.) still preferable.

The author gratefully acknowledges the aid he received from many sources in preparing this and previous editions. Some of those who were of material help are Lewis C. Ward, George F. Ong, A. J. Tassin, Wallace Taggart, and G. G. Thommen of ITT World Communications (Mackay Radio); C. C. Pitts, G. P. Aldridge, and Ralph Scott of RCA; George P. Stafford and Robert Dean of Alden Electronic and Impulse Recording Equipment; John Davis of Sonoma County Communications; Al Stoney of Atcomm Inc. of Santa Rosa, CA; Steve Pinch of KSRO; Joseph Perez of KFTY-TV; Loren Peterson, K6EDV, of Peterson Marine Electronics; Donald Johnson, W6AAQ; and the late Emery L. Simpson, friend and educator. Other instructors who aided in many ways are friend and computer author Jefferson C. Boyce, W7INR; Edwin Pollock, W6LC; Oliver J. Ruel, Thomas J. Bingham, Jr., Thomas G. Vavrina, Michael A. Bochnack, Louis G. Gross, and those instructors at the many schools the author visited during the preparatory stages of the revision who helped materially in determining the structure of this fifth edition.

As always, the reviewing help and ever patient understanding of a greatly appreciated and loving wife, Dorothy, W6ECU, has made this and all other editions possible.

Robert L. Shrader, W6BNB

NOTE

Radio license-type questions have been included in this edition as a supplement following the index. There are two groups of questions. The first group should aid in passing tests for the Commercial General Radiotelephone license, the three Commercial Radiotelegraph licenses, all industry-developed certification tests, and will aid in studying for all of the Amateur Radio tests. The second group of questions is specifically for the subjects included in the Amateur Radio tests, Elements 2 through 4*b*.

1

Current, Voltage, and Resistance

The objective of this first chapter is to familiarize the reader with some of the basic physical theories that underlie any study of electricity and electronics, specifically voltage, current, and resistance. Resistors and color codes are discussed briefly, as are some of the methods of making satisfactory electric circuit connections. All of the checkup questions should be mastered before moving to the next section.

1-1 ELECTRICITY

The complicated electronic systems involved in modern-day missiles, communications satellites, nuclear power plants, radio, and television, and even up-to-date automobiles, require service technicians who understand the functioning of electric and electronic circuits. It is the principal goal of this book, however, to outline the basic theory related to electronic communication, particularly those systems that must be operated or serviced by technically trained personnel.

The term "electronic" infers circuits including either the first of the electronic devices—vacuum tubes—or the newer solid-state devices such as diodes and transistors, as well as integrated circuits (ICs). The term "electric" or "electrical" is usually applied to systems or circuits in which electrons flow through wires but which involve no vacuum tubes or solid-state devices. Actually, many modern electrical systems are now using electronic devices to control the electric current that flows in them. But what is electric current? What makes such a simple thing as an electric lamp glow? It is easy to pass the problem off with the statement, "The switch connects the light to the power lines." But what does connecting the light to the power lines do? How does energy travel through solid copper wires or through space? What makes a motor turn, a radio play? What is behind the dial that allows you to pick out one radio station from thousands of others operating at the same time?

There are no simple single answers to any of these questions. Each question requires the understanding of many basic principles. By adding one basic idea to another, it is possible to answer, eventually, most of the questions that may be asked about the intriguing subjects of electricity, electronics, and radio.

When the light switch is turned on at one point in a room and a light suddenly glows, energy has found a path through the switch to the light. The paths used are usually copper wires, and the tiny particles that do the moving and carry the energy are called *electrons*. These little electrons are important to anyone studying electronics and radio, since they are usually the only particles that are considered to move in electric circuits.

To explain what is meant by an electron, it will be necessary to investigate more closely the makeup of all matter. The word "matter" means, in a general sense, anything that can be touched. It includes substances such as rubber, salt, wood, water, glass, copper, and air.

Water is one of the most common forms of matter. If a drop of water is divided in two and then divided again and again until it can be divided no longer and still be water, this smallest particle is known as a *molecule*. The water molecule can be broken down into still smaller particles, but these new particles will not be water.

Physicists have found that there are three particles making up a molecule of water: two *atoms* of hydrogen (H) and one atom of oxygen (O), as shown in Fig. 1-1. Oxygen, at normal temperatures, is one of several gases that constitute the air we breathe. Hydrogen is also a gas in its natural state; it is found in everyday use as part of the gas used for heating or cooking. If a gaseous mixture containing 2 parts of hydrogen and 1 part of oxygen is ignited, a chemical reaction in the form of an explosion takes place. The residue of the explosion will be water (H_2O) droplets. Atoms can chemically combine into molecules.

It has been determined that an atom is also divisible, being made up of at least two types of particles: electrons and *protons*. Both are electrical particles, but neither one is divisible. All molecules of all matter of the universe are composed of such electrical proton-electron pairs.

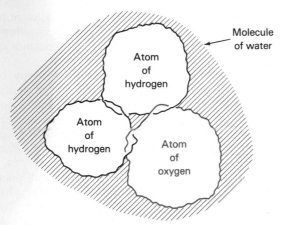

FIG. 1-1 Two atoms of hydrogen and one atom of oxygen, when chemically interlocked, form one molecule of water.

1-2 ELECTRONS AND PROTONS

Electrons are the smallest and lightest of particles. They are said to have a *negative* charge, meaning that they are surrounded by some kind of invisible *field of force* (Fig. 1-2) that will react in an electrically negative manner on anything that is elec-

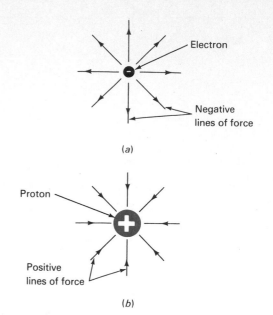

FIG. 1-2 Electrostatic lines of force: (*a*) outward from a negative charge; (*b*) inward toward a positive charge.

trically charged and brought within the limits of the field. The electric field may be represented pictorially as being composed of outward-pointing *lines of force*. Whether lines of force actually exist is not known, but they are used for explanatory purposes.

Protons are about 1800 times as massive as electrons and have a *positive* electric field surrounding them. The positive field is represented by inward-pointing lines. Theoretically, an electron has exactly as many outward-pointing lines as a proton has inward-pointing lines. The proton is exactly as positive as the electron is negative; each has a unit electric charge.

In theory, negative lines of force will not join other negative lines of force. In fact, they repel each other, tending to push each electron away from every other electron (Fig. 1-3*a*). Positive lines of force do the same thing (Fig. 1-3*b*).

When an electron and a proton are far apart, as illustrated in Fig. 1-4, only a few of their lines of force join and pull together. The contracting pull between the two charges is therefore small. When brought closer together, the electron and proton are able to link more of their lines of force and will pull together with greater force. If close enough, all the lines of force from the electron are joined to all the lines of force of the proton and there is no external field. They form a neutral, or uncharged,

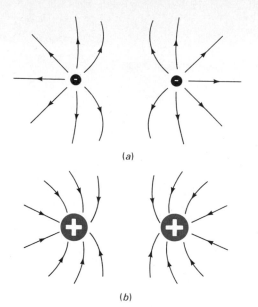

(a)

(b)

FIG. 1-3 Lines of force of like (similar) direction repel.

group. One or more such neutral atomic particles, known as *neutrons*, exist in the nucleus of all atoms heavier than hydrogen.

The fact that electrons repel other electrons, protons repel other protons, but electrons and protons attract each other follows the basic physical law: *Like charges repel; unlike charges attract*.

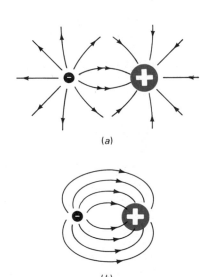

(a)

(b)

FIG. 1-4 Unlike lines of force attract. (a) Little interlinkage if remote. (b) Complete interlinkage when close, forming a neutral group.

Because the proton is about 1800 times heavier than the electron, it seems reasonable to assume that when an electron and a proton attract each other, it will be the tiny electron that will do most of the actual moving. Such is the case. It is the electron that usually moves in electricity. Small though it is physically, the field near the electron is quite strong.

If the field strength around an electron at a distance of one-millionth of a meter is a certain amount, at two-millionths of a meter it will be one-quarter as much; at four-millionths of a meter it will be one-sixteenth as much; and so on. If the field decreases as distance increases, the field is said to vary *inversely* with distance. Actually, it varies inversely with the distance squared.

Since the electric-field strength of an electron varies inversely with the distance squared, the field strength a millimeter (one-thousandth of a meter) or so away might be very weak.

When an increase in something produces an increase in something else, the two things are said to vary *directly* rather than inversely. Two million electrons on an object produce twice as much negative charge as one million electrons would. The charge is directly proportional to the number of electrons.

The fields surrounding electrons and protons are known as *electrostatic fields*. The word "static" means, in this case, "stationary," or "not caused by movement." The presence of a field indicates a storage of energy in it.

When electrons are made to move, the result is *dynamic electricity*. The word "dynamic" indicates that motion is involved.

To produce a movement of an electron, it will be necessary to have either a negatively charged field to push it, a positively charged field to pull it, or, as normally occurs in an electric circuit, both a negative and a positive charge (a pushing and pulling pair of forces).

1-3 THE ATOM AND ITS FREE ELECTRONS

There are more than 100 different kinds of atoms, or *elements*, from which the millions of different forms of matter found in the universe are composed. The heaviest elements are always radioactive and unstable, decomposing into lower-atomic-weight atoms spontaneously.

The simplest and lightest atom is hydrogen. An atom of hydrogen consists of one electron and one

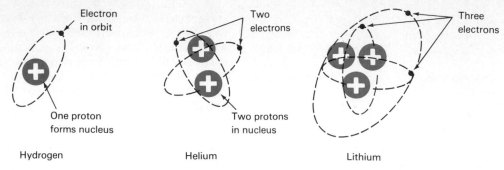

FIG. 1-5 Simplified atoms. Hydrogen and helium have one electron orbit layer. Lithium has two orbit layers.

proton (Fig. 1-5). In one respect this atom is similar to all others: the electron whirls around the proton, or *nucleus*, of the atom, much as planets rotate around the sun. Electrons whirling around the nucleus are termed *planetary*, or *orbital*, electrons.

The next atom in terms of weight is helium, having two protons and two electrons. The third atom is lithium, with three electrons and three protons, and so on (Fig. 1-5).

Some of the common elements, in order of their atomic weights, are:

1. Hydrogen (H)	28. Nickel (Ni)
2. Helium (He)	29. Copper (Cu)
3. Lithium (Li)	30. Zinc (Zn)
6. Carbon (C)	32. Germanium (Ge)
8. Oxygen (O)	47. Silver (Ag)
11. Sodium (Na)	79. Gold (Au)
13. Aluminum (Al)	82. Lead (Pb)
14. Silicon (Si)	88. Radium (Ra)
26. Iron (Fe)	92. Uranium (U)

Most atoms have a nucleus consisting of all the protons of the atom plus one or more neutrons. The remainder of the electrons (always equal in number to the number of nuclear protons) are whirling around the nucleus in various layers. The first layer of electrons outside the nucleus can accommodate only two electrons. If the atom has three electrons (Fig. 1-5), two will be in the first layer and the third will be in the next layer. The second layer is completely filled when eight electrons are whirling around in it. The third is filled when it has eighteen electrons.

Some of the valence electrons in the outer orbit, or *shell*, of the atoms of many materials such as copper or silver exist in a slightly higher *conduction level* and can be dislodged easily. These electrons travel out into the wide-open spaces between the atoms and molecules, and they may be termed *free*

electrons. Other electrons in the outer orbit may resist dislodgment and may be called *bound* electrons. Materials consisting of atoms (or molecules) having many free electrons will allow an easy interchange of their outer shell electrons. The materials are known as *conductors*. Atoms in materials with only bound electrons will hinder any electron exchange. These are known as *insulators*.

When a conductor is heated, greater energy is developed in the free-moving electrons. The more energy they have, the more the electrons resist orderly movement through the material. The material is said to have an increased *resistance* to the movement of electrons through it.

1-4 THE ELECTROSCOPE

An example of electrons and electric charges acting on one another is demonstrated by the action of an *electroscope* (Fig. 1-6). This device consists of two very thin gold or aluminum leaves attached to the

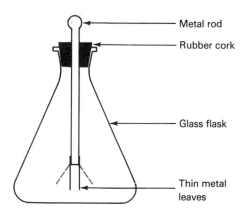

FIG. 1-6 An electroscope indicates the presence of either a negative or positive charge.

bottom of a metal rod. To prevent air currents from damaging the delicate leaves, rod and leaves are encased in a glass flask. The rod projects through a rubber cork.

To understand the operation of the electroscope, it is necessary to recall these facts:

1. Normally an object has a neutral charge.
2. Like charges repel; unlike charges attract.
3. Electrons are negative.
4. Metals with free electrons are conductors.

Normally the electroscope rod has a neutral charge, and the leaves hang downward parallel to each other (Fig. 1-6). Rubbing a piece of hard rubber with wool causes the wool to lose electrons to the rubber, charging the rubber negatively. When such a negatively charged object is brought near the top of the rod, some of the free electrons at the top of the rod are repelled and travel down, away from the negatively charged object. Some of these electrons force themselves onto the leaves. Now the two leaves are no longer neutral but are slightly negative and repel each other, moving outward, shown by the dotted lines. When the charged object is removed, electrons return up the rod to their original areas. The leaves again have a neutral charge and hang parallel.

Since the charged object did not touch the electroscope, it neither placed electrons on the rod nor took electrons from it. When electrons were repelled to the bottom, making the leaves negative, these same electrons leaving the top of the rod left the top positive. The overall charge of the rod remained neutral. When the charged object was withdrawn, the positive charge "induced" into the top of the rod pulled the displaced electrons up to it. All parts of the rod were neutral again.

If a positively charged object, such as a glass rod vigorously rubbed with a piece of silk, is brought near the top of the electroscope rod, some of the free electrons in the leaves and rod will be attracted upward toward the positive object. This charges the top of the rod negatively because of the excess of free electrons there. Both leaves are left with a deficiency of free electrons and, being positively charged, repel each other and move outward.

If a negatively charged object is *touched* to the metal rod, a number of excess electrons will be deposited on the rod and will be immediately distributed throughout the electroscope. The leaves spread apart. When the object is taken away, an excess of electrons remains on the rod and leaves, and the leaves stay spread apart. If the negatively charged rod is touched to any large body that can accept free electrons, such as a person, a large metal object, or the earth, the excess electrons will have a path by which they can leave the electroscope and the leaves will collapse as the charge returns to neutral. The electroscope has been discharged ("grounded").

If a positively charged object is touched to the metal rod, the rod will lose electrons to it and the leaves will separate. When the object is taken away, the rod and leaves still lack free electrons, they are positively charged, and the leaves remain apart. A large neutral body touched to the rod will lose some of its free electrons to the electroscope, discharging it, and the leaves will hang down once more.

The electroscope demonstrates the free movement of electrons (known as *electron current*) that can take place through metallic conductors when electric pressures are exerted on free electrons.

1-5 THE BIG THREE IN ELECTRICITY

Without calling them by name, the discussion thus far has touched on the three elements always present in all operating electric circuits:

- *Current.* A progressive movement of free electrons along a wire or other conductor produced by electrostatic lines of force.
- *Electromotive force.* The electron-moving force in a circuit that pushes and pulls electrons (current) through the circuit.
- *Resistance.* Any opposing effect that hinders free-electron progress through a conductor in a circuit when an electromotive force is attempting to produce a current in the circuit.

Changes in the values of any one of these "big three" will produce a change in the value of at least one of the others. Note the interrelationship of these in the simple electric circuit below.

1-6 SIMPLE ELECTRIC CIRCUIT

A simple electric circuit consists of (1) some sort of electron-moving force, or *source*, such as a dry cell, or battery; (2) a *load*, such as an electric light; (3) connecting wires; and (4) a control device. A method of both picturing and diagraming a very simple circuit is shown in Fig. 1-7.

A schematic diagram is far simpler to draw and read than a picture diagram. For this reason, schematic diagrams are used in electronics as much as

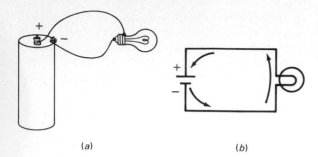

FIG. 1-7 (*a*) Lamp connected across a dry cell. (*b*) A schematic diagram of the same circuit.

possible. It is important to observe the diagrams that are shown and to practice drawing them until they can be reproduced rapidly and correctly.

Although the wires connecting the source of electromotive force to the load may have some resistance, it is usually very small in comparison with the resistance of the load and is ignored in most cases. A straight line, then, in a schematic diagram is considered to connect parts electrically, but it does not represent any resistance in the circuit.

In the simple circuit shown, the cell produces the electromotive force that continually pulls electrons to its positive terminal from the lamp's filament and pushes them out the negative terminal to replace the electrons that were lost to the load by the pull of the positive terminal. The result is a continual flow of electrons through the lamp filament, connecting wires, and source. The special resistance wire of the lamp filament heats when a current of electrons flows through it. If enough current flows, the wire becomes white-hot and the lamp glows.

The addition of a switch in series with one of the connecting wires of the simple circuit affords a means of controlling the current in the circuit (Fig. 1-8). When the switch is closed, electrons find an uninterrupted path in the circuit and flow through

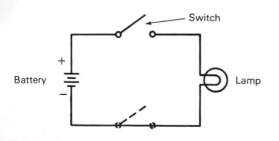

FIG. 1-8 A simple circuit. The battery is the source, the lamp is the load, and the switch is the control device.

the lamp. When the switch is opened, the electromotive force developed by the battery is normally insufficient to cause the electrons to jump the switch gap in the form of an arc and the electron flow in the circuit is interrupted. The lamp cools and no longer glows.

Since the only purpose of the switch is to interrupt or close the circuit, it may be inserted anywhere. It is shown in the upper connecting wire but will give the same results if placed in the lower one.

1-7 CURRENT

A stream of electrons forced into motion by an electromotive force is known as a current. The atoms in a good conducting material such as copper are more or less stationary, but one or more free electrons of the outer ring are constantly flying off at a high rate of speed. Electrons from other nearby atoms fill in the gaps. There is a constant aimless movement of billions of electrons in all directions at all times in every part of any conductor.

When an electric force (from a battery) is impressed across a conductor it drives some of these aimlessly moving free electrons away from the negative terminal toward the positive. It is unlikely that any one electron will move more than a fraction of an inch in a second, but an energy flow takes place along the conductor at approximately 186,000 miles per second (mi/s) or 300,000,000 meters per second (actually 299,792,462 m/s).

A simple analogy of energy flow can be illustrated with automobiles parked in a circle (Fig. 1-9). All the automobiles are parked bumper to bumper except cars 1 and 2, which are separated by a few feet. Nothing is happening in the circuit. The driver in car 1 steps on the gas for an instant, producing mechanical energy, and his car is propelled forward, striking car 2. The force of the impact transfers energy to car 2, which in turn transfers energy to car 3, and so on. An instant later the energy is transferred to the back bumper of car 17, and it is propelled forward, striking the back of car 1. Energy has traveled completely around the circuit in a very short space of time, and yet none of the cars except cars 1 and 17 may have moved more than a few centimeters. In an electric circuit the electrons are somewhat similar to the cars. By moving suddenly in one direction, electrons can repel other electrons. These repel others farther along, and so on. The energy transfer of an

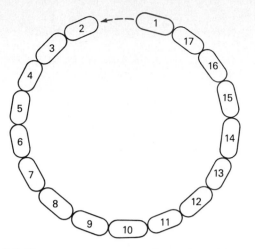

FIG. 1-9 How energy can travel faster than the energy carriers.

impulse is very rapid, but the *drift* of the electrons themselves is relatively slow.

A source of electric energy does not increase the number of free electrons in a circuit; it merely produces a concerted pressure on loose, aimlessly moving electrons. If the material of the circuit is made of atoms or molecules that have no free electrons, the source cannot produce any current in the material. Such a material is known as an *insulator*, or a *nonconductor*.

The amount of current in a circuit is basically measured in *amperes*, abbreviated A or amp. An ampere is a certain number of electrons passing a single point in an electric circuit in one second. Therefore, an ampere is a *rate of flow*, similar to gallons per minute in a pipe.

The quantity of electrons used in determining an ampere (and other electrical units) is the *coulomb*, abbreviated C or coul. An ampere is one coulomb per second. A single coulomb is about 6,250,000,000,000,000,000 electrons. Such a large number is more easily expressed as 6.25×10^{18}, which is read verbally as "6 point 25 times 10 to the eighteenth power." The 10^{18} means the decimal place in the 6.25 is moved 18 places to the, *right*. This method of expressing numbers is known as the *powers of 10* and is handy to use when very large or very small numbers are involved. An example of a small number is 42.5×10^{-7}. In this case the 10^{-7} indicates the decimal point is to be moved 7 places to the *left*, making the number 0.00000425. When very large and very small numbers are multiplied, the powers of 10 are added algebraically; that is, if both powers are negative numbers, they are

added and the sum is given a negative sign. If both powers are positive numbers, they are added and the sum is given a positive sign. If one power is negative and the other positive, the smaller is subtracted from the larger and the sum is given the sign of the larger.

EXAMPLE: Multiply 3.2×10^{14} by 4.5×10^{-12}. When multiplied, $3.2 \times 4.5 = 14.4$. The powers of 10, that is, 10^{14} and 10^{-12}, together equal 10^2. The answer is then 14.4×10^2, or 1440.

The unit of measurement of current is the ampere, but the unit of measurement of electrical quantity, or charge, is the coulomb. One coulomb passing a point in 1 second means an average of 1 ampere of current is flowing.

▌ Test your understanding; answer these checkup questions.

1. What is the name of the moving particle of electricity? _____
2. What is the unit formed when atoms combine chemically? _____
3. What electric charge does a proton have? _____ An electron? _____ An atom? _____ A molecule? _____ A neutron? _____
4. What type of field forms between positive and negative bodies? _____
5. How do like charges interact? _____ Unlike charges? _____
6. What particles are found in an atomic nucleus? _____ _____
7. Where in conductors are valence electrons? _____
8. What is the name of the device which indicates a positive or negative charge is being held near it? _____
9. What are the "big three" in electricity? _____ _____ _____
10. What are the four minimum parts of any electric circuit? _____ _____ _____ _____
11. What is the rate of electric-energy flow in meters per second? _____
12. What are materials that have no free electrons at room temperature called? _____ That have many free electrons? _____
13. Is an ampere a quantity or a rate of flow? _____
14. How many electrons are there in a coulomb? _____
15. What does 1.86×10^5 miles per second represent? _____

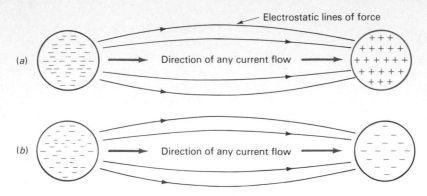

FIG. 1-10 Direction in which energy or current will flow (*a*) between negative and positive bodies and (*b*) between highly negatively and less negatively charged bodies.

1-8 ELECTROMOTIVE FORCE

The electron-moving force in electricity, variously termed *electromotive force* (emf), *electric potential, potential difference* (PD), *difference of potential, electric pressure,* and *voltage* (V), is responsible for pulling and pushing of the electric current through a circuit. The force is the result of an expenditure of some form of energy to produce an electrostatic field.

An emf exists between two objects whenever one of them has an excess of free electrons and the other has a deficiency of free electrons (Fig. 1-10*a*). Should the two objects be connected by a conductor, a discharge current will flow from the negative body to the positive one.

An emf also exists between two objects whenever there is a difference in the number of free electrons per unit volume of the objects (Fig. 1-10*b*). If both objects are negative, current will flow from the more negatively charged object to the less negatively charged object when they are connected together. There will also be an electron flow from a less positively charged object to a more positively charged object.

The electrostatic field, the strain of the electrons trying to reach a positive charge, or to move from a more highly negative to a less negative charge, or to move from a less positive to a more positive charge, *is* the emf in electricity. When a conducting material is placed between two points under electric strain, current flows.

The unit of measurement of electric pressure or emf, is the *volt* (V). A single flashlight dry cell produces about 1.5 V. A wet cell of a storage battery produces about 2.1 V.

A volt can also be defined as the pressure required to force a current of one ampere through a resistance of one ohm. (The *ohm* is the unit of measurement of resistance, to be discussed in Sec. 1-12.)

An emf can be produced in many ways. Some of the more common methods are:

1. Chemical (batteries)
2. Electromagnetic (generators)
3. Thermal (heating the junction of dissimilar metals)
4. Piezoelectric (mechanical vibration of certain crystals)
5. Magnetostriction (filters and special energy changers called transducers)
6. Static (laboratory static-electricity generators)
7. Photoelectric (light-sensitive cells)
8. Magnetohydrodynamic (MHD, a process that converts hot gas directly to electric power)

1-9 A BATTERY OR CELL IN A CIRCUIT

In the explanations thus far, "objects," either positively or negatively charged, have been used. A common method of producing an emf is by the chemical action in a battery (Chap. 28). A brief outline of the operation of a battery is given here.

Consider a flashlight *cell*. Such a cell (two or more cells form a *battery*) is composed of a zinc can, a carbon rod down the middle of the cell, and a black, damp, pastelike chemical *electrolyte* between them (Fig. 1-11). The zinc can is the nega-

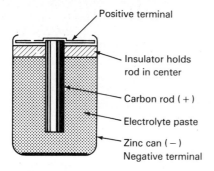

FIG. 1-11 Cross section of a common dry cell.

tive terminal. The metal cap connected to the carbon rod is the positive terminal. The active chemicals in such a cell are the zinc and the electrolyte.

The materials in the cell are selected of such substances that electrons are pulled from the outer orbits of the molecules or atoms of the carbon terminal chemically by the electrolyte and are deposited on the zinc can. This leaves the carbon positively charged and the zinc negatively charged. The number of electrons that move is dependent upon the types of chemicals used and the relative area of the zinc and carbon electrodes. If the cell is not connected to an electric circuit, the chemicals can pull a certain number of electrons from the rod over to the zinc. The massing of these electrons on the zinc produces a *backward* pressure of electrons, or an electric strain, equal to the chemical energy of the cell, and no more electrons can move across the electrolyte. The cell remains in this static, or stationary, 1.5-V charged condition until it is connected to some electric load.

If a lamp is connected between the positive and negative terminals of the cell, the 1.5 V of emf starts a current of electrons flowing through the lamp. The electrons flowing through the wire start to fill up the deficient outer orbits of the molecules of the positive rod. The electron movement away from the zinc into the lamp begins to neutralize the charge of the cell. The electron pressure built up on the zinc, which held the chemical action in check, is decreased. The chemicals of the electrolyte can now force an electron stream from the positive rod

through the cell to the zinc, maintaining a current of electrons through the lamp and battery as long as the chemicals hold out.

Note that as soon as the lamp begins to carry electrons, the electrolyte also has electrons moving through it. This motion produces an equal amount of current through the whole circuit at the same time. This point is a very important one to understand. There are no bunches of electrons moving around an electric circuit like a group of racehorses running around a track. A closed circuit is more like the racetrack with a single lane of automobiles, bumper to bumper. Either all must move at the same time, or none can move.

In an electric circuit, when electrons start flowing in one part, all parts of the circuit can be considered to have the same value of current flowing in them instantly. Most circuits are so short that the energy flow velocity, 300,000,000 m/s, may be disregarded for the present.

When a circuit is broken by the opening of a switch, electron progress comes to a sliding halt (Fig. 1-12). The cell chemicals keep pumping elec-

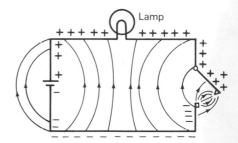

FIG. 1-12 Electrostatic lines of force across an open circuit. The greatest concentration occurs at the sharp points of the switch.

trons into the wire at the negative terminal until this half of the circuit attains a 1.5-V charge of electrons when compared with the positive half-circuit terminal. When this occurs, the electron charge, or strain, across the open switch equals the chemical strain produced in the cell and all electron progress in the circuit ceases. The circuit is charged, and electrostatic lines of force are developed as illustrated.

Note that the greatest concentration of lines of force appears at the sharpest point of the switch. The sharper this point the more concentrated the electric field becomes there. If the field is intense enough to heat the air and *ionize* it (Sec. 1-10), a visible *spark* will be seen in the air when the switch is opened. A spark that holds continually between two points is called an *arc*.

If a wire in a very high voltage circuit has a sharp point on it, the concentration of lines of force developed at that spot may pull electrons out of the wire into the air (or out of the air into the wire), forming a *corona*, or *brush discharge* around the point.

1-10 IONIZATION

When an atom loses an electron, it lacks a negative charge and is therefore positive. The electronless atom in this condition is a *positive ion*.

In most metals the atoms are constantly losing and regaining free electrons. They may be thought of as constantly undergoing ionization. Because of this, metals are usually good electric conductors.

Atoms in a gas are not normally ionized to any great extent, and therefore a gas is not a good conductor under low electric pressures. However, if the emf is increased across an area in which gas atoms are present, some of the outer orbiting electrons of the gas atoms will be attracted to the positive terminal of the source of emf and the remainder of the atom will be attracted toward the

negative. When pressure increases enough, one or more free electrons may be torn from the atoms. The atoms are then ionized. If ionization happens to enough of the atoms in the gas, a current flows through the gas. For any gas at a given pressure and temperature there is a certain voltage value that will produce ionization. Below this voltage, the number of ionized atoms is small. Above the critical value, more atoms are ionized, producing greater current flow, which tends to hold the voltage across the gas at a constant value. In an ionized condition the gas acts as an electric conductor.

Examples of ionization of gases are lightning, neon lights, and fluorescent lights. Ionization plays an important part in electronics and radio.

1-11 TYPES OF CURRENT AND VOLTAGE

Different types of currents and voltages are dealt with in electricity. In this book the following nomenclature will be used:

- *Direct current (dc)*. There is no variation of the amplitude (strength) of the current or voltage. Obtained from batteries, dc generators, and power supplies (Fig. 1-13a).

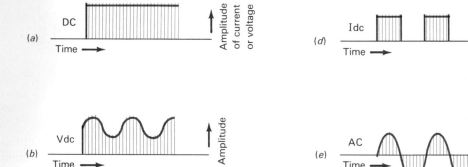

FIG. 1-13 Different forms of voltage or current. (*a*) Direct current. (*b*) Varying dc. (*c*) Pulsating dc. (*d*) Interrupted dc. (*e*) Alternating current. (*f*) Damped ac.

- *Varying direct current (vdc)*. The amplitude of the current or voltage varies but never falls to zero. Found in many transistor and vacuum tube circuits (Fig. 1-13b).
- *Pulsating direct current (pdc)*. The amplitude drops to zero periodically. Produced in rectifier circuits (Fig. 1-13c).
- *Interrupted direct current*. Current or voltage starts and stops abruptly ("square wave"). Produced by vibrators, choppers, and special circuits (Fig. 1-13d).
- *Alternating current (ac)*. Electron flow reverses (alternates) periodically and usually changes amplitude in a more or less regular manner. Produced in ac generators, oscillators, some microphones, and radio in general. This is the usual house current (Fig. 1-13e).
- *Damped ac*. This is alternating current which dies out in amplitude. Produced by spark-type oscillators and inadvertently in many circuits as they make and break (Fig. 1-13f).

1-12 RESISTANCE

Certain metals, such as silver and copper, have many free electrons flying aimlessly, at all times, through the spaces between the atoms of the material. Other metals, such as nickel and iron, have fewer free electrons in motion. Still other materials, such as glass, rubber, porcelain, mica, and quartz, have practically no interatom free-electron movement. When an emf is applied across opposite ends of a copper or silver wire, many free electrons progress along the wire and a relatively high current results. Copper and silver are very good conductors. When the same emf is applied across an iron wire of equivalent size, only about one-sixth as much current flows. Iron may be considered a fair conductor. When the same emf is applied across a length of rubber or glass, no electron drift results. These materials are insulators. Insulators are used between conductors when it is desired to prevent electric current from flowing between them. (Semiconductors, which conduct under certain circumstances, are discussed in Chap. 9.)

Silver is the best conductor, and glass is one of the best insulators. Between these two extremes are found many materials of intermediate conducting ability. While such materials can be catalogued as to their conducting ability, it is more usual to think of them by their resisting ability. Glass (when cold) completely resists the flow of current. Iron resists

much less. Silver has the least resistance to current flow.

The resistance a wire or other conducting material will offer to a current depends on four physical factors:

1. The type of material from which it is made (silver, iron, etc.)
2. The length (the longer, the more resistance)
3. Cross-sectional area of the conductor (the more area, the more molecules with free electrons, and the less resistance)
4. Temperature (the warmer, the more resistance, except for carbon and other semiconductor materials)

A piece of silver wire of given dimensions will have less resistance than a similar iron wire. It is reasonable to assume that if a 1-ft piece of wire has 1 *ohm* (unit of measurement of resistance), 2 ft of the same wire will have 2 Ω. (Greek letter omega, Ω, the symbol for ohms. The Greek alphabet is shown in Appendix A.)

On the other hand, if a 1-ft piece of wire has 1 Ω, two pieces of this wire placed side by side will offer twice the cross-sectional area, will conduct current twice as well, and therefore will have half as much resistance (Fig. 1-14). (The *cross-*

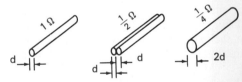

FIG. 1-14 Resistance of a single wire, two wires in parallel, and a wire with twice the diameter.

sectional area is the area seen when a wire is cross-sectioned, or cut in two.) A wire having twice the diameter (d) of another wire will have four times the cross-sectional area ($A = \pi \times$ radius *squared*, $= \pi r^2$) and therefore one-fourth the resistance.

If the resistance of a conductor is directly proportional to its length, but inversely proportional to its cross-sectional area, what would be its resistance if made 10 times as long and if its cross-sectional area were doubled?

$$R = \frac{L}{A} = \frac{10}{2} = 5 \text{ times original } R$$

A round wire of 0.001-in. diameter is said to have 1 circular mil (abbreviated cmil or cir mil) of cross-sectional area. The word *mil* means $1/1000$ of an inch. A round wire of 2-mil diameter has twice the radius and, by the formula above, has 4 times the cross-sectional area, or $A = 4$ cmil. A 3-mil-diameter wire has 9 cmil, and so on. The number of circular mils in any round wire is equal to the number of thousandths of an inch of diameter *squared*.

The number of circular mils is considered when determining how much current a wire may pass safely. When current flows through any wire, heat is produced in the wire. If too much heat is produced, the insulation on the wire may be set on fire or the wire may even melt.

It has been found that a copper wire, having a diameter of 64 mil, or $64^2 = 4096$ cmil, will allow 4.1 A to flow through it in a confined area without overheating. This represents about 1000 cmil/A. Therefore it may be assumed that any copper wire may carry 1 A for every 1000 cmil of cross-sectional area. In some applications, when a highly heat-resistant insulation is used, it may be possible to use wire with 500 cmil/A or less. The wire will heat considerably more than it would with 1000 cmil/A, but it cannot destroy the ruggedized insulation at the temperature that will be developed. A 64-mil wire that carries 4 A safely in confined spaces may carry 15 A or more in free air, where it can rapidly dissipate heat that has developed in it.

The unit of measurement of resistance is the ohm. For practical purposes 1 Ω may be considered to be the resistance of a round copper wire, 0.001 in. in diameter, 0.88 in. long, at 32° Fahrenheit (32°F). (Metrically this is: diameter = 2.54 mm, length = 2.235 cm, at 0°C. See Sec. 1-14.)

The *specific resistance* of a conductor is the number of ohms in a 1-ft-long 0.001-in.-diameter round wire of that material. The specific resistance of several common materials at room temperature is listed in Table 1-1.

An aid in recalling the order of resistance of five common conductor materials is to remember how they go down the "scail" (misspelling of the word "scale"), where the letters of "scail" indicate silver, copper, aluminum, iron, and lead.

Materials such as german silver and nichrome are alloys of two or more metals and are used in the construction of resistors because their resistance changes very little if they become heated (have a zero temperature coefficient). When wire made of these substances is wound on a tubular ceramic form, the result is a *wirewound resistor*, as shown in Fig. 1-15. These resistors are usually covered with a hard, vitreous protective coating.

TABLE 1-1	THE SPECIFIC RESISTANCE OF CONDUCTORS	
Conductor		**Specific resistance, Ω/mil ft**
Silver		9.75
Copper		10.55
Aluminum		17.30
Nickel		53.00
Iron		61.00
Lead		115.00
German silver		190.00
Nichrome		660.00

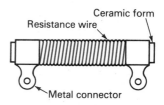

FIG. 1-15 Wirewound fixed resistor.

A *carbon resistor* is made from powdered carbon mixed with a clay binding material and baked into a small, hard cylinder with a wire attached to each end (Fig. 1-16). The value of resistance depends on the percent of carbon in the mixture used. Carbon resistors range in value from a fraction of an ohm to several million ohms. Size has no bearing on resistance value, affecting only the power (heat) or "wattage" rating of the resistor. Common ratings are 0.1, 0.25, 0.5, 1, and 2 W (watts). A 0.5-W resistor is about 0.125 in. in diameter and 0.5 in. long. Lower-power resistors are smaller, and higher-power resistors are larger. Carbon resistors

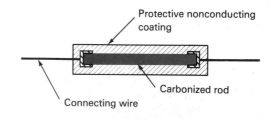

FIG. 1-16 Construction of a fixed-type carbon resistor.

have a positive temperature coefficient of resistance (resistance increases when carbon resistors are heated).

A *carbon-film resistor* consists of a glass or ceramic rod coated with a carbonized layer, which in turn is covered with a ceramic nonconductive coating. A *metal-film resistor* has a metal such as nichrome deposited on a glass or ceramic form. A spiral groove is cut with a lathe or laser beam through the deposit to produce a spiral-wound resistor, which is then covered with a protective layer.

The symbol for a *fixed* (nonvariable) *resistor*, either carbon or wirewound, is shown connected across a battery (Fig. 1-17).

FIG. 1-17 Symbol of a resistor across a battery.

The symbols used for variable resistors, called *rheostats*, are shown in Fig. 1-18. A rheostat has two connections, one to one end of the resistor and a sliding arm that moves along the length of the resistor. Rheostats may be either wirewound or carbon.

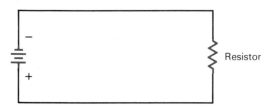

FIG. 1-18 Various symbols used to indicate a rheostat. Note that there are only two terminals on a rheostat.

The symbol for a *potentiometer* ("pot"), a rheostat with connections at both ends plus the sliding contact, is shown in Fig. 1-19. Potentiometers are used in most cases to select a desired proportion of the total voltage across them. They are used as "voltage dividers." Pots can be used as rheostats.

An *adjustable* resistor is a wirewound resistor with a sliding contact that can be locked into position when the desired value of resistance is found. Its moving contact can be tightened by a machine screw to make it immobile (Fig. 1-20). A resistor of this type may be partly covered with a vitreous coating, but part of the wires will be bare to allow the slider to make contact.

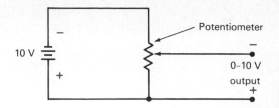

FIG. 1-19 Symbol of a potentiometer used as a voltage divider.

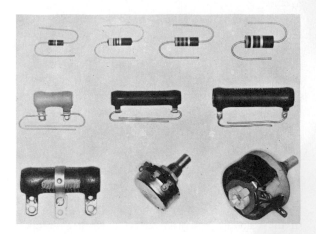

FIG. 1-20 Resistors. Top row, 1/2-W carbon, 1-W wirewound, 1-W carbon, and 2-W carbon. Middle row, 5-W wirewound, 10-W wirewound, and 20-W wirewound. Bottom row, wirewound adjustable resistor, carbon-type potentiometer, and wirewound rheostat in the OFF position.

1-13 COLOR CODES

Resistors and other components are often color-coded rather than having their resistance values printed on them. Technicians must know this code. Two methods have been used to color-code resistors. One is the three-stripe method; the other, the body-end-dot method. The body-end-dot resistors are no longer manufactured, but older equipment may contain some of them. In both methods the same color code is used. The colors and their meanings are:

Brown = 1	Blue = 6
Red = 2	Violet = 7
Orange = 3	Gray = 8
Yellow = 4	White = 9
Green = 5	Black = 0

To read the resistance of a color-coded resistor, as in Fig. 1-21, start with the stripe nearest the end

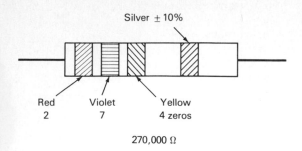

Silver ± 10%

Red — 2
Violet — 7
Yellow — 4 zeros

270,000 Ω

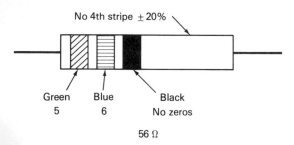

No 4th stripe ± 20%

Green — 5
Blue — 6
Black — No zeros

56 Ω

FIG. 1-21 Color-coded resistors.

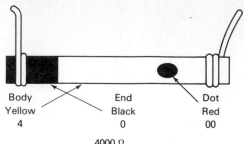

Body Yellow 4
End Black 0
Dot Red 00

4000 Ω

FIG. 1-22 Old-fashioned body-end-dot resistor.

of the resistor. The first stripe is the first number. The second stripe is the second number. The third stripe, the "10 multiplier," indicates the number of zeros following the second number.

Resistors with values of less than 10 Ω have a multiplier, or third stripe, of gold or silver. Gold indicates that the first two numbers are to be multiplied by 0.1. (Yellow, violet, gold = 4.7 Ω.) A silver third stripe indicates multiplication by 0.01. (Green, blue, silver = 0.56 Ω.)

Three-stripe resistors have a "tolerance" of 20%. (Its actual value may vary 20% up or down from its indicated value.) If a silver fourth stripe is added, it indicates the resistor has a tolerance of 10%. A gold fourth stripe represents a tolerance of 5%. A red fourth stripe represents a tolerance of 2%. The ohmic values are often printed on 1% and 2% resistors.

When resistors have a double-width *first* stripe, the resistor is wirewound rather than carbon. One is shown in Fig. 1-20.

The old body-end-dot coding (Fig. 1-22) is read in that order. If no dot is visible, the dot and body number are assumed to be the same.

Although resistors can be specially ordered in any values, the values considered as off-the-shelf or standardized are shown in Appendix B.

Test your understanding; answer these checkup questions.

1. What must exist when an emf is produced between two points? _____
2. Electric current flows in what direction between a positive and a negative body? _____ Greatly positive and lesser positive bodies? _____ Greatly negative and lesser negative? _____
3. In what unit is emf measured? _____ Current? _____ Resistance? _____
4. What may develop at a sharp point in high-voltage circuits? _____
5. If an atom loses an electron, what is it then called? _____
6. Under what condition may gas be a good conductor? _____
7. List the six types of current. _____ _____ _____ _____ _____ _____
8. List five common materials in order of their descending conductivity. _____ _____ _____ _____ _____
9. Of what materials are resistors commonly made? _____ _____
10. Does the physical size of a resistor represent power-dissipating capabilities or resistance value? _____
11. How many contacts on a carbon resistor? _____ A rheostat? _____ A potentiometer? _____ An adjustable resistor? _____
12. Which resistor color-coding method is no longer used? _____
13. On a color-coded resistor, what does the first stripe indicate? _____ The second? _____ The third? _____ The fourth? _____
14. The first stripe on a color-coded resistor is double width. What does this indicate? _____
15. A 40-Ω resistance is desired. What is the closest off-the-shelf value with a tolerance of 1%? _____ 5%? _____ 10%? _____

16. What is the value in ohms of a yellow-violet-brown-silver striped resistor? _____ A green-gray-gold striped resistor? _____

1-14 THE METRIC SYSTEM

Scientific measurements are more and more being given in the metric system. It is a multiple-of-10 system (similar to the United States dollar-dime-cent-mil monetary system). The basic metric units of measurement are:

- Length: *Meter*. Approximately 39.37 in. All other units of length are multiples or submultiples of 10 of the meter.
- Volume: *Liter*. A centimeter (cm) is $1/100$ meter (0.3937 in.). A cubic centimeter (cm³) is a cube with all sides 1 cm long. 1000 cm³ is a liter (1.06 qt).
- Weight: *Gram*. The weight of 1 cm³ of water at 4° Celsius or centigrade (4°C), or 0.353 oz, avoirdupois weight.

Volts, amperes, ohms, etc., may also use metric-based prefixes. The prefixes in general use in electronic and radio work are listed in Table 1-2.

Some examples of these prefixes are:

- pF (picofarad): A trillionth of a farad, or 1 pF or 1 $\mu\mu$F
- ns (nanosecond): A billionth of a second, or 1 ns or 1 mμs

TABLE 1-2	METRIC-BASED PREFIXES AND POWERS OF TEN		
Atto (a)	= quintillionth of	= 10^{-18} times
Femto (f)	= quadrillionth of	= 10^{-15} times
Pico (p), or $\mu\mu$	= trillionth of	= 10^{-12} times
Nano (n), or mμ	= billionth of	= 10^{-9} times
Micro (μ)	= millionth of	= 10^{-6} times
Milli (m)	= thousandth of	= 10^{-3} times
Centi (c)	= hundredth of	= 10^{-2} times
Deci (d)	= tenth of	= 10^{-1} times
	unity	= 10^{0} = 1
Deka (da)	= ten times	= 10 times
Hecto (h)	= hundred times	= 10^{2} times
Kilo (k)	= thousand times	= 10^{3} times
Mega (M)	= million times	= 10^{6} times
Giga (G), or kM	= billion times	= 10^{9} times
Tera (T)	= trillion times	= 10^{12} times

- μV (microvolt): A millionth of a volt, or 1 μV
- mA (milliampere): $1/1000$ A (0.001 A)
- mm (millimeter): $1/1000$ m (0.001 m)
- cm (centimeter): $1/100$ m (0.01 m)
- dB (decibel): $1/10$ B (bel) (0.1 B)
- kW (kilowatt): 1000 W
- km (kilometer): 1000 m, 0.64 mile
- MHz (megahertz): 1,000,000 Hz, also known as 1,000,000 cps (cycles per second)
- GHz (gigahertz): 1,000,000,000 Hz

It is often necessary to convert one metric-prefixed value to another. A method of doing this is first to arrange the two values in ratio form, larger to smaller. Then multiply this ratio by the powers-of-10 difference between the two values.

EXAMPLE: How many hertz are there in 5.23 kHz? If kilo = 10^3, the ratio of the two values would be 5.23 × 10^3 to 1, which is 5230 Hz.

EXAMPLE: How many microvolts are there in 2.4 kV? If kilo = 10^3 and micro = 10^{-6}, in ratio form the problem would read 2.4 × 10^3 divided by 1 × 10^{-6}. The difference between 10^3 and 10^{-6} is 10^9. Therefore, there must be 2.4 × 10^9, or 2,400,000,000 μV in 2.4 kV.

1-15 WIRE SIZES

Most wire used in electronics and radio is made of copper. It may be either *hard-drawn* (stiff) or *soft-drawn* (pliable). It is manufactured in various sizes, with or without an insulation coating on the wire. Some of the insulating materials used are enamel, silk, glass fibers, cotton, rubber, varnish, and various plastics.

Table 1-3 lists some of the more commonly used American Wire Gauge (AWG) copper wire sizes and information regarding them.

Wires in air are air-cooled and will carry more current without overheating. For example, a No. 12 wire will carry safely only 6.5 A in a confined space, as in a transformer, but as much as 20 A in open air.

1-16 MAKING LOW-RESISTANCE CONNECTIONS

The addition of resistance to an electric circuit reduces the amount of current that can flow. Loose or oxidized (rusted) electrical connections may act as

TABLE 1-3	COPPER WIRE TABLE		
AWG No.	Diameter, mils	Ω/1000 ft, room temperature	Current-carrying capacity at 1000 cmil/A, as in transformers
0	325	0.1	90 A
8	128	0.641	16.5 A
10	102	1.02	10.4 A
12	81	1.62	6.5 A
14	64	2.58	4.1 A
16	51	4.09	2.6 A
18	40	6.51	1.6 A
20	32	10.4	1.0 A
22	25	16.5	640 mA
24	20.1	26.2	400 mA
26	15.9	41.6	250 mA
28	12.6	66.2	160 mA
30	10.0	105	100 mA
32	7.95	167	63 mA
34	6.3	265	40 mA

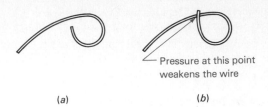

FIG. 1-23 (*a*) Properly looped wire to fit around a machine screw and under a nut. (*b*) Improperly looped wire.

Two wires may be scraped clean and then twisted tightly together, as in a Western Union splice (Fig. 1-24*a*). However, over a period of time such wires may corrode and a resistance joint can result. To prevent this and to give added strength, the joint may be covered with a protective layer of solder.

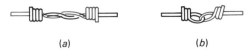

FIG. 1-24 (*a*) Properly made splice between two wires. (*b*) Very poor method of connecting two wires.

resistances and often result in improper operation of a circuit. When equipment is constructed or repaired, the technician must make sure that all connections are tight and have low resistance.

When a wire is scraped clean and looped around a machine screw and a nut is tightened down on it, a good, low-resistance electrical connection is made between the wire and the machine screw. Greater nut pressure produces more contacting surface between wire and nut and less resistance. The looped end should be fitted snugly around the screw (Fig. 1-23*a*) in such a direction that the tightening of the nut tends to close the loop rather than open it. The wire should not be overlapped on itself (Fig. 1-23*b*), as pressure exerted by the nut may squeeze the wire at the point of overlap to half thickness and weaken the joint.

Solder is an alloy of tin and lead having the capability of melting at a relatively low temperature. All solders, when going from solid to liquid, change into an intermediate *plastic* state at a temperature of about 360°F. The type known as 60/40 solder, which is 60% tin and 40% lead, becomes liquid at about 370°F. A less desirable 50/50 solder requires a temperature of about 420° before it changes from plastic to liquid and flows freely.

For electronics work solder is made up into thin, hollow wires. The hollow space is filled with rosin or some other nonacidic *flux*. The job of the flux is to melt before the solder and flow over the connection to be soldered. The hot flux reacts with the metal, cleaning it somewhat, which allows the hot solder to adhere more firmly to it. When cold, the rosin loses its ability to react with the metal but may absorb water from the air and therefore should be removed. Although acid-core solder and flux are useful when soldering sheet metal, they should never be used in electrical work. Acid solder joints may produce an undesirable voltage-generating cell between solder, acid, and metal that can cause noise in electronic equipment, rectification (Chap. 9), higher resistance, and a weakening of the soldered joint.

When two wires are to be soldered together, (1) the surfaces of both wires must be clean, (2) the two wires should make good mechanical contact, and (3) both wires must be heated before solder is applied. The hot soldering iron should be applied so that both wires receive heat simultaneously. Solder and flux are then held against the point where the two wires touch. As soon as hot solder flows evenly over the hot metal surfaces, the iron and solder are removed. The point must be held motionless until the molten solder passes from liquid through plastic to solid form; otherwise, the solder may crystallize and lose its holding ability.

There are many types of soldering irons. A common, 25- to 250-W (watt) type is shown in Fig. 1-25. For small, transistor work a 35-W iron may be satisfactory. For larger work a 50- to 100-W soldering iron or soldering gun may be more desirable. For heavy-duty work on large wires or in the open air, a 250-W iron may be required. It is imperative that the tip of a soldering iron be *tinned*. A tinned surface has a layer of fresh solder applied to a cleaned portion of it. The tinned area can be recognized by its shiny surface in comparison with the brown, oxidized copper, untinned area. A tinned surface will transfer heat many times faster than an oxidized surface.

Solder that has been heated and cooled several times at a connection oxidizes and crystallizes. It no longer holds properly. If the solder is in this condition, it should be melted from the joint and new solder substituted.

There are other modern means of joining wires to special connector lugs in electric circuits. Crimped connections are produced with a pressure tool that physically crimps a wire and lug together. There is also a special wire-wrapping tool that twists the wire around a square terminal lug so tightly that the connection is considered equal to a soldered joint and may be superior if the joint is subject to vibration.

1-17 PRINTED CIRCUIT BOARDS

All methods of producing a printed circuit (PC) board result in narrow copper-strip conductors laid down on thin insulating sheets of a phenolic or Fiberglas board. The components, or parts, to be interconnected are then soldered to the copper strips.

One method of producing a PC board is to start with a small oblong of $1/16$-in.-thick phenolic insulator sheet with a thin copper plating on one side. The desired circuit wires are drawn on the copper with a special *resist ink*. The board is then immersed in a ferric chloride solution until all the copper except that under the resist ink has been etched (eaten away) chemically. After this, the board is washed and the resist ink removed, leaving a series of copper conductor lines on the phenolic sheet. These may then be silver-plated or plated with solder. Holes are drilled through the copper strips and base. Connector wires from the required electronic components are then slipped into the holes and soldered to the copper lines.

Each connection can be soldered separately, which is rather slow. If all the components are fed through the holes from the insulated side of the board, all the connections can be *wave-soldered* at once by moving the copper-connection surface of the PC board over a wave of molten solder. The wave is developed over a cylinder rotating just under the surface of a small tub full of molten solder.

One of the difficulties with printed circuits stems from the difference of expansion of the copper and the phenolic boards. Temperature changes can loosen the copper strips from the phenolic backing, or hairline cracks can open up across the strips, breaking the electrical connections. This may occur during the original soldering, if heated during operation, or during unsoldering to remove and replace faulty components. Bending the boards may develop similar troubles. While resistance meters (ohmmeters) can be used to determine if a circuit has broken open, visual examination with a powerful magnifying glass will often be simpler.

To remove components from a printed circuit, a very small soldering iron must be used to prevent

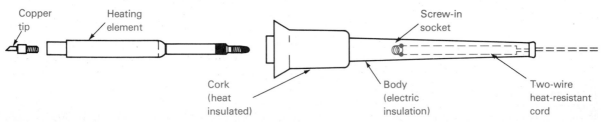

Copper tip   Heating element   Screw-in socket   Cork (heat insulated)   Body (electric insulation)   Two-wire heat-resistant cord

FIG. 1-25 A soldering iron, 25- to 50-W types.

overheating of the PC board. A pulling tension is applied to one end of the lead of the component being removed while a 25- to 35-W soldering iron tip is applied to the soldered contact. As the solder melts, the component wire pulls free of the PC board hole. If the hole remains plugged with solder, it may be heated and a stainless-steel wire, sharp lead pencil, or other thin device to which solder will not adhere can be run into the hole to clear it. A far better method is to use a special small soldering iron called a *desoldering tool* (Fig. 1-26), having a hollow tip to which is coupled a rubber bulb. The bulb is squeezed, forcing air out the hollow tip. The hot tinned tip is then placed around the wire to be desoldered until the solder melts. When the bulb is released, solder is sucked up into the hollow iron, freeing the component lead of solder and allowing the lead to be extracted. Squeezing the bulb again forces the molten solder out of the hollow tip, readying the iron for use again.

Another desoldering method is to heat a soldered joint with an iron and touch a pure copper-wire braid to the joint. The solder will be pulled up the braid by capillary action, removing all the solder from the joint.

One commercial process of producing printed circuit boards is *photoetching*. The original circuit is drawn on paper and photographed. The negative is then used to transfer, by photographic methods, an etch-resistant image of the wiring onto a photographic-surfaced copper-clad phenolic board. The rest of the method is as described above. Other methods used are silk screening, chemical deposition, and vacuum distillation.

▌Test your understanding; answer these checkup questions.

1. How many hertz in a kilohertz? _____ In a megahertz? _____
2. How many milliamperes in 2.45 A? _____ In 0.0358 A? _____
3. How many microamperes in 29 mA? _____ In 7.2 A? _____
4. How many amperes in 450 mA? _____ In 75 μA? _____
5. How many milliwatts in 850 μW? _____ In 5.6 W? _____
6. From Table 1-3, how many ohms does 100 ft of No. 10 copper wire have? _____ No. 20? _____
7. Using 1000 cmil/A, what size wire is required for a 4-A flow? _____
8. Using 400 cmil/A, what size wire is required for a 1-A flow? _____ A 4-A flow? _____
9. At what temperature does the best solder melt? _____
10. What is the reason for soldering a wire splice? _____
11. What is the duty of solder in PC board connections? _____
12. What is the duty of solder fluxes? _____
13. With what may circuit connections be drawn on PC boards before they are chemically treated? _____
14. What wattage iron would be used on an antenna wire outdoors? _____ On a PC board? _____ On indoor electrical wiring? _____
15. What is the coating of solder on the tip of a hot soldering iron called? _____
16. What method of soldering can make a hundred soldered connections at once on PC boards? _____
17. What is the best device to use to remove solder from PC board connections? _____

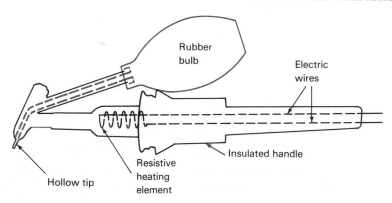

FIG. 1-26 Desoldering tool required with PC boards.

1. What is the name of the moving particle in electricity?
2. What is the name of the smallest particle of water?
3. What is the name of the smallest particle of oxygen?
4. What is the relative charge of an electron compared to a proton?
5. What kind of field surrounds electrons or protons?
6. Is there a difference in weight between electrons and protons, and if so how much?
7. What is the name of the central part of an atom?
8. In general, does a hot or cold conductor have more resistance?
9. What does a conductor have that an insulator does not?
10. What does an electroscope demonstrate?
11. In electricity, what does the "big three" refer to?
12. What are the four basic parts of a simple electric circuit?
13. Why is 300,000,000 m/s important?
14. What is the basic unit of measurement of electric current?
15. What is 42,500 in powers of 10?
16. What is 0.000053 in powers of 10?
17. Express the unit of quantity of electricity in powers of 10.
18. What is the basic unit of measurement of electric pressure?
19. What is meant by "emf"?
20. List eight methods of generating an emf.
21. What is an electrolyte?
22. Why are sharp points not desirable on high-voltage circuits?
23. When a spark holds for a time, what is it called?
24. What is another name for brush discharge?
25. What is an atom that loses an electron called?
26. List three types of ionized gases.
27. What are the meanings of dc; vdc; pdc; ac?
28. What is a damped ac?
29. In general, how much resistance does a conductor have in comparison with an insulator?
30. What four things determine the resistance of a conductor?
31. What is the new resistance of a 5-Ω conductor if it is stretched until it is twice as long and has a cross-sectional area of two-thirds the original?
32. How many circular mils are there in a wire having a diameter of 10 mils?
33. List five common metals by their specific resistivity.
34. Why are nichrome and german silver wires used?
35. How does a potentiometer differ from a rheostat?
36. What is the resistance of a resistor marked green, gray, green, silver?
37. What is meant by the tolerance of a resistor?
38. What are the basic metric units of length, volume, and weight?
39. Give the power of 10 for the following: pico; nano; micro; milli; kilo; mega; giga.
40. How many mV are there in 3.5 kV? μV in 2 V?
41. Which is larger, a No. 8 or a No. 30 AWG wire?
42. Which would require the larger wire to carry 1 A, a wire in a confined area or one in the air, and why?
43. When wires are spliced, what does solder do?
44. When solder is used on PC boards, is it a means of carrying current?
45. What does rosin do in soldering?
46. What are two methods of removing solder from PC boards?

2

Direct-Current Circuits

In this chapter several of the laws and theorems that apply to electric circuits will be discussed and problems will be solved using them. These are Ohm's law, Watt's law, Kirchhoff's laws, and Thevenin's theorem.

2-1 MATHEMATICS FOR ELECTRICITY

This is not intended to be a textbook on mathematics. It may be well, however, to point out some basic mathematical operations that can be used in working electrical problems involving formulas we will be using. These operations are a form of simple algebra. Only a few will be given. If you are familiar with simple algebraic manipulations, you might wish to omit this and continue with Sec. 2-2.

When you work with algebraic formulas, there are certain operations to remember:

1. The sign for addition is +; the sign for subtraction is −; the sign for multiplication is ×, parentheses ( ), or brackets [ ].

$$+2 + 2 = +4$$
$$-3 - 4 = -7$$
$$+5 - 2 = +3$$
$$+3 - 5 = -2$$
$$3 \text{ times } 4 = 3 \times 4 = (3)(4) = 3(4)$$
$$(A)(B) = A(B) = AB$$

2. Any number (or letter) multiplied by 1 is unchanged, and therefore the 1 may be dropped.

$$1(3) = 3 \qquad 1(A) = 1A = A$$

3. Any number (or letter) divided by 1 is equal to the number (or letter), and therefore the 1 may be dropped.

$$\frac{4}{1} = 4 \qquad \frac{A}{1} = A$$

4. Any number (or letter) when multiplied by itself is equal to the number (or letter) squared.

$$2 \times 2 = 2^2 \qquad V(V) = V^2 \qquad VV = V^2$$

5. Any number (or letter) divided by itself is equal to 1.

$$\frac{4}{4} = 1 \qquad \frac{F}{F} = 1 \qquad \frac{XQZ}{ZXQ} = 1$$

ANSWERS TO CHECKUP QUIZ ON PAGE 18

1. (1000)(1,000,000) **2.** (2450)(35.8) **3.** (29,000)(7,200,000) **4.** (0.45)(0.000075) **5.** (0.85)(5600) **6.** (0.102)(1.04) **7.** (No. 14) **8.** (No. 24)(No. 18) **9.** (360°F) **10.** (Hold wires and protect from corrosion) **11.** (Hold components and carry current) **12.** (Clean surfaces to be soldered) **13.** (Resist inks, adhesive tapes) **14.** (250 W)(25–35 W) (50–100 W) **15.** (Tinned surface) **16.** (Wave) **17.** (Desoldering tool)

6. If both sides of an equation are multiplied by the same number (or letter), equality still holds. For example,

$$2 = 2$$

If multiplied by 4:

$$4(2) = 4(2) \quad \text{or} \quad 8 = 8$$

Another example:

$$6 = 2(3)$$

If multiplied by 4:

$$4(6) = 4[2(3)] \quad \text{or} \quad 24 = 24$$

With letters:

$$X = AC$$

If multiplied by B:

$$BX = BAC$$

7. If both sides of an equation are divided by the same number (or letter), the equation will still be correct. For example,

$$6 = 2(3)$$

When divided by 2:

$$\frac{6}{2} = \frac{2(3)}{2} \quad \text{or} \quad 3 = 3$$

With letters:

$$V = IR$$

When divided by R:

$$\frac{V}{R} = \frac{IR}{R} \quad \text{or} \quad \frac{V}{R} = I$$

8. If both sides of an equation are squared, the equation will still be correct.

$$6 = 2(3) \qquad 6^2 = [2(3)]^2$$
$$6^2 = 2^2 3^2 \qquad 36 = 36$$

With letters:

$$A = BC \qquad A^2 = (BC)^2$$
$$A^2 = B^2 C^2$$

9. If the square root is taken of both sides of an equation, the equation will still be correct.

$$16 = 2(8) \quad \sqrt{16} = \sqrt{2(8)} \quad \text{or} \quad 4 = 4$$
$$6^2 = 3(12) \quad \sqrt{6^2} = \sqrt{3(12)} \quad \text{or} \quad 6 = 6$$
$$V^2 = PR \quad \sqrt{V^2} = \sqrt{PR} \quad \text{or} \quad V = \sqrt{PR}$$
$$A = BC \quad \sqrt{A} = \sqrt{BC}$$

10. A negative number on one side of the equation becomes a positive number when moved to the other side. (No sign in front of a number or letter indicates it is positive.)

$$7 + 2 = 9$$
$$7 = 9 - 2$$
$$B + C = A$$
$$B = A - C$$
$$B - C = A$$
$$B = A + C$$

When rearranging formulas involving simple fractions, a first step may be to *cross-multiply*. This means to multiply the top of the fraction on one side of the equation by the lower numbers of the fraction on the opposite side of the equals sign. The same is done to the other halves of the fractions. These two answers are set as equal to each other.

$$4 = \frac{8}{2} \quad \text{may be written} \quad \frac{4}{1} = \frac{8}{2}$$

When cross-multiplied,

$$\frac{4}{1} \diagup\!\!\!= \frac{8}{2}$$

becomes $\quad 2(4) = 1(8) \quad$ or $\quad 8 = 8$

With a letter formula,

$$I = \frac{V}{R} \quad \text{or} \quad \frac{I}{1} = \frac{V}{R}$$

When cross-multiplied,

$$\frac{I}{1} \diagup\!\!\!= \frac{V}{R}$$

becomes $\quad 1(V) = I(R) \quad$ or $\quad V = IR$

By cross-multiplying it has been determined what V equals. A further possible step is to determine what R equals. To do this, divide out the unwanted letters on one side of the equation, leaving only the desired letter. For example,

$$V = IR$$

To find what R equals, divide out the I from both sides:

$$\frac{V}{I} = \frac{IR}{I}$$

The I's on the right-hand side cancel each other, leaving

$$\frac{V}{I} = R \quad \text{or} \quad R = \frac{V}{I}$$

These two operations, cross-multiplying and dividing out the unwanted, can be used in a surprising number of electrical problems. (If trained in mathematics, you may know other methods.)

It is recommended that the reader use a pocket calculator to do the mathematical work required in basic electronics. The calculator should have at least add, subtract, multiply, divide, sine (SIN), cosine (COS), tangent (TAN), logarithm (LOG), square root ($\sqrt{}$), inversion (1/X), polar/rectangular (P/R) function keys, and some memory.

If a pocket calculator is not available, a slide rule with scales, A, B, C, CI, D, K, L, S, and T can give reasonably quick answers correct to the 3rd significant digit. If a calculator or slide rule is not available, the appendix of this book has tables for sine, cosine, and tangent used in many ac-circuit problems, as well as log tables for dB problems.

■ Test your understanding; answer these checkup questions.

1. If $P = VI$, then $V = $ _____ $I = $
   _____
2. If $Q = X/R$, then $X = $ _____ $R = $
   _____
3. If $Z = X^2/R$, then $R = $ _____ $X = $
   _____
4. If $FL = 1/FC$, then $1 = $ _____
   $C = $ _____ $L = $ _____
   $F = $ _____
5. If $3 - A = L/X$, then $X = $ _____ $L = $
   _____ $A = $ _____ HINT: $3 - A$ either is $+3 + -A$ or can be considered as the unit $(3 - A)$.
6. If $2(B - C) = Q/Z$, then $Q = $ _____ $Z = $
   _____ $B = $ _____ $C = $ _____
   HINT: $2(B - C)$ is $+2B - 2C$.
7. What is the square root of 525? _____ Of 10,000? _____
8. What is the square root of 1000? _____ Of 10? _____
9. What is the square root of 0.05? _____ Of 0.5? _____

2-2 OHM'S LAW

Wherever electric circuits are in use, voltage, current, and resistance are present. It is interesting to see how the theory of more complex circuit operation unfolds by starting with a simple circuit and slowly adding one step to another. To the beginner, each step may appear understandable enough, but remembering it and, more important, learning when

to apply it are the secrets to success in the study of electric circuitry. Once readers comprehend something of the physical nature of current, voltage, and resistance, they are ready to use this knowledge to learn when, where, how, and why these factors may be applied to electric circuits.

A change in current can be produced by changing either the voltage or the resistance in the circuit. An increase in voltage will increase current. Therefore, voltage and current are *directly* proportional to each other.

An increase in resistance in a circuit *decreases* current. Therefore, resistance and current are *inversely* proportional to each other.

These two facts can be condensed into one statement, known as Ohm's law: *Current varies directly as the voltage and inversely as the resistance.*

Ohm's law is a simple statement of the functioning of an electric circuit. It can be expressed mathematically as

$$I = \frac{V}{R}$$

where I = intensity of current (in amperes, A)
   V = emf (in volts, V)
   R = resistance (in ohms, Ω)

Ohm's law might also be expressed as

$$\text{Amperes} = \frac{\text{Volts}}{\text{Ohms}}$$

By multiplying both sides of the equation above, $I = V/R$, by R, the Ohm's-law formula becomes

$$I(R) = \frac{V(R)}{R} \qquad V = IR$$

By dividing both sides of this last equation by I, the Ohm's-law formula becomes

$$\frac{V}{(I)} = \frac{IR}{(I)} \qquad R = \frac{V}{I}$$

These three variations of the Ohm's-law formula make it possible to determine (1) the current value if the voltage and resistance are known, (2) the voltage in the circuit if the current and resistance are known, and (3) how much resistance is in the circuit if the voltage and current are known.

An understanding of this law and an ability to use it are quite important. Electronic job test questions for civil service work and other testing are certain to involve the application of Ohm's law in several different ways.

The so-called magic triangle or circle (Fig. 2-1)

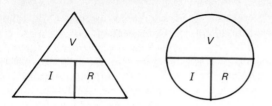

FIG. 2-1 Aids to remember the three Ohm's-law formulas.

may help in learning Ohm's law. If the symbol for the unknown is covered, the mathematical formula to solve for this letter is shown by the position of the other two symbols. For example, cover I; it is necessary to divide V by R. Cover V; multiply I by R. Cover R; divide V by I.

2-3 USING OHM'S LAW

The following examples illustrate the use of Ohm's-law formulas in determining the functioning of simple electric circuits.

In the circuit of Fig. 2-2, if the voltage V of the battery is 10 V and the load resistance R is 20 Ω, what is the value of the current I in the circuit?

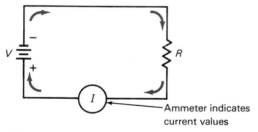

Ammeter indicates current values

FIG. 2-2 Current in a circuit is directly proportional to emf and inversely proportional to resistance.

Solution: Using Ohm's law,

$$I = \frac{V}{R} = \frac{10}{20} = 0.5 \text{ A}$$

In the same type of circuit, if the ammeter reads 4 A and the resistance is known to be 30 Ω, what value of emf must the source have?

Solution:

$$V = IR = 4(30) = 120 \text{ V}$$

In the same type of circuit, the ammeter reads 3 A and the source voltage is known to be 150 V. What is the load-resistance value?

Solution:

$$R = \frac{V}{I} = \frac{150}{3} = 50 \text{ }\Omega$$

An electric shock of more than 15 mA (15 milliamperes or 0.015 A) flowing through the body is considered dangerous to human life. What current will flow through a person having body contact resistance of 2200 Ω across 110 V?

$$I = \frac{V}{R} = \frac{110}{2200} = 0.05 \text{ A or 50 mA}$$

If the resistance in a circuit is increased 5 times and the voltage is tripled, what is the final current value?

$$I = \frac{V}{R} = \frac{3V}{5R} = \frac{3}{5} \text{ as much}$$

▌ Test your understanding; answer these checkup questions.

1. What is the basic formula for Ohm's law? _____ What are the two derivations of this formula? _____ _____
2. What current flows in a circuit with 100 V and 1000-Ω resistance? _____
3. What voltage is required to produce 2 A of current through 60 Ω? _____
4. What resistance will limit current flow to 4 A in a circuit having a 200-V supply? _____
5. A lamp connected across 120 V is found to have 3 A flowing through it. What is its resistance? _____
6. A relay coil having 315 Ω is across 6.3 V. What current value will it draw? _____
7. A 5000-Ω resistor in a receiver has 5 mA flowing through it. What voltage-drop is developed across it? _____
8. The resistance of a circuit remains the same, but the current through the resistor suddenly triples. What has happened to the circuit voltage? _____
9. If the voltage applied to a circuit is doubled but the resistance remains unchanged, what will the current value do? _____
10. R triples and V doubles, what will be the new current value? _____

2-4 POWER AND ENERGY

Electric pressure, or emf, by itself can do no work. A battery develops an emf, but if no load is connected across it, no current flows and no electrical work is accomplished.

When a conductor is connected across a source of emf, a current of electrons is developed. The current represents movement. The product of the pressure and the movement (volts and amperes) does accomplish work. The unit of measurement of the rate of doing work, or the unit of measurement of power, is the *watt* (W): 1 V causing 1 A to flow in a 1-Ω resistor produces 1 W of power. In formula form,

$$P = VI$$

where P = power (in watts, W)
V = emf (in volts, V)
I = current (in amperes, A)

EXAMPLE: What is the power input to a transmitter having an input voltage of 2000 V and an input current of 0.5 A?

$$P = VI = 2000(0.5) = 1000 \text{ W}$$

The Ohm's-law formula states: $V = IR$. If this is true, then (IR) can be substituted for V in the power formula:

$$P = VI \qquad P = (IR)I \qquad \text{or} \qquad P = I^2R$$

EXAMPLE: What is the heat dissipation, in watts, of a resistor of 20 Ω having a current of $\frac{1}{4}$ A passing through it?

$$P = I^2R = 0.25^2 \times 20 = 0.0625(20) = 1.25 \text{ W}$$

From Ohm's law again, $I = V/R$. By substituting V/R for the I in the basic power formula.

$$P = VI \qquad P = V\left(\frac{V}{R}\right) \qquad \text{or} \qquad P = \frac{V^2}{R}$$

EXAMPLE: What is the minimum power-dissipation rating of a resistor of 20,000 Ω to be connected across a potential of 500 V?

$$P = \frac{V^2}{R} = \frac{500^2}{20,000} = 12.5 \text{ W}$$

These three formulas to determine power in an electric circuit are undoubtedly as important to

ANSWERS TO CHECKUP QUIZ ON PAGE 22

1. $(P/I)(P/V)$ **2.** $(QR)(X/Q)$ **3.** $(X^2/Z)(\sqrt{RZ})$ **4.** (F^2LC) $(1/F^2L)(1/F^2C)(\sqrt{1/LC})$ **5.** $[(L/(3 - A)][(3 - A)X](3 - L/X)$ **6.** $[2Z(B - C)][Q/2(B - C)](C + Q/2Z)(B - Q/2Z)$ **7.** $(22.9)(100)$ **8.** $(31.6)(3.16)$ **9.** $(0.224)(0.707)$

ANSWERS TO CHECKUP QUIZ ON PAGE 23

1. $(I = V/R)(R = V/I, V = IR)$ **2.** (0.1 A, 100 mA, or 100,000 μA) **3.** (120 V) **4.** (50 Ω) **5.** (40 Ω) **6.** (0.02 A or 20 mA) **7.** (25 V) **8.** (It has tripled) **9.** (Double) **10.** (Two-thirds the original)

know as the three Ohm's-law formulas, and may be considered as Watt's-law formulas.

When a current of electrons flows through a conductor, the conductor always becomes warmer. Some of the power in the circuit is converted to heat and is lost. If a perfect conductor could be found, it would carry current without such a heat loss. However, even the best of conductors has some resistance, so there will always be some heat loss in electric circuits. Note that the main factors in the conversion of electric power to heat are the current and the resistance. The power formula $P = I^2R$ will always give true power values.

Not all power in electricity is converted into heat. In a radio receiver, some power is converted into sound waves, but some heat will be developed in the radio in the process of this conversion. With transmitters, power is changed into radio waves in the air. When current flows through the resistance-wire filament of an electric light, the filament becomes so hot that it glows brightly and is also radiating energy in the form of light. The power formula will give the total amount of power being consumed. The light energy is a small percent of the total. Fluorescent lights utilize a method of producing light energy other than by a hot filament. Less heat is required to produce the same amount of light. Such lights are more efficient because of their lower percent of heat loss.

The basic unit of power measurement is the watt. For smaller quantities the *milliwatt* (0.001 W) may be used. For larger quantities the *kilowatt* (1000 W) may be used. Another unit of power is represented by 746 W, called a *horsepower*. If an electric motor were 100% efficient, 746 W of power fed to it would produce the equivalent of one mechanical horsepower of twisting force (*torque*).

The terms "power" and "energy" are sometimes used synonymously. Actually, these two terms do not mean the same thing. *Power* is the ability to do work and is measured in watts. *Energy* is usually computed by multiplying the amount of power by the length of *time* the power is used. One watt of power working for one second is known as a *watt-second*, or as a *joule* of energy.

If a 100-W lamp is turned on for 1 second (1 s), it uses 100 joules (100 J) of energy. During the time it is working, it is dissipating 100 W of power. If the light is left on for 10 s, it consumes 1000 J of energy, but while it is working it is still dissipating only 100 W of power. Electric power companies may produce power, but they sell energy. Instead of using wattseconds, they use the larger basic units, the *watthour* (number of watts times the number of hours) or the *kilowatthour* (watts times hours divided by 1000). Every establishment buying electric energy has a kilowatthour meter measuring how many kilowatthours (kWh) of energy flow in the power lines. If electricity costs 5 cents per kilowatthour, a 1-kW lamp may be operated for 1 hour (1 h) for 5 cents, or a 100-W lamp may be operated for 10 h for the same cost.

Power, by the formula $P = VI$, includes time, since the ampere, I in the formula, is a coulomb per second. An ampere can be expressed as

$$A = \frac{Q}{T}$$

where A = current (in amperes, A)
Q = electron quantity (in coulombs, C)
T = time (in seconds, s)

For example, if 10 C moves through a circuit in 2 s, the average current is 5 A.

If power equals volts times amperes and amperes equals Q/T, then power must equal volts times Q/T. In formula form,

$$P = VI = V\left(\frac{Q}{T}\right) = \frac{VQ}{T} = VQ/T$$

This formula tells us power is equal to volts times coulombs per second.

The unit of energy was given as the wattsecond, or joule, and it is usually computed as power times time. This brings up an interesting fact. If energy equals power times time and the power formula $P = VQ/T$ is multiplied by time to give energy (E_n), the equation for energy will then be

$$E_n = \frac{VQT}{T} = VQ$$

Since time cancels out, energy is something that may do work if given a chance but has no reference to time itself. Although normally measured for convenience as power times time, energy is timeless. The energy formula with the time canceled is $E_n = VQ$, or joules equals volt-coulombs. A power company actually sells volts of pressure times the number of coulombs it delivers.

Pressure times quantity, or *volt-coulombs*, equals the energy that is available. Pressure times movement, or *volt-amperes* (VI), equals work done, or power.

In problems involving energy, the wattsecond, watthour, or kilowatthour is used. For example, to determine the energy consumed by a radio receiver drawing 60 W of power for 20 h, power times time is employed, using hours as the time unit. In this case, 60 W times 20 h equals 1200 Wh of energy, or 1.2 kWh, or 4,320,000 Ws, or joules.

2-5 OTHER POWER FORMULAS

The power formula $P = VI$, or Watt's law, states in mathematical form: *The power is directly proportional to both voltage and current.* Since the current increases if the voltage increases, doubling the voltage will also double the current, and the power will increase fourfold.

The power formula $P = I^2R$ states in mathematical form: *The power is directly proportional to the resistance and also to the current squared.* If the current is doubled, the power dissipation is equal to 2 squared, or 4 times as much.

The power formula $P = V^2/R$ says: *The power is directly proportional to the voltage squared and inversely proportional to the resistance.* If the resistance is kept constant, doubling the voltage will produce 4 times the power.

From the three basic power formulas, other useful formulas involving power can be derived. From $P = VI$, it is possible to solve for V by dividing both sides of the equation by I:

$$P = VI \qquad \text{or} \qquad \frac{P}{(I)} = \frac{VI}{(I)} \qquad \text{or} \qquad V = \frac{P}{I}$$

where V = emf, in V
P = power, in W
I = current, in A

EXAMPLE: A resistor rated at 50 W and 100 mA (0.1 A) will stand how much voltage across it without becoming excessively hot? From $V = P/I$, the maximum voltage is equal to 50/0.1, or 500 V.

From the formula $P = VI$, when both sides are divided by V, the result is the formula

$$P = VI \qquad \text{or} \qquad \frac{P}{(V)} = \frac{VI}{(V)} \qquad \text{or} \qquad I = \frac{P}{V}$$

EXAMPLE: How much current will flow in a television set that is rated at 240 W when connected across 120 V? From the formula $I = P/V$, the current is equal to 240/120, or 2 A.

From the formula $P = I^2R$, dividing both sides by R and then taking the square root of both sides results in the formula

$$\frac{P}{(R)} = \frac{I^2R}{(R)} \quad \text{or} \quad \sqrt{\frac{P}{R}} = \sqrt{\frac{I^2R}{R}} \quad \text{or} \quad I = \sqrt{\frac{P}{R}}$$

EXAMPLE: What is the maximum rated current-carrying capacity of a resistor marked "5000 Ω, 200 W"? From the formula $I = \sqrt{P/R}$, the current is equal to $\sqrt{200/5000}$, or $\sqrt{0.04}$, or 0.2 A.

From the formula $P = I^2R$, dividing both sides by I^2 results in the formula

$$P = I^2R \quad \text{or} \quad \frac{P}{(I^2)} = \frac{I^2R}{(I^2)} \quad \text{or} \quad R = \frac{P}{I^2}$$

EXAMPLE: A radio receiver rated at 55 W draws 2 A from the line. The effective resistance is $R = P/I^2$, or $55/2^2$, or 13.75 Ω

From the formula $P = V^2/R$, multiplying both sides by R and then taking the square root of both sides results in the formula

$$P(R) = \frac{V^2(R)}{R} \quad \text{or} \quad \sqrt{V^2} = \sqrt{PR}$$

or $\quad V = \sqrt{PR}$

EXAMPLE: What is the maximum voltage that may be connected across a 10-W 1000-Ω resistor? From the formula $V = \sqrt{PR}$, the voltage is equal to $\sqrt{10(1000)}$, or $\sqrt{10,000}$, or 100 V.

From the formula $P = V^2/R$, cross-multiplying (or multiplying both sides by R) and then dividing both sides by P results in the formula

$$R = \frac{V^2}{P}$$

EXAMPLE: What is the resistance of a 3-W 6-V lamp? The resistance is equal to 6^2 divided by 3, or $36/3$, or 12 Ω.

2-6 POWER DISSIPATION IN RESISTORS

Resistors, whether they are carbon or wirewound, have a resistance and a power rating. The power rating indicates how much heat the resistor is capable of dissipating under normal circumstances. If cooled by passing air across it, the resistor may be capable of considerably greater power dissipation. If enclosed in an unventilated area, it may become excessively hot and burn out when dissipating its rated power or less. Usually, the power rating required is computed by one of the power formulas and a resistor of twice the computed power rating is employed. If it is computed that a resistor in an operating circuit must be capable of dissipating at least 5 W, a 10-W resistor would be used. If the resistor is tightly enclosed, a rating three or four times the computed value might be required.

Instead of rating resistors by power dissipation, they might be rated in current-carrying ability. Some wirewound resistors carry resistance, power, and current ratings. When the current rating is not given, the rearranged power formula $I = \sqrt{P/R}$ can be used. For example, a 100-Ω 1-W resistor will carry safely

$$I = \sqrt{\frac{P}{R}} = \sqrt{\frac{1}{100}} = 0.1A \text{ or } 100 \text{ mA}$$

▌ Test your understanding; answer these checkup questions.

1. A receiver across a 120-V line has 0.75 A flowing through it. How much power is being used? _____

2. A 420-Ω resistor has 30 mA flowing in it. How much power is producing heat in the resistor? _____

3. A 120-V electric iron has 36 Ω resistance. How much heating power does it produce? _____
4. How many milliwatts are there in a kilowatt? _____

5. How much energy is used in 30 days by a 120-V clock having an internal resistance of 5000 Ω? _____

6. If electricity costs 4 cents a kilowatthour, how much does it cost to operate a TV set 24 h if it draws 2 A when across 120 V? _____
7. If 0.5 C passes one point in a circuit in 0.01 s, what is the average current value? _____
8. How much power is developed when 100 V forces 80 C through a point in a circuit in 0.5 s? _____ How much energy is this?

9. How much energy is being used by a transmitting station drawing 40 A frpm a 440-V power line in 2 h of operation? _____
10. Across how many volts must a 600-W heater be connected if it is drawing 5 A? _____
11. A 25-W emergency light draws how many amperes from a 6.3-V storage battery? _____
12. How much current is a 1-W 2500-Ω resistor capable of carrying without overheating? _____
13. A 100-W lamp draws 0.9 A. What is the resistance value? _____

14. A 25-W 500-Ω resistor can be safely connected across how much voltage? _____
15. How many ohms of resistance does the ordinary 75-W house lamp have when operating at its rated 120 V? _____
16. A 1700-Ω resistor must carry 8 mA. What should the power rating of the resistor be? _____

2-7 FUSES

To protect circuits from damage caused by accidental overloads or *short circuits*, fuses are installed in a series with one of the lines carrying current from the source to the load. A fuse should carry current with little or no voltage loss to the circuit. This requires a fuse with low resistance in circuits carrying high current. Low-current circuits may have fuses with several ohms of resistance.

A fuse is placed in such a position that all the current flowing through the circuit to be protected must flow through the fuse (Fig. 2-3). If a short

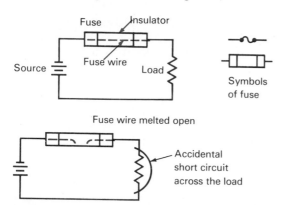

FIG. 2-3 The fuse wire melts open if excessive current is made to flow in the circuit.

circuit develops across the load, the current from the source flows through the fuse and the low-resistance short circuit. This produces a heavy enough current to melt the special low-melting-point fuse wire, interrupting the excessive current flow and protecting the source from damage. Without a fuse, a "short" may cause the connecting wires of a circuit to become hot enough to ignite the insulation on the wires and start a fire.

Fuses are rated for current-carrying ability and also for maximum voltage of the circuit in which they are used. High-current fuses use relatively heavy fuse wire and are recognizable by their relatively large diameter. Low-current fuses may be made quite small with delicate fuse wire. Low-

voltage-circuit fuses may be physically short, but fuses for high-voltage circuits are long. This prevents any high voltage that appears across the burned-out section of the fuse from jumping the gap and striking an arc of current, which would prevent the fuse from "open-circuiting." The greater length results in better insulating properties of the fuse *after* it burns out.

Fuses are available in such ratings as 100 A, 30 A, 15 A, 1 A, ½ A, ¼ A, and 1/32 A. "Slow-blow fuses" are made to withstand short-duration overloads due to current surges, such as occur when a motor is started, but they will burn out after a short interval of time if the current does not decrease. They are not suitable when fast protection is important. A "chemical" fuse is an ultra-slow-blow fuse that will carry a slight overload current value for an appreciable time before opening.

2-8 MEASURING DEVICES

Six types of measuring devices in general use are the voltmeter, ammeter, wattmeter, watthour meter, ohmmeter, and oscilloscope. These are explained in greater detail in Chap. 13. At this point meters will only be shown in their usual positions in simple electric circuits. The symbol for a meter is a circle with a letter in it to indicate the type of meter.

A *voltmeter* measures the difference of potential, or the emf, *across* a circuit. It is always connected across the difference of potential to be measured (Fig. 2-4).

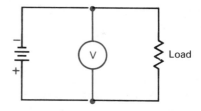

FIG. 2-4 A voltmeter is always connected *across* the circuit to be measured.

An *ammeter* is connected *in series* with the line carrying the current to be measured. It indicates the number of electrons, in coulombs per second, flowing through it. The meter in Fig. 2-5 is measuring the total current of the circuit. To measure the current in any one of the three branches, the meter must be moved to points *X, Y,* or *Z* shown. Since an ampere is a relatively large current value in electronic circuits, milliammeters or microammeters

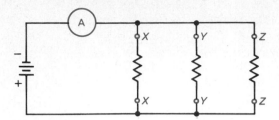

FIG. 2-5 Ammeters are connected in series with a circuit. One ammeter reads the sum of three branch currents.

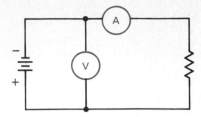

FIG. 2-7 Voltmeter times ammeter readings give the power in watts demanded by the load.

are frequently used. (A milliampere is 0.001 A, and a microampere is 0.000001 A.)

A *wattmeter* measures power. It is a voltmeter and ammeter combined in such a way that it shows the product of the voltage and current on its scale. It is therefore connected across the difference of potential and also in series with the line carrying the current. The wattmeter may have three or four terminals, whereas most other meters have only two. Figure 2-6 shows a three-terminal wattmeter

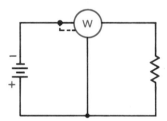

FIG. 2-6 A wattmeter is connected in series with and also across the load of a circuit.

with the necessary connection, if a fourth terminal is used, shown in dotted lines.

In many applications, a separate voltmeter and ammeter are used instead of a wattmeter (Fig. 2-7). The voltmeter value multiplied by the ammeter value gives the power value in watts.

A *watthour meter* measures electric energy. It is actually an electric motor geared to an indicator similar to the hand of a clock. How far the indicator rotates depends on the current flowing, the voltage across the circuit, and the length of time the

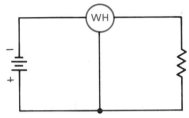

FIG. 2-8 A watthour meter is also connected in series with and across the load.

current flows. Figure 2-8 shows a three-terminal watthour meter connected in a simple circuit.

An *ohmmeter* is a sensitive ammeter plus an internal battery, both contained in a small plastic case. It can be used only when the resistance being measured is in a dead circuit, that is, when the resistance being measured has no current flowing through it. The ohmmeter is usually a portable meter with flexible leads (Fig. 2-9).

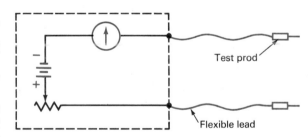

FIG. 2-9 An ohmmeter consists of a milliammeter calibrated in ohms, a battery, a calibrating resistor, and test prods.

An *oscilloscope* is an electronic device with a small TV-like picture tube in it (Chap. 13). It can show the shape of voltages with respect to time as they vary in strength, or even if they reverse in direction. It is usually connected across circuits as a voltmeter is.

ANSWERS TO CHECKUP QUIZ ON PAGE 26

1. (90 W) **2.** (0.378 W) **3.** (400 W) **4.** (1,000,000 mW) **5.** (2074 Wh) **6.** (23 cents) **7.** (50 A) **8.** (16,000 W)(8000 J, or Ws) **9.** (35.2 kWh) **10.** (120 V) **11.** (3.97 A) **12.** (0.02 A) **13.** (123.4 Ω) **14.** (112 V) **15.** (192 Ω at 120 V) **16.** (2 × 0.109, or ¼ W)

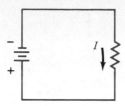

FIG. 2-10 A simple circuit has only one load.

2-9 TYPES OF CIRCUITS

A source of emf with a single load, as shown in Fig. 2-10, is a *simple circuit*.

A source of emf with two or more loads connected across it in such a way that there is only one current path through the whole circuit, as in Fig. 2-11, is called a *series circuit*.

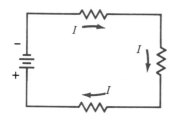

FIG. 2-11 A series circuit has two or more loads in series.

A source of emf with two or more loads connected across it in such a way that each branch has only its own current flowing through it, as in Fig. 2-12, is termed a *parallel circuit*.

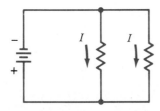

FIG. 2-12 A parallel circuit has loads in parallel.

When speaking of paralleling resistors or loads, it may be said that they are connected in *shunt*. If something is connected across the terminals of something else, it may be said that the first is shunted across the second.

When a group of loads is connected in a more or less mixed and complex group of series and parallel circuits, the whole group may be said to be connected in *series-parallel*.

2-10 SERIES CIRCUITS

The same amount of current flows through all parts of a series circuit. More electrons can never flow into a resistance than flow out the other end. The electrons forming the current may lose energy in the form of heat while moving through a resistor, but electrons themselves are not lost.

In working with series circuits, the sums of the unit values are simply added. For example, Fig. 2-13a shows a series-resistance diagram with a total resistance of 150 Ω. Figure 2-13b shows a series of batteries with a total voltage of 600 V.

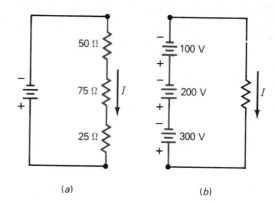

(a)   (b)

FIG. 2-13 (a) These resistors in series total 150 Ω. (b) These batteries in series total 600 V.

Resistors can be connected in series when it is desired to have a greater resistance and thereby a smaller current.

Batteries are connected in series, positives to negatives (Fig. 2-13b), when it is desired to produce maximum voltage. However, a battery can develop only a certain value of current. By connecting batteries in series, the sum of all the voltages of all the batteries can be obtained, but the maximum current possible through the circuit is no greater than the greatest current that the weakest battery can deliver. If one of the batteries is capable of passing 1 A through it, another 2 A, and the third 3 A, the maximum current the three batteries in series can pass without damage to any is only 1 A. This series combination will result in 600 V with a maximum current capability of 1 A. According to Ohm's law ($I = V/R$), if the resistor across the three batteries has 600 Ω resistance, the current in the circuit will be 1 A. If less than 600 Ω is used, the weakest battery will be overworked, may overheat, and the voltage across its terminals will drop.

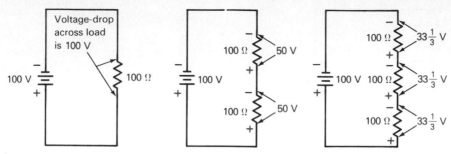

FIG. 2-14 Voltage-drops in series circuits.

In Fig. 2-14, the voltage of all batteries is 100 V, and each of the resistors has a value of 100 Ω. Note how the voltage is distributed in the three circuits and how the sum of the voltage-drops across the resistors always equals the battery voltage of 100 V. Can you see why a series circuit might be considered a "voltage-divider" circuit?

If the emf across each resistor is considered to be a *voltage-drop*, or loss of voltage, then the sum of all the voltage-drops (considered as minus values) around the circuit, when added to the source voltage (a plus value), gives an algebraic sum (a plus and a minus sum) of zero volts in the circuit. This is stated in Kirchhoff's voltage law: *The algebraic sum of all the voltages plus the voltage-drops in a series circuit is zero.*

To find the total resistance of a group of series resistors, the formula is simply

$$R_{total} = R_1 + R_2 + R_3 + \cdots$$

If all the resistors are equal in value, the value of any one can be multiplied by the number of resistors to give the total. Five 25-Ω resistors in a series present 5(25), or 125 Ω.

2-11 OHM'S LAW IN SERIES CIRCUITS

The use of Ohm's law was previously discussed as it applied to a circuit made up of a source of emf and a load. When more complex circuits are used, Ohm's law may still be employed, but additional factors must be considered. Complex circuits, in this instance, mean parallel, series, or series-parallel types.

There are three important rules regarding the use of Ohm's law in working problems:

VOLTAGE RULE. It is possible to determine the voltage across any particular known R in a group of resistances, if the *current through that particular R* is known, by $V = IR$.

CURRENT RULE. It is possible to determine the current through any particular known R in a group of resistances, if the *voltage across that particular R* is known, by $I = V/R$.

RESISTANCE RULE. It is possible to determine the resistance of any one part of a circuit, if the voltage across *that part* and the current flowing through *that part* are known, by $R = V/I$.

These seem simple, but it is difficult when first starting electronics to determine where Ohm's law can and cannot be used. Consider two impossible problems: In Fig. 2-15 it is impossible to find the

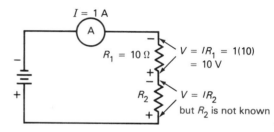

FIG. 2-15 Insufficient data given to compute unknown circuit values.

voltage across the second resistor because none of the three rules can be applied to any one part of the circuit.

In Fig. 2-16 it is impossible to determine R_1 because none of the rules can be applied to any one part of the circuit. The source voltage is not across R_2 but across R_1 and R_2 in the series.

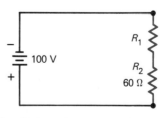

FIG. 2-16 Insufficient data given to compute unknown circuit values.

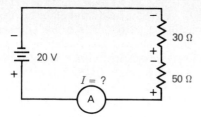

FIG. 2-17 Adequate data given to compute circuit values.

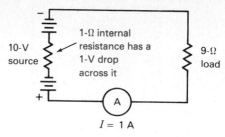

FIG. 2-18 In some cases the internal resistance in the source may have to be considered.

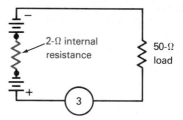

FIG. 2-19 This source has an output of 150 V with the 50-Ω load, but 156 V with no load.

A problem that can be computed is shown in Fig. 2-17. Being a series circuit, the current value in all parts must be equal. Since the total resistance is equal to the sum of all the resistors, the total resistance across the battery is 30 + 50, or 80 Ω. (The source "sees an 80-Ω load.") Once the total resistance is known, the current value in the circuit can be found by Ohm's law:

$$I = \frac{V}{R} = \frac{20}{80} = 0.25 \text{ A}$$

The voltage-drop developed across either of the resistances by the current flowing through it can be found by applying Ohm's law to that part. For the 30-Ω resistor,

$$V = IR = 0.25(30) = 7.5 \text{ V}$$

For the 50-Ω resistor,

$$V = IR = 0.25(50) = 12.5 \text{ V}$$

The voltage-drops across the resistances are considered voltage losses. The voltage of the source is considered a gain voltage. The sum of −7.5, −12.5, and +20 is zero volts (Kirchhoff's voltage law). The sum of the voltage-drops around a series circuit must always equal the source emf.

If the source has internal resistance, the voltage loss across the internal resistance must be considered. Suppose a 10-V source has 1 Ω internal resistance and is connected across a 9-Ω load (Fig. 2-18). With 10 Ω and 10 V, the current is 1 A, which through the 1-Ω internal resistance produces a 1-V loss. Therefore the 10-V source actually produces only 9 V across its terminals and across the 9-Ω resistance load. The other volt of pressure is lost inside the source. If no current is flowing through the source, however, no *voltage-drop* is developed across the internal resistance and the terminal voltage is 10 V. This is important to understand!

If a circuit carrying 3 A has an internal resistance of 2 Ω in the source and a 50-Ω load (Fig. 2-19),

what is the terminal voltage of the source? This problem can be analyzed and worked two ways. One solution is to consider that the terminal voltage of the source is the voltage-drop across the 50-Ω resistor, since the two are directly connected. The voltage across the 50-Ω resistor when 3 A flows through it is equal to V = IR, or 3(50), or 150 V.

Another solution for this problem is to consider the total resistance in the circuit as equal to 50 Ω plus 2 Ω, or 52 Ω. The current is 3 A. From Ohm's law, the total voltage in the circuit is V = IR, or 3(52), or 156 V. The voltage-drop developed across the 2-Ω internal resistance is V = IR, or 3(2), or 6 V. This 6 V does not appear outside the source and must be subtracted from the total voltage present in the circuit. This gives 156 V less 6 V, or 150 V across the terminals of the source. On the other hand, if the 50-Ω resistor is disconnected, the source has an unloaded terminal voltage of 156 V.

Sometimes internal resistance may be disregarded. A 6-V storage battery has an internal resistance of 0.01 Ω. What current will flow when a 3-W 6-V lamp is connected across it, as in Fig. 2-20? It is assumed that the storage battery has 6 V with no load. The resistance of the lamp is determined by using the rearranged power formula $R = V^2/P$, or $6^2/3$, or 12 Ω. The total resistance of the circuit is the load plus the internal resistance of

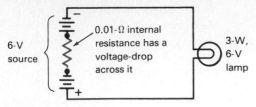

FIG. 2-20 Internal resistance in this circuit is negligible in comparison with the load resistance value.

the source, or 12.01 Ω. The current, according to Ohm's law, is $I = V/R$, or 6/12.01, or 0.4996 A. By most meters this would read 0.5 A. (The loss of voltage across the internal resistance, called *internal drop*, is only about 0.004996 V.)

It is possible to solve for missing values in some series problems. In Fig. 2-21, what is the total

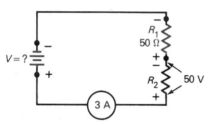

FIG. 2-21 The sum of the voltage-drops across R_1 and R_2 equals the source voltage.

impressed emf across the circuit? Two factors regarding R_1 are known: $R_1 = 50$ Ω, and $I = 3$ A. The voltage across it can therefore be determined as $V = IR$, or 3(50), or 150 V. This voltage is in series with the 50 V across R_2, resulting in a total of 200 V across the two resistors, which must therefore be the source voltage.

In Fig. 2-22, a vacuum tube has a filament rated at 0.25 A and 5 V and is to be operated from a 6-V battery. What is the value of the necessary series resistor? The filament can be considered a resistor

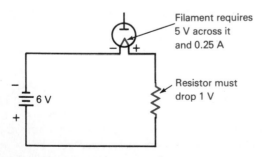

FIG. 2-22 Circuit required to drop the voltage of the source to a usable value across a load.

requiring 0.25 A flowing through it to develop a 5-V drop across it. A 6-V battery directly across the filament will cause too much current to flow. To prevent this, a resistance with a value sufficient to drop 1 V across it when 0.25 A flows is required. According to Ohm's law, its resistance should be $R = V/I$, or 1/0.25, or 4 Ω. The minimum power dissipation for this resistor is equal to $P = VI$, or 1(0.25), or ¼ W. In practice, a 4-Ω ½-W resistor should be used. A 1- or 2-W resistor would be satisfactory but would be physically larger and more expensive.

A keying relay coil (Fig. 2-23) has a resistance of 500 Ω and is designed to operate on 125 mA. If

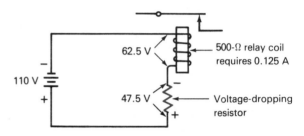

FIG. 2-23 Voltage-dropping resistor required to prevent excessive voltage across a relay coil.

the relay is to operate from a 110-V dc source what value of resistance should be connected in series with the relay coil? The relay coil can be considered as a resistor in this problem. A current value of 125 mA is equal to 125/1000, or 0.125 A. The voltage required to produce this current value through the coil is found by Ohm's law $V = IR$, or 0.125(500), or 62.5 V. The emf available is 110 V, or 47.5 V too much. To drop 47.5 V, a resistance of $R = V/I$, or 47.5/0.125, or 380 Ω, is required. The minimum power dissipation is $P = VI$, or 47.5(0.125), or 5.94 W. Any 380-Ω resistor with a power rating of 10 W or more would be satisfactory.

If two 10-W 500-Ω resistors are in series, what are the power-dissipation capabilities of the combination? Since these are similar resistors, they will stand the same amount of current. With the maximum current for their rating, or $I = \sqrt{P/R}$, or 0.141 A, each will produce 10 W of heat, resulting in 20-W maximum safe dissipation from the two.

If a 10-W 500-Ω resistor and a 20-W 500-Ω resistor are in series, the 10-W resistor will have a maximum safe current of 0.141 A, but the 20-W resistor will have a maximum safe current of $I = \sqrt{P/R}$, or 0.2 A. However, the limiting factor in the circuit is the 0.141 A. The 10-W resistor will dissipate 10 W, but the 20-W resistor will be held to 10 W of dissipation because only 0.141 A should be flowing through it. The total safe power dissipation for the two resistors in series is only 20 W.

When resistors are connected in series, the maximum safe current for each must be determined before the limiting current can be judged. The highest wattage rating does not always indicate the greatest safe current rating. For example, a 20-W 2000-Ω resistor has a maximum current value of only 0.1 A, which is lower than the 0.141-A capability of a 10-W 500-Ω resistor.

> Test your understanding; answer these checkup questions.

1. Four 37.5-Ω resistors are in series across 120 V. What is the voltage-drop across one of them? _____
2. A 30-Ω, a 60-Ω, and a 150-Ω resistor are in series across a 24-V battery. What voltage-drop appears across the 60-Ω resistor? _____ How much current flows in the 150-Ω resistor? _____
3. A 50-Ω, a 90-Ω, and an unknown-value resistor are in series across a 60-V generator with $^1/_3$ A flowing through it. What is the voltage-drop across the unknown resistor? _____ What is its resistance? _____
4. A 12.6-V automobile battery is across a 1.5-Ω headlight lamp. If the battery has 0.14 Ω internal resistance, what is the current value? _____ What is the loaded terminal voltage of the battery? _____
5. A 500-Ω relay coil operates with 0.2 A. What resistance must be connected in series with it if operation from a 110-V line is required? _____
6. A 6-V mobile receiver draws 36 W. If it is to operate from a 12-V battery, what value resistor must be used in series with it to maintain the 36-W receiver dissipation? _____
7. A 5000-Ω 20-W resistor and a 1000-Ω 5-W resistor are in series. What is the maximum voltage that can be applied across them without exceeding the wattage rating of either resistor? _____
8. A 6.3-V tube filament (consider it a resistor) requires 300 mA. What resistance is required in series with it if operation is across a 110-V line? _____ What power would be dissipated in heating the resistor? _____

2-12 CONDUCTANCE

An interesting point regarding the study of electricity is the necessity, at times, of observing the same thing from two viewpoints for a better understanding of the whole. One example of this is the resistance versus the conductance of a circuit.

In an operating electric circuit, there must always be a source of emf and a load. In the simplest circuit, the load may be a single resistor. As explained before, the greater the resistance in the load, the less the current in the circuit. However, the very fact that some current is flowing indicates that the resistance is not infinite (immeasurable, or endless) but is actually a conductor to some degree. It may be a very poor conductor, or it may be a fairly good conductor. In any event, the greater its conducting ability, or *conductance*, the less its resistance value. Conversely, the less its conductance, the greater its resistance. Conductance and resistance refer to the same thing but from opposite viewpoints. They are said to be *reciprocals* of each other.

The meaning of "reciprocal" is, roughly, "mathematically opposite." In stating that one thing is the reciprocal of another, or that it varies inversely as the other, the two are placed on opposite sides of an equals sign and one of them expressed as a fraction by putting a 1 over it.

R stands for resistance in ohms. G stands for conductance in siemens, S (formerly *mho*, which is *ohm* spelled backward, symbol $\mho$). Since R and G have opposite meanings, they may be expressed:

$$R = \frac{1}{G} \quad \text{or} \quad G = \frac{1}{R}$$

If a resistance has a value of 2 Ω, then its conductance value is 1 over 2, or ½ S or $\mho$. If the R value is raised to 3 Ω, the conductance value becomes ⅓ S. As the resistance is increased from 2 to 3 Ω, the conductance value decreases from ½ to ⅓ S.

Since $R = 1/G$, Ohm's law can be expressed in terms of conductance by using $1/G$ in place of R in the three formulas:

$$V = IR = I\left(\frac{1}{G}\right) \quad \text{or} \quad V = \frac{I}{G}$$

$$I = \frac{V}{R} = \frac{V}{1/G} = V\left(\frac{G}{1}\right) \quad \text{or} \quad I = VG$$

$$R = \frac{V}{I} \quad \text{or} \quad \frac{1}{G} = \frac{V}{I} \quad \text{or} \quad GV = I \quad \text{or} \quad G = \frac{I}{V}$$

EXAMPLE: What is the conductance of a circuit if 6 A flows when 12-V dc is applied?

$$G = \frac{I}{V} = \frac{6}{12} = \frac{1}{2} \text{ S (or mho)}$$

2-13 PARALLEL RESISTANCES

Conductance is a fitting preliminary to the subject of parallel resistors. Understanding conductance makes it possible to see that when two 10-Ω resistors are connected in parallel across a source of emf, as in Fig. 2-24, the conductance of the circuit

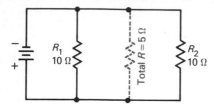

FIG. 2-24 Two 10-Ω resistors in parallel present an equivalent resistance of 5 Ω to the source.

is greater and therefore the total resistance must be less. Two parallel 10-Ω resistors provide a conductance value twice that of one resistor and therefore a resistance value of one-half of 10, or 5 Ω (not 20 Ω, as in series circuits). This apparent adding of resistances to a circuit and obtaining a resultant resistance less than any of the resistances may be confusing unless it is seen from the conductance viewpoint.

The total conductance G_t of a circuit is equal to the sum of all the conductances in parallel across the circuit, or

$$G_t = G_1 + G_2 + G_3 + \cdots$$

Since any single conductance value is equal to the reciprocal of its resistance value, the formula for the total conductance G_t of a parallel circuit is also

$$G_t = \frac{1}{R_1} + \frac{1}{R_2} + \frac{1}{R_3} + \cdots$$

Substituting the reciprocal of the total resistance R_t for the total conductance gives the formula

$$\frac{1}{R_t} = \frac{1}{R_1} + \frac{1}{R_2} + \frac{1}{R_3} + \cdots$$

If this equation is made into a pair of fractions by placing a 1 over both sides, it becomes

$$\frac{1}{1/R_t} = \frac{1}{1/R_1 + 1/R_2 + 1/R_3 + \cdots}$$

The rule for simplifying compound fractions is: Invert the lower fraction and multiply it by the upper. In this case the left-hand part of the equation can be simplified to

$$\frac{1}{1/R_t} = 1\left(\frac{R_t}{1}\right) = R_t$$

The complete formula to solve for the total resistance of any number of parallel resistances is

$$R_t = \frac{1}{1/R_1 + 1/R_2 + 1/R_3 + \cdots}$$

This formula can be solved in one of two ways: by using fractions or by using decimals.

EXAMPLE: If 5-, 3-, and 15-Ω resistors are in parallel, what is the total resistance?

$$R_t = \frac{1}{1/5 + 1/3 + 1/15}$$

To add the fractions $1/5$, $1/3$, and $1/15$, they must be expressed in their lowest common denominator: $1/5$ equals $3/15$, and $1/3$ equals $5/15$. Thus

$$R_t = \frac{1}{3/15 + 5/15 + 1/15}$$

$$= \frac{1}{9/15} = 1(15/9) = 15/9 = 1\,2/3\ \Omega$$

The problem can be solved by expressing the fractions $1/5$, $1/3$, and $1/15$ in their decimal equivalents by dividing: 5 into 1 (0.2); then 3 into 1 (0.3333); then 15 into 1 (0.0667); then substituting in the formula

$$R_t = \frac{1}{1/5 + 1/3 + 1/15}$$

$$= \frac{1}{0.2 + 0.3333 + 0.0667}$$

$$= \frac{1}{0.6} = 1.67 = 1\,2/3$$

When there are only two parallel resistances, the formula can be rearranged to read

$$R_t = \frac{R_1 R_2}{R_1 + R_2}$$

EXAMPLE: What is the total resistance of a parallel circuit consisting of one branch of 10-Ω resistance and one branch of 25-Ω resistance?

$$R_t = \frac{R_1 R_2}{R_1 + R_2} = \frac{10(25)}{10 + 25} = \frac{250}{35} = 7.14\ \Omega$$

If there are three resistances in parallel, the total of two of them can be computed by this formula and the answer considered as R_1. By using this resistance value and the third resistance as R_2, the formula can be used to solve for the total of the three resistances. The same procedure can be employed to determine the value of any number of parallel resistances.

Another formula to compute three resistors in parallel is

$$R_t = \frac{R_1 R_2 R_3}{R_1 R_2 + R_2 R_3 + R_1 R_3}$$

When all the resistances in parallel are equal in value, such as five 100-Ω resistances, the total resistance is most easily determined by dividing the resistance value of one resistor by the number of resistors. Five 100-Ω resistors in parallel present 100/5, or 20 Ω to the source.

A quick check on answers to problems involving parallel resistances is possible by noting that the answer must always be a lower value than the lowest of the parallel resistances. Also, when one resistance is about 10 times another and the two are in parallel, the total resistance will be about 10% less than the lower. If one resistance is 100 times a second resistance, it may often be possible to disregard the first entirely, since it will affect the total resistance by less than 1%.

When two 10-W 500-Ω resistors are in parallel, what are the total power-dissipation capabilities of the combination? Since they have similar power and resistance ratings, they will stand the same maximum value of voltage across them, $V = \sqrt{PR}$, or $\sqrt{10(500)}$, or 70.7 V. With the two resistors in parallel across 70.7 V, each will dissipate 10 W, giving a total 20 W of heat.

Since two parallel resistors across a source split the source current between them, can you see how a parallel circuit might be considered a "current divider"?

In parallel circuits the controlling and limiting factor is the voltage, whereas in series circuits it is the current. With parallel resistors it is necessary to determine what maximum voltage each will stand to produce enough current through it to make it dissipate its maximum rated power. The lowest maximum voltage that a group of resistors will stand determines the highest voltage that may be applied across the parallel group.

EXAMPLE: A 10,000-Ω 100-W, a 40,000-Ω 50-W, and a 5000-Ω 10-W resistor are connected in paral-

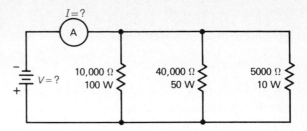

FIG. 2-25 To determine the maximum allowable current, the maximum safe source voltage must first be found.

lel (Fig. 2-25). What is the maximum voltage that may be applied across the circuit? What is the maximum total value of current through the parallel combination which will not exceed the wattage rating of any of the resistors?

For the 10,000-Ω 100-W resistor R_1:

$$V = \sqrt{PR_1} = \sqrt{100(10,000)}$$
$$= \sqrt{1\ 00\ 00\ 00} = 1000 \text{ V}$$

For the 40,000-Ω 50-W resistor R_2:

$$V = \sqrt{PR_2} = \sqrt{50(40,000)}$$
$$= \sqrt{2\ 00\ 00\ 00} = 1414 \text{ V}$$

For the 5000-Ω 10-W resistor R_3:

$$V = \sqrt{PR_3} = \sqrt{10(5000)}$$
$$= \sqrt{5\ 00\ 00} = 224 \text{ V}$$

The maximum allowable voltage across the circuit must not exceed 224 V, or the 10-W resistor will be drawing more than its rated current. With 224 V the 10,000-Ω resistor draws

$$I = \frac{V}{R_1} = \frac{224}{10,000} = 0.0224 \text{ A}$$

With 224 V the 40,000-Ω resistor draws

$$I = \frac{V}{R_2} = \frac{224}{40,000} = 0.0056 \text{ A}$$

With 224 V the 5000-Ω resistor draws

$$I = \frac{V}{R_3} = \frac{224}{5000} = 0.0448 \text{ A}$$

The total current value in the circuit is the sum of the three separate branch currents, or

$$I_t = IR_1 + IR_2 + IR_3 = 0.0728 \text{ A}$$

Kirchhoff's current law states: *The total current entering a point in a circuit will equal the total current leaving that point.* This is another way of saying that the sum of currents to a parallel group equals the current from the source.

2-14 SERIES AND PARALLEL BATTERIES

If two 9-V batteries, each capable of 0.1 A maximum safe current through them are connected in series, as shown in Fig. 2-26, the circuit will be capable of producing 18 V at 0.1 A and therefore 1.8 W of *power output.*

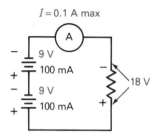

FIG. 2-26 Two 9-V 100-mA batteries in series are capable of 18 V at 100 mA.

The same two 9-V batteries connected in parallel (Fig. 2-27) are capable of only 9-V output, but each allows 0.1 A to flow through it, permitting the load to draw a total of 0.2 A of current from the two. Maximum safe *power output* is still 1.8 W.

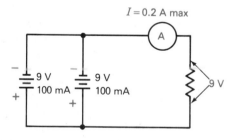

FIG. 2-27 Two 9-V 100-mA batteries in parallel produce 9 V at a maximum of 200 mA.

For maximum voltage output, batteries must be connected in series, negative to positive, as shown.

For maximum current output, batteries must be connected in parallel, but all such batteries must have the same voltage. A 9-V battery cannot be paralleled with a 12-V battery, or the higher-voltage battery will soon run down, discharging through the lower-voltage battery. Care must be taken to connect negative terminal to negative and positive terminal to positive when paralleling batteries.

Test your understanding; answer these checkup questions.

1. What is the conductance of a circuit having two 300-Ω resistors and one 500-Ω in parallel? _____
2. What is the conductance of a circuit having 250 V and a current of 50 mA? _____
3. What is the lowest common denominator of the fractions $^2/_{30}$, $^7/_{60}$, $^{11}/_{90}$? _____ Of $^8/_{25}$, $^9/_{15}$, and $^3/_{10}$? _____
4. What is the effective resistance of three parallel resistors of 1000 Ω, 2000 Ω, and 3000 Ω? _____
5. A 240-Ω and a 180-Ω resistor are in parallel across a 100-V source. What is the total circuit current? _____
6. A 55-Ω resistor and a 23-Ω resistor are in parallel. The current through the source is 2.5 A. What is the source voltage? _____
7. A 4000-Ω resistor and a 3672-Ω resistor are in parallel across a 500-V supply. How much current flows in the 4000-Ω resistor? _____ How much flows in the 4000-Ω branch if the 3672-Ω branch is disconnected? _____
8. A 75-Ω resistor and a 100-Ω resistor are in parallel. The total current through them is 3 A. How much current will flow in the 100-Ω resistor if the 75-Ω resistor is disconnected? _____
9. A 400-Ω 10-W resistor and a 1500-Ω 50-W resistor are in parallel. What is the maximum voltage that can be applied across this circuit without exceeding the wattage rating of either resistor? _____ What is the maximum total current that can flow in the combination and not exceed the wattage rating of either resistor? _____
10. A 12.6-V tube filament and a 6.3-V 0.3-A tube filament with a series resistor are in parallel across 12.6 V. What is the value of the required resistor? _____
11. Draw a diagram by which three resistors of equal value produce a total resistance of one-third of one unit. _____
12. A power company charges 7 cents per kWh. What is the cost of operating three 120-V lamp bulbs in parallel for 24 h if each has an internal resistance of 100 Ω? _____

2-15 COMPLEX DC CIRCUITS

There are countless circuit configurations that involve resistors in series and in parallel. To attempt to give examples covering all is obviously impossible, but by applying the basic principles involved in series and in parallel circuits it should be possible to solve most circuit problems. Remember: (1) All parts connected in series have the same current flowing through them. (2) All parts connected in parallel have the same voltage across them.

A relatively complex looking circuit is shown in Fig. 2-28. By examination it can be rationalized into a single resistance value. Resistors R_1 and R_2

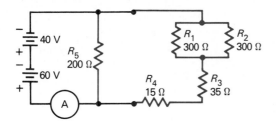

FIG. 2-28 An easily solved, relatively complex circuit.

are connected in parallel. Two 300-Ω resistors in parallel present 150 Ω of resistance. In series with this 150-Ω resistance are two other resistances, R_3 and R_4, totaling 50 Ω. To the source, this one branch presents a total of 150 plus 50, or 200 Ω.

There are two branches across the source, one made up of R_5 alone and the other of R_1, R_2, R_3, and R_4. Both branches present 200 Ω of resistance each and are in parallel with each other. Two 200-Ω parallel resistances present 100 Ω. Therefore, the source of 40 plus 60 V, or 100 V, sees a total load resistance of 100 Ω. With a total emf of 100 V and a total resistance of 100 Ω, the ammeter will read $I = V/R$, or 100/100, or 1 A.

The current through the series-parallel branch is $I = V/R$, or 100/200, or 0.5 A.

With 0.5 A flowing through it, R_4 will have a voltage-drop across it of $V = IR$, or 0.5(15), or 7.5 V. Similarly, the drop across R_3 will be 17.5 V. The voltage across R_1 will be the voltage-drop across R_1 and R_2 in parallel, or that produced by 0.5 A flowing through 150 Ω, or 75 V. Note that the voltage across both R_1 and R_2 is the same voltage, not two equal voltages. The current in R_1 (or R_2) is $I = V/R$, or 75/300, or 0.25 A.

Consider the circuit shown in Fig. 2-29. Two resistors of 30 and 15 Ω are connected in parallel. In series with this combination is a 12-Ω resistor, and in parallel with the total combination is connected a 22-Ω resistor. The total current through the combination is 5 A. What is the current value in the 15-Ω resistor? The current through neither

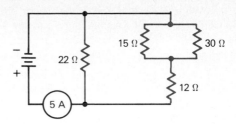

FIG. 2-29 To determine the current in the 15-Ω resistor, the source voltage must first be found.

branch is known, nor is the source voltage known. By determining the resistance of the series-parallel branch, however, both unknowns can be determined.

The 15-Ω and 30-Ω resistances in parallel equal 10 Ω. The total resistance of the series-parallel branch is 10 plus 12, or 22 Ω.

The resistances of the two branches are now known to be 22 Ω and 22 Ω, which in parallel equals 11 Ω. A 5-A current through 11 Ω computes to $V = IR$, or 5(11), or 55 V as the source voltage.

With 55 V across the 22-Ω series-parallel branch, the current through it is $I = V/R$, or 55/22, or 2.5 A. If 2.5 A flows through the 10-Ω parallel group, the voltage across it must be $V = IR$, or 2.5(10), or 25 V. With 25 V across the 15-Ω resistance, the current through it must be $I = V/R$, or 25/15, or 1.67 A.

In this case it is necessary to solve many circuit values in order to apply Ohm's law, $I = V/R$, to the single resistance in question. Had the source voltage been given, the first 22-Ω branch could have been disregarded, as it has no effect on the current flowing through the other branch.

Figure 2-30 is an example of an apparently highly complex circuit. Observation shows, how-

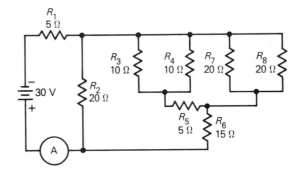

FIG. 2-30 A complex-looking circuit that may be computed mentally.

ever, that relatively simple steps can solve the total resistance. The current through the ammeter should be determinable without pencil and paper. Check these steps:

1. R_3 and R_4 can be computed as a parallel group.
2. The answer in step 1 can be computed with R_5 as a series group.
3. R_7 and R_8 can be computed as a parallel group.
4. The answers in steps 2 and 3 can be used to compute a parallel group.
5. The answer in step 4 can be added to R_6 as a series circuit.
6. The answer in step 5 can be computed with R_2 in parallel.
7. The answer in step 6 and R_1 can be computed in series, which is the effective resistance seen by the source.
8. The current is 2 A.

The diagrams in Fig. 2-31 illustrate four methods of connecting three similar resistors to produce vari-ous total resistances. Three 1-Ω resistors in series (diagram a) result in a total resistance of 3 Ω. Three 1-Ω resistors in parallel (b) result in a total resistance of ⅓ Ω. Two in series shunted across the third (c) result in ⅔ the resistance of one alone. Two parallel resistors in series with the third (d) result in a total resistance of 1½ times the resistance of one resistor alone.

The way a circuit is diagramed often names the circuit. For example, the series-parallel circuit of Fig. 2-32 might be termed a *pi-type circuit* after the Greek letter π (pi). It would be computed as a series-parallel circuit. The current in the 15-Ω leg would be $I = V/R$, or 60/15, or 4 A. The second

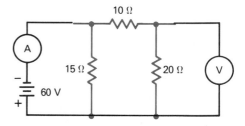

FIG. 2-32 A series-parallel group of resistors laid out as a π-network circuit.

leg consists of 10 Ω plus 20 Ω, or 30 Ω, of resistance, and would draw $I = V/R$, or 60/30, or 2 A. The total source current is 4 A plus 2 A, or 6 A.

With 2 A flowing through the 20-Ω resistor, the voltage-drop across it would be $V = IR$, or 2(20), or 40 V. Similarly, the voltage-drop would be 20 V across the 10-Ω resistor.

The power dissipation in the 15-Ω leg would be $P = VI$, or 60(4), or 240 W. The power dissipated by the 20-Ω resistor would be 2(40), or 80 W. The power converted into heat in the 10-Ω resistor would be 40 W. The total power dissipation of this pi-type circuit would be 240 + 80 + 40 = 360 W.

2-16 MATCHING LOAD TO SOURCE

Ranking high among electrical concepts is matching the load to the source. To produce maximum power in any load, it is necessary that the load resistance equal the internal resistance of the source.

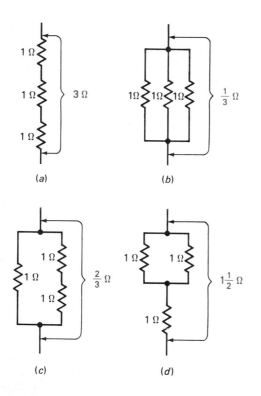

FIG. 2-31 Four possible circuit configurations for three similar resistances.

ANSWERS TO CHECKUP QUIZ ON PAGE 36
1. (0.00867 S) **2.** (0.0002 S) **3.** (180)(150) **4.** (545 Ω)
5. (0.973 A) **6.** (40.5 V) **7.** (125 mA)(0.125 A) **8.** (1.29 A)
9. (63.2 V)(0.201 A) **10.** (21 Ω) **11.** (Three in parallel)
12. (72.6 cents)

The diagram in Fig. 2-33 shows a 100-V source with 5-Ω internal resistance, R_1, and a load with a resistance, R_L, of 5 Ω also. With the load and internal resistance equal, the voltage in the circuit is 100 V, the total resistance is 10 Ω, and the current is

$$I = \frac{V}{R_t} = \frac{100}{10} = 10 \text{ A}$$

The power delivered to the load in this case is

$$P = I^2 R_{L_1} = 10^2(5) = 500 \text{ W}$$

However, if the load resistance mismatches the source and is 15 Ω, the current in the circuit is

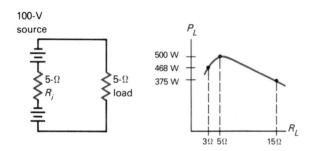

$$I = \frac{V}{R_t} = \frac{100}{20} = 5 \text{ A}$$

In this case, the power delivered to the load with mismatch is only

$$P = I^2 R_{L_2} = 5^2(15) = 375 \text{ W}$$

If the load resistance mismatches the source and is 3 Ω, the current in the circuit is

$$I = \frac{V}{R_t} = \frac{100}{8} = 12.5 \text{ A}$$

The power delivered to the load with this mismatch is only

$$P = I^2 R_{L_3} = 12.5^2(3) = 468.75 \text{ W}$$

These figures can also be used to demonstrate an important fact. When the load matches the source, half the power is dissipated in the source and half in the load. The total power in the example above is 1000 W. The *efficiency* of the circuit, the ratio of output to input powers, is equal to 500/1000, or 0.5, usually expressed as 50%.

With the 15-Ω load, the power dissipated in the source is 125 W. The total power is 125 W in the source and 375 W in the load, or 500 W. This results in an output efficiency of 375/500, or 75%.

With the 3-Ω load, the power dissipated in the source is 781.25 W. The total power is 781.25 W in the source and 468.75 W in the load, or 1250 W. This results in an output efficiency of 468.75/1250, or only 37.5%.

Matching source to load may produce maximum power output in the load, but mismatching with a higher resistance load gives better output efficiency. Theoretically, 100% efficiency can occur only with an infinite-resistance load, in which case there is no power being fed to the load.

When speaking of matching or mismatching a source to a load, the term *impedance* is usually employed rather than resistance. Impedance is a general term indicating the opposition to current flow particularly in circuits in which ac is flowing. The term may be used properly in place of resistance in dc circuits also. Impedance is discussed in Chap. 7, where ac circuits having inductance or capacitance are explained.

2-17 KIRCHHOFF'S LAWS

Application of Kirchhoff's laws can be used to solve complex circuits that are not readily solved by Ohm's law. First, consider the simple circuit shown in Fig. 2-34. The 10-V battery pressure will reduce

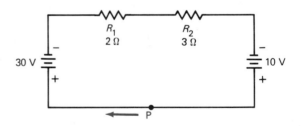

FIG. 2-34 Simple Kirchhoff circuit.

the 30-V battery pressure in the circuit to a total of only 20 V. With 20 V and 5 Ω, the circuit current will be $I = V/R$, or 20/5, or 4 A.

Consider this circuit another way. Starting at point P, and progressing clockwise around the circuit, we come first to a 30-V source of voltage, which can be given a + value because it adds voltage to the circuit. Next is a resistor across

FIG. 2-33 (a) Load matches source R. (b) Graph of mismatched and matched load powers.

which there will be a voltage-drop of $V = IR$ (a loss of voltage), which will therefore be given a negative or $-IR$ value. This is followed by another resistor with another $-IR$ value. The final component is a 10-V source, but since its pressure is reversed in direction from the higher 30-V source, it will also be given a -10-V value. We can therefore express this Kirchhoff's voltage law (the algebraic sum of all voltages plus voltage-drops equals zero) circuit as

$$30 + (-IR_1) + (-IR_2) + (-10) = 0$$

or, removing the parentheses

$$30 - IR_1 - IR_2 - 10 = 0$$

substituting known resistance values

$$30 - I2 - I3 - 10 = 0$$

or $\quad 30 - 2I - 3I - 10 = 0$

solving for I,

$$30 \qquad\qquad - 10 = 2I + 3I$$
$$20 = 5I$$
$$\frac{20}{5} = I = 4 \text{ A as above}$$

The two-loop circuit in Fig. 2-35 cannot be computed by simple Ohm's law alone. But it can be

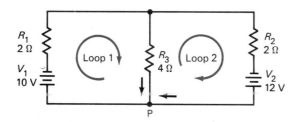

FIG. 2-35 Kirchhoff's-law circuit.

solved by computing the two loops separately. Starting at point P for loop 1 gives the following:

$$+10 - I_1R_1 - I_3R_3 = 0$$

substituting known values

$$+10 - 2I_1 - 4I_3 = 0$$
$$+10 \qquad\quad - 4I_3 = 2I_1$$

dividing both sides by 2

$$5 \qquad\quad - 2I_3 = I_1$$

The expression for the loop-1 current is $I_1 = 5 - 2I_3$. This is not yet a numerical value for I_1, just an algebraic expression for it.

The loop-2 current expression, regardless of current direction, is solved similarly, again starting at point P:

$$- I_3R_3 - I_2R_2 + 12 = 0$$
$$- 4I_3 - 2I_2 + 12 = 0$$
$$- 4I_3 \qquad\quad + 12 = 2I_2$$
$$- 2I_3 \qquad\quad + 6 = I_2$$

The expression for the loop-2 current is therefore $I_2 = 6 - 2I_3$.

Point P sees two currents approaching it, $5 - 2I_3$ and $6 - 2I_3$. Since both have the common I_3 term, it is possible to add the expressions for currents approaching and leaving point P (Kirchhoff's current law)

$$I_1 + \quad I_2 \qquad = I_3$$
$$5 - 2I_3 + 6 - 2I_3 = I_3$$
$$5 + \quad 6 \qquad = I_3 + 2I_3 + 2I_3$$
$$11 \qquad = 5I_3$$
$$\frac{11}{5} \qquad = I_3 = 2.2 \text{ A}$$

The R_3 current is now known to be 2.2 A. The voltage-drop across R_3 must be $V = IR$, or 2.2(4), or 8.8 V.

Since the current through R_1 was determined to be the expression $5 - 2I_3$, then I_{R_1} must be $5 - 2(2.2)$, or $5 - 4.4$, or 0.6 A. The voltage-drop across R_1 must be $V = IR$, or 0.6(2), or 1.2 V.

Similarly, the loop-2 current is $6 - 2I_3$, or $6 - 2(2.2)$, or $6 - 4.4$, or 1.6 A. The IR drop across R_2 is then 3.2 V.

The power dissipated by the resistors can be determined by any of the three power formulas.

2-18 THEVENIN'S THEOREM

Thevenin's theorem says, *Any complex resistive circuit may be replaced by an equivalent circuit consisting of a single voltage source in series with a single resistor.* Consider the voltage-divider and battery circuit in the two-terminal "black box" in Fig. 2-36. A voltmeter between points A and B would indicate 5 V because the two 5-kΩ resistors are equal in value. If a zero-resistance ammeter is connected from A to B, it should read $I = V/R_1$, or 10/5000, or 0.002 A. If the letter "d" (or Greek letter Δ) means "change in," then it may be said that when the ammeter is connected, the dV from A to B becomes $d5$ V (from 5 down to 0 V), and the dI is $d0.002$ A (from 0 A with no ammeter to 0.002 A with the ammeter connected). Using Ohm's law, $R = dV/dI$, or $d5/d0.002$, or 2500 Ω is the effective resistance. This means that as far as

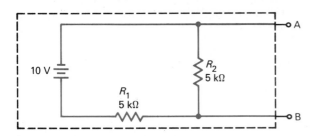

FIG. 2-36 Black-box circuit.

the outside world is concerned, inside the black box there is apparently a 5-V battery in series with a 2500-Ω resistor.

The complex circuit shown back in Fig. 2-35 can also be solved by using a form of Thevenin's theorem. If a voltage-resistor network is opened at a point, and the voltage is measured (or computed) at the opening, other values of the circuit can be determined.

In Fig. 2-37a, if the middle branch of this network is opened and a voltmeter is connected across

be rationalized that 12 V minus 10 V, or 2 V, is all that is at work in the series circuit with the R_3 branch open. With $R_t = 2\ \Omega + 2\ \Omega$ and $V = 2$ V, the current in the circuit will be $I = V/R$, or 2/4 of 0.5 A. The voltage-drop across R_2 will be $V = IR$, or 0.5(2), or 1 V. Therefore, the voltage across the voltmeter will be 12 − 1, or 11 V. This can be called the Thevenin voltage, V_{th}. The polarity markings on the meter would indicate the top terminal to be the negative and the bottom the positive. The circuit is redrawn in Fig. 2-37b without either of the sources, but inserting the 11-V Thevenin voltage, V_{th}, (dashed) in such a direction as to make current flow according to the polarity markings on the meter (− to +). The circuit is now a reasonably simple series-parallel circuit of 4 + 1, or 5 Ω across 11 V, producing $I_{R_3} = V/R$, or 11/5, or 2.2 A, the same as was determined to be flowing through R_3 by Kirchhoff's law.

Either of the other two branches could be opened and the V_{th} for that branch could be measured (or computed). By redrawing the circuit without the sources but with V_{th} inserted in the opened branch, the current through and the voltage-drop across that branch's resistor could be computed.

Thevenin's theorem states that any complex network may be replaced by a single source of voltage in series with a single constant resistance. There is also Norton's theorem, which states that any complex network may be replaced by a constant-current source in parallel with a single resistance. A 1-A constant-current source would feed 1 A to a 1-Ω load (by having 1 V), and if the load is changed to 10 Ω the source would change its voltage output to 10 V to maintain the 1-A load current.

Test your understanding of the operation of both the Thevenin voltage and the Kirchhoff's laws methods by computing (1) I through and (2) V across the headlamps in Fig. 2-38 when the switch

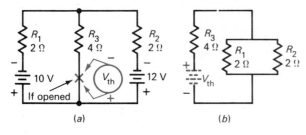

FIG. 2-37 Determining the Thevenin voltage.

the opening, it will read some voltage value. Since the voltages and resistances are known here, it can

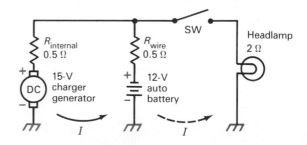

FIG. 2-38 Automobile lighting and charging system.

is closed. What is the battery charging rate (current flowing through it − to +) (3) with the switch open, and (4) with the switch closed? The chassis ground symbols at the bottom indicate zero ohms resistance between all of them. (5) Is V_{th} developed across the open switch?[1]

| Test your understanding; answer these checkup questions.

1. A 200-Ω resistor and a 300-Ω resistor are in parallel. In series with them is a 180-Ω resistor. The whole combination is across a 50-V generator. What current flows through the generator? _____ Through the 180-Ω resistor? _____ Through the 200-Ω resistor? _____ What voltage appears across the 300-Ω resistor? _____

2. A 40-Ω resistor and a 60-Ω resistor are in parallel. In series with them are two other resistors of 20 Ω and 30 Ω. The whole combination is connected across a 1.5-V dry cell. What is the current in the dry cell? _____ What is the current in the 60-Ω resistor? _____ What is the voltage-drop across the 20-Ω resistor? _____

3. A component requires 6.3 V across it with a current of 0.3 A. A second component requires 12.6 V at 0.15 A. The two components are connected in series. What is the value of the resistor that must be connected across the 12.6-V component to allow it to operate properly when in series with the 6.3-V component? _____ What is the value of the resistance that must be connected in series with the combination to allow operation from a 110-V source? _____

4. Component A requires 6.3 V at 1.2 A. Component B requires 6.3 V at 0.15 A. Component C requires 6.3 V at 0.6 A. Component D requires 12.6 V at 0.15 A. Draw a diagram of the most economical method of connecting these four components across a 12.6-V battery. _____ What are the resistance value and the wattage rating of the required resistor? _____

5. A 10-Ω and a 25-Ω resistor are in parallel, and are in series with a 20-Ω and a 30-Ω parallel pair. If there is 5 V across the 10-Ω resistor, what is the current through and the voltage-drop across each component in the circuit? _____
_____ _____ _____
_____ _____ _____
_____

6. In Fig. 2-34, if the R values are doubled and the V values are halved, what is the I_t? _____ What would it be if one of the sources were reversed in polarity? _____

7. In Fig. 2-35, if all Rs are halved and all Vs are doubled, what would be the I_3 value? _____

8. In problem 7, what would be the Thevenin voltage? _____

---

[1][(1) 6 A, (2) 12 V, (3) 3 A, (4) 0 A, (5) Yes]

DIRECT-CURRENT CIRCUIT QUESTIONS

1. State the three basic formulas for Ohm's law.
2. What resistance does a transistor have if 50 mA flows through it when there is 8 V across it?
3. What voltage is required to force 150 mA through a 270-Ω resistor?
4. What current flows through a 23-Ω component when across 6.3 V?
5. If R doubles and V triples, what will the new I value be?
6. What is the power input to a transistor drawing 4.5 A from a 13-V power supply?
7. What is the one difference between power and energy?
8. What is another term for joule?
9. How many watthours must be paid for if a TV set draws 1.8 A from a 120-V line if it runs 5 h a day for 25 days?
10. If electricity costs $0.06 per kWh, how much does it cost to run the TV set in problem 9?
11. Which indicates power, volt-coulombs, kWh, or volt-amperes?
12. What are the three formulas to find dc power?
13. What is the minimum power rating for a 330-Ω resistor across 12.6 V?
14. A 5-kΩ potentiometer is to have no more than 10 mA flowing through it. What wattage rating should it have?
15. A transmitter draws 504 W from its 12.3-V power supply. What is its effective resistance?
16. What are the three ratings that wirewound resistors may carry?
17. How much current will an 82-Ω $1/2$-W resistor safely carry?
18. What does it tell you if an ohmmeter indicates the resistance of a fuse is 10 Ω?
19. What does it tell you if a $1/3$-in.-diameter fuse is 4 in. long?
20. What type of fuse is made to withstand short-duration overloads? Longer-duration overloads?
21. List the six types of basic meters mentioned.
22. In what kind of circuit does the same current value flow through all components?

23. A 5-Ω and a 20-Ω resistor are connected end-to-end across a 3-V battery. What are the battery current and the voltage-drops across the two resistors?
24. "The algebraic sum of all the voltages plus the voltage-drops around a series circuit is zero" expresses what law?
25. Do question 23 answers prove question 24?
26. An electron current is flowing from the top of a resistor to the bottom. Which is the "negative end"?
27. A 24-V unloaded power supply has $^1/_2$-Ω internal resistance. If across a 5-Ω load, what is the current value through and the voltage across the load?
28. When is internal source resistance of little consequence?
29. It is required to operate a 12-V receiver drawing 0.86 A from a 20-V power supply. What values of R and P should the voltage-dropping resistor have?
30. If a 47-Ω 2-W resistor is in series with a 24-Ω $^1/_2$-W resistor, what is the maximum I that can flow safely through them? What could be the maximum voltage across the two?
31. What are two possible units of measurement of conductance?
32. What is the conductance of a soldering iron if it has 0.25 A flowing when across 120 V?
33. A 10-Ω 2-W and a 5-Ω 10-W resistor are in parallel. What is the maximum safe circuit V- and P-dissipation possible?
34. The formula $V_t = IR_1 + IR_2$ is a statement of what law?
35. What is the total circuit voltage and current available if two 12.6-V 8-A batteries are in parallel?
36. A 20-Ω R_1 is in series with a parallel 40-Ω R_2 and 60-Ω R_3. If across 100 V, what is the I through and V across each resistor?
37. A π-network has a 50-Ω input leg, a 60-Ω output leg, and a 30-Ω top resistor. What are the resistance values seen by the source and by the output load?
38. What is gained by matching load to source impedances?
39. A 20-V source (negative up) with a 2-Ω internal R_1, a 30-V source (negative up) with a 5-Ω internal R_2, and a 10-Ω load R_3 are all in parallel. What is the current in all resistors by Kirchhoff's law?
40. Work the same problem as in problem 39 using Thevenin's theorem. What is the Thevenin voltage and what are the three currents?

3

Magnetism

This chapter presents the basic theories and terms regarding magnetism that will be used later in explaining inductors, transformers, ac machinery, and various other devices.

3-1 MAGNETISM AND ELECTRICITY

Any wire carrying a current of electrons is surrounded by an unseen area of force called a *magnetic field*. For this reason, any study of electricity or electronics must consider magnetism.

Almost everyone has had experiences with magnets or with pocket compasses at one time or another. A magnet attracts pieces of iron but has little effect on practically everything else. Why does it single out the iron? A compass, when laid on a table, swings back and forth, finally coming to rest pointing toward the North Pole of the world. Why does it always point in the same direction?

These and other questions about magnetism have puzzled scientists for hundreds of years. It is only comparatively recently that theories to answer questions about magnetism have been developed.

Electronic apparatus such as relays, circuit breakers, earphones, loudspeakers, transformers, chokes, magnetron tubes, television tubes, phonograph pickups, tape and disk recorders, microphones, meters, vibrators, motors, and generators depend on magnetic effects to make them function. Every coil in a radio receiver or transmitter is utilizing the magnetic field that surrounds it when current is flowing through it. But what is a *magnetic field*?

3-2 THE MAGNETIC FIELD

An electron at rest has a negative *electrostatic* field of force surrounding it (Chap. 1). When energy is imparted to an electron to make it move, a new type of field develops around it, at right angles to its electrostatic field. Whereas negative electrostatic lines of force are considered as radiating outward from an electron, the *electromagnetic field* of force develops as a ring around a *moving* electron, at right angles to the current path (Fig. 3-1). (Moving protons also develop magnetic fields around them.)

Electrons orbiting around the nucleus of an atom or a molecule produce an electromagnetic field around their path of motion. In most cases, this magnetic field is either balanced or neutralized by

ANSWERS TO CHECKUP QUIZ ON PAGE 42

1. (0.167 A)(0.167 A)(0.1 A)(20 V) **2.** (0.0203 A)(8.12 mA)(0.406 V) **3.** (84 Ω)(303 Ω) **4.** (Component D across line, Component A and Component C in series across line, Component B and a resistor in series across line)(14 Ω) (Either 2 × 2.84 W or 5 W OK) **5.** (5 V, 0.5 A; 5 V, 0.2 A; 8.4 V, 0.42 A; 8.4 V, 0.28 A) **6.** (1 A)(2 A) **7.** (6.8 A) **8.** (22 V)

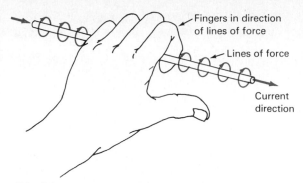

FIG. 3-1 With the thumb in the direction of current, the left-hand fingers indicate the magnetic-field direction.

the magnetic effect of any proton movement in the nucleus, or the movement of one orbital electron is counteracted by another orbital electron whirling in an opposite direction. In almost all substances the net result is little or no external magnetic field.

An electric conductor carrying current along a wire produces a magnetic field around the conductor. The greater the current, the more intense the magnetic field. Normally, the field strength around a current-carrying conductor varies inversely as the distance from the conductor. At twice the distance from the conductor the magnetic-field strength is one-half as much.

To indicate the presence of a magnetic field around a wire, circular lines may be drawn, as in Fig. 3-1. Note that the lines of force are given direction by arrowheads drawn on them. The arrowheads do not mean that the lines of force are moving in this direction but only that a relative polarity is present in them. The direction of these lines is determined in relation to the direction of the electron movement, or current flow. In the illustration, the electron flow is indicated to be from left to right. If the left hand grasps a current-carrying conductor with the thumb extended in the direction of the current, the fingers will indicate the accepted field direction of the magnetic lines of force. This is known as the *left-hand magnetic-field rule*. If the current flows in the opposite direction, the left hand must be turned over so that the thumb points in the direction of the current. The direction of the lines of force will now be opposite to that shown.

If the direction of the lines of magnetic force is known, the same rule can indicate the current direction in a conductor (*left-hand current rule*).

When a current in a conductor increases, more electrons flow, the magnetic-field strength increases, and the field extends farther outward.

In many electric circuits the current flows through a coil of several turns of wire. Figure 3-2 illustrates the magnetic field as it might be set up around a single turn. Note that the indicated field

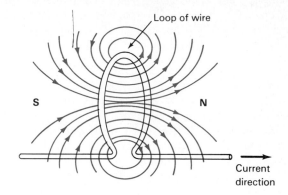

FIG. 3-2 By looping a conductor, magnetic lines of force are concentrated in the central core area of the loop.

direction conforms to the left-hand rule and that the greatest concentration of lines of force appears at the center, or the *core*, of the turn.

When several turns of wire are formed into a coil, the lines of force from each turn add to the fields of the other turns and a more concentrated magnetic field is produced in the core of the coil (Fig. 3-3). The more current and the more turns,

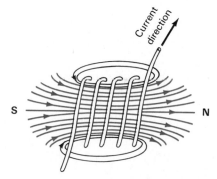

FIG. 3-3 By forming a coil with several loops, the field is concentrated and an electromagnet is produced.

the stronger the magnetic field of such an *electromagnet*. The direction of the field of force can be reversed by reversing the current direction or by reversing the winding direction.

At one end of the coil the field lines are leaving, and at the other end they are entering. When a coil or piece of metal has lines of force leaving one end of it, that end is said to have *north* polarity. The end with the lines entering is the *south* pole. However, each line of force is actually a complete loop and has no north or south points on it.

The terms "north" and "south" indicate *magnetic* polarity, just as "negative" and "positive" indicate *electrostatic* polarity. They should not be used interchangeably. The negative end of a coil is the end connected to the negative terminal of the source and does not refer to the north or south magnetic polarity of the coil.

A different *left-hand coil rule* can be explained by Fig. 3-3. If a coil is grasped by the left hand with the fingers pointing in the direction of current flow, the thumb indicates the *north* end of the coil and also the direction of lines of force in the core. Conversely, if the direction of the lines of force through the core is known, the same left-hand coil rule can be applied to determine the current direction in the coil.

All magnetic lines of force are complete loops and may be considered somewhat similar in their action to stretched rubber bands. They will contract back into the circuit from which they came as soon as the force that produced them ceases to exist.

Magnetic lines of force never cross each other. When two lines have the same direction, they will oppose mechanically if brought near each other. This is illustrated in Fig. 3-4, which shows a cross-

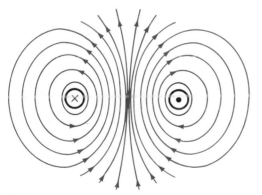

FIG. 3-4 Similar-direction magnetic lines of force repel.

sectional view of two wires with current flowing in opposite directions in them. Where they are adjacent, the lines from both wires have a similar and upward direction and repel each other. The two wires tend to push apart. The little cross in the wire represents the rear view of a current-direction arrow, indicating the current to be flowing away from the viewer. A dot represents the point of a current-direction arrow approaching the viewer.

Magnetic lines of opposite direction or polarity are attracted to each other. The loops surrounding wires carrying current in the same direction are opposite in polarity where they are adjacent (Fig. 3-5). Such loops can attract each other and join into

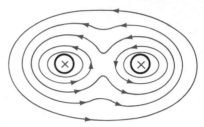

FIG. 3-5 Opposite-direction magnetic lines attract and may join.

single large loops encircling both of the wires carrying the current. The two wires will now be subjected to a common contracting force that tends to pull them together physically.

3-3 FLUX DENSITY, *B*

The complete magnetic field of a coil is known as the flux and is usually denoted by the Greek letter ϕ (phi). If a current flowing through a certain coil produces 100 lines of force in the core of the coil, it may be said that the core has a flux of 100 lines (ϕ = 100 lines, or 100 *maxwells* or Mx).

Flux density of a magnetic field is the number of lines per square inch (in English units) and is indicated by the letter B. In the example above, the core has a total of 100 lines. If the cross-sectional area of the core is 2 in.^{2}, the flux density B is 100/2, or 50 lines/in.^{2}, or 7.75 *gauss* or G. 1 G = 1 Mx/cm^2 = 6.45 lines/in.^{2}

Strong laboratory-produced magnetic fields have over 2,000,000 lines/in.^{2}. A toy magnet may have 500 to 750 lines/in.^{2}.

A magnetic field is one way that energy can be stored for use at a later time.

3-4 MMF, $\mathscr{F}$, AND FIELD INTENSITY, *H*

The energy that produces magnetic lines of force is contained in the movement of electrons (current). The *magnetomotive force*, or *mmf*, is the force that produces the flux in a coil. Magnetomotive force may be computed by multiplying current flow in amperes by the number of turns (ampere-turns, or A · t) in a coil, or

$$\mathscr{F} = NI$$

where $\mathscr{F}$ = mmf, in A · t
N = number of turns in coil
I = current, in A

A 20-turn coil with 0.5 A flowing in it has an mmf of 10 NI, or 12.6 *gilberts* (Gb), since 1 Gb equals 0.796 A · t.

Past the ends of the coil, the magnetic field starts spreading and flux density begins decreasing. Within the core itself, lines of force are relatively straight and parallel and present a nearly constant flux density. A practical unit of measurement of the magnetizing force working on the column of material forming the core of the coil is determined by dividing the total mmf by the core length in inches. This reduces the total mmf to a unit for comparison and computation. Thus, a 3-in. coil with 60 A · t has a standardized magnetizing force, or field intensity, of 60/3, or 20 A · t/in. This magnetizing force is represented by H. In this case, H = 20 A · t/in., or 9.9 *oersted* (Oe), since 1 Oe or 1 Gb/cm equals 2.02 A · t/in.

To illustrate the direct relationship of flux lines to A · t, a 100-turn choke coil has 1 A of current flowing through it (100NI), which produces 500 flux lines. If the number of turns is reduced to 60 and the current to 0.7 A, what is the number of flux lines? The total NI = 60(0.7) = 42 A · t. Setting up the ratios, 100/42 = 500/X, and solving for the unknown, X = 500(42/100) = 210 lines.

3-5 PERMEABILITY, μ

When a coil of wire is wound with air as the core, a certain flux density will be developed in the core for a given value of current. If an iron core is slipped into the coil, a very much greater flux and flux density will exist in the iron core than was present when the core was air, although the current value and the number of turns have not changed.

With an air-core coil, the air surrounding the turns of the coil may be thought of as pushing against the lines of force and tending to hold them close to the turns. With an iron core, however, the lines of force find a medium in which they can exist much more easily than in air. As a result, lines that were held close to the turns in the air-core coil are free to expand into the highly receptive area afforded by the iron. This allows lines of force that would have been close to the surface of the wire to expand into the iron core. Thus, an iron core allows a greater flux density, B, in itself,

although no more magnetizing force (NI) may have been developed than in an air-core coil. The iron core merely brings the lines of force out where they can be more readily used and concentrates them.

When comparing the ability of different materials to accept or allow lines of force to exist in them, air may be considered as the standard of comparison. A magnetizing force H of 20 A · t/in. will produce a flux density B of 20 lines/in.^{2} in an *air-core* coil. By substituting a core of fairly pure iron, the same 20 A · t/in. may produce a flux density of perhaps 200,000 lines/in.^{2} in the core.

The ratio of B to H, or B/H, expresses the ability of a core material to accept, or be *permeated* by, lines of force. The permeability of most substances is very close to that of air, which may be considered as having a value of 1. A few materials, such as iron, nickel, and cobalt, are highly permeable, with permeabilities of several hundred to several thousand times that of air. (Note that the word "permeability" is a derivation of the word "permeate," meaning "to pervade or saturate," and is not related to the word "permanent.")

Permeability is represented by the Greek letter μ (mu, pronounced "mew"). In formula form

$$\mu = \frac{B}{H}$$

where μ = permeability (no unit)
B = flux density, in lines/in.^{2}
H = magnetizing force, in A · t/in.

In the preceding example, the iron has a permeability equal to B/H, or 200,000/20, or 10,000. In practice, this is not particularly high. Pure iron approaches a permeability of 200,000. Special alloys of nickel and iron may have a permeability of 100,000. Alloying iron makes it possible to produce a wide range of permeabilities. A cast iron can be made that has almost unity (1) permeability. Most stainless steels exhibit practically no magnetic effect, although some may be magnetic.

Any substance that is not affected by magnetic lines of force and is reluctant to support a magnetic field is said to have the property of *reluctance*, the symbol of which is $\mathscr{R}$. Air, vacuum, and most substances have nearly unity reluctance, while iron has a very low reluctance. In formula form,

$$\mathscr{R} = \frac{\mathscr{F}}{\phi}$$

In electric circuits the reciprocal of resistance is called *conductance*. In magnetic circuits, the reciprocal of reluctance is called *permeance*, symbol $\mathscr{P}$. Thus $\mathscr{P} = 1/\mathscr{R}$, and $\mathscr{R} = 1/\mathscr{P}$. Air has unity permeance, whereas iron has a high permeance.

Permeability may also be considered as permeance per unit volume. Thus permeability is lines of force (ϕ) per ampere-turn (NI) per cubic inch. When speaking of a magnetic circuit, "permeance" is used. "Permeability" is used when discussing how magnetic materials behave.

▌Test your understanding; answer these checkup questions.

1. Under what condition does an electron have a magnetic field? _____
2. When a preponderance of electrons are rotating around a nucleus in the same direction, what does the atom then form? _____
3. Do arrowheads on magnetic lines of force indicate movement in that direction? _____
4. Which magnetic pole has lines pointing into it? _____ Which has lines pointing out? _____
5. When the fingers of the left hand are around a coil in the direction of current flow, to what does the thumb point? _____
6. When current ceases in a coil, what happens to the magnetic-field lines? _____
7. Where in a current-carrying iron-core coil is the magnetic field most intense? _____
8. What does a cross on the cross-sectioned end of a current-carrying wire indicate? _____
9. If two parallel wires have similar-direction current flowing in them, what physical effect will be produced? _____
10. What is the symbol for magnetic flux? _____ In what English unit is it measured? _____
11. What is the symbol for flux density? _____ In what English unit is it measured? _____
12. What is the symbol for magnetomotive force? _____ In what English unit is it measured? _____
13. What is the symbol for field intensity? _____ In what English unit is it measured? _____
14. What is the permeability symbol? _____ The formula? _____
15. What is the μ of air? _____ Of pure iron? _____
16. What is reluctance? _____ What is its symbol? _____ Its formula? _____
17. What is the reciprocal of reluctance? _____ Its symbol? _____
18. Where do magnetic lines go when they collapse back into a wire? _____

3-6 THE ATOMIC THEORY OF MAGNETISM

The discussion here will be a condensed version of the atomic theory of magnetism.

From atomic theory it is known that an atom is made up of a nucleus of protons surrounded by one or more electrons encircling it. The rotation of electrons and protons in most atoms is such that the magnetic forces cancel each other. Atoms or molecules of the elements iron, nickel, and cobalt arrange themselves into magnetic entities called *domains*. Each domain is completely magnetized.

Groups of domains form in crystals of magnetic material. The crystals may or may not be magnetic, depending on the arrangement of the domains in them. Investigation shows that while any single domain is fully magnetized, the external resultant of all the domains in a crystal may be a neutral field.

Each domain has three directions of magnetization: *easy, semihard,* and *hard*. If an iron crystal is placed in a weak field of force, the domains begin to line up in the easy direction. As the magnetizing force H is increased, the domains begin to roll over and start to align themselves in the semihard direction. Finally, as the H is increased still more, the domains are lined up in the hard direction. When all the domains have been lined up in the hard direction, the iron is said to be *saturated*. An increase in magnetizing force will then produce no more magnetic change in the material.

The result of this action can be seen in the *BH* curve for a piece of iron shown in Fig. 3-6. The

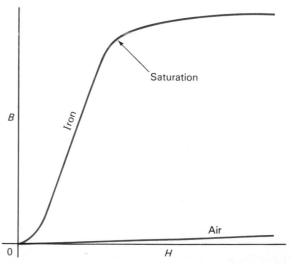

FIG. 3-6 Magnetization or *BH* curves for iron and air.

graph shows that an increase of H produces an increase of B. At low values of magnetizing force the iron produces relatively little flux density. The permeability, or B/H, is relatively low under this condition. As the magnetization is increased, the flux density increases rapidly in respect to the increase of magnetizing force, resulting in a large B/H ratio, or a high value of permeability. As saturation is approached, further increase in H produces little increase in B. This general curve form is to be expected of all magnetic materials, although the steepness of the curve changes with different permeabilities. The BH curve of air does not bend, as indicated.

3-7 FERROMAGNETISM

Substances that can be made to form domains are said to be *ferromagnetic*, which means "iron-magnetic." The ferromagnetic elements are iron, nickel, and cobalt, but it is possible to combine some nonmagnetic elements and form a ferromagnetic substance. For example, in the proper proportions, copper, manganese, and aluminum, each by itself being nonmagnetic, produce an alloy which is similar to iron magnetically.

Materials made up of nonferromagnetic atoms, when placed in a magnetic field, may weakly attempt either to line up in the field or to turn at right angles to it. If they line themselves in the direction of the magnetic field, they are said to be *paramagnetic*. If they try to turn from the direction of the field, they are called *diamagnetic*. There are only a few diamagnetic materials. Some of the more common are gold, silver, copper, zinc, and mercury. All materials which do not fall in the ferromagnetic or diamagnetic categories are paramagnetic. Most materials are paramagnetic.

Ferromagnetic substances will resist being magnetized by an external magnetic field to a certain extent. It takes some energy to rearrange even the easy-to-move domains. Once magnetized, however, ferromagnetic substances may also tend to oppose being demagnetized. They are said to have *retentivity*, or *remanence*, the ability to retain magnetism when an external field is removed.

As soon as the magnetizing force is released from a magnetized ferromagnetic substance, it tends to return at least part way back to its original unmagnetized state, but it will always retain some magnetism. This remaining magnetism is *residual*

magnetism. Paramagnetic and diamagnetic materials always become completely nonmagnetic when an external magnetizing force is removed from them.

3-8 THE HYSTERESIS LOOP

When a magnetizing force (H) is applied to a piece of completely demagnetized ferromagnetic material, the flux density (B) rises as shown by the dashed curve X in Fig. 3-7. If the magnetizing force (H) is

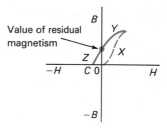

FIG. 3-7 Portion of hysteresis loop developed when H is applied, is removed, and coercive force applied.

removed from the ferromagnetic material, the flux density decreases but drops back only part way to zero, as shown by the curve Y. The opposition of the domains to roll back to the unmagnetized state is known as *hysteresis*. The magnetism left in the metal is residual magnetism. For soft iron the residual magnetism will be small, while for steels and iron alloys it will be considerably greater. With the latter materials the B value will not drop so far toward zero.

If a magnetizing force in the opposite direction is now applied, a certain value of this $-H$ will be necessary to counteract the residual magnetism and bring the Z portion of the curve down to point C. This point represents zero flux density, or complete demagnetization.

The opposite-direction magnetizing force, or $-H$, required to demagnetize the material completely is known as the *coercive force* and is indicated by the length of the line 0 to C.

When an iron core has a coil of wire wrapped around it and an alternating current is made to flow through the coil, a curve called a *hysteresis loop* (Fig. 3-8) may be drawn to indicate the flux-density variations under changing magnetizing forces. The dotted line shows the magnetization when the unmagnetized iron was first subjected to the force. The remainder of the loop indicates the B that results when the H is alternately reversed.

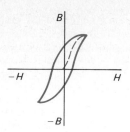

FIG. 3-8 Full hysteresis loop of an iron core produced by one cycle of ac in the coil surrounding the core.

As current through the coil increases in one direction, the H increases, producing an increase in B. As the current decreases, the H decreases, resulting in a lessening of B. As the current reverses, the H increases in the opposite direction, the domains are rolled over, and the B follows in the opposite direction.

With a full cycle of ac a hysteresis loop enclosing a given area is developed. A soft iron will be represented by a slim loop (Fig. 3-9a). Steel results in a broader loop (Fig. 3-9b). Special iron alloys, such as alnico, have a loop approaching a rectangle in shape (Fig. 3-9c).

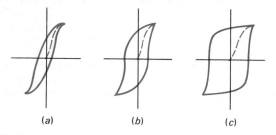

(a)   (b)   (c)

FIG. 3-9 Hysteresis loops for (a) soft iron, (b) steel, (c) alnico-type materials.

An expenditure of energy is required to reverse the magnetism, or realign the domains, in a piece of ferromagnetic material. The energy lost due to hysteresis appears as heat in the magnetic material. For this reason, when transformers or relays using alternating current are made with iron cores, it is

necessary to select metals with narrow hysteresis loops and with little *hysteretic loss*. Even so, the core of a transformer or ac relay always heats a little because of loss of energy due to hysteresis. New fast-cooled molten iron alloys can be made to form an "iron glass" ribbon which has a remarkably small hysteresis loop area and can result in transformers and relays having almost no hysteretic losses. Such *Metglas* has a random atomic structure rather than the crystalline structure of iron or other metals.

3-9 PERMANENT AND TEMPORARY MAGNETS

A ferromagnetic substance that holds magnetic-domain alignment well (high retentivity) and has a broad hysteresis loop is used to make permanent magnets. One of the strongest permanent magnets is a combination of iron, aluminum, nickel, and cobalt called *alnico*. It is used in horseshoe magnets, electric meters, headphones, loudspeakers, radar transmitting tubes, and many other applications. Some magnetically hard, or permanent-magnet, materials are cobalt steel, nickel-aluminum steels, and other special steels.

Ferromagnetic metals that lose magnetism easily (low retentivity) make temporary magnets. They find use in transformers, chokes, relays, and circuit breakers. Pure iron and Permalloys (*perm* from "permeable," not "permanent") are examples of magnetically soft, temporary-magnet materials. Iron oxides, such as manganese ferrite, usually insulators, are used for magnetic cores in many applications. These are called *ferrite* cores.

3-10 MAGNETIZING AND DEMAGNETIZING

One method of magnetizing a ferromagnetic material is to coil wire around the material and force a direct current through the coil. If the core material has a high value of retentivity, it will become a permanent magnet. If the core is heated and allowed to cool while subjected to the magnetizing force, a greater number of domains will be swung into alignment and a greater permanent flux density may result. Hammering or jarring the core while under the magnetizing force also tends to increase the number of domains that will be aligned.

A less effective method of magnetizing is to stroke a high-retentivity material with a permanent magnet. This will align some of the domains of the material and induce a relatively weak permanent magnetism.

If a permanent magnet is hammered, many of its domains will be jarred out of alignment and the flux density will be lessened. If heated, it will lose its magnetism because of an increase in molecular movement that upsets the domain structure. Strong opposing magnetic fields brought near a permanent magnet may also decrease its magnetism. It is important that equipment containing permanent magnets, such as some earphones and microphones, be treated with care. The magnets must be protected from physical shocks, excessive temperatures, and strong alternating or other magnetic fields.

When objects such as screwdrivers or watches become permanently magnetized, it is possible to demagnetize them by slowly moving them into and out of the core area of a many-turn coil in which an alternating current (ac) is flowing. The ac produces a continually alternating magnetizing force. As the object is placed into the core area, it is alternately magnetized in one direction and then the other. As it is pulled away slowly, the alternating magnetizing forces become weaker. When it is finally out of the field completely, the residual magnetism will usually be so low as to be of no consequence.

3-11 THE MAGNETIC CIRCUIT

The magnetomotive force, the reluctance of the magnetic material used, and the flux developed are often likened to the electric circuit, in which electromotive force applied across resistance produces current. The similarities are:

V = electromotive force (emf) $\mathscr{F}$ = magnetomotive force (mmf)

R = resistance $\mathscr{R}$ = reluctance

I = current ϕ = flux

In the magnetic circuit the idea that the total flux is directly proportional to the mmf and inversely proportional to the reluctance of the material used can be expressed by the *Ohm's law for magnetic circuits formula*

$$\phi = \frac{\mathscr{F}}{\mathscr{R}}$$

where ϕ = flux, in lines

$\mathscr{F}$ = mmf, in A · t

$\mathscr{R}$ = reluctance (no unit)

Since the permeability graph of all magnetic materials is curved, or *nonlinear* (meaning "not a

straight line"), the reluctance value of magnetic materials will also be nonlinear under varying values of H. Unless permeability curves for the metal being investigated are available, accurate magnetic-circuit computations are not possible.

3-12 ENGLISH, CGS, MKSA, AND SI UNITS

Magnetism has been explained in English (inches, square inches, ampere-turns) and cgs (cm, cm²) terms. However, there is another system that is used. The English and the cgs, or centimeter-gram-second, were used for many years before the rationalized mksa, or meter-kilogram-second-ampere, system with its SI units (Système International d'Unités) was developed. Table 3-1 illustrates the relationship between the magnetic terms used in these three systems. All three systems are still in use.

▌ Test your understanding; answer these checkup questions.

1. What does "ferromagnetic" mean? _____
   Into what basic entities do molecules of ferromagnetic substances form? _____
2. In what category is a substance that tries to turn at right angles to a magnetic field? _____ That tries to line up in the direction of a magnetic field? _____
3. When an increase of H can produce no more B, what has been reached? _____
4. Which would have no retentivity, paramagnetic, diamagnetic, or ferromagnetic materials? _____ Which would have no hysteresis loop? _____
5. Is $+H$, $-H$, $+B$, or $-B$ required to produce coercive force in a material? _____
6. What shape hysteresis loop must a material have to be a good relay core? _____ A permanent magnet? _____
7. What are iron oxides used in magnetic cores called? _____ Would they have a high or low resistance value? _____
8. What happens to a permanent magnet in the core of a coil carrying a heavy alternating current? _____ Why might it heat? _____
9. What is the Ohm's-law formula for magnetic circuits? _____
10. In the English magnetic system what is the unit of measurement for ϕ? _____ For $\mathscr{F}$? _____ For H? _____ For B? _____
11. In the cgs system what is the unit of measurement

TABLE 3-1 MAGNETIC UNITS

Term and symbol	English units		Cgs units		Mksa (SI) units, rationalized
Magnetic flux (ϕ)	1 LINE OF FORCE	=	1 MAXWELL (Mx)	=	10^{-8} weber (Wb)
	10^8 lines of force	=	10^8 Mx	=	1 WEBER
Magnetomotive force ($\mathscr{F}$)	1 AMPERE-TURN (NI)	=	1.26 gilbert (Gb)	=	1 AMPERE-TURN
	0.796 A $\cdot$ t	=	1 GILBERT	=	0.796 A $\cdot$ t
Magnetizing force (H)	1 AMPERE-TURN/INCH	=	0.495 oersted (Oe)	=	39.4 A $\cdot$ t/m (NI/m)
	2.02 A $\cdot$ t/in.	=	1 OERSTED, or		
			1 Gb/cm	=	79.6 A $\cdot$ t/m
	0.0254 A $\cdot$ t	=	0.0126 Oe	=	1 AMPERE-TURN/METER
Magnetic flux density (B)	1 LINE/INCH^2	=	0.155 gauss (G)	=	0.155×10^{-4} Wb/m^2
	6.45 lines/in.^{2}	=	1 GAUSS, or a Mx/cm^2	=	10^{-4} Wb/m^2
	6.45×10^4 lines/ in.^{2}	=	10^4 G	=	1 TESLA (T), or 1 Wb/m^2

for ϕ? _____ For $\mathscr{F}$ _____ For H? _____ For B? _____

12. In the mksa system what is the unit of measurement for ϕ? _____ For $\mathscr{F}$? _____ For H? _____ For B? _____

3-13 PERMANENT-MAGNET FIELDS

When a piece of magnetically hard material is subjected to a strong magnetizing force, its domains become aligned in the same direction. When the magnetizing force is removed, many of the domains remain in the aligned position and a permanent magnet results. A north pole is where the direction of the magnetic lines of force is outward from the magnet (Fig. 3-10) and a south pole is where the direction of the lines is inward.

It has been explained that the lines of force surrounding current-carrying wires oppose each other if the lines have the same direction. They attract each other if they are of opposite directions. This is true of the fields of permanent magnets. When sim-

ilar poles are held near each other, lines of similar direction are opposing, tending to push the magnets apart physically (Fig. 3-11a). Unlike poles held near each other produce a physically attracting effect because the lines of force from both magnets join as long, contracting loops (Fig. 3-11b).

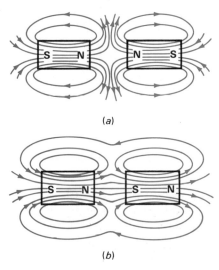

(a)

(b)

FIG. 3-11 (a) Fields with like lines repel and push magnets apart. (b) Unlike lines attract, join, and pull magnets together.

3-14 MAGNETIC-FIELD DISTORTION

Under normal circumstances, a permanent magnet made into bar shape and suspended in the air will

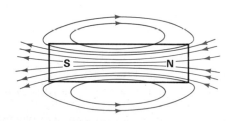

FIG. 3-10 Magnetic poles are determined by the direction of lines of force: out of north, into south.

have a field surrounding it somewhat similar to that shown in Fig. 3-10.

When a piece of highly permeable material such as pure iron and a piece of low-permeability material such as copper are placed in the field (Fig. 3-12), the field pattern becomes distorted. The cop-

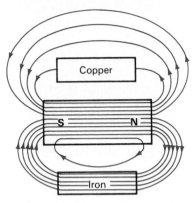

FIG. 3-12 Distortion of a bar-magnet field by iron and lack of distortion by low-permeability material.

per, having practically the same permeability as air, has almost no effect on the magnetic lines of force. The iron, being thousands of times more permeable than air, is a highly acceptable medium in which the lines of force may exist, and most of the lines on that side and some of the lines from the other side of the magnet move into the iron. This produces a distorted field pattern with highly concentrated fields between the magnet and the piece of iron. This ability to produce a concentrated field by using iron to close the magnetic gap between north and south poles of magnets is common in electronics.

If a magnet is completely encased in a magnetically soft iron box, all its lines of force remain in the walls of the box and there is no external field. This is known as magnetic *shielding*. Shielding may be used in the opposite manner. An object completely surrounded by an iron shield will have no external magnetic fields affecting it, as all such lines of force will remain in the permeable shield.

3-15 THE MAGNETISM OF THE EARTH

Enough of the ferromagnetic materials making up the earth have domains aligned in such a way that the earth appears to be a huge permanent magnet [field strength of approximately 70 microteslas (μT) on the surface]. The direction taken by the lines of force surrounding the surface of the earth is inward

at a point near what is commonly known to be the North Pole of the world and outward near the earth's South Pole. This results in a rather confusing set of facts. The North Geographical Pole of the earth is actually near its South Magnetic Pole, and the South Geographical Pole is near the North Magnetic Pole (Fig. 3-13).

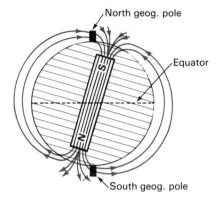

FIG. 3-13 Representation of the magnetic field of the earth.

The familiar magnetic navigational compass consists of a small permanent magnet balanced on a pivot point. The magnetic field of the compass needle lines itself up in the earth's lines of force. As a result, the magnetic north end of the compass needle is pulled toward the earth's South Magnetic Pole because unlike poles attract each other. This means that when the "northpointing end" of the compass points toward the geographical north, it is actually pointing toward the South Magnetic Pole of the earth.

3-16 USING A MAGNETIC COMPASS IN ELECTRICITY

A small magnetic pocket compass moved to different points in a field produced by a permanent magnet will indicate the magnet's field (Fig. 3-14). In the same manner, the direction of the lines of force surrounding a coil carrying a direct current can be checked. Note the indication given by the compass needle when inside the coil (Fig. 3-15): the north end of the compass now points to the north end of the coil. Outside, the compass lines up in the lines of force, pointing toward the south end of the coil.

If the current direction in the coil is reversed, the compass will reverse. When an ac is flowing in the coil, the compass can not follow rapid field rever-

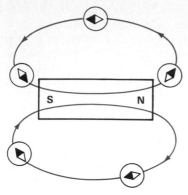

FIG. 3-14 Compasses can indicate the contour of a magnetic field. (Dark ends are north.)

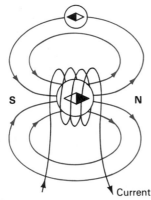

FIG. 3-15 Compass indications inside and outside a current-carrying coil.

sals and will give no indication. If the ac is strong enough, the compass may be demagnetized and become useless.

The direction of current in a wire can be determined with a compass. Current flowing along a wire sets up lines of force around it. If a compass is brought near the wire, the needle will swing at right angles to the current and in the direction of the lines of force surrounding the wire. The north pole of the compass indicates the direction of the lines of force around the wire (Fig. 3-16). By

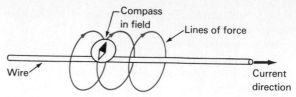

FIG. 3-16 A compass indicates the direction of the magnetic field around a current-carrying wire.

applying the left-hand current rule, the direction of current can be determined. If the current direction is reversed, the direction of the lines of force will reverse and the compass needle will swing around 180°. As in a coil, the compass will not indicate if the current is alternating rapidly.

3-17 ELECTRONS MOVING IN A MAGNETIC FIELD

Electrons will travel in a straight line in a magnetic field provided they are moving in a direction parallel to the lines of force.

When electrons are propelled across lines of force, they move in a curved path. Figure 3-17

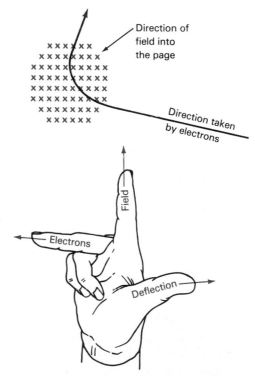

FIG. 3-17 Direction taken by electrons moving through a magnetic field and illustration of the right-hand motor rule.

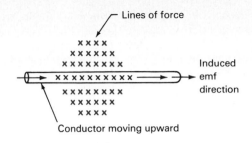

Lines of force

× × × ×
× × × × × ×
× × × × × × × ×
× × × × × × × × × × → Induced
× × × × × × × × × × × × → emf
× × × × × × × × direction
× × × × × ×
× × × ×

Conductor moving upward

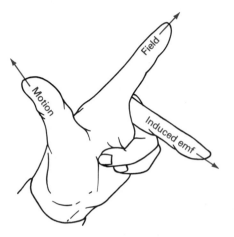

Field

Motion

Induced emf

FIG. 3-18 Direction of induced emf when a conductor moves through a magnetic field and illustration of the left-hand generator rule.

illustrates the path that will be taken by electrons moving into the magnetic field shown. The crosses represent the rear view (feathers) of arrows indicating lines of force pointing into the page.

The direction taken by the electrons as they curve through a magnetic field can be determined by using the *right-hand motor rule*. This states: With the thumb, first, and second fingers of the *right* hand held at right angles to each other, the first finger pointing in the direction of the magnetic field, the second finger in the direction of the electron movement, then the thumb will indicate the

deflection of the electron. Check the electron path taken in the illustration by this rule.

If a wire is carrying the electrons through the magnetic field, an upward pressure on the electrons tends to move the wire in an upward direction. This is the principle by which electric motors operate, discussed in Chap. 27.

Free electrons moving through a vacuum but in magnetic deflecting fields are found in television picture tubes.

3-18 GENERATING AN EMF

When a wire is moved parallel to lines of force, no effect is produced on free electrons in the wire. However, when a conductor is moved upward through a magnetic field (Fig. 3-18), an electron flow is induced in the conductor. This current direction may be determined by using the *left-hand generator rule*. This three-finger rule is somewhat similar to the right-hand motor rule. It may be stated: With the thumb, first, and second fingers of the *left* hand held at right angles to each other, if the first finger is pointed in the direction of the magnetic field and the thumb is pointed in the direction of the conductor's motion, then the second finger indicates the direction of induced emf (or current) in the wire.

An explanation of the induction of a current in a conductor (Fig. 3-19) may be given as follows: When a wire moves into a magnetic line of force (*a*), the line is stretched around the wire (*b*), rejoins itself behind the wire (*c*), and the little loop developed around the wire collapses inward (*d*). The free electrons in the wire move at right angles to the inward-collapsing magnetic loop (*e*), and are forced along the wire. The induced current in the wire shown, due to the collapsing loop, is outward from the page. Check this with the left-hand generator rule. (HINT: The collapsing-field motion is inward, which is the same as an outward motion of the conductor.)

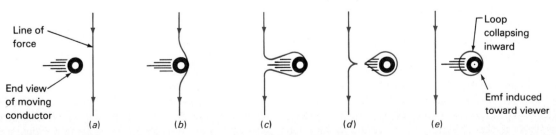

Line of force

End view of moving conductor

(a) (b) (c) (d) (e)

Loop collapsing inward

Emf induced toward viewer

FIG. 3-19 How an emf is induced into a wire cutting through a line of force.

While the explanation is given in terms of an induced electron *current*, it is usually considered to be an emf that is induced in a conductor moving across a magnetic field. Both statements are correct, since producing a difference of potential between the two ends of the conductor requires some of the free electrons in the wire to be moved to one end from the other end. This is considered a charging current. If a load is connected across the ends of the conductor, many more electrons will be forced into movement. Induced currents produce expanding magnetic fields around them in a direction that opposes the original moving magnetic field. This is an application of *Lenz' law,* which states: Any induced current must be in such a direction that its own magnetic field will oppose the magnetic field that produced the induced current. Physical or mechanical energy will now be required to move the conductor through the field. With no electrical load across the conductor, almost no energy is required to move it. This is the basic theory of the electric generator, which is explained in Chap. 27.

The factors that determine the amplitude, or strength, of the induced emf are:

1. Strength of the magnetic field (number of lines per square inch)
2. Speed of the conductor motion across the lines of force
3. Number of conductors connected in series (as in a generator)

These factors can be simply expressed by: The greater the number of lines cut per second, the higher the voltage induced. *When magnetic lines are cut at a rate of 10^8 (or 100,000,000) lines per second, an average emf of 1 V is produced.*

3-19 MAGNETOSTRICTION

In some electronic and radio communication equipment the property of ferromagnetic materials to expand or contract when they are subjected to a magnetizing force is used. This effect is called *magnetostriction. Nickel constricts* when under a magnetizing force; *iron expands.* A part of the magnetic energy stored in such metals is convertible to mechanical energy in this way.

The maximum contraction or expansion occurs at magnetic saturation. The direction of the magnetizing force has no effect on the direction of strain. Nickel wrapped with a coil of wire will contract regardless of the direction of current fed into the coil.

When the magnetizing force is removed, the magnetostrictive material springs back to its normal shape or size. The mechanical energy is converted back into magnetic energy and can be converted further to electric energy. If the current used as the magnetizing force is continuously pulsating or alternating, the material will continuously vibrate mechanically.

3-20 RELAYS

A relay is a relatively simple electromagnetic device. It may consist of a coil or solenoid, a ferromagnetic core, and a movable armature on which make and break contacts are fastened. Figure 3-20 illustrates a relay used to close a circuit when

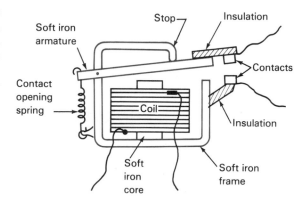

FIG. 3-20 Functional parts of a SPST NO dc relay.

the coil is energized. It is known as a single-pole single-throw (SPST), normally open (NO), or "make-contact" relay.

The core, the U-shaped body of the relay, and the straight armature bar are all made of magnetically soft ferromagnetic materials having high permeability and little retentivity. One of the relay contacts is attached by an insulating strip to the armature, and the other to the relay body with an insulating material. The contacts are electrically separated from the operational parts of this particular relay. A spring pulls the armature up and holds the contacts open.

When current flows in the coil, the core is magnetized and lines of force develop in the core and through the armature and the body of the relay. The gap between the core and the armature is filled with magnetic lines trying to contract. These contracting lines of force overcome the tension of the spring and pull the armature toward the core, closing the relay contacts. When the current in the coil is stopped, the magnetic circuit loses its magnetism and the spring pulls the armature up, opening the contacts.

Relays are useful in remote closing and opening of high-voltage or high-current circuits with relatively little voltage or current flow in the coil. Three basic relays are shown in Fig. 3-21.

In some cases it may be required to change the operating voltage of a relay coil.

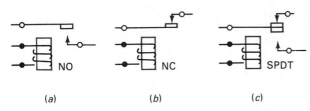

(a)   (b)   (c)

FIG. 3-21 Schematic symbols of simple relays. (a) Normally open or make-contact relay. (b) Normally closed or break-contact relay. (c) Single-pole double-throw relay.

EXAMPLE: A 6-V relay is to be used on 12 V. The magnetizing force to attract the armature is equal to the ampere-turns of the coil. If the coil has 6 Ω and 300 turns the current will be $I = V/R$, or 6/6, or 1 A. The ampere-turns required is 1 A times 300 turns, or 300 NI. A resistance must be added in series with the coil to limit the current to 1 A when a 12-V source is used. The resistor must have a 6-V drop across it when 1 A is flowing through it. Therefore, $R = V/I$, or 6/1, or 6 Ω. However, the heat developed in the resistance is a complete waste. It may be more desirable to purchase a 12-V coil for the relay or to wind one. To wind one, the size of the wire in the 6-V coil must be determined. A wire having twice the resistance per unit length or 70% of the diameter must be used. This is obtained by using a wire with a gauge three units higher. For example, No. 27 wire has twice the *cross-sectional area* of No. 30 and therefore half the resistance. No. 30 has a diameter of 10 mils, which is 70% of the 14.2-mil diameter of No. 27 wire. When a coil is wound with the thinner wire to the same size as the original coil, it will have four times the resistance and twice as many turns. With 12 V, ½ A will flow through the 600 turns. This represents 300 NI, the required magnetizing force.

If a 12-V relay coil is to be changed to operate on 6 V, a wire 1.4 times the diameter of the original should be used.

Relay contacts are usually made of silver or tungsten. Silver oxidizes but can be cleaned by using a very fine file or burnishing tool, abrasive paper, or a piece of ordinary letterhead paper rubbed between two contacts. If the contacts are pitted by heavy currents, they may be smoothed, but the original shape of the contacts should be retained to allow a wiping action during closing and opening to keep them clean.

Vacuum relays have all their working parts in an evacuated bulb. Their tungsten contacts will stand extremely high voltages and currents without sparking or deteriorating.

Reed switches are a form of relay used in many modern electronic circuits. They consist of two iron reeds that overlap about 0.1 inch (100 mil) in an evacuated glass tube (Fig. 3-22). When current

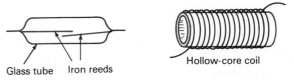

Glass tube   Iron reeds   Hollow-core coil

FIG. 3-22 The two components of a reed switch or relay.

flows through a hollow-core coil placed around the tube, both iron reeds are magnetized. The induced N polarity of one reed is attracted to the induced S end of the other, and an electric circuit can thereby be closed. Such reed relays are very fast acting and are quite small.

▌ Test your understanding; answer these checkup questions.

1. Does a magnetic shield capture lines of force or repel them? _____
2. Which would shield a permanent-magnet field best, copper, lead, glass, iron, or aluminum?
   _____
3. The end of a compass needle that points to the geographic North Pole of the earth has what magnetic polarity? _____
4. What do the thumb and first two fingers indicate in the left-hand generator rule? _____
   _____ _____ In the right-hand motor rule? _____ _____
   _____
5. "Induced currents always produce expanding magnetic fields around their conductors in a direction that opposes the original magnetic field" is an expression of what law? _____

6. How many lines of force must be cut per second to produce an average of 1 V? _____
7. What is the name of the effect whereby a ferromagnetic material changes size or shape when magnetized or demagnetized? _____
8. What would DPDT NC stand for when speaking of relays? _____
9. Why is silver a good material for relay contacts? _____ Why is tungsten? _____
10. What magnetic unit is important when determining how much pulling force a coil will have on a relay arm? _____
11. What is the advantage of using vacuum relays? _____ Reed? _____

MAGNETISM QUESTIONS

1. What is required to produce a magnetic field?
2. What is the space relationship between electrostatic and electromagnetic lines of force?
3. Using the left-hand magnetic-field rule with a wire, what does the thumb indicate? The fingers?
4. In a coil, where is the magnetic field most intense?
5. What effect does coiling a wire have on its magnetic field?
6. Does coiling a wire develop additional lines of force?
7. Where is there a north pole when current flows in a straight wire?
8. To what pole do arrowheads on lines of force point?
9. What effect do two parallel wires have on each other if they have similar-direction currents flowing in them?
10. What happens to magnetic fields when current ceases in a wire?
11. With the left-hand magnetic-field coil rule, what does the thumb indicate? The fingers?
12. What is another term meaning magnetic line of force?
13. What is the unit of measurement of flux density, B?
14. What is mmf, and how is it computed?
15. In what unit is magnetizing force or field intensity, H, measured?
16. What is the advantage of a high-permeability material?
17. What metal accepts lines of force best? Does it hold them best?
18. What does the ratio B/H express?
19. What do low-reluctance materials do in a magnetic field?
20. What is the reciprocal of reluctance, and what is a formula for it?
21. Does iron have a high or low reluctance?
22. What are magnetic domains?
23. What are the three directions of domain magnetization?
24. What is iron said to be when at or past the knee of its BH curve?
25. What are substances that form domains called?
26. What are paramagnetic substances? Diamagnetic?
27. What is remanence?
28. What is hysteresis?
29. What is coercive force?
30. What does a wide hysteresis loop indicate?
31. What kind of hysteresis loop should relay cores have?
32. Are ferrites conductors or insulators?
33. For what are hard magnetic materials used?
34. What effect does heat have on a permanent magnet?
35. What are the three formulas for the magnetic Ohm's law?
36. What are the names of the three magnetic-unit systems?
37. What effect do like magnetic poles have on each other?
38. Why does iron and not copper make a good magnetic shield?
39. What is the magnetic polarity of the North Geographic Pole?
40. How might you determine the north pole of a coil carrying a direct current?
41. To what does the thumb, first finger, and second finger of the right-hand motor rule point?
42. To what does the thumb, first finger, and second finger of the left-hand generator rule point?
43. What law states, "Any induced current must be in such a direction that its own magnetic field will oppose the magnetic field that produced the induced current"?
44. What factors determine the amplitude of an induced emf?
45. What is the effect called when a magnetic material expands or contracts when magnetized?
46. Is an NO, an NC, or an SPST relay a "make-contact" type?
47. Would it be possible to have an SPST relay? A DPDT? A 3P2T? A 3P3T?
48. What determines the pulling power of a relay coil on its armature?
49. Should relay contacts be flat or rounded?
50. What are the advantages of reed relays?

4
Alternating Current

This chapter presents some of the basic concepts and terminology regarding alternating current used in all chapters that follow. Topics included are: frequency; cycles; alternations; peak, effective, and average values; and the frequency spectrum.

4-1 METHODS OF PRODUCING EMF

Some of the various methods by which an electromotive force can be produced were listed in Sec. 1-8. These can be further separated into methods of producing unidirectional pressure, called direct current (dc), and the generation of alternating-direction current (ac).

There are seven methods of producing dc: (1) *chemical* (in batteries, Chap. 28); (2) *magnetic* (generators, Chap. 27); (3) *thermal* (heated thermocouple junction, Sec. 13-18); (4) *photoelectric* (conversion of light energy into movement of electrons, Sec. 9-4); (5) *friction* (produced by rubbing two substances together, as tires rolling on a road, walking across wool rugs, or stroking a cat); (6) *magnetohydrodynamic* (MHD, converting heat energy in hot gases directly to electric energy); and (7) *piezoelectric* (by exerting pressure on a crystal). Note that all these are merely methods of converting energy of one form to electric energy.

Although alternating current has been mentioned only briefly so far, from this point on it becomes more important. Some methods by which it can be produced are *magnetic* (mechanical motion of a wire across alternate-direction magnetic lines induces alternating emf in the wire, Chap. 27); *magnetostrictive* (mechanical vibration of ferromagnetic materials induces an alternating emf in a wire coiled around the material, Sec. 3-19); and *piezoelectric* (mechanical vibration of quartz or rochelle salt crystals produces an alternating emf between two metal plates on opposite sides of the crystal, Sec. 11-12). Transistors or vacuum tubes operating with coils, capacitors, and resistors may form *oscillator* circuits which can generate ac of various forms.

4-2 A BASIC CONCEPT OF ALTERNATING CURRENT

A generator developing an emf that alternately forces electrons through a circuit in one direction, stops them, and then moves them in the opposite direction is called an *alternator*. It produces alternating voltages, which in turn can produce alternating currents in a circuit.

When it comes to doing electrical work, such as rotating motors, lighting lamps, and so on, either dc or ac can be used. In some cases it may be easier to do the job with one; in other cases, with the other.

One advantage of ac is the ease with which its

voltage can be stepped up or down by a transformer, which is a relatively simple piece of equipment. When electric power is transported over long distances, much less power is lost in heating the power lines themselves if the voltage is high, because with higher voltages less current is required to produce the same amount of power at the far end of the line. Power loss in a line is computed by the power formula $P = I^2R$. Since power loss is proportional to the current squared, anything that will lessen the required current will greatly lessen the power lost in heat. Doubling the voltage in a system reduces the required current to one-half for the same power. Since power loss is proportional to the current squared, one-half squared equals one-quarter of the power loss in the resistance of the wires. Power companies transport electric energy over long distances at potentials of 120,000 to more than 500,000 V.

Alternating current or voltage can be produced in many forms. Figure 4-1a represents current flowing

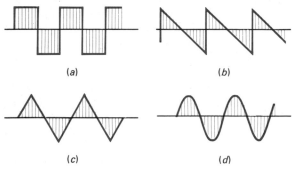

FIG. 4-1 Four possible ac waveshapes: (a) square wave, (b) sawtooth, (c) triangular, (d) sinusiodal.

in one direction at a constant amplitude for a period of time, immediately reversing to the opposite direction of flow in the circuit for a period of time, again reversing to the first direction, and so on. This is a *square-wave* form of ac.

Figure 4-1b shows the current rising to a peak value instantaneously, slowly decreasing to zero, then slowly increasing in the opposite direction. This is a *sawtooth* ac waveform.

Figure 4-1c shows the current rising linearly to a peak and then dropping to zero linearly again. This is a triangular waveform.

Figure 4-1d shows current approaching its peak slowly. If the current increases as the sine of the angle (explained in Sec. 4-3), it is known as *sine-wave, or sinusoidal,* ac. Sine-wave ac is considered to be the perfect waveform. It is the type normally used by power companies and generated by radio transmitters. When ac is mentioned, it is considered to be sinusoidal unless specified to be otherwise.

It is possible to make dc vary in a sinusoidal, square, sawtooth, or triangular manner.

4-3 THE AC CYCLE

The electron flow in an ac circuit is continually reversing. Each time it reverses, it is said to be *alternating*. Two of these alternations result in a *cycle*. Figure 4-2 shows two cycles or four alternations of ac.

The curved line represents the amount of current flow (or voltage) in a circuit. It indicates that current begins to flow at the point of time marked 0°, increases until the 90° point is reached, and then decreases in strength to the 180° point. This completes one-half of a cycle, or one *alternation*. The current now reverses direction and increases in strength or magnitude to the 270° point and then decreases to zero current at the 360° point. It has completed one cycle, or 360°. The start of a new cycle has been reached. While a cycle may also be considered as starting at a 90° point and moving through to the next similar 90° point, normally a cycle is considered to start at 0°.

The horizontal line in Fig. 4-2 on which the degrees are marked is known as the *time line*. It represents time progressing to the right.

A complete cycle is considered to be 360°, a half cycle to be 180°, and a quarter cycle to be 90°. A point on the time line one-third of the way from 0° to 90° is represented as 30°. A point two-thirds of the time from the zero point to the maximum is said to be 60°. The maximum points are normally at 90° and 270°.

A point 30° past 90° is 120°. The magnitude of a sinusoidal current, or voltage, at 120° will equal that of the current, or voltage, at the 60° point. This can be seen in the illustration.

If the cycle is a true sine wave, the magnitude of the current at the 30° point will be exactly one-half of the maximum value. It will have the same value at the 150°, 210°, and 330° points. Mathematically,

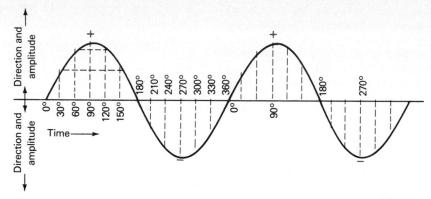

FIG. 4-2 Two cycles of sine-wave ac with magnitudes indicated every 30 electrical degrees.

the sine of 30° is 0.5, or one-half of the maximum value, whatever the maximum happens to be. The sine values of angles in degrees or to tenths of a degree can be found in Appendix C. At this time, the sines of only four angles, 0°, 30°, 60°, and 90°, will be used.

The sine of 60° is 0.86. Therefore the amplitude of the cycle at 60° is 0.86 of the maximum value. The magnitude of the current at the 120°, 240°, and 300° points will be the same as at 60°.

The sine of 0°, 180°, and 360° is zero. On the other hand, the sine of 90° and 270° is 1.0, called either the *maximum* or the *peak* value.

A circle may also be used to represent a complete sine-wave cycle. In electrical work, the circle (cycle) is considered as starting at the right-hand side and rotating counterclockwise, as in Fig. 4-3.

On the circle, 90° is actually one-quarter of a cycle from the start, 180° is one-half of a cycle, and so on. This method of indicating the number of

electrical degrees, or the angle from 0°, is quite graphic.

The arrow from the center of the circle to the zero point in Fig. 4-3 may be termed a *vector* arrow. A quarter cycle represents 90° rotation of the vector arrow. A half cycle represents 180° rotation of the vector. The position of the vector in relation to the zero direction indicates an angle from zero. This can be called an *angular vector*, since it is indicating an angle.

An example of a vector arrow used as a *magnitude vector* is shown in the cycle of Fig. 4-4. In this case the length of the vector arrows varies as the strength of the current varies. These vectors indicate not angles but the magnitude of the current, or voltage, at the angle in degrees along the time line at which they are drawn. Their direction indicates the relative direction of the current in the circuit. They show that the current is flowing in one direction in the circuit during the first half cycle and in the opposite direction during the second half cycle.

Vector arrows are handy devices to represent

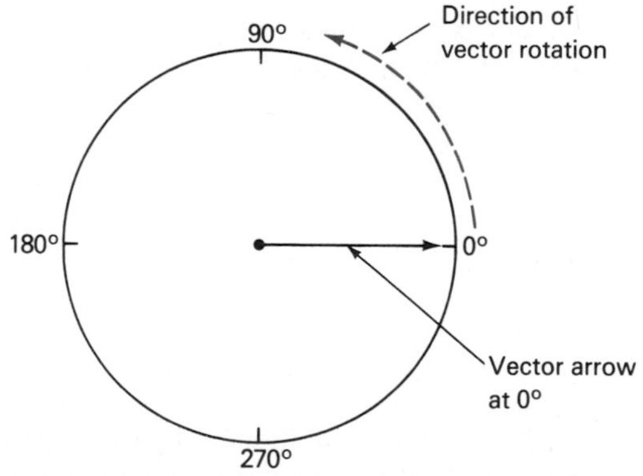

FIG. 4-3 Rotating vector method of indicating an ac cycle of 360° is a circle.

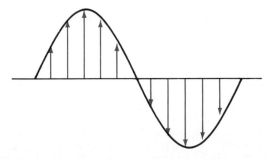

FIG. 4-4 Amplitude vectors indicating instantaneous magnitudes every 30° of one cycle.

quantities, angles, or directions. The magnitude vectors are visually indicating the *instantaneous* values. If a vector arrow in a sine wave is drawn at the peak, the instantaneous value is the maximum value. If the vector is drawn at the 30° point on the time line, the instantaneous value is 0.5 of the maximum voltage (0.5 V_{max}). The vector arrow at 60° will show an instantaneous voltage value of 0.86 V_{max}. An instantaneous value indicates the magnitude of a voltage (or current) at a particular time interval from 0°.

Rotating vectors can indicate the angle of an instantaneous value of current, or voltage, in a cycle. The distance between the point of the rotating vector arrow and the zero vector indicates the instantaneous strength (Fig. 4-5).

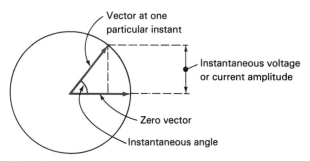

FIG. 4-5 Angle between starting point and vector determines instantaneous voltage or current value.

While this book uses 360° as one cycle, some engineering texts may use *radian* measure. One radian is the angle produced by a vector arrow tip rotating a distance equal to its own length. This angle is equivalent to 57.3°. From circular measure, 2π radius = circumference (π = 3.14). From this, 2π radians = 2(3.14) 57.3°, or 360°. Thus π rad = 180°, and 2π rad = 360°. Therefore, 3 cycles = 3(2π rad) = 6π rad, or 6(3.14) rad = 18.84 rad. Converting this from radian to degree measure, 6(3.14)57.3° = 1080°.

For more exact pocket calculator work, π = 3.141593, and 1 radian = 57.29578°. For a beginner, accuracy to 3 or perhaps 4 significant figures should suffice. It indicates proper procedure.

4-4 PEAK AND EFFECTIVE AC VALUES

A 10-A dc and a 10-A *peak* ac are graphed to the same scale in Fig. 4-6. The dc represents a constant 10-A value. The ac rises to a peak of 10 A but is at this value for only an instant. Then it drops to zero, reverses, increases to 10 A in the opposite direc-

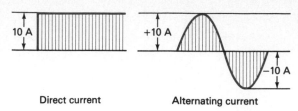

Direct current   Alternating current

FIG. 4-6 Comparison of a 10-A dc with a 10-A peak ac.

tion, and then drops to zero again. Although it is equally effective in doing work on both half cycles, during most of the cycle the ac has a value less than the constant dc and will be unable to produce as much heat or accomplish as much work.

Power being proportional to either V^2 or I^2 ($P = V^2/R = I^2R$), if all the instantaneous values of a half cycle of sine-wave current (or voltage) are squared and then the average, or mean, of all the squared values is found, the square root of this mean value will be 0.707 of the peak value. This *root-mean-square*, or rms, value represents how effective a sinusoidal ac will be in comparison with its peak value.

The *effective* value of a sine-wave ac cycle is also equal to 0.707 of the peak value. Comparing dc and ac, the 10-A peak ac will be as effective at producing heat and work as 7.07 A of dc. It may be said that the effective value of an ac is its *heating* value. Peak and effective dc values are the same for unvarying dc.

To determine a peak value of ac that will be as effective as a given dc, it is necessary to multiply the effective value given by the reciprocal of 0.707 (1/0.707), which is 1.414. For an ac to be as effective as 10 A of dc, it must be 10 × 1.414, or 14.14 A, at its peak.

The peak-to-peak (p-p) ac value is twice the peak value—the value between the maximum negative and maximum positive half cycles of any ac wave.

Peak and effective factors of 1.414 and 0.707 are applicable only when the current and voltage are sine waves. With square-wave ac the effective factor will be essentially 1.0. With most sawtooth waveshapes the factor is usually a little less than 0.5. The ac produced by microphones responding to voice sounds may have an effective value of only 0.3 of the peak value.

Today the FCC tends to rate radio transmitter output in peak power values, termed *peak envelope power* (PEP). What would be the effective power value? First, consider Fig. 4-7a, a half wave of sinusoidal ac. The effective ac V or I values are 0.707 of the peak value. If the peak voltage is 10

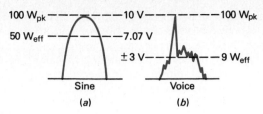

FIG. 4-7 (a) Peak and effective power levels for sine-wave ac. (b) For voice ac.

V, then the effective is 7.07 V. Using the dc or effective ac power formula $P = V^2/R$, and a resistance of 1 Ω for simple calculations, the peak power would be $10^2/1$, or 100 W. The effective, or the normally considered value would be $7.07^2/1$, or 50 W, *exactly half* of the peak power value. Similarly, using $P = I^2R$, and 10-A peak, the peak power would be $10^2(1)$, or 100 W. The effective power would be $7.07^2(1)$, or 50 W, again half the peak.

If an ac voice peak is measured to be 10 V across a 1-Ω resistor, regardless of its waveform, the peak power is $10^2/1$, or 100 W, but what would the effective value be? According to the voice wave shown in Fig. 4-7b, the effective power (contained in 70.7% of the wave area) would be at a voltage level of about 30% of the peak, indicating an effective power value of about $P = V^2/R$, or $3^2/1$, or only 9 W. With square-wave ac the peak and the effective powers would be the same, unless the corners of the square wave were slightly rounded, in which case the effective would be slightly less than the peak power.

When problems are given, it is assumed that the effective value is being employed unless stated otherwise. Thus, when the power company states it is furnishing 120 V ac, it means 120 V_{eff}, which is 120 × 1.414, or nearly 170 V peak.

| Test your understanding; answer these checkup questions.

1. What kind of voltage (ac or dc) is produced by friction? _____ A crystal? _____ Chemical action?_____ Heat? _____
2. If all that is specified is that ac is used, what waveshape would the ac be assumed to have? _____
3. How much are power line losses decreased if power is carried at 10 times the normal voltage? _____
4. What is the main advantage of ac? _____
5. At what voltages do you think power might be carried between the mountains and cities? _____ Along a city block to homes? _____
6. How many degrees in an alternation? _____ In a cycle? _____
7. At how many points in a normal cycle will the instantaneous voltage equal the voltage at 15°? _____ At 90°? _____ At 300°? _____
8. According to Appendix C, what is the sine of 15°? _____ 88°? _____ 160°? _____ 220.5°? _____ 321.8°? _____
9. Does a vector arrow normally rotate counterclockwise (CCW) or clockwise (CW)? _____
10. If sine-wave ac at maximum is 20 V, what is its amplitude at 40°? _____ At 135°? _____ At 292.3°? _____
11. What can vector arrows represent? _____ _____
12. How many degrees in one radian? _____ In 2π rad? _____
13. A current at 120° is 3 A. What is it at 265°? _____
14. A 350-V peak ac is as effective as what value of dc? _____
15. A filament of a vacuum tube requires a 0.4-A dc current to heat it. What peak ac is required? _____ What rms value? _____
16. A sinusoidal radio carrier ac reads 300 V effective across a 50-Ω line. What is the effective carrier power? _____ The PEP? _____
17. A voice ac reads 200 V peak across a 50-Ω line. What is the PEP? _____ An approximate P_{eff}? _____

4-5 THE AVERAGE VALUE OF AC

An ac cycle has a peak, an effective, and also a less frequently used value called the *average*. This is the value obtained when 180 instantaneous values, from 0° to 180°, each separated by 1°, are added and then the average is found by dividing the sum by 180. The average value for a sine wave is 0.636 of the peak value. For dc, the average, rms, and peak are all the same.

The "average" value may be confused with "effective," because it seems that the average value should represent how much work an ac cycle might do. As pointed out previously, however, it is the effective, or rms, value that does this. To summarize, for sine-wave ac:

AC peak value = 1.414 × effective value
AC effective value = 0.707 × peak value
AC average value = 0.636 × peak value

We will disregard the average value temporarily. Later, in Chaps. 10 and 13, in the study of power supplies and meters, when ac is changed to pulsating dc, the average value will be used.

4-6 FREQUENCY

The number of times an alternating current goes through its complete *cycle per second* is known as its *frequency*. The international unit of measurement of frequency is the *hertz*, abbreviated Hz. The English unit is cycles per second, abbreviated cps (1 Hz = 1 cps). This text will use hertz, although cps could be used correctly.

To simplify terminology, 1000 Hz is called a *kilohertz*, abbreviated kHz, and 1,000,000 Hz is called a *megahertz*, abbreviated MHz. Previous corresponding units were kc (kilocycles/s) and Mc. A frequency of 3,500,000 Hz can be expressed as 3500 kHz (previously 3500 kc), or as 3.5 MHz (previously 3.5 Mc), or as 0.0035 GHz.

The vibration rate of sound waves in air may also use the term *frequency*. When *middle C* is played on a musical instrument for example, an air disturbance with a frequency of 262 Hz is set up. The lowest tone that can be heard by human beings is about 15 Hz. The highest audible, audio, or sonic tones are usually 15–20 kHz. A microphone is a device or *transducer* that can change sound waves to an equivalent-frequency ac.

Frequencies that can produce sound waves that can be heard by humans are considered to be *audio frequencies* (AF). Frequencies that can be fed to antennas and will radiate electromagnetic and electrostatic waves are considered to be *radio frequencies* (RF).

The list of frequencies in Table 4-1 indicates the terminology that may be used for different ac frequencies. Note the overlapping of the frequencies from 10,000 to 1,000,000 Hz. The letters in parentheses are commonly used abbreviations.

At power frequencies, materials such as fiber,

cambric, cotton, some types of glass, black rubber, and impure Bakelite are satisfactory insulators. At higher radio frequencies, mica, low-loss hard rubber, special porcelains, Isolantite, Mycalex, polystyrene, steatite, plastics, and special glasses have lower losses and are to be preferred.

4-7 PHASE

There are three ways in which ac currents, emf's, or waves can differ. These are (1) amplitude, (2) frequency, and (3) phase. The *amplitude* means the peak height of the ac wave. The *frequency* is the number of cycles per second. The *phase* is the number of electrical degrees one wave leads or lags another. Figure 4-8a illustrates two ac waves graphed on the same time line. The solid-line wave starts upward before the dotted-line wave and is therefore leading the dotted line. The waves are out of phase, or out of step, by approximately 45°. If they were in phase, they would go to maximums at the same instant and to zero at the same instant.

The two waves shown in Fig. 4-8b start in phase, drop out of phase, and then return in phase again. To change phase this way, one or both of the waves must change frequency slightly. These waveforms may represent two voltages, two currents, or a voltage and a current. The two waves in Fig. 4-8c are 180° out of phase.

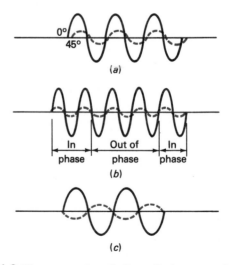

FIG. 4-8 Three examples of phase displacements between two waves. (*a*) Solid line leads dotted line by 45°. (*b*) Dotted line falls out of and back into phase. (*c*) Two waves 180° out of phase.

TABLE 4-1   AC FREQUENCY TERMINOLOGY AND RANGE

Terms used	Frequency limits
Subaudible frequencies	0.001 to 15 Hz (cps)
Audio frequencies (AF)	15 Hz to 20 kHz (ck)
Power frequencies	10 Hz to 1 kHz
Voice frequencies (VF)	250 to 3000 Hz
Video frequencies	0 to over 4.5 MHz (Mc)
Supersonic or ultrasonic frequencies	20 kHz to over 2 MHz
Extremely low frequencies (ELF)	30 to 300 Hz
Superlow frequencies (SLF)	0.3 to 3 kHz
Very low frequencies (VLF)	3 to 30 kHz
Low frequencies (LF)	30 to 300 kHz
Medium frequencies (MF)	300 kHz to 3 MHz
High frequencies (HF)	3 to 30 MHz
Very high frequencies (VHF)	30 to 300 MHz
Ultrahigh frequencies (UHF)	300 to 3000 MHz (3 GHz)
Superhigh frequencies (SHF)	3 to 30 GHz
Extremely high frequencies (EHF)	30 to 300 GHz (0.3 THz)
Heat or infrared	0.3 to 430 THz
Visible light, red to violet	430 THz to 1 kTHz
Ultraviolet	1 to 60 kTHz (6×10^{16} Hz)
X-rays	6×10^{16} to 3×10^{19} Hz
Gamma rays	3×10^{19} to 5×10^{20} Hz
Cosmic rays	5×10^{20} to 8×10^{21} Hz

4-8 OTHER AC CONSIDERATIONS

In some electronic circuits, a sinusoidal ac input signal becomes distorted and the output contains *harmonics* as well as the fundamental input frequency. A harmonic is a whole-number multiple of a frequency. That is, the first three harmonics of 1000 Hz are 1 kHz, 2kHz, and 3 kHz. Note that the fundamental and the first harmonic are the same thing, and that harmonics develop in 1, 2, 3, 4, 5, etc., steps.

We sometimes use the term *octave*. If an ac signal has a frequency of 1 MHz, then one octave higher is 2 MHz. Two octaves above 1 MHz is 1,000,000 × 2 × 2, or 4 MHz. Three octaves above 1 MHz is 1,000,000 × 2 × 2 × 2, or 8 MHz. Octaves increase in a 1, 2, 4, 8, 16 rate. While octaves always fall on harmonic frequencies, in our example the 3rd, 5th, 6th, and 7th harmonics are not involved in the octaves.

The *wavelength* of a cycle is the distance between successive positive (or negative) peaks of any wave. It is computed by dividing the wave velocity by its frequency. If a sound wave travels at a rate of 1100 ft/s, the wavelength of a 550-Hz sound wave is

$$\lambda = \frac{v}{f} = \frac{1100}{550} = 2 \text{ ft}$$

where λ = wavelength in velocity units
f = frequency
v = velocity

Wavelength is discussed in relation to antennas in Chap. 20.

The *period* of a wave is the time that it takes to complete one cycle of that frequency. For example, a 5 kHz ac goes through its cycle 5000 times a second. The period of such a wave is simply the reciprocal of 5000, or 1/5000, or 0.0002 of a second. It takes 0.2 millisecond (ms) to progress from the beginning of one of the cycles to the beginning of the next.

How fast a vector will rotate depends on the frequency of the ac involved. For example, with a frequency of 1 Hz, the vector will rotate 360°/s. The time to rotate 180° would be 180°/360° of a second, or 0.5 s. To rotate 45° would require 45°/360°, or 0.125 s. How long would it take for a vector to rotate 90° if the frequency is 1 MHz? This period of time can be computed by using 90°/360° divided by 1,000,000, or 0.25/1,000,000, which is 0.000 000 25 s, or 0.25 μs.

▌Test your understanding; answer these checkup questions.

1. What is another term for rms, or root-mean-square? _____ For hertz? _____
2. Most ac voltmeters read in rms values. What would a 50-V ac voltmeter indicate when across 28-V sinusoidal rms? _____ 50-V sinusoidal peak? _____ 100-V sinusoidal peak-to-peak? _____ 15-V square wave? _____
3. What is the average value of 50-V peak sinusoidal ac? _____ 2.5-A rms ac? _____ 8-V p-p ac? _____
4. If voltages above the time line are positive, which of the following would be negative? Voltages at 210°, 87°, 160°, 359°, 30°, 175°? _____

5. How much more than the average value is the effective value? _____
6. Which of the following materials make good insulators for power frequencies but not for high radio frequencies: mica, Mycalex, polystyrene, steatite, cotton, Isolantite, rubber? _____
7. A 10-V peak and a 5-V peak 60-Hz ac generator are in series. What will be the peak voltage if these two sources are in phase? _____ 180° out of phase? _____
8. What type of transducer changes sound waves to equivalent-frequency ac? _____ Changes ac to equivalent-frequency sound waves? _____
9. If two ac voltages do not reach their peaks at the same time, what are they said to be? _____
10. What harmonic of 1.5 MHz is 3 MHz? _____ 4.5 MHz? _____ 0.75 GHz? _____ 1500 kHz? _____
11. What is the distance between successive negative peaks called? _____ How would this be related to frequency? _____
12. How long does it take a 1-MHz cycle to rotate 360°? _____ 45°? _____
13. If velocity is 300,000,000 m/s, what is the wavelength of a 500 kHz ac? _____
14. What is the period of a 20-kHz ac wave? _____

ALTERNATING CURRENT QUESTIONS

1. What are four methods of generating ac?
2. What is the proper name for an ac "generator"?
3. What is a main advantage of ac over dc?
4. What are the four basic forms of ac waves?
5. What is one-half of an ac cycle called?
6. How many degrees in one cycle? One alternation?
7. When an ac cycle is graphed, in which direction does time progress?
8. What is the amplitude of a sine wave 30° from maximum? 60° from maximum? 90° from maximum? 180° from maximum?
9. When a circle represents an ac cycle, in which direction is its angular vector arrow considered to rotate?
10. How many degrees represent a quarter cycle? Half cycle? Three-quarter cycle? One cycle?
11. One radian is how many degrees? How many radians in a half cycle? In a full cycle?
12. To how many volts dc is 10-V effective ac equivalent in working ability?
13. What is another term meaning "effective" ac?
14. What is the effective value of 15-V peak ac?
15. Why would a power graph of a circuit having sine-wave ac voltage not graph as a sine wave?
16. What is the peak value of 110 V_{rms}? Its p-p value?
17. For a 30-V peak square-wave ac, what is the rms value? The average value? The p-p value?
18. What is the PEP delivered to a receiver drawing 50 W from a 120-V power line?

19. What is the PEP of a sinusoidal emission having a measured 70 V_{eff} into a 50-Ω load?
20. The average value is what fraction of the peak voltage value? Of the effective value?
21. What is another term meaning hertz?
22. How many kHz in 4.56 MHz?
23. How many GHz in 32.8 MHz?
24. Middle C represents what frequency?
25. What is the approximate range of audio or audible frequencies for young people? For old people?
26. What frequencies are considered to be radio frequencies (RF)?
27. What are the three ways in which ac waves may differ?
28. In what units is ac phase measured?
29. If two 180° out-of-phase 200-V_{rms} voltages are in series and are fed to a 50-Ω load, what power would be dissipated? If the voltages were 90° out of phase? If in phase?
30. What is the 5th harmonic of 800 kHz? The 5th octave above 800 kHz? The 5th octave below 800 kHz?
31. What is the wavelength of C below middle C produced by an organ?
32. What is the period of a 500-kHz wave?
33. How long would it take a vector to rotate 270° if the frequency is 1000 Hz?

5

Inductance and Transformers

This chapter discusses self-inductance of coils and chokes and basic transformer theory in terms of mutual inductance. Power transformers and their various ratios and losses are outlined. The idea of inductive reactance and its computations are developed along with the phase relations of inductive circuits.

5-1 INDUCTANCE

Coils of wire were mentioned in Chap. 3 when the electromagnetic effect produced by current flowing through them was considered. An equally important aspect of the operation of a coil is its property to oppose any *change* in current through it. This property is called *inductance*.

When a current of electrons starts to flow along any conductor, a magnetic field starts to expand from the center of the wire. These lines of force move outward, through the conducting material itself, and then continue into the air. As the lines of force sweep outward through the conductor, they induce an electromotive force (emf) in the conductor itself. This induced voltage is always in a direction opposite to the direction of current flow. Because of its opposing direction it is called a *counter emf* (cemf), or a *back emf*. The direction of this self-induced counter emf can be verified by using the left-hand generator rule (Sec. 3-18).

The effect of this backward pressure built up in the conductor is to oppose the immediate establishment of maximum current. It must be understood that this is a temporary condition. When the current eventually reaches a steady value in the conductor, the lines of force will no longer be in the process of expanding or moving and a counter emf will no longer be produced. At the instant when current begins to flow, the lines of force are expanding at the greatest rate and the greatest value of counter emf will be developed. At the starting instant the counter emf value almost equals the applied, or source, voltage (Fig. 5-1a).

The current value is small at the start of current flow. As the lines of force move outward, however,

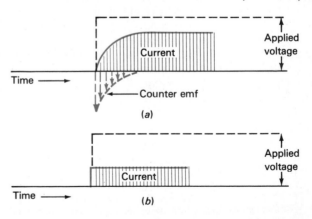

FIG. 5-1 Current, counter emf, and applied voltage plotted on a time base for (*a*) an inductive circuit and (*b*) a resistive circuit.

the number of lines of force cutting the conductor per second becomes progressively fewer and the counter emf becomes progressively less. After a period of time, the lines of force expand to their greatest extent, all counter emf ceases to be generated, and the only emf in the circuit is that of the source. Maximum current can now flow in the wire or circuit, since the inductance cemf is no longer reacting against the source emf.

The current is said to rise in an *exponential* manner:

$$i = I(1 - \epsilon^{-tR/L})$$

where ϵ = 2.718
   t = time in s from start
   R = ohms
   L = henrys
   i = instantaneous current
   I = maximum current

The $-tR/L$ is the "exponent."

If it were possible to produce current by applying a voltage across a wire and not produce a counter emf, then Fig. 5-1b would represent the action of the current. The figure shows a current reaching the maximum value instantly—just as soon as the voltage is applied. This is essentially true for a purely resistive circuit having short connecting wires.

5-2 SELF-INDUCTANCE

When the switch in a current-carrying circuit is suddenly opened, an action of considerable importance takes place. At the instant the switch breaks the circuit, the current due to the applied voltage should cease abruptly. With no current to support it, the magnetic field surrounding the wire should collapse back into the conductor at a tremendously high rate, inducing a high-amplitude emf in the conductor. Originally, when the field built outward, a counter emf was generated. Now, with the field collapsing inward, a voltage in the opposite direction is being induced. It might be termed a *counter-counter emf*, but is usually known as a *self-induced*

emf. The self-induced emf is in the direction of the applied source voltage. Therefore, as the applied voltage is disconnected, the voltage due to self-induction tries to establish current flow through the circuit in the same direction and aiding the source voltage. With the switch open it would be assumed that there is no path for the current, but the induced emf immediately becomes great enough to ionize the air at the opened switch contacts and an arc of current appears between them. Arcing lasts as long as energy stored in the magnetic field exists. This energy is dissipated as heat in the arc and in the circuit itself.

With purely resistive circuits involving low current and short wires, the energy stored in the magnetic field will not be great and the switching spark may be insignificant. With long lines (large inductance values) and heavy currents, inductive arcs several inches long may form between opened switch contacts on some power lines. The heat developed by arcs tends to melt the switch contacts and is a source of difficulty in high-voltage high-current switching circuits.

Note that, regardless of any change of current amplitude or direction in a conductor, the induced emf's oppose the current *change*. With a steady, unvarying direct current, there is no change of current and no opposition develops. When a varying dc is flowing, the counter emf opposes any increase of source voltage. As the source voltage decreases, the self-induced emf opposes the decrease. With alternating current, the constant state of change results in a continual opposing or reacting action. From this comes the definition: *Inductance is the property of any circuit to oppose any change in current, and in which energy is stored in the form of an electromagnetic field.*

The unit of measurement of inductance is the *henry*, defined as the amount of inductance required to produce an average counter emf of one volt when an average current change of one ampere per second is under way in the circuit. Inductance is represented by the symbol L in electrical problems, and henrys is indicated by H.

5-3 COILING AN INDUCTOR

It has been indicated that a piece of wire has the ability of producing a counter emf and therefore has a value of inductance. Actually, a small length of wire will have an insignificant value of inductance by general electrical standards. One henry represents a relatively large inductance in many circuit applications, where millihenrys and microhenrys (Sec. 1-14) are more likely to be encountered. A straight piece of No. 22 wire 1 meter long has about 1.66 μH. The same wire wound around an iron nail or other high-permeability core may produce 500 or more times that inductance.

Even without the iron core, a given length of wire will have much greater inductance if wound into coil form. Consider Fig. 5-2, showing two

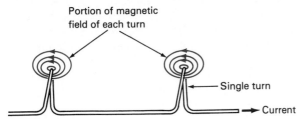

Portion of magnetic field of each turn

Single turn

Current

FIG. 5-2 Fields from two widely separated turns do not intercouple.

loops of wire separated by enough distance so that there is essentially no interaction between their magnetic fields. If the inductance of the connecting wires is neglected, these two loops, or *turns*, have twice the inductance of a single turn.

When the two loops are wound next to each other (Fig. 5-3) with the same current flowing,

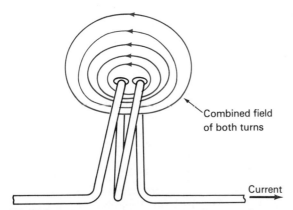

Combined field of both turns

Current

FIG. 5-3 Fields of close-wound turns intercouple.

there are twice the number of magnetic lines of force cutting each turn. With 2 turns, 4 times the counter emf is developed. If 3 turns are used, 3 times the number of lines of force cut 3 turns, so 9 times the counter emf is developed. The inductance of a coil varies as the number of turns squared (n^2). However, the length of the coil is also going to enter into the exact computation of the inductance of a coil. If the turns are stretched out, the field

intensity will be less and the inductance will be less. The larger the radius or diameter of the coil, the longer the wire used and the greater the inductance. In single-layer air-core coils with a length approximately equal to the diameter, a formula that will give the approximate inductance in microhenrys is

$$L = \frac{r^2 n^2}{9r + 10l}$$

where L = inductance, in microhenrys (μH)
r = radius of coil, in inches
n = number of turns
l = length of coil, in inches

Using metric measurements, essentially the same formula is

$$L = \frac{r^2 n^2}{24r + 25l}$$

where r = radius, in centimeters (cm)
l = length, in cm

For more accurate and detailed formulas to compute inductance of various coils having differing permeability cores, refer to radio or electrical engineering handbooks.

The inductance of straight wires alone is encountered in antennas, in power lines, and in ultrahigh-frequency equipment. In most electronic applications where inductance is required, space is limited and wire is wound into either single-layer or multi-layer coils with air, powdered-iron-compound (ferrite), or laminated (thin iron sheet) cores. The advantage of multilayer coil construction for high values of inductance becomes obvious when it is considered that, while 2 closely wound turns produce 4 times the inductance of 1 turn, the addition of 2 more turns closely wound on top of the first 2 will provide almost 16 times the inductance. *Ferrite* materials like zinc- or manganese-iron oxides have magnetic properties but are insulators.

In many applications, coils are wound around a ferrite cylinder that can be screwed into or out of the core space of the coil. This results in a controlled variation of inductance, maximum when the iron-core "slug" is screwed into the coil and minimum with it backed out.

A special type of coil is the *toroid*, which consists of a doughnut-shaped ferrite core, either single-layer-wound as shown in Fig. 5-4, or multi-layer-wound. Its advantages are high values of inductance with little wire, and therefore little resistance in the coil. Furthermore, all the lines of

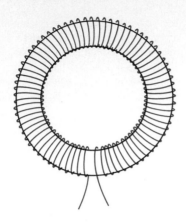

FIG. 5-4 Coil wound on a toroidal form.

force are in the core and almost none outside (provided there is no break in the core). As a result it requires no shielding to prevent its field from interacting with outside circuits. Two toroids can be mounted so close that they nearly touch, and there will be almost no interaction between them. If a secondary is wound over the first winding, a toroidal transformer (Sec. 5-15) results.

5-4 *LR* CIRCUIT TIME CONSTANT

The time required for the current to rise to its maximum value in an inductive circuit after the voltage has been applied will depend on both the inductance and the resistance in the circuit. With a constant value of resistance in a circuit, the greater the inductance, the greater the counter emf produced, and the longer the time required for the current to rise to maximum.

With a constant value of inductance in a circuit, the more resistance, the less current that can flow. The less current, the less possible counter emf to oppose the source emf and the less time required to reach a maximum current value.

The time required for the current to rise to 63% of the maximum value (called the *time constant*) can be determined by

$$T_c = \frac{L}{R}$$

where T_c = time, in seconds (s)
L = inductance, in henrys (H)
R = resistance, in ohms (Ω)

According to this formula, a 10-H coil with 10-Ω resistance will allow current to rise to 63% of max-

imum in one second. In the next second, the current will rise 63% of the remaining amount toward maximum, and so on. (If a coil could be produced with zero ohms of resistance and be connected across a power supply of infinite current capabilities, theoretically the current would never reach a maximum value.) A time equivalent to five times the time-constant formula value results in a current within 1% of maximum. This is usually considered as the maximum value. A circuit with zero inductance and only resistance will reach its maximum current value instantly (Fig. 5-1b).

The time-constant formula also indicates the time required for current to decrease by 63% from maximum in an inductive circuit if the voltage decreases abruptly.

5-5 ENERGY IN A MAGNETIC FIELD

Current flowing in a wire or coil produces a magnetic field around itself. If the current suddenly stops, the magnetic field held out in space by the current will collapse back into the wire or coil. Unless the moving field has induced a voltage and current into some external load circuit, all the energy taken to build up the magnetic field will be returned to the circuit in the form of electric energy as the field collapses.

The amount of energy in joules that is being stored in a magnetic field at any instant can be determined by the formula

$$E_n = \frac{LI^2}{2}$$

where E_n = energy, in joules (wattseconds)
L = inductance, in H
I = current, in A

5-6 CHOKE COILS

The ability of a coil to oppose any change of current can be used to smooth out varying or pulsating types of current. In this application an inductor is known as a *choke coil*, since it chokes out variations of amplitude. For radio-frequency (RF) ac of perhaps 100 kHz to 500 MHz, an air-core coil may be used. For lower-frequency circuits greater inductance is required. For this reason iron- or ferrite-core choke coils are found in AF, lower RF, and power-frequency applications. Several types of inductors are shown in Fig. 5-5.

An iron-core choke coil will hold a nearly constant inductance value until the core material be-

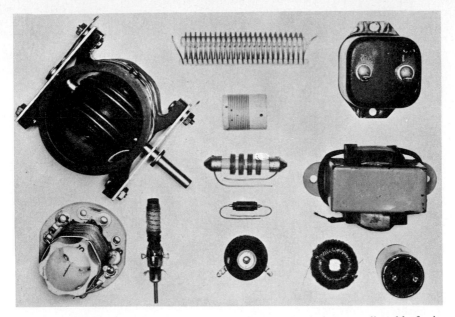

FIG. 5-5 Inductors. Left, from top, variable air-core, fixed air-core, adjustable ferrite-slug coil. Middle group, various air-core coils. Right group, iron-core inductors: from top, shielded audio choke, unshielded power choke, toroid, miniature shielded high-inductance choke.

comes saturated. When enough current is flowing through the coil to saturate the core magnetically, variations of current above this value can produce no appreciable counter emf's and the coil no longer acts as a high value of inductance to these variations. To prevent the core from becoming magnetically saturated, a small air gap may be left in the iron core. The air gap introduces so much reluctance in the magnetic circuit that it becomes difficult to make the core carry the number of lines of force necessary to produce saturation. The gap also decreases the inductance of the coil. An air-core coil cannot be saturated.

Figure 5-6 shows the symbol for an air-core coil

or choke, an iron-core coil or choke, and a simplified picturization of a laminated iron-core choke.

Small beads of ferrite material can be slipped over circuit wires and will act as RF chokes for frequencies above a few megahertz.

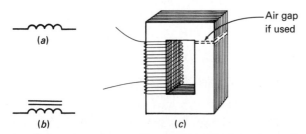

(a)

(b)   (c)

FIG. 5-6 (a) Air-core coil symbol. (b) Iron-core coil symbol. (c) Simplified construction of an iron-core choke coil.

Test your understanding; answer these checkup questions.

1. Does a 1-m wire have any inductance if ac is flowing in it? _____ If dc is flowing in it? _____

2. In a switched dc circuit, which is greater, the counter emf or the self-induced emf? _____

3. What effect causes switch or relay contacts to develop pits and mounds on them? _____

4. How much inductance is in a circuit if 2 V of counter emf is developed in 0.5 s? _____

5. What is the property of a circuit that opposes any change in current called? _____ That opposes current? _____

6. What is the advantage of using toroid coils? _____

7. What is the L of a coil in μH having a diameter of 1.5 in., a length of 2 in., and 50 turns? _____ In mH? _____

8. A 5-H coil is in series with a 100-Ω resistance, a switch, and a battery. How long will it take for

the current to reach 63% of I_{max}? —————
Approximately I_{max}? —————

9. How much energy is contained in the field of a 7-H choke coil when 500 mA is flowing through it? —————

10. What relative-frequency circuits might use choke coils with air cores? ————— Iron cores? —————

11. To prevent saturation of the core of a choke coil, should the core be continuous or have an air gap? —————

12. If a 10-turn coil has a second layer of 10 turns wound over the first, what should be the total inductance? —————

5-7 MUTUAL INDUCTANCE

Any coil has a value of inductance, or self-inductance. It has 1 H of inductance if an average current change of 1 A/s produces an average counter emf of 1 V in it.

If one coil is placed near a second, it will be found that alternating or varying currents in the first produce moving magnetic fields that will induce voltages in the second coil. The farther apart the two coils are, the fewer the number of lines of force that interlink the two coils and the lower the voltage induced in the second coil (100,000,000 lines/s cutting across one turn induce 1 V).

When an average current change of 1 A/s in the first coil can produce moving fields that will induce an average of 1 V in the second coil, the two coils are said to have a *mutual inductance* of 1 H, regardless of the inductance values of the two coils themselves.

The mutual inductance can be increased by moving the two coils closer together or by increasing the number of turns of either coil.

In power *transformers*, two coils are so arranged that almost all the lines of force on the first coil cross the turns of the second coil. A large mutual inductance value results.

When *all* the lines of force from one coil cut *all* the turns of a second coil, *unity coupling* exists, and the mutual inductance may be found by the formula

$$M = \sqrt{L_1 L_2}$$

where M = mutual inductance, in H
L_1 = inductance of coil 1, in H
L_2 = inductance of coil 2, in H

This formula assumes 100% coupling between the two coils. If all the lines from the first coil do not cut all the turns of the second, M is determined by

$$M = k\sqrt{L_1 L_2}$$

where k = percent of lines that couple

EXAMPLE: If a 2-H coil has 84% of its lines of force cutting across a 4.5-H coil, what mutual inductance exists? By substituting in the formula above,

$$M = k\sqrt{L_1 L_2} = 84\%\sqrt{2(4.5)}$$
$$= 0.84\sqrt{9} = 0.84(3) = 2.52 \text{ H}$$

5-8 COEFFICIENT OF COUPLING

The degree, or closeness, of coupling can be expressed as a percent, as above, although the term *coefficient of coupling* is to be preferred. For examples, 100% coupling is equivalent to a coefficient of coupling of 1.0, or *unity*; 95% is equivalent to a coefficient of coupling of 0.95; and so on.

The coefficient of coupling between two coils can be computed from the rearrangement of the mutual-inductance formula

$$k = \frac{M}{\sqrt{L_1 L_2}}$$

The answer obtained with this formula is always a decimal value, unless the coefficient is unity.

EXAMPLE: The mutual inductance between two coils is 0.1 H, and the coils have inductances of 0.2 and 0.8 H. What is the coefficient of coupling?
By substituting in the formula,

$$k = \frac{M}{\sqrt{L_1 L_2}} = \frac{0.1}{\sqrt{0.2(0.8)}}$$
$$= \frac{0.1}{\sqrt{0.16}} = \frac{0.1}{0.4} = 0.25 \text{ or } 25\%$$

Coils with relatively high coefficients of coupling may be said to be *tightly* coupled. With low values of coupling they are said to be *loosely* coupled. What is tight and what is loose will vary in different applications. In power transformers the coefficient may exceed 0.98, while in some radio circuits a coefficient as low as 0.01 is all that may be required.

5-9 INDUCTANCES IN SERIES

Electric circuits often have two or more inductances in them. Whether the magnetic fields of the two coils interlink or not determines the effective amount of inductance presented to the circuit by the coils.

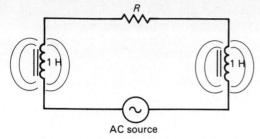

FIG. 5-7 Two uncoupled inductances in a series circuit.

Figure 5-7 shows two 1-H coils and a resistor in series across an ac generator, called an *alternator*. Since the two coils are widely separated, no interlinkage of fields occurs and the total inductance in the circuit is simply 2 H (any slight inductance in the connecting wires can be neglected). The formula for uncoupled inductances in series is

$$L_t = L_1 + L_2 + L_3 + \cdots$$

where L_t = total inductance, in H
$L_1, L_2, \cdots$ = separate inductances, in H

Note the similarity of this series-inductance formula to that of series resistors.

Figure 5-8 shows two 1-H coils and a resistor in series across an alternator. The two coils are close

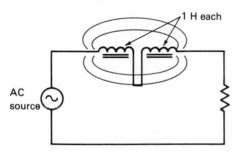

FIG. 5-8 Two series inductances placed to intercouple their fields.

enough that the lines of force from one coil interlink the other. Now, the mutual inductance will affect the total-inductance value. If the coils are wound in the same direction, the emf induced from one to the other will be *in phase*, or additive, and the total-inductance value will be more than the simple sum of the two inductance values alone. The effective inductance of two in-phase series-connected inductors can be determined by using the formula

$$L_t = L_1 + L_2 + 2M$$

where M = mutual inductance, in H

If the two coils were unity-coupled and each had 1 H of self-inductance, the total-inductance value would be 4 H.

If two coils are wound in opposite directions and coupled, the induced emf in the coils will be in opposition, or *out of phase*, and will tend to cancel each other, resulting in less effective inductance. The formula for total inductance in this case is

$$L_t = L_1 + L_2 - 2M$$

If the two coils were unity-coupled and each had 1 H of self-inductance, the total-inductance value would be zero henrys. The two coils would have completely canceled each other's inductance.

5-10 INDUCTANCES IN PARALLEL

In some circuits two or more inductors may be connected in parallel as in Fig. 5-9.

If the two inductances have 1 H each, the resultant inductance will be 0.5 H (no interaction of their

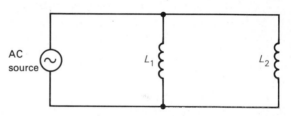

FIG. 5-9 Two parallel inductances, fields uncoupled.

fields is assumed). This inductance is computed by using a formula similar to the parallel-resistance formula:

$$L_t = \frac{L_1 L_2}{L_1 + L_2}$$

If three or more inductances are in parallel, a formula similar to the parallel-resistance formula may be used:

$$L_t = \frac{1}{1/L_1 + 1/L_2 + 1/L_3}$$

(Since coils are rarely connected in parallel and intercoupled, the more complicated formulas for such circuits will not be presented.)

5-11 SHORTING A TURN IN A COIL

There are several methods of reducing the inductance of a coil. Examples are taking turns off the coil, stretching the coil out until it has a greater

Connector shorts
out one turn

FIG. 5-10 Coil with one shorted turn.

length, using a less permeable core, or shorting one or more turns of the coil as in Fig. 5-10.

If one turn of a coil is shorted, there is one less turn in the coil and the coil has a little less inductance. If dc is flowing through the coil, there will be relatively little change in the magnetic field around the coil. However, when ac flows through the coil, the expanding and contracting fields from adjacent turns cut the shorted turn and induce an emf in it. In a shorted turn even a small emf induces a relatively high-amplitude current in the turn. Since the emf induced in the shorted turn is a counter emf, it produces a current in the shorted turn in an opposite direction to that flowing in the remainder of the coil. This results in a counteracting field, partially canceling the field of the coil. The inductance of the coil is reduced by much more than would result from cutting off one turn. A shorted turn may become noticeably warm or hot, depending on how much current is induced in it.

In some radio applications in which a variation of inductance is desired, a shorted turn in the form of a loop of wire or a brass disk may be brought near a coil, effectively reducing the inductance of the coil. A brass screw turned into the core of the coil has the same effect. The closer a shorted turn is to the center of a coil, the more it decreases the inductance.

5-12 INDUCTIVE REACTANCE

It has been explained that dc flowing through an inductor produces no counter emf to oppose the current. With varying dc, as the current increases, the counter emf opposes the increase. As the current decreases, the counter counter emf opposes the decrease. Alternating current is in a constant state of change, and the effect of the magnetic fields is a continual induced voltage opposition to the current.

This reacting ability of the inductance against the changing current is called *inductive reactance*.

Inductance is the property of a circuit to oppose any change in current and is measured in henrys. Inductive reactance is a measure of how much the counter emf in the circuit will oppose current variations, and it is measured in *ohms*.

It may be considered that the inductance of a coil does not change. Whether it is used in a dc circuit, a 60-Hz circuit, or a 10,000-Hz circuit, or whether it lies unused on a shelf, a 1-H inductance has a value of 1 H. Its property has not changed. When dc is flowing through a 1-H coil, there will be no opposition to current flow except for any ohmic resistance in the coil. In a low-frequency 60-Hz circuit, the magnetic field builds up and collapses relatively slowly and relatively little counter emf may be developed to oppose circuit current. In a higher-frequency circuit, the magnetic field moves more rapidly and produces more counter emf, which opposes the current more. The amount of opposition, or reactance, is directly proportional to the frequency of the current variation.

The value of counter emf induced can be determined by

$$v = L \frac{di}{dt}$$

where v = counter emf
L = inductance, in H
di = change in current
dt = change in time, hence frequency

A 1-H coil will produce a certain value of opposition to a 60-Hz ac. A 2-H coil will have twice the counter-emf-producing capability and oppose the current change twice as much. Therefore, reactance is also directly proportional to the inductance value.

When the inductance in henrys is multiplied by the frequency in cycles per second and this is multiplied by $2 \times \pi$ (Greek letter pi), an inductive-reactance value results that is similar to a resistance value in ohms. The formula is

$$X_L = 2\pi f L$$

where X_L = inductive reactance, in Ω
π = 3.14 (for simplicity)
f = frequency, in Hz
L = inductance, in H

The Greek letter ω (omega) is often used to indicate $2\pi f$. (This is known as the angular velocity of the cycle and indicates how fast the current is changing.) Thus, $X_L = \omega L$.

As an example of how much inductive reactance is presented by a circuit composed of a source of ac and an inductance, consider the circuit in Fig. 5-11,

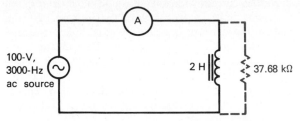

100-V, 3000-Hz ac source   A   2 H   37.68 kΩ

FIG. 5-11 Current is determined by the inductive reactance, voltage, and frequency in the circuit.

in which the source has a frequency of 3000 Hz and the inductance is 2 H. By substituting in the inductive-reactance formula,

$$X_L = 2\pi fL = 2(3.14)3000(2)$$
$$= 6.28(6000) = 37,680 \ \Omega$$

The 2-H coil presents 37,680 Ω of opposition to a 3000-Hz ac and will limit the amplitude of the current in the circuit exactly as much as if a 37,680-Ω resistance were used instead. (The lower-case Greek letter ω is often used to indicate reactive ohms, with the upper-case Ω used for resistive ohms. We will use Ω for both conditions.) Reactance can be substituted for resistance in Ohm's-law formulas but only if the circuits are purely reactive. Ohm's law for reactive circuits states: The current in a reactive circuit is directly proportional to the voltage and inversely proportional to the reactance. In formula form,

$$I = \frac{V}{X} \qquad V = IX \qquad X = \frac{V}{I}$$

where I = current, in A (usually rms)
V = emf, in V (usually rms)
X = reactance, in Ω (or ω)

The general reactance symbol X is used in these formulas rather than X_L because there is another reactance, known as *capacitive reactance*, with the symbol X_C (Chap. 6). The Ohm's-law formulas apply to either X_L or X_C. To specify inductive reactance only, the symbol X_L should be used.

If the ac generator in Fig. 5-11 produces an emf of 100 V, the circuit current by Ohm's law is

$$I = \frac{V}{X} = \frac{100}{37,680} = 0.00265 \text{ A or } 2.65 \text{ mA}$$

In practice, Ohm's-law formulas for reactive circuits should be used only if the reactance value is

more than 10 times the resistance. If there is an appreciable proportion of resistance in the circuit, Ohm's law for ac circuits (Sec. 7-2) must be used.

If two uncoupled inductive reactances are in series, the total reactance is

$$X_{LT} = X_{L_1} + X_{L_2}$$

If two uncoupled inductive reactances are in parallel, the total reactance may be computed by using the formula

$$X_{LT} = \frac{X_{L_1}X_{L_2}}{X_{L_1} + X_{L_2}}$$

The inductive reactance formula can be rearranged to solve for inductance if X_L is known. Thus

$$X_L = 2\pi fL \qquad \text{and}$$

$$L = \frac{X_L}{2\pi f} \qquad \text{or} \qquad L = \frac{X_L}{\omega}$$

Test your understanding; answer these checkup questions.

1. If an average 2 A in one coil induces 0.5 V into an adjacent coil, what is the M value? _____
2. What is the coefficient of coupling expressed in decimals for a k of 3.5%? _____
3. What is the mutual inductance of two unity-coupled coils of 4 and 9 H? _____
4. What is the M value of a 4-H and a 5-H coil having a coefficient of coupling of 0.85? _____
5. What is the inductance of a 3-H and a 4-H coil in series, fields aiding, and 40% coupled? _____
6. What is the L value when a 5-H and an 8-H coil are in series, fields opposing, if the k value is 0.75? _____
7. What is the inductance if a 0.5-H an a 0.2-H coil are in parallel? _____
8. What happens to the L value if a brass slug is moved into the core area of a coil? _____ If a ferrite slug is used? _____
9. What part of a reactance formula indicates cyclic consideration? _____
10. What is the reactance of a 10-H coil to 120-Hz ac? _____ A 5-mH choke to 1000 kHz? _____
11. What is the X_L of a 600-μH coil to 4 MHz? _____
12. What inductance is required to present 500 Ω of reactance to a 10-kHz ac? _____
13. At what frequency does a 0.04-H coil have a reactance of 3 kΩ? _____
14. How much I flows through a 2-H choke when connected across a 120-V 60-Hz power line? _____

15. An RF choke coil develops a 200-V drop across itself when a 5-mA 2-MHz ac flows through it. What reactance does it present to this frequency? _____

16. How much voltage-drop will occur across an audio choke coil having 5 H when 150 mA of 40-Hz ac flows through it? _____

17. A zero-R choke coil limits current through it to 0.5 A when across a 150-V, 60-Hz line. What is its: X_L? _____ L? _____

5-13 PHASE RELATIONS WITH INDUCTANCE

When a coil is connected across an ac generator, as in Fig. 5-12, the current in the coil will not rise to a peak at the instant that the voltage attains a peak value. Theoretically, the current in the coil will lag

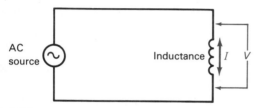

FIG. 5-12 A purely inductive circuit.

the source voltage by 90° (assuming negligible resistance in the circuit).

When such an inductive circuit is first turned on, the current and voltage start out in phase as a varying dc. (The starting operation is quite complex and will not be dealt with here.) After a few voltage cycles, however, the circuit current begins to alternate and then settles down into what is known as a *steady-state* ac and continues to operate in this manner until something in the circuit is changed. All explanations will be for steady-state conditions.

Figure 5-13 represents one cycle of steady-state ac flowing through a purely inductive circuit. The

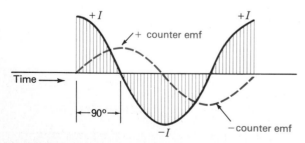

FIG. 5-13 Phase relations of the current and the counter emf in an inductor.

current cycle under consideration is from a maximum positive current value to a maximum negative value and back to a maximum positive value.

As the current increases through an inductor, the magnetic field increases, exactly in step. At the instant the current reaches its maximum positive value, its *rate of change* is zero. This means maximum field strength but zero magnetic-field *movement* and therefore zero counter emf due to moving magnetic fields, indicated by the dotted curve. Thus, maximum current and zero counter emf occur at the same instant.

As the current diminishes toward zero, the magnetic field collapses inward toward the center of the coil. As the current nears the zero point, its rate of change is very rapid, producing maximum induced positive voltage at the zero-current instant.

The current then starts to increase in the negative direction, producing an opposite-polarity magnetic field expanding as the current increases. This reversed-polarity expanding field produces a voltage of the same *polarity* as the original contracting field produced. This is a significant point. Reversing the field polarity and at the same time reversing a contracting motion to an expanding motion constitutes a double reversal. A double reversal produces the same, or original-direction, induced emf. (If you reverse direction twice, you are still going in the original direction.) Therefore, as the current reverses, the induced emf is still in the same direction (positive) and continues to be developed as long as current is changing. As the current reaches maximum in the negative direction, its *rate of change* decreases to zero. At this instant there is zero induced voltage in the coil again.

As the current drops from maximum negative toward zero, its rate of change increases, developing an induced emf again, this time in the negative direction. When the current reaches zero, the induced voltage is again at a maximum but at a maximum negative value.

As the current swings up from zero to maximum in the positive direction, the induced voltage drops from maximum negative to zero once more. This completes one full cycle of current and induced voltage.

By reference to the figures it can be seen that the current and induced voltage in the coil are constantly a quarter of a cycle, or 90°, out of phase.

Since the induced voltage is a counter emf (counter to the source voltage), the source voltage must be 180° out of phase with it. The three curves in Fig. 5-14 show the phase relations of the current,

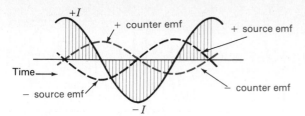

FIG. 5-14 Phase relations between current, counter emf, and source emf in an inductor.

the counter emf or induced voltage, and the source voltage in a purely inductive circuit. The current lags 90° behind the source voltage.

5-14 PHASE WITH BOTH *L* AND *R*

When inductance and resistance are in series in a circuit, the number of degrees the source voltage and the circuit current will be out of phase depends upon the relative resistance and reactance values.

The circuit in Fig. 5-15 has its resistance and inductive reactance values equal. Remember: The

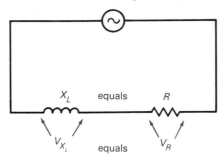

FIG. 5-15 When X_L and R are in series and equal, the voltage-drops across them are equal.

current in all parts of this circuit is exactly the same at any single instant, since it is a series circuit.

When the current through the resistance is at a maximum value, the voltage-drop across it is at a maximum. The voltage across and the current through a pure resistance are always exactly in phase.

The current in the coil leads the induced or counter emf by 90°. The applied voltage across the coil is 180° out of phase with the induced emf and therefore leads the current by 90°.

There is only one current in the circuit, but the source sees two voltage-drops in series, one across the resistor (in phase with the current) and the other across the coil (and leading the current by 90°).

Since the resistance and reactance are equal, the source sees two voltage-drops of equal magnitude but 90° out of phase. As a result, the source voltage sees a resultant voltage-drop equal to itself but 45° out of phase and leading the circuit current. This is graphed in Fig. 5-16.

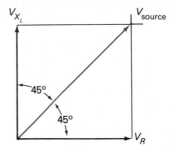

FIG. 5-16 Vector representation of the source voltage, voltage across an inductive reactance, and voltage across the resistor.

If there is more R than X_L, the phase difference will be closer to zero degrees. If there is more X_L, the phase will be closer to 90°. (This is discussed in Sec. 7-3.)

5-15 TRANSFORMERS

One of the common components, or parts, used in electricity, electronics, and radio is the transformer. The name itself indicates that the device is used to transform, or change, something. In practice a transformer may be used to step up or step down ac voltages, to change low-voltage high-current ac to high-voltage low-current ac, or vice versa, or to change the impedance of a circuit to some other impedance in order to transfer energy better from a source to a load.

In its simplest form, a transformer consists of a *primary* wire and a *secondary* wire laid side by side (Fig. 5-17). The only parts of the primary and secondary circuits to be considered are the portions lying parallel to each other. When the source is producing an alternating voltage, an alternating cur-

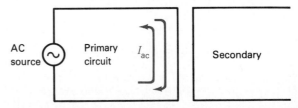

FIG. 5-17 Basic transformer with ac flowing in the primary.

rent will be developed in the primary wire, as indicated by the arrows, producing expanding and contracting alternating magnetic fields around the primary wire. These fields induce a counter emf in the primary, which attempts to counteract the source voltage and thereby limit the primary current value. In practical transformers the primary coil has a sufficient number of turns (inductance and inductive reactance) to produce a counter emf almost equal to the source voltage and, consequently, when unloaded has very little primary current.

In addition to the counter emf induced into the primary circuit by self-induction, the expanding and contracting fields from the primary cross the secondary wire and induce an ac emf in it. According to the left-hand generator rule, by electron current theory (Sec. 3-18), if the current in the primary flows downward, the induced emf in the secondary will be upward and 180° out of phase with the primary emf (Fig. 5-18).

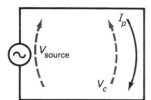
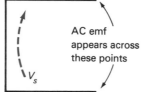

FIG. 5-18 Basic transformer voltages and currents with no load across the secondary.

There is no load shown across the secondary circuit. With no secondary load, no current flows in the circuit, although a voltage is developed across it. If the secondary has a voltage induced in it but no current, no power is developed in the secondary and the current in the primary will be the same as though there were no secondary. (Actually, some electrons flow in the turns of the secondary to produce the + and − charges that result in the secondary voltage.)

When a load resistor is connected across the secondary of a transformer (Fig. 5-19), several things occur. Step by step these are:

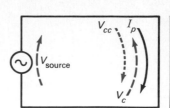
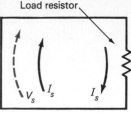

FIG. 5-19 Currents and voltages in a transformer when a load resistor is connected across the secondary.

1. The source emf, V_{source}, produces primary current, I_p.
2. Primary current produces counter emf, V_c, in the primary.
3. Primary current also produces an induced secondary emf, V_s.
4. V_s produces current I_s through the secondary and its load.
5. I_s produces a magnetic field expanding outward from the secondary.
6. Expanding fields from the secondary induce a counter counter emf, V_{cc}, in the primary (opposite to the original counter emf in the primary and in the same direction as the source voltage). Mutual inductance is now present.
7. V_{cc} partially cancels the counter emf of the primary.
8. Cancellation of primary counter emf allows the source emf to send more current through the primary.

Therefore, when a load is connected across the secondary, the primary current increases. This results in feeding more electric power to the primary to be converted to magnetic energy, which is transferred in turn to the secondary and reconverted to electric energy and power in the secondary.

Consider transformer loading another way. By Lenz' law, I_{sec} flows in such a direction that its magnetizing action opposes the magnetizing action of I_{pri}. Thus, any increase in I_{sec} cancels primary flux, reduces primary cemf, and increases I_{pri}.

5-16 TRANSFORMER CONSTRUCTION

Transformers are constructed with a primary coil and one or more secondary coils. A second secondary may be termed the *tertiary* (meaning "third") winding, and a third secondary may be termed the *quaternary* (meaning "fourth") winding. It is, however, more usual to designate the windings by their function or voltage. Thus, a transformer may be

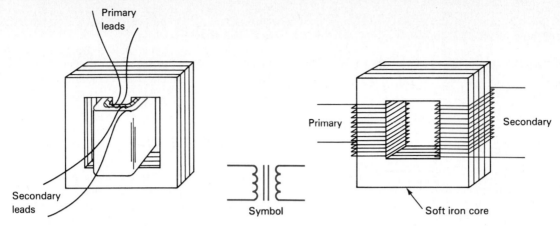

FIG. 5-20 Two basic methods of constructing iron-core transformers and symbol for both.

said to have a primary, a high-voltage secondary, a 5-V winding, etc.

Figure 5-20 illustrates two of many possible methods of constructing a power- or audio-frequency transformer and the symbol for such transformers. Winding the secondary coil over the primary coil is probably the more frequently used method.

In power transformers, there are usually several hundred turns on the primary and an equivalent number of turns on the secondary if it is desired to produce a secondary voltage equal to the voltage applied across the primary. If a greater secondary voltage is desired, more turns will be wound on the secondary than on the primary.

Transformers for higher frequencies use less iron in their cores. If the frequency is in the RF range, either air or ferrite cores may be used. Figure 5-21 illustrates a possible RF transformer and its symbol. The core is often some nonconducting material hav-

ing the same permeability as air. One or both coils may be "tuned" (Sec. 8-7).

5-17 EDDY CURRENTS

To produce a transformer of high efficiency and with a minimum number of turns, the primary and secondary are wound on a core of iron or other high-permeability material. As a result, when a transformer is in operation, intense moving magnetic fields are produced in the core. These fields induce circulating currents in the core material because iron is a fairly good electric conductor. Such whirlpool-like currents are known as *eddy currents*, and they can produce a considerable I^2R power loss in the form of heat in the core. Figure 5-22 indi-

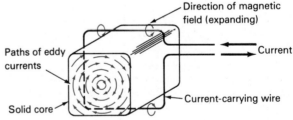

FIG. 5-22 Eddy currents induced in a solid iron core.

cates the path of eddy currents in a solid core wound with a single turn of wire in which the current is increasing (left-hand generator rule).

Eddy currents are decreased in strength by using many thin sheets rather than a solid block of iron. Each separate sheet must be coated with an insulating scale, varnish, or plastic coating. Any one eddy-current path is limited to the thickness of the sheet (Fig. 5-23). Limiting the length of the path

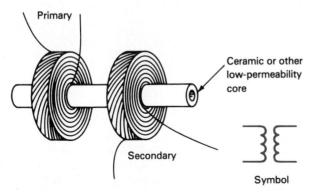

FIG. 5-21 One possible air-core transformer construction and symbol.

limits the amplitude of the eddy currents and holds the I^2R heat loss in the core to a minimum. Slicing the core into thin sheets is known as *laminating* the core. The thinner the laminations, the less the eddy-current loss. The cores in Fig. 5-23 are shown laminated.

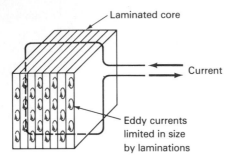

FIG. 5-23 Eddy currents are limited in amplitude by laminating the core.

As frequency is increased, magnetic flux movement and eddy-current losses both increase. It has been found advantageous to use ferrite cores in higher-frequency applications. All magnetic particles of ferrites are insulated from adjoining particles, which prevents eddy currents from developing. Ferrites are nonconductors. *Air-core transformers* have no eddy current losses.

5-18 HYSTERESIS

If iron is in an unmagnetized state, its domains are not arranged in any particular manner. When a magnetizing force is applied to them, the domains rotate into a position in line with the magnetizing force. If the magnetizing force is reversed, the domains must rotate into an opposite position. In rotating from one alignment to the opposite, the domains must overcome a frictional *hysteresis*, or resisting, effect in the substance. In some materials the resisting effect is very small; in others hysteresis is appreciable. The energy converted into heat overcoming hysteresis is known as *hysteretic loss*.

Hysteresis occurs in only the iron cores of transformers. As frequency is increased, the alternating magnetizing force may no longer be able to magnetize the core completely in either direction. Before the core becomes fully magnetized in one direction, the opposite magnetizing force will begin to be applied and start to reverse the rotation of the do-

mains. The higher the frequency, the less fully the core magnetizes.

Transformers operated on low-frequency ac may not have much hysteresis, but the same cores used with a higher frequency have more hysteresis and will be less efficient. Air-core transformers have no hysteretic losses.

5-19 COPPER LOSS

Iron-core transformers are subject not only to eddy-current and hysteresis losses in the core but also to a *copper loss* which occurs in the copper wire of the primary and secondary. The current flowing through whatever resistance exists in these windings produces heat. The heat in either winding, in watts, can be found by the power formula $P = I^2R$. For this reason the copper loss is also known as the I^2R loss. The heavier the load on the transformer (the more current that is made to flow through the primary and secondary), the greater the copper loss.

With one layer of wire wound over another in a transformer, there is a greater tendency for the heat to remain in the wires than if the wires were separated and air-cooled. Increased temperature causes increased resistance of a copper wire. As a result it becomes necessary to use heavier wire to reduce resistance and heat loss in transformers than would be required for an equivalent current value if the wire were in open air (cooled) during operation.

5-20 EXTERNAL-INDUCTION LOSS

Another loss in a transformer is due to external induction. Lines of force expanding outward from the transformer core may induce voltages and therefore currents into outside circuits. These currents flowing through any resistance in an outside circuit can produce a heating of the external resistance. Power lost in heating outside circuits represents a power loss to the transformer, since the power is not delivered to the transformer secondary circuit. In a well-designed transformer, the amount of power lost in this fashion is very small. Voltages induced into nearby wires of certain types of amplifying circuits can, however, produce undesirable voltages in these circuits, even though the power loss to the transformer is negligible.

Any lines of force from the primary of a transformer that do not cut secondary turns and that induce no emf into the secondary or outside circuits are considered leakage flux (not a power loss).

5-21 TRANSFORMER VOLTAGE RATIO

One of the main uses of transformers is to step up a low-voltage ac to a higher voltage. This can be accomplished by having more turns on the secondary than on the primary. The transformer in Fig. 5-24 has a single wire for the primary and two

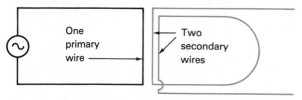

FIG. 5-24 Basic transformer with a 1:2 step-up voltage and turns ratio.

wires in series in the secondary. Each of the secondary wires will pick up an equal voltage, since both are being cut by the same number of lines of force from the primary. If two 1-V induced emf's in the secondary are in series, this results in an output of 2 secondary volts.

As long as the coefficient of coupling is high in a transformer, the no-load voltage ratio will equal the turns ratio. If the primary is wound with 500 turns and the secondary with 1000 turns, the secondary voltage will be twice any voltage applied across the primary. The fact that the voltage ratio is equal to the turns ratio can be expressed by

$$\frac{T_p}{T_s} = \frac{V_p}{V_s}$$

where T_p = primary turns
T_s = secondary turns
V_p = primary voltage
V_s = secondary voltage

If the primary has 200 turns and the secondary has 40 turns, with 100 V connected across the primary, the secondary voltage can be determined by rearranging the preceding formula to

$$V_s = \frac{V_p T_s}{T_p} = \frac{100(40)}{200} = \frac{4000}{200} = 20 \text{ V}$$

5-22 TRANSFORMER POWER RATIO

There is no step-up of power in transformers. It is possible to step voltage up or down, but the basic ratio of power into the primary to power out of the secondary is 1:1. Actually, because of losses in a transformer, less power will always be drawn from

the secondary than goes into the primary.

Power transformers are constructed to handle a certain number of watts, or voltamperes (VA). For example, a 100-V primary, 500-W (properly 500-VA) transformer will have a primary wound with wire that will carry only enough current to produce 500 VA in the primary. By the power formula $P = VI$, or $I = P/V$, it can be computed that 5 A is all the primary wire will be required to carry, regardless of whether the transformer is going to step the secondary voltage up or step it down. The primary will be wound with the thinnest wire that will carry 5 A without excessive heating (Table 1-3). If more than 500 W is drawn by the secondary load, the primary will be called on to carry more than 5 A, will become overly hot, and may burn the insulation on the wires or melt the wire itself. The secondary also will be wound with a wire that will approach its safe heating limit when 500 W is being drawn from the secondary.

If the transformer has a step-up ratio of 1:4, the secondary voltage will be 4 times 100 V, or 400 V. Inasmuch as the limiting factor is the 500 W into the primary, the secondary can be called upon to deliver only 1.25 A at 400 V. Otherwise, the power delivered by the secondary will be more than 500 W and either the primary or the secondary winding may overheat and fail.

To protect transformers from overloads, fuses or overload relays are connected in the primary circuit. A 5-A fuse in series with the primary of the above transformer would burn out if more than the maximum safe current of 1.25 A were drawn by the load across the transformer secondary.

Test your understanding; answer these checkup questions.

1. Where in time is the current peak in relation to the voltage peak in a purely inductive circuit? _____ In a purely resistive circuit? _____
2. At what points in a sine wave is the rate of change at maximum? _____ At zero? _____
3. A 10-Ω R is in series with a 20-Ω X^L across an ac source. Are V and I leading, lagging, or in phase in the R? _____ In the X^L? _____ In the source? _____
4. If a transformer secondary is unloaded, is any counter counter emf induced in the primary? _____
5. Is a ceramic-cored transformer classed as an air-core type? _____

6. What is decreased by laminating transformer cores? _____

7. What is the term that indicates the inability of a core material to reverse magnetic polarity completely? _____

8. Does an increase in load on a transformer increase the eddy-current loss? _____ The hysteretic loss? _____ The copper loss? _____ The external-induction loss? _____

9. What ratio in any transformer is always about 1:1? _____

10. What gauge wire would be used for a primary winding that must carry 5 A? _____ 1.25 A? _____

11. An unloaded transformer has a 4:1 step-down ratio and a k of 0.8. With 100 V across the primary, would the voltage across the secondary be 25 V, less than 25 V, or more than 25 V? _____

12. Short-circuiting the secondary winding of a transformer has what effect on the inductance of the primary? _____ On the current of the primary? _____

13. Would increasing the gauge size of the wire used in a transformer affect the output voltage? _____

5-23 TRANSFORMER CURRENT RATIO

A step-up transformer may produce more voltage across the secondary than is applied across the primary, but the secondary current will have to be proportionately less than the primary current. This was indicated by the 500-W 1:4 ratio step-up transformer previously mentioned. With 100 V across the primary, the secondary produces 400 V. With a load that draws 1.25 A (a 320-Ω resistor, for example) connected across the secondary, the primary will be called upon to draw 5 A of current from the source. This represents a primary-to-secondary step-down of current from 5 to 1.25 A.

Secondary current is inversely proportional to the turns ratio. This can be expressed by formula as

$$\frac{T_p}{T_s} = \frac{I_s}{I_p}$$

where T_p = primary turns
T_s = secondary turns
I_p = primary current
I_s = secondary current

The turns, voltage, current, and impedance (Z, Chap. 7) ratios can be expressed:

$$\frac{T_p}{T_s} = \frac{V_p}{V_s} = \frac{I_s}{I_p} = \frac{Z_p}{Z_s}$$

5-24 TRANSFORMER EFFICIENCY

It will always be found that more power is fed to the primary of a transformer than is delivered by the secondary to a load. The difference between the input power and the output power is the sum of all the *power losses* in the transformer.

The ratio of the output power to the input power is the efficiency of the transformer. The factors that determine efficiency are the copper, eddy-current, hysteretic, and external-induction losses. The output/input power ratio always results in a decimal number less than 1.0. In practice, efficiency is given in percent rather than in the decimal equivalent. It is only necessary to multiply the decimal by 100 to determine the percent. The formula for percent efficiency is

$$\text{Percent efficiency} = \frac{P_o}{P_i} \times 100$$

where P_o = power output
P_i = power input

If the overall efficiency of a transformer is known, the primary power times the percent efficiency is the secondary power, or $(\%)(P_i) = P_o$.

EXAMPLE: If a power transformer has a primary voltage of 4400 V, a secondary voltage of 220 V, and an efficiency of 98% when delivering 23 A of secondary current, what is the value of primary current? In this case, 23 A at 220 V, or 5060 W, represents 98% of the power being fed into the primary. By rearranging the formula,

$$(\%)(P_p) = P_s$$

$$P_p = \frac{P_s}{\%} = \frac{5060}{0.98} = 5163 \text{ W}$$

By using the power formula $P = VI$ the primary current can be found:

$$P_p = V_p I_p$$

$$I_p = \frac{P_p}{V_p} = \frac{5163}{4400} = 1.17 \text{ A}$$

Power transformers are always warm to the touch when operating, due to internal losses. In some cases, it becomes necessary to air-cool transformers to keep them from overheating and damaging the insulation on the windings. Some transformers are built into oil-filled cases. The oil helps to insulate the internal wiring and prevent moisture from forming on the insulation, which might result in a breakdown, and also carries heat from the windings to the outer case to be dissipated into the air.

5-25 AUTOTRANSFORMERS

An *autotransformer*, or *autoformer*, consists of a single winding with one or more taps, Fig. 5-25.

If a 100-V source of ac is connected between points *A* and *C* and there are 100 turns between

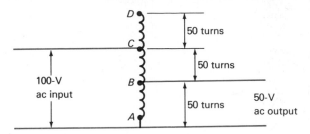

FIG. 5-25 Autotransformer or autoformer.

FIG. 5-26 Transformers. Left, iron-core; from top, power, variable autotransformer. Middle, unshielded, shielded, and two transistor audio transformers. Right, air-core, top two with shields removed, and transistor IF transformer.

these points, an emf of 1 V will be developed across each of the turns, as well as induced in each of the turns from *C* to *D*. If a load is connected across points *A* and *B*, as shown, it will be across 50 V. If connected across *A* and *C*, it will be across 100 V. If connected across *A* and *D*, it will be across 150 V. Thus, an autotransformer can be used as a voltage step-down or step-up device.

If the load is connected between *A* and a tap that can be adjusted to any turn between *A* and *D*, any desired voltage up to 150 in 1-V steps can be developed across the load. Such autotransformers are made and sold under trade names such as Powerstat and Variac.

A disadvantage of autotransformers is the common connection between primary and secondary. For safety, it is desirable to have primary and secondary circuits isolated from each other electrically.

If isolation is not a factor, any common transformer can be connected as an autotransformer (primary and secondary windings connected in series) and more or less output voltage can be obtained than the turns ratio of the transformer would normally give.

5-26 PRACTICAL TRANSFORMER CONSIDERATIONS

In electronics, there are many types of transformers in use. Three common types are power, AF, and RF transformers (Fig. 5-26).

A power transformer is normally made to operate across a 120- to 480-V ac line. It is a heavy iron-encased piece of equipment. The resistance in the primary winding ranges from a fraction of an ohm to possibly 5 Ω. The inductive reactance of the primary winding acts to limit the primary current to a low value when connected across an ac power line. If connected across a similar-voltage dc line, the low resistance will allow excessively high current to flow in the primary. The primary will overheat and burn out, or the line fuses will blow. Care must be taken not to connect the primary of a power transformer across a dc power line.

Power transformers are made to operate on one particular frequency, usually 50 or 60 Hz. In most cases, such transformers will operate fairly well on any frequency between 40 and 70 Hz. If the frequency is much too high, however, the inductive reactance of the primary will prevent the primary from drawing sufficient power. There will be more iron in the core than is necessary, and hysteresis and eddy-current losses will be excessive. If the frequency is too low, the primary will have insufficient reactance and too much primary current will flow, producing considerable copper loss. The transformer may start to smoke. There will not be enough iron in the core, and the transformer will not be capable of its rated power output.

If a turn of either the primary or the secondary of a power transformer shorts out for some reason, a high current will be induced in the turn, producing excessive heat in the transformer, not only because of the shorted turn's heating but also because of the cancellation of the inductance of the primary by the magnetic field set up by the shorted turn. Cancellation of the inductance materially decreases the inductive reactance of the primary, and excessive primary current flows.

Audio transformers are also iron-cored, are usually smaller than power transformers, and may be connected in series with a relatively high-resistance transistor or vacuum tube across a source of dc. The resistance of the transistor or vacuum tube limits the primary current to a safe dc value and prevents the primary from burning out.

Radio-frequency transformers are normally air-core transformers and are made to operate across RF ac directly or in series with either a transistor or a vacuum tube and a source of dc.

| Test your understanding; answer these checkup questions.

1. Would a voltage step-up transformer be considered a current step-up or step-down transformer? _____

2. A transformer has a V_p of 120 V, a V_s of 24 V, and an efficiency of 96% when delivering 4 A to a load. What is the I_p value? _____

3. It is desired to produce 1000 V by using a 120-V source. How many turns should the secondary have if the primary has 240 turns? _____

4. A transformer is to supply 12.6 V from a 120-V line. How many secondary turns are required if there are 300 primary turns? _____ To supply a 4-A load at 600 cmil/A, what gauge wire should be used on the secondary? _____ On the primary? _____

5. A 120-V primary has a 240-V secondary and a 12-V tertiary. Disregarding losses, if the tertiary load draws 5 A and the secondary 0.1 A, what current flows in the primary? _____

6. If a choke coil has a tap near its center, what type of transformer might it be? _____ Would it be a step-up or a step-down type? _____

7. What might occur if a 110 V dc is connected across a 120-V primary power transformer? _____

8. What difficulties occur if the ac frequency fed to a power transformer is too low? _____ Too high? _____

9. What type of core is usually used with an RF transformer? _____ An AF transformer? _____

10. Under what condition is dc applied safely to the primary of a transformer? _____

INDUCTANCE AND TRANSFORMERS QUESTIONS

1. What component opposes any change in current?
2. When a magnetic field is expanding from a wire, what is the name of the voltage it generates in the wire?
3. When dc is connected across an inductance, in what way is the current amplitude said to rise?
4. In what direction is a self-induced emf generated?
5. If a dc current is turned on and off in an inductive circuit, when is a spark developed at the switch?
6. Does inductance oppose a rise in current? A fall?
7. What are the symbols used for inductance and its unit of measurement?
8. How is the inductance value of a coil related to the number of turns?
9. What is the inductance of an air-core coil 0.5 in. in diameter, 0.5 in. long, and having 25 turns?
10. What is a ferrite slug?
11. What is the name of a doughnut-shaped core?
12. How long does it take for a dc current to rise to essentially its maximum value in a 50-mH coil having 30-Ω R?
13. How much energy is stored in an 80-mH inductor when 1.5 A is flowing through it?
14. What does a choke coil choke?
15. What frequency chokes use iron cores?
16. How many lines of force must cut across a conductor in half a second to induce 2 V?
17. If all of the lines from a 5-H coil cut all of the turns of a 10-H coil, what mutual inductance exists?
18. In problem 15, if only 40% of the lines interlinked, what would be the M value?
19. What is the coefficient of coupling if two 2-H coils have a mutual inductance of 0.5?
20. How may "tight coupling" be otherwise expressed?
21. The formula to determine total inductance of two or more chokes in series is similar to what other previous formula?
22. What is the formula to determine total inductance of two inductors if their mutual inductance is known?
23. A 0.5- and a 0.8-mH coil are in parallel. What is the total inductance if no intercoupling exists?
24. If a brass disk or core is run into a coil, what effect does this have on the inductance value? If a ferrite core is used?
25. What is the ability of a coil to react against variations of current through it called? In what unit is it measured?

| ANSWERS TO CHECKUP QUIZ ON PAGE 81

1. (Lags behind)(In phase) **2.** (As time line is crossed)(At peaks) **3.** (In phase)(I lags)(I lags) **4.** (No) **5.** (Yes) **6.** (Eddy currents) **7.** (Hysteresis) **8.** (No)(No)(Yes)(No) **9.** (Power) **10.** (No. 12)(No. 18) **11.** (Less than 25 V) **12.** (Reduces L)(Increases I greatly). **13.** (Theoretically no but with heavy I larger voltage-drops occur with small wires).

26. If a change of 0.5 A occurs in 0.02 s in a 200-mH coil, what voltage is being induced?
27. What is the formula for inductive reactance?
28. To the 3rd significant figure, what X_L does a 750-mH coil have to 1000 Hz? To 4 MHz?
29. If a coil has 2700-Ω reactance to a 60-Hz ac, what is its inductance?
30. An ac circuit contains a 90-V 500-Hz source and a 3-H choke as the load. What value of current flows?
31. What is the phase of the current in relation to the voltage in any purely inductively reactive circuit?
32. Why is counter emf zero when ac current reaches maximum in a coil?
33. A 500-Ω X_L and a 400-Ω R are in series across an ac source. Across which is the greater voltage-drop developed? Which has the greater current in it?
34. What effect does a transformer secondary have on the primary if there is no secondary load?
35. Is the secondary voltage of a transformer developed similarly to the primary counter emf?
36. If a transformer has a third winding, what may it be called?
37. What types of cores may be used in RF transformers?
38. How can eddy currents be reduced in iron-core transformers?
39. What type of transformers have no eddy-current losses?
40. What is the frictional loss caused by domain movement called?
41. What property of a conductor causes copper loss?
42. What determines the voltage ratio of a transformer?
43. In a transformer is it possible to step up voltage? To step up power? To step up current?
44. If a transformer has V_p = 120 V, V_s = 14 V, I_s = 10 A, and efficiency = 95%, what is the I_p value?
45. A center-tapped choke is what kind of a transformer? What is the disadvantage of such a transformer?
46. What is the disadvantage of using too high a frequency ac on a power transformer?
47. What are the disadvantages of too low a power line frequency on a power transformer?

6

Capacitance

This chapter discusses the various types of capacitors, the factors determining capacitance, and losses in capacitors. Capacitive reactance is explained, as are capacitors in series and in parallel, as well as V and I phase relationships, and color codes.

6-1 THE CAPACITOR

One of the most used parts in radio and electronics is the *capacitor*, previously called a *condenser*. A capacitor has the ability to hold a charge of electrons. The number of electrons it can store under a given electric pressure is a measure of its *capacitance* (sometimes called *capacity*). Two separate metallic plates with a nonconducting substance sandwiched between them, as in Fig. 6-1, form a

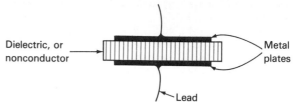

FIG. 6-1 A capacitor consists of two conductive plates separated by a nonconducting dielectric.

simple capacitor. Symbols for capacitors are shown in Fig. 6-2.

Old   New   Variable

FIG. 6-2 Symbols used to denote fixed and variable capacitors.

It is important to understand the current and voltage changes that take place in a circuit in which a capacitor is connected. Figure 6-3a shows a simple circuit with a battery, switch, resistor, and capacitor in series. Figure 6-3b is a graph of the current and voltage changes that will take place when the switch is closed. With the switch open, the capacitor will be assumed to have zero charge; that is, both plates have the normal number of electrons and protons in the molecules of the metal.

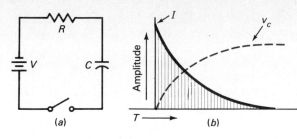

FIG. 6-3 *RC* circuit with graph of current flowing into a capacitor and voltage developing across it after the switch is closed.

When the switch is closed, the electric pressure of the battery begins to force electrons into the top plate from the negative terminal and pull others out of the bottom plate toward the positive terminal. As the electron difference is developed between the two plates, electrostatic lines of force appear in the region between the plates.

At the instant the switch is closed, there is no opposing emf in the capacitor and the amplitude of the current is determined only by the resistance in the circuit. As time progresses, more electrons flow into the capacitor and a greater opposing emf is developed in it. The difference between the source and the opposing emf becomes less. The opposing emf across the capacitor continually increases, and the charging current continually decreases. When the opposing emf equals the source emf, there will no longer be any charging current flowing into the capacitor. The voltage across the capacitor will be at a maximum and equal to the source voltage. The voltage across the capacitor rises "exponentially" and can be determined by

$$v_C = V(1 - \epsilon^{-t/RC})$$

where v_C = instantaneous emf
 V = source emf
 ϵ = 2.718
 t = time in s from start
 R = ohms
 C = farads

The $-t/RC$ is the exponent in question.

A capacitor that will store a difference of 1 coulomb (1 C or 6.25×10^{18} electrons) when an emf of 1 V is applied across it has the capacitance value of 1 *farad* (1 F). A total of 2 V across this capacitor would store 2 C. Capacitance is measured in farads, microfarads (millionths of a farad), nanofarads (billionths of a farad or 10^{-9}) or in picofarads (trillionths of a farad or 10^{-12}). Microfarad is

abbreviated μF (uf, μf, μfd, ufd, and mfd may also be seen at times). A micromicrofarad ($\mu\mu$F) is the same as a picofarad, abbreviated pF. Nanofarad is abbreviated nF. It is important to be able to convert from microfarads to nanofarads and picofarads. For instance,

$$1\ \mu\text{F} = 1{,}000{,}000\ \text{pF}\ (\mu\mu\text{F})$$
$$0.005\ \mu\text{F} = 5000\ \text{pF}$$
$$0.00004\ \mu\text{F} = 40\ \text{pF or } 40{,}000\ \text{nF}$$
$$250\ \text{pF} = 0.00025\ \mu\text{F or } 0.25\ \text{nF}$$
$$1\ \mu\text{F} \times 10^{-6} = 1\ \text{pF}$$
$$1\ \text{pF} \times 10^{6} = 1\ \mu\text{F}$$
$$1\ \text{pF} \times 10^{3} = 1\ \text{nF}$$

The time required for a capacitor to attain a charge is proportional to the capacitance and resistance in the circuit. The *time constant* of a resistance-capacitance circuit is

$$T_C = RC$$

where T_C = time, in s
 R = resistance, in Ω
 C = capacitance, in farads (F)

The time in the formula is that required to charge the capacitor to 63% of the voltage value of the source. It is also the time required for a charged capacitor to discharge 63% when connected across the value of resistance used in the formula. (Note the difference in the *RC* time-constant formula compared with that of an *RL* circuit. Sec. 5-4.) The time required to bring the charge to about 99% of the source voltage is approximately five times the time computed by using the time-constant formula.

If there is no resistance in the circuit, charging time is zero and a capacitor will charge (or discharge) instantaneously.

A 4-μF capacitor and a 1-MΩ (1-megohm) resistor have a time constant of 0.000004 $\times$ 1,000,000 (or $4 \times 10^{-6} \times 1 \times 10^{6}$) or 4 s. The capacitor will charge almost completely in 4×5 time constants, or 20 s.

Inductance was defined as *the property of a circuit to oppose any change in current*. Capacitance may be defined as *the property of a circuit to oppose any change in voltage*. As discussed above, in a dc circuit a capacitor develops an opposition emf from a zero value up to the source value. When a resistor is connected across a charged capacitor that has been disconnected from the charging source, the opposing emf developed in the capacitor during the charging period discharges, driving current through the resistor. In a varying dc or ac circuit, in which the emf is continually varying in amplitude, a capacitor will be charging and discharging continually and will continually oppose any source emf variations.

The current value flowing into a capacitor is computed by

$$I_C = C\left(\frac{dv}{dt}\right)$$

where d signifies "change in."

When a resistor and a capacitor are in series across an ac source, the charging of the capacitor will alternate at the source frequency. However, with resistance in series, the ac voltage across the capacitor will always be less than the source emf; how much less will depend on the time constant of the R and C being used. With a very long time constant the ac voltage across the capacitor will be very small.

6-2 CAPACITANCE FACTORS

The factors that determine the capacitance are the area of the plates exposed to each other, the spacing between the plates, and the composition of the nonconducting material between the plates.

Two plates, each with an area of 1 in.^{2}, when separated by 0.001 in. of air, produce a capacitance of 225 pF. If each plate area is doubled (spacing remaining 0.001 in.) the capacitance doubles, to 450 pF. Capacitance is directly proportional to the plate areas.

If the spacing of the two 1-in.^{2} plates is increased to 0.002 in., the path of the electrostatic lines of force between the negative plate and the positive plate is twice as great, resulting in only half as intense an electrostatic field and only half as much capacitance. Capacitance is inversely proportional to the spacing between plates.

The nonconducting *dielectric* material determines the concentration of electrostatic lines of force. If the dielectric is air, a certain number of lines of force will be set up for a given emf value. Other materials offer less opposition to the formation of electrostatic lines of force in them. For example, with one type of paper instead of air, the number of electrostatic lines of force between the plates may be twice as great. Such a capacitor will have twice as much capacitance and will have two times as

many electrons flowing into and out of it with the same applied source emf. The paper is said to have a *dielectric constant*, or *specific inductive capacity*, twice that of air. Capacitance of a capacitor is directly proportional to the dielectric constant.

A formula to determine the capacitance of a two-plate capacitor is therefore

$$C_{\text{pF}} = \frac{0.225\ KA}{S}$$

where C = capacitance, in pF
K = dielectric constant
A = area of one of the plates, in in.^{2}
S = spacing between plates, in in.

This formula is for a two-plate capacitor. For greater capacitance, plates can be stacked on top of one another and separated with strips of dielectric material. A 3-plate capacitor has twice the plate area exposed, as shown in Fig. 6-4, and twice the

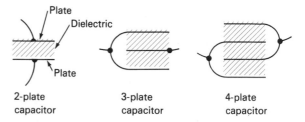

2-plate capacitor   3-plate capacitor   4-plate capacitor

FIG. 6-4 A 3-plate capacitor has twice the capacitance of a 2-plate capacitor. A 4-plate capacitor has 3 times the capacitance of a 2-plate capacitor.

capacitance. A 4-plate capacitor has three times the plate area and capacitance. The formula for multiplate capacitors is

$$C_{\text{pF}} = \frac{0.225\ KA(n-1)}{S}$$

where n = number of plates in capacitor

The approximate dielectric constant, or specific inductive capacity, of some common dielectric materials is given in Table 6-1.

The dielectric constant of solid dielectric materials may decrease with an increase in frequency. The molecules of the dielectric do not have sufficient time to conform to the rapidly changing electrostatic lines of force that they must support. If the lines of force cannot be fully developed in the dielectric molecules, the dielectric constant and the capacitance will be less. Thus, a 0.1-μF paper capacitor may have 0.1 μF at

TABLE 6-1 DIELECTRIC CONSTANTS

Material	Dielectric constant
Vacuum	1
Air	1.0006
Rubber	2–3
Paper	2–3
Ceramics	3–7
Glass	4–7
Quartz	4
Mica	5–7
Porcelain	6–7
Water	80
Barium titanate	7500

1 MHz but will have considerably less at 100 MHz. Mica is less affected by frequency. Air and vacuum dielectrics are not noticeably affected. If a dielectric material expands when heated, the plates will be forced apart and the capacitance value will decrease.

6-3 DIELECTRIC LOSSES

Almost all the energy stored in the electrostatic field of a capacitor is converted into some other form of energy when the capacitor is discharged. However, two losses occur in the dielectric itself.

Electrons on the negative plate of a charged capacitor may find a high-resistance path to the positive plate through the dielectric (or over the surface of the capacitor) forming a leakage to the other plate. Therefore leakage current is a loss $(P = I^2R)$ to the capacitor.

Another dielectric loss is due to hysteresis and is indicated by heat in the capacitor. It is caused by the friction of the molecules of the dielectric material as they are changed from one strained position to the opposite by any reversing of electrostatic lines of force. Hysteresis is normally significant only when an alternating current produces rapid charging and discharging of the capacitor. It increases as the frequency of the ac increases. For this reason, many capacitors operate satisfactorily at lower, but not at very high, frequencies. Vacuum, air, mica, and ceramic dielectric capacitors have little leakage or hysteresis. Paper may have considerable leakage and hysteresis, particularly if the dielectric has moisture in it.

An interesting and significant phenomenon is produced in some high-capacitance solid-material dielectric capacitors. If charged by a dc voltage they can be disconnected from the charging source and will remain charged. A wire touched across the capacitor terminals will discharge the capacitor, usually with an audible and visible electric spark. If the wire is left disconnected for a short time and then touched across the discharged capacitor again, another spark is produced, indicating that the dielectric had not released all the stored energy on the first discharge. During the charged period dielectric molecules some distance from the capacitor plates capture some leakage electrons. These take time to leak back to the plates through the high resistance of the dielectric after the capacitor has been discharged. This *dielectric absorption* may not be very significant when the capacitor operates in a dc circuit, but in high-frequency ac or varying dc circuits it decreases the effectiveness of the capacitor.

6-4 WORKING VOLTAGE AND DIELECTRIC STRENGTH

One rating of capacitors is the *working voltage*. This is the maximum voltage at which the capacitor will operate without leaking excessively or arcing through. Sufficient leakage through the dielectric over a period of time can produce a carbonized path across the dielectric, and the capacitor will act as a conductor. In such a case it is said to be *burned out*, or *shorted*.

A burned-out capacitor should not be confused with an *open* one. An open capacitor has lost its storage ability by the breaking off of a wire lead internally or, in the case of electrolytic capacitors, because the electrolyte has dried out.

The *working voltage* is usually rated as a dc value. A capacitor may be rated to work at 600 V on dc circuits; but when used in power-frequency ac circuits, its *effective ac working voltage* will be about one-half the dc rating. As the frequency of the ac is increased, the working voltage of the capacitor decreases, particularly when the frequency rises above a few megahertz. Any dielectric heating will decrease the breakdown voltage value.

The *dielectric strength*, or number of volts that a dielectric material will stand per 0.001 in. of dielectric thickness, varies considerably with different materials. Air's dielectric strength is about 80 V; Bakelite, about 500 V; glass, 200 to 300 V; mica, about 2000 V; untreated paper, a few hundred volts; waxed or oiled paper, 1000 to 2000 V; rubber, 400 V; ceramics, 80 to 200 V; Teflon, 1500 V; practical vacuums may exceed 10,000 V.

6-5 ENERGY STORED IN A CAPACITOR

When a capacitor is charged and then disconnected from the charging source, it retains its difference of electrons between the two plates and the dielectric molecules are still under the stress of electrostatic lines of force. If the charged capacitor is connected across a light bulb, for example, the excess electrons on the negative plate will flow through the bulb to the positive plate until the electron inequality between plates no longer exists. When the two plates have an equal number of electrons, the capacitor will no longer have any charge and all current will cease.

While moving through the light bulb, the electrons liberate the energy of their motion in the form of heat. In this case the light bulb may flash for an instant and then go out. The amount of energy stored by the electrostatic field in the dielectric of a capacitor can be computed by the formula

$$E_n = \frac{CV^2}{2}$$

where E_n = energy, in wattseconds or joules
C = capacitance, in F
V = voltage, in V

6-6 CAPACITOR CHARGE

The *charge* of a capacitor is the difference in number of electrons on the two plates. Since this difference involves a quantity of electrons, the unit of quantity of charge is the coulomb. In formula form, the quantity of charge in a capacitor is

$$Q = CV$$

where Q = charge, in coulombs
C = capacitance, in F
V = voltage, in V

EXAMPLE: If a 2-μF capacitor is across 10 V dc, the electron difference between the positive and negative plates will be

$$Q = CV = 0.000002(10)$$
$$= 0.00002, \text{ or } 2 \times 10^{-5} \text{ coulomb}$$
In electrons $= (6.25 \times 10^{18})(2 \times 10^{-5})$
$$= 12.5 \times 10^{13} \text{ electrons}$$

It can be reasoned that if a 0.1-μF capacitor is charged by a 125-V source, an electron difference will be developed of $Q = CV$, or 0.0000001(125), or 0.0000125 coulomb between the plates. If the charged capacitor is disconnected from the source,

it still retains the electron difference on its plates (assuming no leakage). If a similar but uncharged capacitor is connected across the charged one, electrons flow from the charged to the uncharged capacitor. Since both are of equal capacitance, each will have half the electron difference, or 0.00000625 coulomb. A capacitor losing half its electron charge will have only half the voltage across it. Each of the two parallel capacitors now has 62.5 V across itself. Nothing has been lost. If the two capacitors are reconnected in series, the total voltage-drop across them will be 125 V and the same number of electrons will still be in storage in the capacitors.

▌Test your understanding; answer these checkup questions.

1. What is the old term meaning capacitor? _____

2. If a 0.1-μF capacitor and an 8-MΩ resistor are connected across 550 V dc, how long does it take for the capacitor to charge to 63% of the source value? _____ To 99%? _____

3. What is the time constant of a 75-kΩ resistor and a 0.2-μF capacitor? _____

4. What device has the property of opposing any change in current through it? _____ Any change in voltage across it? _____

5. Is the capacitance directly or inversely proportional to plate area? _____ Plate spacing? _____ Dielectric constant? _____

6. A seven-plate air-dielectric capacitor with 3.5-in.² plates separated by 0.002 in. would have what capacitance in μF? _____ In pF? _____

7. What is another name for specific inductive capacity? _____

8. What dielectric material mentioned would make the highest-capacitance capacitor? _____ The lowest-capacitance capacitor? _____

9. What are the names of three losses present in a solid-dielectric-type capacitor? _____

10. What is the approximate ac working voltage of a capacitor in comparison with its dc rating? _____

11. What dielectric mentioned has the greatest dielectric strength? _____

12. A 4-μF capacitor is disconnected from a 600-V dc supply. How much electric energy will a person receive if he grabs hold of the two terminals? _____ How many electrons will pass through his flesh? _____ Might this kill him? _____

13. An uncharged 8-μF capacitor is connected across the charged capacitor in question 12. What voltage will appear across the two capacitors in parallel? _____

6-7 ELECTROLYTIC CAPACITORS

A capacitor was represented as a flat metal plate, dielectric, metal plate, dielectric, metal plate, etc. Mica and air capacitors are made in this manner. If the dielectric is a flexible solid, capacitors can be made of two long sheets of aluminum foil separated by long strips of the dielectric material. The whole capacitor is then rolled into a relatively small tubular form with a wire lead from one end connected to one of the foil plates and a wire lead from the other end connected to the other foil plate.

An *electrolytic* capacitor consists of an aluminum-foil positive plate immersed in a solution called an *electrolyte* (ionizable solution capable of carrying current). The aluminum foil is the positive plate, and the electrolyte is the negative "plate," if a liquid can be called a plate. To make an electrical connection to the liquid, another aluminum foil is placed in the solution. To prevent the two foils from touching each other, a piece of gauze is placed between them. The aluminum positive plate and the negative foil in contact with the electrolyte are then subjected to an electric potential to *form* electrochemically an oxidized film on the positive plate. It is this thin oxide film between positive plate and electrolyte that is the dielectric. The thickness of the dielectric film depends upon the forming potential. If formed with a 450-V potential, the capacitor should be used at, or slightly under, this voltage. Electrolytic capacitors formed at 450 V but used on 300-V circuits may re-form to the lower voltage, which results in a thinner film and higher capacitance value. Electrolytic capacitors are rolled into tubular shape and sealed to make them airtight.

If electrolytic capacitors are connected across a circuit with reversed polarity (positive plate of capacitor to negative terminal of the circuit), the film deforms, the capacitor becomes a good conductor, high current flows, heat is developed, the electrolyte boils, and the capacitor may explode. Care must be taken when connecting *polarized* electrolytics into circuits that proper polarities are observed. This means that the average electrolytic capacitor cannot be used in ac circuits. (A special "non-polarized" electrolytic capacitor can be used during starting periods in ac electric motors.)

Although physically small and relatively inexpensive, electrolytic capacitors dry out over a period of time and lose capacitance. They have a small leakage current when in operation, which tends to raise

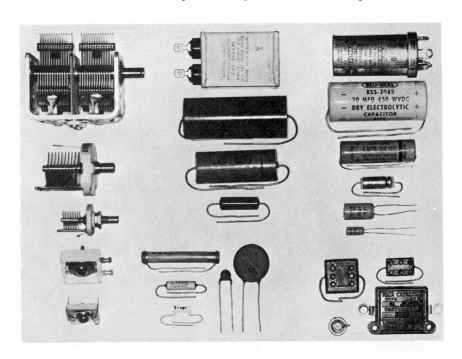

FIG. 6-5 Capacitors. Left column, variable and adjustable. Top center, paper. Bottom center, ceramic. Top right, electrolytics. Bottom right, mica.

the power factor (Sec. 6-10) of the capacitors. When used on voltages above the rated values, the leakage current increases, tending to dry out the capacitors. *Tantalum* electrolytics operate at higher temperatures than are possible with the older types.

6-8 VARIABLE CAPACITORS

Two types of capacitors can be adjusted or varied. *Adjustable* capacitors are usually constructed of two or more flat metal plates separated by sheets of mica or plastic installed on oblong ceramic holders, as at lower left in Fig. 6-5. The plates are so bowed that they normally hold themselves apart somewhat. Turning a machine screw presses the plates together, thereby increasing the capacitance. Adjustable capacitors may be called padders or trimmers. They are commonly available in capacitances from a few to 1000 or more pF, with a working voltage of 100 to 600 V.

Variable capacitors have *stator*, or stationary, plates and *rotor*, or rotatable, plates. When the shafts (Fig. 6-5) are rotated, the rotor plates mesh into the spaces between stator plates without touching. This varies the exposed plate areas and thereby the capacitance. The dielectric of these capacitors is usually air, although some employ sheet mica, plastic, or a ceramic material. There are also special vacuum-dielectric variable capacitors. Variable capacitors in radio receivers have very little spacing between plates. Those in radio transmitters or high-power high-frequency generators may have up to 1 in. or more of spacing between plates, depending on the voltages encountered in the circuit in which they are used. Symbols for variable capacitors are shown in Fig. 6-6.

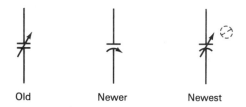

| Old | Newer | Newest |

FIG. 6-6 Variable or adjustable capacitor symbols. Dashed circle denotes a screwdriver-adjustable capacitor.

6-9 CAPACITOR TYPES

There are many types of capacitors in general use. Some are listed below with a brief statement as to dielectric leakage, fixed or variable type, working voltages, capacitance values, and frequencies.

1. *Vacuum-dielectric*. Practically no leakage. Made in both fixed and variable types. Used in 5000- to 50,000-V service. Capacitances of 5 to 250 pF. Efficient to well over 1000 MHz. Used mostly in transmitters.

2. *Air-dielectric*. Little leakage except through insulators. Made fixed, adjustable, and variable. Used in low- and high-voltage applications, receivers, and transmitters. Wide variety and capacitance range for both fixed and variable. Rarely much more than 400-pF capacitance. Variable air capacitors are used as tuning elements in receivers, transmitters, and other high-frequency equipment.

3. *Mica-dielectric*. Little leakage except through material which encases the plates and dielectric. Made fixed and adjustable. Working voltages from 350 to several thousand volts. Capacitances from 1.5 pF to 0.1 μF. Used in RF circuits up to more than 300 MHz, although efficiency drops off over 10 MHz. Fixed types are used as RF bypass capacitors, etc. Adjustable types used as padders or trimmers. Silver-mica capacitors are within 5% of their rated capacitance values and hold constant capacitance under adverse conditions.

4. *Ceramic-dielectric*. Low leakage. Fixed flat, round "disc," or tubular, and adjustable types. Capacitances from 1.5 pF to 0.01 μF for fixed types and up to 100 pF for adjustable types. Approximately 500-V working voltage. Useful up to more than 300 MHz.

5. *Paper-dielectric*. Paper impregnated with oil, wax, polychlorinated biphenyls (PCB), or ester is used. Relatively low leakage when new. When moisture seeps in, leakage becomes very high and the dielectric carbonizes at low voltages. Fixed types range from 10 pF to 10 μF. Working voltages are from 150 to several thousand volts. Efficient up to 1 or 2 MHz. Above this they become less effective due to dielectric fatigue and hysteresis.

6. *Plastic-dielectrics*. Various sheet plastics (Mylar, polystyrene, etc.) used in place of paper dielectrics. Tubular forms. Better for higher frequencies than paper. Range up to 2 μF and from 200 to 600 V.

7. *Electrolytic*. Some leakage, particularly if used on voltages over the rated value. Fixed types only. Range from a few μF to 50,000 μF or more. Working voltages from 6 to about 700 V. Normally polarized. Dry out and lose capacitance. Limited life expectancy. Useful only in dc circuits or in circuits where the dc component exceeds the ac component. Tantalum electrolytics are the best type.

6-10 POWER FACTOR IN CAPACITORS

Capacitors may be tested to find their *power factor* (a percentage rating of losses). With little or no leakage or losses, a capacitor has a low power factor. With no losses, the power factor is zero. As leakage develops through the capacitor, the power factor increases. A power factor of 1 denotes all leakage and no capacitive effect at all. Most capacitors when tested for power factor will indicate very close to zero. Electrolytic capacitors may read small power-factor values and still be usable. If their power factor increases above 0.1 or 0.2, it is assumed that they should be replaced.

6-11 INDUCTANCE IN CAPACITORS

In all capacitors there is a small value of inductance from the counter emf developed in the leads of the capacitor and the current traveling along the plates. In most low-frequency applications this inductance can be ignored. At frequencies over about 10 MHz the inductance of long leads may become detrimental to circuit operation.

Just as there is inductance in capacitor leads, so there is distributed capacitance in coils. There is a small value of capacitance between adjacent turns of a coil and from one end of the coil to the other. For this reason, it is impossible to produce pure inductance or pure capacitance.

6-12 CHANGING VARYING DC TO AC

When a capacitor is in a series circuit consisting of a source of voltage, a switch, an ammeter, a capacitor, and a load (Fig. 6-7), current will flow in the circuit under certain conditions.

If the source of emf is a battery or other dc supply, when the switch is closed, electrons will flow in the circuit until the capacitor is charged to the source voltage. The ammeter will respond with a momentary indication and then fall back to zero and remain there. From then on, no current will

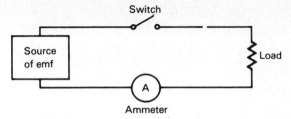

FIG. 6-7 A capacitor, load, ammeter, and switch across a source of emf (ac, dc, or varying dc).

flow in the circuit. It may be said that a series capacitor blocks the continuous flow of unvarying dc in a circuit.

If the source produces ac, the emf from the source will constantly be changing, the capacitor will constantly be charging and discharging, and an alternating current will flow through the meter. A capacitor may be used to complete a circuit if the source is producing ac. It may be considered a conductor for ac, but it must be understood that electrons do not pass across the dielectric.

If the source produces a varying dc emf, the periodic increase and decrease of emf results in a continual charging and partial discharging of the capacitor. Since the charging and discharging current flows back and forth through the meter, the ammeter will have an alternating current flowing through it. A capacitor in series with a circuit having a varying dc source results in an alternating current flow. The capacitor is said to block the dc but pass the *ac component*. Figure 6-8 shows a

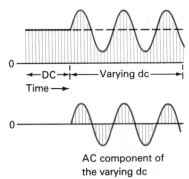

FIG. 6-8 Varying dc source emf and the resultant ac component if a capacitor is in series with the circuit.

varying dc and the ac component of the variation. This is an important point. Many transistor circuits have a varying dc flowing in them. When a load is coupled to a *varying dc* source by a capacitor, only the variation, the ac component, is transferred to the load. AC, not varying dc, flows in the load.

6-13 CAPACITIVE REACTANCE

How effective a capacitor may be in allowing ac to flow depends upon its capacitance and the frequency used. The greater the capacitance, the more electrons required to charge it to the source-voltage value. The smaller the capacitor, the fewer electrons required to bring it to full charge. It is possible to control the current flow in an ac circuit by changing the capacitance, somewhat as current can be controlled by varying the resistance in a circuit. The actual ac resistance effect of a capacitor is known as its *capacitive reactance*, which is measured in ohms, and can be determined by inserting the capacitance and frequency in the formula

$$X_C = \frac{1}{2\pi f C}$$

where X_C = reactance, in Ω
  f = frequency, in Hz
  C = capacitance, in F

EXAMPLE: The reactance of a 0.005-μF capacitor to a frequency of 1000 kHz may be found by substituting in the formula:

$$X_C = \frac{1}{2\pi f C} = \frac{1}{6.28(1,000,000)0.000000005}$$

$$= \frac{1}{6.28(0.005)} = \frac{1}{0.0314} = 31.8\ \Omega$$

If C is in microfarads, the formula to use is

$$X_C = \frac{10^6}{2\pi f C}$$

The X_C formula can be rearranged to solve for C if X_C is known. Thus

$$X_C = \frac{1}{2\pi f C} \quad \text{and} \quad C = \frac{1}{2\pi f X_C}$$

The reactance of a capacitor is inversely proportional to the frequency. If the reactance is 31.8 Ω at 1000 kHz, it will be one-half as much, or 15.9 Ω, at 2000 kHz and one-tenth as much at 10,000 kHz. This inverse proportion can be expressed by

$$\frac{f_1}{f_2} = \frac{X_{C_2}}{X_{C_1}}$$

where f_1 = one frequency
  f_2 = another frequency
  X_{C_1} = capacitive reactance at frequency f_1
  X_{C_2} = capacitive reactance at frequency f_2

This formula can be used to solve problems such as: What is the reactance of a capacitor at the frequency of 1200 kHz if its reactance is 300 Ω at 680 kHz? By substituting the given values,

$$\frac{f_1}{f_2} = \frac{X_{C_2}}{X_{C_1}}$$

$$\frac{1200\ \text{kHz}}{680\ \text{kHz}} = \frac{300}{X_{C_1}}$$

$$1200 X_{C_1} = 300(680)$$

$$X_{C_1} = \frac{204,000}{1200} = 170\ \Omega$$

The same answer will be obtained if the 300-Ω reactance and 680 kHz are used in the capacitive-reactance formula rearranged to solve for the capacitance. The unknown reactance is then determined by using this capacitance at 1200 kHz.

With the capacitive-reactance value known, it is possible to use Ohm's law for reactive circuits, as explained in Sec. 5-12. Reactance is substituted for resistance in the dc Ohm's law. The formulas are

$$I = \frac{V}{X_C} \qquad V = I X_C \qquad X_C = \frac{V}{I}$$

6-14 CAPACITORS IN PARALLEL

Capacitors are often connected in parallel to obtain greater capacitance (Fig. 6-9a). Plates A and B in the first capacitor have a certain area and are separated by a dielectric. The other capacitor has similar plate areas, similar dielectric, and therefore the same capacitance. If plates A and C and plates B and D are connected, as in Fig. 6-9b, the result will

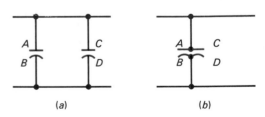

(a)   (b)

FIG. 6-9 Two similar parallel capacitors have twice the capacitance of one alone.

be equivalent to a capacitor with twice the plate area and twice the capacitance.

The formula for computing the capacitance of two or more capacitors in parallel is simply

$$C_t = C_1 + C_2 + C_3 + \cdots$$

where C_t = total capacitance
C_1 = one capacitor
C_2 = second capacitor
C_3 = third capacitor

EXAMPLE: Capacitors of 1, 3, and 5 μF are connected in parallel:

$$C_t = C_1 + C_2 + C_3 = 1 + 3 + 5 = 9 \ \mu F$$

Care must be taken when connecting two or more capacitors in parallel. The highest voltage a group of parallel capacitors will stand will be determined by the capacitor with the lowest working-voltage rating. If one capacitor has a working-voltage rating of 50 V and a second has a rating of 40 V, no more than 40 V should be used across the circuit if the two are in parallel.

█ Test your understanding; answer these checkup questions.

1. What is the main advantage of electrolytic capacitors? _____ What are the main disadvantages? _____ _____ _____
2. What would be the material of the dielectric in an electrolytic capacitor? _____ Of the nonmetallic plate? _____
3. What is the dielectric material of an improved type of electrolytic capacitor? _____
4. What are the three most likely dielectric materials used in adjustable and variable capacitors?
_____ _____ _____
5. Does a capacitor have any inductance? _____ An inductor any capacitance? _____ A resistor any capacitance or inductance?
_____
6. Will a capacitor pass dc? _____ AC? _____ Varying dc? _____
7. What is the capacitive reactance of a capacitor having 0.2 μF in a 15.9-kHz circuit? _____
8. What is the reactance of a 0.01-μF capacitor to a 3-kHz ac? _____
9. What is the capacitive reactance of a capacitor having 400 pF in a 3.8-MHz circuit? _____
10. If a capacitor has 1500 Ω at a frequency of 8 MHz, what reactance will it have to 400 kHz?
_____
11. If a 3-μF capacitor is across an 800-Hz signal generator producing 10 V rms, what value of ac current flows? _____
12. If a C and an R are in series across a 100-V ac line, how much voltage will be developed across

the capacitor if the X_C is 300 Ω and the current is 200 mA? _____
13. A 400-V power supply has a 450-V 40-μF capacitor and a 150-V 80-μF capacitor in parallel across its output. What is the total capacitance? _____ What will happen when the power supply is turned on? _____
14. Required, a capacitor that has 1-Ω X_C at 1 MHz. What C value is needed? _____

6-15 CAPACITORS IN SERIES

There are many cases when two or more capacitors are connected in series (Fig. 6-10). If two 4-μF capacitors are in series, the total capacitance is 2 μF. The circuit is acting as one capacitor with the

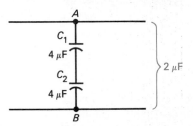

FIG. 6-10 Series capacitors increase dielectric thickness across a line and result in less total capacitance.

top plate connected to point A and the bottom plate connected to point B. Between these two points there is twice the dielectric spacing of one capacitor. Whenever the spacing between plates is increased, the capacitance decreases (Sec. 6-2).

The capacitance of two or more *similar* capacitors in series is equal to the capacitance of one divided by the total number of capacitors. Thus, four 10-μF capacitors in series present a total capacitance of 10/4, or 2.5 μF. Since capacitive reactance is inversely proportional to capacitance [$X_C = 1/(2\pi fC)$], the reactance of two equal series capacitors will be twice the reactance of one.

Two useful formulas for computing unequal capacitors in series are given below. The first is for two series capacitors. The second can be used for any number of capacitors in series. Note the similarity to parallel-resistance formulas (Sec. 2-15).

$$C_t = \frac{C_1 C_2}{C_1 + C_2}$$

$$C_t = \frac{1}{\dfrac{1}{C_1} + \dfrac{1}{C_2} + \dfrac{1}{C_3} + \cdots}$$

EXAMPLE: The total capacitance of three series capacitors of 5, 3, and 7 μF, is

$$C_t = \frac{1}{1/5 + 1/3 + 1/7}$$

$$= \frac{1}{21/105 + 35/105 + 15/105} = \frac{1}{71/105}$$

$$= \frac{105}{71} = 1.48 \ \mu F$$

The C_t of three capacitors in series can also be determined by the formula

$$C_t = \frac{C_1 C_2 C_3}{C_1 C_2 + C_2 C_3 + C_1 C_3}$$

The voltage-drop across any of the series capacitors (C_1 for example) can be found by using the voltage-divider formula

$$\frac{V_{c_1}}{V_t} = \frac{C_t}{C_1} \qquad V_{c_1} = \frac{V_t C_t}{C_1}$$

Since two similar capacitors in series present twice the dielectric spacing across the circuit, the working voltage of two such capacitors in series will be doubled. If each has a rating of 500 V, the two in series will stand 1000 V.

When capacitors in series are connected across a difference of potential, the sum of the voltage-drops across each of them will always equal the source voltage. Furthermore, the value of the voltage-drop across a particular capacitor in a series group will be inversely proportional to the ratio of its capacitance to the total, or directly proportional to its reactance. For example, a 1-μF and a 2-μF capacitor are in series across a 300-V source. There will be 200 V across the 1-μF capacitor and 100 V across the 2-μF. This will be true whether the capacitors are across ac or dc (if leakage is not present in the dc case).

It is usually desirable to equalize the dc voltage across capacitors of supposedly equal value when in series. This can be done by connecting resistors of equal value across them (Fig. 6-11).

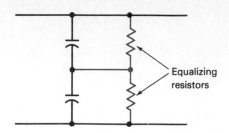

FIG. 6-11 Equalizing resistors across capacitors in series.

When an emergency arises or if many low-voltage capacitors are on hand, it is possible to connect several in series and use them across a high-voltage circuit. For example, if it is desired to have 1.5 μF of capacitance across a 1600-V circuit and a number of capacitors rated at 400 V and 2 μF each are available, four 2-μF 400-V capacitors in series produce a capacitance of 0.5 μF capable of standing 1600 V. Three parallel groups of four capacitors in series (twelve capacitors) will give the desired 1.5-μF capacitance suitable for the 1600-V circuit. Equalizing resistors across the capacitors would be desirable.

As with inductive reactances, two capacitive reactances in series have a total value of

$$X_{C_t} = X_{C_1} + X_{C_2}$$

Two capacitive reactances in parallel have a total value of

$$X_{C_t} = \frac{X_{C_1} X_{C_2}}{X_{C_1} + X_{C_2}}$$

6-16 PHASE RELATIONS WITH CAPACITANCE

In Chap. 5 it was explained that alternating current and voltage will be out of phase in a circuit in which inductance is present. In an inductive circuit the current may lag the voltage by as much as 90°. In an ac circuit containing series capacitors, the current and voltage may also be out of phase by as much as 90° but the *current leads the voltage*.

The steady-state voltage and current relations of a capacitive circuit can be explained by referring to Fig. 6-12, which shows a capacitor across a source of ac emf. A graph of the ac emf is indicated by the dashed line. The current is represented by the solid line. The voltage varies and alternates from maximum in one direction, point 1 on the graph, to maximum in the opposite direction, point 4, and back to the original value, point 5.

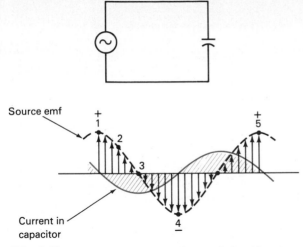

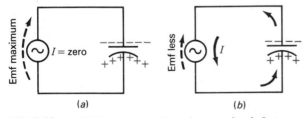

FIG. 6-12 Voltage and current phase relationships in a purely capacitive steady-state ac circuit.

As the source emf increases toward a maximum, more and more electrons are forced into the capacitor. At the instant of maximum pressure, point 1, the capacitor is charged; there will be no electrons moving into the capacitor or in motion at any place in the circuit. The condition that exists is maximum pressure but no current (Figs. 6-12, 6-13a).

FIG. 6-13 (a) With source emf maximum, circuit I stops. (b) As source emf decreases, I flows out of capacitor against diminishing source emf.

The emf is at a maximum value for only an instant and then begins to fall off. As the electric pressure becomes less, all the electrons that had been forced into the capacitor as it was charged comprise an opposing emf greater than the decreasing source emf and current begins to move out of the negative plate of the capacitor into the source in a direction opposite to the source pressure (Fig. 6-13b and point 2 of Fig. 6-12).

The source emf continues to decrease, and the current flows out of the capacitor in a direction opposite to the emf, until the emf reaches zero and reverses. At the instant the voltage reaches zero, the number of electrons moving from one plate to the other attains a maximum value (point 3, Fig. 6-12). From now on, as the emf increases toward maximum (in the negative direction of the graph), the capacitor is charging. When the source emf reaches maximum, the capacitor reaches maximum charge and the number of electrons moving into it will be zero. The condition is again maximum source voltage and no current flow in the circuit (Fig. 6-12, point 4).

It should be noted that at point 3 on the graph the current is at a maximum in the negative direction, but the voltage does not reach maximum in the negative direction until 90° later. Therefore, the current leads the voltage in a purely capacitive circuit by 90°.

The applied voltage and the current will be 90° out of phase only if there is no resistance in the circuit. With any series resistance the difference in source phase will be less than 90°. The phase difference decreases as the proportion of resistance to capacitive reactance increases. When the circuit has a small value of capacitive reactance and a proportionately large resistance value, the circuit is predominantly resistive and the source phase angle is small. If $R = X_C$ in a series R and C circuit, V lags I by 45°.

6-17 SELECTING CAPACITORS

When replacing or purchasing capacitors, several factors should be considered:

1. *Working voltage.* If the circuit in which the capacitor is to be used is a 35-V circuit, purchase a capacitor with a working-voltage rating at least 10 to 20% higher than 35 V.
2. *Capacitance.* Replace with a capacitor having as nearly the same capacitance as possible.
3. *Type of dielectric.* For RF, mica-, air-, vacuum-, or ceramic-dielectric capacitors are suitable. For AF ac, mica-, ceramic-, paper-, or plastic-dielectric capacitors are suitable. For dc filter circuits, electrolytic or paper capacitors are suitable.
4. *Physical size.* Ceramics are usually smaller than mica and paper equivalents. Electrolytics are much smaller than paper.
5. *Cost.* Probable cost, per microfarad, in ascending order: electrolytic, ceramic, paper, plastic, mica, air, vacuum.
6. *Variable, adjustable, or fixed.* As similar to the original as possible.
7. *Temperature.* Capacitors in confined areas may

become overheated and burn out. In particular, it is not advisable to overheat paper or electrolytic capacitors. Electrolytics dry out near hot transistors, tubes, and resistors.

8. *Temperature coefficient*. Some capacitors have a positive temperature coefficient (increase capacitance with an increase of heat); others have a negative temperature coefficient, and still others have a zero temperature coefficient. This fact is important when constant capacitance values are required, as in oscillator circuits.

6-18 CAPACITANCE COLOR CODE

Fixed capacitors may be marked with their capacitance and working voltage. The markings will be either in printed numbers or in colors, using the same number-color code employed with resistors (Sec. 1-13).

The simplest code is the three-dot code, used with 500-V mica capacitors, with a tolerance (guaranteed value) of 20% (Fig. 6-14).

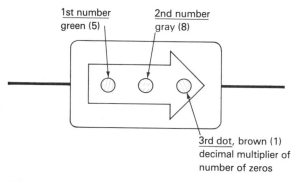

FIG. 6-14 A 3-dot EIA 580-pF mica capacitor.

In the six-dot Electronic Industries Association (EIA) code (Fig. 6-15), reading clockwise from the top left, white indicates mica; the next two dots indicate capacitance numbers; and the fourth dot is the capacitance multiplier, or number of zeros to follow the numbers. Capacitance is in picofarads. The fifth dot is the tolerance (Table 1-2, Sec. 1-13). The sixth dot is the temperature coefficient in parts per million (ppm), increasing as the numbers diminish from 4. The voltage ratings are 500 V up to about 500 pF and 300 V for greater capacitances.

The military standard is similar to the six-dot EIA, except that a black first dot indicates a mica dielectric and silver a paper dielectric.

Some ceramic capacitors are cylindrical and are marked with a series of colored dots or bands. An

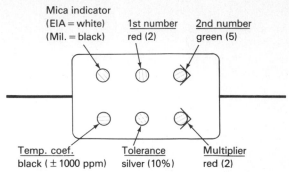

FIG. 6-15 A 6-dot EIA mica capacitor having 2500-pF, 10% tolerance, ±1000 ppm TC.

example of a 3800-pF 10% tolerance capacitor is shown in Fig. 6-16.

The *temperature coefficient* is the degree by which the capacitor will change its capacitance with a change in temperature. If the capacitor does not change its capacitance at all, it has a zero coefficient. If it increases capacitance with increased

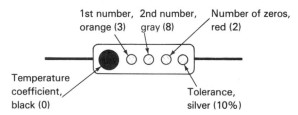

FIG. 6-16 A 3800-pF, 10% tolerance, 0-TC ceramic capacitor.

temperature, it has a positive coefficient. If the coefficient is −150, the capacitance will decrease by 150 ppm per degree Celsius increase in temperature. A list of color-coded temperature coefficients for ceramic capacitors is given in Table 6-2.

TABLE 6-2	TEMPERATURE COEFFICIENTS
Color	TC/°C
Black	0
Brown	−33
Red	−75
Orange	−150
Yellow	−220
Green	−330
Blue	−470
Violet	−750
Gray	+30
White	+500

6-19 TOUCH SWITCHES

An application of series-parallel capacitance is the *touch switch* (Fig. 6-17). The switch consists of two metal plates, *B* and *C*, cemented against the back of a piece of plastic or glass, and a touch

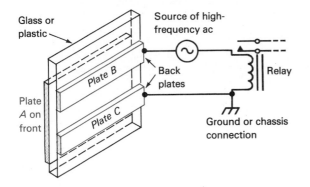

FIG. 6-17 Touch switch consists of 2 plates separated from a third by a glass or plastic dielectric sheet.

plate, *A*, on the front side. This combination produces two plastic or glass dielectric capacitors in series, one formed by plates *A* and *B*, the other by plates *A* and *C*. If each capacitor has 20 pF, the series total is 10 pF, which is in series with the high-frequency ac source and the relay. Circuit current is very low. If someone touches the touch plate, he or she acts as a ground return circuit and adds perhaps 50 pF across the *A*-to-*C* capacitor, resulting in 20 pF in series with 70 pF, or about 16 pF total series-parallel circuit capacitance. With increased capacitance more ac current flows through the relay coil, closing its contacts. The relay con-

tacts can activate power circuits to operate an elevator motor and turn on lights, for example.

▌ Test your understanding; answer these checkup questions.

1. What is the total capacitance when six 4-μF capacitors are connected in series? _____
2. What is the total capacitance when a 2-, a 3-, and a 4-μF capacitor are in series? _____
3. What is the voltage-drop across a 2-μF capacitor in series with a 5-μF capacitor across 100 V? _____
4. Why are resistors usually connected across capacitors in series? _____
5. What is the phase relation between *V* and *I* in a purely capacitive ac circuit? _____
   In a purely resistive circuit? _____
   In an equally resistive and capacitively reactive circuit? _____
6. If a capacitor increases capacitance when warmed, what temperature coefficient does it have? _____
7. Would a mica-dielectric capacitor operating at 1 MHz have more, the same, or less capacitance when operating in a 100-MHz circuit? _____
8. On which plate of an electrolytic capacitor does the dielectric form? _____
9. What might be two causes of electrolytic capacitors exploding? _____ _____
10. In a purely capacitive circuit, what is the current amplitude when the voltage amplitude is at 270°? _____ At 180°? _____
11. Why should capacitors be kept away from hot resistors, transistors, etc.? _____
12. What is the *C* value of a capacitor marked with dots of: Red, green, and yellow? _____
    Blue, gray, and brown? _____
    Black, green, blue, and black? _____
13. A color-coded tubular capacitor has the following dots: violet, yellow, white, brown, and silver. What do you know about it? _____

CAPACITANCE QUESTIONS

1. What is the old name for capacitor? Is it ever used any more?
2. Name four parts of a simple capacitor.
3. In a dc circuit, what determines how long it takes a capacitor to reach full charge?
4. What is the name of the rate of charging of a capacitor?
5. What is the basic unit of measurement of capacitance, and what other smaller units are used?
6. How long will it take a 0.01-μF capacitor to charge to 63% of full charge if a 470-kΩ resistor is in series across a 12-V circuit? How long to reach full charge?

7. What is another term meaning dielectric constant? How is capacitance affected if dielectric constant is increased?
8. What is the capacitance of a 2-plate barium titanate capacitor if the plates are 0.5 in.^{2} and dielectric thickness is 0.001 inch?
9. What are three possible dielectric losses in a capacitor?
10. If in a 15-V dc circuit, what working voltage ratings should nonelectrolytic capacitors have? In 15-V ac circuits?
11. How much energy is stored in a 10-μF capacitor

across 500 V dc? Could this produce a lethal shock?

12. How much energy is stored in a 20,000-μF capacitor across 12 V dc? Would this produce a lethal shock? Why?

13. What is the electron difference, or charge, across a 0.05-μF capacitor across 12.3 V dc?

14. What are the main advantages of electrolytic-type capacitors? Disadvantage?

15. What working voltage should be used when using electrolytic capacitors in dc circuits?

16. What is a capacitor called if its capacitance is changed by compressing plates together? By rotating rotor plates in or out of stator plates?

17. What dielectrics are used for variable capacitors? For adjustable capacitors?

18. List seven types of capacitor dielectrics.

19. Why is a low power factor capacitor usually desired?

20. Do all capacitors have inductance? Do all inductors have capacitance?

21. What can be the advantage gained by coupling a varying dc signal to a load through a capacitor?

22. What is the ac-resistance effect of a capacitor called?

23. How many ohms does a 0.2-μF capacitor have in a 500-Hz circuit? In a 4-MHz circuit?

ANSWERS TO CHECKUP QUIZ ON PAGE 99

1. (0.667 μF) **2.** (0.923 μF) **3.** (71.4 V) **4.** (To equalize voltage across capacitors) **5.** (90°)(0°)(45°) **6.** (+) **7.** (Somewhat less) **8.** (+ plate) **9.** (Connected with incorrect polarity, used on ac) **10.** (Zero)(Maximum) **11.** (Change capacitance, electrolytics dry out) **12.** (250,000 pF)(680 pF)(56 pF) **13.** (TC = −750 ppm, 490 pF, 10% tolerance)

24. If the reactance of a capacitor is 5 kΩ to a frequency of 700 kHz, what is its capacitance?

25. Express Ohm's law for purely capacitive or inductive ac circuits.

26. If a capacitor is lossless, is the formula $P = I^2X$ meaningful? Why?

27. Two similar capacitors in parallel have how much total capacitance? How much total X_C?

28. Two similar inductors in parallel have how much total inductance? How much total X_L?

29. What is the total capacitance if a 3-, 4-, and 5-μF capacitor are in parallel? If in series?

30. Is the voltage-drop across one capacitor in a series group directly or inversely proportional to its C value? To its X_C value?

31. A 1- and a 3-μF capacitor are in series. If a 5-kΩ equalizing resistance is across the 1-μF capacitor, what value equalizing resistance should be across the 3-μF capacitor?

32. What is the phase relationship of the current in respect to the voltage in a purely capacitive ac circuit? How does doubling the frequency change the phase?

33. At the instant of ac voltage maximum across a capacitor, why is current at zero?

34. List eight factors that may have to be considered when selecting a capacitor for a circuit.

35. A 6-dot EIA color-coded capacitor is marked clockwise: White, blue, gray, brown, gold, violet. What does this tell you?

36. What temperature scale is used with capacitor temperature coefficients?

37. Why is high-frequency ac used with the touch-switch shown?

7

Alternating-Current Circuits

This chapter develops methods of computing currents, voltages, impedances, phase angles, power values, and power factors for inductors, capacitors, and resistors in various series, parallel, and series-parallel circuits.

7-1 EFFECTS OF *L*, *C*, AND *R*

In preceding chapters it was pointed out that *inductance alone* in a circuit has the property by which it (1) opposes any change in current, (2) produces an electromagnetic field around itself, (3) tends to limit ac flow in the circuit, (4) passes dc without attenuation, and (5) produces a phase difference of 90°, with the current lagging the voltage of the circuit.

Capacitance alone in a circuit has the property by which it (1) opposes any change in voltage, (2) produces an electrostatic field between its plates, (3) tends to limit ac flow in the circuit, (4) blocks dc flow, and (5) produces a phase difference of 90°, with the current leading the voltage.

Resistance alone in a circuit limits the current that can flow at a given voltage but produces no phase difference between the current flowing through and the voltage across the resistance.

7-2 *L* AND *R* IN SERIES

When an inductance and a resistance are in series across a source of alternating emf (Fig. 7-1), an alternating current flows in the circuit. If resistance were alone in the circuit, the current and the volt-

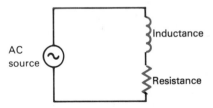

FIG. 7-1 *L* and *R* in series across an ac source.

age would be in phase, all the power in the circuit would be converted to heat, and the formula used to compute *V, I,* and *R* would be the Ohm's-law formula $I = V/R$.

If inductance were alone in the circuit, the current and voltage would be 90° out of phase, all the power in the circuit would be expended in producing a magnetic field around the coil during one-half of the ac cycle, and then all the power would be returned to the circuit again during the other half of the cycle. No power would actually be lost in heat. The Ohm's-law formula used to compute *V, I,* and *X* in reactive circuits would be $I = V/X_L$.

With resistance and inductance in series in an ac circuit it is often necessary to compute current, voltages, impedance, phase angle, voltamperes (apparent power), true power, and power factor.

If the inductance value and the ac frequency are

known, the inductive reactance of the coil in ohms can be determined by $X_L = 2\pi f L$ (Sec. 5-12).

Since the load in Fig. 7-1 is not purely resistive, the total opposition will not be the resistance value alone. The opposition of the load is not purely reactive either and will not be the reactance value alone. Instead, a value that is a resultant of the resistance and reactance will have to be determined. This will be the *impedance (Z)* of the circuit, which is also measured in ohms.

While reactance acts as an opposition and in some ways may be likened to resistance, it must be considered to be setting up its opposition at right angles to the opposition of resistance rather than opposing in the same direction. This can be shown by a vector arrow diagram (Fig. 7-2a). The vector diagram indicates, by the relative lengths of the vector arrows, a circuit in which the R in ohms is less than the X_L in ohms. The actual ac opposition will be something more than either X_L or R values alone, but how much?

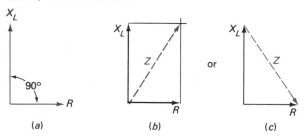

(a)   (b)   or   (c)

FIG. 7-2 (a) X_L and R vectors plot at right angles. (b) and (c) The third side of the triangle is the Z value.

To determine the actual value of opposition, or impedance, lines may be drawn parallel to the resistance vector and to the reactance vector, forming a parallelogram (Fig. 7-2b). The length of the dashed Z vector from the point of origin diagonally across the parallelogram indicates the impedance value of the circuit. How can this be determined?

The triangle in Fig. 7-2c, consisting of the R, Z, and X_L vectors, is a "right" triangle, since the R and X_L vectors are at 90° or at right angles. It is possible to determine the numerical value of the Z side of the right triangle, if the R and X_L lengths are known, by using the pythagorean theorem, which states: *The square of the hypotenuse value of a right triangle is equal to the sum of the squares of the values of the other two sides.* (The hypotenuse is the side opposite the right angle, the Z side in this case.) The pythagorean theorem can be stated in formula form:

$$Z^2 = R^2 + X_L{}^2$$

By taking the square root of both sides of the equation,

$$\sqrt{Z^2} = \sqrt{R^2 + X_L{}^2}$$
$$Z = \sqrt{R^2 + X_L{}^2}$$

This formula can be used in a problem such as the following:

The resistance in a series RX circuit is 3 Ω, and the inductive reactance is 7 Ω. What is the impedance? Figure 7-3 illustrates the circuit.

$$Z = \sqrt{R^2 + X_L{}^2}$$
$$= \sqrt{3^2 + 7^2} = \sqrt{9 + 49} = \sqrt{58} = 7.62\ \Omega$$

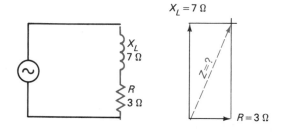

FIG. 7-3 Schematic and vector diagrams of the problem: What is the impedance if R is 3 Ω and X_L is 7 Ω?

With the impedance value it is possible to solve problems involving Ohm's law in ac circuits by substituting the Z value for the R in the dc Ohm's-law formulas. The formulas in Table 7-1 are for use in dc circuits (or in ac circuits having negligible reactance), in reactive circuits with negligible resistance, and in ac circuits in which R and X are both involved.

In the example problem above, the impedance was 7.62 Ω. If a current of 10 A is flowing through the circuit, the voltage of the source must be

$$V = IZ = 10(7.62) = 76.2\ V$$

TABLE 7-1 OHM'S-LAW FORMULAS		
For dc or resistive ac circuits	For purely reactive ac circuits	For circuits having both R and X
$I = V/R$ $V = IR$ $R = V/I$	$I = V/X$ $V = IX$ $X = V/I$	$I = V/Z$ $V = IZ$ $Z = V/I$

What is the impedance of a coil if its resistance is 5 Ω and 0.3 A flows through it when 110 V at 60 Hz is applied across it? In this case both current and voltage are given. Using Ohm's law for ac circuits, $Z = V/I = 110/0.3 = 367$ Ω. Neither the 5-Ω resistance nor the frequency of 60 Hz is needed to solve for the impedance.

7-3 PHASE ANGLE WITH L AND R

The current through a resistor will always be in phase with any voltage across it. If V increases, so does I, and exactly in phase. It is said that the ac V and I phase angle is 0°.

The current flowing in an inductor lags the voltage across it by 90°. If the inductor itself has negligible or zero resistance, the phase angle of the V and I of the coil will be 90°.

When both R and X_L are present in a circuit, the phase angle of the circuit as a whole will be neither 90° nor 0° but some intermediate value. This will be the phase angle that is seen by the source as it looks into the whole circuit connected across it. The number of degrees will be equal to the angle formed between the Z and the R sides of a Z, R, X triangle of the circuit. This phase angle is indicated by the Greek letter θ (theta). Figure 7-4 illustrates a

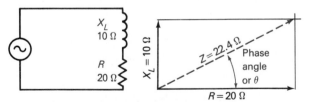

FIG. 7-4 The phase angle is always formed by the R and Z sides of the vector triangle.

circuit having 10-Ω inductive reactance and 20-Ω resistance. The impedance value is found by

$$Z = \sqrt{R^2 + X^2} = \sqrt{20^2 + 10^2} = 22.4 \ \Omega$$

One method of determining the number of degrees of the phase angle is to draw to scale the R, X_L, Z vector diagram of the circuit and with a protractor measure the angle between R and Z.

Another method of determining the phase angle is to compute the numerical value of the ratio of the R and Z sides. This ratio of R to Z, or R/Z, has been given the name cosine. In Fig. 7-4, the cosine of R/Z is equal to 20/22.4, or 0.89286.

Trigonometric tables (Appendix C) will indicate the angle associated with the cosine value of

0.89286. By searching through the cosine ("cos") values in the tables, the number 0.89259 is nearly equal to 0.89286 and is indicated as being equal to 26.8°. (This is found much faster and more accurately by pocket calculator to be 26.76550057°!) The angle 26.8° is accurate enough for general use. The angle by which the current lags the voltage in this inductive circuit is 26.8°. The circuit has a phase angle of 26.8°.

Although the phase angle can be found by using the cosine ratio R/Z, it can also be found by using the ratio of sides X/Z, which is known as the sine ("sin" in tables) value. The angle associated with the sine value will be the same phase angle. The ratio of sides X/R is known as the tangent ("tan"), and may also be used to determine the θ. Remember:

$$\cos \theta = \frac{R}{Z} \qquad \sin \theta = \frac{X}{Z} \qquad \tan \theta = \frac{X}{R}$$

In a circuit having equal sides of R and X the vector arrows are equal in length, the impedance will be 1.414 times the R (or X_L) value, and the phase angle will be half of 90°, or 45° (Fig. 7-5). Check this by using cos, sin, and tan formulas.

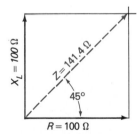

FIG. 7-5 Vector diagram when R and X_L are equal.

7-4 VOLTAGE VECTOR ADDITION

The same 10-Ω inductive reactance, 20-Ω resistance, 22.4-Ω impedance circuit, plus an ammeter indicating a 10-A current are shown in Fig. 7-6.

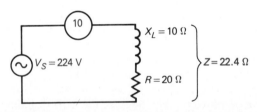

FIG. 7-6 A 224-V ac source will develop 100 V and 200 V across the X_L and R of this series circuit.

If 10 A flows through a reactance of 10 Ω, the voltage-drop across the reactance will be

$$V_{X_L} = IX_L = 10(10) = 100 \text{ V}$$

If 10 A flows through a resistance of 20 Ω, the voltage-drop across the resistance will be

$$V_R = IR = 10(20) = 200 \text{ V}$$

The sum of the two voltage-drops, the reactive 100 V plus the resistive 200 V, would seem to equal 300 V, but the voltage across the reactance will lead the current through it by 90° and the voltage across the resistance is in phase with the current through it. These two *voltages*, being 90° out of phase, must be added vectorially (Fig. 7-7),

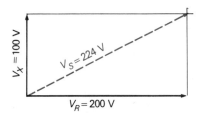

FIG. 7-7 Vector diagram of the source voltage plotted with the reactive and resistive voltages.

using a method similar to that used to determine Z when R and X are known. The reactive and resistive voltages are plotted 90° out of phase, and the resultant is the source-voltage value V_S.

Using the pythagorean theorem, the formula for determining the resultant or source voltage is

$$V_S^2 = V_R^2 + V_X^2$$
$$V_S = \sqrt{V_R^2 + V_X^2}$$

where V_S = source voltage
V_R = voltage across R
V_X = voltage across X

By substituting the values in the problem above,

$$V_S = \sqrt{200^2 + 100^2} = \sqrt{40{,}000 + 10{,}000}$$
$$= 224 \text{ V}$$

Here is a case in which 200 and 100 V equals 224 V, but only if they are added *vectorially* and at right angles.

The total voltage may also be determined by using the Ohm's-law formula

$$V = IZ = 10 \times 22.4 = 224 \text{ V}$$

An important point here is the understanding that the source voltage represents a *resultant* of the reactive and resistive voltage-drops across the circuit,

and vice versa. Also, in an ac circuit involving L, C, and R in series, the sum of all the voltage-drops in the circuit always adds up to more than the source voltage unless they are added vectorially.

When two voltages (or currents) are in phase, they can be added vectorially, head to tail, in the same direction (Fig. 7-8a). This would apply when

(a)   (b)

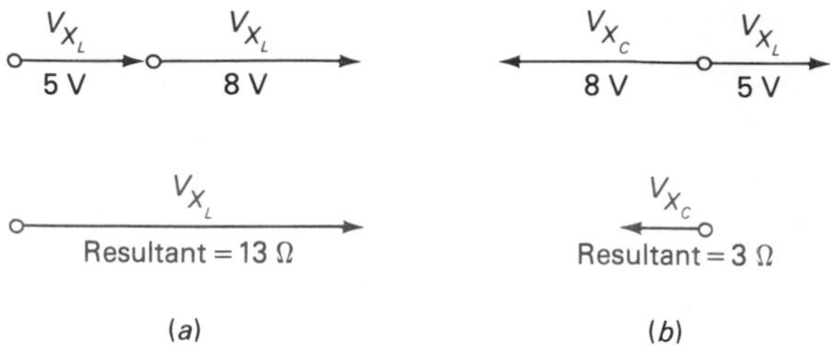

FIG. 7-8 Vector addition of (a) two voltages in phase and (b) two voltages 180° out of phase.

the voltage-drops across two coils in series are added. The resultant voltage is the simple sum of the voltages. If two voltages (or currents) are out of phase by 180°, they must be added vectorially in opposite directions, starting at the same reference point (Fig. 7-8b). This would apply when the voltage-drop across a coil is added to the voltage-drop across a capacitor when they are in series. The resultant voltage will be the difference of the two voltages and will carry the sign of the larger.

Test your understanding; answer these checkup questions.

1. In what unit is impedance measured? _____
2. State the pythagorean theorem in the impedance formula form. _____
3. What is probably being solved when the partial equation $\sqrt{a^2 + b^2}$ is seen? _____
4. A 100-V ac circuit has a 10-Ω R and a 12-Ω X_L in series across it. What is the Z value? _____ The I_R value? _____ The V_R value? _____
5. In question 4, what is the cosine ratio in symbol letters? _____ The cos value? _____ The phase angle of the circuit? _____
6. If a series ac circuit has an impedance of 25 Ω and 10 Ω of X_L, what is the sine ratio in symbols? _____ The sin value? _____ The θ value? _____
7. If the reactance in question 6 were capacitive, do you think the sin value would be the same? _____ The θ value? _____
8. A series ac circuit has an X_L of 50 Ω and a R of 20 Ω. What is the tangent ratio in symbols?

_____ The tan value? _____
The θ? _____

9. Could the formula $Z = \sqrt{R^2 + X^2}$ be used to solve R and X in parallel? _____
10. A series ac circuit has 20 V across the resistor and 30 V across the reactor. What is the source-voltage value? _____ The tan value? _____ The phase angle? _____
11. Why is the pythagorean theorem not used with I_R and I_X in series circuits to determine the source current? _____
12. What is the resultant voltage in a series circuit if $-V_R = 75$ V and $+V_R = 55$ V? _____ If $V_X = 75$ V and $V_R = 55$ V? _____ If $V_{X_L} = 75$ V and $V_{X_C} = 55$ V? _____

7-5 APPARENT AND TRUE POWERS

In the circuits described thus far, the only part of the circuit actually using power is the resistor. A resistor loses power in the form of heat whenever current flows through it. All the current flowing through a coil produces energy that is stored in the magnetic field around the coil. When the current alternates, the magnetic field collapses and returns all its energy to the circuit. Therefore, there is no loss of power in the inductance itself (if it has negligible resistance in its turns and no coupling to any external circuit).

The same series circuit as shown in Fig. 7-6, but with a voltmeter included, is shown in Fig. 7-9. All

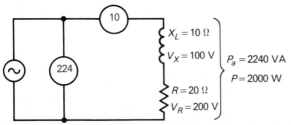

FIG. 7-9 Apparent power is the value computed when meter voltage is multiplied by meter current.

known values are indicated. The *apparent power* (P_a) can be determined by using the visually apparent values shown by the voltmeter (224 V) and by the ammeter (10 A). Since power can be determined by $P = VI$, or $P = VA$, the apparent power of the circuit is 224(10), or 2240 VA (not watts).

The power lost in heat is usually computed by the formula $P = I^2R$, in this case $10^2(20)$, or 100(20), or 2000 W. This is called the *true power* (P) of the circuit. P is always less than P_a.

The true power of this simple circuit can also be determined by using either of the formulas $P = V^2/R$ and $P = VI$ if the voltage across the resistor is used, but the formula $P = I^2R$ should be employed if possible. Wattmeters always indicate true power values.

The use of the term *voltamperes* is preferred by many to the term "apparent power." In a purely reactive circuit, there may be volts forcing amperes to flow, but if all the energy stored in the field of the reactance is returned to the circuit on the second half of each cycle, there is only an apparent power (VA) but no actual loss of power.

7-6 POWER FACTOR WITH L AND R

When the ratio of true to apparent powers, or P/VA, is computed, a decimal number between zero and 1.0 results. The ratio of true/apparent power in a circuit is known as the *power factor* (pf) of the circuit.

$$pf = \frac{\text{true power}}{\text{apparent power}} = \frac{I^2R}{VA}$$

In the circuit used as the example thus far, this ratio is

$$pf = \frac{2000 \text{ W}}{2240 \text{ VA}} = 0.89286$$

Note the reappearance of the number 0.89286. This will be recognized as the cosine value determined by dividing the resistance by the impedance. Besides representing the cosine of 26.8°, this decimal figure 0.89286 is the circuit's power factor. Since the ratio of true to apparent power and the ratio of resistance to impedance produces the same decimal figure, either ratio can be used to determine the power factor of a series circuit. The usual methods of determining power factor are

$$pf = \frac{P}{VA} = \frac{I^2R}{VA} = \frac{R}{Z} = \cos \theta$$

where θ is the VI phase angle.

By algebraic rearrangement, the true power can be computed by multiplying the apparent power by the power factor, pf (not pF, which is picofarad!):

$$P = VA(pf) \quad \text{or} \quad P = VA(\cos \theta)$$

The pf is in reality a comparison of the amount of power a circuit is apparently using and what it is actually using. It is often expressed as a percent. Thus, a pf of 0.89286 is also a pf of 89.286%.

Power factor is important when wiring a circuit. Consider an electric circuit having a pf of 0.5, a source voltage of 100 V, and an impedance of 20 Ω. According to Ohm's law the current in the circuit is $I = V/Z$, or 100/20, or 5 A. An ammeter in the circuit will read 5 A. Apparent power is the product of source voltage and current through the source, in this case 100(5), or 500 VA. Since true power is apparent power times pf, or VA(cos θ), the load in the circuit must be receiving 100(5)(0.5), or 250 W. The wires of the circuit must carry the current of 5 A that is shown by the ammeter instead of only the 2.5 A normally required when 100 V produces 250 W of power in a resistive load.

Why is 5 A flowing when only 2.5 A should produce the power? The answer lies in the fields developed in the reactors of the circuit. Energy is required to build up the field of a reactor but this energy is returned to the circuit when the field collapses. The energy is carried in the form of current from the source and is returned as current to the source. At the same time the load is constantly demanding energy and therefore current. The wires must carry the reactive field-building current too.

Apparent power does not mean fictitious power. Apparent power has its effect in electric circuits. Engineers must so construct circuits that true power approaches apparent power (high pf), to allow the use of smaller wire to lessen costs.

If the reactance causing a low pf in a circuit is inductive reactance, it is possible to raise the pf by adding some capacitive reactance to the circuit to counteract the inductive reactance, leaving the circuit more nearly resistive (higher pf).

Power factor may be referred to as being either *leading* or *lagging*. A lagging pf indicates the *current* is lagging the voltage in the circuit (inductive circuit). A leading pf means the *current* is leading the voltage (capacitive circuit).

If a 220-V line delivers 100 W at 80% pf to a load, what is the phase angle between the line current and line voltage and how much current flows in the line? First, the cosine of the phase angle is the pf, or cos θ = pf = 80% = 0.80. From a table of trigonometric functions or a calculator, 0.80 is found to be the cosine of 36.9°,

which is the phase angle between the current and voltage. (It is impossible to tell whether the current is leading or lagging the voltage from the information given.)

Second, in practice, when power is mentioned, true power is understood. Therefore the 100 W in the problem must be true power. Since true power equals voltamperes times power factor

$$P = VA(pf) = VA(\cos \theta)$$

then $$VA = \frac{P}{\cos \theta} = \frac{100}{0.80} = 125 \text{ VA}$$

The voltampere value represents the product of what would be read by a voltmeter across the load and an ammeter in series with the load. When the apparent-power formula $P = VA$ is rearranged to read $A = P/V$, the desired current value is

$$A = \frac{P}{V} = \frac{125}{220} = 0.568 \text{ A}$$

7-7 C AND R IN SERIES

A circuit composed of an ac source, a capacitor, and a resistance in series is shown in Fig. 7-10. If the capacitance value and the ac frequency are known, the capacitive reactance of the capacitor in ohms can be computed by $X_C = 1/2\pi fC$ (Sec. 6-13).

FIG. 7-10 Series *RC* circuit across an ac source.

The capacitive-reactance value in ohms can be used in conjunction with the resistance for ac computations in very much the same way as inductive reactance and resistance are used.

As with an X_L and R in series, the capacitive-reactance (X_C) vector value is also plotted at right angles to the resistance vector (Fig. 7-11). Note that whereas the X_L vector was always shown pointing upward, the X_C vector points downward, a 180° difference in direction. This difference indicates that X_L and X_C vectors will tend to cancel each other.

The vector sum of the capacitive reactance and resistance is the impedance value and is determined by using the pythagorean formula:

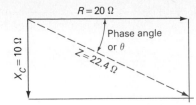

FIG. 7-11 Vector diagram of R, X_C, and the resultant Z.

$$Z = \sqrt{R^2 + X_C^2}$$

If the impedance value in ohms is known, voltage or current values can be computed by using Ohm's-law formulas for ac circuits:

$$I = \frac{V}{Z} \qquad V = IZ$$

In a series circuit the current is the same in all parts of the circuit at any instant. In a purely capacitive circuit the voltage across the capacitor lags the current by 90°. In a resistive circuit the voltage and current are in phase. In a circuit made up of both X_C and R, the phase angle between V and I will be something between 0° and 90°. The phase angle, as in inductive circuits, may be solved for graphically or by determining the angle represented by the ratio of R/Z. The ratio of the resistive voltage-drop to the source voltage (V_R/V_S) will be proportional to the ratio of resistance to impedance and may also be used to determine phase angle.

A capacitive circuit has a *leading power factor* (current leads). The pf can be found by:

1. Dividing R by Z (pf = R/Z)
2. Dividing true power by the apparent power (pf = I^2R/VA)
3. Finding the cosine of the angle between current and voltage of the circuit (pf = cos θ)

Since the ratio of the voltage-drop across the resistance to the voltage-drop across the source is proportional to the ratio of resistance to impedance, power factor in a series circuit is also

$$\text{pf} = \frac{V_R}{V_S}$$

The computations of impedance, phase angle, and power factor are performed the same way, whether the circuit has a coil and resistor in series, or a capacitor and resistor in series.

A leading power factor in a capacitively reactive series circuit can be corrected (raised) by adding inductance to the circuit.

■ Test your understanding; answer these checkup
■ questions.

1. Under what two conditions might a coil lose power? _____ _____
2. What power is determined by multiplying ac ammeter times ac voltmeter values? _____
3. What power is registered by a wattmeter? _____
4. In what unit is apparent power measured? _____ True power? _____
5. What is the ratio of true to apparent power called? _____
6. Does the cosine value of an ac circuit indicate true power, VA, or pf? _____
7. What is the phase relationship of the voltage in a circuit having a lagging pf? _____
8. A 120-V line delivers 500 W at 90% pf. What is the phase angle? _____ The line current? _____
9. What is the effective reactance in a series circuit having 12-Ω X_C and 20-Ω X_L? _____ 40-Ω X_C and 40-Ω X_L? _____ 60-Ω X_C and 35-Ω X_L? _____
10. A 300-Ω R and 100-Ω X_L are in series across a 50-V ac source. What is the Z value? _____ The VA value? _____ P value? _____ Pf? _____ θ? _____ Power lost in heat? _____ I value? _____
11. A series RL circuit has a Z of 141 Ω to 500-Hz ac. An ohmmeter shows the total resistance as 100 Ω. The source voltage, V_S, is 120 V. What is the reactance value? _____ The VA? _____ _____ P? _____ θ? _____ V_L? _____ V_R? _____

7-8 L, C, AND R IN SERIES

A circuit containing a coil, capacitor, and resistor in series across a source of ac is shown in Fig. 7-12. If the inductance, capacitance, and frequency are known, the reactances can be found by applying the usual reactance formulas:

$$X_L = 2\pi f L \qquad \text{and} \qquad X_C = \frac{1}{2\pi f C}$$

When drawing a vector diagram of R, X_L, and X_C in series, the dissimilar reactances must be plotted

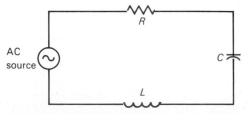

FIG. 7-12 Series RCL circuit across an ac source.

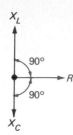

FIG. 7-13 Inductive and capacitive reactances plot 180° out of phase but 90° from resistance.

in opposite directions, each 90° from resistance. Figure 7-13 illustrates a large value of X_L and a smaller value of X_C. The vector diagram shows that X_C and X_L are each 90° from R but in such directions that they tend to cancel each other. If there is more X_L than X_C, as shown, the resultant, or net reactance, will be X_L. The net reactance is obtained by subtracting the smaller value from the larger. The complete impedance formula for a series ac circuit in which X_L is greater than X_C is

$$Z = \sqrt{R^2 + (X_L - X_C)^2}$$

In cases in which X_C is greater than X_L, the formula is

$$Z = \sqrt{R^2 + (X_C - X_L)^2}$$

EXAMPLE: What is the Z of a series circuit consisting of an R of 4 Ω, an X_L of 9 Ω, and an X_C of 6 Ω? Circuit and vector diagrams are shown in Fig. 7-14.

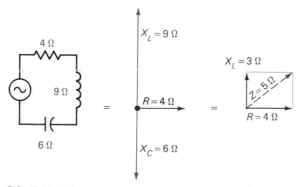

FIG. 7-14 Schematic and vector diagrams of 4-Ω R, 9-Ω X_L, and 6-Ω X_C in series.

$$Z = \sqrt{R^2 + (X_L - X_C)^2} = \sqrt{4^2 + (9 - 6)^2}$$
$$= \sqrt{16 + 9} = \sqrt{25} = 5$$

If the source produces 50 V, the current, by Ohm's law would be

$$I = \frac{V}{Z} = \frac{50}{5} = 10 \text{ A}$$

Conversely, if the circuit current is known to be 10 A, the voltage of the source, according to Ohm's law for ac circuits, is

$$V = IZ = 10(5) = 50 \text{ V}$$

In a series circuit the current is the same in all parts of the circuit. At any instant when the current reaches its peak value in the resistance, the current is also at a peak value in the inductance and capacitance. The voltages across the components may be out of phase with the current, but the current in all parts of the circuit, including the source, has the same value at any instant.

The voltage-drop across each part of the circuit above, with a current of 10 A, will be

$$V_R = IR = 10(4) = 40 \text{ V}$$
$$V_{X_C} = IX_C = 10(6) = 60 \text{ V}$$
$$V_{X_L} = IX_L = 10(9) = 90 \text{ V}$$

The power factor of the problem above is probably most simply determined by the ratio R/Z, or 4/5 which equals 0.80. The phase angle is determined by $\cos \theta = 0.80 = 36.9°$.

If a series circuit has capacitance and inductance but has *zero resistance*, the impedance formula eliminates the R value and becomes

$$Z = \sqrt{(X_L - X_C)^2} = X_L - X_C$$

With no resistance, the phase angle will be 90°. The power factor will be R/Z, or 0/Z, which is zero. With a power factor of zero there is no loss of power in the circuit. The circuit is completely reactive and appears as either a purely capacitive or purely inductive circuit to the source, depending on which reactance is greater.

In any series LCR circuit the total impedance will always exceed or equal the resistance value.

In the special case in which X_L equals X_C, a condition known as *resonance* occurs. When the impedance formula for a series circuit is applied to a series-resonant circuit, $Z = R$, or

$$Z = \sqrt{R^2 + (X_L - X_C)^2} = \sqrt{R^2 + 0^2} = R$$

7-9 ACCURACY OF COMPUTATIONS

When the resistance in a series circuit is more than 10 times the net reactance, for most practical purposes the reactance could be disregarded. Consider the following:

A series circuit has a resistance of 100 Ω and a reactance of 10 Ω. The impedance is

$$Z = \sqrt{R^2 + X^2}$$
$$= \sqrt{100^2 + 10^2} = \sqrt{10,100} = 100.5 \ \Omega$$

The 100.5-Ω impedance is within 0.5% of 100 Ω. Most meters used for voltage, current, and resistance measurements are guaranteed to be accurate only within 1 to 3% of full scale. This is one reason why mathematical accuracy to the *third or fourth significant figure* is all that is required in general work. (15,273 = 15,300 is correct to the third significant figure, and 15,270 is to the fourth.)

Since the circuit above acts almost as a pure resistance, the source current and voltage will be nearly in phase and pf will approach unity (1).

If the values were reversed, that is, the reactance were more than 10 times the resistance, for most purposes the circuit might be considered to be completely reactive. (Simplifications of this type should be applied with caution.)

The knowledge that any series circuit has a resistance value much larger than its total reactance value should indicate, even before a problem is worked, what the approximate impedance, phase angle, and power factor should be. If computed answers do not correspond to generalized theory, there is a good possibility that errors were made in the mathematical solving of the problem.

7-10 PROVING PROBLEMS

It is possible to work a problem in which values of an equation are squared and an incorrect answer results. For example:

A series circuit has 7-Ω R, 8-Ω X_L, and 13-Ω Z. What is X_C? Figure 7-15 shows the diagram of the circuit. By the series-circuit formula:

$$Z = \sqrt{R^2 + (X_L - X_C)^2}$$
$$Z^2 = R^2 + (X_L - X_C)^2$$
$$Z^2 - R^2 = (X_L - X_C)^2$$
$$\sqrt{Z^2 - R^2} = X_L - X_C$$
$$+ X_C + \sqrt{Z^2 - R^2} = X_L$$

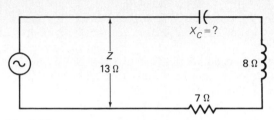

FIG. 7-15 Diagram of circuit in problem.

$$X_C = X_L - \sqrt{Z^2 - R^2} = 8 - \sqrt{13^2 - 7^2}$$
$$X_C = 8 - \sqrt{120} = 8 - 10.95$$
$$X_C = -2.95 \ \Omega$$

Should the answer -2.95 Ω be substituted in the original formula to prove the answer, the impedance works out to be 8.7 Ω instead of the specified 13 Ω. *Something is wrong.* It must be remembered that the answer is -2.95.

When the formula indicates $X_L - X_C$, this should be computed as $X_L - (-X_C)$, or $X_L + X_C$, which proves as 13 Ω of impedance.

A vector diagram of the problem would have shown that an X_C greater than the X_L value would be required to produce the 13-Ω Z and that the formula with a greater X_C would be indicated as the one to use. Always draw a vector representation of an ac circuit before working it.

7-11 THE j OPERATOR

The letter j is used as a means of notation, or as a means of labeling the quantity in front of which it is placed. It is known as the j *operator*.

The j operator in front of a number indicates that the quantity is 90° (no more and no less) out of phase with something else. It is considered as a vector rotation of 90°. Numbers with a j operator in front of them may be called *imaginary* numbers.

A positive j indicates a normal, *counterclockwise* 90° rotation of the vector. It is possible to express "A 3.7-A current in some capacitive circuit is leading the voltage 90°" by merely writing $j3.7$ A. The notation $j2$ V indicates that some 2-V vector is leading its current vector by 90°, which would be true in an inductive circuit.

A negative j $(-j)$ indicates a *clockwise* rotation of the vector by 90°. "A 4.2-A current in some inductive circuit lags the voltage by 90°" may be expressed by $-j4.2$ A. The notation $-j83$ V indicates a capacitively reactive voltage of 83 V lagging the current by 90°.

The j indicates 90°, and the polarity sign prefixing it indicates the direction of vector rotation.

When drawing vectors to represent reactance values, it is standard practice to show the X_L values 90° ahead of the zero value, or upward. The X_C value is shown 90° behind the zero value, or downward. Therefore a j operator before a reactance value such as $j276\ \Omega$ indicates 276-Ω X_L, while $-j75\ \Omega$ indicates 75-Ω X_C.

A j operator is not used in front of a Z value because the Z infers some R and X and therefore less than 90° phase shift.

▌ Test your understanding; answer these checkup questions

1. When X_L, X_C, and R are in series, what formula is used to determine Z? _____
2. When X_L and X_C are known but R is zero, what is the formula to determine Z? _____ What is the value of θ? _____
3. What is the special term used when X_L equals X_C in a series circuit? _____ What is the formula for Z in this case? _____
4. In most practical applications, how much greater should the X value be than the R to allow R to be disregarded? _____ What percent Z error does this produce? _____ What θ is this? _____
5. A series ac circuit has $14\ \Omega$ of X_L, $6\ \Omega$ of R, and $6\ \Omega$ of X_C. What is the Z value? _____ Pf? _____ θ? _____
6. A series ac circuit of 12-Ω R, 15-Ω X_L, and 40-Ω X_C has 5 A flowing in it. What is the value of Z? _____ V_S? _____ Pf? _____ θ? _____ V_L? _____ V_C? _____
7. A 100-W 115-V lamp is in series with a 355-Ω X_L and a 130- X_C across 220 V. What is the lamp R? _____ Z of this circuit? _____ I? _____ Pf? _____ θ? _____
8. A potential of 110 V is applied to a series circuit containing 25-Ω X_L, 10- X_C, and 15-Ω R. What is the Z? _____ Pf? _____ θ? _____
9. A series circuit having 5-Ω R, 25-Ω X_C, and 12-Ω X_L has 10 A flowing. What is the voltage across the L? _____ C? _____ R? _____ What is the source voltage? _____
10. What is the Z value if a series circuit contains $j40\ \Omega$, $-j50\ \Omega$, and $15\ \Omega$? _____ What is θ? _____

7-12 PARALLEL AC CIRCUITS

Methods for solving series ac and parallel ac circuits are quite different. Considerable confusion may result from a carry-over of procedure between the two types. To try to prevent this and at the same time introduce two new electrical terms, an entirely different method will be employed to compute the impedance of parallel circuits. Basically, this method is similar to the method of computing resistors in parallel by converting the resistance values into their equivalent reciprocal values, their conductances, and then adding the conductances in siemens, S (formerly mhos, Ω). The letter used to denote conductance is G. The reciprocal of the total conductance is the total resistance, or

$$G_t = \frac{1}{R_1} + \frac{1}{R_2} + \frac{1}{R_3} = \frac{1}{R_T}$$

and

$$R_T = \frac{1}{1/R_1 + 1/R_2 + 1/R_3}$$

This last formula is valid for an ac circuit as well as a dc circuit as long as there is only resistance in the circuit. If any reactance is present the formula will not hold true. However, a variation of this formula can be used.

To solve for the impedance, power factor, phase angle, and so on in circuits having resistance and reactance in parallel, two new terms are used. The first is *susceptance*, B, the reciprocal of reactance, or $1/X$, measured in siemens. The second is *admittance*, Y, the reciprocal of impedance, or $1/Z$, also measured in siemens. Therefore:

$$B = \frac{1}{X} \qquad Y = \frac{1}{Z}$$

Susceptances are simply added to determine the total value of parallel reactances. The susceptance of a 20-Ω X_L and an 8-Ω X_L in parallel is

$$B_t = B_1 + B_2$$

$$B_t = \frac{1}{X_{L_1}} + \frac{1}{X_{L_2}} = \frac{1}{20} + \frac{1}{8}$$

$$= \frac{2}{40} + \frac{5}{40} = \frac{7}{40}\ \text{S}$$

If the total B is equal to $7/40$ S, then the total X_L will be the reciprocal of B_t or $40/7$, or $5.71\ \Omega$ of X_L.

If the two reactances are of different types (Fig. 7-16), the total B will be the *difference* between the two susceptances. When 20-Ω X_C and 8-Ω X_L are in parallel the susceptance is now

$$B_t = \frac{1}{8} - \frac{1}{20} = \frac{5}{40} - \frac{2}{40} = \frac{3}{40}\ \text{S}$$

The reciprocal of B_t is $40/3$, or $13.3\ \Omega$. Since the

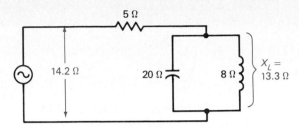

FIG. 7-16 Two parallel reactances with a resistor connected in series with them.

current through the 8-Ω inductor would be the greater, the parallel group acts as a 13.3-Ω X_L.

With the 5-Ω resistor connected in series with the parallel group, the circuit impedance can be solved as a 5-Ω R in series with a 13.3-Ω X_L, or

$$Z = \sqrt{5^2 + 13.3^2} = 14.2 \ \Omega$$

with an inductive (lagging) power factor.

7-13 L AND R IN PARALLEL

To solve for the total impedance of R and X_L in parallel, it is important to understand that the series impedance formula $Z = \sqrt{R^2 + X_L^2}$ cannot be applied. However, a similar type of formula can be used, provided *reciprocal* values are substituted. The reciprocal of Z is Y; the reciprocal of R is G; and the reciprocal of X is B. Therefore, the formula to solve for the admittance of a *parallel* R and X_L is

$$Y = \sqrt{G^2 + B_L{}^2} \qquad \text{or} \qquad Y = G + jB$$

Impedance, the reciprocal of admittance, Y, can be found by dividing Y into 1. For the circuit of Fig. 7-17

$$Y = \sqrt{G^2 + B^2} = \sqrt{0.1^2 + 0.1^2}$$
$$= \sqrt{0.01 + 0.01} = \sqrt{0.02} = 0.1414 \ \text{S}$$
$$Z = \frac{1}{Y} = \frac{1}{0.1414} = 7.07 \ \Omega$$

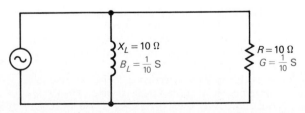

FIG. 7-17 Parallel RL circuit across an ac source.

This formula is solving a right triangle having G, B, and Y sides, similar to an RXZ right triangle of a series circuit, Fig. 7-18.

In series circuits the phase angle is the angle between the R and Z sides. In parallel circuits, the phase angle is the angle between the 1/R and 1/Z (G and Y) sides. If G and Y are known, the phase angle of the circuit is equal to the cosine G/Y. In

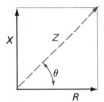 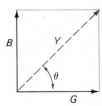

FIG. 7-18 Series-circuit X, R, and Z vectors compared with parallel-circuit B, G, and Y vectors.

Fig. 7-17, G/Y is equal to 0.100/0.1414, or 0.707. From a table of trigonometric functions, this angle, the phase angle, is found to be 45°. The angle should have been recognized as 45°, because both the B and G vectors are equal.

With the impedance of the circuit known, Ohm's law can solve either current or voltage. If the source voltage in Fig. 7-17 is 20 V, the source current or I_t is

$$I_t = \frac{V}{Z} = \frac{20}{7.07} = 2.83 \ \text{A}$$

The individual branch currents are

$$I_R = \frac{V}{R} = \frac{20}{10} = 2 \ \text{A}$$

$$I_{X_L} = \frac{V}{X_L} = \frac{20}{10} = 2 \ \text{A}$$

As with the vector diagram of B and G in a parallel circuit, it is found that a vector diagram of the branch *currents* will also plot in such a way that the phase angle, and therefore the power factor, can be determined from it. Figure 7-19 shows the current vector diagram of the circuit described above, using the resistive current and the inductive-reactance current, which together produce a resultant 2.83-A source current.

In parallel R and X circuits, where R and X are

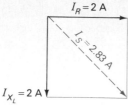

FIG. 7-19 In parallel circuits the currents of the branches can be added vectorially.

not present in the same branch, the circuit may also be solved without the use of G, B, and Y by assuming a convenient source voltage and solving for the separate branch currents. By solving the current vector triangle of I_R, I_X, and I_S, the cosine I_R/I_S will be the power factor and the cosine of the phase angle. Increasing or decreasing the assumed source voltage will have no effect on the ratio of the currents and therefore will give the same θ and pf. Phase angle can also be determined from power factor using P/VA, or from $\cos \theta = G/Y$, or $\sin \theta = B/Y$, or from $\tan \theta = B/G$.

EXAMPLE: What is the impedance of a circuit having an X_L of 50 Ω in parallel with an R of 25 Ω? Assume a convenient source voltage, such as 100 V, as in Fig. 7-20.

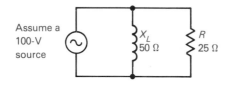

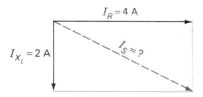

FIG. 7-20 Any convenient source voltage can be assumed when computing impedance of a parallel circuit if branch currents are vectored.

$$I_R = \frac{V_S}{R} = \frac{100}{25} = 4 \text{ A}$$

$$I_{X_L} = \frac{V_S}{X_L} = \frac{100}{50} = 2 \text{ A}$$

$$I_S = \sqrt{I_R^2 + I_{X_L}^2} = \sqrt{4^2 + 2^2}$$
$$= \sqrt{16 + 4} = \sqrt{20} = 4.48 \text{ A}$$

Then $\quad Z = \dfrac{V_S}{I_S} = \dfrac{100}{4.48} = 22.3 \ \Omega$

7-14 C AND R IN PARALLEL

Capacitance and resistance in parallel are computed in the same way as inductance and resistance, except that the susceptance vectors, or the reactive current vectors, are drawn in opposite directions.

▌ Test your understanding; answer these checkup questions.

1. What letter symbol represents $1/R$? _____ $1/X$? _____ $1/Z$? _____
2. What term is represented by the letter symbol B? _____ Y? _____ G? _____
3. What is the X_t of a 20-Ω X and a 30-Ω X if both are similar reactances and in series? _____ Are similar reactances and in parallel? _____ Are opposite reactances in series? _____ Are opposite reactances in parallel? _____
4. Which two sides of an impedance triangle form the phase angle in a series circuit? _____ Which triangle sides form θ in a parallel circuit? _____
5. A 50-V source has a parallel 40-Ω R and a 25-Ω X across it. What is the admittance of the circuit? _____ The Z? _____ The I_L? _____ The I_R? _____ The P? _____
6. What is another way of expressing $Y = G + jB$? _____ $Y = G - jB$? _____
7. If the currents in a resistive branch and a parallel capacitive branch are known, what formula is used to solve for I_t? _____
8. If G, B, and Y are known, how is pf determined? _____
9. Voltages can be vectored in series circuits. What can be vectored in parallel circuits? _____
10. A 120-V source delivers 200 W at 90% pf to a parallel R and X_L. What is the θ? _____ I_S? _____ Z? _____ VA? _____
11. A 1-μF capacitor is across a 530-Ω electric light operating on a 110-V 60-Hz ac line. What is the Z_t? _____ Pf? _____ θ? _____ I_R? _____ I_S? _____

▌ ANSWERS TO CHECKUP QUIZ ON PAGE 110

1. $[Z = \sqrt{R^2 + (X_C - X_L)^2}]$ **2.** $(Z = X_L - X_C)(90°)$ **3.** (Resonance)$(Z = R)$ **4.** (10 times)(0.5%)(84.3°) **5.** (10 Ω)(0.620)(53.2°) **6.** (27.7 Ω)(139 V)(0.43)(64.5°)(75 V)(200 V) **7.** (132 Ω)(261 Ω)(0.842 A)(0.506)(59.6°) **8.** (21.3 Ω) (0.707)(45°) **9.** (120 V)(250 V)(50 V)(139 V) **10.** (18 Ω) (33.7°)

7-15 C, L, AND R IN PARALLEL

With C, L, and R in parallel, the same voltage is across all branches. Current in the inductive branch will lag the source voltage by 90°. Current in the capacitive branch will lead the source voltage by 90°. Therefore, the two reactive currents will tend to cancel each other insofar as the source is concerned. If these reactive currents happen to be equal, the only current the source must supply is to the resistive branch. This special condition, in which $X_L = X_C$, is known as *parallel resonance* (Chap. 8).

When X_L and X_C are not equal in a parallel circuit (Fig. 7-21), the circuit can be solved by the

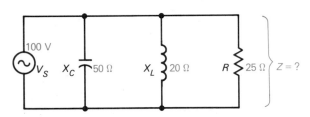

FIG. 7-21 Parallel *RLC* circuit across an ac source.

admittance (or the current-vector) method. The X's are converted to their B values and the R to its G value. For the example circuit,

$$X_C = 50\ \Omega \qquad B_C = {}^1/_{50} = 0.02\ \text{S}$$
$$X_L = 20\ \Omega \qquad B_L = {}^1/_{20} = 0.05\ \text{S}$$
$$R = 25 \qquad G = {}^1/_{25} = 0.04\ \text{S}$$

Since X_L and X_C are plotted 180° out of phase, the B_L and B_C are also plotted in opposite directions. As a result, the net susceptance of the circuit will be the difference between B_L and B_C. In this case, the net, or B_t, is

$$B_t = 0.05 - 0.02 = 0.03\ \text{S}$$

If the reciprocal of the difference of the *reactances* (50 − 20 = 30) is used, the net value will be ${}^1/_{30}$, or 0.0333 S, which is incorrect.

The net susceptance is 0.03 S, while the conductance is 0.04 S. By substituting these values in the admittance formula,

$$Y = \sqrt{G^2 + B^2} = \sqrt{0.04^2 + 0.03^2}$$
$$= \sqrt{0.0016 + 0.0009} = \sqrt{0.0025}$$
$$= 0.05\ \text{S}$$

$$Z = \frac{1}{Y} = \frac{1}{0.05} = 20\ \Omega$$

With this impedance value and a source emf of 100 V, the current of the whole circuit is

$$I = \frac{V}{Z} = \frac{100}{20} = 5\ \text{A}$$

Power and power factor can be determined as in series circuits. The true power is the power dissipated in the resistance, usually found by the formula $P = I^2R$, where I is the current flowing through the resistance. For Fig. 7-21

$$I_R = \frac{V}{R} = \frac{100}{25} = 4\ \text{A}$$
$$P = I^2R = 4^2(25) = 400\ \text{W}$$

The apparent power of the whole circuit can be found by multiplying the source voltage by the source current $P_a = \text{VA} = 100(5) = 500\ \text{W}$.

In series circuits the power factor is equal to R/Z. In parallel circuits pf $= G/Y$. In this circuit

$$\text{pf} = \frac{G}{Y} = \frac{0.04}{0.05} = 0.8000$$

In series circuits the power factor is the ratio of resistive to source *voltages*. In parallel circuits, it is the ratio of resistive to source *currents*, or pf $= I_R/I_S$. In the example, pf $= 4/5$, or 0.8.

In either series or parallel circuits the power factor will be the ratio of the true to the apparent power, or pf $= P/\text{VA}$, or in this example, 400/500, or 0.8. With parallel circuits the power factor may also be found by pf $= Z/R$.

The phase angle of this circuit is the angle represented by the cosine value of the power factor, as determined by any of the methods given above, or 36.9° (from the table of trigonometric functions).

In a parallel circuit, when the net X value is more than 10 times the R value, for most practical purposes the X can be disregarded. It will have little effect on the value of current flowing, the impedance of the circuit, the phase angle, or the power factor.

Current-vectoring can be used to solve parallel ac circuits. For example, what is the total circuit drain from the source in Fig. 7-22? The current values through the three branches are

$$I_R = \frac{V}{R} = \frac{115}{30} = 3.83\ \text{A}$$
$$I_{X_L} = \frac{V}{X_L} = \frac{115}{17} = -j6.76\ \text{A}$$
$$I_{X_C} = \frac{V}{X_C} = \frac{115}{19} = j6.05\ \text{A}$$

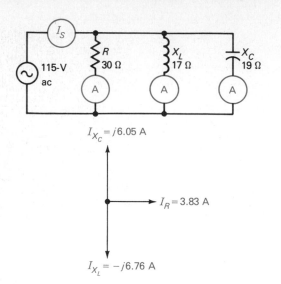

$I_{X_C} = j6.05$ A

$I_R = 3.83$ A

$I_{X_L} = -j6.76$ A

FIG. 7-22 The source current can be determined by vectoring the three branch currents.

The net current of the $-j6.76$ and the $j6.05$ A is $-j0.71$ A.

The resistive current is 3.83 A, and the net inductively reactive current is 0.71 A (Fig. 7-23).

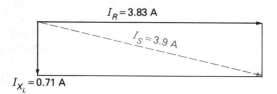

$I_R = 3.83$ A

$I_S = 3.9$ A

$I_{X_L} = 0.71$ A

FIG. 7-23 Resultant vector diagram after the reactive current values have been subtracted.

The pythagorean theorem can be used to solve for the hypotenuse of the right triangle, I_S

$$I_S = \sqrt{I_R^2 + I_X^2} = \sqrt{3.83^2 + 0.71^2}$$
$$= \sqrt{14.7 + 0.504} = \sqrt{15.2} = 3.9 \text{ A}$$

The impedance can be found by Ohm's law:

$$Z = \frac{V}{I} = \frac{115}{3.9} = 29.5 \text{ } \Omega$$

The power factor is the ratio of the I_R to the I_S, or pf $= I_R/I_S$, or 3.83/3.9, or 0.98205. From tables or a calculator the phase angle is found to be 10.87°.

True power is $P = I^2R$, or $3.83^2(30)$, or 440 W. Apparent power is 115(3.9), or 448.5 VA.

7-16 COMPLEX PARALLEL-SERIES CIRCUITS

The admittance formula $Y = \sqrt{G^2 + B^2}$ can also be applied to more complex parallel-series ac circuits, such as Fig. 7-24.

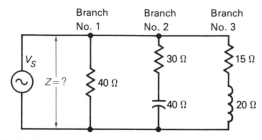

FIG. 7-24 A more complex parallel-series circuit that can be solved by the admittance method.

In branch 1, composed of R only, the G value is simply the reciprocal of the resistance, or

$$G = \frac{1}{R}$$

In the two other branches both R and G are in series. To find the G component, a formula that takes both R and X into consideration must be used. The special formula for this is

$$G = \frac{R}{(\sqrt{R^2 + X^2})^2} = \frac{R}{R^2 + X^2}$$

Similarly, to compute the B component of a branch that has both R and X in series, the special formula is

$$B = \frac{X}{(\sqrt{R^2 + X^2})^2} = \frac{X}{R^2 + X^2}$$

The admittance of the whole circuit can be determined by using the general formula

$$Y = \sqrt{G_t^2 + B_t^2}$$

In this formula, the G_t value is the sum of all the branch G's, regardless of whether the branch is inductive or capacitive. (In a purely reactive branch the G value is zero.)

The B_t is the *difference* between the B_L and B_C of the different branches. (In a purely resistive branch the B value is zero.)

To solve for Z_t (by $1/Y_t$) in Fig. 7-24, the expanded admittance formula would be

$$Y_t = \sqrt{(G_1 + G_2 + G_3)^2 + (\pm B_1 \pm B_2 \pm B_3)^2}$$

in which B_L is assigned a plus value and B_C a minus value.

The suggested steps are:

Branch 1:

$$G_1 = \frac{1}{R} = \frac{1}{40} = 0.025 \text{ S}$$

$$B_1 = 0 \text{ S}$$

Branch 2:

$$Z_2 = \sqrt{R^2 + X^2} = \sqrt{30^2 + 40^2}$$
$$= \sqrt{900 + 1600} = 50 \text{ }\Omega$$

$$G_2 = \frac{R}{R^2 + X^2} = \frac{30}{900 + 1600} = \frac{30}{2500}$$
$$= 0.012 \text{ S}$$

$$B_2 = \frac{X}{R^2 + X^2} = \frac{40}{900 + 1600} = \frac{40}{2500}$$
$$= 0.0016 \text{ S}$$

Branch 3:

$$Z_3 = \sqrt{R^2 + X^2} = \sqrt{15^2 + 20^2}$$
$$= \sqrt{625} = 25 \text{ }\Omega$$

$$G_3 = \frac{R}{R^2 + X^2} = \frac{15}{225 + 400}$$
$$= \frac{15}{625} = 0.024 \text{ S}$$

$$B_3 = \frac{X}{R^2 + X^2} = \frac{20}{225 + 400}$$
$$= \frac{20}{625} = 0.032 \text{ S}$$

The sum of all G's (G_t):

$$G_t = G_1 + G_2 + G_3$$
$$= 0.025 + 0.012 + 0.024 = 0.061 \text{ S}$$

The sum of all B's (B_t):

$$B_t = B_1 - B_2 + B_3$$
$$= 0 - 0.016 + 0.032 = 0.016 \text{ S}$$

Solving for impedance:

$$Y = \sqrt{G_t^2 + B_t^2} = \sqrt{0.061^2 + 0.016^2}$$
$$= \sqrt{0.003721 + 0.000256}$$
$$= \sqrt{0.003977} = 0.06306 \text{ S}$$

$$Z = \frac{1}{Y} = \frac{1}{0.06306} = 15.86 \text{ }\Omega$$

The power factor can be found by using the formula pf $= P/\text{VA}$ and employing any convenient source voltage. The true power is then the sum of all three I^2R power values in the circuit. The apparent power is the power indicated by the product of the voltage across the circuit and the current from the source. Power factor is also the cosine of the phase angle.

7-17 RECTANGULAR AND POLAR NOTATIONS

The solving of complicated ac circuit impedance, phase angle, and, from them, the currents, voltage-drops, VA, P, and pf can be accomplished by the admittance method explained above. The pythagorean theorem formula that is used involves computations of values such as R (or G) and X (or B) at right angles, as in Fig. 7-18. This is known as *rectangular notation*. If $Z = \sqrt{R^2 + X^2}$ is a form of rectangular notation, then the *complex number* (containing a j operator) used in the expression $Z = R + jX$ has a similar meaning and is also a form of rectangular notation.

When a vector, such as a Z vector, is rotated forward, or counterclockwise (CCW), from a zero-degree direction (far right usually), the angle between its position and zero represents the phase angle of the V and I in the circuit. The length of the vector, or *phasor*, arrow indicates the value of the impedance. For example, in a series RL circuit, if $R = 3 \text{ }\Omega$ and $X_L = 4 \text{ }\Omega$, then $Z = 5 \text{ }\Omega$ (from $Z = \sqrt{R^2 + X^2}$) with a phase angle of 53.1° (by $\tan \theta = X/R$). When this information is written $Z = 5 \underline{/53.1°} \text{ }\Omega$, it is read as "an impedance of 5 Ω with a phase angle of 53.1°" and is said to be in *polar notation* form. For a circuit having a 10-Ω R and a 10-Ω X_C, in rectangular form $Z = \sqrt{10^2 + 10^2}$, which computes to the polar information $Z = 14.1 \underline{/-45°} \text{ }\Omega$. The negative sign indicates the circuit is capacitively reactive with the current leading the voltage. There is really nothing new in this except how the circuit Z and θ are expressed.

The trigonometric formulas, discussed briefly in Sec. 7-3, which may be utilized in converting from polar to rectangular notation, or vice versa, are

$$\cos\theta = \frac{R}{Z} \qquad \sin\theta = \frac{X}{Z} \qquad \tan\theta = \frac{X}{R}$$

Using the simple 3-4-5 RXZ triangle in which $Z = 5\underline{/53.1°}\ \Omega$, the impedance vector is 5 Ω long and is at a CCW angle of 53.1° (Fig. 7-25). To

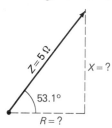

FIG. 7-25 Polar representation of a circuit having an impedance of 5 Ω and a phase angle of 53.1°.

convert this to rectangular notation, determine the R and X values. To find R if Z and θ are known, use $\cos\theta = R/Z$, rearranging to solve for R,

$$R = Z(\cos\theta) = 5(\cos 53.1°) = 5(0.600) = 3\ \Omega$$

Then, to solve for the X value, the formula that contains X and Z, or $\sin\theta = X/Z$, is used. Rearranging to solve for X,

$$X = Z(\sin\theta) = 5(\sin 53.1°) = 5(0.800) = 4\ \Omega$$

Note how simply total impedance of three series ac networks can be found by adding their complex number rectangular expressions algebraically, as

$$
\begin{aligned}
Z_1 &= 10 - j30 && (R \text{ and } X_C)\\
Z_2 &= 15 + j20 && (R \text{ and } X_L)\\
Z_3 &= 0 + j60 && (\text{only } X_L)\\
\hline
Z &= 25 + j50
\end{aligned}
$$

The whole series circuit has the equivalent Z of a 25-Ω R and a 50-Ω X_L in series. Try converting this to polar notation. Your answer should be $Z = 55.9\underline{/63.4°}\ \Omega$.

It is fairly complicated to divide or multiply rectangular notation complex numbers, but polar notation values can be divided or multiplied easily. If it is desired to divide

$$\frac{15\underline{/20°}\ \Omega}{3\underline{/-25°}\ \Omega} = \frac{15\underline{/20° + 25°}\ \Omega}{3} = 5\underline{/45°}\ \Omega$$

The rule is to divide the impedances but move the phase angle from below the line to above the line, change its sign, and then add the two algebraically. In this case, the final polar expression is $5\underline{/45°}\ \Omega$.

To multiply polar numbers, multiply the impedances and add the angles algebraically. For example, $10\underline{/40°}\ \Omega$ times $5\underline{/-10°}\ \Omega$ is simply $50\underline{/30°}\ \Omega$.

What are the parameters that can be determined from the circuit shown in Fig. 7-26a by utilizing

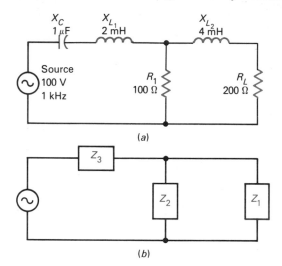

(a)

(b)

FIG. 7-26 (a) Schematic diagram of problem in text. (b) Block diagram of three impedances to be computed.

polar and rectangular notations? This circuit consists of a 100-V 1000-Hz source coupled to a 200-Ω resistor load through a capacitor in series with an RL "T network."

The load R_L and the 4-mH coil form a series impedance network which in turn is in parallel with the 100-Ω R_1. The capacitor and the 2-mH coil form a series impedance network. The three lumped impedances are indicated in block form in Fig. 7-26b. First, it is necessary to determine the values of X_C, X_{L_1}, and X_{L_2}. By the reactance formulas (to the third significant figure) the reactances are

$$X_C = \frac{1}{2\pi fC} = 159\ \Omega$$

$$X_{L_1} = 2\pi fL = 12.6\ \Omega$$

$$X_{L_2} = 2\pi fL = 25.1\ \Omega$$

Solving the Z_1 series network first,

$$Z_1 = \sqrt{R_L{}^2 + X_{L_2}{}^2} = \sqrt{200^2 + 25.1^2} = 202\ \Omega$$

Using the tangent formula,

$$\tan \theta = \frac{X}{R} = \frac{25.1}{200} = 0.1255$$

$$\theta = 7.2°$$

$$Z_1 = 202 \,\underline{/7.2°}\ \Omega$$

Solving for Z_1 and Z_2 together by the general parallel impedance formula,

$$Z_t = \frac{Z_1 Z_2}{Z_1 + Z_2} = \frac{(202 \,\underline{/7.2°})(100 \,\underline{/0°})}{(202 \,\underline{/7.2°}) + (100 \,\underline{/0°})}$$

$$= \frac{(202)(100) \text{ and } 7.2° - 0°}{(200 + j25.1) + (100 + j0)}$$

$$= \frac{20,200 \,\underline{/7.2°}}{300 + j25.1}$$

$$= \frac{20,200 \,\underline{/7.2°}}{301 \,\underline{/4.8°}} = 67.1 \,\underline{/7.2° - 4.8°}$$

$$= 67.1 \,\underline{/2.4°}\ \Omega$$

Converting to rectangular notation from $67.1 \,\underline{/2.4°}\ \Omega$,

$$X_{L_{1,2}} = Z(\sin \theta) = 67.1(\sin 2.5°) = 2.81\ \Omega$$
$$R_{1,2} = Z(\cos \theta) = 67.1(\cos 2.5°) = 67.0\ \Omega$$
$$Z_{1,2} = 67 + j2.8$$

Solving for Z_3 (two pure reactances),

$$Z_3 = \begin{cases} 0 - j159 \\ 0 + j12.6 \\ 0 - j146.4 \end{cases}$$

Solving for Z_t, the total of Z_3 and $Z_{1,2}$,

$$Z_t = \begin{cases} Z_3 = 0 - j146.4 \\ Z_{1,2} = 67 + j2.8 \end{cases}$$

$$= \quad\quad 67 - j143.6$$

$$= 157 \,\underline{/-65°}\ \Omega$$

Solving for I_t by Ohm's law,

$$I_t = \frac{V}{Z} = \frac{100}{158} = 0.633\ \text{A}$$

Then

$$V_{X_C} = I_t X_C = 0.633(158) = 100\ \text{V}$$
$$V_{X_1} = I_t X_1 = 0.633(12.6) = 7.98\ \text{V}$$
$$V_{Z_{1,2}} = I_t Z_{1,2} = 0.633(67.1) = 42.5\ \text{V}$$
$$I_{R_1} = \frac{V_{R_1}}{R_1} = \frac{42.5}{100} = 0.425$$

$$I_{R_L} = I_{Z_1} = \frac{V_{Z_{1,2}}}{Z_1} = \frac{42.5}{202} = 0.21\ \text{A}$$

$$V_{R_L} = I_{R_L} R_L = 0.21(200) = 42\ \text{V}$$

$$P_t = I_{R_1}^2 R_1 + I_{R_L}^2 R_L$$

$$= 0.425^2(100) + 0.21^2(200)$$

$$= 18.1 + 8.82 = 26.9\ \text{W}$$

$$\text{VA} = 100(0.633) = 63.3\ \text{VA}$$

$$\text{pf} = \frac{P_t}{\text{VA}} \quad \text{or} \quad \cos(-65°) = 0.424$$

$$= 42.4\%$$

$$P_{R_1} = V_{R_1} I_{R_1} = 42.5(0.425) = 18.1\ \text{W}$$
$$V_{X_2} = I_{X_2} X_2 = 0.21(25.1) = 5.27\ \text{V}$$

Test your understanding; answer these checkup questions

1. When R, X_L, and X_C are in parallel, to solve for Z what is the first operation? _____ The second? _____ What formula is then used? _____

2. When R, X_L, and X_C are in parallel, to solve for source current, what is the first operation? _____ The second? _____ The formula to use? _____

3. A 300-Ω X_L, a 100-Ω X_C, and a 400-Ω R are in parallel across 120 V ac. What is the Z? _____ The pf? _____ Is the pf leading or lagging? _____ What is the I_S? _____ The θ? _____

4. A 100-V ac generator has a 70-Ω X^L, a 90-Ω X^C, a 600-Ω R and a 400-Ω R all in parallel across it. Which branch dissipates the most power? _____ What is the Z? _____ VA? _____ P? _____ pf? _____ θ? _____

5. A 50-Ω R, a 100-Ω X^L, and a 25-Ω X^C are in parallel. Assume a 100-V source to make current computations simple. What is the I^S value? _____ By Ohm's law the Z value? _____ pf? _____ θ? _____ If 200 V had been used, which answers would have been different? _____

6. Draw a diagram of a 50-Ω resistor across a source of 100 V with a 90-Ω X^C in series with a 40-Ω R across the source. Using the admittance method, what is G^t? _____ B^t? _____ Y? _____ Z? _____ VA? _____ P? _____ pf? _____ θ? _____

7. A series 20-Ω R and 30-Ω X_C are in parallel with a series 40-Ω R and 80-Ω X_L. Using $Z_T = Z_1 Z_2/(Z_1 + Z_2)$, what is the polar Z_t value? _____ The rectangular Z_t value? _____ Circuit pf? _____

1. What component develops a 0° *V-I* phase difference in it? A *V* lag of 90°? A *V* lead of 90°?
2. Express the pythagorean theorem two ways, using electrical symbols.
3. What is the impedance of a series ac circuit having 24-Ω *R* and 32-Ω *X*?
4. In problem 3, if the source *V* is 100 V: What is the circuit *I*? The voltage-drop across *R*? Across *X*?
5. A 50-Ω *R* and 65-Ω X_L are in series across 20 V. What is the *Z*? The I_{source}? V_{X_L}? V_R? θ? Is θ leading or lagging?
6. A coil and resistor are in series across an ac line. The voltage measured across *R* is 25, and across *L* is 15. What is the line voltage? The *RL* phase angle?
7. What are two ways an inductor can lose power in a circuit?
8. Which will always be greater, VA or true power?
9. What are the true and apparent power if a 10-Ω *R* and a 50-Ω X_L are in series across a 12-V ac line?
10. In problem 9, what is the pf? θ?
11. If *P* = 175 W and θ is 45°, what is VA?
12. If VA = 400 W and θ is 15°, what is *P*?
13. When wiring a circuit should the wire size depend upon *P* or VA? Why?
14. How can a high pf be lowered?

15. What kind of circuit has a leading pf?
16. A 14-Ω *R* and a 21-Ω X_C are in series across a 30-V ac line. What is the *Z*? *I*? V_R? V_{X_C}? *P*? VA? pf? θ?
17. If *R* and *C* are in series, with V_R = 86 V and V_S = 120 V, what is the pf? θ?
18. A 12-Ω X_L, a 9-Ω *R*, and an 18-Ω X_C are in series across 40 V. Is the circuit inductive or capacitive? Why?
19. In problem 18, what is the circuit *Z*? *I*? V_{X_L}? V_{X_C}? *P*? VA? pf? θ?
20. A series circuit has a 40-Ω *R*, 50-Ω X_L, and 90-Ω *Z*. What is the X_C value?
21. How could you represent: "A 7.2-A current lags its voltage by 90°," using the *j* operator? "A 32-V voltage leads its current by 90°"? "A capacitive reactance of 46 Ω"?
22. What is the total reactance when a 75-Ω X_L is in parallel with a 40-Ω X_C?
23. A 10-Ω X_L and a 20-Ω X_C in parallel are in series with a 20-Ω *R* across 100 V. What is the total *X*? *Z*? I_R? V_{X_C}? V_{X_L}? I_{X_C}? I_{X_L}?
24. In problem 23, how can both reactive currents be greater than the series resistance current?
25. By admittance, what is the impedance of a 250-Ω *R* and a 175-Ω X_L in parallel?
26. A 20-Ω *R* and a 25-Ω X_C are in parallel across a 100-V source. What is I_S? *Z*? pf? θ?
27. A 50-Ω *R*, a 40-Ω X_C, and a 30-Ω X_L are in parallel across a 15-V source. What is I_R? I_X? I_S? *Z*? pf? θ?
28. In problem 27, what is B_C? B_L? *G*? B_t? *Y*? *Z*?
29. What is the *Y* and *Z* of a 3-branch parallel circuit having a 100-Ω X_L and a 100-Ω *R* in series in branch No. 1, a 150-Ω *R* as branch No. 2, and a 20-Ω X_C and 120-Ω *R* in series in branch No. 3?
30. Convert *Z* = 26 − *j*31 to polar notation.
31. Convert *Z* = 42 ∠ 62° Ω to rectangular notation.

8

Resonance and *LC* Filters

This chapter discusses resonant and antiresonant circuits, *Q* of *LCR* circuits, basic decibel computations, and bandwidth of *LC* circuits. Special filter circuits are developed by combining *L*, *C*, and *R* in different configurations.

8-1 RESONANCE

Resonant circuits are the basis of all transmitter, receiver, and antenna operation. Without resonant circuits there would be no radio communication.

Brief mention was made of series and parallel resonance in preceding chapters. When the inductive reactance (X_L) of a coil equals the capacitive reactance (X_C) of a capacitor in a circuit, a condition known as resonance occurs. Figure 8-1 illus-

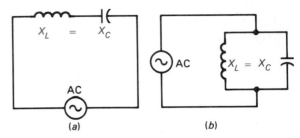

FIG. 8-1 (*a*) A series-resonant circuit across an ac source. (*b*) A parallel-resonant circuit across an ac source.

trates a series-resonant and a parallel-resonant circuit.

Since resonance is the condition where X_L equals X_C, the formula for resonance is

$$X_L = X_C \quad \text{or} \quad 2\pi f L = \frac{1}{2\pi f C}$$

$$\text{or} \quad \omega L = \frac{1}{\omega C}$$

where X_L = inductive reactance, in Ω
X_C = capacitive reactance, in Ω
f = frequency, in Hz
L = inductance, in H
C = capacitance, in F
$\omega = 2\pi f$

The middle formula shows that the inductive reactance is directly proportional to frequency, and that capacitive reactance is inversely proportional to frequency. With any given coil and capacitor, as the frequency increases, the reactance of the coil increases but the reactance of the capacitor decreases. At some frequency, the two reactances will be equal in value. At that one frequency the condition of resonance occurs. At all other frequencies the circuit shown in Fig. 8-1*a* is merely a series ac circuit, and Fig. 8-1*b* is merely a parallel ac circuit.

To determine the frequency at which a coil and capacitor will resonate, the resonance formula may be rearranged to solve for f:

$$2\pi fL = \frac{1}{2\pi fC}$$

$$2\pi fL(2\pi fC) = 1$$

$$4\pi^2 f^2 LC = 1$$

$$f^2 = \frac{1}{4\pi^2 LC}$$

$$f = \frac{1}{2\pi \sqrt{LC}}$$

By dividing the 2π portion into the 1, the frequency formula can be simplified to

$$f = \frac{0.159}{\sqrt{LC}} \quad \text{or} \quad f = \frac{0.159}{\sqrt{L}\,\sqrt{C}}$$

These formulas can be used to determine the resonant frequency of any LC circuit. For example, if inductance is 150 μH and capacitance is 160 pF, what is the resonant frequency?

$$L = 0.000\ 150\ \text{H} = 15 \times 10^{-5}\ \text{H}$$

$$C = 0.000\ 000\ 000\ 160\ \text{F} = 16 \times 10^{-11}\ \text{F}$$

$$f = \frac{0.159}{\sqrt{LC}} = \frac{0.159}{\sqrt{15 \times 10^{-5} \times 16 \times 10^{-11}}}$$

$$= \frac{0.159}{\sqrt{240 \times 10^{-16}}} = \frac{0.159}{15.5 \times 10^{-8}}$$

$$= \frac{0.159 \times 10^8}{15.5} = 0.01026 \times 10^8$$

$$= 1{,}026{,}000\ \text{Hz},\ 1026\ \text{kHz, or}\ 1.026\ \text{MHz}$$

According to the formula, the frequency of resonance is inversely proportional to the square root of either L or C. Thus, increasing L 4 times results in a lowering of the frequency to one-half of the original. Similarly, increasing C 4 times will also result in a frequency one-half of the original.

Any time the LC *product* is increased 4 times, by quadrupling L, by quadrupling C, by doubling both L and C, or by any other means, the frequency will be one-half of the original.

Any time the LC product is decreased to ¼ by any means, the frequency will be doubled.

As long as the LC product remains the same, the frequency will remain the same. For example, in a 1000-kHz circuit, if L is halved and C is doubled, the LC product has not changed and the frequency remains 1000 kHz.

It is possible to rearrange the resonance formula another way to solve for the inductance needed to resonate with a given capacitance, or the capacitance needed to resonate with a given inductance,

to form a resonant circuit at some desired frequency:

$$2\pi fL = \frac{1}{2\pi fC}$$

$$4\pi^2 f^2 LC = 1$$

$$C = \frac{1}{4\pi^2 f^2 L} = \frac{1}{(2\pi f)^2 L} = \frac{1}{\omega^2 L}$$

$$L = \frac{1}{4\pi^2 f^2 C} = \frac{1}{(2\pi f)^2 C} = \frac{1}{\omega^2 C}$$

where $\omega = 2\pi f$

For example, to determine the capacitance to shunt across a 56-μH coil to resonate at 5000 kHz

$$C = \frac{1}{4\pi^2 f^2 L} = \frac{1}{4(9.86)(5 \times 10^6)^2 (56 \times 10^{-6})}$$

$$= \frac{1}{39.4(25 \times 10^6)56} = \frac{1 \times 10^{-6}}{39.4(25)56}$$

$$= \frac{0.000001}{55{,}160} = 0.000\ 000\ 000\ 018\ \text{F}$$

$$= 0.000\ 018\ \mu\text{F} = 18\ \text{pF}$$

While a given coil and capacitor will resonate at essentially the same frequency whether connected in series or in parallel, a series-resonant circuit behaves in many ways opposite to a parallel-resonant circuit.

8-2 SERIES RESONANCE

The series ac circuit shown in Fig. 8-2 can be classed as series-resonant because the inductive reactance equals the capacitive reactance at the frequency of the source.

In any series circuit, whether resonant or not, the same value of current flows in all parts of the circuit at any one instant. However, the voltage across the capacitor is 90° behind the circuit current, while the voltage across the coil is 90° ahead

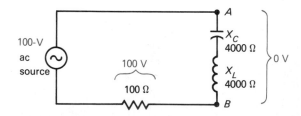

FIG. 8-2 The voltage-drop between A and B is zero. The whole source voltage appears across R.

of the current. The current through any resistance is in phase with the source voltage.

Figure 8-3 is a vector diagram of the voltages and current in Fig. 8-2. The V_{X_L} and the V_{X_C} are 180° out of phase. At resonance ($X_L = X_C$) the

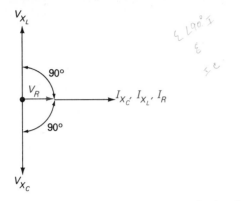

FIG. 8-3 Vector diagram of the phase and amplitude of the current and voltages in a series-resonant circuit.

Capacitors are usually considered to have negligible series resistance in them. However, coils may have considerable resistance in the wire with which they are wound. This resistance in the coil itself is usually treated as an external resistor. In the circuit of Fig. 8-2, if the coil having 4000 Ω of reactance also has 100 Ω of resistance in its wire, the diagram as drawn is a proper method of indicating the circuit, but showing internal resistance as R_i.

In the common electronic circuit of Fig. 8-4, the condition of *series* resonance is not apparent. The

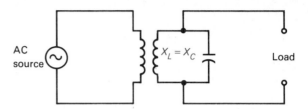

FIG. 8-4 The secondary is a series-resonant circuit

reactive voltages, being exactly equal and opposite, cancel each other completely insofar as the source is concerned. Therefore, between points A and B in Fig. 8-2 there is zero volts, although there is a voltage-drop across the X_C equal to that across the X_L. The full source voltage will be across the resistor.

If the source sees the coil and capacitor together as having a zero voltage-drop across them, it sees them as a perfect conductor, or as zero impedance. If the reactances are not exactly equal (a nonresonant condition), the voltages do not exactly cancel and the source sees the two reactances as having a resultant voltage-drop across them and therefore as having some value of reactance or impedance.

Theoretically, if a series LC circuit has no resistance and is connected across a source of ac to which it is resonant, it presents zero reactance, zero resistance, zero impedance, and infinite current!

The current-limiting factor in a series-resonant circuit is the resistance. In Fig. 8-2, with a source of 100 V, a resistance of 100 Ω, and 4000-Ω reactances, the reactances cancel, leaving the source looking at the 100-Ω resistance. The current in the circuit is $I = V/R = 100/100 = 1$ A, and the voltage-drop across each reactor is $V = IX = 1(4000) = 4000$ V! A series-resonant circuit is considered to be purely resistive. Its impedance value is its resistance value. Voltages much greater than the source voltage may be developed across each reactor.

transformer secondary coil has a capacitor across it with a reactance equal to the reactance of the secondary, forming a resonant circuit. At first glance it appears to be parallel-resonant. The primary, however, is inducing an ac emf into each turn of the secondary coil. An emf is not being induced across the ends of the coil. Theoretically, the secondary may be considered to have a source of ac inserted in series with its turns (Fig. 8-5). Since the induced

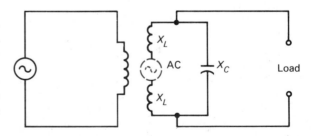

FIG. 8-5 Effectively, emf is induced in series with the secondary inductance turns.

emf is in series with the coil turns, the circuit is classed as a series-resonant circuit. Any load on the secondary, however, is in parallel with the capacitor and coil. Any emf developed across the resonant circuit produces current in the load.

If the primary coil (Figs. 8-4 and 8-5) is brought to resonance by connecting a proper value of capacitance across it, the primary will be a parallel-resonant circuit, since the ac voltage is being fed across it and not induced into its turns.

Tuned transformers are used in radio receivers, transmitters, and electronic circuits to select a desired frequency when many frequencies may be present. Figure 8-6 shows an antenna connected to

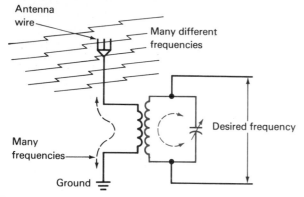

FIG. 8-6 A resonant circuit accepts signals at its resonant frequency but rejects all others.

the primary of a transformer. The antenna circuit is completed by connecting the lower end of the primary to ground (earth). Radio signals in the air pass across the antenna wire, inducing radio-frequency ac voltages in the antenna-to-ground circuit. Thousands of different-frequency signals are being induced into the antenna simultaneously. The problem is to pick out only the desired frequency.

A secondary coil can be loosely coupled to the primary, and a *variable* capacitor connected across its terminals. By varying the capacitance, it is possible to tune the series-resonant secondary circuit over a band of frequencies. At any frequency where the X_L of the coil equals the X_C of the capacitor, the secondary will appear as a low-impedance circuit to this frequency, and as a result this one frequency produces a significant current in the secondary. With a high-amplitude current flowing, a relatively high-amplitude voltage at the resonant frequency will be developed across the reactances.

To any frequency other than the resonant frequency, the series-resonant circuit will have greater impedance and will oppose the flow of ac at that frequency, resulting in lesser currents and reactive voltages for frequencies off resonance.

It has been pointed out that the impedance of a series circuit at resonance is equal to the resistance value of the circuit, but what impedance will a resonant circuit have to frequencies other than its resonant frequency?

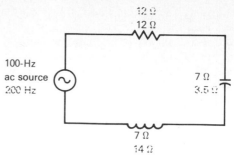

FIG. 8-7 Doubling the frequency doubles the X_L but halves the X_C.

The circuit shown in Fig. 8-7 illustrates a series-resonant circuit with a 12-Ω R, a 7-Ω X_C, and a 7-Ω X_L across a 100-Hz source. The impedance of the circuit is equal to the value of R alone, or 12 Ω. If the frequency is doubled, to 200 Hz, what is the impedance of the circuit?

Since X_L is directly proportional to frequency, when the frequency is doubled, X_L will double to 14 Ω. The X_C, being inversely proportional to frequency, becomes one-half its original value, or 3.5 Ω. The R value remains the same for all frequencies. The impedance is now

$$Z = \sqrt{R^2 + (X_L - X_C)^2}$$
$$= \sqrt{12^2 + (14 - 3.5)^2} = \sqrt{144 + 110.25}$$
$$= \sqrt{254.25} = 15.9 \ \Omega \text{ inductive (lagging pf)}$$

At 50 Hz, the circuit again presents 15.9 Ω impedance, but is capacitive, and has a leading pf.

Test your understanding; answer these checkup questions.

1. What is the basic formula for resonance? _____ To determine f if L and C are known? _____ To determine L if f and C are known? _____
2. A 2-H L and a 0.01-μF C are in series. What is the resonant frequency? _____ What is f if they are in parallel? _____
3. What is the resultant f if the LC product is increased 10 times? _____ Decreased to one-half? _____
4. What C value is needed to resonate a 2-H coil to 3 kHz? _____
5. What L value is needed to resonate a 70-pF C to 5 MHz? _____
6. Does the Z of a series-resonant circuit increase or decrease if the f value is increased? _____ If the f is lowered? _____

7. Is a resonant circuit a resistive or a reactive load on a source? _____
8. Across what component in a resonant circuit is the source voltage-drop value developed? _____
9. What might happen if a low-resistance resonant circuit were connected across a power line?
_____
10. A series circuit consisting of a 6.5-Ω R and equal 175-Ω inductive and capacitive reactances is across 260 V ac. What is the Z of the circuit? _____ I? _____ V_L? _____ V_C? _____ V_R? _____
11. In a series circuit having equal values of R, X_L, and X_C of 11 Ω, if f is reduced to 0.411 of its value at resonance, what is the new X_L? _____ X_C? _____ Z? _____
12. If the values of both L and C are doubled, how does this affect the resonant frequency of the circuit? _____

8-3 PARALLEL RESONANCE

A coil and capacitor connected as in Fig. 8-8 form a parallel ac circuit. If X_C and X_L have the same reactance to the frequency of the ac, the circuit is known as a *parallel-resonant* or, more correctly, an *antiresonant* circuit.

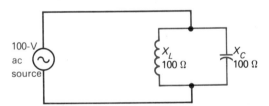

FIG. 8-8 Parallel-resonant or antiresonant circuit.

If the capacitor is temporarily disconnected, leaving the 100-Ω reactance coil across the 100-V source, according to Ohm's law for ac circuits the current in the coil will be

$$I = \frac{V}{X_L} = \frac{100}{100} = 1 \text{ A}$$

If the 100-Ω-reactance capacitor is reconnected across the coil, 1 A will flow in the capacitor also. Placing the capacitor across the coil in this circuit does not change the current in the coil even though a resonant circuit is formed (provided there is no resistance anywhere).

In a parallel-resonant circuit, the same voltage is across both the coil and the capacitor. In the inductive branch, however, the current lags the source voltage by 90°, and in the capacitive branch the current leads the source voltage by 90°. Since the two currents are 180° out of phase, at any instant that current is flowing down through the coil an equal current must be flowing up into the capacitor (Fig. 8-9).

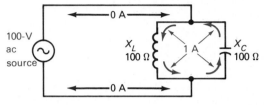

FIG. 8-9 With no resistance in an antiresonant circuit, current oscillates in it but the source current is zero.

As seen by the source, 1 A is flowing down through the coil and 1 A is flowing up through the capacitor at the same time. Since it is impossible for current to flow in two directions at once in either of the source lines, there must be zero current from the source but 1 A flowing from one reactance to the other. The electrons that make up the current in the reactances move from the top plate of the capacitor down through the coil and up to the lower capacitor plate. When the source voltage alternates, the electrons retrace their path back up through the coil to the top plate of the capacitor. This *LC* circulating current of 1 A flows between the reactances, but no current flows into and out of the source! Energy is first stored in the coil as an electromagnetic field, and then in the capacitor as an electrostatic field.

Because the source is supplying no current, it should be possible to disconnect the source and the current should continue to *oscillate* back and forth between capacitor and coil indefinitely. With no resistance or losses in the circuit, this would be true. The ability of a resonant circuit to sustain electron oscillation is known as *flywheel effect*, because of its similarity to the action of a mechanical flywheel, which, once started, tends to keep going until stopped by friction or other losses.

Since the source voltage is across the resonant circuit, but no current flows in the source, the parallel-resonant circuit impedance must be

$$Z = \frac{V_S}{I_S} = \frac{V_S}{0} = \infty \text{ (infinite) } \Omega$$

Both infinite-ohms impedance and ceaseless oscillations are impossible. The circuits shown have neglected resistance as well as inductive losses to external circuits. There is no coil or capacitor that does not have some series-resistant value. Figure 8-10 shows the circuit redrawn in a more practical form, including resistances in each branch. With

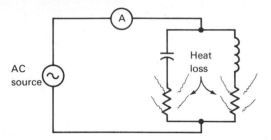

FIG. 8-10 Losses in the antiresonant circuit demand current from the source.

the same source voltage, less current will flow in the branches because of increased impedance. The impedance of each branch can be now computed as $Z = \sqrt{R^2 + X^2}$.

Current flowing through the resistances heats them. The heat represents a power loss. If the source is disconnected, the energy of the electrons will almost immediately be dissipated in heating the resistors and the electrons will cease to oscillate. With a power loss in the circuit, the source must feed power into the circuit to make up for the loss. To feed the power into the circuit, current will have to be fed to it. Therefore, enough source current must flow to all parallel-resonant circuits to make up for the losses in the circuit and maintain the voltage across the reactances.

With pure reactances and zero resistance in the circuit, the source current is zero. With resistance in series with either branch, source current increases, and reactive currents decrease.

As a generalization, the following applies to resonant and antiresonant circuits with little resistance:

RESONANT CIRCUITS
- The impedance across the circuit is low (equals any series resistance).
- The voltage-drop across the circuit is low.
- The current flow from the source is high (theoretically infinite if $R = 0$).

ANSWERS TO CHECKUP QUIZ ON PAGE 122

1. $(X^L = X^C)$ [$1/(2\pi \sqrt{LC})$ or $0.159/\sqrt{LC}$] [$1/(4\pi^2 f^2 C)$] 2. (1124 Hz)(Same) 3. (0.316 original)(1.414 times original) 4. (0.0014 μF) 5. (14.5 μH) 6. (Increase)(Increase) 7. (Resistive) 8. (Resistor) 9. (Melt LC circuit, blow fuses, start fire, etc.) 10. (6.5 Ω)(40 A)(7000 V)(7000 V)(260 V) 11. (4.52 Ω)(26.8 Ω)(24.9 Ω) 12. (Halves it)

- The voltage-drop across either reactance is equal to $V = IX$ and may be considerably greater than the source voltage.
- The circuit acts as a purely resistive (zero-reactance) load to the source and therefore has a power factor of 1.
- The phase of the current and voltage, as seen by the source, is 0°, or in phase.

ANTIRESONANT CIRCUITS
- The impedance across the circuit is high.
- The voltage-drop across the circuit is equal to the source voltage (less if there is any line R).
- The current from the source is low.
- The current through each reactance is equal to $I = V/X$ and will be greater than the source current.
- The circuit acts as a purely resistive (zero-reactance) load to the source and therefore has a power factor of 1.
- The phase of the source current and voltage is 0°, or in phase.
- The phase of the current in each reactance is 90° from the source voltage.
- Adding a little resistance to either branch lowers the impedance of the whole circuit.

The impedance of a parallel-resonant circuit can be determined in several ways:

1. When both the source voltage and the current are known, impedance can be found by

$$Z_p = \frac{V_s}{I_s}$$

2. The product of the series impedance of both legs of the circuit ($Z_C = \sqrt{R^2 + X_C^2}$ multiplied by $Z_L = \sqrt{R^2 + X_L^2}$), divided by the total *series* impedance of the coil, capacitor, and any resistance in the two parallel branches [$Z_s = \sqrt{R^2 + (X_L - X_C)^2}$], will give the parallel impedance of the circuit. The formula is

$$Z_p = \frac{Z_C Z_L}{Z_s}$$

where Z_p = impedance of parallel circuit
Z_C = impedance of capacitive leg
Z_L = impedance of inductive leg
Z_s = series impedance of the two legs

Since the impedance of a *series-resonant* circuit equals the resistance in the circuit, the antiresonant formula may also be given as

$$Z_p = \frac{Z_C Z_L}{R_s}$$

where R_s = total resistance of the two legs in series

3. When the reactance value is more than 10 times the resistance of the inductive branch, a simple formula that will give an approximate antiresonant impedance value is

$$Z_p = \frac{X_L{}^2}{R}$$

4. The circuit may be treated as a parallel ac circuit and the admittance computed by the parallel-circuit formula $Y = \sqrt{G^2 + B^2}$. The reciprocal of the admittance is the impedance.
5. The circuit may be treated as a parallel ac circuit. Assume a convenient source voltage, compute the currents that would flow in the two legs, and then determine the circuit current by using the formula

$$I_s = \sqrt{(I_{Z_L} \cos\theta + I_{Z_C} \cos\theta)^2 \atop + (\pm I_{Z_L} \sin\theta \pm I_{Z_C} \sin\theta)^2}$$

The total impedance is then determined by Ohm's law, $Z_p = V_X/I_s$.

Consider the parallel-resonant circuit in Fig. 8-11, which has resistance in series with the 100-V

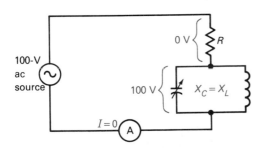

FIG. 8-11 At resonance the voltage-drop across the tuned circuit is maximum; across R, minimum.

source lines. When the capacitor is varied, there should be some value of capacitance at which X_C to the source frequency is equal to the X_L offered by the coil. At this resonant condition, the source current should drop to nearly zero (negligible resistance in L and C is assumed). With little current through the resistor R, there will be little voltage-drop across it, and almost the full 100-V source voltage will appear across the parallel circuit. The

flywheel current oscillating back and forth between L and C will be at a maximum.

When the capacitance is varied to any other value, the circuit will no longer be resonant. The reactive currents will no longer cancel each other. The circuit will no longer present as high an impedance to the source ac, and current will flow in the source lines. Current flowing through the resistor R will produce a voltage-drop across it. If there is a 10-V drop across the resistor, the voltage across the resonant circuit will be 90 V. The further the circuit is tuned from resonance, the greater the source current and the voltage-drop across R, and therefore the less the voltage across the parallel circuit. The less voltage appearing across the parallel circuit, the less flywheel current that will circulate in it. Maximum voltage across a parallel circuit, when resistance is in series with it, will occur at resonance. Maximum circulating current in the LC circuit will also occur at resonance. (This explains why resonance of the collector or plate tuning circuit of an RF amplifier stage in a transmitter is indicated by minimum collector or plate current.)

Test your understanding; answer these checkup questions.

1. Give the basic formula for antiresonance? _____
2. If the C is removed from an antiresonant circuit, how does this change the I_{X_L} if there is zero line R? _____
3. Does maximum or minimum current flow in the line feeding a resonant circuit? _____ An antiresonant circuit? _____
4. What are electron alternations in an antiresonant circuit called? _____
5. What must be the Z of an antiresonant circuit that draws 0.2 A from a 120-V line? _____
6. If the Z of the C branch of an antiresonant circuit is 2.5 kΩ, the Z of the L branch is 2 kΩ, and the series Z of the two branches is 1.5 kΩ, what is the Z of the parallel-resonant circuit? _____
7. A parallel-resonant circuit having 380-Ω reactances and zero R is in series with a 24-Ω R across 120 V. What is the voltage-drop across R? _____ Across the LC circuit? _____ The I value in the LC circuit? _____ The X value across the C? _____
8. What is the antiresonant frequency of a 500-pF C and a 150-μH L with a 10-Ω R in it? _____
9. What C is required to tune a 40-μH coil to a frequency of 8 MHz? _____
10. What L is required to tune a circuit with 0.0005 μF to 450 kHz? _____

11. A variable capacitor has a minimum C of 20 pF and a maximum C of 300 pF. If it is connected across a 100-μH coil, what are the highest and the lowest values of f at which it will be resonant? _____

12. An antiresonant circuit has an 800-Ω X_L and no R. If it is across 35 V ac, what is the I value in the LC circuit? _____ In the source? _____

8-4 THE Q OF A CIRCUIT

A term often applied to ac circuits in which inductance and capacitance are involved is Q. The symbol Q can be considered to mean "quality." A coil with no resistance or other losses would be a perfect inductor and would have an infinitely high Q. Since a coil without losses is not possible, the Q of a coil will always have some finite value. Q is a ratio of reactance to resistance, or losses, or

$$Q = \frac{X_L}{R} = \frac{2\pi f L}{R}$$

From the formula, an inductor should have a higher Q if used in higher-frequency circuits (provided losses in it do not increase proportionally).

As an example, at 6 kHz a certain coil has a 1000-Ω X_L and a 20-Ω R. Its Q is X_L/R, or 1000/20, or 50. At 12 kHz the X_L increases to 2000 Ω. If the resistance remains essentially the same, the Q should be 2000/20, or 100.

At higher frequencies, however, electrons flowing in a wire or coil travel nearer the surface of the wire. Thus, only a small portion of the conductor is actually carrying current. The lessening of the usable cross-sectional area results in an effectively higher resistance of the same wire to higher-frequency ac. This increased resistance, known as *skin effect*, is one cause of lower Q in a coil. Skin effect can be decreased by (1) using larger wire; (2) silverplating the wire used, since silver is the best conducting material; (3) using fewer turns but increasing the permeability of the core by using powdered-iron cores; and (4) using Litzendraht ("Litz") wire, an insulated multistrand wire. Several thin strands have more surface for a given wire diameter than does a solid wire. (Litz wire is effective only up to about 1 MHz.)

Capacitors also have a value of Q. As with coils, the Q is the ratio of the capacitive reactance to the effective resistance of the capacitor. This resistance may be in the leads, may be skin effect, or may be the result of internal losses of the capacitor. The formula is

$$Q = \frac{X_C}{R} = \frac{1/(2\pi f C)}{R} = \frac{1}{2\pi f C R}$$

From the formula, the Q of a capacitor should be halved when it is used at twice the frequency. Thus the Q of an LC circuit should decrease somewhat if tuned to higher frequencies.

When resistance is in series with any reactance, an increase in resistance produces a lower Q. When a resistor is connected across a coil or capacitor, however, the effective Q of the circuit will vary directly with the value of the resistance. A high shunting resistance means a high Q; a low parallel resistance means a low Q. Thus

$$Q_p = \frac{R}{X_L} \quad \text{or} \quad Q_p = \frac{R}{X_C}$$

The dielectric materials of capacitors normally have extremely high resistance, in the hundreds or thousands of megohms. If the dielectric leaks even a little, however, the ratio of resistance across the capacitor decreases tremendously, lowering the Q of the capacitor as well as the Q of any resonant circuit in which it is used.

A shunt resistor is often connected across a parallel LC circuit to lower its Q. This makes the circuit less sensitive to any one frequency. A parallel-resonant circuit with a loading resistor shunted across it to lower the Q is shown in Fig. 8-12. The

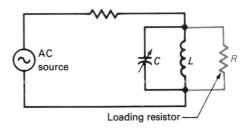

FIG. 8-12 A loading resistor across a tuned circuit lowers Q and broadens the frequency response.

lower the value of the parallel resistance, the greater the proportion of resistive current flowing in the circuit. With low resistance the source sees the LCR circuit more as a resistance than as a resonant circuit and as having a low impedance. Because of the resistance, current of all frequencies will flow from the source to the circuit. The higher the Q of a resonant circuit, the proportionally greater current that will flow in it at the resonant frequency (Sec. 8-6). Q formulas for series- and parallel-resonant circuits are

$$Q_s = \frac{\sqrt{L/C}}{R} \qquad Q_p = \frac{R}{\sqrt{L/C}}$$

When an antiresonant circuit is loaded by a resistor across it, the Q is computed by

$$Q = \frac{R}{X}$$

where R = effective load resistance

Since the Q of a coil is usually much lower than the Q of a capacitor, it is the controlling factor in the Q of LC circuits. Some of the methods of attaining a higher-Q coil are (1) using larger wire, or wire of better conducting material, to reduce skin effect; (2) spacing the wires of the coil by a distance approximately equal to the diameter of the wire used; (3) for single-layer coils, using a coil with a length slightly greater than its diameter; (4) using low-loss core materials to decrease eddy-current and hysteresis losses; and (5) insulating the wire with a material that has low dielectric hysteresis loss. This increases the Q of the capacitance that exists between adjacent turns in a coil, known as *distributed capacitance*.

Generally, the frequency of a tuned circuit is determined by L and C values,

$$f = \frac{1}{2\pi \sqrt{LC}}$$

The Q of resonant circuits used in communications may be about 5 to 15 in radio transmitters, 25 to 200 in RF tuned circuits in receivers, or several hundred in specially designed high-Q filter circuits. By using regeneration in transistor circuits, it is possible to obtain a resonant circuit with a Q over 10,000.

8-5 DECIBELS

One of the important mathematical tools of communications is the *decibel* (*deci-* = $^1\!/_{10}$), abbreviated dB. This is a measurement of the ratio of one power to another. An audio amplifier can increase the power of an AF ac signal. If it can increase a 5-W input to a 50-W output, it has increased the signal power 10 times, or 1 *bel* (B), or 10 dB. The same amplifier will also increase a 0.5-W input to a 5-W output, still a gain of 10 dB. If the volume control is turned up, the 0.5-W input signal may be amplified to a 50-W output. The hundredfold increase is a gain of two 10-times gains, or 20 dB. This will be recognized as a logarithmic increase by those familiar with logarithms. For those not familiar with them, a brief explanation is given here.

In the equation $10^2 = 100$, the 2 is called the *exponent*, the 10 is called the *base*, and 100 the *number*. The exponent 2 can also be called the *logarithm* of the number 100 to the base 10. This is written $\log_{10} 100 = 2$ and is read: "The logarithm to the base 10 of the number 100 is 2." Since the logarithms in general use commonly employ a base of 10, it is customary to express an equation as merely $\log 100 = 2$ and say, "The log of 100 is 2."

The equation $10^3 = 1000$ may be expressed: $\log 1000 = 3$. The logarithm of any number between 100 and 1000 will have to be between 2 and 3, or 2 plus some decimal fraction. For example, $\log 500$ happens to be 2.6990. A logarithm such as 2.6990 is composed of two parts. The whole number (2.) is the *characteristic*, and the decimal-fraction part (6990) is the *mantissa*.

The characteristic is determined by finding between which $\log_{10}$ the number falls (Table 8-1).

Note that the characteristic is always 1 less than the number of whole number digits. The number 435 has a characteristic of 2; 86 has a characteristic of 1; but for numbers less than 1 the characteristic is one more than the number of zeros to the right of the decimal point, so 0.05 has a characteristic of -2; and so on.

The mantissa, or decimal-fraction part of the logarithm, may be found in a table of common logarithms (Appendix D) or with pocket calculators.

EXAMPLE: The log of the number 8450 = ? The characteristic is 1 less than the four digits in the number, or 3 in this case. Therefore, log 8450 = 3.+? The mantissa is found in the tables in the 84 line under the 5 column and is 9269. Thus, log 8450 = 3.9269.

TABLE 8-1	RANGE AND CHARACTERISTICS OF NUMBERS	

Range of numbers	Characteristic
0.001–0.009999	-3
0.01–0.09999	-2
0.1–0.9999	-1
1–9.999	0
10–99.99	1
100–999.9	2
1000–9999	3

Check the following logarithms for characteristic and mantissa:

$$\log 23 = 1.3617$$
$$\log 15,500 = 4.1903$$
$$\log 629 = 2.7987$$

From bel = $\log_{10} P_1/P_2$, the number of decibels change between two power values will be

$$dB = 10 \log_{10} \frac{P_1}{P_2}$$

How many decibels gain does an amplifier have if it produces 40-W output with an input of 0.016 W? By the formula,

$$dB = 10 \log \frac{40}{0.016}$$
$$= 10 \log 2500 = 10(3.3979) = 33.9 \text{ dB}$$

Similarly, how much output power will be produced by an amplifier capable of 25-dB gain if fed an output of 0.001 W? From the formula

$$dB = 10 \log \frac{P_{\text{out}}}{P_{\text{in}}} = 10 \log \frac{P_o}{0.001}$$
$$25 = 10 \log \frac{P_o}{0.001}$$
$$2.5 = \log \frac{P_o}{0.001}$$

This states that the log of the number $P_o/0.001$ is 2.5, or

$$\log \frac{P_o}{0.001} = 2.5000$$

The 5000 is the mantissa. By searching through the mantissas in the tables, 5000 is found in the 31 line and 6 column. The 2 is the characteristic and indicates three whole numbers in the logarithm. Therefore,

$$\frac{P_o}{0.001} = 316.0$$
$$P_o = 316(0.001) = 0.316 \text{ W}$$

For exact decibel computations the logarithmic formulas should be used, but a fairly accurate approximation is possible by applying one or more of the following ratios, particularly when logarithmic tables or pocket calculators are not available:

1 dB = power gain of 25%	(actually × 1.26)
3 dB = power gain of 2	(approximately)
6 dB = power gain of 4	(approximately)
10 dB = power gain of 10	(exactly)
20 dB = power gain of 100	(exactly)

While the power ratios given above are stated as gains, they may also be stated as losses: −3 dB indicates a loss to half power, etc.

The 1-, 3-, and 10-dB approximations can be used to solve the problem involving the amplifier with 0.001-W input and 25-dB gain:

Since 10 times the power equals 10 dB, the 0.001-W input power increased by 10 dB represents 0.01 W. A second 10-dB increase (total of 20 dB) represents 0.1 W. The remaining 5 dB can be computed: 3 dB of this represents twice the power at 20 dB, or 0.2 W. One decibel more (total, 24 dB) represents 0.2 + ¼ of 0.2, or 0.2 + 0.05, or 0.25 W. One decibel more (a total of 25 dB) represents 0.25 + ¼ of 0.25, or 0.25 + 0.063, or 0.313 W. Compare this with the 0.316 W obtained by logarithmic computation.

A simpler, "in the ball park" approximation of the last 5 dB would be to say it was slightly less than 6 dB, or a little less than 0.4 W.

Inasmuch as the decibel is used as a unit of measurement in systems in which power decreases to a threshold value before it reaches an absolute zero, it is necessary to make an arbitrary selection of some power value and assign it as "zero" dB. Power values more than the chosen zero level are considered as +dB; those less than the zero level are indicated as −dB.

In the past, several reference points have been established as zero levels. Today, 0.001 W, or 1 mW, is being accepted generally as the standard, although 0.006 W is also used. The term *dBm* signifies dB with a zero level of 1 mW. A postfix k indicates kilowatt, and μ or u indicates 1 microvolt as the zero level, or base. The VU (volume unit) is another dB unit having 1 mW as the zero level, but in a 600-Ω impedance circuit (Chaps. 13 and 23).

The decibel has been explained as a ratio of two powers. Since power is proportional to both voltage (V^2/R) and current (I^2R), it is possible to compute decibel ratios between input and output voltages or currents. The basic formulas are

$$dB = 10 \log \frac{V_o{}^2/R_o}{V_i{}^2/R_i}$$

$$dB = 10 \log \frac{I_o{}^2 R_o}{I_i{}^2 R_i}$$

where V_o = output voltage
V_i = input voltage

As an example, the dB gain of an amplifier feeding 0.1 V to an 800-Ω input circuit to produce a 5-V output in a 4000-Ω load is

$$dB = 10 \log \frac{V_o{}^2/R_o}{V_i{}^2/R_i} = 10 \log \frac{5^2/4000}{0.1^2/800}$$

$$= 10 \log \frac{25/4000}{0.01/800} = 10 \log \frac{0.00625}{0.0000125}$$

$$= 10 \log 500 \qquad = 10 \times 2.69897$$

$$= 27 \text{ dB}$$

For transistor amplifiers the power values above the fraction line would be the larger output values. For resistive pads or other power-loss networks the larger input values should be used above the fraction line.

In the special case in which the input and output resistance or impedance values are equal, these values cancel mathematically and the formulas are

$$dB = 10 \log \frac{V_o{}^2}{V_i{}^2} \qquad or \qquad dB = 10 \log \frac{I_o{}^2}{I_i{}^2}$$

Rather than working with squared values, the formulas may be simplified to

$$dB = 20 \log \frac{V_o}{V_i} \qquad and \qquad dB = 20 \log \frac{I_o}{I_i}$$

As an example, the voltage across a 500-Ω input circuit of an amplifier is 0.2 V. The output is 5.6 V across a 500-Ω load. The gain of the amplifier in decibels is

$$dB = 20 \log \frac{5.6}{0.2} = 20 \log 28$$

$$= 20(1.4472) \qquad = 28.94 \text{ dB}$$

▌ Test your understanding; answer these checkup questions

1. What word does the letter Q stand for? _____ What is the formula for Q of a coil? _____ Of a capacitor? _____
2. Would a coil have a greater Q at 1 or at 10 MHz? _____ If the Q is lower at 10 MHz, what is the probable cause? _____

3. Which would normally have the higher Q, a C or an L? _____ Which would have greater effect on the Q of an LC circuit? _____
4. What is the name of the capacitance between turns of a coil? _____
5. What is the log of 420? _____ 27? _____ 42,600? _____ 0.135? _____ 0.00423? _____
6. An amplifier has a 53-dB gain and a 10-W output. What input signal power does it have? _____
7. A 2-W signal is fed to a 600-Ω resistor network which loses some of the power fed to it. How many dB must it lose to have an output of zero VU? _____
8. If a microphone is rated at -65 dBm, how many dB gain must an amplifier have to bring its output to 0.001 W? _____ 10 W? _____ 18 W? _____
9. An amplifier is fed a 2-mV signal and has a 0.025-V output. How many dB gain does it have if $Z_i = Z_o$? _____
10. An amplifier is capable of 20-dB gain. If the input signal is 0.0045 A, what is the I_o if $Z_i = Z_o$? _____
11. Is it possible to have a dB gain in a voltage step-up transformer? _____ Why? _____
12. A 5-V signal to a 200-Ω input produces what dB gain or loss if the V_o is 20 V across 10,000 Ω?
13. The output of an amplifier having a voltage gain of 30 dB is 25 V. What is the V_i if $Z_i = Z_o$? _____
14. What would 3 dBk indicate? _____
15. Current tends to travel on the surface of conductors at higher frequencies. What is this called? _____

8-6 BANDWIDTH

In electronics, there are many cases in which it is desired to limit the passage of ac to one particular frequency or, more practically, to a small band of frequencies. It has been stated (Sec. 8-2) that a series-resonant circuit in series with an ac line will pass current of the frequency of resonance very well but will *attenuate* (decrease or reduce) current of frequencies either higher or lower. The further from the resonant frequency, the greater the attenuation (Fig. 8-13). How wide a band of frequencies is being passed by this series-resonant circuit? The band of frequencies being passed at absolute maximum is rather small. A much wider band of frequencies is being passed between half-amplitude points. According to the curve, a very wide band of frequencies is being passed between low-amplitude

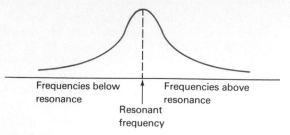

FIG. 8-13 Maximum current flows in a series circuit at resonance; less at other frequencies.

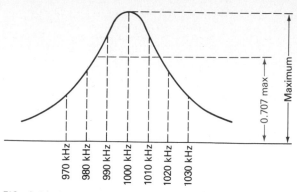

FIG. 8-14 Curve of a 30-kHz bandwidth circuit.

points. Just where should the bandwidth be measured?

One method of measuring bandwidth is to measure the width of either the voltage or the current response curve between points at 0.707 of maximum. Since power is proportional to either voltage or current *squared*, the 0.707 point is also the 0.5-power point $(0.707^2 = 0.5)$, or down 3 dB. Thus, bandwidth is normally measured between half-power points, or -3-dB points.

The higher the Q of a resonant circuit, the narrower its bandwidth. This is expressed by the half-power bandwidth formula

$$BW = \frac{f_o}{Q}$$

where f_o = frequency of resonance
$\quad$ BW = bandwidth

Figure 8-14 illustrates the curve of a circuit resonant at 1000 kHz. The two points on the curve that are 0.707 of maximum are at 985 and 1015 kHz, respectively. In this circuit the bandwidth is 30 kHz. If the same circuit were tuned to resonate at 1020 kHz, its bandwidth would remain essentially the same, 30 kHz. To represent the circuit the curve would have to be moved to the right along the page about ⅜ in. When a receiver is tuned, its response curve is being shifted up or down the frequency spectrum.

8-7 BANDWIDTH OF TRANSFORMERS

An important circuit is the tuned transformer, in which both primary and secondary are tuned to the same frequency. The bandwidth of a tuned transformer will vary with degree of coupling, basic Q of both primary and secondary circuits, and load on the circuit. Determination of the bandwidth of such circuits is complex. Only a few fundamental ideas will be presented.

In Fig. 8-15 if all the lines of force from the primary cut all the secondary turns, the condition of

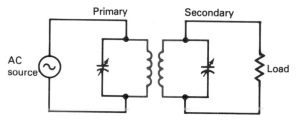

FIG. 8-15 A double-tuned transformer.

unity coupling exists and the coefficient of coupling k equals 1 (Sec. 5-8). In such a condition the transformer will pass practically all frequencies equally well and represents the extremity of broadness. Tuning the circuits would have very little effect on the frequency response.

In a more practical form, if the coefficient of coupling is equal to, possibly, 0.001, with each circuit tuned to the same frequency, and assuming an equal Q for both primary and secondary circuits, the frequency response of the circuit will follow the *universal resonance curve* (Fig. 8-16). This is the relative frequency response curve of series-resonant circuits (when the Q is over 5), or of parallel-

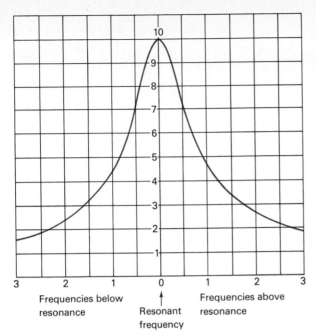

FIG. 8-16 Universal resonance curve.

resonant circuits, or of two tuned and coupled circuits.

The universal resonance curve illustrates, for example, the current amplitudes in a series-resonant circuit if a variable frequency is applied to the circuit. When the applied ac is 2 frequency units below resonance, the circuit will develop only about 25% of the current that it will when the frequency is at resonance. If the frequency is raised 2 units above resonance, the circuit current drops to about 25% of the resonant value.

If the Q of the circuit is raised, the peak-amplitude current can be developed with less source voltage and the response at the quarter-current point will be less than 2 units from resonance. If the Q is doubled, the quarter-current frequency may be 1 unit above and below resonance for this particular circuit and the bandwidth is halved. The curve for the high-Q circuit will still have the same relative shape, but the frequency units would be further out from the resonant frequency.

This same curve also indicates the relative rise in impedance for a parallel-resonant circuit when the circuit is subjected to a variable frequency. At resonance, the peak impedance is reached.

A transformer with both primary and secondary tuned to the same frequency has two resonant circuits, each one attempting to select its resonant

frequency and rejecting all other frequencies. The response curve for such a circuit, with a low coefficient of coupling, will be the product of the two curves separately. What would have been the quarter-amplitude point for one circuit alone will be ¼ × ¼, or ¹/₁₆, the response with both circuits. The two tuned circuits (with similar Q's) when loosely coupled, produce a bandwidth approximately one-fourth that of either coil alone.

As coupling is increased, the curve of the primary current begins to dip at the resonant frequency due to a resistive effect coupled into it from the secondary (Fig. 8-17).

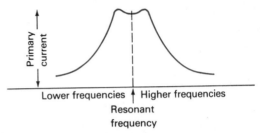

FIG. 8-17 Primary current dips at resonance if coupling is tight.

As soon as the primary-current dip occurs, the curve of the secondary current no longer continues to rise with increased coupling as sharply as when the primary had no dip, and the secondary current peak flattens (Fig. 8-18).

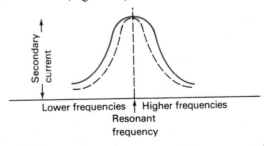

FIG. 8-18 Secondary-current curve flattens as coupling increases.

As the coupling between primary and secondary is increased, the top of the secondary-current curve flattens more and more until an increase in coupling no longer raises the secondary-current peak amplitude. This degree of coupling is known as *critical coupling*. The coefficient of coupling k may be in the region of 0.01. The bandwidth at critical coupling is wider than when the two circuits are loosely coupled, but the amplitude is maximum. If all coils are high Q, the bandwidth may still be relatively narrow. For minimum bandwidth, the coils are coupled to less than the critical value. In applications where a flattened-peak response is desired, circuits will be coupled to the critical value or slightly over.

When coupling is increased past the critical point, the peak current in the secondary at resonance drops and two peaks appear. The two circuits are now said to be over-coupled, and a secondary-current curve dip results (Fig. 8-19). The peaks of

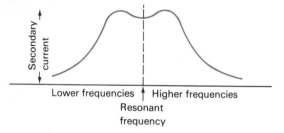

FIG. 8-19 Tight coupling develops two shoulders on the secondary-current curve, broadening the response.

the two "shoulders" remain at about the same amplitude as at critical coupling, but the bandwidth of the curve is broader. This double-humped response, sometimes called *split tuning*, is the result of over-coupling or detuning of primary or secondary.

8-8 FILTERS

Many circuits in radio and electronics use *filters*. A filter may be considered to be a combination of capacitors, coils, and resistances that will pass or impede certain frequencies.

The transformer back in Fig. 8-15 is actually a type of *bandpass-filter circuit*. It passes the frequency to which it is tuned and a few adjacent frequencies but attenuates those higher and lower.

The *shape factor* of a filter is the ratio of its bandpass 60 dB down from the midband value to its bandpass 6 dB down. The steeper the skirts of its curve, the smaller the shape factor (Fig. 8-20).

A *bandstop filter* is used to attenuate one frequency or a small band of frequencies and pass all others. A *wave trap* is an elementary form of bandstop or *notch* filter. It can be used in the antenna or other circuits of a receiver to prevent undesired signals from interfering with a desired signal. Figure 8-21 shows two wave traps, one a parallel-resonant circuit in series with the antenna and the other a series-resonant circuit between antenna and ground terminals of a receiver. The parallel-

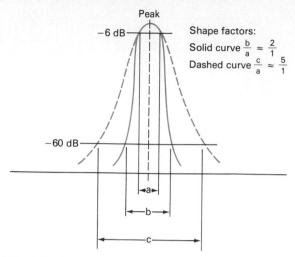

FIG. 8-20 Shape factors of two bandpass filters, one of 2:1 and the other of 5:1.

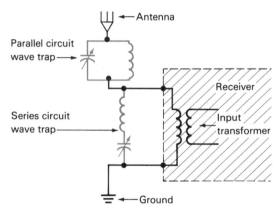

FIG. 8-21 Two possible positions for wave traps at the input (antenna-to-ground) circuit of a receiver.

resonant circuit offers high impedance to the frequency to which it is tuned and relatively low impedance to all other frequencies. The high impedance reduces current flow at the frequency of resonance into the receiver input winding. This one frequency is not received. The series-resonant circuit offers low impedance to any signals at its resonant frequency, effectively short-circuiting the antenna-to-ground circuit of the receiver at this frequency and preventing this frequency from being received. Usually a single wave trap is adequate, but for greater attenuation of an interfering signal both may be used.

Figure 8-22a shows a bandstop or band-elimination filter. The parallel-resonant-frequency LC circuit tends to prevent the unwanted frequency

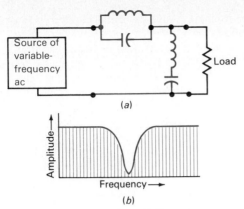

(a)

(b)

FIG. 8-22 (a) Bandstop filter. (b) Response curve.

from traveling to the load. The series-resonant circuit shorts out any of it that does pass through the antiresonant circuit. Figure 8-22b shows the response curve of a bandstop filter. If the frequency of resonance of the two wave traps are slightly different, the bandwidth at the bottom of the curve will be wider and a greater band of frequencies will be greatly attenuated.

An example of a *low-pass filter* using a single coil in series with a line between a source and load and a capacitor across the line is shown in Fig. 8-23a. The inductive reactance of the coil will oppose higher frequencies more than it will oppose low frequencies. The capacitive reactance of the capacitor presents a better path across the line for higher frequencies than does the load. In this way,

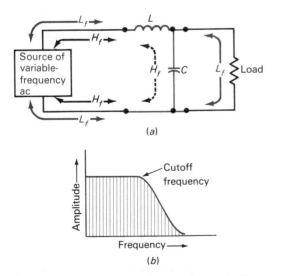

(a)

(b)

FIG. 8-23 (a) Constant-k low-pass filter. (b) Response curve.

both reactances are acting to attenuate the high frequencies but allow the low frequencies to pass to the load.

This *single-section* low-pass filter is the simplest of the *constant-k filters*. They derive their name from the fact that the product of the X_L times the X_C is constant at all frequencies. For example, at a certain frequency X_L may be 400 Ω and X_C may be 100 Ω. The product of the two is 40,000. At twice the frequency X_L will be 800 Ω and X_C will be 50 Ω. The product is still 40,000, the k constant for this filter.

A constant-k low-pass filter circuit will pass all frequencies up to the *cutoff frequency* (Fig. 8-23b). Past this the L and C rapidly attenuate all frequencies. L and C values can be computed by the formulas

$$L = \frac{R}{\pi f_c} \qquad C = \frac{1}{\pi f_c R}$$

where L = inductance, in H
C = capacitance, in farads, F
R = impedance, Ω (both source and load)
f_c = cutoff frequency

Simpler filter theory assumes source and load impedances to be equal. If not, the filter may not have the desired frequency characteristics.

The higher the Q of the reactances used, the sharper the cutoff. For still sharper cutoff characteristics, two or more sections are used.

Since the load is looking into a capacitor which has the input current flowing through it, the load sees a circuit with V lagging the I by 90°.

A simple constant-k *high-pass filter* is shown in Fig. 8-24a. The capacitor in series with the line passes the high frequencies to the load, while the coil across the line forms a better path than does the load for low frequencies, preventing low frequencies from appearing in the load. The cutoff frequency (Fig. 8-24b) is approximately the series-resonant frequency of C and L. The C and L values for the single-section high-pass constant-k filter of Fig. 8-24 can be determined by the formulas

$$L = \frac{R}{4\pi f_c} \qquad C = \frac{1}{4\pi f_c R}$$

The constant-k filters shown are "unbalanced," meaning one of the lines could be grounded. The high-pass filter in Fig. 8-25 is balanced because the filter elements are similar on both sides of the line, and the center of the reactances across the line can be connected to ground, as indicated.

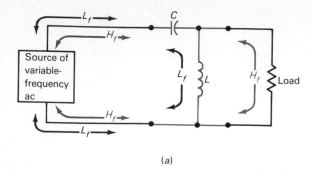

(a)

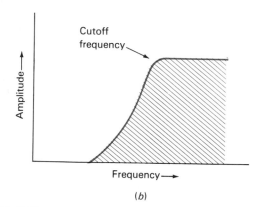

(b)

FIG. 8-24 (*a*) Constant-*k* high-pass filter. (*b*) Response curve.

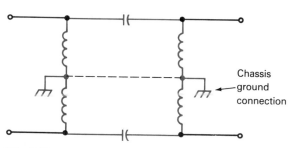

FIG. 8-25 A balanced constant-*k* high-pass filter circuit.

Low-pass filters are used in electronic power supplies to pass dc but not variations of current or voltage. They may also be used in voice-frequency circuits where only frequencies up to perhaps 3 kHz are to be passed. They can be employed between a transmitter and an antenna to prevent higher frequencies (such as harmonics) from appearing in the antenna.

High-pass filters with the proper cutoff frequency can be used in audio-frequency circuits or between a TV receiver and its antenna to prevent nearby lower-frequency signals from interfering with TV reception.

For a sharper cutoff than a constant-*k* filter gives, an *m-derived* filter is used. This filter can be recognized by the parallel- or series-type resonant circuit in series with or across the line (Fig. 8-26). These

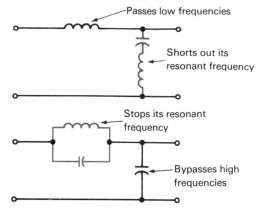

FIG. 8-26 Two possible single-section *m*-derived low-pass filter configurations.

wave-trap circuits produce essentially infinite attenuation of the frequency to which they are tuned and therefore zero transmission of that frequency along the line. (Filters with series-resonant elements in *series* with a line or parallel-resonant elements *across* the line do not attenuate the frequency of resonance. They fall in a constant-*k* category.)

The *m* represents a ratio of the frequency of infinite attenuation (zero output) to the cutoff frequency, or $m = f_\infty/f_{co}$, and will be something between 1 and 0. An *m*-derived filter with an *m* value of 1 has the same attenuation or transmission curve as a constant-*k* filter and has no point of infinite attenuation. In such a case, the tuned circuit involved has either negligible capacitance or negligible inductance. As the *m* value decreases, a point of infinite attenuation develops. As the *m* value decreases further, the point of infinite attenuation begins to approach the cutoff frequency, producing an increasingly steep attenuation (Fig. 8-27). The lower the *m* value, however, the less the attenuation of frequencies beyond the point of infinite attenuation. With an *m* value approaching zero, the resonant portion operates as a wave trap (as shown dashed). In practice an *m* value of about 0.6 is a good compromise between steep cutoff and reduced transmission past infinite attenuation.

Two single-section high-pass *m*-derived filter circuits are shown in Fig. 8-28.

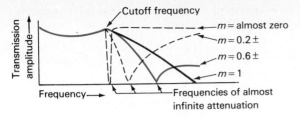

FIG. 8-27 Frequency response curves of an *m*-derived low-pass filter with different *m* values.

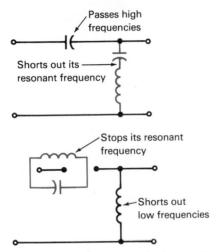

FIG. 8-28 Two single-section *m*-derived high-pass filters.

Practical filters are usually made up into two or more sections. There are three basic section configurations, *T, L,* and *π* types, named for their appearance when diagramed (Fig. 8-29). A balanced-T type might be termed an H type.

A composite filter consists of two or more separate sections coupled together. Figure 8-30 shows a constant-*k L* section coupled to an *m*-derived *L* section to form a low-pass filter with maximum attenuation at about 1000 Hz and operating from a 600-Ω source into a 600-Ω load. It is important that

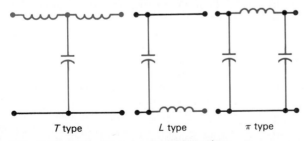

FIG. 8-29 Basic filter-circuit configurations.

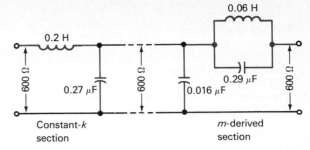

FIG. 8-30 Low-pass composite filter with one constant-*k* and one *m*-derived section.

the separate sections match each other in impedance. Figure 8-31 shows the filter as it might actually be constructed. (See also Chap. 17 for information on sideband filters.)

For further information on filters the reader is referred to the author's text *Electrical Fundamentals for Technicians* or to radio or electronic engineering handbooks.

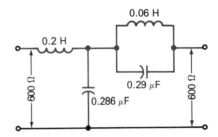

FIG. 8-31 Same low-pass filter as in Fig. 8-30 as it would actually be constructed.

Test your understanding; answer these checkup questions

1. At what amplitude is bandwidth (BW) of a tuned circuit usually measured? _____ What is another method of expressing BW? _____
2. Why would it be useless to try to tune a power or audio frequency transformer? _____
3. What would be the BW represented by Fig. 8-16 if the frequencies were in kHz? _____ If in MHz? _____
4. Does the curve of Fig. 8-16 represent *I* versus *f, V* versus *f, Z* versus *f,* or all of these? _____
5. Which dips first with increased coupling, the I_p or the I_s? _____
6. Maximum BW without secondary dip is produced by what degree of coupling? _____
7. What are the four basic types of filters? _____ _____ _____ _____

8. Of what is a wave trap made? _____ What basic type of filter is it? _____
9. What would be the C and L values of an unbalanced single-section constant-k 500-Hz low-pass (LP) filter for use in a 1000-Ω line? _____ For a similar high-pass (HP) filter? _____
10. What type of filter would be used between an antenna and a TV receiver to prevent interference with the picture by local lower-frequency radiations? _____ By local higher-frequency radiations? _____

11. What type of filter gives sharper cutoff response than a constant-k type? _____
12. What is the m value of a high-Q wave trap? _____
13. What are the four basic configurations found in LCR filters? _____ _____ _____ _____
14. What is the shape factor of a filter with a bandpass of 2.8 kHz at 6 dB down if its bandpass is 6.5 kHz at 60 dB down? _____

RESONANCE AND *LC* FILTER QUESTIONS

1. Give three variations of the resonance formula.
2. What is the resonant frequency of a 25-mH coil and a 0.002-μF capacitor?
3. If a resonant frequency is computed to be 5.673 MHz, why might it be best to indicate the answer as ± 5.673 MHz?
4. If the LC product in the resonance formula is halved, what frequency change will occur?
5. What C is used with a 15-μH L to resonate at 7 MHz?
6. What type of a circuit is resonant? Antiresonant?
7. A 300-Ω X_C and a 40-Ω R are in series in a series-resonant circuit across 18 V. What is the voltage-drop across the R? The X_L?
8. What is the phase relationship of the voltages across the L and C of a series-resonant circuit? Of the currents?
9. If a capacitor is across the secondary of a transformer is the secondary circuit series- or parallel-resonant?
10. Why are variable capacitors used in radio-receiver-tuned circuits?
11. What is an antiresonant circuit?
12. In an antiresonant circuit what is the phase relationship of the branch currents? Of the branch voltages?
13. A 500-Ω-reactance L and C are in parallel, and in series with a 20-Ω R across a 12 V ac. What is the voltage across R? X_C?
14. What is the impedance across a low-loss resonant circuit? An antiresonant circuit?
15. What is the power factor of a resonant circuit? Antiresonant circuit?
16. An antiresonant circuit has a branch with a 250-Ω X_C and a 50-Ω R. It is in parallel with an X_L branch having 80-Ω R. What is the total Z of the circuit?
17. In problem 16, if the R values were 10 and 20, what would be the circuit Z? Why the great difference?
18. What might the current in the branches of an antiresonant circuit be called?
19. If an antiresonant LC circuit has zero R, how long might electrons oscillate in it?
20. What is the Q of an inductor having 500-Ω X_L if it has a 50-Ω R? If 5-Ω R?
21. What effect causes a coil to have greater resistance at higher frequencies?

22. How can the Q of an antiresonant circuit be lowered?
23. In what two terms might an increase in AF power from 1 W to 10 W be expressed?
24. In the logarithm 3.1536, what is the 1536 part called? The 3?
25. What is the log of 8? Of 24? Of 356? Of 0.47?
26. How much of a dB gain does an amplifier have if it has 15-W output and the input is 0.24 W?
27. What is the power output of an amplifier capable of 44 dB if the input is 0.0005 W?
28. Generally, what power level is considered 0 dB?
29. What is the dB gain of an amplifier having 0.12 V fed across a 500-Ω input circuit and delivering 15 V across a 40-Ω output?
30. What is the gain of an amplifier having equal input and output impedance if V_o is 45 V and V_i is 0.6 V?
31. Where on a bandpass curve is bandwidth measured?
32. What is the bandwidth relationship considering LC circuit frequency and Q?
33. Why are iron-core transformers never tuned?
34. Which produces narrow bandwidth in a tuned transformer, loose or tight coupling?
35. In Fig. 8-16, if the frequencies represented are in megahertz, what is the bandwidth of this circuit?
36. What develops a dip at the center frequency of a tuned transformer?
37. What are measurement points to determine the shape factor of a filter?
38. What type of wave trap is generally used in series with an electric line? In parallel with the line?
39. What kind of filter would be used to allow only frequencies from 300 to 4000 Hz to be transmitted?
40. It is desired to prevent a frequency of 10 kHz from passing along an electric line. What kind of filter should be connected across the wires?
41. If all frequencies above 4000 Hz should not pass, what type of filter would be used?
42. How can sharper cutoff characteristics be attained with filters? What are the disadvantages of these?
43. What does an m-derived filter have that a constant-k does not?
44. How is a filter "balanced"?
45. What is a common m-value for a filter?
46. What are some of the shapes taken by filters?

9

Active Devices

This chapter describes devices which come under the general heading of diodes, transistors, integrated circuits, and vacuum tubes. It describes their makeup, what makes them operate, and, very briefly, some of the basic rectifying and amplifying circuits in which they will be found. A few gaseous control devices are also discussed.

9-1 ACTIVE DEVICES

The components discussed so far—resistors, inductors, and capacitors—are known as *passive devices*. They do not change the waveshape of ac signals applied to them. Diodes, transistors, and vacuum tubes may rectify, amplify, and alter the waveshapes of ac signals fed to them. (Sometimes some diodes may be considered passive devices.) The great advances made in electronics and radio are due to the ability of *active devices* to oscillate, amplify, and change state (turn on or off). The original active devices were crystal detector diodes. These were followed by vacuum tubes. Since about 1950 a wide variety of solid-state transistors and integrated circuits composed of diodes and transistors have appeared. The advantages of transistors over vacuum tubes are their small size, the small amount of power required to operate them, and their reasonable fabrication cost.

Vacuum tubes today are mostly found in older equipment that is still operating satisfactorily and which is worth maintaining, in TV-type cameras and monitor tubes, and in high-power circuits having power outputs of over perhaps 500 W of RF ac. Each year transistors are being developed to handle more and more ac power, but it will still be some time before transistors can produce RF ac power of 10,000 W and more. There are many applications of vacuum tubes in which the power output greatly exceeds 100 kW, or even 1 megawatt (MW). Vacuum tubes are far less affected by electromagnetic pulses (EMP) caused by lightning and nuclear explosions than are semiconductor devices. Although solid-state devices are the more important to understand today, every technician should also be familiar with at least the basic vacuum devices and their operation.

9-2 DOPED SEMICONDUCTORS

A solid-state device consists of a semiconductor substance such as silicon, germanium, or gallium arsenide, treated in such a way that electron flow through it can be controlled. The physical makeup of a semiconductor should be understood. (Note that electron current, not "conventional" current, is used in all explanations in this text. Electrons flow in a direction opposite to the arrows in semiconductor symbols.)

The outer-orbit, or *valence*, electrons of some atoms, such as metals, can be detached with relative ease at almost any temperature and may be

137

called *free electrons*. These valence electrons are able to move outward from their normal outer-orbit level into a *conduction level* or *band*, from which they can be dislodged easily. Such materials make good electrical conductors. Other substances, such as glass, rubber, and plastics, have no free valence electrons in their outer conduction bands at room temperatures and are good insulators. A few materials have a limited number of electrons in the conduction level at room temperatures and are called *semiconductors*. The application of energy in the form of *photons* (small packets of light or heat energy) to the valence electrons moves some of them up into the conduction band, and the semiconductors then become better electrical conductors. Energy of some form is required to raise semiconductor electrons to a conduction level. Conversely, if an electron drops to the valence level from a higher conduction level, it will radiate energy in some high-frequency form such as heat, infrared light, ultraviolet, and, if the fall is great enough, x-rays.

Semiconductors such as silicon (Si) and germanium (Ge) have four outer-ring electrons. Crystals of these can be laboratory-grown in pure form, called *intrinsic* (I). A perfectly formed semiconductor crystal lattice (Fig. 9-1a) acts more like an insulator than a conductor at room temperatures. If, during its manufacture, a silicon crystal has about one in a million atoms of an impurity, called a

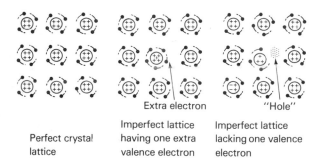

Extra electron   "Hole"

Perfect crystal lattice	Imperfect lattice having one extra valence electron	Imperfect lattice lacking one valence electron
(a)	(b)	(c)

FIG. 9-1 Nucleus and outer ring electrons of atoms in a lattice of (a) I germanium, (b) N germanium, (c) P germanium.

dopant, added to it, the resulting crystal is imperfect (Fig. 9-1b). It has one in a million atoms in the lattice with an apparently excess outer-ring electron not being tightly held. If an electrostatic field is developed across such a doped silicon, the doped semiconductor is about 1000 times better as a conductor than the undoped intrinsic form. Doped silicon with such relatively free electrons is known as *N silicon*, and is a reasonably good conductor at room temperatures. (To form *N germanium*, arsenic can be used as the *donor* dopant.)

When silicon is doped with boron, which has three valence electrons, the crystal lattice is again imperfect (Fig. 9-1c). In this case there is an area, or *hole*, in the lattice that apparently lacks an electron. While the hole may not be actually positive, at least it is an area in which electrons might not be repelled by a negative charge. This positive-appearing semiconductor material is called *P silicon*. When an electrostatic field is impressed across a P-type semiconductor, the hole areas act as stepping stones for electron travel through the material. It can be said that *hole current* flows in a direction opposite to electron current because as an electron moves from one hole to a second, the holes are opened up from the second area to the first. Note that both N and P silicon crystals have zero electric charge, because both have an equal number of electrons and protons in all of their atoms. (One possible *acceptor* dopant used to produce P germanium is gallium.)

Doped silicon has more resistance than germanium, but it is useful in higher-voltage applications, it does not change its resistance as much when heated, and it can withstand greater temperatures without its crystalline structure being destroyed.

9-3 SOLID-STATE DIODES

When N-doped and P-doped semiconductor materials are grown together to form a single long crystal (Fig. 9-2a), a solid-state diode results. Some of the relatively free electrons in the N-type material move into the more or less positively charged holes in the adjacent P-type material, developing an area at the *junction* which is actually slightly negative on the P side of the junction and slightly positive on the N side. This produces a barrier to any further electron flow of about 0.6 V with silicon and 0.2 V with germanium diodes.

The "bias" battery in Fig. 9-2b has its negative terminal to the N-type semiconductor and its posi-

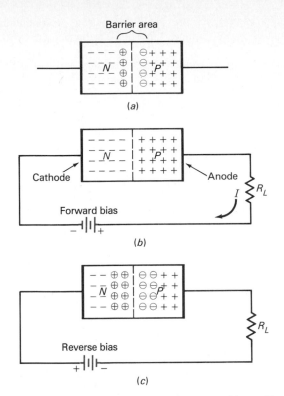

(a)

Cathode ← Anode

R_L

I

Forward bias

(b)

R_L

Reverse bias

(c)

FIG. 9-2 Junction diodes. (*a*) Barrier area with no bias. (*b*) Forward bias neutralizes barrier. (*c*) Reverse bias increases barrier area.

tive terminal to the P-type. The negative-to-positive electrostatic field developed by the battery across the junction overcomes the junction barrier voltage, and current flows through the *forward-biased* junction. It is necessary to add a current-limiting resistor to prevent excessive current flow in the semiconductor because of its low resistance in the forward-biased condition. Because of this low resistance, the voltage-drop across a conducting solid-state diode is very small.

The circuit in Fig. 9-2*c* has the bias connected in reverse. The negative pole of the battery drives electrons into the holes of the P-type material, while the positive pole of the battery pulls some electrons from the N-type material. The barrier area of the junction widens and no current flows. In this condition the diode has very high resistance. (There are always a few free electrons in P-type materials and holes in N-type materials, particularly at a junction. As a result of these *minority carriers*, there is always a small backward or leakage current in solid-state devices.) If the reverse voltage is excessively high, the barrier may break down and

reverse current will flow. This is called *zener effect* if the emf value at breakdown is less than about 5 V and *avalanche* if it is more than about 5 V. This is not a normal operating condition for most semiconductor small-signal or power-supply-rectifier diodes, and it may cause lattice damage, ruining the diode.

The reverse-voltage breakdown effect, however, is used in special *zener diodes*. Curves for three diodes are shown in Fig. 9-3*a*. The solid line repre-

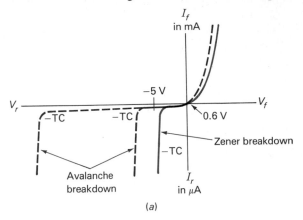

(a)

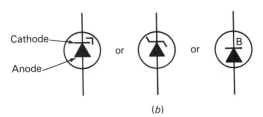

(b)

FIG. 9-3 (*a*) Zener and avalanche curves. (*b*) Symbols used for zener diodes.

sents the actions of a silicon diode that has a low reverse-voltage breakdown [$V_{(BR)r}$]. In the forward-biased direction current starts flowing when V_f exceeds 0.6 V. In the reverse direction, V_r, the diode breaks down at about 4 V. If V_r were increased a few volts more, the diode might be destroyed unless manufactured to operate at high reverse currents or unless a resistor were in the circuit to limit current flow. The circuit across which the diode is connected will not increase in voltage over the 4 to 5 V of the zener effect. For this reason, zener diodes are used as shunt (parallel) voltage-regulating devices. Zener diodes with less than a 5-V breakdown will have a negative temperature coefficient (TC) of resistance, but with breakdown values over 5 V they will have a positive TC. (In the forward-biased direction all diodes have a negative TC.) The dashed lines might also represent characteristics of normal solid-state diodes as well as zeners. Figure 9-3b shows symbols used to designate zener or breakdown diodes.

A *voltage-variable capacitor*, or varactor, utilizes the variation of barrier width in a reverse-biased diode. Since the barrier of a diode acts as a nonconductor, a diode forms a capacitor when reverse-biased, with the N material as one plate, the P material as the second plate, and the junction as the dielectric. If the reverse-bias voltage is increased, the barrier (dielectric) widens, effectively separating the two capacitor plates and reducing the capacitance. The frequency of resonance of an *LC* circuit can be varied by using a varactor across a coil.

A *tunnel diode* is a heavily doped germanium or gallium arsenide diode with radically different VI characteristics. Because of the heavy doping, the diode acts as a good conductor when reverse-biased and has no zener effect, as shown by the solid line in Fig. 9-4a. The heavy doping allows the valence and conduction levels of the N- and P-area atoms to overlap under zero-bias conditions. With a small forward-bias voltage the diode acts as a conductor and electrons "tunnel" through the barrier area with the speed of light. As V_f increases, I_f increases linearly up to about 0.05 V. At this emf value the electrostatic field across the junction begins to develop a barrier. For a further small increase in forward-bias voltage the I_f decreases sharply. This part of the operation curve represents a *negative-resistance* effect (opposite to Ohm's law) in which an increase of *V* results in less *I*. At a little more than 0.2 V the junction begins to behave as a normal forward-biased junction (dashed curve). The negative-resistance effect is used in oscillators (Chap. 11) and in microwave amplifiers (Chap. 22).

When a doped semiconductor crystal is formed against a metal conductor, the heavily occupied conduction band of the metal and the lightly occupied conduction band of the semiconductor average out to be a diode junction which requires almost no voltage to produce conduction but which does not conduct under reverse-voltage conditions. Furthermore, since there are no minority carriers in the junction, current turn-on and turn-off can occur instantaneously. (In any all-solid-state diodes, minority carriers must be swept out of the junction before complete turn-off occurs.) These fast-acting diodes are called *Schottky* or, because they require so little energy to carry electrons across the junction, *hot-carrier diodes* (HCDs).

9-4 LIGHT-FREQUENCY DIODES

A *light-emitting diode* (LED), also known as a solid-state lamp (SSL), utilizes the fall of an electron from the conduction level to the valence level to develop an energy release in the form of heat or light. An electron moving across any PN junction moves to a hole area. This can allow a nearby conduction electron to fall to its valence level, radiating energy. In common diodes and transistors made from germanium, silicon, or gallium arsenide, this electromagnetic radiation is usually at a heat frequency, which is lower than light frequencies. With gallium arsenide phosphide the radiation occurs at red light frequencies. Gallium phosphides produce still higher frequency yellow through green radiations. Gallium nitride radiates blue light. An LED in a biased circuit is shown in Fig. 9-5a. Starting with the bottom gold contact, there is a gallium arsenide layer, an N-type gallium arsenide phosphide layer, and then a very thin P-type GaAsP layer. Current passing through this P GaAsP layer develops red light radiations. About 5 mA of cur-

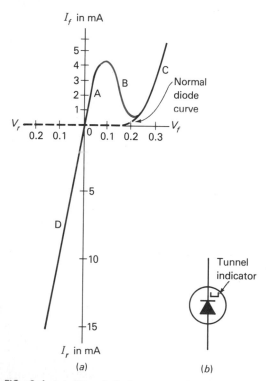

FIG. 9-4 (a) Tunnel-diode curve. (b) Symbol of tunnel diode.

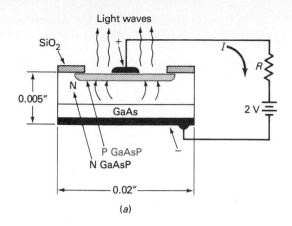

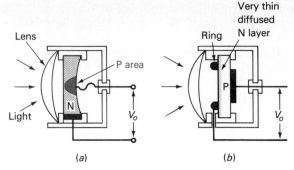

FIG. 9-6 Two forms of photovoltaic diodes.

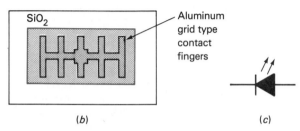

FIG. 9-5 (a) LED cross section, (b) top view looking into light-emitting surface, and (c) symbol.

rent flow in the direction shown produces a weak red glow. A current of 20 mA results in quite a bright glow. Maximum safe current may be in the 100-mA region, depending on how the unit is made. At about 20 mA, the life expectancy is about 100,000 h. To prevent a maximum glow at only the tiny positive electrode, this contact is made in the form of fingers of aluminum spread out across the P GaAsP surface (Fig. 9-5b), resulting in a broader, more visible area of glow. Since this is a diode, reversing the bias battery will produce a barrier at the junction, no current will flow, and no glow will appear. A resistor is always connected in series with an LED to limit current to a desired value.

LEDs are taking the place of the pilot lamps previously used as visual indicators in equipment. An LED plus a resistor can be used on any voltage from 2 to 10 V. It uses much less current than a pilot lamp.

There are several photodiodes and photosensitive devices. Photodiodes convert photons to electric emf. In Fig. 9-6a an N silicon chip has a P silicon diffused into it until the front of the N silicon is so thin that light can penetrate to the PN junction. Photon energy lifts electrons from the valence level of the P silicon and injects them above the conduction level of the N silicon. As long as light strikes

the junction, the diode converts photon energy to emf, up to about 0.5 V, across the diode. Figure 9-6b indicates a metal ring connector held against a thin diffused N layer on a P substrate with a contact against the back of the P layer. Light striking the PN junction develops a movement of electrons into the P silicon. These devices are called photovoltaic cells or diodes.

A *photojunction*, or *photoresistive*, device may not be a diode; it may be a semiconductor material that increases its conductance (reduces its resistance) when struck by photons of light or heat. Photoresistive devices are slow acting.

A PIN-type silicon photodiode-type photoresistor is shown in Fig. 9-7. From the bottom, it consists of a gold electrode, a thin N Si layer, a thick I Si layer, a thin P Si layer, and finally a gold electrode making contact with the P Si. The silicon dioxide (SiO_2) is an insulating or passivating (protecting) material. Photons of heat or light passing through the thin P Si layer strike atoms in the intrinsic layer and produce free electrons and holes. The electrons move to the positive bias potential, and electrons

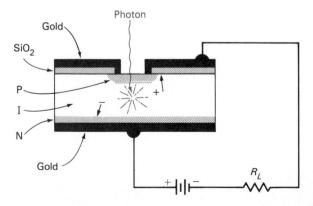

FIG. 9-7 PIN photodiode.

fill the holes from the negative potential. The stronger the light, the more electron-hole pairs developed and the more current that can flow. These are not true diodes, but are very fast acting photoresistors.

Without the light aperture PIN diodes are used as a variable resistance by controlling the forward bias across them (Chap. 22).

Mounting an LED facing a photodiode cell in a tiny lighttight enclosure produces an *opto-isolator*. When current flows in the LED, it illuminates the photodiode cell and produces current in it. An opto-isolator allows signal transfer without coupling wires, capacitors, or transformers. It can couple digital (on-off) or analog (variable) signals.

A variety of diodes are shown in Fig. 9-8.

Test your understanding; answer these checkup questions.

1. Where must valence electrons be in conductors? _____
2. What are small packets of light or heat energy called? _____
3. When impurities are added to a semiconductor crystal lattice, what is the process called? _____
4. Is P germanium positively, negatively, or neutrally charged? _____ N Si? _____
5. Heating a semiconductor has what effect on its conductance? _____
6. To conduct, is the P material of a diode made positive or negative? _____

7. What is the junction barrier potential difference using Ge? _____ Si? _____
8. What current do minority carriers produce? _____
9. How does forward-biasing affect junction resistance? _____
10. What is another name for a varactor? _____ Would it normally be forward- or reverse-biased? _____
11. What are two materials used with tunnel diodes? _____ _____
12. What is different about the characteristic curve of a tunnel-type diode? _____
13. What makes Schottky diodes high-speed types? _____
14. What is an LED? _____ Would it rectify? _____
15. Which determines color in an LED, the material used or the current amplitude? _____
16. What frequency radiation is produced by a forward-biased Ge or Si diode? _____
17. Which are fast-acting devices: photodiodes, photoresistive devices, PIN devices? _____
18. What might be considered as the "majority carrier" in N Si? _____ In P Ge? _____

9-5 JUNCTION TRANSISTORS (BJTs)

The basic and original transistor can be thought of as two diode PN junctions constructed in series (Fig. 9-9). From the bottom, there is a contact plate formed against an N-type *emitter* element (E). Next to this is a thin P-type *base* element (B), with a metal electrode attached to it, forming the first PN *junction*. A second N-type *collector* element (C) is

FIG. 9-8 Various diodes. The cathode end may be marked with a dot, stripe, or arrowhead, or be the threaded end.

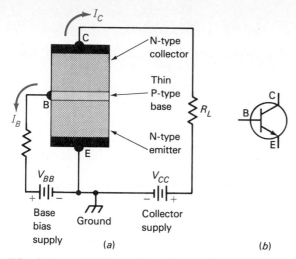

FIG. 9-9 (*a*) NPN-transistor circuit showing currents flowing out of both base and collector and (*b*) symbol.

added, with a contact on it forming the second PN junction. This produces an NPN transistor.

With the base element disconnected, regardless of the polarity of the collector supply, no current can flow in the emitter-collector (EC) circuit (disregarding a small minority current) because one of the PN junctions would always be *reverse-biased*. However, with the circuit connected as shown, the emitter-base (EB) junction is *forward-biased* so some electrons move from emitter to base. Since a semiconductor has relatively few usable carriers, electrons must detour throughout the thin base layer to find a path to the base contact. In so doing, some of them move through the base-collector junction area, aided by the higher collector voltage. This junction becomes a conductor with carriers in it, allowing emitter-collector current to flow. Increasing and decreasing EB current results in increasing and decreasing EC current. The significant feature about a transistor is that a small variation of EB current can control 50 to 150 times as much EC current flowing through its *load* (R_L). Thus, the transistor is an ideal control and amplifying device. A junction transistor of this type can be called a *bipolar junction transistor* (BJT) to differentiate it from a field-effect transistor (FET) discussed later. Note that $I_E = I_B + I_C$ at all times.

Transistors are fabricated in a variety of ways. The original transistors had wires held against a semiconductor and were called *point-contact* types. Since then alloy, rate grown, diffused mesa, passivated planar, epitaxial base, overlay, and many

other types have been developed. Some methods of construction may provide higher-power operation. Others, such as the overlay type, are better at higher-frequency turn-on and turn-off. One type that was developed for high-frequency operation had two connections to the base and was called a *tetrode* (4-element) transistor. It is rarely seen.

The BJT discussed so far is the NPN type. However, transistors operate essentially the same if made in a PNP form. The electrical differences are the bias voltages used in the EC and in the EB circuits, which are reversed from those used in NPN devices.

A *phototransistor* is a very sensitive exposed-junction transistor that operates as a photoresistor. When light is allowed to fall on the exposed EB junction, EC current flows. The brighter the light the less the EC resistance and the more I_C current that can flow.

The "ground" symbol in Fig. 9-9 indicates that this part of the circuit would be connected to a common terminal, or to the metal chassis or outer rim area of the PC board on which the circuit is built. This symbol is said to be a chassis ground indicator. It is usually connected to the negative terminal of V_{cc} if NPN BJTs are used.

9-6 COMMON-EMITTER CIRCUITS

The basic transistor amplifier circuit is the common-emitter (CE), shown in Fig. 9-10*a* with an NPN transistor and in Fig. 9-10*b* with a PNP BJT. The notable differences between the two circuits are reversed arrow directions in the symbols besides reversed currents due to reversed bias and collector supply polarities. This is a *common-emitter* amplifier because the emitter is a part of both base and collector circuits.

With a medium value of base bias voltage, a medium value of I_B will flow in the input circuit and a medium I_C will flow in the output or collector circuit (called class A operation). In a representative transistor, a current variation of 20 μA in the base circuit might produce a 2000-μA (2-mA) change in the collector circuit, or a current gain of 100 times. A small audio ac coupled into the base circuit becomes a much larger audio current variation in the output.

Current gain or amplification is the ratio of collector current change occurring in the collector circuit *load* compared with the change in base current that is causing the I_C change, or

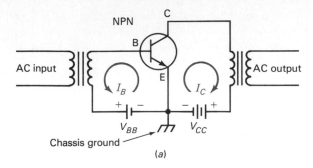

(a)

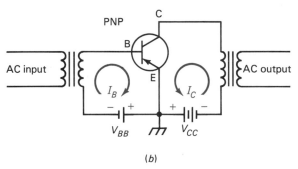

(b)

FIG. 9-10 Basic transistor amplifiers. Note supply polarities (a) with NPN and (b) with PNP transistors.

$$A_i = \frac{\Delta I_C}{\Delta I_B}$$

where Δ = change in
A_i = current amplification

Beta (Greek letter β) is an expression of the change in the collector current of a transistor compared with the change in the base current when there is *no load* in the collector circuit and the emitter-collector voltage is held constant, or

$$\beta_{ac} = h_{fe} = \frac{\Delta I_C}{\Delta I_B} \bigg|_{V_{EC} \text{ constant}}$$

The *dc beta* is the ratio of I_C compared with the I_B value that produces it, and it is roughly similar to the ac beta above. In formula form it is

$$\beta_{dc} = h_{FE} = \frac{\Delta I_C}{\Delta I_B} \bigg|_{V_{EC} \text{ constant}}$$

Producible current gains in amplifiers will range around 70% of the published beta values with load resistors up to perhaps 10 kΩ.

The *alpha cutoff frequency* of a transistor is the frequency at which its current gain drops 3 dB from its gain at 1000 Hz (common-base, Sec. 9-9).

It is important to understand that a transistor must have the base-emitter circuit forward-biased to allow any collector current to flow. With zero or a reverse base bias, no I_C can flow. Note that a reverse bias is a negative voltage for an NPN transistor. Conversely, a PNP transistor is reverse-biased with a positive potential on its base. To simplify explanations, most circuits will be shown with NPN transistors (which is also similar to vacuum-tube circuitry) with the understanding that the circuits will operate with PNP transistors if the supply polarities are reversed.

The circuits above are not practical because the voltage required to forward-bias a germanium transistor is slightly over 0.2 V and slightly over 0.6 V for silicon. No common batteries have such emf values. With transistors it is easier to work with bias currents rather than voltages. Thus a certain low-power transistor might have a bias of 50 μA to produce a medium I_B and I_C flow. Another, higher-power transistor might require 10 mA of I_B to produce a medium value of I_C for that transistor. Although germanium transistors have less resistance, silicon types are used more frequently.

One method of biasing a base circuit, called *fixed bias*, is shown in Fig. 9-11. The dashed line indicates the collector current path through the transistor, load, and power supply. The dotted line represents the base-biasing current path into the emitter, out of the base, through the base-biasing

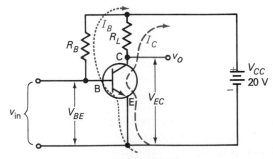

FIG. 9-11 Fixed bias in a common-emitter amplifier, showing base and collector current paths.

resistor, and to the power supply. The value of biasing resistance is varied until the collector-emitter voltage, V_{EC}, is just half of the power-supply voltage, or 10 V in this case. An ac input signal to the base circuit can now vary the voltage-drop across the *load*, R_L, from a quiescent (no-signal) 10 V up to almost 20 V when the base is heavily forward-biased by the signal, and then down to nearly 0 V when the base voltage drops to the EB junction barrier voltage of 0.3 V for germanium, or 0.6 V for silicon transistors. The R_B value is usually in the range of 200,000 Ω for low-power transistors and a few thousand ohms for power transistors. Some useful formulas that can be used with common-emitter transistors are:

$$I_C = \frac{0.5 V_{CC}}{R_L} \qquad I_B = \frac{I_C}{\beta} \qquad R_B = \frac{V_{CC}}{I_B}$$

$$\text{Voltage gain} = \frac{\beta^2 R_{\text{out}}}{R_{\text{in}}} \qquad \text{Power gain} = \frac{\beta^2 R_L}{R_{EB}}$$

This simple biasing arrangement is unstable thermally. If the transistor warms due to a rise in ambient (surrounding) temperature or due to current flow through it, the I_C value increases. The higher the current gain of the transistor, the greater this current instability. Obtaining base bias current by connecting R_B to the bottom of the load resistor (Fig. 9-12) improves thermal stability by a factor of

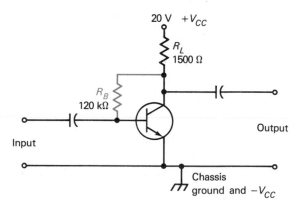

FIG. 9-12 BJT amplifier with self- or collector-bias.

2 or more. The value of R_B is then about half of the fixed-bias value. This is called *self-bias* or collector bias. It produces a collector form of *degeneration*, or out-of-phase feedback.

When a resistor is connected emitter-to-ground (R_E, Fig. 9-13), a positive-going signal on the base increases not only I_C through R_L and I_B, but also I_E

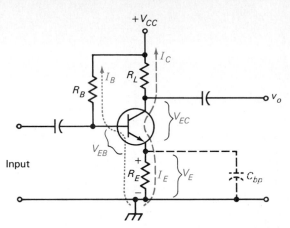

FIG. 9-13 Adding R_E improves dc thermal stabilization and adds degeneration.

through R_E. The increased voltage-drop across R_E reduces V_{EC}, and the gain of the stage is reduced. This is called degeneration, and results in less output from the stage. However, the greater the degeneration, the greater the thermal stability, and usually the less distortion produced in the stage. To reduce ac signal degeneration, a bypass capacitor C_{bp} is added across R_E. If this capacitor is large enough (X_C is less than $^1/_{10}$ R_E at the lowest signal frequency), rapid signal variations do not change its charge materially and no degeneration of the signal from this part of the circuit is produced. Also, since the BE resistance changes with temperature, adding R_E decreases base-to-ground percent of resistance variation considerably. Note the labeling of the various voltages in this stage. Other symbols often seen are V_{CBO}, the voltage between collector and base with the other element (emitter) open-circuited, and I_{CBO}, the minority carrier or leakage current between C and B with E open.

By far, the most used bias system with transistors is the "voltage-divider" type (Fig. 9-14a) because of good thermal stability and β independence. By ratio, the voltage-drop across R_S should be about 10 k/50 k of 20 V, or 4 V. (Since the I_B through R_B will be only $^1/_{100}$ of I_C in a 100-β transistor, it will be disregarded to simplify this explanation.) Because the V_{EB} of a silicon transistor is about 0.6 V, when the base is at +4 V, the emitter must be at about +3.4 V above ground ($-V_{CC}$). With 3.4 V across 500 Ω, I_{R_E} must be $I = V/R$, or 3.4/500, or 6.8 mA. Disregarding the small I_B again, I_C must be equal to I_E, or also 6.8 mA. Since R_L is 1500 Ω and I_C is 6.8 mA, then $V_{R_L} = IR$, or 1500(0.0068), or 10.2 V. The emitter-to-collector voltage must be

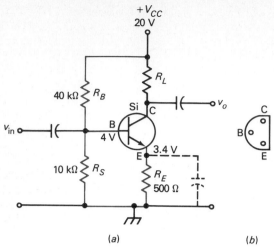

+V_{CC}
20 V

40 kΩ R_B

R_L

Si C

B
4 V

v_o

3.4 V

E

10 kΩ R_S

R_E
500 Ω

v_{in}

C

B

E

(a)   (b)

FIG. 9-14 (a) Voltage-divider or emitter bias. (b) A common base view of pin arrangement of transistors.

$20 - (10.2 + 3.4)$, or 6.4 V. If equal V_{EC} and V_{R_L} values are desired, R_L must be decreased in value somewhat.

The addition of R_S increases thermal stability. The lower the resistance of the voltage-divider network, the better the thermal stability but the lower the input impedance and the greater the loss of signal due to power dissipated in these resistors. The dashed bypass capacitor is used to prevent signal degeneration while maintaining long-term or dc thermal stability.

When transistors are made to plug into a socket, the pin arrangement, looking at the bottom of the transistor (or socket), is often as shown in Fig. 9-14b. However, refer to basing diagrams in transistor manuals to be sure.

A special heat-sensitive resistor made of selenium, silicon, or metal oxides, having a negative temperature coefficient (TC) and called a *thermistor*, can be physically attached to the transistor and electrically connected in series with R_S for further thermal stability. As transistor and thermistor warm, the transistor I_C tends to increase, but the thermistor resistance decreases. This reduces the forward bias on the transistor, lowering the increase in I_C. The $-$TC characteristic of a forward-biased diode can be used in this circuit by adding the diode, connected in a forward-biased direction, in series with R_S. Many solid-state diodes can be used as thermistors.

Two important considerations when selecting a BJT are: (1) The *cutoff frequency*. This is the frequency at which the gain, or beta, of a common-

emitter circuit (or alpha of a common-base circuit) drops to 0.707 (-3 dB) of the gain at 1 kHz. This is considered the *bandwidth* of the device. (2) The *gain-bandwidth product* is the frequency, f_{co}, at which the common-emitter beta drops to zero ($f_{co} = 0$ dB gain).

Test your understanding; answer these checkup questions.

1. Of what type material is the base of an NPN transistor made? _____ The collector of a PNP type? _____
2. How is the EC resistance affected when the EB junction is forward-biased? _____
3. What does BJT stand for? _____
4. What method of fabrication was used with the first transistors? _____
5. What is the layer which protects junctions in transistors called? _____
6. What is the name of one type of high-frequency transistor? _____
7. In a class A amplifier what value of I_C flows? _____ I_B? _____
8. What is the formula for current gain in a BJT? _____
9. What is the formula for beta? _____
10. What is the frequency at which the gain of a transistor drops 3 dB from its 1000-Hz gain called? _____
11. Name three types of biasing circuits used in transistor amplifiers. _____ _____ _____
12. Increasing degeneration decreases what undesirable effects in transistor amplifiers? _____
13. What symbol indicates minority carrier current between B and C with E open? _____
14. What type of biasing arrangement is probably the most common with transistor amplifiers? _____
15. What is a thermistor? _____ Must a diode be forward- or reverse-biased to act as a thermistor? _____
16. Would a thermistor be used in series with R_B, R_E, or R_S? _____
17. If a BJT were plugged into its amplifier circuit socket in reverse (E to C, C to E, and B to B), would the circuit amplify? _____

9-7 BJT CHARACTERISTIC CURVES

The diagram in Fig. 9-15a is a simple bipolar junction transistor resistance-load amplifier with fixed bias. The characteristic curves shown in Fig. 9-15b could be supplied by the manufacturer of the transistor or could be shown on a curve tracer, or could be plotted by hand. The lowest solid line

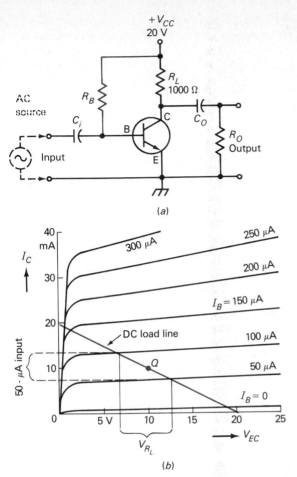

(a)

(b)

FIG. 9-15 (a) CE amplifier. (b) Load line and Q point with a 50-μA peak-to-peak input signal. V_{R_L} is output voltage.

represents the collector current as V_{EC} (or V_{CC} if the curves are developed with no load) is varied from 0 to 25 V, with I_B held at zero. The next curve is the I_C with V_{EC} variations, but with I_B held at 50 μA, and so on.

When using a dc load line what bias should be selected for an R_L of 1000 Ω and a V_{CC} of 20 V? With these conditions known, two points can be marked on the graph. (1) 20 V for V_{EC} when the transistor is at zero bias and no I_C flows. (2) Maximum $I_C = V/R$, or 20/1000, or 20 mA when the transistor is forward-biased to I_C saturation. A line drawn between these two points is the *dc load line*. A no-signal or quiescent point, Q, has been chosen close to the center of the load line, in this case at 75 μA. Resistor R_B can be found by Ohm's law, considering the voltage-drop across it to be 20 −

0.6, or 19.4 V. Thus, $R_B = V/I$, or 19.4/0.000075, or 260,000 Ω.

An ac input signal of ±25 μA (Δ50 μA) peak value produces a varying dc I_C of about 5 mA peak, or a current gain of about 100. The voltage variation across the load with this input signal will be about 5.5 V. (The V_{EB} variation would be only about 0.05 V.) The impedance of the transistor is found by Z = $\Delta V/\Delta I$, or 5.5/0.005, or 1100 Ω. Since the Q point is approximately half of the V_{CC} value, it might have been expected that the transistor and the load impedances would be nearly equal.

The load resistor, R_L, the output coupling capacitor, C_O, and the output resistor, R_O, form what is known as a *resistance coupling* network to deliver the output signal to the next stage, whatever it might be.

9-8 POWER BJTs

Most transistors are tiny, low-power devices operating with fractions to a few milliamperes of I_C (Fig. 9-16). When power is required to drive loudspeakers or radiate radio-frequency energy, transistors must be capable of at least hundreds of milliamperes to many amperes of I_C. AF and RF transistors are different devices. Lower-frequency power transistors are more or less standard types but made larger, resulting in more junction capacitance. Since C_j prevents high-frequency circuits from working properly, RF transistors must be specially designed overlay, VMOS, etc., types discussed later.

A basic requirement in power junction transistors is to remove the heat developed in the device when I_C flows. The EB junction is forward-biased and therefore has a low resistance or impedance value. The CB junction, on the other hand, is reverse-biased and has considerably greater resistance. Thus, with I_E and I_C essentially equal (except for a small I_B in the I_E), from the power formula, $P = I^2R$, the higher-resistance collector junction will develop greater power in heat. For this reason the collector in power transistors is constructed with a means of leading heat to the outside air to allow it to be dissipated. The collector is usually fastened to a relatively large metal body that can radiate heat developed in the collector junction. Any metal cooling device, such as fins clamped to the collector connection, is called a *heat sink*. A transistor that can safely dissipate only 1 W of power without a heat sink might safely dissipate 10 W with an efficient heat sink with cool air passing over it.

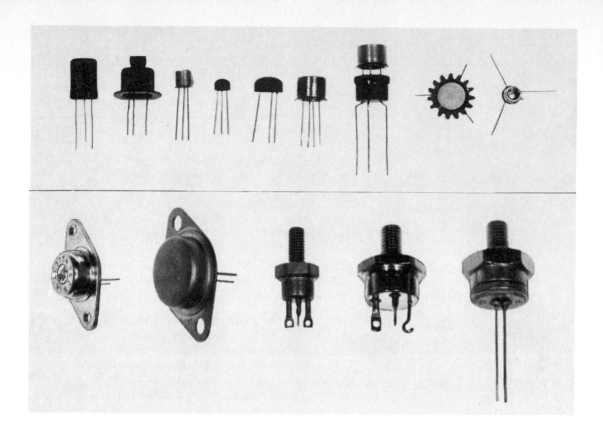

FIG. 9-16 Top, from left, six low-power transistors, BJT in a socket, BJT with heat sink around it, and a phototransistor. Bottom, SCRs and power transistors.

Figure 9-17a is a simple BJT power amplifier. It is similar to previous amplifiers except that it has a transformer and loudspeaker as the load in the collector circuit. If additional degeneration is required, a 2- to perhaps 10-Ω emitter resistor might be added at point X.

The characteristic curves (Fig. 9-17b) represent a transistor with a 20-W collector dissipation rating. The dashed curve indicates that operations must be to the left side of this line, since it follows a 20-W

value. (At $V_{EC} = 20$, $I_C = 1$, so $P = 20$ W. At $V_{EC} = 5$, $I_C = 4$, so $P = 20$ W, etc.)

Normally a dc load line is drawn from V_{EC} to I_C at its maximum possible value. With a low-resistance transformer primary as the load, the maximum I_C would be very high, resulting in an almost vertical load line, as indicated at 6-V V_{EC}.

When an ac signal is fed to the base-emitter circuit, the collector current will vary in the transformer primary (which has some value of ac impedance). If the transformer has a 1 : 1 turns ratio, it also has a 1 : 1 impedance ratio. If a 2-Ω load is connected across the secondary, the primary impedance will look into the 2-Ω load and present this impedance to the transistor. Thus, "ac" current, (actually vdc) flowing in the transformer should be $I = V/Z$, or using 6-V V_{EC}, 6/2, or 3 A. Using 3-A and 6-V V_{EC}, a fundamental ac load line can be drawn, as shown with dashes.

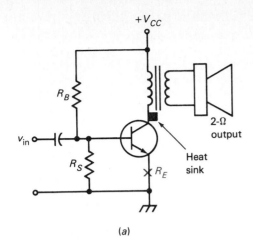

(a)

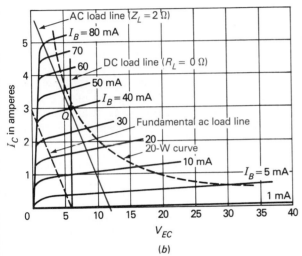

(b)

FIG. 9-17 (a) Power transistor in an AF amplifier circuit. (b) DC and ac load lines for one type of power transistor.

The curves show that a bias of 40-mA I_B will produce approximately 3 A with a 6-V V_{EC}, which is safely to the left of the 20-W curve. This is a possible quiescent or Q value. An ac load line (solid line) is drawn through Q exactly parallel to the dashed fundamental load line.

A sinusoidal input signal that drives the base from its 40-mA quiescent value down to zero, up to 80, and back to 40 mA is an 80-mA peak-to-peak varying input signal. From the load line, a change of I_B from 80 to 0 mA produces a I_C change from 5 to 0 A. The voltage-drop across the transistor is from 2-V V_{EC} (intersection of load line and I_B = 80 mA) to the 12-V V_{EC} point (the sum of the 6-V V_{EC} plus the inductive transformer primary voltage that

is developed) or 10 V. The impedance of the transistor must be $Z = V/I$, or 10/5, or 2 Ω.

Since the load impedance matches the 2 Ω of the transistor, maximum power output will result. The actual power being produced across the 2-Ω load can be found by $P = VI$, but effective values must be used, not these peak-to-peak values. If the p-p load voltage is 10 V, the peak value is 5 V and the effective value is 0.707(5), or 3.5 V. The effective value of the p-p 5 A is 0.707(2.5), or 1.77 A. Using these values, the output ac power is

$$P_o = VI = 3.5(1.77) = 6.2 \text{ W}$$

The power dissipated in the collector of the transistor with no signal is the dc input power, or

$$P_{dc} = I_C V_{EC} = 3(6) = 18 \text{ W}$$

The efficiency of the stage, the ratio of the ac output to the dc input power is

$$\text{Eff} = \frac{P_o}{P_{in}} = \frac{P_{ac}}{P_{dc}} = \frac{6.2}{18} = 0.34 \text{ or } 34\%$$

Note the distortion being produced. An increase of 40 mA of I_B increases I_C by 2 A, but a decrease of 40 mA decreases I_C by 3 A. Had the I_B curves been parallel and equally spaced, there would have been no distortion and the efficiency of the stage operating over the center of the load line (class A operation) would have been 50%, the theoretical class A maximum.

When the Q point is about half of maximum rated I_B, as in the transformer-load output circuit, and without a very efficient heat sink, internal transistor heating may occur. This can result in an increasing I_C, an upward shift of the Q point (crossing the maximum dissipation curve), and a devastating *thermal runaway*: more heat, less semiconductor resistance, more current, more heat, and so on. When the Q point is about half of V_{CC} in resistance-load stages, self-heating is usually self-correcting if dc degeneration is used, and thermal runaway is not likely to happen at normal ambient temperatures.

Excessive heat from too much collector current can almost instantaneously rupture a junction of a transistor. When repairing printed circuits (Secs. 1-16 and 1-17) into which transistors are soldered, care must be observed when removing or soldering in transistors, as heat can travel up the leads and damage a junction. Small heat sinks can be clipped to the leads, or air can be passed over the transistor leads while they are being soldered.

The significant characteristics of transistors can be found in transistor handbooks, which usually specify the maximum or minimum ratings and characteristics of the transistors under certain operating conditions. A few of these are listed for a low-power transistor in Table 9-1. The letter meanings are: T = temperature, J = junction, A = ambient, P = power, t = time, d = delay, r = rise, f = fall, and O indicates the third element is open-circuited. Delay, rise, and fall times refer to how fast the collector current rises and falls when a square-wave pulse is fed to the transistor input.

TABLE 9-1	BIPOLAR TRANSISTOR CHARACTERISTICS		
Maximum ratings			
Collector-to-base voltage	V_{CBO}	40 V	
Collector-to-emitter voltage			
$\quad V_{BE} = -0.5$ V	V_{CEV}	40 V	
$\quad R_{BE} = 10{,}000\ \Omega$	V_{CER}	20 V	
Emitter-to-base voltage	V_{EBO}	15 V	
Collector current	I_C	200 mA	
Transistor dissipation			
$\quad T_A$ up to 25°C	P_T	150 mW	
$\quad T_A$ up to 100°C	P_T	85 mW	
Temperature range			
$\quad$ Operating (junction)	$T_{J(opr)}$	−60 to 100°C	
$\quad$ Storage	T_{STG}	−60 to 100°C	
Lead soldering temperature			
$\quad$ (10 s)	T_L	230°C	
Characteristics			
Base-to-emitter voltage			
$\quad I_B = 10$ mA			
$\quad I_C = 200$ mA	V_{BE}	1.5 max V	
Collector-cutoff current			
$\quad V_{CE} = 20$ V			
$\quad R_{BR} = 10{,}000\ \Omega$	I_{CER}	50 max μA	
$\quad V_{CE} = 40$ V			
$\quad V_{BE} = -0.5$ V	I_{CEV}	50 max μA	
$\quad V_{CB} = 40$ V, $I_E = 0$	I_{CBO}	40 max μA	
Emitter-cutoff current:			
$\quad V_{EB} = 1$ V, $I_C = 0$	I_{EBO}	5 max μA	
Static forward-current			
$\quad$ transfer ratio,			
$\quad V_{CE} = 0.5$ V, $I_C = 30$ mA	h_{FE}	60 to 180	
Small-signal forward-current			
$\quad$ transfer ratio cutoff			
$\quad$ frequency			
$\quad V_{CB} = 6$ V, $I_C = 1$ mA	f_{hFE}	5 min MHz	
Output capacitance	C_o	20 max pF	
Turn-on time	$t_r + t_d$	1 max μs	
Storage time	t_s	0.7 max μs	
Fall time	t_f	0.7 max μs	

9-9 COMMON-BASE CIRCUITS

The previously discussed *common-emitter*, or *grounded-emitter*, circuit might be considered the standard transistor amplifier. However, there are two other basic circuits. One is the *common-base*, or *grounded-base* amplifier. A simplified version using batteries and transformers is shown in Fig. 9-18. The two batteries are in series and produce

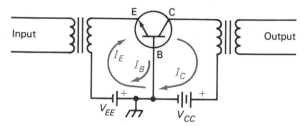

FIG. 9-18 Common-base amplifier.

the emitter-collector current that flows through input and output circuits. The base, tapped up the battery, is forward-biased, allowing I_C to flow. The input circuit current, I_E, is composed of I_C plus I_B. The ratio of common-base I_C to I_E is known as the *alpha* (Greek letter α) of the transistor. In formula form it is

$$\alpha = \frac{I_C}{I_E}$$

Since I_E is always greater than I_C, alpha is always less than unity in junction transistors. Alpha ranges from about 0.95 to 0.98. The smaller the I_B required to control the I_C the more nearly equal I_E and I_C will be and the higher the alpha.

Alpha cutoff frequency in a common-base circuit is the frequency at which the gain drops 3 dB from its gain at 1 kHz.

An input signal between E and B that adds to the forward-bias will increase both I_E and I_C further. The opposite-polarity input signal decreases the forward bias and I_E and I_C decrease. The phase characteristics of common-base amplifiers have input and output currents in phase, preventing self-oscillation (Chap. 11) of such amplifier stages, even at high frequencies. The input impedance is normally quite low, less than 200 Ω. The output load impedance can be anything from a few ohms to over 50 kΩ.

Although the CB amplifier has no current gain, it can have a voltage gain. Assume a 100-Ω input impedance and a 1000-Ω output impedance. With essentially the same current flowing through both circuits the voltage-drop across the output load will be 10 times that fed into the input. There is also considerable power gain in a CB amplifier. The I_C versus V_{CB} curves of a CB circuit are nearly parallel and equally spaced, producing a minimum of distortion.

In the resistance-coupled common-base amplifier shown in Fig. 9-19, the base is forward-biased with

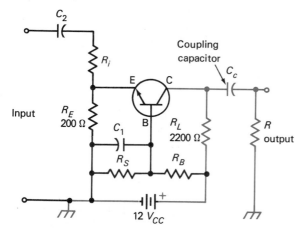

FIG. 9-19 Resistance-coupled common-base amplifier.

a voltage-divider network, R_S-R_B. The bias voltage is held constant by C_1. Since the emitter-base voltage varies very little during operation of a transistor, the voltage-drop across R_E remains essentially constant. The input signal *current*, via C_2 and R_i, takes a path through emitter, base, collector, V_{CC}, to ground. Whatever small current variation develops through the EB junction produces the amplifying action in the transistor.

9-10 COMMON-COLLECTOR CIRCUITS

The third basic transistor amplifier circuit is the *common-collector*, or *emitter-follower*, shown in Fig. 9-20. This circuit is often used as either an impedance converter or an isolation stage. Although it has both current and power gain, an emitter follower always has less voltage output than input. The input impedance seen by the stage ahead is approximately R_L times beta, shunted by R_B, since as far as ac is concerned, R_B goes to ground through the power supply. (With a 100-β transistor in the circuit shown this would be 600 × 100, or 60,000 Ω shunted by 54,000 Ω, or 28,420 Ω input impedance.) The output impedance is usually slightly less than R_L, although if a voltage-divider biasing network is used with values of resistances

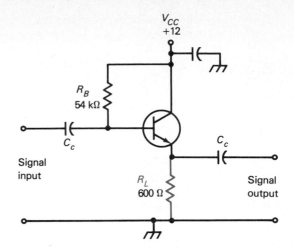

FIG. 9-20 Common-collector or emitter-follower amplifier.

in the few thousand ohms, the output impedance may appear in the 20- to 100-Ω range.

The emitter, or output, voltage follows the input signal almost exactly. The stage has nearly 100% degeneration, and there is little or no possibility of unwanted oscillations or distortion.

Table 9-2 shows some rough approximations of impedances, gains, and phase relationships of the basic amplifiers, common-emitter (CE), common-base (CB), and common-collector (CC).

TABLE 9-2	PARAMETER APPROXIMATIONS		
		Type of circuit	
Parameter	CE	CB	CC
Input Z	1000 Ω	60 Ω	40,000 Ω
Output Z	40,000 Ω	200,000 Ω	1000 Ω
Voltage gain	500	800	0.96
Current gain	20	0.95	50
Power gain	10,000	760	48
Phase in/out	180°	0°	0°

Test your understanding; answer these checkup questions.

1. Where is the Q point on a load line for class A operation? _____
2. Why does the BC junction heat more than the EB? _____
3. What would cooling fins on a transistor be called? _____
4. What is the rms value of a 100-V p-p? _____

5. Is the power output from a transistor amplifier normally ac or dc? _____ The power input? _____
6. What is the formula for percent efficiency of an amplifier? _____
7. What is the maximum possible class A amplifier efficiency? _____
8. What will be the result of unchecked thermal runaway? _____
9. What is another term meaning the same as common emitter? _____ Common base? _____ Common collector? _____
10. Is the alpha of a transistor always more or less than unity? _____ The beta? _____
11. Does a CE amplifier have voltage gain? _____ Current gain? _____ Power gain? _____
12. Does a CB amplifier have a V gain? _____ I gain? _____ P gain? _____
13. Does a CC amplifier have V gain? _____ I gain? _____ P gain? _____
14. If CB circuit I_C versus V_{CB} curves are parallel and equally spaced, what does this indicate? _____
15. What are two main uses of CC circuits? _____ _____

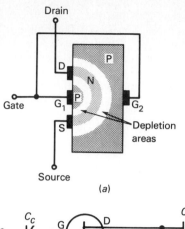

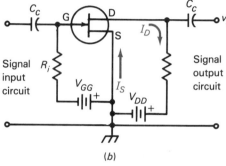

FIG. 9-21 (*a*) Diffused N-channel JFET showing source (S), drain (D), and gate (G) terminals. (*b*) Simple battery-biased JFET amplifier.

9-11 JUNCTION FETs

Unipolar *field-effect transistors* (FETs) and bipolar junction transistors (BJTs) are entirely different devices. Actually, FETs are solid-state amplifying devices that have operating characteristics quite similar to those of triode vacuum tubes. There are three basic types of FETs, the junction (JFET), and two *metal-oxide semiconductor* types known as MOSFETs. One type is called a *depletion*, the other an *enhancement* MOSFET.

The essentials of a JFET are illustrated in Fig. 9-21*a*. A P-type substrate has an N-type area diffused into it (indicated in color), on top of which is diffused another P-type area. Normal barrier or depletion areas develop at the two PN junctions (white areas). The operating drain current (I_D) of an N-channel JFET amplifier (Fig. 9-21*b*) is produced by the supply, V_{DD}. The two P-type areas are connected together and are called the *gate*, G. With no potentials connected to the gate, the *source-drain* current, I_D, would be relatively high.

By connecting a reverse-bias voltage, V_{GG}, between gate and source, the depletion areas are increased, reducing the volume of the N channel, which tends to pinch off the drain current. If sufficient reverse gate bias is used, the I_D can be pinched off completely. Half of the pinch-off bias value results in a medium I_D and class A operation.

An input signal *voltage* (not current as in BJTs) added across gate-to-source varies the effective bias, resulting in a variation of I_D through the load and an amplified voltage-drop across the load resistor.

JFETs have an *amplification factor*, or μ. This is the ratio of the change in gate-source voltage necessary to produce a constant drain current if the drain-source voltage is changed, or

$$\mu = \frac{\Delta V_{DS}}{\Delta V_{GS}} \bigg|_{I_D \text{ constant}}$$

The μ is not the voltage gain achievable in a practical circuit. A JFET with a μ of 50 might operate with voltage amplifications (A_v) of 20 to 30. A higher load-resistance value in the drain circuit or a greater V_{DD} will both increase the A_v.

Another JFET parameter is its *dynamic drain resistance*, or r_D. The dc or ohmic resistance of the N channel of a JFET might read from a few hundred ohms with no bias to essentially infinite ohms at pinch-off. The source-drain ac or varying dc impedance to varying signal voltages is more significant. Dynamic drain resistance (impedance) is the ratio of drain-source voltage change to drain current change produced by the voltage change, assuming the gate bias voltage is being held constant, or

$$r_D = \frac{\Delta V_{DS}}{\Delta I_D} \bigg|_{V_{GS} \text{ constant}}$$

The r_D of JFETs will range from a few thousand to several hundred thousand ohms. Matching r_D to the load impedance produces maximum output power in the load.

The *transconductance*, or g_m, of a JFET is its ability to vary the output circuit current, I_D, when an input-voltage variation is applied (also called forward transadmittance, y_f or g_f). In formula form

$$g_m = \frac{\Delta I_D}{\Delta V_{GS}} \bigg|_{V_{DS} \text{ constant}}$$

This can be expressed in siemens, S (formerly mhos), or, since it is always a small decimal number, in microsiemens (μS or μmhos).

The relations between μ, r_d, and g_m are

$$\mu = r_D g_m \qquad g_m = \frac{\mu}{r_D} \qquad r_D = \frac{\mu}{g_m}$$

Although it is normal for a current, I_{SG}, to flow from source to gate when the gate is made positive, there may also be a little leakage current, I_{GSS}, from gate to source when the gate is more negative than the source (N-channel JFETs). It is usually disregarded in basic considerations.

Just as there are three basic methods of connecting BJT amplifier circuits (Secs. 9-6, 9-9, 9-10) there are three equivalent methods of connecting JFETs in amplifier circuits. These are the common-source, common-gate, and common-drain circuits, illustrated in the 3-stage amplifier of Fig. 9-22. Transistor Q_1 has its source bypassed to ground by C_1, making it a *grounded-* or *common-source* circuit. The input to Q_2 is to its source, with the gate grounded, making this a *grounded-* or *common-gate* circuit. The input to Q_3 is to its gate, but the drain is bypassed to ground by C_2, making this a *grounded-* or *common-drain* circuit. The Q_1 circuit has high-input and relatively high-output impedances. The Q_2 circuit has relatively low-input and relatively high-output impedance. The Q_3 circuit has high-input impedance and relatively low-output impedances. All three circuits can produce an amplified power output signal when compared to the input signal power. Q_1 and Q_2 can also amplify the input signal voltage, but the output signal voltage

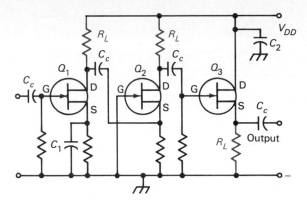

FIG. 9-22 A three-stage JFET amplifier using CE, CG, and CD circuits.

of Q_3 will always be less than the input voltage. In general, the resistance following a coupling capacitor should be equal to or greater than the resistance at the input side of the capacitor. (See also Sec. 9-26.) Note placement of output load resistors for the stages.

Only an N-channel JFET has been discussed. P-channel JFETs are also used. Supply polarities are reversed, as is the arrow direction of the gate in the schematic diagram symbol.

9-12 MOSFETs

An FET formed as in Fig. 9-23a, with the gate insulated from the silicon N channel by a thin layer of silicon dioxide, produces an insulated-gate FET (IGFET), or a metal-oxide semiconductor FET (MOSFET). The P-type base substrate lead, B, is either connected to ground or internally to the source.

When an input voltage is applied, negative to gate and positive to source, its electrostatic field extends into the N channel, reducing its ability to carry current by repelling channel electrons, thus depleting the channel of its carriers. This is a *depletion* MOSFET. Sufficient reverse bias can pinch off the I_D completely. Conversely, a forward bias can increase I_D. As a result, this device can operate with medium I_D (class A) with no bias at all, which simplifies circuitry. Unlike JFETs (and VTs) a forward bias does not produce any input circuit current (source to gate in JFETs) due to the insulation between gate and channel. MOSFETs have very high input circuit impedance. Because the insulation between gate and channel is very thin, it can be punctured by voltages of 20 V or more. To prevent this, most modern FETs incorporate in them re-

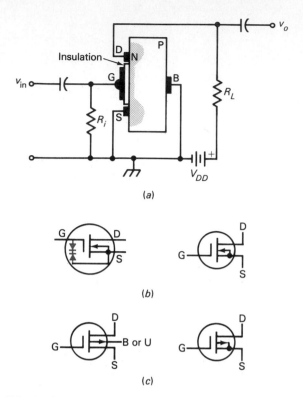

(a)

(b)

(c)

FIG. 9-23 (a) N-channel depletion MOSFET amplifier. (b) Symbols of N-channel MOSFETs, one with diode-protected gate. (c) P-channel MOSFETs.

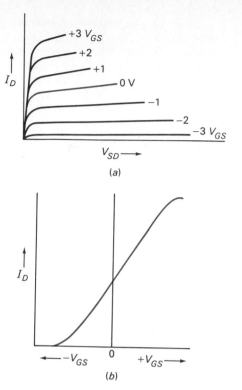

(a)

(b)

FIG. 9-24 (a) $V_{DS}I_D$ and (b) $V_{GS}I_D$ transfer characteristic curves for N-channel depletion MOSFET.

versed series diodes between source and gate to limit excessive electrostatic or other gate voltage buildup. These may not always be shown on symbols. Symbols of N- and P-channel MOSFETs are shown in Fig. 9-23b and c.

A set of transfer characteristic curves for a depletion MOSFET is shown in Fig. 9-24a. The same general information is contained in the V_{GS} versus I_D curve in Fig. 9-24b. Zero bias is close to the center of both sets of curves.

Another MOSFET is formed as in Fig. 9-25a, with its N areas almost touching. With no bias there will be essentially no source-drain current flowing. If a forward bias (+ with this N-channel device) is used, it attracts negative minority carriers

from the P area into the channel, providing carriers to support a source-to-drain current. The more forward bias the more I_D (Fig. 9-24b). This is known as an *enhancement* MOSFET because the channel requires carrier enhancement to produce current flow. Even with forward bias, no gate current can flow through the gate insulation. The input impedance may be in the hundreds or thousands of megohms. In the symbols of MOSFETs, the space between gate and channel represents gate insulation.

Although only N-channel MOSFETs have been discussed, P-channel devices are also used. Their supply polarities must be reversed. The arrowheads on the symbols are also reversed, as shown in Fig. 9-25d.

If the gate contact of a MOSFET is split into two parts and each half has a lead fastened to it, a *dual-gate* MOSFET results. Both gates can affect drain current amplitude.

A complementary metal-oxide semiconductor (CMOS) device has two MOSFETs, an N-channel and a P-channel, working together to produce a well-balanced electronic system.

One of the important transistor developments for

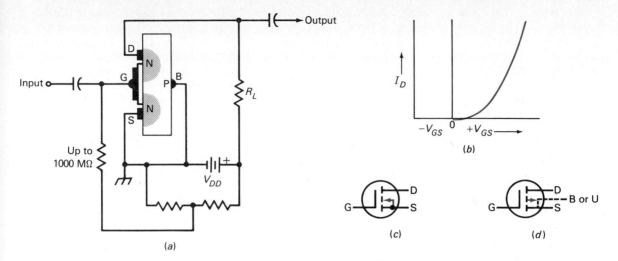

FIG. 9-25 (a) N-channel enhancement MOSFET amplifier. (b) $V_{GS}I_D$ curve. (c) Symbol. (d) P-channel enhancement MOSFET symbol.

audio or radio frequency amplification and for switching is the VMOS (*vertical* metal-oxide semiconductor), or *power MOSFET*. It is a silicon enhancement form of MOSFET, but with a "V" cut down into its chip, on which is formed the insulated gate. Instead of current flowing along a channel beneath a flat insulated gate, the source is at the top of the chip, and current flows vertically downward alongside the V to the drain area at the bottom. The VMOS features high current density; will stand relatively high source-drain (SD) voltages, producing higher powers than other FETs; has high transconductance; and is very linear (low distortion). For audio and other linear work, the protective internal series-diode barrier voltage must not be exceeded. For radio frequency work it may be exceeded somewhat, since distorted waveforms are not as critical for this service. With no forward bias on the gate there may be almost no SD current, requiring a forward bias *voltage* for class A operation. When driven with a maximum positive or saturation voltage on the gate, the voltage-drop between drain and source reaches almost zero volts. Unlike bipolar transistors, drain current *decreases* with increased device temperature, preventing thermal runaway. The symbol is similar to that of a MOSFET.

9-13 UNIJUNCTION TRANSISTORS (UJTs)

A bar of N-type semiconductor with contacts on both ends and a small P-type area alloyed into it (Fig. 9-26a) forms a double-contact diode called a

unijunction transistor (UJT). The ohmic resistance between the *cathode, K*, or base 1, and the *anode, A*, or base 2, may be about 10,000 Ω. The emitter or gate, E or G, is about 60% up the bar from base 1. The emitter acts as a voltage-divider tap on a fixed resistor, with E in the diagram at a potential of 60% of 30 V, or at 18 V. When V_{EE} is increased by moving the potentiometer arm upward, at about 18.6 V (silicon UJT) the PN diode becomes forward-biased and emitter current flows. As I_E starts, holes are injected into the N area between B· and E, and this area becomes a good conductor. As I_E increases, the voltage-drop between B· and E de-

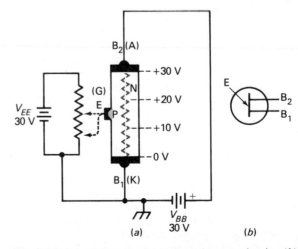

FIG. 9-26 (a) Unijunction transistor in test circuit. (b) Symbol.

creases, which is opposite to Ohm's law. For this reason the UJT is said to exhibit negative resistance, an effect similar to regeneration (Chap. 11), greatly speeding the increase of I_E to maximum and the fall of the B-E resistance to minimum.

When V_E is decreased to a low value, a point is reached at which the PN diode reverse-biases, and I_E decreases to zero as fast as it increased to maximum. This transistor is used in certain types of oscillator and pulse circuits, as well as to fire SCRs. The symbol of a UJT is shown in Fig. 9-26b.

9-14 SILICON CONTROLLED RECTIFIERS (SCRs)

Figure 9-27 illustrates one of several NPNP, or multiple-layer semiconductor, devices. This one is called a *silicon controlled rectifier* (SCR). Since there are three junctions, regardless of any reason-

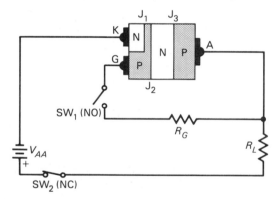

FIG. 9-27 SCR in a dc circuit.

able potential connected across it, there will be at least one junction which will be reverse-biased, limiting current flow to some negligible value. The basic current path is from V_{AA} to the cathode, K, through the device to the anode, A, through the load, and back to the source. (*Cathodes* give off electrons; *anodes* collect electrons.)

Assuming V_{AA} is some value below the forward breakover voltage, $V_{f(BO)}$, of the device, negligible current should flow. If the normally-open (NO) gate-electrode switch, Sw₁, is closed, the first P-type segment, acting like the base of an NPN BJT, is forward-biased through R_G and R_L, allowing current to flow through junction J_2. Since J_3 is forward-biased already, current flows through the SCR, K to A, and through the load. The gate (G)

switch, Sw₁, can now be opened, but current will continue to flow because J_2 loses control of its current carriers. Even reverse-biasing the gate will not stop cathode-anode current. It is necessary to reduce V_{AA} to almost zero or open Sw₂ to stop current flow. No current will flow if the source potential is reversed, even with Sw₁ closed. Therefore, with an ac source, the SCR rectifies ac to pulses of dc.

An ac lamp-dimming circuit operated by controlling the triggering voltage phase of an SCR is shown in Fig. 9-28. During the half cycle in which

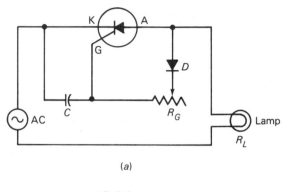

(a)

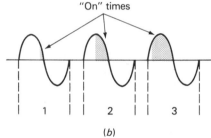

(b)

FIG. 9-28 (a) RC phase-control circuit produces (b) almost 180° of controlled ON time.

the SCR can fire, capacitor C charges through R_L and R_G until it builds up enough voltage to fire the SCR. If R_G is a high resistance, C charges slowly and the SCR is fired late, resulting in only a few degrees of current (cycle 1). If R_G is a medium value, C charges faster and current flows for perhaps 90° (cycle 2). With a low value of R_G, the capacitor charges almost instantaneously, the SCR fires immediately, and almost 180° of current flow is produced. The diode, D, prevents a reverse voltage from being applied to G when the ac polarity reverses.

One rating given SCRs is the maximum safe peak forward voltage (PFV). This is somewhat greater than the forward breakover voltage rating, and if exceeded, will usually damage the SCR. If reverse-direction voltage exceeds the *reverse breakdown voltage*, the SCR will go into thermal runaway and be destroyed.

If two similar SCRs with similar triggering circuits are connected in reverse, back to back, the circuit will conduct in both directions and both halves of each cycle will power the load circuit.

Another multilayer semiconductor device called a *silicon controlled switch* (SCS) not only has the two main electrodes and the gate of an SCR but also a *turn-off gate* which, when biased properly, can shut off the conduction between the main electrodes at any time. An SCR can only be shut off by reducing the ac voltage to a very low level.

9-15 TRIACS AND DIACS

Another breakdown semiconductor, or *thyristor* device, is the *triac*. It is a two-way SCR (Fig. 9-29*a*). Across the triac, the layers from terminals T_2 to T_1 are PNPN. When the source emf attempts to produce current flow downward through R_L, it is opposed by junction J_1, which is reverse-biased. When the emf reverses, J_2 prevents current flow. As T_2 starts to become positive, C begins to charge (shown dotted) through R and R_L. The voltage across C produces a forward bias for the NP junc-

tion next to T_1. The gate acts as a P-type base of an NPN transistor. A positive gate injects carriers across J_2, and current can flow across the top of the triac from T_1 to T_2. Once conducting, there is essentially no voltage-drop across the triac, and C discharges. The higher the resistance of R, or the greater the capacitance of C, the longer the delay of voltage buildup across C to fire the triac, controlling the average current flow.

As the source ac reverses, C starts charging with a negative polarity toward the gate. When the gate NP junction is biased with a negative voltage, carriers are injected into the adjacent junction and current flows across the bottom of the triac. The triac acts as two SCRs back to back, in reverse direction, both directions being under the control of one gate. This can be termed a *full-wave* control circuit, since it operates on both ac half cycles. The triac symbol is shown in Fig. 9-29*b*.

Another thyristor device, a *diac*, is a three-layer semiconductor (Fig. 9-30*a*), with equal doping in

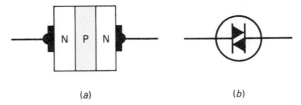

(a)   **(b)**

FIG. 9-30 (*a*) Diac. (*b*) Symbol.

both N layers (BJTs have heavy doping of the emitter). As an ac voltage develops across it, the diac reaches an avalanche point and breaks into conduction. The avalanche voltage is between 25 and 35 V in either direction. Diacs can be used to delay the firing of an SCR until the control voltage rises to about 30 V.

9-16 LIGHT-ACTIVATED SCRs

A light-activated SCR (LASCR), is similar to a light-activated NPN transistor. It is constructed with a glass lens on one side of the metal case surrounding the semiconductor layers (Fig. 9-31*a*). When light falls on the junctions, the device latches ON and cathode-anode current can flow A to K. The diagram of Fig. 9-31*b* is a possible circuit that will turn on a LASCR when it is light-activated. The two arrows pointing to the symbol indicate light and are always part of LASCR or other light-activated device symbols.

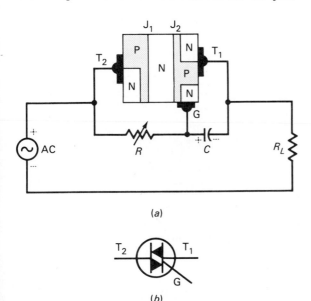

(a)

(b)

FIG. 9-29 (*a*) Triac and full-wave phase-control circuit. (*b*) Triac symbol.

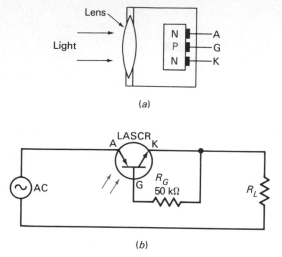

FIG. 9-31 (*a*) LASCR. (*b*) Light-operated LASCR circuits.

9-17 INTEGRATED CIRCUITS

When transistors were developed and took over essentially all of the low-power control and amplifying duties previously handled by vacuum tubes, the size and power requirements of electronic circuits decreased greatly. Then, when methods were developed to fabricate dozens to hundreds of thousands of transistors and diodes on tiny semiconductor chips in what are called *integrated circuits* (ICs), electronic equipment became very much smaller.

An IC may contain only a few MOS or BJT transistors and a few diodes to accomplish some single-circuit action. Or it may have under 100,000 active devices (large-scale IC, or LSI circuit) and form the heart of a highly complex microprocessor having over 100,000 units (very large-scale IC, or VLSI circuit) and be in a still more complex computer. ICs may have from 4 to 40 or more connector leads extending outward from them (Fig. 9-32). They may be plugged into special IC sockets, or may be soldered directly into specially drilled printed circuit boards.

It is possible to produce a variety of transistors in one IC chip, as well as diodes, resistors, and even low-capacitance-value capacitors. An IC by itself is rarely a complete circuit, but, when ICs are used in conjunction with external coils, capacitors, resistors, and transformers, they greatly reduce the

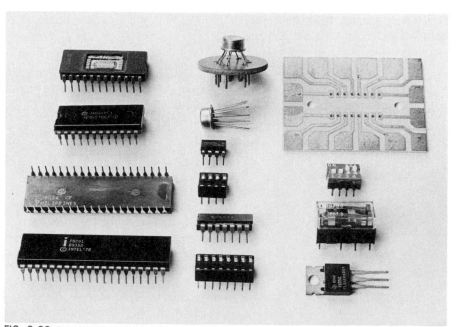

FIG. 9-32 Integrated circuits. Left column, top views of 24-pin ICs, at top showing actual chip at center with leads out to pins; bottom and top views of a 40-pin LSI DIP IC. Center column, round IC made to plug into a 9-pin round socket, round 8-lead IC, 8-pin bridge rectifier IC, 8-pin IC socket, 14-pin IC, 16-pin IC socket. Right column, experimental socket or IC adapter, 4-pole switch, PC board relay for a 16-pin DIP socket, voltage regulator IC.

size of communication receivers, test equipment, computers, etc.

Since ICs are so small, they are not able to handle much power because of the heat that would be developed in them. They perform all of their operations at a very low power level. If it is required that high power be developed, the output of the IC may be fed to external higher-powered transistor or vacuum tube amplifiers or control devices. Our discussions will be more on the circuits that might be found in ICs rather than the endless number of ICs themselves.

There are two basic types of ICs. One is a *digital* type used in computers and other digital circuits. All such devices must be able to change from fully on to fully off in fractions of a microsecond (in a few nanoseconds). The other type is the *linear* IC, having transistors with linear characteristic curves that allow them to be used in audio amplifiers. The number of circuit possibilities with ICs are endless. IC handbooks illustrate some of the standard combinations of circuits available off the shelf. In symbol form an IC is represented usually by an oblong block with numbered connections on two sides. They are termed *dual-in-line packages* (DIP).

| Test your understanding; answer these checkup questions.

1. A JFET is similar in operation to what other device? _____
2. Why is the JFET supply labeled V_{DD}? _____
3. In N-channel JFETs, at what forward emf and polarity would I_G begin to flow in silicon devices? _____ In germanium P-channel JFETs? _____
4. Under class A operation does I_G flow in a JFET? _____ Does I_B flow in a BJT? _____
5. What is the JFET pinch-off bias value? _____
6. What is the formula for μ for a JFET? _____
7. What is the formula for r_D in JFETs? _____ For r_D in MOSFETs? _____
8. What is the formula for g_m in JFETs? _____
9. In JFETs what is increased to pinch off the channel? _____
10. Why would most MOSFETs have a relatively low insulation-layer voltage tolerance? _____
11. For class A operation, which FET requires forward bias? _____ Which no bias? _____ Which reverse bias? _____
12. Which FET would have the lowest Z_{in}? _____
13. In UJTs, what is produced once the emitter voltage is raised to the start of I_B? _____ What effect does this have? _____
14. For what are UJTs used? _____

15. How many layers are there in an SCR? _____ What must be done to start current in an SCR? _____
16. Why are SCRs usually used in ac circuits? _____
17. What is the name of a single-unit solid-state two-way SCR? _____
18. Name three thyristor-type devices. _____
19. Which SCR is turned on by high-frequency radiations? _____
20. What are the four types of devices listed as being built into ICs? _____ _____ _____ _____
21. Could MOSFETs be built into ICs? _____ FETs? _____
22. What are the two types of ICs? _____ _____
23. How many microseconds (μs) are there in 45 nanoseconds (ns)? _____
24. If connected upside down in a circuit (D for S, S for D) would a JFET amplify? _____ Would a MOSFET? _____
25. Identify the symbols shown in Fig. 9-33 on a separate piece of paper.
26. Draw diagrams of the following circuits and check against those shown in the chapter. NPN CE

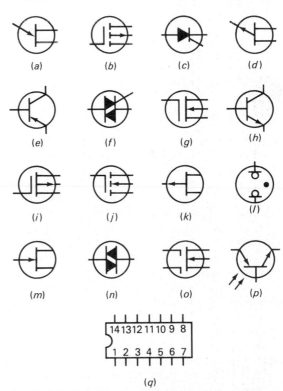

FIG. 9-33 Identify these symbols.

amplifier with fixed bias. NPN CE power amplifier with voltage-divider bias. NPN CB resistance-coupled amplifier. NPN emitter-follower. N-channel JFET amplifier. N-channel depletion MOSFET amplifier. N-channel enhancement MOSFET amplifier. SCR with RC phase control in ac circuit.

9-18 VACUUM DIODES

An evacuated glass bulb, or *envelope*, containing a metallic *cathode* that gives off electrons when hot and an *anode*, or *plate*, that accepts electrons when it is made positive, form a vacuum tube (VT) diode (Fig. 9-34). In this circuit the 5-V winding of the

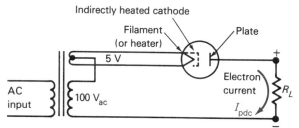

FIG. 9-34 VT diode used to rectify ac to pulsating dc.

transformer heats a tungsten, thoriated tungsten, or oxide-coated nickel wire-filament-type cathode. A 100 V ac from the other transformer winding is in series with the load resistor (R_L) and the diode. When the filament wire is hot it gives off a cloud of electrons, called a *thermionic emission*, that forms a *space charge* in the vacuum space between filament or cathode and plate.

During the ac half cycle that the plate is made positive in respect to the filament, space charge electrons are drawn across the tube to the plate, then flow through R_L, and back through the 100-V secondary to the filament again. On the other half cycle, when the plate is toward the negative end of the 100-V secondary, the negative plate repels electrons and no current flows in the diode. In this way the ac is converted to pulsating dc (pdc) through the load resistor. (A semiconductor diode produces the same result without requiring a filament winding or any cathode heat loss.)

Because the filament wire might heat and cool at the ac power frequency, resulting in a varying space charge, most tubes have an oxide-coated metallic cathode sleeve mounted around the filament (indicated dashed). It is the coated cathode that liberates the electrons. The filament, now known as a *heater*, indirectly heats the cathode sleeve. Since it requires several seconds to warm and cool the sleeve, the cathode surface temperature, with 60 Hz

ac, remains constant, developing a constant space charge. Heater cathodes are desirable because the heating can be done with commonly available power ac. Wire filaments may require dc to heat them.

Whereas a single semiconductor diode may only stand a reverse voltage (anode negative, cathode positive) of a few to a few hundred volts, vacuum diodes can be made which will withstand many thousands of *inverse peak volts* without an internal flashback or arc. However, several solid-state diodes in series will provide a high inverse peak voltage rectifier system also.

It is important to use the correct voltage across a filament or heater to obtain optimum tube life. If a tube is rated at 6.3 V, it will have a certain life expectancy in hours. If operated at 7 V, it will have a greater filament emission because of increased filament current, but cathode molecules boil off of the wire, shortening the life of the filament, and usually forming a darkened area on the inside of the glass envelope. Lower filament voltage will produce less emission but can increase the life expectancy of tubes when used in low-power applications. Note that the feed to the filament winding is into the center tap. This prevents the 5 V ac of the filament winding from being added to the 100 V ac of the other secondary so that the plate potential is 100 V ac, not 105 or 95 V.

Diodes act as one-way conductors and have a very low resistance. With only a few volts across a diode in a low-impedance circuit, enough current may flow to damage the diode. Practical circuits involving diodes always have some limiting resistance in the circuit. Any increase in *plate voltage* (V_p) across a tube produces an increase in the plate current (I_p) through it.

A high degree of vacuum is known as a *hard* vacuum. A vacuum tube that is gassy for any reason is known as a *soft* tube. Tubes may become soft because of leaks through the envelope, or because of unintentional overheating of the metallic elements in the tube, causing them to liberate gas molecules which are always present in metals. This usually produces a blue haze between cathode and anode of a gassy tube.

9-19 VACUUM TRIODES

When a vacuum diode has a wire meshwork built into it between cathode and plate, the meshwork is called a *grid*. A simplified cross-sectional view of such a *triode*, or 3-element, tube is shown in Fig.

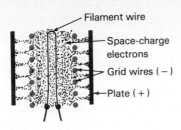

FIG. 9-35 Simplified cross section of a triode tube, showing relative element placement.

9-35. When the filament is heated, the space charge forms around it. Since the grid is usually dc-biased to some negative value, it tends to repel space-charge electrons and holds them close to the filament. The positive charge of the plate tries to reach in through the negative field set up by the grid wires and pull electrons over to the plate. If the grid is highly negative, or if its wires are close together, the plate potential may not be strong enough to pull electrons to the plate. Conversely, with widely spaced grid wires it requires a very high negative grid bias to stop all current flow to the plate. Closely spaced grid wires, or grid wires that are placed close to the filament, result in a tube with a high amplification capability. Widely spaced grid wires, or grid wires that are mounted relatively far from the cathode, result in tubes with low amplification values. Because of its ability to control plate current, the grid is known as a *control grid*.

The diagram of Fig. 9-36 shows a triode tube connected in an audio frequency (AF) amplifier circuit. The triode has two separate circuits. One is the input or grid circuit, consisting of the cathode (K), the bias or C battery, an input AF ac signal source (microphone), and the grid (G). The other circuit, the plate circuit, consists of the cathode, the plate (P), the output load (transformer primary),

and the plate supply, or B battery. Voltage changes in the grid circuit can change the plate circuit current, but changes in the plate circuit have no significant effect on the grid circuit. If a battery is used to heat the cathode, it is known as an A battery.

The amount of plate current, or I_p, is determined by the value of V_{bb} and also by the value of V_{cc}. The microphone is developing AF ac. This is added to the bias voltage of the C battery in the grid circuit, producing a vdc voltage which causes the I_p to vary at the AF ac rate.

The grid is as effective a controlling device as the gate of an FET. In the circuit shown the grid bias is -8 V. The plate circuit supply is $+200$ V. An I_p of 3 mA will be assumed. When V_{bb} is increased by 90 V, to 290 V, it is found that the I_p increases 1 mA. Then, returning to a V_{bb} of 200 V, it is found that reducing the grid bias voltage, V_{cc}, by 3 V also results in an I_p increase of 1 mA. This indicates that a 3-V change in the grid circuit is just as effective in controlling I_p as a 90-V change of plate circuit voltage is. This ratio of 90:3, or 30, is the *amplification factor*, or μ, of the tube, similar to the μ of FETs. By formula

$$\mu = \frac{dV_p}{dV_g}$$

where $d = $ "change in"

Actually, V_p is the voltage between cathode and plate, but may be the same voltage as V_{bb} or B+ when the plate load is a low-resistance transformer winding.

This triode has a μ of 30, but the voltage variation, or output voltage across the plate circuit load (transformer primary), will not be 30 times the signal voltage applied to the grid circuit. Actual voltage amplifications (A_v) or *gains*, equal to one-half to two-thirds of the μ value are normal. This triode should amplify an ac input signal by 15 to 20 times. The formula to compute voltage amplification is

$$A_v = \frac{\mu R_L}{R_L + R_p}$$

If 1 V of AF ac is developed by the microphone, this might produce a gain of 17, or a signal of 17 V across the plate load. If the transformer has a step-up ratio of 3:1, a 3×17, or 51-V ac signal output should result. This is an overall, or whole *stage* gain of 51.

If resistance coupling (Sec. 9-7) had been used as the output network, the gain of the stage would

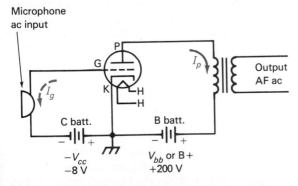

FIG. 9-36 Triode VT connected as an AF amplifier.

only be about 17 instead of 51. With resistance coupling output it would require two stages to provide a gain of at least 51. However, two 17-gain stages in *cascade* (one following the other) would provide a voltage amplification, or gain, of $A_v = 17 \times 17$, or 289. Because of the cost of transformers, it is more economical to add an extra resistance-coupled stage, particularly with transistors.

Plate current *saturation* occurs when the plate is positive enough to accept all of the space-charge electrons. If the grid is made sufficiently positive, it can also produce plate saturation. Any further positive potential on the grid can produce no higher I_p.

Except in special cases, the grid is not driven into the positive region, as this would produce grid current (I_g from G to K) which could affect the stability of the bias voltage if there is any resistance in the grid circuit.

Tubes are not made to withstand continual saturation currents. Such high I_p values would overheat the plate and melt a hole in it. Maximum plate power *dissipation* values are less than half the saturation value.

9-20 CHARACTERISTIC CURVES

How much will a variation of the V_g affect the I_p in a particular triode tube? To determine this, the manufacturer's V_g versus I_p curve (Fig. 9-37) can be used. If this curve is plotted with no load resistance in the plate circuit, it is called a *static* characteristic curve. If a load is used, a *dynamic* characteristic curve results.

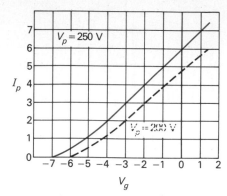

FIG. 9-37 Static curve of I_p versus V_g with a constant V_p.

This graph is read: The V_p is constant at 250 V. With 0 volts on the grid (no bias), $I_p = 6$ mA; with −2 V on the grid, $I_p = 4$ mA; with −5 V, $I_p = 1$ mA; with −6 V, $I_p = 0.5$ mA. This tube is completely cut off by applying a −7-V bias to its grid, provided 250 V is used on the plate. With a lower V_p, less bias is required to attain I_p cutoff (dashed curve).

The negative voltage value required to produce complete I_p cutoff for any triode may be found by dividing the V_p by the μ of the tube:

$$V_{co} = \frac{V_p}{\mu}$$

where V_{co} = cutoff bias voltage
V_p = plate voltage
μ = amplification factor of tube

If the same formula is rearranged, the μ of the tube, the curve of which is shown, would be approximately $\mu = V_p/V_{co}$, or 250/7, or 35.7. This is a "medium-μ" triode. Low-μ tubes have amplification factors between 3 and about 20, whereas high-μ tubes range from about 50 to 150. Note that with different plate voltages the μ value changes somewhat. For the 200-V plate voltage curve, the μ of this tube decreases to 33.3. The μ of a VT is the same idea as the μ of an FET.

It is usually not desirable to operate a tube in the bent portion of the curve, near cutoff. Figure 9-38 shows a curve of a class-A biased amplifier with two input signals of different amplitudes. The first results in I_p variations having the same waveshape as the input signal because all the operation is under a straight portion of the curve. The negative half of the second, higher-amplitude input signal operates in the bent region of the curve, which

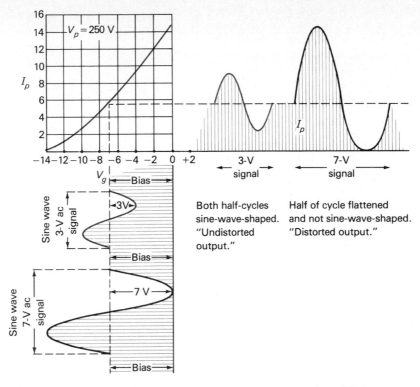

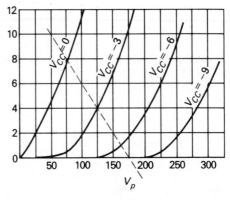

FIG. **9-38** Grid-voltage variations must be under straight portion of $V_g I_p$ curve to produce undistorted I_p variations.

produces a flattening of the plate-current waveshape. The voltage-drop across the plate load with the higher-amplitude signal will not have the same waveshape as the voltage applied to the grid. The tube is distorting the signal.

Another form of graph is the $V_p I_p$ family of curves. This plots V_p against I_p with the grid bias constant for each curve (Fig. 9-39). The dashed line is called a *load line*, and has been drawn through a point where a chosen plate voltage of 125 V intersects with the -3-V bias curve and produces 4 mA of I_p. If a +3-V input signal is applied to the grid, it should reduce the grid voltage to zero and increase the I_p to perhaps 8 mA. The load line is drawn from the $V_{cc} = 0$ and 8-mA-I_p point through the quiescent or no-signal-input point where $V_{cc} = -3$ and $I_p = 4$, and is extended down until it intersects the $I_p = 0$ line. The load-line intersection points on the three curves show that with a 6-V ac peak-to-peak (p-p) signal input, the peak output voltages across the plate load would be from about 80 V to about 163 V_p, for a total p-p output of 83 V. The result is a stage voltage gain of 83/6, or

13.8, with this tube operated under these conditions.

When a resistor (5- to 100-kΩ) is used as the load in an amplifier circuit, the plate current is reduced greatly, and the dynamic curves plot out much flatter. As the grid is driven less negative by a signal voltage, I_p increases, the voltage-drop

FIG. **9-39** $V_p I_p$ curves for different values of grid bias. Possible load line shown dashed.

across R_L increases, and the instantaneous plate voltage (v_p, P to K) is reduced. Thus, with an input signal the v_p will be constantly shifting although the V_{bb} would not be. It is this varying voltage-drop across R_L that is the amplified output voltage from such a resistance-coupled output circuit. Actually, when a transformer is the load, the v_p is also varying due to the counter emf developed across the primary when the i_p in this winding is changing.

Note how similar a triode VT is to the operation of a triode type JFET. Grid and gate, cathode and source, and plate and drain all have the same functions. Even with BJTs, grid and base, cathode and emitter, and plate and collector are similar in function. N-channel FETs have the same power supply and bias polarities. NPN transistors have similar power-supply polarities, but require forward bias for the gate circuit in most circuits.

9-21 OUTPUT IMPEDANCE

The power supply and the series-active device (transistor or VT) appear to the load in the output circuit as a dc source when there is no signal input. With an ac input signal the supply and device appear to the load as a varying I_p dc source with an ac component. The impedance of this cathode-plate circuit for a VT (SD for an FET) can be found by Ohm's law by

$$Z_p \quad \text{or} \quad r_p = \frac{\Delta V_p}{\Delta I_p} \quad \text{or} \quad \frac{\Delta v_p}{\Delta i_p} \quad \text{or} \quad \frac{dv_p}{di_p}$$

The uppercase letters indicate either dc, or in this case, rms ac values. The lowercase letters indicate instantaneous, or usually peak values. The Greek Δ or the italicized d indicate "small change in."

The plate impedance of a triode can be determined from a V_pI_p family of curves such as Fig. 9-39. Starting at the intersection of the load line and the $V_{cc} = -3$-V curve, for a dV_p of 25 V (100 to 125 V) there is a dI_p of 0.002 A (2 to 4 mA). The impedance from the formula would be dv_p/di_p, or 25/0.002, or 12,500 Ω. For maximum power output the load must appear to the active device plus power supply to have 12,500 Ω also. However, for best undistorted voltage output the load impedance should be 2 or 3 times the Z_p. Different load impedances will be obtained if different parts of the same curve are used, because the curve is not completely linear.

Test your understanding; answer these checkup questions.

1. In Fig. 9-33, what would be the peak voltage value of the pulses developed across R_L? _____
2. What is the cloud of electrons around a hot VT cathode called? _____
3. Is a tube that has a heater, cathode, and anode a diode or triode? _____
4. Where is a grid placed in a VT? _____
5. What are the names of the three elements of a triode? _____ _____ _____
6. How is the control grid constructed to produce a high-amplification VT? _____
7. Which circuit in a triode controls the other? _____ What might be considered a third circuit? _____
8. What are two formulas to compute amplification factor? _____ _____
9. How can stage gain exceed the μ of the tube used in an amplifier? _____
10. Why should AF amplifiers not operate in the bent portion of the V_gI_p curve? _____
11. What indicates the output signal voltage when using a V_pI_p family of curves? _____

9-22 TRANSCONDUCTANCE AND POWER

As with FETs, one of the factors by which a VT can be judged is its *transconductance*, or *mutual conductance* (g_m). As with the FET formula

$$g_m = \frac{di_p}{dv_p}$$

As an example, in Fig. 9-38 the di_p along the 100-V V_p line for a dV_{cc} from -3 to 0 V is a dI_p of about 9.5 mA. Substituting in the formula, $di_p/dv_p = d0.0095/d3$, or 0.003167 S, or 3167 μS.

Another formula for transconductance is

$$g_m = \frac{\mu}{r_p}$$

and from this, $r_p = \mu/g_m$, and $\mu = g_mr_p$.

When a high-voltage amplification is required a high-μ triode or FET should be used. To produce a large current variation, in the primary of a transformer for example, the active device should have a high g_m. Practical g_m values range from about 500 to over 15,000.

Computation of the power output of an amplifier can be made by measuring the voltage developed across the load and the current flowing through it.

The product of these two values will be the power developed in the load ($P_o = V_L I_L$). If the resistance of a load is known, the common formulas $P = I^2 R$ or $P = V^2/R$ may be used.

A fairly accurate formula for the output power of a triode device is

$$P_o = \frac{\mu^2 V_g^2 Z_L}{(Z_L + Z_p)^2}$$

where P_o = output power, in W
μ = amplification factor
V_g = rms signal voltage on grid
Z_L = load impedance, in Ω
Z_p = plate impedance of tube, in Ω

9-23 SECONDARY EMISSION

The *primary emission* in a vacuum tube is the electron emission from the cathode. When the plate is highly positive, it accelerates electrons traveling toward it so much that, when they strike, each electron may dislodge one or more electrons. These electrons moving from the plate out into the vacuum form a *secondary emission*.

The effect of secondary emission may be great enough to produe a small negative space charge (cloud of electrons) near the plate, interfering with normal electron flow to the plate.

If the control grid of a triode is driven positive, some of the secondary-emission electrons may be attracted to that element instead of to the plate. This is undesirable, since only those electrons that flow through the plate-circuit *load* can produce usable output.

The grid of a triode is normally held negative and does not attract any secondary electrons. For this reason, secondary electrons in triodes are eventually returned to the plate and can flow through the load.

9-24 TETRODE TUBES

To obtain a higher amplification in a triode, either the plate can be backed away from the cathode, or a finer grid meshwork can be used, necessitating relatively higher V_p to maintain I_p. Also, in triodes a relatively high *interelectrode capacitance* exists between plate and grid because of their proximity to each other. This capacitance can prevent the tube from amplifying properly at RF frequencies.

To overcome some of the difficulties that arise

when the μ of a triode is high, and to reduce grid-plate capacitance, the *tetrode*, or 4-element tube, was developed. In the tetrode, the grid is kept close to the cathode and the plate is moved outward to increase the μ. Then, between grid and plate, a second *screen grid* is added (Fig. 9-40).

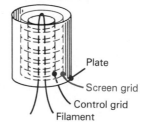

FIG. 9-40 Relative positions of the elements in a filament-type tetrode tube.

The screen grid is connected to a static, or unvarying, positive dc potential of 100 V or more. The positive field of the screen grid draws space-charge electrons from the cathode through the control grid wires. This puts the space-charge electrons in such a position that the plate potential can attract them. Thus, relatively high plate current and high amplification are possible. The screen grid is dynamically (ac) connected to the cathode with a *bypass* capacitor (Fig. 9-41). The diagram of a tetrode

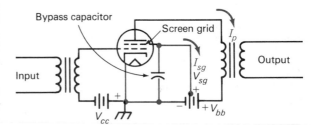

FIG. 9-41 Screen-grid circuit shown in colored lines.

circuit is the same as that of a triode except for the addition of the screen grid and its bypass capacitor.

Since the control grid and plate are far apart, interelectrode capacitance between them is greatly reduced. More important, bypassing the screen grid to cathode develops an electrostatic shield between plate and grid. With a higher g_m and less plate-to-grid interaction due to capacitance, a tetrode produces more stable operation and amplification.

Triode tubes are rated by their μ values, but since the dc voltage applied to the screen grid (V_{sg})

can control the amplification of the tube, tetrodes (and pentodes, Sec. 9-25) are rated in g_m only.

The positive screen grid has the effect of accelerating the electrons on their way to the plate. As a result, the secondary-emission current from plate to screen becomes rather high. In fact, if the V_p falls below the V_{sg}, secondary-emission current from the plate may exceed the current to the plate.

Tetrodes have very high plate-impedance values, ranging from 50,000 to 100,000 Ω. This presents difficulties in load-impedance matching, but only if a transformer is used as the load.

As long as the V_p is greater than the V_{sg}, the I_p in a tetrode is nearly independent of the V_p. Doubling the V_p will raise the I_p very little, whereas in a triode doubling the V_p should double the I_p. However, grid-voltage variations, either control grid or screen grid, can control I_p. The control grid is perhaps 50 times as effective as the screen in controlling I_p in tetrodes.

The disadvantage of the secondary-emission electrons returning to the screen grid in the original tetrodes may be overcome in two ways. First, the screen and control grids can be so placed in the tube that the screen-grid wires are in the electron shadow of the control grid (Fig. 9-42). Most of the

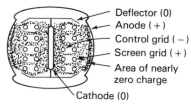

FIG. 9-42 Placement of cathode, grids, deflector, and anode in a beam-power tetrode.

electrons leaving the space charge deviate outward from the negative control grid and do not converge again until they have passed the screen grid. This forms strong streams of electrons flowing to the plate. Second, the tube has two deflector plates at cathode potential, which help to beam electrons to the plate.

The beaming results in many electrons hitting the plate and some secondary emission. Secondary-emission electrons, however, are swept backward

toward the plate again by the advancing beam of electrons. The net result is an area near the highly positive plate where there is virtually no charge, which lowers the speed of electrons flowing to the plate, thereby reducing secondary emission.

The shading of the screen grid is not complete. Some I_{sg} flows, but it is usually less than one-tenth of the I_p value.

Practically the only tetrodes used today are beam-power types, usually made for operation in RF circuits, and specially designed to reduce interelectrode capacitance so that little or no *neutralization* (Chap. 15) is required. The symbol in diagrams is the same as for other tetrodes.

9-25 PENTODE TUBES

Secondary-emission electrons moving from plate to screen grid in a tetrode is undesirable. The addition of a zero-charged third grid, called a *suppressor grid*, between screen and plate decreases the velocity of the electrons approaching the plate. Any secondary emission that does occur has insufficient energy to move back across the zero-charged suppressor grid field. In this way secondary-emission current from plate to screen grid is stopped. This 5-element tube is known as a *pentode*. Placement of the elements in a pentode is shown in Fig. 9-43.

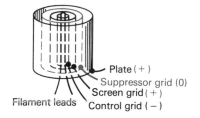

FIG. 9-43 Relative positions of pentode elements.

The only difference between diagraming pentode and tetrode amplifiers is the inclusion of the suppressor grid, which is normally connected to the cathode, often internally by the manufacturer. A pentode tube in an amplifier circuit is shown in Fig. 9-44a.

Besides having higher gain than tetrodes, pentodes have better shielding between plate and control grid, requiring no neutralization when used in high-frequency circuits. Pentodes find use as voltage amplifiers, as in receivers. Plate impedances of pentodes vary from about 100,000 to over 1,000,000 Ω, which makes load matching difficult with transformers. Compare the pentode V_pI_p

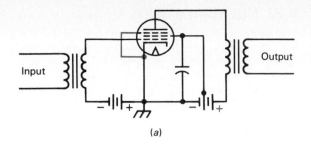

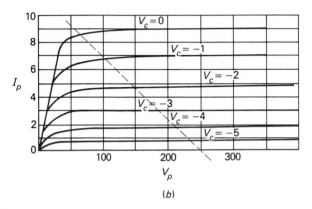

(b)

FIG. 9-44 *(a)* Suppressor grid circuit in color lines. *(b)* $V_p I_p$ curves for a pentode and dashed load line.

curves (Fig. 9-44*b*) with those of a triode (Fig. 9-38). The dashed line is a possible load line centered on $-V_{CC} = -2$ V, V_p of 250 V, and $I_p = 4.7$ mA.

For some circuits a tube is required that does not drop its I_p sharply to cutoff, as most tubes do. Less sharp cutoff can be produced by close-spacing the control-grid wires at top and bottom, but wide-spacing them near the center (Fig. 9-45). With no bias, electrons move freely to the plate between all grid wires. With a little negative bias, the closely spaced wires cut off the plate current through them, but the widely spaced wires allow electrons to flow between them. A very high bias is required to cut

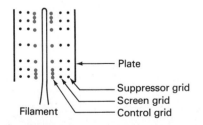

FIG. 9-45 Spacing of control and other grid wires in a remote-cutoff pentode tube.

off plate current completely in such *remote-cutoff, variable-μ,* or *supercontrol* tubes. The $V_g I_p$ curves for both sharp-cutoff and remote-cutoff tubes are shown in Fig. 9-46. (Using remote-cutoff pentodes for AVC is described in Chap. 18.)

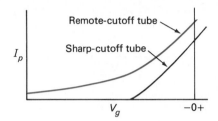

FIG. 9-46 Comparison of $V_g I_p$ curves of sharp- and remote-cutoff pentodes.

In the earlier days of radio, tubes were large. In some circuits one tube interacted with others because of capacitance between them. It was often necessary to place an aluminum shield around each tube to prevent this interaction. Later, metal tubes were developed and the metal envelope was used as the shield. More recent miniature tubes are so small that they may not interact in many circuits. If they do, small metal shields are placed around them and are grounded as in earlier days. Interstage shields may also be required in transistor systems.

9-26 BASIC VT CIRCUITS

Just as BJTs can be operated in common-emitter, common-base, and common-collector circuits, VTs can be operated in common-cathode, common-grid, and common-plate circuits. The input and output impedances and voltage phases are similar for both VTs and transistors.

The common-grid circuit is known better as a *grounded-grid* circuit (Fig. 9-47). The input impedance is relatively low (a few hundred ohms) but the output impedance is relatively high. A difficulty in this circuit is feeding the heater-cathode (or filament), since the cathode is above ground potential.

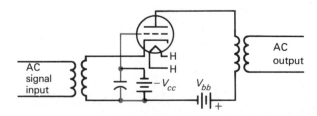

FIG. 9-47 Grounded-grid VT amplifier.

If the voltage between heater and cathode is more than a few volts, the heater insulation may break down and short out the input voltage.

The common-plate is known as a grounded-plate, or usually as a *cathode-follower* (Fig. 9-48). The

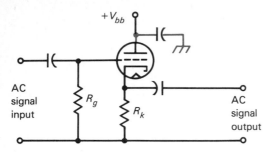

FIG. 9-48 Cathode-follower amplifier.

input impedance can be very high. The output is somewhat less than the value of the load resistor, which may be about 600 Ω. There is no voltage gain in this circuit, but power gain is possible. The circuit finds considerable application as an impedance-changing device, from a high-input to a relatively low-output impedance.

Whereas the input-output voltage phase change of common-cathode amplifiers is 180°, both grounded-grid and cathode-followers have input and output circuit voltages in phase, which is an advantage sometimes.

9-27 MULTIUNIT TUBES

The pentode has the greatest number of grids of the normal amplifying tubes. However, there are tubes having more than three grids. A hexode (6-element) tube has four grids. A heptode (7-element) has five grids. An octode (8-element) has six grids. These tubes are used in special circuits having functions other than mere amplification. In a pentagrid tube (Fig. 9-49), the cathode, G_1, and G_2 are used in conjunction with one frequency. The cathode and G_4 are used in conjunction with a second frequency. The current flowing from cathode to plate will have components of both frequencies in it, producing a third *difference-frequency* component and a fourth *sum-frequency* component, in the I_p.

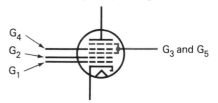

FIG. 9-49 Symbol of a pentagrid converter tube.

To conserve space in equipment, multiunit tubes are used (Fig. 9-50). Some of these are:

- *Twin, or duodiodes*. Two diodes in one envelope with common or separate cathodes, such as a seven-pin miniature type 6AL5.
- *Twin triodes*. Two triodes with either a com-

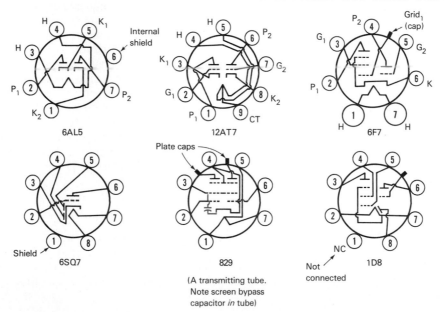

FIG. 9-50 Pin connections (bottom view) of several types of filament and heater-cathode tubes.

mon cathode or separate cathodes, such as a 12AT7 with a nine-pin noval base.

- *Triode-pentodes.* A triode and a pentode with a common cathode, such as a 6F7 with a grid connection at the top of the tube.
- *Duplex-diode-triodes.* Two diodes and one triode with a common cathode, such as a 6SQ7 with an octal base.
- *Twin tetrodes.* Beam-power tetrodes with common cathodes and screen grids, such as an 829 with plate leads coming out of the top.
- *Diode-triode-power pentode.* A diode, triode, and pentode using a common filament, such as a 1D8, battery-portable tube.

Other multiunit tubes have triple diodes, triple triodes, etc.

9-28 TUBE RATINGS AND DATA

If the maximum ratings of a vacuum tube are exceeded, damage or malfunction may occur. Tube ratings are found in manufacturers' manuals. Some examples are:

- *Control grid-to-plate capacitance*, in pF, if high, causes undesirable feedback of energy.
- *Input capacitance*, in pF, of grid and all other elements to cathode.
- *Output capacitance*, in pF, plate to cathode.
- *Heater voltage*, dc or rms ac value for normal cathode heating.
- *Maximum dc plate-supply voltage*, if exceeded, may produce arc-over.
- *Maximum peak positive-pulse voltage*, the probable plate-cathode arc-over voltage.
- *Maximum negative-pulse plate voltage*, if exceeded, may cause arc-over in the tube.
- *Maximum screen-grid voltage*, if exceeded, may overheat screen wires or plate.
- *Maximum peak negative control-grid voltage*, if exceeded, may cause internal arc-over.
- *Maximum plate dissipation*, watts of heat plate can withstand safely.
- *Maximum screen-grid dissipation*, watts of heat screen-grid wires can withstand safely.
- *Maximum dc cathode current*, total grid, screen-grid, suppressor grid, and plate currents the cathode can continually deliver.
- *Maximum peak cathode current*, maximum short-term cathode current.
- *Maximum control-grid circuit resistance*, if exceeded, may result in loss of control by the grid over other tube element currents due to reverse grid current.

9-29 RECEIVING AND TRANSMITTING TUBES

Diode, triode, tetrode, and pentode tubes are used in both transmitters and receivers. The difference between a transmitting and a receiving tube is generally in physical size and ruggedness of element construction. Receiving tubes are small and usually have all the element connections attached to metal pins at the base of the tube.

Transmitting tubes are larger and have heavier filaments for greater electron emission. Larger plates accommodate greater I_p and dissipate greater power in heat. The plate connection is usually brought out the top of the tube to provide lower grid-plate capacitance and maximum insulation between the plate and other elements. Plate voltages may be 500 to 10,000 V.

In some transmitting tubes the metal plate is the outside envelope of the tube, with fins attached to it (Fig. 9-51). A forced draft of air rapidly dissipates heat developed on the plate. Such tubes are said to be *air-cooled*. Heat on the plate of small tubes must be *radiated* from the plate and through the envelope, resulting in less rapid dissipation of heat.

Some high-power tubes are cooled by passing cold water over their externally constructed anodes. The resulting warmed water is recirculated by a pump through a cooling system. If water temperature increases excessively, or if water pressure drops, the voltage to the plate is automatically removed to protect the tube from overheating. Distilled water must be used in such a system. If it becomes impure, its resistance lowers, which is shown by a meter, indicating replacement is necessary. A water-cooled tube having fins tends to boil the water coolant and is known as a *vapor cooling*

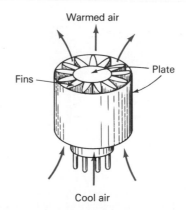

FIG. 9-51 Air flowing over metal fins attached to external plate dissipates heat of plate.

system. The steam is condensed, cooled, and returned to the system.

Older tubes operated up to about 30 MHz. Newer miniature and subminiature tubes operate up to 500 MHz. Above this, specially constructed *lighthouse* triodes are used. They have short leads (little inductance) and the spacing between cathode and plate is small, allowing *transit time* (time taken by electrons to move from cathode to plate) to be minimized. Special VHF and UHF tubes, such as the *nuvistor*, are useful up to 1000 MHz.

A few of the thousands of different kinds of vacuum tubes that have been produced are shown in Fig. 9-52. Cathode-ray, magnetron, and other special tubes are discussed in other chapters.

Improper operation of a tube may be indicated by excessive or insufficient plate, screen grid, or control grid voltages or currents. Cathode surfaces may lose emission, or filaments may burn out. Gas may develop in the tube. Internal elements may heat, warp, and short together. Internal element welds may be jarred loose and open-circuit that element or short it to another element.

Excessive I_p may be caused by excessive V_{sg}, V_p, insufficient or positive grid bias, or gas in the tube.

Insufficient I_p may be caused by low V_p or V_{sg}, excessive negative grid bias, open control or screen-grid circuits, low cathode emission, or gas in the tube.

Filament or heater failure may be due to loss of surface emission, jarring, or excessive V_f.

A blue glow in a tube indicates gas. Internal sparking indicates excessive voltages, loosened elements, or parasitic ac oscillations.

The majority of high-power triodes and tetrodes today have air-cooled anodes, with ceramic insulation between the cathode and grid, grid and screen grid, and screen grid and plate. Few are made now with glass insulation.

9-30 GASEOUS TUBES

If a diode tube is evacuated and then some liquid mercury or an inert gas is inserted, a gaseous diode is produced. Such gas-filled tubes are called soft tubes, although the word "soft" here does not indicate faulty operation.

In gaseous diodes, cathode electrons are emitted

FIG. 9-52 Electron tubes. Top, from left, three octal (8-pin) metal tubes, a loctal glass tube, three 7- or 9-pin glass miniature tubes, two subminiature tubes, and a nuvistor high-frequency tube. Bottom, vacuum phototube, glass octal amplifier tube, gaseous VR tube, two low-power lighthouse tubes, air-cooled anode lighthouse tube, and air-cooled tetrode transmitting tube.

as in vacuum tubes, but they do not go directly to the plate. They are attracted toward the plate but soon strike a gas molecule. With mercury vapor, if the difference of potential between cathode and plate exceeds 15 V, electron velocity is sufficient to dislodge one or more electrons from gas molecules. This splits the gas molecule into a positive ion (molecule minus one electron) and a free electron. The relatively heavy positive ion starts toward the filament, and the free electron toward the positive plate. The positive ion moves only a short distance before an electron from the cathode fills in the hole produced by the freed electron, neutralizing the ion. It is no longer attracted to positive- or negative-charged elements.

The free electrons either travel directly to the plate or ionize another molecule as they move toward the positive element. This electron movement from ion to ion is continued, resulting in a current of electrons from filament to plate.

An advantage of mercury vapor diodes is their efficiency. Since electrons travel only short distances, they cannot accelerate and strike the plate with great force. Thus, little heat dissipation is produced at the plate. These tubes require little ventilation, carry relatively high currents, and have a relatively constant voltage-drop across them (Chap. 10). One gaseous diode used argon gas, had a coiled filament wire, a carbon plate, was known as a Tungar bulb and was used in battery-chargers. Gaseous tubes glow blue with mercury vapor, and purple, pink, or orange with other gases.

If a cylindrical third "grid" element is inserted between cathode and plate (Fig. 9-53a) so that an electrostatic charge on it will affect the ionization potential of the gas between cathode and anode, it is possible to control plate current in the tube. If the grid is made negative in respect to the cathode, the negative electrostatic field produced in the region between cathode and grid will counteract the positive charge of the plate, resulting in lower electron velocities and less tendency to ionize the gas. It will now be necessary to increase the positive potential on the plate to produce ionization. The value of the negative charge on the grid determines

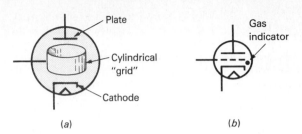

FIG. 9-53 (a) Cylindrical "grid" and (b) symbol of a thyratron.

the ionization potential of the gas. Once conducting, these tubes will not de-ionize until the plate potential drops below the extinction voltage of the gas used.

Figure 9-53b is a symbol of a 3-element *gaseous triode*, or *thyratron* (operates somewhat similar to an SCR). There are also 4 element gaseous control tubes. The dot in the symbol of a tube indicates gas.

Test your understanding; answer these checkup questions

1. What are two formulas to determine Z_p or r_p? _____ _____ To determine g_m? _____ _____
2. A 15-μ tube with a 5-kΩ Z_p has a 2-V rms input signal and is operating into a 10-kΩ load. What is the power output? _____
3. Through what must electrons flow to produce usable output power from a triode? _____ From a tetrode? _____
4. What does a screen grid reduce? _____ What two things does it increase? _____ _____
5. What is the only tetrode used today? _____
6. What does a suppressor grid suppress? _____
7. What is a 5-element tube called? _____
8. What are two other names for remote-cutoff tubes? _____ _____
9. With VTs, what is the meaning of octode? _____ Of heptode? _____ Of hexode? _____
10. In what two basic ways do transmitting tubes differ from receiving tubes? _____ _____
11. To what is a thyratron similar? _____ What would it be used for? _____

ACTIVE DEVICES QUESTIONS

1. What are the three basic passive devices? The three active devices?
2. What is the basic limitation of semiconductors in comparison with vacuum tubes? Their basic advantages?
3. In what direction do electrons flow in respect to the arrowheads shown on semiconductor symbols?
4. What are materials called that have a limited number of free electrons in the conduction level at room temperature?

5. If electrons drop to the valence level, what happens?
6. What is the name for semiconductor crystals that have perfect lattices?
7. What type of silicon has a free electron in its lattice? What type lacks an electron at some point in the lattice?
8. What makes a semiconductor hole move?
9. What is the electric charge of an N-type semiconductor? Of a P-type?
10. What type of material is used in most semiconductor devices?
11. Which area is the cathode of an NP diode? What relative potential is applied to the cathode to make current flow?
12. If reverse current flows in a diode, what carries the current?
13. In the forward-biased condition, what TC does a diode have? In the reverse-biased direction, if $V_{(BR)r}$ is over 5 V?
14. What diodes make use of their reverse breakdown voltage?
15. Why does a voltage-variable capacitor diode have less capacitance if V_r is increased?
16. What is different in the current characteristic of a tunnel diode? How is this produced?
17. What is another name for a fast-acting Schottky diode?
18. What does LED mean? SSL?
19. Why are LEDs not made with silicon or germanium?
20. Why is a resistor always in series with an LED?
21. What is the useful output of a photovoltaic diode?
22. Which operates faster, a photojunction or a PIN device?
23. What is inside a lighttight opto-isolator?
24. Name the thinnest element in a BJT. Name the other two elements.
25. If a voltage is applied between E and C of a BJT, why does no I_C flow? What must be done to make it flow?
26. Will a BJT operate if E and C leads are reversed?
27. Name the first type of transistor developed. Is it a BJT?
28. What takes the place of the forward bias of a BJT in a phototransistor?
29. What potential is applied to the collector of an NPN transistor? Of a PNP transistor?
30. What is the current gain of a BJT if a $d2$-mA I_B produces a $d0.16$-A I_C in a common-emitter circuit?
31. How do A_i and h_{fe} differ?

32. What are the two dc current paths through an NPN BJT?
33. Which might be considered the better, fixed or self-bias? Why?
34. Draw a diagram of a BJT common-emitter amplifier using voltage-divider bias.
35. What is a heat-sensitive resistor having a $-$TC called?
36. Where is the cutoff frequency of a common-emitter BJT?
37. What is the gain-bandwidth product of a common-emitter BJT?
38. What can be determined with a dc load line of a BJT?
39. Using the curves of Fig. 9-17, if the load were 3 Ω and V_{CE} is 10 V, what would be the quiescent value? What approximate total BE current excursion could be tolerated? What I_C excursion would this produce? What would be the power output? The dc input power?
40. The formula $\alpha = I_C/I_E$ is used with what type of amplifier circuit? What does it indicate? Can it exceed 1? Can the power gain exceed 1?
41. What is the frequency called at which the gain of a CB amplifier drops 3 dB below its 1000-Hz value?
42. Draw a diagram of a resistance-coupled common-base amplifier.
43. What is another name for a common-collector amplifier? What are two uses for it?
44. Draw a diagram of a common-collector amplifier.
45. Which amplifier (CE, CB, or CC) has: Highest Z input? Lowest Z input? Greatest voltage gain? Greatest power gain? 180° signal phase inversion?
46. What are the three basic types of FETs?
47. What are the names of the three terminals on a JFET?
48. What is the amplification-factor formula for a JFET?
49. What is the dynamic-drain-resistance formula for a JFET?
50. What is the transconductance formula for a JFET?
51. What are the three types of JFET amplifier circuits that correspond to the CE, CB, and CC circuits of BJTs?
52. Draw a diagram of a 3-stage amplifier having a CS, a CG, and a CD stage.
53. What other type of JFET is used besides the N-channel? How does it differ?
54. What is the main difference between JFETs and MOSFETs?
55. Does dc gate current ever flow in a JFET? In a MOSFET?
56. What type of MOSFET requires reverse gate bias to stop I_D flow? What type does not?
57. What type of MOSFET allows two different input signals to control the I_D?
58. Would a MOSFET amplify if its S and D connections were reversed?

59. How does N-channel MOSFET circuitry differ from P-channel?
60. What is a CMOS device?
61. What is a VMOS device? What are its uses and advantages?
62. What are the uses of a UJT?
63. Draw a diagram of an N-channel enhancement MOSFET amplifier.
64. What is an SCR? How is it turned on? Turned off?
65. Draw a diagram of an SCR used to vary the current through a load. What determines the effective load current value?
66. How does an SCS differ from an SCR?
67. What is a triac?
68. What does a diac do?
69. What is a LASCR and what causes it to conduct?
70. What is the determining point between LSI and VLSI?
71. What is the device called that has many transistors and diodes in it and usually performs many functions?
72. What are the names of the parts of a semiconductor diode? Of a VT diode?
73. What VT device operates like a low-power FET? What are its element names?
74. Draw a diagram of a triode stage capable of amplifying the weak ac voltages from a microphone.
75. What is the function of an A battery? Of a B battery? Of a C battery?
76. What is the formula for the μ of a triode? The A_v of a stage?
77. What is the cutoff bias voltage for a 20-μ triode having a V_p of 2500 V?
78. What may happen to a VT if operated in a saturated condition?
79. From the graph of Fig. 9-39, what would be the approximate quiescent bias voltage if V_p is 125 V and I_p maximum is to be 8 mA?
80. What is the P_o of an active device if it has a μ of 30, V_{in} rms of 4 V, R_L of 5 kΩ, and Z_{out} of 3 kΩ?
81. What is a 4-element VT called? What is the 4th element called? Why might it collect secondary emission?
82. What type of screen-grid tube is used mostly in transmitters?
83. What is the name of a 5-element tube? The name of the 5th element, and what does it reduce? Where will these VTs usually be found?
84. Draw a diagram of a basic tetrode amplifier.
85. Do characteristic curves of tetrodes and pentodes resemble those of BJTs or FETs?
86. Into what three basic circuit configurations can BJTs be connected? FETs? VTs?
87. What semiconductor device does the same work as a pentagrid converter?
88. Draw a diagram of a grounded-grid VT amplifier and a cathode-follower amplifier.
89. Name three ways in which heat is removed from the plate of high-power VTs?

10

Power Supplies

This chapter is a study of basic half-wave and full-wave rectifier circuits and the use of *C* and *L* as filters. Solid-state, high-vacuum, and mercury-vapor rectifier circuits are used. Voltage-multiplying circuits, as well as shunt- and series-regulating circuits are also explained. Switching and 3-phase type power systems are discussed.

10-1 POWER SUPPLIES

Power to operate electronic equipment may be obtained from a variety of sources. Batteries can produce a dc emf by chemical action. Photons of heat or light from the sun can be converted to dc electric energy by photocells. Fuel cells combine hydrogen and oxygen gases in an electrolyte to produce a dc emf. A fossil-fuel engine or a fall of water can rotate dc or ac generators (Chap. 27).

The dc sources are often able to operate electronic equipment directly, although some means of regulating or maintaining a constant emf under changing load conditions may be necessary. The most generally available energy, alternating current, must be changed (rectified) to a pulsating dc, which in turn must be smoothed (filtered) to a nonvarying voltage. The resultant dc may also require voltage regulation to operate electronic circuits properly.

Generally speaking, the term "power supply" usually means a rectifier-filter system that converts ac to pure dc. There are many different power-supply circuits that may be employed to do this. The basic components used for the simpler circuits are transformers, rectifiers, resistors, capacitors, and inductors. More complex regulated supplies may add transistors or triodes as voltage-sensing and -controlling devices, plus zener diodes or VR tubes to establish reference voltages.

10-2 HALF-WAVE RECTIFICATION

The simplest rectifier circuit is the half-wave circuit (Fig. 10-1). The ac input produces an alternating emf in the secondary of the transformer, which attempts to force current through the secondary circuit, first in one direction and then in the opposite,

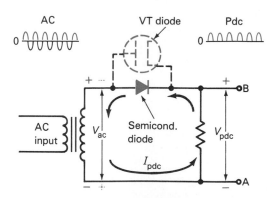

FIG. 10-1 Solid-state (or vacuum-tube) half-wave rectifier circuits produce pulsating dc.

alternately. Without the rectifier, ac would flow through the load resistance. With the rectifier, current can flow in one direction only. Although the transformer-secondary *voltage* may be alternating, *current* flows in it only during one half of each cycle. This produces a pulsating dc in the circuit as shown. The voltage-drop across points A and B is pulsating and has essentially the same voltage value as one half cycle of the ac from the transformer secondary.

A half-wave circuit has disadvantages. Only half of each ac cycle is used. The average current is equal to only 0.318 (half of 0.636, Sec. 4-5) of the peak-current value. It is more difficult to *filter* smooth than full-wave rectification.

Diodes used in rectifier circuits can be solid-state, vacuum, or mercury-vapor types. The basic rectifying circuits are all similar, but with vacuum and mercury diodes an additional filament winding is required on the power-supply transformer. With the exception of older equipment or high-power transmitters, which may use vacuum or mercury-vapor diodes, electronic power supplies today use solid-state silicon rectifiers.

If dc pulses with a peak value of 100 V are required, the secondary of the transformer must have a peak ac voltage output of 100 V (actually, silicon rectifiers with their 0.6-V voltage-drop or barrier voltage would require a peak voltage of 100.6 V, with 100.3 V for germanium).

10-3 FULL-WAVE RECTIFICATION

To utilize both half cycles of the power-frequency ac, most power supplies use a *full-wave rectifier* system. There are two such circuits: one is a *bridge* rectifier, and the other, used in the past with vacuum or mercury diodes, is a *full-wave center-tap* rectifier.

Bridge circuits require four diodes, connected as shown in Fig. 10-2. When the power transformer polarity is as shown by the solid + and − signs, electrons are pushed out of the negative end of the transformer, through diode A, up through R_L, through diode B and are pulled into the top terminal of the transformer, which is positive at this time. This half of the ac cycle produces one pulse of dc upward through the load resistor. (This is not a "power supply" yet because it has no filtering added.)

On the next half cycle of ac, the polarities on the transformer are reversed, shown dotted. Now electrons are pushed out of the top of the transformer,

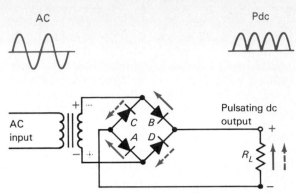

FIG. 10-2 Bridge-type full-wave rectifier circuit.

through diode C, again up through R_L, through diode D, and pulled into the positive end of the transformer, which is now the bottom terminal. Since the second half cycle also produces a second pulse of dc up through the resistor, power is delivered to the load on both halves of the ac cycle. If dc pulses with peak values of 100 V are required, the transformer secondary must furnish a 100 V peak ac (actually 101.2 V with silicon diodes).

A full-wave center-tap rectifier circuit is shown in Fig. 10-3. It is shown with vacuum tubes to

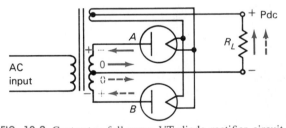

FIG. 10-3 Center-tap full-wave VT diode rectifier circuit.

illustrate a filament circuit, but operates otherwise the same with solid-state diodes. With the polarities indicated by solid + and − signs, electrons are pushed out of the relatively negative center tap, up through R_L, through diode A, and are pulled into the positive or top terminal of the transformer secondary. On the next half cycle (dotted signs), electrons are pushed out of the center tap again, up through R_L again, through diode B, and are pulled into the now positive bottom terminal of the transformer. Each power line ac cycle is thus changed into two similar-direction pulses of dc in the load circuit. If 100 V peak dc is required, the transformer *secondary* must provide *two* 100-V peak windings, or a total secondary voltage of 200 V peak. This rectifier requires only one filament winding with VT diodes, whereas a bridge rectifier requires three windings, one for diode A in Fig. 10-2, one for diode C, and one for both B and D.

Solid-state (and mercury) diodes have a relatively constant voltage-drop across them when carrying current. VT diodes increase the voltage-drop across them with any increase in load current. In many commercial high-voltage power supplies using VT diodes, a high-voltage transformer and a separate filament transformer are used to allow the filaments to be brought up to temperature before applying the high voltage to the plates of the rectifiers.

10-4 CAPACITIVE FILTERING

The output voltage of a rectifier is never smooth. Since the voltage required in most electronic applications must have an unvarying characteristic, it will be necessary to smooth the pulsations by *filtering*. The most common method is capacitive filtering.

Figure 10-4 shows a capacitor connected across the output of a half-wave rectifier circuit with no load.

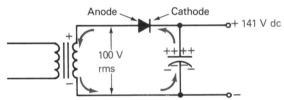

FIG. 10-4 The first filter capacitor charges to the peak ac voltage.

During that half of the ac cycle when the rectifier anode is negative, no current flows in the circuit. On the next half cycle, the top of the transformer becomes positive, pulls electrons off the top plate of the capacitor through the rectifier, and drives electrons onto the lower plate of the capacitor. This charges the capacitor to the peak value of the ac voltage. With 100 V rms ac, the capacitor charges to 1.414×100, or 141 V dc.

On the next half cycle (anode negative), the current cannot push back through the rectifier; so the capacitor remains charged at 141 V.

When the next positive half cycle occurs, the capacitor is already charged to the peak value, so nothing happens in the circuit. The voltage across the filter capacitor remains an unvarying 141 V dc.

During the charging half cycle in Fig. 10-5a, the current in the transformer secondary has two components. One charges the capacitor; the other flows through the load resistor.

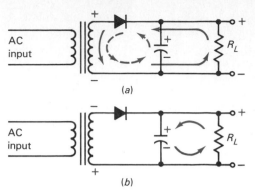

FIG. 10-5 (a) Load current comes from transformer. (b) Load current comes from charged filter capacitor.

During the noncharging half cycle, the transformer does nothing. Any load current flowing now must come from the electrons stored on the plates of the capacitor. The capacitor discharges through the resistor (Fig. 10-5b), still moving current upward through the load. If the capacitor is large, it can hold sufficient electrons to keep current flowing through the resistor during all of the noncharging half cycle. As it discharges, however, the voltage across the capacitor and resistor decreases. During the next charging half cycle, the capacitor is recharged to full voltage and the transformer drives the peak value of current through the resistor again. The voltage across the load resistor will vary as shown in Fig. 10-6a.

FIG. 10-6 Effect of increasing the load on the output waveshape. (a) Light load. (b) Heavy load.

A high-resistance load will discharge the capacitor slowly, as in Fig. 10-6a. If a heavy load (low resistance) is across the output, the capacitor will be discharged rapidly by the load, causing considerable variation of voltage between cycles, as in Fig. 10-6b. Additional capacitance will be required to counteract the rapid drop-off. This simple type of capacitive filtering across the output of a bridge rectifier, using several thousand microfarads of capacitance, is all that is needed for most transistorized equipment if followed by a voltage regulator circuit (Sec. 10-24). If the power supply and its attached equipment is insulated from the outside world completely, this type of supply may operate directly from the power line to provide about 150 V dc.

When the ac voltage reverses and applies an *inverse voltage* across a power-supply diode it is important that the diode be able to withstand such an inverse emf. If a diode is rated at 100-V peak inverse voltage (PIV) and a 50 V_{rms} ac is applied to its half-wave rectifying circuit using capacitive filtering, the diode may fail. The inverse voltage will rise to twice the ac peak value (the value to which C charges), or to $50 \times 1.414 \times 2$, or to 141.4 V!

Power-supply *ripple* (dc variation) can be reduced greatly by using the electronic *capacitance multiplier* shown in Fig. 10-7. The load sees an output capacitance equal to C times the beta (β) of the transistor. In this case the effective filter capacitance, C_{eff}, is 200(125), or 25,000 μF, which should be quite effective in reducing any dc ripple under normal load conditions.

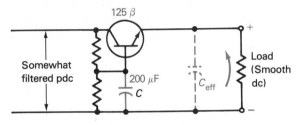

FIG. 10-7 Capacitance multiplier circuit.

10-5 TWO VOLTAGES FROM ONE RECTIFIER

A bridge rectifier circuit using a center-tapped transformer (Fig. 10-8) can produce two different positive voltages, one twice the other. As far as R_{L_1} is concerned, the circuit appears to be a bridge rectifier producing 200-V dc output. As far as R_{L_2} is concerned, it sees diodes A and C as two rectifiers of a center-tapped full-wave circuit that provides a 100-V output across it. The negative terminal is common to both rectifier circuits. If all diodes were

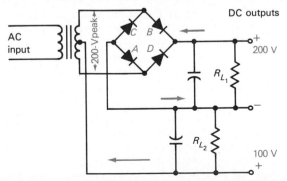

FIG. 10-8 Two dc voltages from one bridge rectifier.

reversed, the common terminal would be positive and the two output voltages would be -100 and -200 V dc.

Diodes A and C are carrying current for both of the output circuits. The additional current flow would have to be considered when selecting the diodes to be used.

10-6 INDUCTIVE FILTERING

When inductance alone is connected in series with a rectifier circuit (Fig. 10-9a), a smoothing action results.

Inductance has the property to oppose any change in current. Pulses of current through a choke coil build up a magnetic field around it, taking energy from the circuit to produce the field. As the pulse decreases, the magnetic field collapses and returns energy in the form of current to the circuit, thereby tending to hold the current constant. When pulsating dc from a rectifier passes through the circuit in which a choke coil is placed, the pulse amplitude will be lessened and the dropping off of the pulse will result in a lengthening of the pulse (Fig. 10-9c). It is impractical to produce a steady dc by using only inductance, but it is possible to use L and C together and produce substantially pure dc with reasonably inexpensive parts.

While series inductance is not often used in solid-state power supplies, its rounding effect on a pulse waveform can prevent proper transmission of square-wave pulses in many electronic circuits.

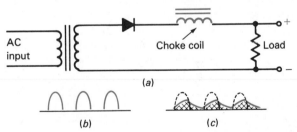

FIG. 10-9 (*a*) Inductor in series with pulsating dc and load. (*b*) Pulses if inductor were shorted. (*c*) Pulses with inductor in circuit.

■ Test your understanding; answer these checkup questions.

1. What polarity does an anode have if there is a 200-V inverse voltage across it? _____
2. Which of the common types of rectifiers has the lowest forward resistance value? _____ The highest? _____

3. How would anodes dissipate heat from vacuum rectifiers? _____ Solid-state rectifiers? _____
4. What is the average value of a 200-V rms sine-wave ac if half-wave rectified? _____ If full-wave rectified? _____
5. Which rectification might tend to permanently magnetize the core of a power supply transformer? _____ Why? _____
6. Name two circuits which produce full-wave rectification? _____ _____ Which one may require three filament windings? _____
7. Why are most modern rectifier circuits bridge-type? _____
8. What is required with a bridge rectifier circuit to obtain two voltages, one twice the other? _____
9. What components produce shunt filtering? _____ Series filtering? _____
10. What type of filtering smooths the leading edge of pulses? _____ The trailing edge? _____
11. In a half-wave filtered supply what feeds current to the load during the conduction half cycle? _____ During the nonconduction half cycle? _____
12. A 100-V rms secondary feeds a rectifier and a filter capacitor with no load. What is the voltage across C if the rectifier is a VT? _____ A silicon diode? _____
13. Draw diagrams and waveforms of half-wave, full-wave bridge, and full-wave center-tap rectifiers using solid-state diodes. Repeat using vacuum diodes.

10-7 CAPACITIVE-INPUT FILTERING

There are many variations of filter circuits. A common circuit for unregulated high-voltage power supplies is the capacitive-input type (Fig. 10-10). This should be recognized as a low-pass pi-type

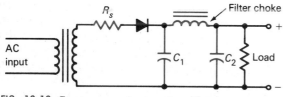

FIG. 10-10 Common π-type capacitive-input power-supply filter circuit.

filter (Sec. 8-8). When used to filter the output of a rectifier, it discriminates against any voltage and current variations but passes steady dc (zero frequency).

When looking at the filter from the source of power (the transformer), the first part of the filter seen is the capacitor C_1. This is the *input capacitor* of the filter. It charges to the peak-voltage value of the input ac and discharges somewhat during the

noncharging half cycle. During the noncharging interval, the input-capacitor current flows through the load resistor and through the choke coil. The choke opposes any change in this current, resulting in a fairly smooth dc flow through the load. Added to this, the output capacitor C_2 further tends to hold the voltage constant across the load.

Practical values of inductance might be 0.1 to 0.5 H for high-current, low-voltage transistor power supplies, used mostly only if the supplies are unregulated. For high-voltage relatively low-current VT circuits, the inductance might be in the 5- to 20-H range. The value of the output capacitance is usually double that of the input. An input capacitor tends to determine the output voltage of the supply, but an output capacitor acts more as a storage tank of electrons for the load.

Under medium loads, the output of a capacitive-input filter system is roughly 85% of the ac peak-voltage value. Power supplies having outputs of more than 450 V use paper-dielectric capacitors. Below this voltage either paper or electrolytic capacitors are suitable. Electrolytics are always used in low-voltage power supplies. When selecting input capacitors for power supplies, a working voltage (WV) of at least 10% over the peak ac voltage value is desirable (450 WV for 400 V peak ac from the transformer). With electrolytics, the peak rectified ac voltage is the WV rating to use.

Low-voltage high-current supplies for transistorized equipment may use filter capacitors of 500 to 20,000 μF. When the supply is turned on, the first surge of charging current with such high-value capacitors may damage a solid-state rectifier. To prevent this a few ohms of series "surge" resistance (R_s in Fig. 10-10) may be added to limit surge current.

10-8 INDUCTIVE-INPUT FILTERING

The first parts seen by the source when looking into the filter of Fig. 10-11 are a coil L_1 and a capacitor C_1 in series. This is known as an *inductive-input* filter. If the load is disconnected, the pulsating dc from the rectifier will charge C_1 and C_2 to the peak value of the transformer after a few pulses. When a load is connected, however, current flows through L_1 and L_2. Since the current through L_1 will be pulsating (has an ac component), a reactive voltage-drop will occur across this choke coil ($V = IX_L$). C_1 can no longer charge to the peak value but will usually fall to about 65% of the peak. As a result, the voltage across C_2 may never be greater than this

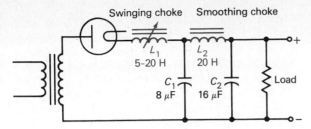

FIG. 10-11 An inductive-input high-voltage power-supply filter circuit for a gaseous diode.

value and may be somewhat less. Because the current through L_2 is essentially unvarying dc, there is very little reactive voltage-drop across it, although there will be a resistive voltage-drop across the resistance of the wire in the choke ($V = IR$). The inductive-input filter circuit will usually have better voltage regulation, but it will have less output voltage than a capacitive-input filter with the same power transformer. The input choke is usually a swinging choke (Sec. 10-11). Any other chokes are *smoothing* chokes. When an inductive-input filter is used, no surge resistor R_s is needed with any type of rectifier, regardless of how large a filtering capacitor is used. Mercury-vapor rectifiers always require inductive-input filters.

10-9 *RC* FILTERS

If the load on a supply is constant (as with class A amplifiers), adequate filtering may be produced by using a resistor in place of the choke coil of a π-type filter (Fig. 10-12).

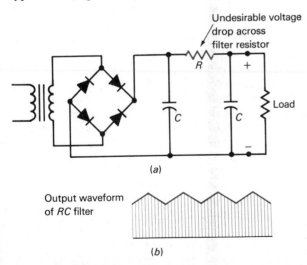

(a)

Output waveform
of *RC* filter

(b)

FIG. 10-12 (a) Full-wave power supply with an *RC* filter. (b) Output waveform under heavy load.

Disadvantages of this type of filtering are the voltage-drop across the filter resistor. The resistance value may be 10 to 100,000 Ω, depending on the load current, and the variation of the output voltage if the load changes.

An *RC* filter is used in very low current drain applications, such as in the high-voltage circuit of cathode-ray tubes (CRT) or TV picture tubes, which may require little voltage regulation. This is also a common type of filter for power supplies that incorporate output voltage regulation systems.

10-10 FILTER CHOKES

Some factors to be considered regarding iron-core-type filter chokes are listed here.

The winding of a choke coil must be insulated from the core to withstand voltages developed between the winding and the usually grounded core.

To prevent fluctuating fields of the choke from inducing an ac into nearby wires, the winding may be shielded in an iron case.

If the windings of chokes are of small-diameter wire and many turns, the inductance may be great, but the resistance may be high, resulting in a dc voltage-drop across the choke. Increasing current flow increases the voltage-drop across the choke, resulting in poor power-supply voltage regulation. The greater the inductance in a choke, the more reactance to varying currents and the greater choking action it will have. It becomes a problem of how much inductance versus resistance versus physical size.

Since capacitance passes ac or pulsating dc, a choke with a high value of *distributed capacitance* across it will choke down the variations by its inductance but pass them by its distributed capacitance. One counteracts the other. Chokes must have low distributed capacitance.

The current that can be carried by a choke is determined by the size of the wire used and ventilation of the choke windings. With poor ventilation a relatively small current may cause enough heat accumulation to produce deterioration of the insulation on the wires and to increase wire resistance, which compounds voltage-regulation difficulties.

If the variation of current in a choke is of high amplitude, eddy currents may develop in the core, resulting in a heat loss. The cores of chokes are laminated with low-retentivity steel to decrease hysteresis and eddy-current losses. If the laminations are not securely bolted together, they may vibrate when varying dc flows through the windings. Except for an audible buzzing sound, the operation of the choke is not impaired.

Choke coils react against current variations only if their cores are not magnetically saturated. To prevent saturation at a relatively low current, the core is constructed with an interruption in the continuity of the magnetic-iron circuit. To increase the reluctance of the core, an air gap will be found in all cores of *smoothing*-type filter chokes. This gap may be only 1 or 2 mm wide, perhaps with a piece of cardboard in it, but the gap prevents the core from becoming saturated at normal current levels.

Whenever possible, the metal cases of chokes and transformers are connected to ground potential or to the metal chassis. This prevents personnel from receiving electric shocks and the metal cases from picking up an electrostatic charge.

10-11 SWINGING CHOKES

High-voltage power supplies may use a *swinging* choke as the input inductor to improve voltage regulation. A swinging choke has little or no air gap in its core. Because of this the magnetic-iron core begins to saturate at a medium-current value. When low current is flowing through it, the choke has high inductance and filters effectively. With high current it saturates and has less inductance. Thus, with a light load and little current, the swinging choke has a high reactance and develops a large voltage-drop across it. When the load increases, the choke saturates and has less reactance and less voltage-drop across it. The voltage of the power supply tends to remain more constant under varying loads, improving the voltage regulation of the supply.

A swinging choke might have inductance values that swing from 5 H with a heavy load (high current) to 20 H with a light load.

10-12 VACUUM RECTIFIERS

Vacuum diodes may still be found in vacuum-tube equipment power supplies and in high-voltage supplies for high-powered transmitters.

ANSWERS TO CHECKUP QUIZ ON PAGE 177

1. (Neg) **2.** (Solid-state)(VT) **3.** (Radiation, convection, conduction)(Heat sink) **4.** (89.9 V)(179.8 V) **5.** (Half-wave) (Load current flows one direction only) **6.** (Bridge, center-tap)(Bridge) **7.** (Usually solid-state; no filament windings) **8.** (Center tap on transformer) **9.** (C)(L) **10.** (L)(C and L) **11.** (Transformer)(Filter C) **12.** (141.4 V)(140.8 V) **13.** (See chapter illustrations)

High-voltage vacuum rectifiers are made in single units, with filament leads coming out at the base and the plate connection at the top of the tube.

Low-voltage vacuum rectifiers are usually constructed as duodiodes, with two plates and two parallel filaments in one envelope. The filament and plate leads all terminate at pins in the tube base.

Some of the significant points regarding high-vacuum rectifiers are listed below.

The voltage-drop across the tubes will vary directly with the current flowing through them.

These tubes dissipate considerable heat, due to high filament temperatures and radiation from the plate. Adequate ventilation must be provided, particularly in high-power supplies.

They can be used with either inductive- or capacitive-input filter circuits.

Low-power filament type diodes can be operated as soon as the filament is turned on (no warm-up period). With indirectly heated cathodes there is a warm-up period of about 10 s.

Any light blue or purple haze between filament and plate of a high-vacuum rectifier indicates unwanted gas. In some cases the tube may operate for a long period of time with some gas in it. In other cases, when the gas leak is more rapid, the tube may require immediate replacement. Any gassy tubes bear watching.

10-13 MERCURY-VAPOR RECTIFIERS

Mercury-vapor diodes (Chap. 9), if found today, are single-diode, high-voltage, high-current types, with filament connections at the bottom and a plate connection at the top. They may be found in transmitters and other high-power applications.

The ionization potential of any gas depends on several factors: gas used, pressure of the gas, electrode size, and whether a hot cathode is one of the electrodes. As pressure is reduced, the ionization-voltage value increases until in a vacuum tube (no gas) ionization does not occur under normal operating conditions.

Ionized mercury vapor affords a low-resistance path between cathode and plate and will support relatively high current flow with little heating. Such a tube depends on ionization of the gas rather than on electron emission from a hot filament to produce current flow. The result is greater current through the tube, less power to heat the filament, less heating of the plate, and greater overall efficiency.

When a mercury-vapor tube is at operating pressure, about 15 V between filament and plate will

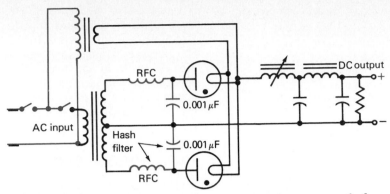

FIG. 10-13 A complete high-voltage mercury-vapor tube power supply for a transmitter.

cause heavy ionization and a large current flow. The potential required to produce an undesired current backward through the tube (plate to hot cathode) is several thousand volts. By holding the gas pressure within limits, a rectifier tube with a 15-V ionization emf one way can have an *inverse ionization voltage* of several thousand volts in the other direction. However, if overheated, gas pressure may increase, inverse ionization voltage will fall below the peak voltage of the transformer, and a flashback, or arc-back, will occur in the tube.

Mercury-vapor tubes must be warmed for 15 to 20 s and must not be overheated. They operate satisfactorily between about 20 and 70°C.

Mercury-vapor tubes must not work into a capacitive-input filter. In such a circuit, when the alternating emf of the transformer increases from zero to +15 V, current suddenly starts flowing. This sudden, steep-sided wave of current represents the equivalent of a very high frequency pulse. The reactance of the capacitor is very low, practically a short circuit, for such a high frequency. As a result, during this instant, a relatively high current flows in the transformer-tube-capacitor circuit and produces double ionization and deterioration of the filament.

Even with an input choke, a slight current surge due to distributed capacitance across choke, coil, and transformer may be developed at every instant of ionization. This surge in the mercury *plasma* (ionized gas) sets up a damped oscillatory wave, similar to a spark-transmitter emission, in the transformer-tube-filter circuit. A low-intensity wideband signal may be picked up by nearby receivers, audio-frequency (AF) amplifiers, tape recorders, etc., as a disturbing *hash* (buzzing sound). To stop this, 5- to 20-mH RF chokes may be connected in each plate lead of the rectifier tube plus 0.001- to

0.005-μF mica RF filter (bypass) capacitors.

Mercury-vapor tubes pass relatively high current (0.5 A for a type 866). When more current must be passed, two or more tubes may be connected in parallel, but *equalizing resistors* must be connected in series with each tube. They range in value from 0.5 to 5 Ω.

A complete full-wave mercury-vapor power supply with swinging choke input filter, hash filters, and a means of heating the filaments before the high-voltage ac is applied is shown in Fig. 10-13.

10-14 120 OR 240 VOLTS TO ONE TRANSFORMER

Many power supplies are constructed to operate from either 120-V or 240-V ac lines. To maintain the same secondary voltage output from the transformer the primary is wound for 240-V operation, but it is split *exactly* in the center. When used on 240 V ac, the two halves of the primary are connected in series, producing the desired output from the secondary (perhaps 25 V ac). When used on a 120-V line, the two halves of the primary are connected in parallel, again producing 25 V from the secondary. The power (VA) output of the transformer will be the same with either connection. Such a switched primary is shown in Fig. 10-14.

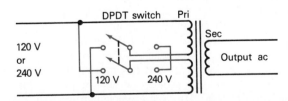

FIG. 10-14 Split-primary transformer and switch to enable operation from either 120 or 240 V ac.

1. What is the input component of a π-type filter? _____ The output? _____
2. What swings in a swinging choke? _____
3. Where is a swinging choke used? _____ What is the name of the other type of filter choke? _____
4. What are two disadvantages of using an *RC* filter? _____ _____ Advantages? _____ _____
5. What kind of capacitance should a choke not have? _____
6. What is the disadvantage of having resistance in a filter choke? _____
7. With mercury-vapor diodes: What is damaged by double ionization? _____ What is their forward ionization potential value? _____ Inverse ionization potential value? _____
8. Why do high-power rectifier tubes use separate filament transformers? _____ _____
9. What components are used in a hash filter? _____
10. How long a warm-up period should mercury-vapor diodes have? _____
11. What is the approximate loaded output voltage of a *C*-input filter in a full-wave power supply if the peak ac is 1000 V? _____ If the filter is *L*-input? _____
12. What type dielectric filter capacitor is used in 1000-V supplies? _____ In 12-V supplies? _____
13. What type of choke has a gap in its core? _____
14. What filter configuration is never used with mercury-vapor rectifiers? _____
15. Draw a diagram of a power supply with a π-type filter. With an *RC* filter. With mercury-vapor tubes.

10-15 RIPPLE FREQUENCY

When filters are designed for power supplies, the pulse frequency is important. The higher the ripple frequency, the easier it is to filter. For example, the power frequency may be 60 Hz. A half-wave 60-Hz rectifier produces 60 pulses, or ripples, per second and requires a certain amount of *C* (and *L*) to filter the pulses adequately. A full-wave 60-Hz rectifier produces a ripple frequency of 120 pps and requires only half as much *C* (and *L*). In aircraft, frequencies up to 800 Hz produce a full-wave ripple frequency of 1600 Hz, requiring much smaller and lighter filter components.

When 3-phase ac is investigated (Sec. 10-29), it is found that a half-wave 60-Hz, 3-phase rectifier has a ripple frequency of 180 Hz. A full-wave 60-Hz, 3-phase rectifier has a ripple frequency of 360 Hz. Because the pulses overlap each other, rectified 3-phase ac is in the form of varying instead of pulsating dc and is much easier to filter. For these reasons 3-phase ac is very often used in broadcasting stations and other high-powered applications.

10-16 BLEEDER RESISTORS

High-voltage power supplies always use a *bleeder* resistor across the output of the supply. It serves two purposes. *First*, it bleeds off the charge of the capacitors when the power supply is turned off. (It is possible to turn off a transmitter with an open bleeder and several hours later receive a lethal shock from the charge still left in unbled filter capacitors.) *Second*, the bleeder resistor aids in holding the voltage output more constant. According to the graph in Fig. 10-15, with no current

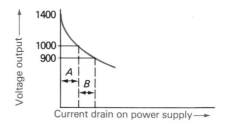

FIG. 10-15 Effect of using bleeder resistor on output voltage.

drain the power supply has a 1400-V output. If a current drain of the magnitude indicated by line *A* is drawn from the supply, the output voltage is reduced to 1000 V, a 40% voltage change. By connecting a bleeder resistor which draws a current of the magnitude of *A* across a power supply, an equal additional load equal to *B* drops the output voltage to 900 V, only an 11% load voltage variation.

Regulated low-voltage solid-state power supplies may not use bleeders.

10-17 DETERMINING REGULATION

When a power supply is operating into a load, it will have a certain output voltage. If the load is removed, the output voltage will increase. The percent of voltage *increase* is considered the *regulation* of the power supply. For example, a power supply delivers 1000 V to a telegraph transmitter with the key down and 1200 V with the key up. The regulation is the ratio of the difference of voltages to the full-load voltage. The percent of voltage regulation can be found by the formula

$$\% \text{ Reg} = \frac{(V_{nl} - V_{fl})100}{V_{fl}}$$

$$= \frac{(1200 - 1000)100}{1000} = \frac{200(100)}{1000} = 20\%$$

where V_{nl} = no-load voltage
V_{fl} = full-load voltage

The same formula can be applied to batteries, motor-generators, transformers, etc., to determine regulation. The lower the percent, the better the regulation.

The regulation formula can be rearranged to compute the full-load voltage. If the no-load voltage is 140 V and the regulation is 15%, what is the full-load voltage?

$$\% = \frac{(V_{nl} - V_{fl})100}{V_{fl}}$$

$$\%(V_{fl}) = 100V_{nl} - 100V_{fl}$$

$$\%(V_{fl}) + 100V_{fl} = 100V_{nl}$$

$$V_{fl}(\% + 100) = 100V_{nl}$$

$$V_{fl} = \frac{100V_{nl}}{\% + 100} = \frac{14,000}{115} = 121.7 \text{ V}$$

The regulation formula can be rearranged to compute the no-load voltage also. If the full-load voltage is 240 V, and regulation is 11%, what is the no-load voltage?

$$\frac{(V_{nl} - V_{fl})100}{V_{fl}} = \%$$

$$(V_{nl} - V_{fl})100 = \%(V_{fl})$$

$$V_{nl} - V_{fl} = \frac{\%(V_{fl})}{100}$$

$$V_{nl} = \frac{\%(V_{fl})}{100} + V_{fl} = \frac{11(240)}{100} + 240$$

$$= 26.4 + 240 = 266.4 \text{ V}$$

If a power-supply voltage decreases when the load across it demands more current, it is acting as a high-impedance device. A varying load current results in a varying power-supply output voltage. Thus, variations of current of one circuit connected to a common supply produce variations of voltage

in other stages connected to the same supply. This is one form of interstage feedback and can result in weakened output, distorted amplification, or oscillation of the system. Well-regulated power supplies are said to have low impedance, from

$$Z = \frac{dV}{dI}$$

where d = a change in

If V does not change when I changes, the Z value must be low.

These conditions in an unregulated power supply may cause an increase in regulation percentage:

1. Resistance in the wires of any choke or transformer produces internal voltage-drops. As I increases, output voltage decreases.
2. Using vacuum rather than semiconductor or mercury-vapor diodes.
3. Using C-input filtering instead of L-input.
4. Failure to use a swinging choke as the filter input.
5. Insufficient capacitance in the filter circuit.
6. Too high or too low a bleeder-resistance value.
7. Using half-wave rectification instead of full-wave.

10-18 FILAMENT CIRCUITS

A variety of methods may be found to heat the cathode of VT rectifiers. A standard power-supply transformer has a primary, a high-voltage secondary, and a filament winding. Most low-power rectifier tubes use filamentary cathodes. However, some have a heater-cathode.

In high-voltage, high-power systems the tubes usually have heavy filaments or heaters which heat slowly. This requires a separate filament transformer with switching circuits to allow the filaments to be turned on some time before the ac is switched to the high-voltage transformer primary. Such switching can be done manually, but in commercial equipment it is usually done with a timing relay. With heavy filament tubes, it is often advisable to start filament heating at lower voltages, slowly increasing the operating voltage with a variable autotransformer to prevent fracturing the filament by sudden thermal expansion.

In transformerless receivers or television sets, known as ac/dc equipment, all of the tubes of the system must have similar-current heaters, which are connected in series, and across the 120-V ac line, usually in series with a voltage-dropping resistor.

CAUTION: This equipment can be *dangerous to service* because one side of the power-supply circuit is always connected to the hot wire of the power lines. It is not isolated from ground as when a power transformer is used. (There are some solid-state sets with series transistors across a rectified power line, which are also dangerous to service!)

10-19 VACUUM DIODE POWER SUPPLIES

The circuit shown in Fig. 10-16 represents a common type of power supply used to provide the necessary 100- to perhaps 500-V V_{bb} or B+ used in the plate circuits of older VT equipment. It shows

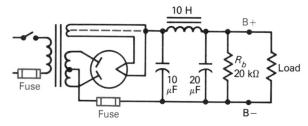

FIG. 10-16 Most common VT power-supply circuit.

two possible places for a fuse. Only one would normally be used. Less expensive supplies connect the filter to one side of the filament as shown. It would be better to use the center tap of the filament winding, shown dashed. Although a single double-diode tube is indicated here, two separate vacuum diodes might be used. If mercury-vapor diodes are used, they require an additional input swinging choke, and the tube symbol would show a dot in the envelope(s). R_b is a bleeder resistor (Sec. 10-16).

Higher voltage supplies, 500 to 10,000 V or more, often use a bridge rectifier. If solid-state rectifiers are not used, a separate filament transformer is needed to provide the three separate filament windings required by such a tube-rectifier bridge circuit.

10-20 POWER-FACTOR COMPENSATION

Power-transformer primaries have inductance. With inductance in an ac circuit, current and voltage are out of phase. If current lags (or leads) the voltage, there is a lessening of the power factor in the circuit. A low power factor means less secondary power output per ampere of current in the primary. With 60-Hz ac, power factor may be disregarded in most power supplies. With 400- to 800-Hz ac, the power factor may decrease considerably. To overcome the inductive effect of the transformer, an 8- to 20-μF capacitor may be connected in series with the primary. The current lead of the capacitor counteracts the current lag of the inductance, raising the power factor.

▌ Test your understanding; answer these checkup questions.

1. What is the inverse peak voltage on a rectifier if a transformer has 500 V rms in a full-wave center-tap circuit? _____ Bridge circuit? _____ Half-wave with π-type filter? _____
2. What is the ripple frequency of 600-Hz 1-phase ac if half-wave-rectified? _____ If bridge-rectified? _____ If bridge-rectified using 3-phase ac? _____
3. What diode radiates the most heat? _____
4. What is reduced by using the center tap of the filament winding of a power supply? _____
5. Name two places in a center-tap full-wave circuit where a circuit breaker might be used. _____ _____
6. When is pf compensation usually required for power-supply transformers? _____
7. Name two reasons for using bleeders on high-voltage power supplies. _____ _____
8. What resistance bleeder might be used across a 5-kV 0.5-A output power supply? _____ What should be its power rating? _____
9. A power supply has 12 V loaded. What is its regulation if the no-load V_o is 13 V? _____ If it is 12.1 V? _____
10. A transformer has 11.1% regulation with 900 V as the loaded output. What is the unloaded output? _____
11. Draw diagrams of a full-wave center-tap mercury-vapor power supply. A vacuum diode power supply.

10-21 VOLTAGE-MULTIPLIER CIRCUITS

It is possible to rectify an ac voltage and produce a dc voltage under load almost twice the ac rms value by using a full-wave *voltage-doubler* circuit. In Fig.

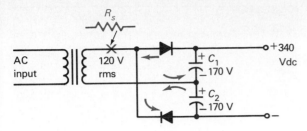

FIG. 10-17 A full-wave voltage-doubler circuit.

10-17, when the top of the transformer secondary is positive, current flows through the top rectifier and charges capacitor C_1. When the cycle reverses and the bottom of the secondary is positive, C_2 is charged through the lower rectifier. The output dc is across the two charged capacitors in series. The larger the value of the two capacitors, the better the regulation and the higher the output voltage under loaded conditions. Values in the range of 40 μF to more than 500 μF for each capacitor are used. The output voltage of the voltage-doubler circuit ranges from twice the peak value of the applied ac at no load to about the peak value under heavy-load conditions. With 120-V rms ac input, this would produce a no-load voltage of almost 340 V (from 120 $\times$ 1.414 $\times$ 2), but with a heavier load an output of perhaps only 150 to 200 V.

With solid-state rectifiers, a 2- to 5-Ω surge resistor (R_s at point X) or small choke should be connected in series with the rectifiers to limit surge currents.

A *half-wave* voltage doubler circuit is shown in Fig. 10-18. When the top power line is negative,

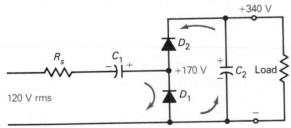

FIG. 10-18 A solid-state half-wave or cascade voltage doubler.

the input capacitor charges through D_1 to the peak voltage of the line (+170 V). When the line voltage reverses, the output capacitor, through D_2, is connected to the positive line voltage in series with the charged input capacitor. The output capacitor charges to almost twice the line-voltage peak value. Both capacitors can be electrolytics.

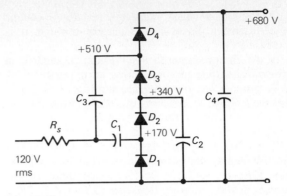

FIG. 10-19 Voltage-quadrupler circuit.

The same basic circuit, by adding D_3 and C_3 (Fig. 10-19), produces a voltage tripler. Adding D_4 and C_4 results in a quadrupler. As in Fig. 10-17, D_1, C_1, D_2, and C_2 act to charge C_2 to nearly 340 V. When the upper input line is negative, the third capacitor is in series with D_3 and the charged C_2. When C_3 charges, it adds another 170 V, for a total of nearly 510 V. When the upper line reverses to positive, C_4 charges through D_4, adding another 170 V, for a total unloaded voltage of nearly 680 V. With any load at all, the output voltage drops drastically. Voltage reduction depends on the capacitance values used and the current demand of the load that tends to discharge the capacitors.

Voltage multipliers can be used to produce high-voltage low-current for cathode-ray tubes or other special devices.

10-22 SHUNT-REGULATING DIODES

The constant voltage-drop across a zener diode, or a gas-filled tube when the gas is ionized, can be used as a means of shunt-regulating the dc output of a power supply. As previously explained (Chap. 9), when a voltage is increased across a resistor in series with a reverse-biased zener diode (Fig. 10-20), a voltage value will be reached where the diode breaks down and passes current. The diode

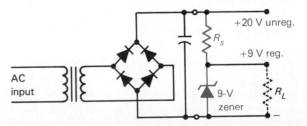

FIG. 10-20 Zener-diode shunt regulator supply.

will maintain a nearly constant breakdown voltage value across it even if the current through it is increased.

In the circuit shown, a 9-V breakdown zener is in series with a resistor, R_s, across an unregulated 60 V dc from a rectifier-filter circuit. If R_s has 22 Ω, and the voltage-drop across it is 11 V (from $20 - 9 = 11$), the current through both R_s and the zener must be $I = V_s/R_s$, or 11/22, or 0.5 A.

If a 45-Ω load (the dashed R_L) is added across the 9-V zener, the current through the load will be $I = V/R$, or 9/45, or 0.2 A. Since there is still current flowing through the zener ($0.5 - 0.2 = 0.3$ A), the voltage-drop across it is still 9 V. The total current has not changed, so the voltage-drop across R_s must still be 11 V.

If the load changes to 36 Ω, the load current becomes $I = V/R$, or 9/36, or 0.25 A. This leaves a similar 0.25 A flowing through the zener, but there is still 9 V across the zener and R_L. If the load increases and provides 18-Ω resistance, the current through R_L is then $I = V/R$, or 9/18, or 0.5 A. Now the load is taking the whole half ampere and the zener has no current through it. If the load becomes 8 Ω, for example, the total resistance across the 20-V supply is now $8 + 22 = 30$ Ω. The current flow is $I = V/R$, or 20/30, or 0.667 A. The voltage across R_L is now $V = IR$, or 0.667(8), or 5.336 V. Increasing the load has resulted in the voltage-drop across the load falling from 9 V down to 5.336 V. With this much load the zener is unable to maintain its regulated 9-V value. The circuit is known as a *shunt regulator* because the load is shunted across the regulating device, in this case a zener diode. It is advisable to have a minimum of about 5 mA flowing in a zener diode to maintain good regulation.

Zener diodes are selected by voltage breakdown (2.5 to 200 V), maximum current rating (10 mA to 2.4 A), and power rating (0.5, 1, 5, and 10 W). Another rating is their impedance, from which the actual variation of the zener voltage can be approximated under changing load conditions. A listing of four 10-V zeners is shown in Table 10-1.

The Ohm's-law formula for impedance can be stated as $Z = V/I$, or as $Z = dV/dI$. Consider the

ANSWERS TO CHECKUP QUIZ ON PAGE 184

1. (1414 V)(1414 V)(2828 V) **2.** (600 Hz)(1200 Hz)(3600 Hz) **3.** (Vacuum) **4.** (60-Hz ripple) **5.** (Primary)(Center-tap lead) **6.** (High-frequency ac) **7.** (Discharge)(V reg.) **8.** (40 kΩ+)(1250 W) **9.** (8.33%)(0.833%) **10.** (1000 V) **11.** (See chapter illustrations)

TABLE 10-1	ZENER CHARACTERISTICS		
Volts	**Watts**	I_{max}	**Z**
10	0.5	0.050 A	60 Ω
10	1	0.095 A	8 Ω
10	5	0.480 A	3 Ω
10	10	0.880 A	3 Ω

1-W zener in Table 10-1. It has an impedance of 8 Ω, its nominal rated voltage is 10 V, and from breakdown to maximum current the dI is 0.095 A. Rearranging the formula to solve for dV,

$$Z = \frac{dV}{dI}$$

$$dV = Z(dI) = 8(0.095) = 0.76 \text{ V}$$

Assume that the voltage at breakdown will be the nominal 10-V rating minus about half of 0.76 V, or 9.24 V. At maximum zener current the zener voltage will be about 0.38 above 10 V. It can be seen that the zener does not hold its load voltage at exactly 10 V with a widely varying load current. However, the line voltage may vary perhaps 10% and have *transients* (voltage spikes) of over 300 V, but the zener and its load voltage may vary imperceptibly.

A gaseous device called a VR (voltage-regulator) tube may be found in 75- to 150-V low-current circuits. They can be used in the same way as a shunt-regulating zener diode (Fig. 10-21). A VR

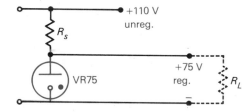

FIG. 10-21 VR-tube regulator circuit. Dot indicates gas.

tube is a gas-filled glass envelope containing a cylindrical metal cathode with a thin wire running down the center of the cylinder. If filled with neon gas and if the emf across it is brought up to about 100 V, the gas ionizes. If it is assumed that a dropping resistor is in series with the VR tube and the dc source, the voltage across the tube drops to about 75 V and remains there. Since many more electrons move from a larger area electrode to a smaller one in a gaseous device (making the VR a diode), the cathode and anode in the symbol are just the opposite of the actual operation of the device. VR tubes require about 5 mA to hold them ionized, and they will safely pass about 40 mA. Therefore, they can be used to regulate up to about 35 mA of load current variation. Their voltage variation from 5 to 40 mA may be in the 5- to 10-V range. They are known as VR75 (or OA3), VR90 (OB3), VR105 (OC3), and VR150 (OD3) types. Two VRs (or two zeners) in series will regulate to the sum of their voltage ratings. Normally, the unregulated supply should have a voltage about 50% greater than the regulated-voltage value.

Zener diodes look like other solid-state diodes. The low-power units are axial-lead types, that is, small plastic cylinders with a wire lead extending out each end. The 10-W units are stud-mount types, metal-cased, with a threaded machine screw anode. The cathode lead is a rod, flattened at the end with a hole in it. The stud is made to bolt to a heat sink. VR tubes (Fig. 9-52) are glass types with connecting pins protruding from the base of the tube.

10-23 SHUNT-REGULATED SUPPLIES

Whereas a volt of power-supply variation may not seriously degrade the operation of many high-voltage VT circuits, most transistor circuits require low-voltage, high-current power supplies that must reduce output variations to less than 20 mV at worst. More often, the variations or ripple must be reduced to less than 1 mV. To do this, more complex circuits than simple zener diode circuits may be necessary.

Figure 10-22a shows a shunt-type regulator with the regulating transistor and the load in parallel (shunt). If the load increases, the output voltage sags somewhat. This reduces the forward bias of Q_1 (provided by voltage divider $R_B R_S$), and its collector current decreases, allowing this amount of current to go to the load. The output voltage tends to return to its original value.

If the shunt regulator can be made more sensitive to output voltage changes by amplifying the variations, the regulation will be better. The circuit in Fig. 10-22b uses a zener to hold the base-collector of Q_1 constant. If the load increases and the output voltage sags, the base voltage of Q_1 becomes less positive due to the decreased voltage-drop across R_Z. This decreases the I_C of amplifier Q_1, which decreases the I_C of amplifier Q_2, in turn decreasing the current through R, and the voltage-drop across it, increasing the voltage output. The sensitivity

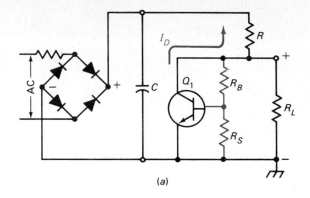

(a)

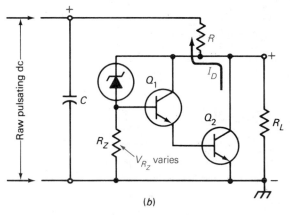

(b)

FIG. 10-22 (a) Simple transistor shunt regulator. (b) Shunt regulator with error-signal amplifier.

may be 100 times that of the single-transistor shunt regulator.

If the unregulated dc has a little ripple in it, the regulator circuit action filters it smooth.

10-24 SERIES-REGULATED SUPPLIES

The current-regulating device that results in a regulated voltage output can be connected in series with a power-supply line.

One of the simpler transistor series-regulator circuits is shown in Fig. 10-23a. The bridge rectifier and capacitor, C, produce a reasonably well-filtered unregulated 20 V dc. With no load connected, no I_C flows through Q_1. Its base is held at +12 V above ground by the zener diode. When a load, R_L, is connected, collector current, I_C, flows through Q_1. Since Q_1 is normally a silicon device, under a light load there is a voltage-drop of 0.6 V between emitter and base, which leaves +11.4 V as the load voltage. If the load increases, I_C increases and the

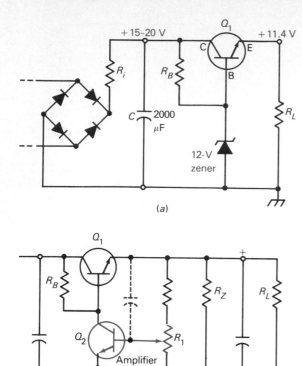

FIG. 10-23 (*a*) Simple transistor series regulator. (*b*) Series regulator with error-signal amplifier.

EB voltage increases slightly. This represents a greater forward bias for Q_1, it passes more current to the load, and the output voltage tends to rise. The change in voltage caused by a change in load current can be called an *error voltage* or signal. The correction of the error voltage in most circuits does not quite return the output voltage to its former value. If this error voltage can be amplified, the voltage regulation can be improved. Note that Q_1, called a *pass transistor*, is functioning as an *emitter-follower* amplifier (Sec. 9-10), R_L is its load, with the base bias held constant by the zener. With no voltage variation on the base, there should be no voltage variation across the load of the emitter follower.

The circuit in Fig. 10-23*b* uses a second amplifier transistor, Q_2, to sense the error voltage and feed an amplified error signal to the control transistor, Q_1. Resistor R_B with Q_2 and D_z in series forms a voltage-divider bias circuit for Q_1. If transistor Q_1

has its bias changed, its conduction also will change. Potentiometer R_1 and the resistors in series with it form a voltage-divider biasing circuit for Q_2. Moving the arm of R_1 will vary the emitter-collector resistance of Q_2, which in turn will vary the bias on Q_1.

Assume R_1 is adjusted to produce an output voltage of 15 V. The zener diode holds the Q_2 emitter at a constant voltage value. If the load is increased, the output voltage should decrease slightly. With the error voltage becoming less positive (less forward bias for Q_2), the EC resistance of Q_2 increases. This allows the base of Q_1 to become more positive (more forward-biased), and the series transistor conducts more, reestablishing the output voltage to nearly the original value. Since Q_2 is acting as an amplifier for the error voltage, the regulation of this circuit is quite good. (More complex regulators may include more amplification and circuits to counteract temperature variations.)

The capacitor shown dashed has a value of about 0.1 μF. It aids in reducing any residual ripple that might appear in the output and prevents voltage "hunting" and pickup of stray emf's.

Since short-circuiting the output of a power supply may damage the rectifiers and the series control device, some form of overload protection should be provided. This might take the form of a fuse or overload relay. An electronic overload circuit is shown in Fig. 10-24*a*. Q_1 is a silicon transistor (EB = 0.6 V). Two silicon diodes in series and a 1-Ω resistor form the overload circuit. The diodes are forward-biased whenever any load current flows through R_1. However, the voltage-drop across them must exceed 1.2 V before they will conduct. A load current of 0.5 A produces 0.5 V across R_1, which, in series with the 0.6-V EB, is still not enough to make them conduct. As soon as the load current reaches 0.6 A the diodes begin to conduct, holding the base of Q_1 at 1.2 V above the output voltage. The diodes in series with R_B form a voltage-divider bias circuit. As the load increases, the Q_1 bias (V_{R_1} − 1.2 V) decreases, and not enough current can flow through Q_1 to damage it.

The whole Fig. 10-24*a* discrete component circuit (enclosed in dashed lines) could be built into IC form, Fig. 10-24*b*, with only three external connections: (1) the unregulated + input at upper left; (2) the regulated + output at upper right; and (3) the lower negative terminal, common to both input and output. Note the 1-μF capacitors always used at the input and output terminals.

If the pass transistor in a regulator circuit short-

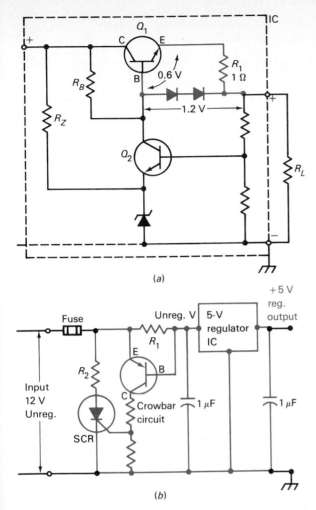

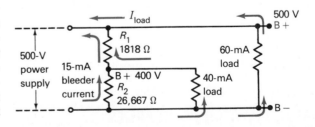

FIG. 10-24 (*a*) Overload protection for series-regulated power supply. (*b*) Crowbar-type protection circuit.

circuited, the high unregulated voltage would be applied to the load, would draw high current, and might damage load circuit components. To prevent this, a *crowbar*-type circuit of some type can be used. For example, if current drain on the unregulated supply line increases over normal, the voltage-drop across R_1, in Fig. 10-24*b*, becomes high enough to forward-bias the transistor into conduction. Current then flows in the two voltage-divider collector resistors. This turns on the SCR and immediately blows the fuse. R_1 determines the voltage-drop that forward-biases the transistor. R_2 must be low enough to burn out the fuse immediately but not damage the SCR.

If a 3-terminal 15-V regulator IC has a 500-Ω resistor across its output with a second 500-Ω re-

sistor connected from the negative IC terminal to ground, the ground to +15-V output terminal will now provide a regulated 30 V. The two resistors form an equal-value series voltage divider. There must be 15 V across the top one—therefore, there must be 15 V across the lower resistor, totaling 30 V.

Figure 10-25 illustrates the similarity of solid-state and VT series-type voltage regulators.

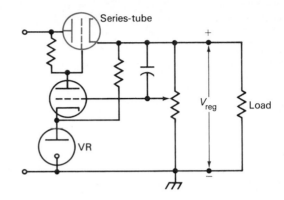

FIG. 10-25 Series-tube regulator.

10-25 VOLTAGE DIVIDERS

In some applications it is necessary to have two different unregulated load voltages. A *voltage-divider* network of resistors can develop a lower voltage from a higher voltage of a power supply. The voltage divider may also serve as a bleeder. Regulation is poor with voltage dividers.

EXAMPLE: A 500-V power supply is used in a 2-stage transmitter. One stage requires 500 V at a current of 60 mA. The other requires 400 V at 40 mA. The bleeder current is to be 15 mA. What resistors will the voltage-divider circuit require? See Fig. 10-26.

FIG. 10-26 Voltage-divider circuit feeds different voltages to two loads.

R_1 and R_2 in series form both the voltage-divider and the bleeder circuit.

The voltage across R_2 is 400 V at a current of 0.015 A. According to Ohm's law, $R_2 = V/I$, or 400/0.015, or 26,667 Ω.

The current through R_1 is the sum of the 0.015-A bleeder current plus the 0.04 A of the 400-V load circuit, or 0.055 A. The voltage-drop across R_1 is 100 V. Its resistance, by Ohm's law, is $R_1 = V/I$, or 100/0.055, or 1818 Ω. The wattage for the resistors is usually twice the computed I^2R, VI, or V^2/R values.

10-26 COPPER-OXIDE AND SELENIUM DIODES

If a sheet of lead and a sheet of copper covered with copper oxide are pressed together, current will flow more readily from copper to oxide than in the other direction. Such a unit is called a *copper-oxide rectifier*. They will stand only a limited inverse peak voltage. In high-voltage applications, many units must be connected in series for high inverse-voltage capabilities. They have little barrier voltage and are used as meter rectifiers.

When iron is coated with selenium, a selenium rectifier is produced. They are sometimes manufactured in a full-wave bridge circuit. Four rectifiers are mounted on a single rod and interconnected in such a manner that when ac is fed to the proper pair of leads, dc is obtained from the other two leads. Figure 10-27 shows a bridge rectifier with each rectifier numbered and then the physical interconnections of such a rectifier group.

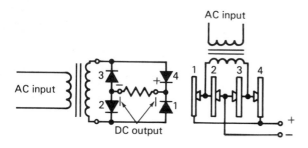

FIG. 10-27 Bridge circuit used with copper-oxide meter rectifiers and one method of stacking and connecting units.

10-27 SWITCHING SUPPLIES

Up to this point power supplies have converted ac to dc. There are many cases in which low-voltage

dc, 12 V from an automobile battery, for example, must be converted to 20 to 80 V dc for power transistors or to several hundred volts for VT circuits.

The dc-to-dc converter or *switching*-type power supply in Fig. 10-28 uses two transistors which alternately pulse current through the two halves of the 40-turn primary. This is stepped up by the transformer, *bridge-rectified*, and filtered to provide, in this case, 300 V.

If 12 V is applied to the forward-biased BJTs, one of them, perhaps Q_1, begins to draw more I_C than the other, and greater current starts flowing downward through the 40-turn primary. This induces voltages into both 5-turn secondaries. The upper secondary voltage forward-biases Q_1 still more and its I_C rapidly rises to saturation. At the same time, the voltage induced into the lower secondary reverse-biases Q_2 and no I_C flows in Q_2.

When the I_C of Q_1 reaches saturation, the core of the transformer no longer has a magnetic flux building up in it and the induced voltages in the 5-turn secondaries drop to zero. The 25-μF capacitor of Q_1, which has been positively charged by the upper 5-turn secondary, begins to discharge through the $R_B R_S$ bias resistors. A reversed current is developed in the upper 5-turn secondary, and by induction Q_2 becomes forward-biased. This drives an I_C pulse upward through the lower half of the 40-turn pri-

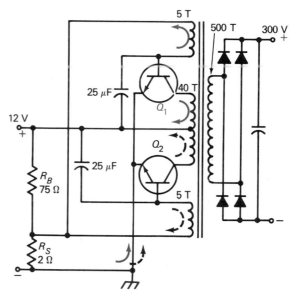

FIG. 10-28 Transistor-type dc-to-dc converter.

mary, reverse-biasing Q_1 and cutting off its I_C, but driving Q_2 into I_C saturation. The Q_2 upward-flowing I_C pulse induces a high voltage into the high-voltage secondary of opposite polarity from the voltage induced when Q_1 was functioning. The net result is a continuing oscillation of saturated I_C pulses being switched up and down in the primary inducing a square-wave ac into the secondary. Since the induced ac is flat-topped, it is easier to filter when rectified than rectified sine-wave ac would be. The whole BJT circuit is a push-pull Armstrong oscillator (Chap. 11). The frequency of the ac developed depends on the values of the two capacitors, the biasing resistances, and the inductance of the coils involved. The higher the frequency of the ac, the less filter necessary and the lighter and smaller the transformers. These circuits may develop and rectify 500 Hz to 100 kHz ac. The transformer is often an iron-cored toroid.

If no rectifiers are used on the secondary circuit in Fig. 10-28, the converter becomes a dc-to-ac circuit and is known as an *inverter*. Inverters are often made to operate at 60 Hz.

One of the first switching power supplies was the vibrator type. Instead of having two transistors, it used an electromagnetic vibrator. When the mechanical arm vibrated between upper and lower contacts it fed dc to upper and lower halves of a transformer primary, inducing ac into the secondary. If dc output was required, the ac was fed to a rectifier-filter circuit.

Some modern switching-regulated supplies (Fig. 10-29a) use a series or pass transistor which is switched on and off at a given rate by a transistor square-wave ac generator, called an oscillator (Chap. 11). To maintain a constant output voltage, an amplifier senses any lowering of the power-supply output under load and lengthens the on-time of the switching oscillator, resulting in more energy being fed to the filter capacitor, C, and therefore a return to normal of the output voltage.

In other switching supplies (Fig. 10-29b) the lowering of the output voltage under load is sensed by an amplifier. Its output voltage is fed to a voltage-controlled oscillator (VCO). As the amplifier voltage changes the VCO output ac frequency, the oscillator frequency approaches the resonant frequency of an LC circuit, and the output voltage of the LC circuit increases. This increased ac voltage amplitude fed to a switching transistor drives it harder, enabling it to feed more energy to the filter capacitor, C, forcing the supply voltage to return to the normal voltage output.

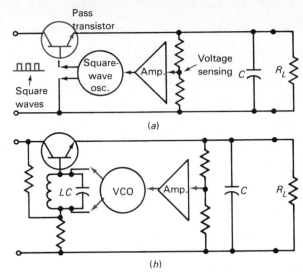

FIG. 10-29 Switching supplies in semi-block diagram form. (*a*) By controlling ON time and (*b*) by controlling VCO frequency.

Test your understanding; answer these checkup questions.

1. What determines the regulation of voltage multipliers? _____
2. Solid-state rectifiers operating into capacitive filters usually require what other component? _____
3. How many diodes are required in a voltage tripler? _____
4. What would be a resistance to use between a 20-V unregulated dc and a 12-V zener feeding a 25-mA load? _____ What power rating should the resistor have? _____ The diode? _____
5. Within about what voltage range would a 5-W, 10-V zener regulate a d95-mA current? _____ A d9.5-mA current? _____
6. What is a shunt regulator connected across? _____
7. With what is a series regulator in series? _____
8. What four voltages will VR tubes regulate? _____ _____ _____ _____
9. What is the disadvantage of using high unregulated voltages for shunt regulators? _____
10. What is the minimum current for proper ionization of a VR tube? _____ Maximum VR current? _____
11. What current change occurs in the dropping R_s of a zener or VR tube when I_L increases? _____
12. How can transistor regulators be made to improve their regulation? _____
13. Is an emitter-follower type regulator a series or shunt type? _____

14. Why does a pass transistor have to have a heat sink? _____

15. Does a crowbar circuit protect the load, the unregulated power supply, or both? _____

16. Does a voltage divider provide a high or low percent regulation to a load? _____

17. What solid-state diodes were used prior to silicon and germanium? _____

18. What are the advantages of using high frequencies with switching supplies? _____ _____

_____

19. Draw diagrams of the following circuits: voltage doubler (two types). Voltage quadruplers. Zener-type shunt-regulated power supply. VR regulator circuit and load. Simple transistor shunt regulator (two types). Simple series regulator. VT series regulator. Series regulator that might be used in a regulator IC. Crowbar circuit. Transistor-switching power supply.

10-28 THREE-PHASE POWER

For high-power high-voltage supplies in AF or RF stages of radio transmitters, 3-phase (3-ϕ) power has several advantages over single-phase:

1. The ripple frequency is three times that of single-phase. A 60-Hz single-phase full-wave rectifier has a ripple frequency of 120 Hz. A 60-Hz 3-ϕ full-wave rectifier has a ripple frequency of 360 Hz. Half-wave 3-ϕ with its 180-Hz ripple is easier to filter than full-wave single-phase.

2. The pulses of rectified current in either full- or half-wave 3-ϕ circuits overlap. The current value never drops to an instantaneous value of zero, making the 3-ϕ rectified current still easier to filter. Power is present in the unfiltered circuit at all times.

3. The output voltage of a 3-ϕ system may be 73% higher than the turns ratio of the transformers would indicate.

Three-phase ac is produced by special ac generators, properly termed *alternators*. In effect, they are three single-phase alternators in one. A single-phase alternator has two leads coming from it (Fig. 10-30). A 2-ϕ alternator would have four leads, and a 3-ϕ alternator six leads.

A simple 3-ϕ alternator may have three separate pickup coils that rotate between electromagnetic-field poles. In each pickup coil a single-phase ac is induced. These coils are set on the revolving part, called a *rotor*, or *armature*, in such a way that the voltages induced in the different coils are 120° apart. The 3-ϕ alternator in Fig. 10-30 would be capable of supplying three separate 1-ϕ circuits, or the coils can be combined in one of two methods to produce a three-wire 3-ϕ circuit (Fig. 10-31). The first diagram shows the Y connection (also known as *wye*, or *star*). All three pickup coils are connected together at one point. The other ends of the coils form the leads that are brought out of the alternator. Any two output leads will carry voltages from two of the coils in series. Since the voltages in the coils are 120° out of phase, there will never be a time when the two voltages will be at a maximum together. With 100 V as the peak in all coils, there will never be a time when 200 peak V will appear across any two leads. By trigonometry, the peak voltage can be computed to be 173 V between any two legs of a Y-connected, 100-V per coil, 3-ϕ alternator. The voltage across any output line of a Y-connected alternator is the voltage of one of its pickup coils times the factor 1.73.

In Δ-connected (delta) circuits (Figs. 10-31 and 10-32), any two lines leading from the machine are directly connected across a pickup coil. If the coil has 100 V induced in it, there will be 100 V across the line. The other two coils have voltages induced in them, but these will be out of phase, with a

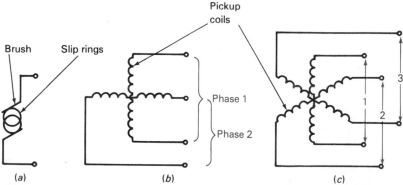

FIG. 10-30 (*a*) Two terminals from a single-phase alternator; (*b*) four from a 2-phase alternator; and (*c*) six from a 3-phase alternator.

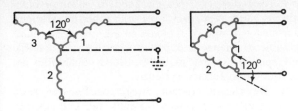

FIG. 10-31 Alternator pickup coils or transformer secondaries connected in Y and in delta (Δ).

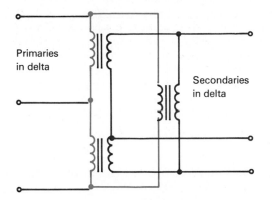

FIG. 10-32 Transformers connected Δ-Δ.

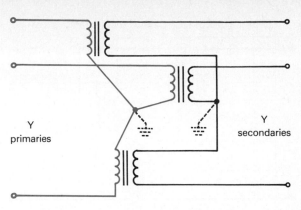

FIG. 10-33 Transformers connected Y-Y.

connected with primaries in Δ and secondaries in Y (Fig. 10-34), each output phase will have 1.73 times the voltage of the primary phases. If the transformers have a 1:10 step-up ratio, the secondary-line voltage will be 17.3 times the primary voltage.

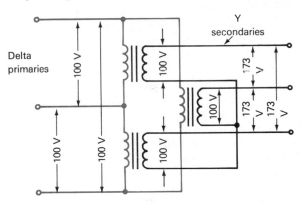

FIG. 10-34 Transformers connected Δ-Y.

resultant of 100 V at the instant the first coil attains its peak value. This parallels all voltages at all times and in such phases as to produce 1.73 times as much current as any one pickup coil alone might carry.

A Y-connected alternator produces higher voltage at lower current. The very same alternator if Δ-connected produces lower voltage output but greater current. The power output is the same.

Three-phase power lines usually have three wires, although the center, or *neutral*, connection on a Y circuit may be grounded (dashed, Fig. 10-33).

When either a single-core 3-ϕ transformer or three separate transformers terminate in a three-wire line, there are five methods of connecting the windings: (1) Δ primary, Δ secondary—termed Δ-Δ; (2) Y-Y; (3) Δ-Y; (4) Y-Δ; and (5) open Δ (which requires only two single-phase transformers).

Figure 10-32 illustrates three transformers with primaries and secondaries connected in Δ. Figure 10-33 illustrates three transformers with primaries and secondaries in Y.

If the transformers used have a 1:1 ratio, there will be no step-up or step-down in either the Δ-Δ or the Y-Y connections. With the same transformers

Using the same transformers connected with primaries in Y and secondaries in Δ (Fig. 10-35), each output phase will have a voltage equal to the reciprocal of 1.73, or 0.578, times the input- or primary-phase voltage. However, the current for each phase of the output can be 1.73 times the current in a primary phase.

With balanced loads on a 1:1 transformer 3-ϕ system the following apply:

Delta-connected:

$$V_{\text{line}} = V_{\text{phase}}$$
$$I_{\text{line}} = 1.73 I_{\text{phase}}$$
$$I_{\text{phase}} = I_{\text{line}}/1.73$$

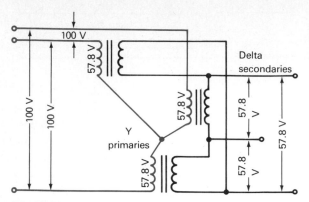

FIG. 10-35 Transformers connected Y-Δ.

Wye-connected:

$$V_{\text{line}} = 1.73 V_{\text{phase}}$$
$$V_{\text{phase}} = V_{\text{line}}/1.73$$
$$I_{\text{line}} = I_{\text{phase}}$$

For either delta or wye:

$$\text{VA} = 1.73 V_{\text{line}}(I_{\text{line}})$$
$$\text{W} = 1.73 V_{\text{line}}(I_{\text{line}})\ \text{pf}$$
$$\text{pf} = \text{W}/\text{VA}$$

It is possible to use only two transformers in a 3-ϕ system. This configuration is known as an open-Δ circuit. It produces 3-ϕ ac by using the three secondary wires, as indicated in Fig. 10-36. It is equivalent to a 3-transformer-Δ circuit with one

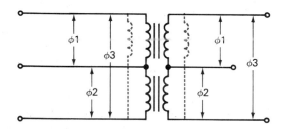

FIG. 10-36 Two transformers connected open-Δ. (Transformer windings dotted in color.)

of the transformers open-circuited in either or both primary and secondary, or with one transformer disconnected (shown dotted). In an emergency it is possible to change a Δ power system to an open Δ and operate at 58% of the full-load capabilities of the three-transformer system.

A Y-connected output circuit can power one, two, or three single-phase circuits and still operate 3-ϕ motors, etc., from the same transformers at the same time (Fig. 10-37). Only the secondaries of the transformers are shown.

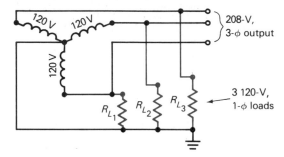

FIG. 10-37 A 208-V 3-ϕ Y with grounded neutral can also supply three 120-V 1-ϕ lines.

In some cases one phase winding of a 240-V Δ-connected power transformer may be center-tapped and used as the neutral for two relatively low-power 120-V single-phase circuits.

Computations of power become quite involved in 3-ϕ circuits. However, if reasonably well-balanced, with each phase taking approximately one-third of the load, the total power can be determined by measuring the power in any two phases separately. If the power factor is more than 0.5, as is usually the case, the total power of the three phases is equal to the *sum* of the two power readings taken. If the power factor is less than 0.5, the total power is equal to the *difference* between the two readings. Figure 10-38 shows how two single-phase watt-meters can be connected to read the total power in

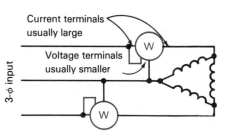

FIG. 10-38 Two 1-ϕ wattmeters connected to read the power in a 3-ϕ system.

a 3-ϕ system. There are also 3-ϕ wattmeters that consider all phases and indicate the true power of the whole circuit.

Proper phasing (connections) of transformers in either Δ or Y circuits is absolutely necessary.

10-29 THREE-PHASE POWER SUPPLIES

When the three phases of 3-ϕ ac are plotted against time, they form the pattern shown in Fig. 10-39. This can be resolved into three separate single-

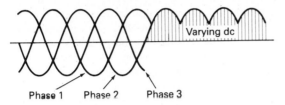

FIG. 10-39 Three-phase ac and varying dc ripple when half-wave-rectified.

phase sine waves separated by 120°. The shaded lines indicate the resultant voltage or current when a 3-ϕ ac is half-wave-rectified. The pulses never drop to zero, making the waveform a varying rather than a pulsating dc, and the equivalent of three ripples per ac cycle.

A half-wave-rectified 3-ϕ power-supply circuit is shown in Fig. 10-40. The 1:10 ratio transformers are connected Δ-Y. This circuit does not give the 1.73-voltage gain as a half-wave rectifier, because no two phases are in series across the load at any time.

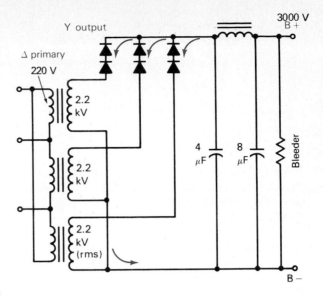

FIG. 10-40 Three-phase half-wave Y-connected power supply. Number of diodes depends on their inverse voltage ratings.

A full-wave 3-ϕ bridge-rectifier circuit is shown in Fig. 10-41. The power transformers are connected Δ-Y. The neutral connection cannot be grounded if B$-$ is. The dashed lines indicate one current path at one instant of maximum voltage in one phase. Two of the secondary windings are in series across the load. This results in a 1.73-voltage step-up over the turns ratio of the transformers used, and the equivalent of 6-ϕ pulses.

The waveform of full-wave-rectified 3-ϕ ac is

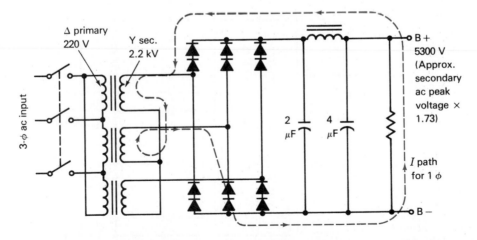

FIG. 10-41 Three-phase full-wave Y-connected power supply showing current path during the peak of one of the phases.

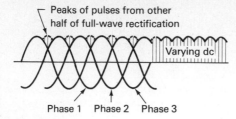

FIG. 10-42 Three-phase ac and varying dc ripple when full-wave-rectified.

shown by the shaded lines in Fig. 10-42. Because of the small amplitude variation, these pulses are very easily filtered and may need little or no filter choke.

If two full-wave 3-ϕ rectifier circuits, one connected delta and the other star, are in series, the two sets of rectified pulses will be 30° out of phase. This results in a 12-ϕ power supply (Fig. 10-43). Such a supply is twice as easily filtered as the 6-ϕ or normal full-wave-rectified 3-ϕ supplies. Only about 5 μF of capacitance is needed (no filter choke at all). The output voltage will be the sum of the two peak-rectified voltages. The delta windings will have to have a 1.73 times higher turns ratio than the Y windings.

10-30 TROUBLESHOOTING POWER SUPPLIES

An important phase of a technician's duties is to be able to locate and repair equipment failures when they occur or are about to occur.

SOLID-STATE POWER SUPPLIES. Little warning of malfunction is given with this type of power supply. There may be some preliminary heating of components and perhaps some smoke before the fuse, overload relay, or circuit breaker goes out. Fuses may wear out and open after years of operation for no reason. Try a new fuse in such cases. However, there is usually trouble when a fuse opens. It may be a short circuit in the primary wiring, a shorted transformer, welded-together diodes, shorted filter capacitors, shorted-to-ground filter chokes, defunct active devices, or a short in the equipment the power-supply feeds.

The idea of troubleshooting is to localize the difficulty as rapidly as possible. First, examine the equipment carefully, looking for signs of heated parts, loose metal pieces across connecting wires, etc. Then, if nothing seems wrong, check the load on the power supply first. An ohmmeter across the output of a power supply feeding semiconductor circuits should read relatively higher resistance one way and lower the other. If a zero or near-zero resistance appears across the load terminals, disconnect the load. If zero resistance shows across the load, this is probably where the trouble is. If the load reads higher resistance one way than the other, the trouble is probably in the power supply. Zero ohms across the power supply alone indicates it is malfunctioning.

First checks should be of the filter capacitors for shorts. An ohmmeter should show a charging of the capacitors by first showing zero ohms resistance, which slowly changes to higher values over a pe-

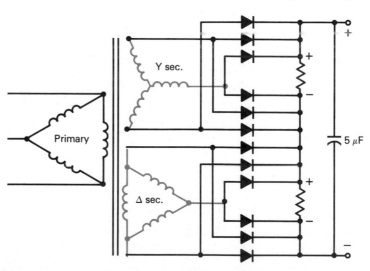

FIG. 10-43 A 12-ϕ rectifier power supply.

riod of a few seconds. If charging is not observed, disconnect one lead of the capacitor and check this component itself. If the capacitor checks shorted, replace it and try the power supply again. Test rectifier diodes next. They should all show much higher resistance one way than the other. If not, replace them and try the power supply again. Continue checking resistors, capacitors, diodes, and transistors, looking for higher resistance one way than the other with semiconductors and shorts or opens with other components.

If a pass transistor shorts emitter to collector it will apply full unregulated voltage to the load circuit, often damaging load circuit components unless the fuse opens immediately. An open zener reference-voltage diode will heavily forward-bias the pass transistor, possibly burning it out, and it may also cause load damage.

If nothing wrong is found, disconnect the rectifiers from the transformer and try turning on the supply with only the transformer in the circuit. If the fuse blows again, either the transformer is internally shorted or there is some unseen short circuit in the wiring related to the transformer. If the power supply is disconnected from the load and operates with its rated voltage output, the trouble is probably in the load equipment.

TUBE-TYPE POWER SUPPLIES. There are often several possible indications of power-supply failure. Light gray wisps of smoke may curl up from under a chassis. Sparks may be seen in a rectifier tube. The hissing of an electric arc-over may be heard. Rectifier plates may turn red-hot. Mercury-vapor tubes may turn a brilliant blue-white. A purple or blue haze may appear between filament and plate in a vacuum diode. Meters in associated equipment drop to zero or indicate excessive values. Transformers or chokes may hum ominously or start to smoke. Circuit breakers or overload relays may snap open; fuses may burn out; lights may appear on panels; alarm bells may ring.

With transmitters, the operator should (1) immediately shut down all the equipment or (2) note the stages that indicate normal operation and then shut down the equipment and start looking for the trouble in the first stage that indicates improper operation.

In power supplies, tubes age and lose emission rather slowly, usually indicated by a slowly decreasing output current, or the filament of a tube may burn out without warning. Vacuum tubes may become gassy and develop a purple or blue glow between filament and plate. If the tube develops a rapid gas leak, it will suddenly turn milky and the filament will burn out immediately, depositing a white oxide layer on the inside of the envelope. Visual examination of the power supply may tell the operator which tube is at fault.

Smaller receiving-type rectifier tubes can be tested on a tube tester. Transmitting tubes can be tested by heating the filament of a *good tube* to a normal degree and connecting a dc source, an ammeter, and a variable resistance in series with the plate circuit of the tube. The resistance is adjusted until the maximum rated operating plate current flows through the tube. When a suspected tube is tested in place of the good tube, the plate current should be at least 80% of that of the tube known to be good.

A short circuit in equipment usually produces dramatic indications (red plates, internal sparking in the tube, blown fuses). A short circuit may be produced when a filter-capacitor dielectric arcs over and carbonizes. This results in a low resistance through the capacitor and a heavy current in the transformer-rectifier-filter circuit. The first components to check are the filter capacitors. Chokes and filament transformers are usually installed in the positive high-voltage lead with their cores and metal cases grounded. A short circuit between winding and core will cause heavy current in the transformer-rectifier-filter circuit.

If there is a suspected short to ground, an ohmmeter reading between ground and the positive terminal will show a very low resistance. If there is no short circuit, the ohmmeter reading should be the value of the bleeder resistor, usually between 10,000 and 40,000 Ω. (Be sure the equipment is off and capacitors are discharged before using an ohmmeter for any tests.)

Sometimes a filter capacitor does not completely short out but develops a high resistance across it. It is said to have developed a *leak*. A leaking capacitor can be checked by connecting an ohmmeter across it. Any good capacitor of 0.1 μF or more will produce a kick of the ohmmeter needle when first tested. This is the charging current flowing into the capacitor from the ohmmeter battery. If a paper dielectric capacitor registers less than about 10 MΩ (10 megohms) after a second or two, it may be considered as leaking. An electrolytic will read low resistance the first instant it is tested but should register more than 100,000 Ω after a few seconds. If an electrolytic capacitor is tested with the ohmmeter leads reversed, a different resistance value will usually be obtained. This is normal.

If one is available, a *capacitor analyzer* will give a reading of the capacitance, leakage, and power factor and show whether the capacitor is open or shorted. This is better than an ohmmeter test.

To test a capacitor alone in a power supply, one of its leads must be disconnected from the circuit and a meter connected across it; otherwise, the whole filter circuit is being tested.

It is also possible that a short in the load circuit will produce the same symptoms in the power supply as a short in the power supply itself. Disconnecting the power supply from the load and checking both with an ohmmeter will make it possible to localize the trouble more closely.

Sometimes turns in a power transformer short together because of insulation breakdown. Heavy primary current will flow whether the shorted turns are in the primary or secondary, and fuses or overload relays will go out. The transformer may smoke. If it continues to heat or smoke when all parts are disconnected from it, a new transformer is required.

If one-half of the secondary of a center-tapped transformer in a full-wave rectification system burns out, the equipment may operate but at reduced power because of lower voltage output when the circuit operates as a half-wave power supply.

Sometimes a short circuit is an instantaneous thing caused by arcing across a moistened surface or by an insect crawling between two points of high potential in the equipment. A sudden line surge can produce an overload that will burn out fuses or open circuit breakers. As a result, when a piece of equipment suddenly ceases operating and it is desirable that it be placed in operation as soon as possible, it may be expedient to replace the fuse or close the circuit breaker and try momentarily turning on the equipment again. If there is nothing seriously wrong, the piece of equipment will operate normally. If not, the fuse or circuit breaker will go out again and it will be necessary to shut down to locate the trouble. In the first case, it would be well to examine the equipment closely for burns or signs of arcing the next time it is turned off, cleaning it thoroughly at the same time.

After a transmitter has been turned off for servicing, the operator should *always* touch the positive terminals of all power supplies with a grounded flexible wire before servicing the equipment. If such a wire is not available, an insulated-handle screwdriver held across the filter-capacitor terminals will discharge such capacitors. CAUTION: *The operator should make sure he or she is not grounded* when doing this.

If a piece of equipment using a power transformer in the power supply is plugged into a power line carrying 120 V dc instead of the required 120 V ac, the inductance of the primary limits the current flow with ac, but it has no opposition to dc. An excessively high current flows through the primary, burning out the primary or a fuse immediately.

The metallic shields and cores of power transformers and filter chokes should be grounded to prevent them from picking up a static-electricity charge which may be dangerous to personnel. Ungrounded equipment in transmitters may allow high-frequency RF ac to find its way into it and cause damage to internal insulation.

Test your understanding; answer these checkup questions.

1. Why is 3-ϕ rectified ac easy to filter? _____
2. What are the two basic methods of connecting 3-ϕ circuits? _____ _____
3. Using 1:1 ratio transformers and 100-V 3-ϕ ac inputs, what are the output voltages if the transformers are connected: Y-Y? _____ Δ-Y? _____ Y-Δ? _____ Δ-Δ? _____ Open Δ? _____
4. By what percent may 3-ϕ ac step up voltage? _____
5. By what percent may 3-ϕ ac step up current? _____
6. What 3-ϕ circuit may have four wires? _____
7. What 3-ϕ system uses only two transformers? _____
8. What 3-ϕ voltage and connection will provide three 120-V single-phase lines? _____ _____
9. How many 1-ϕ wattmeters are required to measure total power in a 3-ϕ balanced system? _____
10. What kind of current results from using a 1-ϕ bridge rectifier? _____ A 3-ϕ half-wave rectifier? _____
11. How many filament windings are required to supply tube rectifiers in a 3-ϕ half-wave rectifier? _____ A bridge rectifier? _____
12. What might be the first thing to do when a power supply ceases to operate? _____ Next? _____ Next? _____
13. If a nonoperating power supply shows no obvious burns, what might be the next thing to do? _____ Next? _____
14. Before servicing high-voltage power supplies what two precautions should the technician always take? _____ _____

15. Under what conditions does a smoking power transformer not indicate trouble in the transformer?

_____

16. Draw diagrams of transformers in Δ-Δ. Transformers in Y-Y. Transformers Δ-Y. Transformers Y-Δ.

Transformers in open Δ. 208-V 3-ϕ Y to supply three 120-V 1-ϕ lines. Wattmeters to read total power in 3-ϕ lines. Half-wave 3-ϕ power supply. Full-wave 3-ϕ power supply.

POWER SUPPLIES QUESTIONS

1. In what form is electrical energy most available to us?
2. In most cases, what is a "power supply" in electronics?
3. Why is a simple rectifier not usually suitable as an electronic power supply?
4. What are two reasons why half-wave rectification is not used in very many power supplies?
5. What are the names of the two full-wave rectifier circuits?
6. Which requires a higher-voltage transformer for an equal output voltage, a bridge or a full-wave center-tap circuit?
7. Which requires more filter, a high- or a low-resistance load?
8. What is power supply output voltage variation called?
9. Draw a diagram of a semiconductor power supply having capacitive filtering and: half-wave rectification; full-wave bridge rectification; full-wave center-tap rectification with a capacitance multiplier.
10. Using a 20-V rms ac secondary transformer, what would the peak-inverse voltage rating have to be for the diode(s) used in a capacitively filtered: half-wave supply? full-wave bridge supply? full-wave center-tap supply?
11. If a bridge circuit is used to provide two voltage outputs, what is the ratio of these voltages?
12. What effect would series inductance have on the rising wavefront of a square-wave pulse? On the falling waveshape?
13. What type of filter capacitors are used in 5- to 200-V power supplies? 200- to 500-V supplies? 500- to 5000-V supplies?
14. In what one case must inductive input filtering be used?
15. Draw a diagram of a bridge rectifier with C-input filtering.
16. Why is RC filtering not used with heavy current supplies?
17. List nine factors which might have to be considered for power supply or other choke coils.
18. What is the advantage of using a choke with an air gap in its iron core? What is such a choke called?
19. Draw a diagram of a mercury-vapor full-wave rectifier power supply.
20. Draw a diagram of a DPDT switch used to switch a

split primary transformer for operation on 120 or 240 V ac.
21. What is the relationship between ripple frequency and required amount of filter?
22. What are two reasons for using bleeders on high-voltage power supplies? Why are these not valid reasons for 20-V supplies?
23. What is the voltage regulation of a power supply that has 14-V output with no load and 13-V output with a load? If it has 12 V under load?
24. List seven ways to improve regulation of an unregulated-type power supply.
25. Why is full voltage sometimes not applied across high-power tube filaments immediately?
26. Draw a diagram of a common VT power supply.
27. Why is a large-value capacitance sometimes connected in series with the primary of a high-frequency power transformer?
28. Draw a diagram of a full-wave voltage-doubler power supply circuit.
29. In the full-wave voltage-doubler circuit, what is the peak inverse voltage across either of the diodes?
30. If a surge-current choke is used in the voltage-doubler circuit, approximately what inductance might it have?
31. Draw a diagram of a half-wave voltage tripler.
32. If a power-supply load is in parallel with a regulating device, what type of regulator is it called?
33. It is desired to have a 10-V zener regulate a 70-mA load from an unregulated 18-V supply that drops to 17 V under load. What is the required R_s? The no-load zener current? What wattage zener would be used?
34. Does a zener diode protect its load against high line voltage transients?
35. Why is a 2-transistor shunt regulator better than a 1-transistor circuit?
36. Draw a diagram of a 2-transistor shunt voltage regulator.
37. Why should the unregulated voltage not be too much more than the output voltage of a zener shunt regulator?
38. Why is less filter capacitance required if the output of a power supply is regulated?
39. What does the BJT in the base circuit of a pass transistor do?
40. Draw a diagram of a 2-BJT series-regulator circuit.

41. Draw a diagram of an overload protection circuit in a series regulator.
42. Against what does a crowbar circuit protect?
43. Draw a diagram of a crowbar-type overload protection circuit for a regulator IC.
44. In Fig. 10-26, if the power supply is a regulated 12 V and the second voltage is to be 5 V at 25 mA, with a bleeder current of 20 mA, what are the R and P values of the resistors?
45. What accomplishes the switching in the basic switching power supply described?
46. What was one of the first switching-type power supplies?
47. What are two other methods of regulating output voltage in switching power supplies?
48. What is the ripple frequency of a full-wave 1-ϕ power supply from a 60-Hz line? Of a full-wave 3-ϕ?
49. What would polyphase ac be?

50. How many phase degrees between peaks of 3-ϕ ac?
51. What is another name for Y-connected alternators?
52. What is the voltage ratio of 1:1 transformers if connected in Y-Y? Δ-Δ? Y-Δ? Δ-Y?
53. What 3-ϕ circuit usually has a ground connection?
54. When only two transformers are used in a 3-ϕ system, what is the circuit called?
55. Draw a diagram of Y-connected transformers. Is this the same as a Δ-Y circuit?
56. Draw a diagram of open-Δ transformers.
57. How many 1-ϕ wattmeters are required to measure power in 3-ϕ circuits? How many 3-ϕ wattmeters are required?
58. Draw a diagram of a 3-ϕ half-wave Y-connected power supply.
59. Draw a diagram of a 3-ϕ full-wave Y-connected power supply.
60. Draw a diagram of a 12-ϕ rectifier power supply.
61. If a fuse blows what should be done first?
62. List three important steps in localizing trouble.
63. How is a power-supply transformer tested?
64. In transmitters, how is trouble detected?
65. In VT equipment, what is checked and in what order?
66. In power supplies, what is a good check-out procedure?
67. After turning off high-power transmitting equipment, what should always be done before checking parts or circuits?

ANSWERS TO CHECKUP QUIZ ON PAGE 198

1. (Pulses overlap) **2.** (Wye, or Y) (Delta, or Δ) **3.** (100) (173) (57.8) (100) (100) **4.** (73%) **5.** (73%) **6.** (Grounded neutral Y) **7.** (Open Δ) **8.** (208 V) (Y) **9.** (2) **10.** (Pulsating dc) (Varying dc) **11.** (1) (3) **12.** (Check fuse) (Examine) (Replace fuse and try) **13.** (Disconnect load and try) (Ohmmeter tests) **14.** (Pull plug) (Short HV filters) **15.** (If still connected to rectifier) **16.** (See chapter illustrations)

11

Oscillators

This chapter explains the operation of some of the many different types of oscillator circuits used today with semiconductor devices and vacuum tubes. This includes Hartley, Colpitts, TPTG, ECO, crystal, *RC*, and PLL oscillator circuits.

11-1 TYPES OF OSCILLATORS

An oscillator circuit produces alternating current from a fraction of a watt to possibly thousands of watts. When high-power power-frequency ac (10 to 1000 Hz) is required, various types of electromagnetic alternators may be used. For higher frequencies, in the audio- and radio-frequency ranges, transistor or tube oscillator circuits are employed.

The first type of oscillator used to generate HF (high-frequency) ac was the spark circuit. It produced a damped (decaying) ac waveform up to more than 2 MHz. Another early HF oscillator was the Poulsen arc. It developed a constant-amplitude ac up to about 500 kHz. Still other types of early RF generators were the Alexanderson and Goldsmith alternators for frequencies below 50 kHz. Then came the triode tube, now capable of generating ac up to more than 3 GHz.

Since the 1930s, the magnetron tube, klystron tube, traveling-wave tube (TWT), and backward-wave-oscillator (BWO) tube have extended the vacuum-device oscillating frequency up to 50 GHz and more. Modern transistors can produce ac above 10 GHz, while tunnel and other special diodes can produce ac at frequencies over 100 GHz (Chap. 22).

11-2 SHOCK EXCITATION

As a means of explaining the operation of oscillator circuits involving coils and capacitors, the shock-excitation, or flywheel, theory will be used.

If the switch in Fig. 11-1 is closed for a very brief instant and then opened, electrons from the

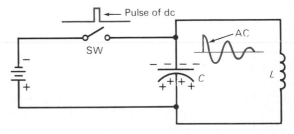

FIG. 11-1 Basic *LC* oscillator. Quick closing and opening of the switch charges the capacitor.

battery (1) flow to the top plate of the capacitor and charge it negative and (2) are pulled from the bottom plate, charging it positive. The inductance of the coil prevents any material current flow through it for the instant that the switch is closed. As the switch opens, electrons deposited on the top plate of the charged capacitor start to move toward the

201

positive plate, downward through the coil. The battery has *shock-excited* the coil-capacitor circuit, and the circuit starts into operation, using the pulse of energy from the battery as the motivating power.

The current of electrons flowing through the coil causes a magnetic field to expand outward (Fig. 11-2a), inducing a counter emf in the coil which prevents the capacitor from discharging immediately.

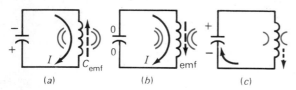

FIG. 11-2 (a) Capacitor discharges through the coil, producing a magnetic field. (b) Collapsing field discharges capacitor. (c) Continued collapsing recharges capacitor to opposite polarity.

As the capacitor discharges, it reaches a point where there are the same number of electrons on both plates (Fig. 11-2b). With no emf across the coil, there should be no current in it and nothing to hold the field out around the coil. The field collapses back inward, inducing a downward emf in the coil. This forces free electrons in the wire of the coil and from the top plate of the capacitor to flow down through the coil to the bottom plate, charging the bottom plate negative and the top plate positive (Fig. 11-2c).

Now the capacitor is charged again, this time with an opposite polarity, but the *amplitude* of the charge is exactly the same. This charged condition now causes the current to reverse itself in the circuit, swing back up through the coil, and recharge the capacitor to a polarity the same as originally developed in it by the shock excitation of the battery. One cycle of ac has been produced.

If there were no losses in the circuit this ac current would *oscillate* back and forth, indefinitely, with a constant amplitude for each cycle. The result would be a perpetual ac generator. Because there are always losses in circuits, each succeeding half cycle has an amplitude less than that of the one before. In a short time the ac is damped out entirely (Fig. 11-3). The less the resistance or losses, the longer it takes the current to damp out.

The oscillation of electrons back and forth in an *LC* (inductance-capacitance) circuit is known as *flywheel effect*. Most sine-wave-generating oscillators utilize this effect.

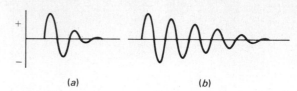

FIG. 11-3 AC damps out (a) rapidly in low-Q circuits and (b) slowly in high-Q circuits.

The explanation of shock excitation assumed that the capacitor received the original pulse of energy. It is also possible to shock-excite an *LC* circuit by inducing a pulse of current into the coil, which can also produce ac oscillation in the circuit.

If analyzed, electrostatic energy in the capacitor is alternately converted into electromagnetic energy around the coil and then back to electrostatic energy in the capacitor. Energy in an *LC* circuit oscillates from electrostatic to electromagnetic form.

The spark oscillator originally used for radio communication inserted a spark gap in series with an *LC* circuit and then keyed a low-frequency ac voltage across the capacitor (Fig. 11-4). When the

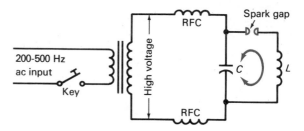

FIG. 11-4 Spark-gap oscillator. Low-frequency ac charges *C*, ionizes air of gap, and allows damped HF ac to oscillate in the *LC* circuit.

emf reached a voltage high enough to ionize the air between the electrodes of the gap, a spark flashed across, heating the air. The ionized hot air acted as a conductor, allowing the charge that had built up in the capacitor to produce oscillations of a damped or dying-out type at the natural frequency of resonance of the *LC* circuit. The radio-frequency chokes (RFC) prevented the high frequencies from feeding back into the power-frequency transformer.

11-3 ELECTRONIC *LC* OSCILLATORS

The basic theory of the generation of sine-wave ac of constant amplitude and frequency in transistor and vacuum-tube oscillator circuits might be outlined as follows:

1. A tuned *LC* circuit is shock-excited into oscillation by any small voltage.
2. The ac voltage from the oscillating circuit is amplified by some type of active device.
3. Amplified ac energy is fed back to the original tuned circuit by either inductive or capacitive coupling.
4. The energy is fed back in phase to aid the oscillating energy in the *LC* circuit and must be strong enough to overcome all *LC* circuit losses.
5. The feedback keeps the shock-excited *LC* circuit oscillating at its resonant frequency.
6. The oscillator circuit draws all its oscillating energy from the dc power supply of the active device.
7. Ac power produced in the oscillating *LC* circuit can be taken from it by either inductive or capacitive coupling (Figs. 11-5, 11-7).

11-4 ARMSTRONG OSCILLATORS

An Armstrong, or inductive-feedback, oscillator circuit is shown in Fig. 11-5*a* and *b*.

When the switch is closed (Fig. 11-5*a*), a surge of electrons begins to flow through the *tickler coil* producing an expanding magnetic field around it. The field cuts across the turns of the coil of the *LC* circuit, inducing a voltage in it. This induced voltage shock-excites the tuned circuit, which starts to oscillate at the *LC* frequency.

As the *LC* circuit starts to oscillate, the ac voltages developed across it are fed to the amplifying device, producing a relatively high-amplitude *variation* of tickler current. Expanding and contracting magnetic fields around the tickler coil induce ac into the *LC* circuit, keeping the circuit oscillating.

Note that the tickler coil and the *LC* circuit form a transformer, with the tickler coil as the primary. All the ac energy in the *LC* circuit is the result of induction from the tickler.

As long as the two coils are oriented in such a manner as to produce an aiding effect between them, oscillation of the *LC* circuit will continue to increase in strength until maximum oscillation power is reached (almost immediately). The maximum amplitude is determined by several factors—power-supply voltage, degree of coupling between tickler and *LC* circuit, *Q* of the *LC* circuit, amount of bias developed, and value of capacitance from tickler to ground.

If the tickler coil were wound in the improper direction, the voltages induced into the *LC* circuit by the tickler-coil current would be out of phase

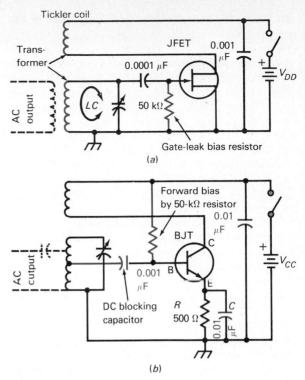

FIG. 11-5 Armstrong oscillators. (*a*) JFET with inductive-output coupling. (*b*) NPN BJT with capacitive-output coupling.

with oscillations in the *LC* circuit, the circuit would not produce sustained oscillations, and no ac would be generated.

All *LC* circuits have some losses (resistance, or induced energy into some external circuit). The energy fed back to the *LC* circuit must be sufficient to overcome all losses to allow the *LC* circuit to oscillate at a constant amplitude and not damp out. Actually, considerably more feedback than this minimum value is used.

The in-phase feedback effect capable of producing oscillation is known as *regeneration*. Out-of-phase feedback will prevent oscillation and is known as *degeneration*. If it is desired to make an electronic circuit oscillate, it will be necessary to introduce regeneration in it. If it is desired to prevent it from going into oscillation, it may be necessary to introduce some degeneration into it, or to *neutralize* it.

In the Armstrong circuit there are several methods by which the ticker-to-*LC* circuit feedback can be controlled: (1) by varying the tickler-to-*LC* circuit coupling, (2) by varying the dc power-supply voltage (and therefore the tickler-coil current), and

(3) by varying the tickler-to-ground capacitance, which completes the ac path from tickler circuit to source or emitter. The greater this capacitance, the greater the ac component flowing in the tickler coil.

The operation of an NPN BJT Armstrong circuit (Fig. 11-5b) is essentially the same as with a JFET (or a triode). Note that the LC circuit is tapped down to a lower impedance point on the coil to match the relatively low base-emitter impedance of the BJT better. Note also that the lower-impedance transistor circuits use larger circuit capacitors than higher-impedance FET (or VT) circuitry does. Changes in V_{cc} will affect the frequency of the oscillator considerably.

The parallel RC network in the emitter circuit is a stabilizing circuit to prevent *thermal runaway*. Such a stabilizing network is found in many bipolar transistor circuits.

11-5 OSCILLATOR BIAS

An N-channel FET (or VT) oscillator circuit requires a relatively high negative dc bias voltage to allow it to operate efficiently. If a battery were used to supply the bias voltage, the circuit could oscillate but might not be self-starting. A satisfactory bias voltage can be developed with *gate-leak* bias (grid-leak for VTs). The oscillator will always be self-starting because no bias voltage is developed until the circuit actually begins to oscillate. Figure 11-5a shows a shunt-resistor bias circuit. Figure 11-6

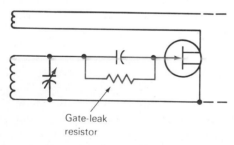

Gate-leak
resistor

FIG. 11-6 Series-type gate-leak bias circuit.

shows series-resistor gate-leak bias, so named because electrons trapped on the gate produce a biasing voltage-drop as they leak through the resistor from gate to emitter.

When the power-supply switch is closed in Fig. 11-5a, output circuit current begins to flow through the tickler coil. A magnetic field expands, inducing a voltage into the LC-circuit coil, and electrons start oscillating in it. On the positive half of the first induced LC cycle, the gate is driven positive, fur-

ther increasing the drain current, which still further increases the voltage induced into the LC circuit. This regenerative action continues until I_D reaches maximum. The magnetic field of the tickler coil can no longer expand, and induction into the LC circuit ceases. Electrons in the LC circuit now begin to move in the opposite direction. The positive voltage at the gate drops off and is driven negative during the negative half of the cycle.

While the gate is being driven positive, it picks up electrons from the source. These electrons can not leak back to the source through the coil because of the 0.0001-μF blocking capacitor and the high-resistance bias resistor (25 to 100 kΩ). The electrons that are trapped on the gate form a negative charge that does not leak off completely even during the negative half cycle of LC oscillation because of the relatively long RC time constant of this bias circuit. The electron charge on the gate over a whole cycle of oscillation averages enough to produce the desired high (class C) bias. The I_D pulses flow only during a small part of the positive cycle on the gate. The oscillator operates quite efficiently (50 to 70%).

The higher the biasing resistance, the higher the average bias voltage and the greater the efficiency of the circuit. If the bias is too high, drain-current pulse energy is reduced and the oscillator will not deliver much ac power output. Bias capacitors vary from 0.00005 μF for high-frequency circuits to perhaps 0.001 μF for low-frequency oscillators. Too much capacitance may cause low-frequency RC oscillations to occur at the same time the LC circuit is oscillating at a high frequency. Such a low-frequency *parasitic oscillation* is normally undesirable.

A BJT Armstrong oscillator without any forward bias can oscillate but will not be self-starting. Without forward bias, closing the switch in the collector circuit does not produce any I_C flow in the tickler coil to excite the LC circuit into oscillation. If forward-biased with a 50-kΩ dropping resistor (Fig. 11-5b), the circuit will start oscillating as soon as the switch is closed. Compare the bias resistor connections for JFET (or triode) and BJT circuits.

> Test your understanding; answer these checkup questions.

1. List the three early types of high-power HF ac generators. _____ _____ _____
2. In what type of circuit does flywheel effect occur? _____
3. What are trains of ac that decay called? _____

4. When electrons rotate back and forth in an *LC* circuit, what is it termed? _____
5. In an Armstrong oscillator, what type of feedback is used? _____ What is the output circuit coil called? _____
6. Why must a transistor Armstrong oscillator be forward-biased? _____ Why is this not true of JFET or VT Armstrongs? _____
7. In Fig. 11-5, what kind of BJT is used? _____
8. If the tickler coil of an Armstrong oscillator is rotated 180°, what would be controlled? _____
9. What term is used to indicate in-phase feedback? _____ Out-of-phase feedback? _____
10. Why are *LC* tank circuits often tapped down when BJTs are the active devices? _____ What other method might produce the same effect? _____
11. How may thermal runaway in transistor oscillators be prevented? _____
12. If an oscillator oscillates at two frequencies at the same time, what is the undesired oscillation called? _____
13. With grid-leak bias does grid current flow when the circuit is oscillating? _____ If the circuit is not oscillating because of excessive coupling to a load? _____

11-6 TUNED-INPUT TUNED-OUTPUT OSCILLATORS

The Armstrong is an inductively coupled oscillator. Energy from the output circuit is induced into the *LC* circuit by transformer or inductive coupling. A VT tuned-plate tuned-grid (TPTG) circuit is an example of a capacitively coupled oscillator.

Why these oscillators work is the reason for many high-frequency amplifiers *not* operating satisfactorily. When it is understood how this circuit oscillates, it will be evident why an amplifier using the same general circuitry will sometimes oscillate instead of amplifying as it should.

In Fig. 11-7a, capacitor C_{gp}, shown dotted, is usually not needed. (In Fig. 11-7b, C_{cb} may be required.) The C_{gp} of the tube elements themselves is normally sufficient to couple back enough ac energy capacitively from the plate circuit to the grid circuit to produce sustained oscillations of electrons in the tuned L_1C_1 circuit, which in turn keeps L_2C_2 oscillating.

The ac grid circuit of Fig. 11-7a consists of the grid, the grid-leak capacitor, the *LC* circuit, and the cathode. The dc grid circuit consists of the grid, the grid-leak resistor, and the cathode.

The ac plate circuit consists of the plate, its *LC* circuit, the bypass capacitor C_{bp}, and the cathode.

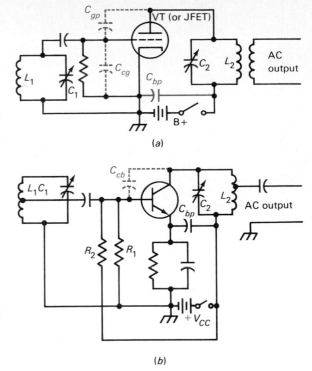

(a)

(b)

FIG. 11-7 TPTG-type oscillators (a) with a triode and (b) with a BJT.

The dc plate circuit consists of the plate, the *LC* circuit, the B battery, the switch, and the cathode. When drawing diagrams, always check for complete grid and plate (gate and drain, base and emitter) circuits for both ac and dc continuity.

The TPTG oscillator must have its grid and plate *LC* circuits tuned to approximately the same frequency. When the switch is closed (Fig. 11-7a), a sudden surge of I_p begins to flow in the plate circuit. This shock-excites the plate *LC* circuit into oscillation at a frequency determined by the values of its inductance and capacitance. When L_2C_2 starts oscillating, an ac voltage is developed across it. This ac voltage is divided across the bypass capacitor C_{bp}, the cathode-grid capacitance of the tube, C_{cg}, and the grid-plate capacitance of the tube, C_{gp}.

The bypass capacitor has relatively low reactance, and very little of the total ac voltage-drop is across it. As a result, practically all the ac voltage across the plate *LC* circuit is between cathode and plate in the tube. If C_{cg} and C_{gp} were equal, about half the L_2C_2 ac would be developed across the grid-cathode circuit. In Fig. 11-7a, the grid *LC* circuit is connected between cathode and grid through the grid-leak capacitor and is therefore subjected to a considerable fraction of the L_2C_2 ac emf. This voltage, fed to the grid LC circuit, forces it into oscillation. However, if the two circuits are not adjusted to approximately the same frequency, the two ac voltages may be sufficiently out of phase to prevent circuit oscillation.

To change frequency in a TPTG oscillator, the grid and plate circuits should be tuned simultaneously to keep them in proper phase and the circuit in optimum oscillation. Actually, the plate circuit must be tuned to a frequency slightly higher than that of the grid to prevent both circuits from being completely resistive (oscillation would stop).

When used to generate lower-frequency ac, the TPTG circuit may not have enough energy fed back from plate to grid circuit by the interelectrode capacitances of the tube. A capacitor of 5 to 10 pF may have to be connected between grid and plate, C_{gp} in Fig. 11-7a. This explanation would have used "gate" and "drain" if a JFET had been used, because there is little basic difference in the operation of a JFET and a triode VT.

Because of the low impedance of transistor junctions, an output-to-input capacitor (C_{cb} in Fig. 11-7b) may be needed in BJT tuned-base tuned-collector (TBTC) circuits. Note also, in the circuit shown, how a voltage-divider network (R_1 and R_2) is used to forward-bias the base instead of using a simple voltage-dropping resistor from $+ V_{CC}$ to base.

11-7 SERIES AND SHUNT FEED

The route taken by the dc in either the output or input circuit determines the type of *feed* used in the circuit. There are two possible means of feeding, *series* and *shunt* (parallel). (1) If dc flows through an active device, an LC circuit, and a dc source in series, the circuit is series-fed. (2) If an active device, a tank circuit, and a dc source are all in parallel, the circuit is shunt-fed.

Figure 11-8a illustrates series-fed gate and drain circuits. Figure 11-8b illustrates shunt-fed base and collector circuits. It is not necessary to use the same type of feed in both input and output circuits.

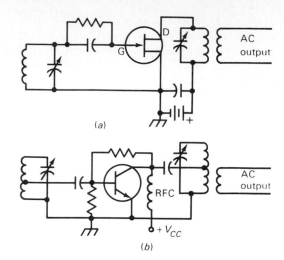

FIG. 11-8 (*a*) Series-fed TGTD gate circuit. (*b*) Shunt-fed TBTC with shunt-fed collector circuit.

When any oscillator is said to be series-fed, it usually means that the output circuit is series-fed.

Figures 11-9*a, b,* and *c* all illustrate shunt-fed Armstrong oscillators. The radio-frequency chokes

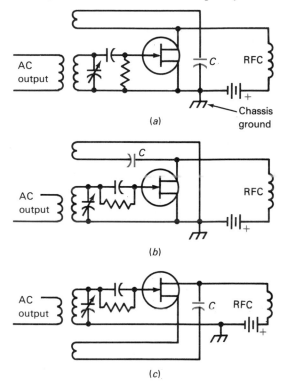

FIG. 11-9 Armstrong oscillators. (*a*) Shunt-fed, (*b*) both coils at ground potential, and (*c*) tickler below LC circuit.

(RFC) have sufficient inductance and little enough distributed capacitance to exhibit high impedance to RF ac or RF varying dc. An RFC is usually not needed in series-fed circuits, although it is sometimes used to keep RF ac out of the power supply.

Series feed of *LC* output circuits puts them at high dc voltages with attendant insulation and safety problems in high-power applications.

11-8 HARTLEY OSCILLATORS

A widely used *variable-frequency oscillator* (VFO) is the Hartley (Fig. 11-10). It is an inductively coupled circuit; I_D variations in the drain part of the tank coil produce induced voltages in the gate half

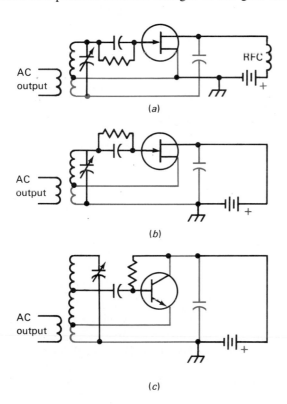

FIG. 11-10 (*a*) Incorporating tickler and *LC* circuit in one tuned circuit changes an Armstrong to a Hartley. (*b*) Common variation to ground one side of tuned circuit. (*c*) BJT Hartley.

of the coil, which are in-phase, or regenerative, and produce sustained oscillations of the tank circuit. However, the circuit will oscillate if the gate and drain halves of the tank circuit are isolated from each other, indicating that part of the regenerative

effect must result from the capacitive coupling through the *LC* tuning capacitor.

The Hartley oscillator may be series- or shunt-fed. It has a center-tapped coil and one tuning capacitor. The tap will be closer to the gate end of the coil for maximum power output and to the drain end for best frequency stability.

Figures 11-9*c* and 11-10 show the evolution of the Hartley from an Armstrong oscillator. All diagrams are practical oscillator circuits. The circuit used will depend on the requirements. If it is desired to keep the rotor of the tuning capacitor at ground potential to reduce *hand capacitance* (detuning the oscillator frequency when a hand is brought near the knob on the rotor shaft of the tuning capacitor), the Fig. 11-10*b* circuit would be used.

Triodes, tetrodes, pentodes, BJTs, or FETs can be used in these circuits. Compare the FET in Fig. 11-10*b* with the tetrode in Fig. 11-11. The bypass

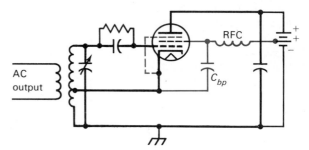

FIG. 11-11 Tetrode Hartley circuit (pentode if suppressor grid is used).

capacitor C_{bp} holds the screen grid and cathode at the same ac potential above ground. To prevent the power supply from ac-grounding the cathode, a high-*Z* RFC (or resistor) must be added between screen grid and the dc supply. For a pentode the dashed suppressor grid circuit is added.

In all FET (BJT) Hartley circuits, the gate (base) couples to one end of the *LC* circuit, the drain (collector) to the other end, and the source (emitter) to the center tap on the coil. Check for this also in the Colpitts oscillators described next.

11-9 COLPITTS OSCILLATORS

A frequently used VFO is the *Colpitts*. It is similar to a shunt-fed Hartley except that, instead of the tank coil being center-tapped, two capacitors in series center-tap the *LC* circuit (Fig. 11-12).

The Colpitts oscillator is normally shunt-fed in both input and output circuits. Without C_2, the dc supply voltage would be across the tuning capaci-

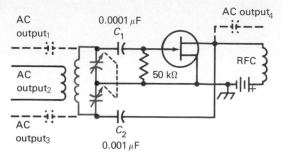

FIG. 11-12 Colpitts oscillator with four possible points and methods of obtaining output ac.

tors. If the rotor and stator plates ever shorted, damage would result to the power supply or the RFC.

Since the two tuning capacitors in the Colpitts circuit are in series, each will be twice the value of the tuning capacitor used in an equivalent LC circuit of a Hartley oscillator. Such large capacitances across the internal capacitances of the active device mean that small changes of internal capacitances that may occur in the device as it warms will have little effect on the frequency of oscillation of the LC circuit. Internal capacitance changes have only about half as much effect on the frequency of oscillation in the Colpitts as they have in the Hartley with any given LC tank-circuit coil.

In any VFO, the greatest frequency stability is usually at the lower-frequency end of the tuning range, because at this end the shunting capacitance of the tuning circuit is greatest and the ratio of possible internal capacitance change to tuning capacitance is highest.

As in the Hartley, the center tap of the tank circuit is not usually in the center. If the gate-end tuning capacitor has the greater capacitance, the greater voltage-drop will appear across the drain-end capacitor and the greater power output will be produced. Better frequency stability may result if the drain-end capacitor has the greater capacitance.

While a variable capacitor is used in explaining the tuning of oscillators, fixed capacitors may be used and a ferrite (or brass) slug can be screwed into and out of the core of the LC circuit coil to tune it. This is known as permeability tuning. A *permeability-tuned oscillator* (PTO) usually uses some form of Colpitts circuit. A PTO is noted for its frequency stability.

Tetrodes or pentodes can also be used in a Colpitts circuit. Try diagraming a pentode Colpitts circuit and then a MOSFET circuit.

11-10 ULTRAUDION OSCILLATORS

The ultraudion oscillator circuit in its usual form is a series-fed Colpitts VT circuit used in very high frequency (VHF) or ultrahigh frequency (UHF) applications. The cathode is brought to the approximate center of the LC circuit capacitance by the interelectrode capacitances of the tube (Fig. 11-13).

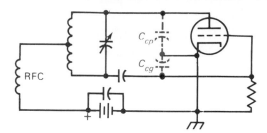

FIG. 11-13 Ultraudion oscillator circuit.

Since the circuit is used mostly at frequencies of over 100 MHz, the coils are quite small and the interelectrode capacitances are large enough to serve as a Colpitts-like voltage-dividing network across the LC circuit. Although shown connected to the "cold" (center) point on the coil, the circuit will operate with the RFC connected to either the grid or plate end of the coil. If a JFET were used, small capacitors would be added where C_{cp} and C_{cg} are shown, and the circuit would be a Colpitts oscillator.

11-11 ELECTRON-COUPLED OSCILLATORS

The true electron-coupled oscillator (ECO) is a combination of one of the previously described VFOs plus the amplification possible in the plate circuit of a tetrode or pentode (Fig. 11-14a). It is an oscillator and amplifier stage in one tube. The amplification results in greater power output and, more important, tends to isolate the oscillator from external circuits (prevents frequency changes due to variations in any circuits following the oscillator). The oscillating section will oscillate even if the output circuit, L_2C_2, is disconnected.

The oscillating circuit is made up of the cathode, grid, and screen grid acting as the anode of the oscillator circuit. When oscillating, the grid voltage varies at the oscillation frequency, developing I_p pulses at the same frequency. The only connection between the output coil in the plate circuit and the oscillating grid circuit is by electron pulses passing the screen grid on their way to the plate; hence the name *electron-coupled oscillator*. The oscillator

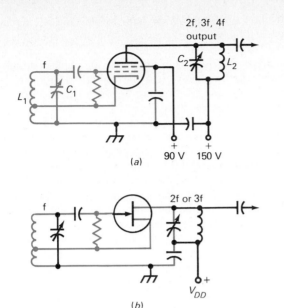

FIG. 11-14 (a) Hartley ECO. (b) JFET ECO-like Hartley.

does not rely upon electron coupling, but the amplifying section does. Changing the output LC circuit tuning or loading should produce little effect on the frequency of oscillation. This is highly desirable.

The plate-circuit load is shown as a tuned LC tank circuit. When output is desired on the fundamental frequency of the oscillator, the LC circuit may be replaced with an RFC, or perhaps a 2- to 10-kΩ resistor. If maximum output is desired on a *harmonic* frequency (2, 3, or 4 times the oscillator frequency), the output tank must be tuned to the desired harmonic.

The ECO-like circuit in Fig. 11-14b will provide output at 2f or 3f. The capacitor of the output LC circuit effectively completes the ac circuit from drain to ground allowing the oscillator to operate as a Hartley at frequency f. Ac output will be developed in the output LC circuit if it is tuned to 2f or 3f, provided the bias voltage is high enough.

Test your understanding; answer these checkup questions.

1. What produces the feedback in a TPTG oscillator's circuit? _____
2. From which LC tank of a TPTG is ac power taken? _____ What are the two coupling methods used? _____ _____
3. To provide proper phasing for oscillation, which tank of a TPTG should be at the higher frequency? _____

4. Why would the grid tank usually determine the oscillation frequency of a TPTG? _____
5. What type of feed has current flowing through the tank coil? _____ Through the RFC? _____
6. In what way is a Hartley oscillator similar to a TPTG? _____ To an Armstrong? _____
7. Where on a Hartley tank is the tap for greatest power output? _____ Why? _____
8. If you reach toward an oscillating LC circuit and it changes frequency, what causes this? _____
9. Basically, how does a Colpitts differ from a Hartley? _____
10. Why might Colpitts stability be better than Hartley? _____
11. Is it possible to have a series-fed Hartley? _____ A series-fed Colpitts? _____
12. What is the name of the VHF oscillator that uses interelectrode capacitances as the ac voltage divider? _____
13. What does ECO mean? _____ Can an ECO be used with BJTs? _____ Triodes? _____ Pentodes? _____
14. What are three advantages of an ECO over simple Hartley or Colpitts-type circuits? _____ _____ _____

11-12 CRYSTAL OSCILLATORS

All modern communications equipment uses quartz *crystal* oscillators because they will not drift more than a few hertz from the frequency for which they are ground. A variable-frequency or "self-excited" oscillator may drift considerably.

A quartz crystal may look like a piece of thin frosted window glass cut into ¼- to 1-in. squares. To operate as an oscillator, crystalline quartz must be cut in thin slices and ground smooth. Such quartz crystals have peculiar properties. If a crystal is held between two flat metal plates and the plates are pressed together, a small emf will be developed between the two plates, as if the crystal became a battery for an instant. When the plates are released, the crystal springs back to its original shape and an opposite-polarity emf is developed between the two plates. In this way, mechanical energy is converted to electric energy by the crystal. Also, when an emf is applied across the two plates, the crystal will distort its normal shape. If an opposite-polarity emf is applied, the crystal will reverse its physical distortion. In this way, electric energy is converted to mechanical energy by the crystal. These two reciprocal actions of a crystal are known as the *piezoelectric effect*. Man-made lithium tantalate, lead zirconate, and lead titanate crystals may be superior to natural quartz crystals in some ways.

If a crystal between metal plates is shock-excited by either a physical stress or an electric charge, it will vibrate mechanically at its natural frequency for a short while and at the same time produce an ac emf between the plates. This is somewhat similar to the damped electron oscillation of a shock-excited LC circuit. A vibrating crystal will produce an ac emf much longer than an LC circuit will when shock-excited, because the crystal has a much higher Q (fewer losses) than an LC circuit.

A tuned-gate tuned-drain (TGTD) crystal oscillator substitutes a crystal for an LC tank in the gate circuit (Fig. 11-15). In this circuit the crystal is

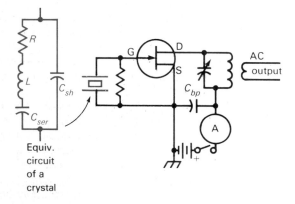

Equiv. circuit of a crystal

FIG. 11-15 Crystal oscillator, TGTD-type, and equivalent circuit of a crystal in its holder.

operating as a high-Q parallel-resonant circuit. No gate-leak blocking capacitor may be needed, since the crystal is an insulator and does not short out the resistor as an LC coil would.

When the switch is closed, the LC tank in the drain circuit is shock-excited into oscillation by a sudden surge of I_D. The ac developed across this LC circuit is fed back to the top crystal plate by internal capacitance (C_{DG}), and to the bottom plate of the crystal by C_{bp}. The crystal starts vibrating and working as an ac generator on its own. The emf generated by the crystal develops I_D variations

in the LC circuit. With both crystal and LC circuit oscillating and feeding each other in proper phase, the whole circuit oscillates as a very stable ac source. The LC circuit must be tuned slightly higher in frequency than the crystal to produce the required phase relationship between the two circuits to sustain oscillations.

With no oscillation of the circuit, no bias will be developed across the gate-leak resistor and the I_D will rise. It is possible to tune this type of oscillator by watching a milliammeter in the drain circuit. A decrease of I_D as the LC circuit is tuned is an indication that the circuit is oscillating and developing bias. The harder the crystal oscillates, the greater the bias and the lower the I_D.

When tuning a crystal stage, as the LC tank is *increased* in frequency, the circuit suddenly breaks into strong oscillation and the I_D suddenly drops to minimum as the resonant frequency of the crystal is reached. However, if the LC circuit is *decreased* in frequency while tuning, the I_D gradually decreases to the minimum value and then pops up to a maximum as the circuit stops oscillating. This is a tuning characteristic of all TGTD circuits.

A minimum I_D may indicate strongest oscillation of a crystal (also greatest crystal heating), but it does not necessarily indicate the optimum operating condition. For most satisfactory operation the I_D should be about 20% above minimum. This will also allow the circuit to be immediately self-starting (important in radiotelegraph transmitters). As the LC circuit is adjusted, the frequency of oscillation may vary as much as a kilohertz.

The frequency of oscillation of a crystal is determined by its material, thickness, physical size, angle of cut, pressure on the plates, type of circuit, and temperature. It is also possible to vary the frequency of oscillation by a few hertz by connecting a small variable capacitor across it or a small variable inductance in series with it. (Too large a capacitance or inductance will stop the crystal from oscillating.) Replacement of an oscillator's active device usually changes the oscillation frequency somewhat. Even if a manufacturer indicates a certain frequency of oscillation for a crystal, it is possible that the crystal may oscillate a few hertz high or low, depending on the circuit used.

Low-frequency crystals may require a small capacitor between drain and gate to increase feedback sufficiently to produce oscillation. Excessive feedback capacitance can result in crystal fracture if power supply voltages over 100 V are used.

Although *thickness* vibration of thin crystals has

been used in the explanations, a crystal, if cut from quartz at the correct angle, will vibrate corner to corner (*shear*) or end to end (*longitudinally*). The last-named mode results in a much lower frequency of oscillation when using the same-sized crystal. Crystals are often silver-plated on the two flat surfaces, with a connecting wire soldered to the middle of each silvered surface. Crystals may also vibrate *flexurally* (by bending back and forth), *torsionally* (in a twisting movement), or in two modes at the same time.

Although crystals are quite stable in frequency, changes in dc supply voltage may shift the oscillation frequency. Some form of well-regulated oscillator power supply is advisable.

11-13 CRYSTALS AND TEMPERATURE COEFFICIENTS

The manner in which a crystal is sliced from raw quartz will determine its oscillating frequency, its frequency stability, and its temperature coefficient and even name the type of crystal.

Oscillating crystals are cut from six-sided quartz crystals. End views of two such crystals are shown in Fig. 11-16. The dashed lines indicate the *X* and

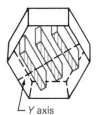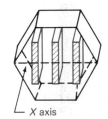

FIG. 11-16 Angles at which *Y*-cut and *X*-cut crystals are sliced from the raw crystalline quartz.

Y axes. The *Y* axis is from one flat face to the opposite flat face of the crystal. The *X* axis is from one corner to the opposite corner. A crystal sliced out of the quartz at right angles to the *Y* axis is a *Y-cut* crystal. If cut at right angles to the *X* axis, it is called an *X-cut* crystal.

When crystals are cut at angles other than *X* and *Y*, they are given other names, such as *AT, BT, CT, Z*, etc. Some operate best on high frequencies, others, on low. Different cuts have different temperature coefficients.

The following should make clear the meaning of *temperature coefficient* (TC) of a crystal.

1. If a change in temperature produces a large variation in frequency, a crystal has a *high TC*.
2. If a change in temperature produces a small variation in frequency, a crystal has a *low TC*.
3. If a change in temperature produces no variation in its oscillation frequency, a crystal has a *zero TC*. (No crystal has a true zero TC.)
4. If an increase in temperature increases its oscillating frequency, a crystal has a *+TC*.
5. If an increase in temperature decreases its oscillating frequency, a crystal has a *−TC*.

Some crystals may have a +TC when operated at about 20°C, nearly zero at about 50°, and a −TC at temperatures above 60°. Such crystals, if their temperature can be held at about 50°, will operate as though they had a zero TC.

As an example of TC with a crystal:

A 600-kHz crystal, calibrated at 50°C and having a TC of −20 parts per million (ppm) per degree Celsius, will oscillate at what frequency when its temperature is 60°C?

First, the "−20 parts per million per degree" can be read as "a −TC of 20 Hz/°C/MHz." Since 600 kHz is 0.6 MHz, the expression above can be stated "−20 Hz/°C/MHz × 0.6." The change in temperature is 10°. The change in hertz is then

$$-20 \times 10 \times 0.6 = -120 \text{ Hz}$$

By increasing 10° in temperature, the crystal loses 120 Hz from its 600,000-Hz calibration. It oscillates at 599,880 Hz, or 599.88 kHz. (If the crystal had a similar but +TC, the 120 Hz would have been added instead of subtracted.)

In many transmitters, the oscillator may be operating on a certain frequency and the output may be on some integral multiple, or *harmonic*, of the fundamental frequency. For example: If a transmitter uses a 1000-kHz crystal with a TC of −4 cycles/°C/MHz and the crystal temperature increases 6°, the crystal will decrease its frequency of oscillation −4 × 6, or −24 Hz. If the output of the transmitter is to be 4 times the fundamental, or 4000 kHz, the output frequency will decrease 4 × −24, or −96 Hz. The output frequency will be 4,000,000 less 96 Hz, or 3,999,904 Hz.

11-14 CRYSTAL HEATER CHAMBERS

If a crystal can be kept at a constant temperature and its oscillator circuit has a constant-voltage regulated power supply, it should maintain a constant frequency. A thermostatically controlled heated chamber can provide the necessary constant temperature of operation for the crystal. To assure that a hot day does not allow the crystal chamber to overheat, it is necessary to maintain the chamber at some above-normal temperature, such as 130°F (55°C). This can be done with reasonably simple circuits.

Figure 11-17 illustrates a 130°F *bimetallic-element* temperature-controlled crystal "oven." The

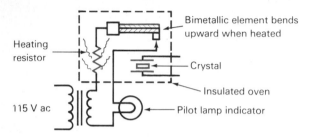

FIG. 11-17 Temperature-controlled oven using a bimetallic element.

two strips, of different metals and welded together, have dissimilar thermal expansion coefficients. When heated, the strip bends because the metal with the greater thermal expansion expands more than the other. The element shown bends upward when heated. If the temperature in the oven is less than 130°F, the element completes an electric circuit across the ac voltage, through the heating resistor and the pilot lamp. The current that flows heats the resistor and raises the temperature of the chamber. When the air in the chamber is heated to 130°F the bimetallic element bends upward enough to open the contact at the right-hand end, turning off the heater. The chamber now starts to cool again. The pilot lamp blinking on and off indicates that the chamber circuit is operating properly.

Figure 11-18 illustrates a temperature-controlling system using a BJT. A special mercury thermometer has two connections, one to the base of the mercury column, and a metallic contact at the 130°F point. When the chamber is cool, each positive half cycle of the ac from the two transformer windings feeds a positive or forward-biasing voltage to both the base and the collector of the BJT. This results in an I_C flowing through the heater resistor. When the chamber is heated to 130°F, the mercury switch closes. The BJT base is now grounded (zero bias), even when the collector is being positively biased, cutting off the I_C through the heater resistor. With no

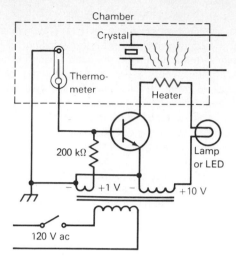

FIG. 11-18 A thermometer-type temperature control chamber system.

heater current, the chamber begins to cool again. The pilot lamp blinking slowly indicates system operation.

Figure 11-19 illustrates a thermocouple-controlled crystal chamber. A thermocouple (Chap. 13) produces a dc voltage and current when heated. If the chamber is cold, there is no voltage from the ther-

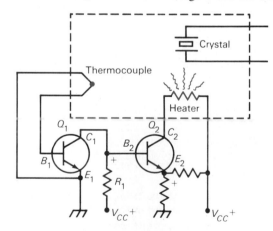

FIG. 11-19 Thermocouple-controlled crystal oven. Variation of R_1 will determine operating temperature.

mocouple, B_1 of Q_1 has no bias, and the resistance between C_1 and E_1 is very high. Thus, B_2 is forward-biased through R_1, and collector current flows through the heater resistor, warming the chamber. As it heats, the thermocouple warms, producing a voltage which forward-biases B_1. With forward bias, the C_1E_1 resistance is reduced, bringing B_2 closer to ground potential. This reduces the forward bias on Q_2, and its collector current decreases, reducing the heating of the chamber. The amplifying ability of the transistors produces a highly sensitive thermostat.

In broadcast stations crystals are always in temperature-controlled chambers, maintained within ±0.1°C (±0.2°F). With low-temperature-coefficient crystals, the requirement is only ±1°C. Most broadcast stations leave their crystal ovens running 24 hours a day to aid frequency stability.

11-15 CRYSTAL HOLDERS

Several types of holders are used for crystals. One consists of a hollowed-out block of insulating material in which two flat metal plates and the crystal are mounted (Fig. 11-20a). A spring against one

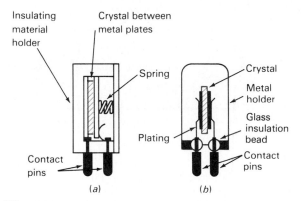

FIG. 11-20 (a) Original type of crystal holder. (b) Modern form.

plate holds together the sandwich formed by the crystal and the two plates. An electrical connection is made to each plate and to base pins. Sometimes the metal plates have raised corners touching the crystal. These air-gap mountings are used with high-frequency crystals.

Instead of pressing plates against its flat surfaces, a crystal may be plated with a thin layer of silver on both sides. The crystal is supported in a metal can by two steel wires that make connections to the two plated surfaces, terminating in pins at the bottom (Fig. 11-20b). The metal cans may be filled with inert gas to prevent oxidation of the silver plates.

Unplated crystals will not oscillate if they are dirty. To clean them, wash with soap and water or with a cleaning solvent. The crystals should not be touched with the bare hands but should be handled with clean, lint-free cloths and by the edges rather

than by the flat surfaces. The metal plates must also be cleaned.

Crystals are very fragile and will fracture easily. If excessive regeneration is used in an oscillator circuit, the crystal may fracture itself. If one edge or surface of a crystal is ground down a little, the frequency of oscillation will increase. Grinding is done on a flat surface such as a piece of plate glass, using a mixture of very fine carborundum dust and water as the grinding compound. The frequency of unplated crystals can be lowered a few hundred hertz by marking an X from corner to corner on one or both flat surfaces with solder or lead.

11-16 OTHER CRYSTAL CIRCUITS

Besides the basic tuned-input tuned-output crystal oscillator described, crystals will operate in other circuits. The *Pierce* oscillator (Fig. 11-21) uses a

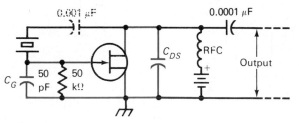

FIG. 11-21 Triode Pierce crystal oscillator. Capacitive voltage divider (in color) is across the crystal.

crystal as the LC circuit in a shunt-fed *ultraudion* oscillator. It requires no tuned circuits. In VT circuits the 0.001-μF capacitor is inserted to prevent dc voltage strain across the crystal. It is not needed in low-voltage BJT or FET circuits. Note that C_G and C_{DS} make this a Colpitts oscillator.

In Fig. 11-22, the crystal is used in place of the coil in a Colpitts-type oscillator. The two voltage-dividing capacitors across the crystal control the degree of coupling between drain and gate (collector and base, or plate and grid) and should be as small as possible.

Crystals are extremely thin when ground to resonate at frequencies above 10 MHz. For higher frequencies it may be desirable to use a lower-fundamental-frequency crystal and employ frequency-multiplying oscillators with the output plate-tank circuit tuned to a harmonic of the crystal frequency. Figure 11-23 illustrates a VT crystal ECO employing the first three elements as a triode Pierce oscillator. The harmonic output tank can be tuned to 2, 3, or 4 times the crystal

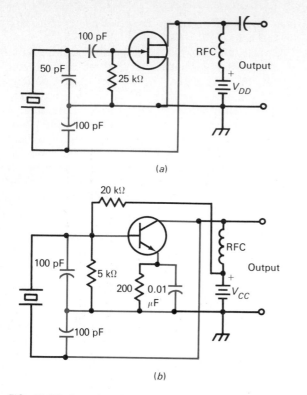

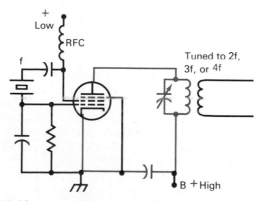

FIG. 11-22 Crystal Colpitts oscillators: (*a*) JFET and (*b*) BJT.

FIG. 11-23 Pierce ECO crystal oscillator capable of even- or odd-harmonic output.

It is possible to make almost any crystal oscillate in three layers. Such a *third-mode* oscillation produces a harmonic output very near 3 times the *series*-resonant frequency of the crystal. These *overtone* oscillator circuits resemble some self-excited *LC* oscillators with the crystal acting as the coupling device to complete the oscillator circuit. In Fig. 11-24*a* the crystal is completing the circuit to

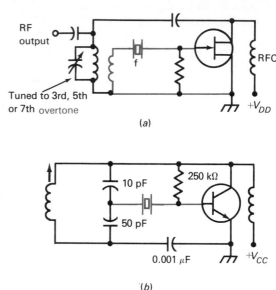

FIG. 11-24 Odd-harmonic overtone VHF oscillators: (*a*) Armstrong-type and (*b*) BJT-Colpitts type.

the feedback (gate tickler) coil in a tuned-drain Armstrong oscillator. The *LC* circuit must be tuned to the 3rd, 5th, or 7th overtone. No fundamental frequency will be generated. Thus, overtone and harmonic output crystal oscillators are not the same. Figure 11-24*b* is a BJT Colpitts-type overtone oscillator.

The Butler crystal oscillator is described in Sec. 25-5.

In Fig. 11-25, two plated lead zirconate or titanate crystals are laminated to opposite sides of a pliable piece of sheet metal. The crystals are developed with molecular alignments to make one expand and the other contract when similar polarity voltages are applied to the outer contact plates, forcing the unit to bend. It will bend in the opposite direction if the potentials are reversed. Such piezoelectric benders can be used as fast-operating relays, vibrators, low-frequency oscillators, fans, etc. If mechanically driven back and forth, such benders can develop ac voltages that can be used in

frequency. The output will be an exact multiple of the oscillation frequency of the crystal.

If a crystal is used to couple the output of a 2-stage amplifier to the input of the first stage, the circuit will oscillate at the series-resonant frequency of the crystal, which is slightly lower than the parallel-resonant frequency produced in TPTG type crystal oscillator circuits.

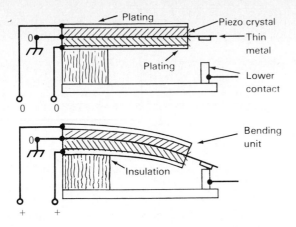

FIG. 11-25 Piezoelectric bending unit as a SPST relay. If negative potentials were applied it would bend upward.

various ac, or if rectified, in various dc power applications. Almost anything that can be done electromagnetically (by current) can be done electrostatically (by voltage) with piezoelectric devices, often with greater efficiency.

▌ Test your understanding; answer these checkup questions.

1. What kind of oscillators are not VFOs? _____
2. To what does a quartz crystal convert mechanical energy? _____ What is this effect called? _____
3. To what is an oscillating crystal equivalent electrically? _____
4. What is the relative I_p value when a crystal oscillator is oscillating? _____ The $-V_g$? _____
5. Why is it advantageous to have a separate regulated oscillator power supply? _____
6. What TC does a crystal have if an increase in temperature increases its frequency? _____ Has no effect on frequency? _____
7. What type crystals are sliced parallel to the sides of the raw crystal? _____
8. A 4-MHz crystal at 50°C with a +10 Hz/°C/MHz has what frequency at 60°C? _____
9. Is a harmonic ever anything other than a whole-number multiple of the fundamental? _____
10. Why are crystals often kept in temperature-controlled chambers? _____ Would this benefit LC tanks also? _____ Whole oscillator circuits? _____
11. What are three devices used to keep crystal chambers at a constant temperature? _____ _____ _____
12. To what LC oscillator circuit is a Pierce similar? _____

13. What are two advantages of using a crystal ECO?
14. Is an overtone the same as a harmonic? _____
15. Which overtones can be produced by crystal oscillators? _____
16. What is the first harmonic frequency of a 10-MHz crystal? _____

11-17 SOME HIGHER FREQUENCY OSCILLATORS

Generating low-frequency ac presents no serious difficulties. The higher the frequency, however, the less power output and the poorer the frequency stability of oscillation. Audio frequencies, up to about 20 kHz, may use laminated-iron-core coils and capacitors for the LC tank circuit. For frequencies up to several megahertz either air-core or powdered-iron-core coils may be used. Up to several hundred megahertz air-core coils, sometimes with brass slugs in the core area, are employed. In the UHF and superhigh frequency (SHF) range resonant-line tanks, hairpin tanks, coaxial tanks, and resonant cavities are required. The higher the frequency, the less inductance and capacitance required to produce a resonant circuit. At about 300 MHz a 4-in. piece of wire bent into a hairpin and connected between input and output leads of an active device to form a Colpitts-type oscillator may provide all the L and C needed to act as the resonant circuit. Above this frequency the usual LC-circuit idea may not be practical.

An electric impulse travels at 186,000 miles, or 300,000,000 meters, per second. A 300-MHz ac completes its cycle in $\frac{1}{300,000,000}$ s. In this time, from the formula $\lambda = v/f$, or *wavelength = velocity/frequency*, the 300-MHz ac will travel 300,000,000/300,000,000, or 1 m (39.37 in.).

If a 300-MHz generator is connected to a load by a long pair of wires (Fig. 11-26), the voltage values along the line at one instant might be as shown. Note that points half of a wavelength apart are 180°

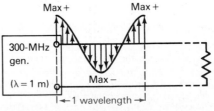

FIG. 11-26 Instantaneous voltage distribution of an ac having a frequency of 300 MHz.

out of phase. This is the end-to-end phase relationship of a resonant *LC* circuit. Therefore, it should be expected that a half-wavelength (19.685-in. or 50-cm) wire connected between base and collector of a BJT should form a resonant Colpitts or Hartley tank at 300 MHz. When bent into a hairpin shape, the end-to-end capacitance of the wire, plus the interelectrode capacitance of the active device added to the inductance of the connections, requires a much shorter hairpin to resonate at this frequency. If a small tuning capacitor is added to tune the circuit, the hairpin length must be still shorter.

Figure 11-27 shows a TPTG-type hairpin tank oscillator. It should be understood that the quarter-

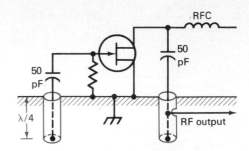

FIG. 11-28 Quarter-wavelength coaxial tanks in a TGTD oscillator.

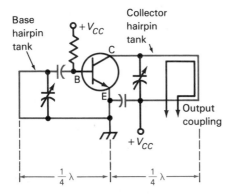

FIG. 11-27 UHF hairpin tank oscillator.

wavelength (¼ λ) indications represent the electrical, not the physical, length of the base and output circuit hairpins. Hairpin tanks are not practical at low frequencies. One wavelength at 5 MHz is v/f, or 300/5, or 60 m, or 196.85 ft long.

A variation of the hairpin tank circuit is the *coaxial* tank, a thin conductor running up the center of a λ/4 (quarter-wavelength) copper tubing. The conductor is connected to the sealed end. A TGTD circuit using coaxial tanks is shown in Fig. 11-28. Such tank circuits isolate the oscillating electrons from outside effects. Electrons oscillate up and down the central wire and *inner* surface of the

copper tubing (not on the outside). RF ac output is tapped off through a hole in the coaxial tank, as shown.

At still higher frequencies, UHF and SHF (Chap. 22), a *resonant cavity* is used. As a simplified explanation, many hairpin circuits of the same dimensions laid side by side and soldered together form a single wide hairpin. If the open sides of the hairpin are closed over with a sheet of metal, a metallic cavity (Fig. 11-29) is produced. Electrons

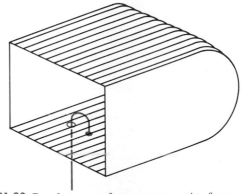

FIG. 11-29 Development of a resonant cavity from many parallel hairpin tanks.

oscillate back and forth inside the cavity, setting up alternating magnetic fields in the cavity. A loop or hook can provide a means of extracting energy from or of feeding energy to a cavity.

The higher the frequency, the greater the skin effect and the higher the effective resistance of a conductor. High-frequency circuits use large-area conductors such as resonant cavities. Since current travels only on the skin of conductors at high frequencies, it is desirable to silver-plate conductors to increase their *Q*.

11-18 DYNATRON AND TUNNEL-DIODE OSCILLATORS

In all the oscillators thus far, inductive or capacitive feedback of energy from output to input circuit is used to produce sustained oscillations. Dynatron and tunnel-diode oscillators operate on an entirely different *negative resistance* principle. According to Ohm's law, an increase in emf in a circuit produces an increase in current. If an increase of emf in a circuit results in less current, the circuit is said to be exhibiting a negative resistance effect.

A negative resistance effect can be developed by operating the grid of a triode at a positive potential higher than the plate potential. With the grid at a constant positive value, the I_p will increase as the V_p is brought up from zero to a certain point, X in Fig. 11-30a. Then, as the V_p is increased further, the I_p begins to decrease because of secondary-emission electrons from the plate traveling to the grid. (If more electrons are leaving the plate as secondary emission than arrive on it, current flows from plate to grid in the tube.) If the V_p is increased further, the plate is able to attract the secondary-emission electrons more than the grid, and the I_p will increase again. The downward slope of the $V_p I_p$ curve, X to Z, indicates the range of plate voltages over which a negative resistance effect is exhibited by the tube.

When the switch is closed, the LC circuit is shock-excited. Ac voltages developed across it are added to the V_p. If the V_p is a value near Y on the curve, the negative resistance effect will be operating in such a phase as to oppose any resistance effect of the LC circuit and the circuit will function as a generator of sustained LC oscillations. Frequency stability of dynatron oscillators is good, but power output is low. The circuit is rarely used, but the *dynatron principle* occurs in several electronic devices. Elements acting in this manner are called *dynodes*.

Compare the basic dynatron circuit with the basic solid-state tunnel-diode oscillator, Fig. 11-30b, in which an LC circuit is across a diode biased to the midpoint of the negative resistance portion of its curve. As long as an LC circuit is across a source exhibiting negative resistance (provided tank circuit losses are not excessive), LC oscillations should result. The bias-voltage (V_b) point for germanium tunnel diodes is about 0.15 V. The tunnel diode may be tapped down the LC circuit instead of being across the whole coil as shown.

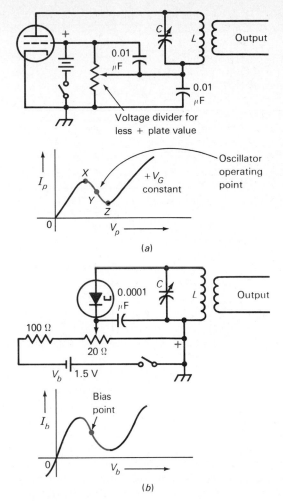

FIG. 11-30 Negative resistance curve oscillators. (*a*) Triode dynatron and (*b*) tunnel diode.

11-19 AUDIO OSCILLATORS

The oscillators described thus far have been RF types. The ac generated is presumably well above the audible frequencies. An audible-frequency ac may be desired for a code-practice circuit or as an AF ac source. An Armstrong oscillator using an iron-core transformer is shown in Fig. 11-31. The gate-leak resistor and capacitor help to determine oscillation frequency by their RC time constant.

A Colpitts-circuit AF oscillator using a choke coil is shown in Fig. 11-32. Note the use of a resistor instead of an AF choke in the collector circuit. When only a little audio power is desired, it

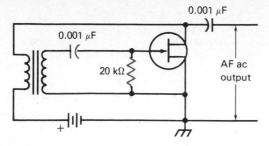

FIG. 11-31 Armstrong audio frequency oscillator.

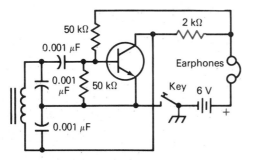

FIG. 11-32 Code-practice circuit using a BJT in a Colpitts-type oscillator circuit.

is more economical to use the resistance to obtain the required impedance across which the AF ac voltage-drop is developed.

11-20 RC OSCILLATORS

All oscillators discussed so far generate sine-wave ac, produced by flywheel effect in an oscillating LC circuit. There are many applications for other waveshapes, such as sawtooth or square-wave. The circuits used to produce these waveshapes usually rely on charge and discharge times of capacitors in conjunction with resistors.

A simple RC oscillator (Fig. 11-33) employs a resistor, capacitor, and neon bulb in a *relaxation*

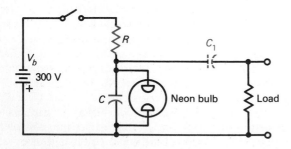

FIG. 11-33 Neon bulb relaxation oscillator.

oscillator circuit. When the switch is closed, the time required to develop a charge in the capacitor C depends on the V_b and R values in series with C. When the voltage across C rises to about 90 V, the ionization potential of neon, the neon acts as a good conductor and discharges C rapidly to about 75 V. At this voltage the neon deionizes, and the capacitor begins to charge slowly again. Thus the voltage across C repeatedly rises slowly from 75 to 90 V and falls rapidly to 75 V, producing a sawtooth varying dc waveshape across capacitor and load (Fig. 11-34a). If a capacitor C_1 is added

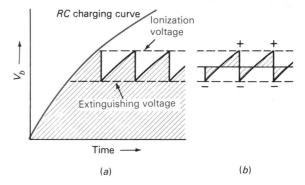

FIG. 11-34 (a) RC charging curve and varying dc produced across bulb in oscillator. (b) Sawtooth ac wave if C_1 is used.

in series with the load resistor, the waveshape remains the same but the charge-discharge action of the output capacitor converts the current in the load to sawtooth ac (Fig. 11-34b). The power-supply voltage should be at least twice the ionization potential of the neon.

SCRs can be used in the relaxation oscillator circuit of Fig. 11-35. Variation of the gate bias controls the firing potential of the SCR and therefore the frequency of the output wave. Thyratrons can be used in similar circuits.

True sine-wave ac has no harmonic energy output, but sawtooth ac is rich in harmonics. A 1-kHz

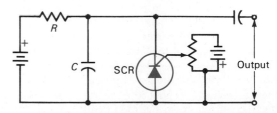

FIG. 11-35 SCR relaxation oscillator. Frequency can be controlled by R, C, or bias voltage.

sawtooth wave may contain weak but usable harmonic energy every 1 kHz up to several hundred thousand hertz.

11-21 MULTIVIBRATOR OSCILLATORS

One of the many possible relaxation oscillators is the *multivibrator*. It consists of two active devices connected in *RC*-coupled circuits such as those in Fig. 11-36. Common-source, -emitter, and -cathode

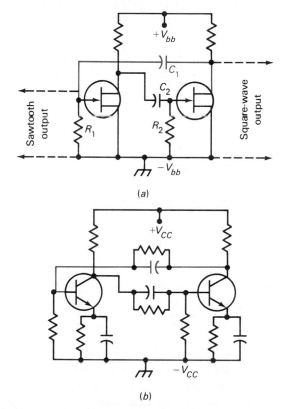

(a)

(b)

FIG. 11-36 (*a*) JFET multivibrator. (*b*) NPN BJT multivibrator.

amplifiers have 180° phase change from the input to the output of any stage, or from the input of one stage to the input of the next. If the input of the first device suddenly goes more negative, the input of the second goes more positive, and its output goes more negative (less positive). If the output of the second device is fed back to the input of the first, the feedback voltage will be in phase (180 + 180 = 360° = 0°) and will increase the effect of the originating voltage change at the input of the first device. This in-phase feedback produces a slow discharging of coupling capacitors C_1 and C_2

through resistors R_1 and R_2 and then a rapid charging through the low-resistance output circuit resistors when the device ahead starts to conduct. A sustained relaxation oscillation is produced. The voltage waveform developed across the input resistors as they discharge the coupling capacitor is sawtooth. The voltage waveform across the output resistors as their devices are alternately held at cutoff and then full conduction is square-wave. Frequency of oscillation depends on the *RC* time constant of the coupling capacitors and their associated *input* resistors. Unequal *RC* constants produce cycles of ac in which one alternation is longer than the other. If an *LC* circuit is substituted for one of the output load resistors, an essentially sine-wave voltage can be developed across the *LC* circuit.

In the BJT multivibrator (Fig. 11-36*b*) the resistors across the coupling capacitors are required to forward-bias the bases. *Switching* transistors with low storage times, and abruptly rising collector-current curves are suitable for oscillators and other class C (input bias voltage sufficient to cut off output current) circuits.

Multivibrators are used to produce sawtooth or square-wave ac, to generate a fundamental frequency with many harmonics, and to develop voltages to gate or switch electronic circuits on and off (Sec. 12-10).

11-22 PARASITIC OSCILLATIONS

Parasitics are unwanted (parasite) oscillations that may occur in almost any type of circuit, oscillator, amplifier, receiver, transmitter, etc. As an example, a transmitter is operating on 6 MHz, but nearby receivers can hear signals from it every 100 kHz from about 4 to 8 MHz. The transmitter tubes may overheat from overload, or tuning may be erratic. In all probability one of its stages is oscillating at 100 kHz in some way. This is a low-frequency parasitic oscillation.

In another case a 6-MHz transmitter is operating, but tuning is erratic. It may be found that one of the stages is oscillating around 100 MHz. This would be a high-frequency parasitic oscillation.

Sometimes, in audio amplifiers, undesired feedback will produce an oscillation at an audible, subaudible, or supersonic frequency. These are all forms of parasitic oscillations.

One possible method of producing a parasitic oscillation at a frequency lower than the operating frequency is illustrated in Fig. 11-37, showing a TGTD oscillator (which is similar to a basic RF

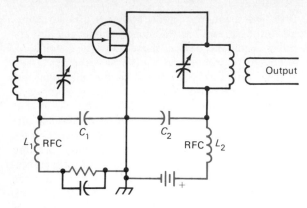

FIG. 11-37 TGTD oscillator (or amplifier) circuit in which low-frequency parasitic oscillations are possible.

amplifier stage). If L_1C_1 happens to resonate at the same frequency as L_2C_2, the circuit may oscillate at this particular frequency and at the same time work at the frequency of the two tank circuits. This would be a lower-frequency parasitic. RF chokes have greater inductance than the coils in the tuned circuits have, and the bypass capacitors have greater capacitance than the tuning capacitors. The parasitic oscillation can be stopped by using a resistor instead of the grid RFC, by using a smaller-inductance RF choke in the plate circuit, or a smaller-value bypass capacitor in the grid circuit.

TGTD-type parasitic oscillations are not the only type that may be developed. It is possible to have unwanted Hartley, *RC*, Colpitts, and even dynatron parasitic oscillations develop in electronic circuits.

High-frequency parasitic oscillations may exist in an amplifier or oscillator and not be detected. TV receivers make excellent detectors for any parasitic oscillations or their harmonics if they fall on TV channels. Figure 11-38a shows in color the elements of a stage that determine the normal frequency of operation. Figure 11-38b is the same circuit but the colored lines show a possible parasitic oscillation circuit having a frequency determined by two λ/4 hairpin tanks. The tuning capacitors act as bypass capacitors across the ends of the λ/4 lines. Should both the grid and plate circuits happen to have approximately the same-length leads, the circuit may produce oscillations at a VHF or UHF frequency at the same time that it is operating on LF or HF.

VHF parasitics may be stopped by winding a half-dozen turns of wire around a 50-Ω 1-W carbon resistor and inserting this *parasitic choke* in either the input or output leads of the circuit. Parasitic chokes should be installed as close to the terminals

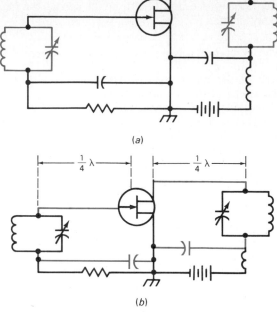

FIG. 11-38 (*a*) TGTD may oscillate at *LC* circuit frequency and (*b*) as hairpin-tank VHF parasitic oscillator at the same time.

of the transistor or tube as possible. A 20- to 300-Ω resistor alone, or one or two *ferrite beads* around the input or output leads, will often produce the same results as parasitic chokes. (A ferrite bead is shown in Fig. 11-44.) Changing lead lengths can often stop parasitic oscillations.

11-23 INDICATIONS OF OSCILLATION

When an RF oscillator is first turned on, several methods can be used to determine whether the circuit is oscillating.

RECEIVER. A nearby radio receiver will indicate by a change of sound when an oscillator is tuned across the frequency to which it is set. With beat-frequency oscillator (BFO) circuit of the receiver turned on, the oscillator will be heard as a whistle.

BIAS VOLTAGE. The presence of any dc bias voltage in an FET, VT, or BJT oscillator usually indicates that the circuit is oscillating. A change of bias voltage will be produced if oscillations are stopped by shorting the capacitor of an *LC* circuit.

OUTPUT CIRCUIT CURRENT. When a circuit is oscillating, bias voltage is developed, reducing I_D, I_C, or I_p. High current may be an indication of nonoscillation of a circuit. (Some BJT circuits increase I_C with oscillation.)

RF INDICATOR. A flashlight lamp or LED in series with a loop of insulated wire will glow when the loop is coupled to the oscillator tank coil if the oscillator produces 100 mW or more of RF power. A sensitive RF thermogalvanometer or a 0- to 1-mA meter with a solid-state diode across it can be used in place of the lamp.

OSCILLOSCOPE. An insulated-wire loop coupled to the oscillator tank coil and to the vertical input of the oscilloscope will show the presence of RF energy as a horizontal band on the screen.

NEON LAMP. A neon lamp will glow if touched against the *LC* circuit of an oscillating circuit, provided the oscillator is generating 90-V peak or more in the tank circuit.

LEAD PENCIL. A dangerous method with even medium-powered RF circuits is the use of a wooden pencil with soft lead. When touched to an oscillating tank circuit, it will produce a spark if there is more than a few watts of RF energy.

ELECTRONIC COUNTER. If oscillations are strong enough, a two- or three-turn loop can be used to couple ac voltage from an *LC* circuit of an oscillator to an electronic counter.

For Audio-Frequency Oscillators. The presence of bias is a reliable indication. A tone in earphones in series with two 0.0005-μF capacitors connected from ground to either input or output element gives a good indication. The output of the oscillator can be connected to an oscilloscope for a visual indication.

11-24 OSCILLATOR STABILITY

A main requirement for an oscillator, whether crystal or variable-frequency (VFO) types, is usually frequency stability. Factors for this are:

1. *Constant dc supply voltage.* Use a separate oscillator power supply or use regulated voltages in a multistage system.
2. *Low active device current.* The less current the less heating of the devices involved and of the components in the circuits.
3. *Low power output.* Loose coupling keeps device currents low and tank Q high, tending to hold frequency constant.
4. *Rigid mechanical structure.* Vibration of any oscillator part produces frequency variation.
5. *Buffer stage.* Oscillators should be followed by a buffer stage to prevent output stages from affecting the oscillator.

6. *Heavy coil wire.* Heavy wire for the oscillator coil results in higher Q (particularly if silver-plated), less contraction and expansion due to heating, and less chance of vibration.
7. *Drafts.* Changing temperature of oscillator-stage parts because of air drafts produces a drift in the frequency.
8. *Temperature-control chamber for crystals.*
9. *Proper tapping of coils.* For the Hartley, move tap to point on coil where best stability results. For the Colpitts, proper tuning-capacitor ratios are required.
10. *High C/L ratio.* The more capacitance in the tuning circuit, the less effect external and internal parameters have on frequency.
11. *Tapping down coil.* Tapping down the coil, as in Fig. 11-5*b*, produces higher Q, and results in circuit changes having less effect on the frequency of resonance.
12. *Shielding parts.* The shielding of parts prevents air currents, hand capacitance, and humidity from affecting tubes, transistors, coils, resistors, and capacitors.
13. *Correct bias.* Important factors are the values of biasing resistances and capacitances, usually found by trial.
14. *Active device used.* Some FETs, VTs, and BJTs change too much while operating and will never make stable oscillators.
15. *Negative TC capacitors.* Most *LC* circuits decrease frequency when they warm. Use of a $-$TC capacitor as part of the *LC* circuit compensates and holds frequency more constant.

Some other difficulties regarding stability are discussed in Sec. 16-6.

11-25 IC OSCILLATORS

There are countless digital or linear ICs which have amplification possibilities. They can be used to develop oscillators of any ac waveform desired. For sine-wave ac, an IC may be shown with an external *LC* circuit plus external resistors and capacitors that form any of the oscillators described previously or with a few resistors and capacitors, any one of a variety of phase-shift oscillator circuits can be formed. For square, triangular, or sawtooth waveshapes, resistors and capacitors will be added to produce some form of multivibrator circuit.

Figure 11-39 is a diagram of a 14-pin integrated circuit connected as a parallel resonant crystal oscillator and amplifier. The triangles inside the IC indicate amplifiers of some sort. Note the few ex-

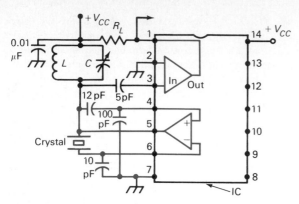

FIG. 11-39 IC crystal oscillator and amplifier (capacitances in pF).

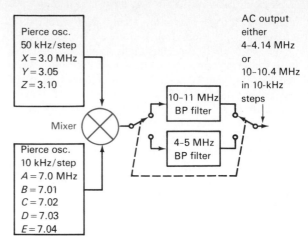

FIG. 11-40 Frequency synthesis with crystal oscillators.

ternal parts needed to provide oscillator output. The lower amplifier with the crystal and tuned circuit form the oscillator. The upper amplifier couples amplified ac to the output. The blank pins are connected to other internal amplifier circuits not involved in this particular circuit. Pin 7 is usually ground, and 14 is usually V_{CC} for a 14-pin IC.

In schematic diagrams ICs are usually shown as an oblong or a circle with pin numbers around it. The designer has determined the internal circuitry of the IC but, since it is usually quite complex, does not bother to indicate internal circuitry in the diagram. Even the two triangle-amplifiers shown in Fig. 11-39 might not be shown. To determine the internal circuits used it is necessary to refer to an IC data book available from the IC manufacturer.

11-26 FREQUENCY SYNTHESIS

It is possible to *synthesize* stable frequencies in steps by using two or more crystal oscillators. Figure 11-40 is a block diagram of one oscillator with a 3-MHz crystal (Z) and a second oscillator with a 7-MHz crystal (A), both feeding into a *mixer* circuit (diode, transistor, or VT, Sec. 18-12). The output of any mixer will be 4 frequencies; the original two (3 and 7 MHz), the *sum* of the two (10 MHz), and the *difference* of the two (4 MHz). If the 10-kHz-per-step crystals, A through E, are switched into their oscillator, and with the X crystal being used in the other oscillator, the output ac through the 9–11 MHz bandpass filter will be 10 to 10.04 MHz in 10-kHz steps. By switching in the next crystal, Y, into its oscillator, the output frequencies can then be switched from 10.05 to 10.09 MHz. With the third crystal, Z, the output signals will range from 10.1 to 10.14 MHz. Thus, with 6 crystals (X plus A

through E) 5 output signals can be synthesized. The addition of the 7th crystal (Y) allows 10 synthesized frequencies, the 8th crystal (Z) provides 15 synthesized frequencies, and so on. By switching the 3–5 MHz bandpass filter in place of the 9–11 MHz filter, 15 other output frequencies are possible from 4 to 4.14 MHz. The filters allow only the desired mixer resultant frequencies to pass into the output.

11-27 PHASE-LOCKED LOOPS

A highly important circuit is the phase-locked loop, or PLL. Its use as a means of synthesizing stable RF ac is explained here; how it can detect FM is explained in Sec. 19-4; its effect in a horizontal sweep AFC circuit, Sec. 24-19; and its use in a communication receiver, Sec. 25-12. It can be used as an AM detector, for frequency-shift keying, as an FM modulator, and in many other applications.

How a PLL circuit can lock a voltage-controlled oscillator (VCO) to a reference frequency can be explained by the basic circuit of Fig. 11-41a. A stable (crystal) reference oscillator ac and the ac from an oscillator whose frequency can be varied by the dc voltage applied to it (an oscillator with a voltage-variable capacitor across its LC circuit, for example) are both fed to a *phase detector* or *comparator* circuit. If the frequency and phase of the reference and VCO signals are equal, the dc output voltage from the phase detector will be some particular value. This dc is fed through a low-pass filter to a dc amplifier and to the voltage-input circuit of the VCO. If the VCO tries to shift off frequency, the phase detector develops a change in

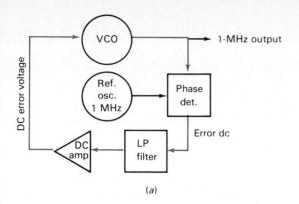

(a)

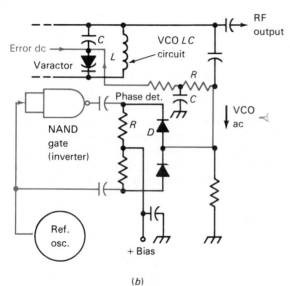

(b)

FIG. 11-41 (a) Basic PLL system. (b) Possible phase detector coupled to VCO tank circuit with voltage-variable capacitor.

the dc "error" voltage which corrects the VCO frequency and locks it to the reference frequency again. The VCO frequency is corrected not only to exactly the frequency of the reference ac but to the same phase. Once locked in, whatever the reference oscillator frequency does the VCO will do.

The phase detector might be a diode-resistor circuit as in Fig. 11-41b. The reference ac from a crystal oscillator can be fed to the bottom of the resistor-diode (RD) circuit. Feeding this voltage through a NAND gate (Sec. 12-7) reverses its phase, providing an equal amplitude but opposite phase ac for the top of the RD circuit. The VCO ac is fed to the center tap of the two diodes. The center tap of the two resistors is biased to a value which allows the diodes to conduct properly and is bypassed to

ground. If the reference ac and the VCO ac are exactly on frequency and in phase, the dc output voltage is some particular value. If the VCO tries to change frequency the voltage at the center tap between the diodes develops an error signal. This is passed through the low-pass RC filter (then usually to a dc amplifier) and to the VCO voltage circuit, which forces the VCO to lock into the reference frequency again. In this circuit, the VCO output can be only the reference frequency.

The circuit of Fig. 11-42 is a PLL system in which the output ac from the VCO locks into some

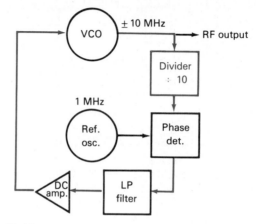

FIG. 11-42 PLL has output 10 times reference oscillator frequency.

chosen frequency and maintains this frequency with the stability of the 1-MHz reference oscillator. If the TTL (transistor-transistor-logic) divider IC (Sec. 12-14) divides the ±10-MHz VCO frequency by 10, the VCO will lock into a 10-MHz output, provided the VCO free-running or natural oscillation frequency is sufficiently close to 10 MHz. If the divider divides by 10.001, the VCO locks into a 10.001-MHz output. Thus the divider and the reference frequency determine the output frequency of the VCO.

The circuit of Fig. 11-43 is a possible PLL synthesizer. It has two TTL dividers; one divides a fixed number of times for relatively large frequency changes, but the other is a variable TTL down counter for small frequency changes. How many times it divides the ac frequency fed to it is determined by the combination of high and low digital voltages applied to its control terminals by the switching circuit IC.

A whole PLL circuit except for the crystal and dividers can be incorporated in one IC.

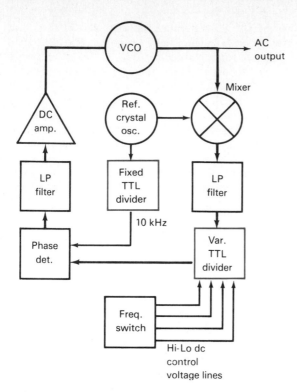

FIG. 11-43 PLL synthesizer in 10-kHz steps.

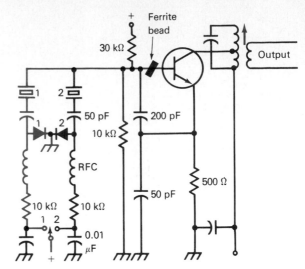

FIG. 11-44 Diode crystal switching circuit. Output circuit is slug-tuned.

11-28 DIODE SWITCHING AND TUNING

In many VHF transmitters and receivers it is necessary to shift rapidly from one specific frequency to another. This may be accomplished by using several crystals, but only coupling one of them at a time to the oscillator circuit. To control the coupling of the crystals, *biased diodes* are used. When a diode is not forward-biased, it is essentially an open circuit. When forward-biased, it is essentially a closed circuit. The diagram of Fig. 11-44 shows how a simple rotary (or push-button) switch can feed a forward bias to selected diodes to allow them to couple desired crystals to the oscillator circuit. If the switch is in the position 1, diode 1 is forward-biased, coupling crystal 1 to the Pierce-type transistor-oscillator circuit. Moving the switch to position 2 results in crystal 2 being electronically coupled to the oscillator. To allow trimming each crystal exactly to frequency, small adjustable capacitors are usually connected across each crystal.

The same diode system can be used to switch capacitors instead of crystals across tuned circuits to electronically tune the *LC* circuits involved. If varactor diodes (Sec. 9-3) are used, varying the

bias will vary the effective capacitance, and an electronically variable-tuned circuit results. Thus, tuning is accomplished with a potentiometer rather than with a capacitor. It is particularly desirable when remote tuning is required, since the dc control circuit can be any distance from the circuit being tuned. The voltage regulation of the bias must be very good.

11-29 SIGNAL GENERATORS

Testing electronic systems often requires the application of a known frequency ac signal at a known amplitude to a system input, and checking for its appearance at different parts of the system. Electronic equipment that provides such signals are known as audio-frequency (AF) signal generators, radio-frequency (RF) signal generators, and function (various waveshape) generators.

A simple form of AF generator may develop sine-wave, sawtooth or triangular, and square-wave ac at selectable frequencies from 20 to over 20,000 Hz. The output voltage may be variable from perhaps 10-V rms down to zero. The usual circuit is a 2-device *Wien bridge* 360° feedback oscillator, which produces very low distortion sine-wave ac. An amplifier increases the output to the desired level. To produce square-wave ac the sine-wave ac may be highly amplified but is then limited in amplitude by a shunt diode. More sophisticated AF signal generators cover a wider bandwidth and may use electronic squaring circuits, such as a Schmitt trigger circuit, for square-wave output.

RF signal generators use a variable-frequency oscillator, such as a Hartley or Colpitts, with a single variable capacitor but six to ten different size coils to provide oscillations from perhaps 20 kHz to 100 MHz. Besides developing continuous wave (CW) signals, most of these oscillators will have some means of providing amplitude modulation (AM) or frequency modulation (FM) (Chaps. 17 and 19) of the output to make the signal audible in receivers. For VHF and UHF bands the *LC*, linear-tank, or cavity-tank-type oscillators cover from perhaps 50 to 2000 MHz. Microwave signal generators use a variety of types of oscillators to cover perhaps 2 to 20 GHz, with "doublers" to produce 20- to 40-GHz signals. Besides CW, these systems may incorporate FM and pulse-modulation circuits in them.

Synthesized signal generators are switch- or push-button-tuned from 0.01 Hz to over 1 GHz with the stability of their 10-MHz crystal clock oscillator (Sec. 13-25), and they may have CW, AM, and FM outputs.

Function generators develop a variety of waveforms, such as sine-wave, square-wave, and triangular-wave, positive and negative pulses of varying widths and shapes, and ramp or sweep (slow-rise fast-fall) voltages. Function generators operate in the frequency range of a fraction of a hertz to over 10 MHz. They may be *LC* oscillators or synthesized systems.

Most engineering-type signal generators today are manufactured to operate into either a 50- or a 600-Ω load. Their output may be adjustable from about 10 V down to zero.

Test your understanding; answer these checkup questions.

1. List four types of resonant tanks used in UHF and SHF. _____ _____ _____ _____

2. An *LC* tank represents what wavelength? _____
3. What type of VT makes the best dynatron oscillator? _____ Could BJTs or FETs be used? _____
4. What solid-state device mentioned has a negative resistance effect? _____
5. What is the bias point for all negative resistance oscillators? _____
6. What waveform(s) is produced by neon lamp *RC* oscillators? _____
7. From what part of a multivibrator is a square wave available? _____ A sawtooth wave? _____
8. If the grid resistance of one VT of a multivibrator is 10 times the second R_g, what is the output wave at the first VT plate? _____
9. Why is the output of a multivibrator rich in harmonics? _____
10. What are two possible methods of producing a lower-frequency parasitic oscillation? _____
11. What can be done to stop higher-frequency parasitic oscillations? _____
12. What are some signs of parasitic oscillation? _____
13. List seven devices that can be used to indicate oscillation in an RF oscillator. _____ _____ _____ _____ _____ _____ _____
14. List 12 factors that aid frequency stability of an oscillator. _____ _____ _____ _____ _____ _____ _____ _____ _____ _____ _____ _____
15. What are two methods of synthesizing frequencies? _____ _____
16. What circuits are involved in a basic PLL circuit? _____
17. List the frequencies developed by oscillators in the AF category. _____ RF. _____ VHF-UHF. _____ Microwave. _____
18. What generates the signals in the less costly signal generators? _____ In the more costly? _____

OSCILLATORS QUESTIONS

1. List six devices or methods that are used to generate ac.
2. What effect is produced by shock-exciting an *LC* circuit?
3. If a high-*Q* *LC* circuit is shock-excited by a pulse of dc, what is the result? If the *LC* circuit has low *Q*?
4. What is required to enable an *LC* circuit to oscillate continuously at a constant amplitude?
5. What is the name of the oscillator circuit that uses inductive feedback?
6. Why is the *LC* circuit tapped in many BJT oscillator circuits?
7. Why would reversing tickler leads prevent oscillations?
8. What is another term meaning out-of-phase feedback? In-phase feedback?

9. How is the gate-leak resistor connected in a series-type biasing circuit?
10. If an oscillator circuit oscillates at two frequencies at the same time, what is the undesired oscillation called?
11. What type of output circuit feed always requires an RFC?
12. Draw diagrams of a BJT and a JFET Armstrong oscillator.
13. Draw diagrams of BJT and JFET tuned-input tuned-output oscillators.
14. Name three capacitive feedback-type oscillators.
15. What oscillator circuit using a JFET (or VT) must have its LC circuit tapped?
16. Is a Hartley oscillator an inductive or capacitive feedback type?
17. Draw diagrams of a BJT and a JFET Hartley oscillator in which one terminal of the tuning capacitor is grounded.
18. How does the Colpitts differ from the Hartley circuit?
19. How could Fig. 11-12 be made series-fed?
20. Why would gate-leak bias never be used with MOSFET oscillators?
21. Why is a JFET ECO not truly an ECO circuit?
22. Draw a diagram of a JFET ECO circuit.
23. What are the reciprocal effects present in any material that exhibits piezoelectric effects?
24. What seven factors determine the frequency of oscillation of a piezoelectric crystal?
25. Draw a diagram of a TGTD-type JFET crystal oscillator.
26. In what five directions or ways may crystals oscillate?
27. How is an X-cut crystal sliced from the crystal?
28. What is the most desirable TC of a crystal?
29. A crystal has a +TC of 6 Hz/°C and operates at 4,999,844 Hz when at 20°C. What will be its frequency if warmed to 45°C?
30. What are the three types of crystal ovens described?
31. Draw diagrams of the three types of crystal ovens in question 30.
32. Crystal oscillator circuits, if not of the TGTD-type, are forms of what oscillators?

33. Draw diagrams of a JFET Pierce oscillator, a JFET Colpitts crystal oscillator, and a BJT crystal Colpitts.
34. What oscillator circuit can make a crystal oscillate in three or five layers?
35. Name four uses of voltage-driven piezoelectric benders.
36. What can piezoelectric benders produce if vibrated mechanically?
37. In formula form, express the relationship of frequency, wavelength, and velocity.
38. What type of tuned circuits are used in UHF and SHF oscillators?
39. Draw a diagram of a UHF hairpin tank oscillator.
40. What is a tunnel diode said to be exhibiting when operated in the downward portion of its $V_b I_b$ curve?
41. How might the LC circuit of an AF oscillator differ from that of an RF oscillator?
42. What controls the frequency of oscillation of an SCR relaxation oscillator?
43. What is the waveform at the gate of a JFET multivibrator? At the drain?
44. If the sources of a JFET 100-kHz multivibrator are disconnected from ground and connected to the opposite ends of a 100-kHz LC circuit having its center tap grounded, what waveform ac would be developed in the LC circuit?
45. What are the two types of parasitic oscillations discussed?
46. List eight possible methods of determining if HF oscillations are present.
47. List 15 possible factors affecting oscillator stability.
48. In the IC diagram, Fig. 11-39, what type of oscillator circuit does this appear to be? Does anything indicate where the power-supply voltages are connected to the amplifiers?
49. In Fig. 11-40, how many crystals would be required to synthesize 25 different frequencies?
50. What are the five circuits required in a basic phase-locked loop?
51. In Fig. 11-41a, if the reference oscillator drifts 8 Hz, how much will the VCO drift?
52. In the PLL, why is the LP filter used?
53. Would a PLL operate without the dc amplifier?
54. Draw a diagram of a PLL that can change the frequency of the VCO over a relatively narrow band of frequencies.
55. Draw a diagram of a diode crystal switching circuit that will select one of three different frequency crystals.
56. Why does the diode crystal switcher select one crystal only?
57. What oscillator circuit is used in most audio-frequency signal generators? What might be used in RF signal generators?
58. What type of internal oscillator system would be used in a push-button frequency selecting signal generator?
59. Most signal generators today have what voltage output capabilities? What output impedances?

ANSWERS TO CHECKUP QUIZ ON PAGE 225

1. (Resonant line) (Hairpin) (Coaxial) (Cavity) **2.** (Half wave) **3.** (Tetrode) (No) **4.** (Tunnel diode, unijunction) **5.** (Middle of negative portion of curve) **6.** (Sawtooth) **7.** (Output) (Input) **8.** (Narrow pulse) **9.** (Nonsinusoidal waves) **10.** (RC charge-discharge) (RFCs input and output) **11.** (Parasitic chokes, ferrite beads, or resistors at elements) **12.** (Irregular tuning, excessive heating, received signals on wrong frequencies) **13.** (See Sec. 11-23) **14.** (See Sec. 11-24) **15.** (Crystals) (PLL circuits) **16.** (VCO, ref osc, phase det, LP filter, dc amp) **17.** (20–20,000 Hz) (20 kHz–100 MHz) (50–2000 MHz) (2–40 GHz) **18.** (Oscillators) (PLLs)

12

Digital Fundamentals

This chapter presents some of the more important ideas regarding digital numbers and simpler logic gates and circuits. Boolean logic is introduced. Simplified logic circuits, registers, counters, LED and LCD displays, A/D converters, and some microprocessor information is presented.

12-1 DIGITAL DEVICES

A rapidly developing area in communications is in *digital logic gate* control, processing, and computer circuits.

General-purpose BJTs, FETs, and VTs are linear input-vs-output devices. That is, if an input voltage increases 1 mV, the output current may increase 1 mA, and a 2-mV input change should produce a 2-mA output change. Such devices can amplify input voltage variations linearly, with minimal output waveshape distortion.

When a BJT is used to turn a circuit on or off, as a switch would, it is only necessary that the base be fed enough forward bias to drive the collector into saturation. This results in a very low voltage-drop (± 0.2 V) between emitter and collector. Rise and fall times (t_r and t_f) should be as nearly instantaneous as possible. Linearity is not a consideration.

Digital circuits are all involved in rapidly rising and lowering voltages, usually ranging from 0 to 2 V as the low, and 3 to 6 V as the high values for TTL (transistor-transistor-logic) 7400 series devices. For CMOS devices, like the 3000 series, the low levels are 0 to 4 V, and the highs are 9 to 15 V. Such two-level, or *binary* circuits may be used in counters, frequency meters, indicators, transmitters, receivers, computers, controls of various types, etc. New applications for digital circuits are constantly being developed.

12-2 BINARY NUMBERS

The common decimal counting method in daily use is based on 10 digits, 0 through 9. To use this method in electronic systems would require circuits that would respond to 10 discrete voltage or current levels, which would be difficult. The one used in electronics is the simplest one possible, the *binary* system. It has two states or levels, *on* and *off*. Besides on-off, it may be called a high-low, a 1-0, or a true-false system.

Decimal counting starts at 0 and moves up to 9 before starting a second column with 10. The 2-digit binary system starts at 0 and moves only to 1 before starting a new column. Note how counting in binary (indicated by the subscript 2) progresses, compared to decimal:

0 =	0_2		5 =	101_2
1 =	1_2		6 =	110_2
2 =	10_2		7 =	111_2
3 =	11_2		8 =	1000_2
4 =	100_2		9 =	1001_2

and so on. You can probably deduce the progression pattern from an examination of these few numbers. In order to handle the binary number 1001_2, four on-off circuits are needed, the first to handle the first 1, two more to handle the two zeros, and a fourth to handle the last 1. This means that handling binary numbers requires many simple on-off circuits in a string, or in *series*, to allow the high or low (5-V or 0-V) levels to be stored in them so that they can be extracted and operated on when required. Binary-type devices employed are gates, flip-flops, registers, and memories. These may use BJTs, diodes, FETs, or combinations of these, plus other special semiconductor devices made up into simple or very complex ICs. Monitor screens are the only vacuum devices used.

Since only two values are used, 0 and 1, the circuitry for binary circuits is basically uncomplicated. However, because so many digits will be used (for example, $256_{10} = 1\ 0000\ 0000_2$), long strings of simple circuits will be required to be able to handle all the digits.

The weights of the binary numbers compared to decimal numbers are:

Decimal 512 256 128 64 32 16 8 4 2 1
                1   1   1 1 1 1 1 1 1 1

(or binary 11 1111 1111 = decimal 1023)

EXAMPLES:

$4_{10} = 1_2$ in the 3rd position $= 100_2$

$16_{10} = 1_2$ in the 5th position $= 10000_2$

$128_{10} = 1000\ 0000_2$

$5_{10} = 4_{10} + 1_{10} = 100_2 + 1_2 = 101_2$

$6_{10} = 4_{10} + 1_{10} + 1_{10} = 100_2 + 1_2 + 1_2 = 110_2$

$72_{10} = 64_{10} + 8_{10} = 1000000_2 + 1000_2$
$= 1001000_2$

$80_{10} = 64_{10} + 16_{10} = 1000000_2 + 10000_2$
$= 1010000_2$

12-3 BINARY ARITHMETIC

Adding binary numbers is relatively simple. For example, using binary notation, $0_2 + 0_2 = 0_2$. Also, $0_2 + 1_2 = 1_2$, and $1_2 + 1_2 = 0_2 + c$, where c is a carry of a 1_2 overflowed to the next left column. Thus

$$1_2 + 1_2 = \begin{array}{r} c \\ 1 \\ +1 \\ \hline 10_2 \end{array} = 10_2$$

Check the five binary addition problems below. The decimal equivalents are shown below the binary problems (the subscript 2s are left off for clarity).

	c	cc	c	$cccc$
0101	0011	0110	1001	1111
+0010	+1010	+0111	+1100	+1111
0111	1101	1101	10101	11110
5	3	6	9	15
+2	+10	+7	+12	+15
7	13	13	21	30

To add three or more binary numbers, the first two are added using simple adder circuits, then this sum is added to the third number, and so on.

Subtracting binary numbers first requires "2s complementing" the number to be subtracted. Ones complementing is simply reversing the 1s and 0s. Twos complementing means first changing the 0s to 1s and 1s to 0s and then adding 1. For example, 101_2 when 2s complemented is first 1s complemented to 010_2, and then by adding 1 is 2s complemented to 011_2. The subtraction process consists of *adding* the larger number and the 2s complemented smaller number. Thus, in decimal, $7 - 5 = 2$. In binary this is $111_2 - 101_2 = ?$ First, 2s complementing the smaller value, 101, gives 011. Adding the two numbers, using carries,

$$\begin{array}{r} ccc \\ 111 \\ +011 \\ \hline 1010 \end{array}$$

The *overflow*—any 1 in the column higher than the most significant digit in the problem—is dropped in subtraction. This drops the first 1 in the answer above, leaving 010_2, which is equal to decimal 2— the correct answer. Thus, subtracting is only a form of adding, and is easily handled with digital *hardware* (devices and circuits).

Multiplying in binary is handled as in decimal. To multiply 7×3 it is only necessary to add 7 three times ($7 \times 3 = 7 + 7 + 7 = 21$). Thus, multiplication may be a form of multistep addition. (A shift and add system is actually used.)

Dividing is a form of subtraction, since to divide 24 by 6 the steps are $24 - 6 - 6 - 6 - 6 = 0$. Thus $24/6 = 4$ (steps). Binary hardware may compute division as subtractions, which, in turn, are forms of addition. Binary circuits might handle the problem 3210/10 by subtracting 10 from 3210 a total of 321 times to get the answer 321. But, if it only

228 • CHAPTER 12 DIGITAL FUNDAMENTALS

takes 1 μs per operation, it requires only 0.000 321 s to produce the answer!

12-4 THE AND GATE

Digital systems are full of *gates*. A gate is an electronic circuit which will function in some desired manner when one or more high- or low-voltage signal(s) are fed to its inputs. Consider the AND-type gate formed by two BJTs in series in an emitter-follower configuration, Fig. 12-1a.

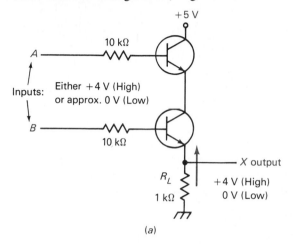

(a)

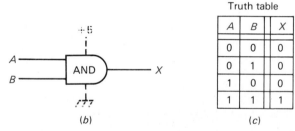

(b)

A	B	X
0	0	0
0	1	0
1	0	0
1	1	1

Truth table

(c)

FIG. 12-1 (a) A possible AND gate. (b) Symbol. (c) Truth table.

With the two input lines A and B either at ground potential or floating, there is no forward bias on either BJT and neither can conduct. As a result there is no current through R_L, and the output is at ground potential or 0. If input line A goes high (+5 V applied to it), the top BJT is forward-biased, but the lower BJT is still a nonconductor, and the output X remains at 0. If just the lower input is driven high, the upper BJT is an open circuit, only a little I_{EB} flows through R_L, and the output remains low (at about 0.4 V). If both inputs are fed +5 V, both BJTs will conduct, current will flow through R_L and a positive voltage will appear at the output

(perhaps 4.6 V). This is an AND circuit because both input A *and* input B must be high for the output to be *high*. The symbol for a 2-input AND gate is shown in Fig. 12-1b. The "truth table" of a 2-input AND gate is shown in Fig. 12-1c, where the output X for the four possible conditions of inputs A and B is shown, "A AND B" can be written $A \cdot B$ or AB.

If there were three BJTs in series, a 3-input AND gate would result. Since a transistor is an amplifier, low-power input signals will allow the output of the gate to drive several other low-power input TTL gates. The capability of an amplifying-device gate to produce adequate "fanout" to drive several other gates is usually quite important. (Resistor-diode gates have no power amplification capabilities.) Besides BJTs, gates are made with JFETs and MOSFETs. It is interesting to note that two SPST switches in series form an AND gate.

These explanations will be given for TTL-type gates using BJTs in their circuits. Such units are usually designated as being 7400 series. Another type of construction is the complementary metal-oxide semiconductor, or CMOS. These are not as fast as the TTL gates but they use much less current and power to operate, employing up to 15 V for the power supply, whereas TTLs use only 5 V. A third construction is the emitter-coupled logic, or ECL, which also uses bipolar transistors. These have the highest speed of operation, but they require more power for very high speed operation. ECL units are designated as 1600 series. A CMOS designated as 7400C means a unit that can be used to replace a TTL gate directly.

An example of a TTL AND gate is the 7408, a quad 2-input AND (four separate AND gates in one) IC.

12-5 THE OR GATE

Another important digital circuit is the OR gate. Figure 12-2a illustrates two BJTs in parallel in a common-emitter circuit. With both inputs A and B low, neither BJT is forward-biased and the output X is low. If either (or both) input A or input B is driven with a +5-V signal, that BJT will conduct, current will flow through R_L, and the output X will be nearly +5 V (or high). This is a 2-input OR gate. With four BJTs in parallel, the circuit would form a 4-input OR gate. The symbols for 2- and 3-input OR gates are shown in Fig. 12-2b. The truth table for a 2-input OR gate is shown in Fig. 12-2c.

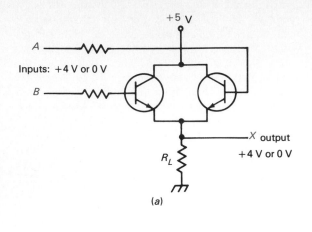

Inputs: +4 V or 0 V

X output
+4 V or 0 V

(a)

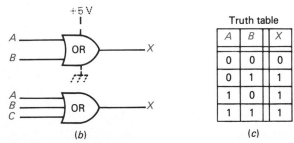

Truth table

A	B	X
0	0	0
0	1	1
1	0	1
1	1	1

(b) (c)

FIG. 12-2 (a) Possible OR gate. (b) Symbols for 2- and 3-input OR gates. (c) Truth table.

Two SPST switches in parallel can function as an OR gate. A OR B can be written $A + B$.

A type 7432 TTL IC is an example of a quad 2-input OR gate.

12-6 THE INVERTER

In many cases in digital systems a high signal must be inverted to form a low signal, or vice versa. This can be done using a common-emitter BJT circuit, shown in Fig. 12-3a. With the input fed a low signal (grounded), or if it is open, the BJT cannot conduct, there will be no current flow through R_L, and the output X will be high (+5 V). Conversely, if the input is fed a +5-V (high) signal, the BJT is forward-biased and is saturated. Enough current now flows through R_L to produce a voltage-drop of almost 5 V across it. As a result, the output is low and is inverted from the input. The symbol of an inverter is shown in Fig. 12-3b. The triangle indicates an amplifier and the "bubble" (little circle) indicates inversion. The truth table for an inverter is shown in Fig. 12-3c. Inverted or NOT A is written $\bar{A}$.

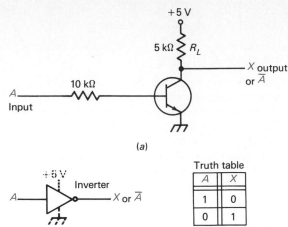

Truth table

A	X
1	0
0	1

(b) (c)

FIG. 12-3 (a) Possible inverter. (b) Symbol. (c) Truth table.

A 7404 TTL IC is a hex (6-unit) inverter. A CD4009 is a hex CMOS inverter IC.

12-7 THE NAND GATE

If an inverter is coupled to the output of an AND gate, the combination is called a not-and, or NAND gate. Whereas two high-input signals to an AND gate produce a high output, two high-input signals to a NAND gate produce a zero or low output. The symbol for a NAND gate is the same as an AND gate with the addition of a bubble at the output, to indicate inversion, Fig. 12-4a. The truth table for a NAND gate is shown in Fig. 12-4b.

A 7400 is a quad 2-input NAND gate. The CD4011 is a good 2-input CMOS NAND gate.

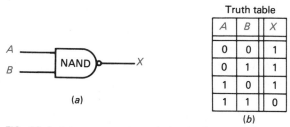

Truth table

A	B	X
0	0	1
0	1	1
1	0	1
1	1	0

(a) (b)

FIG. 12-4 (a) NAND gate symbol. (b) Truth table.

12-8 THE NOR GATE

When an inverter is coupled to the output of an OR gate, the combination is a not-or or NOR gate. With an OR gate, a high signal to any input will produce a high output. With a NOR gate, a high signal to any input produces a low output. The symbol for a NOR gate is a bubble added to the OR gate symbol, as in Fig. 12-5a. The truth table for a NOR gate is shown in Fig. 12-5b.

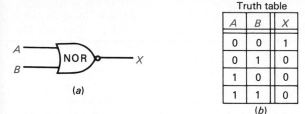

Truth table

A	B	X
0	0	1
0	1	0
1	0	0
1	1	0

(a)

(b)

FIG. 12-5 (a) NOR gate symbol. (b) Truth table.

A 7202 is a quad 2-input NOR TTL gate IC. The CD4001 is an example of a quad 2-input CMOS NOR gate IC.

12-9 THE EXCLUSIVE-OR (XOR) GATE

By connecting two inverters, two AND gates, and an OR gate as in Fig. 12-6a, an exclusive-OR (XOR) gate results. With this gate, if either A or B input goes high, the output will go high. But, unlike an OR gate, if both A and B inputs go high, the output will not go high. Note that if A is high and B is low, No. 2 AND gate enables the OR gate and output X is high. If both A and B inputs are high, neither AND gate is enabled (due to the inverters) and the output X from the OR gate is not driven high. The symbol for an XOR gate and its truth table are shown in Fig. 12-6b. The symbol for an exclusive-NOR (XNOR) gate and its truth table are shown in Fig. 12-6c. Whereas the OR symbol is +, the exclusive-OR symbol is $\oplus$.

An XNOR gate can form a simple bit *comparator*. It has a high output only if both bits are the same, both 0 or both 1. (A *level comparator* that compares a rising voltage with a reference voltage is a more complicated device.)

An XOR gate can also invert or "complement" any *bit* (*b*inary dig*it*). It can change a 0 to 1, or a 1 to 0. If the A input is held at 1 (Fig. 12-6b), any bit fed to B will be inverted at X (truth table). With 4 XORs in four lines, the binary word 1001 becomes 0110 at the X's if the A inputs are high.

A type 7486 TTL IC is an example of an exclusive-OR gate.

12-10 FLIP-FLOP CIRCUITS

A flip-flop (FF) circuit is somewhat similar to a multivibrator. It has many applications in digital

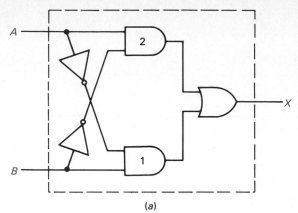

(a)

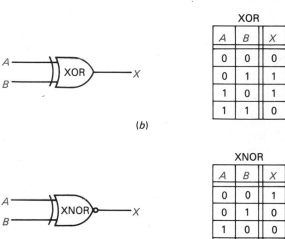

XOR

A	B	X
0	0	0
0	1	1
1	0	1
1	1	0

(b)

XNOR

A	B	X
0	0	1
0	1	0
1	0	0
1	1	1

(c)

FIG. 12-6 (a) Exclusive-OR gate. (b) XOR symbol and truth table. (c) XNOR symbol and truth table.

circuitry. A basic FF using discrete BJTs is shown in Fig. 12-7a. If the set (S) input is driven high for an instant, BJT_1 saturates, and its collector, indicated as $\overline{Q}$ ("not-Q"), is lowered to nearly ground potential. In this condition $\overline{Q} = 0$. If the collector of BJT_1 goes to ground, BJT_2 has no forward bias, no collector current flows in it, and point Q goes high (Q = 1). Now, Q = 1 and $\overline{Q}$ = 0, and the circuit *latches*, or holds in this condition. It remains latched until the reset (R) input is driven high for an instant. Then BJT_2 saturates, Q = 0, and $\overline{Q}$ = 1. The circuit remains latched until S is driven high again. Thus, a FF can store a voltage condition indefinitely and act as a memory cell. The symbol for an RS (reset-set) FF is shown in Fig. 12-7b. In ICs an RS FF may be made with two NAND gates, Fig. 12-7c.

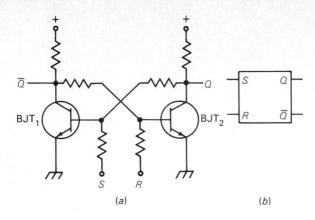

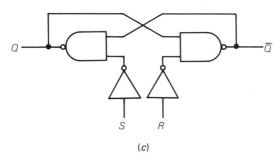

FIG. 12-7 *RS* FFs (*a*) with BJTs, (*b*) symbol, and (*c*) with NAND gates.

By adding an inverter and two AND gates to the *S* and *R* inputs, an *RS* FF becomes a *clocked D* (data) *FF*, Fig. 12-8*a*. The truth table is shown in Fig. 12-8*b*.

The clock input may be either a square wave or narrow pulse of dc from an oscillator, perhaps at 1 MHz. All circuits driven by the same clock pulses operate simultaneously, provided they are enabled by some other circuitry. If they are "inhibited" they will ignore the clocking pulses. For example, if the base of BJT$_1$ in Fig. 12-7*a* were grounded, no set or clocking signal would actuate the FF. The FF would be inhibited.

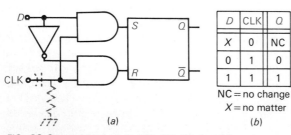

D	CLK	*Q*
X	0	NC
0	1	0
1	1	1

NC = no change
X = no matter

(*a*)   (*b*)

FIG. 12-8 (*a*) Clocked *D* FF. (*b*) Truth table.

In the clocked *D* FF of Fig. 12-8*a*, if *D* is made high before the clock pulse arrives, the upper AND gate is not enabled until the pulse is applied, at which time the *Q* output goes high. If *D* is low when the clock pulse arrives, the lower AND circuit will be enabled and *Q* will go low (and $\overline{Q}$ goes high). The truth table of Fig. 12-8*b* indicates the condition of *Q* under the various conditions of *D* and CLK. Such a circuit can be used to steer digital voltage levels from one part of a system to another at a desired clock instant.

If the CLK signal is fed in through a small capacitor in series with the line, with a low value of resistance to ground (Fig. 12-8), the FF is called an edge-triggered FF, because the leading edge of the clock pulse will be very short and triggers the FF rapidly.

One flip-flop that *toggles* (changes state with every input pulse) is the *RST* FF, Fig. 12-9. If *Q* is

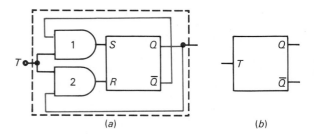

FIG. 12-9 *RST* FF that toggles.

low, then $\overline{Q}$ is high, and $\overline{Q}$ connected to the top input of AND-1 gate is high. When a high pulse is fed through *T* to the lower input of AND-1, this gate output and *S* both go high, so *Q* goes high. On the next pulse, AND-2 has both inputs high and $\overline{Q}$ goes high. With each input pulse the FF toggles.

By adding two more OR circuits, *preset* (P) and *clear* (CLR) controls are formed, Fig. 12-10*a*. This *D*-type FF can be cleared (*Q* = 0), or one bit of data can be clocked into it, or it can be preset high, whichever is desired. The symbol is shown in Fig. 12-10*b*.

One of the most versatile flip-flops is the *JK* FF shown in Fig. 12-11. The AND gates in this circuit are 3-bit (input) types.

If both *J* and *K* inputs are low, the *Q* output remains in its last state, regardless of clock pulses. If *J* is made high and *K* is low, *Q* will be set by the clock (*Q* = 1). If *J* = 0 and *K* = 1, *Q* will be reset by the clock (*Q* = 0). Also, if both *J* and *K* are high, each clock pulse will toggle (reverse) both

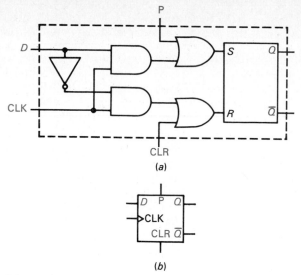

(a)

(b)

FIG. 12-10 (a) Clocked D FF with P and CLR controls. (b) Symbol.

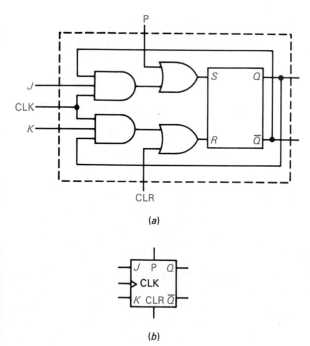

(a)

(b)

FIG. 12-11 (a) JK FF. (b) Symbol.

outputs. The symbol of a JK FF is shown in Fig. 12-11b.

A type 7474 TTL IC is a dual D-type edge-triggered FF. The 7476 is a dual JK amplitude-triggered FF. A CD4013 is a CMOS dual D-type FF.

Test your understanding; answer these checkup questions.

1. Would a slide rule be a digital or an analog device? _____
2. What does TTL mean? _____ What are the high- and low-voltage ranges of TTL devices? _____ _____
3. How does a machine multiply 6 by 4? _____ Divide 88 by 22? _____
4. High A and B inputs drive the output X high. What type(s) of gate might this be? _____
5. High A and low B drive X high. What type(s) of gate might this be? _____
6. True A and false B drive X false. What type(s) of gate might this be? _____
7. True A and B drive X false. What type(s) of gate might this be? _____
8. What does a "bubble" at the output of a gate symbol indicate? _____
9. What type of digital gate forms a simple comparator? _____
10. What circuit is used to store a 1 or 0 voltage condition? _____ In what two broad categories might such a circuit be used? _____ _____
11. What does a "clock" generate? _____
12. What type of flip-flop is the most versatile? _____
13. Of what two circuits might a NAND gate be composed? _____
14. In reference to digital FFs, what is the meaning of: R? _____ S? _____ P? _____ CLR? _____ Q? _____ $\overline{Q}$? _____

12-11 REGISTERS

A register is a single digital word storage device. It can consist of a series of FFs which will hold binary words or numbers until they are wanted. An example of a 4-bit binary word might be 1101_2. A register consisting of 4 FFs could have this 4-bit binary word clocked into it. Each FF of the register would hold one of the bits of the word. A register holds a word until it is to be used.

A possible register is shown in Fig. 12-12. The 4 bits of the word 1101_2, in the form of high and low voltages are indicated on the top *bus* (parallel conductors). They can be loaded and clocked into the D inputs of the FFs through the top four 2-input AND gates when the clock/load AND gate inputs are both high. The output of the four Q-outputs of the FFs deliver the digital word to the lower 4-wire bus whenever the lower AND gates are enabled by simultaneous high inputs to the clock/enable AND

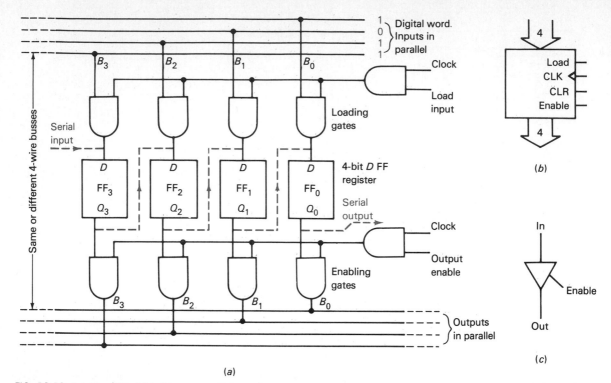

FIG. 12-12 (*a*) Possible 4-bit binary parallel word register. Dashed lines indicate connections for serial in/out register. (*b*) Symbol. (*c*) 3-state switch symbol.

gate. The symbol for a 4-bit register is shown in Fig. 12-12*b*.

Instead of the four AND gates, four *3-state* gates or switches might be used. The three switching states are *open, closed,* and *floating*. The symbol for a 3-state switch is shown in Fig. 12-12*c*.

In registers of this type the clocking signal may be appearing every microsecond, but if the loading or enabling signal is not present, there is no input to or output from the FFs. Registers might be 4, 8, 12, 16, 32, 64, or any number of bits wide. The register shown is a 4-bit parallel loading type. There are also serial loading types (dashed lines) in which a bit is loaded into the first FF, then moves into the second on the next clock signal, and so on (Sec. 12-21).

A 4-bit register of the type described could be developed from two 7408 quad AND gates, plus a 7475 quad *D*-type FF, plus a half of another 7408 for the loading and enabling gates.

12-12 MEMORIES

The 3-bit FF register of Fig. 12-13 could be considered a 3-bit-word *memory cell*. It could hold one of eight different 3-bit words, 000 to 111. If only FF_0 is fed a high when the input loading switches are closed, the stored digital word would be 001. Any time that the three output switches are closed, the output word to the 3-wire bus would be 001. If the FFs are reset and then input switches 1 and 2 reprogram highs to the FFs, at any time the output switches are closed the output word on the bus would be 110. The input switches represent the loading gates, and the output switches represent the enabling gates. Such a memory group can store a 3-bit digital word that can be recalled (output) at any time. If reset, the word is lost, and the output would be 000. A memory bank made up of groups of such programmable cells is called *random access memory*, or RAM.

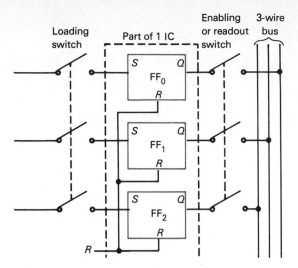

FIG. 12-13 Rudimentary switch loading and enabling 3-bit RAM memory cell.

If the memory cells are fabricated by the manufacturer to have no input data ports but with the different cells either constructed or preprogrammed to have high or low output voltages, the cells are said to form a *read-only memory*, or ROM. In Fig. 12-14, two 3-bit memory cells are coupled to a

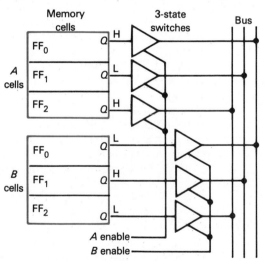

FIG. 12-14 Two ROM memory cell groups, both capable of feeding the bus if enabled.

common 3-wire bus through 3-state switches. If the A memory cells 3-state switches are fed a high clocking pulse, the bus will receive the data word (voltages) HLH, or 101, during this pulse time. If the B memory cells 3-state switches are pulsed, the

bus will be fed the data word LHL, or 010. It is necessary to develop some form of steering or addressing system so that only the desired set of memory cells are activated at any given instant. An address number may be keyed into a system through a decimal number key pad. This number must be converted to digital, and the digital number will be made to activate only the 3-state switches of the memory cell having that digital number.

A 3-bit address word can decode eight output words, but a 4-bit address word could decode sixteen output words. A 5-bit address could produce 32 words, an 8-bit address 256 words, and a 16-bit address could select any one of 65,536 data words.

A RAM system is more complex than a ROM. A 4-bit RAM must be able to: (1) have a certain memory cell selected by an address word fed to the address bus from a *memory address register*, or MAR, (2) load a high/low series of values into the selected memory cell, (3) hold the high/low digital word indefinitely, and (4) enable the output of the memory cell to feed the high/low series of values into a 4-bit data bus for transfer to some register or device.

For simplicity, only simple FFs have been used in describing memories. Actually there are many different types of memories. Some are simple switch circuits, some programmable ROMs (PROMs), or perhaps magnetic cores, discs, tapes, or bubbles, or charge-coupled devices (CCDs). The switch and magnetic types are nonvolatile and will retain their data if power is turned off. Volatile memories will lose their data and must be reprogrammed if the power goes off. CCDs must be refreshed every few microseconds. A single IC chip can easily store 65,536 (known as 64k) bits or more.

A 74S200 is a 256-bit memory TTL IC. A 4116 CMOS IC is a 16,384-bit memory chip.

12-13 BINARY CODED DECIMAL

The term *binary coded decimal* (BCD) is often encountered. To explain its significance, let us say that a series of binary switches is used to activate a circuit to produce a frequency of 591 kHz. Since decimal numbers require 10 digits, it will be necessary to use more than three binary switches to produce the full 0 to 9 of the decimal system in binary (111 = only 7). Thus, to switch in a number such as 591, a series of three sets of four switches is required. They would be switched as shown in

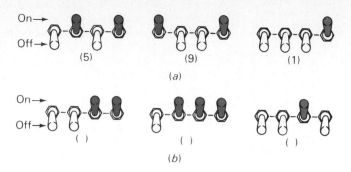

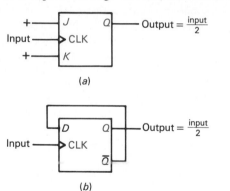

FIG. 12-15 (a) Binary-coded decimal switches set for 591. (b) What is the decimal number indicated in b?

Fig. 12-15a. What BCD number is represented by Fig. 12-15b?

12-14 DIVIDERS OR COUNTERS

The JK and the D FFs connected as in Fig. 12-16a and b operate as divide-by-2 counters of pulses fed to the CLK inputs. In Fig. 12-16a, remember that

FIG. 12-16 Divide-by-2 counters using (a) a JK FF and (b) a clocked D FF.

if J and K inputs are both high, the output, Q, toggles with each input pulse. That is, for every two clocking pulses there will be only one repeat of the Q level. In Fig. 12-16b, by feeding back the $\overline{Q}$ to the D input, a D-type FF does the same as a JK FF. If two JK FFs are cascaded (Q of the first coupled to CLK of the second) the output will equal the input divided by 4. With 3 JK FFs the output will equal the input divided by 8. With 4 JK FFs the output will equal the input divided by 16.

A binary counter that will register from 0000 to 1111 is shown in Fig. 12-17a, using JK FFs. With the first clocking pulse, A goes high (0001). The

second pulse drives B high (0010). The third pulse drives A high again (0011). The fourth pulse drives C high and both B and A low (0100). This continues until all DCBA outputs are high (1111, or 15). The next pulse clears all outputs and DCBA reads 0000.

To divide by only 10, it will require at least 4 FFs but with some connect-back circuit to interfere with the normal 16 count. This is so that the counter will stop counting (start at 0 again) when the tenth pulse has been fed to it. One possible divide-by-10 or *modulus-10* (10 states before repeating) circuit is shown in Fig. 12-17b. The binary coded decimal output will be from the D-C-B-A connections. For the first clock pulse the DCBA output will be 0001, for the second, 0010, for the third, 0011, and so on up to the ninth, 1001. The tenth produces 0000 output, plus generating a carry of 1 into the next modulus-10 decimal counting unit (DCU) circuit by utilizing the dropoff of the high voltage from the D output as it changes to 0. The D output is high for only the numbers 8 and 9 (1000 and 1001). A whole DCU can be made up into a single tiny IC.

12-15 LED READOUTS

While the high DCBA outputs of the BCD counter of Fig. 12-17 might be used to illuminate four lamps, to read the decimal value it would be necessary to know the binary code. It is easier to read if the BCD outputs D, C, B, and A are made to control a 7-segment or 7-bar LED decimal number display (Fig. 12-18).

If the BCD code fed to DCBA is 0001, the gate circuits in the IC will connect segments b and c to ground through the 330-Ω current-limiting resistors.

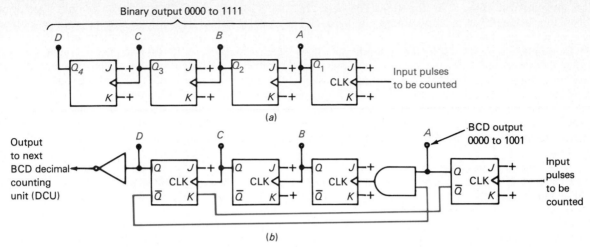

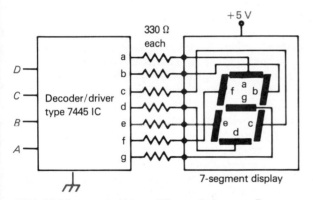

FIG. 12-17 (a) Binary counting unit. (b) Decimal counting unit.

FIG. 12-18 Decoder/driver IC coupled to a 7-segment display unit.

The unusual feature of an LCD is its *nematic* (light-polarity-shifting) liquid. A reflective type of LCD has several layers (Fig. 12-19). The first is a glass polarized filter. Next to this is a glass plate with the 7-segment bars printed on it, but with such

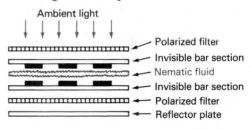

FIG. 12-19 Constituent layers of a reflective LCD.

The current flow through these two segments will light them to form a "1" to the viewer. If the BCD code is 0011 (3), the IC will ground segments a, b, g, c, and d. The viewer will see a good representation of a 3. In this way all the decimal numbers 0 through 9 can be displayed. If a display is to show an 8-digit decimal number, then eight modulus-10 counter ICs, eight BCD-decoder/driver ICs, and eight 7-segment displays are required.

12-16 LIQUID CRYSTAL DISPLAYS (LCDs)

Another bar-segment type of display is the LCD. These displays require only microamperes to operate them, whereas LED and ionized gas (neon) displays require many milliamperes, and segmented incandescent displays may require amperes.

a thin metallic layer that they are invisible. The nematic liquid is between the metalized surface and a second set of invisibly thin metalized segments on another plate of glass. Against the back of the last printed segment glass is a second polarized filter. Finally, there is a reflecting surface at the bottom of the stack.

Ambient light is always of mixed polarity. When it shines through the first polarized filter, it becomes unipolar (let's say vertically polarized). It passes through the various layers of the stack, reflects back out of the top surface, and the observer sees an illuminated background. If an ac voltage (about 5 V at about 30 Hz) is applied to any two segments that are located one above the other, the nematic liquid rotates the polarity of the light 90° in this area. As a result no light can pass through the bottom polarized filter to be reflected. The observer sees this segment area as returning no light and therefore it appears black.

In another type of LCD the bottom polarized filter is rotated 90° from the top one and therefore no background light reaches the reflecting surface and no light is returned to the observer. Now, if any segments are ac actuated, they rotate the incident light 90° and the segments appear light (or whatever the color of the reflecting surface happens to be).

The 7-segment numbers are driven with circuits similar to the 7-segment drivers of LEDs. All of these reflective LCDs only show figures when there is incident light to be reflected.

If the reflective surface is made transparent and a light is allowed to shine through it from the back (or edge), the observer again sees a lighted background. When the segments are ac-driven, the nematic liquid rotates the polarized light 90° and this light cannot pass through the front polarized filter. The segments again appear black to the observer. Such LCDs can be read in the dark.

LCDs turn on rather rapidly, although not nearly as fast as LEDs, but they tend to turn off slowly (0.1 s). When the LCD is at a temperature below freezing, the turn-off time begins to lengthen and even slow changing numbers, like seconds on a watch, tend to become latent images and blur together. Specially developed fast operating (0.05 s) back-lighted LCDs are used in small flat-screen TV sets (Sec. 24-26).

12-17 A/D CONVERSION

A changing voltage, as from a microphone, is said to be an *analog* voltage. In many modern systems it becomes necessary to sample an analog wave, convert the sampled voltage levels to relative binary numbers, and then transmit this digital information. At the receiving end the digital numbers can be converted back to analog form if required.

One possible scheme to convert *analog-to-digital* (A/D) is illustrated in Fig. 12-20. The analog voltage is shown with 19 sampling times. The voltage values at approximately these times must be "quantized" into relative digital values (numbers).

The heart of this system is the 10-MHz pulse generator and the divide-by-100 square-wave-output circuit. At sampling time T_1, the 100-kHz pulse enables AND gates A, B, and C. Enabling gate A feeds dc to the amplifier, and it begins passing the analog voltage in the amplifier at that time to one input of the level comparator. At the same time enabling gate B feeds dc to a ramp voltage generator. It begins to generate a linearly increasing voltage which is fed to the other input of the comparator. The comparator output remains 0 until the ramp voltage builds up to the value of the analog voltage being sampled. At this instant the comparator output becomes 1. Enabling gate C passes 10-MHz clock pulses through AND gate D which is still being enabled by a 1 from an inverted 0 out of the comparator. Thus, clock pulses are being fed into the binary counter, and it starts counting these 10-MHz clock pulses in binary (not BCD in this case), as 0000, 0001, 0010, 0011, 0100, etc.

These numbers appear as a 6-digit binary number at the output (B_5 to B_0). When the ramp voltage equals the sampled analog voltage, the comparator output changes to 1, the inverter applies a 0 to AND gate D inhibiting it, and counting stops. The output of the counter represents the relative amplitude of the analog voltage at T_1 but expressed as a digital number. When the 100-kHz pulse drops to 0, *inverter* 2 applies a 1 to enable the output AND gates, feeding the sampled total number onto a 6-wire bus. Shortly thereafter, but delayed by the *propagation delay time* of inverters 3 and 4, the binary counter is cleared to all zeros.

The output numbers are only latched onto the bus for the propagation delay time before the counter is cleared and its output returns to all zeros. Since T_2 is seen to be higher than T_1, the binary output number from the counter for sample T_2 will be a greater value than the binary number counted at T_1.

Single-chip A/D ICs are available.

12-18 D/A CONVERSION

The previous section described how an analog voltage might be quantized into relative digital values (A/D). In that circuit the analog voltage output is represented by up to a 6-digit binary number. This allows 2^6, or 64, discrete sampled values of the analog voltage.

To convert the A/D output to a digital-to-analog (D/A) signal requires different circuitry. To simplify the explanation it will be assumed that any high value (any "1") bit in an input digital number is +10 V and any low value (0) is at ground potential. To convert the digital number values to relative voltages, a resistive ladder network (Fig. 12-21) might be used.

Without going into the mathematics of the cir-

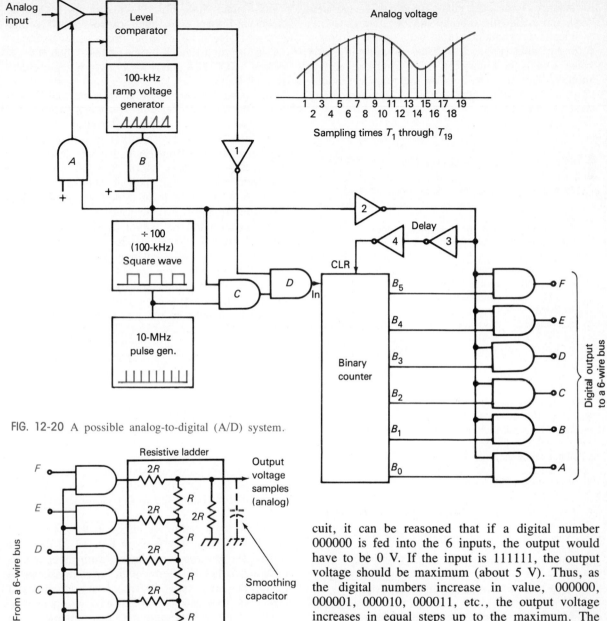

FIG. 12-20 A possible analog-to-digital (A/D) system.

FIG. 12-21 A possible digital-to-analog (D/A) system using a resistive ladder.

cuit, it can be reasoned that if a digital number 000000 is fed into the 6 inputs, the output would have to be 0 V. If the input is 111111, the output voltage should be maximum (about 5 V). Thus, as the digital numbers increase in value, 000000, 000001, 000010, 000011, etc., the output voltage increases in equal steps up to the maximum. The output voltage at any one instant is determined by the value of the digital number on the bus at the enabling time.

The input to the ladder must be clocked on while the digital number is on the bus. The result is a series of short-duration voltages at the output terminal that vary as the original analog voltage varied. A capacitor across output-to-ground would charge and discharge as the peaks increase and decrease, producing a smoother analog output voltage than instantaneous sampled pulse values and result in a wave more like the original analog voltage.

Telephone companies often use systems of binary coding of analog speech voltages to transmit analog voice signals in binary or BCD from one city to another. This is termed PCM (pulse code modulation). It is used in other areas of communication, such as satellite, etc.

▌ Test your understanding; answer these checkup
▌ questions.

1. How many digital words does a register hold? _____ As explained, it consists of what kind of circuits? _____
2. If a register has eight input lines, how many output lines will it have? _____
3. In question 2, how many bus lines will be required to feed to and from it? _____
4. What is the meaning of RAM? _____ ROM? _____ PROM? _____ EPROM? _____
5. Are addresses associated with registers, memories, or FFs? _____
6. A 1-of-16 decoder is fed 1010. Which memory cell will it enable? _____
7. Name a volatile memory. _____ A memory that must be refreshed. _____ What memories are nonvolatile? _____
8. Using BCD, 375 would be expressed as _____ .
9. What does BCD 1000 0101 0111 0010 equal in decimal? _____
10. What modulus does a single *JK* FF represent? _____
11. What does LED mean? _____ DCU? _____
12. How many small lighted segments does it require to show all digits from 0 to 9? _____
13. What is the main advantage of LCDs over other displays? _____
14. What is accomplished by using an A/D converter? _____ A D/A converter? _____
15. What is a resistive ladder used to convert? _____
16. What is an example of an analog voltage? _____

12-19 BOOLEAN LOGIC

The means of mathematically describing and simplifying complex binary-logic gate systems is called *Boolean logic* or *algebra*. While it can become quite complicated, there are a few basic points that should be understood.

The simple truth tables involving A, B, and X, shown in previous sections, express some of the ideas of Boolean logic. In an inverter, for example, if A is the input and X is the output, then $A = \overline{X}$ (spoken "NOT X") and $\overline{A} = X$, or A is the complement of X.

The special symbols used in Boolean logic are + for OR, the symbol · (or any sign meaning multiply) is for AND, and $\oplus$ is for exclusive-OR.

$A \cdot B = X$ is read "A AND $B = X$," and is the logic term for a 2-input OR gate (Fig. 12-2).

$A + B = X$ is read "A OR $B = X$," and is the logic term for a 2-input OR gate (Fig. 12-2).

$\overline{A + B} = X$ is the logic term for a NOR gate in which the output level is inverted *after* inputs A and B are ORed (inputs 1s, output 0)(Fig. 12-5).

$\overline{A} + \overline{B} = X$ is the logic term for an OR gate in which the input levels are inverted *before* they are fed to the OR gate. It is an OR gate being fed by two inverters and is indicated by two little circles or "bubbles" at the inputs. It might be called a NOTED OR gate, or a bubbled OR gate (Fig. 12-22*a*).

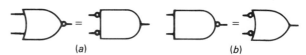

FIG. 12-22 Symbols showing De Morgan's two logic laws. (*a*) First law. (*b*) Second law.

$\overline{AB} = X$ represents a NAND gate, whereas $\overline{A} \cdot \overline{B} = X$ is an AND gate in which the levels have been inverted before being fed to the gate (a NOTED AND gate).

Boolean logic can prove that $\overline{A + B} = \overline{A} \cdot \overline{B}$ and that $\overline{AB} = \overline{A} + \overline{B}$. The gates in Fig. 12-22 illustrate the two *De Morgan* logic laws. From them it may be deduced that all digital logic operations can be developed using only NAND, NOR, and inverter gates. (An AND gate is a NAND gate plus an inverter, etc.) This greatly simplifies the construction of digital IC devices.

Examples of Boolean notation with accompanying basic logic circuits are shown in Fig. 12-23. When the inverter is the last device, the NOT sign is over the whole Boolean expression. If the inputs are NOTED, only those letters are NOTED. Check the logic circuits and their Boolean expressions that are shown. The circuit in Fig. 12-23*j* is called a *half adder* (HA). It will add only two binary digits, but if both are 1s the sum is a 0, but a carry of 1

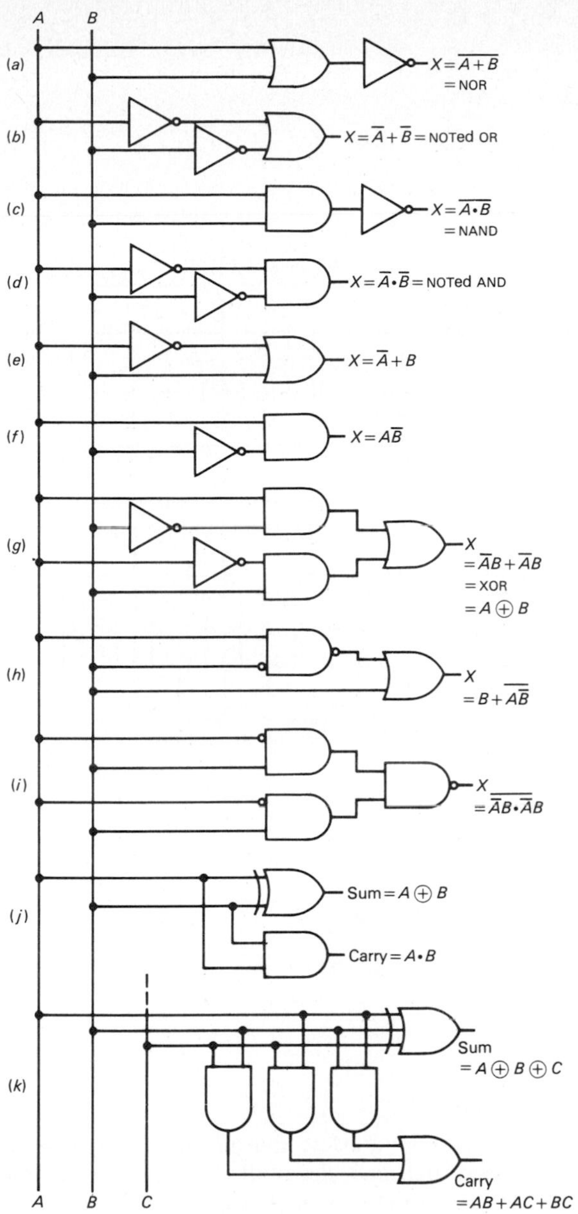
A
B
(a)
X = $\overline{A+B}$
= NOR
(b)
X = $\overline{A}+\overline{B}$ = NOTed OR
(c)
X = $\overline{A \cdot B}$
= NAND
(d)
X = $\overline{A} \cdot \overline{B}$ = NOTed AND
(e)
X = $\overline{A}+B$
(f)
X = $A\overline{B}$
(g)
X
= $\overline{A}B+A\overline{B}$
= XOR
= $A \oplus B$
(h)
X
= $B+\overline{A\overline{B}}$
(i)
X
= $\overline{\overline{A}B \cdot \overline{A}B}$
(j)
Sum = $A \oplus B$
Carry = $A \cdot B$
(k)
Sum
= $A \oplus B \oplus C$
Carry
= $AB+AC+BC$
A
B
C

FIG. 12-23 Some logic circuit combinations and their Boolean logic expressions.

results, giving 10_2 or binary 2. The *full adder* (FA) is shown in Fig. 12-23*k*. It adds 3 bits, 2 digits, and a carry, and can also produce a carry.

To add two 3-bit binary numbers, a half adder and two full adders are required (Fig. 12-24). If the number $A_2A_1A_0$ is 011, and $B_2B_1B_0$ is 010, sum S_0 from the HA is 1 with no carry (c). The first FA

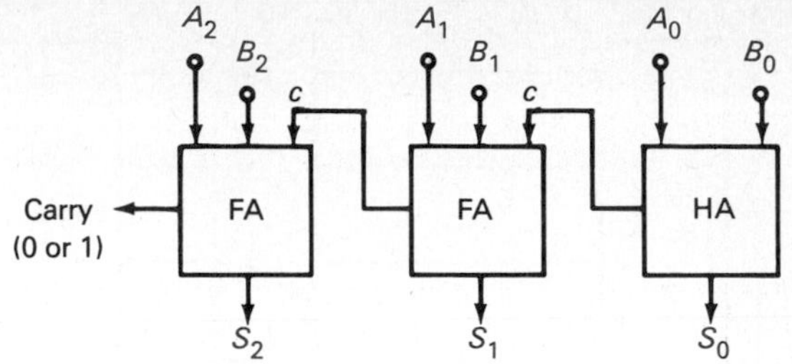
A_2
B_2
A_1
B_1
A_0
B_0
c
c
Carry
(0 or 1)
FA
FA
HA
S_2
S_1
S_0

FIG. 12-24 A 3-number binary counter using a half adder and two full adders. Numbers to be added (*A* and *B*) at top, sum (*S*) at bottom.

adds A_1 and B_1 to produce a sum S_1 of 0 with a carry into the second FA. The carry gives a S_2 of 1, for a total binary sum of 101. Thus 3 + 2 = 5. To add two 16-bit binary numbers the adder will require one HA and 15 FAs.

12-20 A KEYBOARD MATRIX

When a keyboard key is pressed it must develop a digital word that can then be made to produce some desired action. A keyboard matrix circuit can be made to work with diodes, software, or by scanning. A relatively simple scanning idea is shown in Fig. 12-25*a*. The color and black matrix represents crossed sets of vertical and horizontal wires below the keys on a keyboard. If the "A" key is pressed it pushes wires 2 and *W* together. Pushing the "B" key shorts wires 3 and *X* together, and so on. Note how a separate digital code word might be developed by activating each of the 16 keys in this circuit.

The 100-kHz multivibrator (MV) feeds square-wave pulses up to the 4-output binary-coded counter (Sec. 12-13). The counter starts counting pulses as soon as all circuits have been reset, cleared, or preset where indicated. Pulses are also fed to both a 17-count and to one of two 4-count ring counters. A starting high on the CLR/PRE line presets FF_1 and FF_5 of the scanning ring counters with a high at D_1 and D_5, but clearing D_2, D_3, D_4, and D_6, D_7 and D_8 to lows. The system is ready to scan the matrix.

Basically, the first ring counter is going to scan or test the four horizontal wires sequentially from the bottom for any connections first to wire *W*, then to wire *X*, then to wire *Y*, and finally to wire *Z*. The first MV pulse drives FF_1 to develop a high on Q_1, which drives the lower input of the AND-1 gate high, and also clocks FF_5 to develop a high on Q_5

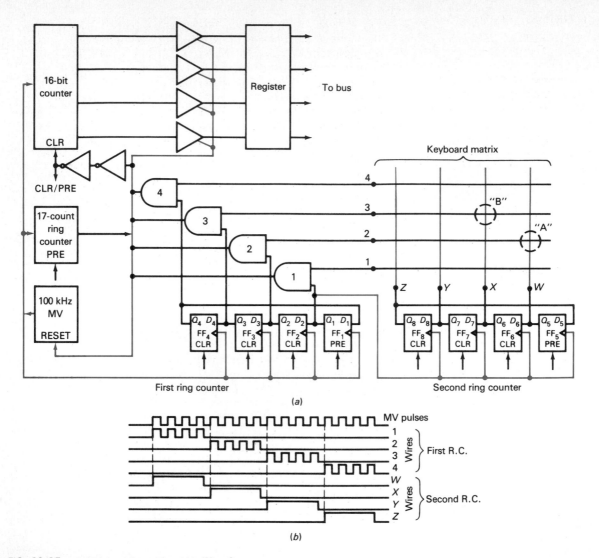

FIG. 12-25 (a) Keyboard matrix. (b) Waveforms.

and wire W. If no key is pressed, none of the four horizontal wires will ever contact any of the vertical wires, and there can be no high AND-gate output to load any counted numbers through the 3-state gates into the register to be held there. After 16 unsuc-

cessful scanning pulses, the 17-pulse ring counter develops a high that clears all circuits, allowing the MV to start another scanning-pulse cycle.

If wire W and wire 2 are connected by pressing the "A" key, the first MV pulse clocks Q_1 high. This places a high on the lower input of AND-1, which drives Q_5 and therefore wire W high. Since there is no contact between wires W and 1, nothing happens. The second MV pulse clocks Q_2 and the lower input of AND-2 high. Now, the high still on wire W is connected to the upper input of AND-2 feeding a high to the 3-state switches to load the 2 counts developed in the counter into the keyboard register. The same AND-2 high output resets the

MV and is delayed by propagation through the two inverters before it clears or presets all other circuits. (A delay line might have been used.) Note that the CLR/PRE occurs after the count number is loaded into the register. The system is now ready to start another scanning-pulse cycle.

Even a rapid pressing of a key will produce at least a 30-millisecond closure, allowing a minimum of 3 scanning cycles and the clocking of the same number, 2 in this case, into the register. (If only one loading is desired, circuitry can be added which will hold off the multivibrator until the output of the register is clocked onto a bus.) If a second letter key is pressed before the register is cleared, the first stored letter will be lost. To prevent this the register can be clocked into a series of first-in-first-out memory cells, called a *buffer*. If the buffer has 16 memory cells it will be able to store up to 15 digital key numbers before it would be overrun by a next key pressing.

If wire X and wire 3 are connected by pressing the "B" key, during the first 4 pulses, while the AND gates are waiting for highs for their upper inputs, there can be no AND-gate action. With the 5th pulse, Q_1 is driven high again, clocking Q_6 high, making wire X high. Since there is no contact between wire X and wire 1, there is no AND-gate action. On the 6th pulse Q_2 and the lower input of AND-2 go high. Since there is no contact between wires X and 2, there is no AND-gate action. However, on the 7th pulse Q_3 and therefore the lower input of AND-3 are both high. Now the high of wire X is fed to the upper input of AND-3 and it feeds an output high to the 3-state switches, loading a 7-count (0111) into the register before resetting, clearing, or presetting the system circuits again. It

can be reasoned that pressing 16 different keys will produce 16 different digital numbers (0000 to 1111). While this 4×4 matrix will only produce 16 digital words, an 8×8 matrix is capable of 64 different digital words. This is more than enough for the 26 letters of the alphabet, 10 numbers, many punctuation marks, and any desired machine functions (space, line feed, carriage return, shift-up for capitals, shift-down for lowercase letters, etc.).

12-21 PARALLEL-TO-SERIAL DATA

The binary word stored in the register of Fig. 12-25 is in parallel form. It might be the word 1011, for example. To transmit this by radio or telephone line it must be sent as 4 separate on-off signals (on, off, on, on). It is necessary to convert the parallel word to serial form, 1, 0, 1, 1. This can be accomplished by using a shift-right 4-*JK*-FF *parallel-serial shift register*, Fig. 12-26. When the CLK_1 line goes high for an instant, the 1011 word in the keyboard register is clocked into the 4 *JK* FFs through the AND gates. The speed of the CLK_2 line pulses determines the rate at which the serial pulses will be transmitted out of the serial register. On the first CLK_2 pulse the state of FF_1 (a high) is fed out of the Q_1 serial output port, perhaps to a radio transmitter, and all remaining states shift one FF to the right. Now FF_1 is holding a low, FF_2 a high, and FF_3 a high. (What the contents of FF_4 are does not matter as only 4 CLK_2 pulses will be used until after another CLK_1 pulse is developed.) On the next CLK_2 pulse the Q_1 serial output is a low, and the states of the next two FFs shift to the right again. On the third CLK_2 pulse the output is a high. On

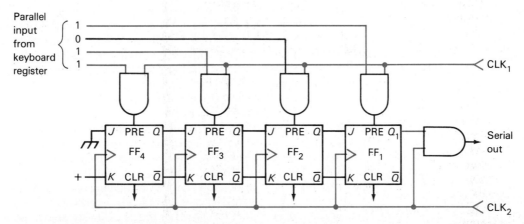

FIG. 12-26 Parallel-serial shift register.

the fourth CLK$_2$ pulse the output is again high, and the word 1011 has been serially transmitted.

If this digital code were to be transmitted by radio, the system could be shown by a block diagram (Fig. 12-27*a*). At a distant receiving location

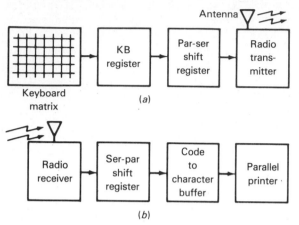

FIG. 12-27 (*a*) Keyboard-to-transmitter block diagram. (*b*) Receiver-to-printer block diagram.

the signals could be picked up, amplified, passed through a *serial-parallel shift register*, fed to a code-to-character converter, and the character could actuate a parallel printing machine (Fig. 12-27*b*). In this way letters typed at one place could be printed at a distant point. The parallel 4-bit register in Fig. 12-12 can be used as a shift register by utilizing the dashed lines. It might also be used as a serial-in parallel-out, or a parallel-in serial-out register.

12-22 MICROPROCESSORS

A microprocessor (μP) is a tiny digital chip, often made up into a 40-pin (or more) DIP form, that can process digital information. There are many different types of microprocessors. Some are dedicated, or specially manufactured to do one job. Others are nonspecialized and may be used in a variety of systems to accomplish various results. They are all composed of thousands of gates and several registers, with one or more buses in them. Some "architectures" need no external memories or other subsystems, but most operate into and from external memory chips and other devices. Although the complete operation of a μP is far too involved for an explanation here, there are some salient points that can be discussed.

A basic μP, showing more or less standard input and output circuits, plus an external clock signal

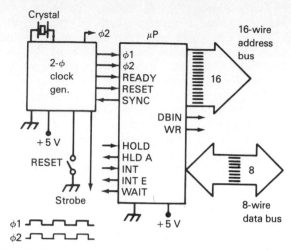

FIG. 12-28 A basic microprocessor and an external 2-ϕ clock generator.

generator, is shown in Fig. 12-28. The 2-phase (2-ϕ) clock generates square-wave pulse-pairs derived from a square-wave oscillator at a frequency of perhaps 1 MHz. For each cycle of the oscillator two pulses, offset in time, are developed. The first, $\phi1$, can be used to enable or ready a bus and register system, and the second, $\phi2$, can be used to transfer or clock digital words from one register or circuit to another. As long as the first pulse is held off from a register or circuit, the second cannot load or enable the register. Controlling the input or output of a register can determine whether it will be used to input or output data or instructions from or to some other part of the system.

An expanded look at a possible μP plus some external devices is shown in Fig. 12-29. A 16-wire address bus is connected to both the ROM and RAM memory banks, and possibly to one or more external devices. The address word from the program counter (PC) is fed to the memory address register (MAR), then out the address bus, is decoded at the memory (or input-output, I/O, device), and a single memory cell (or I/O device) is selected. When data-bus input (DBIN) outputs a 1, the selected word in memory is able to be clocked into the bidirectional 8-bit data bus, moving to either an internal instruction register or to the accumulator register via an 8-bit internal bus in the μP, whichever is enabled at that time. The data bus is made bidirectional by having two back-to-back 3-state switches, forming a "transceiver" in each bus line, as shown in Fig. 12-30. By enabling the write gate, data can be written into any memory cell or

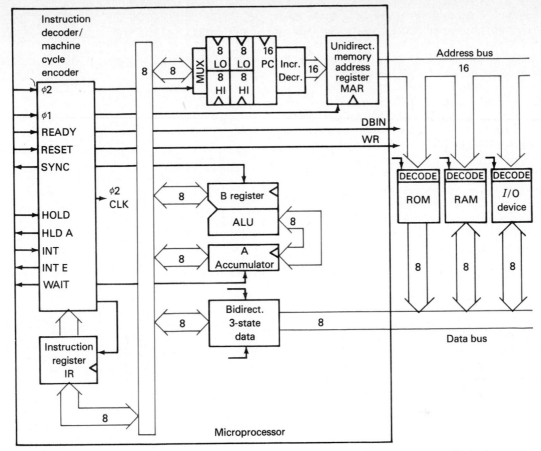

FIG. 12-29 Basic architecture of a microprocessor coupled to ROM, RAM, and an I/O device.

location that is being addressed and open to loading at the time. Conversely, by enabling the DBIN gate the contents of a memory cell can be guided to any µP register that is open to loading at that time. If neither gate is enabled, the 3-state switch is floating and the data bus can be used to communicate data between memories and I/O devices.

The heart of the µP is the instruction-decoder/machine-cycle-encoder. Instruction words from ROM or RAM are clocked into the instruction register via the 8-bit internal bus, which in turn clocks

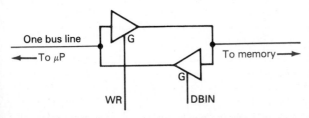

FIG. 12-30 Bidirectional 3-state switch transceiver.

the word into the instruction decoder. From the instruction word's information, the decoder determines which register, buffer, memory, or device to load or enable. On the next clock cycle it sends load/enabling high or low voltages wherever necessary. The program counter obtains the first program word address from ROM or RAM and starts the program into operation. Each time a new program step is performed the PC increments one number higher in the address number it is holding, so that the next system cycle will enable the next higher program step and memory cell in the program section of the memory. The PC with 16 data bits is being fed from the internal 8-bit bus through a multiplexer (MUX) to pairs of 8-bit registers. For high address numbers it is possible for the multiplexer to first load one of the register pairs with the low (LO) half of the address number. Then on the next clock it can load the high (HI) half of the address number into the PC and deliver both to the memory address register.

The orderly step-by-step action of a main program can be interrupted (INT) and a subroutine program can be instituted. This stops the main program until the subroutine is completed, whereupon the main program resumes by going back to the next program address number it has been holding in another internal register called the "stack."

When the μP is used to compute mathematically, its program will involve the arithmetic and logic unit (ALU), which performs both arithmetic and logic operations. The results of arithmetic operations are addition, subtraction, division, and multiplication. Digital values clocked into both the accumulator and the B register are fed into the ALU for computation and are stored in the A register. The logic operations that can be performed are AND, OR, NAND, NOR, XOR, complement A, or complement B. With these operations many different types of decisions can be made, which is the power of such a computer system.

Some of the actions that can be taken by the μP are: READY: 1 = μP working, 0 = μP stopped, used by I/O devices. RESET: 1 = All internal registers cleared. SYNC: Used while fetching an operation code. HOLD: Disables both buses. HOLD A: Acknowledges the hold. INT: Stops program execution for a subroutine. INT E: 1 = Interruption still in force. WAIT: μP not ready for data input.

12-23 A MICROCOMPUTER SYSTEM

A μP or CPU (central processing unit) is the heart of a microcomputer system. A small microcomputer, as in Fig. 12-31, might be used for a communication terminal. The μP shown has been manufactured to handle 8-bit digital words, which might represent binary data, encoded binary data, an encoded instruction, or an encoded character. In the microcomputer a particular word could be stored in, and could be retrieved from, memories, registers, or flip-flop latches in 2-potential form, such as 0-5-0-0-5-5-0-5 V. It is the job of the μP to move such 8-bit voltage/no-voltage words (8 bits = 1 *byte*) from place to place within itself; to add them, subtract them, compare them, and eventually to deliver the results to the outside world.

Since any data being handled is in 8-bit bytes, an 8-wire bus running the length of the μP will allow parallel interconnections between the various registers, enabling latches, decoders, etc., that make up the μP. This internal bus is also connected to the external microcomputer system, but through a con-

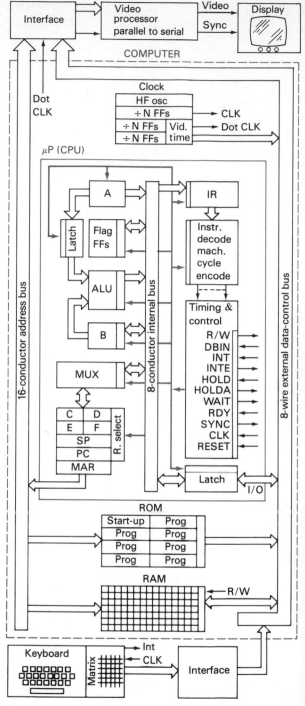

FIG. 12-31 Microprocessor computer with two interfaces.

trolled input/output (I/O) latch or buffer. Each section of the μP has an electronic 1- or 2-way gate to provide a means of connecting to or from the bus.

Many of the sections of the μP are registers. These may have eight 3-state switches as *loading* and *enabling* controls to allow digital bytes to be loaded into them, or to enable, at a desired instant, the byte voltage levels to be passed on to other sections of the μP or to external (peripheral) devices. Three of the registers shown, SP (stack pointer), PC, and MAR, can hold 16 bits of binary information.

Two memory banks are used, possibly having hundreds to thousands of addressable 8-bit memory cells. The read/write (R/W) or random access memory (RAM) is used to store data words in specific cells identified by numerical addresses expressed in up to 16-bit binary words, which allows 65,536 ("64K") specifiable addresses. When one cell's address is activated, the 8-bit data in this cell is released to the 8-wire bus for delivery to some other section of the computer, or to peripheral devices.

A read-only memory (ROM) has programs of data fixed in it that cannot be erased or cleared. It is the brain of the microcomputer. Programs of bytes in ROM tell the μP what to do, how to do it, when to do it, and then where to put the results. When the computer is first turned on, for example, the 5-V dc power-supply voltage begins to rise. When a 1-V level is reached, a Schmitt-trigger circuit develops a pulse that activates another circuit to send an address to the ROM where a fixed "start-up" program exists. This program may consist of several operations that cause all the registers and RAM memories to be cleared and to ready the μP, and peripherals for operation.

The first few thousand address numbers are often assigned to ROM, with RAM being assigned only higher-numbered cell addresses.

The controlling of the interchange of bytes in a computer is based on clock pulses. Each part of a program requires a separate machine cycle. A machine cycle consists of two movement operations, *fetch* and *execute*. During fetch, the address in the program counter (PC) register is moved into the memory-address register (MAR) at the first clock pulse. On the next clock pulse a cell in either ROM or RAM is addressed and the data at that address is moved through the latch to the accumulator (A) or to the instruction register (IR). On the next pulse the PC is incremented to the next higher-number address. During the execute portion of the cycle, the remaining steps necessary to carry out the program are taken, keyed into action by clock pulses.

Computers must be given a program of steps that are to be performed in sequence to produce a desired result. Once given a "RUN" signal, a computer will move through the steps so fast that the answer on the display may appear to be instantaneous.

Instructions may be 1, 2, or 3 bytes in length. When there is more than one byte, the first is called the operation code, or op code. The other instructions may be data bytes, a device-code byte, or two-address bytes. The 8-bit op code is broken up into 2- to 4-bit *fields* which tell the μP where to deliver and what to do with the next byte or bytes. The device code indicates which internal or external device to input from, or output to. Many instructions are stored in ROM, but others may be written into the RAM and stored. Data can be addressed and loaded into another part of the RAM and can be erased as desired. Care must be taken to address the memory cells properly. If a data byte is addressed when an instruction byte should have been, the machine does not know what to do with the word. The programmer must play by machine rules, always programming instructions first and then the data to be worked on.

To handle 16-bit words the MUX can first load the LO bytes and then the HI bytes into the general-purpose registers (C,D,E,F), with LO going to D and F, and then HI going to C and E. The stack pointer (SP) is a 16-bit register loaded by the MUX. It stores the address of the next *word* to be used from a group of words previously stacked into one area of memory. The program counter (PC) is a 16-bit register that stores the next *instruction* that will be executed in a routine or program.

The command lines that feed timing and control voltages to and from the μP are: *read* or *write* into external devices, depending on logic levels 1 or 0 (R/W); *data-bus input* control from external devices (DBIN); interrupt routine to μP (INT); interrupt enable to external devices (INTE); hold command from external devices (HOLD); hold acknowledge to external devices (HOLDA); wait (WAIT); ready indication from external devices (RDY); synchronizing pulse to external devices (SYNC); clock pulse (CLK); reset (RESET).

When instructions are received from external devices, or ROM, they may be clocked first into the instruction register. Data is then clocked by the instruction-decoder/machine-cycle-encoder into applicable registers, latches, flags, memories, and multiplexer on succeeding clock pulses. The flag FFs are set either to 1 or 0 by the ALU, or if the flags are set externally, enable the ALU whenever it must add, subtract, etc.

Interfaces are the means by which the world external to the *computer* can input or receive information to or from the computer. A keyboard, display system, cassette or disk memories, Morse code generators, Morse code decoders, teleprinter code generators, teleprinter code decoders, etc., are examples of communication devices that may interface with a computer.

To write programs into memory, a keyboard can be interfaced to the data bus as shown. When a key is pressed it connects together two of the crossed wires of an 8 × 8 matrix to develop an 8-bit word. (In smaller computers, panel switches may be used.) This byte can be decoded to a 16-bit ROM address which contains a data byte in some other desired code, such as the 7-bit, 128 character/function *American Standard Code for Information Interchange* (ASCII) used between computers, or the 5-bit, 60-character/function *Baudot* code used between teleprinters (Sec. 12-24). The keyboard matrix output is a parallel 8-bit word. When a character word which represents an address for ROM is ready at the output of the keyboard, an interrupt (INT) signal is sent to the μP's latch circuit. As soon as any machine cycle already in operation is concluded, the μP halts momentarily and directs the keyboard address word onto the external bus. The μP passes this address to the ROM, and the ASCII character data there is moved to the video interface for processing and display, or in other applications keyboard words may be moved to registers or RAM for computation operations.

Readout from a computer can be a video display or a printout on paper using some form of printer, or both. Consider the video system in the block diagram. A 12-in. or larger TV display tube is generally used. Its horizontal and vertical sawtooth oscillators and amplifiers are brought into sync by timing signals from the clock section. Characters (or graphics) are developed by dots on the screen. Each character is formed by selectively illuminating a block of perhaps 35 dots, five horizontally and seven vertically, with one blank dot between character blocks and five blank sweeps between character lines. With 64 characters per line, for instance, and 16 lines to the whole "page" on the screen, there is a total of nearly 74,000 dot positions. To write a letter "I," the horizontal sweep starts with the electron beam blanked out. When the beam gets to the block locations where the top of the I is to be, it unblanks and writes three dots

(dots number 2, 3, and 4), then blanks for the fifth dot, and the sweep continues across the screen. On the next five sweeps across, only dot 3 of this character block is written. On the seventh sweep dots 2, 3, and 4 are written and the I is complete. To write a letter "O" the illuminated dots would be numbers 2, 3, and 4, then five sweeps of dots 1 and 5, and then on the bottom or seventh sweep, dots 2, 3, and 4 are written again. Each character has its own format (program) of dots.

Besides the horizontal and vertical sync pulses the video timing circuit must generate a dot signal so that the video processor can key dots on the screen at the correct times, controlled by the ASCII code from the ROM. (There are several other methods of developing a display.)

The programs in ROM are specialized to produce desired movements of bytes through the system. When programs are developed to use other capabilities of the computer, they are typed in via the keyboard and stored in RAM. These RAM programs can be transferred to tape cassettes, or magnetic disks, or any other form of *nonvolatile* (program does not erase if power goes off) memory-storage devices via interfaces off of the data bus. Later they can be played back into the computer RAM to be used in programs as desired by the programmer.

There are two ways that RF equipment and computers can interfere. The computer clock signal, being a high-frequency (1–5 MHz) square wave, has many harmonics that can be picked up by nearby MF, HF, VHF, or UHF receivers. If the computer equipment is not well shielded, nearby RF fields from transmitters can also cause added signals or wipe out digital bits (glitches) in the circuits, or may even burn out ICs or other components.

12-24 PULSE AND DATA CODES

Information can be transmitted by a variety of methods. Speech and music use microphones and operate with analog waveforms. Morse code is a form of pulse transmission in which the length of the pulses helps identify the sound of letters to the operator. The sound and length of each character are different from all others. For either Baudot or ASCII codes (Fig. 12-32) all letters, numbers, punctuation marks, or keyboard functions require the same length of time to transmit. For example, the 5-pulse (7-element) Baudot code characters, when sent at 61.33 words per minute (usually

ASCII					
@	100	0000	∅	011	0000
A	100	0001	1	011	0001
B	100	0010	2	011	0010
C	100	0011	3	011	0011
D	100	0100	4	011	0100
E	100	0101	5	011	0101
F	100	0110	6	011	0110
G	100	0111	7	011	0111
H	100	1000	8	011	1000
I	100	1001	9	011	1001
J	100	1010	:	011	1010
K	100	1011	;	011	1011
L	100	1100	<	011	1100
M	100	1101	=	011	1101
N	100	1110	>	011	1110
O	100	1111	?	011	1111
P	101	0000	SP	010	0000
Q	101	0001	!	010	0001
R	101	0010	''	010	0010
S	101	0011	#	010	0011
T	101	0100	$	010	0100
U	101	0101	%	010	0101
V	101	0110	&	010	0110
W	101	0111	'	010	0111
X	101	1000	(	010	1000
Y	101	1001	)	010	1001
Z	101	1010	*	010	1010
[	101	1011	+	010	1011
\	101	1100	,	010	1100
]	101	1101	—	010	1101
∧	101	1110	.	010	1110
—	101	1111	/	010	1111

BAUDOT					
A	00011	—	Q	11101	1
B	11001	?	R	01010	4
C	01110	:	S	10100	BELL
D	10010	$	T	00001	5
E	10000	3	U	11100	7
F	10110	!	V	01111	;
G	01011	&	W	11001	2
H	00101	STOP	X	10111	/
I	01100	8	Y	10101	6
J	11010	'	Z	10001	"
K	11110	(	CR	00010	
L	01001	)	LF	01000	
M	00111	.	SU	11011	
N	00110	,	SD	11111	
O	00011	9	SP	00100	
P	01101	∅			

FIG. 12-32 Above, ASCII code in binary. Below, Baudot code. CR = carriage return, LF = line feed, SU = shift up, SD = shift down (LTRS), SP = space bar. For serial transmission the right-hand 1 or 0 is sent first.

called 60 wpm), take 163 milliseconds (ms) per character. When sent at a 100 wpm data rate, they require 100 ms for each character.

The speed of transmission of pulse-type codes is usually stated in *bauds*. A baud is a pulse or *bit per second* (b/s). The *baud rate* is the number of b/s of the transmitted code. This can be determined by taking the reciprocal of the time of the shortest pulse that will be transmitted. The shortest pulse width of a Baudot code character at 60 wpm is 22 ms. Therefore, the baud rate is 1/0.022, or 45.45 bauds.

ASCII is basically a 7-bit ("7-level") code. An eighth bit, called a *parity bit*, may be added to serve as an error check. If the letter A in Fig. 12-32 is transmitted, it has an even number of 1s in it (2). If C is sent it has an odd number of 1s (3). If the A character has a "0" parity bit added to it, developed by a 7-input XNOR gate, it will total an even number of 1s, which will not activate an 8-input XOR receiving gate. If C has a "1" parity bit added, it will have an even total and will not activate the XOR gate either. If all characters have 1 or 0 parity bits added to make them total an even number of 1s, then if a noise impulse changes a character bit it will total as an odd number and will activate an XOR gate and an error will be indicated (the machine may leave a space where the letter should be). If a start and a stop pulse are added, the code becomes a 10-bit code. If two stop pulses are used in mechanical machine circuits, it will become an 11-bit code. At 100 wpm the shortest pulse will be 10 ms, and the baud rate will be 1/0.01, or 100. The baud rate and wpm would be the same.

For Morse code (Sec. 31-3) the standard word is PARIS. Counting dots as a 1-bit length, between-dot-dash spaces as 1 bit, dashes as 3 bits, spaces between letters as 3 bits, then PARIS, plus a 7-bit between-words space after it, equals 50 bits. If PARIS is transmitted in 1 second (60 wpm), the pulse rate will be 50 b/s. However, unlike Baudot and ASCII, which consider no space between bits, Morse has equal *mark* (dot on) and *space* (time between dots) times. Instead of the Morse dot being 1/50, or 0.02 s, it will be half of this, or 0.01 s at 60 wpm, and the baud rate will be 1/0.01, or 100 baud. The maximum square-wave frequency of a 100-baud string of dots would be 50 Hz. If it is assumed that the first 10 harmonics of the fundamental are required to reproduce a square-wave shape, an amplifier or filter with a minimum bandpass or bandwidth of 10(50), or 500, Hz would be

TABLE 12-1 WPM VS. BAUDS

Baudot		ASCII		Morse	
Wpm	Baud	Wpm	Baud	Wpm	Baud
				15	25
				30	50
61.33	45.45			60	100
76.67	56.88			75	125
100.00	74.2	100	100	100	167
		300	300		
		10,000	10,000		

required to pass the signal with no distortion. With strong signals less bandwidth may be satisfactory.

Table 12-1 illustrates some wpm vs. baud rates of the three codes.

12-25 MORSE CODE WITH A COMPUTER

To use a computer system to transmit Morse code (Sec. 12-24) by radio at a desired speed requires additional hardware and a special RAM "software" program (or a complete ROM program). When a key is pressed on the keyboard, a parallel digital word (usually in ASCII) is produced, which must be converted via a software program to a coded digital word that will produce serial dot-dash-space outputs from a serial register. The coded digital word is in parallel form and can be loaded into a series of buffer registers. Assume a cascaded (one after another) 3-register system. The first two registers are parallel-in parallel-out types. The third is a parallel-in serial-out shift register. The digitally coded Morse-character letter loads into the first register. If there is nothing in the second register, the letter is transferred to the second register on the next clock signal. If there is nothing in the shift register, the letter immediately transfers into it. The coded letter in the shift register comes out as 1-unit dots, 3-unit dashes, or 1-, 3-, or 7-unit spaces at a rate determined by the manually variable-speed clock used to control the output Morse code speed. The serial output can operate a keying relay to key a transmitter. However, with only three buffer registers, the operator can only type two letters ahead of the letter being transmitted.

To receive Morse code by radio and display it on a computer screen requires different hardware and another software program. Serial-type dot-dash-space Morse characters being received are first loaded into a serial-in parallel-out register that is being clocked at a rate that is synchronized with the received code speed. When the 3-dot-length space used between letters is recognized, the serial register clocks the Morse letter into a parallel register which then converts the Morse letter information into a computer-understood (ASCII) coded word by a self-contained program. The computer word is then fed to the display circuits and is shown on the screen as a character.

Similar systems but using different programs can be used to transmit and receive teleprinter (RTTY) signals. The computer's own ASCII words can also be transmitted and received, but must be converted to serial form and slowed down in a buffer-register for transmitting, and then be reconverted from serial back to parallel to be displayed at the receiving end.

12-26 SOME GATE USES

Inverter, NOR, and NAND gates, separately or in ICs, are actually amplifiers. A NOR gate may be shown as in Fig. 12-33a, but it must be remembered that it always has positive and negative power-supply connections, and should be thought of as shown in Fig. 12-33b. Furthermore, if two NOR or NAND gate inputs are tied together as in Fig. 12-33c, the circuit becomes an inverter.

If gates amplify they can be used as oscillators, amplifiers, etc. In Fig. 12-33d, two inverters (NOR or NAND gates) are connected as a crystal oscillator. An *LC* circuit in place of the crystal will produce a variable-frequency oscillator (VFO). The waveshape can be varied from sine to square by varying *R*. In Fig. 12-33e two NOR gates using a telegraph key and an 8- or 16-Ω loudspeaker form an audio-frequency oscillator, useful for code practice.

The 2-inverter circuit in Fig. 12-33f is a *Schmitt trigger*. If a rising sine wave or other analog voltage at the input reaches a trip point, the two amplifiers suddenly reinforce each other because of feedback through R_f and the output snaps to fully on (logic 1). When the input analog voltage decreases to a trip point slightly lower than the turn-on voltage (the difference is called the *hysteresis*), the circuit snaps off (logic 0). Thus an analog voltage cycle can be converted to a square-wave output. Schmitt triggers may be used as *debouncing* circuits to make sure that when a switch closes there is only one closing, not several rapid bounces,

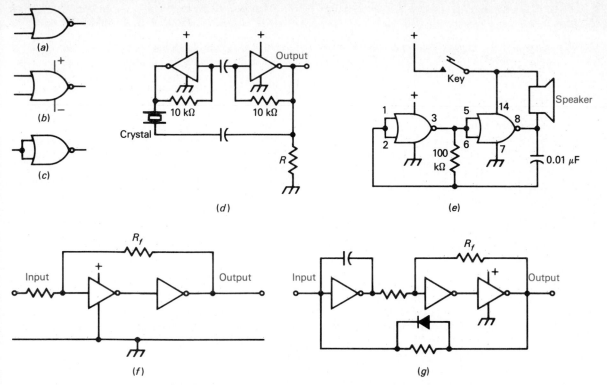

FIG. 12-33 (a) NOR gate, (b) showing power supply leads, (c) connected as an inverter, power leads deleted. (d) Crystal oscillator. (e) A 1-kHz AF code practice oscillator using a CMOS 4000 quad NOR IC. (f) Schmitt trigger. (g) VCO, no supply or ground lines shown on two gates.

which is common for many switches. The Schmitt trigger with added inverter and diode (Fig. 12-33g) forms a voltage-controlled oscillator (VCO) capable of operating up to at least 4 MHz without LC components and much higher with them.

To test if high (1) or low (0) voltages are present on operating BJT, FET, or IC digital TTL circuit devices, a *logic* probe can be used (Fig. 12-34). The ground clip is clipped to a ground connection of the circuit to be tested, and the +5-V lead is clipped to the +5-V point of the system power supply. In this circuit, if the probe is touched to any point in a digital circuit that is high, Q_1 becomes forward-biased and current flows through R. This drives the input of inverter A high, and its output goes low. LED H is now between a low-voltage point through the inverter's internal circuits and +5 V, so it glows. If the probe is touched to a point that is low, inverter B now has a low input and a high output. Inverter C has a high input and a low output, which leaves LED L between a low

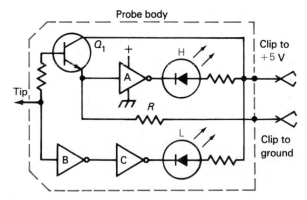

FIG. 12-34 Logic probe to indicate circuit conditions of 1, 0, or active pulsing at terminals of a working digital system.

and +5 V, so it glows. If the probe is touched to a point which is being pulsed by digital pulses, both LEDs will glow. LED H might be red and LED L might be green to allow easy identification.

Test your understanding; answer these checkup questions.

1. What type of logic circuit is expressed by $A = \overline{X}$? _____ By $\overline{AB} = X$? _____ By $A + B = X$? _____ By $AB = X$? _____ By $\overline{A} + B = X$? _____

2. Express the two De Morgan laws in logic symbols. _____ _____

3. What do the De Morgan laws tell about simplifying logic circuitry? _____

4. What does DIP mean? _____ CPU? _____ μP? _____ I/O? _____ R/W? _____

5. What is a bit? _____ A byte? _____

6. What would a 4-MHz crystal oscillator be used for in microcomputers? _____ What waveshape would its usable output have? _____

7. What are names of 6 types of storage devices in a microcomputer? _____ _____ _____ _____ _____ _____

8. What is the width of the internal bus in the microcomputer? _____ Of the address bus? _____

9. What are the two operations in a machine cycle? _____ _____

10. What does PC mean? _____ IR? _____ LO? _____ HI? _____ MAR? _____ SP? _____

11. What does the MUX do? _____

12. What do the fields in an op-code byte do? _____

13. What happens to programs in ROM if the power is lost temporarily? _____ In RAM? _____

14. What does the program counter store? _____ The stack counter? _____

15. What must be outputted from a keyboard before a letter byte from the matrix can be transferred to the ROM? _____

16. What is a volatile memory? _____ Name two such memory storage devices. _____

17. In Fig. 12-23, what is the output (1 or 0) if circuits (a) through (k) have input levels of $A = 1$, $B = 0$, $C = 0$? _____ $A = 0$, $B = 1$, $C = 1$? _____ $A = 1$, $B = 1$, $C = 1$?

18. What would be the required bandwidth of an amplifier to pass 300-baud ASCII? _____ 74.2-baud Baudot? _____

19. What types of ICs might be used in Fig. 12-25? _____

DIGITAL FUNDAMENTALS QUESTIONS

1. What is an amplifier said to be if it amplifies a signal without distorting it?

2. What are the "low-" and "high-" voltage ranges of TTL devices? Of CMOS devices?

3. What is the decimal number 2 in binary? 4? 8? 16? 25? 33?

4. How many binary on-off circuits are needed to handle the binary number equivalent of 8_{10}? For 15?

5. What are the digital and decimal sums of $0101 + 1010$? $0010 + 1010$? $1001 + 1111$?

6. What are the digital and decimal answers to these subtraction problems: $1010 - 0100$? $111 - 110$? $10011 - 1010$?

7. What is the decimal answer if these binary numbers are multiplied: 0110×0010? 1011×0011? 10110×0100?

8. What is the decimal answer if these binary numbers are divided: $11000/1100$? $10010/0110$?

9. In an AND gate, if A and B are both high what is X?

10. What type of logic is the fastest? Slowest? Uses least power?

11. What does "quad AND gate" mean?

12. In an OR gate, if A and B are high what is X?

13. With an inverter, if A is high what is X?

14. With a NAND gate, if A and B are high what is X?

15. What two circuits might make up a NAND gate?

16. In a NOR gate, if either A or B are high what is X?

17. With an XOR gate, if A and B are high, what is X? If only A is high?

18. With an XNOR gate, if A or B is high, what is X?

19. What gate can be used as a simple bit comparator?

20. Draw symbol diagrams of AND, OR, Inverter, NOR, XOR, and XNOR gates.

21. What does "RS" flip-flop mean? For what can it be used?

22. What is an inhibited FF?

23. Draw a diagram of a BJT RS FF.

24. Draw a symbolic diagram of a FF that toggles.

25. How is a JK FF made to toggle?

26. What is a register?

27. Diagram a 4-bit parallel operating register.

28. Why is an AND gate useful as an enabling gate?

29. What are internal parallel interconnecting lines called?

30. What are the minimum connections to a single BJT RAM FF?

31. What are the minimum connections to a single BJT ROM FF called?

32. What would a memory be called if it could be permanently programmed by the user?

33. Draw a diagram of two 3-bit ROMs coupled to a 3-wire bus.

34. What is a volatile memory?
35. What would be the high/low switch settings to represent the BCD number 739?
36. If two *JK* FFs are cascaded, with all *J*'s and *K*'s high, what will the output frequency be if the input is sinusoidal 1 MHz? What will be the output waveform?
37. Draw a diagram of a 4-bit binary counter using *JK FFs*.
38. Draw a diagram of a 4-bit BCD output DCU.
39. Draw a diagram of a decoder/driver IC coupled to a 7-segment LED display unit.
40. List the layers in a reflective LCD.
41. What is the advantage of reflective LCDs over LEDs? Disadvantage?
42. Why are analog signals often changed to digital for transmission, and then the received digital data is changed back to analog?
43. What is PCM and where might it be used?
44. Most quantized analog audio ac is quantized to 128 values. What bus width would be required to carry this data?
45. For what is a resistive ladder used in PCM systems?
46. What are the Boolean algebra symbols for OR, AND, and exclusive-OR?
47. What is the Boolean logic term for a 2-input AND gate? OR gate? NOR gate with inversion at output? OR gate with inversion at the inputs?
48. Draw a symbol diagram of the following circuits: NOR, OR, NAND, $A + B$, AB, XOR.
49. What circuits are required to add two 3-bit numbers?
50. How many digital words can a 4 × 4 matrix develop? What matrix would provide all letters and digits?
51. Would data transmission inside a computer or in any single IC be parallel or series type? Why?
52. When is it necessary to change parallel data to serial? To change serial to parallel?
53. Would all electronic printers be parallel data types?
54. Draw a diagram of a parallel-serial shift register.
55. What are some of the internal circuits in a microprocessor?
56. Is the μP or CPU in Fig. 12-28 a CMOS, ECL, or TTL type? What indicates this? What kind of clock oscillator does it have?
57. If the 3-state switches are floating, what can the data bus in a computer be used for?
58. In Fig. 12-28, how many memory addresses could be activated? What does $\phi1$ do? What does $\phi2$ do?
59. How does a microcomputer differ from a microprocessor?
60. What memory data will never be lost if the power line fails?
61. What is a byte?
62. How are characters made visible on a CRT? Graphics?
63. What is the basic pulse code used with computers? What are two other codes that can be handled by computers, and how?
64. What is the relationship between baud or wpm and bandwidth?
65. What is the baud rate of ASCII at 300 wpm?
66. What must be done to keyboarded Baudot code to enable it to be transmitted over two wires or radio? Morse code? ASCII?
67. What is the ASCII code for "T"?
68. Why can gates in ICs be used as oscillators?
69. Draw a diagram of a crystal oscillator using inverters. A 1-kHz code practice oscillator using NOR gates. A Schmitt trigger using inverters.
70. What does a Schmitt trigger circuit do?
71. What does a logic probe tell you?

13

Measuring Devices

This chapter presents information of importance to technicians about analog and digital measuring devices, including voltmeters, ammeters, ohmmeters, wattmeters, watthour meters, bridges, electronic voltmeters, dip meters, counters, frequency meters, and oscilloscopes

13-1 ANALOG AND DIGITAL METERS

The original, and still widely used, measuring devices are known as *analog meters*. They are recognized by their moving pointer and operate by the repelling of internal magnetic fields. The newer *digital meters* (Secs. 13-31 and 13-32) use LED or LCD readouts and operate by timing how long is required for a ramp voltage to build up from zero (or decay) to a triggering point. Analog meters for the most part are current-operated devices. Digital meters are activated mostly by voltages.

Analog meters are superior for indicating varying values of voltage, current, or power, but are usually limited to accuracies of 1 to 3% of their full-scale value. Digital meters are often accurate to less than 1%, and may be in the 0.01% range, but are not used for varying voltages, such as voice or music signals.

Analog meters vary in size from the 1-in. type used in portable equipment, through the 2- to 4-in. *panel* meters, up to *switchboard* meters measuring as much as 12 in. across, and made for distant viewing. Digital meters may range from 3- to 10-unit displays. The LED or LCD display numbers may range from 0.2 in. to over 0.5 in. in height, depending on how far they are made to be read. Portable digital meters generally use LCDs rather than LED displays to conserve battery power. Meters operating from public utility power lines will use LED, neon, or incandescent displays.

13-2 DC METERS

The basic dc meter is known as a *moving-coil* or *d'Arsonval* meter. This one type of meter can be used as a dc ammeter, milliammeter, micro-ammeter, voltmeter, or ohmmeter, and with rectifiers it can indicate alternating currents and voltages. Its symbol is a small circle with a V, A, mA, etc., inside it.

The moving-coil meter is an electromagnetic device consisting of the following parts:

1. A horseshoe-shaped permanent magnet
2. A round iron core between the magnet poles
3. A rotatable mechanism which includes
   a. A lightweight coil
   b. A pointer attached to the coil
   c. Two delicate spiral springs to return the pointer to zero position
   d. Two precisely ground bearings
4. A calibrated paper or metal scale
5. A metal or plastic case

A simplified drawing of the working parts of the moving-coil meter is shown in Fig. 13-1. The coil

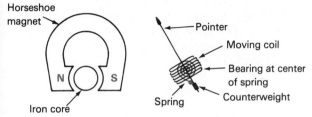

FIG. 13-1 D'Arsonval meter. The moving-coil assembly has been removed from its place between the magnet poles.

and pointer are shown away from their normal position in the circular slots between the magnet poles and the iron-core piece. The iron core does not rotate, but the coil assembly rotates in the spaces between the core and the magnet

The distance between the magnet and the soft-iron core is made quite small to reduce the reluctance of the path of the lines of force from north to south pole. This provides a strong field in which the coil moves, reduces leakage lines of force from the magnet, and increases the sensitivity of the meter. It also decreases interaction between the lines of force and other outside magnetic fields.

Each of the two spiral bronze springs is connected electrically to an end of the rotating coil. The other ends of the springs are brought to insulated terminals on the magnet and pole-piece assembly. Besides being used to zero the indicating needle, the springs provide the path by which current is fed to the moving coil.

When current flows through it, the coil becomes an electromagnet with a north pole at one end and a south pole at the other. If the current direction develops an electromagnetic north pole at the upper end of the coil in the illustration, there will be a magnetic attraction between the north end of the coil and the south pole of the magnet. At the same time the south end of the coil will be attracted to the north pole of the magnet. This rotates the coil assembly (and the pointer) clockwise against the spring tension. If the current is small, the springs will not allow much rotation. The greater the current flow, the farther the pointer and coil assembly will rotate.

Since the coil is constructed of many turns of very fine wire, care must be taken never to feed excessive current through it. The full-scale deflection current will not injure the coil, but a 100% or greater overload may burn out the coil or take the temper out of the springs or burn them out, or the very delicate aluminum pointer may be bent against the bumper at the end of the scale.

When coil current ceases, the springs return the pointer to a zero setting. Reversing the current through the coil moves the pointer off scale to the left. This may not damage the meter mechanism, but it may bend the pointer.

On most meters an adjustment screw is brought out at the front. By rotating this screw, more or less torque, or twisting effort, can be placed on one of the centering springs. With this adjustment the pointer can be accurately set to a zero point on the scale when no current is flowing.

Well-balanced meters will read the same if held horizontally or vertically. Others may require a zero readjustment if their operating position is changed. This can sometimes be corrected by adjusting the position of the counterweight (Fig. 13-1) to balance the coil and pointer assembly more accurately. Additional *quadrantal* counterweights are often used at right angles to the indicator needle.

Meters are delicate. They must be handled gently and should not be subjected to strong external magnetic fields. They may hold satisfactory accuracy for 60 years or more.

Meters are made to be mounted on either steel (iron) or nonmagnetic panels (aluminum, plastic, etc.). If mounted on an improper panel, the permanent-magnet field strength will be affected and incorrect indications may result.

Meters are considered most accurate above half-scale. Many ac meters are difficult to read below one-third scale.

There are other dc meters operating on different electromagnetic principles, but they are not in wide use. Some meters use a thin steel *taut band* instead of a pivot, jewel, and springs, which results in a very fine movement.

13-3 LINEAR AND NONLINEAR SCALES

A moving-coil meter usually has good scale *linearity;* that is, if a current flow of 4 mA moves the pointer tip through an arc of 40°, 2 mA will move the pointer tip through an arc of 20°, 1 mA will result in a 10° deflection, etc.

Meters other than d'Arsonval types often have a nonlinear scale. The scale divisions near the zero value (on the left side of the scale) are usually crowded together and widen toward the higher values. Alternating-current and current-squared meters are examples of such scales (Fig. 13-2). If 2 mA

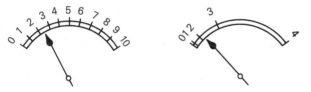

FIG. 13-2 A linear (left) and a nonlinear meter scale.

produces a 10° deflection, but 4 mA produces 40°, the meter is said to have a *current-squared* response.

One special application meter, made so the pointer rests at center scale when no current flows through it, is a *galvanometer*. Other meters may have the pointer resting at the right-hand end of the scale with no current flowing.

13-4 DC AMMETERS

A 0–1 mA meter can be used to read high current values by connecting a resistor across the meter. Circuit current divides, part going through the meter and part going through the *shunt* resistor (Fig. 13-3).

If the coil of a 0–1 mA meter has 25 Ω resistance and a 25-Ω resistor is connected across it, half of any current will pass through the meter and the other half through the resistance. If 1 mA is flowing in the circuit, ½ mA flows through the meter. Only half as much electromagnetic effect will be developed in the meter coil, and only half deflection of the meter pointer will result. The meter will

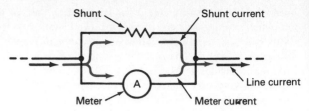

FIG. 13-3 Part of the line current flows through the shunt of an ammeter.

now read full deflection when 2 mA flows in the line. For correct meter indications it will be necessary to replace the 0–1 mA scale with a 0–2 mA scale.

By using the correct shunt resistance and scale calibration, it is possible to make a 0–1 mA meter read full-scale deflection for any desired value above its basic 0–1 mA reading.

A shunt can be connected to the external contacts of the meter, but is usually found inside the case and is an integral part of the meter.

Most metals heat when current passes through them, and their resistance increases. To retain accuracy, an ammeter must use shunts that do not change resistance under a change of temperature. Special *zero temperature coefficient of resistance alloys,* such as constantan, manganin, etc., are used.

If the current flow is more than the full-scale value of an ammeter, it is possible to connect another ammeter in parallel with it. The readings may not be equal if the sensitivies of the meters are not the same, but the circuit current will be the sum of the two readings.

If the accuracy of the ammeters on hand is not known, a more accurate value may result by connecting two ammeters in series and using the average of the two readings.

13-5 COMPUTING SHUNT RESISTANCES

If a milliammeter with a 0–1 mA movement is available, it is possible to convert it to a full-scale 0–10 or 0–100 mA meter by connecting different shunt resistors across it.

EXAMPLE: It is desired to make a 0–1 mA meter with 25-Ω internal resistance read 0–10 mA full-scale. The shunt will have to carry 0.009 A, and the meter 0.001 A. Being in parallel, the meter and the shunt will have the same voltage across them. The current in any leg of a parallel circuit is inversely proportional to the resistance. Therefore, $1/9$ of 25 Ω is required for the shunt, or 2.78 Ω. Whatever the meter reads must now be multiplied by 10. A 0.6 on the meter now indicates 6 mA.

If the meter is to be used as a 0–100 mA meter, the shunt must be $1/99$ of 25 Ω, or 0.253 Ω. The scale graduations must now be multiplied by 100. A reading of 0.74 indicates 74 mA.

Ammeters have low resistance and must not be connected across a source of potential, or excessive current may flow through them.

It is possible to determine the current in a circuit by connecting a voltmeter across any resistance in series with the circuit. The current is found by applying Ohm's law, $I = V/R$. This is the theory on which digital ammeters operate.

13-6 METER SENSITIVITY

A sensitive meter is one that requires very little current to produce full-scale deflection of the pointer.

If anything is done to a meter that makes it necessary to use more than normal current to obtain full-scale deflection, the meter is said to be desensitized. Shunting a meter may make the meter operate as if it were less sensitive, but the *movement sensitivity* (coil and magnet) has not been changed.

The movement sensitivity of meters in electronic equipment and testing apparatus varies widely. The more common dc meters have full-scale sensitivities of 0–50 μA, 0–200 μA, 0–500 μA, 0–1 mA, and 0–5 mA, to mention a few. The sensitivity of a voltmeter is expressed in *ohms per volt* (Sec. 13-9).

13-7 DAMPING

Meters with no shunt resistors across them sometimes have very lively moving pointers. Those with very low resistance shunts have very slow moving pointers. The slowing of the pointer movement is known as *damping*. Some damping is desirable to prevent the pointer from oscillating back and forth when the current through the meter is changed a little. If a current is suddenly fed to a meter and the pointer moves up past the correct reading, the meter is underdamped. If it comes to the correct reading rapidly but does not overshoot, it is *critically damped*. If overly damped, it will rise slowly and will not indicate rapid variations adequately.

When current flows through an ammeter, the coil moves. As it moves across the lines of force of the magnet, a counter emf is developed in it. This induced voltage develops a counter-current through coil and shunt. The counter-current bucks the current flow through the meter coil, preventing the coil and pointer from swinging upward as rapidly as they normally might. As the coil slows, the counter emf drops off. The meter is damped in both upward and downward motions. The lower the resistance of the shunt, the greater the damping effect.

Damping is aided by using an aluminum form for the moving coil. The metal form acts as a shorted turn. When the coil movew in the magnetic field, it induces a current in the shorted turn, setting up a counter field that tends to oppose the movement of the shorted turn and coil.

A third method of producing damping on larger meters utilizes small aluminum paddles attached to the coil assembly. The motion of the paddles through an enclosed air chamber prevents rapid movement of the coil and pointer.

▌ Test your understanding; answer these checkup questions.

1. How are d'Arsonval meters adjusted to exact zero? _____
2. How is current led into and out of the moving coil of a d'Arsonval meter movement? _____
3. What must be adjusted if a meter changes its zero position when rotated? _____
4. Can external magnetic fields affect the indications of moving-coil meters? _____
5. Of what are meter pointers usually made? _____
6. At what part of their scale are meters assumed to be most accurate? _____
7. To what might quadrantal counterweights be attached? _____
8. What returns the taut-band meter's point to zero? _____
9. Which would usually have the more linear scales, dc or ac meters? _____
10. What is it that constantan and manganin have that makes them good shunt resistors? _____
11. Why should ammeters never be connected across a line? _____
12. What is the shunt value needed to make a 50-Ω 0–1 mA meter into a 0–50 mA meter? _____
13. What are two methods of expressing the sensitivity of a meter? _____ _____
14. What do pointers on underdamped meters do? _____ On overdamped meters? _____
15. What are three ways of producing damping in meters? _____ _____ _____

13-8 ELECTROSTATIC VOLTMETERS

A voltmeter is an instrument that will indicate the difference of potential across a circuit. It is always connected across the difference of potential.

An electrostatic voltmeter is constructed in the form of a variable capacitor with a pair of stationary metal plates and a pair of light, balanced metal plates that rotate on a central axis (Fig. 13-4). A

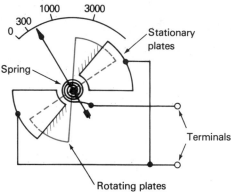

FIG. 13-4 An electrostatic voltmeter.

pointer is attached to the rotating plates. A spiral spring returns the pointer to zero.

When the plates of the meter are connected across a dc source of voltage, the positive- and negative-charged plates are attracted to each other. The rotating plates swing toward the stationary plates, overcoming the tension of the spring, and rotate the pointer across the scale.

Electrostatic voltmeters can be used to measure either dc or ac emf values from about 50 V to several thousand volts. The scale is not linear at the lower voltage readings but can be made reasonably linear at the higher-scale readings by shaping the plates. These are laboratory meters and are used to measure high voltages. The electrostatic voltmeter requires no current flow through it to produce deflection, although it does require electrons as an initial charging current. When used to measure ac,

its capacitive reactance results in an alternating current flow to its plates, although no electrons actually flow through the meter.

13-9 D'ARSONVAL VOLTMETERS

The usual dc voltmeter consists of a sensitive d'Arsonval milliammeter with a resistance in *series* with the meter. A dc ammeter has a shunt, but a voltmeter has a high-value series resistor, called a *multiplier,* normally installed inside the case.

One meter movement used in voltmeters is a 0–1 mA meter with an internal resistance of about 25 Ω. If this meter is connected across a 100-V dc source, the current through it will be $I = V/R$, or 100/25, or 4 A, which would burn it out. To limit the current to 0.001 A, the meter circuit must have $R = V/I$, or 100/0.001, or 100,000 Ω. The meter has 25 Ω resistance in its coil. To this must be added 99,975 Ω (nominally 100,000 Ω). The meter with this multiplier resistance will give full-scale deflection when across 100 V, half-scale deflection across 50 V, and so on.

If it takes 100 V to produce a full-scale deflection with a 100,000-Ω multiplier, a 200,000-Ω multiplier will require 200 V to produce full-scale deflection. For this reason, a 0–1 mA meter is said to have a *sensitivity* of 1000 Ω/V.

A 0–50 μA meter (0.00005 A), being 20 times as sensitive as a 0–1 mA meter has a sensitivity of 20,000 Ω/V. Since the resistance of the moving coil of the meter is such a small percent of the total multiplier-resistance value, it would be disregarded. To produce a 0–150 V meter, a 0–1 mA meter with a 150-kΩ multiplier or a 0–50 μA meter with a 3-MΩ multiplier could be used.

The ohms/volt sensitivity of a meter will be the reciprocal of its full-scale *current* reading in μA. Thus, for a 100-μA meter

$$\text{Sensitivity} = \frac{1}{100 \times 10^{-6}} = \frac{1}{10^{-4}} = 10,000 \ \Omega/\text{V}$$

To measure a voltage known to be more than the full-scale voltage of any meter on hand, two voltmeters can be connected in series. The sum of the two voltage readings will be the voltage across the circuit. Each meter indicates the voltage-drop existing across itself.

If the accuracy of the voltmeters on hand is not known, two voltmeters can be connected in parallel across the circuit. The average of the two readings should be a fairly accurate value.

13-10 VOLTMETERS IN HIGH-RESISTANCE CIRCUITS

A low-sensitivity voltmeter may give correct readings when measuring in low-resistance circuits, but may give inaccurate indications when used to measure voltages in high-resistance circuits. In Fig. 13-5

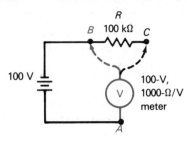

FIG. 13-5 The voltmeter reads 100 V between A and B but only 50 V between A and C.

if a 0–100 V 1000 Ω/V meter is across the 100-V source of emf, A to B, it will read 100 V. Since no current is flowing through the 100,000-Ω resistance R, there is no voltage-drop across it and there must be 100 V between points A and C. But if the meter is connected between points A and C, the meter will indicate only 50 V because the current is now flowing through the 100,000-Ω resistor R and through the meter with its multiplier resistance of 100,000 Ω. With a source of 100 V across 200,000 Ω, the current will be $I = V/R$, or 0.0005 A, and the meter reads only half-scale. The meter is actually reading the correct voltage across its terminals, but there is now a 50-V drop across resistor R also.

If a 100-V 20-kΩ/V meter is substituted for the 1-kΩ/V meter in Fig. 13-5, it will read differently. Across A and B it will show 100 V, but when across A and C, it will read about 95 V instead of 50 V. The multiplier resistance of this 100-V meter is 2 MΩ. The multiplier plus resistance R totals 2.1 MΩ. The current through the combination is $I = V/R$, or 0.000 047 6 A, or 47.6 μA. The meter, being a 50-μA meter, will read 47.6/50 × 100 V, or 95.2 V, an error of only about 5%.

It can be seen that the more sensitive the meter used, the more accurate the voltage readings will be when high-resistance circuits such as AVC circuits in receivers are measured. It is possible that even the 20,000-Ω/V meter might not give accurate enough readings in very high resistance circuits. This is where digital and electronic voltmeters excel. An input resistance of at least 10 MΩ can be expected with such voltmeters. They often utilize op amps (Sec. 14-20) to amplify the input resistance.

13-11 OHMMETERS

A standard piece of test equipment used in electronics is an ohmmeter. With this meter it is possible to read directly the value of a resistor, the amount of resistance in a coil, or the value of resistance in a circuit, or to make continuity tests on transistors, vacuum tubes, capacitors, transformers, or entire circuits.

An ohmmeter can be a relatively simple piece of equipment (Fig. 13-6) composed of a moving-coil

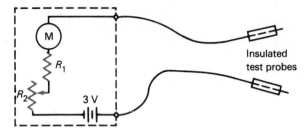

FIG. 13-6 The basic ohmmeter circuit.

meter, a low-voltage battery (1½ or 3 V), a fixed resistor R_1, and a rheostat R_2.

The meter with resistors R_1 and R_2 forms a 3-V voltmeter. When the two test probes are held together, the meter reads full-scale. The *probes touching each other represent a zero-resistance connection. Therefore, this point is marked 0 Ω on the meter scale (Fig. 13-7).

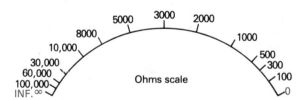

FIG. 13-7 Most ohmmeter scales have zero ohms at the right-hand end of the scale and infinite ohms at the left.

If the meter in Fig. 13-6 is a 0–1 mA meter with 25 Ω internal resistance, the total of R_1, R_2, and the meter must equal 3000 Ω for full-scale deflection with the 3-V battery. If the probes are touched across a 3000-Ω resistor, there are 6000 Ω in the circuit and the current is ½ mA. The meter deflects to a half-scale reading. If the probes are held across 1000 Ω, the total resistance in the circuit is 4000 Ω and the meter will deflect to three-quarters scale. Deflection can be computed by

$$D = \frac{R_m(100)}{R_m + R_x}$$

where D = percent of deflection
R_m = resistance of multiplier
R_x = resistance of unknown resistance

If a 60,000-Ω resistance is measured, the meter deflects only one-twentieth of the full scale. Thus, resistance values crowd together at the high-resistance end, with infinite resistance being equal to the zero-deflection setting (Fig. 13-7). This particular meter will not read values above about 60,000 Ω accurately, nor will it give satisfactory readings of resistances under 100 Ω.

Rheostat R_2 is made variable to compensate for battery aging. Dry cells have about 1.5 V when new but about 1.3 V when they age. It is necessary to hold the test probes together and adjust the rheostate to 0 Ω before making resistance tests to assure correct resistance readings.

A multirange ohmmeter is shown in Fig. 13-8. The meter and resistors R_1 and R_2 form a 3-V

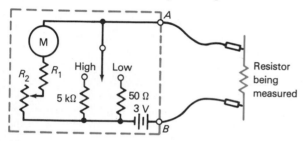

FIG. 13-8 Basic circuit of a multirange ohmmeter.

voltmeter as before. Test probes are connected to terminals A and B. When the probes are held together, the voltmeter is across the 3-V battery, and rheostat R_2 is adjusted to a 0-Ω reading. If the probes are touched across a 50-Ω resistor with the high-low switch in the low position, a 100-Ω voltage divider is formed. The meter is across only 1.5 V and will read half-scale. Fairly accurate readings can be obtained from 0 to about 500 Ω with this setting. If the switch is in the high position, a 5000-Ω resistance across the probes will give approximately half-scale deflection. Reasonably accurate readings can be expected up to about 100,000 Ω.

By using a 50-μA meter instead of the 0–1 mA meter and several values of fixed-range resistors instead of only two, fairly accurate measurements can be made from 1 Ω to over 1,000,000 Ω. The meter must be rezeroed for each range.

CAUTION: The current through ohmmeters on the low-resistance settings may be 100 or more mA. Milliammeters, microammeters, transistors, ger-

manium diodes, or circuits which will not stand this much current through them must not be measured or tested with an ohmmeter on its low range. Also, when using an ohmmeter, make sure that the circuit being tested has no current flowing in it. Any voltage in the circuit being tested will produce incorrect resistance indications, or may burn out some ohmmeters.

13-12 VOLT-OHM-MILLIAMMETERS

A handy piece of equipment is a *volt-ohm-milliammeter* (VOM), also known as a *multimeter*. This is usually a relatively sensitive dc meter in a small box with a battery, switch, and several terminals (Fig. 13-9). If connection is made across the proper

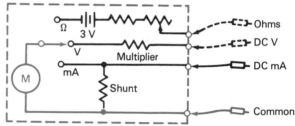

FIG. 13-9 A basic volt-ohm-milliammeter (VOM).

terminals and the switch is set to the correct position, the single meter can be used as a voltmeter, an ohmmeter, or a milliammeter. The meter face will be marked with three separate scales.

VOMs may also have bridge-type solid-state rectifiers to enable them to indicate ac voltages. A separate ac volts scale will appear on the face of the meter. Usually several dc and ac voltmeter, milliammeter, and ohmmeter ranges can be selected.

13-13 ELECTRONIC VOLTMETERS

Sensitive dc voltmeters using 20–50 μA movements load down high-resistance circuits very little, giving reasonably accurate readings. However, an *electronic voltmeter* (EVM) using JFETs (or MOSFETs) may have an input impedance in the millions of ohms and may measure voltages quite accurately in very high resistance circuits. A simple EVM circuit is shown in Fig. 13-10.

Consider the circuit first without any input voltage across the probes at the left so no current flows through R_1 to R_4. Assume the zeroing resistor R_z is centered. The bias voltage developed across R_b will be the same for both JFETs, developing equal I_D

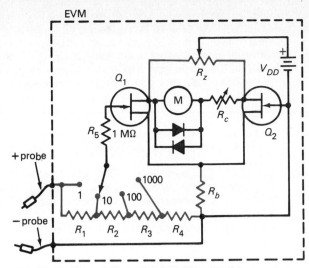

FIG. 13-10 Electronic voltmeter using JFETs.

values through Q_1 and Q_2, and therefore equal voltage-drops across both sides of R_z. As a result the voltage across the meter circuit, drain to drain, is zero, and the meter pointer rests at zero. If it is not at zero, R_z can be adjusted to zero the meter. (Some EVMs have automatic zeroing circuits in them.) It will be assumed that the meter scale has a linear scale reading to full scale.

If a 1-V emf is connected across the probes (+ to the top probe) and the switch is moved to the "times 1" position, Q_1 will now be forward-biased. I_D increases through R_s, developing more bias across it, reverse biasing Q_2 more than before. Now Q_1 has more current and Q_2 has less. The result is a greater voltage-drop across the left half of R_z and less across the right half, developing a voltage difference across the meter and its calibrating resistor, R_c. The meter indicator should now point to 1. If it does not, R_c is adjusted until the meter is calibrated to read exactly 1.

If the switch is moved to the 10 position, and if R_1 is 9 MΩ while the sum of R_2, R_3, and R_4 totals 1 MΩ, only one-tenth of the voltage being measured will be across the input to Q_1. The meter will read one-tenth scale, or 0.1 V. Since the switch is on the "times 10" range, the technician operator knows that the voltage shown on the meter is actually 10 times the indicated value. Similarly, when the switch is moved to the 100 position, the actual voltage across the probes will be 100 times that shown on the meter scale. If the meter now reads 0.34, the voltage beween the probes must be 34 V. The meter shown could be used to measure dc up

to 1000 V. (To measure higher voltages, regular voltmeters are preferred.)

The $V_G I_D$ curves of all JFETs bend somewhat, which would result in nonlinear scales on the meter. Since the two JFETs are in opposite sides of the circuit, their bent portions cancel, and all scale markings are equally spaced (linear).

The two diodes are to protect the meter from any excessive voltages being developed across it. If they are silicon diodes, whenever the voltage-drop across the meter exceeds 0.6 V, the diodes represent very low resistances to any higher currents than that necessary to develop the 0.6 V. These diodes also help to protect the meter from any voltages developed by stray RF fields from nearby transmitters, for example.

Should a voltage be applied across the probes that exceeds the bias voltage of Q_1, current will begin to flow through R_5, developing a reverse bias across it, protecting the meter from excessive current.

If the probes are connected in reverse across the voltage to be measured, a double-pole double-throw (DPDT) polarity-reversing switch (Fig. 13-11) can

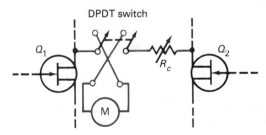

FIG. 13-11 A DPDT polarity-reversing circuit as used in an EVM.

be thrown and the meter will then read correctly. (Many EVMs use automatic polarity selection.)

For a still higher input impedance and greater sensitivity to very low voltages, an EVM may have a MOSFET dc amplifier connected ahead of its two-JFET "balanced *bridge* circuit" (Sec. 13-24).

An EVM can also have internal circuits to read resistance values. A simple two-range ohmmeter circuit is shown in Fig. 13-12. A separate ohms scale is required on the meter face.

To measure current, an internal 1-Ω resistor can be switched across the probes. When a current of 1 A flows through it, a voltage-drop of 1 V develops across it and the meter reads 1 A.

To allow EVMs to read ac voltages, a solid-state diode probe can be added in place of the normal dc

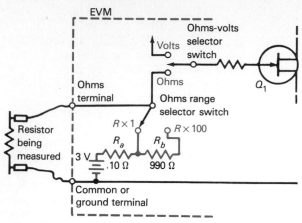

FIG. 13-12 Ohmmeter connections for an EVM.

voltmeter probe (Fig. 13-13). The ac input voltage is rectified, and the capacitor C_1 charges to the peak voltage of the ac. R_1 and R_2 form a voltage-dividing network that reduces the peak value to 0.707 of the peak, or the rms value, which is then indicated by the EVM pointer. For minimum barrier voltages, selenium or copper oxide diodes may be used for low-frequency ac, but Schottky diodes can be used for any frequency ac.

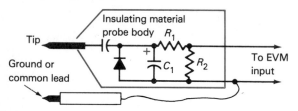

FIG. 13-13 An ac probe for an EVM.

The original electronic voltmeters used VT triodes and were called *vacuum-tube voltmeters,* or VTVMs. EVMs may sometimes be referred to as VTVMs although they have no VTs in them, of course.

EVMs are particularly handy because they are portable and have little battery drain. They do not require the power supply always found in VTVMs.

▌ Test your understanding; answer these checkup questions.

1. What is the name of the resistor connected in series with voltmeter movements? _____
2. Which is more sensitive, a 0–50 μA or a 0–1 mA meter? _____

3. What is the Ω/V sensitivity of a 0–2 mA meter? _____ Of a 0–25 μA meter? _____
4. What would be the value of the multiplier used with a 50-μA-movement 300-V meter? _____
5. Could a 100-V voltmeter with 0–1 mA movement and a 100-V voltmeter with 0–50 μA movement be used in series across 125 V? _____ Why or why not? _____
6. A 100-V 0–1 mA voltmeter reads 100 V when across one of two 100,000-Ω series resistors across a power supply. What is the power-supply voltage? _____
7. If a 20,000-Ω/V meter with 5 kΩ internal resistance is used in an ohmmeter with a 3-V battery, what internal resistance is required in the meter to produce proper zeroing? _____
8. On a simple ohmmeter where is the 0-Ω graduation? _____ On a multirange ohmmeter? _____
9. On what range on multirange ohmmeters is the meter most likely to be dangerous to equipment being tested? _____
10. What does VOM mean? _____
11. Does switching to a higher full-scale range produce more, less, or the same sensitivity with an EVM? _____ With a d'Arsonval voltmeter? _____
12. What component prevents meter and device damage in an EVM? _____
13. What is gained by using balanced JFETs in an EVM? _____
14. Would the calibrate or the zero-adjust control be on the front panel of an EVM? _____
15. In what kind of circuits are "low-power" ohmmeters particularly useful? _____
16. What meter requires no current flow through it to produce deflection? _____ Does it read ac, dc, or both? _____
17. What is required to convert a dc EVM to an ac-reading EVM? _____

13-14 AC METERS

A dc meter will deflect in one direction or the other, depending on the direction of the current flowing through it. With a 1-Hz ac, it will swing back and forth at a 1-Hz rate. With 10-Hz ac, it may attempt to swing back and forth 10 times per second, but because of damping and inertia the pointer cannot move fast enough and will only vibrate a little above and below zero. With more than about 20 Hz, the needle will not move at all. It is possible to increase the ac through such a meter until the meter burns out and still have no deflection or indication at any time. Thus, a dc meter alone is not suitable for ac-circuit measurements. An electrostatic meter will give an indication with either dc or ac. Most ac calibrations are in effective values (0.707 of the peak value if the waveform is sinusoidal). A half-wave rectifier probe, as shown with the EVM, may be used.

13-15 THE RECTIFIER METER

In radio and electronic circuits, where the frequency involved is not much above the audible-frequency range (20,000 Hz), a moving-coil meter with a bridge rectifier is frequently used. Four rectifiers are connected in a full-wave bridge-rectifying circuit to form an ac voltmeter (Fig. 13-14). The bridge rec-

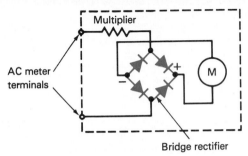

FIG. 13-14 Bridge-rectifier type ac voltmeter.

tifier converts ac into pulsating dc and produces a deflection equivalent to 0.636 (the average value) of the ac peak.

It is standard practice to calibrate the face scales of ac meters in effective values. Thus, the peak-voltage value of any *sine-wave* ac being read is the meter reading times 1.414.

The rotation of the moving coil in a *dc meter* is always proportional to the *average* value of the current or voltage being measured. When a voltmeter is connected across a 6-V battery, the meter reads 6 V. If a 2 V peak ac is added in series with the 6 V dc, the voltage varies from 4 to 8 V alternately, but the average is still 6 V. This is the value the meter will indicate.

The effective value of a sine-wave cycle of ac is 0.707 of the peak value. The average value is 0.636 of the peak. To convert from the effective value to the peak value, the multiplying factor 1.414 (reciprocal of 0.707) is used. To convert from average to peak, the factor 1.57 (or 1/0.636) is used.

If a 10-V *dc* meter is to be used to read *ac* voltage by employing a full-wave bridge rectifier, an indication of 10 V on the dc scale will actually be the average value of the dc pulses flowing through it, or 0.636 of some peak-voltage value. To

find the peak value when the average is known, the multiplying factor 1.57 is used: $1.57 \times 10 \text{ V} = 15.7 \text{ V}$. Therefore, the 10-V dc scale reading indicates 15.7 V peak ac. The effective value of a 15.7-V peak is 0.707 of it, or 11.1 V. This is 1.11 times the average 10-V value shown on the meter. Thus, when using a dc-calibrated meter and a full-wave rectifier to measure ac, the dc scale reading must be multiplied by 1.11 if the effective ac value is desired.

If a half-wave rectifier is used, the average value is half the full-wave value of 0.636, or 0.318. Now a 10-V reading on the meter indicates $10 \times 1.57 \times 2$, or 31.4 V peak. Therefore, to determine the effective ac value, the conversion factor to use if a half-wave rectifier is employed with a dc meter is 2.22 times the dc scale reading. (All rectifiers are nonlinear near their zero-voltage points. As a result, neither of these theoretical factors, 1.11 for full-wave and 2.22 for half-wave rectifiers, will give exact values.)

Meters using copper oxide rectifiers may not be accurate at higher frequencies because of the capacitance across the rectifier units. This can be partially overcome by using Schottky (or perhaps germanium) diodes, but the inductive reactance of the moving coil increases with frequency and the meter reads lower as frequency increases

D'Arsonval meters with self-contained rectifier units are normally calibrated in effective ac values at 60 Hz.

13-16 PEAK-READING METERS

In audio work the waveform of the ac is rarely sine wave. As a result, the values shown on the meters are rarely 0.707 of the peak values. If the peak value is required, the normal ac meter is not satisfactory. A *peak-reading* voltmeter is required that rectifies the ac being measured (Fig. 13-15),

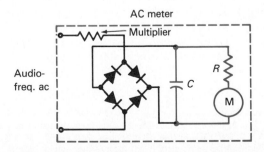

FIG. 13-15 Peak-reading ac voltmeter. M is a sensitive dc or EVM to indicate voltage across C.

charges a capacitor C with the pulsating dc obtained, and then measures the voltage across the capacitor. The meter may be either a sensitive moving-coil type with a rectifier or a VTVM. The high resistance of the meter circuit slowly leaks off the charge of the capacitor. The indicator rises rapidly as the capacitor charges but hangs at or near the peak reading for a period of time. The less sensitive the meter, the less resistance it has and the quicker the capacitor discharges through the meter.

13-17 DB AND VU METERS

The volume-unit (VU) meter is an ac peak-reading *voltmeter* of the rectifier type used to monitor the amplitude of the AF program signals in broadcast or other audio circuits. It must be connected across a 600-Ω impedance line to maintain accuracy of its *power*-level calibrations.

In the past, different services used different values of power as the reference, or 0-dB value. Some used 500 Ω and others 600 Ω as standard line impedances. Some used 0.006 W as the 0 reference point, some used 0.0125 W, and others used 0.001 W. All such meters may be calibrated in decibel units (Sec. 8-5) with the zero-power-level indication at about center scale. If the power in the line is more than the reference value, it produces a higher voltage across the line and the meter reads to the right of the zero mark. If the power is less than the reference value, the indicator does not reach the 0-dB mark.

If a reference power of 0.006 W is used, a +10-dB reading on the meter indicates that 10 times the reference power, or 0.06 W, is being delivered to the 600-Ω load. Similarly, a 20-dB reading indicates 100×0.006, or 0.6 W, etc. On the other hand, a −10-dB reading on the meter indicates $^1/_{10}$ of 0.006, or 0.0006 W, while a −20-dB reading indicates 0.00006 W is operating in the line, etc.

In 1940, a standard power-level meter was adopted. It is a rectifier-type meter having specified pointer ballistics (damping, etc), and scale markings. It is calibrated with 0.001 W as the 0-dB reference level when across a 600-Ω impedance line.

This is the standard VU meter for all audio services. The standard VU meters are built with identical characteristics, whereas dB meters may not have the same scale markings, damping, or zero levels. A VU meter usually has a higher degree of damping than a dB meter. A dB meter is used in applications where the power level is not changing rapidly. In circuits carrying speech and music the power level is always changing and VU meters are used.

The unit of measure dBm indicates decibel values referenced to 1 mW (0.001 W). VU readings could be dBm values also.

There are two scale markings used in VU meters (Fig. 13-16). The VU meter is constructed to oper-

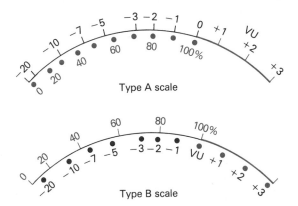

FIG. 13-16 VU meter scales are graduated in plus or minus volume units and 0% to 100% modulation.

ate with a 3600-Ω external series multiplier resistor (Fig. 13-17). When the power in the line being measured is higher than the +3 VU at the high end of the meter scale, three properly selected resistors

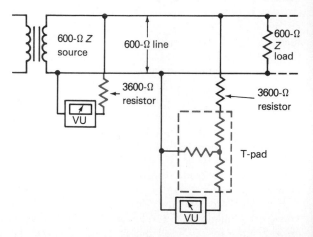

FIG. 13-17 A VU meter across a 600-Ω line and another VU meter with an attenuation pad to desensitize it.

forming a T-type *attenuation pad* can be switched into the meter circuit to desensitize the meter by a predetermined number of VU (dB) to prevent it from indicating off scale. If the attenuation pad desensitizes the meter by 8 VU, any reading given by the meter will be 8 VU lower than the true value. An indication of -11 VU represents an actual line power of -3 VU if an 8-VU attenuator pad is used.

The 3600-Ω external resistor is not installed inside the meter so that an attenuation pad can be added in the meter circuit if required.

13-18 THERMOCOUPLE AMMETERS

A thermocouple ammeter consists of a dc moving-coil meter connected across a thermocouple junction (Fig. 13-18). Current flowing from A to B or from B to A produces heat at the junction. The junction is heated whether the current through it is ac or dc and the amount of heat produced is independent of frequency.

The junction is composed of two dissimilar metals welded together. When the joint is heated, different values of electron activity are developed in the two metals. This results in a dc emf between them and an electron movement from one to the other. The heated junction becomes a thermal dc generator. The current developed is small, but with a sensitive meter and high enough junction temperature satisfactory deflections are produced. The greater the current through the resistance of the junction, the warmer it becomes and the greater the dc developed across the meter. A thermocouple ammeter's indicator needle moves slowly because of the time required to heat the junction.

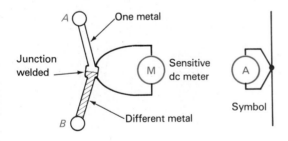

FIG. 13-18 Thermocouple ammeter (dc meter across a thermocouple junction) and symbol.

Thermocouple meters are calibrated in rms ac values. If calibrated at 60 Hz, they will be accurate up to 20 MHz or more. In dc circuits thermocouple ammeters may read slightly high or low, depending on the current flow direction through them, due to a small dc voltage-drop developed across the junc-

tion. This emf is in series with that produced by the heated junction and may either aid or buck it. The average of two readings, one with current flowing in one direction and one with current in the opposite gives a correct value.

The range of a thermocouple ammeter can be varied in several ways. Different junction metals will have different electron-activity capabilities and will produce different dc emf's with the same heat. A resistance in series with the meter will reduce the meter current, decreasing the sensitivity of the meter. A shunt across the junction will reduce the current flowing through the junction and will also reduce heating. Deflection is essentially current squared with heat-actuated meters.

The main use of thermocouple ammeters is to measure the RF current flowing in the antennas of radio transmitters. In the past "hot-wire" ammeters were also used to measure RF.

13-19 ELECTRODYNAMOMETERS

The electrodynamometer, or dynamometer, is somewhat similar to the dc moving-coil meter except that it has no permanent-magnet field. Instead, it has two wire wound *field coils* that produce an electromagnetic field when current flows through them. The pointer is attached to a moving coil that is returned to the zero position by a pair of spiral springs (represented by the wavy lines, Fig. 13-19).

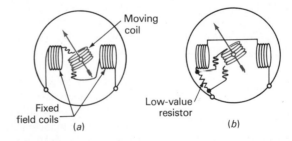

FIG. 13-19 (*a*) Low-current and (*b*) high-current dynamometer connections.

Two methods of connecting the field and moving coils of an electrodynamometer are shown. The series connection is used for voltmeters and low-current ammeters. The circuit with the moving coil across a low-value series resistance is used in high-current ammeters. In either case, the currents in the moving and stationary coils are in phase. When current flows through one, it also flows through the others, producing a magnetic pulling effect that rotates the pointer against the tension of the springs.

Because the polarity of both the fixed- and the rotary-coil magnetic fields reverses with a reversal of current, the rotational pull is always in the same direction. For this reason, the meter will indicate on either ac or dc. Although dynamometers can be made with sensitivities of a few milliamperes, they are generally rather insensitive (a few ohms per volt) and find greatest use in power-frequency circuits where power drain is of little importance.

In the electrodynamometer, the same current flows through the field and moving coils. Doubling the current in one coil results in a doubling of the current in the other and a quadrupled magnetic effect. Thus, deflection is proportional to the *current squared*. For example, if the pointer deflects 1 cm when 1 mA flows, it will deflect 4 cm with 2 mA, 9 cm with 3 mA, and so on. If these nonlinear current values are squared and the squared markings are also placed on the meter scale, the squared markings will be linear (Fig. 13-20).

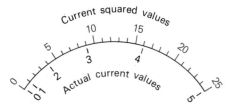

FIG. 13-20 Current-squared-meter face. I^2 and actual I values shown.

Dynamometers calibrated on dc will hold calibration reasonably well for ac up to about 500 Hz. At higher frequencies increasing reactance of the coil decreases the current flow through the meter and introduces errors.

Any power-frequency ac meter may use a step-down transformer when measuring potentials higher than 1000 V. When measuring heavy currents, a step-down *current transformer* can be connected in series with the line. The meter is connected across the secondary.

Test your understanding; answer these checkup questions.

1. Why is bridge rather than single-diode rectification used for ac meters? _____
2. A 10-V voltmeter is across a 5 V dc having a 3-V ac component. What value will it indicate? _____
3. What will a dc meter with a bridge rectifier read if across 10 V sinusoidal ac? _____
4. Under what condition would 10 V rms shown on a meter not indicate 14.14 V peak? _____

5. What is the average value if an ac voltmeter reads 10 V effective? _____
6. Why do many ac voltmeters read low at frequencies above 20 kHz? _____
7. What might be the effect if a very sensitive meter were used as a peak-reading meter? _____
8. Across what impedance line will VU meters indicate accurately? _____ Is this true of dB meters? _____
9. What power is in the line if a VU meter indicates +30? _____ −20? _____ How is it possible for a VU meter to read a +30 level? _____
10. Are VU meters dB meters? _____
11. What do the two scales on a VU meter indicate? _____
12. If the thermocouple part of a thermocouple meter were used as a thermometer, what would have to be done to the meter scale? _____
13. Would a thermocouple ammeter read accurately on 5 Hz? _____ Would the pointer vary at this rate? _____ Why? _____
14. What thermal ammeter is rarely seen anymore? _____
15. Basically, how does a dynamometer differ from a d'Arsonval movement? _____
16. What meter(s) might normally have current-squared calibrations? _____ .

13-20 REPULSION-TYPE METERS

The repulsion-type *moving-vane* meter consists of a coil of wire and, inside the coil, two *vanes* made of thin sheets of highly permeable, low-retentivity iron (magnetizes and demagnetizes easily). One vane is stationary and the other rotary (Fig. 13-21). The

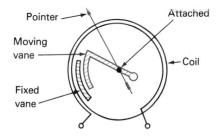

FIG. 13-21 Iron-vane type of repulsion ac meter.

pointer is attached to the moving vane and is returned to zero by a spiral spring (not shown).

When the coil is energized, it magnetizes both vanes by induction. Since they are lying in the same plane, the two vanes are magnetized with like polarity and repel each other. The moving vane is repelled from the fixed vane and against the tension of the spring, regardless of the current direction. It has low sensitivity, but can be used as a voltmeter (or ammeter) for power-frequency ac, or for dc. If the meter is used as an ammeter, the coil is made of a few turns of heavy wire. As a voltmeter, the coil has many turns of fine wire. By shaping the vanes it is possible to produce fairly linear graduations over most of the scale.

An *inclined-vane* (Thompson) meter has its iron vane set on an inclined plane inside the coil and fastened to the pointer. When magnetized, the vane tries to line itself in the lines of force, rotating the indicator needle against a return-spring tension. It will operate with either ac or dc.

A purely ac meter is the *inclined-coil,* or inclined-loop, meter. AC fields flowing in the external coil induce ac into an internal shorted coil, or loop. The induced current in the loop produces a magnetic field of its own, in opposition to the external-coil field, rotating the loop and the indicator attached to it against the return-spring tension.

13-21 WATTMETERS

The value of power in a circuit can be determined by multiplying the voltage by the current. If an electrodynamometer type of meter is connected with its field coils in series with the line (Fig. 13-22), all

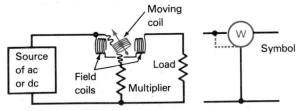

FIG. 13-22 Wattmeter and symbol. Dotted connection is used if meter has a fourth terminal on it.

the current to the load passes through the field coils and produces a magnetic field proportional to the current. If the moving coil and a resistor are connected as a voltmeter across the line, the magnetic field around the moving coil is developed proportional to the voltage across the circuit. In one meter are represented both current and voltage effects. Increasing the current value increases the deflection of the pointer. Increasing the voltage across the line results in a greater current through the moving coil, a stronger magnetic field around it, and a greater scale deflection. Increases of either current or voltage, or both, increase the power in the circuit and the scale deflection of the meter. The meter may be

calibrated in watts or kilowatts. It may be used with dc or low-frequency ac.

A wattmeter always indicates the *true power* in an ac circuit. If line current and voltage are out of phase, the current-carrying field coils and the voltage coil automatically allow for this and power-factor correction (Sec. 7-6) is not necessary.

13-22 WATTHOUR METERS

Electric energy is measured in watthours, watt-seconds, or joules. Energy-measuring meters are known as watthour meters or kilowatthour meters. They operate on a principle somewhat similar to that of a wattmeter. Instead of moving an indicator needle, however, the current and voltage fields rotate the armature of an electric motor. The motor rotation gear-drives indicator hands that rotate like clock hands. A thousand watts operating for a minute will rotate the motor for a minute, gear-driving the hands through a small arc. The same power operating for an hour will move the indicator hand through an arc 60 times as great. There are usually four indicator hands, each reduction-geared 10 times the preceding indicator. The first reads kilowatthours; the second tens of kilowatthours; the third hundreds of kilowatthours; and the fourth, thousands of kilowatthours. The symbol of a watthour meter is shown in Fig. 13-23.

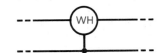

FIG. 13-23 Symbol of a watthour meter.

13-23 AMPERE-HOUR METERS

Ampere-hour meters are energy-indicating instruments used mostly in battery-charging circuits. When a battery is being *discharged,* current flows out of its negative terminal and into its positive. To *recharge* a battery, it is necessary to reverse the current flow through it by using an emf greater than that of the battery. This forces current into the negative and out of the positive terminals (Fig.13-24). If an ampere-hour meter is in series with a discharging battery, current flows through the meter in one direction. When the battery is charging, current flows through the meter in the opposite direction, reversing its pointer rotation.

The ampere-hour meter is a small mercury-pool motor. The direction of motor rotation depends on

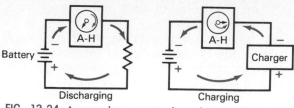

FIG. 13-24 Ampere-hour meter in a battery-discharging and in a battery-charging circuit.

the current direction through it. The speed of rotation depends on current strength. The motor is geared to a rotating pointer. During battery discharge, the indicator needle rotates clockwise. During battery charging, the indicator needle reverses, moving counterclockwise. The more current flowing, the farther the indicator needle rotates in a given time. The position of the pointer indicates the state of charge of the battery.

13-24 BRIDGES

Any discussion of measuring instruments must include some bridge-type devices. The Wheatstone bridge (Fig. 13-25) can be used to measure resist-

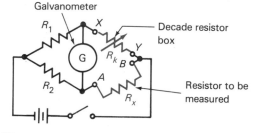

FIG. 13-25 Wheatstone bridge for visual null indications.

ance values accurately from a fraction of an ohm to millions of ohms. The unknown resistance R_x is connected between points A and B, and a calibrated variable resistance R_k is connected betweeen points X and Y. If the four resistances are proportional so that R_1 is to R_2 as R_k is to R_x, the voltage-drops across R_k and R_x will be equal. With no difference of potential across the meter, it will read zero. If the resistances are not proportional, the meter will

indicate some value, plus or minus. The formula for the balanced circuit is

$$\frac{R_1}{R_2} = \frac{R_k}{R_x} \quad \text{or} \quad R_x = \frac{R_k R_2}{R_1}$$

In Fig. 13-25, if it is known that R_1 is 5 Ω, R_2 is 5 Ω, and R_k is 50 Ω when the meter is zeroed, R_x must be

$$R_x = \frac{50(5)}{5} = 50 \ \Omega$$

Thus, if R_k is adjusted until the galvanometer reads zero, R_x will be exactly equal to R_k.

A bridge circuit requires a sensitive galvanometer with a zero setting in the *center* of the scale.

The theory of proportional balance to produce a null indication is used in other bridges. There are many different types capable of measuring impedance, reactance, frequency, capacitance, and inductance. Space limits this discussion to one other type, an inductance bridge (Fig. 13-26). This bridge

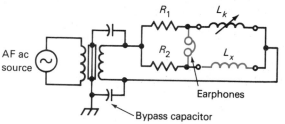

FIG. 13-26 Inductance bridge gives aural null indication.

uses an ac source, such as an audio-frequency oscillator transformer-coupled to the bridge. When R_1 is to R_2 as L_k is to L_x, the difference of ac potential across the earphones will be zero and no tone will be heard. If the proportions are not correct, the oscillator tone will be heard in the earphones.

The variable resistances (inductances or capacitances) used to balance bridge circuits are available in *decade boxes* ("deca" means "ten"). A decade box might have five rotary switches; the first switch selects, for example, one of nine 0.1-Ω resistors in series (from 0.1 to 0.9 Ω). The second switch has resistors in 1-Ω units up to 9 Ω. The third switch has resistors in 10-Ω units up to 90 Ω. The fourth switch has resistors in 100-Ω units up to 900 Ω. The fifth switch has resistors in 1000-Ω units up to 9000 Ω. By proper switch selections any value of resistance from 0.1 to 9999.9 Ω can be selected.

Resistor decades are very accurate. Capacitor and inductor decades are less accurate. Try to diagram a resistance decade box.

13-25 FREQUENCY TOLERANCES

It is quite often necessary to measure the frequencies of transmitters, oscillators, signal generators, received signals, power line ac, and so on. Simple meters are made which can give fairly close indications of many frequencies, but to determine ac frequencies accurately a more complicated system is usually required.

How close a transmitter must operate to its assigned frequency is known as its *frequency tolerance,* normally set by the FCC. An AM broadcast band (535–1605 kHz) transmitter must maintain its assigned frequency within ±20 Hz. This is approximately 20 Hz per 1,000,000 Hz, or 20 parts per million (ppm). This can be expressed as either 20/1,000,000, or within 0.00005, or 0.005% of the assigned frequency.

An FM broadcast band (88–108 MHz) transmitter must maintain its assigned frequency within ±2000 Hz. This is approximately 2000 Hz per 100,000,000 Hz, which is also 0.005%.

In general, fixed land transmitters operating between 50 and 450 MHz must maintain their frequency within 0.0005%; from 450 to 512 MHz, within 0.00025%; and from 800 to 900 MHz, within 0.00015%. Mobile stations above 450 MHz are usually allowed twice the tolerance of fixed stations (0.0005% instead of 0.00025%). For frequencies above 1 GHz the tolerance of all transmissions is usually 0.03%.

Consider the following example of frequency tolerance. A radio station has an assigned frequency in the 8-MHz band with a frequency tolerance of 20 Hz. If the oscillator operates at one-fourth of the output frequency, what would be the maximum permitted deviation of the oscillator which would not exceed the tolerance? If the oscillator frequency drifts 1 Hz, at 4 times this frequency the drift would be 4 Hz (at the operating frequency). Thus, the maximum drift of the oscillator without exceeding the tolerance would be 20/4, or 5 Hz.

As an example of tolerance of fire department (police, taxi, etc.) transmitters, the operating frequency might be 154.31 MHz and the tolerance 0.0005%. (0.0005% has a coefficient of 0.000005.) The tolerance is therefore 154,310,000(0.000005), or 771 Hz. All transmitters must operate at all times within 771 Hz of 154.31 MHz. However, if the measuring device has a tolerance of only twice the transmitter tolerance, or 0.00025% in this case, the transmitting equipment must operate within 771/2, or within 385 Hz of 154.31 MHz according to the measuring device.

Most electronic oscillators drift one way or the other to some extent. The direction of drift should first be determined. The oscillator should be set when cold at a frequency within tolerance, so that when the oscillator drifts, it will drift past its assigned frequency to settle down at some frequency as near as possible to the assigned frequency.

Simple frequency-measuring devices will be discussed first, followed by comparison-type systems, and finally by electronic counters.

13-26 POWER-LINE FREQUENCY METERS

Frequency meters to measure power-line ac frequency are usually of two types, vibrating reed and electrodynamic. A vibrating-reed type may indicate from only 58 to 62 Hz, It might be composed of nine vibrating iron reeds, having natural periods of vibration of 58 Hz, 58.5 Hz, 59 Hz, and so on. An electromagnet excited from the circuit being measured produces an alternating field at the frequency of the current being measured. The reeds are placed in this alternating magnetic field. If the frequency is 60 Hz, the reed tuned to 60 Hz falls into resonance with this frequency and vibrates vigorously, while adjacent reeds vibrate less. Observation of the reeds indicates the frequency of the ac. If two adjacent reeds vibrate at the same amplitude, the frequency of the ac is halfway between the frequency of the two reeds.

The induction- or electrodynamic-type frequency meters utilize a principle of balancing an indicator needle at the center of the scale by using the magnetic field of a resistive circuit and the magnetic field of an inductively reactive circuit to oppose each other. If the frequency increases, the current through the reactive circuit decreases while the current through the resistive circuit remains the same. This pulls the needle toward the resistive-circuit side of the scale (indicating a higher frequency). A lower frequency produces greater field strength of the reactive circuit, pulling the indicator toward the reactive-circuit side of the scale. These frequency meters are connected across the line like a voltmeter.

13-27 ABSORPTION WAVEMETERS

The simplest, least accurate RF measuring device is an absorption wavemeter. It consists of a coil and a variable capacitor in parallel, usually with some form of indicator (Fig. 13-27). The variable capaci-

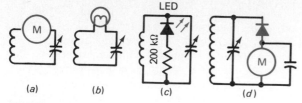

FIG. 13-27 Indicators used with wavemeters. (*a*) Thermocouple milliammeter. (*b*) Flashlight lamp. (*c*) LED. (*d*) Dc microammeter and rectifier.

tor may have a dial calibrated in frequency attached to it.

When a wavemeter is held close to the *LC* tank circuit of an oscillator or low-powered RF source, energy is absorbed by the meter if it is tuned to the same frequency as the tank, and the indicator will provide a maximum indication. The wavemeter should be held as far from the tank circuit as possible. The looser the coupling, the more accurate the indication given. When a self-excited oscillator is measured, a shift of frequency will occur if the coupling to it is too tight. With the wavemeter adjusted to the resonant point, the frequency of the transmitter (and wavemeter) is indicated on the dial. Figure 13-28 illustrates a possible wavemeter dial calibrated to read from 3 to 5 MHz.

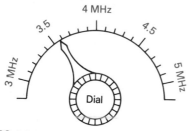

FIG. 13-28 Dial with scale calibrated directly in frequency.

Hand capacitance (detuning due to the presence of a hand changing the dielectric constant of the space surrounding a capacitor or inductor) will cause less detuning of a circuit being measured if the wavemeter is coupled to the *cold* (close-to-ground) end of the tank circuit.

Absorption wavemeters seldom have a frequency accuracy better than 0.05%. They are used to determine an approximate fundamental frequency, being relatively unresponsive to harmonic output.

A wavemeter may have several plug-in coils in order to cover a wide band of frequencies. In this case an arbitrary scale from 0 to 100° is used on the dial. A *calibration chart*, or graph, is used with

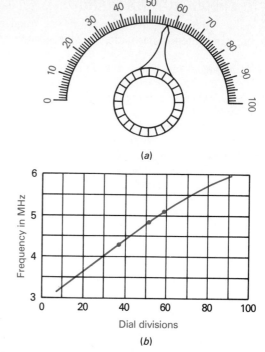

FIG. 13-29 (*a*) Linear 0 to 100 unit scale and (*b*) calibration chart to convert dial reading to frequency.

each plug-in coil. With the graph, arbitrary dial indications can be converted to frequencies. Figure 13-29 shows an arbitrarily marked dial scale and a frequency-versus-dial-division calibration chart. The curve on the chart is used to convert dial-division readings to frequency. According to the indication shown on the dial (fifty-seventh division), the frequency of resonance of the wavemeter from the graph is approximately 5.1 MHz. Note that the line on this graph is not linear. With special straight-line-frequency variable capacitors, a nearly straight curve can be produced.

If a wavemeter has an error proportional to frequency and is accurate within 200 Hz at a frequency of 1000 kHz, it will be accurate within 400 Hz at a frequency of 2000 kHz. If a frequency-measuring device is accurate to 20 Hz when set at 1000 kHz, its error when set at 1250 kHz can be determined by setting up the ratio, 1250 is to 1000 as X is to 20 Hz:

$$\frac{1250}{1000} = \frac{X}{20}$$

$$1000 \ X = 25,000$$

$$X = 25 \ \text{Hz possible error}$$

When parallel wires are used as a linear wavemeter, they are called *lecher wires*. They are sometimes used in the 100–900 MHz range. Cavity-type microwave wavemeters are discussed in Sec. 22-6.

13-28 DIP METERS

To measure the frequency of resonance of a parallel L and C when they are not in an operating circuit, a *dip meter* can be coupled to them. A dip meter may consist of a JFET oscillator, usually a Colpitts, with a microammeter in series with the gate leak resistor (Fig. 13-30). When the stage is oscillating, some I_G

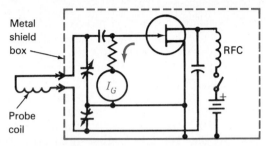

FIG. 13-30 Basic dip-meter circuit.

flows. When the oscillating tank probe coil is coupled to an external LC circuit tuned to the same frequency, the external circuit will absorb energy from the oscillating tank, weaken its oscillation, and lessen its I_G. If this oscillator is coupled to an LC circuit having an unknown frequency of resonance, the I_G dips as the oscillator is tuned past the resonant frequency of the LC circuit.

The lower the degree of coupling between the dip meter and the circuit being tested, the more accurate the results. A dip meter is not designed to have great accuracy. To cover a wide band of frequencies, dip meters usually have plug-in probe coils. There is a separate chart for each plug-in coil, calibrated in dial divisions vs. frequency. A dip meter is valuable when constructing, revamping, or testing receiving and transmitting LC circuits and resonance of antennas.

Most dip meters will act as wavemeters when the active device in them is turned off.

The original dip meters used VTs and were called *grid-dip meters*.

▌ Test your understanding; answer these checkup questions.

1. Name five types of ac voltmeters. _____ _____ _____ _____ _____

2. Which of the meters of question 1 will not indicate with dc applied to it? _____
3. In a wattmeter to what is moving-coil current proportional? _____ Field-coil current? _____
4. How is power factor determined if volts, amperes, and watts of a circuit are known? _____
5. What kind of device is a watthour meter? _____
6. Where might a watthour meter be found? _____
7. How many indicators does a watthour meter usually have? _____
8. What kind of device is an ampere-hour meter? _____
9. For what are ampere-hour meters generally used? _____
10. Are the graduations of a current-squared meter linear? _____ Are the graduations of squared current-squared values linear? _____
11. For what is a Wheatstone bridge used? _____
12. What is a center-zero meter called? _____
13. What is used in a Wheatstone bridge to serve as the known resistance value? _____
14. Is a VTVM a bridge circuit? _____
15. Is a VOM ohmmeter a bridge circuit? _____
16. Why might AF ac and earphones make a good null indicator when used in bridges? _____
17. What is the percent tolerance of 200 ppm? _____
18. What is the tolerance for mobile transmitters operating in the 450–515 MHz band? _____
19. What is the maximum drift that a 100-MHz oscillator may experience without exceeding a 0.0005% tolerance? _____
20. What are the two types of power-line frequency meters? _____ _____
21. What three components make up a wavemeter? _____ _____ _____
22. What are three types of wavemeters? _____ _____ _____
23. What is measured by dip meters? _____ What dips? _____

13-29 FREQUENCY STANDARD SIGNALS

The National Bureau of Standards (NBS), Fort Collins, Colorado, has a cesium-beam atomic *primary frequency standard* oscillator from which it develops a 2.5-MHz-frequency ac accurate to 1 cycle in 10^{14}. NBS station WWVB transmits one RF carrier continuously on 60 kHz. Another NBS station, WWV, transmits RF carriers continuously on 2.5, 5, 10, 15, and 20 MHz, accurate to at least two parts in 10^{11}. One of these frequencies should be receivable at any time of day in any part of North America.

The WWV carrier is modulated by 5 cycles of 1000-Hz ac every second, which sounds to the listener like a "tick." The tick is skipped on the twenty-ninth and fifty-ninth seconds of each minute. Six seconds before each minute a man's voice announces what the next minute will be in UTC (universal time coordinated, also known as GMT, or Greenwich mean time). Besides the 1000 Hz of each tick, 100-, 440-, 500-, and 600-Hz tones are transmitted at some time during each hour. Weather and other advisories are also transmitted at different times during the minute or hour.

Another NBS station, WWVH, in Kekaha, Kauai, Hawaii, transmits on 2.5, 5, 10, and 15 MHz. It makes its voice announcement 14 seconds before each minute, using a woman's voice. During the hour and minute it transmits some 60- and 1200-Hz signals, plus advisories.

By comparing a local 10-MHz oscillator signal on a receiver with one of the carrier frequencies of the NBS stations, it is possible to check the local oscillator frequency to 1 part in 10^7. To do this, one of the NBS carriers is tuned in on a radio receiver. At the same time an approximately equal-amplitude signal from the local oscillator is fed to the receiver. The difference between the two frequencies appears as an output beat signal that wavers in amplitude a certain number of times per second. This waver frequency is the difference between the local oscillator and the NBS carrier frequency.

A *secondary frequency standard* is an oscillator which when checked against a primary standard can be used to produce accurate output signals every 1, 10, 100, 1000, 10,000, or 100,000 Hz all across the usable frequency spectrum (up to perhaps 1000 MHz). An example of a reasonably simple secondary frequency standard would be a 100-kHz crystal oscillator synchronizing a 100-kHz multivibrator oscillator as shown in block diagram form in Fig. 13-31. The multivibrator, which develops square

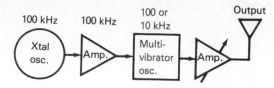

FIG. 13-31 Block diagram of a simple secondary frequency standard.

and sawtooth waves, has an output rich in harmonics. Every 100 kHz across the spectrum a weak signal should be heard from it on nearby receivers. As long as a harmonic of the 100-kHz crystal oscillator is synchronized with a received WWV signal, the multivibrator output signals will be calibrated and will be exactly 100 kHz apart. If the output amplifier is tunable, 100-kHz harmonics can be peaked in strength at any desired frequency.

If the *RC* constants of the multivibrator are changed to provide a natural frequency of oscillation of 10 kHz, this can also be held in synchronization by the 100-kHz crystal oscillator. The multivibrator now develops output signals every 10 kHz across the spectrum.

The frequency of a received signal on a receiver can be determined fairly accurately if the signal is compared with the closest two 10-kHz markers received from a local secondary standard. If it is assumed that there is a linear dial-versus-frequency curve on the receiver, an approximation of any signal can be made by noting the dial reading of one 10-kHz marker signal, the received stations carrier, and the next 10-kHz marker. If the receiver has a BFO and a whistle is heard for markers and received signal, the beat tone between the closest marker and the received signal can be compared with the same tone from an audio signal generator, and an exact frequency above or below a known marker can be determined.

How far a received signal is from a 10-kHz marker can also be determined by using a variable-frequency oscillator having straight-line frequency-versus-dial calibrations. Such a *transfer oscillator* is brought to zero beat with the signal, and then to zero beat with the two marker signals on each side. By mathematical ratio, how far the signal frequency is above the lower marker can be found. For example, if

Lower marker = 4010 kHz = 50° on dial
Signal   = 58° on dial
Upper marker = 4020 kHz = 64° on dial

Let

Lower marker to signal $= f_1 = f_x$
Lower to upper marker $= f_2 = 10$ kHz
Lower marker to signal on dial $= d_1 = 8°$
Lower marker to upper on dial $= d_2 = 14°$

Then, frequency 1 is to frequency 2 as dial 1 reading is to dial 2 reading:

$$\frac{f_1}{f_2} = \frac{d_1}{d_2} \quad \text{or} \quad f_1 = \frac{(10 \text{ kHz})(8)}{14} = 5.714 \text{ kHz}$$

Therefore the beat signal is 5.714 kHz above the lower marker of 4010 kHz, or at a frequency of 4015.714 kHz.

Digital frequency synthesizers make excellent secondary frequency standards. They can generate any ac frequency desired by operating the frequency selector switches. If their 10-MHz master oscillator is checked against WWV to 0.00001%, all the synthesized frequencies will be accurate to this degree. An unknown frequency compared to a synthesizer output can indicate the unknown frequency within 1 Hz. Radio receivers using PLLs for their local oscillators may give digital indications to within 10 Hz or better at any place on the dial.

13-30 HETERODYNE FREQUENCY METERS

The term "heterodyne" means the mixing together in some nonlinear circuit two different frequencies to produce a beat or difference frequency. (A fourth or sum frequency will also be developed, but is not used in this case.)

A simple form of *heterodyne frequency meter* might consist of a well-shielded Hartley dual-gate MOSFET oscillator with an accurately calibrated dial, earphones in the output circuit, and a short antenna wire (Fig. 13-32). (This actually forms a single-device *radio receiver*.)

The lower gate is used in the Hartley oscillator circuit to generate one frequency, while the upper gate brings in an external signal.

The oscillator tunes from 1 to 2 MHz and has a dial accurately calibrated in frequency. Assume a nearby transmitter is transmitting on 1.400 MHz. As the oscillator dial is tuned across 1.4 MHz, there will first be heard a high-pitched beat whistle that decreases in frequency until it is so low that it can not be heard (zero beat). If the tuning is continued, the tone will again increase in frequency on the other side of zero beat until it becomes too high

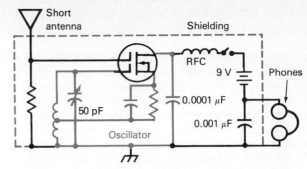

FIG. 13-32 Possible simple heterodyne frequency meter.

to hear again. If the lowest tone that is audible to the human ear is 20 Hz, this leaves 40 Hz of unheard beat signal. However, the center of the zero-beat area can be closely approximated, and the dial is read at that point. In this case the dial should read 1.40 MHz. If not, either the transmitter is off frequency or the frequency meter dial calibrations are not correct.

If a local transmitter were on 2.8 MHz, a zero beat would be heard at the same spot on the dial because now the second harmonic of the oscillator mixes with the incoming transmitter signal to produce a beat signal in the MOSFET output circuit.

With a simple circuit such as this, hand capacitance might shift the oscillator frequency. A strong received signal might force the oscillator frequency of oscillation to fall into synchronism with the incoming frequency, making the zero beat very broad. Actual heterodyne frequency meters will be more likely to be as shown in block form in Fig. 13-33. (In block diagrams a triangle indicates an amplifier, a circle represents an oscillator, and a

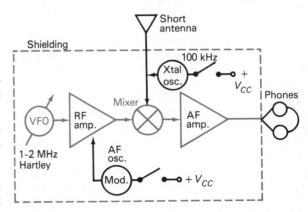

FIG. 13-33 Block diagram of a practical heterodyne frequency meter.

circle with a cross in it is a mixer, although some block diagrams may use all oblongs and label their functions.)

In this circuit the oscillator is followed by an amplifier which will amplify all frequencies fed to it and also isolates the oscillator from the antenna, preventing hand capacitance effects. The output of this stage feeds into a mixer (such as a dual-gate MOSFET). The mixer is also fed incoming signals from the short antenna as well as from a 100-kHz *checkpoint* crystal oscillator signal. Heterodyne or beat signals from the mixer are amplified and fed to earphones. If the oscillator-amplifier is to be used as an RF signal generator, it is sometimes advisable to modulate the Hartley signal with an audio tone to make it easier to identify when picked up by a receiver. This modulation (Sec. 17-10) is provided by the AF oscillator when its power switch is closed.

After the variable oscillator has warmed up for 30 min or so, the checkpoint oscillator can be turned on and one of its harmonics should be brought to zero beat with WWV on a receiver tuned to 2.5, 5, 10, or 15 MHz. When synchronized with WWV, all of the 100-kHz harmonics will be exactly on frequency and can be used to check the variable oscillator dial calibrations every 100 kHz.

A simple 180° dial would not provide very accurate readings of frequency. A *vernier dial* is better (Fig. 13-34). The vernier scale does not move, but the tuning dial rotates below it. Note that there are ten vernier scale marks for nine dial marks. In Fig. 13-34*a* the vernier arrow appears to be centered between 28 and 29°. This is borne out by checking the one vernier mark that lines up with a dial mark, namely, number 5. Therefore the dial must be set on exactly 28.5°. In Fig. 13-34*b* the vernier and dial marks line up at 3, indicating a dial reading of 68.3°. If the dial had 100 graduations, the vernier scale would allow reading accurately to 1000 graduations.

In most heterodyne meters the dial is not graduated in frequency but in arbitrary numbers. In such cases a frequency-versus-dial chart or calibration curve is developed using a secondary standard to determine the proper frequency, or dial, settings. Then, when received signals are brought to zero beat, the dial reading is noted, and the chart is used to determine what the dial setting should be converted to for frequency.

Suppose an absorption-type wavemeter indicates that the approximate frequency of a transmitter is 500 kHz. At the same time the transmitter signal

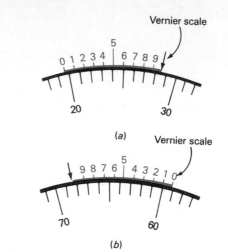

FIG. 13-34 Vernier scale on a dial reading (*a*) 28.5 and (*b*) 68.3.

produces a zero beat on an accurately calibrated heterodyne frequency meter at a dial reading of 374.1. The frequency-meter calibration chart indicates the following dial readings versus frequencies:

Dial	Frequencies
371.5° =	499.6 or 999.2 kHz
376.0° =	499.8 or 999.6 kHz

Since the absorption wavemeter indicates the approximate frequency to be 500 kHz, the dial reading of 374.1 must indicate a frequency between 499.6 and 499.8 kHz, rather than one between the harmonic frequencies 999.2 and 999.6 kHz. To find the exact frequency from the given information, it is necessary to interpolate:

$$b \left\{ a \left\{ \begin{array}{l} 371.5° = 499.6 \text{ kHz} \\ 374.1° = \text{the signal} \\ 376.0° = 499.8 \text{ kHz} \end{array} \right\} c \right\} d$$

By setting up the ratio and solving for the unknown difference,

$$\frac{a}{b} = \frac{c}{d}$$

$$\frac{2.6}{4.5} = \frac{c}{0.2}$$

$$4.5c = 0.2(2.6)$$

$$c = \frac{0.52}{4.5} = 0.1156 \text{ kHz}$$

Therefore, the unknown frequency is 0.1156 kHz greater than 499.6 kHz, or 499.7156 kHz.

Once the checkpoint oscillations indicate the dial calibrations are correct, the meter can also be used as an accurate signal generator (emitter) to check the dial calibrations of receivers. Some of the variable-frequency oscillator signal, as well as the 100-kHz crystal signal, leaks into the antenna and is radiated. If the meter is set to 1.5 MHz but a receiver picks up the signal at 1.510 MHz, the receiver must be 10 kHz off calibration. With the same 1.5-MHz meter setting, the receiver should pick up meter harmonics every 1.5 MHz, or at 3.0 MHz, 4.5 MHz, 6.0 MHz, etc.

Maximum heterodyne signal will be produced when the two signals being mixed have the same strength at the mixer.

When active devices have to be replaced in frequency-meter oscillators it may be necessary to recalibrate the dial for maximum accuracy.

Because of mechanical play in most dials, called *backlash*, frequency-meter dials should always be brought to zero beat while tuning in the same direction (clockwise, for instance).

13-31 DIGITAL COUNTERS

Digital flip-flops can be connected together to count pulses (Secs. 12-10 and 12-14). If the pulses run through an AND gate (Sec. 12-4) which is gated on for exactly 1 s, the number of counts will be the frequency of pulses.

Figure 13-35 is a simplified electronic counter used as a frequency meter. At the bottom, the positive half cycle of the ac to be measured is limited in amplitude, rectified, and shaped into very narrow positive pulses by a Schmitt trigger circuit. These are fed to the AND gate, *A*.

A 1-MHz oscillator, or generator, feeds its ac output to a divide-by-1,000,000 (6-decade) divider (Sec. 12-14). The 1-Hz output toggles a *JK* FF (Sec. 12-10) to produce a precise 1-s positive square-wave timing pulse followed by a precise 1-s space. The positive timing pulse is used to enable the AND gate for exactly 1 s. During this time the narrow pulses are able to pass through the gate and be counted by the decimal counting units (DCUs).

If the input ac frequency is 527 Hz, the three DCUs will arrive at totals of 5, 2, and 7 in binary-

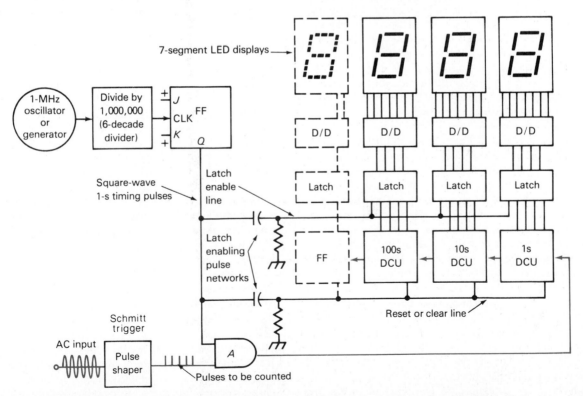

FIG. 13-35 Block diagram of digital counter frequency meter with 7-segment LED display.

coded decimal (BCD) just as the timing pulse drops off. The BCD codes developed in each DCU are fed to the input of the three "latches," which are inhibited during the timing pulse. As the timing pulse drops off, it does two things. (1) It sends a reset or clear pulse developed by the differentiating *RC*-diode network to the DCUs, clearing them for counting when the timing pulse goes high again. (2) It very briefly enables the latches allowing them to transfer the BCD codes for the numbers 5, 2, and 7 to the decoder-drivers (D/D, Sec. 12-15) and to the displays, which then indicate 5, 2, and 7.

After the short differentiated pulse, the latches return to an inhibited state, but the decoder-drivers continue to drive the displays until the next time the timing pulse drops off. At that time the enabled latches feed the new DCU outputs to the decoders. If the frequency has changed, a new number will be displayed. (Without the latches the displays would blur, as numbers are counted until the drop-off of the timing pulse. The indication would be steady for a second until the next run-up blur would occur.)

With only three DCUs the highest frequency that can be displayed is 999 (a 3-digit meter). If a simple FF (dashed) is added after the 100s DCU, the overflow (10th pulse) from this DCU could trip the FF and illuminate a 1 in the 4th display. The highest number now possible is 1999 (called a 3½-digit meter). If a 4th DCU is used instead of the FF, a 4-digit meter is produced, and so on. With six DCUs a frequency of 999,999 could be displayed. To measure higher frequencies, such as 100 MHz, the speed of counting for the 1s, and possibly for the 10s, DCUs is very high. Special high-speed DCUs are needed.

When the frequency being measured is too high for the meter, the overflow output from the most significant digit DCU can be used to illuminate an "overflow" indicator LED, advising that the display is reading incorrectly.

A possible error with some digital frequency meters is a 1-cycle ambiguity. This is due to the chance that the 1-s timing pulse will either just accept or just miss the first narrow pulse to be counted. If a synchronizing circuit is added to allow enabling of the AND gate only at the rise of one of the input pulses, the 1-cycle ambiguity is eliminated.

To use a counter to count slow events, the 1-s pulse can be disconnected from the AND gate at Q and a +5-V potential can be switched to the upper AND gate and to the latch-enable line. All impulses

will be counted and displayed until the switch is opened.

13-32 MEASURING FREQUENCIES WITH COUNTERS

The development of digital frequency counters has made accurate frequency measurements relatively simple.

To measure the frequency of an oscillator or a transmitter, a small link-coupling coil can be connected to the counter input. The counter is turned on and the link is slowly coupled to the ac source until just enough signal (usually less than 1 V) is developed to produce uninterrupted counting. Too much voltage due to excessively close coupling can damage the input circuitry of a counter. For measuring at a distance of a few yards, a tuned circuit with a small antenna can be connected to the input of the counter. However, if interference pulses of any kind are picked up, such as automotive ignition, they will either be added to the RF being measured or may blot out some of the cycles that should be counted, resulting in erroneous readings.

When measuring the frequency of a received signal, a *transfer oscillator* can be used. The transfer oscillator output is coupled to the counter input. The signal is tuned in, the transfer oscillator is adjusted to an exact zero beat as heard on the receiver, and the frequency of the transfer oscillator is read off on the counter while the signal is still at zero beat.

Counters may have top frequencies of 100 kHz to over 18 GHz. With special heterodyning circuits, they can read up into the hundreds of gigahertz. Assume a 14-MHz signal is to be measured but the only available counter has a 10-MHz top frequency. If a stable 10-MHz oscillation (from the master 1-MHz crystal signal multiplied by 10) is heterodyned against the 14-MHz signal, the resultant 4-MHz difference frequency can be read on the counter. In this case the signal frequency is 10 + 4, or 14 MHz. Thus, a 20-MHz transfer oscillator would allow measurements between 20 and 30 MHz, a 30-MHz oscillator would allow measurements between 30 and 40 MHz, and so on.

Transmitters and receivers using frequency synthesizers (Sec. 11-26) as their oscillators may have better frequency tolerances than most measuring devices (other than counters) provided their master crystal oscillators are on frequency. However, if a synthesizer malfunctions, it could develop an improper output frequency. This would normally be considerably off the desired frequency and counters or other frequency-measuring meters would show that a malfunction was present.

13-33 DIGITAL METERS

A digital panel meter (DPM) may be made to indicate voltage, current, resistance, etc. The measured values are displayed by a series of DCUs similar to those used in counters. D'Arsonval meters may be superior for monitoring varying signals and may have long, reliable lives, but digital displays can give better repeatability, readability, and accuracy of constant amptitude values. Voltage- or analog-to-digital (A/D) counts can be produced by several methods: single-ramp, dual-ramp, and voltage-to-frequency, for example.

A single-ramp meter has a circuit that generates a dc ramp voltage (Fig. 13-36) a few times per sec-

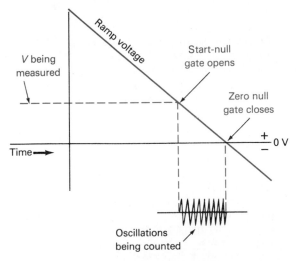

FIG. 13-36 Digital voltmeter (DVM) ramp voltage. While counting gates are open, internal oscillations are counted.

ond. Both the ramp voltage and the voltage to be measured are fed to a null detector or comparator circuit. When the ramp voltage decreases to a value equal to the voltage being measured, the null detector gates a high-frequency oscillator into operation. Every cycle of the oscillator is counted by a series of DCUs and displayed by 7-segment LEDs, LCDs, neon Nixie tubes, etc. When the ramp voltage reaches zero, another null detector circuit stops the oscillator and the last digital number displayed represents the voltage being measured. Low-input voltages allow the oscillator to run for only a few

cycles. Higher voltages allow more cycles to be generated, counted, and displayed.

A dual-ramp DPM uses an *operational amplifier* (op amp) IC with an amplifying capability of several hundred thousand; but by output-to-input feedback the net amplification is reduced to perhaps 10. Any distortion produced in such amplifiers is reduced by almost the feedback ratio. As a result, the op amp output *current* will be a faithful replica of the input *voltage* (to be measured). In this way the op amp acts as a voltage-to-current converter. The current, proportional to the input voltage, is fed to an integrator circuit which is gated on for a short period of time. The integrator circuit current builds up at a rate determined by the amplitude of the current fed to it by the converter (T, Fig. 13-37). At

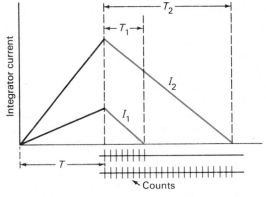

FIG. 13-37 Dual slopes of two different voltages are being measured in a DPM.

the end of time T, *the integrator circuit cuts itself off and* (1) the current starts to decay (I_1 or I_2), and (2) a high-frequency oscillator is fed into decimal counting units. When the current decays to zero, a null detector stops the DCUs and the numbers displayed represent the voltage. A low voltage produces a small-rise current, I_1. A higher voltage produces a high-rise current, I_2. Since both decay at the same rate, I_1 shuts off the DCUs before I_2 and therefore indicates less voltage.

A voltage-to-frequency DPM also employs an op amp as a voltage-to-current converter. The output current from the op amp is used to charge a small capacitor (C, Fig. 13-38). When the capacitor voltage builds up to the zener voltage V_Z, a comparator or null detector circuit forward-biases the transistor across C, discharging C immediately. The bias drops off the transistor and C starts recharging again. The greater the voltage being measured, the greater the charging current, and therefore the more

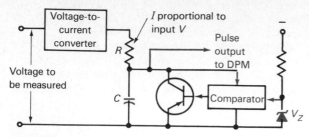

FIG. 13-38 Voltage-to-frequency converter for a DPM.

times per second the charge-discharge pulses are generated. These pulses are fed to DCUs and counted. By gating the comparator circuit on and off over the proper interval, the digits shown on the DCUs indicate the proper voltage.

Most DPMs have a hold circuit that allows the voltage display to be held from a fraction of a second to any desired time before it recycles again. One advantage of these meters is their BCD output which allows recording of readings or the feeding of any number of remote displays.

For current measurements the voltage across a shunt resistor in a circuit is read. For resistance readings the DPM is used in place of the analog meter in the usual ohmmeter circuits. For added sensitivity, dc amplifiers (op amps usually) may be used ahead of the analog section. Only by using ICs can such complicated systems be manufactured in small cases. To enable ac-powered meters to operate in circuits that are above ground potential, they may be isolated by transformer coupling from the line or by opto-isolating devices.

Digital meters are available in 3, 3½, 4, 4½, etc., digit types.

Digital VOMs, or multimeters (DMMs), may be found in hand-held probe form. The LCD readout is on the side of the probe body, along with the range and function switches. A common-connection ("ground") clip lead is the only lead other than the probe.

Sophisticated modern DMMs use a microprocessor and memory to store and activate programs, such as complete calibration of all functions by comparing them against the voltage of an internal lithium (constant-voltage) battery. All analog inputs are converted to digital values, which are then counted and displayed. The digital values can also be fed to external computers which in turn can provide printout. Displays may be of voltage or current, either dc or true rms (regardless of ac waveform up to perhaps 100 kHz), and resistance. All functions and ranges can be calibrated automati-

cally by mathematical programs in memory. Display may be 12 or more alphanumeric (letter or number) LED or LCD characters to display numbers, or to show words from the external computer. All ac inputs are converted to true rms and then to the equivalent dc before being fed to the A/D section and display. If an internal calibration is suspect, it can be compared with a laboratory standard, and the internal programs will change the meter display to the correct values.

Test your understanding; answer these checkup questions.

1. What are the carrier frequencies of WWVB? _____ WWV? _____ WWVH? _____ Which uses a woman's voice? _____

2. What is used as the primary standard at WWV? _____

3. Name the four stages of a secondary standard? _____ _____ _____ _____

4. To measure the frequency of a received signal, what kind of oscillator is used? _____

5. What must be used in conjunction with a secondary standard when measuring frequency? _____

6. Name two methods of determining the AF beat between a signal and a marker. _____

7. Over how many dial divisions does a vernier scale extend? _____

8. How is backlash prevented? _____

9. When is AF modulation used with heterodyne meters? _____

10. How long should heterodyne frequency meters be warmed before use? _____ Against what is the crystal frequency checked? _____

11. Can frequencies above the oscillator range of a heterodyne meter be measured? _____ Below? _____

12. If 5160 kHz = 86° on the dial, 5150 = 68°, and the signal is at 72°, what is the signal frequency?

13. Name the circuits between the AND gate and the LED display in a digital counter. _____ _____

14. When measuring the frequency of a received signal with a counter, what else must be used? _____

15. Why should coupling to a frequency-measuring coupler be increased slowly? _____

16. In a counter, which acts faster, the least- or the most-significant-digit DCU? _____

17. What does an overflow warning light indicate in a frequency counter? _____

18. Could DPM-type devices be used as bench meters? _____

19. What are the no-feedback gain possibilities of op amps? _____ With feedback? _____
20. What is an advantage of using feedback?
21. How many counts would a 3-digit meter display? _____ A 4½-digit meter? _____
22. What circuit can be used as a linear voltage-to-current converter? _____
23. How can digital meters be made to indicate current? _____
24. If the analog section of a digital meter is isolated by a transformer from the display and power-supply section, how might the analog section obtain its operating power? _____
25. What is meant by a ramp voltage? _____

13-34 OSCILLOSCOPES

An extremely useful measuring device is the *oscilloscope*, or the *oscillograph*. It can present an instantaneous visual indication of voltage excursions that no other measuring device can show. The indicator is known as a cathode-ray tube (CRT), which is somewhat similar to the picture tube in a television receiver, the basic difference being in the type of deflection used.

A CRT consists of an *electron gun*, four deflection plates, a fluorescent screen, an Aquadag coating, and a glass envelope (Fig. 13-39). The electron

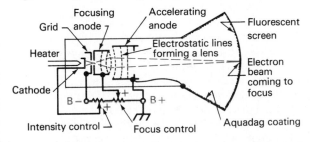

FIG. 13-39 Electron gun of an electrostatic-deflection cathode-ray tube. (Deflection plates not shown.)

gun consists of a *heater cathode* to emit electrons; a small metal cylinder called the *grid*, which has a hole in one end; and two other metal cylinders, one the *focusing anode* and the other the *accelerating anode*. The grid is negatively charged with respect to the cathode by being connected to a more negative point on a voltage-divider resistor across a power supply of 600 to 2000 V B+ or more.

The negative charge on the grid would normally prevent electrons from passing through the one small hole in it, except that the focusing anode on the other side is at a relatively high positive poten-

tial and attracts them. Some of these electrons strike the focusing anode and move to the power supply, but most of them pass on through the focusing anode and into the *electrostatic lens* that is formed at the open end of this anode.

Because of the difference in potential between the two anodes, the electrostatic lines of force in this area bend. These lines of *equipotential* across the opening between the anodes form an electrostatic lens. Electrons moving through this lens converge at the screen, as light beams converge and focus when passed through a glass lens. By varying the voltage difference between the first and second anodes (focus control), the configuration of the lens can be changed, allowing the electron beam to focus as a spot smaller than the head of a pin on the fluorescent screen.

The accelerating anode, besides aiding in focusing the beam, increases the speed of the electrons by its high potential. The more rapidly the electrons travel, the brighter the spot produced when they strike the fluorescent screen at the back of the face of the tube.

The intensity of the spot produced on the screen is basically controlled by the potential of the grid. The more negative the grid, the fewer electrons that can be pulled through its hole and the less intense a spot produced on the screen. Spot brightness can be controlled by changing the grid bias with the intensity control potentiometer. This may also have a slight effect on focus.

The electron beam can be deflected up or down by two *vertical deflection plates* in the neck of the tube (Fig. 13-40). If the lower deflection plate is connected to the second anode (ground) and a positive potential in respect to it is applied to the upper

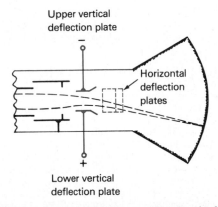

FIG. 13-40 Deflection plates in an electrostatic-deflection CRT.

deflection plate, the electrons in the beam will be attracted toward the upper plate and the spot on the screen will move upward. How high the spot moves is determined by the magnitude of the voltage on the deflection plates. If the upper plate is made negative in respect to the lower plate, the electron beam will be deflected downward a distance proportional to the voltage applied. Applying an ac voltage to the two plates produces a vertical line on the screen representing the peak-voltage value. A transparent graph with centimeter and millimeter spacing between lines is attached to the face of the tube. If the sensitivity of the deflection plate is 0.5 mm/V, for example, and the deflection of a pulsating dc is 6 mm, the peak voltage applied to the plates must be 12 V. The direction of deflection from the center of the screen indicates the polarity of the voltage being measured. Since there is no horizontal deflection, voltage variations will trace a vertical line. With an ac voltage applied to the deflection plates, the line on the screen will extend equally above and below the center of the screen. Oscilloscopes are excellent peak-to-peak-voltage indicating devices.

A CRT also has a pair of *horizontal deflection plates,* shown dashed, at right angles to the vertical deflection plates. Since these plates are closer to the screen, they will deflect the beam less and as a result have a lower sensitivity. Voltages applied to the horizontal plates deflect the spot horizontally.

The *Aquadag* coating, a conductive layer on the inside of the tube, is connected to the accelerating anode. Electrons striking the phosphorescent-fluorescent painted screen on the inner face of the tube cause a bright spot wherever they hit. These electrons bounce back as a secondary emission, are attracted to the positively charged Aquadag coating, and from there move to the positive terminal of the power supply.

The connections to the elements are usually made to pins at the cathode end of the tube. These pins

fit into a 7- to 14-pin socket, depending on the tube type. There are many different types of cathode-ray tubes, having different face sizes and shapes, sensitivities, persistence of illumination, colors of screens, numbers of anodes, filament voltages, etc. Oscilloscope tubes may have round or oblong screens, of 1-in. to over 12-in. diameter or width. (Flat-screen devices are mentioned in Sec. 24-26.)

There are many uses for oscilloscopes. In practically any electronic circuit the oscilloscope can show what is occurring to circuit voltages.

13-35 FREE-RUNNING OSCILLOSCOPES

The original and a few of the simpler modern oscilloscopes use a free-running sawtooth ac oscillator for the horizontal-deflection voltage. When the sawtooth voltage is applied to the horizontal-deflection plates of the CRT, the spot is moved slowly (relatively) across the screen, left to right, and then is snapped back so fast that no trace is left. A 1-Hz sawtooth ac will produce a spot that moves across the screen in 1 s, disappears, but reappears immediately at the left side to move across the screen again. If this sweep ac is over 20 Hz, the spot appears as a line, due to the persistence of vision of the human eye. The persistence of illumination of most CRT tube screens is less than 1 ms and may be as little as 40 μs.

One application of an oscilloscope is to view some signal voltage waveform. A simple oscilloscope system that will do this is shown in Fig. 13-41 in block diagram form. The cathode-ray tube is represented as a round screen with horizontal- and vertical-deflection plates (HDP and VDP) shown outside the CRT for clarity.

There are four separate sections of a simple oscilloscope: CRT, sweep circuitry, signal amplification circuitry, and power supply (not shown).

The *phase-inverter* circuits convert the output signal from the signal amplifier to equal-amplitude positive-going and negative-going ac voltages. This allows an ac signal voltage to be centered on the CRT. The centering controls allow both vertical and horizontal centering, or any other movement of the display desired. The gain controls are adjusted by the operator to the desired signal size and sweep width.

Consider what is displayed when a 100-Hz sine-wave ac is applied to the vertical-deflection plates and a 100-Hz sawtooth ac is applied to the horizontal-deflection plates. The signal voltage will be considered as starting with zero volts on the VDPs.

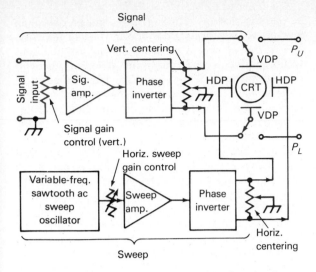

FIG. 13-41 Block diagram of a simple free-running oscilloscope.

The sawtooth sweep at this instant is at maximum positive on the left HDP (maximum negative on the right HDP), placing the spot at the far left edge of the screen at the start.

1. In $1/400$ s the horizontal sweep moves the spot one-fourth of the distance to the right. At the same time the sine wave has risen to a maximum potential (positive on top VDP) and has pulled the spot upward, perhaps an inch, depending on the magnitude of the voltage, tracing the first quarter of a sine wave on the screen (Fig. 13-42a).
2. During the next $1/400$ s the horizontal sweep moves the spot to a position one-half of the total distance to the right (Fig. 13-42b). At the same time, the sine wave has fallen to zero volts, allowing the spot to fall back to the starting level. The second quarter of a sine wave has been traced.
3. During the next $1/400$ s the horizontal sweep moves the spot to a position three-fourths of the total distance to the right (Fig. 13-42c). At the same time, the sine wave increases to a maximum negative and the spot moves in a downward direction, tracing the third quarter of a sine wave on the screen.
4. In the next $1/400$ s the horizontal sweep moves as far to the right as it is going (Fig. 13-42d), and then snaps back to the starting point. During this time, the sine-wave voltage drops back to the starting level, tracing the final quarter of the sine wave on the screen. One cycle is completed, and the next one starts retracing immediately.

With the cycles occurring 100 times per second, to the eye a sine wave appears to be standing still on the screen. If the ac applied to the vertical plates is nonsinusoidal, the display will be nonsinusoidal. In this way, the waveshape of any ac or varying dc applied to the vertical plates is made visible. It is only necessary to synchronize the horizontal-sweep frequency with the frequency of the wave being applied to the vertical plate to make the pattern stand still on the screen. It is also possible to use the sweep at some submultiple of the signal frequency and stop the motion of the figure displayed. For example, if the sweep is 50 Hz and the signal is 100 Hz, the sweep makes only one-half of its total excursion by the time the signal completes its first cycle. As a result, two cycles of the signal voltage will be shown across the screen. If the signal and sweep voltages are not synchronized, the display moves to the right or left across the screen.

If the sweep frequency (or the sweep rate) is known accurately, it is possible to measure the frequency of an ac by stopping the display of it on an oscilloscope, counting the number of cycles shown, and multiplying the cycles by the sweep frequency.

Some oscilloscopes have switches to permit direct connections to the VDPs (P_U and P_L on the diagram). This allows signals to be fed directly to the deflection plates without passing through the amplifiers, which may not amplify at very high

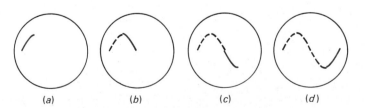

FIG. 13-42 Developing a cycle on an oscilloscope screen.

frequencies. Signals up to several hundred megahertz can be displayed with direct connections. Since there are so many cycles per second, the display is not of separate cycles, but an illuminated band of closely packed waves across the screen. However, variations of the signal strength will show as variations of the band height, which is one way of monitoring amplitude modulation in radiotelephone transmissions (Chap. 17).

Patterns developed by two different ac cycles fed to the VDPs and HDPs of a CRT are known as *lissajous figures*. If two sine waves of equal frequency are used, a stationary single circle, oval, or loop will be formed on the screen. If the ratio of frequencies is exactly 2:1, there will be two stationary loops formed, etc.

13-36 TRIGGERED OSCILLOSCOPES

The usual and more sophisticated type of oscilloscope uses a triggered sweep. It is usually accurate enough to measure voltages and times. With no input signal, no sweep voltage is developed, and the cathode of the CRT is maintained so highly positive in respect to the grid that no electrons can form a beam to the screen. When a signal is applied to the input amplifier, it is amplified and fed in two directions: to a synchronization amplifier and to a delay line (Fig. 13-43).

The input signal voltage is further increased by the sync amplifier. Either the rising positive or the rising negative portion of the signal cycle is used to key a circuit to develop a sharp trigger pulse of voltage, T. The trigger pulse is fed to two circuits. One produces a slowly increasing dc *ramp* voltage, T to R, which is used to sweep the spot horizontally

across the screen. The other circuit produces an *unblanking* bias voltage at time T to bring the cathode of the CRT to a potential that allows electrons to flow through the electron gun. The ramp generator and the unblanking circuit are interconnected in such a way that when the ramp returns to zero volts, time R, the unblanking circuit again produces a blanking voltage on the cathode that stops the electron beam.

During the generation of the trigger, ramp, and unblanking voltages, the input signal is being fed through a delay line that delays the *signal* perhaps 150 ns before it is further amplified and fed to the VDPs. This is time enough for the developing of the trigger and the starting of the ramp voltage so that the beginning of the signal cycle that produced the trigger can be displayed on the CRT.

If the amplifiers are single-ended (one of the two output lines is at ground potential), it will be necessary to use phase inversion to feed the deflection plates. In many cases, phase inversion is produced at the beginning of the amplification system and all signals are then fed through direct-coupled double-ended (push-pull) amplifiers to prevent distortion of high-frequency and very low frequency signals. A 20-MHz oscilloscope should be able to show separate cycles of ac at this frequency, as well as be able to show variations that take several seconds for a single cycle. The sweep voltage times may be either continuously variable, or variable in switched steps, such as 1, 2, 5, and 10 μs, 1, 2, 5, and 10 ms, and 1, 2, 5, and 10 s per centimeter division on the screen. With calibrated sweeps, frequency can be determined. For example, with a 1-μs/cm sweep a 1-MHz ac should take up exactly 1 cm across the screen.

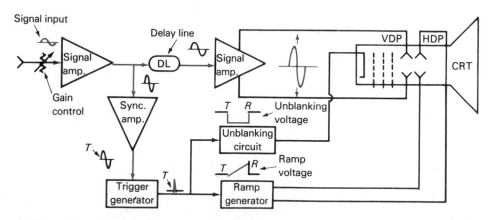

FIG. 13-43 Block diagram of a simple triggered-sweep oscilloscope.

There are a variety of other possibilities for these scopes. Dual-trace scopes show two displays on one screen by alternating a sweep line above the center and a sweep line below the center and applying two different signals to the two lines to allow comparison of two signals (such as an input and output signal of an amplifier) at the same time. Some scopes feature variable-persistence, or *storage,* that allows the CRT to retain a selected display for fractions of a second, seconds, hours, or days if need be. When the frequency to be viewed is higher than the capabilities of the vertical amplifiers, use may be made of a *sampling* system whereby an additional unit ahead of the scope input samples different parts of repetitive waveforms and delivers the samples at a frequency that the amplifiers can handle and display. Sampling scopes can display signals well up into the gigahertz range.

13-37 SPECTRUM ANALYZERS

An important adaptation of the oscilloscope is the *spectrum analyzer,* or *signal analyzer,* capable of showing relative spacing of transmitter carriers, their sidebands, harmonics, etc. A simplification of such a system is shown in Fig. 13-44. Basically, it

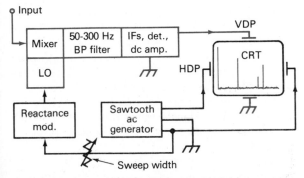

FIG. 13-44 Block diagram of a simplified spectrum analyzer.

consists of a very narrow bandwidth (1–300 Hz) IF filter and an electronically tuned local oscillator (LO) "receiver" system (Chap. 18), a sawtooth ac sweep generator, a CRT, and a reactance-type modulator (Chap.19) to sweep-tune the local oscillator of the receiver.

Sawtooth ac applied to the HDPs sweeps the CRT spot completely across the bottom of the screen, left to right. The same sawtooth ac is fed to a reactance-type modulator to shift the frequency of the local oscillator of the receiver, tuning the receiver from a low to a high frequency in step with

the trace being swept across the CRT. Controlling the amplitude of the sawtooth ac fed to the reactance modulator determines the frequency span that the LO will sweep. If the receiver is set for a minimum frequency of 7 MHz and the sawtooth ac sweeps the LO 300 kHz, all signals between 7 and 7.3 MHz that are received will produce separate vertical lines on the CRT. For example, if there is a signal at 7.1 MHz, one at 7.2 MHz, and one at 7.25 MHz, there will be three vertical lines projecting upward from the bottom sweep line, as in the illustration. (The vertical line at the far left edge, at the beginning of the sweep line, is a calibrating start signal of known amplitude developed on every sweep.) The amplitude of the vertical received signal lines indicate the relative strength of them. The horizontal lines on the face *graticule* of analyzer CRTs are marked off in either decibels or voltage units to give usable indications of signal strengths. The low-amplitude vertical lines shown along the sweep base line are caused by noise and are referred to as *grass.*

On the screen (Fig. 13-44) a vertical calibration line extends apparently up to the top of the screen. If the screen graticule is calibrated in 100 dB units, bottom to top, the first received signal appears to be about 30 dB down from the maximum 100-dB level. The second received signal appears to extend up only to a point about 70 dB from the top. Therefore, this second signal can be said to be about 40 dB below the first signal in amplitude. The third signal is about 45 dB down. With an accurately calibrated graticule, the decibel values can usually be estimated ±1 dB. The sweep can be adjusted so that horizontal graticule graduations can read frequency separations between signals within perhaps 0.2% (about 1.4 kHz for a 7–7.3 MHz band).

If the sweep of the LO is only 40 kHz, an amplitude-modulated carrier signal can be centered on the CRT and all of the sidebands being generated can be seen for 20 kHz on both sides of the carrier indication. If the sweep is set for a 1–50 MHz presentation, any harmonics or spurious signals generated by a local transmitter on 3.8 MHz, for example, could be seen up to 50 MHz. The signal would appear at the 3.8-MHz point on the base sweep line, the 2nd harmonic at 7.6 MHz, etc. The true bandwidth of transmitters, filters, etc., can be determined with a spectrum analyzer when set to a narrow frequency sweep.

To check the spectrum purity of any piece of equipment, its output may be connected directly into the output of the analyzer with coaxial cable and through a resistive attenuating network. In this way, no other signals can confuse the display.

Modern spectrum analyzers may use digital synthesizing and PLL circuits to produce accurate and programmable frequency sweeps for high-accuracy measurements.

13-38 TEST PROBES

Electronic test equipment usually has a ground lead that attaches to the ground circuit on the chassis of the equipment under test and a "hot lead," or probe, that is touched to the point under test.

Probes may range from simple wires up to highly complex, compensated, amplifying devices. A simple probe may consist of an insulated wire terminated with an insulating-material handle with a contact tip at the end (Fig. 13-45a). Inside the handle a 1–100 kΩ isolation resistor may be inserted. These probes are adequate for relatively low-impedance circuits involving frequencies up through the AF range (20 kHz).

For higher frequencies a compensated probe may be used (Fig. 13-45b). The internal capacitance of the flexible coaxial cable (dashed capacitor) has a continually lowering reactance as frequency increases. The result would be a constantly decreasing signal reaching the oscilloscope as higher frequencies were tested. To compensate for this, a 3–30 pF capacitor across a series input resistor is added. At dc the two capacitances are not effective, and the probe is a 10:1 reduction probe. The coaxial capacitance bypasses the ac signal more as frequency increases, but the reactance of the compensating capacitor is also decreased, feeding more signal to the coaxial line, maintaining the signal to the test equipment at the same amplitude. The circle with a line across it is the symbol of a screwdriver adjustment for the capacitor. The probe is calibrated by watching a square-wave ac on a scope. The capacitor is varied to show the sharpest corners possible on the square-wave signal. A ground clip may be added at the base of the handle. To prevent hand capacitance effect, the inner wall of the probe body may be metallic and grounded to the outer sheath of the coaxial cable ("shielding").

The probe in Fig. 13-45c is a half-wave rectifier that converts RF ac to pulsating dc, which is filtered to dc by the capacitance of the coaxial cable. This probe presents a dc voltage to the test instrument that varies as the circuit ac varies in amplitude.

A *signal tracer* is a specialized probe. It consists of a small cylinder with a short probe extending out one end. To this probe is connected a diode detector, amplifier, small loudspeaker, and battery. If touched to an electronic circuit that has voice- or music-modulated signals present in it, the signals become audible. It is used for testing receivers (Sec. 18-30).

13-39 ACTIVE DEVICE TESTS

One of the most important tests or measurements that a technician must make is to determine if active devices are operating properly. An ohmmeter is sometimes all that is readily available. With the device unenergized an ohmmeter alone can give some very definitive information.

With solid-state diodes, and the ohmmeter set to $R \times 1000$, there should be low resistance one way and high resistance the other. If not, the diode is either shorted or open.

With BJTs, using $R \times 1000$ on the ohmmeter, E-B should read high resistance one way and low resistance the other. The same with B-C. E-C

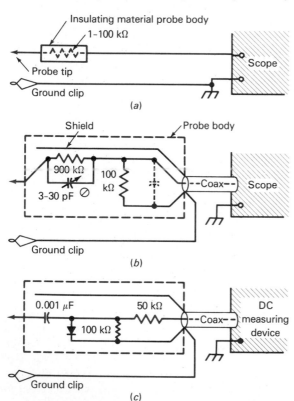

FIG. 13-45 Simple probes: (a) low frequency, (b) compensated HF, (c) HF-SHF ac meter probe.

should read high resistance in both directions. The higher-powered the device, the lower the resistance values.

With JFETs, the same results should be obtained as with BJTs except the channel reads relatively low resistance both ways.

With MOSFETs, an ohmmeter should not be used between gate and channel, as it may cause insulation breakdown.

With VTs, filament or heater connections should show low resistance. There should be no continuity at all between any two elements (k-g, g-p, p-sg, etc.) in a good tube.

A *transistor tester* can be used to test for shorts, gain, and leakage of BJTs, JFETs, MOSFETs, and SCRs. Some will test the device while it is in circuit; others require removal of the device before tests are valid. Some indicate relative gain and leakage on a meter; others are connected to an oscilloscope and show dynamic curves of the amplifying ability of the device when they apply a small audio voltage to the input circuit.

There are two types of *vacuum-tube testers*. One ties together all elements except the cathode and charges them positive, checking how much cathode current is given off at normal heater voltage. It can also test for interelement shorts and for leakage current if there is any gas in the tube. A better tester applies a small audio ac to the grid-cathode circuit and indicates the transconductance (g_m) of the tube on a meter scale while the plate potential is applied to the plate. With all tube testers the tube must be removed from the equipment and plugged into the appropriate socket in the tester.

■ Test your understanding; answer these checkup questions.

1. List the component parts of a CRT. _____
2. What is an advantage of using a high potential on the last anode of a CRT? _____

3. What is controlled by varying the potentials between anodes? _____ By varying grid bias? _____
4. Which deflection plates control the beam more? _____
5. To which plates are signal voltages usually applied? _____ What is applied to the other plates? _____
6. What is the function of the Aquadag? _____ Its potential in respect to ground? _____ In respect to the cathode? _____
7. What determines the number of cycles shown on a scope for a given frequency? _____
8. What is a block diagram symbol of an amplifier? _____ Of an oscillator? _____
9. Why is phase inversion used in scopes? _____ What types of amplifiers do they follow? _____
10. What will be seen with sawtooth 500 Hz on the HDPs and sine-wave 2 kHz on the VDPs? _____ If the sawtooth is 501 Hz? _____ If 499 Hz? _____
11. How can high-frequency ac be displayed if the amplifiers cannot handle the frequency? _____
12. In a triggered scope, what starts the sweep voltage? _____ What determines the sweep speed? _____ Why must HDPs use a blanking voltage? _____ Why is a delay line used? _____
13. Why are transistor amplifiers desirable in scopes? _____
14. What is a CRT that can hold a display called? _____
15. What are two ways of measuring frequency on a scope? _____ _____
16. What is improved in compensated oscilloscope probes? _____
17. Is it possible to make meaningful tests with an ohmmeter on BJTs? _____ JFETs? _____ VTs? _____
18. What is the best type of test for VTs? _____ JFETs? _____
19. Can transistors be tested while still in a circuit? _____ VTs? _____
20. What device can show fundamental and harmonics? _____

MEASURING DEVICES QUESTIONS

1. For what are analog meters superior? For what are digital meters superior?
2. What type of meters use LCDs?
3. Of what does the rotating mechanism of a d'Arsonval meter consist?
4. What are the major components of a d'Arsonval meter?
5. Why are meter magnet poles and the iron-core piece placed close together?
6. Why are counterweights used on meter pointers?
7. If springs are not used in a d'Arsonval meter, what does it use?
8. What is a meter called if its zero indication is at center scale?
9. What type of meter is most likely to have a nonlinear scale?
10. Why does a shunted meter have a higher full-scale reading than if it is not shunted?

11. Why would copper coils not make good meter shunts? What is used?
12. A 300-Ω, 0–50 μA meter is to be used as a 0–1 mA meter. What shunt value should be used?
13. How are digital meters made to read current values?
14. In what value is the sensitivity of a dc ammeter usually expressed? Of a dc voltmeter?
15. List three means by which meter movements are damped.
16. Where is an electrostatic voltmeter usually found?
17. How can a d'Arsonval ammeter be converted to a voltmeter?
18. A 100-μA meter movement has 200-Ω internal resistance. What would be its full-scale voltage rating? What multiplier would be necessary in order to make it a 0–1 V voltmeter? A 0–100 V meter?
19. Why are high-sensitivity voltmeters used in high-resistance circuits?
20. Why is a fixed resistor always included in an ohmmeter?
21. What resistance value is usually at the right-hand full-scale point of an ohmmeter? Left-hand?
22. Draw a diagram of a two-scale ohmmeter.
23. When using an ohmmeter, what should usually be done before making a resistance measurement?
24. Of what does a VOM consist?
25. What does EVM mean? Why is the meter of an EVM shunted by two diodes?
26. Draw a diagram of a simple four-range EVM using JFETs. What would be the values of the four input resistors if the meter has 1-, 10-, 100-, and 1000-V ranges and if the top one is 9 MΩ? What is the meter's best sensitivity? Worst?
27. Draw a diagram of a simple ac probe for an EVM.
28. In EVMs, why are two JFETs used instead of one?
29. If a 100-V dc meter is across 200 V ac, what will it read? Will this damage it?
30. A 0–100 V dc voltmeter using a bridge rectifier across an ac line reads 47 V. What is the actual ac peak value? The rms value?
31. If the meter in problem 30 used a half-wave rectifier, what would be the actual peak and rms values?

ANSWERS TO CHECKUP QUIZ ON PAGE 285

1. (Electron gun composed of cathode, grid, and anodes, VDPs, HDPs, screen, Aquadag, envelope) **2.** (Accelerates electrons, bright display) **3.** (Focus)(Brightness) **4.** (First, or VDPs) **5.** (VDP)(Sweep voltage) **6.** (Collects electrons)(Usually low)(High +) **7.** (Sweep frequency or speed) **8.** (Triangle)(Labeled square or circle) **9.** (Allow centering of display)(Single-ended) **10.** (Four sine cycles)(Cycles move right)(Move left) **11.** (Connect to VDPs directly, sampling) **12.** (Trigger V)(Slope of ramp)(Prevent dot on screen with no signal)(Allow display all of starting cycle) **13.** (Good HF capabilities) **14.** (Storage CRT) **15.** (Lissajous)(Calibrated sweep) **16.** (Square waveforms) **17.** (Yes)(Yes)(Yes) **18.** (g_m)(Same) **19.** (Sometimes)(Rarely) **20.** (Spectrum analyzer)

32. How could a bridge-rectifier dc meter be made to indicate peak audio voltages relatively well?
33. Is a VU meter a dB meter? How does it differ from a dB meter?
34. What is the 0-dB power level for most dB meters?
35. Why do VU meters come with a separate 3600-Ω resistor?
36. What are the names of the two types of VU scales?
37. What type of meter is generally used to indicate RF ac in transmitter antennas? Is it an ac or dc movement?
38. How does an electrodynamometer differ from a d'Arsonval meter?
39. Does a d'Arsonval meter have a linear or nonlinear scale? An electrodynamometer?
40. Name three meters that use the electrodynamometer principle.
41. Name two repulsion-type meters. Where might they be likely to be found?
42. Name the two circuits in a wattmeter. Which is connected in series with the line? Which is across the line?
43. Why does a wattmeter always read true power?
44. What does a watthour meter measure? Where are these meters usually found?
45. In what application is an ampere-hour meter usually found? What kind of device is it?
46. In a Wheatstone bridge, if R_1 is 100 Ω, R_2 is 1000 Ω, and R_k is 650 Ω, what is the unknown resistance?
47. How many different resistance values could a bridge with a 6-decade box measure?
48. What are two other ways of expressing a frequency tolerance of 10 ppm?
49. How much frequency drift might occur in a 250-MHz transmitter and still be within a tolerance of 0.0005%?
50. Which of the two power-line frequency meters mentioned in the text would probably be more accurate?
51. What is the simplest but least accurate RF frequency meter? What is its main use? In what frequency bands might it be used?
52. What dips in a dip meter? What does a dip meter measure?
53. Who transmits standard frequency and time signals in the United States? On what frequencies?
54. What is a secondary frequency standard?
55. For what is a transfer oscillator used?
56. What modern type of system makes an excellent secondary frequency standard? How could it be made a primary standard?
57. How is hand capacitance reduced in a heterodyne frequency meter?
58. What are four advantages of the more advanced heterodyne meter over the simple one?
59. A dial marked off in 250 units can read how many values if a vernier scale is used on it?
60. What must be added to an electronic counter to make

a frequency meter out of it? What determines the meter's accuracy?

61. How many DCUs and FFs would be required to measure 150 MHz?

62. How might the frequency of a self-excited oscillator be measured with a counter or frequency meter? Of a Pierce crystal oscillator?

63. Draw a block diagram of a digital counter, or frequency meter.

64. What does a digital voltmeter actually count?

65. List the operating parts of an oscilloscope CRT.

66. Where is the potential difference that focuses electrons in a CRT? What controls the beam intensity?

67. Draw a block diagram of a simple free-running oscilloscope.

68. In an oscilloscope, to what is the sawtooth ac fed? The amplified input signal?

69. What do the phase inverters do in a scope?

70. What is the desirable beam return interval in a free-running scope?

71. What determines the starting of a sawtooth ac in a triggered scope? Where is it developed?

72. In a triggered scope to what may the unblanking voltage be fed?

73. With a 5-μs/cm sweep on a 10-cm-wide CRT, how many cycles would be shown by a 0.2 MHz ac?

74. What type of scope allows retention of a waveform for long periods of time? What type can show SHF ac cycles?

75. Draw a block diagram of a triggered scope.

76. Which of the following cannot be shown by a spectrum analyzer: signal amplitude, frequency, width of signals, ac waveform, harmonics, adjacent frequency signals?

77. Draw diagrams of a HF compensated scope probe. A HF ac meter probe.

78. For what is a signal tracer used?

79. What test meter can be used to make rudimentary tests on BJTs, JFETs, and MOSFETs? What is preferred? What is best to use to test VTs?

14
Audio-Frequency Amplifiers

This chapter describes the operation of amplifiers of audible-frequency signals. These may be voltage or power amplifiers, solid-state or vacuum-tube, single-ended or push-pull, and classes A, AB, or B. The various types of coupling and distortions developed in these circuits are discussed. The effects of inverse feedback are described. Special solid-state circuits, such as the Darlington, differential, complementary symmetry, and operational amplifiers are explained.

14-1 AUDIO FREQUENCIES

Chapter 9, "Active Devices," outlined the principles of amplification. Variations of voltage or current from small fractions of a cycle per second up, or a hertz, to thousands of millions of hertz can be amplified. For variations below the physiological limit of audibility (about 16 Hz) direct-coupled stages are usually required. Direct, capacitive, or transformer coupling can be used between amplifier stages for frequencies up to 1 GHz or so. Above this, specialized equipment is used (Chap. 22).

Any frequency that can be heard when applied to earphones or loudspeakers is called an audible or audio frequency (AF). These are generally considered to be from 16 to 20,000 Hz. Electromagnetic waves having frequencies above 10 kHz are known as radio frequencies (RF). *Subaudible frequencies* are those from a fraction of a hertz to 16 Hz. *Supersonic frequencies* range from 20 kHz up to several megahertz.

If a system can handle only the frequencies between about 250 and 3500 Hz, it will do a reasonably good job of amplifying voltages developed in a microphone by the human voice. This voice frequency (VF) range is what telephones must respond to and amplify. Music signals transmitted through such speech-type systems leave a lot to be desired. If the amplifiers are engineered to pass all frequencies from 50 to 5000 Hz equally well, music (and voice) is reproduced quite well. But for true *high fidelity* (hi-fi, or low distortion), audio amplifiers must be able to amplify equally well all frequencies from about 20 to 20,000 Hz. In this chapter the fundamentals of amplifiers of only audible-frequency signals will be discussed.

14-2 VOLTAGE VERSUS POWER AMPLIFIERS

The ac signals developed by microphones, turntables, detector stages of receivers, etc., may be in the microvolt or microampere to millivolt or milliampere range. To enable such weak signals to drive the input of a power amplifier adequately, one or more low-level (low-power) amplifiers may be required.

The first stage of an amplifying system must be a low noise type because all following stages will be amplifying any noise (usually considered 5-kHz and higher pulses) that it generates. Internal noise can be from random electron motions in the input conductors or resistors, from electron-hole combinations in semiconductor devices, or from uneven cathode emission in vacuum tubes.

An AF power amplifier develops AF ac power into a load of some form. It may be a few milliwatts into earphones, 100 mW to 2 W into loudspeakers, up to hundreds of watts into some stereo or rock-band amplifier loudspeakers, or tens of thousands of watts into an AM broadcast station modulation system.

Physically, low-level devices are relatively small and operate at very low current and power levels. Power-amplifying devices are relatively larger, may involve amperes of current, and may develop enough heat to require special cooling methods for their collectors, drains, or plates. Many low-level stages may operate from the same low-voltage power supply. High-power amplifiers often require a separate power supply for each of their high-level stages.

14-3 DEVICES USED

The basic operation of the devices used in AF amplifiers is discussed in Chap. 9. These devices are the bipolar junction transistor (BJT), the junction field-effect transistor (JFET), the insulated-gate FET (MOSFET), and the various vacuum tubes (VTs).

One of the important operating parameters of any of these devices is the value of the bias voltage or current developed in the input circuit that sets or limits the output circuit current to a proper value. When an ac or varying dc signal voltage or current is applied to the input of an amplifying device, the output current varies up and down from the no-signal value, producing an ac or varying dc output power in the load greater than that of the input signal. This is *amplification*.

While the basic function of all amplifier circuits is to amplify power, in some cases it is only necessary to amplify the input ac signal voltage. Such stages are known as *voltage amplifiers*.

A basic common-source, or grounded-source voltage-operated-device AF amplifier circuit using an N-channel JFET is shown in Fig. 14-1a. For normal amplification, called class A, with no input signal between gate and ground, a reverse-bias voltage-drop is developed across the self-bias source resistor, R_S, selected to produce a drain-source voltage of about half of $+V_{DD}$. Assume the voltage-drop, V_S, across R_S is a 0.5-V biasing value.

If a 0–0.1 V peak-to-peak (p-p) sine-wave signal is applied to the input circuit and the amplifica-

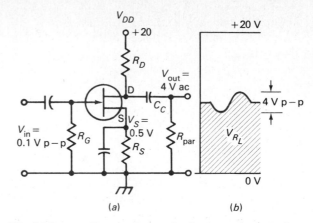

FIG. 14-1 (a) JFET amplifier. (b) Output voltage variation across R_L.

tion factor (μ) of the JFET is 60, the actual stage gain may be about two-thirds of this, or 40. The output-signal-voltage drain to ground (or across R_D since $+V_{DD}$ is always heavily bypassed to ground) should be 0.1(40), or a p-p 4 V varying dc (Fig. 14-1b). This should develop a 4 V p-p ac when coupled by C_c to the parallel resistive load R_{par}.

If the input signal is increased to 0.5 V p-p, the output will be 0.5(40), or 20 V p-p, which is the maximum because this is the total value of the power-supply voltage. If the input is increased over 0.5 V p-p, both the positive and negative peaks of the output signal voltage will become flattened and badly distorted. Actually, the input signal would have to be held down to about 0.4 V p-p to prevent some distortion from being developed, because the $V_G I_D$ curve is not linear, curving more at its low I_D end.

Usually, the higher the value of the drain load resistance, or the higher the power-supply voltage, the closer the stage voltage gain will be to the device's μ value.

In an amplifier stage the resistance of the device and the load resistance are in series across the power supply. The ratio of R_L times μ of the device, to the total ac-circuit R_L (where R_L is R_d and R_{par} computed in parallel) plus the device resistance gives the stage gain; that is,

$$A_v = \frac{\mu R_L}{R_L + R_d}$$

where A_v = voltage gain of the stage
R_d = device resistance
R_L = total load resistance
μ = device amplification factor

Consider the case of an amplifier having $\mu = 30$, $R_d = 5\ k\Omega$, and total $R_L = 10\ k\Omega$. The stage gain should be

$$A_v = \frac{30(10{,}000)}{10{,}000 + 5000} = \frac{300{,}000}{15{,}000} = V\ \text{gain of } 20$$

The gain of an amplifier is how much the input voltage is increased in the output, and may also be expressed in decibels. If *similar input* and *output impedances* are assumed, an amplifier with a voltage gain of 40 has a decibel gain of 20 log (V_o/V_i), or 20 log 40, or 20(1.60), or 32 dB. A pentode resistance-coupled stage may have a voltage gain of over 250, or a decibel gain of 20 log 250, or 20(2.398), or 48 dB. If an AF amplifier is made up of a first high-gain stage having 30 dB, a second lower-gain stage of 20 dB, and a power-output stage having 15 dB gain, the total amplifier gain is 65 dB.

An N-channel JFET (or VT) can usually stand a little gate (grid) current. If there is any current that flows through any input circuit resistance during the positive half cycle of the input signal, there will be a voltage-drop across this resistance. Such a voltage-drop will be a reverse emf and will result in a clipping of the positive peak in the output signal.

A basic current-operated-device AF amplifier circuit using an NPN BJT is shown in Fig. 14-2. In

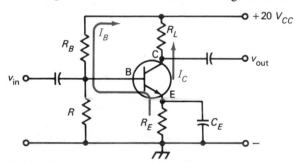

FIG. 14-2 Audio amplifier stage using a BJT.

this circuit, bias current is drawn up through R_E, through the emitter to the base, and through R_B to $+V_{CC}$. This forward-biasing current allows EC current to flow through R_L. The resistances are selected so that with no input signal the voltage-drop across R_L equals the EC voltage-drop. Resistors R and R_E are added to help to stabilize the circuit thermally. Capacitor C_E ac-grounds the emitter and prevents signal degeneration which would occur across R_E. When an ac signal is applied between base and ground (actually B to E through C_E), the collector current varies, producing a varying voltage-drop

across R_L. This is the output signal from the stage, and is usually 15 to 30 dB greater than the input signal.

The three devices mentioned so far, the JFET, BJT, and VT, will be the only ones used in most explanations. The depletion and enhancement MOSFET-type devices operate much the same as the JFET except that since they have insulated gates, there can never be any gate current. Enhancement MOSFET bias circuits may differ from those of JFETs when they use either zero bias or forward bias. When zero bias is required, no source-bias resistor is used. When a forward bias is necessary, two series resistors, such as R_B and R in Fig. 14-2, will be connected from V_{DD} to the gate and from the gate to ground (source).

14-4 A CLASS A AMPLIFIER

With AF amplifiers there are three basic "classes" of operation, A, AB, and B. These classes are based on how much bias is used in the circuit. A class A amplifier having a resistor load, as in Fig. 14-1, is said to be biased to class A if essentially half the voltage value of the power supply appears as a voltage-drop across the load resistor. If this happens with -1 V of bias, then when the bias is -2 V, the output circuit current should decrease to nearly zero, or to output circuit current "cutoff," resulting in almost no voltage-drop across the load resistor. If the bias decreases to 0 V, the output circuit current should increase so that there is now a voltage-drop across the load resistor of almost the whole power-supply potential.

When a transformer is used as the output coupling device in a stage (rather rare), T_2 in Fig. 14-3, there is relatively little resistance in the primary winding and the drain potential will be almost

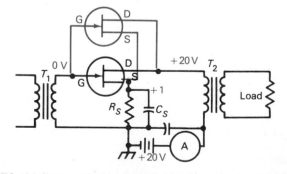

FIG. 14-3 Transformer-coupled JFET stage. A second JFET in parallel to double power output is shown in color.

the same as the power-supply voltage. However, when a positive signal emf is applied to the input gate by T_1, the output circuit current in the primary of T_2 increases, developing a counter emf across the primary. Since the lower terminal of the primary is held at ac-ground potential by the bypass capacitor across the power supply, the counter emf that is developed partially cancels the drain voltage D to ground. When the signal goes more negative than the bias, the output current decreases and the induced voltage in the primary adds to the drain voltage. The peak-to-peak ac voltage developed across the primary of an output transformer under strong signal conditions may almost equal the power-supply voltage value. If a JFET is rated at 100 V maximum, it should never be operated transformer-coupled with more than 50 V of V_{DD}.

When amplifiers are operated with resistors as the output circuit load, they are considered to be voltage amplifiers. When they are operated with transformers, they are usually power amplifiers. Power output from an amplifier is determined by $P = V_L I_L$.

The product of the no-signal output circuit current times the power-supply voltage must not exceed the dissipation rating of the device or burnout may result.

Class A JFET or VT stages should never be driven hard enough to draw input circuit current, and therefore will require essentially no power to drive them. The output circuit current varies equally above and below the *static* or no-signal value so rapidly that it makes minimum demands on power-supply voltage regulation. An ammeter in the output circuit should show no movement with or without signal input.

If the bias voltage is too high or too low, the input signals may drive past the linear limits of the characteristic curve and distortion will be developed. With small signal input values, the bias can be rather far from the optimum value and no noticeable effect will be produced. It is only when maximum possible output signal current or voltage excursions are produced that the exact bias value is really important.

The optimum bias voltage for an N-channel JFET or triode VT is usually about 0.6 of the voltage between zero bias and the negative value required to produce output circuit cutoff. The cutoff bias voltage can be found by dividing the drain or plate voltage by the μ of the device. For example, in Fig. 14-3, if the V_{DD} value is +40 V and the JFET has a μ of 30, the bias voltage for class A opera-

tion would be approximately $-V_G = 0.6V_D/\mu$, or 0.6(40)/30, or -0.8 V.

To produce 0.8 V across R_S with an I_D of 5 mA, by Ohm's law, $R = V/I$, or $R_S = -0.8/0.005$, or the bias resistor should be 160 Ω.

The capacitor C_S across R_S prevents a 180° out-of-phase, or degenerative, voltage from being added in the source-gate circuit which would partially cancel the input signal. To make the capacitor effective at canceling degeneration, its reactance should be one-tenth the resistance value. From this can be derived the basic bypass or coupling capacitance formula

$$C = \frac{1,600,000}{f_L R}$$

where C = coupling capacitance, μF
f_L = lowest frequency used
R = resistor involved

In the circuit above, the required bypass capacitor for a 100-Hz lowest frequency would be $C = 1,600,000/100(160)$, or 100 μF.

(This same formula can be used to compute the coupling capacitance C_c to R_{par} in Fig. 14-1.)

14-5 PARALLEL AND PUSH-PULL CLASS A

If a single transformer-coupled output device can produce 1 W of undistorted AF output in a load, but more power is needed, two similar devices can be connected in parallel, as shown in color in Fig. 14-3. The same input signal voltage applied to two parallel-connected devices will produce twice the current in the output primary and therefore twice the output power. Doubling the total drain current also doubles the source current, so that the bias resistor must be reduced in value to one-half to retain the same class A bias voltage for the two devices. Any number of active devices can be connected in parallel to produce higher output powers.

Another method of doubling the power output is to connect the two active devices in push-pull (Fig. 14-4) using transformers for simplicity. In this circuit if the top device Q_1 is being fed a positive input signal as indicated, at the same time the lower device Q_2 will be fed a negative. Q_1 and Q_2 are being fed 180° out of phase. The drain current of Q_1 increases through the upper half of the output transformer's primary while the Q_2 drain current is decreasing in the lower half of the primary. Both of these effects act to induce an upward voltage in the secondary of T_2. Both devices are aiding in devel-

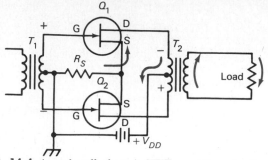

FIG. 14-4 A push-pull class A JFET amplifier circuit.

oping half the current in the load. The result will be twice the output power.

Since both source currents are flowing through R_S, the voltage-drop developed across it will be the bias for both JFETs. (Both gates return to the negative end of the R_S voltage-drop.) A bias filter capacitor may not be needed to prevent degeneration in this circuit because as Q_1 source current increases, Q_2 source current decreases a similar amount, and a constant bias potential results.

There is another advantage of push-pull operation besides producing twice the power output. In all devices, given the same small input signal, if the bias value is low, there will be slightly more output current change than if the bias value is high (near cutoff). If output current is plotted against bias voltage (Fig. 14-5), the resulting curve always has

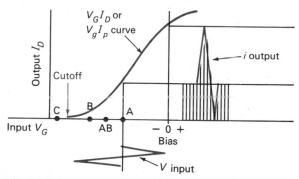

FIG. 14-5 JFET $V_G I_D$ characteristic curve. Unequal output current peaks due to the bend in the curve.

a slight bend. Because of this bend all single-device stages distort input signals to some extent, thus developing harmonics in the output. By using push-pull operation, when one device is operating near cutoff, the other is operating near zero bias. In the output transformer primary, one device is developing a stronger half cycle when the other is developing a weaker similar half cycle. The result is a more linear amplification at all times, and a reduction of even-order (second, fourth, etc.) harmonics.

Distortion percentage may be reduced from perhaps 5% down to less than 1% by using push-pull class A circuits instead of single-ended class A circuits.

Still another advantage of push-pull is its ability to reduce low-frequency "hum" resulting from an inadequately filtered power supply. If the power-supply voltage is varying upward at one instant, it will be rising in both of the push-pull devices at the same time. This tends to induce similar current increases in the opposite halves of the transformer primary, canceling this common-mode hum voltage in the secondary.

With push-pull AF circuits there may be three classes of operation, A, AB, and B, each with its own specific bias value.

As explained previously, bias for a JFET (or VT) is approximately $-V = V_D/\mu$, where V_D is the source-to-drain voltage and μ is the amplification factor of the device. For class A the bias formula would be approximately

$$-V_A = \frac{0.6V_D}{\mu}$$

For class AB

$$-V_{AB} = \frac{0.8V_D}{\mu}$$

For class B

$$-V_B = \frac{0.95V_D}{\mu}$$

14-6 CLASSES B AND AB

Only a single device is necessary in a class A amplifier to produce minimal distortion, particularly if the amplifier is operating over only a small portion of the characteristic curve (minimal bend). If it were to be biased to class B, which means that the bias is increased almost to the point of output circuit current cutoff (Fig. 14-5), only half of the input signal would produce output circuit current. The distortion would be intolerable. However, if two JFETs (VTs or BJTs) are connected in push-pull and biased to class B, during one-half of the input cycle one device will be developing current into one-half of the output transformer primary (Fig. 14-4). On the next half cycle the first device will be cut off but the other device will be developing primary current in the opposite direction in the other half of the primary. As a result both halves of the input cycle are producing output current and power in the output load. One device amplifies the positive half of the input signal cycle, and the other amplifies the negative. Since the two output pulses do not occur at the same time, they cannot cancel bent-curve distortion. Class B operation should always be more distorted than push-pull class A operation.

Two class B biased devices in push-pull should be able to produce at least twice as much power output as when they are in class A, provided the input signal voltage is twice as great. Whereas a class A AF amplifier may only operate at about 25% efficiency, a class B AF amplifier may operate at about 50% efficiency because each half cycle of input signal operates over twice as much of the curve. If input gate (or grid) current can be tolerated (no resistance in the input circuit), it may be possible to have the input signal drive past the zero-bias point on the curve and develop still higher-amplitude output current pulses. The efficiency may be increased to perhaps 60% in this manner. As soon as input circuit current is made to flow, however, the demands on the stage driving the push-pull amplifier increase. Now the driving stage must feed not only a signal voltage (at essentially no current) to the input circuit, but also current and therefore power ($P = VI$) to the input circuit. Thus, high-efficiency class B JFET and VT stages must be driven by power amplifiers, but lower-efficiency stages (classes A and AB) may be driven by voltage amplifiers.

Because with no input signal there is almost no output circuit current, the bias resistor in Fig. 14-4 cannot be used to bias a class B stage. A separate well-regulated voltage supply is required, perhaps using a zener diode regulator or even a separate bias supply.

If the bias of a push-pull AF stage is adjusted to a value halfway between class A and class B operation, it is said to be biased to class AB. For JFETs and VTs there are two ways of operating class AB: If no input circuit current ever flows, the operation is known as class AB_1; if the peaks of the driving signal produce input circuit current, the operation is known as class AB_2. Being intermediate between classes A and B, the power output may not be as great as class B, but the distortion might be less than that of class B, approaching that of class A. With class AB_1, if heavily bypassed, the bias resistor shown in Fig. 14-4 might be used, but zener diode regulation would be preferable. For class AB_2, zener regulation or a separate bias supply is required.

The subscripts 1 and 2 are sometimes used with classes A and B also, and have the same meaning. A_1 or B_1 means no input current is to flow. A_2 and B_2 indicates input current is expected to flow at the peaks of the driving signal. Actually, "class B" operation usually assumes some input signal current flow (B_2).

Class A operation maintains a constant average output circuit current with or without signal. Classes AB and B have low current values with no signal and higher current with any signal. This can place quite a demand on the power supply to maintain a constant voltage output. An ammeter in the drain, collector, or plate circuit will increase whenever signals are being amplified, although more with class B than with class AB. Because I_D (I_C, I_p) flows all of the time in a class A amplifier, such an amplifier is said to have an *operating angle* of 360°. Class B has about 200°, and class AB has an operating angle of about 280°.

If the bias voltage is at a value past output circuit current cutoff, the stage is in class C. Since part of the input voltage will not be affecting output circuit current, the output signal will be badly distorted. Class C is never used in AF amplifiers, but is used in RF power amplifiers (Chap. 15).

▌ Test your understanding; answer these checkup questions.

1. What is the range of audio frequencies? _____ Subaudible frequencies? _____ Supersonic frequencies? _____
2. What is the range of voice-communication AF? _____
3. What stage of an amplifier must generate the least noise? _____
4. What two factors determine the no-signal current in an amplifier device? _____ _____
5. With an ac input signal, what kind of voltage is developed across the output load if it is a resistor? _____ A transformer? _____
6. What happens to the output waveform if the input circuit is greatly overdriven? _____ What is the result of this called? _____
7. What is the usual class in which amplifying devices operate? _____
8. What stage gain can be expected from a device having a μ of 50? _____
9. What is developed in the input circuit of a JFET or VT stage if input current flows and there is resistance in the circuit? _____
10. Name an amplifying device which will never have any input current flowing in it. _____

11. In what class must all single-ended AF amplifiers operate? _____

12. In what class(es) may push-pull AF amplifiers operate? _____

13. What three classes of amplifiers require no power to drive them? _____ _____ _____

14. What is meant by a static V, I, or P value? _____

15. What are two ways of expressing the gain of an amplifier? _____ _____

16. Is correct bias more critical with low- or high-input signals? _____

17. What is the formula to determine a workable bypass capacitor value to be used across a source or cathode self-bias resistor? _____

18. What would be the bypass capacitance for a 500-Ω self-bias resistor for hi-fi response? _____ For reasonably good music response? _____

19. What is the stage gain of a triode-type device having a μ of 60, a R_d of 10 kΩ, and a Z_L of 30 kΩ? _____

20. Draw diagrams of a single-ended JFET class A AF amplifier. A parallel-device JFET class A AF amplifier.

14-7 PHASE INVERTERS

All single-ended AF stages must be class A and may be driven by voltage amplifiers. They can use transformer interstage coupling to produce a stepping up of the signal voltages, but this is expensive. Transformers may radiate alternating magnetic fields that can induce voltages into other circuits. External magnetic fields from 60-Hz power transformers may induce hum voltages into AF transformer windings, which are then amplified by all following amplifiers. For push-pull stages, center-tapped transformers can be used to feed the two devices 180° out of phase. These are usually necessary in class B and class AB$_2$ stages.

Rather than using a center-tapped coupling transformer to feed a push-pull class A or class AB$_1$ amplifier, a *phase-inverter* stage can be used (Fig. 14-6). The input to the phase inverter is from a single-ended stage. In this circuit R_S, the JFET, and R_D are all in series and will have exactly the *same current value* flowing through them at all times. If R_S and R_D are similar in resistance value, any voltage-drops across them will be equal. With no signal input, both points A and B will be at ground potential as far as ac is concerned. With equal voltage-drops across R_S and R_D, D will be as much less positive than $+V_{DD}$ as S will be more positive than ground. Inasmuch as $+V_{DD}$ is held at ac ground

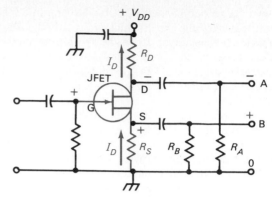

FIG. 14-6 A paraphase or split-load phase inverter.

potential by bypass capacitance, points D and S are 180° out of phase with each other (D is less positive, S is more positive). With a positive-going input signal to the gate, I_D increases. The voltage-drops across both resistors increase equally. D becomes less positive; S becomes more positive. This feeds a more negative voltage (less positive) to point A, and a more positive voltage to point B. If A and B are connected to the input of a push-pull class A amplifier, the inputs to the two push-pull devices will be driven 180° out of phase, just as was done by a center-tapped input transformer.

This *split-load* or *paraphase* phase inverter has no gain. The voltages at A and B will be somewhat less than the signal voltage fed to G. There may be essentially no distortion produced by phase inverters. On the other hand, AF transformers tend to drop off in signal response at lower frequencies because of insufficient iron in the cores. Distributed capacitance in the windings attenuates the higher frequencies. Nonlinearity of the magnetic circuit BH curve also produces some output distortion from a transformer. However, an AF voltage step-up is possible when transformer coupling is used.

Phase inverters of this type cannot be used to drive class B amplifiers if the amplifier input circuits draw current, as such current flowing through the R_A and R_B resistors would develop voltage-drops across them which would tend to cancel the driving voltage peaks due to degeneration.

14-8 EARPHONES

To change AF ac or AF variations of dc into sound waves, a *transducer* is required. Either an earphone or a loudspeaker can be used. Magnetic earphones may be low-impedance (5- to 600-Ω) types, medium-impedance (1000- to 3000-Ω) types, or high-impedance (20-kΩ or higher) types.

Common earphones operate on an electromagnetic principle. A varying current produces a varying-strength magnetic field of the internal coil. The varying magnetic field attracts and releases a thin iron diaphragm. Vibration of the diaphragm sets up airwaves which are recognized as sound by the human ear.

Figure 14-7 illustrates a basic earphone. When no current is flowing through the coils, the permanent

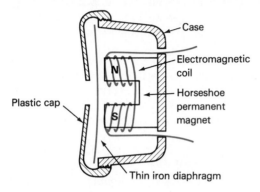

FIG. 14-7 Construction of an electromagnetic earphone.

magnet attracts the diaphragm and holds it in a strained position, bent slightly inward. If current flows through the coils, it will either add to the magnetism of the magnet and pull the diaphragm farther inward, or counteract the magnetic field and allow the diaphragm to swing outward, depending on current direction. If the current is varying or alternating, the diaphragm will swing back and forth at the frequency of the current variations or alternations. The diaphragm vibrations cause airwaves or sound to come out through the hole in the plastic cap.

Medium-impedance earphones have many turns of fine wire around the magnetic core. With this impedance they can be connected directly in the output circuit of many amplifiers, but care must be taken to connect them in such a way that current will flow in a direction that will keep the permanent magnets from becoming demagnetized. Earphone cords are often marked with a red tracer to indicate which wire should be connected to the positive end of the power supply. When earphones are connected across the secondary of an output transformer, they are fed ac and the polarity of the leads does not matter. If earphones ever lose their permanent magnetism, each *half cycle* of AF ac applied

to them produces one complete vibration, producing extreme distortion.

Low-impedance earphones have fewer turns of heavier wire and operate with higher current from a step-down output transformer or from low-Z transistor circuits. They are not made to operate directly in the output circuit of high-Z amplifier circuits, although they will produce weak sounds if connected there.

A source- (emitter-, cathode-) follower circuit may be used to convert high impedance to low impedance (Fig. 14-8). Input signals vary the I_D,

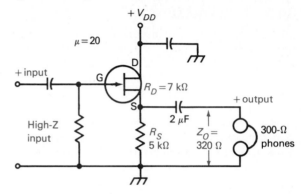

FIG. 14-8 A source-follower amplifier used for impedance reducing or as an isolating circuit.

developing an output voltage across the load. The voltage gain can be found from

$$A_v = \frac{\mu R_S}{R_D + R_S(1 + \mu)}$$

and is always less than 1.

The output impedance, Z_o, is approximately

$$Z_o = \frac{R_D R_S}{R_D + \mu R_S}$$

or about 320 Ω in this case.

Unlike common-source (-emitter, -cathode) amplifiers, which change the signal phase 180°, a source-follower produces no phase shift. This type of circuit is useful as an isolating circuit because any changes that may occur in the load do not affect the input circuit.

Earphones should be handled carefully. A dented diaphragm will cause weak signal output or distortion if the diaphragm touches the core during its vibrations. A mechanical shock, as from dropping, demagnetizes the magnet core, causing weakened output and distortion.

Besides electromagnetic earphones, there are

piezoelectric, or crystal, earphones. These are very high-impedance (20,000 Ω) types. They operate from high-impedance output circuits. Inside the earphone is a piezoelectric crystal attached firmly to a thin diaphragm. The two sides of the flat crystal are metal-coated. When an audio voltage is applied to the two opposite metalized surfaces of the crystal, the crystal bends, pushing the diaphragm inward or outward, depending on the polarity of the audio voltages being applied. Crystal earphones are quite sensitive and have excellent fidelity.

14-9 LOUDSPEAKERS

In early-day radio, loudspeakers were similar to high-impedance magnetic earphones but were larger and had heavier wire in them to handle greater plate currents. A horn, or large diaphragm area, was used to allow a larger mass of air to vibrate, producing louder sounds.

Most modern loudspeakers operate on a permanent-magnet field principle. They are known as p-m, or *dynamic*, speakers (Fig. 14-9). When cur-

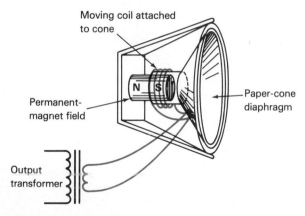

FIG. 14-9 Components of a p-m loudspeaker.

rent flows through the coil attached to the diaphragm assembly, the coil becomes an electromagnet. The coil will now be either attracted inward toward the magnet or repelled outward away from it, depending upon the direction of the current

in the coil and its magnetic polarity. Since the coil is attached to the diaphragm, any movement of the coil carries the diaphragm back and forth with it, producing the air vibrations necessary to make sound. The ends of the coil are attached to two points on the paper diaphragm. Flexible leads from these points carry the ac signal currents to the coil from the output transformer. The impedance of these loudspeakers varies from 3 to 100 Ω or more.

Electrodynamic loudspeakers are similar to p-m speakers except that the central core is an iron temporary magnet. On this core is a fixed field coil, which, when energized, can produce a stronger field than that of the permanent magnet in a p-m speaker.

Electrostatic loudspeakers are very high impedance devices and might ·be likened to an air-dielectric capacitor with one plate free to move. Varying high voltages attract and release the movable-plate diaphragm, making it vibrate.

14-10 COUPLING METHODS

The three basic methods of coupling AF ac signals into and out of an amplifier stage are transformer coupling, resistance (resistance-capacitance, or *RC*) coupling, and impedance coupling (Fig. 14-10).

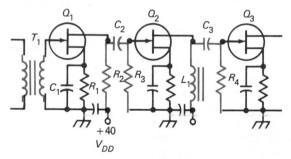

FIG. 14-10 Inductive, resistance, and impedance interstage coupling.

(Direct coupling, discussed later, feeds a *varying dc* to the input of the next stage.)

Either AF ac or AF varying dc in the primary of T_1 feeds AF ac to the gate of Q_1. This is *transformer*, or *inductive, coupling*. Amplified varying dc flowing through R_2 develops an AF varying dc voltage across it, which is coupled to the gate of Q_2 through C_2 as an ac voltage across R_3. This is known as *resistance coupling*. Varying dc from the drain of Q_2 flowing through the choke coil L_1 develops an ac (counter emf and induced emf) across the choke, which is fed through C_3 as an AF ac across R_4 to the gate of Q_3. This is known as *impedance coupling* (generally used to feed power to a high-impedance load, such as a high-impedance loudspeaker, although it can be used as an interstage coupling, particularly in RF stages).

The resistance values of the input circuit resistors, R_3 and R_4, are normally at least twice the resistance of the output load impedances, R_2 and L_1, at the lowest frequency to be amplified.

For Q_1, the load resistance R_2 might be 50 kΩ. The voltage-drop across it should be about half of $+V_{DD}$, or 20 V with no signal input. The current through R_1, Q_1, and R_2 must be $I = V/R$, computed either as 40/100,000 or 20/50,000, or 0.4 mA. If the Q_1 μ is 50, the bias voltage should be (from Sec. 14-4) $-V = 0.6V_D/\mu$, or 0.6(20)/50, or -0.24 V. R_1 can be determined by Ohm's law, $R = V/I$, or 0.24/0.0004, or 600 Ω. Capacitance C_1 would be approximately (from Sec. 14-4) $C_S = 1,600,000/200(60)$, or 13 μF. R_3 should be twice 50 kΩ, or 100 kΩ. Coupling capacitance C_2 can be determined in the same way as C_1, or $C_c = 1,600,000/200(100,000)$, or 0.08 μF. A little more capacitance will not be detrimental to circuit operation and may increase low-frequency response, but too much coupling capacitance may cause oscillation.

If L_1 has 10 H of inductance, for a low frequency of 100 Hz the inductive reactance (essentially the impedance if minimal resistance is assumed) would be $X_L = 2\pi fL$, or 6.28(100)(10), or 6280 Ω. R_4 might be 12,500 Ω or more.

Since T_1 is operating into a class A input (no current flowing), it may be advantageous to load the secondary with a 10–100 kΩ resistor across it for flatter response and more stable operation.

Inductive coupling can multiply the input signal to the stage it feeds by its turns ratio. Resistance coupling can deliver peak ac voltages to the next stage at not quite half the $+V_{DD}$ value. Impedance coupling can produce counter emf and induced emf p-p voltages nearly equal to the $+V_{DD}$ value.

▌Test your understanding; answer these checkup questions.

1. What is the fundamental advantage of parallel over single-ended operation? _____
2. List three advantages of push-pull over parallel operation. _____ _____ _____
3. What class of operation utilizes over twice the characteristic curve length of class A_1? _____
4. What class of amplifier will produce no variation of the pointer of an ammeter in its output circuit when signal is applied? _____
5. What is the operating angle of an amplifier biased to class A? _____ Class AB? _____ Class B? _____ Class C? _____
6. What three classes of AF amplifiers may be driven by a phase inverter? _____ _____ _____
7. What are two names for the phase inverter circuit shown? _____ _____
8. Why may common earphones lose sensitivity if dropped? _____
9. What would be the output impedance of a cathode follower if $\mu = 20$, $R_p = 10$ kΩ, and $R_k = 1$ kΩ? _____ The stage gain? _____
10. What type of earphones have very high impedance values? _____
11. An earphone magnet becomes completely demagnetized. What is heard if it is fed a 500-Hz signal? _____
12. What are the three significant parts of a p-m speaker? _____ _____ _____
13. What is the impedance of p-m speakers? _____
14. Name the three ac interstage coupling circuits. _____ _____ _____
15. What two coupling methods might pick up hum voltages? _____ _____
16. In Fig. 14-10, what would be the coupling capacitance value of C_2 if R_3 is 100 kΩ and the lowest frequency is 100 Hz? _____
17. Draw diagrams of push-pull class A JFETs. Push-pull class B JFETs. The three forms of ac coupling circuits.

14-11 MATCHING IMPEDANCES WITH A TRANSFORMER

When the maximum power output is desired from an ac source, the load impedance must equal the source impedance. While this goal cannot always be attained in practice, it can be approached. If a 4-Ω loudspeaker is connected directly into a 4000-Ω-output circuit, such a mismatch will produce very little signal output.

Transformers can step up or step down voltages, depending on the turns ratio of the primary and the secondary. A 3:1 ratio transformer either steps up or steps down any applied ac voltage 3 times.

Transformers can also be used to convert from one impedance value to another. As might be expected, a 1:1 ratio transformer will have the same impedance in the secondary as in the primary. When the turns ratio is anything other than unity, the *impedance ratio* of primary and secondary will be equal to the *turns ratio squared*:

$$\left(\frac{T_p}{T_s}\right)^2 = \frac{Z_p}{Z_s} \quad \text{or} \quad \frac{T_p}{T_s} = \sqrt{\frac{Z_p}{Z_s}}$$

where T_p = number of primary turns
T_s = number of secondary turns
Z_p = primary impedance
Z_s = secondary impedance

The turns ratio (T_p/T_s) of a transformer to match a source impedance of 500 Ω to a load of 10 Ω is

$$\frac{T_p}{T_s} = \sqrt{\frac{Z_p}{Z_s}} = \sqrt{\frac{500}{10}} = \sqrt{50} = 7.07:1$$

The impedance ratio for a transformer with a 4:1 turns ratio can be found by the same formula:

$$\frac{Z_p}{Z_s} = \left(\frac{T_p}{T_s}\right)^2 = \left(\frac{4}{1}\right)^2 = \frac{16}{1}$$

The answer, 16:1, is the impedance ratio between primary and secondary. If the primary is looking into an impedance of 16,000 Ω, the secondary will appear as a 1000-Ω source for any 1000-Ω load.

To match a 4000-Ω output circuit to a 4-Ω speaker, the transformer turns ratio for maximum power output is

$$\frac{T_p}{T_s} = \sqrt{\frac{Z_p}{Z_s}} = \sqrt{\frac{4000}{4}} = \sqrt{1000} = 31.6:1$$

In AF amplifiers maximum power output produces some distortion. It is usually better for the load not to match but to have an impedance at least twice the output impedance of the device.

14-12 COMPLEMENTARY SYMMETRY

The use of an AF transformer to couple the output circuit of an amplifying device to a loudspeaker is costly, adds greatly to the weight of the equipment, takes up considerable space, and can add to the distortion of signals. By using the low impedance of two matched BJTs it is possible to drive 8- or 16-Ω speakers directly, using a complementary-symmetry circuit somewhat like the basic circuit

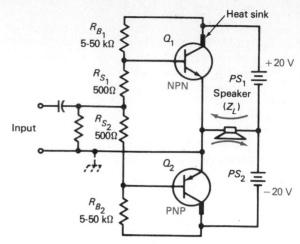

FIG. 14-11 Complementary-symmetry amplifier requiring two power supplies.

shown in Fig. 14-11. The NPN and PNP power transistors are selected to have identical characteristics, R_{B_1} and R_{S_1} bias Q_1, and R_{B_2} and R_{S_2} bias Q_2. If both biases are equal, similar I_C values try to flow through L, resulting in zero current in the speaker load. When a positive half cycle of ac is applied to the input, I_C increases in the NPN but decreases in the PNP and current flows to the left in the load. On the negative half cycle, the PNP is forward-biased and current flows to the right through the speaker. An ac input produces an ac in the load. The difficulties with this circuit are the two power supplies and deciding on what point to ground.

Another complementary-symmetry-type amplifier using only one power supply is shown in Fig. 14-12. Again, matched NPN and PNP power transistors are used. If properly biased, the midpoint in the circuit, M, will be at +20 V. Capacitor C charges to this potential through the speaker. A positive-going signal to Q_1, as indicated, decreases the EC resistance of Q_1, increasing the forward bias on the PNP. This decreases the PNP EC resistance also, discharging C downward through the load. As the collector of Q_1 becomes less positive, the base of Q_2 is made more negative, tending to cut off Q_2. Point M can decrease almost to zero if the positive input signal is strong enough. When the input signal reverses polarity, Q_3 is driven toward cutoff and Q_2 is forward-biased. Point M increases in potential and capacitor C charges toward +40 V, pulling current up through the load.

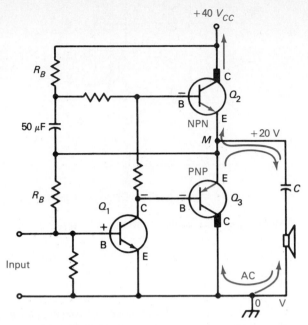

FIG. 14-12 Complementary-symmetry amplifier using only one power supply.

14-13 TYPES OF DISTORTION

When the waveshape of the output signal from an amplifier varies in any respect, other than in amplitude, from the waveshape of the signal fed into the amplifier, the amplifier is distorting the signal. There are four basic forms of distortion:

1. *Amplitude*. If a 3-V sine-wave signal fed into an amplifier produces a 30-V output signal, but a 6-V input produces only a 40-V output, the amplifier is not amplifying all signal amplitudes properly. This is amplitude distortion. The output waveshape of the 6-V input signal must be flattened on the peaks. This same flattening of the peaks can be produced by adding to the fundamental signal just the correct number and amplitudes of harmonics of the signal frequency. Since the amplifier is emitting the fundamental plus all the harmonics, amplitude distortion is also known as *harmonic distortion*. It is usually caused by overdriving the active device or working on a nonlinear part of its curve.
2. *Frequency*. If a 3-V 100-Hz signal is fed to an amplifier and produces a 30-V output voltage but a 3-V 5000-Hz input produces only a 20-V output, the amplifier is not amplifying all frequen-

cies equally well. This is known as *frequency distortion*. In this case, if a curve of frequency-versus-output amplitude is drawn, it will vary up and down and is said to be nonlinear (not a straight line), or not *flat*. A flat amplifier will amplify all frequencies equally well.

3. *Phase*. If a 500-Hz signal shifts phase 180° in passing through an amplifier but a 5000-Hz signal shifts 175°, the amplifier is producing *phase-shift distortion*. This differing amount of phase shift for different frequencies is caused by inductances and capacitances in the amplifier circuit working in conjunction with resistances. Phase distortion is not bothersome in AF amplifiers since the ear does not distinguish such distortion. In circuits amplifying TV, video, square, or complex waves, phase distortion produces very undesirable effects.
4. *Intermodulation*. When two different-frequency signals are added in a nonlinear or overdriven circuit, they intermix or intermodulate, producing sum and difference frequencies of the original two frequencies. These products appear as dissonant high- or low-frequency sounds or noises in AF amplifiers.

Distortion is usually related to the bias-vs.-output current curve. If the bias used produces operation over a nonlinear part of the curve, or results in operation too close to either the cutoff point or the other end of the curve, the output current will be distorted.

Some possible causes of distortion in AF amplifiers are:

1. Overdriving an input circuit
2. Leaking coupling or bypass capacitor
3. Open or shorted resistor or capacitor
4. Improper bias value
5. Improper matching of load impedance
6. Undesired interstage coupling of signals
7. Transformer core saturation
8. Poorly regulated power supply
9. Insufficient filter in power supply
10. Malfunctioning active device

14-14 INVERSE FEEDBACK

Signals fed into audio amplifiers are usually distorted somewhat as they pass through each stage, particularly when audio transformers are used. It is possible to decrease the distortion produced in a stage by feeding a small part of the output signal back into the input circuit 180° out of phase. This may be called *out-of-phase, inverse, degenerative,* or *negative feedback*. Any variations of the signal waveform produced in the stage and fed back out of phase will be materially reduced when reamplified out of phase. This results in a less distorted but somewhat weaker output signal. Distortion already in an input signal cannot be corrected by inverse feedback, of course.

A JFET circuit with *inverse-voltage feedback* is shown in Fig. 14-13*a*. The feedback circuit is shown

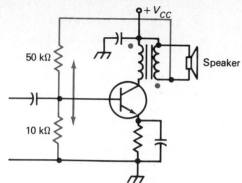

FIG. 14-14 Inverse-voltage feedback from speaker output.

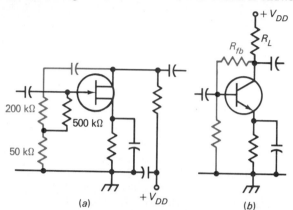

FIG. 14-13 Inverse-voltage feedback circuits (*a*) with a JFET and (*b*) with a BJT.

in color. A signal applied to the input is amplified and appears across the output load. Since one end of the load is bypassed to ground, all of the output signal appears between drain and ground. Connected across this output signal is a dc blocking capacitor and the voltage-divider network made up of the 200- and 50-kΩ resistors. One-fifth of the output-signal ac will appear across the 50-kΩ resistor, so 20% of the output is being developed in series with the gate circuit. Since the output-voltage signal of this form of amplifier stage is always 180° out of phase with the input signal, the part of the output signal across the 50-kΩ resistor represents an out-of-phase voltage being added to the input circuit.

The BJT circuit in Fig. 14-13*b* uses a high resistance R_{fb} as both feedback element and biasing resistor. If connected to the top of R_L, no feedback would result.

Another inverse-voltage feedback circuit takes the feedback voltage from the secondary of the output transformer (Fig. 14-14). If the voltage fed back to the grid is found to be regenerative and raises the gain, or causes the stage to oscillate, it can be made degenerative by reversing either the primary or the secondary connections of the output trans-

former. The dots indicate points of similar phased ac voltages.

Inverse-current feedback utilizes the inverse voltage developed across an unbypassed source- or cathode-bias resistor, or emitter stabilizing resistor, to provide degeneration and reduction of distortion. The BJT stage (Fig. 14-15) has no bypassing of the

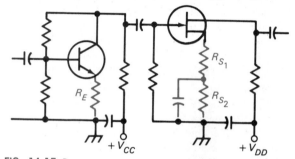

FIG. 14-15 Inverse-current feedback circuits.

emitter resistor, which may produce excessive inverse feedback. The JFET stage has only part of its source-biasing resistor bypassed, reducing the degree of inverse feedback. If the bypass capacitor is across the whole resistor, there is considered to be no inverse feedback. However, at very low frequencies the reactance of the capacitor may be so high that it is not effective, so lower frequencies may be progressively reduced in amplitude.

The voltage gain of an amplifier at the mid-frequency of its response curve with inverse feedback can be determined by

$$A_f = \frac{A}{-\beta A + 1}$$

where A_f = amplification with feedback
A = amplification without feedback
$-\beta$ = decimal fraction of output voltage fed back to input

As an example, an amplifier has a gain of 20. If 1 V is fed to its input, 20 V should appear across the output load. If a $-\beta$ of 0.05 of the output is fed back into the input circuit, the total gain of the amplifier is

$$A_f = \frac{20}{0.05(20) + 1} = \frac{20}{2} = \text{gain of 10}$$

Without feedback this amplifier might have dropped to half-voltage output at some lower frequency. With this value of feedback, at the same frequency the output may drop only 10%. Thus, inverse feedback widens the bandwidth of an amplifier. Also, instead of developing a 45° phase shift at any original half-voltage frequency, the resultant phase shift will be about 27°. Therefore, inverse feedback reduces phase shift in an amplifier.

Voltage feedback increases the input impedance of a stage and reduces the output impedance. Conversely, current feedback reduces input and increases output impedances.

If the gain of an amplifier without feedback is 10 and with feedback is 8, the feedback factor β can be determined by rearranging the formula given above to

$$\beta = \frac{1}{A_f} - \frac{1}{A} = \frac{1}{8} - \frac{1}{10} = \frac{1}{40} = 0.025$$

It is possible to use a feedback loop that includes several stages. This decreases distortion developed in all stages. However, with transformer *interstage* coupling, a phase reversal may occur at some frequency in the audio range, and increased distortion or oscillation due to regeneration at that one frequency may result.

14-15 CONTROLLING VOLUME

The usual method of controlling volume in JFET and VT amplifiers is to use a voltage divider (potentiometer) in the input circuit (Fig. 14-16a). In resistance-coupled circuits the potentiometer resistance should be twice the load resistance of the device ahead. When the sliding contact is at the top of the potentiometer, the gate (or grid) receives full output voltage from the first stage. When halfway down, it receives half voltage, and at the bottom, no signal.

The potentiometer across the secondary of an interstage coupling transformer should have a resistance high enough to reflect the proper impedance back to the primary. It may range from 1 to 500 kΩ.

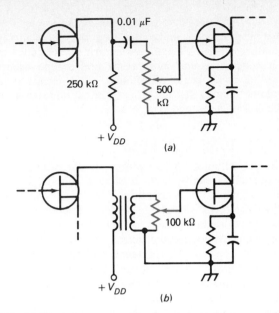

(a)

(b)

FIG. 14-16 Volume controls for (a) resistance-coupled stages and (b) transformer-coupled stages.

In BJT AF circuits the base connection to the biasing resistors must not be varied or the bias value will change. For this reason the volume-control potentiometer is placed in the collector circuit ahead (Fig. 14-17).

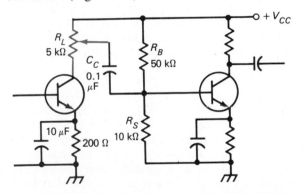

FIG. 14-17 Volume control used in BJT AF amplifiers.

Potentiometers have different resistance tapers. A *linear-taper* potentiometer has the same resistance change for a given rotation angle of the arm regardless of whether the arm is at the lower end, the middle, or the top end. When starting from zero signal and increasing the control slightly, the audio signal will rapidly increase at first. Above the midpoint there seems to be relatively little increase in volume. To overcome this undesirable feature, potentiometers are available with logarithmic *audio tapers*. A small angle of rotation at the low end of the potentiometer moves the arm over a relatively small resistance change. As the arm is continued in rotation, more and more resistance change occurs for the same angle of rotation. This results in a volume control that has a less critical adjustment at the low-volume end.

14-16 TONE CONTROLS

Theoretically, an AF amplifier should amplify all frequencies equally well. It should amplify 50 Hz just as much as it amplifies 500 or 15,000 Hz. When correctly designed, modern amplifiers actually have such a linear or flat response. If any capacitance is connected across the input or output circuits, however, its reactance will form a path to ground for the signal voltage. Since the reactance of a capacitor varies inversely as the frequency, the same capacitor will bypass the high frequencies to ground more than it will the low frequencies.

The higher the impedance of the circuit across which a capacitor is placed, the more effect the capacitor will have on the loss of high frequencies. For example, a 0.0002-μF capacitor across a 100,000-Ω circuit will effectively decrease the higher audio frequencies. It takes 0.001 μF to produce the same effect across a 20,000-Ω circuit.

An amplifier in which the high frequencies are attenuated will sound to the listener as though the low frequencies are being highly amplified. Many people prefer such an increased *bass* (low-frequency) response.

Random noises in amplifiers originate from thermal agitation of electrons in the first-amplifier-stage. These and other noises contain more energy at frequencies above 5 kHz. A capacitor across an amplifier output can reduce such noises.

While a variable capacitor across the input circuit would effectively decrease and control *treble* (high-frequency) response, it is not very practical. Possible treble-attenuating tone controls are shown in Fig. 14-18. With maximum resistance in series with the capacitor, there is practically no capacitive effect. With zero resistance, the capacitor is directly across the circuit, attenuating the high frequencies. Since the output circuit impedances are much less than the input, capacitance values in this circuit must be greater. BJT amplifiers require even larger capacitance values because of their much lower impedance values.

The dual tone controls of Fig. 14-19 are capable of boosting or attenuating either the bass or the tre-

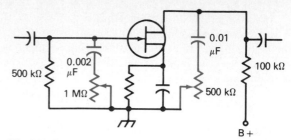

FIG. 14-18 Shunt-capacitance tone controls for JFETs or VTs. (Multiply C and divide R values by about 20 for BJTs.)

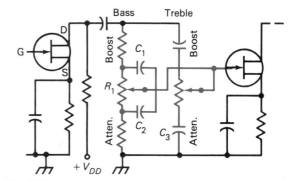

FIG. 14-19 A bass-treble tone-control unit.

ble to the gate of the second JFET. When the arm of the bass control is at the bottom of R_1 (attenuate), high frequencies are coupled to the gate by C_1. With the arm up (bass boost), C_2 shunts the high frequencies. When the treble-control arm is down, the gate is shunted by C_3, which attenuates the high frequencies at the gate. With the arm up, C_3 is not as effective, and maximum high frequencies are coupled to the gate.

Low-pass, high-pass, or bandpass *LC* audio filters are often used in amplifiers. They are not adjustable but have sharp cutoff characteristics. "Active" filters may also be used (Sec. 14-21).

14-17 MILLER EFFECT

A high-gain resistance-coupled JFET amplifier may have a poor high-frequency response. This may be due to *Miller effect*, an electronic capacitance that develops across the input circuit of grounded-source-type circuits. The value of this capacitive effect is approximately

$$C_m = C_{GS} + (A)C_{GD}$$

where C_m = Miller-effect capacitance, pF
$\quad\quad C_{GS}$ = gate-source capacitance, pF

A = amplification of stage

C_{GD} = gate-drain capacitance, pF

As an example of the effective capacitance across the input circuit of a JFET in which C_{GS} is 4.2 pF, A is 12, and C_{GD} is 3.8 pF,

$$C_m = C_{GS} + (A)C_{GD}$$
$$= 4.2 + (12)3.8 = 49.8 \text{ pF}$$

The JFET itself may have only 4.2 pF C_{GS}, but it may operate as though more than 50 pF is across the input circuit and is bypassing high-frequency signals to ground.

Miller effect varies with the stage gain. High-gain stages produce high effective capacitive effects. Inverse feedback (Sec. 14-14) decreases Miller effect.

Since the gain of transistors is so very dependent on voltages, currents, etc., Miller effect can become a problem, particularly in RF circuits which must be held to a certain frequency.

14-18 DIRECT COUPLING

The coupling capacitors in resistance and impedance coupling and transformers in inductive coupling may not pass frequencies below 50 Hz too well. To improve low-frequency response a direct-coupled *Loftin-White*-type circuit may be used (Fig. 14-20). Current flowing out of the negative terminal

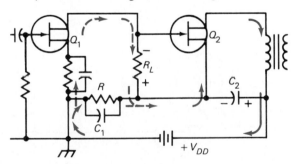

FIG. 14-20 Loftin-White-type amplifier.

of the power supply divides, the major portion flowing through resistor R and a little flowing through voltage amplifier Q_1 and resistor R_L. These two currents join and as one current flow through power amplifier Q_2, through the primary of the output transformer, and to the power supply.

The I_D of Q_1, flowing through R_L, produces a voltage-drop across it. Since the negative end of the resistor is at the top and the positive is at the bottom, this potential acts as bias for Q_2.

When a signal voltage is fed to the gate of Q_1, its I_D varies, producing a varying potential across R_L. This varying dc potential is directly coupled to the gate-source circuit of Q_2, causing the I_D of Q_2 to vary. Varying current in the primary of the output transformer produces ac output power in the secondary.

C_1 must be large enough to hold the voltage across R constant. C_2 must be large enough to hold the voltage constant from the source of Q_2 to $+V_{DD}$. Effectively, there are two power supplies in series. The one actual power supply must supply enough voltage for the two stages in series. With two separate power supplies, or by using zener diodes in place of C_2 and R, this circuit is capable of coupling between the two stages with little distortion over the audio spectrum and down to dc. Thus, a *dc amplifier* may be either a direct-coupled one, or may be used to amplify relatively slow dc variations.

▌ Test your understanding; answer these checkup questions.

1. In what value resistor will maximum power be dissipated if the source is 38 Ω and a 5:1 step-down coupling transformer is used? _____
2. What transformer ratio matches a 600-Ω AF line to a 16-Ω speaker? _____
3. What is the main advantage of using complementary symmetry in an AF amplifier? _____
4. What is another name for amplitude distortion? _____
5. A nonlinear response curve indicates what type of distortion? _____
6. What type of distortion can be caused by internal L, C, and R in an amplifier's circuit? _____
7. What distortion results from mixing two frequencies in a nonlinear circuit? _____
8. List 10 possible causes of distortion. _____
9. What are the names of the two degenerative feedback systems? _____ _____
10. What are three advantages of negative feedback? _____ _____ _____
11. What effect does inverse voltage feedback have on the input circuit impedance? _____
12. What is the gain of a stage having a normal gain of 40 if 10% negative feedback is added? _____
13. Why is a logarithmic taper pot better for an AF gain control? _____
14. Where should the pots be set for hi-fi if a bass-treble tone-control circuit is used? _____
15. Why would Miller effect be greater when the plate-load resistance value is high? _____

16. If power-supply voltage is increased, would Miller effect be increased? _____
17. What is a Loftin-White circuit? _____ Its main advantage? _____
18. Draw diagrams of a complementary-symmetry amplifier feeding a loudspeaker. Inverse-voltage feedback. Inverse-current feedback. A bass-treble tone control. A Loftin-White amplifier.

14-19 TYPICAL TRANSISTOR AMPLIFIERS

The three basic types of BJT amplifier circuits—common-emitter, common-base, and common-collector—were described in Secs. 9-6, 9-9, and 9-10. These basic circuits are often combined to produce some desired result.

There are many electronic circuits that could have used VTs but never did because of the difficulty of heating the cathodes. The two-stage amplifier in Fig. 14-21 is an example. In this circuit an

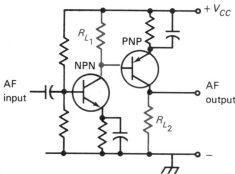

FIG. 14-21 One form of direct-coupled two-stage amplifier.

NPN and a PNP amplifier are direct-coupled. Using an above-ground-positive power supply, the PNP has to be connected "upside down" to allow I_C to flow through it. Current variations in R_{L_1} produce amplified variations in R_{L_2}.

Another dc amplifier is shown in Fig. 14-22. BJT Q_2 receives its forward bias through Q_1. A positive-

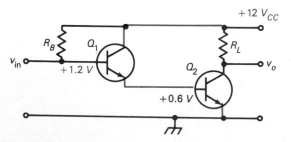

FIG. 14-22 Another form of transistor dc amplifier.

going input signal on the base of Q_1 decreases its EC resistance in proportion to the beta of Q_1. This increases the forward bias on Q_2, and it amplifies the input change still further. The total gain is approximately $\beta\beta R_L$.

An extensively used amplifier is the *Darlington pair* (Fig. 14-23). It is similar to the amplifier in

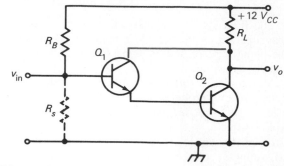

FIG. 14-23 Darlington pair amplifier. R_s may be used for greater stability.

Fig 14-22 except that the collector lead of Q_1 is connected to the opposite end of R_L. This introduces degeneration, and the circuit is thermally more stable as the devices warm during operation.

A *differential amplifier* (Fig. 14-24) responds to the difference between two signals. Assume a gain

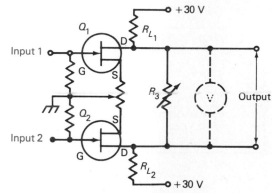

FIG. 14-24 Basic differential amplifier using JFETs.

of 10 for each JFET. If input 1 goes +1 V and input 2 goes +1 V also, the currents in both R_{L_1} and R_{L_2} increase like amounts and there is no voltage difference across the output terminals. If input 1 goes +1 V, but input 2 goes to only +0.5 V, the current in R_{L_2} increases only half as much as that in R_{L_1}, and the difference of the two voltage-drops is 10 − 5, or 5 V. The differential amplifier amplifies the 0.5-V difference to 5 V. The potentiometer is used to balance the two circuits to zero dc output voltage with no input voltage. The gain of the amplifier is controlled by R_3. With zero resistance the output must be zero; with high resistance the gain is high. For better thermal stability differential amplifiers may use some form of inverse feedback. Any signal fed in phase to both inputs is known as a *common-mode signal*, and is canceled in the output. If the internal circuits are not perfectly balanced, there will be some output dc voltage, known as *offset*.

A series-connected low-noise amplifier pair that has its first device react to input voltages and its second device react to current variations in the first is called a *cascode* amplifier (Fig. 14-25a). The Q_2 gate is held at a fixed positive value by the zener diode, D, with its source at a higher positive potential (biasing the gate negative). If a positive-going signal is fed to the Q_1 gate, current increases

through Q_1, R_{L_1}, Q_2, R_{L_2}, and the power supply. The positive-going Q_1 gate pulls its drain voltage down to a less positive value. If the source of Q_2 is more negative, its fixed potential gate must now be relatively more positive, further increasing the current flow through R_{L_2}. Being unbypassed, R_{L_1} introduces degeneration into the circuit, thermally stabilizing the whole stage. The amplitude stability and bandwidth are good up to UHF with JFETs and BJTs and to at least 400 MHz with ICs.

The balanced cascode circuit of Fig. 14-25b has better thermal stability if it is assumed that the devices on both sides of the circuit are similar. If the base of Q_2 is grounded, the amplifier can be used as either a balanced single-ended cascode amplifier or a cascode phase inverter if both outputs are used. When voltages are applied to the bases of both Q_1 and Q_2, the circuit operates as either a differential or a push-pull cascode amplifier. The zener diode (D) holds the bases of Q_3 and Q_4 at a constant potential. The dashed rheostat would act as a gain control. Potentiometer R_1 is used to balance the circuit.

The three diagrams in Fig. 14-26 illustrate three solid-state *cascade* amplifiers (one following the other). The first is *RC*-coupled; the others are direct-coupled. All show 180° feedback circuits. The voltage-drops across the diodes assure proper bias values.

14-20 OPERATIONAL AMPLIFIERS

One of the most versatile of all electronic circuits is the operational amplifier, or *op amp*. It is a two-input amplifier with very high gain (A_v may exceed 10^6). One input amplifies its signal, producing an inverted output (out of phase). The other input will amplify equally well, but its output is in phase with its input. Operating open loop (no feedback) the inputs have a very high impedance to ground (0.5–1000 MΩ), and the output is a very low impedance to ground, nearly the optimum characteristics of an amplifier. In many applications op amps require a positive and a negative power supply, although the negative terminal can be grounded and a single supply can be used in some cases.

A simplified op amp circuit is shown in Fig. 14-27a. If it has an A_v of 10^6, a 1-μV ac signal should be able to develop a 1-V output with perhaps 3% distortion. The feedback impedance, Z_f, usually holds the gain to less than 100. This would reduce distortion to less than 0.0003%, which is negligible. Because of the large feedback factor, the ac

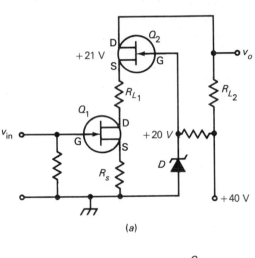

(a)

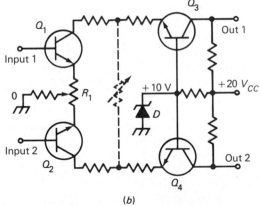

(b)

FIG. 14-25 Cascode amplifiers. (a) Single-ended with JFETs. (b) Balanced or push-pull form with BJTs.

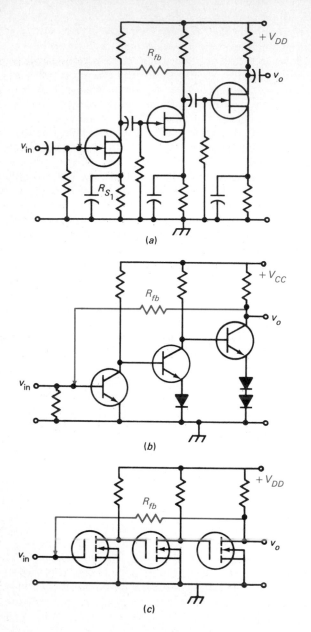

FIG. 14-26 Three-stage amplifiers with feedback. (*a*) *RC*-coupled JFETs. (*b*) Direct-coupled BJTs. (*c*) Direct-coupled enhancement MOSFETs.

1. (1.52) **2.** (6.12:1) **3.** (No transformer) **4.** (Harmonic) **5.** (Frequency) **6.** (Phase) **7.** (Intermodulation) **8.** (See text) **9.** (*V*)(*I*) **10.** (Reduce distortion)(Wider BW)(Reduce phase shift) **11.** (Increase) **12.** (8) **13.** (Spreads lower values) **14.** (Midpoints usually) **15.** (Greater amplification) **16.** (Yes, more gain) **17.** (DC amplifier)(Amplifies low frequencies) **18.** (See chapter illustrations)

voltage-drop and impedance from the *summing junction*, SJ, and ground is so small that it can be neglected. Any input signal will produce its voltage across Z_i, and the output signal will be developed between the output terminal and SJ, since it is essentially at ground potential. The gain of the amplifier will be V_{out}/V_{in}, or Z_f/Z_i. Thus, if both Z_i and Z_f are 1000 Ω, the gain of the amplifier is forced to be 1, or unity; if Z_i is 1 kΩ and Z_f is 10 kΩ, the gain is 10; if Z_i is 1 kΩ and Z_f is a 100-Ω resistor plus a 100-kΩ rheostat in series, an amplifier with a variable gain from 0.1 to 100 results. The peak output voltage is limited to about 50% of the V_{CC} value.

To mention a few of the many uses of op amps:

1. If two separate input impedances are fed to the SJ, the output voltage will be the *algebraic sum* of the two inputs, and the circuits can be used in an *analog* (nondigital) *computer*.
2. If Z_f is a capacitor and Z_i is a resistor, the circuit acts as a precision integrator, developing a linearly increasing ramp voltage.
3. If Z_f is a resistor and Z_i is a capacitor, the circuit can be used as a precision *differentiator* to develop sharp voltage peaks from square-wave input signals.
4. If Z_i is a resistor and Z_f is a diode, the circuit becomes a logarithmic amplifier (one that amplifies according to a logarithmic curve rather than a linear curve).

Simpler op amps usually consist of a differential input amplifier feeding a high-gain direct-coupled amplifier, followed by a complementary-symmetry single-ended output amplifier. Either or both of the differential amplifier inputs (the +, or "noninverting"; or the −, or "inverting") can be used to amplify a signal. If two signals are fed to the two inputs, the difference between the two will be the amplified output. Op amps are manufactured in IC form and are made up in a variety of packaging forms. The symbol of an op amp is shown in Fig. 14-27*b*. If the triangular amplifier symbol has only one input shown, it may or may not be an op amp.

Improved op amps have JFET or MOSFET input stages to reduce noise, increase input impedance, and widen bandwidth materially.

Three basic forms of op amp circuits are shown in Fig. 14-28. In Fig. 14-28*a* the input voltage is applied to the noninverting input and the gain is determined by the ratio of Z_i/Z_f. In Fig. 14-28*b* the signal is fed to the inverting input and the gain is determined by the feedback ratio of Z_f/Z_i. In Fig.

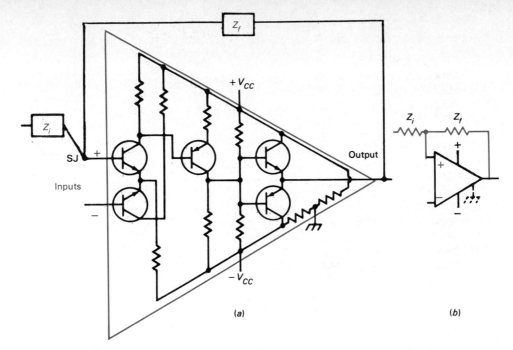

(a)

(b)

FIG. 14-27 (a) Basic op amp circuitry with feedback. (b) Symbol.

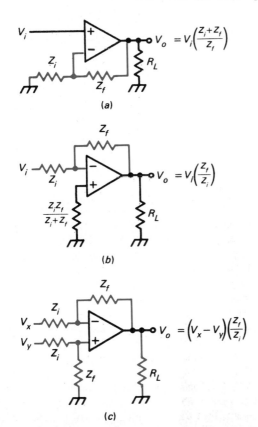

(a)

$$V_o = V_i\left(\frac{Z_i + Z_f}{Z_f}\right)$$

(b)

$$V_o = V_i\left(\frac{Z_f}{Z_i}\right)$$

(c)

$$V_o = (V_x - V_y)\left(\frac{Z_f}{Z_i}\right)$$

FIG. 14-28 Basic op amp circuits with feedback.

14-28c the op amp is being used as a differential amplifier and amplifies the difference between the two input voltages.

Besides the circuits mentioned, op amps are useful in many other specialized circuits, such as D/A converters, Schmitt triggers, etc.

Low-power IC op amps may have an output in the 1.5-mW range, with a total consumption of perhaps 300 to 500 mW. Other power-type op amps may provide 5 to 10 W of reasonably undistorted output power at a low enough impedance to drive a loudspeaker directly.

Without any feedback, an op amp may have a very high gain, perhaps 100 dB, at low frequencies. Because of internal capacitances its gain drops off so that at perhaps 1 MHz it has only unity gain. If a little feedback is added, the gain may be reduced to a maximum of perhaps 40 dB, but the *closed-loop* (with feedback) gain may now be flat to about 10 kHz before it begins to drop off. If the feedback is increased to provide only 20 dB of closed-loop gain, the amplifier may be flat to perhaps 60 kHz. By decreasing the gain to 10 dB the amplifier may be flat to perhaps 500 kHz. However, an op amp with a higher *slew rate* (output voltage change per microsecond) can provide more amplification than this and it can provide it over a wider bandwidth.

The output impedance of an op amp is very low, approaching 0 Ω, particularly with feedback being used. A common load resistor value to use might be about 2000 Ω.

14-21 AN ACTIVE BANDPASS FILTER

Among its many applications, an op amp can be used to produce bandpass or other filters without using any inductors, only resistors, capacitors, and the op amp. In Fig. 14-29, together, R_1 and C_1

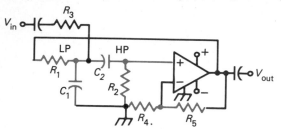

FIG. 14-29 A simple active bandpass AF filter.

form a basic RC low-pass filter. Together, R_2 and C_2 form a basic RC high-pass filter. Two such circuits in cascade, if overlapping in frequency, will form a bandpass filter. The active Q of the filter will be the ratio of the center frequency f_c in Hz, to the bandwidth BW in Hz. If the center frequency is to be 750 Hz and the desired BW is chosen to be 250 Hz, the Q will be 750/250, or 3. The ratio of R_5/R_4 is found by $Q - \sqrt{2}/Q$. A convenient value for C_1 and C_2 in AF filters should be chosen in the range of 0.001 and 0.1 μF. Thus,

f_c = center frequency, in kHz
$C_1 = C_2$, in μF
$R_1 = R_2 = R_3$, in kΩ

Choose

$$C_1 = C_2 = 0.01 \ \mu\text{F}$$

Determine

$R_{1, 2, 3}$, in kΩ by $\quad \dfrac{\sqrt{2}}{2\pi f_c C}$

Compute

$$Q = \frac{f_c}{\text{BW}}$$

Determine

$\dfrac{R_5}{R_4}$ ratio by $\quad \dfrac{Q - \sqrt{2}}{Q}$

For the filter above, the component values would be $C_1 = C_2 = 0.01 \ \mu$F and $R_1 = R_2 = R_3 =$ 29.7 kΩ. The R_5/R_4 ratio is 2.53, so if R_4 is selected to be 10 kΩ, then R_5 would be 25.3 kΩ. The

input and output capacitors are not critical and might be something between 0.02 and 0.1 μF.

A single-section, or "pole" (an op amp and its components), filter such as this does not have steep drop-off of its skirts, particularly to the high-frequency side. By adding a second pole in cascade the drop-off will be twice as steep. Practical filters may use as many as three or four poles.

Besides the bandpass filter explained, active RC filters can also be developed to have only high-pass or only low-pass characteristics. Because common op amps drop off in gain at higher frequencies, the RC active filters are not practical for frequencies over a few kilohertz unless high-slew rate op amps are used. By using an inductance in parallel with a circuit capacitor, a practical active bandpass LC filter can be developed for use up to 1 MHz or more.

14-22 DECOUPLING AF STAGES

The phase of any signal passing through a common-source-type amplifier stage is changed 180°. As the gate of a first JFET in an amplifier becomes less negative, the gate of the next becomes more negative. Follow the step-by-step action in Fig. 14-30:

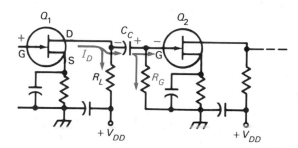

FIG. 14-30 As the first stage gate is driven positive the second stage is driven negative.

1. Q_1 gate goes less negative (more positive).
2. I_D increases.
3. The voltage-drop across R_L ($V_{R_L} = I_D R_D$) increases because I_D increased.
4. The top of R_L is now closer to ground potential (more negative than previously).
5. The drain side of C_c is now more negative.
6. When the left side of C_c becomes more negative, electrons are driven out of the right side onto the gate of Q_2, and some discharge slowly to ground through the high resistance of R_G.
7. Electrons now on Q_2 gate charge it negatively, or 180° from Q_1.

The drain currents of two JFETs in cascade will also be 180° out of phase. As the Q_1 current increases, the Q_2 decreases. Any coupling, either capacitive or inductive, between the gates of two adjacent stages, or between the drains of two adjacent stages, will tend to cancel each other. This inverse or *degenerative* feedback (Sec. 14-14) decreases the overall amplification of the cascade amplifier but also decreases any tendency for the stages to oscillate.

Feedback from the third- to the first-stage output circuits in a cascade amplifier is twice 180°, or 360°, and is in phase, or *regenerative*. The stages may form an oscillatory circuit. Regenerative feedback is rarely desirable in audio amplifiers, since it increases noise in the circuit, increases distortion, may cause erratic operation of the stages, or may produce oscillations. To decrease this effect, it has been found that decoupling the output load of each of the stages using a common power supply will prevent current variations in one stage from producing voltage variations in other stages.

Figure 14-31 illustrates output circuit decoupling as applied to a BJT amplifier. A similar *RC* circuit is used with JFET and VT amplifiers. When the base voltage varies up and down because of an input signal, the collector current through both R_L and decoupling resistor R_d would be expected to vary up and down. However, with bypass-filter capacitor C_{bp} having a relatively large value, the $R_d C_{bp}$ combination acts like an *RC* low-pass power-supply filter to hold the voltage across C_{bp} constant. As a result there is essentially no changing load on

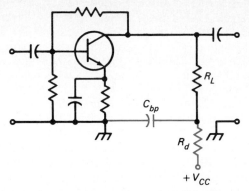

FIG. 14-31 Output circuit decoupling in a BJT amplifier.

the power supply and therefore little chance of varying its output voltage, which might affect other stages in a regenerative or degenerative way. The value of R_d may be from 50 to 1000 Ω, depending on how much voltage-drop can be tolerated across it. The value of C_{bp} may be from 0.01 to several microfarads, depending on the R_d and I_C values and on the signal frequency.

Interaction between stages in a multistage amplifier can be reduced by using decoupling circuits to prevent feedback via a common power supply, by shielding electrostatically or electromagnetically between stages, and by proper placement of circuit components.

14-23 A GENERAL-PURPOSE TRANSISTOR AMPLIFIER

Figure 14-32 illustrates a possible simple, discrete-device transistor amplifier capable of perhaps 10 W

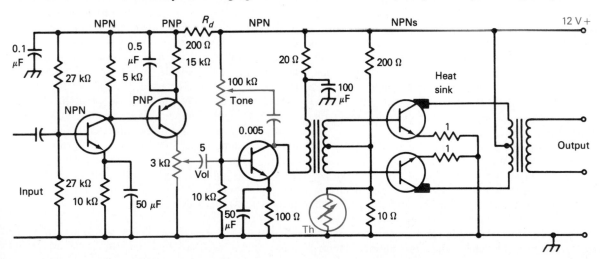

FIG. 14-32 A discrete parts four-stage transistor AF amplifier with two direct-coupled stages, tone and volume controls.

if the power transistors have adequate heat sinks but only about 1 W if they do not.

The first two stages are directly coupled and use an NPN and a PNP transistor. The resistors in these stages must be chosen so that both transistors are forward-biased to their optimum class A points. A positive-going signal to the NPN base increases the stage I_C. This produces an amplified negative-going (180° out-of-phase) potential at the NPN collector. Since the base of the PNP is directly coupled to the NPN collector, the negative-going potential forward-biases the PNP base, and also increases its I_C. The increase of current through the 3-kΩ resistor develops an amplified positive-going signal at the top of this resistor. A negative-going input signal reduces the currents in both stages.

The third stage is an NPN common-emitter power amplifier that should be capable of at least 0.5 W output to drive the following push-pull stage adequately. The volume control is at the input of the driver stage. This stage also has a simple tone control incorporated in its voltage-divider bias network. The stage has an emitter-stabilization RC circuit to protect it against thermal runaway if I_C flow heats the transistor. The current in this class A stage should be approximately 100 mA.

The fourth stage has push-pull NPNs biased to class AB or B for high power output. Note the low value of emitter-stabilizing resistors to allow maximum emitter-collector current to flow in the transistors. These resistors are so low in value that it is not feasible to bypass them. A dc milliammeter in the center-tap lead of the output transformer should read only a few milliamperes with no signal and perhaps 1500 mA or more with full-power output.

With zero bias on the base of a transistor, I_C should be zero (biased beyond cutoff, or to class C). In the output stage, therefore, some forward bias must be provided to prevent *crossover distortion* when the ac signal switches conduction from one transistor to the other. The 10- and 200-Ω voltage-divider resistors provide this very small forward bias.

When transistors become warm, they become better conductors and their collector current increases. The *thermistor* (Th) is a negative-temperature-coefficient resistor that is physically attached to the transistors. As the temperature of the transistors increases, the thermistor resistance decreases, thereby decreasing the forward bias, reducing collector current, and preventing thermal runaway.

Rather than the simple power output amplifier circuit shown, it would probably be more usual to find some type of complementary-symmetry output amplifier stage, since semiconductor devices are now less expensive than transformers. Furthermore, complete low-power audio amplifiers are now made up into single small ICs and would incorporate more sophisticated stages in them. Only external volume and tone control potentiometers and bypass capacitors are needed to complete such systems.

To develop about 10-W AF output, about 30 W of dc power-supply power would be required.

14-24 A GENERAL-PURPOSE VACUUM-TUBE AMPLIFIER

Figure 14-33 represents a possible four-stage VT amplifier capable of amplifying weak signals from a source, such as a microphone, up to 10 W of undistorted power, or adequate audio power for a 40- by 80-ft room. It illustrates how VT forms of the circuits described might be used.

The first stage is a *contact-potential-biased* (10-MΩ grid resistor) 6AU6 pentode *preamplifier* (stage before the volume control) with a tone control across its output circuit and a 10-μF 20-Ω plate-circuit decoupling network.

The second stage is one-half of a 12AU7 twin-triode amplifier with a volume (gain) control in the input circuit, inverse-current feedback in the cathode circuit, and plate-circuit decoupling. If the dotted circuit is added, the output stage, phase inverter, and speech amplifier are all included in an inverse-voltage feedback loop.

The third stage is a split-load phase inverter with plate-circuit decoupling, using the other half of the dual-triode 12AU7.

The fourth stage is a push-pull 6BQ5 class A power amplifier capable of 10 W of undistorted output. If the V_{bb} is raised to 300 V, the output should exceed 15 W (an increase of about 2 dB).

The resistance-coupled stages should draw about 1 mA of I_p each, the phase inverter about 2 mA, the power amplifiers about 62 mA of I_p and 7 mA of I_{SG} with no signal, rising to a total of about 73 + 17, or 90 mA, under full output. At 6.3 V_f, the 6AU6 and the 12AU7 draw 0.3 A each and the 6BQ5s draw 0.75 A each. Thus the power supply must furnish 250 V dc at 90 mA (22.5 W), 6.3 V ac at 2.1 A for heaters (13.23 W), plus 10 W or more for a rectifier filament, or over 46 W to produce 10 W or less of audio power.

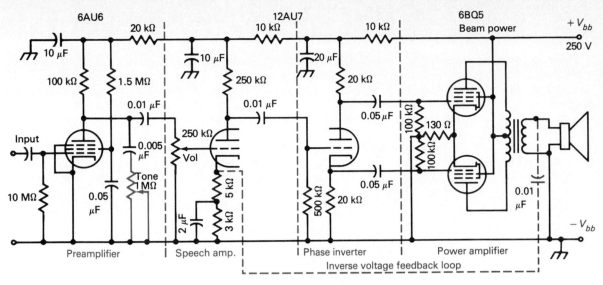

FIG. 14-33 A discrete parts four-stage 10-W AF amplifier using VTs.

Test your understanding; answer these checkup questions.

1. In Fig. 14-21, if the input signal is going more positive, what is the output signal doing? _____

2. In Fig. 14-22, with germanium BJTs, what would be the voltage value at the bottom of R_B? _____ At the bottom of R_L? _____

3. In the Darlington pair, if the input voltage is going more positive, what is the output voltage doing? _____

4. A 0.6-V 1000-Hz and a 0.5-V 1000-Hz ac are fed to the inputs of a 10-gain differential amplifier. What is the output if the signals are in phase? _____ If out of phase? _____

5. Why are the two signals in question 4 not common-mode signals? _____

6. What is a cascade amplifier? _____

7. What is the advantage of a cascade amplifier? _____

8. For how many things can the balanced cascade amplifier be used? _____

9. What is the relative impedance from SJ to ground in an op amp? _____ Of the output to ground? _____

10. How is flat wideband operation produced with op amps? _____

11. What is the output voltage change per microsecond of an op amp called? _____

12. At what frequencies does an op amp have its greatest gain? _____

13. Basically, what two types of amplifiers does an op amp provide? _____ _____

14. A 300–3300 Hz bandpass active filter using $C = 0.02\ \mu F$ should have what resistor values? _____

15. For steep skirts, how many poles should an active bandpass filter have? _____

16. In Fig. 14-32, why is there no bypass across R_s in the p-p stage? _____ What does the thermister do? _____ Why must the power supply have good regulation? _____ What three coupling methods are used? _____ _____ _____

17. In Fig. 14-33, which stage has greatest gain? _____ What class is the output stage? _____ Which stage has current feedback? _____ In which stage is the gain control? _____

18. Draw diagrams of a direct-coupled two-stage BJT amplifier. A Darlington pair amplifier. A JFET differential amplifier. A JFET cascade amplifier. A three-stage dc amplifier. The three basic op amp amplifier circuits. A basic active bandpass filter. The BJT stage with decoupling.

14-25 VT AMPLIFIERS

Amplifiers have been discussed mostly in transistor terms. There are still millions of pieces of vacuum tube equipment working too well to be discarded. Lately, replacement VTs have become very costly. In many cases vacuum rectifiers are being replaced by solid-state power-supply diodes with high inverse-voltage ratings. In some amplifier and oscillator applications JFETs can replace triodes or pentodes with only minor circuit changes and adjustments.

There are some features of high-power vacuum tubes which are completely different from transistors and should be understood when servicing VT equipment. A major difference between VTs and transistors lies in their filament or heater systems. A few VTs use only filaments, usually heated by ac. To reduce 60-Hz hum in amplifiers it is necessary to center-tap the filament transformer winding and use this as the grid and plate return point (Fig. 14-34a). The two sides of the filament are usually

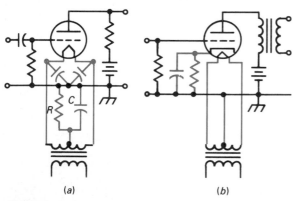

(a)   (b)

FIG. 14-34 (a) Circuit for filament-type VTs. (b) Grounding the center-tap for cathode-type VTs.

bypassed to the grounded point. If used, the R and C shown will provide self-bias in a filament-type tube. An alternative method of returning grid and plate circuits to the filament center is to use a center-tapped resistor of about 50 Ω across the filament and return to this point. Even when a heater-cathode is used, it is often necessary to center-tap the ac heater circuit and ground it to reduce hum to a minimum (Fig. 14-34b).

If the insulation between heater and cathode breaks down and allows ac to flow to the cathode and to ground, a strong 60-Hz hum may be developed across any self-bias resistor being used. Other

causes of hum are drying out of electrolytic power-supply filter capacitors, which results in insufficient filtering and 120-Hz hum (with full-wave rectification). If one rectifier diode in a center-tap full-wave power-supply circuit becomes ineffective, a 60-Hz hum may develop. Audio transformers, chokes, and conductors in an amplifier can have a 60-Hz ac induced in them from stray power-supply fields. To prevent hum voltages from being induced in a metal chassis, all ground connections of each stage should be brought to a common ground point. If filament wires carrying 60-Hz ac are twisted together, their adjacent but opposite magnetic fields tend to cancel each other and induce much less ac emf into nearby wiring.

In VT systems voltage amplifiers are often pentode resistance-coupled stages. Figure 14-35 illus-

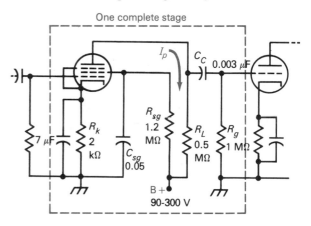

FIG. 14-35 Pentode resistance-coupled amplifier with representative component values.

trates a class A pentode resistance-coupled voltage amplifier. The plate-load resistor R_L may range from 25 kΩ to as much as 1 MΩ. The higher its value, the higher the gain of the stage. (With a 6AU6 tube, 250 kΩ produces a voltage gain of 280, 500 kΩ gives a gain of 360, and 1 MΩ gives a gain of 450.)

The grid resistor R_g is usually twice the value of the R_L it follows. The screen-grid dropping resistor R_{sg} is several times the R_L value. The cathode resistor R_k ranges from 1 to 10 kΩ, depending on V_p and R_L values.

The value of the coupling capacitor C_c depends on the R_L and R_g values. With a 250-kΩ load a capacitor of 0.01 μF may be required. With a 1-MΩ load a 0.002-μF capacitor may suffice. The cathode capacitor ranges from about 10 μF across a 1-kΩ resistor to about 1.5 μF across a 10-kΩ resistor. Screen-grid bypass capacitors are usually between 0.03 and 0.5 μF. Capacitance values given are for reasonably linear amplification down to 100 Hz.

The interstage coupling capacitor C_c of VT resistance-coupled stages is often troublesome. The voltage difference across this capacitor is relatively high. As it ages, it may become *intermittent*, operating correctly part of the time and either shorting or opening during the remainder of the time. The amplifier works normally part of the time and then either distorts or decreases output. If its dielectric breaks down, a positive bias is applied to the grid, which produces distortion. If this capacitor opens, the signal transferred across it drops to almost zero. In transistor amplifiers the coupling capacitor is an electrolytic type and may dry out in time.

A capacitor that often shorts is the screen-grid bypass. If it shorts, V_{sg} drops to zero and I_p ceases. Any resistor in the screen circuit will heat and possibly burn out. If a resistance in an amplifier begins to heat, it will rarely be the fault of the resistance. Look for a shorted capacitor, a shorted tube, or two wires touching and shorting. Charred resistors should be replaced, as heat may change their resistance or cause intermittent or noisy operation.

Shielded cables, such as microphone, pickup head, or speaker cables, should be grounded at one end only, usually as close as possible to the metal chassis, or through the ground connection of a PC board, never at both ends. This prevents pickup of undesired *ground-loop* hum voltages.

Unlike some types of transistors, the bias in a VT is always negative, and may be developed by battery, by cathode resistor (self-bias), by a voltage divider across the power supply with the cathode returned to the desired voltage above the negative terminal, by contact potential, by the trapping of electrons on the grid of high-μ tubes when the grid is driven positive, or by capture of electrons directly from the space charge, and their slow leakage back to ground through a high-resistance grid resistor (Fig. 14-33).

14-26 SERVICING AMPLIFIERS

When an audio amplifier or the AF amplifier section of a receiver or transmitter ceases to function properly, it is up to a technician to determine the trouble and correct it. The following are a few servicing suggestions.

There are several levels of trouble that might occur in an amplifier system: (1) no signal output, (2) weak signals, (3) distorted signals, and (4) bursts of noise.

Equipment that may be used to test AF amplifiers includes VOM or EVM, AF signal generator, oscilloscope, signal tracer (a high-gain, usually battery-powered portable amplifier with an input probe and an output speaker), transistor tester, VT tester, earphones, and clip leads.

NO SIGNAL OUTPUT.   Jar the amplifier to check for any loose connection. Check the line-cord plug at the wall receptacle. Check the fuse or overload relay. Check for proper power-supply voltage. Check volume control for noise when adjusted. If the background noise increases with increased gain setting, check input connections to the amplifier. Touch ohmmeter on $R \times 1$ across the loudspeaker and listen for a click. If heard, short the input circuit of the power amplifier to ground for a brief instant and listen for a click. If heard, short each input circuit to ground and listen for clicks. The stage input at which clicks are not heard may be the inoperative stage. Check all transistors or VTs with an ohmmeter or with a transistor or VT tester. With a VOM voltmeter check for V_{DD}, V_D, V_G, etc., at each stage. With malfunctioning ICs external components can be checked, but replacement of the IC is often necessary.

WEAK SIGNALS.   Feed a 1000-Hz sinusoidal ac to the input of the final stage first. Reduce this signal for lower-level stages. Any stage where a marked decrease in signal is apparent may be the malfunctioning stage. (Remember, phase inverters and common drain-type amplifiers have no voltage gain.) Refer to pin connections for ICs.

DISTORTED SIGNALS.   Feed a weak 1000-Hz sinusoidal ac to the input of the first stage. Check with a signal tracer or oscilloscope for progressively stronger undistorted signals at the output of each higher-level stage. If a stage is found where the signal becomes aurally or visually distorted, even when the signal input is decreased, it is possibly the malfunctioning stage. If no signal tracer or oscilloscope is available try earphones in series with a 0.001- to 0.01-μF capacitor in place of the signal tracer. Usual causes of distortion are bad active devices, shorted or leaking coupling or bypass capacitors, improper bias voltages, open or changed-value resistors, improper power-supply voltages, and insufficient power-supply filter. Dried-out filter or decoupling circuit capacitors can allow an amplifier to break into oscillation and produce a hum- or whistle-type output. Make tests of each stage for proper bias voltages and output circuit voltages with a VOM or EVM. Refer to pin connection diagrams for ICs.

BURSTS OF NOISE. These are usually caused by leakage current in coupling or bypass capacitors, malfunctioning active devices, or broken and arcing resistors or contacts. Localize the trouble by shorting the input of each stage, starting at the last or power amplifier. If the trouble is in the power supply, shorting the last stage input may not stop them.

Active devices should be checked with a transistor or VT tester. If none is available, a VOM may be used. For BJTs, with power supply off, base to either emitter or collector should show low resistance on the $R \times 10$ or $R \times 100$ ohmmeter setting one way, and very high resistance with the ohmmeter leads reversed. If this is not obtained, unsolder or remove the BJT and test it out of the circuit. Compare its ohmmeter readings with a similar BJT known to be good. Do not use $R \times 1$ setting when testing semiconductors.

For JFETs the channel should read perhaps 150 Ω on the $R \times 10$ ohmmeter setting. Gate to either source or drain should read high resistance one way and low resistance the other with the ohmmeter on $R \times 100$ or $R \times 1000$. If not, remove from circuit and test and compare. JFETs can be tested with most transistor testers.

For MOSFETs, the channel may be checked with an ohmmeter for continuity, but gate to source or drain has an easily ruptured insulator and may not stand the voltage of an ohmmeter that has a 1.5- or 3-V battery in it. MOSFETs can be tested with modern transistor testers.

For VTs, if no tube tester is available, there should be ohmmeter continuity between filament or heater leads only, and infinite resistance between all other elements. An ohmmeter only tells if there are interelement shorts or if the heater is open. Operating filaments may be determined by feeling the tube. If turned on but cold, there must be no heater current.

Capacitors should be checked on a capacitor checker for capacitance value and leakage. An ohmmeter on $R \times 1$ MΩ may indicate leakage or a shorted dielectric, or if 0.001 μF or larger, should show a charging current when first touched across the ohmmeter probes, and should then return to an infinite reading.

Resistors should be checked out of circuit with an ohmmeter and should be within 10% of their rated values. If they show brownish signs of overheating,

replace them, but first find what caused the overheating. It may be a leaking capacitor, shorted active device, or improper bias on the related active device. Resistors should never overheat by themselves. Over a long period of time carbon resistors may increase their resistance as much as 100%.

Transformers can usually be checked for opens or interwinding shorts with a VOM. Shorted internal turns cause distortion and excessive heating. Usually a smoking AF transformer indicates an active device that is not biased properly or is shorted. A smoking power transformer indicates too heavy a load due to an improper operating amplifier, shorted filter or other across-the-line capacitors, or shorted internal turns in one of the windings. If an output circuit transformer heats with a light load on its secondary, it usually indicates internal shorted turns, requiring a new transformer.

Test your understanding; answer these checkup questions.

1. With 60-Hz power, if there were an insufficient power-supply filter, what would be developed if full-wave rectification were used? _____ If half-wave? _____ If an ac heated filament were not center-tapped? _____
2. What is gained by twisting heater leads together? _____
3. What capacitors in an amplifier are most likely to break down? _____ _____ _____
4. What may be developed if both ends of shielded cables are grounded? _____
5. What are the four levels of trouble in an AF amplifier? _____ _____ _____ _____ _____
6. List steps to take if an amplifier has no signal output. _____
7. List steps to take if an amplifier produces weak signals. _____
8. List steps to take if an amplifier's signals are distorted. _____
9. List steps to take if an amplifier has noisy output. _____
10. Is a BJT probably good or bad if an ohmmeter shows low R E-C? _____ Low R C-E? _____ Low R E-B one way? _____ Low R B-C one way? _____
11. Is a JFET probably good or bad if an ohmmeter shows low R S-D both ways? _____ High R S-G one way? _____ High R D-S? _____ Low R D-G both ways? _____
12. Is a VT probably good or bad if an ohmmeter shows infinite R across the filament? _____

Medium R h-k? _____ Low R k-p? _____
High R k to all elements? _____

13. What two test instruments can be used to check capacitors? _____ _____

14. If a resistor burns out, what should be done before replacing it? _____

15. What AF transformer fault cannot be checked with an ohmmeter? _____

16. Draw a diagram of a filament-type triode AF amplifier. A one-stage pentode AF amplifier.

AUDIO-FREQUENCY AMPLIFIERS QUESTIONS

1. What is the range of frequencies included in AF? VF? Subaudible frequencies? Supersonic frequencies?
2. Which stage of an AF amplifier system must have the lowest input noise level device?
3. A JFET has a μ of 50, 1-V bias, a 13-kΩ R_G, a 20-kΩ R_D, a 40-kΩ R_{par}, a 10-kΩ Z_D, and a 0.2-V p-p input signal. What is the V gain? dB gain?
4. Diagram two parallel JFETs transformer-coupled.
5. What two devices may operate with reverse-input dc?
6. Under what conditions is optimum bias required?
7. What is the optimum class A bias value of a JFET having a V_D of 30 V if it has a μ of 45?
8. For a 250-Ω biasing resistor, what should be the bypass value across it for a VF amplifier?
9. Draw a diagram of a push-pull AF amplifier.
10. What effect would paralleling a second amplifying device across a first have on the output transformer?
11. What approximate percentage of cutoff bias value is used to bias a stage to class A? Class B? Class AB?
12. What are the advantages of using two similar devices in push-pull rather than in parallel?
13. What is the operating angle of class B? AB? A?
14. What type of phase inverter is described? Draw a diagram of it.
15. Which develops more distortion to a class A p-p stage: transformer coupling or a phase inverter?
16. What is the output Z of a source-follower having a 200-Ω R_S, a 50-μ, and a 15-kΩ Z_D?
17. What are the two types of earphones described?
18. Name three types of loudspeakers.
19. Name three ac interstage coupling circuits. DC.
20. Which types of coupling could be used to couple signal voltage to the next stage? To couple power?
21. In a resistance-coupled stage, if μ is unknown, how might the biasing resistance be found?
22. Why might impedance coupling provide more output voltage than resistance?
23. Diagram a three-stage amplifier using transformer, R, and Z interstage coupling circuits.
24. A 35-Ω Z_C of a BJT is to feed a 4.2-Ω speaker. What turns-ratio transformer is needed? In which winding does more current flow?
25. What is the advantage of complementary symmetry power output stages?
26. Draw a diagram of a simple split- or double-supply complementary-symmetry AF output circuit.
27. Draw a diagram of a single-power-supply complementary-symmetry AF output circuit.
28. What are four forms of distortion in AF amplifiers?

Which is least objectionable to a listener?
29. List ten possible causes of AF distortion.
30. What are three terms for "out-of-phase" feedback?
31. Diagram a two-stage JFET amplifier with both inverse voltage and current feedback.
32. An amplifier has a V gain of 25. If a $-\beta$ of 0.07 is used, what would its gain be?
33. What is the feedback factor if gain without feedback is 30 and with feedback is 20?
34. Diagram volume controls in JFET and BJT circuits.
35. Draw a diagram of a bass-treble tone-control circuit.
36. What is the formula to determine Miller effect in JFETs? In BJTs?
37. What is the result of Miller effect on the output signals of AF amplifiers?
38. Diagram a Loftin-White-type circuit using JFETs.
39. Draw a diagram of a common-source to source-follower dc amplifier.
40. Draw a diagram of a Darlington-pair amplifier.
41. Diagram a basic JFET differential amplifier.
42. Draw a diagram of a two-JFET cascade amplifier.
43. Draw a diagram of a three-stage dc amplifier using enhancement MOSFETs.
44. What gain can be expected from an op amp with no feedback? What determines its actual amplification?
45. Besides AF amplification, what might an op amp be used for?
46. An op amp has a Z_f of 8 kΩ and a Z_i of 400 Ω. If it is fed a 3-mV input signal, what is the output?
47. What might be the output power of low-power op amps? Of high-power op amps?
48. What is the output Z of an op amp with feedback?
49. What is the advantage of a single-pole active filter? The disadvantage?
50. Diagram an active bandpass filter using an op amp.
51. What might be the component values of an active bandpass filter having a bandpass of 700–1000 Hz?
52. What would be a proper term to use for a three-cascaded-stage op amp bandpass filter?
53. Why are decoupling circuits used in AF amplifiers?
54. Why are heat sinks used on power amplifier devices?
55. What type of device is usually found in the power output stage of a VT AF amplifier?
56. What is the frequency of audible hum if a power supply uses full-wave rectification? Half-wave?
57. Why are 60-Hz current-carrying wires often twisted?
58. What may cause a microphone to pick up hum?
59. What are the four basic levels of trouble that may occur in an AF amplifier?

15

Radio-Frequency Amplifiers

This chapter first discusses low-level radio-receiver type RF amplifiers and then, at greater length, transmitter *RF power amplifiers*. Unneutralized stages are described first, followed by neutralization methods with 3-element devices. Parallel, push-pull, and push-push amplifiers with various forms of coupling are considered. Special requirements for high-power amplifier stages are discussed, as are amplifiers for VHF and UHF, and troubleshooting.

15-1 RF AMPLIFIER FREQUENCIES

In this chapter two basic types of high- or radio-frequency (RF) amplifiers will be discussed: (1) RF voltage, low-level, or small-signal amplifiers and (2) RF power amplifiers. The low-level amplifiers are found in radio receivers, or in small-signal applications in transmitters. RF power amplifiers are used in transmitters or special applications where high-power RF ac is required. Because most high-power amplifiers use VTs, these devices are used in many explanations, but BJT or VMOS devices can be used up to at least 500-W output.

The RF spectrum is considered to be from 10 kHz to 300 GHz, although the most common communications are below 1 GHz.

The fundamental difference between AF and RF amplifiers is the bands of frequencies they are expected to amplify. AF amplifiers amplify a major portion of the AF spectrum (20 to 20,000 Hz) equally well. RF amplifiers amplify only a relatively narrow portion of the RF spectrum, attenuating all other frequencies. Radiotelegraphic code transmissions (called CW, or A1A) require receiver amplifiers of 200- to perhaps 1000-Hz bandwidth. Single-sideband (SSB, or J3E) voice transmissions require about a 3-kHz bandwidth. RF carrier waves amplitude-modulated (AM, or A3E) by voice frequencies require a 6-kHz bandwidth. Standard broadcast band AM for music requires at least 10 kHz. Frequency-modulated (FM, or F3E) voice emissions use a 20-kHz bandwidth, and FM broadcast stations have a bandwidth of 200 kHz. Television transmissions (TV, or C3F) use a 6-MHz bandwidth. Radar may use a bandwidth up to 10 MHz. Even the bandwidth of radar emissions represents only a small portion of the whole RF spectrum.

RF amplifiers, unless broadbanded, are tuned to a desired frequency, amplifying only that frequency and a relatively few frequencies on both sides. AF amplifiers may amplify frequencies equally well from perhaps 300 to 800 Hz for CW, 250 Hz to 3 kHz for voice, and 20 Hz to 20 kHz for music.

(Special *wideband*, or *video*, amplifiers are usually directly coupled types and may be flat from dc to 5 MHz or higher. These are used in oscilloscopes, TV, and radar, discussed in later chapters.)

In general, amplifiers with tuned *LC* circuits accept the resonant frequency and a few frequencies above and below. At 50 kHz *LC* circuits tend to have high *Q* and narrow bandwidth, 0.5 to 3 kHz. At 500 kHz the bandwidth may be 5 to 30 kHz, again depending on the *Q* of the *LC* circuits. At 5 MHz the bandwidth may be 20 to 150 kHz. Thus, for applications requiring amplification of a wide band of frequencies, signals can be converted to high-frequency RF circuits to be amplified. If only a narrow band of frequencies is to be amplified, signals may be converted to lower RF circuits to be amplified. However, by utilizing the high-*Q* capabilities of piezoelectric crystals, very narrow bandwidths can be obtained for any radio frequencies up to about 15 MHz with crystal filters.

At high frequencies the *Q* of coils decreases because of skin effect and increased core losses. The dielectric losses in capacitors increase. Interaction between stages in receivers or transmitters increases. Shielding becomes more necessary and more difficult to achieve. Wiring between parts of a circuit must be made shorter. Equipment must be made more compact. Insulating materials, such as Bakelite, paper, cambric, black rubber, and cotton, usually satisfactory at lower frequencies, are found to have too much loss at higher RF. Special insulating materials, such as mica, steatite, Isolantite, ceramics, and special plastics, must be used. In general, at higher frequencies circuits become more difficult to manage. New, smaller parts and more advanced circuitry, however, are widening the usable spectrum. Before 1940, little commercial use was being made of frequencies above 100 MHz. At present, frequencies above 100 GHz (microwaves) are in daily commercial use.

The amplifiers in this chapter operate in the 500-kHz to 300-MHz range.

15-2 LOW-LEVEL RF AMPLIFIERS

A low-level, or small-input-signal, RF amplifier can use transistors, linear ICs, or VTs; is biased to class A; and may be involved with ac signals in the microvolt or nanoampere range. Both input and output loads may be resonant circuits in small-input-signal RF and intermediate-frequency, or IF (Chap. 18), amplifiers. In RF power amplifiers the output may be the only circuit that is tuned.

The basic RF amplifier is similar to that of the basic AF amplifier except that it normally has an *LC* circuit tuned to a desired frequency feeding the input, and another *LC* circuit tuned to the same frequency in, or coupled to, the output circuit. This may be recognized as a description of a regenerative capacitive feedback (TDTG, TPTG) oscillator. To prevent such an amplifier from becoming an oscillator it is necessary to introduce some degenerative or inverse feedback into the circuit. In this application degenerative feedback is known as *neutralization*. A neutralization circuit is shown in color in Fig. 15-1.

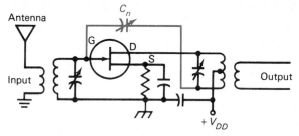

FIG. 15-1 JFET RF amplifier with possible neutralizing circuit in color.

The solid-line part of the circuit is a common-source-type JFET-tuned input and output RF amplifier, biased to class A by the source *RC* self-bias network. Without C_n in the circuit, the capacitance between gate and drain, both internal in the JFET and external due to proximity of input and output wiring, would produce a tuned-input tuned-output oscillator. Since the output circuit is center-tapped, when the top part of the drain tuned circuit is ac-positive, the bottom will be equally ac-negative. If the capacitance of the neutralizing capacitor C_n is adjusted to equal the total G-D capacitance, the gate will be receiving equal and opposite voltages from the drain circuit, and the circuit effectively has no feedback; it cannot oscillate, only amplify.

In this circuit the antenna is picking up some desired-frequency RF wave and inductively coupling it to a tuned *LC* circuit. When the input *L* and *C* are tuned to this frequency, a maximum voltage for this signal will be fed to the gate and will appear as an amplified signal in the drain tuned circuit. The amplified signal is inductively (but might be capacitively) coupled to the next stage.

With careful placement of components, operation of this basic circuit up to a few hundred kilohertz might be possible, but for MF, HF, and higher-frequency operation, neutralization will normally be required. If only half of the source resistor *R* is bypassed by C, degeneration will be introduced, which should reduce "instability" (tendency to self-oscillate) and increase the possible frequency of operation. In general, the lower the input and output load impedances the less likely that an RF or IF amplifier will go into self-oscillation for a given value of feedback capacitance.

If the output load circuit is not tuned near the frequency of the input circuit, or is an untuned circuit, the amplifier should not oscillate, and should be usable without neutralization through the HF range at least.

A BJT RF amplifier is shown in Fig. 15-2. Since both the base-emitter and the collector-emitter im-

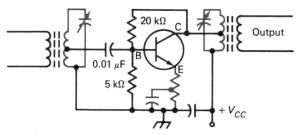

FIG. 15-2 RF amplifier using a BJT.

pedances are quite low, both LC circuits are tapped down to provide a better low-impedance match. The emitter stabilizing resistor is only partially bypassed to provide some degeneration. The forward-bias network is also connected to the collector to provide some degenerative feedback to the base. The dashed lines in the transformers indicate ferrite-type, possibly toroidal, cores on which the coil turns are wound.

Note that the curved part of capacitor symbols is always connected to ground, or to the lowest-impedance, or closest-to-ground, part of the circuit. The curved line indicates the outer foil plate if a capacitor is a spiral-wound type, or the stator connection or screwdriver adjustment part of a variable or adjustable capacitor. This reduces hand capacitance or intercomponent capacitive coupling and is important in RF amplifiers.

An excellent RF amplifier can result by using a dual-gate MOSFET (Fig. 15-3). These high-Z devices have minimal internal capacitance and provide good gain with little instability. The dashed lines with an arrowhead indicate an adjustable ferrite core (as does an arrow through a coil) with which the fixed-capacitor LC circuit is tuned to resonance. The RF input signal is applied to one of the dual gates G_1 of this enhancement MOSFET. The other

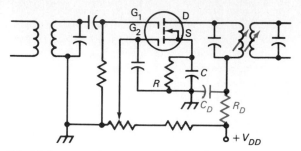

FIG. 15-3 RF or IF amplifier using a dual gate MOSFET.

gate is receiving its positive bias from a potentiometer between $+V_{DD}$ and ground. The pot can also operate as a variable gain control for the I_D and the amplification of the stage. If oscillations are produced in this circuit, it may be necessary to center-tap the drain LC circuit and add a neutralization capacitor back to G_1. Neutralization becomes more necessary as the frequency of operation increases. The circuit shown might be stable at 10 MHz but requires neutralization at perhaps 20 MHz or higher. The RC network source to ground develops class A bias across it when I_D flows.

Note the inclusion of an output decoupling circuit C_D-R_C (Sec. 14-22) in this stage. A similar circuit is usually found in all RF amplifiers, but was left out previously to simplify the basic circuits.

15-3 NONREGENERATIVE RF AMPLIFIERS

The grounded- or common-source (-emitter, -cathode) circuit has the greatest amplification, in the range of 30 dB, but it is subject to self-oscillation. The grounded- or common-drain (-collector, -plate) circuit, also known as the source- (emitter-, cathode-) follower circuit, does not reverse the phase of the input signal in the output and should not self-oscillate. This is also true of the grounded- or common-gate (-base, -grid) circuit.

A common-drain or source-follower small-signal RF amplifier is shown in Fig. 15-4. The gate-ground impedance of this circuit is quite high, but the output LC circuit is tapped down to match the low-impedance source-ground circuit better. The RC network in the source circuit provides the class A self-bias voltage. The gain of such amplifiers may be in the range of 15 dB.

A common-gate RF amplifier is shown in Fig. 15-5. The input LC circuit is tapped down because the source-gate circuit has relatively low impedance, whereas the drain-ground circuit has relatively

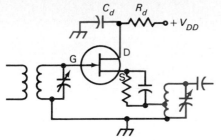

FIG. 15-4 Common-drain or source-follower JFET RF amplifier.

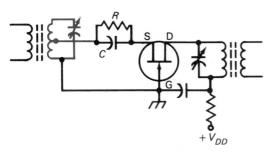

FIG. 15-5 Common-gate or grounded-gate RF amplifier using a JFET.

high impedance. Again, the *RC* network provides the class A bias when I_D flows through it. The diode formed by G and D is in the nonconduction direction, so no current flows from gate to drain (electron current only flows backward against an arrowhead in symbols). The gain of these amplifiers is usually in the 15-dB range.

Low-signal RF amplifiers are available in a variety of IC forms. They are capable of perhaps 60 dB (1 mV input produces 1 V output) gain at 10 MHz and 30 dB at 100 MHz. These ICs are essentially wideband linear amplifiers that require only a few external resistors, capacitors, and inductors. Figure 15-6 illustrates a simple IC used as an RF amplifier. A diagram of the actual BJTs and resistors formed in the manufacture of the IC is shown in Fig. 15-6*a*. The socket connections are shown in Fig. 15-6*b*. The diagram of an RF amplifier using the actual components is shown in Fig. 15-6*c*. How the circuit might be drawn in IC diagraming is shown in Fig. 15-6*d*. This IC is a differential wideband amplifier. The circuit shown uses its emitter-coupled differential capabilities. It could also be connected as a single-ended or as a cascode (Sec. 24-14) amplifier. The IC will also operate as a low-signal AF voltage amplifier. It has many other capabilities.

15-4 PENTODE RF AMPLIFIERS

Vacuum-tube triodes have sufficient g-p capacitance to assure self-oscillation if both input and output circuits are tuned to the same frequency. Pentodes and tetrodes were developed to reduce g-p capacitance to such a small value that neutralization should not be necessary. Consequently, almost all VT small-signal RF amplifiers use pentodes connected in circuits somewhat similar to those in Fig. 15-7.

The input signal is fed to the first grid, the control grid, which varies the plate current. The plate voltage is commonly in the +250-V range. The second grid, the screen grid, operates at a positive potential of about 100 V, and draws one-tenth to one-third the current of the plate. Resistor R_2 is selected to provide a 150-V voltage-drop for the screen grid. The bypass capacitor C_2 must be large enough to hold the screen-grid potential steady when the control grid is varying the electron stream to plate and screen grid. The third grid, the suppressor grid, draws no current and is usually connected to the cathode, although better isolation of plate and control grid is produced by grounding the suppressor, at the cost of some reduction in I_p and stage gain. R_1 and C_1 form the self-bias circuit. R_3 and C_3 form a plate-circuit decoupling network to isolate this stage from other stages that might otherwise intercouple via the power supply. The heater circuit is not shown, but the heater leads are usually connected to a 6- or 12-V winding of the power-supply transformer. The gain of a pentode stage can exceed 50 dB.

To decrease instability in pentode stages, the connections shown in color should be as short as physically possible, and should all be brought to a single ground point on the chassis or PC board. Such a reduction of return-lead length to ground is important in all circuits operating at higher frequencies. Input and output circuit "hot" leads (heavy lines) must be kept well separated to reduce g-p capacitance and output-to-input circuit feedback. In Fig. 15-7 the grid circuit itself is not tuned, but is inductively coupled to a tuned circuit. As far as the grid circuit is concerned, its input is tuned and it will oscillate if sufficient g-p capacitance is present. The same is true for the plate or output circuit. The plate circuit coil may be untuned, but if closely coupled to a tuned circuit, it acts as if a tuned circuit was actually the output load. The same is true for transistor circuits also.

Note the use of the symbol letter B+, which was

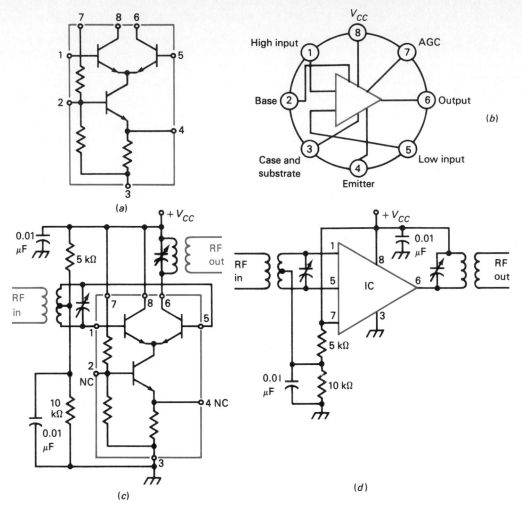

FIG. 15-6 Simple RF differential wideband IC amplifier. (*a*) Internal structure. (*b*) Base or socket connections. (*c*) Application in an RF amplifier. (*d*) As the circuit might be diagrammed. (NC = not connected.)

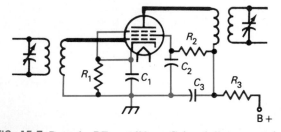

FIG. 15-7 Pentode RF amplifier. Colored lines must be short leads. Heavy leads must be kept apart.

the original symbol. It will still be found in some phases of radio, such as television servicing, for example.

15-5 RF POWER AMPLIFIERS

The remainder of this chapter pertains mostly to RF amplifiers in *high-power* applications. These amplifiers are designed to take an RF ac input signal, amplify it, and produce enough power output to drive another, higher-power RF amplifier, or excite a transmitting antenna, or operate other loads (nuclear research, etc.). Since present transistors are not capable of high-power output (over perhaps 500 W) at high frequencies, much of the discussion will be in VT terms.

While a small-signal RF amplifier is basically class A in operation, distorting the signal very little, an RF power amplifier may be biased from 1.2

to 4 times I_p cutoff, allowing it to operate as a high-efficiency class C amplifier. Ten watts applied by the power supply to a class A stage (transistor or VT) may produce only about 2.5 W of ac output (25% efficiency). The same device, biased to class C, may be able to produce 6 W output (60% efficiency). To do this, however, it is necessary to change the sine-wave input signal to pulses of current and then rely on the flywheel effect of the output circuit tuned coil and capacitor to restore the sine-wave shape to the output signal ac.

In many cases where class C amplifiers are used, class A or B could be used but with lower efficiency. To amplify RF ac signals *without distortion*, class A, AB, or B is used.

Bias for the different classes of operation of a triode (or JFET) can be determined by the formulas

Class A $\quad = 0.6$ cutoff $\quad = \dfrac{0.6V_p}{\mu}$

Class AB $= 0.8$ cutoff $\quad = \dfrac{0.8V_p}{\mu}$

Class B $\quad = 0.95$ cutoff $\quad = \dfrac{0.95V_p}{\mu}$

Class C $\quad = 1.2$ to 4 cutoff $= \dfrac{1.2V_p}{\mu}$ to $\dfrac{4V_p}{\mu}$

As an example, a triode has a V_p of 1250 V and a μ of 25. Its bias voltages might be approximately

Class A $\quad = \dfrac{0.6(1250)}{25} = 30$ V

Class AB $= \dfrac{0.8(1250)}{25} = 40$ V

Class B $\quad = \dfrac{0.95(1250)}{25} = 47.5$ V

Class C $\quad = \dfrac{1.2(1250)}{25}$ to $\dfrac{4(1250)}{25}$

$\qquad = 60$ to 200 V

Because I_p is more dependent on V_{sg} than on V_p, for transmitting *tetrodes* the cutoff bias is approximately one-fifth of the V_{sg} voltage, or $V_{co} = 0.2V_{sg}$. Class C bias ranges from about $0.2V_{sg}$ to $0.8V_{sg}$.

Plate power input is the product of plate voltage and current, as read from a voltmeter across the power supply and a plate-current meter, or $P_i = V_pI_p$ (or V_DI_D, or V_CI_C). If a cathode-current meter is used instead of a plate milliammeter, the I_g and I_{sg} must be subtracted from the cathode-current reading to determine the I_p.

A common-emitter transistor circuit with no forward bias has no I_C and is in class C. Any *RC* emitter-stabilization circuit produces reverse bias for the base when a signal is applied and emitter current begins to flow through it. This places the stage further into class C operation when driven.

▍Test your understanding; answer these checkup questions.

1. In what frequency bands are the following:
   1 kHz? _____ 50 kHz? _____
   500 kHz? _____ 4 MHz? _____
   50 MHz? _____ 400 MHz? _____
   Which is not a radio frequency? _____
2. What range of frequencies are required to reproduce voice signals adequately? _____
   AM broadcast music? _____ FM broadcasts? _____
3. What is the main difference between small-signal RF and AF amplifiers? _____
4. What is an inverse feedback circuit in an RF amplifier called if it prevents self-oscillation? _____
5. To what class of operation are low-signal RF amplifiers biased? _____
6. Why might ferrite toroidal cores be preferable in RF amplifiers? _____
7. Which should go to the higher-Z point, the straight or the curved line of a capacitor symbol? _____
8. Name the three different types of JFET amplifier circuits _____ _____ _____
   Which may oscillate? _____ Which has the greatest gain? _____
9. What external additions would always be used with single-frequency RF amplifier ICs? _____
10. What active device has the highest gain for low-signal RF amplifiers? _____
11. Which grid controls I_p most, the first, second, or third? _____ Which has the least effect on I_p? _____
12. With no bias in its circuit, to what class is a BJT operating? _____ A JFET? _____ A depletion MOSFET? _____ An enhancement MOSFET? _____ A VT? _____

15-6 BIAS-SUPPLY CLASS C RF POWER AMPLIFIERS

A beam-power tetrode RF power-amplifier circuit with power-supply bias is shown in Fig. 15-8. The I_p cutoff bias is -50 V. Since the grid is biased to -100 V, the tube is biased to class C. With no RF ac input, the bias holds the I_p and I_{sg} to zero. This

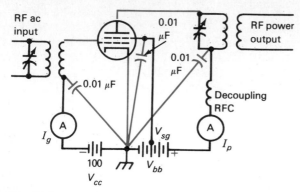

FIG. 15-8 Beam power tetrode RF amplifier. Colored leads must be short and use a central ground point.

condition is indicated on the $V_g I_p$ curve of Fig. 15-9 as *condition 1*. The curve shows the cutoff bias point as -50 V, the working bias value of -100 V, and the zero bias point. The resultant I_p during condition 1 is zero.

In *condition 2*, the RF input induces a 50-V peak in the grid-circuit coil. The ac is added to the -100-V bias, producing, in effect, a varying dc bias, alternately -50 to -150 V, but still no I_p flows because the grid voltage is not driven up to cutoff.

In *condition 3*, the RF input induces a 100-V peak into the grid coil. This, added to the -100-V bias, produces a dc bias that varies from zero to -200 V. During the portion of the input-signal cycle when the bias is under the $V_g I_p$ curve, I_p flows as a steep, narrow pulse, less than a half cycle wide. The I_p flows for less than 180° of the input cycle, for 120° in this case. Thus far, the grid has not been driven into the positive region, and therefore I_g does not flow in the grid circuit. This is a possible class C stage condition, but is not the normal one.

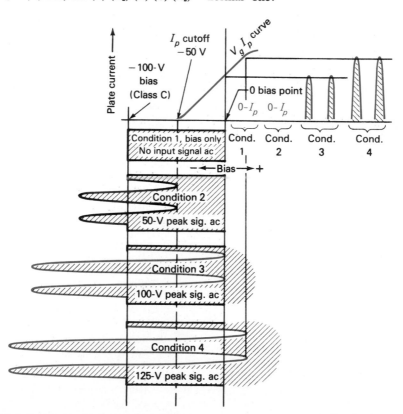

FIG. 15-9 $V_g I_p$ curve and four possible I_p conditions with class C amplifier operation. (JFET would be similar.)

Condition 4 is the usual class C condition. The RF source induces a voltage with a peak value greater than the bias value. This drives the grid into the positive region for a portion of each cycle, causing a pulsating dc I_g to flow from the cathode to the grid, down through the grid coil and the ammeter, and through the bias battery in a direction opposite to normal current flow for a battery. The I_p peaks are considerably higher in amplitude and exist for more than 120°.

Plate current heats the plate of a tube. The plate-dissipation rating of a tube expresses how much average heat the plate is capable of dissipating without overheating. The use of high bias voltages results in narrow I_p pulses. Thus, class C amplifiers have a relatively long duration between pulses, allowing the tube to rest for a major portion of each cycle. During the time the tube is working it can be driven hard, and still its average plate dissipation may not exceed the plate-dissipation rating. Thus, class C amplifiers are high in efficiency. Class B amplifiers, biased to cutoff, have I_p flowing for a full half of the input cycle, rest less, and are therefore less efficient. Class A amplifiers have I_p flowing all the time and tend to overheat their plates unless the V_p is held to a low value.

If the plate circuit power supply is 3000 V and the plate ammeter reads 2 A, the *dc power input* is $P_i = V_p I_p$, or 3000(2), or 6 kW. If the device is operating as a class C amplifier at 60% efficiency (70% is about maximum), the *RF power output* is 6000(0.6), or 3.6 kW, delivered to an antenna or other load.

The VMOS or power FET is similar in operation to a triode VT except that no gate current ever flows. A VMOS is limited in single-device operation to between 100 and 200 W input. (Some VTs produce over 1 MW RF output!) Being a triode device, a VMOS may require neutralization at higher frequencies if used in a common-source circuit. It requires forward biasing like a BJT to produce any I_D flow in class A or B operation (Fig. 15-10*a*).

Bipolar transistors have much lower impedances than the FETs and VTs, and range in power up to over 150 W output in the HF range. A possible BJT class C NPN common-emitter RF amplifier is shown in Fig. 15-10*b*. Since the input impedance is only a few ohms, the 10-Ω resistor tends to desensitize the device to some extent and may make neutralization unnecessary provided the output is loaded heavily (to I_C saturation). If not loaded heavily enough and not driven hard enough, the

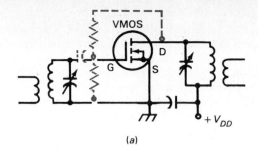

(a)

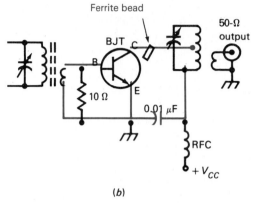

(b)

FIG. 15-10 (*a*) VMOS class C RF power amplifier. Dashed colored components for class A, AB, or B. (*b*) BJT class C amplifier.

operating beta value may increase and the stage may break into self-oscillation. Most transmitting amplifiers use a form of π-network tuning and coupling (Sec. 15-29) rather than the simpler LC circuits used in these explanations. Many modern RF amplifier circuits use a 50-Ω output impedance value. A ferrite bead is shown around the collector lead to suppress parasitic oscillations and help stabilize the circuit. A ferrite bead may be used around the base lead also. The leads shown in color must be as short as possible.

A VT RF power amplifier may be biased to class A or AB but be driven into the positive grid region and draw I_g. While not the accepted class A_1 or AB_1 operation, it is used in RF power amplifiers and results in increased stage efficiency.

As in low-level RF stages, all bypass capacitors should return to a single ground point. Some special high-frequency VTs may have two (or more) leads to the cathode. By grounding both of them, the cathode-to-ground inductance is halved, which helps to stabilize the circuit.

Screen-grid current varies in step with the I_p, although at a lower amplitude. With regard to the screen grid in an RF amplifier: (1) Use short leads on the bypass capacitor; (2) do not exceed screen-grid voltage specifications or I_p will increase excessively; and (3) low I_p may increase I_{sg} drastically.

If the bypass capacitor can maintain the bottom of the LC circuit at essentially RF ac ground potential, the additional filtering or decoupling of an RFC may be unnecessary.

RFCs may be constructed as long single-layer coils, but may exhibit parallel-resonant (high-impedance) or series-resonant (low-impedance) effects at certain frequencies. The resonant peaks are harmonically related. To overcome such undesirable effects, RF chokes may be made up of three to five separate, *universal-wound* (layers wound on top of other layers) *pies*, with all pies connected in series (Fig. 15-11). This produces a

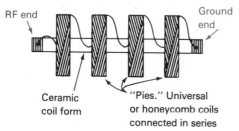

FIG. 15-11 RF chokes are usually constructed in pies.

high impedance to the relatively wide band of frequencies for which the choke is manufactured, and greatly reduces end-to-end distributed capacitance. The lead nearest the core should be connected to the RF ac point. Circuits may sometimes operate better if an RFC is reversed. In many HF, VHF, and UHF circuits ferrite beads around connecting wires act as RFCs.

15-7 TUNING THE OUTPUT CIRCUIT

The I_p flowing in a class C tube (I_D for FETs, I_C for BJTs) occurs as short-duration pulses of dc which appear to the output LC circuit more as a distorted form of ac than as a dc. Consequently, the tuned circuit sees the device and power supply as a source of quasi ac energy. The LC circuit is a parallel-resonant circuit connected across an ac source, presenting a high impedance to the pulses. When the LC circuit is resonant, little I_{dc} flows. The I_{dc} meter will dip to a minimum value.

If the output LC circuit is detuned from the frequency of the I_{dc} pulses, it presents a lower impedance to the pulses and allows a greater pulsating current to flow through source and LC circuit. The

I_{dc} meter indicates a higher value and makes a good indicator of output LC circuit resonance.

In a parallel circuit at resonance, a maximum number of electrons oscillate back and forth between the coil and capacitor of the LC circuit. Because they are oscillating at the same frequency as the incoming pulses from the source, little energy is required from the source pulses to maintain oscillations, and the ac peak voltage across the LC circuit is nearly equal to the voltage of the power supply. At resonance it requires little energy (little I_{dc}) to produce this ac voltage value in the LC circuit. Thus maximum RF current is circulating in the LC circuit when I_{dc} is at a minimum.

In a VT the plate is taking a minimum number of electrons from the space charge of the tube when the plate load is tuned to the resonant frequency, and the grid is free to attract a maximum number during its positive swing. As a result, the I_g meter rises somewhat when the plate circuit is tuned to resonance. The I_g falls off as the plate circuit is detuned or if the I_p increases for any reason. As a result, an I_g meter may indicate plate-circuit resonance. I_{sg} also increases, sometimes dangerously, if I_p decreases greatly.

15-8 COUPLING LOAD TO AMPLIFIER

The RF ac energy oscillating in the output LC circuit can be coupled into an antenna, into the input circuit of another RF amplifier, or into some other kind of load circuit. If the coupling between the output LC circuit and the circuit that follows is very loose, little energy will be drawn from it. In this case, the output circuit current I_{dc} dip, when the LC circuit is tuned to resonance, will be very deep. For example, if the off-resonance I_{dc} maximum is 200 mA, the resonant value may dip to perhaps 20 or 30 mA (and I_{sg} may rise dangerously in VTs).

If the coupling between the output tank and the following circuit is increased, the tank circuit will have more energy drawn from it. This lowers the parallel-circuit impedance of the tank circuit, increasing the I_{dc}. The I_{dc} dip will no longer be as great. The off-resonance value may still be 200 mA, but the value at resonance may now be 100 mA. If the load is too tightly coupled to the tank circuit, it may be found that no I_{dc} dip will occur.

How much coupling should be used? The best answer is to consult a manual giving the operating characteristics of the device being used. Adjust the electrode voltages to the recommended values. Adjust the bias and drive to produce the recommended I_{dc}. Then, starting with loose coupling, keep increasing the coupling and dipping the I_{dc} by tuning the output circuit until the I_{dc} at the dip is the value recommended in the manual. Readjust all voltages and currents to the recommended values.

In an emergency adjusting the coupling until the I_{dc} at the dip is about 75% of the off-resonance maximum should result in fairly efficient operation. If the off-resonance value is 200 mA, the coupling should be increased until the maximum dip reads 150 mA. Proper operation of a class C RF amplifier rarely results with less than a 20% dip in I_{dc} from the off-resonance value.

A more accurate method of determining the proper coupling to the output circuit is to use some means of measuring the output power or of obtaining a relative indication of it. An RF ammeter in series with a noninductive resistive load (Fig. 15-12) will serve as a *dummy load* and an indicator

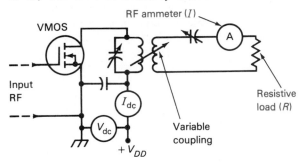

FIG. 15-12 RF power output is determined by RF current squared times load resistance value ($P = I^2R$). Dc power input is $P_i = V_D I_D$.

of the relative power output of an amplifier circuit. (An RF wattmeter may also be used.)

With loose coupling to the load circuit, both the dc input power at resonance and the RF ac output power will be low. As the coupling is increased, both dc input and RF output values increase. A degree of coupling at which the RF output power will show no further increase will be reached. Further increase of coupling will usually result in a decrease in RF output but an increased input power. This indicates that optimum coupling has been passed. The coupling should be reduced to the lowest value of dc power that produces almost maximum RF output. At this degree of coupling a condition of approximate impedance match between the output impedance of the amplifying device and the tank-circuit impedance occurs, resulting in maximum power being developed in the load.

It is usually possible to get a still greater power output by increasing the C and decreasing the L of the tank circuit, or vice versa, and running through the tests again. When maximum RF output results with as little dc power input as possible, optimum coupling has been attained. Care must be taken not to operate a transistor or tube at current values higher than those recommended by the manufacturer of the device.

Exact values for capacitance and inductance of the LC tank, degree of coupling, and load impedance are not easily determined. Fortunately, a rather wide variation of L and C values from the optimum may often produce satisfactory output.

Circuit efficiency is the ratio of RF power output to the dc power input, or

$$\text{Percent efficiency} = \frac{P_{rf}}{P_{dc}} \times 100$$

Of the total dc power input, some is dissipated as heat in the active device, some is dissipated as heat in the LC tank and wiring of the circuit, and the remainder is the RF power output. For example, if the dc power input is 1000 W and the RF power output is 650 W, plate dissipation is the difference, or 350 W. The efficiency of the stage must be $(650/1000) \times 100$, or 65%.

Maximum output power is obtained when the load impedance matches the source impedance. The higher the load impedance in comparison with the plate impedance, the greater the plate-circuit efficiency, but the lower the power output. A transmitter should not be adjusted with highest efficiency as the only consideration.

15-9 CLASS C BIASING

Figure 15-13 shows a grid-leak biased class C tetrode RF amplifier stage. The grid-leak resistor value is selected to produce an average bias voltage in the class C range.

Grid-leak (gate- or base-leak with JFETs or BJTs) bias can be used only if input circuit current flows. It cannot be used in receiver or other small-signal class A RF amplifiers because no input current flows in these circuits. It is developed by I_{dc} flowing through a leak-back resistor when the input is driven into the positive region by an input signal. This requires a power amplifier as the driver stage to produce the I_{dc}.

The value of the bias voltage can be determined by Ohm's law. If the average I_g shown on the

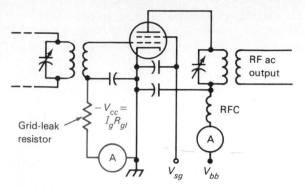

FIG. 15-13 A common tetrode RF power amplifier with grid-leak bias.

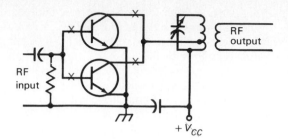

FIG. 15-14 Class C RF power amplifier with parallel BJTs.

meter, is 12 mA and the resistance is 10,000 Ω, the average grid bias voltage is $-V_{CC} = I_g R_g$, or 0.012(10,000), or 120 V. If this is less than desired, the coupling from the RF ac source can be tightened or the value of the grid-leak resistance may be changed.

There is one danger with grid-leak bias. It is normally used in high-power class C RF amplifiers with relatively high V_p. If the driver stage suddenly ceases operating for any reason, the RF amplifier is left with no input signal, zero bias, and high V_p. This will produce dangerously high I_p and I_{sg} that may damage the tube unless some precaution is taken. Many class C amplifiers use either a separate bias supply, or cathode-resistance bias in addition to the grid-leak bias. If the drive is suddenly interrupted, the fixed bias will be sufficient to limit the I_p and I_{sg} to safe values.

Another safety device is an *overload relay* in series with the output circuit. If excessive I_{dc} flows through the coil of the relay, the relay arm is pulled in, trips a latch, and shuts off the power-supply voltages.

15-10 PARALLEL OPERATION

To produce twice the RF power output possible from one transistor or VT, two or more may be connected either in parallel or in push-pull.

A parallel RF amplifier circuit (Fig. 15-14) has all similar elements connected together. Parasitic chokes, ferrite beads, or losser resistors can be connected as close as possible to the device's input and output terminals (colored Xs) to discourage high-frequency parasitic oscillations.

To test an RF amplifier stage for parasitic oscillations, the RF drive is removed, and any bias voltage is turned off. The power-supply voltage is adjusted low enough to limit the electrode currents to values that will not exceed their dissipation rating. If the output circuit I_{dc} varies when the output LC circuit is tuned across resonance, the circuit is unstable and may oscillate.

Transistors made for VHF or UHF operation may give parasitic trouble when used in HF systems.

Two devices in parallel lower the impedance as seen by the LC tank circuit. This may require lowering the value of inductance and raising the capacitance of the tuned circuit in order to match device and load impedances. More oscillatory current will now be flowing through the coil, which may require a coil made with larger wire.

Two devices in parallel will have twice the input and output capacitances, which may be a disadvantage for VHF and UHF operation. Two parallel devices in class C require twice the grid driving power of a single device.

Test your understanding; answer these checkup questions.

1. What percent of cutoff bias is used for class A? _____ AB? _____ C? _____
2. What would be the output waveform if narrow pulses of dc were fed to a class C AF power amplifier? _____ RF power amplifier? _____
3. In a power-supply biased class C stage is it possible to have an input signal and no output signal? _____ In a class B stage? _____
4. What is the approximate angle of output current flow in a properly operating class A stage? _____ In a class B stage? _____ In a class C stage? _____
5. When a class C stage has grid current flowing, would this charge or discharge a battery producing the bias? _____
6. Why can a VT operate with higher power-supply voltage in class C than in class A? _____
7. Can I_G flow in a class C JFET? _____ In a class C VMOS? _____

8. What does a ferrite bead do in an RF amplifier? _____

9. Why may circuits operate better if the RFC is reversed? _____

10. Does a dip in I_p indicate LC circuit resonance with VTs? _____ Does a dip in I_D, with FETs? _____ Does a dip in I_C, with BJTs? _____

11. When I_p is dipped, what does the I_g do? _____ The I_{sg}? _____

12. Why are V_{bb} and V_p the same in an RF power amplifier? _____

13. Where "L" means antenna, $I_p = 0.78$ A, $V_p = 3200$ V, $I_L = 5.7$ A, and $R_L = 50$ Ω, what is the stage efficiency? _____

14. With $V_p = 2500$ V, $I_p = 1.2$ A, $I_g = 15$ mA, and $R_{g1} = 10,000$ Ω, what is the bias value? _____

15. If grid-leak bias alone is used, what may happen if the driving stage output ceases? _____

16. Two active devices in parallel have 5000 Ω internal impedance each. What should the Z_L be for maximum P_o? _____

17. Why is parallel operation not good in VHF or UHF circuits? _____

15-11 PUSH-PULL OPERATION

RF amplifier tubes or transistors can be connected in push-pull to produce twice the output power obtainable from one alone. Figure 15-15 shows a

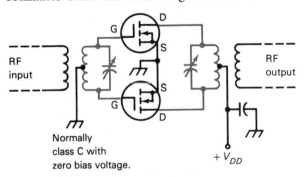

FIG. 15-15 Push-pull class C VMOS RF power amplifier.

Normally class C with zero bias voltage. $+V_{DD}$

pair of VMOS transistors in push-pull. An advantage of push-pull is the cancellation of even-order (2nd, 4th, etc.) harmonic energy in the output. Each gate must be fed the same-amplitude signal voltage, the LC tank circuit must be accurately center-tapped, the coupling to the output should be taken equally from both devices, and they must be evenly matched.

The interelectrode capacitances across the tuned circuits are in series. This results in only half the

capacitance across the tuned circuits that would be present with one device alone and one-quarter the capacitance of parallel operation, a decided advantage at higher frequencies.

Note that multiple ground indications simplify diagraming. It is assumed that the reader understands that each stage has a single grounding point.

15-12 TYPES OF FEED

It is possible to feed I_C from a BJT through the coil of the tuned LC tank circuit (series feed, Fig. 15-16a), or the I_C can be led off through an RF choke (Fig. 15-16b). In the latter case, the reactive

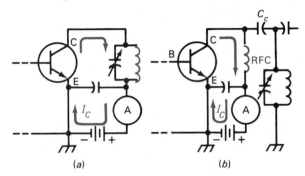

FIG. 15-16 (a) Series-fed and (b) shunt-fed RF amplifier output circuits.

ac voltage-drop across the RFC, developed by pulsating dc flowing through it, is coupled to a tuned circuit by a coupling capacitor C_c (parallel or shunt feed).

In either series- or shunt-fed circuits, as the LC circuit is tuned to the frequency applied to the input the I_{dc} will dip. With shunt feed, one side of the tuning capacitor is grounded, a mechanical advantage. Also, there are no high dc potentials on the coil or capacitor. (A π-network coupling circuit has the same advantage.)

15-13 COUPLING RF AMPLIFIERS

The two stages in Fig. 15-17 are inductively or transformer-coupled. With tuned inductive coupling between stages, the circuits will require very loose coupling and may have to be placed farther apart than expected. It may be preferable to close-couple the two coils and *tune only one* of them.

The two-transformer method of coupling shown in Fig. 15-18 is known as *link coupling*. The drain coil is the primary of a step-down transformer, and the gate coil is the secondary of a step-up trans-

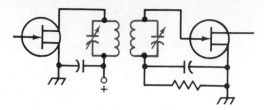

FIG. 15-17 Tuned transformer coupling between two RF amplifiers.

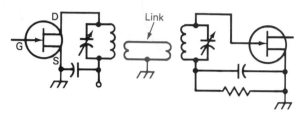

FIG. 15-18 Link coupling between two tuned circuits.

former. The first transformer ratio reduces the impedance of the link line to a low value. The second steps it up again. Energy can be carried long distances with little loss, particularly if coaxial cable is used between the link loops. The intercoupled circuits need not be physically close together, as with other forms of coupling. Link coupling reduces stray capacitive coupling, thereby reducing transmission of unwanted harmonic frequencies that might be generated in the drain circuit. This is aided by grounding one side of the link line.

A form of capacitive or impedance coupling is shown in Fig. 15-19. The degree of coupling can

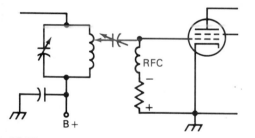

FIG. 15-19 Capacitive coupling to an RF amplifier input.

be changed either by varying the coupling capacitor or by the tap on the coil. The closer the tap to the ground end of the coil, the less the coupling. The RF choke may not be necessary in the grid circuit. This form of coupling tends to transfer harmonics generated in the driving stage to the next input circuit. Since the coupling capacitor has high B+ on one side and negative bias on the other, it is under quite a voltage strain, particularly when RF ac is added. If it shorts out, it applies high B+ to the grid of the next tube, which can damage that tube. If it opens, the associated grid-leak bias falls to zero, resulting in dangerously high I_p and I_{sg} in this tetrode.

A fourth method of coupling RF amplifiers is by a π-network circuit (Sec. 15-29).

15-14 TRIODE-TYPE RF AMPLIFIERS

Triode-type devices such as VMOSs, BJTs, and VTs are used in RF power amplifiers, although tetrodes and pentodes have the advantage of greater sensitivity; require less RF drive in grounded-source, -emitter, or -cathode circuits; and may not require *neutralization*. If an RF amplifier in a transmitter is not properly neutralized, it may generate many spurious signals, interfering with other radio services. The I_{dc} may not vary smoothly as the output circuit is tuned through resonance (a common indication of an improperly neutralized stage), or there may be RF output when the driving ac is withdrawn.

There are several ways by which neutralization may be introduced into a stage to prevent it from oscillating. In vacuum-tube terms, some of these are (1) plate neutralization, (2) grid neutralization, (3) direct neutralization, (4) inductive neutralization, and (5) use of a losser resistor.

The general theory of the first four methods is somewhat the same. Any ac emf fed back output to input by interelectrode capacitance is counteracted by feeding back an equal but opposite phase voltage, resulting in neutralization.

A losser resistor consists of a resistor in series with the gate, base, or grid lead (Fig. 15-20). The loss of energy in the losser resistor plus the out-of-phase, or degenerative, voltage developed across it when any current flows through it adds enough loss to the input circuit to counteract small values of regenerated energy. A losser resistor may discourage HF or VHF parasitic oscillations and also produce degeneration at the frequency to which the amplifier is tuned. The losser-resistor value should

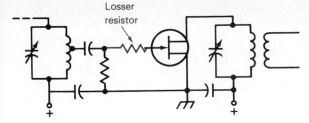

FIG. 15-20 Placement of a losser resistor in the gate circuit of an RF amplifier.

be as low as possible to suppress oscillations without too much loss of gain. Values may range from 50 to more than 1000 Ω. Ferrite beads around input and/or output circuit leads may work better than a losser resistor.

15-15 HAZELTINE NEUTRALIZATION

When neutralization is used in the output circuit of a VT RF amplifier, it is usually known as "plate neutralization," but is properly termed a *Hazeltine balance circuit*. Because triode-type transistors in RF amplifiers have such a small value of interelectrode capacitance, they may often operate satisfactorily without any neutralization. This is not true in the case of VT triodes in grounded-cathode circuits.

As discussed in Sec. 15-2, if C_n in Fig. 15-21 is made equal to the interelectrode capacitance (shown

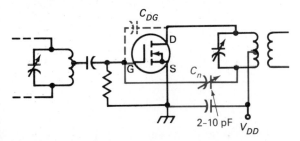

FIG. 15-21 Hazeltine-neutralized RF amplifier.

dashed) between gate and drain and if the coil is center-tapped, C_n will feed an equally negative voltage back to the gate when the top of the coil is feeding a positive voltage back to the input. Under this condition the two feedbacks cancel, the stage is neutralized, and it cannot oscillate, only amplify.

The neutralizing capacitance is a variable capacitor in order that the proper capacitance can be found by trial. Actually, it is not necessary to center-tap the output coil exactly. If the tap is closer to the bottom of the coil, a smaller neutraliz-

ing voltage will be developed across the lower part of the coil. This can be overcome by adjusting the neutralizing capacitor to a higher value of capacitance.

Actually, to produce oscillation of an RF amplifier, the output circuit must be tuned slightly higher in frequency than the input circuit.

Instead of center-tapping the tuned circuit, a *split-stator* capacitor may be used across the tuning coil (Fig. 15-22). This is a dual variable capacitor

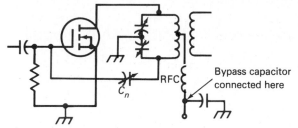

FIG. 15-22 Hazeltine neutralization with a split-stator capacitor.

with a common rotor connection for both stator sections. It center-taps the capacitive half of the LC circuit. Center-tapping the LC circuit in this manner also establishes a central ground-potential point, allowing the opposite ends of the coil to have equal potentials but opposite polarities. Either method of center-tapping the tuned circuit operates equally well, but only the C or L should be center-tapped, not both. If a split-stator capacitor is used, the coil should not be bypassed to ground at the center, or the circuit forms two tuned circuits, one made up of the top capacitor and top half of the coil and the other of the bottom capacitor and bottom half of the coil. The center of the coil should be connected to the power supply through an RF choke.

15-16 RICE NEUTRALIZATION

Another way of neutralizing a triode RF amplifier is to center-tap the input LC circuit (Fig. 15-23). Amplified energy in the output circuit will feed back to the top of the input circuit coil via the inter-

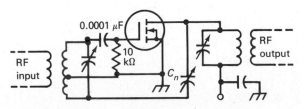

FIG. 15-23 Rice-neutralized triode RF amplifier.

electrode capacitance. The neutralizing capacitor C_n, if properly adjusted, will feed the same value of energy back to the bottom of the input circuit coil. Now the same voltage is being fed into both ends of the input coil at the same time, canceling any effective feedback of energy. The stage is neutralized. The input circuit may use a split-stator capacitor instead of a center-tapped coil to develop a ground potential at the center of the LC circuit. Such circuits are called *Rice balance* or *grid neutralization circuits*.

15-17 NEUTRALIZING BY RF INDICATOR

A method of neutralizing a low-powered triode RF amplifier stage is:

1. Turn off the power-supply voltage to the stage to be neutralized.
2. Couple to the tank circuit some device that will indicate the presence of RF ac in the circuit, such as a loop of insulated wire in series with a sensitive RF thermogalvanometer, a loop and flashlight lamp, a loop and an oscilloscope, a loop and a milliammeter with a germanium diode across it, any form of wavemeter having an RF-indicating device in it, or a neon lamp held to the "hot" (ungrounded) end of the tank coil.
3. Make sure the stage is being excited by RF ac.
4. Rotate the neutralizing capacitor to maximum capacitance to deneutralize the circuit.
5. Tune the tank circuit for maximum RF ac indication. (The tank circuit is now being fed RF via the neutralizing capacitor and should give a strong RF indication when the LC circuit is tuned to resonance.)
6. Rotate the neutralizing capacitance to zero RF ac indication. Remove the RF indicator!

The stage should now be neutralized and ready to operate as soon as the power-supply voltage is applied.

This method of neutralizing can be used with either Hazeltine- or Rice-connected transistor RF power amplifiers.

15-18 NEUTRALIZING BY I_g METER

A method of neutralizing higher-powered triode RF amplifiers by I_g meter only is to:

1. Turn off the power-supply voltage for the stage and apply drive to the input circuit.

2. Tune the output tank capacitor through resonance and watch the grid milliammeter. If the I_g fluctuates at all during the tuning, the stage is not neutralized.
3. Keep tuning the output circuit back and forth through resonance and adjusting the neutralizing capacitor until the I_g does not change as the output circuit is tuned through the resonant frequency. The stage is now neutralized.

If there is no meter in the grid circuit, an RF voltmeter can be connected across the input circuit of either VT or transistor amplifiers to act as the input circuit indicator.

15-19 NEUTRALIZING PUSH-PULL STAGES

The neutralizing circuit used with push-pull RF amplifiers (Fig. 15-24) is a combination of both Rice and Hazeltine circuits. Note that C_{n_1} is acting as a

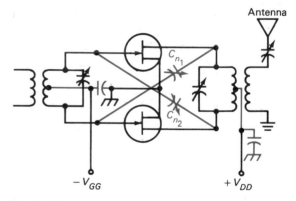

FIG. 15-24 Neutralized push-pull RF amplifier feeding an antenna.

Hazeltine neutralizing capacitor for the top JFET and a Rice neutralizing capacitor for the bottom. C_{n_2} is a Rice neutralizer for the top JFET and a Hazeltine neutralizer for the bottom. Both neutralizing capacitors are adjusted at the same time to maintain equal-capacitance values for each.

Because both input and output circuits are center-tapped and are assumed to be evenly balanced electrically, neutralization of push-pull stages tends to produce a neutralized condition over a relatively wide band of frequencies. Single-ended stages often require reneutralization if the frequency of operation is changed materially.

15-20 DIRECT NEUTRALIZATION

A completely different method of producing neutralization is the *direct* method, used mostly in broadcast stations. It operates on the theory that a parallel-tuned circuit has a very high impedance to the frequency to which it is tuned. If the correct value of inductance is selected and connected as L in Fig. 15-25, a parallel-resonant circuit is formed

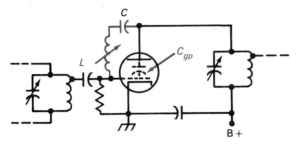

FIG. 15-25 Direct neutralization.

by L and C_{gp}. If this circuit resonates at the desired frequency of operation, it forms a very high impedance to this frequency and no ac energy is fed back from plate to grid circuit.

If the coil L were used without capacitor C, it would place B+ on the grid and the stage would not operate. To prevent this, blocking capacitor C must be added in series with the coil. It is selected to have less than one-tenth the reactance of C_{gp} to the operating frequency, so that the L and C_{gp} circuit resonates as if it were not there. Direct neutralization is effective only if the L and C_{gp} are tuned to the frequency of operation. A new value of inductance is required if another frequency of operation is used.

15-21 INDUCTIVE NEUTRALIZATION

Another method of neutralizing an RF amplifier is *inductive neutralization* (Fig. 15-26). To neutralize any RF amplifier it is possible to couple a pickup loop loosely to the output circuit. The energy picked up by this loop can be inductively coupled into the input circuit by a second loop. If either loop is turned over, the phase of the inductive feedback will be reversed. With the loops coupled one way, the stage may be regenerative and will oscillate. If one loop is turned over, the feedback becomes degenerative. Then, varying the coupling of either loop can produce neutralization.

The loops should be coupled to the *cold*, or

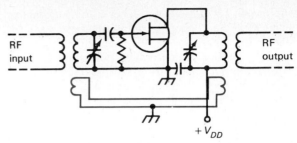

FIG. 15-26 Inductive neutralization of an RF amplifier.

ground-potential, ends of tuned circuits, as shown. Any load should be coupled to the cold end rather than to the hot end of any LC circuit to prevent detuning it. Couple to the center for push-pull circuits.

15-22 NEUTRALIZING OTHER DEVICES

In general, tetrode and pentode tubes made for use in grounded-cathode RF circuits will need no neutralization at frequencies below 20 to 50 MHz. Above this, particularly if the input and output circuits are not adequately isolated and shielded, neutralization may be necessary. The very small grid-plate capacitance in such tubes makes necessary very little neutralizing capacitance. For example, with the grid circuit below the chassis level and the plate circuit above, it may only be necessary to run a stiff wire from one end of a center-tapped grid coil up through a hole in the chassis to a position near the plate of the tube to produce enough capacitance to neutralize the tube.

Some high-frequency tetrodes are constructed in such a way that they may operate up to 500 MHz without neutralization.

When connected in grounded-grid or common-plate circuits, tubes should not require neutralization any more than grounded-gate, -drain, -grid, or grounded-base or -collector circuits do.

15-23 FREQUENCY MULTIPLIERS

Oscillator circuits may have good frequency stability when operated on lower frequencies, but at higher frequencies the stability may not be good. It is possible to operate an oscillator at a lower frequency and feed its output to frequency-multiplier stages. Such stages will produce an output two, three, four, or five times the frequency fed to its input. The stability, as well as the frequency of the output, is always an exact multiple of that of the oscillator. This results in a better output stability than if the oscillator were used to generate the higher frequency directly. Frequency multipliers are found in most FM transmitters (Chap. 19).

Basically, a multiplier stage is a class C RF power amplifier, but with its output circuit tuned to resonate at a frequency some whole-number multiple of that fed to its input.

The VMOS amplifier circuit (Fig. 15-27) might be expected to be a frequency multiplier because it lacks neutralization. Since the output is not tuned to the same frequency as the input, the stage cannot break into self-oscillation and needs no neutralizing. A *frequency doubler* usually has an RF output power about half that of a normal amplifier. Its efficiency may range between 30 and 40%.

When the output circuit is tuned to resonate at a frequency three times that applied to the input, the stage operates as a *tripler*, emitting energy at a frequency exactly three times the input. The power output from a tripler is less than from a doubler. Still higher multiplications produce progressively less power output. It is unusual to have a multiplier operating at more than the fifth harmonic.

Maximum output is produced in a doubler by using a high value of class C bias, usually two to four times cutoff, and driving hard enough to produce high-amplitude, narrow-width I_D pulses. Each pulse excites the output LC circuit. The electrons are driven down through the LC circuit but oscillate back up, then down, and then up again by flywheel effect before the next I_D pulse arrives. In this way each pulse produces two cycles of nearly sine-wave ac in the output LC circuit. In the case of a tripler,

the LC circuit must use the flywheel effect of a high-Q circuit to allow three complete flywheel cycles to be formed for each I_D pulse. Each succeeding unexcited oscillation, or ac cycle, decreases in amplitude, resulting in a lowered average output power from a doubler and still less from triplers.

The conditions for the best power output on a harmonic frequency from a multiplier stage can be summarized as: High bias voltage, high driving voltage, high supply voltage, high-Q tank circuit, and loose coupling to the load.

The opposite conditions are required to reduce harmonic output from an RF amplifier: As little bias as possible while still operating efficiently, as little drive as possible with good efficiency, loose coupling between output tank and load, and a low L/C ratio in the output tank.

A frequency doubler with high efficiency is the *push-push* circuit (Fig. 15-28). The gate circuits are in push-pull, and the drains are in parallel. The stage is biased to class C by the RC network. The output circuit is tuned to twice the frequency of the input

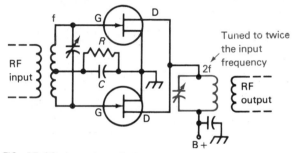

FIG. 15-28 A push-push frequency doubler.

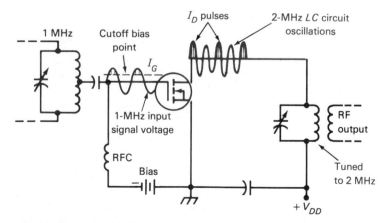

FIG. 15-27 A frequency-multiplier, or -doubler, stage indicating input and output signal voltages and currents.

circuit. The two gates are excited 180° out of phase.

When the gate of the top JFET is positive, an I_D pulse flows downward in the output LC circuit, starting an oscillation at the LC circuit frequency. During this time the gate of the lower JFET is highly negative, and this JFET is inoperative. Between half cycles of the excitation voltage there is a period of time when neither JFET is passing current and the flywheel effect of the tank circuit allows electrons to oscillate back up through the tank coil. When the cycle in the gate circuit reverses, the gate of the bottom JFET is driven positive and a I_D pulse again flows downward in the LC circuit, again exciting the tank. In the push-push circuit only one half cycle of each tank-circuit oscillation depends on flywheel effect. In single-device doublers, the flywheel effect must operate by itself for three half cycles before another excitation pulse occurs. This is the reason the push-push circuit can produce RF output at twice the input frequency with almost the same efficiency as a single-ended RF amplifier.

A push-push stage needs no neutralization. It will not operate as a tripler but will produce output on even-order (2nd, 4th, etc.) harmonics. A push-push VT stage has a relatively high plate-to-cathode capacitance because the two plates are in parallel, making VHF and UHF operation difficult. Transistors in push-push operate up into the microwave region.

Unlike push-push, a *push-pull* class C stage (Fig. 15-24 without the neutralizing capacitors) will not operate as a doubler but makes an excellent tripler, and produces output on odd harmonics (3rd, 5th, 7th, etc.).

The first harmonic is the fundamental frequency. A harmonic is a whole-number multiple of some frequency. When a fundamental is multiplied by 2, the answer is the second harmonic. Thus, the 2nd harmonic of 800 kHz is 1600 kHz, the 3rd is 2400 kHz, etc.

To develop output power on high frequencies, several frequency multipliers can be used in *cascade*. The output frequency of three doublers in cascade is $2 \times 2 \times 2$, or 8 times the fundamental. If the output frequency of three doublers in cascade is 16,840 kHz, the input frequency to the first doubler is 16,840/8, or 2105 kHz.

If a tripler and two doublers are used with an input of 1 MHz, the output will be $3 \times 2 \times 2$, or 12 MHz. The same output will result if the arrangement is doubler, doubler, tripler, or doubler, tripler, doubler.

■ Test your understanding; answer these checkup questions.

1. What harmonics are lessened by using push-pull? _____ Push-push? _____
2. Which operates better in VHF circuits, push-push or push-pull? _____
3. Would dc current have to be flowing through an RFC in a series- or in a shunt-fed RF amplifier? _____
4. What type of feed has no dc potential on its LC components? _____
5. What are the four types of RF ac coupling mentioned? _____ _____
6. What type of coupling does not discriminate against harmonics? _____
7. Where might losser resistors be used to advantage in VT circuits? _____ In FET circuits? _____ In BJT circuits? _____
8. Should the input circuit be tuned a little higher or lower in frequency to produce TPTG-type oscillation? _____
9. For what is a split-stator capacitor used in an RF amplifier? _____
10. What are other names for Rice neutralization? _____ For Hazeltine neutralization? _____
11. Why must an amplifier have no output circuit dc applied to it when it is being neutralized? _____
12. An RF indicator in the input circuit of a stage being neutralized dips when the output circuit is tuned across resonance. Is the stage neutralized? _____
13. In what service might direct neutralization be found? _____
14. Which circuit would never be neutralized: single-ended, push-pull, or push-push? _____
15. What is the approximate efficiency of a single-ended doubler? _____ Of a push-push stage? _____
16. What type of bias is used to produce only class C operation? _____
17. What multiplier circuit produces odd harmonics best? _____ Even-order harmonics? _____
18. For a 4 MHz ac, what is its 4th-harmonic frequency? _____ Its 4th-octave frequency? _____

15-24 DETERMINING L AND C FOR RESONANT CIRCUITS

When tuned circuits are discussed, what size coil and what value capacitor should be used? Actually, there is only one value of L and C that will give best output for a given set of circuit conditions. Since many of the factors in a circuit are variable, an accurate determination is rather involved. An old rule of thumb for HF circuits can be used to determine a workable capacitance. With the resonance formula an inductance value can be computed. In the earlier days of radio, frequency of RF ac was not employed as much as it is today. *Wavelength* was used instead. A wavelength is the distance that an electric impulse, or wave, will travel in the time it takes to complete one cycle of the ac being considered (Fig. 15-29). For example, consider a

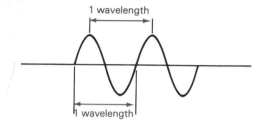

FIG. 15-29 Two possible measurements of the wavelength of an ac cycle.

frequency of 1,000,000 Hz. Its *period* (1/f) is 0.000001 s. It takes this much time to complete one cycle. In this time, the electric impulse, or wave, will travel 300 meters (m) through space. The velocity of electric impulses, or radio waves, is 300,000,000 m/s. A wave having a frequency of 1 MHz has a wavelength of 300 m.

The formulas to convert wavelength to frequency or frequency to wavelength are

$$\lambda = \frac{V}{f} = \frac{300,000,000}{f} = \frac{300,000}{f_{kHz}} = \frac{300}{f_{MHz}}$$

$$f_{Hz} = \frac{V}{\lambda} = \frac{300,000,000}{\lambda}$$

where V = velocity of propagation, m/s
 f = frequency, Hz
 λ = wavelength, m

The rule of thumb mentioned above is: For a high-impedance resonant circuit, use 1 pF of capacitance for each meter of wavelength. For a 1-MHz *LC* circuit, 300 pF is a possible *C* to be used. What

capacitance could be used in a tuned circuit operating on a frequency of 6000 kHz? The wavelength is

$$\lambda = \frac{300,000,000}{6,000,000} = 50 \text{ m}$$

A possible *C* to use is 50 pF. This is unlikely to be the value of capacitance for optimum operation of the tuned circuit. It is merely a simple starting approximation. In some circuits twice the value may be better; in others half may be better. For low-impedance BJT RF circuits 100 times the rule of thumb might be best.

To determine the *L* required to resonate the 50-pF *C* to 6000 kHz, the resonance formula (Sec. 8-1) can be used:

$$L = \frac{1}{4\pi^2 F^2 C}$$

$$= \frac{1}{4(9.86)(36 \times 10^{12})(50 \times 10^{-12})}$$

$$= \frac{1}{70,900} = 14 \ \mu\text{H}$$

If the coil must have 14 μH, what size and how many turns should it have? Assume an RF coil with diameter and length approximately equal. In Sec. 5-3, a formula is given for such a coil:

$$L = \frac{r^2 n^2}{9r + 10l}$$

The required coil might be made in many dimensions. For this problem, a 1-in. diameter and a 1-in. length will be tried. How many turns will be needed? Rearranging to solve for turns *n* yields

$$n = \sqrt{\frac{L(9r + 10l)}{r^2}} = \sqrt{\frac{14[9(0.5) + 10(1)]}{0.5^2}}$$

$$= \sqrt{\frac{203}{0.25}} = \sqrt{812} = 28.5 \text{ turns}$$

From a copper-wire table in a handbook, it is found that a No. 19 enamel-insulated wire will wind about 27 turns to the inch. Therefore, a 1-in.-long coil wound with No. 19 enameled copper wire on a 1-in.-diameter form will produce an inductance of approximately 14 μH. This coil, when shunted with a 50-pF capacitor, should form a resonant circuit at about 6000 kHz. (Distributed capacitance and other circuit capacitances will lower the frequency of resonance considerably.)

If the diameter of No. 19 wire is too small for the current that will be flowing, it will be necessary to increase the dimensions of the coil until the desired wire size can be accommodated.

A frequency doubler has an input frequency of 1000 kHz. The output circuit inductance is 60 μH. What output capacitance is necessary for resonance? Since the stage is a doubler, the output frequency is 2000 kHz. Using 60 μH and solving for C with the resonance formula rearranged yields

$$C = \frac{1}{4\pi^2 F^2 L} = \frac{1}{4(9.86)(4 \times 10^{12})(60 \times 10^{-6})}$$
$$= \frac{10^{-6}}{39.4(240)} = 0.000106 \ \mu\text{F} = 106 \ \text{pF}$$

This is the total capacitance, including device and incidental circuit capacitances, needed to resonate a 60-μH coil to 2000 kHz.

15-25 CENTER-TAPPING FILAMENTS

Many high-power transmitting tubes use directly heated cathodes. The heating of these filaments is usually by 60-Hz ac. As explained in Secs. 9-18 and 14-25, to produce as little hum as possible, the filament circuit is center-tapped (and grounded). Both grid- and plate-circuit returns are made to the center tap (Fig. 15-30). In RF stages two bypass

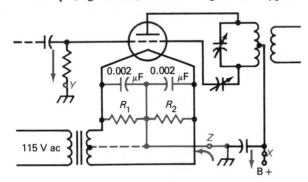

FIG. 15-30 Center-tapping the filament circuit of an RF amplifier triode.

capacitors must be connected from the filament terminals to ground with leads as short as possible. The center tap of the filament can be developed by using two equal-value resistors, R_1 and R_2 (10 to 50 Ω), or the center of the filament winding can be grounded (dotted line). Since the capacitors complete the RF ac circuit from filament to ground, the length of the filament wires to the transformer is not critical.

A milliammeter inserted at point X will read I_p. If it is at point Y, it will read I_g. If it is inserted at point Z, it will read the sum of both I_p and I_g, called the *cathode current*, since both currents flow

to the filament center tap. In a tetrode it would read the sum of I_p, I_g, and I_{sg}. In a pentode it would also read any suppressor current.

15-26 GROUNDED-GRID RF POWER AMPLIFIERS

When the input signal is fed to the grid-cathode (gate-source, or base-emitter) circuit with the grid at ac ground potential (Fig. 15-31), a grounded-grid

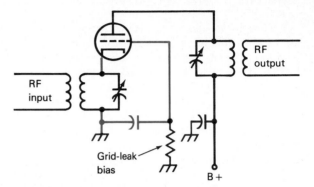

FIG. 15-31 A grounded-grid class C RF power amplifier.

circuit is produced. As the cathode is driven more positive by the input signal, the grid at the other end of the input tank becomes more negative than the cathode. As the grid becomes negative in respect to the cathode, the I_p decreases and the plate becomes more positive. Thus, the tops of the input and output tanks are positive at the same time, or in phase, not 180° out of phase, as in grounded-cathode amplifiers. With this phase relation no regenerative feedback occurs and *no neutralization is required*. This is a widely used circuit.

The input impedance of a grounded-grid circuit is relatively low, requiring a high C/L ratio in the input tank. Such a stage requires several times as much power to drive it, but most of the driving power is available as output because input and output circuits are in series. Since the input and output are in series, the output signal will be proportional to the input voltage plus the μ of the tube ($\mu + 1$). The gain of a grounded-grid stage is

$$A = \frac{(\mu + 1)R_L}{R_L + R_p}$$

With the grid ac grounded by the grid-leak capacitor, during one-half of the input cycle the grid is driven more positive than the cathode, and grid current flows. Current flowing down through the grid-leak resistor results in enough bias to produce class C operation for VTs (JFETs, BJTs).

In some RF power amplifiers it is necessary that the output circuit reproduce any variations in RF amplitude that are fed to the input circuit without changing, or distorting, the variations. Under this condition the amplifier must be a linear device and must be biased to class A, AB, or B. To operate the grounded-grid amplifier of Fig. 15-29 as a class A, AB, or B stage, it is necessary to substitute a low-impedance bias supply for the grid-leak resistor.

Although an indirectly heated cathode tube is shown for simplicity in the basic diagram, some high-power tubes use directly heated cathodes. One method of feeding filament current to the filament without allowing RF ac to appear in the filament transformer is to make the cathode coil of hollow copper tubing and feed both filament leads through the hollow center, bypassing the filament terminals to the top and bottom of the cathode coil, as in Fig. 15-32. Another possibility is to use an RF choke in

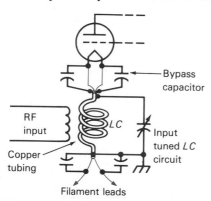

FIG. 15-32 A method of feeding the filament in a grounded-grid power amplifier.

each filament lead and bypass both filament terminals to the top of the cathode coil. The filament leads at the bottom of the copper tube coil and at the bottom of the RF chokes are also bypassed to ground.

Although a triode RF power amplifier is shown, tetrode and pentode RF power amplifiers may also use a grounded-grid circuit.

15-27 VHF AND UHF AMPLIFIERS

In AF stages the length of component leads, placement of parts, shielding, and physical size are not too important. In the HF range (3 to 30 MHz),

these factors begin to affect operation. Capacitor leads must be short; coils and parts must be placed in positions that will not induce unwanted voltages from one to the other; interstage shielding becomes important; input and output leads must be kept apart; neutralization may be required.

In the VHF range (30 to 300 MHz), the factors mentioned become increasingly important. Interelectrode capacitances of tubes shunting LC circuits may lower the resonant frequency excessively. *Transit time*, the time required for electrons to move from the grid area to the plate, may be longer than one-half of a cycle of the signal frequency, preventing operation of circuits. Special active devices must be used. LC circuits are supplanted by hairpin, coaxial, or cavity tanks. Grounded-grid-type circuits are favored over grounded-cathode. Connecting wires are larger to obtain higher Q and less loss, and may be silver-plated to reduce skin effect. Parts are placed as close together as possible to reduce lead length. Wiring goes as directly as possible, component to component, with no length loss in right-angled lead dress. Capacitors must be mica or other low-loss dielectric types. Resistors tend to become capacitive and pass VHF or UHF ac across them. Because of reduced effective capacitance, push-pull circuits are better than single-ended ones.

Bipolar transistors are available that can be used up to 5 GHz. Low-power JFETs can be used to 1 GHz. VMOS can be used up to perhaps 200 MHz with power inputs of 200 W.

15-28 TRANSISTOR RF POWER AMPLIFIERS

Whereas VT RF power amplifiers may produce hundreds of kilowatts of RF power, HF transistor RF amplifiers may not exceed 200 W. VHF and UHF amplifiers usually run less than 100 W. However, there are far more transistorized transmitters and receivers in operation than VT equipment.

Generally speaking, discrete transistor RF amplifier circuitry and VT circuitry (single-ended, push-pull, etc.) are similar. There are an unlimited number of possible circuits. Figure 15-33 shows four cascaded stages that illustrate some of the variations possible in HF BJT circuits.

The first stage is a common-base circuit (similar to grounded-grid) with no base-emitter bias. It operates in class C without bias and needs no neutralization.

The second stage is a common-emitter circuit, again with no bias and operating in class C. The neutralizing circuit in dashed lines should be recognized as being a Hazeltine balance type of neutralization. Without any neutralization and with some added reverse bias, or with some "base-leak" bias, this stage could act as a frequency multiplier.

The third stage is also a common-collector circuit, again with no forward bias in the base circuit. When the base is driven with a signal, pulsating emitter current flows through the RFC, developing an ac voltage-drop across it. Positive half cycles of the input signal will draw electrons from the emitter down through base-leak resistor R_b to ground, which develops reverse bias. The high bias allows the circuit to operate efficiently as a frequency multiplier.

The fourth stage is a push-push frequency doubler, biased to class C by the 2- and 100-Ω resistors that place a positive potential on the emitter. Push-push and push-pull stages require transistors with very short storage times or they will not function as high-frequency multipliers.

With no input signal, collector current is zero in all of these circuits. With any driving signal I_C develops. As the collector loads are tuned to resonance, I_C values should dip to minimums.

Variation of the output circuit tuning in BJT amplifiers reflects considerable impedance variation back into the input, detuning the driving stage somewhat.

Whereas VT transmitters may have many grid and plate circuits metered in high-power equipment, transistorized systems usually do not. They may have an output signal indicator (SWR meter, RF voltmeter, etc.), but for checking separate stages "test points" (TP) are brought out somewhere on the chassis. Test points may be connected directly to points in the circuit, as shown in Fig. 15-33, or they may be decoupled from circuit points through 1-kΩ resistors. (With a 20-kΩ/V meter the voltage-drop across a 1-kΩ resistor is considered negligible.)

Assume a 12-V power supply is being used for this amplifying system. With the oscillator stage off there is no signal passing through any of the class C stages, so there is no I_C. There is therefore no voltage-drop across any of the decoupling resistors R_d. Test points 1, 3, 5, and 7 should read 12 V; TP$_2$ and TP$_4$ should read 0 V; TP$_6$ should read $2/_{102}$ of 12 V, or +0.24 V.

If the oscillator is turned on, TP$_1$, TP$_3$, TP$_5$, and TP$_7$ should all read a fraction of a volt less than 12 V if the stages are close to being in tune. If not, with an RF indicator (oscilloscope, VOM or EVM with an RF diode probe) TP$_2$ should show a peak of RF ac when the CB stage is tuned to resonance, and the TP$_1$ voltage should *peak* slightly. Reading dc voltage at TP$_4$ should show a peaking when the CE stage is resonated, the TP$_3$ voltage should peak slightly, and the TP$_5$ voltage should dip because more I_C is now flowing in the CC amplifier stage. TP$_6$ should indicate a positive peak of voltage when the CC amplifier is resonated. When the push-push stage is resonated, I_C decreases and the TP$_7$ voltage should peak with the TP$_6$ voltage dipping slightly. If the output load is increased or tuned to resonance to accept more power, I_C should increase, TP$_7$

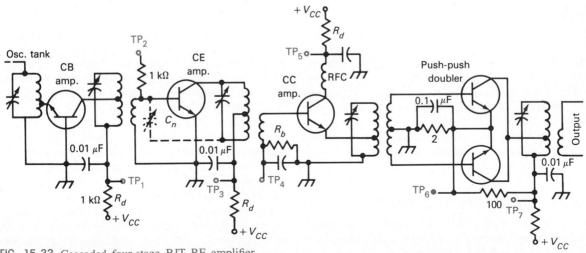

FIG. 15-33 Cascaded four-stage BJT RF amplifier.

should decrease, and TP_6 should increase its voltage reading.

If the stages are nearly in tune, all stages can be trimmed to resonance by watching the TP_7 voltage. As each stage is tuned, it drives the following stages harder, the push-push stage draws more I_C, and TP_7 should show a further dip in reading. But when the push-push stage is tuned, TP_7 should peak slightly, indicating its I_C dip.

In BJT LC circuits the C/L ratio is always high, being four to six times that of FET or VT circuits, unless the coil is tapped down. The tap point may be from one-tenth to one-third of the way up the coil, although in diagrams it is often shown as a mid-tap.

Many things might go wrong with the transistorized system shown, and perhaps only minimal notable difference might result. Any of the decoupling resistors could short out or increase in resistance and not affect circuit operation greatly. The 2-Ω resistor or its bypass capacitor in the push-push stage could short out, or the 100-Ω resistor could open with only a slight reduction of output, although I_B and I_C might increase excessively. The biasing resistor or its bypass capacitor in the CC stage could short out and reduce output only slightly, but I_B might now increase enough to burn out the EB junction. (Note that emitter RC networks are usually not found in class C RF *power* amplifiers.) Shorting of the output circuit bypass capacitors may excessively heat or burn out the decoupling resistor associated with it. If the RFC in the CC stage shorted, the collector would be bypassed to ground and there would be little or no RF output. If it opened, the EB junction would act like a diode and feed rectified ac to the output tank circuit, and a very weak output signal would result from the stage. The CC-EB junction might burn out and short without stopping all RF feed to the push-push stage. If one of the push-push BJTs opened, the output power would be halved and the TP_7 voltage would increase. If any BJT shorted, its output circuit TP voltage would drop to zero or to a very low reading. If a BJT open-circuits, there is usually minimal RF output, although there might be a little due to interelectrode capacitance.

15-29 PI-NETWORK OUTPUT

Probably the most frequently found output tuning and coupling circuits for HF and VHF transmitters is the *pi* (π) *network*, shown with a power FET

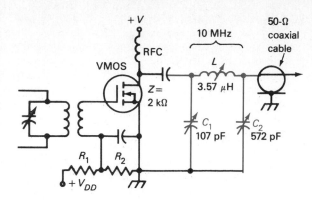

FIG. 15-34 A π-network output circuit.

(VMOS) in Fig. 15-34. It is a shunt-fed circuit, with no I_D flowing in the tuned circuit components. C_1 and C_2 are in series across the tuning coil L. The ratio C_1/C_2 determines what fraction of the total LC circuit impedance is coupled to the 50-Ω load and what fraction is seen by the active device. C_2 is the output impedance-matching component, and C_1 is used to bring the LC circuit to resonance. The greater the capacitance of C_2, the lower the output impedance. If the circuit is to be used on different frequencies, L may have to be variable, because for any one frequency there is only one inductance value which will produce optimum output when C_1 and C_2 are properly tuned. Note the forward bias to produce class A, AB, or B operation, developed by the R_1R_2 voltage divider. This FET will operate in class C with no forward bias, similar to a BJT.

The π network can also be used for interstage coupling and impedance conversion. (If C_2 is smaller than C_1, the input impedance will be lower than the output impedance.) While shown in a common-source circuit, it can be used with common-drain and common-gate circuits. It is also used with BJTs and VTs in any of their circuit configurations. Although it might be connected as a series-fed circuit, it is rarely found that way.

Computing the values of C_1, L, and C_2 is rather rigorous. The C and L values given in Fig. 15-34 are for a 50-Ω output load impedance and a 2000-Ω device impedance at a frequency of 10 MHz. At 10 times the frequency the C and L values would all be one-tenth of those shown. However, if the device impedance is one-tenth of 2 kΩ, the capacitances would both be 10 times greater, but the L value would be one-tenth that shown. From this information, values for any frequency and device impedance could be approximated. (Note that these values are only for a 50-Ω load.)

Test your understanding; answer these checkup questions.

1. What is the wavelength of 5 MHz? _____ What C value might be used for a 5-MHz LC circuit? _____ What L value? _____
2. What is reduced by center-tapping the filament circuit of a directly heated RF amplifier tube? _____
3. What currents make up the cathode current of a beam-power tetrode RF power amplifier? _____
4. Why would grounded-grid input circuits have large values of tuning capacitance? _____
5. In what way does the RF power driving a grounded-grid stage differ from that of a grounded-cathode type? _____
6. What is the main electrical difficulty when constructing a grounded-grid RF power amplifier? _____
7. Would transit time affect operation more in 1000-kHz or in 500-MHz circuits? _____
8. Why are UHF and microwave inductors and tanks often silver-plated? _____
9. Why do two active devices in push-pull require more tuning C than a single-ended stage? _____
10. With high-power VT transmitting systems how is stage operation known? _____ With transistorized transmitters? _____
11. When the I_C of a stage increases through a decoupling resistor R_d, what voltage change would its TP indicate? _____
12. In Fig. 15-33, with the circuit operating, what would the TP_5 voltage do if TP_3 bypass C shorted? _____ If the 2-Ω resistor opened? _____ If TP_4 bypass C shorted? _____ If the oscillator stopped oscillating? _____ If RFC open-circuited? _____ If RFC shorted? _____
13. Give three uses of an L and C π network. _____ _____ _____
14. What type of FET is shown in the π-network diagram? _____

15-30 TROUBLES IN HIGH-POWER TRANSMITTERS

When something goes wrong with an RF power amplifier, servicing personnel must consider all the evidence given by meters, smoke, heat, and whatever else can be seen or heard. The meters alone in the high-powered section of radio transmitters can help to localize the difficulty.

The plate-circuit milliammeter indicates normal or improper operation by its reading. If it reads *low*, there are several possibilities (check the reasoning behind these): (1) Power-supply voltage has dropped off. (2) Load circuit is not tuned or is not functioning properly. (3) Bias voltage may be too high. (4) There may be loss of RF drive if the circuit uses BJTs, or is power-supply biased. (5) With VTs, the filament voltage may have decreased or the cathode may have lost its emission. (6) Screen potential may be low. (7) Plate-circuit bypass capacitor may have shorted.

If the plate-circuit milliammeter indicates *higher* current than normal, some possibilities are: (1) Output circuit is detuned from a minimum current value. (2) Load is too tightly coupled. (3) There is loss of or insufficient bias. (4) With VTs, V_p or V_{sg} is too high. (5) There is lack of RF drive if grid-leak bias only.

The grid milliammeter is also a good indicator of the operation of VT stages. If it reads *low* it may indicate (1) too much bias; (2) low filament voltage, or the tube losing its emission; (3) lack of RF grid drive; or (4) an excessive I_p.

If the grid milliammeter reads too *high*, some reasons might be (1) loss or decrease of bias; (2) too much RF drive; (3) plate or screen circuits open or shorted to ground; or (4) load decoupled from the plate circuit.

While capacitors, resistors, transistors, and tubes are the most likely components to break down, meters, switches, transformers, and coils can oxidize and open if not burn out due to the shorting of some other part.

In a VT amplifier with grid-leak bias, if either the RF drive ceases or the I_p drops to zero, electrons driven off the hot cathode will flow through the grid-leak resistor, forming a small current to ground. The voltage-drop thereby developed is called *contact-potential bias*, and can help in analyzing what may be wrong with a transmitter. When I_p increases, the contact current decreases.

15-31 VT TROUBLESHOOTING PRACTICE

The TPTG oscillator, link-coupled to a neutralized triode amplifier (Fig. 15-35), represents a possible transmitter using unregulated power supplies. What would the different meters read (VH for very high, H for higher than normal, N for normal, L for lower than normal, VL for very low, R for reversed reading, and 0 for zero) if the faults listed on page 340 occurred? (Answers on page 342.)

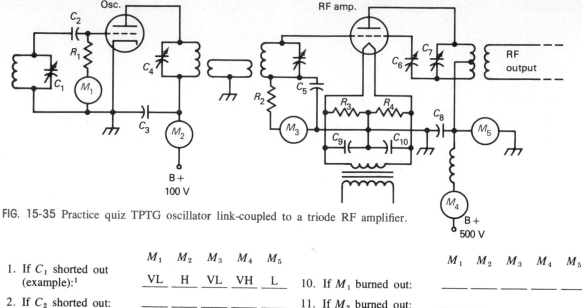

FIG. 15-35 Practice quiz TPTG oscillator link-coupled to a triode RF amplifier.

	M_1	M_2	M_3	M_4	M_5
1. If C_1 shorted out (example):[1]	VL	H	VL	VH	L
2. If C_2 shorted out:					
3. If C_3 shorted out:					
4. If C_4 shorted out:					
5. If C_5 shorted out:					
6. If C_6 shorted out:					
7. If C_7 shorted out:					
8. If C_8 shorted out:					
9. If C_9 shorted out:					

	M_1	M_2	M_3	M_4	M_5
10. If M_1 burned out:					
11. If M_2 burned out:					
12. If M_3 burned out:					
13. If M_4 burned out:					
14. If M_5 burned out:					
15. If RFC shorted:					
16. If R_1 burned out:					
17. If R_2 burned out:					
18. If R_3 burned out:					
19. If oscillator filament burned out:					
20. If amplifier filament burned out:					

[1]M_1, no oscillation, contact-potential current only; M_2, no bias, high I_p; M_3, no drive, contact current only; M_4, no bias, high I_p; M_5, heavily loaded power supply, voltage drops.

RADIO-FREQUENCY AMPLIFIERS QUESTIONS

1. In what two basic ways do RF amplifiers differ from AF amplifiers?
2. In what range of frequencies are the RF amplifiers of this chapter expected to operate?
3. Name the out-of-phase feedback circuit that prevents a small-signal RF amplifier from oscillating.
4. Draw a diagram of a possible 5-MHz small-signal RF amplifier using a JFET. Using a BJT.
5. Draw a diagram of a dual-gate MOSFET RF amplifier.
6. Name two nonregenerative RF amplifier circuits using JFETs. BJTs. VTs.
7. In what two shapes may IC symbols be shown?
8. Draw a diagram of a simple RF differential wideband IC amplifier.
9. Draw a diagram of a pentode RF amplifier.
10. Why is B+ used sometimes with VTs instead of the V_{bb} symbol?
11. A triode with a μ of 20 has a V_p of 3200 V. What voltage would bias it to class A? Class AB? Class B? Class C?

ANSWERS TO CHECKUP QUIZ ON PAGE 339

1. (60 m) (60 pF) (17 μH) **2.** (Hum) **3.** (I_g, I_{sg}, I_p) **4.** (Low-Z circuits) **5.** (Appears in output) **6.** (Feeding cathode) **7.** (500 MHz) **8.** (Reduce skin effect, raise Q) **9.** (C's are in series) **10.** (Meters) (TPs) **11.** (Lowered) **12.** (Go to 12 V) (Raise) (0 V) (12 V) (12 V) (Decrease) **13.** (LP filter) (LC circuit) (Z matching) **14.** (Enhancement)

12. A P-channel JFET with a μ of 30 has a power-supply voltage of 50 V. What voltage will bias it to class A? Class AB? Class B? Class C?

13. How do the waveshapes of the input signal and the output circuit device current compare in class A? Class B? Class C?

14. Without any gate bias, in what class of operation would a VMOS operate?

15. Draw a diagram of a class C RF power amplifier using a VMOS. Using a BJT.

16. Should an RFC present a resonant or an antiresonant effect to the RF ac involved?

17. What indication does the output circuit dc meter give when a stage is tuned to resonance?

18. How should an RF output stage be adjusted for optimum output if RF output and dc input can be measured?

19. An RF power amplifier has a power supply of 300 V and draws 150 mA when delivering 0.8 A to a 50-Ω load. What is the stage efficiency?

20. What class of operation can use grid-, gate-, or base-leak biasing? What devices cannot use it?

21. When devices are connected in parallel in RF amplifiers, what may have to be added to input and output leads?

22. Why should devices be evenly matched and all circuits accurately center-tapped in push-pull amplifiers?

23. Draw a diagram of push-pull VMOSs in a class C RF amplifier. In a class A RF amplifier.

24. What type of RF output circuit feed must use an RFC?

25. Name the two types of inductive coupling used in RF amplifiers.

26. When a losser resistor is in the gate circuit of an N-channel JFET, what polarity voltage does it develop at the gate?

27. What would be the proper term for a "drain-neutralization" circuit?

28. What is a split-stator capacitor?

29. What would be the proper term for a "gate-neutralization" circuit?

30. When neutralizing with an RF indicator, should the circuit be tuned for zero or maximum indication?

31. Why is the C always added to the L in direct neutralization?

32. What type of tuned circuit does a push-push multiplier have in its output? In its input?

33. What are two major requirements when using an RF amplifier as a frequency doubler?

34. Which would work better as a tripler, a push-push or a push-pull circuit? As a quadrupler?

35. Draw a diagram of a JFET push-push circuit.

36. What might be the L value that would resonate with the C of a 5-MHz rule-of-thumb LC circuit for a high-Z JFET or VT circuit? For a BJT circuit?

37. How many turns are required of a 1-in.-long and 1-in.-diameter coil to produce 10 μH?

38. What is needed in an RF VT filament circuit that might not be required in an AF filament circuit?

39. What is the advantage of using a grounded-grid circuit for an RF power amplifier stage?

40. What might be the maximum gain of a grounded-grid triode with a μ of 15 and an R_p of 1000 Ω?

41. List 10 factors that are important in VHF and UHF circuit construction.

42. Draw a diagram of a four-stage BJT RF amplifier, including a CB, a CE, a CC, and push-pull output stages.

43. In Fig. 15-33, why do TP_1, TP_3, TP_5, and TP_7 peak slightly when the stages involved are tuned to resonance?

44. Why could TP_7 be used to tune all stages and the load?

45. Draw a diagram of a VMOS RF power amplifier using π-network output coupling. What would the component values be for operation at 5 MHz?

46. In Fig. 15-34, if a BJT with 200-Ω Z_D were used, what would the component values be?

47. What determines the S-D impedance of an active device in an RF amplifier?

48. Under what conditions would contact-potential bias exist alone?

49. Is it possible to have grid- or gate-leak bias if there is no output circuit I_{dc}?

16

Basic Transmitters

This chapter outlines the basic systems used to provide the carrier frequency power for all types of transmitters, whether they are keyed radiotelegraph, amplitude-modulated, or frequency-modulated. The main subject of this chapter is radiotelegraph (CW) systems, especially vacuum-tube applications since these devices are commonly used in the higher-power or final amplifier stages of transmitters. Various types of on-off keying are discussed, as is frequency-shift keying. Very simple transmitters through the MOPA and heterodyne systems are explained. Possible transmitter troubles and their correction are outlined.

16-1 RADIO TRANSMITTERS

Thus far, study has been confined to basic devices and circuits, such as transistors, vacuum tubes, oscillators, AF amplifiers, RF amplifiers, power supplies, and meters. This and subsequent chapters will be concerned with complete electronic *systems*. For example, a radiotelegraph transmitter may consist of an oscillator, one or more low-level RF amplifiers or multiplier stages, an RF power amplifier, a power supply, a means of keying the system, and an antenna. Such transmitters may operate in the frequency range of 10 kHz to well over 200 MHz.

The transmission of intelligence in a radiotelegraph system is accomplished by *keying* the system. If a telegraph key is inserted in the power-supply circuit of a simple oscillator (Fig. 16-1), the oscillator generates ac when the key is held down and

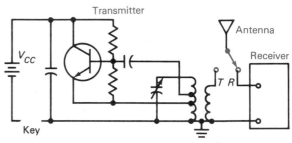

FIG. 16-1 Single-stage Hartley oscillator transmitter, receiver, and transmit-receive (TR) changeover switch.

ANSWERS TO PRACTICE QUIZ ON PAGE 340

	M_1	M_2	M_3	M_4	M_5
2.	O	H	VL	VH	L
3.	L	VH	VL	VH	L
4.	VL	H	VL	VH	L
5.	L	H	O	H	L
6.	H	L	R	VH	L
7.	H	L	L	VH	L
8.	L	H	H	VH	O
9.	N	N	N	N	N
10.	O	O	VL	VH	L
11.	L	O	VL	VH	L
12.	H	L	O	O	H
13.	L	H	H	O	O
14.	N	N	N	N	O
15.	N	N	N	N	N
16.	O	O	VL	VH	L
17.	H	L	O	O	H
18.	N	N	L	L	N
19.	O	O	VL	VH	L
20.	H	L	O	O	H

none when the key is up. By operating the key according to some dot-dash code, it is possible to transmit words and information.

The first radio transmitters were *spark* oscillators (Sec. 11-2) coupled to long-wire antennas. They produced trains of *damped* RF ac waves (decreasing series of cycles), previously called *spark* or *type B emissions*, which could be radiated into space by the antenna. Manual operation of a telegraph key was the only method of sending code with these transmitters. Their damped emission had a basic *carrier* transmitting frequency but contained many energy components called *sidebands* on both sides of the carrier, resulting in a wide transmission bandwidth.

Another early method of generating RF was by using Alexanderson or Goldsmith alternators. These machines could generate sinusoidal ac up to about 50 kHz but were bulky and could not be keyed rapidly. They were not used much after 1930. An electric arc has a negative-resistance effect and, if in series with an *LC* circuit, can produce a keyed carrier (A1A, Appendix E-2). *Arc* transmitters generated a purer waveform than the spark, and were used for high-power RF transmission, but were limited to an upper frequency of about 500 kHz.

As the ability to use higher and higher frequencies increased, the more adaptable vacuum tube outmoded all other types of transmitters. A preponderance of high-power transmitters use VTs in the output stages. For low-power applications, transistor transmitters are in general use. In most cases transistors are used for the low-power circuitry of a high-power transmitter and the high-level stages use VTs.

16-2 SINGLE-STAGE TRANSMITTERS

The simple Hartley oscillator coupled to an antenna wire produces a workable transmitter. By using high-powered tubes, a single-stage transmitter could produce as much as 100,000 W of RF. However, its frequency stability would not be good. That is, when the key was pressed, the plate potential would rise from zero to the voltage of the power supply. As the supply voltage of an oscillator changes, the frequency changes somewhat. Therefore, during the make and break of keying, the oscillator would not be frequency-stable, producing chirping sounds in a receiver and sidebands on both sides of the carrier frequency. The result is a wide-frequency-spectrum signal, which is undesirable.

The instability of an oscillator increases as the power and frequency increase. Even if a keyed oscillator might produce a reasonably stable emission at 500 kHz, at 5 MHz the chirp produced might be intolerable.

The closer, or *tighter*, the coupling between an oscillator and an antenna, the less stable the oscillator will be. Variations of antenna characteristics such as tuning, varying the height and swinging the antenna wire, will all affect the frequency generated by the oscillator.

If the dc power supply shown were changed to an ac source, such as a 500-Hz alternator, the circuit would be known as a self-rectified transmitter. It would transmit only during the positive half cycles of the power supply ac, would have a very broad emission, and would be classified as A2A. It would be illegal for anything other than a distress transmission.

16-3 KEYING RELAYS

The system in Fig. 16-1 represents a *simplex* method of communication. The same antenna is used for both transmitting and receiving. The operator sends a message, then switches the antenna manually from the transmit to the receive position and listens for the other station.

A keying relay (Fig. 16-2) is often used. It is a fast-acting double-pole double-throw (DPDT) type switch. With the key up, the relay coil is not energized and both relay arms are in the up position, connecting the antenna to the receiver. Incoming signals can be heard. When the operator presses the key, the electromagnet relay coil pulls both

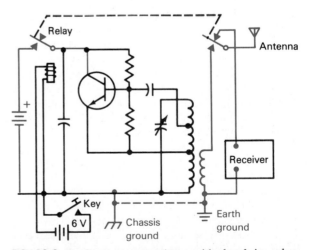

FIG. 16-2 Single-stage transmitter with break-in relay. The chassis is usually connected to earth ground.

relay arms down. This connects the antenna to the transmitter and the power supply to the oscillator, producing an RF ac carrier that energizes the antenna and transmits energy.

This type of simplex operation is known as *full break-in*, or QSK (Appendix G), because whenever the key is up, the transmitting operator can hear the other station. A distant operator who wishes to break in on the sending holds his or her key down long enough for the local operator to hear the tone being produced between the dots and dashes, which tells the operator to stop sending. The distant operator then explains why the break was made. Most modern manual commercial radiotelegraphic communication systems use break-in. *Duplex*, or *simultaneous* transmission in two directions, is possible by using two different frequencies, usually separated by at least a few hundred kilohertz.

While the key is down, the RF ac carrier level is continuously the same. Hence the term for such a code transmission is CW (continuous wave), also indicated by the letter-number designation A1A, meaning simple *amplitude on-off* changes.

With the key held closed, a transmitter radiates an RF ac emission (NØN), but unless it is keyed (A1A) or voice-modulated (A3E) with information of some kind, it is technically not a "signal."

16-4 THE MOPA

To develop high-power RF ac with good frequency stability it is necessary to go to a system of two or more stages. The oscillator stage determines the frequency of operation, and the RF amplifier stage or stages develop the high-power output. For maximum power output and best efficiency, class C operation of the amplifier is employed. Such a 2-stage transmitter, known as a *master oscillator power amplifier* (MOPA), is shown in block diagram form in Fig. 16-3a. It consists of a Hartley oscillator capacitively coupled to a neutralized VMOS amplifier inductively coupled to a tunable "Marconi" antenna, as shown in Fig. 16-3b.

Since the B− is being keyed, both oscillator and amplifier are keyed on and off at the same time. No provisions are made for break-in.

If the oscillator is left running at all times and only the B+ to the amplifier is keyed, a greatly improved frequency stability will result. However, the oscillator signal may block out a local receiver tuned to its frequency. If the amplifier is properly neutralized, any such oscillator signal leak-through (called a *backwave*) will be minimized.

DC meters in a transmitter should be bypassed with 0.01-μF capacitors. This prevents RF ac from flowing through the thin wires and hairsprings of the meter and burning them out. DC meters are never in RF-hot parts of circuits (drain, collector, plate leads, etc.). The RF ammeter in the antenna circuit must not be bypassed, of course, or it will not indicate.

Neutralization of the amplifier is important. If it is improperly neutralized, not only will there be a strong local signal when the key is open if the amplifier alone is keyed, but the oscillator frequency will pull (shift) as the amplifier or antenna is tuned. The amplifier may break into strong self-oscillation and emit spurious signals far from the operating frequency. The frequency of oscillation can be determined with a receiver or frequency meter. The output code signals can be monitored with a local receiver, or the key may also activate an AF oscillator connected to a loudspeaker to enable the operator to monitor the dots and dashes being sent.

In explanatory diagrams batteries are shown instead of actual voltage-regulated power supplies. If such power supplies had insufficient filter, the resulting varying dc would produce a variation of output amplitude at 120 Hz (A2A emission) which would be heard as a humming sound in the receivers.

16-5 TUNING AN MOPA

The tuning of an MOPA transmitter, such as Fig. 16-3, to operate properly on a desired frequency involves several steps. The procedure outlined below is a possible one. To protect PA circuits, use low amplifier power when tuning a transmitter.

1. Disconnect the amplifier B+. Decouple the antenna coil. Adjust the interstage coupling capacitor to minimum capacitance.
2. Decrease PA output by opening the Hi-Lo drain switch and close the key, energizing the oscillator. M_1 might read between 5 and 20 mA. M_2 and M_3 will both read zero.
3. Select the desired frequency on a receiver or frequency meter and rotate the oscillator tuning capacitor C_1 until the signal is heard.
4. Couple energy to the amplifier by increasing C_2 until M_2 reads $\pm 10\%$ of that required for normal operation. M_1 should increase a little. Retune the oscillator if necessary.
5. Neutralize the amplifier if necessary (Secs. 15-15 to 15-18).

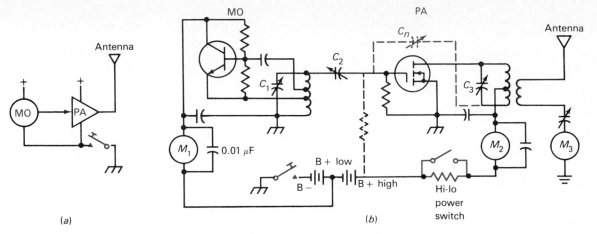

FIG. 16-3 MOPA transmitter. (*a*) Block diagram and (*b*) schematic diagram.

6. Connect the amplifier B+ and immediately tune C_3 for minimum I_{dc} on M_2.
7. With *very loose* coupling between antenna and amplifier tank, tune the antenna by adjusting the antenna capacitor for a peaking of I_{dc} on M_2 or for maximum *RF* current by M_3. (The antenna is now tuned to resonance and usually needs no further adjustment.)
8. Increase antenna coupling. This will increase M_2 and M_3, indicating that the antenna is taking RF power from the amplifier.
9. Close hi-lo switch for high power.
10. Redip M_2 by adjusting C_3 and readjust the antenna coupling until the desired I_{dc} occurs with the tank circuit at resonance (I_{dc} dip). A manual will indicate what the proper I_D operating value is for a given V_D.
11. Recheck the frequency once more, and the transmitter is ready to operate.

16-6 FREQUENCY STABILITY

Self-excited oscillator circuits, such as Hartley and Colpitts, may have poor frequency stability and may not hold a constant frequency of oscillation (Sec. 11-24). Some possible causes of an *abrupt frequency variation* in an oscillator are (1) power-supply voltage changes; (2) loose connections in the oscillator, amplifier, or antenna circuits; (3) poor soldered connections in the oscillator or in the following stages; (4) poor connections between tube or transistor base pins and sockets; (5) faulty capacitors, resistors, tubes, or transistors; (6) poor electric contact between the bearing surfaces of variable capacitors; (7) loose shield cans, and (8) parts

which, when heated, expand and either make or break a contact.

Some causes of a *slow drift of frequency* in a self-excited oscillator are: (1) heating and expansion of the oscillator coil due to RF ac flowing in it; (2) heating of any capacitors in the oscillator or the stage following it; (3) heating of the oscillator tube or transistor causing a change in electrical characteristics; (4) dc and RF ac heating of resistors, causing them to change values slightly; (5) aging, which may cause the characteristics of an active device to change, resulting in a slight frequency change.

Slow frequency changes are decreased or eliminated by using a crystal oscillator instead of a self-excited oscillator. Where the *frequency tolerance* (number of cycles per second that a transmitter may operate off the assigned frequency) is small, it may be necessary to keep the crystals in temperature-controlled chambers (Sec. 11-14).

Frequency variation due to variation of the power-supply voltage can be eliminated by using a voltage-regulated power supply.

Modern multifrequency transmitters rely on phase-locked loops to develop their oscillator frequencies. Stability is then dependent on one very stable low-power crystal reference oscillator.

All oscillators have a warm-up period, during which some frequency variation will occur. With temperature-controlled crystals, the period is the time required for the oven and crystal to attain their operating temperature. With self-excited oscillators, half an hour to several hours may be required, but correct-value temperature-coefficient capacitors can counteract drift due to component heating.

A sudden hot or cold draft striking any self-excited oscillator stage may cause a change of several hundred hertz in frequency, but only a few hertz with a crystal.

16-7 DUMMY ANTENNAS

To prevent a transmitter from emitting a signal and interfering with someone else on the frequency during tests, the transmitter should be coupled to a *dummy* (*artificial, phantom*) antenna (Fig. 16-4).

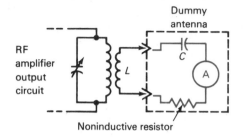

FIG. 16-4 A plug-in dummy antenna circuit.

The capacitor of the dummy antenna should have a reactance value equal to the reactance of the coupling coil being used. The noninductive resistor (not made in coil form) should have a resistance similar to the impedance of the antenna (often 50 Ω). With the dummy antenna coupled to the amplifier output circuit, the resistor should dissipate, in heat, the amount of power the antenna will radiate as a radio carrier.

The power *output* of the transmitter can be approximated by $P = I^2R$, using the dummy antenna ammeter value and the resistance of the noninductive resistor. When the antenna is connected in place of the dummy [and using the same amplifier *input* power ($P = V_{dc} I_{dc}$)] the antenna should be radiating the same power as that computed with the dummy circuit.

Tuning of any transmitter, with or without a dummy load, should be done at reduced power-supply voltages if possible to prevent inadvertent overloads during tune-up.

16-8 AMPLIFIER KEYING

One method of keying a radio transmitter is to open and close the output circuit dc with a telegraph key operated relay. The relay can be inserted in series with the negative end of the power supply (at X, Fig. 16-5), or in series with the positive terminal

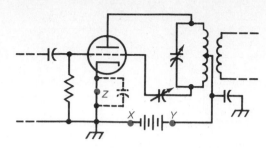

FIG. 16-5 Keying at X or Y is plate-circuit keying; at Z, cathode keying.

(at Y). If the relay is in series with both output and input circuits, between cathode (emitter, source) and ground (at Z), the stage is *cathode-keyed*. The cathode must then be bypassed to ground.

Test your understanding; answer these checkup questions.

1. In what frequency range might radiotelegraph transmitters operate? _____
2. What were three methods of generating carriers before VTs were used? _____ _____ _____
3. What is the main disadvantage of a single-tube transmitter? _____
4. Why would a swinging antenna change the frequency of a simple transmitter? _____
5. What is the name of the operation that uses an antenna changeover switch? _____ What can be added to make this full break-in? _____
6. If A1 is CW, what would AØ (A-zero) be? _____ When might AØ be used? _____
7. What is the main advantage of an MOPA over a single-tube transmitter? _____
8. What are two indicators for neutralization that might be used with the MOPA? _____
9. Plate tank resonance of a PA is shown by what I_p indication? _____ What I_g indication?
10. Why might heating a fixed capacitor change its capacitance? _____ A variable capacitor? _____
11. What are two other terms used for "dummy" antenna? _____ _____
12. For what two reasons are dummy antennas used? _____ _____
13. What three formulas might be used with dummy antennas to determine power output of a transmitter? _____ _____ _____
14. If both grid and plate circuits are keyed simultaneously, what is this keying called? _____ What would a comparative BJT stage-keying be called? _____ A JFET stage? _____

15. Draw diagrams of a single-stage CW transmitter with TR switch. A single-stage CW transmitter with full break-in. An MOPA with capacitively tuned antenna. A dummy load. An RF amplifier showing a key in three places that could be used for keying.

16-9 SHAPING A CW SIGNAL

When the key of a code transmitter is closed, RF ac is radiated. The RF ac of the letter "S" (three dots) can be represented as in Fig. 16-6a. The waveshape of the dots is square. Figure 16-6b shows the sharp

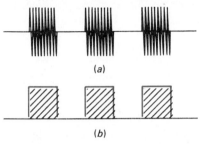

FIG. 16-6 (a) Three keyed bursts of RF ac form the telegraphic letter "S." (b) How the letter "S" appears in the detector circuit of a receiver.

corners of the square wave, as it looks at the receiver after being rectified by the detector.

Transmitting a square-wave signal produces sidebands, or frequencies on both sides of the carrier frequency. The reason for this can be understood by considering what constitutes a square wave.

Figure 16-7 shows, first, a sine wave, second, the result of adding the third harmonic, and finally, the result of adding up to the ninth harmonic.

By starting with a fundamental frequency, 10 Hz, for example, and adding odd-order harmonics up to the 9th in proper phase and amplitude, it is possible to produce a fairly square 10-Hz waveform. If the first 25 odd-order harmonics are added in the proper

proportions, an excellent square wave can be produced. The more odd-order harmonics involved, the squarer the corners of the square-wave signal. By working backward with this theory, any square wave must be a fundamental plus many odd harmonics of that fundamental frequency. (Even-order harmonics produce triangular-shaped waves.)

If a series of square-wave dots are transmitted, the signal must be made up of the dots plus many harmonics of the *dot frequency* (not to be confused with possible harmonics of the RF carrier frequency). If the dots are made at a speed of 10 per second, they form a 10-Hz fundamental keying frequency. If the dots have a square waveshape, there may be up to 50 or more strong harmonics present. This means that harmonics will be present for 10 × 50, or 500 Hz on both sides of the carrier. If there are 20 dots per second, harmonics will be generated out to 1000 Hz or more on each side of the carrier. The amount of RF spectrum space required is directly proportional to the speed of the sending. Shaping the transmitted square-wave signals by rounding them off represents the presence of fewer harmonics (see also Sec. 12-24).

The generation of sidebands by keying may be exaggerated by variations of the power-supply voltage of the transmitter. When the key is pressed, a heavy load is suddenly applied to the power supply and its output voltage may sag and rise several times in a fraction of a second. This *transient* voltage imposes an added waveshape on the transmitted signal. The mixture of the transient with the normal waveshape can produce high-order harmonics and objectionable sidebands hundreds of kilohertz on both sides of the carrier. These sidebands can produce clicking sounds in receivers tuned far from the frequency of the transmitter.

The rounding of the corners of a CW signal is said to produce *soft* keying. *Hard* keying has a sharp-cornered square waveform. Hard keying can be softened by the use of a *key-click filter*, shown in heavy lines in Fig. 16-8. When the key is

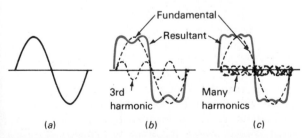

FIG. 16-7 Formation of a square-wave cycle. (a) Fundamental. (b) Addition of 3rd harmonic. (c) Addition of many odd harmonics.

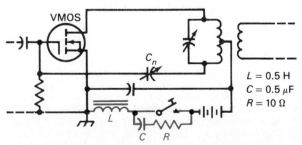

$$L = 0.5\ \text{H}$$
$$C = 0.5\ \mu\text{F}$$
$$R = 10\ \Omega$$

FIG. 16-8 LCR key-click filter in an RF amplifier.

closed, the inductance L opposes the rise of current, rounding off the sharp *make* corner of the signal. When the key is opened, C allows current to flow into it until it becomes charged to the power-supply potential. R decreases C-discharge sparking at the key contacts when the key is closed again. L, C, and R values vary with I_{dc} values. A characteristic waveshape with such a filter is shown in Fig. 16-9.

In general, if sparking occurs across any electrical contacts, a capacitor and resistor in series may be connected across the contacts to decrease the sparking.

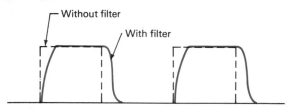

FIG. 16-9 Effect of key-click filter on normal sharp corners of keyed dots of RF ac waves.

16-10 SERIES-DEVICE KEYING

One method of keying a radiotelegraph transmitter is to key an active device connected in series with an RF amplifier stage, as shown in Fig. 16-10. The

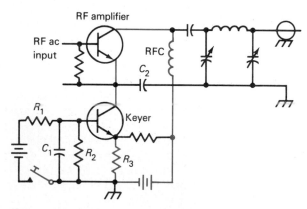

FIG. 16-10 Series-device keying circuit.

keyer device in this case is the lower BJT, but might be a VMOS, an FET, or a VT. With the key open the keyer BJT is reverse-biased well into nonconduction by the voltage-divider network formed by R_3 and R_4. (The base is at ground potential through R_2 and emitter is *positive* with respect to ground and the base.) No current can flow to the upper RF amplifier BJT. When the key is closed the keyer BJT is forward-biased and current can now flow to the RF amplifier BJT which then amplifies the RF fed to its base-emitter circuit. The amplified RF ac is fed to an antenna or other load through the π-network tuner. C_2 bypasses or RF-grounds the RF BJT emitter. R_1, C_1, and R_2 form a key-click filter circuit to soften the output CW signals.

16-11 BLOCKING-BIAS KEYING

If the bias of an RF amplifier is made sufficiently negative to prevent I_p regardless of how high the driving RF ac voltage may be, the amplifier stage cannot operate. The blocking of I_p by applying a high negative charge to a grid is known as *grid-block*, *blocked-grid*, or *grid-bias* keying. (With VMOSs or JFETs it is blocked-gate, and with BJTs it is blocked-base keying.) One such keying circuit requiring a separate bias supply is shown in Fig. 16-11.

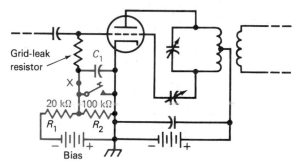

FIG. 16-11 Blocked-grid keying with a bias supply.

When the key is open, R_1 and R_2 form a voltage divider across the bias supply. The voltage-drop across R_2 must be high enough to cut off I_p, preventing any RF output.

When the key is closed, all the bias supply voltage is across R_1 with none across the key or in series with the grid circuit. The only bias in the grid circuit is that developed by the grid-leak resistor, which allows the stage to operate normally.

Capacitor C_1 is an RF bypass, holding the bottom of the grid-leak resistor at RF ground. If key clicks are produced, a resistance inserted in the circuit at point *X* may reduce them.

If the bias voltage should fail for any reason, the stage will emit a signal even with the key open.

Another circuit in which a JFET gate is blocked with the key open but which requires no separate bias supply is shown in Fig. 16-12. With the key

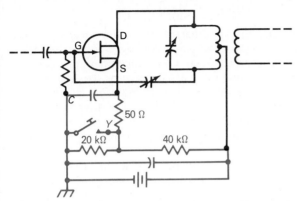

FIG. 16-12 Blocked-gate keying requiring no bias supply.

open, the 20- and 40-kΩ resistors form a voltage divider across the power supply. One-third of the power-supply voltage is on the gate to block it. When the key is closed, the 20-kΩ resistor is shorted out and the only bias in the gate circuit is that developed across the gate-leak resistor. The full voltage of the power supply is now between source and drain.

Another type of emission can be produced by connecting a chopper-wheel-type device (Fig. 16-13) into Fig. 16-12 at point *Y*. If the metal

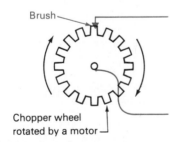

FIG. 16-13 A chopper wheel electrically makes and breaks at the brush.

chopper makes and breaks contact 500 times per second and the key is held down for $\frac{1}{10}$ s to transmit a dot, the carrier will be turned on and off 50 times. This results in an interrupted CW (A2A) signal, composed of pulses of nearly square-wave RF ac having a "modulation" frequency of 500 Hz.

It would be legal only for distress transmissions. A more modern method of producing A2A is by adding a 500-Hz ac in series with an RF amplifier power supply to produce a 500-Hz amplitude-modulated signal (Chap. 17). Such signals not only transmit a single-frequency carrier, but also RF signals 500 Hz on each side of the carrier. In the case of distress communication reception there is less chance of missing a carrier signal if there are three dispersed signals being transmitted. Also, all types of receivers will reproduce this signal form.

Transistor RF stages may be keyed with simple OR, AND, or inverter gates.

16-12 FREQUENCY-SHIFT KEYING (F1B)

The usual radio receiver is sensitive to changes in signal amplitude. When a carrier of a transmitter is keyed on and off, as when sending dots and dashes, the receiver should produce a tone for each dot or dash and be quiet in between. Unfortunately, both man-made and natural static occur as changes in amplitude, producing noises in the receiver when the transmitter is off. If the received signal is strong, it will be well above this noise level and the dots and dashes can be easily distinguished. When signals are only slightly stronger than the noise level, instantaneous crashes of static between dots and dashes confuse reception of the code.

As long as there is even a weak carrier, the receiver is quieted somewhat. Therefore, if the carrier can be left on at all times, some of the effect of background noise can be overcome. If the carrier is on all the time, how can dots and dashes be sent? One method is to shift the carrier back and forth in frequency. When no code is being sent, the carrier is on one frequency, called the *space* frequency. When a dot or dash is transmitted, the carrier is shifted to the other *mark* frequency. This is *frequency-shift keying* (FSK, F1A, or F1B, Appendix E-2). If the mark and space frequencies are relatively close (60 to 800 Hz), the receiver remains in a quieted condition, since it can receive both signals on one setting of the dial. By making the receiver sensitive to frequency changes only, rather than to amplitude changes, an improvement of reception results.

There are a variety of circuits that can frequency-shift an oscillator under the control of a telegraph key, teleprinter, or digital on-off gating circuit. The circuit as shown in Fig. 16-14 will shift from a key-up higher-frequency space signal to a key-down lower-frequency mark signal. It does this by shift-

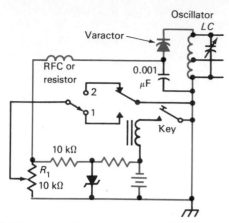

FIG. 16-14 One possible varactor-type FSK circuit.

ing the bias on a voltage-variable capacitor (varactor) across the tuned circuit of a self-excited oscillator. Regardless of the position of the arm of potentiometer R_1, with key-up the varactor reverse bias is half of the zener voltage, developing some capacitance across the oscillator LC circuit. If the arm of R_1 is at center position, with key-down the relay arm shorts out the bottom half of R_1, lowering the reverse bias voltage on the varactor, raising its capacitance value, and lowering the frequency of oscillation. Variation of the arm on R_1 varies the amount of frequency shift. Common values used for FSK are 60, 100, 170, 425, and 850 Hz.

If the switch is moved to position 2, the mark-space frequency shifts are reversed.

A similar circuit can be used to remotely control any self-excited oscillator. If the relay and switch are removed and the arm of R_1 is grounded, the arm position controls the bias on the varactor and therefore the oscillator frequency.

A reactance-type modulator circuit (Chap. 19, Fig. 19-14) across an oscillator will shift its reactance if its bias voltage is keyed. The frequency of a crystal can be shifted a few hundred hertz by keying a small capacitor across or in series with it, or keying a small inductance in series with it. Two separate crystals can be keyed on and off by using the circuit shown in Fig. 11-44. The divide-by circuitry of a PLL synthesizer can be keyed to provide FSK output.

Another form of keying a constantly-on carrier is *phase-shift keying* (PSK). In PSK the RF carrier can be fed directly to a transmitter RF stage, then by relay or transistor it can be coupled to the same RF stage but through a common-emitter-type amplifier to shift its phase 180°. At the receiver the shift in phase is detected and is made to shift the conductivity of a circuit to produce a keyed condition. PSK is often used in data transmission, as in satellite communications.

16-13 BUFFER AMPLIFIERS

Keying the amplifier of an MOPA results in a decided improvement over keying a simple oscillator. More nearly perfect CW operation is possible if a *buffer* amplifier is added between the MO and the PA. A buffer does two things: (1) amplifies the weak output of the MO enough to drive the input of the PA and (2) prevents any interaction between the MO and the antenna circuit, which might result in frequency instability.

The oscillator and buffer stages should be considered as a frequency-determining pair. While the buffer might be keyed and produce good CW, it is preferable to key a stage following the buffer, as in Fig. 16-15a. In this circuit a Pierce oscillator is capacitively coupled to a VMOS buffer, which is link-coupled to a power tetrode PA stage, which is π-network- (L, C_t, and C_z) coupled to an antenna. The PA stage is blocked-grid-keyed. The V_{cc} of -100 V is sufficient to hold the PA far enough in I_p cutoff so that there is no power output at all with the key up. When the key closes, the voltage divider formed by R_1 and R_2 lowers the bias to a normal class AB, B, or C operating value for the input ac from the buffer, and the PA functions normally. With the voltage values shown, the stage might develop 600 W of RF ac output with about 1000 W of dc input. This assumes the VMOS stage can develop enough driving power. If not, another VMOS *driver* stage might have to be added between the buffer and the PA. (Note zener diode limiters to prevent overdriving the VMOS gate.)

A procedure for tuning a multistage transmitter such as this might be:

1. Start with the key open and V_{bb} and V_{sg} off.
2. Turn on V_{cc} and V_{DD} and tune M_1 to a dip in I_D, indicating resonance of the VMOS output LC circuit.
3. Close the key and tune the PA grid LC circuit to a peak reading of M_2. Retune the buffer output if necessary and open the key.
4. Set the impedance adjusting capacitor C_z to maximum capacitance (minimum impedance and coupling to the antenna).

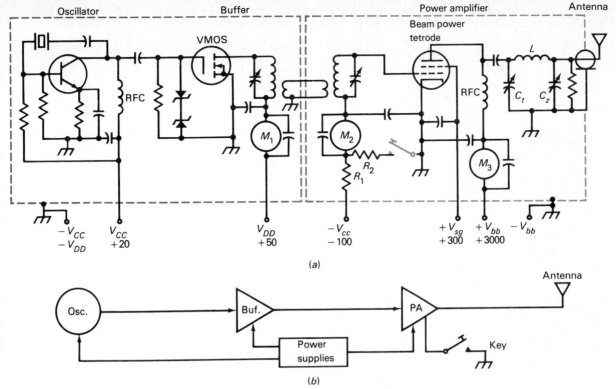

FIG. 16-15 (*a*) Schematic diagram of an improved MOPA transmitter. (*b*) Block diagram.

5. Turn on V_{bb} and V_{sg}.
6. Close the key and adjust the tuning capacitor C_t to an I_p dip shown by M_3. This should result in a lower I_p than the recommended operating value.
7. Decrease C_z slightly and redip C_t. This should increase the M_3 reading at the I_p dip.
8. Repeat operation 7 above until the I_p at the dip is the desired value. Send the word "test" and sign the station call.

If this tuning might interfere with operations on the frequency, during tune-up use a dummy load having the same impedance as the antenna. When the frequency is clear connect the antenna and trim C_t and C_z until the desired I_p is developed. Send the word "test" and sign the station call.

If any buffer, driver, or PA stage(s) require neutralization, neutralize them before output circuit power is applied to them.

The PA should always be shielded from the oscillator-buffer section to prevent RF interaction, feedback, oscillation, or harmonic radiation.

Figure 16-15*b* illustrates a block diagram of the same transmitter, including the power supply.

16-14 A CLAMP CIRCUIT

A grid-leak-biased amplifier *following a keyed stage* has no bias when there is no RF excitation to its grid. Without bias, the stage will draw excessive I_p and damage the tube. Three methods of protecting such a grid-leak-biased stage are:

1. The amplifier may be biased with a bias supply to limit key-up current to some safe value.
2. The stage may use cathode-resistor bias as a safety bias when no RF excitation is applied.
3. A *clamp tube* may be used in screen-grid amplifiers (Fig. 16-16).

The clamp tube should have a high transconductance, a low cutoff-bias value, and a large plate. With no excitation to the RF amplifier, there is no grid-leak bias and the clamp tube, which also derives its bias from the grid leak, has a low plate resistance, passing a high value of current through itself and the screen dropping resistor R_{sg}. This produces a large voltage-drop across R_{sg}, a low voltage across the clamp tube, a low screen voltage for the RF amplifier, and therefore a low amplifier I_p.

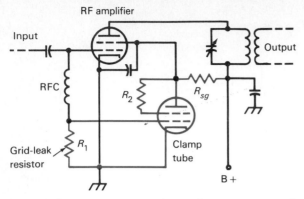

FIG. 16-16 Clamp circuit used on a beam-power tetrode RF amplifier when a preceding stage is keyed.

With RF excitation grid-leak bias is produced, biasing the clamp tube to cutoff and stopping its I_p. This reestablishes a normal V_{sg} for the amplifier, and it operates as though the clamp tube were not in the circuit. Resistor R_2 prevents excessive clamp tube I_{sg}.

16-15 TRANSMITTERS WITH MULTIPLIERS

The lower the frequency of oscillation, the more stable the oscillator and the less drift likely. To take advantage of this, an oscillator is often operated at a lower frequency than the PA frequency. For example, it is desired to transmit on a frequency of 4 MHz. If the oscillator operates on 1 MHz stably, the 1-MHz signal can be fed into a pair of doubler stages (Sec. 15-23). The output of these stages will drive the PA at the desired 4-MHz frequency with greater stability than if the oscillator had been on 4 MHz (Fig. 16-17). Since the PA frequency is not

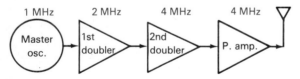

FIG. 16-17 Block diagram of a transmitter with an output on 4 MHz and an oscillator on 1 MHz.

the same as that of the oscillator, there is less interaction between stages. Shielding between multiplier stages may not be required.

Transmitters in the VHF band (30–300 MHz) may use three or more multiplier stages. A possible system to transmit a signal on 162 MHz is shown in block diagram form in Fig. 16-18. A crystal is the only commercially acceptable oscillator to main-

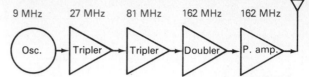

FIG. 16-18 Block diagram of a system to produce 162-MHz output.

tain frequency stability. However, crystal-based frequency synthesizers (Sec. 11-26) are used in many transmitters that must change to several different working frequencies.

If the frequency of the oscillator stage in a multistage transmitter changes for any reason, the multiplier stages multiply the change. Suppose a 1-MHz crystal is used in a transmitter operating on an assigned frequency of 8 MHz with three doubler stages. If the *fundamental frequency* crystal changes 5 Hz, the first doubler changes 10 Hz, the second changes 20 Hz, and the third changes 40 Hz. A 5-Hz oscillator change produces a 40-Hz output signal change.

If the crystal has a temperature coefficient (TC) of −4 Hz/MHz/°C and the temperature of the crystal increases 6°C, the output frequency will *decrease* by 4 Hz (−4 Hz indicates a negative temperature coefficient) times 1 (1 MHz), times 2, times 2, times 2 (three doubler stages), times 6 (temperature increase in degrees), or 192 Hz. If the output frequency had been 8,000,000 Hz, increasing the temperature of the crystal 6°C would have decreased the output by 192 Hz, to 7.999808 MHz. If the crystal had a +TC of +4 Hz/MHz/°C, the rise in temperature would have increased the frequency to 8.000192 MHz.

Test your understanding; answer these checkup questions.

1. How is keying speed related to bandwidth? _____
2. Does a key-click filter result in soft or hard keying? _____
3. What is reduced by rounding off square-wave emissions? _____ _____
4. What is a baud? _____
5. A key-click filter uses what components? _____
6. What is a transient voltage? _____
7. Why is A2 used for distress messages? _____ _____
8. What is keyed in FSK or F1 emissions? _____
9. What is the range of F1 shifts used? _____

10. What are the two duties of a buffer amplifier? _____ _____

11. Why would pentodes make better buffers than triodes? _____

12. In Fig. 16-15, what are the names of the three types of coupling used? _____ _____ _____ What type of keying is used? _____

13. How can a grid-leak-biased tetrode RF amplifier be protected from excessive I_p if a driving stage is keyed? _____ _____ _____

14. What is gained by operating an oscillator on some subharmonic of the transmission frequency? _____

15. A 1-MHz oscillator has 10-Hz FSK. What is the output shift if its signal is fed through three doublers? _____

16. Draw diagrams of a key-click filter. A series-tube keying circuit. A blocked-grid keying circuit. A remote oscillator tuning circuit. A three-stage blocked-grid CW transmitter. An RF amplifier with clamp tube. A block diagram of an RF system to produce 162-MHz output.

16-16 REDUCING UNWANTED EMISSIONS

No RF amplifier emits a pure sine-wave RF ac. Since any deviation from the pure sine wave represents harmonic content, RF amplifiers can be expected to transmit some harmonic energy. Such harmonic (multiples of the carrier frequency) energy will fall on frequencies used by other radio services, causing interference. When a harmonic or parasitic spurious signal falls on TV frequencies, it is known as *television interference* (TVI).

There are many ways of decreasing harmonic and parasitic radiations. Figure 16-19 illustrates some of them. Numbered points are described.

1. The use of link or any form of inductive coupling discriminates against passing higher frequencies, whereas capacitive coupling will pass higher (harmonic) frequencies better than the lower fundamental.

2. A high-Q circuit tuned to the fundamental in the grid circuit decreases the possibility of transferring harmonics generated in the driving stage through the final amplifier.

3. The coupling to the grid circuit should be held to a low enough value to prevent driving to I_p saturation at the peak of the RF voltage (high harmonic output).

4. The bias voltage should be as low as possible, to produce I_p flow as near 180° (class C) of the RF cycle as possible without decreasing stage efficiency. Classes B, AB, and A produce progressively fewer harmonics.

5, 6, and 7. Small mica capacitors (10 to 50 pF) connected to ground from grid, screen, and plate bypass harmonics.

8. A small coil and capacitor wave trap tuned to any VHF being interfered with tends to decrease harmonic transfer to the tank circuit. A few turns of wire around a 2-W 50-Ω carbon resistor may be used as a parasitic choke, or ferrite beads around the lead may be used.

9. The tank circuit should have a reasonably high Q to produce good flywheel effect.

10. A *Faraday shield* may be installed between PA tank and antenna. It is a gridwork of parallel

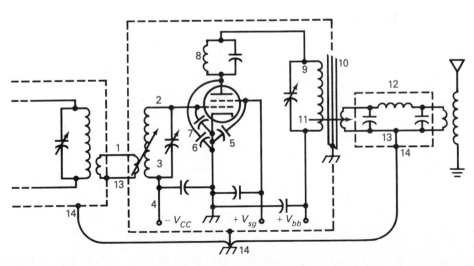

FIG. 16-19 Methods of decreasing harmonics or parasitics from a beam-power tetrode RF amplifier.

wires, not connected together at the top but soldered together at the bottom and grounded. This *electrostatic shield* decreases capacitive coupling, thereby decreasing harmonic transfer to the antenna circuit.

11. As loose a coupling as possible should be used between PA tank and antenna and still maintain output power. Overcoupling lowers tank-circuit Q, raising harmonic output.

12. A low-pass filter with a cutoff below the second harmonic of the transmitter may be added to the antenna feed line. All frequencies above its cutoff will be attenuated, and will not be delivered to the antenna to be transmitted. The filter, transmission line, and PA must all be completely shielded to make the filter effective. A more complex filter is preferable to the simple one shown.

13. One wire of a link-coupling pair should always be grounded, or a coaxial line with a grounded shield should be used.

14. All stages, as well as the low-pass filter, should be separately shielded and then connected to a common ground point on the chassis.

Not shown in this illustration, but particularly effective in reducing even-order harmonics, is a well-balanced push-pull stage. A π-network output tuning system tends to discriminate against harmonics, too.

Harmonics are one of several types of spurious signals generated by transmitters. Others are parasitic oscillations, or beat signals caused by two ac frequencies mixing in some nonlinear circuit, or the results of overmodulation (Sec. 17-9). Stopping parasitics requires chokes, ferrite beads, circuit rearrangement, etc. Stopping beats requires re-engineering, better bypassing or shielding, or better filtering of the output.

While a receiver having a calibrated "S" meter can give a rough idea of harmonic strengths, an accurately calibrated signal-strength meter can measure how far down, in decibels, any specific

harmonic is from the measured carrier level. A *spectrum analyzer* is the best method of checking harmonics and parasitics (Sec. 13-37).

Spurious transmitted signals often show up as cross-hatching (narrow lines) across a TV screen. If the carrier is on some lower frequency, a high-pass filter across the input terminals of the receiver may help. When RF signals interfere with the audio of TV, FM, or broadcast receivers, or stereo equipment, it is often the first audio stage of the equipment picking up the RF signal and detecting it that is at fault.

16-17 SHIELDING AND PROTECTIVE DEVICES

Shielding stages, circuits, or parts has been mentioned. Shields must (1) shield one circuit *electromagnetically* from other circuits and (2) shield one circuit *electrostatically* from other circuits.

As a simple case of shielding, consider the two coils in Fig. 16-20. The dotted lines surrounding

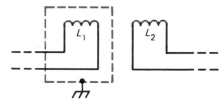

FIG. 16-20 Coil L_1 is shielded from L_2 by a metal box.

the first coil indicate that this coil is encased in a metal container or shield. Assume, first, that there is no shield present. Coils 1 and 2 are mutually inductive. Ac in the first will induce ac into the second because of expanding and contracting magnetic fields. If a metallic box is placed around the first coil, almost no current is induced into the second coil. The expanding and contracting magnetic fields induce an ac in the metal shield. The induced current produces a field, according to Lenz' law, opposite in phase to that of the field that produced it. If the shield were a perfect conductor, the strength of the opposing field would equal the inducing field and there would be zero resulting field to move outward toward the second coil. Shielding would be complete.

Unfortunately, there is no perfect conductor. Therefore shielding of magnetic fields can never be perfect. However, silver is near enough to a perfect conductor to produce very good shielding. Copper and aluminum are only a little less effective. Most shielding is aluminum except in circuits operating at frequencies over perhaps 100 MHz, when silver-plated copper or brass may be used.

Again, consider the two coils without the shield. Each is a conductor, and the air between them is a dielectric, forming a capacitor. Ac energy can travel from the first coil to the second via this capacitance. Actually, energy is transferred from one coil to the other by both electromagnetic and electrostatic lines of force. When the metallic shield is placed around the first coil, there is capacitance between the coil and the shield and also between the shield and the second coil. If the shield is connected to ground, the electrostatic energy flows back and forth between the first coil, the shield, and ground and none reaches the second coil. Electrostatic shielding is complete with most metals.

What is true of shielding between the two coils is also true between two or more RF stages, between an RF amplifier and an antenna, and between an RF and an AF stage.

Shielding can decrease or prevent degeneration, regeneration, oscillation, distortion, parasitics, instability, hand capacitance (detuning), and other undesired effects. Communications equipment is encased in metal to provide solid physical construction as well as good shielding.

Properly constructed transmitters and receivers have no high-voltage-carrying leads external to the equipment. All exposed metal parts are engineered in such a way that they can be connected to the chassis and grounded, thereby preventing operating personnel from receiving shocks. Grounding the metal parts of a transmitter also prevents the accumulation of a static charge on them. As a further protection to personnel, *interlock switches* may be incorporated in doors to interrupt all high-voltage circuits automatically if the doors of the transmitter are opened. All possible controls to adjustable parts are brought outside, making it unnecessary for the operator to reach inside the cabinet, or housing, to tune or adjust the circuits. When fuses are used, they are usually available from outside the equipment. When electromagnetic overload relays or circuit breakers are used instead of fuses, the restore mechanisms are usually available from outside. The operator determines the operation of high-powered equipment by meter readings alone. The meters found on a VT transmitter are power-line voltmeters and wattmeters, filament voltmeters, plate-current meters, plate-voltage meters, grid-current meters, and either an antenna ammeter or voltmeter.

16-18 TRANSMITTER SYSTEMS

Modern transmitters might be classified as to whether they operate on one, on several, or in bands of frequencies.

AM broadcast, TV, FM broadcast, and some fixed point-to-point stations are examples of transmitters assigned to operate on one frequency only. Mobile police, fire, etc., may operate on one or several relatively nearby frequencies, using crystal-controlled oscillators. The RF sections of such transmitters are tunable but, once tuned, are not expected to be adjusted for long periods of time. Mobile equipment is constructed to operate from 12 V dc and is as light and rugged as possible. Mobiles usually operate on assigned frequencies from 30 to 900 MHz, using between 10- and 100-W output.

Shipboard radiotelegraph or radiotelephone, aircraft, and many point-to-point transmitters are examples of systems that operate on certain assigned frequencies during one time of the day and on other frequencies at other times of the day to communicate over the same or different distances. The original self-excited oscillators used in these services have been supplanted by crystal-type oscillators. Modern shipboard transmitters use synthesizers to develop the many assigned frequencies with the necessary 50-Hz stability.

The only service using self-excited, or variable-frequency, oscillators in transmitters is the amateur group. Even this service uses many synthesized oscillators.

Since the amateur has been assigned bands of frequencies such as 3.5–4 MHz, 7–7.3 MHz, etc., the transmitter may use a variable-frequency oscillator to permit operation on any frequency within the band limits. Crystal-controlled oscillators may be used, but they restrict the amateur's ability to shift to any desirable frequency. The amateur transmitter should be capable of operating on several bands at the choice of the operator and on any frequency in these bands.

16-19 A TRANSISTOR CW TRANSMITTER

Transistorized CW transmitters are usually capable of 1 to 100 W, depending on the transistors used. Circuitry can be simple. Figure 16-21 illustrates a possible single-frequency all-BJT crystal oscillator transmitter capable of 10 W by using heat sinks and a 12-V battery, or of more than 50 W output with a V_{CC} of 30 V or more.

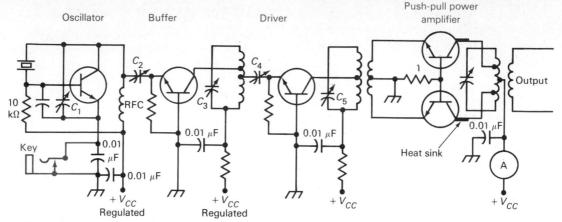

Oscillator Buffer Driver Push-pull power amplifier

FIG. 16-21 A transistor CW transmitter capable of 1 to 50 W, depending on transistors, V_{CC} value, and heat sinks used.

Capacitive coupling is shown in low-level stages. Coupling is varied by using variable capacitors. Inductive coupling is used to the push-pull common-base stage. Many dual-transistor stages are parallel, although push-pull is shown. Keying can be accomplished by inserting a key into the closed-circuit jack in the oscillator emitter circuit. Common-base circuits are used for stability. If one of the stages is to operate as a frequency multiplier, it should be changed to a common-emitter circuit.

Capacitors C_2 and C_4 are variable because coupling between stages may be critical. C_1 should be adjusted for best keying characteristics.

The meter can be used to tune all stages. It will dip when the PA stage is resonated and will increase as the driver and buffer stages are resonated. Some form of RF indicator (oscilloscope, etc.) may also be used, or test points may be brought out to allow tuning each stage separately.

16-20 HETERODYNE VFO TRANSMITTERS

Modern transmitting systems often utilize the *heterodyne, translation,* or *mixing* principle to develop the desired basic frequency to be amplified and transmitted.

A simplified heterodyne system that might be used to cover three different bands of frequencies, 3.5–4, 7–7.5, and 14–14.5 MHz, is shown in Fig. 16-22. Switches SW_1 and SW_2 are shown in position to operate on the 3.5–4 MHz band. A 9-MHz Pierce crystal oscillator feeds RF ac to the upper gate of a dual-gate MOSFET. A Hartley variable-frequency oscillator (VFO) capable of tuning be-

tween 5 and 5.5 MHz is feeding RF ac to the lower gate of the MOSFET. As a result, with the key down, the I_D of the MOSFET *mixer* stage contains two components, one 9 MHz, and the other perhaps 5 MHz. The output of the mixer contains four components: (1) 9 MHz; (2) 5 MHz; (3) the difference of the two inputs, or 4 MHz; and (4) the sum of the two inputs, or 14 MHz. Since only the frequencies in the 3.5–4 MHz band are required at this time, the output LC circuit can be tuned to 4 MHz, and only this difference-frequency is passed on to the following amplifiers (not shown). If the VFO is tuned to 5.5 MHz, the difference-frequency will be $9 - 5.5 = 3.5$ MHz, the other end of the desired band of frequencies. The output LC circuit should be trimmed to maximum output for this frequency (if not broadbanded), as should all following RF amplifiers. Since the key is in the mixer stage, neither oscillator circuit is being keyed, resulting in good frequency stability when keying. Equally important, with the key up there is no oscillator running on the transmitting carrier frequency.

For operation on the 7–7.5 MHz band, both SW_1 and SW_2 are switched to the next-lower position. Note that these switches are ganged, shown by dashed lines. When one is adjusted the other is automatically switched. This switches in a 2-MHz crystal and shorts out part of the inductance of the LC circuit so that it will resonate at the required higher frequency. If the VFO is tuned to 5 MHz, the sum of this frequency and the 2-MHz crystal output is 7 MHz, one end of the desired band. If the VFO is tuned to 5.5 MHz, the output from the

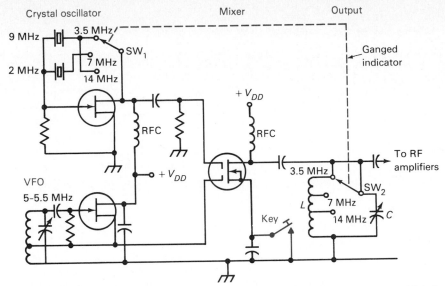

FIG. 16-22 Heterodyne-type VFO using a dual-gate MOSFET mixer stage which is keyed.

mixer is 5 + 2.5, or 7.5 MHz, the other end of the desired band. The *LC* circuit must be trimmed to this new frequency, *unless* a 7–7.5 MHz bandpass (or a 7.5-MHz low-pass) filter is used in place of the *LC* circuit. This is a preferred way of handling such a system, making amplifier tuning unnecessary. If the transmitter operates on three bands there would be three separate bandpass filters that might be switched across the output circuit of the mixer (and three more for each following amplifier).

To operate on the third band, 14–14.5 MHz, the switches are moved to the lowest position. Again the crystal frequency is 9 MHz and the VFO is 5–5.5 MHz. If 9 and 5 MHz are used, the heterodyne frequency is now the *sum* of 9 + 5, or 14 MHz, one end of the desired band. With 9 + 5.5 MHz the output is 14.5 MHz, the other end of the desired band.

To develop other 500-kHz bands of frequencies, it is necessary only to switch in other crystal frequencies to heterodyne against the 5–5.5 MHz VFO to cover those bands also.

The output power of a heterodyne or mixer stage is usually quite low, in the range of perhaps 0.1 dBW (dB with a reference zero of 1 W). To produce a 30 dBW (1000-W) output, a possible system might be as that shown in Fig. 16-23. If the BJT stage is capable of 10 dB gain its output would be 0.1 × 10, or 1 W, or 1 dBW. If the VMOS stage has 13 dB of gain its output would be 10 dB + 3 dB, or 1 W × 10 × 2, or 20 W, or

13 dBW. If the beam-power tetrode has a gain of 17 dB, its output would be 20 W + 10 dB + 3 dB + 3 dB + 1 dB, or 20 W × 10 × 2 × 2 × 1.25, or 1000 W, or 30 dBW.

To vary the output power of a transmitter the plate voltage of the final amplifier is usually varied. A common method is to use a variable autotransformer (Sec. 5-25) controlling the input to the primary winding of the high-voltage power supply feeding the PA, as shown in Fig. 16-23. It would also be necessary to feed the screen-grid power-supply transformer from the same autotransformer so that the plate voltage could never be less than the screen-grid voltage, or the screen might be fed through a dropping resistor from the high-voltage $+V_{bb}$. Degree of coupling of PA to antenna also affects output power.

To produce maximum efficiency and the highest power output the amplifier stages would be operated in class C. To reduce possible harmonic output (or when amplifying modulated signals, discussed in following chapters) the amplifiers are normally operated in class A, AB, or B. Since current is flowing at all times in class A, AB, and B amplifiers, the first stage develops low-amplitude "white" noise impulses which are then amplified by all following stages. The result is a constant-level noise emission being radiated on the transmitting frequency when the key is open. White noise can be eliminated by using a full break-in relay system, or biasing the PA to class C when the key is opened.

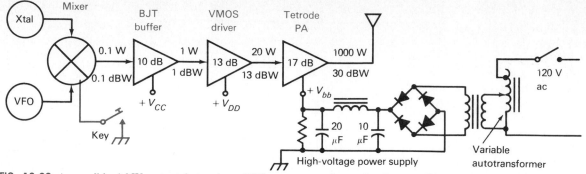

FIG. 16-23 A possible 1-kW output heterodyne VFO-type transmitter, showing possible stage gains and a means of varying the output power.

| Test your understanding; answer these checkup questions.

1. What type of coupling discriminates against harmonic transmission? _____
2. Why does a high-Q circuit discriminate against harmonics? _____
3. What class of bias should produce least harmonics? _____
4. What type of field does a Faraday shield stop? _____
5. What type of circuit reduces even-order harmonics? _____
6. Which is more difficult to shield, electromagnetic or electrostatic fields? _____
7. What is the purpose of an interlock switch? _____
8. What are the only types of transmitters that now use VFOs? _____
9. In Fig. 16-24, name the oscillator used. _____ In what class is the buffer and driver operating? _____ What is the purpose of the 1-Ω resistor in the PA? _____
10. In Fig. 16-22, name the crystal oscillator circuit. _____ Is it series- or shunt-fed? _____ Name the VFO circuit. _____ How is it fed? _____ Name the heterodyne circuit device. _____ How is it fed? _____
11. In Fig. 16-22, what might be used in place of the output LC circuit? _____
12. In Fig. 16-22, how could operation be produced on 5–5.5 MHz? _____
13. Why might a resistor with a switch across it sometimes be included in the dc output circuit of RF amplifiers? _____
14. How many watts are represented by 40 dBW? _____
15. What class VT RF amplifier might produce white noise? _____
16. Draw diagrams of a tetrode RF amplifier with link-coupling input and output, coupled to an antenna

through a low-pass filter. A four-stage BJT crystal-oscillator CW transmitter. The translation section of a heterodyne CW transmitter. A block diagram of a heterodyne CW transmitter with means of varying the power output.

16-21 INDICATIONS OF TROUBLE

Some obvious indications of trouble, particularly in high-power transmitting equipment, is the hiss or crackle of an electric spark jumping, the blue curl of smoke, the flicker of a flame, the smell of burning insulation, the loud humming of a transformer, the lack of filament glow in tubes, the failure of the keying relay to close when the key is pressed, the clank of an overload circuit breaker opening, or the sounding of a trouble alarm. Less obvious but just as significant are the indications of meters on the panel of a VT transmitter or the cessation of the sound from a keying monitor. While high-power stages may use VTs, low-level RF stages of almost all high-power radio transmitters use transistors. Some of the problems that can arise with transistor stages were discussed in Sec. 15-28. Much of the following information, apparently referring only to high-power VT stages, can apply to transistorized circuits also. Some possible meter indications of circuit malfunctions are listed here. Where plate and grid currents are indicated, collector and base currents, or drain and gate currents, might be read into the statements.

OSCILLATORS. I_p Increased. Stage probably not oscillating; inoperative crystal; detuned plate circuit; shorted grid circuit; shorted tube.

I_p Decreased. The next stage may not be accepting RF power; oscillator may be faulty; V_f or V_p may be low.

I_p Zero. The power supply may not be functioning; meter may be burned out; oscillator filament may be open; poor connection to the plate, screen, or control grid; fuse may be burned out.

RF AMPLIFIERS. I_g Increased. Loss of bias supply voltage; excessive RF drive from the preceding stage; shorted grid leak; or low I_p. When the plate is taking no electrons, the grid is free to collect more of them than normal. Thus, low V_p or V_{sg}, decreased coupling to next stage, or open plate or screen circuits may all result in increased I_g.

I_g Decreased. Decreased excitation to grid; driving circuit detuned; low filament voltage; low filament emission; increased I_p. (As I_p increases, I_g decreases. Therefore, anything that produces high I_p, detuned plate circuit, high V_{sg}, or tight coupling to next stage produces a lowered I_g.)

I_g Zero. A burned-out filament; no filament voltage; resistor or RF choke in grid circuit burned out; insufficient excitation if power-supply bias used; no excitation with grid-leak bias; open meter.

I_p Increased. Loss of bias voltage; loss of RF excitation from previous stage if grid-leak biased; detuned plate circuit; increased coupling to the next stage; a soft, or gassy, tube; high V_{sg}.

I_p Decreased. Low plate voltage; decreased coupling to plate-circuit load; low filament emission; low V_f; low V_{sg}; excessive bias on grid; open in the antenna or other load circuit.

I_p Zero. No plate voltage; burned-out fuse in power supply; shorted filter capacitor; open RFC or power-supply choke; open overload relay in plate circuit; no V_f; burned-out filament; zero V_{sg}; open in the plate circuit.

ANTENNA. Meter Increased. Power-supply voltage increased, or faulty meter.

Meter Decreased. Usual indication something is wrong in the transmitter; low or excessive coupling to the antenna; detuned output circuit; low power-supply voltage; low input excitation; excessive bias on PA; low filament emission; low V_f; insufficient or excessive V_{sg}; loss of amplifier bias.

Meter Zero. No RF output from PA; antenna or coupling circuit faulty; meter is burned out. It is possible to have the antenna-meter moving coil burn out and still have the transmitter emitting normally. The meters in the final amplifier will read properly in this case.

A PINNED METER. When a meter is suddenly *pinned* (the pointer driven to the stop pin past the maximum scale indication), there is usually a shorted part in the stage; bypass capacitor shorted; RFC short to ground; variable capacitor short to ground; active device may develop a short circuit of internal elements. If no fuse, circuit breaker, or overload relay in the circuit, the meter may burn out before the power supply can be turned off. With the transmitter completely off, an ohmmeter may be used to determine what shorted.

If the RF excitation to an amplifier is removed but the antenna meter continues to indicate the presence of RF, either the amplifier is improperly neutralized and is self-oscillating, or the meter moving-coil and pointer may be stuck.

An *overload* relay carrying I_p in its coil can be made to latch open if the current becomes excessive, protecting the stage.

Time delay relays open only after a greater-than-normal current has been flowing for a period of time. The time delay can be by some mechanical means, such as an oil or air dashpot, or a thermal device which snaps open when heated by an excessive current through it.

Recycling relays restore a number of times automatically before latching open. They are designed for transmitters in which sometimes, for no accountable reason, a spark will jump a gap and momentarily cause a surge of high current in the circuit. The overload relay opens, the spark extinguishes, and the relay closes. If the spark occurs again, or if the fault does not clear itself, the overload falls out again. The relay will latch open after the second or third restoration. Lightning striking on or near a transmitting antenna will often cause such an instantaneous interruption.

16-22 EMERGENCY REPAIRS

When something malfunctions in a transmitter, a qualified or licensed operator must determine what the fault is and then make the necessary repairs. When a faulty part is found it is best to use an identical replacement. Unfortunately, identical replacements are not always immediately available, and emergency repairs must be made. This sometimes tests the ingenuity of the operator or technician.

Most stations are required to have replacement tubes, circuit boards spare parts, and testing devices on hand for their radio equipment. Sometimes only a VOM may be available. Between voltmeter readings and continuity tests with an ohmmeter, however, a surprising number of troubles can be located. Some of the steps that can be taken with transmitting equipment are listed below.

TUBES. Vacuum tubes wear out. An indication

of loss of filament emission is a permanent decrease of I_g and I_p. Excitation to the next stage decreases. If the tube has a glass envelope, a blue or purple haze may be seen between filament and plate, indicating gas. The grid, screen grid, or plate may become red-hot. If the envelope develops a crack in it, air will rush in and a white coating forms on the inside of the glass as the filament wire burns out. Loose elements may heat, sag, and short to adjacent elements, causing sparks and high-current indications on meters. Often pin terminals of tubes oxidize and develop a poor contact. If the tube is pulled out and then put back, the equipment may work normally again. The terminals of the tube and the socket contacts should be lightly sanded with fine sandpaper. A continuity check with an ohmmeter will tell if a filament is burned out. An ohmmeter may also be used to test a tube for element-to-element shorts. Receiving-type tubes can be checked in a tube tester, if one is available. Transmitting tubes must have identical replacements to determine if they are faulty. Tubes of somewhat similar construction with similar socket-pin connections can be substituted temporarily.

TUBE SOCKETS. These may short out internally or open-circuit. Because they so seldom become faulty, they may be overlooked when they do. Ohmmeter tests may indicate the difficulty.

TRANSISTORS. BJTs: with an ohmmeter set to $R \times 100$ or $R \times 1000$; conductivity B-E and B-C with ohmmeter leads one way; no conductivity with ohmmeter leads reversed. Same tests with JFETs. MOSFETs and VMOSs: no conductivity G-S or G-D. Same conductivity S-D and D-S. If devices are in circuit, the components may alter readings to some extent. It is preferable to use a transistor tester.

TRANSFORMERS. Power transformers often give trouble. If they become damp, the insulation around the internal wires may ionize and spark across, sometimes starting a fire. RF ac may burn insulation or short-circuit a transformer. If charred insulation is found, it is sometimes possible to scrape all the blackened parts away and coat the wires with an insulating clear lacquer, or household cement. The transmitter may operate under lowered power for long periods of time with such a temporary repair. If RF is suspected in a transformer, bypass primary and secondary windings. It is not unusual to find transformer wires corroded to the extent that they finally open-circuit. If the wires are within reach, it is sometimes possible to scrape them clean and solder them together again. Sometimes a transformer with several taps for different voltages will open-circuit. It may be possible to jumper the open part and use the remainder of the winding operating at reduced power. An ohmmeter can be used to check continuity of windings. A transformer can be checked by an ac voltmeter (in a VOM), or by putting a low-voltage ac across the primary and measuring the secondary voltage. If there is no voltage output, the transformer is probably burned out or shorted. If burned out, it will remain cool; if turns are shorted, it will soon become hot.

RESISTORS. Resistors often burn out. They may become excessively heated over a period of time and increase resistance, or open. Usually they overheat because some other part is faulty. For example, a screen-grid resistor suddenly becomes hot and burns out. It will usually be found that the screen bypass capacitor is shorted, connecting the resistor from ground to V_{bb}. In some cases, replacement of resistors with others of half or twice the resistance of the burned-out part will produce a temporary repair that will be satisfactory. Sometimes it will not.

CAPACITORS. When capacitors are working the dielectric is under constant strain. If there is a weak spot on the dielectric, eventually it may break down. Sometimes the short circuit is intermittent and is hard to find. Electrolytic capacitors can dry, lose their capacitance, but not short out. When tested with an ohmmeter, they will not show the charging current that a good capacitor will. Nor will they show low resistance as a shorted capacitor would. If a mica capacitor in an RF circuit shorts it may be possible to use mylar capacitors. The bearings, or sliding contacts, on variable capacitors may become dirty. They can be cleaned with alcohol, lacquer thinner, or other solvents. A 100% variation in the capacitance of a bypass capacitor may not noticeably effect the operation of a transmitter. In an emergency, CW transmitter power supplies can be operated with little or no filter capacitance.

CHOKE COILS. RFC coils sometimes oxidize and

ANSWERS TO CHECKUP QUIZ ON PAGE 358
1. (Inductive or π network) 2. (Purer sine-wave ac) 3. (Class A) 4. (Electrostatic) 5. (Push-pull) 6. (Electromagnetic) 7. (Protect personnel) 8. (Amateur) 9. (Pierce) (Class C) (Add bias, limit I_C) 10. (Pierce) (Shunt) (Hartley) (Series) (Dual-gate MOSFET) (Shunt) 11. (Bandpass filter) 12. (Remove crystals, tune LC to 5–5.5 MHz) 13. (Tune with low power) 14. (10,000) 15. (Class A, AB, or B) 16. (See chapter illustrations)

open. If the faulty layer, or pie, can be found with an ohmmeter, it can be jumpered and the choke may work satisfactorily. An RFC may be made by winding several hundred turns of wire onto a pencil-sized piece of insulating material or even dry wood. A choke coil in a power supply may burn out or short to the core. If burned out, it may be jumpered, or the primary or secondary of a spare power transformer may be used in its place. If shorted to the core, the core may be loosened from its metal chassis, insulated from the chassis with sheet insulation, and operated temporarily. The core will be "hot" electrically when the power supply is on and must not be touched.

METERS. A dc ammeter moving coil can burn out or oxidize apart. The meter will still carry current through the shunt, and the transmitter continues to operate, but with no current indication by the meter. If a shunt burns out, the moving-coil element always burns out too. In a voltmeter, the series-multiplier resistor can oxidize and open, but the moving coil may not be damaged. A new series resistor of approximately the same resistance will result in fairly accurate indications by the voltmeter. I_p and I_g values are functions of the filament temperature. If the filament circuit voltmeter burns out, it is possible to arrive at an approximation of the correct filament voltage by adjusting the filament control until normal I_g and I_p is indicated, or until the color of the hot filament seems normal. In an RF thermocouple meter, if the thermocouple circuit becomes faulty, RF ac may still flow through the thermocouple, but no dc will be fed to the meter and no indication of RF will be given. If an RF ammeter is the only means of indicating when antenna and final amplifier are correctly tuned, substitute a 150- to 300-W electric light for the meter if it burns out. Care must be taken not to operate the equipment at a power level that will burn out the light. After the transmitter is tuned at low power, the light can be shorted out and the transmitter can be run at full-power output.

ANTENNAS. Any kind of wire will work as an antenna in an emergency. While copper wire is better, iron wire will radiate almost as well, and the difference in reception is not noticeable. No wire is too thick, and any wire that does not burn out is satisfactory for emergency communications. Dry rope; most plastic materials; and dry, oiled, or waxed wood can make satisfactory temporary antenna insulators.

RELAYS. If a relay coil burns out, it can often be taken apart and rewound (Sec. 3-20). If the relay contacts become so badly worn that they are inoperative, it is often possible to file them smooth.

16-23 ELECTRICAL HAZARDS

Electric shock can range from 1 mA for the threshold of feeling, to 5 mA for decided pain, to 10 mA for the beginning of arm paralysis, to 30 mA for stoppage of breathing and possible death, to 75 mA for ventricular fibrillation (discoordinated heartbeats) and probable death. With 4 A the heart may stop but may start again when current ceases. Above 5 A causes tissue burning, but may not be fatal if vital organs are not involved.

A relatively small shock can cause a violent reaction thay may throw a person backward and possibly against some hard object or off a ladder, which may cause an injury.

Some appropriate skin contact resistances are: finger touch, dry, 50 kΩ, wet, 5 kΩ; grasping wire, dry, 15 kΩ, wet, 5 kΩ; holding pliers, dry, 5 kΩ, wet 2 kΩ; foot to wet ground through wet shoe, 5 kΩ, hand in water, 300 kΩ. Thus, standing on damp ground with wet shoes, holding a pair of pliers, and touching a hot wire, the resistance to ground is perhaps 10 kΩ. From Ohm's law, $V = IR$, the voltage to produce a sharp pain = 0.005 × 10,000, or 50 V. If the hand is perspiring, the voltage would be only about 35 V. Across 120 V ac enough current might flow to produce respiratory paralysis because after a short time the resistance of a body contact decreases, increasing the shock current.

The path of the current determines how lethal a shock may be. Across a hand or up one arm may not be lethal, whereas across the body and heart, or involving the brain or backbone areas, shocks may be very dangerous. One should always make sure not to be grounded when working with electrical equipment, tuning transmitters, etc.

If someone is being subjected to electric shock, the electricity should be turned off before touching the person. If this is not possible, the person should be dragged off of the circuit using dry clothing, or pushed away from the circuit with sticks of dry wood or other insulating materials. Be careful not to get in the electric circuit too! If breathing has stopped mouth-to-mouth resuscitation must be given immediately. If there is no carotid (throat) pulse, CPR (cardiopulmonary resuscitation) must be started at once. Everyone should be familiar with CPR and be able to give it. Contact the Red Cross for CPR instructions and training.

High-voltage RF ac, as from an *LC* circuit or the end of a transmitting antenna, even with only a few watts of RF, will produce skin burns rather than the more familiar shock sensation. With higher power, the burns will be deeper and heal slowly. Ends of transmitting antennas must always be high enough to be out of the reach of everyone (see also Sec. 20-33).

▌ Test your understanding; answer these checkup questions.

1. What might a decrease in PA I_g indicate? _____
2. What does a blue haze between filament and plate indicate? _____
3. Does a cherry-red plate indicate excessive I_p? _____
4. How is it determined if a high-power transmitting tube is faulty? _____ A receiving-type tube? _____
5. What might go wrong with a socket? _____
6. What may cause overheating of a power transformer? _____
7. Why are bypass capacitors often connected across power or audio transformers in a transmitter? _____
8. What about a transformer can be tested with an ohmmeter? _____ _____ _____
9. What should be looked for if a resistor burns out? _____
10. If an ammeter burns out, what can be done in an emergency to get the transmitter on the air? _____
11. If a voltmeter burns out, will the ability of the transmitter to function be affected? _____
12. What is the usual indication if a VT crystal circuit stops oscillating? _____
13. If the PA I_p meter suddenly changes but other meters in the transmitter remain normal, where might the trouble be? _____
14. The PA I_p meter reads normally but the antenna ammeter drops to zero. What is probably wrong? _____
15. Buffer and amplifier stage meters in a transistorized transmitter drop to zero. What are two possibilities? _____ _____
16. If the drive to the PA ceases but the antenna ammeter continues to read, what is the probable trouble? _____ _____
17. How much current produces a painful shock? _____

BASIC TRANSMITTERS QUESTIONS

1. What type of transmitter develops a type B emission?
2. What type of emission is developed by an arc transmitter? By an Alexanderson alternator?
3. Why are single-stage transmitters not used?
4. What is simplex operation? Duplex?
5. How can QSK operation be accomplished?
6. What does MOPA mean? What is its advantage over a single-stage transmitter?
7. Draw a diagram of a BJT-VMOS MOPA capable of radio telegraphy coupled to an antenna.
8. In Fig. 16-3, in what class is the VMOS operating? If the dashed resistor were used?
9. In Fig. 16-3, why would M_2 read zero with C_2 at minimum C? What other terms might be used for B−, B+ low, and B+ high?
10. What meter(s) is (are) not bypassed in a transmitter?
11. When tuning an MOPA should low oscillator power be used? For best frequency stability and an exact operating frequency what kind of oscillator would be used?
12. What are the two basic forms of oscillator instability?
13. What is the basic component in a dummy antenna? What other two components may be added?
14. What would be the names of the three possible forms of keying of a BJT stage? Of a VMOS stage?
15. How are CW signals reduced in bandwidth?
16. What kind of harmonics develop square waves on RF signals? Triangular?
17. How are keying speed and bandwidth related?
18. What is hard keying?
19. Draw a diagram of a key-click filter.
20. Draw a diagram of a BJT series keyer circuit.
21. In Fig. 16-10, what component(s) controls the rounding of the make part of the keyed signal? The break part?
22. What is the main disadvantage of the circuit in Fig. 16-11?
23. Name two ways a CW signal could be modulated by a 500-Hz tone.
24. Draw a diagram of a varactor-type FSK circuit. What does setting the switch to position 2 do?
25. What does PSK mean and what angle is used?
26. How can a PLL synthesizer be made to produce FSK?
27. When is a driver stage required in a transmitter?
28. Why should all stages of a transmitter be shielded?
29. Draw a diagram of a three-stage unkeyed radio transmitter using a BJT oscillator, a VMOS buffer, and a beam-power tetrode output.
30. What are three ways to protect a grid-leak biased

stage from damage due to the loss of RF drive?

31. What are some advantages of using multipliers in transmitters?

32. If the 150.42-MHz carrier of a transmitter having four doubler stages drifts 155 Hz, what, if anything, has happened to the crystal frequency?

33. A crystal operates on 9.984 MHz at 20°C, but at 9.983 MHz at 40°C. What is its TC?

34. List at least 12 items which might reduce unwanted emissions in a beam-power tetrode RF power amplifier.

35. How can harmonics and parasitics be best located?

36. If magnetic shielding is not complete by encasing an RF stage in a metal housing, what might be done to improve the shielding?

37. What two types of oscillators are used in commercial-type transmitters? In amateur transmitters?

38. Draw a diagram of a CW transmitter using BJTs having an oscillator, a buffer, a driver, and a push-pull PA.

39. What are advantages of a heterodyne transmitter for CW?

40. How can a VFO heterodyne transmitter be made a single-dial control system?

41. Draw a diagram of a heterodyne-type VFO system that would be usable on three bands of frequencies.

42. Why would a heterodyne VFO be useful as a full break-in transmitting system?

43. If 2 and 6 MHz are fed to the input of a mixer stage, what are the possible output frequencies of this stage?

44. What is the dB gain of two cascaded RF amplifiers if each has 16-dB gain? If the input signal to the first is 0.1 W, what is the output power from the second?

45. What does a recycling relay do?

46. In a high-power radio station, if the output suddenly drops off, what should the operator do first? Second?

47. What current-value shock produces the threshold of feeling? Decided pain? Arm paralysis? Stoppage of breathing? Ventricular fibrillation? Tissue burning?

48. If you make a 10,000-Ω connection across a power supply, how many volts must it have to produce decided pain? Possible arm paralysis?

17

Amplitude Modulation and SSB

This chapter discusses the important fundamentals of amplitude modulation, starting with the historic and simplest forms, progressing through the circuits used in standard broadcast transmitters and in the picture information of television, up to single-sideband suppressed-carrier (SSSC or SSB) transmitters used in many modern two-way voice communication systems.

17-1 MODULATION

The term modulation may be considered to mean variation or shaping. If a direct current is made to vary in amplitude (strength) 500 times per second, it is being *amplitude-modulated* at a 500-Hz rate. In this case, the dc is the *carrier* current, and the variations impressed on it represent modulation. This is the basic principle of the telephone.

If a transmitter emits an RF ac carrier wave and this *carrier* is made to vary in amplitude at a 500-Hz rate, amplitude modulation (AM or A3E, previously A3 emission) is produced. Here, the RF ac is the carrier current, and the variations impressed on it represent the modulation. This is the basic principle of AM radiotelephone, or AM broadcast transmissions.

If the RF carrier is made to vary in frequency (not amplitude) 500 times per second, it is being *frequency-modulated* (FM, Chap. 19). See Appendix E for a listing of emissions.

17-2 WHY THE CARRIER IS MODULATED

A normal radio receiver can be tuned to the carrier frequency of a radio transmitter but will detect no sound from the carrier alone. However, if the carrier is made to vary in amplitude at speech or musical tone rates, the detector stage in the receiver can detect the modulation on the carrier and reproduce the speech or music variations, producing speech or music sounds in its loudspeaker.

There are many methods by which speech or music frequency ac can be made to amplitude-modulate a carrier, but first there are some basic concepts regarding sound and microphones that should be understood.

17-3 SOUND

When a firecracker explodes, a sound is heard. If the firecracker is near, it sounds loud. If far away, it sounds weak. When the firecracker explodes, it suddenly and violently pushes air outward, creating an expanding ball of compressed air all around itself. Compressing molecules of air as it goes, the ball of compressed air moves outward in all directions, at a rate of about 334 m/s (1100 ft/s). If the ball were examined closely, it would be found that just inside, and next to, the compressed-air molecules is a rarefied area of air molecules. This compression with its attendant rarefaction forms a *sound wave*.

As the sound wave travels outward, it is constantly expanding. The mechanical energy it contained when it was confined in a small area near the firecracker rapidly becomes less for any given square inch of the wave front. If the wave travels only a short distance, it strikes the eardrum with considerable energy. If it travels a long distance, it strikes the eardrum with very little energy.

A single sound wave striking the eardrum causes the outer-ear diaphragm to vibrate. This transmits a mechanical vibration into the inner ear, in turn actuating delicate nerve endings that relay electro-chemical impulses to the brain, and the person is made aware of the "sound." By past experience, this nerve-to-brain impulse is recognized as a "bang."

Any musical instrument producing a constant single tone sets up a continuous series of waves at the tone frequency. These waves, striking the diaphragm of the ear, are transferred to the inner ear and energize the nerve endings that are resonant to this particular frequency, making the listener aware of this tone only. Higher-frequency sound waves energize higher-frequency nerve endings, and the *pitch*, or tone, heard sounds "higher."

The greater the energy of a wave striking the eardrum, the stronger the nerve impulse generated and the louder the sound to the listener.

A sound wave has both amplitude (loudness) and frequency (pitch). It also has *quality*, or purity (great or little harmonic content). A sine-wave ac signal generator adjusted to 262 Hz (middle C) coupled to a loudspeaker produces a pure 262-Hz tone. A musical instrument playing a fundamental 262-Hz tone will sound different because it cannot produce a pure vibration. It may produce a fundamental 262-Hz vibration, but many harmonics, or *overtones*, are generated at the same time. The number and amplitudes of the different harmonics produced identify the type of instrument being played. A piano, an oboe, and a saxophone all sound different even though they may all be playing

the same fundamental tone. Two singers sound different because their vocal cords have different dimensions and produce different harmonics.

Most sounds are highly complex, being made up of fundamentals and many harmonics. Figure 17-1a

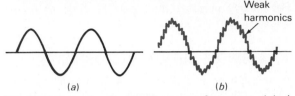

FIG. 17-1 Two signals with the same frequency. (*a*) A pure tone and (*b*) a fundamental with harmonics, or overtones.

represents a pure tone and (*b*) a possible musical or speech tone having the same fundamental frequency plus harmonics.

Sound waves are often represented by sine waves for pure tones and jagged waveforms for speech or musical tones. The half of the drawing above the zero, or time line, may indicate the compression of air molecules. Then that shown below the line will represent the rarefaction part of the sound wave.

More sine-wave ac power is required to modulate a carrier to a maximum extent than is required by most other ac waveshapes. All natural sounds produce a given peak value with a lesser average power. Transmitters engineered to handle sine-wave audio power will be capable of handling any normal sound waveforms. For this reason, explanations are made in terms of sine-wave-modulating ac. Only about half as much average AF power is required to fully modulate a transmitter with speech sounds as to fully modulate with sine-wave tones.

The amount of sound power transmitted to a diaphragm can be measured in microwatts per square centimeter. The lowest audible intensity, or threshold, of sound for the human ear is about 1×10^{-10} μW/cm^2. On the loudness scale, this power level is given a *zero* value. Ten times this sound power is one *bel* louder. Ten times more sound (100 times the threshold value) is again 1 bel louder, or has a value of 2 bels. This is a logarithmic increase. The logarithmic ratio is used because the human ear responds closely to this ratio; that is, an increase in a weak sound of one decibel (one-tenth of a bel), or dB, is barely noticeable to a listener. Actually, this is a power increase of 26% (Sec. 8-5). A very loud sound, when increased by 26%, will also result in a just barely discernible change in sound intensity to a listener.

The ear responds best to sounds between 1000 and 4000 Hz. The response is down about 20 dB at 200 Hz and down about 40 dB at 100 Hz. The older a person is, the less response at the higher frequencies. For teen-agers, sounds of 15 kHz may be down 20 dB, and 18 kHz may be down 40 dB. For elderly persons, there may be little or no response above 12 kHz.

17-4 SINGLE-BUTTON MICROPHONES

A microphone is a device used to convert mechanical sound energy into electric energy of equivalent frequency and relative amplitude.

The original *single-button*, or *carbon-button* microphone is still used in most telephones. It will convert voice sounds into varying electric currents fairly well but will not faithfully reproduce over a wide enough frequency range to be used for high-fidelity music. It is in reality a sound-variable resistor. When its resistance is changed, the current flowing through it varies accordingly.

The construction of a single-button microphone is shown in Fig. 17-2. The metal button is electrically

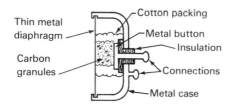

FIG. 17-2 Basic single-button carbon microphone cross-sectioned.

insulated from both the diaphragm and the case. One electrical connection is made to the case and diaphragm, with another to the button. Sand-sized particles of carbon are held between diaphragm and button by a fluffy cotton washer that allows the diaphragm to vibrate to and from the button without allowing carbon particles to fall away. If the diaphragm is moved toward the button, the granules are compressed and the diaphragm-to-button resistance is decreased.

When sound waves beat against the diaphragm, compressions push it inward; rarefactions pull it outward. Thus sound waves cause the resistance of the microphone to vary at the rate of the compressions and rarefactions. High-frequency tones vibrate the diaphragm rapidly; low frequencies vibrate it slowly. Weak sounds cause little vibration. Loud sounds produce wide movements of the diaphragm

and great resistance changes. Pure tones produce an even, smooth inward and outward swing. Harmonic-containing sounds produce a jerky inward and outward swing of the diaphragm.

In the single-button-microphone circuit (Fig. 17-3), the dc emf keeps current flowing through the

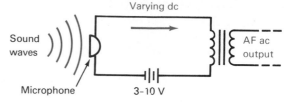

FIG. 17-3 A single-button-microphone circuit.

microphone and transformer primary at all times. Resistance variations, due to sound waves, produce corresponding current variations in the microphone circuit. The vdc in the primary produces an ac in the secondary with a frequency and amplitude proportional to the frequency and amplitude of the sound itself.

If earphones are connected either in place of the primary of the transformer or across the secondary, the electrical variations will reproduce the original sound. This represents a simple one-way telephone circuit.

The single-button microphone is relatively inexpensive, withstands rough handling, and is not subject to deterioration due to heat or cold. It has a relatively high output power in comparison with other microphones. It can be spoken into at close range, allowing the voice to override background noises. It is most sensitive to voice frequencies, between 200 and 3000 Hz.

Sometimes the carbon granules will pick up moisture, stick together, and *pack*. The microphone may have to be shaken or jarred to free the granules. This microphone distorts sounds more than most other types of microphones. When dc flows through the granules, they move very slightly, producing a random variation of current resulting in a constant, weak hissing sound in the output. A current of 5 to 20 mA in the microphone circuit produces satisfactory results. The microphone has an impedance of about 100 to 1000 Ω.

17-5 ABSORPTION MODULATION

The oldest and simplest method of amplitude-modulating a radio carrier wave with voice frequencies is called *absorption*, or *loop*, modulation. An

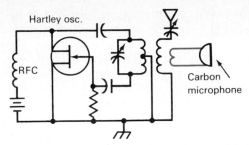

FIG. 17-4 One form of loop, or absorption, modulation.

RF ac source is necessary to provide a constant RF ac carrier to the antenna.

In Fig. 17-4 a single-button microphone is inductively coupled to the antenna circuit. Most of the RF carrier energy being fed into the antenna circuit by the oscillator is radiated, but some is fed into the loop and is dissipated as heat in the resistance of the microphone. The microphone resistance varies at the audio rate of any sound causing diaphragm vibration. As a result, the amount of RF energy being absorbed and dissipated by the microphone is varying.

If the power output of the oscillator is essentially constant but the microphone is absorbing alternately more and less power from the oscillator, the antenna finds itself with alternately less and more power to be radiated. This variation of RF power output at an audio rate from the antenna produces the modulated carrier of the system. Since the power output is being varied in accordance with the strength of the sound waves, the modulation is known as *amplitude modulation* or AM.

The effect of the microphone on the output RF carrier power is shown in Fig. 17-5. This form of modulation is impractical by present standards and is rarely used. However, a variation of the absorption principle using biased PIN diodes across microwave waveguides or across RF or AF circuits is used (Sec. 22-10).

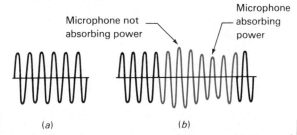

FIG. 17-5 Antenna current (*a*) with no modulation and (*b*) when the microphone is alternately absorbing less and then more power from the circuit.

The absorption loop acts as a partly shorted turn coupled to the oscillator tank. As the microphone resistance changes, the effective inductance of the shorted turn changes, which also modulates the *frequency* of the oscillator. Such simultaneous FM and AM produces unwanted distortion products, and is called *dynamic instability*. The term refers to the carrier and indicates that the carrier *frequency* is not stable under operating conditions.

17-6 SERIES MODULATION

One method of producing AM is by *series modulation*, Fig. 17-6. A continuous RF carrier is provided by an oscillator and RF amplifier. The output of the class C amplifier is fed to an antenna and is radi-

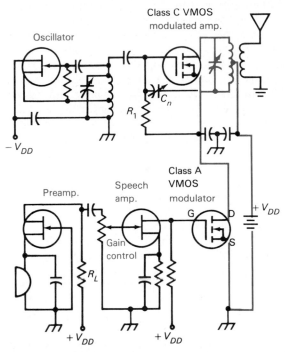

FIG. 17-6 A series-modulation circuit involving two VMOSs.

ated. The RF stage is the *modulated amplifier*. The final AF stage coupled to the modulated amplifier is the *modulator*. Stages preceding the modulator are known as *speech amplifiers*. They amplify the AF emf from the microphone to a value that will drive the gate of the VMOS modulator.

In series modulation, the modulator, the modulated stage, and the power supply are all in series. A variation of the resistance in any part of this series circuit will affect the dc flowing in the RF amplifier and therefore the power output of that stage (see also Sec. 23-18).

The modulator in series modulation is an electronic variable resistor, since an AF voltage variation applied to its gate varies the dc drain resistance of the VMOS (compare with the BJT keying circuit, Fig. 16-10).

When a sound strikes the microphone, the resistance changes, the JFET bias changes, and its I_D varies, developing a varying voltage-drop across R_L. This is amplified by the speech amplifier and applied to the gate of the modulator. When the modulator gate is driven more positive than normal, its drain resistance decreases, the voltage-drop across the modulator decreases, and the RF amplifier is across a greater proportion of the power-supply voltage. More current flows through the RF amplifier, and its RF output increases.

When the modulator gate is driven *more negative*, its drain resistance increases and a greater voltage-drop is developed across the modulator. Less of the total power-supply voltage is across the RF amplifier, and its output decreases. In this way, AF variations of the gate voltage of the modulator vary the RF ac power output of the RF amplifier.

Series modulation has the advantage of not requiring any modulation transformer to limit frequency response. It has the disadvantage of requiring approximately twice the power-supply voltage, since both modulator and modulated amplifier, are in series across the supply.

The audio voltage applied to the gate of the modulator must never be high enough in amplitude to produce I_D cutoff. This places the operation of the modulator in the class A category. Note the method of forward-biasing the modulator VMOS in Fig. 17-6.

17-7 THE MODULATED ENVELOPE

All modulated RF stages are biased to class C. As a result, the drain current will always be narrow pulses of dc that occur during the positive half cycle of the RF ac gate excitation. These pulses of I_D produce a flywheel effect in the drain-tank circuit, resulting in a very nearly sinusoidal RF ac output waveform (Fig. 17-7).

If the modulator suddenly allows more I_D to flow through the RF amplifier, the drain pulses increase in amplitude and the RF flywheel amplitude also increases. Figure 17-8 illustrates the result of flywheel effect on the output ac when varying am-

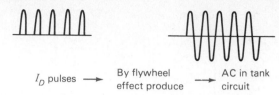

FIG. 17-7 Pulses of I_D by flywheel effect produce sinusoidal RF ac in the LC tank circuit.

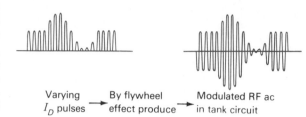

FIG. 17-8 Varying-amplitude I_D pulses produce a modulated RF ac in the LC tank circuit.

plitude I_D pulses are flowing in the drain circuit. Note that both the positive and the negative halves of each RF ac cycle are increased. The diagram of the RF varying in amplitude during modulation illustrates a *modulation envelope*.

Note that in single-ended circuits, flywheel action alone is responsible for the reproduction of the second half of the RF ac cycle. Each cycle of RF of the modulated envelope produced by flywheel action has an almost perfect sine waveshape if a tank-circuit Q of 10 or more is used. If the RF amplifier is push-pull, I_D pulses flow in opposite directions alternately in the tank circuit and aid in producing the second half of the RF cycle.

Test your understanding; answer these checkup questions.

1. What is the letter-number designation of AM voice or music emissions? _____ FM? _____
2. If a broadcast station is transmitting but not modulating, what is heard in a receiver tuned to its frequency? _____
3. What is a radio wave? _____ What is a sound wave? _____
4. Does audible "pitch" refer to frequency, amplitude, purity, or harmonics? _____
5. What is the frequency of middle C? _____ Of one octave above middle C? _____
6. What is the threshold power value for the human ear? _____ What power is 1 bel louder? _____

7. What frequencies are heard best by the human ear? _____
8. What is another name for a single-button microphone? _____ Where are these used most? _____
9. How is a packed single-button microphone cleared? _____
10. What was the first type of modulation, using a SB microphone, called? _____
11. What is dynamic instability? _____
12. In series modulation what name is given to the last AF stage? _____ To the RF stage to which it is coupled? _____
13. List three advantages of series modulation.

_____ _____ _____
14. What is another name for an AF volume control? _____
15. What device can be used to make a modulated envelope visible? _____
16. Draw diagrams of a carbon-button microphone circuit. A loop-modulated oscillator. A series-modulated transmitter. A 100% modulated envelope.

17-8 BASIC PLATE MODULATION

It is possible to produce modulation of a carrier by adding an AF ac in series with the plate (collector, drain) circuit (Fig. 17-9). This is basic plate modulation. Sound waves striking the diaphragm of the

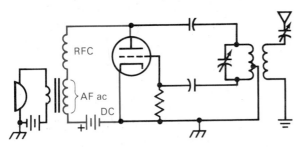

FIG. 17-9 Basic plate modulation of an oscillator.

microphone produce an AF ac voltage in the secondary of the microphone transformer.

On the positive half cycle of the AF ac, the audio emf will be in the same direction as the plate-supply emf, and the two will be additive. As V_p increases, I_p increases, and RF output increases.

On the negative half cycle the audio emf will be opposing the plate-supply emf. The resultant V_p is decreased, I_p decreases, and RF output decreases.

During the positive half cycle of the modulating AF ac, the *positive peak of modulation* of the carrier is produced (Fig. 17-10). During this half cy-

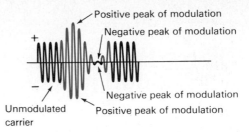

FIG. 17-10 Positive and negative peaks of modulation.

cle, both positive and negative half cycles *of the RF ac* are greater than the unmodulated carrier.

During the negative half cycle of the modulating ac, the *negative peak of modulation* is produced. In this half cycle both positive and negative half cycles of RF ac output are less than the carrier value. (It may seem strange that the negative *peak* should actually be the lowest point as far as the RF ac voltage in the emitted wave is concerned.)

The explanations have been given in terms of voltage but are valid for carrier currents also, since current is directly proportional to voltage. If the carrier voltage is doubled, the carrier current is also doubled. At the instant the voltage and current are doubled, the output power is four times the carrier value. An illustration of *power* at the positive peak of modulation would have to be shown four times as high as the carrier value.

When an ac voltage is used to modulate a self-excited oscillator, the variation of V_p will produce some FM. Furthermore, during high-amplitude negative peaks of the modulating ac, the V_p may drop so low that the oscillator will stop oscillating for a period of time, producing distortion. Thus, full AM of an oscillator is undesirable.

The voltage output of a microphone alone will not be sufficient to produce much modulation. However, by adding audio amplifiers, it is possible to increase the audio power and voltage to produce adequate plate (drain, collector) modulation.

17-9 PERCENT OF MODULATION

The percent of modulation of a carrier wave is determined by how much the carrier voltage or current varies in amplitude. A strong carrier with a low percent of variation produces a weaker response in a receiver than a weaker carrier with a greater percent of modulation.

The variation value is expressed as a *percent of modulation*, with 100% being the highest possible undistorted modulation. It may be found by:

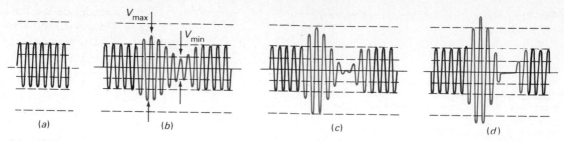

FIG. 17-11 (*a*) An unmodulated carrier. (*b*) A 50% modulated carrier. (*c*) A 100% modulated carrier. (*d*) An overmodulated carrier.

$$\% \text{ mod} = \frac{V_{max} - V_{min}}{V_{max} + V_{min}} \times 100$$

Figure 17-11 illustrates four values of sine-wave modulation of an RF carrier. The unmodulated carrier (Fig. 17-11*a*) represents 0% modulation. A carrier is modulated 50% if its positive peak voltage rises to a value 50% greater than the unmodulated-carrier voltage at the positive peak, and drops 50% at the negative peak of modulation (Fig. 17-11*b*). A carrier is modulated 100% if its positive peak rises to a value twice the unmodulated-carrier maximum and also drops 100% (to zero) at the negative peak (Fig. 17-11*c*). If too much modulating voltage is applied to the modulated stage, overmodulation occurs. The positive peak rises to more than twice the carrier level, and the negative peak drops to zero and remains at zero for a time. The 50 and 100% illustrations show sine-wave modulation. The overmodulated signal (Fig. 17-11*d*) is not a sine wave at the negative peak and therefore will produce undesirable harmonics of the modulating frequency, known as *splatter, buckshot,* or *spurious emissions*.

The results of overmodulation are interference to other radio services on frequencies above and below the carrier frequency as well as a distorted, harsh-sounding transmitted signal. Care must be exercised that no more than 100% modulation of the *negative peaks* occurs in radiotelephone transmitters. As long as positive peaks of overmodulation retain an un-

distorted waveform, they are not responsible for distortion. It is the abrupt arrival of the negative peak at zero output and its abrupt rise from zero that produce the undesirable by-products of overmodulation.

Because the percent of modulation is directly related to the loudness and intelligibility of the signal produced in a receiver, the higher the average percent of modulation without distortion, the more effective the communication can be. To maintain as high a percent as possible, an *audio peak limiter* is usually incorporated in one of the speech amplifiers of commercial radiotelephone equipment. It clips off signal-voltage peaks passing through it which rise above a predetermined value. While this clipping distorts the peaks and generates AF harmonics, the latter can be filtered with low-pass filters. The distortion becomes hardly noticeable, and the emission is not broadened. Excessive limiting, or clipping, of the waveform will produce noticeable distortion.

17-10 PLATE MODULATION

A diagram of a possible broadcast station transformer-type plate modulation is shown in Fig. 17-12. A triode RF amplifier stage is being modulated by adding an AF ac voltage in series with the plate power-supply voltage of the RF amplifier. The power-supply voltage of the RF amplifier is 2000 V, so the AF voltage output of the modulation transformer must be 2000 peak V, positive and negative, to produce 100% modulation. Under this condition, V_p will vary alternately from 4000 to 0 V, producing positive voltage peaks of modulation twice the value of the carrier voltage and negative peaks of zero output.

The RF amplifier stage is grid-leak-biased to class C by RF ac applied to the grid by the driver stage. The output of the RF amplifier is coupled to a Marconi (grounded at one end) antenna.

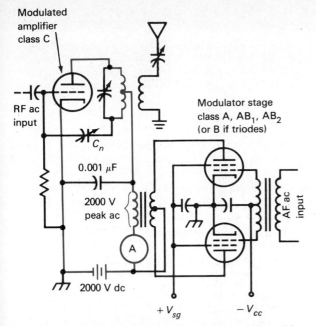

Modulated amplifier class C

RF ac input

C_n

0.001 μF

2000 V peak ac

A

2000 V dc

$+V_{sg}$ $-V_{cc}$

Modulator stage class A, AB$_1$, AB$_2$ (or B if triodes)

AF ac input

FIG. 17-12 A plate-modulated RF power amplifier with a class B push-pull modulator.

The modulator is a push-pull beam power tetrode stage developing the high AF power needed to produce the desired modulation and at the same time reduce even-order AF harmonics.

An important factor in plate modulation is determining how much AF power is required to produce modulation (usually 100%). Assume that a modulated RF amplifier is operating with $V_p = 2000$ V and $I_p = 0.5$ A, shown by ammeter A.

To determine the power required to modulate the RF amplifier to 100%:

1. According to the power formula $P = VI$, the dc power input to the RF stage must be 2000×0.5, or 1000 W.
2. According to Ohm's law $R = V/I$, the resistance of the plate circuit of the RF stage is 2000/0.5, or 4000 Ω.
3. To produce 100% plate modulation, the sum of the instantaneous AF ac peak voltage, plus the dc V_p must equal twice the unmodulated RF amplifier V_p and zero volts, alternately.
4. Twice the voltage on the plate of the RF tube will produce twice the I_p, or 1 A.
5. Twice the voltage and twice the current produce a *peak* power of 4 times the dc power input, or 4 × 1000, or 4000 W.
6. Even at 100% modulation, the I_p meter M will not visibly vary. Therefore the *average* power

being drawn from the RF amplifier power supply must still be 1000 W. It would seem that the modulator must furnish 3000 W to produce the 4000-W peak power. However, consider what the modulator is actually doing.

7. The modulator is feeding its power output through the modulation transformer into the 4000-Ω plate circuit of the RF amplifier.
8. The modulator must develop a 2000-V *peak* ac voltage into the 4000-Ω resistive RF amplifier-plate circuit to produce 100% modulation. How much power will it take to produce 2000-V peak ac across 4000 Ω?
9. The power formula $P = V^2/R$ assumes effective-voltage values. To use this formula the peak-voltage value must be converted to the effective value by multiplying the peak by 0.707. In this case, $2000 \times 0.707 = 1414$-V effective AF ac.
10. By substituting these figures in the formula for power,

$$P = \frac{V^2}{R} = \frac{1414^2}{4000} = \frac{2,000,000}{4000} = 500 \text{ W}$$

11. It requires 500 W of sinusoidal AF power from the modulator to produce a 2000-V peak and 100% plate-modulate an RF amplifier with a plate-power input of 1000 W. (This exact 2:1 ratio is for sinusoidal modulation only.)

Note that the AF power must be delivered into the RF amplifier-plate circuit. Any AF power lost, because of inefficiency of the modulation transformer, will not reach the RF amplifier. As a result, the modulator must feed slightly more than one-half of the dc plate power into its output-transformer primary to produce the required secondary power.

It is possible to apply the same reasoning and formulas to determine the required AF power to produce any percent of modulation. For example:

A peak AF modulating ac equal to 100% of the dc V_p produces 100% modulation.

A peak AF modulating ac equal to 75% of the dc V_p produces 75% modulation.

In order to find the required audio power to plate-modulate an RF amplifier with 2000-V V_p and 0.5-A I_p ($R_p = 4000$ Ω) to 50%:

$$P = \frac{V^2}{R} = \frac{(1000 \times 0.707)^2}{4000}$$

$$= \frac{707^2}{4000} = \frac{500,000}{4000} = 125 \text{ W}$$

Therefore 50% modulation requires only one-quarter of the power required for 100% modulation. This represents a 75% decrease in required audio power between 100 and 50% modulation.

A simpler method of determining the required power to produce a given percent of modulation is

$$P_{af} = \frac{m^2 P_{dc}}{2}$$

where P_{af} = audio power needed
P_{dc} = dc plate-power input
m = modulation percent as a decimal ($50\% = 0.5$)

Check this formula with the 100% and 50% problems explained above.

If a given amount of AF power is available, it is possible to determine the dc power which can be modulated to a desired percent by rearranging the formula above to

$$P_{dc} = \frac{2P_{af}}{m^2}$$

An AF power of 500 W will modulate what RF amplifier dc power input to 50%?

$$P_{dc} = \frac{2P_{af}}{m^2} = \frac{2(500)}{0.5^2} = \frac{1000}{0.25} = 4000 \text{ W}$$

How much modulator dc power input is required to produce a given AF ac power? If the modulator stage is class A, it will be only about 25% efficient. If class B, it may be more than 50% efficient. If the transmitter has an output power of 1000 W and the final RF amplifier is only 50% efficient, the plate input power must be 2000 W. It will require 1000 W of sine-wave AF ac to modulate the 2000 W of dc plate input. The AF power output from the modulator stage will be equal to its plate-power input times the percent of efficiency of the stage, or

$$P_o = P_{in}(\%)$$

where P_o = AF ac power output
P_{in} = dc power input
$\%$ = efficiency of stage

This formula can be rearranged to

$$P_{in} = \frac{P_o}{\%}$$

If the modulator is a class B AF amplifier with an efficiency of 60%, the dc power supply input to the modulator to produce 1000 W of audio ac is

$$P_{in} = \frac{P_o}{\%} = \frac{1000}{0.60} = 1670 \text{ W}$$

In all cases above, modulating ac is assumed to be sinusoidal. Because of the jagged, high-peaked characteristics of speech sounds, their effective power is far below that required to produce a given peak value of sine-wave voltage. It is generally considered that if 500 W of audio will produce 100% sine-wave modulation, half of 500 W, or 250 W of voice-type audio power will produce 100% modulation on peaks. However, with some voices 250 effective watts of audio may not have AF voltage peaks high enough to produce 100% modulation at any instant. Other voices may have many peaks overmodulating the transmitter with less than 250 W effective voice AF ac.

In Fig. 17-12, at the positive peak of modulation the plate potential rises from the 2000-V carrier value to 4000 V. This should produce twice the I_p and therefore four times the RF power output. However, as the V_p is doubled, the I_p may not quite double. As a result, the output power may not be quite four times, and the positive peak will not rise as much as it should. If power-supply bias alone is used in the RF amplifier, this distortion of the positive peak will always occur. With grid-leak bias, when the I_p increases, fewer electrons will be available to flow to the grid, reducing I_g and therefore the bias value. Reducing the bias increases I_p, increasing the positive peak of modulation. Plate-modulated stages should either use power-supply *and* some grid-leak bias or only grid-leak bias.

The value of the RF grid excitation is quite important. A grid-leak-biased amplifier with no RF grid excitation has no bias, and excessive I_p will flow. With little RF grid excitation there will be only a little bias and the signal voltage will operate over only a small portion of the $V_g I_p$ curve of the tube. This will result in low efficiency and low RF ac power output. If the plate circuit is modulated, the grid is not being driven positive enough to lower the impedance of the tube and the positive peaks of modulation will not be developed, although the negative peaks may be. As excitation is increased (without modulation), the RF power output will increase up to a point. The tube is approaching the point of saturation. Further increase of excitation may not materially increase the RF output.

If modulation is applied with low RF drive, low percent of modulation may be produced satisfactorily, but high positive peaks of modulation will not be. The excitation must be increased to allow high-percent positive peaks to be produced linearly. The correct RF excitation will be the minimum value required to produce a positive voltage peak of modulation twice the carrier value and a negative peak just to zero, as indicated on an oscilloscope (Sec. 17-20). The approximate values can be approached by adjusting the stage according to the manufacturer's operating data for the tube used.

If a class C amplifier has a V_p of 1000 V with an I_p of 0.15 A and the modulator tube has a plate impedance of 15,000 Ω, what turns-ratio transformer will match the modulator to the modulated tube? The RF amplifier plate circuit has an impedance of $R = V/I$, or 1000/0.15, or 6670 Ω. From the turns-ratio formula in Sec. 14-11,

$$T_{\text{ratio}} = \sqrt{\frac{Z_1}{Z_2}} = \sqrt{\frac{15,000}{6670}} = 1.5{:}1$$

The primary should have 1.5 times as many turns as the secondary. In practice, any ratio between 1.5:1 and 2:1 would operate satisfactorily.

17-11 PLATE-MODULATING TETRODES AND PENTODES

Tetrodes or pentodes may also be modulated. Triodes must be neutralized, whereas tetrodes and pentodes may not require neutralization below perhaps 30 MHz and usually have higher-power sensitivities (require less RF driving power) and need less bias voltage.

If a modulating voltage is applied to the plate circuit of a tetrode or pentode but a constant voltage is applied to the screen grid, 100% modulation is not possible because I_p is fairly independent of V_p. It is necessary to modulate both plate and screen-grid circuits simultaneously to produce high percent modulation. To do this the modulation transformer may have a secondary and a tertiary winding, producing a high AF ac voltage for the plate circuit and a lower modulating voltage for the screen-grid circuit (Fig. 17-13). The plate- and screen-modulating voltages must be in phase.

The screen grid may be fed a modulating voltage through a voltage-dropping resistor connected to the plate end of the modulating transformer (Fig. 17-14). If the resistor is connected to the B+ end of the transformer, no modulating AF ac will be fed to the screen grid.

A high-inductance, low-resistance choke coil in series with the screen-grid supply lead (Fig. 17-15)

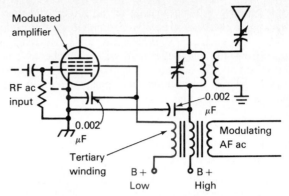

FIG. 17-13 Plate modulation of a tetrode (or pentode) using a modulation transformer with a tertiary winding.

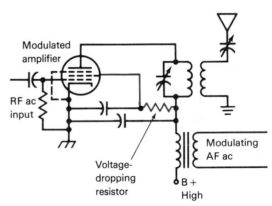

FIG. 17-14 Plate modulation of a tetrode (or pentode) using a voltage-dropping resistor and one power supply.

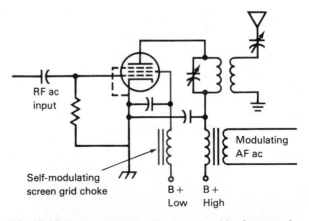

FIG. 17-15 Self-modulating the screen grid of a tetrode (or pentode) while plate-modulating.

will produce self-modulation of the screen-grid voltage. Any increasing positive potential on the plate causes increasing I_p and decreasing I_{sg}. Decreasing I_{sg} allows the magnetic field of the choke to collapse, inducing a more positive voltage on the screen. Thus the screen and plate become more positive at the same time.

▌ Test your understanding; answer these checkup questions.

1. In a single-tube class C amplifier, why is the radiated modulated envelope not made up of varying amplitude dc pulses? _____
2. In a modulated wave is the power in the positive peak equal to the power in the negative peak? _____
3. Why is simple plate modulation not very practical? _____
4. What is the greatest possible modulation value without distortion? _____
5. A plate-modulated RF stage has $V_p = 1$ kV. What modulation is produced if an AF of 2000 V p-p is added to the plate circuit? _____ 500-V peak? _____ 100-V peak? _____ 700-V rms? _____ 1000-V rms? _____
6. What does overmodulation produce? _____ _____
7. Which produces greater interference, overmodulated positive or negative peaks? _____
8. Does using a transmitter audio peak limiter result in louder or in weaker received signals? _____ Why? _____
9. How much AF power is required to sine-wave plate-modulate a 50,000-W transmitter to 100%? _____ To 50%? _____ To 10%? _____
10. In question 9, if the plate modulator stage were 50% efficient, what would be the dc input to it when producing 100% sinusoidal modulation? _____ What class would it probably be? _____
11. What kind of bias should be used on a plate-modulated stage? _____ Why? _____
12. What is the result of low RF excitation to a plate-modulated RF stage? _____
13. What are the three ways of simultaneously modulating the screen and plate circuits of pentode or tetrode tubes? _____ _____ _____
14. Draw diagrams of the basic plate modulation of an oscillator. A 50% modulated envelope. A plate-modulated triode with push-pull pentodes. The simplest form of plate-modulating a tetrode or a pentode.

17-12 OPERATING POWER

A plate-modulated RF amplifier with 2000 V V_p and 0.5-A I_p has a 1000-W input to the plate circuit. It is usually assumed that the maximum RF output from a class C RF amplifier will be about 70% of the input. For 1000-W input, about 700 W of output, or *operating power*, will be produced.

For AM broadcast transmitters licensed at more than 5000 W, the 70% factor may be raised to 85%. A transmitter licensed for 8500-W operating power will require an input power of 10,000 W.

The efficiency of an RF amplifier should remain constant regardless of the modulation percent. A 1000-W input transmitter has a 700-W carrier power whether modulated or not. When it is 100% sine-wave-modulated, the dc plus the audio ac power input is 1000 plus 500 W, or 1500 W. The total RF power output with 100% modulation is 70% of 1500 W, or 1050 W. The carrier is still only 700 W. The other 350 W of RF power is in *sidebands*, two other RF signals generated by the modulation process and emitted when the carrier is transmitted. (One-third of the radiated power is in sidebands at 100% modulation.)

RF power output determination as described above is known as the *indirect* method. The *direct* method of determining RF power output is either by $P_o = I_a^2 R_a$ (where I_a is antenna current at the feed point and R_a is the feed-point impedance value), or by $P_o = V_a^2/R_a$ (where V_a is the RF voltage at the point where R_a is measured).

17-13 SIDEBANDS

The sine wave is the perfect ac waveform. A sine-wave ac RF carrier will have no harmonics or any other frequency components in it. If it has a frequency of 1500 kHz, the only possible emission from it would be a 1500-kHz signal (Fig. 17-16a).

If the illustration of the modulated envelope back in Fig. 17-11 is examined, during the time the carrier is constant in amplitude (not modulated), the RF ac might have a substantially sine waveshape. When modulation is applied, however, each succeeding cycle is higher or lower in amplitude than the one before it. Any progressive increase or decrease in amplitude changes the sine waveshape slightly. The change that occurs when a 2000-Hz sine-wave modulating signal is applied is such that two other RF frequencies, called *sidebands*, are developed, one on each side of the carrier. One of

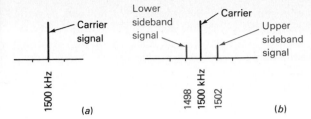

FIG. 17-16 (*a*) A 1500-kHz carrier and (*b*) the sidebands formed when the carrier is modulated by a 2-kHz audio signal.

the sideband signals occurs 2 kHz above and the other 2 kHz below the carrier (Fig. 17-16*b*).

Sidebands are generated by *mixing, beating,* or *heterodyning* one frequency with another in a nonlinear circuit (VT, transistor, diode, Sec. 16-20). The result is always at least four output frequencies: (1) one of the frequencies, (2) the other frequency, (3) the *sum* of the two frequencies, and (4) the *difference* between the two frequencies.

In a modulated RF amplifier-plate circuit, if a 1500-kHz carrier is mixed with a 2-kHz AF from the modulator, the result is: (1) 1,500,000 Hz, (2) 2000 Hz, (3) 1,502,000 Hz, and (4) 1,498,000 Hz. The RF amplifier-tank circuit and the antenna circuit have practically zero impedance to 2000-Hz, so no voltage of this frequency can develop in them and the 2000-Hz frequency is lost. The carrier and the two sidebands are close enough in frequency so that the amplifier-tank circuit, and the antenna will accept and radiate them.

All transmitted intelligence is in the sidebands. If 350 W of RF power is radiated in the sidebands, each sideband will have half, or 175 W. A narrow-bandwidth receiver can tune to the carrier or to either sideband signal alone. With a 2-kHz tone modulation, the receiver will produce no tone from any of the three signals *alone*. A broader receiver will accept all three signals at once and produce a 2-kHz tone when the carrier and sideband frequencies mix or recombine in the nonlinear receiver *detector* circuit to reproduce a 2-kHz resultant. It is also possible to have a sharp receiver pick up the carrier and only one sideband. These two frequencies can mix in the detector to produce a 2-kHz signal, but at only half the power (-3 dB) output.

If the 2-kHz tone modulation is distorted for any reason, harmonics of 2 kHz will be present in the modulated envelope. Instead of confining all the radiated energy within 2 kHz on either side of the carrier,

spurious harmonics of 2 kHz will be produced as additional sidebands far out, above and below the carrier frequency. This results in a broad emission which may produce interference to other radio services as well as distorted-sounding 2-kHz signals in a receiver.

In plate modulation, the AF power output of the modulator produces the sidebands. Modulator AF output power is converted to RF sideband power by mixing them in the nonlinear plate circuit.

$$P_{sb} = P_{af} \times \text{RF amplifier efficiency}$$

The sideband power present in a modulated signal can be determined with the formula

$$P_{sb} = \frac{m^2 P_{car}}{2}$$

If the carrier output of a transmitter is 1000 W, how much power is in the sidebands with 80% sine-wave modulation?

$$P_{af} = \frac{m^2 P_{dc}}{2} = \frac{0.8^2(1000)}{2} = 320 \text{ W in SBs}$$

17-14 BANDWIDTH

The *bandwidth* of an AM transmitter is determined by the highest-frequency audio ac being transmitted, and is the difference in frequency between the furthest removed upper and lower sidebands. A carrier modulated with an 800-Hz audio tone has a bandwidth of 1600 Hz. The bandwidth in Fig. 17-16*b* is 4 kHz.

If the modulating frequency is 3 kHz but the audio is distorted, as from overmodulation, the bandwidth will be determined by the number of harmonics of 3 kHz that are significantly strong. If the fifth harmonic is still relatively strong, the bandwidth is at least $2 \times 15 = 30$ kHz. Near the transmitter, where even weak harmonics are receivable, the signal will appear even wider.

For radiotelegraph transmitters keyed 30 words per minute, the bandwidth should be about 4.2 times wpm, or 4.2(30), or 126 Hz. For speech transmission, where 3000 Hz is the highest frequency to be transmitted, the bandwidth should be 6 kHz. For music transmission, in which the highest frequency to be transmitted is 5000 Hz, bandwidth is 10 kHz. For high-fidelity transmissions, in which the highest frequency is 15,000 Hz, bandwidth is 30 kHz.

17-15 HEISING MODULATION

One type of plate (collector, drain) modulation is the Heising system (Fig. 17-17). The plate currents of both modulator and RF amplifier flow down through the modulation choke. The same V_p is

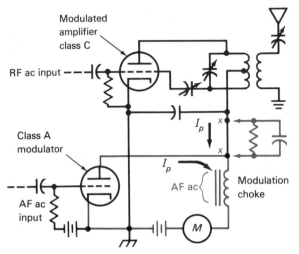

FIG. 17-17 Heising, or constant-current, modulation circuit.

being applied to modulator and modulated amplifier. The modulator is a class A stage and the RF amplifier is class C. The modulation choke has enough inductance (30 to 100 H) to present a very high impedance to any AF current variations that attempt to flow through it. With no modulation, there is a steady average dc flowing through the choke, part being the modulator I_p and part being the RF amplifier I_p.

When the grid of the modulator is driven less negative, the I_p of the modulator increases. The current increasing (downward) through the choke develops a counter emf (upward voltage) across the choke. The counter emf is in series with the power-supply voltage and the plate circuit of the RF amplifier. However, the direction of the counter emf is

opposite to the power-supply voltage and partially cancels the V_p on the RF amplifier. This decreases the RF amplifier I_p and the RF output. The RF amplifier I_p decreases as much as the modulator I_p increases.

If the modulator grid is driven more negative, the modulator I_p decreases, producing a collapsing magnetic field and an induced emf (downward) across the choke which adds to the power-supply voltage for the RF amplifier. This increases RF amplifier I_p and RF ac output. The RF amplifier I_p increases as much as the modulator I_p decreases. The result is a constant current value in the choke and power supply and is why Heising is known as *constant-current modulation*. Meter M should never change.

The peak AF ac voltages developed across the choke coil in a Heising class A modulation system will never equal much more than 80% of the plate supply voltage. If the RF amplifier V_p cannot be doubled at the peaks of modulation, 100% modulation cannot be attained. If a resistor, with an AF bypass capacitor across it, is connected in series with the plate circuit at the points marked x in the diagram, the voltage-drop across the resistor will lower the dc voltage on the RF amplifier plate but not on the modulator plate. If the power supply is 1000 V and the modulator can develop 800-V peak AF ac across the choke, the series resistor should have a value large enough to provide a voltage-drop of at least 200 V to allow 100% modulation.

The capacitor across the resistor should have a low reactance to all audio frequencies in order not to attenuate them. The resistor alone will produce a dc voltage-drop across itself, but the modulating AF will also produce a voltage-drop across it. The capacitor passes the AF to the RF amplifier plate without any AF voltage-drop.

17-16 GRID MODULATION

In grid- (base-, gate-) bias modulation an AF voltage is added in series with the bias supply of the modulated amplifier (Fig. 17-18). Since the bias voltage can control the output power of the RF amplifier, variations of the bias voltage can produce an A3 output. In plate modulation, the modulating ac works into a plate circuit having a definite resistance value ($R = V_p I_p$). In grid modulation if the bias and modulating-signal values are so adjusted that the RF amplifier grid draws no current at any time during modulation, the modulator is then

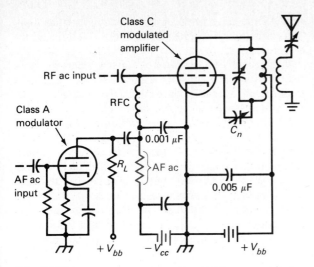

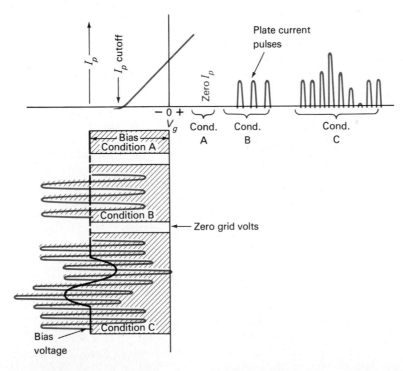

FIG. 17-18 A control-grid, or grid-bias, modulation circuit.

working into a very high load resistance (the value of R_L), and almost no power is required to produce modulation.

A grid-modulated stage is biased to class C and must use a bias supply. Grid-leak bias requires I_g flowing through a grid-leak resistor to produce a bias voltage. With no I_g in grid modulation there would be no grid-leak bias voltage.

Figure 17-19 illustrates the V_gI_p curve of a grid-modulated amplifier (never any I_g). In *condition A*, the bias is adjusted to about 1½ times the I_p cutoff value. There are no RF excitation or modulating voltages, no I_p pulses, and no output.

In *condition B*, the RF excitation voltage is to the point where the peaks are at the midpoint of the V_gI_p curve between zero grid volts and the cutoff value. With medium I_p pulses an output carrier is produced.

In *condition C* the bias voltage is modulated, moving the RF excitation peaks toward and away from the zero grid-voltage line. The I_p pulses vary in amplitude as the modulating voltage changes. Note that the RF excitation voltage does not vary in amplitude, but the whole excitation signal is swung back and forth under the curve by varying the bias voltage, thereby producing the varying-amplitude I_p pulses.

There seems to be an inconsistency in grid modulation. The plate power-supply voltage is constant. At the point of 100% modulation, the amplitude of the I_p pulses is twice what it was with the carrier alone. With the same V_p and twice the current there

FIG. 17-19 How modulation of the grid-bias voltage produces modulated-amplitude I_p pulses.

should be only twice the power output from the tube at 100% modulation peaks. But the positive peak of modulation requires four times the power output (Sec. 17-8). During modulation the excitation voltages are operating over a greater portion of the curve, and the efficiency of operation of the tube increases. When the bias is swung sufficiently to produce I_p pulses of double amplitude, twice the length of $V_g I_p$ curve is being used and the efficiency of stage operation doubles. As a result, at 100% modulation the tube is twice as efficient, and the peak current is twice as much, producing the required fourfold increase of RF power output at the 100% positive peak. Because of this, grid modulation is known as *efficiency modulation*.

During grid (or plate) modulation, the I_p pulses increase as much as they decrease, resulting in a constant average I_p. A plate ammeter does not vary if the modulation is undistorted.

Operating a broadcast station grid-modulated amplifier as explained will produce reasonably undistorted modulation up to about 95%. The stage will operate at only about 20% efficiency when producing the carrier alone, rising to a peak of about 40% under 100% modulation conditions. To increase the efficiency of the amplifier stage, the RF excitation may be raised to the point where the peaks of the RF grid voltage approach the zero grid-voltage point. This operates the tube over a greater proportion of its curve, at about 35% efficiency. When modulated to 100%, the positive peak of modulation will be produced while the tube is operating at about 70% efficiency. Under these conditions, I_g flows during part of the positive peaks of modulation, flattening the peaks. To prevent this type of distortion, transformer coupling must be substituted between the modulator and the grid circuit in Fig. 17-18. A loading resistor may be connected across any grid LC circuit to load equally both halves of the RF excitation cycle. The AF power required for grid modulation is rarely more than 1 or 2 W.

To produce the highest possible power output and the best linearity of modulation, the V_p on the modulated stage should be as high as the tube will stand safely. When resistance-coupled, grid modulation can be linear to over 6 MHz.

Among the disadvantages of grid modulation are the critical adjustments. The degree of coupling to the antenna, the correct L/C ratio in the plate-tank circuit, the bias voltage, the excitation voltage, and the modulating voltage are all more critical than in plate modulation. While grid modulation requires less AF power than plate modulation and has simpler circuits, it requires a higher V_p and is easily overmodulated.

Suppressor-grid modulation occurs when the control grid is fed a constant value of RF ac, the plate circuit has a constant value of high-voltage dc, and the suppressor grid is biased to class A (no suppressor grid current flow). Addition of AF voltages to the suppressor grid circuit through an AF transformer, for example, causes variation of the I_p and modulation of the RF ac output. The control grid may use grid-leak bias, but the suppressor requires either battery or cathode-resistor bias.

Screen-grid modulation is developed when the AF modulating voltage is added in series with the screen-grid power-supply lead. Whereas suppressor-grid modulation is like control-grid modulation, requiring no AF power, screen-grid modulation is modulating the I_{sg} and requires some AF ac power. The control grid may be grid-leak-biased. The suppressor grid is connected to the cathode directly. V_p is high, but V_{sg} is reduced to about 60% of normal class C operation for the tube used. Although screen- and suppressor-grid modulation may not be seen any more, they do illustrate two other methods of producing modulation, or mixing RF and AF to produce sidebands.

17-17 HIGH-LEVEL AND LOW-LEVEL MODULATION

Modulated stages can be divided into three separate categories. These are:

1. *High-level* modulated stages, in which the modulated RF stage is plate- (drain-, collector-) modulated and its output feeds the antenna.
2. *Low-level* modulated stages, in which the modulated RF stage is plate- or grid-modulated and is followed by a linear amplifier that feeds the amplified modulated signal to the antenna.
3. Grid (gate, base) modulation, in which the final RF amplifier is grid-modulated.

Test your understanding; answer these checkup questions.

1. What is the approximate dc input power of a broadcast station licensed for 2-kW operating power? _____ If licensed for 7.5 kW? _____

2. What percent of the total radiated power is in the sidebands with 100% sinusoidal modulation? _____ 50%? _____

3. What are the four output frequencies when 4 MHz and 5 MHz are mixed? _____ When 500 Hz and 8 MHz are mixed? _____
4. If the 500-Hz modulation of a 1-MHz carrier has 3rd harmonic distortion, what is the emission bandwidth? _____
5. What is the approximate bandwidth of a voice-modulated A3 transmitter? _____ Hi-fi A3? _____
6. What are two names of the A3 modulation which uses an AF choke? _____ _____ Why is an *RC* network used in series with the choke? _____
7. What do you think is the main advantage of grid over plate modulation? _____ Disadvantage? _____
8. Under what conditions can grid current be allowed in grid modulation? _____
9. Why is grid modulation called efficiency modulation? _____
10. Is any substantial AF power required for grid modulation? _____ Suppressor-grid modulation? _____ Screen-grid modulation? _____
11. What is meant by high-level modulation? _____ Low-level? _____ What other type is there? _____
12. Draw diagrams of a carrier and sidebands when modulating with 2-kHz tone. Heising modulation with 100% capabilities. A grid-bias modulation circuit. A $V_g I_p$ curve showing V_g and I_p when grid modulating.

17-18 CHECKING MODULATION WITH AN OSCILLOSCOPE

The oscilloscope (Sec. 13-34) provides one of the most satisfactory methods of determining the percent of modulation on a carrier. It may also indicate certain types of distortion. One method shows the modulation envelope. Another produces a trapezoidal modulation figure.

The *modulation envelope* is displayed by feeding a sawtooth ac of 100 to 500 Hz to the horizontal deflection plates (usually from a sawtooth ac-generating circuit in the oscilloscope). RF ac can be picked up by loose-coupling a two- or three-turn loop to the antenna or final-amplifier coil (Fig. 17-20). A short antenna to a tuned *LC* circuit can also be used as the RF pickup device.

With no modulation, the RF ac drives the electron beam up and down as the sawtooth ac drives it left to right. The result is a wide band across the face of the cathode-ray tube. When modulation is applied, the band is modulated. The positive peaks increase the height of the band. The negative peaks

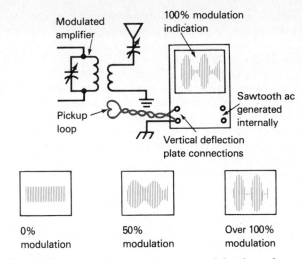

FIG. 17-20 A circuit to produce a modulated-envelope display on an oscilloscope and four possible displays.

decrease the band height. If a free-running sawtooth sweep voltage has a frequency of 200 Hz and the modulation is a sinusoidal 400-Hz signal, the modulated carrier will appear as two sine-wave variations of the carrier band. If the carrier is modulated 100%, the negative peaks will show as a tiny spot on the screen and the positive peaks will have twice the vertical amplitude of the carrier alone. The oscilloscope shows the modulation envelope as in Fig. 17-11. If the transmitter is being modulated by a sine-wave signal, any deviation from the sine waveshape shows on the CRT.

It is possible to check stage by stage with an oscilloscope to determine where distortion first appears. To do this, a 0.01- to 0.1-μF capacitor can be connected to the ungrounded vertical plate, as shown in Fig. 17-21. By touching the capacitor to the input or output terminal of each device in the audio section of the transmitter, an indication will be given of the waveshape at that point. Care must be taken when touching plate-circuit terminals because dangerously high voltages may exist at these

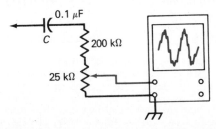

FIG. 17-21 Method of displaying an AF voltage on an oscilloscope.

points. WARNING: With high-power VT equipment turn off the transmitter, fasten the capacitor to the desired point, and then turn on the transmitter to make the test.

The pattern on a CRT will stand still only if the modulating frequency is some exact multiple of the sawtooth sweep frequency. Voice and music produce a jagged, jumping series of waveforms on the face of the tube. If the positive or negative peaks are badly flattened (distorted), it will be apparent to a trained eye.

If short bright horizontal lines are developed on the modulated envelope at the negative peaks of modulation, it is an indication of overmodulation. Precise neutralization of the modulated stage can be accomplished by overmodulating the negative peaks and then adjusting the neutralizing capacitor to a minimum line.

A *trapezoidal display* (Fig. 17-22) is produced by feeding modulated RF directly to the vertical plates,

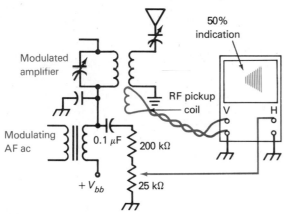

FIG. 17-22 Circuit to produce a trapezoidal modulation display.

and a small fraction of the modulating ac from the modulation-transformer secondary to the horizontal plates of an oscilloscope. With no modulation, the carrier produces a thin vertical line on the CRT. When modulation is applied, the AF drives the

electron beam back and forth and the line widens to the right and left. The negative peaks of modulation are indicated by the side that drops off in amplitude, and the positive peaks by the side that increases. At 100% modulation, the trapezoid forms a point at one side and is twice the carrier amplitude on the other side (a triangle). Figure 17-23 shows

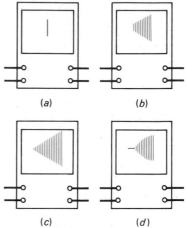

FIG. 17-23 Four possible trapezoidal displays of modulation on an oscilloscope: (*a*) 0%, (*b*) 50 %, (*c*) 100%, and (*d*) overmodulation.

(*a*) an unmodulated carrier, (*b*) a 50% modulated carrier, (*c*) a 100% modulated carrier, and (*d*) an overmodulated carrier. If the slanting sides of the trapezoid are not straight, distortion is indicated. If the positive peaks show a flattening (Fig. 17-23*d*) instead of rising to a sharp point, negative carrier shift (Sec. 17-22) is indicated.

Envelope displays on an oscilloscope show the result of adding sidebands to the carrier but do not show the sidebands themselves. The sidebands and their relationship to the carrier can be shown with a spectrum analyzer (Sec. 13-37). The analyzer can also show spurious signals that may be developed by the modulation process.

17-19 LINEAR AMPLIFIERS FOR A3E

A linear amplifier is one which will amplify without distortion. Linearity can be obtained by operating an amplifier tube over the straight portion of its $V_g I_p$ curve. If operated over a nonlinear section, distorted amplification results. The amplifiers that produce undistorted signals arc the class A, AB, and B. Class C amplifiers are nonlinear because

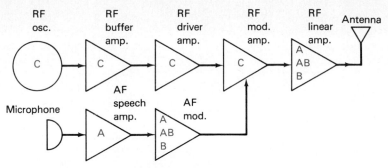

FIG. 17-24 Block diagram of an AM transmitter with low-level modulation and a linear amplifier showing classes of operation.

they are not operated on the straight portion of the V_gI_p curve for the whole of the input cycle.

If the modulated RF stage is not the final RF amplifier in a broadcast transmitter, *all stages* after the modulated stage must be linear amplifiers to enable them to amplify the modulated RF carrier without distorting it. A block diagram of a low-level modulated transmitter is shown in Fig. 17-24. The letters indicate possible classes of operation of the stages. Because class A amplifiers are low in efficiency, classes AB and B are more likely to be used as linear RF power amplifiers.

In audio, a class B amplifier must be a push-pull type to reproduce both halves of the input cycle. In RF amplifiers a class B stage can operate with only one device. The missing half of each RF cycle is reproduced by the flywheel action of the *tuned* output *LC* circuit of the stage (not if a broadbanded amplifier, which must be class A).

Figure 17-25 shows a 100% modulated envelope impressed on the characteristic curve of a class B linear amplifier and the resulting output circuit pulses. The amplifier is excited almost to zero bias by the carrier signal. High positive peaks of modulation fall into the positive bias region and produce input circuit current flow. This requires a bias supply with good regulation and low input circuit resistance.

To produce undistorted modulation from any modulated amplifier it is necessary that it work into a constant load (an antenna in high-level modulation systems). To produce a constant impedance load on the modulated stage, the linear-amplifier input circuit may be loaded with a noninductive resistance (Fig. 17-26).

For higher efficiency a linear amplifier is sometimes biased to class C, but the modulated input signal is made to vary only over the linear portion of the characteristic curve. The RF input signal

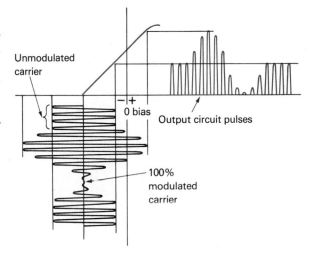

FIG. 17-25 As the input excitation to a linear amplifier is modulated, the I_p pulses are varied in amplitude.

must not be modulated to a high percent. This is an advantage, because it is not difficult to produce a lower percent of undistorted modulation. In Fig. 17-27 the linear amplifier is biased to about 1¼ times cutoff and requires an excitation carrier modulated to only about 70% to produce 100% modulated output RF ac. Since the stage is biased to class C but modulates over the same portion of the curve as if it were biased to class B, it is known as a *class BC* amplifier.

The milliammeter in any linear-amplifier plate circuit should vary only slightly as the percent of modulation changes. To hold distortion to a minimum, all power supplies must be well regulated. A tube used in a linear amplifier can operate with higher V_p than if it were plate-modulated. A tube limited to 2000 V on the plate when plate-modulated may use 2500 V as a linear amplifier.

According to the FCC, the dc plate power input

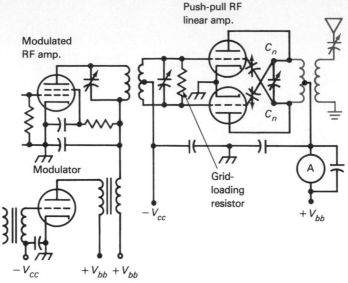

FIG. 17-26 Low-level plate-modulated tetrode driving a push-pull linear amplifier.

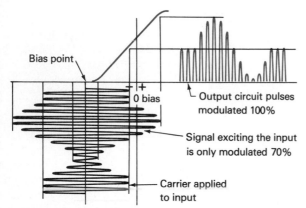

FIG. 17-27 How 100% pulse modulation can be developed by 70% modulated input signal if stage is biased to class BC.

times 0.35 equals the operating power output of a class B linear amplifier, an efficiency of 35%. Compare this with the 0.7 to 0.85 for plate-modulated stages. (Linear amplifiers of single-sideband emissions may operate at more than 60% efficiency.)

A linear amplifier in an AM transmitter is similar to a grid-modulated stage in many respects. The dc I_p meter does not vary. When the excitation voltage is doubled, the I_p pulses double in amplitude, efficiency doubles, and the output increases four times. The output power is proportional to the excitation voltage squared. With increased modulation, the I_p

varies over a wider portion of the $V_g I_p$ curve and the efficiency of the amplifier increases. Whereas the plate of a plate-modulated tube may turn red-hot at a high percent of modulation, a grid-modulated or linear-amplifier plate will cool under modulated conditions because of lessened plate dissipation. (If dc input is constant, plate dissipation must decrease if RF output increases.)

17-20 TUNING AM LINEAR AMPLIFIERS

It is possible to tune a class B linear RF amplifier of an AM transmitter by these steps:

1. Bias the stage to cutoff by applying full operating V_p and observing the I_p as the bias voltage is decreased from more than required for I_p cutoff. (Use no RF excitation while determining the cutoff-bias value.) When the I_p is a few milliamperes, the bias value is satisfactory.
2. Apply a weak unmodulated RF excitation to the grid, and tune the plate circuit to minimum I_p. (Neutralize if necessary.)
3. Increase the RF excitation until some I_p begins to flow. At this value of RF excitation note the I_p. (For example, it might be 2 A.)
4. Decrease the RF excitation until the I_p is half of the above-noted value (1 A). This should be approximately the correct unmodulated carrier-excitation value.

5. Couple the antenna to the amplifier. Check the modulation percent and linearity as shown on an oscilloscope, with a sine-wave signal generator feeding into the speech amplifier, making adjustments on bias, RF excitation, and antenna coupling until essentially undistorted 100% modulation is produced at the desired power output.

17-21 DOHERTY LINEAR AMPLIFIERS

A linear AM amplifier is only about 30 to 35% efficient. The 2-tube Doherty linear (Fig. 17-28) may be 60% efficient. The class B stage operates at all times, but the class C stage operates only during

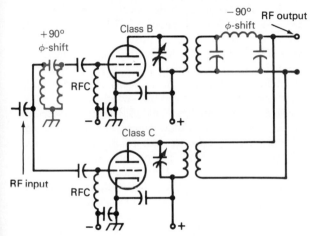

FIG. 17-28 Basic Doherty high-efficiency linear amplifier.

positive half cycles of modulation. With the carrier alone, and during negative halves of the modulated cycle, the class C tube does nothing.

The class B tube, excited to saturation by the carrier, produces a wide swing of I_p (60% efficient) into a plate tank load of twice the tube impedance. This mismatch decreases the power output somewhat, but gives good efficiency. The class B stage produces all the carrier power plus the negative half cycle of modulation.

During the positive half cycle of modulation the class B stage is in a saturated condition and assumably does not increase its output. The class C stage can now go into operation. It feeds RF power into the output circuit. This reflects a lowered impedance back on the class B LC circuit, forcing tube and load impedance to match better, and additional power comes from the class B stage. As a result, at the positive peak of modulation the neces-

sary RF peak power of four times the carrier value is produced by the action of the two tubes.

To produce the necessary impedance matching in the output circuit, the equivalent of a quarter-wave line must be inserted in the output of the class B tube. This shifts the phase of the output RF by 90°, necessitating an opposite change of phase in the grid of the class B stage in order that the RF of both stages be in phase.

These amplifiers are used in broadcast stations where operation is on only a single frequency. Tuning from one frequency to another is difficult. Both phase-shift networks must be tuned.

17-22 CARRIER SHIFT

Distortion of modulation in an AM transmitter may result in positive or negative *carrier amplitude regulation* or *carrier shift* problems.

A meter that will indicate this problem is shown in Fig. 17-29. A dc voltmeter (50-μA meter and

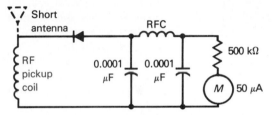

FIG. 17-29 A possible carrier-shift meter.

resistor) is connected across the output of an RF pickup coil and a diode. The output of the rectifier is dc, pulsating at the RF rate. The RFC and capacitors form a low-pass filter for RF but do not smooth AF variations. The pickup coil is coupled close enough to the modulated RF output of a transmitter to give a half-scale meter reading.

With undistorted modulation, the increased positive voltage peaks and decreased negative peaks of modulation should be equal. Since the meter cannot swing fast enough to follow the audio variations of the modulated envelope and since the average remains constant, the meter needle should not change with or without modulation.

If during the positive peak of modulation the RF voltage increases more than it decreases during the negative peak, the average voltage shifts in the positive peak direction (Fig. 17-30a). A positive shift in carrier amplitude regulation is present, and the meter pointer swings upward.

If negative peaks of modulation decrease more than positive peaks increase, the result is a shifting

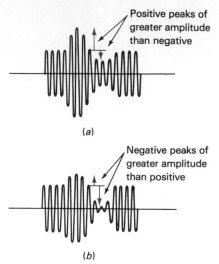

FIG. 17-30 Carrier with (*a*) positive and (*b*) negative carrier amplitude regulation (shift).

of the average voltage in the negative peak direction (Fig. 17-30*b*). Negative carrier shift is present, and the meter swings lower with modulation. This is also known as *downward modulation*.

With carrier amplitude regulation problems, even if the original AF voltage were sinusoidal, the modulated envelope voltage is not sine-wave-shaped. This indicates distorted or *asymmetrical* (nonsymmetrical) modulation.

Up to about 5% of carrier power shift distortion is not noticeable to listeners. If considerable negative carrier shift is present, the signals have a compressed, choked, and weak sound. With negative carrier shift the antenna ammeter will not increase normally or may even decrease. With positive carrier shift the antenna ammeter will increase more than normally.

For voice communication, some positive carrier shift is an advantage, provided the negative peaks are not distorted. A greater proportion of sideband power can be radiated for a given carrier value, and receivers produce louder signals.

17-23 CAUSES OF CARRIER SHIFT

There are many types of distortion that will indicate on carrier-amplitude-regulation meters.

PLATE-MODULATED STAGES. Negative Carrier Shift. Insufficient RF excitation to the grid; power-supply bias only on the modulated stage; distorted AF from the modulator; improper imped-

ance match between the modulator and the modulated stage; improper antenna coupling; low V_f or an old tube in the RF amplifier; excessive V_{sg}; failure to modulate the screen of a tetrode or pentode tube with plate modulation; one of the two tubes of a class AB or B push-pull modulator burns out.

Positive Carrier Shift. Insufficient or no V_p on the modulated RF amplifier (overmodulated condition); too high a setting of the gain control in the speech amplifier.

GRID-MODULATED STAGES. Either form of carrier shift is easily produced if the grid-modulated stage and the modulator are not carefully adjusted. Too high a bias can produce positive carrier shift, as can too little RF grid excitation. Too low a bias can produce negative carrier shift, as can too much RF grid excitation. Excessive AF modulating voltage will usually result in positive carrier shift. A weak tube or low V_f in the modulated stage can produce negative carrier shift. Improper antenna coupling can produce either form of carrier shift, depending upon other circuit factors.

LINEAR AMPLIFIERS OF AM SIGNALS. These are similar to grid-modulated stages. Proper bias and RF excitation values are important. An overexcited grid produces negative carrier shift. If the modulated stage has a certain type of carrier shift in its output, the output of the linear amplifier normally has the same. It is possible to balance out small values of carrier shift as the signal passes through a linear amplifier. If the original modulated signal has a slight negative carrier shift, the linear amplifier can be adjusted to produce a slight positive carrier shift.

If an RF amplifier tuning circuit or the antenna sparks across on positive peaks of modulation, the carrier shift indication will be negative.

17-24 AM AND CW WITH THE SAME DEVICE

If a tube is used for both CW and plate-modulated AM emissions, the carrier output for AM will usually have to be held at 65 to 75% of that used for CW. The differential is to allow for the additional power being handled by the tube under modulated conditions. If a tube is capable of 1000 W in radiotelegraph service, it should be operated at approximately 700 W if plate-modulated, to allow for the 350 W of sideband power that will be added to the carrier under 100% sine-wave-modulation conditions. Running at 700-W carrier, the tube is actually being overdriven if continually operated at full modulation. However, it is unusual to have a transmitter operating at 100% for any period of time. The average modulation for broadcast speech and music will produce far less than 350 W of effective sideband power for a 700-W carrier.

If the same tube is grid-modulated, instead of a 1000-W CW carrier output, only about 200 to 250 W can be expected as the AM carrier power.

Test your understanding; answer these checkup questions.

1. What sweep frequencies could be used on a scope to stop an envelope pattern of 600-Hz modulation? _____

2. Can voice patterns be stopped on a scope? _____

3. Why would trapezoidal patterns be better than envelope for checking speech modulation? _____

4. What percent of modulation must be used when neutralizing an operating transmitter? _____

5. What is the shape of a trapezoidal pattern at 100% modulation? _____

6. What class amplifier is usually used in linear RF amplifiers with AM? _____

7. What is the advantage of biasing an RF linear amplifier to class BC? _____

8. Why are linear-amplifier grid circuits often resistor-loaded? _____

9. Does the efficiency of a linear amplifier increase with percent of modulation? _____ Why? _____

10. What is the advantage of a Doherty linear? _____ To what class is the carrier amplifier biased? _____ The positive peak amplifier? _____

11. Does carrier shift refer to amplitude, phase, or frequency? _____ What is another term meaning the same thing? _____

12. What carrier shift results with insufficient grid excitation to a high-level modulated stage? _____ Grid-modulated stage? _____

13. Does AF peak-limiting produce carrier shift? _____

14. Why must a plate-modulated AM RF amplifier use less V_p than with CW? _____

15. Draw diagrams of a 125% modulated envelope. A circuit to test AF or RF with scope. A circuit to test trapezoidal modulation. A trapezoidal display of 0, 50, 100, 125% modulation. A block diagram of an A3 system using a linear amplifier. A plate-modulated low-level tetrode with linear amplifier. A Doherty amplifier. A carrier-shift meter.

17-25 THE ANTENNA AMMETER WITH AM

If an AM broadcast transmitter is turned on, any antenna ammeter or voltmeter in it will rise to some value and remain there. If the transmitter is modulated, the antenna meter will rise to a higher value than with the carrier alone. The modulation being applied to the RF amplifier appears in the radiated wave as sideband power and increased antenna current.

How much will the antenna current rise for 100% sine-wave modulation? Assume a carrier power of 100 W in the antenna and an antenna resistance of 1 Ω. (Antenna resistance remains constant.)

1. The antenna current, from $P = I^2R$, is

$$I = \sqrt{\frac{P}{R}} = \sqrt{\frac{100}{1}} = 10 \text{ A}$$

2. With 100% modulation, the sideband power will be equal to one-half of the carrier power (Sec. 17-10). The total antenna power is 100-W carrier + 50-W sidebands = 150 W.

3. The antenna current with 150 W is

$$I = \sqrt{\frac{P}{R}} = \sqrt{\frac{150}{1}} = 12.25 \text{ A}$$

4. The antenna current increases from 10 A unmodulated to 12.25 A when 100% modulated. This is an increase of 12.25/10, or 1.225, or a 22.5% increase.

When an antenna ammeter indicates an increase of 22.5% over an unmodulated value with sine-wave modulation, assuming no distortion, 100% modulation must be present. An 8-A unmodulated carrier should read 8 + (8 × 0.225) = 9.8 A at 100% modulation.

For 50% sine-wave modulation, assuming the same 100-W carrier and antenna as above, to what value should the meter rise? Try working the problem. The answer is 10.6 A.

Note that sine-wave-modulating voltages are specified. Pure tones of music are somewhat sinusoidal, but speech is far from sine-wave-shaped. When a transmitter is voice-modulated, the antenna meter may hardly move at all, although some of the peaks of modulation may be 100% or more.

Sometimes an ammeter will move upward with low percentages of modulation but decrease with an increase of modulating voltage. This may indicate nearly normal modulation to a certain percent. Above this, negative carrier shift sets in.

17-26 MAGNETIC MICROPHONES

There are many microphones that fall into the category of magnetic-induction microphones. They operate on the principle of sound waves striking a diaphragm producing a relative movement between a magnetic field and a conductor, inducing a voltage in the conductor. A *dynamic microphone* is similar to a permanent-magnet (p-m) dynamic loudspeaker (Sec. 14-9). It has an essentially flat frequency range of perhaps 60–10,000 Hz. It requires no battery, being a form of ac generator. It can be built light, small, and rugged at a comparatively low cost. It has an output power of perhaps −30 to −80 dBm (1 mW = 0 dBm), depending on sound amplitude.

Figure 17-31 shows a *dynamic microphone* and its circuit. The conical diaphragm has a small coil

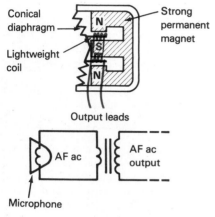

FIG. 17-31 Cross section of a dynamic microphone and the circuit in which it is used.

attached to it but is free to move in and out of the space between north and south poles of the magnet. Vibration of the diaphragm by sound waves moves the coil back and forth across the lines of force, inducing an ac in the turns of the coil. The ac has a

frequency equal to the diaphragm vibration frequency and an amplitude proportional to the extent of the diaphragm vibration.

The moving-coil impedance is only a few ohms. A step-up transformer, usually built into the microphone case, is required to match the 50-, 250-, 600-Ω, or high-impedance input circuits of amplifiers with which the microphone is used.

Sound-powered microphones, used for voice communications in special telephone lines, are dynamic-microphone types. Since high fidelity is not an important factor for voice transmission, it is possible to obtain relatively high power output (0 dBm) by close talking into the microphone. When earphones are connected across them, the resulting signal is quite loud. Such a microphone can also be used as an earphone (as a p-m loudspeaker).

Dynamic microphones must be kept away from ac fields. If the coil is near a power transformer, a hum voltage will be induced in the microphone coil. It is possible to minimize such hum by turning the microphone.

Other microphones operating on the magnetic-induction principle have various names: *variable-reluctance*, *ribbon*, or *velocity* (Fig. 17-32).

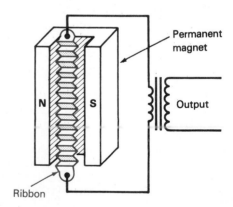

FIG. 17-32 Ribbon microphone. Sound waves vibrate the light aluminum ribbon across the magnetic field, inducing ac into the ribbon.

When it is necessary to use long microphone leads, the lower the impedance of the lines, the less extraneous noise picked up and the less high-frequency attenuation due to capacitance between line conductors. This rule of low impedance for long lines is true for all types of microphones and for audio lines in general.

17-27 CRYSTAL MICROPHONES

The piezoelectric *crystal* microphone is used in public address, amateur communication, and home recording systems, but rarely in broadcasting. It has good frequency response, from about 50 to 10,000 Hz, and relatively good sensitivity, −40 to −60 dBm. It is fairly rugged mechanically, is satisfactory when used as a hand microphone, and may be spoken into at close range.

The crystals used in older microphones are affected by moisture or a temperature of more than 120°F (easily produced by summer sun striking the microphone). Newer developments with ceramics have overcome these difficulties.

Rochelle and other crystals used in these microphones have piezoelectric properties similar to quartz crystals used in oscillators. When the crystal is vibrated mechanically, an ac emf is developed between two foil plates cemented to opposite surfaces of the crystal. The ac will be proportional to both vibration frequency and amplitude.

Sound waves striking the conical aluminum diaphragm of a crystal microphone (Fig. 17-33) will

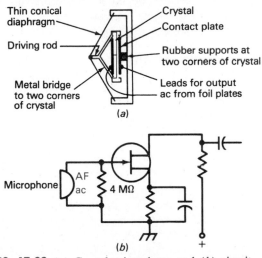

FIG. 17-33 (a) Crystal microphone and (b) circuit.

cause it to vibrate. Vibrations are mechanically transmitted by a small driving rod to the opposite corners of a thin, square crystal, bending the crystal, producing an AF ac between the foil plates. The output leads from the microphone are connected to the two plates.

The output impedance of the crystal microphone is high (several million ohms), necessitating relatively short leads to the input circuit of the micro-

phone amplifier to prevent loss of the higher audio frequencies, or hum pickup. The leads must be well shielded and not more than about 20 ft long.

A new crystal-like microphone is the *electret*. It consists of a thin slice of insulating material with a molecular arrangement which makes it appear to be electrically charged. When a diaphragm movement causes it to vibrate, its molecular arrangement changes and a varying voltage appears across it. It is also a high-impedance microphone.

A *wireless* microphone consists of perhaps an electret microphone mounted on one end of an 8-in. insulating material tube containing a small, low-powered VHF FM transmitter (Chap. 19) operated by a self-contained battery. Its signal can be picked up clearly with a special FM receiver for a distance of 100 meters or more. Wireless telephones use a wireless microphone to communicate to a local control circuit. On another frequency the control circuit transmits incoming voice information as a second FM signal to the wireless handset receiver.

17-28 CONDENSER MICROPHONES

Since the late 1920s, one of the best broadcasting microphones has been the electrostatic, or *condenser* (capacitor), microphone. Its frequency response is good, from about 50 to 15,000 Hz. It has high impedance and its leads must be very short. The first-amplifier stage is constructed in the microphone case for this reason. The cable to the microphone contains not only the output lines from the microphone unit, but also the required power-supply lines for the VT or FET, and for the condenser head. The diaphragm is tightly stretched to prevent it from resonating at any audible frequency.

Figure 17-34 shows the elements of a condenser-microphone head, and a JFET amplifier. When sound waves strike the tightly stretched diaphragm, it vibrates, changing the capacitance between front and back plates according to the frequency and amplitude of the sound waves. Thus, sound waves produce a variation of capacitance.

When the circuit is turned on electrons move down through the resistor R to V_{DD} until there is 200 V across the microphone. Then the current ceases to move, but an excess of electrons remains on the front plate and a deficiency on the back. If the charged capacitor suddenly changes to a greater capacitance by the plates being forced closer together, a current of electrons will flow through the resistor until the charge on the capacitor becomes

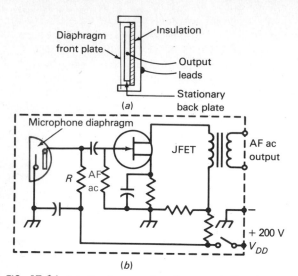

FIG. 17-34 (a) Condenser microphone and (b) circuit.

200 V again. The capacitor-charging current flowing through R produces a voltage-drop across it equal to $V = IR$. As the capacitance varies due to diaphragm vibration, electrons are forced to flow back and forth through R, and an ac emf is developed across it. The emf will vary in frequency and amplitude in accordance with the sound waves striking the diaphragm. A condenser microphone head has a sensitivity of about -90 dBm.

The mounting of a microphone determines its directional characteristics. If the diaphragm points upward, it can receive signals equally well from all directions (omnidirectional). By mounting the diaphragm in a vertical direction and closing the back of the case, the microphone picks up signals best from the front (unidirectional). By leaving both front and back of the diaphragm open (ribbon microphone), it receives signals from both front and back (bidirectional).

When two microphones are fed into the same amplifier, it is possible that signals from one sound source an unequal distance from the two microphones may arrive 180° out of phase, canceling each other. Therefore, it is important that microphones be *phased* properly (correct polarity of microphone leads and proper placement of microphones) when they are used to pick up large orchestras, etc.

17-29 TUNING AN AM TRANSMITTER

A simple but workable *hybrid* (using both transistors and VTs) radiotelephone transmitter is shown in Fig. 17-35. A short description of the operation

of each stage is given. At the start, filaments are on, but power supplies are off.

The oscillator power supply is turned on. The tank circuit of this stage is tuned to a I_D dip, as indicated by meter M_1.

The buffer is neutralized, and then its power supply is turned on. The buffer plate tank is tuned to a dip on meter M_2. The grid tank LC_3 is tuned to maximum I_g on meter M_3. If the I_g is not the desired value, the link coupling to LC_3 can be tightened or loosened, or C_1 can be increased or decreased.

The modulated amplifier is set for low power and its power supply is turned on. The final plate tank is tuned to a dip on M_4.

The antenna is loosely coupled to the final-tank circuit and tuned to resonance, indicated by a peak reading of the RF ammeter M_5, or a peaking of I_p on meter M_4. The modulated amplifier V_p is raised to the operating value, and the coupling is increased until the desired antenna current is produced or until the desired I_p value is shown by M_4.

The *preamplifier stage* is turned on and a sine-wave signal is fed to the input transformer of the preamplifier. The signal-generator output is increased until a change of I_D is noted in meter M_6 and is then decreased slightly. The class A preamplifier is probably not distorting at this signal-input level.

The *speech-amplifier* gain control is turned down, and the stage is turned on. The gain control can be increased up to the point where I_p begins to change, as shown by meter M_7. This is as high as the gain control may be turned with the signal generator set at its present level.

The bias voltage of the modulator stage should be checked. If the modulator is to be operated as push-pull class A, there should be no indication of I_g in meter M_8. It may be necessary to reduce the gain control to a point where no I_g flows. If the modulator is biased for operation as class AB_2 or class B, I_g should flow with the gain control set to its maximum allowable point determined previously.

With the gain control turned down, the modulator is turned on. If the modulators are biased to class A, the I_p in meter M_9 should read the value indicated in the tube manufacturer's data sheet for the type of tubes used. If the modulators are biased to class AB_2, there will be considerably less *static* (no input signal) I_p. If they are biased to class B, there will be very little static I_p.

When the gain control is turned up, the RF am-

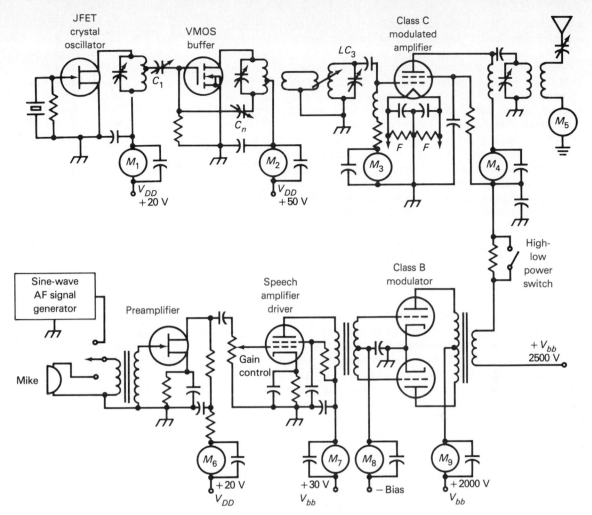

FIG. 17-35 Schematic diagram of a radiotelephone transmitter and meters used to indicate its operation.

meter should rise slightly and all other meters, except possibly the modulator-stage grid and plate meters, should remain steady. If the modulator is biased to class A, neither of its meters should move. However, if it is biased to class AB_2 or B, at a low gain-control setting, no I_g will flow but I_p will increase with modulation. At a high gain-control setting, both the I_p and I_g will rise.

The average I_p of the modulated amplifier should remain constant, with or without modulation, but at high percentages of modulation it will usually drop 1 or 2%.

If an oscilloscope is available, a check on percent of modulation and linearity of the modulated waveform should now be made.

The microphone can be switched on and spoken

into. The antenna current should rise slightly. It will probably be necessary to raise the gain-control setting to produce a high percent of modulation because of the low-amplitude output of microphones. In fact, a second preamplifier or speech-amplifier stage would undoubtedly be needed to bring the weak output of the microphone up to the required level to modulate the transmitter completely.

If a speech amplifier is overdriven, it may cause distortion even if the modulation is not up to 100%. This would show as a flattening of the peaks on the oscilloscope and failure to fully modulate.

With the modulator stage biased to class B, the I_p indication varies directly with the AF output of the stage. Meter M_9 can be used as a relative indication of the percent of modulation. A small I_p increase indicates a low percent of modulation; a large I_p increase indicates a high percent.

When tuning a transmitter, keep in mind that the equipment contains lethal voltages. Touching the antenna leads on high-powered transmitters can result in severe burns or worse. Most commercial transmitters have built-in safety devices to protect personnel and equipment, such as door interlocks that disconnect high-voltage supplies when the doors of the transmitter are opened, and overload relays that automatically shut down the equipment if excessive current flows. In some cases lights flash or bells ring if trouble occurs in any circuit. Relays can be set to sound an alarm if either more or less than normal current flows.

If the modulation transformer suddenly develops one or more shorted turns in either the primary or the secondary, all meter readings will be normal with no modulation. When modulation is applied, a heavy current will be induced in the shorted part, lowering the impedance of the primary, increasing the modulator I_p, and distorting the audio signal. The M_9 indication will increase excessively, the percent of modulation will drop, negative carrier shift will be produced, and the antenna current will not increase normally. The shorted transformer will buzz audibly, become excessively warm, smoke after a short time, and may burn up. The plates of the modulator tubes may become hotter than normal.

Power input can be determined by multiplying $V_p \times I_p$ of the PA stage. Power output can be found by I^2R, using the M_5 reading and the antenna impedance at that point, or with an RF wattmeter in any transmission line used.

An operator who knows the circuits in the equipment can tell by meter readings where any trouble occurs. A transmitter should require only periodic retuning unless it is repaired or retubed.

17-30 MODULATING BJT RF AMPLIFIERS

A BJT stage can be modulated in much the same way as other devices. Figure 17-36 illustrates how the modulating AF might be coupled to the base circuit to produce base modulation, as well as how AF ac might be fed into the emitter circuit to develop emitter modulation.

Because the shape of the V_cI_c curve of a BJT shifts during collector modulation, it is necessary to modulate the collector of the modulated stage as well as feed some modulation to the driver stage collector to produce 100% modulation (Fig. 17-37).

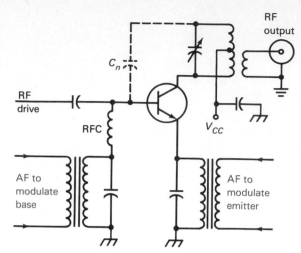

FIG. 17-36 Points where AF may be connected to base and emitter to modulate a BJT stage.

A π-network output coupling circuit is shown. The 10-kΩ resistor is to drain off any dc static charge that might build up on the antenna. An RFC might be used instead. Capacitor C_1 is basically to tune the LC circuit, while C_2 is adjusted to match the impedance of the LC circuit output to the antenna impedance.

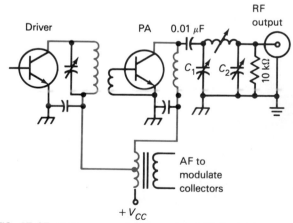

FIG. 17-37 Collector modulation of a BJT for high percent modulation.

17-31 SINGLE SIDEBAND (J3E)

The information to this point has dealt with amplitude modulation to produce a carrier and two sidebands (AM). This emission is used in standard broadcast and shortwave international broadcast stations, aircraft, some industrial communications, some citizens band, some amateur equipment, and the picture information of TV.

Originally, all radiotelephonic communications used AM. Most short-distance communication systems, such as police and fire, have converted to frequency modulation (FM), particularly above 27 MHz. Below this, most radiotelephonic transmissions are now being made with single-sideband suppressed-carrier (SSSC, SSB, or J3E, previously A3J) emissions. Two advantages of SSB over AM are (1) power is saved by not transmitting a carrier and one sideband; and (2) with only one sideband the required bandwidth is halved, allowing twice as many stations to use the available spectrum space. As an example, a 1-kW AM carrier when 100% modulated emits 1.5 kW of RF. If the carrier is deleted, the same information can be transmitted with 500 W of sidebands (SBs), a saving of 67% in power. Since each SB carries the same information, deleting one of them is a further saving (84%), as well as requiring only half as much spectrum space. A 110-W SSB transmitter will do essentially as well for voice communication as a 1000-W AM transmitter will.

Fading signals have always been a problem in long-distance radio communications. Not only do AM signals fade up and down in strength, but the carrier may fade at different rates and times from either sideband. This results in varying signal strengths and a characteristic rolling distortion effect. The result may be voice transmissions that are completely unintelligible at times, regardless of how loud they are, because of carrier and sidebands fading at different times. This can be eliminated by balancing out the carrier at the transmitter, filtering out one sideband, and transmitting only the remaining sideband. Figure 17-38 shows (a) a normal double-sideband A3E signal, (b) the same emission with one sideband filtered out (H3E), and (c) the same signal with one sideband and the carrier removed (J3E). If it is desired to transmit a pilot carrier for receivers to lock onto, instead of canceling the carrier it may be reduced to about 10% of its normal value (R3E emission).

J3E signals produce a muffled noise in the usual AM receiver and cannot be understood because there is no carrier for the sideband signals to beat against to produce audible heterodyne, or beat, frequencies in the receiver. At the receiver it is possible to use an oscillator adjusted to the frequency that the carrier would have had if it had been transmitted. By feeding this frequency into the receiver at the same time as the sidebands are received, the sideband signal can beat against a correct-frequency (local) carrier to produce the audio tones of the original intelligence or sounds. Substituting the nonfading carrier signal at the receiver represents quite a power saving for the transmitter, as well as an improvment in reception of fading signals. Voice transmissions on J3E may fade up and down in strength, and at different instants high or low audio frequencies may predominate, but the rolling distortion or *selective fade* of distant signals is eliminated.

Test your understanding; answer these checkup questions.

1. An antenna ammeter reads 6 A with carrier alone. What should it read with 100% sine AM modulation? _____ 50%? _____
2. If an antenna ammeter decreases with AM modulation, what is indicated? _____
3. What types of microphones have permanent magnets in them? _____ _____ What types do not? _____ _____ _____
4. What types of microphones have a high Z output? _____ _____ Low Z output? _____ _____ _____
5. What microphone has lowest power output? _____
6. In what order would a three-stage transmitter be tuned? _____
7. What is a preamplifier? _____
8. What three meters in Fig. 17-35 might indicate modulation was in progress? _____
9. How does basic plate modulation differ from basic collector modulation? _____
10. Would series modulation also work with VTs and FETs? _____

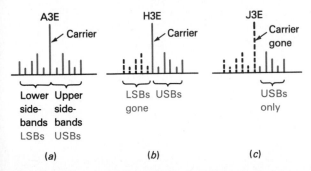

FIG. 17-38 (a) Carrier and sidebands of multitone AM, or A3E, (b) single-sideband with full carrier, or H3E, and (c) single-sideband with suppressed carrier, or J3E.

11. What are two advantages of SSB over AM?
_____

12. Why do fading J3E signals distort less than fading A3E signals? _____

13. What are the letter-number designations of carrier and 1 SB? _____ 1 SB and no carrier?
_____

14. What are five uses of A3 emissions now?
_____ _____ _____
_____ _____ _____

15. About how many decibels of transmitting power are gained by using J3E over A3? _____

16. Draw diagrams of a dynamic microphone and circuit. A crystal or electret microphone circuit. A condenser microphone. A JFET VMOS tetrode transmitter modulated by JFET tetrode push-pull triodes. Base modulaton of BJT. Collector modulation of BJT. Carrier and SBs of A3E, H3E, and J3E emissions.

17-32 FILTER-TYPE SSB TRANSMITTERS

Figure 17-39 is a block diagram of a possible single-sideband transmitter to operate on a (suppressed) carrier frequency of 8 MHz. The 3-MHz crystal oscillator provides the RF carrier to be modulated. The microphone and speech amplifier provide the AF modulating signal (200 to 3000 Hz). The RF and AF are mixed in a special circuit, called a *balanced modulator* (Sec. 17-33), which amplitude-modulates the 3-MHz RF, producing two sets of normal AM sidebands, but cancels or balances out the carrier frequency signal completely. The output of the balanced modulator in this case is

a set of upper sidebands consisting of signals from 3,000,200 to 3,003,000 Hz, and a set of lower sidebands from 2,999,800 to 2,997,000 Hz.

In the diagram the switches are in the USB position. Only upper-sideband frequencies can pass through this narrowband crystal filter. (LSB signals should be attenuated 50 dB or more.) Thus, no carrier and only upper-sideband signals are fed to the mixer. At the same time, another crystal oscillator operating at 5 MHz is fed to the mixer stage to *translate* to 8 MHz. The sum frequency of 3 + 5 MHz is the 8-MHz carrier frequency to be transmitted. Actually, the 5-MHz oscillator is providing a carrier and the 3-MHz sidebands modulate (mix with) this carrier frequency, producing an output that includes its 5-MHz carrier, USBs 3 MHz above 5 MHz (at 8 MHz), and LSBs 3 MHz below 5 MHz (at 2 MHz). The tuned output circuits of the mixer are selective enough to act as a bandpass filter to prevent the undesired products of mixing (the 5-MHz carrier and the 2-MHz LSBs) from being fed to the linear amplifier. Since the 3-MHz carrier was canceled by the balanced modulator, there can be no 8-MHz output from the transmitter without modulation. When modulation is present, upper-sideband information mixes with the 5-MHz oscillator output to produce J3E USB signals having an 8-MHz carrier as their base.

If the filter switch is moved to the LSB position, only the lower-sideband information will pass to the mixer and only a LSB emission will result, still based on an 8-MHz carrier frequency.

By changing the 5-MHz crystal to 4 MHz, the

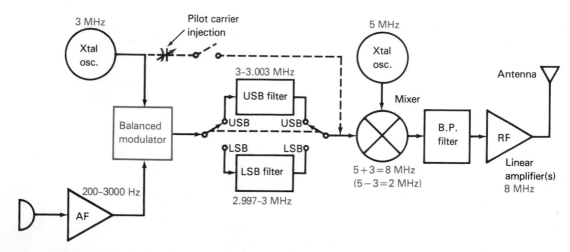

FIG. 17-39 Block diagram of a SSB transmitter using 3-MHz 3-kHz bandpass filters.

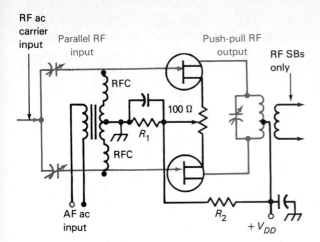

FIG. 17-40 A possible balanced modulator using JFETs.

Figure 17-40 illustrates a possible 2-JFET balanced modulator. In this circuit the RF is fed to the two gates in parallel and the AF gate-modulating voltage is fed to the two gates in push-pull. The RF carrier frequency balances out in the drain circuit because both drains are driven RF-positive at the same time, producing equal and opposite current flows in the drain coil. Since the tuned circuit has zero impedance to AF, there can be no audio output in the drain load either. The results of modulation, namely, the upper and lower sidebands, are not canceled. The output from the tuned circuit consists of two sets of sidebands only, provided the circuit is properly balanced. Balancing may be accomplished by: (1) adjusting the two variable capacitors feeding the gates until the carrier output (as seen on an oscilloscope) becomes zero, (2) adjusting the 100-Ω potentiometer in the source circuit, and (3) adjusting the center tap on the drain tank circuit. Selection of R_1 and R_2 values determines the class of operation, A, AB, B, or C.

Balanced-modulator circuits can be developed with VT or solid-state diodes. VT circuits would be similar to JFET circuits. Figure 17-41 shows three diode-type balanced modulators. The first, a shunt type, requires closely balanced diodes to cancel the carrier adequately. The others have potentiometers to balance out the carrier. The second is called a ring-type balanced modulator, and the third is a modified ring. In each case, AF is fed across the mixer or modulator circuit and the RF carrier is connected, not 180° out of phase, but in what appears to be *quadrature* (90°). This results in cancellation of the carrier and a double-sideband without-carrier (DSB) output. Balanced modulators are available in many ICs.

output frequency of the transmitter would be 3 + 4 = 7 MHz (or 4 − 3 = 1 MHz). Only the output circuits of the mixer and linear amplifier would have to be tuned to the new frequency. Either USB or LSB signals at any desired carrier frequency could be produced with the proper mixer crystal. By using a VFO or frequency synthesizer in place of the mixer-crystal oscillator, transmission can be made on bands of frequencies.

In commercial equipment, a *vestigial*, or *pilot*, carrier may be transmitted (R3E). Although the carrier is canceled at the balanced modulator, a small amount of RF ac from the carrier oscillator can be fed into the mixer input (dashed lines), equivalent to about 10% modulation. At the transmitter the pilot can aid in tune-up. At the receiver the pilot carrier is used to lock the receiver's carrier oscillator on the exact frequency required to produce correct-frequency sideband mixing.

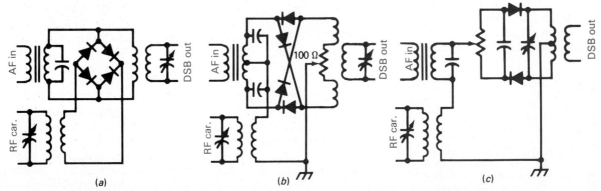

FIG. 17-41 Diode-type balanced modulators: (a) shunt, (b) ring, and (c) modified ring.

17-34 SIDEBAND FILTERS

Bandpass *LC* filters were described in Sec. 8-8, but *LC* types are too broad to be useful as SSB filters at HF because of the relatively low *Q* of the coils. Piezoelectric crystals have *Q* values in the thousands in the HF range. When used as elements in a filter, they produce narrow passbands and steep skirts with good shape factors, but they have some pop-up also. Figure 17-42*a* is a half-lattice crystal

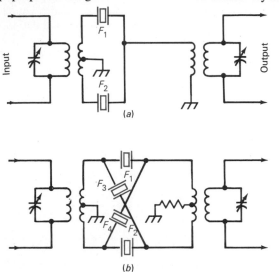

FIG. 17-42 (*a*) Half- and (*b*) full-lattice crystal filters.

filter using two crystals (f_1 and f_2) of 2.8-kHz frequency difference for a 3-kHz passband. It may have a shape factor of 4:1. By using a full-lattice, four-crystal (f_1–f_4) circuit (Fig. 17-42*b*), a shape factor of 3:1 may be attained. Cascading filters improves the shape factor.

Mechanical filters, consisting of tiny thin, machined nickel disks mechanically coupled with wires welded to their edges and with electromechanical (magnetic) coupling devices at both ends, can produce 3-kHz filters in the 100–500 kHz range with shape factors of 2:1.

17-35 PHASE-TYPE SSB TRANSMITTERS

A different method of producing a J3E emission uses two 90° phasing networks. A 90° RF phase-shifting network is shown in Fig. 17-43*a*. With the

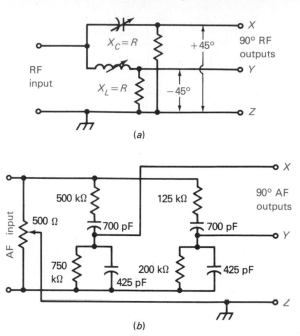

FIG. 17-43 Two 90° phase-shift newtorks: (*a*) single-frequency RF and (*b*) 300- to 3000-Hz AF (approximate values). *C* in picofarads.

reactances equal to the resistances, the current will lag 45° in the inductive circuit and will lead 45° in the capacitive circuit, totaling 90°. The same ac is available as two separate signals (*X* to *Z* and *Y* to *Z*) out of time by 90°. However, this phase relationship holds true for only a single frequency. It can be used to shift the frequency of an RF carrier 90° by tuning the components.

A practical AF phase network, using only resistors and capacitors but capable of holding a constant 90° phase difference from 300 to 3000 Hz, is shown in Fig. 17-43*b*.

In the block diagram, Fig. 17-44, AF from a microphone is amplified and fed into the 90° AF phase-shifting network and then to two balanced modulators. An RF carrier is shifted 90° and is also fed into the same two balanced modulators. The carrier frequency is canceled, but each modulator has upper- and lower-sideband signal output. These are in such phase that, when added in a common circuit, the upper-sideband signals, for example, are in phase but the other sideband signals are now

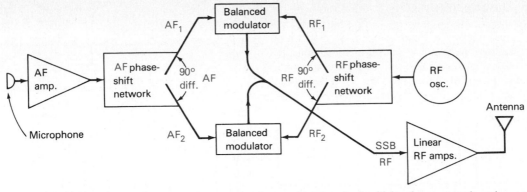

FIG. 17-44 Block diagram of a phasing system that produces a single-sideband suppressed-carrier, or J3E, emission.

180° out of phase (+90° and −90°) and cancel. This leaves only a single-sideband signal to be amplified by the linear amplifier.

By reversing either the AF or the RF leads to the balanced modulators, the opposite sidebands will add and cancel, providing a simple method of switching transmission from USB to LSB.

The phasing type of system has the advantage of not requiring any heterodyning to operate on a specific frequency. Its carrier frequency may be that generated by the RF oscillator. The RF phase-shifting network must be changed if the frequency of operation is changed materially.

17-36 LINEAR AMPLIFIERS FOR SSB

The SSB signal may be generated and converted to the desired transmission frequency at a level well under 1 W. At the transmission frequency it must then be amplified without distortion to the desired power level. Power pentodes, tetrodes, and triodes are used in high-power linear amplifiers in grounded-cathode or grounded-grid circuits. (Power BJT and VMOS transistors are used in linear RF amplifiers up to an output of a few hundred watts.)

RF linear amplifiers are rated by dc power input $(P = V_p I_p)$, or by RF power output $(P = V_p I_p \times$ efficiency of the amplifier, or by antenna power, $P = I_a^2 R_a)$, or by *peak envelope power* (PEP).

Linear amplifiers for SSB emissions are basically the same as those used to amplify AM emissions (Sec. 17-19), except that with SSB there is no RF output when there is no modulation. The output power will be proportional to the sideband signals applied to the input of the amplifier. If biased to class B, there will be almost no I_p with no SSB

drive, and I_p will increase according to the amplitude of the SSB excitation. With class AB_2, there will be more resting current and less average I_p variation when SSB excitation is applied. In class AB_1 there will be still more resting current and little average I_p increase when excitation is applied. With class A operation the resting current will be high and the average I_p may not vary with SSB excitation.

If the $V_p I_p$ meter readings of a class AB_2 SSB linear RF amplifier indicate 1000-W dc input with modulation produced by a single sinusoidal AF tone (assume 2 kHz), the amplifier should produce a single, constant-amplitude RF sideband 2 kHz from the carrier frequency. The sideband should have a peak envelope power of about 700 W, assuming 70% amplifier efficiency. With no modulation the meter values may indicate 200 W or so of static dc input, but the RF output will be zero.

When voice-modulated, the meter values give poor indications of power because of the rapid variations of audio peaks. A satisfactory method of determining the peak envelope power of a SSB emission (also AM, FM, and CW) is to connect an accurately voltage-calibrated oscilloscope across the antenna transmission line. The dc or rms power formula is $P = V^2/R$. If the peak ac voltage (V_{pk}) is used the peak envelope power (PEP) formula is

$$\text{PEP} = \frac{V_{pk}^2}{R}$$

The voltage peaks read on the oscilloscope and the impedance of the transmission line (with SWR of 1:1) can provide a determination of PEP. For example, if the peak (not p-p) SSB envelope voltage indicated by a scope across a 50-Ω line is 200 V, the PEP is $200^2/50$, or 800 W. (The effective voltage values may be only 30 to 50% of the peak.)

The presentation of single-tone SSB modulation on an oscilloscope using a sampling of antenna RF to the vertical plates and a low-frequency horizontal sawtooth sweep of about 300 Hz is a horizontal band (Fig. 17-45a). Proper voice modulation shows

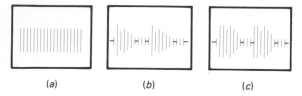

(a)   (b)   (c)

FIG. 17-45 (a) Single-tone SSB emission as seen on a scope. (b) Voice modulation. (c) Overdriven voice modulation.

clear, sharp peaks (Fig. 17-45b). Overdrive flattens the peaks and indicates severe distortion (Fig. 17-45c). Underdrive produces a voice pattern that looks normal but is reduced in amplitude.

A preferred presentation is a two-tone test. This is produced by feeding two constant-frequency sinusoidal AF tones, perhaps 600 and 1500 Hz, into the microphone input. If there is no distortion, the presentation should appear as in Fig. 17-46a. The dashed

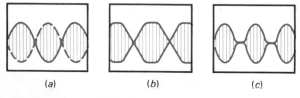

(a)   (b)   (c)

FIG. 17-46 (a) Two-tone modulation test signal. (b) Overdriven two-tone modulation. (c) Excessive bias on linear amplifier.

outline should be sinusoidal. Assuming the SSB generator is operating properly, overdrive of the linear amplifier flattens the peaks (Fig. 17-46b). If the pattern separates at the middle (Fig. 17-46c), it indicates the linear amplifier is operating in the class C region. Parasitics in the linear amplifier produce a fuzzy or ghostlike pattern. The degree of loading on the linear amplifier has a decided effect on the linearity of the output. Insufficient loading flattens the peaks; excessive loading narrows the peaks and lowers the output peak values.

With some linear RF amplifiers stable operation may be difficult to achieve. The circuit must be

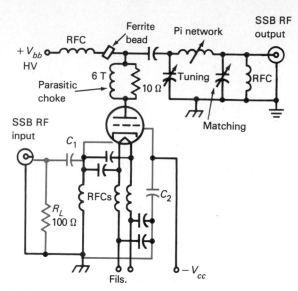

FIG. 17-47 Grounded-grid type linear amplifier used for SSB, AM, or CW transmitters.

completely neutralized. One of the stablest amplifiers is the grounded-grid linear (Fig. 17-47). The term *grounded-grid* means the grid is grounded (bypassed to ground by C_2) for RF, but usually not dc-grounded. Since the RF input and the RF output are in phase, rather than 180° out of phase as in grounded-cathode circuits, the possibility of stray feedback starting oscillation is unlikely. The SSB signal is fed to the noninductive resistive input load resistor, R_L. The RF SSB voltage developed across R_L is fed to the grid-cathode circuit of the triode through capacitors C_1 and C_2. This voltage develops the I_p pulses in the plate circuit. The tuned π network in the output circuit couples the amplified RF to the antenna. R_L is perhaps 100 Ω, and acts as a swamping resistor across the input circuit. The stage operates at class A, AB_1, or AB_2. The gain of such an amplifier is about 10 dB. For a kilowatt output it must be driven by about 100 W of SSB RF. Since input and output circuits are in series through the tube and power supply, most of the input RF appears as output power in the antenna.

With commercial SSB equipment it is usually required that the carrier be reduced at least 40 dB below the maximum PEP value. Thus, a 2-kW PEP transmitter must reduce its carrier to less than 0.2 W.

The clamp-tube circuit back in Fig. 16-16 can be used as a linear amplifier for SSB. With no signal there is no drive to the tetrode or pentode amplifier and no grid-leak bias. The clamp tube clamps the screen voltage to nearly zero. With weak SSB sig-nals applied to the amplifier grid, a little grid-leak bias is produced. This unclamps the clamp tube a little, and the amplifier begins to amplify the input. With strong input signals, the clamp tube is com-pletely unclamped and the amplifier is free to am-plify normally, using grid-leak bias. The clamp tube follows the SSB signal, acting as a *gating* circuit for the amplifier. This linear amplifier does not require well-regulated screen and bias supplies, unlike other types of RF linear stages.

A broadband, untuned transistor RF linear ampli-fier is shown in Fig. 17-48. It has two shunt-fed

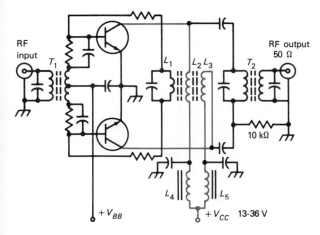

FIG. 17-48 A broadband BJT RF linear amplifier.

BJTs in push-pull to produce 100- to 200-W output. If two additional devices were connected in parallel with these, the output would be doubled. Stage gain is about 15 dB. There are two new items about this circuit. One is the neutralizing or stabilizing net-work made up of L_1 and its resistors, which feed just enough 180° out-of-phase voltage back to the input transformer T_1 to prevent oscillation. The other item is the method of feeding dc to the collec-tors through L_2, L_3, L_4, and L_5. These ferrite-core coils center-tap the collector LC circuit and act as RF chokes through which V_{DD} is fed to the collec-tors. By proper selection of L and C values of ferrite transformers T_1 and T_2, as well as L_1, L_2, and L_3, the amplifier can have equal gain over a band of frequencies such as 10–30 MHz, 40–50 MHz, or 145–170 MHz. To assure no harmonic radiation, a low-pass filter for the band used would follow T_2. The 10-kΩ resistor is a protective static dc dis-charge circuit.

1. What is the designation of SSB with no carrier? _____ With reduced carrier? _____ With full carrier? _____
2. What is the advantage of using crystal filters over *LC* and mechanical filters? _____
3. Why would it not be possible to use a doubler stage to change an SSB signal from a carrier fre-quency of 3–6 MHz? _____
4. What is the result if a balanced modulator is not perfectly balanced? _____
5. If two signals are in quadrature, what is their phase relation? _____
6. With what types of filters does a pop-up occur past the first skirt minimum? _____
7. Which is more desirable, a large or small shape factor? _____
8. In Fig. 17-38 how would it be possible to obtain both USB and LSB emissions on the same carrier frequency with only the USB filter being used? _____
9. What is an advantage in using a phase system for J3E? _____ A disadvantage? _____
10. In a filter system of J3E, how many circuits must be balanced? _____ In a phase system? _____
11. Why might it be difficult to produce music trans-missions with phase-type SSB? _____
12. Why might linear amplifiers with AM excitation be more stable than with J3E operation? _____
13. Why might two-tone tests be better than one-tone or voice if checking with a scope? _____
14. What are two formulas by which PEP might be determined? _____ _____
15. Draw a block diagram of a filter SSB system. Draw diagrams of a 2-JFET balanced modulator. A ring-type balanced modulator. An *m*-derived T-type bandpass filter. Half- and full-lattice crystal filters. An RF 90° phase-shift network. A block diagram of phase system of SSB. Scope presentation of voice-modulated SSB. An undistorted two-tone SSB scope presentation. A grounded-grid linear amplifier for SSB, AM, or CW. A broadband BJT linear amplifier.

17-37 SPECIAL USES OF SSB

In some commercial communications the voice is inverted or scrambled by heterodyning against some frequency such as 3 kHz. The result will be a 3-kHz carrier with USB (3–6 kHz) and LSB (0–3 kHz). Since LF AF becomes HF AF, the 0–3 kHz signals passed through a 2.8-kHz low-pass filter would be unintelligible to a listener. Such an inverted signal can be used as a scrambled modulat-

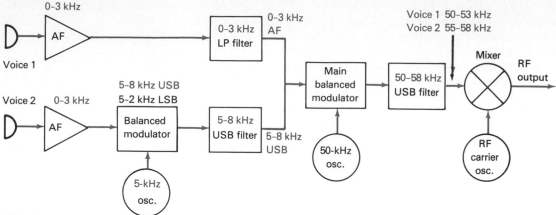

FIG. 17-49 Simple FDM system capable of transmitting two voice signals simultaneously.

ing signal and fed into any transmitter. The process is reversed at the receiver.

One voice signal can be transmitted using the USB and a second can be transmitted simultaneously as the LSB of a single suppressed-carrier emission. This is called B3E, and it is a form of *multiplexing*. Either two separate receivers (one tuned for USB, the other for LSB) or a single receiver having two sideband filters, each feeding a separate detector and AF amplifier, can be used.

Multiplexing ("mux") is the transmission of two or more voice channels on a single carrier, or with telephone systems, into a single pair of wires or into a single coaxial transmission line. Multiplexing may be accomplished by *frequency division* (FDM) or by *time division* (TDM). See also Sec. 21-17.

A simple two-voice FDM radio system is shown in block form in Fig. 17-49. Voice 1 signals are fed through a 0–3 kHz LP filter. Voice 2 signals are first translated up 5 kHz by a balanced modulator and a 5–8 kHz USB filter. Both voice signals are then translated up 50 kHz by a balanced modulator and passed through a 50–58 kHz USB filter. Fi-

nally, this signal is translated up to the desired transmitting frequency. At the receiving end, voice 1 would be audible with any USB receiver by using a 0–3 kHz LP filter after the detector to remove the 5–8 kHz voice 2 signals. A 5–8 kHz filter passes voice 2 voltages to a mixer where they beat with a 5-kHz oscillator to reproduce the voice 2 information. Each channel requires its own AF amplifying system at the receiver. Such a two-voice FDM emission would be categorized as 8K00-J8E-JF (Appendix E-2). Instead of using voice signals, either channel might be used for FSK teleprinter signals, for facsimile, for binary computer data, for telemetry data, etc.

Time-division multiplex is another means of transmitting two or more voice signals without requiring filters. TDM is a synchronized system, usually involving pulse-code modulation (PCM). If an analog voice waveform is chopped at a rate equal to twice the highest expected voice frequency (2 × 4000 Hz = 8000 pps) and these varying amplitude pulses are transmitted along a line, at the far end a listener will hear the 8000/s varying amplitude pulses almost as the original voice sounded. If the chopping were done at twice this frequency, the received signals would only be slightly better. If chopping were at the 16,000-pps rate and only every other pulse were transmitted, the reception would still be good. By chopping another voice channel at the same rate, but accurately sandwiching its pulses in between the first voice pulses, at the receiving end either voice could be heard alone if the receiver circuit were opened and closed in synchronism with the desired channel. A chopping frequency of 32,000 pps would allow four voices to be transmitted with only slight degradation. One of the difficulties with this system is maintaining a sufficiently accurate clock pulse at transmitting and receiving ends to hold the channels in synchronism. Since the pulses are square wave, the bandwidth to transmit TDM is much wider than with FDM.

Rather than using analog AF variations of the various pulses, the variations may be *quantized* (Sec. 12-17) or converted into one of 128 steps or amplitudes. Each step has a binary (off-on, or 0-1) coded amplitude value which is the actual information transmitted. At the receiving end the binary code is decoded and the quantized value is fed to the receiver, reproducing the original voice sound pulses.

A circuit used in many SSB systems is voice-operated transmission (VOX). A simplified VOX diagram is shown in Fig. 17-50. When the micro-

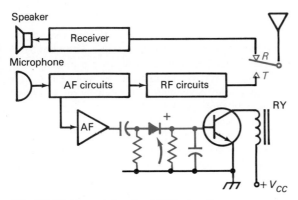

FIG. 17-50 Essentials of a VOX circuit.

phone is being spoken into, some of the AF ac is amplified and rectified, and the resulting dc is used to bias ON a BJT (or VT). When collector current flows in the BJT circuit, its relay is energized and the antenna contacts shift from receiver to transmitter. A second set of relay contacts (not shown) would be used to mute the receiver AF circuits while the transmitter is on. As soon as the speaking stops, the BJT becomes unbiased and the relay transfers the antenna back to the receiver.

17-38 INTERMODULATION

Two sine waves fed into any nonlinear circuit, such as an AF amplifier, RF amplifier, oxidized connection, etc., can mix together, resulting in one signal modulating the other. Such *intermodulation* occurs in AF amplifiers if they are operating nonlinearly, producing distorted output signals.

Two transmitters may intermodulate. For example, transmitter A radiates a signal from antenna A. Some yards away transmitter B radiates a signal from antenna B on some other frequency. If these two transmitters are on the air at the same time, the B signal may induce some of the B energy into the A antenna. The final-amplifier circuit of A now has both A and B energy mixing together, *intermodulating*, or *cross-modulating*. The result is sum and difference frequencies of the B and A carriers, plus sum and difference frequencies of all harmonics that may be present. Many of these frequencies may be radiated by antenna A and cause interference to other services operating on these frequencies. Intermodulation may be going on simultaneously in transmitter B. It can be detected by tunable receivers.

This type of interference can be stopped if a wave trap (Sec. 8-8) tuned to the B frequency is installed in the connecting line between the A transmitter and its antenna. This prevents B energy from appearing and mixing in the nonlinear RF amplifier output circuit of A, and no intermodulation can occur. The antennas themselves may have both A and B energy in them, but antennas are linear R, L, and C circuits and cannot produce sum and difference output. If there is a poor, or oxidized, connection in one of the antennas (or any nearby wiring), however, such a nonlinear joint can mix and produce beat output. An A-frequency wave trap should also be installed in the B-transmitter antenna line.

A receiver picking up two strong signals 10 kHz apart may generate beats in its front end every 10 kHz above and below the two signal frequencies. Usually, only the first beats will be audible in the receiver.

17-39 CITIZENS BAND (CB) RADIO

The CB Radio Service is a two-way, short-distance, voice communications service for personal or business activities. It may also be used for voice paging. CB is *not Amateur Radio* (Chap. 32). CB operations no longer require licenses or call signs. Amateurs must take code and theory tests to get their licenses and calls.

The CB band consists of 40 channels, the first centered on 26.965 MHz, with the last on 27.405 MHz. Channel 9, on 27.065 MHz, is used for emergency communications or for traveler assistance. Most channel centers are separated from the adjacent channel centers by 10 kHz.

CB equipment consists of transmitter-receiver

sets called *transceivers*. The transmitters are limited to 4-W output on AM and 12-W peak-envelope-power (PEP) output on SSB. No external RF power amplifiers may be used with CB transmitters. No internal adjustments should be made to transmitter circuitry except by qualified technicians.

Transmitting antennas are limited in height. The top of the antenna must not exceed 20 ft above the building on which it is erected, or 60 ft above the ground if erected on the ground. Antennas may be simple verticals or may be beam types.

CB transmitters may produce harmonic radiation, which can cause local interference to services at frequencies higher than 27 MHz. For example, the second harmonic of all CB channels falls into the 54–60 MHz range of TV channel 2. The third harmonic of CB channels 1 through 30 falls into the 76–82 MHz range of TV channel 5. To reduce interference to TV receivers, a 28-MHz low-pass filter can be added between the CB set antenna terminal and the antenna feed line. If this is not done, or is not effective, the CB station may have to observe silent periods as determined by the nearest FCC field engineering office.

All CB (police, fire, aeronautical, amateur, etc.) transmitting equipment should be connected with a short wire to an earth ground (ground rod, water pipe into ground, etc.). If the ground connection, or any antenna connection, loosens or becomes oxidized, it may form a semiconductor diode and generate spurious radiated signals that may interfere with any higher- or lower-frequency receivers in the area. It is important that all antenna and ground connections be electrically solid.

Many audio devices, stereo, radio, TV, turntables, tape decks, organs, etc., will pick up CB transmissions, rectify (detect) them in some stage, and reproduce them in the loudspeaker. If the CB transmission is AM, it will be readable, but if it is in SSB, it will be muffled and garbled and unintelligible. This is the fault of the audio device and can be cured only at the device by (1) bypassing power-line leads; (2) using shielded leads to remote loudspeakers and grounding the shield to the chassis (if the chassis is not "hot" and connected to one side of the ac line); (3) adding *RC* or *LC* filters to the input circuit doing the rectifying; and (4) either grounding the chassis of the device, or bypassing it if it is "hot," or shortening the ground lead.

A CB performance tester may be used by servicing technicians. It consists of several test instruments in one package and is usually portable. It may contain many or all of the following: RF signal generators, AF signal generator, SWR meter, RF power meter, dummy load, VOM, and modulation meter.

An AM modulation meter that can be relatively accurate for tone modulation is shown in Fig. 17-51. Carrier RF is picked up by a short antenna.

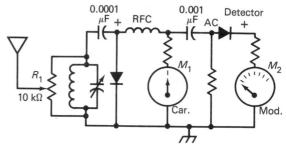

FIG. 17-51 AM percent-modulation and carrier-shift meter.

The *LC* circuit is tuned for a maximum reading on the carrier detector dc voltmeter M_1. The shunt-rectified RF input is adjusted by R_1 to a midscale reading of M_1. When modulation is applied to the transmitter, the percentage of modulation produced can be read directly from the scale of the AF voltmeter M_2. As long as M_1 reads midscale, as when M_2 was calibrated, the modulation percentage should be correct, assuming no negative carrier shift. M_1 also acts as a carrier-shift meter. If it dips when modulation is applied there is negative carrier shift, and the indicated modulation values will not be correct.

17-40 SPEECH PROCESSING

Most AM, SSB, and FM transmitters include some degree of *speech processing*. The goal of all speech processing systems is to raise the average level of the received voice (or music) modulation. With the use of sinusoidal modulating AF ac it is possible to maintain any desired level of modulation from 0 to 100%. Voice signal ac from a microphone is constant in neither frequency nor in amplitude. The characteristic of speech is a low average energy level with peaks two or three times the average voltage value. The peak value depends on the person's voice, or the type of musical instrument used. It is important that these peaks do not overmodulate a transmitting system, or distortion, excessive sidebands, spurious emissions (splatter, buckshot) will be produced on both sides of the channel being used.

Peak clipping is one form of speech processing. The simplest circuit consists of two silicon diodes in parallel and reversed across an audio circuit between two amplifiers (Fig. 17-52). The clipper di-

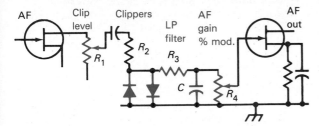

FIG. 17-52 AF clipper and low-pass filter circuit.

odes do nothing to signals up to a few tenths of a volt. When the barrier voltage is reached (silicon = 0.6 V) the diodes begin to draw current, producing a voltage-drop across R_2, preventing the peak voltage from being developed across the volume control R_4. In this way high-voltage peaks are clipped off and are prevented from producing overmodulation. However, since clipping peaks distorts the voice ac waveform, it produces many high-frequency harmonics. These are reduced by the simple R_3C low-pass filter. The gain control before the clipper is used to set the voice level so that only the peaks are being clipped and limited. The output of the filter is peak-limited AF that can be used to modulate a transmitter. Percentage of modulation is set by R_4.

An *automatic level control* (ALC) circuit may be used. If a speaker changes the mouth-to-microphone distance, the sound intensity changes, and the percentage of modulation may vary widely. An ALC system such as that shown in Fig. 17-53 can be used to automatically "ride gain" on the speaker. If the arm of R_2 is at the bottom, there will be no feedback from the output of the second AF amplifier to the gain-controlling bias circuit of the first amplifier. If half of the AF output of the second amplifier is rectified and filtered smooth by D and

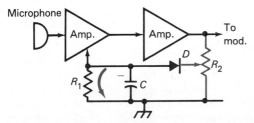

FIG. 17-53 ALC circuit to reduce gain of a preamplifier.

C, a reverse bias is fed back to reduce the gain of the first amplifier whenever signals are being amplified. No bias is developed until the signal output exceeds the barrier voltage of the diode, however. How much reverse bias is fed back is determined by the setting of the contact on R_2. How fast the bias will drop off is determined by the value of R_1 and C. If the speaker backs away from the microphone, the bias dies away and the gain in the first amplifier increases. If the speaker closes in on the microphone, the bias increases and the gain of the first stage is reduced, resulting in a reasonably constant output signal from the system at all times.

If the ALC-type circuit is made to amplify up to a certain voltage value normally but then begins to reduce gain 1 dB for every 2 dB of signal increase, the circuit is called a *compressor*. If a reverse circuit is used at the receiving end and its gain increases 2 dB for every 1 dB of increase of the received signal over a certain level, it is known as an *expander*. When both compression and expansion are used, it is known as a *compander* system. A compander can greatly increase the dynamic audio range, as well as improve the signal-to-noise ratio (S/N, or SNR) of a sound system, whether it is a wire, radio, tape, or disk type.

When linear amplifiers are used in a SSB transmitter, *RF speech processing* can be used. In Fig. 17-54 the modulated RF output from the final am-

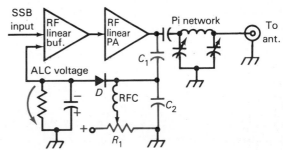

FIG. 17-54 Speech compression circuit in an RF linear amplifier system.

plifier is applied across a capacitive voltage-divider circuit, C_1 and C_2. If the modulation peaks exceed the positive bias voltage level set by the arm of potentiometer R_1, diode D rectifies the RF ac to form an ALC bias voltage. This bias is fed back to control the gain of a lower-level RF linear amplifier. As a result of this feedback, the microphone level can be increased without overdriving the RF amplifiers.

17-41 NARROWBAND VOICE MODULATION

To use less bandwidth for a given voice transmission, the original minimum 5- or 6-kHz bandwidth of A3E signals was cut to about 300–2500 Hz (2.2 kHz) by using SSB. A method of reducing the bandwidth further, to about 300–1600 Hz (1.3 kHz) is called *narrowband voice modulation* (NBVM). It is produced by a circuit added between a microphone and any transmitter.

Speech sounds are generated, fundamentally, in the range of about 300–600 Hz in the human larynx (voicebox). Additional harmonic-type sounds are also developed in the larynx and molded by the aural cavity (throat, nose, and mouth). All together, these develop the complex sounds of human speech. It has been found that frequencies from about 600 to about 1600 Hz do not help the intelligibility of speech materially. But the frequencies from 1600–2500 Hz are needed to give the voice a distinctively human sound. NBVM transmits the 300–600 Hz unchanged but translates the needed 1600–2500 Hz frequencies into a 700–1600 Hz modulating signal, for a total bandwidth of 300–1600 Hz. Modulating with this 300–1600 Hz AF develops SSB sidebands with a width of only about 1.3 kHz. Unfortunately this would be almost unintelligible if received by standard receiving methods. The receiving audio system must have a NBVM converter to retranslate the 700–1600 Hz band of signals back to their proper 1600–2500 Hz values, and feed these plus the 300–600 Hz signals to an amplifier and loudspeaker.

Test your understanding; answer these checkup questions.

1. What method can be used to invert speech? _____
2. What does FDM mean? _____ TDM? _____
3. How are two voices separated in FDM? _____ In TDM? _____
4. What does a VOX circuit do? _____
5. What kind of emission is A8EJ0? _____ J8EJF? _____
6. How is transmitter intermod stopped or decreased? _____
7. What is the maximum power output of a CB A3E emission? _____ J3E emission? _____
8. What is the maximum antenna height for CB? _____
9. What filter may be connected between a CB transmitter and its antenna? _____
10. What else can be indicated besides modulation percentage by the circuit in Fig. 17-50? _____
11. In what two systems in a voice transmitter can speech processing be used? _____
12. What must be done to clipped AF signals to reduce distortion? _____
13. What does the gain control before a clipper determine? _____ After the clipper? _____
14. What does ALC mean? _____ What does such a circuit do? _____
15. What is in a compander system? _____
16. Is SSB voice compression better in the AF or RF section? _____
17. What does NBVM mean? _____ What is the bandwidth of such a SSB system? _____
18. Draw a block diagram of a two-voice FDM system. Draw diagrams of a VOX system. A modulation percentage meter. An AF peak clipper. An ALC circuit. An RF speech compression system.

AMPLITUDE MODULATION AND SSB QUESTIONS

1. What is the emission designation of amplitude-modulated voice or music?
2. What will the normal AM receiver detect from an unmodulated RF ac wave?
3. What makes up one sound wave? To what is this equivalent in electricity?
4. What is the frequency of "middle C" on a piano? Why does this tone sound different on other musical instruments?
5. What sound power represents 1 bel?
6. Who hears high audio frequencies best?
7. Basically, what does a carbon-button microphone do?
8. Where is a single-button microphone used most?
9. What is dynamic instability in a transmitter?
10. Where is a form of absorption modulation used today?
11. What is in series in series modulation?
12. Draw a diagram of a simple series-modulated transmitter.
13. In Fig. 17-6, is R_1 a gate-leak resistor? What type of coupling is used between speech amplifier and modulator? What current flows through the microphone?
14. What is an advantage of series modulation?
15. In RF oscillators and amplifiers, why are drain (collector, plate) current pulses always narrower than a half-wavelength of the frequency used?

16. What happens to the height of the output circuit pulses of an RF amplifier when amplitude modulated? How are they converted to ac in single-ended amplifiers?

17. As seen on an oscilloscope, what is a modulated wave called?

18. In basic plate (collector, drain) modulation, what is added to the plate circuit of the RF stage?

19. Which ac is greater, the carrier level or the negative peak of AM?

20. What is the formula to determine percent of modulation if maximum and minimum voltages are known?

21. What is produced if over 100% of negative peak modulation occurs?

22. What occurs if AM positive peaks are modulated to 120% with negative peaks to only 98%?

23. Draw a representation of an RF carrier modulated 100% by a sinusoidal AF: 50%, 125%.

24. Draw a diagram of a high-powered triode modulated by push-pull beam-power tetrodes.

25. How many watts of sinusoidal AF are required to 100% plate modulate an 85% efficient 50-kW broadcast RF amplifier? What is the dc power input to the modulators if they are 45% efficient?

26. When tetrodes are plate modulated, what else must be done to them?

27. Draw a diagram of a plate-modulated tetrode RF amplifier.

28. What are two other terms essentially synonymous with beating? In what kind of circuits can this occur?

29. In modulation circuits, if a carrier beats with an AF signal what is produced?

30. A 2-kW RF carrier is sinusoidally modulated 100%. How much power is in one sideband? If modulated to only 60%?

31. A carrier is modulated 50% by a 3-kHz AF. What is its bandwidth? If modulated 100%? If it produces four significantly strong harmonics?

32. What is the bandwidth of AM voice modulation? Good music? High-fidelity sound?

33. What is the name of the modulation using a modulation choke? Where was a somewhat similar action seen before?

34. Draw a diagram of a Heising modulation system. Is AF ac or varying dc added to the PA I_{dc}?

35. Draw a diagram of a grid-modulated RF amplifier. Does any I_g flow with 100% modulation?

36. Why can grid modulation be used to modulate the HF video signal of a TV transmitter but plate cannot? What other form of modulation might also be used?

37. What modulation is known as constant current? As efficiency?

38. What is low-level modulation? High-level?

39. How is the modulation *envelope* displayed on an oscilloscope? How is a trapezoidal display produced? Which may show distortion better?

40. Which stages in a broadcast transmitter are normally class C?

41. How much bias is used in class BC operation? How is overmodulation of negative peaks prevented?

42. Draw a diagram of a low-level modulation system, including modulator, modulated amplifier, and linear amplifier. Why is the linear input circuit resistively loaded?

43. By test, how can class B bias be found for an RF amplifier? How can proper grid drive be determined?

44. What is the approximate efficiency of a class B linear RF AM amplifier? The efficiency of a Doherty linear amplifier?

45. When carrier amplitude regulation is present, what has been shifted?

46. What is another term meaning "downward modulation"?

47. What might be the most likely cause of negative carrier shift?

48. How much should an active device be derated when amplitude-modulated in comparison with its operation with CW?

49. A 10,000-W AM broadcast station will have what antenna ammeter increase with 75% sinusoidal modulation? With 75% voice modulation?

50. What part of a radio receiver might be used as a microphone?

51. Name seven types of microphones used in communications.

52. What type of microphone always has the first speech amplifier included in its case? Why?

53. Which microphones have the highest power output? Which has the lowest?

54. What is a hybrid transmitter? Are many used?

55. What is the tuning progression of stages in an AM broadcast-type transmitter?

56. Draw a diagram of a possible hybrid broadcast transmitter.

57. To produce 100% collector modulation, what must usually be done?

58. What are three reasons why SSB is superior to AM for communicating over distances? What are two disadvantages of SSB?

59. What is the designation of an AM voice or music signal? Of a carrier with one SB? Of one SB and no carrier?

60. What is input to a balanced modulator? What is output from it?

61. What is used to develop a SSB signal from the output of a balanced modulator?

62. Draw a block diagram of a filter-type SSB transmitter.

63. Why are mixers and not frequency multipliers used in SSB transmitters?

64. What type of device can be used in balanced modulators?

65. Draw a diagram of a JFET-balanced modulator.

66. Draw a diagram of a diode-balanced modulator.

67. What type of SSB filters might be used in the LF range? In the MF range? In the MF or HF range?

68. Draw diagrams of a half-lattice and a full-lattice crystal filter.
69. Draw a block diagram of a phase-system SSB transmitter. How is shifting from LSB to USB accomplished?
70. In what classes are SSB linear amplifiers operated?
71. If an oscilloscope presentation of a SSB emission shows peaks of 330 V when across a 50-Ω line, what is the PEP?
72. What presentation is given by an oscilloscope of a SSB signal if single-tone-modulated? If two-tone-modulated? If VF modulated?
73. Why are linear SSB RF amplifiers usually grounded-grid types?

74. Draw a diagram of a grounded-grid SSB linear amplifier.
75. Draw a diagram of a broadband push-pull BJT SSB linear amplifier.
76. What do the letters VOX mean? FDM? TDM?
77. Draw a block diagram of a simple FDM system for 2-voice transmission.
78. Draw a block diagram that explains the essentials of VOX.
79. What is it called when RF from two nearby transmitters mix in the first stage of a receiver and produce interference to the desired frequency? How can it be reduced?
80. What type(s) of modulation is used in CB sets?
81. What might a performance tester contain? By whom would it be used?
82. Draw a diagram of an AM modulation-percentage/carrier-shift meter.
83. Draw a diagram of a two-diode AF clipper. What reduces harmonic distortion in the circuit?
84. Draw a block diagram of an AF ALC system. What determines at what level the gain reducing bias starts?
85. What is a compander system?
86. Draw a block diagram of an RF speech compression system around a linear amplifier. What determines at what level ALC starts?
87. Which human voice frequency signals are not transmitted by NBVM? Which are?

ANSWERS TO CHECKUP QUIZ ON PAGE 402

1. (Frequency translation, mixing) **2.** (Frequency division mux)(Time division mux) **3.** (Frequency translation)(Pulse separation, timing) **4.** (Change to transmit whenever voice ac is present) **5.** (Mux)(FDM) **6.** (Filters or wave traps) **7.** (4 W)(12-W PEP) **8.** (60 ft above ground, 20 ft above building) **9.** (Low-pass) **10.** (Carrier shift) **11.** (AF)(RF) **12.** (Filter, low-pass) **13.** (Clipping level) (Mod %) **14.** (Automatic level control)(Maintains more constant modulation level) **15.** (Compressor)(Expander) **16.** (RF) **17.** (Narrowband voice modulation)(1300 Hz) **18.** (See chapter illustrations)

18

Amplitude-Modulation Receivers

This chapter explains the various types of detectors used to demodulate amplitude-modulated signals and then TRF and superheterodyne AM receivers. The variations of system requirements to receive different types of emissions are discussed. Some alignment and troubleshooting techniques are included. Other receiver systems will be found in the chapters on radar, television, and maritime radio.

18-1 RECEIVING RADIO SIGNALS

A receiving system consists of an antenna, in which all passing radio waves induce an RF ac emf, a means of selecting the desired signal, a means of *detecting*, or *demodulating*, the modulation in the signal, and a means of making the detected electrical signal audible.

A wire can serve as an antenna. One particular frequency can be selected by using a resonant circuit (Fig. 18-1). It might be thought that earphones

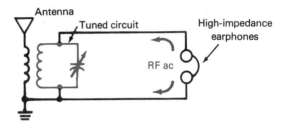

FIG. 18-1 Earphones across an *LC* circuit cannot detect radio signals even though they may be present.

across the tuned circuit would operate as a receiver. If the resonant circuit is tuned to some radio frequency, perhaps 1000 kHz, a voltage of that frequency is being developed across the tuned circuit and a current of that frequency flows through the earphones. However, even if the earphone diaphragms could vibrate at that rate and produce airwaves at a frequency of 1,000,000 Hz, nothing would be heard. The ear is not sensitive to waves in air higher than about 20,000 Hz. If the carrier is modulated, the carrier frequency is still 1000 kHz, the sidebands are a little higher and lower, and all are inaudible. To change the inaudible RF ac to an audible-frequency ac, a *detector* is needed to demodulate the modulation on the carrier.

18-2 DEMODULATING A MODULATED WAVE

If a received RF carrier is amplitude-modulated by voice or music, it is possible to detect the variation of the carrier amplitude with a half-wave rectifier and filter circuit (Fig. 18-2). When the transformer secondary is tuned to the frequency of a local modulated transmitter, RF ac at this frequency is developed in it. A possible modulated RF envelope is illustrated. The earphones and rectifier form a load on the tuned circuit. With the rectifier, only unidirectional pulses can flow through the earphones. The resultant RF pulsating dc is also shown.

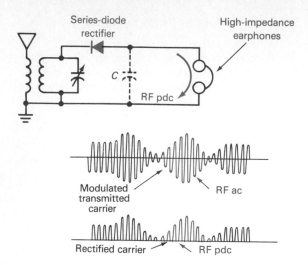

FIG. 18-2 A series-diode detector circuit, a modulated carrier wave, and the rectified (demodulated) carrier.

With an unmodulated carrier, constant-amplitude dc pulses flow through the earphones. This pulls the diaphragm inward a little and holds it there. When the carrier is modulated, the average current of the pulses varies with the modulation and the earphone diaphragm is pulled farther inward or is released to move outward. This controlled vibration of the earphone diaphragm generates airwaves (sound).

The average amplitude of the sinusoidal half-wave pulses is equal to one-half of the average value (one-half of 0.636 V_{max}) or about $V_{max}/3$. The average value can be increased by using a capacitive filter. A 0.0005- to 0.002-μF capacitor charges to the peak value and tends to hold this value until the next RF pulse arrives. As a result, the average current through the earphones is increased and a louder signal is heard.

If capacitor C is too large, it will not discharge rapidly enough and a cycle of high-frequency AF modulation may drop and rise again before the capacitor has had time to discharge. High-frequency audio will not be heard. With correct capacitance, this detector can reproduce AM with good fidelity. Diode detectors are used in essentially all AM receivers.

Diode detectors require no power supply. The incoming signal supplies all the power that appears in the earphones. Only relatively strong or local signals will be audible. While germanium diodes may be used, Schottky hot-carrier diodes are preferable.

Since the sound developed in the earphones of a simple detector is not very loud, the next step is to add an audio amplifier. A *shunt-rectifier* detector resistance-coupled to a JFET amplifier is shown in Fig. 18-3. The RF emf from the tuned circuit is applied across the diode through C_1. When the

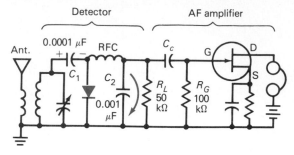

FIG. 18-3 A shunt-type diode detector with an AF amplifier.

anode is driven positive, electrons flow and are stored on the right-hand plate of C_1. On the next half cycle the plate is driven negative and does not draw electrons. Those that had been drawn to C_1 now discharge through the RFC, charging C_2. This capacitor discharges downward through the load resistor R_L. If the carrier is modulated, the charge across C_2 varies with the amplitude of the modulation, producing a varying dc down through R_L. Coupling capacitor C_c and gate resistor R_G are across the varying dc in R_L. Essentially the same amplitude and frequency ac will be developed across R_G as is present in the varying dc of R_L. This signal is fed to the gate of the JFET, producing a much louder signal in the earphones. The RFC is required in this shunt circuit to prevent C_1 and C_2 from forming a detuning voltage-divider capacitance across the LC circuit.

18-3 CRYSTAL DETECTORS

One of the first rectifying demodulators was the *crystal detector*. Rectifier crystals were metallic types, such as galena, iron pyrites, silicon, and carborundum (not quartz- or rochelle-salt-type crystals used in oscillators and microphones).

The crystal was embedded in a lead mounting. A thin, pointed wire, called a *cat whisker*, was pressed against the surface of the crystal. When a sensitive spot was found, considerably more current flowed in one direction than in the opposite and rectification resulted. Crystal detectors were the first junction rectifiers. In Fig. 18-4, if the diode and earphones are tapped part of the way down the

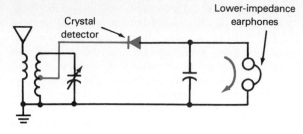

FIG. 18-4 A crystal detector.

tuned circuit, less loading of the tuned circuit results. The circuit has a higher Q, is more *selective* (has narrower bandwidth), and is more *sensitive* (produces more signal voltage from weak signals).

18-4 POWER DETECTORS

With simple rectifying detectors the power of the signal in the earphones is limited to the RF power picked up by the antenna wire. If an amplifying device is used and the rectification is produced in its output circuit, it is possible to detect and amplify in one stage (Fig. 18-5). The potentiometer is

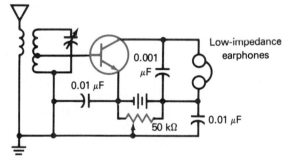

FIG. 18-5 A power detector using a BJT.

adjusted to the point where there is just enough forward bias to produce zero I_C flow in the earphones. Any received signals apply ac to the base-emitter circuit. During the positive half cycles of the RF ac I_C pulses will flow in the collector circuit and through the earphones. Since the BJT is an amplifying device, the pulse amplitude through the earphones will be much greater than would have flowed if the earphones had been in the base circuit, and a greater power output will be produced for a given RF signal input. Local broadcast stations may provide enough power output to operate a 16-Ω loudspeaker. The capacitor across the earphones is to smooth the pulsating dc somewhat and may increase the output signal slightly.

Any active device, if biased just to the point of zero output circuit current (class B), will operate as a power detector. Try diagraming a JFET and a triode power detector.

18-5 LINEAR AND SQUARE-LAW DETECTORS

The graph in Fig. 18-6 represents (a) a plate current that is linear and (b) a plate current that follows a *square law*. In the linear case, each time the grid voltage is doubled, I_p doubles. In the square-law

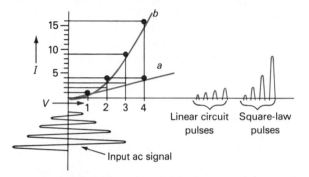

FIG. 18-6 Voltage versus current in (a) a linear rectifying circuit and (b) a square-law rectifying circuit.

case, each time the voltage is doubled, I_p increases four times (the square of V_{in}).

A linear detector will produce an output current waveshape the same as the input voltage waveshape. With a square-law rectifier the waveshape of the output current will differ from the input-voltage waveshape, distortion is produced, but with the same signal input there will be considerably more output from a square-law detector.

A diode is a fairly linear device unless operated close to the zero current point by weak signals, where its curve rounds off in a square-law manner.

18-6 GRID-LEAK DETECTORS

The grid-leak detector (Fig. 18-7) was popular in the early days because it had better sensitivity than a diode. It has a diode rectifier grid circuit and an AF amplifier plate circuit. The grid circuit consists of a tuned circuit (the RF ac source), a rectifier (the cathode and grid), and a resistance as a load with a filter capacitor C across it. Modulated signals produce a varying dc voltage across R and C. Rectified current flows from grid, through the resistor, to cathode, making the grid end of R negative, biasing

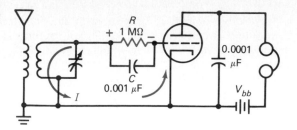

FIG. 18-7 A grid-leak detector circuit.

the triode. With no signal input, there is almost no I_g and almost no bias, I_p is relatively high, and therefore a low V_{bb} (15 to 30 V) is used. With an unmodulated carrier, a bias voltage is developed and the I_p decreases. When the carrier is modulated, the bias varies with the modulation waveform, producing a relatively large I_p variation in the earphones. As a result, the listener hears loud sounds. A JFET can be substituted for the triode.

18-7 REGENERATIVE AND AUTODYNE DETECTORS

To increase the sensitivity and selectivity of the grid-leak detector, a *tickler coil* can be used to regenerate energy from the plate circuit into the grid circuit (Fig. 18-8). Because of the three coils, this is known as a three-circuit tuner detector.

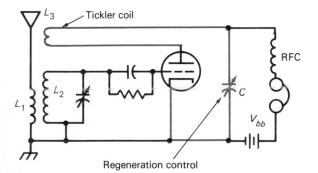

FIG. 18-8 An Armstrong-type regenerative detector.

The tuned circuit of a grid-leak detector would have a fairly high Q except for grid-leak-resistor loading. An RF signal voltage in the grid circuit is amplified and can be fed back *in phase* to the LC circuit by the tickler coil. Some of the signal energy lost due to grid-circuit loading is partially made up. The result is a lessening of losses and a higher Q. With no regeneration (in-phase feedback) the stage operates as a straight grid-leak detector.

As regeneration is increased, the detection efficiency and the signal output increase. If enough regeneration is present, all the losses in the grid circuit will be made up and the LC circuit will begin to oscillate at its natural frequency of resonance as an Armstrong oscillator. The variable bypass capacitor C completes the ac tickler circuit to ground and controls regeneration.

The higher the resistance of the grid leak, the more sensitive and selective the detector. Maximum sensitivity and selectivity, for detection of modulated signals, occur when the regeneration is just under the value required to produce oscillation. If the stage oscillates, it is no longer satisfactory as a detector for AM signals.

As the regeneration and the Q of the LC circuit increase, the coupling requirement from antenna to tuned circuit decreases markedly.

As soon as the regenerative detector breaks into oscillation, it becomes an *autodyne* (self-heterodyning) detector. In this condition it can be used to detect CW, SSB, or FSK emissions. The basis of detection changes from rectification to heterodyne (Sec. 16-20).

If an autodyne detector is oscillating at 1002 kHz and 1000 kHz is fed to it, the difference frequency is 2 kHz. This, plus 1002, 1000, and 2002 kHz appear in the plate circuit. Since the plate circuit has an RF low-pass filter (C and RFC), the three higher-frequency signals do not pass through the earphones. Only the 2-kHz component flows through them. The beat tone can be changed by tuning the LC circuit. If the 1000-kHz incoming signals are in dots and dashes, the audible beat will reproduce the dots and dashes as 2-kHz signals in the plate load.

The autodyne detector has maximum *beat* response when the two signals being heterodyned have the same amplitude. Since two separate frequencies operating in the same circuit tend to synchronize and lock into one frequency of oscillation, a weakly oscillating LC circuit may be forced to synchronize with a strong incoming signal and no beat note will be produced. To prevent such *frequency pulling* it may be necessary to loosen the coupling to the LC circuit, or to increase regeneration until the autodyne circuit is in strong oscillation. The degree of regeneration and oscillation can be controlled by varying capacitor C, the degree of coupling between tickler and LC circuit, or the V_{bb} value.

If a regenerative detector is oscillating and it is tuned across the frequency of a station to which another nearby receiver is set, the detector will produce a whistle on the receiver. Radiation of RF by an autodyne is a distinct disadvantage. Coupled to an antenna, such a detector may radiate a receivable signal for miles.

> **Test your understanding; answer these checkup questions.**

1. What is the circuit that changes modulated RF to AF called? _____
2. Name two transducers that change AF ac to sound waves. _____ _____
3. Why is a diode called an envelope-type detector? _____ Why might it also be thought of as a heterodyne type? _____
4. If the bypass capacitor across the earphones in a diode detector is too large, what is the result? _____
5. Where do crystal detectors obtain the power that vibrates the earphone diaphragms? _____ Where do power detectors? _____
6. Would shunt-type grid-leak bias be possible in a regenerative detector? _____
7. What kind of crystals were used in crystal detectors? _____ In oscillators? _____
8. Why would a JFET power detector be more selective than a BJT power detector? _____
9. What is a plate detector? _____
10. Which produces the louder signal, a linear or a square-law detector? _____ The more distorted signal? _____
11. If earphones were substituted for the grid-leak resistor in a grid-leak detector, why would modulation be audible? _____
12. Why does some regeneration in a grid-leak detector result in louder AM signals? _____ Narrower bandwidth? _____
13. What happens if a regenerative detector starts oscillating while tuned to an AM signal? _____ What detector is it now? _____
14. How can the beat tone in an oscillating regenerative detector be changed? _____ Why might it change if the signal fades? _____
15. How is a regenerative detector adjusted to receive weak CW signals? _____ AM? _____ Why is it not satisfactory in detecting SSB signals? _____

18-8 SUPERREGENERATIVE DETECTORS

A VHF circuit that was used in the past but which is rarely used now is important because its action is sometimes inadvertently developed in electronic circuits. It is the *superregenerative* detector. In this circuit a second oscillation is developed to turn on and then quench at some *supersonic* (20–100 kHz) frequency the oscillations of an oscillating regenerative detector. In block form, a superregenerative circuit has a VHF signal oscillator tuned to the frequency it is desired to receive. In series with its power supply is the supersonic modulating ac (Fig. 18-9).

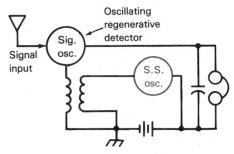

FIG. 18-9 A block diagram of a superregenerative detector circuit.

Each time the quenching oscillator voltage output starts the VHF oscillator, a burst of VHF ac is produced at the signal frequency. Because no signal is received, random noise in the input of the VHF oscillator produces random oscillator starting times and, therefore, random length oscillation bursts. Random energy burst *lengths* result in a loud hissing noise in the detector output.

When a VHF carrier signal is received, the carrier ac forces all detector oscillation bursts to start at a constant time interval, quieting the random noise output. When modulation appears on the carrier, the variation of the amplitude of the signal voltage changes the length of time that the VHF oscillator is turned on, resulting in modulated energy burst lengths and thus a modulated average energy output. This is a form of pulse-width demodulation.

A relatively weak carrier signal will suppress the background noise, and its modulation will produce almost as much output-signal strength as the modulation of a strong carrier. High-amplitude impulse noises, often very strong in the VHF range, are limited in amplitude and will not appear at all if they occur during an RF burst. The detector is broad-tuning even with a high-Q LC circuit. Coupled to an antenna, the superregenerator radiates a broad signal.

A simpler superregenerative detector is the self-quenching circuit shown in Fig. 18-10. The triode oscillates at the VHF signal frequency. The grid-leak R and C set up a supersonic RC relaxation

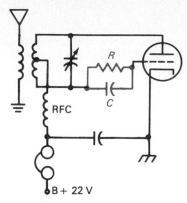

FIG. 18-10 Self-quenching superregenerative detector.

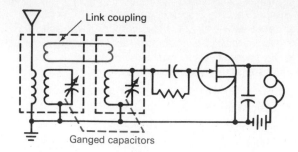

FIG. 18-11 Cascading tuned circuits increases selectivity.

oscillation which produces the quenching effect. Of modern importance, superregeneration sometimes develops in RF amplifiers or oscillators and is characterized by a wide band of spurious signals being generated. It usually disappears when *RC* circuit values are changed.

18-9 TRF RECEIVERS

The next step in receiver development is the tuned radio frequency amplifier (TRF) circuit.

Detectors used alone as receivers lack sensitivity, selectivity, and output audio power.

Selectivity, the ability to tune out all frequencies except the one desired (measured in kilohertz of bandwidth), can be increased by using cascaded loosely coupled tuned circuits (Fig. 18-11). These circuits can be gang-tuned so they will tune to identical frequencies, or *track*, at all times.

In place of the link coupling, a transistor or VT can be used to couple the two tuned circuits. Any loss in signals passing through the tuned circuits is more than made up by such RF amplifying devices.

Figure 18-12 is a block diagram of a TRF receiver. Such receivers use one or two RF amplifiers ahead of the detector. However, the more stages, the greater the difficulty in making them track properly, preventing them from breaking into oscillation, and causing more noise to be generated. Only one or two audio stages are needed to bring the detector output up to loudspeaker volume.

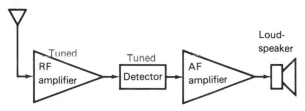

FIG. 18-12 Block diagram of a 3-stage TRF receiver.

The TRF receiver in Fig. 18-13 has a JFET RF amplifier with a variable bias resistor to control the RF gain or sensitivity. R_1 is the minimum value of resistance required to bias to class A. Any additional resistance increases the bias and decreases the sensitivity of the stage. The regenerative detector provides rectification detection for modulated signals when not oscillating and autodyne detection for CW or FSK signals when in oscillation. The audio stage is a common type of volume-controlled amplifier. A single IC chip may contain the RF amplifying, the detection, and the AF amplifying circuitry. It requires only external tuned circuits and a few external resistors, potentiometers, and capacitors.

The detector and RF stages should be shielded (dashed lines) and be in a grounded metal box.

A well-shielded and neutralized RF amplifier ahead of an autodyne or superregenerative detector prevents oscillations in the detector from reaching the antenna and being radiated. Transistorized TRF receivers using JFET or MOSFET RF amplifiers may require neutralization. In VT TRFs pentodes are used as the RF amplifier.

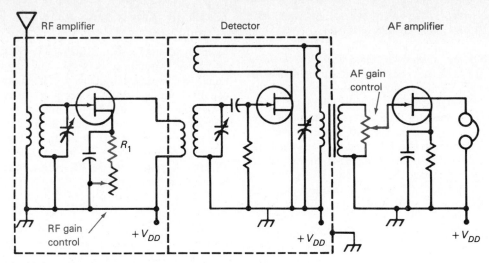

FIG. 18-13 Schematic diagram of a 3-JFET TRF receiver.

TRF receivers are used in simple low-frequency receivers. At HF the sensitivity and selectivity fall off because of lowered tuned-circuit Q. Shipboard emergency receivers and auto-alarm receivers may use a TRF circuit. There are some fixed-frequency microwave TRF receivers also.

18-10 SUPERHETERODYNES

Practically all radio receivers today are super-heterodynes. While this receiver system is more complicated than the TRF, the ability to operate satisfactorily on any frequency with a constant value of selectivity and good sensitivity makes it highly desirable. A block diagram of a simple communications superheterodyne receiver capable of detecting AM, CW, FSK, or SSB signals is shown in Fig. 18-14.

The basic operation of the stages of this super-heterodyne is as follows:

The *RF amplifier* is tuned to 2450 kHz and amplifies any signal on that frequency. The output

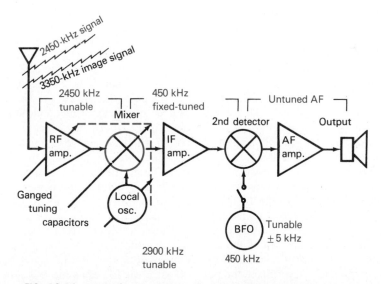

FIG. 18-14 Block diagram of a simple single-conversion-communications superheterodyne receiver.

from this stage is fed to a heterodyne circuit called a *first detector, mixer, or converter*.

The *mixer stage* is also tuned to 2450 kHz, but is also being fed a second signal from the *local-oscillator* (LO) stage. These two frequencies are heterodyned in the mixer stage, where the sum and the difference frequencies are developed in the output. If the LO is tuned to 2900 kHz and the input-signal frequency is 2450 kHz, the difference frequency is 450 kHz, known as the *intermediate frequency* (IF). The signal frequency is *translated* to the IF frequency by the mixer and LO. The IF signal is fed to an IF amplifier. (If the sum frequency of 5350 kHz is used as the IF, the IF is said to be *up-converted*.)

The *IF stage* has both input and output circuits tuned to 450 kHz to amplify only this difference frequency. It rejects the RF signal, the LO signal, and the sum frequency, all of which are present in the output circuit of the mixer. The selectivity of a simple superheterodyne is determined primarily by the number and Q of the tuned circuits in the IF amplifier stages. The 450-kHz IF signal is amplified and fed to the second detector.

The *second detector* may consist of a diode detector to rectify the 450-kHz IF signal. If the signal contains modulation, the output of the diode rectifier is an AF varying dc that is fed to the AF amplifier and then to earphones or loudspeaker.

If the signal is a CW emission, to produce a beat tone the second detector must be changed to a heterodyne detector by switching on the *beat-frequency oscillator*, or BFO. The BFO is tuned a few hundred hertz above or below the 450-kHz IF. By tuning either the LO or the BFO, it is possible to produce any beat tone desired. If the BFO is tuned to the frequency of the incoming carrier, a *zero beat* is produced, and nothing is heard.

To prevent interaction between stages and to prevent interference from undesired signals, all RF, IF, and oscillator stages should be shielded.

It is to be noted that if only an RF amplifier, mixer, and LO are used and the LO is adjusted to within 1 kHz of the RF-mixer frequency, the resulting *heterodyne product* or *synchronous detected* signal will be in the audio instead of in the IF range. Such a simple system is called a *direct conversion* receiver. It would be used for CW or SSB reception, although it would demodulate AM if the LO were tuned to a zero beat with the carrier. To operate more than earphones, the mixer output would be followed by AF amplifiers.

18-11 RF AMPLIFIERS AND S/N RATIO

The first stage in a communications superheterodyne may use a JFET, BJT, MOSFET (Fig. 18-15) or variable-μ pentode. The input circuit is tuned to the signal frequency. The output circuit is usually untuned. With proper parts placement, neutralization should not be required in the HF range. The stage is biased to class A by the source resistor, with additional bias from the automatic-volume-control (AVC) or the AGC voltage (Sec. 18-15). A sensitivity or RF gain control may also be used in this stage.

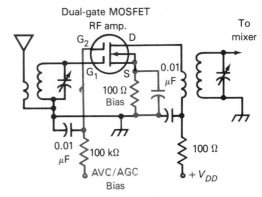

FIG. 18-15 A MOSFET RF amplifier, gain-controlled by AVC voltage.

Strong local signals can block the first RF amplifier, or can overdrive the mixer producing intermodulation distortion (IMD), or undesired beat signals at many places on the dial. Intermod can be reduced by lowering the RF amplifier gain. Bias voltage developed across the source resistor reaches the upper gate through the AVC circuit (not shown), one end of which is grounded.

Any noise generated by the RF amplifier is amplified by all stages that follow it. The RF amplifying device must be selected for minimum noise output (highest signal-to-noise ratio, S/N). *Signal-to-noise ratio* is the ratio of the signal power to noise power developed in an amplifier. If the signal power is two times the power generated by thermal action in the input circuit resistance, the S/N = 2, or 3 dB.

The *noise figure* (NF) of an amplifier is the ratio of the output S/N to the input S/N. This can be expressed as NF = $(S_o/N_o)/(S_i/N_i)$, in decibels. An NF of 6 dB indicates the S/N *power ratio* is 4 times (3 dB + 3 dB) worse at the output than at the input. This is quite good; the amplifier has only two times the noise *voltage ratio* over that present in the input circuit. The receiver has added very little noise to the signal.

The total gain of a communications receiver, known as its *dynamic range*, can be more than 120 dB, from a starting input voltage of about 0.2 μV.

The output of the RF amplifier is fed to the mixer stage.

18-12 MIXERS

The *mixer, converter,* or *first-detector* stage in a superheterodyne receiver has two input signals, RF and local oscillator, or (LO), and one output, the intermediate frequency (IF). In Fig. 18-16a a diode is shown in a *passive* (diode only) single-ended mixer circuit. (The circuit becomes a passive *product mixer* if the dashed diode D_2 is added.) Both the incoming RF signal and the LO signal are mixed in D_1, producing sum and difference IF frequencies. The IF tuned circuit is usually tuned to the difference frequency up through the HF range. Difficulties with this type of mixer are the loss of signal gain in the conversion process, and the lack of isolation of the LO ac from the RF and IF tuned circuits. LO signals can be radiated through the RF tuned circuit or pass into the IF amplifier and desensitize it.

An *active* mixer (Fig. 18-16b) has conversion gain, is isolated to some extent from the RF circuitry, and with a MOSFET has relatively low noise generation.

If RF signals fed into a mixer are too strong, the IF will no longer increase in a linear manner with increased signals, all signals in the mixer passband will sound compressed, *intermodulation distortion* (IMD) will be generated, and the receiver will become desensitized. This can also cause *cross-modulation*, where one modulated signal is heard to ride in on top of another, even though it is not on the same transmitting frequency.

A *double-balanced mixer* (DBM) (Fig. 18-16c) isolates all *ports* (RF, LO, IF) from each other if the inductances are properly center-tapped and the diodes, preferably Schottky hot-carrier-diodes, are matched. A DBM may be either passive as shown or be an active type using BJTs or FETs. These mixers reduce IMD and cross-modulation considerably. They are particularly useful at VHF and UHF.

The local oscillator may be a Hartley, Colpitts,

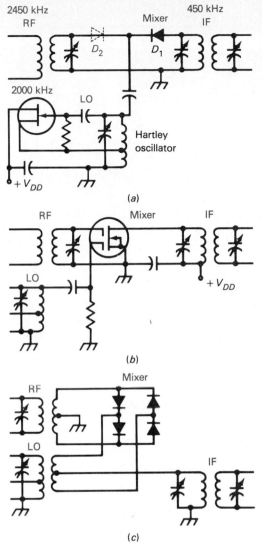

FIG. 18-16 Mixers: (a) passive, (b) active, and (c) DBM.

Armstrong, or synthesizer, or for fixed-frequency applications, a crystal oscillator.

To tune from one station to another it is necessary to track the tuning of the RF amplifier, if any, the mixer, and the LO so that they all tune together to produce the desired intermediate frequency. Consider the mixer-LO (converter) circuit shown in Fig. 18-17. If the IF is to be 450 kHz and it is desired to tune in a station on 800 kHz, the LO will be tuned 450 kHz higher, or to 1250 kHz. For tuning to 900 kHz, the mixer is tuned to 900 kHz and the LO should be gang-tuned (dashed lines) to 1350

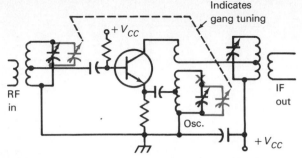

FIG. 18-17 A common BJT AM receiver mixer or converter.

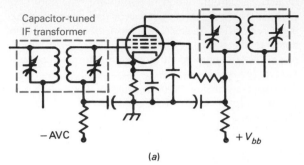

(a)

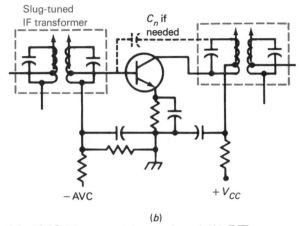

(b)

FIG. 18-18 IF stages: (a) pentode and (b) BJT.

kHz to produce the 450-kHz IF. If both mixer and LO are using similar variable capacitors, as the mixer tunes up 100 kHz it will be found that the higher-frequency LO circuit will tune over a range greater than 100 kHz and the two circuits will not track. For reduction of the capacitance variation so that the mixer tuning range and the LO tuning range do track, a series capacitor can be added at point X in the LO LC circuit. When adjusted to the proper value, the mixer and LO will now track. Another means of tracking is to construct the oscillator capacitor with less plates than the mixer tuning capacitor.

Since mixers may have high noise figures, it may be advantageous to use a tuned RF amplifier ahead of the mixer to raise the signal well above the mixer noise floor.

In older broadcast receivers, special tubes such as the pentagrid converter may be found. Two of the adjacent grids are used as the grid and anode of a triode LO. A third grid will be used as the mixer injection grid. Mixing is accomplished by both sets of grids controlling the single plate current, which feeds through the primary of an IF transformer tuned to the intermediate frequency, thereby rejecting the signal and LO frequencies.

18-13 IF AMPLIFIERS

The superheterodyne is the most practical tunable receiver chiefly because the IF amplifiers do not require tracking. They can be adjusted to the IF and remain that way, regardless of the frequency to which the *front end* (RF amplifier and converter) is tuned. Since the IF is usually a relatively low RF, transistor and pentode amplifier circuits are stable and require no neutralization. In Fig. 18-18 both input and output circuits are tuned, resulting in high gain for the amplifiers and an effective rejection of

all frequencies except those on the IF and a few kilohertz on each side. Except for the tuning of both transformers, primary and secondary, an IF amplifier is similar to an RF amplifier.

IF transformers usually consist of two multilayer coils wound on thin cylinders of insulating material. The coils used to be tuned with two adjustable mica or air capacitors, all enclosed in an aluminum shield can (Fig. 18-18a) Modern IF transformer coils have fixed mica capacitors across them. Ferrite *slugs* are screwed into or out of the cores of the coils to tune them by varying their inductance (Fig. 18-18b). By using low-loss, high-permeability core materials, the coils require only a fraction of the number of turns needed for air-core coils. This can result in higher-Q circuits, narrower bandwidth, greater output, and smaller size.

Several methods are used to produce either a narrower or a variable IF bandwidth. The coupling between primary and secondary of the transformers can be varied from less than critical coupling for a narrow bandwidth to overcoupled for a broad bandwidth. Resistances can be switched in series with the tuned circuits of the IF amplifiers to lower the

Q of the circuits and broaden the bandwidth. In general, the greater the number of tuned circuits in the IF strip, the narrower the passband. Two top-coupled transformers may be used between stages (Fig. 18-19). Tapping down the secondary raises the

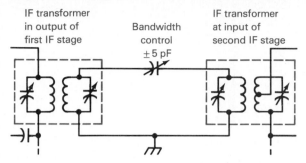

FIG. 18-19 IF selectivity can be increased by cascading loosely top-coupled and tapped-down tuned circuits.

Q, further narrowing the bandwidth. Bandpass filters of various bandwidths may also be switched into the IF strip (Chap. 25).

Shielding is often required between IF amplifier stages to reduce any interaction of electromagnetic fields from one IF transformer to another. Input and output connecting leads must be kept apart physically and be as short as possible.

Some smaller receivers use only one IF amplifier. Most communication receivers have two or more.

For control of the gain of IF or RF amplifiers, their bias may be varied, the screen grid voltage of pentodes may be varied, or a PIN diode across an RF circuit with a variable bias applied to it may be used.

When ICs are used as the IF strip, there will be terminals between which external IF transformers are connected.

18-14 IF FILTERS

Simpler superheterodynes use two stages of IF amplification with tuned coupling transformers. If each transformer has two tuned circuits, there will be six LC circuits all tuned to the same frequency. With 450-kHz IFs, six tuned circuits can provide a 10- to 15-kHz passband with reasonably steep skirts. When narrower bandwidths are required, as in communication receivers, special IF filters similar to those used in SSB transmitters (Sec. 17-34) are used.

If the filter used to couple the mixer to the first IF stage is a well-designed, steep-skirted, 3-kHz or 500-Hz bandpass crystal, or a mechanical filter, the receiver bandwidth is determined by this filter alone. All following circuits may be either broadly tuned, or even untuned, as in some IF ICs. Care must be taken to assure that there is no RF leakage around the filter, or spurious signals may appear in the second detector.

A *monolithic* quartz-type crystal filter is a specially constructed crystal with a large plate on one side and two smaller plates on the other side (Fig. 18-20). The passband of such a crystal filter may be

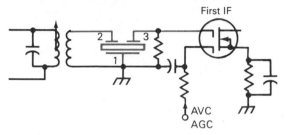

FIG. 18-20 Monolithic IF crystal filter.

about 5 kHz at 450 kHz. Any signal within the passband that is fed to plates 1 and 2 will excite the crystal into mechanical oscillation. The mechanical oscillation of the crystal develops an ac of this frequency between plates 1 and 3, which is passed on to the input of the next circuit.

18-15 AM DETECTION AND AVC

The signal output from the last IF stage is fed to a *second detector*, where it is demodulated and passes on to the AF amplifiers. Consider first a demodulator for an AM emission. Either a simple rectifier-type envelope detector or a power-type detector can be used. The circuit shown in Fig. 18-21 uses a single diode to half-wave rectify the modulated IF ac to AF varying dc, which becomes AF ac on the other side of the AF coupling capacitor. (A full-wave bridge or a voltage doubling rectification circuit may also be used.)

With no input signal there is nothing to rectify except receiver noise. Very little voltage is developed across C_2, and the resistive network R_1, R_2, R_3, and R_4 feeds back almost no bias voltage to the input circuits of the RF and IF amplifiers. These stages are therefore operating at full gain and are ready to receive and amplify signals.

If a weak AM carrier is received, the audio is detected and fed to the AF amplifier. At the same time the top plate of C_1 slowly becomes negatively

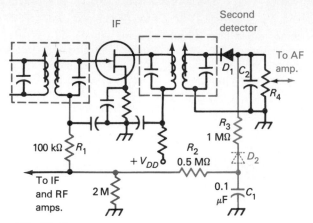

FIG. 18-21 IF, second detector, and AVC circuits.

charged by the rectified carrier ac, because electrons can be driven only to the right through diode D_1. This slight negative charge is fed as a reverse *automatic volume control* (AVC) biasing voltage to the RF and IF amplifiers, reducing their I_D values and hence their gain.

When a strong carrier is received, the negative charge on C_1 is increased, feeding a higher AVC reverse bias to the amplifiers, reducing their gains considerably. With the use of a large C_1 value and relatively high R_1, R_2, R_3, and R_4 values the reverse bias builds up across C_2 slowly. If the received signal is alternately fading up and down in strength, the AVC voltage follows the fading, keeping the detector output signal relatively constant. An AVC circuit is used in all standard broadcast band receivers. C_2 is also an IF bypass to prevent IF ac signals from passing into the AF amplifier.

Delayed AVC, or DAVC, will result if diode D_2, shown dashed, is added. Now AVC voltage cannot be fed to C_1 and the IF and RF amplifiers until the received carrier voltage exceeds the barrier voltage of diode D_2 (0.3 V for germanium diodes). Weak signals will not be attenuated by AVC voltages until they exceed the barrier level of D_2.

18-16 CW-SSB DEMODULATION AND AGC

The signal delivered to the second detector by a CW or SSB signal is not usefully detectable with a diode envelope detector. CW is heard as a series of clicks and changes in background noise. SSB is heard as an unintelligible jumble of sounds that can be made intelligible only by heterodyning the received sidebands with a frequency that is the equivalent of the suppressed carrier of the SSB emission.

This is easily accomplished by mixing a *beat-frequency oscillator* (BFO) ac with the received IF sidebands in a heterodyne, product, or synchronous detector, as shown in Fig. 18-22.

If an SSB signal is to be detected with a receiver having an IF bandwidth of 3 kHz, the BFO should be tuned to one side of the IF passband. As the front end of the receiver is tuned back and forth slowly there should be one setting of the dial where the BFO is supplying the correct frequency to heterodyne with the received SBs to produce intelligible speech output. If not, the BFO should be

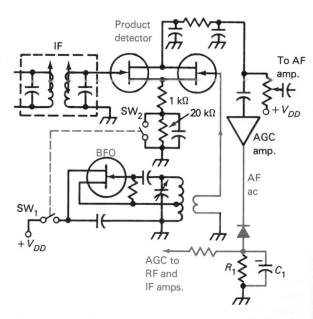

FIG. 18-22 A product or synchronous second detector and AGC for CW or SSB.

shifted to the opposite side of the IF passband, and the front end should be retuned until clear voice output is produced.

Because either upper or lower sidebands may be transmitted, the BFO must be capable of tuning at least 1.5 kHz above and below the center of the IF passband. Rather than the simple Hartley oscillator shown, the BFO may be one of the more stable Colpitts-type oscillators, with an untuned buffer between oscillator and detector to prevent strong received sidebands from affecting the BFO frequency. Many communications receivers use three switchable crystals in the BFO circuit. One crystal provides the proper carrier frequency insertion for USB signals, the second for LSB signals, and the third for CW signals. In SSB receivers the BFO may be called a *carrier oscillator*.

The *automatic gain control* (AGC) voltage developed with the product detector circuit shown amplifies the AF output from the second detector, rectifies it, and rapidly charges capacitor C_1 with a reverse bias voltage, which is then fed to the RF and IF amplifiers. The AGC voltage follows the modulation strength of demodulated signals, reducing the gain with loud signals, and increasing the gain if signals demodulate as weak, or if they fade down.

For CW reception, a preferred bandwidth of the IF filters is about 500 Hz. Since the most desirable tone for receiving codes is approximately 700 Hz, the BFO should be set at a frequency 700 Hz from the center of the IF passband. When the front end is tuned to place the CW signal in the middle of the IF passband, the BFO will beat against the received signals and produce the desired 700-Hz-tone code signal output.

Receivers that have only envelope detectors may switch on a BFO and feed its ac to the second detector to demodulate CW or SSB. Although this may produce demodulation, it is not as satisfactory for SSB as a product detector.

If a CW-SSB receiver must also be used to detect AM emissions, a separate diode-type envelope detector may be switched on in place of the product detector shown. An alternate possibility is to open the ganged switch SW_1–SW_2 shown in Fig. 18-22. This turns off the BFO and adds a power-detector bias resistor for the left-hand JFET of the product detector pair. The receiver output will now be operating with AF AGC rather than AVC. An AVC-type circuit can be switched on in place of the AGC. Finally, if set for SSB reception, a receiver will also demodulate AM signals by tuning to the zero beat of the carrier. As long as the receiver does not drift off zero beat, the AM demodulation will be good; however, only one of the sidebands will be detected.

The values of C_1 and R_1 determine how slowly the AGC will decay. For SSB and slow CW, a slow-decay AGC may be desirable. Added capacitance can be switched in to produce a slower AGC decay action.

The simple circuits discussed here are not the only ones in use. There are many different forms of product and envelope detectors, as well as AVC and AGC circuits. Special ICs may only need external *LC* circuits to form IF, detector, and RF stages. There are other special ICs for IF amplifiers, for different types of demodulators, and for various AF amplifiers, including power output stages.

Strictly speaking, a CW-SSB detector is called a *heterodyne, product,* or *synchronous detector* but is actually a form of envelope detector. In this book only AM demodulators are termed *envelope detectors*.

■ Test your understanding; answer these checkup questions.

1. Regarding superregenerative detectors, what are two advantages? _____ _____ Two disadvantages? _____ _____ In what part of the spectrum were they used? _____
2. Where is superregeneration found today? _____
3. What is meant when it is said that two circuits track? _____
4. What is the term that expresses frequency versus attenuation of an *LC* circuit _____
5. Of what does a TRF receiver consist? _____ Could AVC be used in one? _____
6. List the seven stages in a CW superheterodyne from antenna to speaker. _____
   _____ _____ _____
   _____ _____ _____
7. Why might a triode be better than a pentode as an input RF amplifier? _____ What is the formula for NF? _____
8. Which is better, an NF of 20 or 10 dB? _____
9. What are two other names for a first detector? _____
   _____ _____
10. What circuit accompanies a mixer? _____
11. What circuits are usually used in the LO? _____
    _____
12. What part of a superheterodyne is responsible for the bandpass of the receiver? _____
13. What are three types of filters used in IF strips? _____
    _____ _____ _____

14. What type of second detector is used for AM? _____ For SSB? _____ For CW? _____

15. Is AVC voltage directly or inversely proportional to carrier amplitude? _____ Is it − or + with VTs? _____ With N-channel FETs? _____ With NPN BJTs? _____

16. What is the basic difference between AVC and AGC? _____ Which would be used with AM signals? _____ CW? _____ SSB? _____

17. In DAVC, what is delayed? _____

18-17 A SQUELCH CIRCUIT

A circuit that can operate from the AVC voltage is the *squelch*, or *Q* (quieting), circuit. Tuning from one broadcast station to another with high-gain receivers may produce a loud interchannel noise. To prevent this noise, one can use a squelch circuit such as that shown in Fig. 18-23. The same AVC

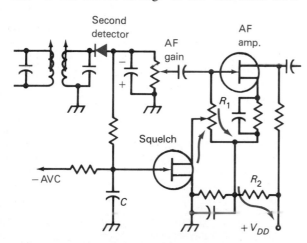

FIG. 18-23 A squelch circuit for the first AF amplifier.

voltage that is fed to the RF and IF stages is fed to a squelch gating JFET. With no negative AVC biasing voltage the source-drain resistance of the JFET is low. Current flows through the JFET, through the lower half of R_1, through R_2, and to $+V_{DD}$. The voltage-drop across the lower half of R_1 is negative at the top, which reverse-biases the AF amplifier well into class C and nonconduction, and it cannot function.

When a carrier is received, a negative AVC voltage is developed across C, which is fed to the gate of the squelch JFET. The negative AVC bias voltage biases the squelch device to class C, removing

its current flow through R_1. The AF amplifier is now free to amplify any modulation detected from the received carrier.

The position of the wiper arm on potentiometer R_1 determines the squelching point. The correct position is where a weak signal will open the squelch. When the signal drops off, the receiver will be squelched and be quiet.

18-18 S METERS

A visual indication of the signal strength of a received signal can be shown on an *S meter* or by a line of LEDs. Any RF or IF amplifier stage being fed an AVC or AGC voltage will undergo reduction of its output circuit dc current in proportion to the AVC voltage it is being fed. A dc milliammeter in the drain, collector, or plate circuit of such a stage will read maximum current with no signal input and less current as the signal increases in strength and the AVC voltage increases. A reverse-reading meter can be calibrated in *S units*. One S unit is often 6 dB, or a gain of twice the voltage. Zero S units is the value of the noise signal present in the receiver with no antenna connected and the receiver input terminated with a resistor equivalent to the RF amplifier input circuit impedance. Any noise voltage that this produces in the output of the receiver is S zero. When this voltage value is doubled by a weak input signal, the signal will be S-1. S meters are calibrated in S units up to S-9, or 54 dB above receiver noise level. With a simple RF or IF output-circuit-current S-meter system it is difficult to set the meter to an accurate zero indication because of drift of current as the active device warms.

A better S meter may first amplify the IF output and then rectify it with a separate diode *D* (Fig. 18-24). The positive dc voltage so developed can

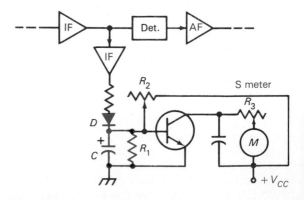

FIG. 18-24 An analog S-meter circuit.

be used to forward-bias a class-B-biased active device, such as the BJT shown. With no received signal, R_2 of the R_1–R_2 voltage divider is adjusted so that the BJT is biased to almost zero collector current flow, as indicated by the meter. An input signal 54 dB above the receiver noise floor should drive the pointer to the S-9 graduation. If not, R_3 is adjusted until the pointer is at S-9. (An accurately calibrated output RF signal generator is required for this adjustment, and V_{CC} must be well regulated.)

A string of LEDs connected along a resistive ladder, as shown in Fig. 18-25, can also be used to

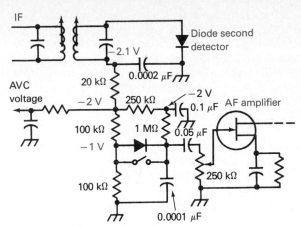

FIG. 18-26 A series-type AF noise-limiter circuit.

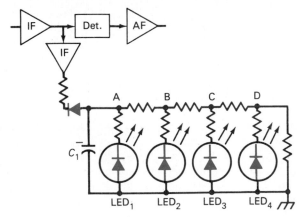

FIG. 18-25 A bar-display signal strength indicator.

indicate levels of signal strength. The IF output signal is picked off as with the S-meter circuit, is amplified, is rectified, and charges capacitor C_1. When point A reaches about -1.5 V, LED$_1$ begins to glow. When the signal is strong enough to develop 1.5 V at point B, LED$_2$ begins to glow. As the signal increases in amplitude the other LEDs progressively begin to glow. The length of the string of illuminated LEDs indicates a relative input signal value. Actual IC bar displays use operation amplifier (op amp) drivers to power the LEDs and may be calibrated in any desired steps, decibels, volts, S units, etc.

18-19 NOISE LIMITERS

Impulse noises, such as automobile ignition, lightning, static, sparking motors, switching heavy-current circuits, and power leaks, can completely cover signals that might otherwise be perfectly readable. All of these are amplitude-type signals and are detected by AM receivers. While noise impulses may be of extremely short duration, they

may have amplitudes 10 to 1000 times the strength of a signal being received.

A circuit used to clip off, or limit, such pulses is the *series noise limiter* (Fig. 18-26). It uses a diode that conducts signals to the AF amplifier if they are less than about 85% modulated. With impulses of greater amplitude the diode stops conducting and no signal is passed to the AF amplifier until the amplitude decreases to 85% again.

The varying dc audio signal of the second detector is developed across the two 100-kΩ resistors in series that form the load for the detector. If the carrier signal produces 2 V negative at the AVC takeoff point, the anode of the limiter diode is then 1 V positive (1 V less negative) in respect to this point. The cathode of the diode is connected to the -2-V point through the 250-kΩ and 1-MΩ resistors. The 0.1-μF capacitor between these two resistors holds an average -2-V charge whether the carrier is modulated or not. (Actually it is not quite -2 V because of the current flowing through the resistors.) Since its anode is more positive than its cathode, the diode conducts. The AF signals developed across the lower 100-kΩ resistor are passed through the conducting diode to the 0.05-μF coupling capacitor to the volume control in the gate circuit of the AF amplifier.

A peak of 100% modulation will produce a -4-V peak of dc at the top of the two 100-kΩ resistors. This makes the anode of the limiter diode -2 V, the same voltage as its cathode (the cathode is obtaining its charge from the 0.1-μF capacitor). The diode ceases to conduct, and no impulse signal of this amplitude or higher can be transferred to the amplifier. The 0.0001-μF capacitor bypasses any diode leakage signal.

A switch across the limiter diode stops the limiting action. This limiter decreases the possible signal-voltage amplitude from the second detector to one-half. With the limiter on, distortion is produced on all signals exceeding about 85% modulation. This limiter will not operate with the BFO on, since the BFO produces so much rectified voltage that the diode conducts all the time.

A shunt noise limiter that will limit impulses over a set voltage value can be useful for CW reception. Two zener diodes can be connected in series, in opposite polarity (Fig. 18-27a), across an

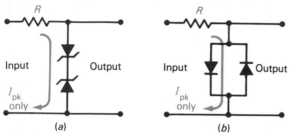

FIG. 18-27 Shunt limiters using (a) zener diodes and (b) silicon diodes.

audio amplifier input circuit. Any noise impulse above the zener voltage causes breakdown of one of the diodes, a current to flow through R with a voltage-drop across it, limiting the input voltage for the duration of the impulse. An opposite-polarity impulse breaks down the other zener. (Zener diodes conduct in the forward direction and range in voltage ratings from 2.5 V to about 200 V.)

Two silicon diodes in parallel and with polarities reversed (Fig. 18-27b) act as limiters across low-voltage circuits such as the primary or secondary of an IF transformer, or across earphones. Silicon diodes begin to conduct at about 0.6 V, germanium at about 0.3 V. When they conduct, they short-circuit

the line across which they are connected for any voltage peaks above their barrier voltage.

When copying CW signals, the supply voltage of the first AF amplifier can be reduced to limit the possible output voltage of the stage.

A noise impulse limiter in the IF strip is called a *noise blanker* (Fig. 18-28). The IF signal from the

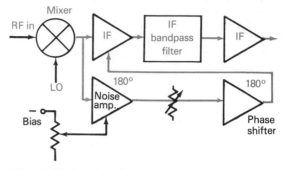

FIG. 18-28 A noise blanker system.

mixer is fed to the IF amplifier and also to another similar stage but biased to class C, called a noise amplifier. To clip all incoming pulses over some particular voltage value, the noise-amplifier threshold control is set to bias the amplifier so no signal up to that value will produce any noise-amplifier output. Pulses over that value will be amplified and reversed in phase. These pulses will be fed back into the IF amplifier 180° out of phase to cancel the pulse as it is being developed in the output of the IF stage. The blanking must be done ahead of the IF bandpass filter, since blanking produces spurious signals. These will be attenuated by the filter and will not reach the second detector.

18-20 AUDIO AMPLIFIERS

The AF amplifiers in a receiver should have a flat 20–15,000 Hz response if high-fidelity music is to be received. Communication receivers may limit the audio response to 200–3000 Hz to decrease noise. Where code or other single-tone reception is required, the AF stages may use a 300–1000 Hz bandpass filter to pass only a narrow band of frequencies. When multiplex transmissions are being received, special bandpass filters may be required.

Most receivers are constructed to operate into loudspeakers having 3- to 16-Ω impedances, or into 600-Ω audio lines. When earphones are plugged in, the loudspeaker may be automatically disconnected.

In many receivers the whole AF section is incorporated into a single IC.

18-21 WAVE TRAPS

When local off-frequency signals produce an undesired response in a receiver, either series- or parallel-resonant-circuit wave traps may be used to advantage (Fig. 18-29).

A *series-resonant* wave trap utilizes the *low* impedance of a series *LC* circuit to its resonant frequency. A series trap connected across a circuit will reduce the formation of any resonant-frequency voltage across it. Such a wave trap may be connected across the antenna to ground circuit of a receiver, from gate to ground, or from output to ground, and sometimes across the power lines where they enter the receiver. As long as the desired signal is not too close to the wave-trap frequency, the trap will not reduce the desired signal.

A *parallel-resonant* wave trap utilizes the *high* impedance developed across a parallel *LC* circuit at its resonant frequency. A parallel wave trap may be connected in series with the antenna lead, in series with an input or output circuit lead in an amplifier, or in series with one or both power lines as they enter the receiver. This introduces so much impedance to the frequency to which the trap is tuned that little energy at this frequency will flow through the trap. A trap in a circuit may detune it somewhat.

A parallel-resonant wave trap may be inductively coupled to any tuned circuit. Signals of the wave-trap frequency are induced into the trap, produce an oscillation in it, and reinduce into the tuned circuit of the receiver a voltage of the same frequency but 180° out of phase (Lenz' law). This effectively cancels or attenuates the signal to which the wave trap is tuned.

The higher the *Q* of the wave trap, the narrower the band of frequencies it will notch out.

18-22 IMAGE FREQUENCIES

Any frequency besides the desired signal that beats with the local oscillator of a superheterodyne and produces a difference frequency equal to the IF is an *image*. For example, if the IF is 450 kHz and the local oscillator is oscillating on 8000 kHz, a signal on either 8450 or 7550 kHz will beat against the oscillator and produce the IF.

The relations between the signal, the oscillator, the IF, and the image frequencies can be expressed in formula form as:

$$f_i = f_o + f_{if} \qquad f_i = f_s + 2f_{if}$$
$$f_i = f_o - f_{if} \qquad f_i = f_s - 2f_{if}$$

where f_i = image frequency
f_o = oscillator frequency
f_{if} = intermediate frequency
f_s = signal frequency

If the receiver above is tuned to a frequency of 7550 kHz, the mixer-tuned circuit usually has a low *Q* and is quite broad. Alone, it may not reject a strong signal on the image frequency of 8450 kHz. The addition of a tuned RF amplifier before the mixer with its added tuned circuit will result in a greater image rejection. A second RF amplifier will reject the image even more. However, at frequencies in the 30-MHz range even two RF amplifiers may not reject images satisfactorily.

The IF of 450 kHz is too low for image-free high-frequency reception. The image signal is only 2 times 450, or 900 kHz, from the signal frequency. By using an IF of 2000 kHz, the image will be twice 2000 kHz, or 4 MHz from the signal frequency. Even a relatively low-*Q* circuit will reject a signal 4 MHz removed.

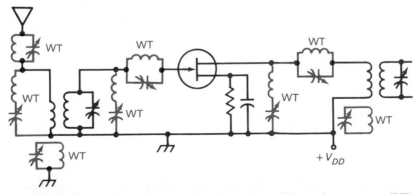

FIG. 18-29 Possible placement of series- and parallel-tuned wave traps (WT) in antenna and RF amplifier circuits.

The problem of image rejection can be solved by using *double-conversion* superheterodynes (Sec. 18-24). The first IF may be 2.5 MHz. After an IF stage at this frequency, a second mixer with a crystal-controlled oscillator converts to a second 450-kHz IF, where it is relatively simple to produce high-Q tuned circuits. Two low-frequency IF stages may produce a very narrow passband. (Narrow bandwidth plus image rejection may be obtained by using a crystal lattice filter at the 2.5-MHz frequency without further conversion.)

If the local crystal oscillator used in conjunction with the second conversion stage is not adequately shielded and isolated, harmonics of this oscillation may produce images or signals on higher-frequency bands. This is somewhat similar to the way in which many broadcast receivers with unshielded oscillators hear higher-frequency amateur or other stations apparently in the broadcast band. Actually, the stations are beating against the second or third harmonics of the local oscillator in the receiver, producing image signals in the improperly shielded set.

18-23 A BROADCAST RECEIVER

The circuit of a possible discrete-component broadcast band AM receiver is shown in Fig. 18-30. It has no BFO, limiter, S meter, squelch, tone control, and so on. The circuit shows only one IF stage, although two would be preferable. Small broadcast receivers seldom have the RF amplifier. A high-Q loopstick L_1 is used in place of a tuned loop antenna. Because of its high Q it "draws in"

signals fairly well from the surrounding space. Mobile and communications receivers use an external antenna and an RF amplifier.

The loopstick tuned by C_1 feeds the RF signal to the gate-source circuit of Q_1. Source-resistor biasing is used. The drain circuit is tuned and acts as the tuned input for the base-emitter circuit of Q_2. The converter BJT acts as both a mixer and an Armstrong oscillator through the use of the tuned circuit in the emitter and the tickler coil in series with part of the IF transformer primary.

Tuning circuits L_1C_1 and L_2C_2 must tune the broadcast band frequencies from 535 to 1605 kHz. At the same time L_3C_3 must tune to frequencies 450 kHz higher, from 985 to 2055 kHz, to assure that any signal being received is converted to the 450-kHz intermediate frequency. Variable air capacitors C_1, C_2, and C_3 are ganged by being mounted on a common rotary shaft.

The primaries of the IF strip transformers are tapped down to better match the relatively low impedance of the collector circuits of the BJTs. This would not be done with JFETs, MOSFETs, or VTs. The secondary of the first IF transformer is tapped down to match the low impedance of the Q_3 base.

The IF amplifier is forward-biased by the AVC voltage-divider network (100-, 5-, and 10-kΩ resistors) to $+V_{CC}$, charging the 10-μF capacitor positive and forward-biasing Q_3 for full gain. When a signal is present, the diode rectifies it and a downward current is developed in the potentiometer. This lowers the positive charge on the 10-μF capacitor and on the IF amplifier base. The gain of the IF amplifier is reduced by the decrease of the forward

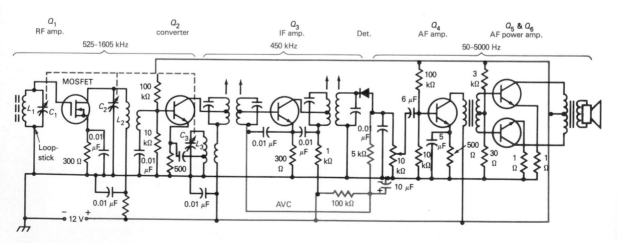

FIG. 18-30 A transistor superheterodyne receiver using a MOSFET RF amplifier and BJTs.

biasing AVC current. To apply AVC to the RF amplifier, a JFET- or VT-type AVC circuit would have to be added.

The first AF amplifier Q_4 is a class A stage driving the class B push-pull power-amplifier output stage Q_5 and Q_6. If the 30-Ω resistor were increased in value, the push-pull stage would be biased to class AB or A.

Most modern receivers use one or more IC chips. There are ICs that contain all the active devices for an RF/IF system, needing only external input- and output-tuned circuits and a few incidental components to be connected to them. Others may be complete IF, second-detector, and AF amplifiers, except for the external-tuned circuits and components. With ICs the possibilities are almost endless.

Test your understanding; answer these checkup questions.

1. How could a TRF receiver be designed to receive SSB signals adequately? _____
2. If a BFO is tunable, over what range of frequencies should it operate? _____
3. What is the BFO in SSB receivers called? _____
4. What does a squelch circuit eliminate? _____ In what part of a receiver is it normally found? _____
5. What would be the bandwidth of an AF filter for AM music? _____ Voice? _____ SSB? _____ CW? _____
6. For what is an S meter used? _____ Will it vary if the receiver is tuned to a local station

transmitting AM? _____ SSB? _____ CW? _____
7. How many dB gain represents one S unit? _____
8. What are circuits that clip off high AF noise peaks called? _____ That clip IF pulses? _____
9. What is used for an antenna in small BC band receivers? _____
10. Will signals be nulled or peaked if a series wave trap is across a circuit? _____ Parallel WT shunts a circuit? _____ Series WT is in series with a circuit? _____ Parallel WT is inductively coupled to a circuit? _____
11. A receiver with an IF of 450 kHz is tuned to 1.5 MHz. What is the probable image frequency? _____ Why might it receive a local signal that is on 3.450 or 4.350 MHz? _____
12. What are two ways in which image response is reduced? _____ _____
13. Does narrowing the IF bandpass reduce images? _____
14. In Fig. 18-30, what type of FET is used? _____ Why use an FET rather than a BJT? _____ What would have to be changed if P-Channel and PNP transistors were used? _____

18-24 DOUBLE-CONVERSION SUPERHETERODYNES

Most of the more sophisticated HF communication (nonbroadcast) superheterodynes use double (or triple) conversion to reduce image response. A block diagram of a discrete parts (no ICs) double-conversion superheterodyne is shown in Fig. 18-31.

For control of the amplitude of the signal into the

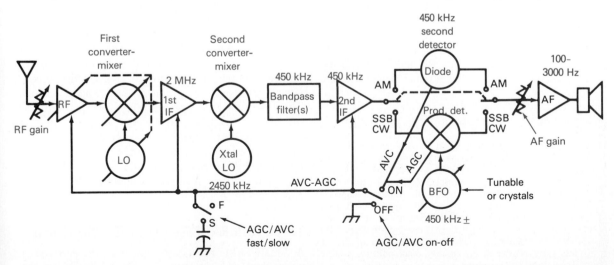

FIG. 18-31 Block diagram of a double-conversion superheterodyne capable of receiving AM, SSB, CW, and FSK.

RF amplifier, a variable or stepped resistive attenuator is used. It may be designed to introduce 20 dB or more of signal attenuation to prevent strong signals from overdriving the front end. AVC or AGC is usually fed to this stage.

The first converter consists of a tuned mixer and a tunable local oscillator to convert the received RF signal into the first intermediate frequency, in this case 2 MHz. The mixer input and the LO are always ganged. The RF stage may be ganged as indicated by the dashed lines, or it may use an RF peaking or tuning control, in which case the RF amplifier may be termed a *preselector*. Many modern receivers use a switchable frequency synthesizer for this local oscillator and a counter to read its frequency and convert that to the received frequency 2 MHz lower.

The 2-MHz IF stage produces amplification of the converted signal but also reduces the possibility of receiving image signals, which would be 4 MHz away from the input tuned circuits in this case. (Some receivers couple directly from the first converter to the second without any amplification at this point.)

The second converter consists of a fixed-tuned mixer at 2-MHz input and 450 kHz at its output, and a Pierce-type crystal LO. It *down-converts* the 2-MHz IF signal to a 450-kHz IF. The 450-kHz signal is passed through a crystal, mechanical, or *LC* filter having the desired bandpass, 6 kHz bandwidth for AM, 2.7 kHz for SSB, and 500 Hz for CW.

The filtered signals pass to the 450-kHz IF amplifier section where they are amplified by two or more broadbanded amplifier stages and are fed to either an AM or a SSB-CW "second detector" (actually the third detector) stage. With the switch in the AM position the diode detector develops AVC which is fed back to the RF and IF amplifiers. In the SSB-CW position the product detector mixes

IF signals with either a tunable, or a switched-crystal BFO to demodulate these emissions. Incorporated in the product detector stage may be an audio-operated AGC-voltage generator that can be used in place of the AVC voltage generated by the AM detector.

The output of either second detector is AF, which passes through a volume control, possibly through an AF peak limiter, and possibly through a 0–3 kHz low-pass AF filter, or a 500–1000 Hz bandpass AF filter for CW. The AF signal is amplified and fed to a loudspeaker or earphones. If a noise blanker is used for noise limiting, it would be added between the second mixer and the IF bandpass filter(s).

The first IF must be a frequency that is not included in the tuning range of the receiver; otherwise the IF section will break into oscillation when the receiver is tuned near that frequency.

In transistorized receivers almost any of the stages might use BJTs, JFETs, or MOSFETs, with BJTs or VMOSs as the final AF stage.

In VT receivers the RF and IF tubes would be variable-μ pentodes for optimum AVC action. Converters might use separate diode, triode, or pentode circuits or special mixer-oscillator "converter" tubes. Low-level AF tubes would be triodes or pentodes. The final AF amplifier is usually a beam-power tetrode.

Transistorized receivers use 9- to 13-V batteries or well-regulated dc power supplies. VT receivers use 100- to 250-V dc power supplies with a 6-V filament winding on the power transformers (unless the filaments are connected in a series string with a resistor in series with them across the 120-V ac line).

When frequency-shift-keyed signals (FSK) are received, two different beat tones will be produced by any CW-type detectors. These must be fed to two narrow-bandpass filters, one to produce the "mark" signal and one to produce the "space" signal. These signals then operate relays or electronic on-off circuits to key teleprinters (RTTYs) or computers. Audio-frequency-shift-keying (AFSK) keys mark and space tones into an AM (or FM) transmitter. The two tones can be demodulated with an envelope detector receiver to operate teleprinters or computers.

18-25 OPERATING SUPERHETERODYNES

A small AM broadcast-band superheterodyne is quite simple to operate. It has an on-off switch, often on the same shaft as the volume control, a tone control, and a tuning knob. It will tune only the *standard broadcast band*, ranging from 535 to 1605 kHz, and demodulate only AM. The tuning knob is rotated to the desired station, the volume is adjusted to the desired level, and the tone control may be varied.

Receivers that monitor a single frequency, such as police, fire, taxicab, and other such services, are crystal-controlled, having an on-off switch, volume control, and a squelch on-off switch with an accompanying noise-threshold adjustment. Operation is as simple as possible.

Communications receivers are much more complex. They may have a power on-off switch, a band-selector switch and a tuning dial, or if synthesized, a series of LO frequency setting switches or a dial-operated divide-by circuit; an RF gain control; an AF gain control; a switch to select the IF filter passband; a switch to select the AF filter passband; a BFO on-off switch; either a knob to tune the BFO, or a switch to select one of three crystal BFO frequencies (for USB, LSB, or CW); an AVC or AGC on-off switch; a switch for fast or slow AVC; an antenna or first RF amplifier trimmer (tuner) control; an S meter; an IF notch filter (a wave trap that tunes across the IF passband); a noise blanker or limiter on-off switch; or a standby on-off switch, used when a local transmitter is in operation, but which leaves the receiver oscillators on continuously to prevent drift.

As the band-selector switch is turned, different-sized coils are simultaneously connected across the RF amplifier, mixer, and local-oscillator tuning capacitors, allowing the receiver to tune over different portions of the radio spectrum. *Low-frequency* receivers will tune from about 15 to 550 kHz in possibly four bands. *All-wave* receivers usually tune from 530 kHz to 30 or 40 MHz in four to seven bands, depending on how wide a frequency coverage is desired on each band. In some receivers, band changing is accomplished by shorting out part of each tuning coil simultaneously. This is not considered particularly good engineering, but it is better than merely tapping down the coils to change frequency, because the unused upper portions may fall into parallel resonance at some frequency and act as a wave trap at that frequency. *Shortwave* receivers tune from about 1.5 to 30 or 55 MHz in four to eight bands.

For broadcast signals, if relatively high fidelity is desired, the bandwidth is set to 10–15 kHz, the AVC is turned on, the RF gain control is turned up full, the audio volume control is advanced to the desired sound level, the tone control is set to what pleases the listener, the noise limiters are normally off, and the BFO is off.

For AM voice signals on the HF bands where sidebands of up to only 3000 Hz are used, the bandwidth should be set to about 6 kHz; AVC is on; RF gain is on full; tone control is set to reduce the high-frequency audio response; the noise limiter is on or off as required; the BFO is off. Loudness is controlled with the AF volume control. If interference appears on one sideband of the desired signal, it is possible to reduce the bandwidth or set the crystal filter to a bandpass of 3 kHz. The receiver is tuned to pick up the carrier and the set of sidebands that are not being interfered with. This is single-sideband-with-carrier reception and is possible on an AM transmission with some loss of signal strength.

To tune in a SSB voice signal, the bandwidth is adjusted to 2.5 or 3 kHz. The AVC is switched off (although long-time constant AGC could be left on, and volume would be controlled with the AF volume control). The AF volume control is set to a midvalue, the AF filters are set for 3 kHz, and the noise limiter and CW filter are turned off. The RF gain control may be used to control the volume of the received signal. The BFO is set to a frequency 1.5 kHz higher or lower than the IF center frequency. The receiver is tuned very slowly and carefully until the voice is readable. If the voice cannot be demodulated, the BFO is switched to the other side of the IF center frequency and the receiver is again tuned. It will be noted that a 200-Hz variation in the setting of the vernier tuning control or of the BFO can make the received signal sound unnatural; 500 Hz off, and the signal is difficult to understand. For this reason receivers used for SSB reception must have good BFO and LO frequency stability.

To tune in CW signals the bandwidth control may be adjusted to a narrow setting. The AVC or AGC is preferably turned off. The audio volume is set to mid-value. The BFO is turned on and adjusted to a frequency about 700 Hz from the center of the IF channel. An AF filter can be adjusted to remove high audio frequencies. An AF noise limiter is turned off, but a noise blanker might be turned on if desired. A 500-Hz AF BP filter may be used if interference is bothersome. The RF gain control is used to control the volume of the received signal. If adjacent frequency signals interfere, the IF filter switch can be turned to a narrower setting.

Frequency-shift keying (Sec. 16-12) requires the

BFO on. Since the carrier is on all the time, the receiver hears none of the background noises that are present when MAB (make and break) keying is used. FSK is used with Morse code and RTTY (radio teleprinter), using either Baudot or ASCII (Fig. 12-24) codes.

When trying to copy CW through interference the RF gain should be as low as possible, to prevent overloading of tuned circuits. The narrower the bandwidth of the IF strip, the less noise present in the output.

To relieve listening fatigue due to continued listening to a single tone while copying code signals, the operator can tune either the BFO or the main tuning dial slightly to change the beat tone of the receiver signal, preferably about half an octave.

18-26 DIVERSITY RECEPTION

To overcome the effects of fading, the AVC circuit was developed. While partially effective, signals can still fade in and then out completely. It has been found that two antennas a few hundred feet apart may have two entirely different amplitude signals in them from the same distant station at any particular instant. It is possible to connect two spaced antennas to separate front ends and IF strips (Fig. 18-32) and feed them to a common second detector, AVC, and AF system. The front end with the stronger signal develops the most AVC, which biases the other one off. Signals that might have been fading severely on a single antenna and receiver may have little signal-strength variation using

such *space diversity* reception. Antennas close together will give a certain amount of diversity reception, but the farther apart the better. To assure that both front ends are tuned to the same frequency, a common local oscillator is employed, as shown.

If a station is transmitting the same code or voice information on two different carrier frequencies, the fading of one frequency will not be the same as the other at a remote receiver. If the receiver has two separate front ends and local oscillators to tune to the two transmitted frequencies, but a common IF, detector, AVC, and AF system, *frequency diversity* is produced. This can also result in a great improvement over single-frequency reception.

18-27 TRANSCEIVERS

A transceiver is a combination transmitter-receiver in one enclosure. Although both units might be tuned separately and be operated on different frequencies, the two sections are usually made to operate on the same frequency at the same time. Commercial and CB transceivers are usually multiple-channel systems with frequencies controlled by crystal or synthesized oscillators. Amateur-band transceivers may use a VFO to allow any frequency in a given band to be used. If the receiver is tuned to 4000 kHz, the transmitter will transmit on that frequency also. The circuitry to produce a system that does this is not simple. There are a variety of methods by which this may be accomplished. A possible SSB system is shown in block form in Fig. 18-33.

The system in heavy lines is a double-conversion superheterodyne receiver. For operation in the 3.5–4 MHz amateur band the incoming signal is amplified and mixed with a 10-MHz crystal oscillator which converts it to some frequency between 6 and 6.5 MHz. This signal is amplified by a broadband IF amplifier and fed to the second receiver mixer. Tuning is accomplished by varying the MASTER VFO feeding the second mixer. The output of this mixer is 1.65 MHz and is fed to the 1.65-MHz IF amplifier. A single 3-kHz crystal filter (1.650–1.653 MHz) can be used for both upper- and lower-sideband reception by selecting one of two crystals in the carrier oscillator.

When transmitting, some of the receiver circuits are used. Starting at the microphone, the voice signals are amplified and then mixed with the desired carrier crystal frequency in a balanced modulator. The output of this stage is two sidebands but no carrier. Both sets of sidebands are amplified by

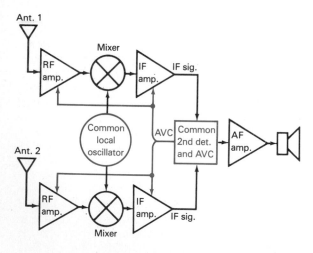

FIG. 18-32 A diversity-reception receiving system.

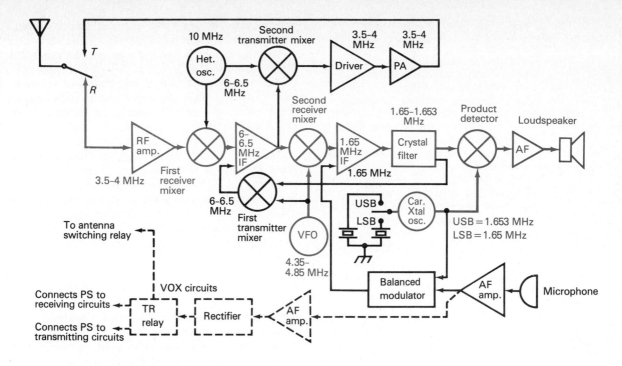

FIG. 18-33 Block diagram of an SSB transceiver.

the 1.65-MHz IF amplifier, but only one will pass through the crystal filter. The sideband that is passed by the filter is fed to the first transmitting mixer, where it is then heterodyned against the 4.35–4.85 MHz MASTER VFO. When the resulting 6–6.5 and 2.7–3.2 MHz signals are fed to the 6–6.5 MHz IF amplifier, only the 6–6.5 MHz SSB signals are amplified and fed to the second transmitting mixer. This signal is mixed with the 10-MHz heterodyne crystal oscillator and is translated to 3.5–4 MHz. The output circuit of the second transmitting mixer is tunable from 3.5–4 MHz, as are the driver amplifier and the power amplifier. The transmitter output is coupled to the antenna by the relay that transfers the antenna from the receive to the transmit position when the equipment is switched to transmit.

The change from receive to transmit may be accomplished by a manual switch that shifts the antenna from receive to transmit and also connects the power supply from receiver circuitry to transmitter circuitry. It may also be accomplished by a voice-operated transmitting (VOX) system. AF picked up by the microphone is amplified, rectified, and made to throw a TR relay that switches the antenna and power-supply circuits. When voice sig-

nals cease, the relay falls back to the receive position and the operator can hear incoming signals. To prevent received sounds from tripping the VOX circuit, an antitrip circuit (not shown) rectifies some of the received AF and uses this dc to buck the VOX dc, thus preventing the VOX TR relay from changing over. Since local voice sounds do not emanate from the receiver loudspeaker, they are not bucked out and the VOX circuits operate.

The frequency selection is accomplished by varying the MASTER VFO, but the receiver RF amplifier, the second transmitter mixer, and the driver are gang-tuned to allow trimming them to frequency for optimum operation. The PA has its own output tuning and coupling controls.

Test your understanding; answer these checkup questions.

1. What services would use crystal LOs? _____
2. What services would use variable LOs? _____
3. What services might use narrowband (200 Hz) AF filters? _____
4. What does a band switch do? _____
5. With what type of emission does mistuning by 200 Hz materially degrade the received signal? _____

6. Why does FSK not require an FM receiver?
_____
7. What system is used to give even better protection against fading than is provided by AVC? _____
8. Would a separate transmitter and receiver in one enclosure be considered a transceiver? _____
9. In the transceiver system described, how many crystal oscillators are required to tune a 500-kHz band? _____ Which stages would probably be trimmed to frequency? _____
10. In the transceiver, how is upper- or lower-sideband operation selected? _____ What stages would be fed AGC? _____ Would an anti-trip circuit be required if earphones were used? _____ What must be the MASTER VFO frequency range to produce 3.5–4 MHz? _____
11. What does VOX mean? _____ What might MOX mean? _____ What is a TR switch or relay? _____
12. In the transceiver, to change to another band, what six things would have to be changed?
_____ _____ _____
_____ _____ _____
13. What does "translated" mean when referring to SBs? _____
14. What are the common circuits in the space diversity receiver shown? _____ What would frequency diversity be? _____ Could a single antenna be used for frequency diversity? _____ For good space diversity? _____

18-28 ALIGNING SUPERHETERODYNES

Tuning the RF, mixer, oscillator, IF, and detector stages of a superheterodyne to their correct frequencies is known as *aligning* the receiver. Fortunately, superheterodynes rarely need realignment. However, if active devices are replaced or repairs are made to circuits, realignment may be necessary. Sometimes a do-it-yourselfer will tighten all the screws in sight, including the RF and IF trimmer capacitors. In this case no signals will be audible, even if the receiver circuits are electrically perfect.

To align a superheterodyne, the technician should have at least a tone-modulated RF signal generator capable of tuning to the IF frequency and across that part of the radio spectrum the receiver is to tune, a small screwdriver made of insulating material, and two 0.0001-μF fixed capacitors. A VTVM or a 20,000-Ω/V meter can be used to measure the AVC voltage during alignment. An ac voltmeter or oscilloscope across the loudspeaker will give a visual indication of signal strength.

The basic procedure is to work from the second-detector input circuit back to the RF input circuits, checking each stage in turn.

To align a small broadcast receiver, the procedure might be as follows:

A tone-modulated RF signal generator is set to the IF, usually 450–465 kHz, and is connected to the input of the last IF stage and to ground (or negative terminal of the power supply) using 0.0001-μF capacitors in series with the signal-generator leads. The detector stage should demodulate this signal and a tone should be heard in the loudspeaker. A VTVM between the AVC line and power supply negative will read the AVC voltage developed across the diode detector load. The RF signal generator should be adjusted to an output as weak as possible and still obtain audible or visual indications of tuning.

The IF transformer between the last IF amplifier and the detector is tuned for maximum audible indication, using the screwdriver to turn the adjusting screws of the IF transformer (insulated hex-wrench for ferrite cores).

The signal generator is moved to the mixer stage's input (or input of the IF stage ahead). A wire is connected across the LO tuning capacitor to prevent it from oscillating. The IF transformer between the mixer and IF stages (or between any two IF stages) is tuned for maximum audible indication. This completes the alignment of the IF section.

The alignment of the mixer and oscillator stages requires tracking them. The various capacitors involved are shown and labeled in Fig. 18-34. After the shorting wire has been removed from the oscillator capacitor, the receiver dial is set to some high-frequency point, such as 1400 kHz. The signal generator is set to the same frequency and coupled to the external antenna terminal. If the signal generator is not heard, the oscillator *trimmer* capacitor is varied until it is. Then the input trimmer capacitor of the mixer is tuned until an audible peak is heard. The receiver is next tuned to a calibrated point at the low-frequency end of the dial, such as 600 kHz, and the signal generator is set to 600 kHz. If the signal generator cannot be heard, the oscillator is not tracking with the dial markings. At the low-frequency end of the dial the oscillator *tracking* capacitor is adjusted until the signal is heard. Then the receiver and signal generator are returned to the 1400-kHz settings, and the oscillator *trimmer* capacitor is readjusted for maximum indication. High- and low-frequency points are rechecked until the signals track with the dial readings (adjust trimmer

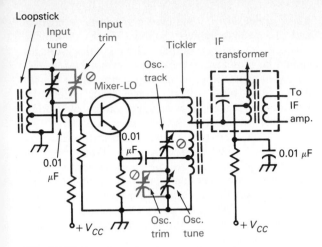

FIG. 18-34 Broadcast-band converter system.

at the high-frequency point, tracking at the low-frequency point). The receiver is now aligned to the dial markings.

In many small BC receivers there is no way to track the mixer and LO calibrated to follow the tuning characteristics of the input circuit. If the LO tuning capacitor has slotted outer plates, it is possible to bend out one or more slotted sections at one end or the other, reducing the capacitance at either the high- or the low-frequency end as desired, to produce tracking.

In communications receivers there are trimmer capacitors across each coil plus a variable ferrite slug changing the inductance of each coil. (Brass slugs lower inductance, acting as shorted turns; ferrite cores increase inductance.) To track circuits having slugs, adjust the *inductance* at the *low-frequency end* of the band and the trimmer *capacitance* at the *high-frequency end*. With receivers having RF amplifiers after the mixer and LO track, the signal is finally fed into the antenna and the RF amplifier is tracked to frequency, using the same method as with the oscillator and mixer. The basic steps are: Track oscillator; track mixer input; track RF amplifier(s).

While the audible-output method is satisfactory for small broadcast receivers, an unmodulated signal generator is required for communication receivers, which are always aligned in their minimum-bandwidth position. With such receivers, the sidebands of a modulated signal can give confusing indications. A peaking of AVC voltage on a VTVM or a peaking of the S meter are possible indications when aligning.

When a tunable single-crystal filter is the input circuit of the IF section, it should be adjusted for minimum bandwidth and the signal generator fed into the mixer stage and varied until a maximum visual indication is obtained. The signal generator is now on the crystal-filter frequency. All IF amplifiers are then trimmed to this frequency.

The aligning of VT superheterodynes is the same as with transistorized systems.

18-29 TROUBLESHOOTING RECEIVERS

One of the requirements of most technical jobs is servicing equipment when it malfunctions. Many books have been written on the subject. Only a few basic ideas can be included here. First, what indicates that the equipment is not operating properly? If it is smoke, the equipment must be turned off immediately. If signals are weak, distorted, or nonexistent, the required steps may vary considerably.

Regardless of the number of jokes made on the subject, in an emergency one of the first steps taken when a receiver suddenly stops working is to deliver a sharp blow with the heel of the clenched fist to each side of the receiver, to the top, and to the front panel. In a remarkable number of cases the receiver will start operating again, but it is not fixed. As soon as possible the equipment should be checked thoroughly. Something may be loose; there may be an oxidized pin connecter somewhere; a screw or solder connection may have loosened. Each transistor, IC, or tube should be worked back and forth gently with the receiver on, to see if the cause of the intermittent operation can be localized. Tapping parts will often help to localize trouble. Receivers usually have parts mounted on printed-circuit (PC) boards, sometimes with transistors in sockets. The usual difficulty with PC boards is a hairline crack or poor solder joint in one of the interpart leads. A magnifying glass may help to locate the trouble.

Fuses should be checked, as should line-cord plug and connections if there is no heating of filaments or dial lights. A fuse often blows out when a receiver is turned on. An old fuse may burn out for no apparent reason. In the case of a blown fuse, replace the fuse and turn on the equipment. If there is something wrong the fuse will burn out again.

Failure of a tube is more often the reason for improper operation of a VT receiver. See if all filaments are lit. A dark filament usually means a burned-out tube. A tube with a burned-out filament will feel cold. Power transistors also should feel warm. If possible, test each tube, noting the physical condition of the tube pins and the socket and surrounding chassis. Clean out any dust. If a tube or transistor tester is not available, substitute each tube or socket-type transistor singly with another, similar device, taking care to replace all devices to their original sockets if not proved faulty. Mixing even the same types of devices in a complex receiver may cause improper operation because of slight variations between supposedly similar devices.

If the steps above have been taken and the equipment still does not operate, some part may have burned out, opened, or shorted. The receiver should be removed from its case and eyeballed. Careful scrutiny may show a burned resistor, signs of excessive heat, bulging fixed capacitors, or loose wires.

In a surprising number of cases, the trouble in a receiver can be localized by using nothing more complicated than a VOM. The voltmeter is used if the receiver is turned on and the power supply is operating. The ohmmeter is used only when the receiver is turned off and the ac line cord is pulled out of the service outlet. An ohmmeter touched across loudspeaker terminals should produce a click. Shorted transistors can often be located with an ohmmeter; they read the *same* low resistance when test probes are reversed across them.

When the power supply is on, its voltage, or a major proportion of it, should be measurable at all collector, drain, plate, and screen-grid points. If a transistor or tube is drawing current, a small fraction of the power-supply voltage should appear across all emitter, source, and cathode resistors.

Many transistorized systems have both a positive and a negative power supply. Each should be checked for proper output voltage. Bias voltages should be indicated at most base terminals, but may not be at gates or grids in amplifiers. There should be class C bias voltage shown with oscillators. Unexpected potentials at an input circuit may originate from leaking coupling capacitors from the stage ahead. Resistors, coils, and transformers may either open-circuit or short-circuit. When they are shorted there will be no voltage-drop across them, but if they are open there may be (proved by an ohmmeter reading). Bypass capacitors may short and have no voltage-drop across them; if open the voltage-drop may be normal, but the circuit may malfunction (oscillate).

If the proper voltages are not known for IC terminals, it is difficult to determine whether they are operating correctly. It may be necessary to replace the IC to find out.

If all measured voltages seem within reason and still no signals can be heard, an audio signal generator can be connected across each audio amplifier in turn, starting at the final amplifier and working back to the second detector. If this does not localize trouble to be in an AF stage, a tone-modulated RF signal generator can be connected to the IF amplifiers and RF circuits, following a procedure of circuit checking similar to alignment, until a stage is found producing little or no output when fed an input signal. If a signal tracer (Sec. 13-39) is available, with a modulated RF signal fed to the RF input, each stage in turn can be checked for modulated RF, IF, or for AF output.

Distortion in the audio amplifiers can be checked with an oscilloscope or even with earphones in series with a $0.005\text{-}\mu\text{F}$ capacitor. When the stage in which the distortion is produced has been reached, it will be visually or audibly evident. It is then necessary to determine which part is faulty.

Sometimes the AVC line grounds. Overdriving of the IF stages occurs, distorting all but the weakest signals. If the detector diode load resistor opens, all signals may be weak and distorted. Volume-control potentiometers often become faulty (sound scratchy), and may open (very distorted AF).

If the power supply has no output voltage, the power switch should be left on and the power-line plug pulled out. An ohmmeter test across the plug should give a reading of 3 to 10 Ω. If a very high resistance reading is indicated, something in the primary circuit (fuse, line cord, or the primary winding) is open.

Secondaries of VT power transformers should show continuity with resistance values of 25 to 350 Ω (less than 5 Ω in transistor supplies). The positive-to-ground circuit should read several thousand ohms, or whatever the bleeder resistance is. If it reads nearly zero, something in the power supply or receiver may be shorted. The output lead of the power supply can be unsoldered from the receiver circuitry and the resistance of the power supply checked. If it shows a relatively high resistance, the receiver circuits must show the low-resistance reading. The section that reads low resistance has the shorted capacitor or part in it. It will be necessary to continue disconnecting circuits from the dc line until the low-resistance circuit is found. Always reverse ohmmeter probes when testing for possible shorts in transistor circuits (semiconductors often act as diodes).

Noisy operation of a receiver can be caused by many things. Sometimes it arises from poor connections and vibration of the equipment. It can be due to intermittent breaking down of the dielectric of coupling or bypass capacitors, or to faulty resistors, transistors, or tubes. If the equipment is battery-operated, noise can be produced by polarization of old, worn-out dry-cell batteries.

In most transistorized equipment the transistors are soldered into printed-circuit boards (Sec. 1-17), which complicates servicing. First, it is difficult to loosen the transistors. Second, if the soldering iron is left on the wire lead of a transistor, heat travels up the wire and deforms the junction to which it is connected and the transistor is destroyed. To help prevent this, the transistor wire lead should be gripped, with a copper heat-sink clip or clamp, between the soldered joint and the transistor so that heat traveling up the wire will be transferred to the clip rather than to the transistor. Sockets are used in some equipment into which the transistors are plugged. Such transistors can be removed easily and tested.

Care must be taken that an electrostatic difference of potential does not build up between transistors in a circuit and the soldering iron. Any difference of potential can be reduced by touching the iron tip to a grounded point on the equipment just before soldering to another point in the circuit. Never solder equipment while it is operating.

Some soldering guns have ac flowing in a hairpin-shaped tip. The magnetic field from this hairpin can induce voltages into low-impedance circuits that can burn out transistors.

Printed-circuit boards are fairly delicate. The printed patterns will pull loose from the backing if they are overheated. Faulty parts can be carefully unsoldered and new units placed in the holes in the board. Often it is quicker and better to clip off a faulty part, leaving ¼ in. of its two leads sticking out from the board. Then solder the replacement part to these two leads.

When printed patterns crack apart, they can be repaired by coating them with solder, although it may be advisable to lay a fine copper wire across the opening and solder it across the joint.

Transistors soldered into PC boards must be unsoldered $1/16$ in. of lead at a time, moving from one lead to the next, using heat-sink clamps and preferably a flow of air across the transistor. Special desoldering irons remove solder from a connection by suction and should be used on PC boards. When the new transistor is replaced, leave a lead length of about ¼ in. above the board connection. Hold leads with a heat-sink clip while soldering them in place.

Power transistors may have their cases bolted to a heat-sink metal, but they often require insulation between case and heat sink. A thin piece of mica coated with special silicon grease to improve heat conduction through it may be used. An ohmmeter check should always be made to make sure the case and heat sink are insulated from each other.

ICs soldered to PC boards can be removed with a desoldering tool. Each pin is freed of solder separately, using minimum tool-to-pin time. Do not hurry. Prevent the IC from heating.

18-30 EMERGENCY REPAIRS

Sometimes there are no spare parts for a receiver that burns out. The operator's problem is to put the receiver into working condition, possibly with a minimum of test equipment.

If the first RF amplifier stage becomes faulty, it is possible to capacitively couple the antenna directly to the next RF amplifier's input or to the mixer input by wrapping an insulated wire around any exposed wiring in the base, gate, or grid circuit. This forms a *gimmick*, which has a few picofarads of capacitance to the wire around which it is connected. It is sometimes possible to remove the faulty transistor or tube and couple the antenna through a 50- to 100-pF capacitor into the plate or collector hole in the socket. The primary of the output RF transformer then acts as an antenna coil. It is still connected to a dc potential, however, and must be treated with caution.

If an RF or IF amplifier transistor or tube becomes inoperative and there are no replacements, the input and output points of this stage can be connected together with a gimmick or small capacitor and signals may be heard.

If the second detector fails, audible signals may be heard by connecting earphones between the emitter, source, or cathode and the AVC end of the second IF amplifier transformer secondary.

If the local oscillator ceases to oscillate, earphones can be connected in series with the mixer output circuit or in series with the last RF stage. These circuits will act as detectors, and strong modulated signals can be demodulated.

If one of the audio amplifiers becomes faulty, it may be possible to jump the signal over this stage with a capacitor of any value between 0.001 and 0.1 μF, or earphones can be connected ahead of the faulty stage.

If a VOM or other test equipment is not available, a continuity checker can be rigged by connecting a 1½-V battery in series with a pair of earphones. When leads from these are connected and disconnected across a coil or the primary or secondary of a transformer, for example, clicks will be heard. If the winding is open, little or no click will be audible. Capacitors of more than 0.002-μF capacitance can also be tested. A capacitor such as a 0.1-μF or larger may click loudly on the first contact, charge to the battery voltage, and produce almost no click if the connection is made to it again immediately. If it continues to click, it is possibly leaking or shorted. The higher the resistance tested with this continuity checker, the lower the amplitude of the click. Tubes can be tested for filament continuity or for shorts between elements this way.

If the earphones have 2-kΩ resistance or more (or have 2 kΩ in series with them), emitter-base and base-collector sections of BJTs can be checked. A click will be heard one way, but not if the test leads are reversed. With JFETs S–D clicks both ways, but G–S should click only one way. With MOSFETs S–D clicks both ways, but *do not check G–S*!

A pair of earphones in series with a 0.0001- to 0.01-μF capacitor can be used to test whether AF signals are present in different stages. When connected from S, E, or k to the AVC connection of an RF or IF transformer, modulated signals should be audible if the receiver is operating up to that point. From the second detector on, the audio signal can be traced to each input and output circuit of the different audio stages, indicating where a loss of signal occurs.

▌Test your understanding; answer these checkup questions.

1. What sections of a superheterodyne are involved in a complete alignment? _____
2. What tools and equipment should be used when aligning? _____
3. What stage in a superheterodyne is aligned first? _____ Last? _____
4. Why is a tone generator best for broadcast receivers when aligning? _____ Why may an unmodulated RF signal generator be better when aligning a communication receiver? _____
5. When aligning the first IF transformer, what circuit should be stopped from operating? _____
6. When aligning the front end, is L or C varied at the high-frequency end of the dial? _____ At the low-frequency end? _____
7. When aligning using AVC, do you tune for a dip or a peak of the voltage? _____ With an S meter? _____
8. What is the first technique to use when servicing an inoperative receiver at the bench? _____ The second? _____ The third? _____ The fourth? _____
9. What polarity and approximate voltage should be read on a VTVM between chassis (ground or B−) and a VT plate? _____ Grid? _____ Cathode? _____ Screen grid? _____ AVC bus? _____
10. What might be approximate VTVM readings between a PNP BJT emitter and its: Base? _____ Ground? _____ Collector? _____
11. What is the best indication of oscillation of an oscillator using a VTVM? _____ How else can it be determined? _____ _____
12. Why are 100-W soldering irons never used on transistorized circuits? _____
13. Why is silicon grease used on power BJT insulation sheeting? _____

AMPLITUDE-MODULATION RECEIVER QUESTIONS

1. Why does an 810-kHz AM signal across earphones produce no audible output?
2. Why might Schottky diodes be preferable to common germanium or silicon diodes for detectors?
3. Draw a diagram of a shunt-rectifier-type AM detector with one stage of AF amplification. How could an AF volume control be added?
4. How can a crystal (diode) detector be made to oper-

ate with narrow bandwidth?

5. Where is the RF ac rectified in a power detector?
6. Does a diode produce linear or square-law output with strong input signals? With weak input signals? Which would be least distorted?
7. Why is a grid-leak or a JFET gate-leak detector actually a power detector?
8. A triode (or JFET) regenerative detector out of oscillation acts like what other detector? What type detector is it when in oscillation?
9. When is a regenerative detector most sensitive to AM signals? To CW signals?
10. How can undesired superregenerative effects in oscillators or amplifiers be cured?
11. What are five advantages of a TRF over a regenerative detector?
12. Why are TRF receivers not used at HF and higher?
13. Diagram a 3-stage TRF using a regenerative detector. What are disadvantages of using a diode detector?
14. Basically, what would be added to a TRF to produce a superheterodyne?
15. To what is the IF always equal in a superheterodyne?
16. Could the LO be below the mixer frequency in a superheterodyne?
17. What is added to a BC superheterodyne to make it a communications receiver? What would be the tuning range of such an added circuit?
18. What is the first detector of a superheterodyne?
19. What is a direct-conversion receiver? What might it be used to detect?
20. Block diagram a simple superheterodyne system.
21. Why must RF amplifier devices be chosen for minimum input noise content?
22. What is the formula for computation of noise figure?
23. What is an adequate dynamic range for a communications superheterodyne? What minimum input signal should it detect?
24. Draw a diagram of a dual-gate MOSFET RF amplifier stage. What controls the gain of this stage?
25. What is the advantage of an active mixer over a passive one? Of using a double-balanced mixer?
26. Draw diagrams of (a) a passive mixer stage, (b) an active mixer stage, and (c) a DBM.
27. What types of local oscillators might be used for tunable superheterodynes? For fixed-frequency superheterodynes?
28. In Fig. 18-17, what oscillator circuit is used in the LO?
29. What is considered to be the "front end" of a superheterodyne?
30. If the LO is at a lower frequency than the mixer, would the LO variable-capacitor plates have to be smaller or larger than those of the mixer?
31. What are two functions of the IF strip?
32. Draw a diagram of a BJT IF amplifier stage. Of a JFET IF stage.
33. Describe four methods of varying bandwidth of IF strips.

34. How can the gain of IF stages be controlled?
35. What are three types of IF filters used? Where in the system are they placed?
36. Draw a diagram of a monolithic crystal filter feeding a dual-gate MOSFET IF amplifier device.
37. What is the meaning of AVC? How does it accomplish its function?
38. What type of detecting circuit is usually used in the second detector of an AM superheterodyne?
39. What is input to and output from the second detector of a superheterodyne?
40. Draw a diagram of a simple AM second detector with AVC. How could DAVC be produced?
41. How is fading counteracted in AM receivers?
42. What is delayed in delayed AVC?
43. What is the meaning of BFO? For what types of emissions is it required? What type of circuits might be used in it?
44. How do AVC and AGC circuits differ?
45. With a 3-kHz bandpass IF system, where is the BFO set for SSB reception? For CW?
46. Draw a diagram of a SSB-CW second detector. How can it demodulate AM?
47. Draw a diagram of a squelch circuit between a second detector and the first AF amplifier. What is its function?
48. Diagram two types of signal strength indicators.
49. Draw a diagram of a series-type noise limiter between a second detector and an AF amplifier. With what kind of signals is it useful?
50. Draw a block diagram of a noise blanker circuit. Why is the noise amplifier biased to class C?
51. What AF bandpass would AM broadcast receivers use? What would AM communications receivers use? SSB receivers? CW receivers?
52. What are the two types of wave traps? Which is connected in series with a line?
53. A receiver has a 2-MHz IF and is tuned to receive an 8.4-MHz signal. What are the possible image frequencies?
54. Draw a diagram of a complete broadcast band AM receiver using all semiconductors.
55. Draw a block diagram of a double-conversion communications superheterodyne. What determines whether AVC or AGC is used?
56. When a band-selector switch is turned, what happens in the receiver?
57. Block diagram a diversity reception system. What is the advantage gained by using such a system?
58. Draw a block diagram of a 3.5–4 MHz SSB transceiver. What should be added to make this a good CW transceiver?
59. Outline the common method of aligning a superheterodyne.
60. Diagram a BJT mixer–LO input stage of a broadcast band receiver and list steps to track its circuits.
61. If a receiver begins to malfunction, what should be done to it first? Second? Third?

19

Frequency Modulation

This chapter presents the fundamentals of frequency and phase modulation as applied to transmitting and receiving systems. While both wideband and narrowband systems are described, additional narrowband circuits are discussed in Chapter 21, "Two-Way Communications," and wideband FM broadcasting is examined in Chapter 23, "Broadcast Stations."

19-1 FREQUENCY MODULATION

Since its inception there has been a constant search for better methods of utilizing radio as a means of communication. By 1930 the superheterodyne had been developed, as had high-powered CW and AM transmitters. Since that time communication equipment has become less bulky and better insofar as sensitivity and selectivity are concerned, but difficulty with man-made and natural static still exists, particularly at frequencies lower than about 10 MHz. Radio broadcasters are usually interested in a primary coverage of only a few miles, so the standard broadcast band of 535–1605 kHz with its 1- to 50-kW stations is satisfactory. However, even in the primary coverage area a lightning storm can make reception of a 50-kW station unpleasant. In order to prevent noises from interfering with reception, it is necessary to make the receiver unresponsive to amplitude variations.

A sinusoidal ac can be changed in only three ways: (1) amplitude, (2) frequency, and (3) phase. The study thus far has been mostly about transmitters and receivers that produce and demodulate amplitude changes. The few short discussions on frequency-shift keying (FSK) indicated that it is possible to transmit intelligence by changing the carrier frequency without changing the carrier amplitude. Frequency modulation (FM) is similar to FSK in that the carrier is made to swing back and forth in frequency, although at an audio rate rather than at a code rate. If the carrier sweeps back and forth 1000 times per second, the carrier is being frequency-modulated at a 1000-Hz rate. When the audio voltage that produces the FM of the carrier is no longer present, the carrier comes to rest at a *center frequency* and remains there until another modulating voltage is applied.

An FM receiver is similar in many respects to an AM receiver. It is a superheterodyne but has a special second detector that demodulates frequency changes instead of amplitude variations. The IF stages are often 10.7-MHz amplifiers, but the last two may also act as amplitude limiters. If the FM detector is made insensitive to amplitude variations, static crashes and other impulse signals are not demodulated and are not audible.

19-2 THE FIVE FIELDS OF FM

There are five major fields in which FM is in use. One is the FM broadcast band, from 88 to 108 MHz, with one-hundred 200-kHz-wide channels, in which wideband FM transmitters broadcast programs to listeners.

A second use of FM is in television. The video signals are amplitude-modulated, but the sound is transmitted by a separate transmitter which is wideband frequency-modulated. Thus, a TV receiver is both an AM and an FM receiver at the same time.

A third use of wideband FM is in the transmission of TV signals, both video and AF, to and from satellites.

A fourth use of FM is in the mobile or emergency services (taxicabs, police, fire, etc.), which transmit voice frequencies only, using narrowband FM.

A fifth use of FM is in the amateur bands. Here again, voice frequencies are transmitted, mostly by narrowband methods.

19-3 BASIC FM CONCEPTS

In AM, the louder the sound striking the microphone, the greater the variation of the strength of the carrier and the stronger the signal developed in the detector stage of the receiver. In FM, the louder the sound, the greater the variation of the carrier frequency from the assigned or center frequency and the stronger the signal developed in the FM detector stage.

If a carrier is made to deviate 75 kHz on each side of a center, or resting, frequency (a total swing of 150 kHz), a signal will be developed at the detector of a receiver that is well above other normal noises that might also be received. This is the maximum carrier excursion set by the FCC for FM broadcast stations, and it is considered *100% modulation*. Actually, any excursion in frequency could have been selected as 100%. If the carrier is 50% modulated, it deviates 0.5 × 75, or 37.5 kHz. Deviation in FM is directly proportional to modulating voltage amplitude.

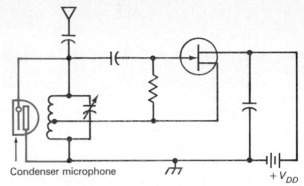

Condenser microphone $+V_{DD}$

FIG. 19-1 A possible method of producing FM.

A simple system by which FM could be developed is a Hartley oscillator with a condenser microphone across the tuned circuit (Fig. 19-1). When sounds strike the diaphragm, it swings in and out, changing the capacitance across the microphone, which changes the frequency of the oscillator. The *amplitude* of the RF ac generated by the oscillator will not change, however.

The louder the sounds striking the microphone, the greater its capacitance change and the farther the carrier swings from its center frequency. If the tone of the sound is 500 Hz, the microphone diaphragm vibrates 500 times per second and the frequency of the oscillator swings higher and lower 500 times per second. If the *same-strength* sound with a frequency of 1000 Hz, strikes the microphone, the diaphragm will vibrate twice as rapidly, producing a carrier that increases and decreases frequency 1000 times per second. If the strength of the tone is the same, the deviation of the carrier from the center frequency is the same. If the sound striking the microphone is weaker, the deviation (and "percent" of modulation) will be less.

In AM, modulation of the carrier produces sidebands. Any single tone used to modulate the carrier produces two sidebands, one on each side of the carrier frequency (Sec. 17-13).

In FM, swinging a carrier from one frequency to another means that no RF ac cycle can be a pure sine wave, since each succeeding cycle is at a slightly higher or slightly lower *frequency* than the one preceding it. This results in sidebands, but instead of only one pair of sidebands for any one tone (as in AM), the number of significantly strong sidebands produced by FM will depend on how far the carrier is swung. The greater the swing, the greater the number of sidebands. Theoretically, an infinite number of sidebands are produced by FM, but only a few may be strong enough to be significant.

With wideband broadcast FM the highest AF that is transmitted is 15 kHz. The widest deviation allowed by the FCC is 75 kHz. The ratio of deviation to audio frequency developing it is known as the *modulation index* (or *deviation ratio* when only the maximums are used), or

$$\text{Modulation index} = \frac{f_{\text{dev}}}{f_{\text{af}}}$$

$$\text{and Deviation ratio} = \frac{f_{\text{dev(max)}}}{f_{\text{af(max)}}}$$

With 75-kHz deviation and 15-kHz audio, the modulation index and deviation ratio is 75/15, or 5. With this modulation index, by Bessel's functions, the number of significant pairs of sidebands that would be developed would be 8 (Fig. 19-2). The furthest significant sideband from the carrier frequency would be 8 × 15, or 120 kHz, producing a total bandwidth of 240 kHz. However, it is unlikely that any tone of this frequency would be broadcast at a 100% modulation level. (The highest fundamental musical tone on a piano or organ is less than 5 kHz.)

A listing of the modulation index values and the number of significant sidebands is shown in Table 19-1.

Narrowband FM is used for VF (200–3000 Hz) communications and is usually limited to 20-kHz bandwidth for commercial services (less for some amateur applications). If a modulation index of 1 is used only three sideband sets would be developed, and the bandwidth for a 3-kHz speech tone would be 2 × 3 × 3000, or 18 kHz. From the modulation index formula, if the modulation index is 1 and the

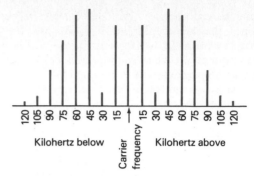

FIG. 19-2 Significant sidebands developed by a 15-kHz tone swinging the carrier 75 kHz above and below the center frequency.

AF is 3 kHz, the deviation of the carrier would be $f_{\text{dev}} = \text{index} (f_{\text{af}})$, or 1(3), or 3 kHz. So, while the total bandwidth might be 18 kHz, the actual carrier deviation would be only plus and minus 3 kHz, or 6 kHz of carrier swing. The FM sidebands extend out past the point where the carrier deviates for all modulation indexes over 0.4.

If the modulation index is held at 0.4 or below, only the first set of sidebands is developed, and with a 3-kHz modulating tone the bandwidth will be 6 kHz, just the same as with amplitude modulation.

The amount of power radiated from a properly tuned FM transmitter is constant, regardless of the percentage of modulation. If the carrier is radiating 1000 W with no modulation, when modulated 100% the sum of all of the sideband powers plus the carrier power will still be equal to 1000 W. Actually, at a modulation index of about 2.4, the carrier power drops to zero and *all of the power* is in the sidebands. At a higher modulation index the carrier again appears.

A midfrequency for voice transmissions is about 1000 Hz. With wideband FM the modulation index for frequencies in this range would be about 75/1, or 75. With narrowband FM the modulation index would be about 3000/1000, or 3, and there would be six sets of sidebands developed, producing a 12-kHz bandwidth emission.

Basically, the bandwidth of any FM transmission will be directly propotional to not the modulation frequency as in AM, but to the amplitude of the modulating AF (modulation percentage).

Since only the frequency of the oscillator is modulated in FM, only a fraction of a watt of audio power is required to produce 100% modulation. Compare this with the 500 W of audio power re-

TABLE 19-1	MODULATION INDEX AND SIDEBANDS
Modulation index	Significant sidebands
0.1–0.4	1
0.5–0.9	2
1.0	3
2.0	4
3.0	6
4.0	7
5.0	8
10.0	14
15.0	20

quired to amplitude-modulate 1000 W of dc plate-power input to an RF amplifier. In AM, the RF sidebands are added to the carrier power. In FM, the carrier power is used to form the sidebands. As a result, the antenna current of an FM transmitter should not change when the carrier is modulated. However, if the tuned circuits of the RF amplifiers of the transmitter or of the antenna have too high a *Q*, their bandpass may be too narrow, the outermost sidebands may be attenuated, and the antenna current may drop off at high percentages of modulation. This may also occur if the stages are not properly tuned.

Some common types of FM emissions are:

- F1A. A carrier shifted or keyed in frequency (FSK) according to Morse code, to be received aurally (previously F1; see Appendix E).
- F1B. A carrier shifted or keyed in frequency to be received mechanically or electronically (previously F1).
- F2A. A carrier shifted or keyed in frequency according to Morse code, tone-modulated, to be received aurally (previously F2).
- F3E. Frequency-modulated carrier, analog-voice or music-modulated (previously F3).
- F3C. Frequency-modulated carrier, modulated with still-picture elements (facsimile, previously F4).
- F3F. Frequency-modulated carrier, modulated by moving picture elements (television, previously F5, used in satellite TV).

When a number precedes the first letter it indicates the bandwidth of the emission: 304HF1B indicates a 304-Hz-wide Baudot or ASCII-type 100-baud, 170-Hz FSK emission to be received mechanically or electronically.

19-4 SLOPE AND PLL FM DETECTION

There are a variety of circuits that will demodulate FM signals. If an AM receiver is detuned to one side or the other of an FM carrier, it will detect any FM on the carrier. This is called *slope detection,* but does not produce acceptable FM detection. True FM detectors are the stagger-tuned and Foster-Seeley *discriminators,* and the ratio, gated-beam and phase-locked loop (PLL) *detectors.*

The basic PLL circuit was described in Sec. 11-27 as it applied to the synthesizing of stable radio frequencies. A block diagram of a PLL circuit adapted to demodulate a frequency-modulated carrier is shown in Fig. 19-3. Assume the center fre-

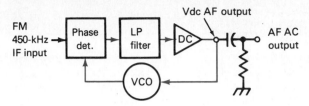

FIG. 19-3 A PLL-type FM detector for narrow-band signals.

quency of the IF passband of an FM receiver is 450 kHz. The frequency-modulated IF signal takes the place of the reference oscillator. The voltage-controlled oscillator, or VCO, should have a free-running frequency close to 450 kHz so that the *lock-in range* is equal on both sides of the frequency and to reduce *lock-in time.* With a 450-kHz carrier arriving at the phase detector, a value of dc error signal will be developed by the detector, which when passed through the low-pass filter and the dc amplifier will lock the VCO frequency to the carrier frequency. If the carrier frequency changes to 451 kHz, the PLL system shifts the VCO to this new frequency by developing the necessary amount of error dc at the phase detector to bring the VCO into a locked condition at the new frequency. If the modulation of the frequency of the carrier alternates up and down at a 1000-Hz rate, then the error dc fed to the VCO will vary at a 1000-Hz rate, which represents the output AF signal from the demodulator. The *RC* network converts the varying dc to an AF ac. In practical applications a single IC containing a phase detector, filter, dc amplifier, and VCO circuit, with a few added external parts, produces the PLL detector system.

19-5 A STAGGER-TUNED DISCRIMINATOR

An FM detector that operates on a theory of bucking voltages is shown in Fig. 19-4*a*. The IF output signal of a superheterodyne receiver, if centered on 10.7 MHz, is fed equally to two *LC* circuits, one tuned 200 kHz above the carrier frequency (10.9 MHz) and the other 200 kHz below (10.5 MHz). The 10.7-MHz voltages induced in both *LC* circuits and rectified by D_1 and D_2 are equal. The current flow develops equal bucking voltages across the equal-value load resistors. Therefore, the output voltage V_o is zero.

If the carrier deviates 50 kHz higher in frequency, the 10.9-MHz *LC* circuit has greater current induced into it and E_{R_1} becomes greater than V_{R_2}. From the current directions shown, this results

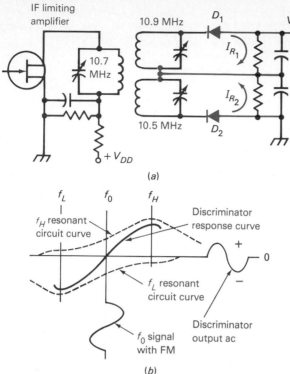

(a)

(b)

FIG. 19-4 (a) Stagger-tuned or balanced discriminator. (b) Resultant characteristic discriminator S curve.

in a negative V_o. If the carrrier deviates 50 kHz low, V_{R_2} exceeds V_{R_1} and V_o becomes positive. In this way frequency variations are converted to negative-positive voltage alternations. FM is converted to AF ac by this *discriminator* (FM detector).

The heavy-line curve in Fig. 19-4b illustrates a discriminator S curve. As long as the FM of the carrier f_0 is beneath the straight portion of the curve, the output AF ac will be a replica of the FM. The dashed lines are the 10.9-MHz (f_H) and the 10.5-MHz (f_L) LC circuit resonant curves, dropping off at the extremities due to the resonance effect of the primary. The primary Q should be half that of the secondaries. If the deviation of a received signal exceeds the limits of the straight portion of the S curve, the detected signal will sound distorted.

Since variations in *strength* of the FM carrier will also produce amplitude variations in V_o, all signals fed to any discriminator should first be limited to a constant amplitude.

The negative- or positive-going output from a discriminator can be used as a detector for FSK signals. A negative-going output signal can forward-bias a PNP transistor into conduction, and the posi-

tive-going output signal can forward-bias an NPN transistor to actuate relays or circuits to key a teleprinter, for example.

▌ Test your understanding; answer these checkup questions.

1. Why was FM first developed? _____
2. To what feature of the modulating tone is FM deviation proportional? _____
3. What are the five basic fields of FM?
   _____ _____ _____
   _____ _____

4. Under what condition would a 1-kHz AF signal produce a single pair of FM sidebands? _____

5. What deviation is considered 100% for FM BC stations? _____ What receiver bandwidth is required to receive this? _____
6. In FM broadcasting, what is the highest required modulating frequency? _____ Lowest? _____

7. What is the ratio of the greatest allowable deviation to the highest modulating frequency called? _____

8. What is the ratio of maximum deviation allowed to the modulating frequency being used called? _____

9. How much AF power is required to plate-modulate a 10-kW PA of an AM transmitter? _____ To modulate a 10-kW FM station? _____
10. What is the letter-number designation for a voice or music FM? _____ FM facsimile? _____ FSK? _____ A keyed FM-tone carrier? _____
11. What is the disadvantage of slope detection of FM? _____

12. What are the two functions of the dc that comes out of the dc amplifier in a PLL-type FM detector? _____ _____

13. Why should discriminators be tuned for a straight characteristic S curve? _____ What would be the advantage of a steep curve? _____
14. Would high-Q coils be more desirable in wide- or narrowband stagger-tuned-type discriminator transformers? _____ Why? _____

19-6 FOSTER-SEELEY DISCRIMINATORS

A well-known FM detector is the Foster-Seeley discriminator, also used in electronics whenever a change in frequency must produce a negative or positive controlling voltage. In Fig. 19-5a it is being used as the second detector of a superheterodyne receiver.

It will be assumed that L_1C_1 and $L_2L_3C_2$ are

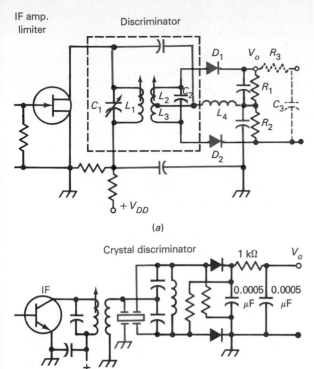

(a)

Crystal discriminator

(b)

FIG. 19-5 (a) Foster-Seeley discriminator. (b) Crystal discriminator.

ations and must be fed by signals that have been limited in amplitude to some relatively low but constant value (Sec. 19-9).

Figure 19-5b is the circuit of a crystal-type discriminator which is similar to the Foster-Seeley in operation but uses a special four-contact IF crystal as the tuned circuits and intercoupling.

19-7 THE RATIO DETECTOR

The FM detector in Fig. 19-6 demodulates FM signals and suppresses amplitude noise impulses without limiter stages ahead of it. In this *ratio detector,* the tuned circuit, the diodes, and resistor

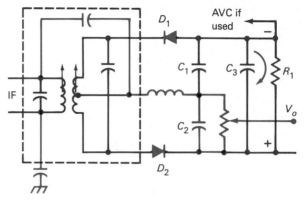

FIG. 19-6 A ratio detector.

tuned to resonate at the center of the IF passband, and that a carrier is being received on that frequency. Since the secondary is center-tapped, equal voltages are developed and rectified by D_1 and D_2, resulting in equal voltage-drops across R_1 and R_2, and zero voltage output V_o. The color lines indicate that L_4 is in parallel with $L_1 C_1$. The voltage-drop across L_4 will act as a reference voltage. When the received carrier shifts in frequency, the L_2 voltage shifts in phase opposite from that of L_3. When added to the reference voltage, these two phase-shifted voltages produce different rectified current values, different R_1 and R_2 voltage-drops, and V_o is no longer zero. When the carrier shifts in the opposite direction V_o reverses in polarity. The greater the deviation of the FM the greater the V_o output. Thus frequency modulation at an AF rate is changed to an V_o ac output that follows the modulating frequency in both amplitude and frequency.

The $R_3 C_3$ low-pass network does two things: it acts as an IF filtering circuit to prevent IF from passing on to the AF stages, and it acts as a de-emphasis circuit (Sec. 19-10).

All discriminators are sensitive to amplitude vari-

R_1 are all in series. (Note that the top diode is reversed from the Foster-Seeley circuit.) Voltages built up across C_1 and C_2 are in series aiding instead of opposing as in a discriminator.

The current through R_1 produces a voltage-drop across it proportional to the average carrier strength being received. The capacitor C_3 has sufficient capacitance (10 μF) to hold the voltage across R_1 constant even if there are instantaneous amplitude variations. The voltage across C_1 plus the voltage across C_2 must always equal the voltage across C_3, from which the circuit derives its name. (Without C_3 the circuit is an AM detector.)

When the received signal is on the center frequency of the tuned secondary, the dc voltages across C_1 and C_2 are equal, as in the Foster-Seeley circuit. If the carrier is deviated lower in frequency, the vector-sum voltage across C_1 will increase and that across C_2 will decrease. When the carrier deviates higher in frequency, C_2 voltage will increase. Thus, frequency modulation at an audio rate produces a varying dc audio signal cross C_2 (or C_1).

The top of C_3 is negative with respect to ground. The stronger the input signal, the more negative it becomes. The voltage may be used as AVC and fed to the grids of RF or IF stages.

19-8 QUADRATURE DETECTORS

One of the circuits incorporated in ICs made for FM receivers is the *quadrature detector* (Fig. 19-7). In this circuit the IF signals feeding the bases of Q_1 and Q_2 are 90° out of phase (in "quadrature")

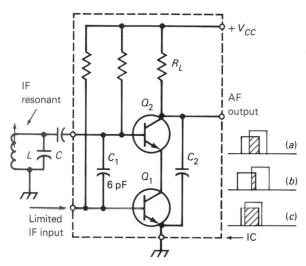

FIG. 19-7 An IC-type quadrature detector.

because of the externally connected high-Q resonant circuit tuned to the carrier frequency of the limited signal coming down the IF strip. The limited (square-wave) IF signal, which will be modulated in frequency, is fed to the base of Q_1, and gates this BJT on and off at the modulated carrier rate. The 6-pF coupling capacitor loosely couples the resonant circuit to the base of Q_2, gating this BJT on and off at the unmodulated carrier rate (the high-Q LC circuit holds its frequency constant, but at a phase 90° from the unmodulated carrier signal). The series-BJTs collector current with no FM is

being gated on and off by two similar frequency ac signals that are 90° out of phase. Because of the 90° difference in gating times, the current through R_L will flow only about half the overlapped gating times, forming a medium-width I_C pulse, as shown in Fig. 19-7a. If the received carrier deviates one way, the gating times will differ more, and the result may be a thinner I_C pulse, as in Fig. 19-7b. If the carrier deviates the other way the gating times differ less and the result is a wider I_C pulse. Since the width of the pulses is varying at the modulation rate, the output energy amplitude varies similarly. When lightly filtered by C_2, the varying dc produced is the AF output signal from the detector.

A 6BN6 tube was developed to limit any IF input signals fed to its first grid. A resonant IF tank connected to the third grid, develops oscillations 90° out of phase with the input IF signal. The resultant I_p pulses vary in width as in the transistorized circuit version, developing the AF output signal. Whereas the tube version requires no IF limiting, the BJT circuit input signal must be limited.

19-9 IF LIMITERS

Except for PLL and ratio detectors, FM detectors will demodulate amplitude variations and noise. It is necessary to use one or, preferably, two *limiter stages* ahead of the detector. These must limit the amplitude of the carrier being received to a constant value, regardless of the strength of the signal. They are actually low-gain IF amplifiers using either input- or output-circuit limiting or both (Fig. 19-8). The limiter shown uses impedance coupling. Transformer and resistance coupling may be used. Re-

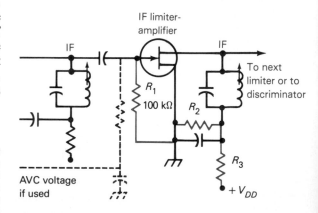

FIG. 19-8 An IF limiter amplifier.

sistor R_1 produces increased bias if the signal increases, decreasing the I_D which tends to hold the output signal to a constant value. The voltage-divider network of R_2 and R_3 lowers the drain voltage to such a value that the JFET has a very limited signal-voltage output. It is preferable to have two limiter stages, since a single stage of this type cannot limit a wide variation of signal-amplitude change. When signals are very weak, the limiters act as low-gain amplifiers without limiting and the discriminator will have noise in its output.

Solid-state diodes or zener diodes acting as shunt-type limiters can be connected across the input circuits of IF stages to make them limit.

While FM receivers using limiter stages may not require AVC, it may be used to prevent front-end overloading by strong local signals. An AVC voltage can be taken from the gate of the *first* limiter stage through a long-time-constant *RC* network, shown dashed in Fig. 19-8.

19-10 PRE-EMPHASIS AND DE-EMPHASIS

Unwanted amplitude variations in electronic circuits are classified as noise. Random electron motion in tubes, transistors, diodes, wires, resistors, or capacitors produces some noise. The warmer the components the greater the noise produced. In the low AF range thermal-noise content is low, but at higher frequencies noise increases. Such noise degrades signals coming through any amplifier. To overcome this partially, pre-emphasis and de-emphasis are used. The FM broadcast transmitter amplifies or emphasizes the higher audio-frequency signals more

than it does the lower. A *pre-emphasis* circuit consists of a series capacitor and a shunt resistor with a time constant of 75 μs ($T = RC$, or $RC = 75 \times 10^{-6}$) in some audio circuit of an FM transmitter system (Fig. 19-18). At 1000 Hz the pre-emphasis is about 1 dB, at 5 kHz about 8 dB, and at 15 kHz about 17 dB. Therefore, FM receivers must have a similar 75-μs time-constant *de-emphasis* circuit (a series R and shunt C, shown dashed in Fig. 19-5) following any discriminator or other FM detector.

19-11 FM RECEIVERS WITH SQUELCH

FM broadcast receivers may be constructed as in Fig. 19-9. These receivers tune 88–108 MHz and have an IF bandwidth of 220 kHz or more.

Solid-state receivers may use discrete components or may use one IC for the RF amplifier, local oscillator and mixer, and one IC for IF, detector, and AF amplifiers, greatly simplifying construction. But inside the ICs there must be circuits which function similarly to those described.

Although pentagrid converters or triode-hexode converters may be used as the mixer and local oscillator, in VT VHF receivers it is preferable to use a separate oscillator to reduce frequency pulling when the mixer is tuned during alignment. The local oscillator usually operates on a lower frequency than the mixer to take advantage of better frequency stability at lower frequencies.

If the receiver in Fig. 19-9 uses *automatic frequency control* (AFC) the dashed circuit will be added to tune the LO to keep the received signal in the center of the IF passband. If the signal happens

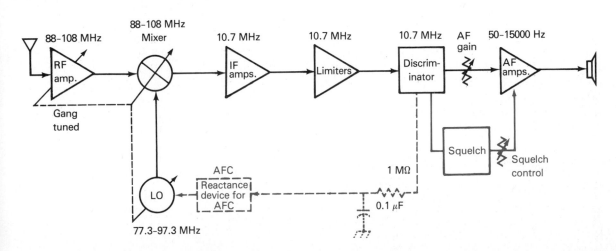

FIG. 19-9 An FM broadcast receiver with squelch and AFC.

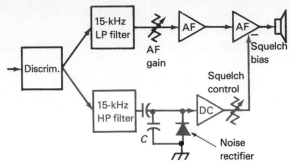

FIG. 19-10 Block diagram of a squelch circuit for broadcast FM.

to be tuned high in the discriminator passband, the average detected output may be slightly positive, and the *RC* circuit will charge *C* positive at the top. The positive voltage applied to the reactance modulator (Sec. 19-13) shifts the frequency of the LO so that the beat frequency will cause the received signal to shift down in the discriminator passband. The result is a constant centering of received signals in the discriminator and undistorted detection. If the receiver drifts the AFC compensates for the drift, within limits of course.

FM broadcast receivers usually have a *carrier squelch* circuit. When there is no signal being received the noise output from the discriminator is very high, producing an undesirable roaring noise from the loudspeaker. The noise signal is made up mostly of higher-frequency components. The output of the discriminator is fed through the AF stages and is also fed to a 15-kHz high-pass filter (Fig.

19-10). Noise frequencies above 15 kHz are rectified and filtered to a dc. The dc voltage is amplified by a dc amplifier and used to bias off one of the AF amplifier stages. When a received carrier is tuned in, it quiets the discriminator noise. There is no longer any higher-than-15-kHz noise to rectify, resulting in the loss of the dc squelching bias and the AF amplifier is free to amplify audio signals detected by the discriminator.

Communication FM receivers are only interested in demodulating VF signals in the 200–3000 Hz range. Figure 19-11 is a block diagram of a fixed-frequency double-conversion communications FM receiver for use in the 162-MHz band. Both of its local oscillators use crystals for fixed-frequency operation. Since no signal frequencies higher than 3 kHz are required, the squelch high-pass filter can have a 4-kHz roll-off frequency. The squelch control is adjusted with no carrier present to the point of just no noise output from the speaker. Note that there is no tuning in this receiver. (All tuned circuits may be screwdriver trimmed, however.) To shift to another received frequency the crystal in the first LO must be changed.

Figure 19-12 is another squelch system. A limiter feeds a discriminator, which feeds its output to an AF amplifier. First, consider the circuit with no carrier. High-amplitude receiver-developed AM noise is present at the output of the limiter and also from the discriminator. D_1C_1 together act as an AM detector, producing a positive voltage on C_1. AF to 3 kHz represents a long-time-constant signal, whereas noise with frequency components above

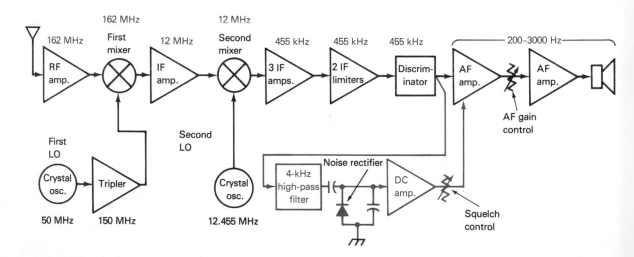

FIG. 19-11 Block diagram of a VHF narrowband FM communication receiver with squelch.

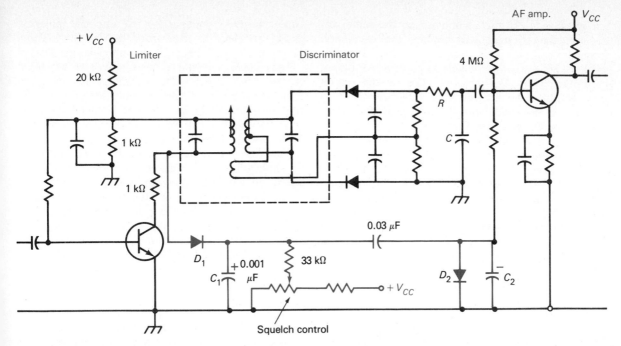

FIG. 19-12 FM limiter, discriminator, squelch, and AF amplifier.

3 kHz develops a high-amplitude short-time-constant positive voltage. If the squelch control is turned down to ground, the positive signals feed through the 0.03 capacitor to another detector, D_2C_2, producing a negative potential that reverse-biases the AF amplifier, completely silencing it. If the squelch control is turned up, it produces a bucking positive voltage that prevents D_1C_1 from rectifying noise, resulting in no noise signal fed to D_2C_2. With no reverse bias the AF BJT is free to amplify the high noise level that comes from the discriminator. There will be a threshold squelch setting that will just stop all noise amplification.

When a carrier is received, it quiets the discriminator and the AM noise detector. With no signal for D_1C_1 to rectify, no reverse bias is developed at the AF BJT. Any modulation detected by the discriminator is amplified by the AF BJT. Due to the short time constant of D_1C_1, any AF modulation on the carrier will not be detected with sufficient signal strength to develop reverse bias in D_2C_2.

VHF reception is subject to much man-made impulse-noise interference from automobile ignition systems, motors, and electric circuits being made and broken. For a given signal strength, the limiter stages in an FM receiver are more effective in eliminating such interference than audio peak limi-

ters in AM receivers. Also, in the VHF region signals are subject to multipath transmission. Signals transmitted from a car may reach the fixed-station receiver from two or more different directions because of reflection from wires and large metal objects. If the *multipath* signals arrive in phase, the signal is strong. If out of phase, the signal is weakened. If the car is moving, the signals may fade up and down rapidly, producing a fluttering effect in the receiver. The AVC circuits in AM receivers cannot follow such a rapid fade, and reception suffers. The FM receiver depends on its limiter stages to keep the signal to the discriminator at a constant level. FM reception of fading signals is not subject to flutter until the signal becomes so weak as not to operate the limiters. Many voice communications at frequencies over 100 MHz are FM; between 30 and 100 MHz they may be FM, SSB or AM, and below 25 MHz most are SSB.

Although FM is authorized in the amateur bands, it is not popular except on the VHF and UHF bands. Few amateurs have HF receivers that will detect FM signals properly.

Two AM stations transmitting on the same frequency will both be detected in an AM receiver and will be heard, even if one signal is 10 times the strength of the other. In wideband FM, if one signal is twice as strong as another on the same frequency, the stronger will capture the oscillation frequency of the discriminator circuit and the weaker signal will not be able to produce AF output in the receiver. This *capture effect* is the reason why aircraft use AM or SSB. Capture effect is lessened as the modulation bandwidth is decreased, being zero at zero deviation (CW).

Because detected noise is directly related to receiver bandwidth, a 3-kHz bandwidth SSB communications receiver may produce a better signal/noise ratio with weak SSB signals than an 18-kHz bandwidth (narrowband) FM receiver can produce, even in the VHF and UHF regions. Newly developed crystals make it possible to maintain the required carrier frequency stability to use SSB in commercial services. ICs can be used for the low-level RF and AF circuits of 12-V dc transmitters, making transceivers quite compact, even with 30- to 60-W output.

19-12 ALIGNMENT OF FM RECEIVERS

FM broadcast receivers use a 10.7-MHz IF to produce a relatively broad (low-Q) 200-kHz bandwidth. FM communication receivers operating in the 25–1000 MHz range may use double conversion, with the first IF at 2–30 MHz, to reject images. The second IF may be near 450 kHz to produce the 18-kHz IF bandwidth required for ±5-kHz deviation F3E.

The alignment of the RF amplifier, mixer, and IF amplifiers in FM receivers is similar to the procedure used in AM receivers (Sec. 18-28). In Foster-Seeley receivers the discriminator and limiters are aligned first. An unmodulated signal generator is adjusted to the center of the IF passband and is fed to the input of the last limiter. This feeds a signal to the discriminator transformer. A sensitive dc voltmeter with an RF choke or a 1-MΩ resistor on the ungrounded probe of a VTVM is connected between ground and the center tap of the two series resistors in the discriminator circuit. The secondary of the discriminator is detuned until an indication of voltage is obtained on the meter. The primary of the discriminator transformer is then tuned for a maximum indication on the meter.

The meter is next connected to read the voltage across both series resistors. The secondary of the discriminator transformer is then tuned to zero volts across the load resistors. The signal generator is shifted 100 kHz above and then 100 kHz below the IF to check the linearity of the discriminator. The

same-amplitude voltage, but of opposite polarity, should appear 100 kHz from the carrier in both directions (use 10 kHz for communications sets).

The signal generator is next moved to the input of the first limiter. The voltmeter is connected across the input of the second limiter, and the output circuit of the first limiter is tuned for maximum voltage indication. If transformer coupling is used between limiters, the input circuit of the second limiter is next tuned, again for maximum voltage indication. The weakest possible signal from the signal generator must be used.

The signal generator is then moved to the input of the last IF amplifier, the voltmeter is moved to the input of the first limiter, and the output circuit of the last IF amplifier is tuned for maximum voltage. If transformer coupling is used between stages, the input circuit of the first limiter is tuned next.

The voltmeter can be left in its last position, and the remainder of the alignment of the IF, mixer, oscillator, and RF stages will follow the pattern explained for AM receivers. If the first limiter indicates maximum limiting, it means that the front-end stages are being overdriven or overloaded.

To align a ratio detector the signal generator is fed to the last IF amplifier input. Both the primary and the secondary of the ratio-detector transformer are tuned for maximum dc voltage across the large capacitor in the detector circuit. The voltmeter may be left across the capacitor to tune the IF transformers also.

Test your understanding; answer these checkup questions.

1. In the Foster-Seeley circuit shown, for what is L_4 used? _____ Could a resistor be used in place of L_4? _____
2. Is the AF output voltage varying dc or ac in a Foster-Seeley circuit? _____ In a stagger-tuned discriminator? _____ In a ratio detector? _____ In a gated-beam detector? _____
3. In question 2, which circuits require limiters ahead of them? _____ Which provide an AVC voltage? _____ Which has its diodes in series? _____
4. To what frequency must the gated-beam quadrature circuit be tuned in an FM BC (broadcast) receiver? _____
5. In what way is the gated-beam detector similar to a superregenerative detector? _____
6. What are the two basic types of limiting that were mentioned? _____ _____

7. With what FM detector(s) would AVC be an advantage? _____
8. What is the reason for using pre-emphasis? _____ How much is produced at 1, 5, and 15 kHz? _____
9. What are the two types of stages in an FM receiver that differ from those in an AM receiver? _____ _____
10. What special circuits are used in a squelch system that can follow changing noise levels? _____
11. Why might FM be better than AM for mobiles? _____ _____
12. What is the order of circuit alignment in an FM receiver? _____

19-13 DIRECT FM AND AFC

There are many methods of producing FM. One direct FM method, the Crosby, utilizes an active device that is made to appear as an inductive or a capacitive reactance across a self-excited oscillator LC circuit (Fig. 19-13).

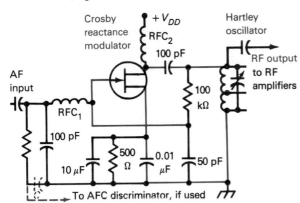

FIG. 19-13 A Crosby reactance modulator across a Hartley tuned circuit.

The 100-kΩ resistor and the 50-pF capacitor across the oscillator LC circuit feed oscillator frequency voltage to the modulator gate, but at nearly a 90° phase *lead* due to the low reactance of the capacitor and the high resistance of the resistor. The phase-shifted voltage fed to the gate produces a modulator-I_D variation through RFC$_2$ that is nearly 90° out of phase with the oscillator tank RF voltage. The oscillator ac is coupled through a 100-pF capacitor to a device that appears to have the same frequency ac but is *lagging* nearly 90°. (Remember, there is a 180° phase shift between gate and drain of a JFET in a common-source amplifier circuit.) The 90° current lag makes the modulator look like

an inductive reactance to the oscillator. Paralleling an inductance across an LC circuit raises the resonant frequency, so the oscillator now operates on a frequency higher than it would normally. When AF voltages are applied to the gate of the modulator, they control the gain of the device, the vector sum of oscillator and modulator RF voltages, and thus the apparent reactance of the modulator. Variation of the reactance causes variation of the frequency of the oscillations of the LC circuit. In this way AF voltages on the gate of the modulator produce FM of the oscillator RF ac. A doubler stage following the modulated stage will double the deviation of the carrier without otherwise affecting the FM.

Since any voltage change in any part of the oscillator or modulator will shift the center frequency, it is necessary to use some method of assuring that the resting-carrier frequency remains constant and on the assigned frequency. A discriminator circuit can be used to return the resting carrier to the center frequency. A block diagram of an FM transmitter using *automatic frequency control* (AFC) is shown in Fig. 19-14. The Hartley oscillator has a center frequency of 5 MHz. The doubler output is 10 MHz. A first tripler would have a 30-MHz output, and a second tripler would have a 90-MHz output to drive the power amplifier. Some of the 10-MHz output is fed to a mixer stage. A 4-MHz crystal oscillator is also fed into the mixer. The difference frequency between 10 and 4 MHz is fed to a 6-MHz IF amplifier and to a discriminator tuned to exactly 6 MHz.

As long as the oscillator remains on 5 MHz, its multiplied output, 10 MHz, will beat against the 4-MHz crystal to produce 6-MHz ac and zero voltage at the discriminator output. If the center frequency of the master oscillator drifts upward, the mixer will be fed a frequency higher than 10 MHz. This beats against the 4-MHz crystal and produces a difference frequency higher than 6 MHz and develops a voltage at the discriminator output. This voltage is fed to the grid of the reactance modulator (dashed line in Fig. 19-13), changing its reactance, which lowers the oscillator frequency. If the oscillator drifts lower in frequency, an opposite-polarity dc voltage is developed at the discriminator, shifting the carrier back to the center frequency again. A reactance tube or transistor AFC system (Fig. 19-9) can be used in FM, AM, and TV *receivers* to keep them tuned to the desired station.

A discriminator alternately develops positive and negative voltages at the modulation frequency. If fed back to a reactance-type AFC circuit in a trans-

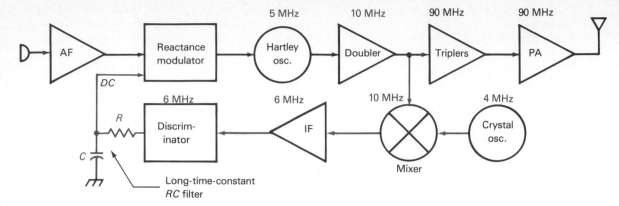

FIG. 19-14 Block diagram of an AFC system to hold a broadcast station on center frequency.

mitter, they would cancel the modulation. To prevent this, a long-time-constant RC circuit is connected between the discriminator output voltage and the modulator grid. The average voltage no longer follows the deviation of the modulation, although it will follow any slow drift of the carrier. Note that this is a form of PLL circuit.

A multiplication of 12 to 24 times the oscillator frequency is usually adequate to produce a 75-kHz deviation with reactance modulators. A multiplication of 18 can be attained with a doubler and two triplers. Thus, a transmitter with a center frequency of 89.1 MHz requires an oscillator with a frequency of 89.1/18, or 4.95 MHz. A deviation of 2 kHz at the oscillator produces a 36-kHz deviation of the final amplifier signal.

A single-ended Crosby circuit may produce a somewhat distorted FM. This can be overcome by using a push-pull reactance modulator.

19-14 FM BY VARACTORS

A *voltage-variable capacitor* diode (varactor, Varicap, etc., see Sec. 9-3), when reverse-biased, will vary its junction capacitance with a variation of bias, producing direct FM (or FSK) of an oscillator

(Fig. 19-15). The two diodes are reverse-biased by a fraction of the +20 V through the RFC. Any AF ac added in series with the bias will modulate the bias, change the capacitance of the diodes, and shift the frequency of the oscillator.

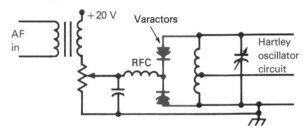

FIG. 19-15 Reverse-biased voltage-variable capacitors used to frequency-modulate an oscillator.

A modulating AF voltage great enough to exceed the linear deviation capabilities of this or any modulation system will produce an emission that will sound distorted. It will develop undesirable sidebands out past the normal bandwidth limits.

19-15 PHASE MODULATION (PM)

Holding a self-excited oscillator on an assigned center frequency may be difficult. A temperature-controlled crystal oscillator will be stable enough, but the frequency deviation produced by reactance modulating a crystal is small and requires many multiplications of the oscillator frequency.

If the crystal-oscillator stage is not modulated but modulation is applied to following stages, the frequency of the emission cannot be directly varied. However, a modulating voltage can shift the phase of the carrier V and I. This results in *indirect FM* called *phase modulation* (PM).

A dc input voltage applied to a reactance modulator shifts the frequency of the oscillator and holds it at the new frequency until the voltage has been changed. In phase modulation the result of shifting the phase will be in an instantaneous frequency change, but the carrier cannot be held at any frequency except the center frequency.

A PM system developed by Armstrong uses a crystal oscillator for frequency stability and a phase-modulated buffer stage. It employed a balanced modulator to cancel the carrier and produce amplitude-modulated sidebands only. These are shifted in phase 90° and fed back to the output of the stage from which the carrier is taken (Fig. 19-16). With no modulation there are no sidebands,

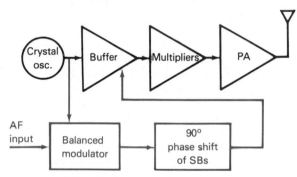

FIG. 19-16 Block diagram of an Armstrong PM system of producing FM.

and the buffer output is that of the crystal. When audio is applied, its amplitude determines the amount of 90° phase-shift energy fed to the output of the buffer. The input of the next stage is fed (1) a carrier and (2) a 90° shifted frequency of varying amplitude. This stage responds to the vector sum of the two phased voltages. With a weak audio signal, little sideband energy is fed to the third stage, there is little phase shift, and the PM is small. With more audio there is greater sideband energy, greater phase shift, and greater PM. Any AM produced in PM (or FM) systems is flattened out by the limiting action of the class C multiplier stages operating into the saturation portion of their curves.

With any method of PM a constant-amplitude, variable-frequency tone capable of producing a 25-Hz deviation of the carrier with an AF of 50 Hz will produce a 7500-Hz deviation with an AF of 15,000 Hz. With PM, deviation is not only directly proportional to modulating audio voltage amplitude, as in FM, but also *directly proportional to the*

audio voltage frequency. To produce equal deviation for all modulating frequencies, it is necessary to use a low-pass RC network in one of the AF amplifier stages. This makes the audio output amplitude inversely proportional to the input frequency. Such a network may consist, for example, of a 500,000-Ω resistor in series with the gate circuit in a resistance-coupled audio amplifier with a 0.1-μF capacitor from gate to source. Then, if pre-emphasis is desired, another network is required to raise the higher-frequency response according to the pre-emphasis curve.

If the maximum low-frequency deviation that can be developed at the modulated stage is only about 25 Hz, it is necessary to multiply this 3000 times to produce 75,000-Hz deviation. For a carrier frequency of 90 MHz, this would require an oscillator frequency of 30 kHz. In actual practice a crystal oscillator of about 200 kHz is used. Its output is fed through a series of multipliers to some frequency in the 30-MHz region. This is heterodyned against a second crystal oscillator and is translated down to about 5 MHz (without losing any frequency deviation) and is then multiplied up to 90 MHz. Many multiplier circuits are required.

19-16 VOICE-MODULATED PM TRANSMITTERS

While the Crosby direct-FM system, the Armstrong phase system, and a few other systems represent some of the methods used in FM broadcast stations, by far the greatest number of frequency-modulated transmitters in use today are found in the millions of taxicabs, police cars, fire engines, pagers, and industrial communication services. These systems are interested only in voice frequency (300–3000 Hz, VF) transmissions. By reducing the deviation ratio and modulating 300 to 3000 Hz only, 5-kHz deviation can be produced using phase modulators with multiplications of only about 12 times. The reduction of the deviation ratio results in a lessened AF output from receivers, but this is partially made up by the lessening of noise with weak signals when the receiver bandwidth is narrowed. Narrowband FM in amateur work is limited to 3-kHz deviation below 29 MHz although above this, wideband FM (75-kHz deviation) is permitted in certain portions of the bands.

A PM circuit is shown in Fig. 19-17. The output of a crystal oscillator is fed across the $C_1R_1C_2$ network and, at the same time, across the C_3L network. The frequency of the ac across R_1 and across L is the same, but because one circuit is an RC

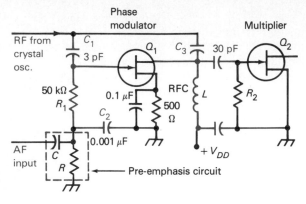

FIG. 19-17 A PM circuit used in many narrow-band FM communications transmitters.

circuit and the other an *LC* circuit, the voltage between the top of *L* and ground will be leading the voltage across R_1. If the phase modulator Q_1 is removed, the phase of the voltage developed across *L* is solely dependent upon C_3. When Q_1 is in the circuit, the out-of-phase voltage on the gate of the modulator produces an amplified out-of-phase current in its drain circuit through *L*. The current that flows through *L* has two out-of-phase components, one due to the voltage fed to the gate, and one due to the voltage from the oscillator. The resultant voltage-drop across *L* is therefore not in phase with either voltage but is at some resultant phase. When an AF ac is fed to the gate circuit of the phase modulator, the I_p still retains an oscillator component, but now the gate circuit component·varies in amplitude, producing a resultant phase shift that varies in accordance with the VF ac. In this way, PM is developed across *L* and is fed to R_2 and the multiplier stage.

Communications transmitters are required to have a means of limiting the percent of modulation in AM or a means of limiting the frequency deviation with FM. In the clipper of Fig. 19-18, current flows through R_2 and R_3 all the time. Current also flows through R_1 and R_3. When a positive audio signal is applied across R_1, the cathode of D_1 approaches the anode potential and current flow decreases. This decreases the current through R_3 and less voltage-drop is developed across it, while more is developed across R_2. Thus, AF ac from Q_1 is passed to the LP filter. When the cathode of D_1 is driven to or exceeds its anode potential (because of high-amplitude audio peak voltages), the diode ceases to conduct and any audio signals above this value do not affect the current through R_3 and therefore

through R_2. The signal is limited, or the peak is clipped off.

On the negative VF half cycle, the cathode of D_1 is made more negative (anode relatively more positive), the current through R_3 increases developing a greater voltage-drop across it, and decreasing the voltage-drop across R_2. When the negative signal reaches a certain value, the current through the diode approaches the maximum. Any further increase in the negative signal produces almost no increase in current in the first diode, nearly full voltage-drop across *R*, and therefore almost no signal variation across R_2. In this way, both the positive and the negative halves of high-amplitude VF signals are limited.

The audio developed across R_2 is fed through a low-pass filter network that does two things. It tends to round off or filter any square-wave-shaped signals developed in the clipping process, thereby reducing higher-frequency audio harmonics that might otherwise be transmitted. It also makes the audio output inversely proportional to the frequency required in PM systems to make them act as direct FM. The setting of the potentiometer arm determines the deviation or percent of modulation of the transmitter.

Crystal-controlled transmitters of this type may gradually drift lower in frequency with age. They may need to be readjusted to the proper center frequency yearly because the silver plating on the crystal increases in weight when oxidized. To allow the crystal to be *warped* to the desired frequency, variable capacitors or inductors may be included in the crystal circuit, to allow the crystal to be tuned 100 or 200 Hz. This produces a kilohertz or more of control at the assigned carrier frequency.

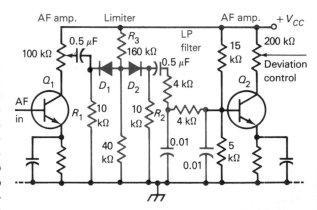

FIG. 19-18 AF limiter, or clipper, circuit with deviation control.

For proper testing of the operation of an FM transmitter, the following devices may be required: RF wattmeter, AF and RF frequency meters, deviation monitor, and audio distortion analyzer. A power supply and a dummy antenna for low-powered or mobile equipment may also be needed. The use of these is discussed in Chaps. 21, 23, and 24.

▌ Test your understanding; answer these checkup questions.

1. What is another name for the reactance-tube modulator? _____ Would it operate with a triode? _____
2. What effect would be produced if a small L were used in place of the 50-pF C in the reactance-tube modulator? _____ If the 50-pF C and 100-kΩ R were interchanged? _____
3. What are the five significant circuits in the AFC system shown? _____ _____ _____ _____ _____

4. Besides the reactance-tube modulator, what is another method of producing direct FM? _____ Could AFC be used with it? _____
5. Back in Fig. 16-14, where might AF ac be applied to produce direct FM? _____
6. In a PM system, where is the PM changed to FM? _____
7. How does the output of a PM stage differ from that of an FM stage? _____
8. What is the advantage of PM? _____ The disadvantage? _____
9. Is AM produced in a reactance modulator? _____ A PM modulator? _____ How is AM eliminated in FM transmitters? _____
10. What was the first BC PM system called? _____
11. Why are limiters used in FM transmitters? _____ In FM receivers? _____
12. In Fig. 19-18 which control would reduce excessive limiting? _____ Which would set maximum deviation? _____

FREQUENCY MODULATION QUESTIONS

1. What are the three ways a sine wave can be changed? Can all of these be used to modulate?
2. What are the fields of wideband FM? Narrowband FM?
3. What deviation is considered 100% modulation in FM broadcasting? 75%?
4. What is the formula to determine modulation index? What is the deviation ratio of broadcast FM?
5. Why is a modulation index of 0.4 important?
6. If a 100-W carrier FM transmitter is modulated 100%, how much power will it radiate then?
7. To what is the bandwidth of an FM emission directly proportional?
8. What is an F3E emission?
9. Draw a block diagram of a PLL FM detector circuit. Is the received signal rectified in this system?
10. What system can be used to demodulate narrowband FM with an AM receiver?
11. Diagram a stagger-tuned discriminator. Could it be used to key a BJT from an FSK signal?
12. Why does a discriminator need a limiter ahead of it?
13. Draw a diagram of a Foster-Seeley discriminator. What is used as the reference voltage?
14. Draw a diagram of a crystal discriminator. To what frequency would the crystal be ground?
15. In what three ways does a ratio detector differ from a Foster-Seeley circuit?
16. Draw a diagram of an IC quadrature detector circuit.
17. In what two ways can an IF amplifier be made to limit the signal it amplifies?
18. What FM radio system uses pre-emphasis? What one uses de-emphasis? What is the net result of using both? Of what are such circuits constructed?

19. Draw a block diagram of an FM broadcast receiver. What circuits are different from an AM broadcast receiver?
20. Can an AM receiver use AFC? A SSB? A CW?
21. Draw a block diagram of a squelch circuit that might be used for VF communications receivers.
22. Draw a diagram of a limiter, discriminator, squelch, and AF amplifier. What is input to and output from the squelch circuit?
23. What is a main advantage of wideband FM?
24. What kind of a signal-generator signal is used to align a Foster-Seeley discrimnator? What are the two steps when tuning the discriminator?
25. How is a ratio detector aligned?
26. Draw a diagram of a Crosby reactance modulator. Is this a direct FM modulator or a phase modulator?
27. Block diagram a system that would keep a reactance-modulated Hartley oscillator on frequency. What would be the stability of the Hartley in this case?
28. Would the same AFC circuit above be used from discriminator to LO of an FM receiver? Explain.
29. Draw a diagram of a two-varactor direct FM circuit.
30. To what is deviation proportional in a PM generator? In an FM generator?
31. What type of PM circuit uses a balanced modulator?
32. Diagram a PM circuit used in VF FM transmitters.
33. How is PM made to be FM?
34. Draw a diagram of an AF limiter-clipper circuit with an RC LP filter in a PM-FM transmitter. What is the function and frequency of the filter?
35. What are some of the instruments used to test communications FM transmitters?

20 Antennas

This chapter discusses the fundamentals of energy radiation through space and various types of antennas used in radio transmission and reception. Transmission lines and their uses in high-frequency applications and in broadcast systems are explained.

20-1 RADIO WAVES

An antenna, or aerial, in one form consists of a piece of wire or other conductor, with insulators at both ends, suspended well above the ground (Fig. 20-1). If the wire is cut to the equivalent of a half wavelength ($\lambda/2$), it acts as an oscillating LC circuit. When excited by some source of RF ac, the free electrons in it will oscillate back and forth from one end to the other. If the wire were a perfect conductor and had a high Q, once shock-excited the electrons might be expected to continue to oscillate back and forth along it indefinitely. However, some of the energy imparted to the electrons is lost in heating whatever resistance is in the wire. A much greater amount of the energy is radiated into space from the wire. It is this energy lost by the antenna into space that is the *radio wave*.

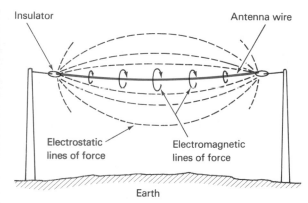

FIG. 20-1 An antenna wire develops electromagnetic fields around it and electrostatic fields end to end.

If an antenna is shock-excited, free electrons in it will be driven to one end, pile up there, and produce a highly negative charge. At the same time the other end of the wire is left with a deficiency of electrons and is positively charged. An electrostatic strain is developed from one end to the other, and an electrostatic field is formed along the antenna (dashed lines in Fig. 20-1).

With nothing to hold the electrons at one end, they reverse their previous motion and start toward the opposite end. As they move, they produce a magnetic field, outward *around* the wire (solid-line loops). The energy that had been stored in the electrostatic field is now transferred to an electromagnetic field. As the potential difference between the ends of the antenna drops to zero, the magnetic field reaches maximum and then begins to contract back into the wire, driving electrons toward the once-positive end of the antenna. This end now has an excess of electrons on it, charging it negative, the other end, having lost electrons, becoming positive. Energy is again stored in an electrostatic field.

This oscillating of electrons and energy in the antenna produces expanding and contracting electrostatic and electromagnetic fields. Parts of these fields travel out so far that they cannot return to the antenna when the fields alternate, are lost to the antenna, and radiate into space.

Actually, neither an alternating static nor a magnetic field can exist alone. If one is present, the other will be also, but at 90° from each other.

If the frequency of the ac exciting the antenna is low, the antenna must be long to prevent the electrons from reaching the end and trying to return before the driving source potential reverses. As a result, low-frequency antennas are long and high-frequency antennas are short. A low radio frequency is 10 kHz and requires a λ/2 antenna 15,000 meters (m) (nearly 9 miles) long. An antenna for the middle of the standard broadcast band, 1 MHz, is 150 m (468 ft) long. For 10 MHz, a λ/2 wire 15 m (46.8 ft) long is required; for 100 MHz, 1.5 m (4.68 ft); and for 10,000 MHz (10 GHz), 0.015 m (0.56 in.). By necessity, antennas of less than a λ/2 are used for low-frequency transmissions. They are *loaded* (tuned) with added inductors until they resonate at the desired frequency.

Radiations of all the lower-frequency (longer-wavelength) radio waves tend to travel along the surface of the earth without attenuation. The *ground wave*, that portion traveling just above the surface of the ground (Fig. 20-2), is usable for hundreds to thousands of miles, day or night.

With higher frequencies, the ground wave weakens as energy from it is more effectively absorbed by the surface over which it is moving. Frequencies in the 500- to 1500-kHz region may have usable ground-wave signals for only 50 to 600 miles.

As frequency is increased above 5 or 10 MHz, the usable ground-wave signal exists for only a few

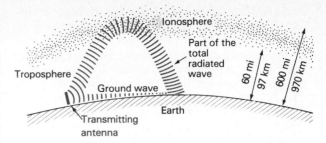

FIG. 20-2 The ionosphere, a layer of ionized air (gas) above the earth, can refract radio waves back to earth.

miles. At these frequencies it is possible to transmit farther by using the *direct wave*. When the receiving and transmitting antennas are within sight of each other, the signal is a direct wave. At high frequencies, where the ground wave is weak, high-flying aircraft may receive direct waves 100 miles or more; satellites, thousands of miles.

While it might be expected that direct-wave or line-of-sight transmissions of VHF, UHF, and SHF signals would only be possible to the horizon, such waves are similar to light waves in some respects. A light wave, with its wavefront parallel to a knife edge, will bend around the edge somewhat as it passes. Radio waves passing over a mountain ridge will do the same. Downward refraction and the resulting dispersion of the wave past the ridge can increase the useful range of these signals by 50% or more.

If static and magnetic fields of energy traveling through the air cross a conductor, an emf at the frequency of the energy wave will be induced in the conductor. The explanations that follow will consider the antenna as a transmitting device. If an antenna transmits maximum signals east and west, it will receive best from east and west also. The radiation resistance of an antenna is the same for transmitting as for receiving. The only differences are the insulation requirements and wire diameter, since transmitting antennas may carry large currents and develop high voltages.

20-2 THE IONOSPHERE

Near the earth the air is rather dense, but from about 60 to 600 miles above the earth the air thins and radiated energy from the sun can ionize the widely spaced air molecules. The different degrees of ionization produced form into several recognizable layers called D, 30–60 miles; E, 60–100 miles; and F, 100–250 miles (splits into F_1 and F_2 during daylight). The ionized atmosphere allows the radiated wave to travel faster through it than in the more dense, un-ionized lower air. As a result, the top part of a wavefront moving into the *ionosphere* speeds up and forges ahead of the lower part of the wavefront and eventually may turn, or *refract*, downward into the *troposphere* again. (In the 50–70 mile range the air is thick enough to cause meteors to disintegrate due to friction and produce ionized paths which can reflect radio waves.)

The lower the frequency of the waves, the less penetrating effect they have and the greater the proportion of them that may be turned back toward earth. The higher the frequency the more penetrating energy the radio wave contains. Signals of 2–50 MHz may be deflected (refracted) or they may penetrate the ionosphere depending on the time of day, the angle at which the wave strikes the ionosphere, and the degree of ionization present. With weak ionization they penetrate; with stronger ionization they may be refracted. With still stronger ionization the wave energy may be totally *absorbed* and dissipated in the ionosphere. During times of sunspot activity (an 11-year cycle) and high *solar-flux-index* values (most noise from the sun), ionization is increased and long-distance transmission may be improved because of the condition of the *F* layer. Aurora borealis–reflected signals have a rapid fluttering sound because of the ionic turbulence produced, and voice communications are poor (CW is the best means of communicating).

Strong solar winds (particles) can cause "sporadic *E*-layer" ionization resulting in unusual 500–1500 mile radio return paths in the 25–60 MHz range. This is called *short skip*. Any *sudden ionospheric disturbance* (SID) can greatly effect radio communications, even shutting them down.

The rotation period of the sun is about 27 days. As a result it can be expected that radio conditions due to sunspot activity will repeat in 27-day cycles.

The *maximum usable frequency* (MUF) is the highest frequency at which ionospheric-refracted signals return to earth with usable strength. MUF varies greatly (between 8 and 30 MHz) with time of day, distance, direction, season, and solar-flux index.

Besides the frequency factor, the angle at which the radio wave enters the ionosphere determines penetration or refraction of the wave. While there may be some reflection of lower-frequency signals traveling directly upward, almost all higher-frequency waves that are transmitted at an angle of nearly 90° above the surface of the earth either penetrate or are absorbed by the ionosphere (Fig. 20-3). At angles less than 90°, there is more refraction. The higher the frequency, the greater the penetration and the lower the angle required to produce refraction. At high frequencies, there may be long distances between the end of the usable ground-wave signal and the reappearance of the reflected *sky* wave. This no-signal area is known as the *skip zone*. At lower frequencies the sky wave often returns to earth in the ground-wave region.

If the sky wave returns to earth and strikes a good conducting surface such as salt water, it can be reflected back upward and take a double hop. Double-hop signals may carry long distances.

Active satellites parked in orbit over the equator receive and transmit at frequencies in the gigahertz range. Signals from and to earth are refracted very little and remain the same night and day. Signals

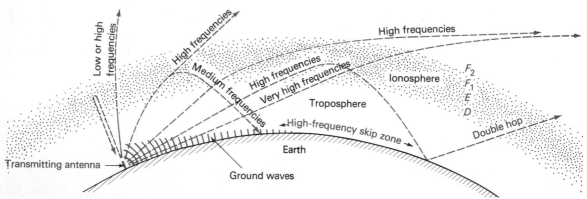

FIG. 20-3 Paths of radio waves in the ionosphere. Low- and medium-frequency waves are affected by the ionosphere more than high frequencies are.

that are received by the satellite are translated to another frequency and relayed back to some other point on earth.

20-3 FADING

If a sky wave is being received, variations in the ionosphere may refract more or less energy to the receiving point at different instants, producing a varying amplitude or fading signal.

When the receiving antenna is within both ground- and sky-wave range, the addition of the two waves may be in or out of phase. When arriving in phase, the waves add to each other, producing a strong signal. When out of phase 180°, they tend to cancel each other. Variations in the ionosphere can change the sky-wave travel distance and therefore the phase relationship of the ground and sky waves. Two sky waves refracted to the same point by two different areas of the ionosphere may arrive in or out of phase.

In the VHF or UHF regions, where the direct wave is used, as in television, an airplane flying overhead will act as a reflector of the direct wave. A receiving antenna below the plane will receive the direct wave from the TV station plus the wave reflected by the airplane. Since the plane is moving, the transmitter-reflector-receiver distance constantly changes and the relationship of the direct and the reflected waves at the receiving point is alternately in and out of phase, producing a continually fading signal. Mobile reception may change amplitude due to the shielding effect when the unit moves behind buildings, hills, thick trees, etc., or due to reflected signals.

20-4 NIGHT AND DAY TRANSMISSIONS

The ground wave remains the same night and day. Only the sky wave changes.

As night approaches, the sun can no longer ionize the atmosphere above the darkened part of the earth, and the ionized layers become thinner. The thinly ionized layers turn sky waves back to earth over a wider arc. Sky waves may return to earth farther away than during the daytime.

In the *VLF range* (10–30 kHz) and the *LF range* (30–300 kHz) there is not much difference between night and day transmissions, although at distances of several thousand miles the nighttime signals will be stronger. Received signals are mainly by ground waves. Services in this range include radio naviga-

tion, time and frequency standards, maritime mobile, and aeronautical communications.

In the *MF range* (300–3000 kHz), out past the ground-wave range, signals improve materially at night, increasing from a few hundred to several thousands of miles. Services in this range include maritime and aeronautical aids and communications, AM broadcasting, fixed and mobile communications, time and frequency broadcasts, and amateur communications.

In the *HF range* (3–30 MHz), during the daytime lower frequencies may return to earth 20 to 500 miles or more away. At nighttime the ionosphere thins and refracted signals return to earth 200 to many thousands of miles away. The higher-frequency signals may return to earth 200 to 5000 miles or more away. At nighttime higher-frequency signals may pierce the ionosphere and not return at all. Communications systems may have to shift from one frequency to another during the day and night to maintain usably strong signals. Services in this range include aeronautical, maritime, land, amateur, and point-to-point communications, plus short-wave BC and time and frequency broadcasts.

In the *VHF range* (30–300 MHz), signals in the lower-frequency portion may sometimes be refracted during the daytime, but rarely at night. These frequencies are not too reliable for long-distance use, but are useful for communications up to 50 miles. Signals above 100 MHz are rarely subject to refraction by the ionosphere, acting more like light waves. They are used for ground-wave, direct-wave, or extraterrestrial satellite communications. Services in this range include land and aeronautical mobile, industrial, amateur, FM and TV broadcasting, radio navigation, space, and meteorological communications.

In the *UHF range* (300–3000 MHz) transmissions are essentially line-of-sight, day and night. Services include TV broadcasting, land and aeronautical mobile, radioastronomy, telemetry, satellite, and amateur communications. (Above about 900 MHz is considered to be *microwaves*.)

In the *SHF range* (3–30 GHz) radio paths are line-of-sight day and night. Services are limited to microwave relays, satellite plus exploratory, and amateur communications for the most part.

The *EHF range* (30–300 GHz) has characteristics essentially the same as those of the SHF range.

Beaming strong VHF, UHF, or SHF signals toward the horizon can produce forward, side, and even backscatter signals when the wave hits discontinuities in the air, troposphere, or ionosphere, or conductive objects on the horizon. This can result in communications well over the horizon, to either side, and sometimes backward. The equator is subject to greatest ionization and can often produce scatter signals. Thus, *transequatorial scatter* or refraction communications can be made over 1000 to 5000 mi, or *backscatter* can provide a communication path from one area into its skip area if both antennas are pointed to the south. Troposcatter is effective up to about 300 to 400 mi.

Normally, the greater the altitude the colder the air. If for any reason a layer of warmer air forms above a colder stratum, a *heat inversion* is present. The two layers have different densities of air and can affect VHF or UHF radio waves enough at times to refract them back to earth at distances of a few hundred miles or less. When the layer is thin, it may act as a *duct*, or pipeline, for VHF to SHF tropospheric signals. Ducted signals may travel hundreds of miles before leaving the duct and returning to earth. Such ducts often form over water areas.

20-5 EFFECT OF LIGHTNING

A bolt of lightning produces RF energy across almost the whole usable radio spectrum, but the percent of energy decreases as the frequency increases. At lower frequencies there is considerable energy, and with good ground-wave transmission, storms hundreds or thousands of miles away can produce considerable interference. The higher frequencies, particularly between 5 and 15 MHz, are subject to local storms, but storms in the skip zones may not be heard. As a result, the higher frequencies are much less subject to lightning-produced radio noise (static). At VHF and above lightning causes little disturbance.

To protect antenna systems from lightning, stations may use arcing points from antenna to ground, overvoltage circuit breakers, antenna ground switches, or grounded-type antennas.

20-6 POLARIZATION OF WAVES

The polarization of a radiated wave is in the direction of the electrostatic field of the antenna. A horizontal antenna (Fig. 20-1) transmits a horizontally polarized radio wave. A vertical antenna (Fig. 20-2) radiates a vertically polarized wave.

TV and FM broadcasts are made with horizontally polarized antennas. A receiving antenna to pick up a maximum signal voltage from the transmitted wave must have the same polarization. As a result, TV and FM receiving antennas have horizontal elements. If a TV receiving antenna is erected vertically, almost no signal will be received from the transmitting station.

Most transmissions below 2 MHz will be vertically polarized. Standard-broadcast-band transmitters (535–1605 kHz) use vertically polarized antennas. The antennas may have horizontal wires, but the vertical radiation component will be greater than the horizontal.

In the frequency ranges from 3 to 30 MHz, under certain conditions, and for a given transmission distance, vertical polarization sometimes operates better than horizontal. Under other conditions, horizontal polarization may be better.

For frequencies above 30 MHz, most communication transmissions, with the exception of TV and FM, are vertically polarized.

Because of reflection from nonvertical objects or ionosphere contour, the polarization of a ground, direct, or sky wave may twist. Thus at times a vertically polarized wave may be received best on a horizontal antenna some distance away. Antennas developed for *circular polarization* have both horizontal and vertical elements (Sec. 20-31).

20-7 THE HALF-WAVE ANTENNA

The wavelength of a radio wave, a sound wave, or any other wave varies inversely as the frequency and is determined by the basic formula

$$\lambda = \frac{v}{f}$$

where λ = wavelength in whatever unit of length is used
v = velocity of the wave in the same unit
f = frequency in Hz

From this are derived three formulas often used in radio

$$\lambda = \frac{300,000,000}{f} = \frac{300,000}{f_{kHz}} = \frac{300}{f_{MHz}}$$

where $\quad \lambda$ = wavelength in m
f = frequency in Hz
f_{kHz} = frequency in kHz
f_{MHz} = frequency in MHz
$300,000,000$ = velocity of radio waves, m/s

The velocity of a radio wave is considered to be

300,000,000 m/s (actually 299,792,462), a little more than 186,291 mi/s or 983,616,000 ft/s.

If a 1-MHz transmitter is exciting an antenna, the radio wave from it will travel 300 m in the time it takes the ac to complete one cycle. A frequency of 1 MHz has a wavelength of 300 m. If used with a 1-MHz radio transmitter, the basic length of an antenna will be one-half of the full 300-m wavelength, or 150 m (492 ft). When an antenna of this length is excited by a 1-MHz RF ac, electrons have just enough time to reach the end of the wire as the source RF ac reverses polarity. At no time should the natural period of electron oscillation be ahead of or behind the phase of the exciting RF.

While this theory is basically correct, it neglects *end effect*. End effect must be considered with any antenna. It may be regarded as a dielectric effect of the air at the end of the antenna that effectively lengthens the antenna. End effect makes a half-wave-antenna wire act as if it were about 5% longer than it actually is. This will produce an interference between the exciting and the oscillating currents, a lessening of the oscillation amplitude, and a corresponding lessening of the radiated field. To overcome end effect and to resonate the antenna, it must be cut to a physical length approximately 95% of the electrically computed $\lambda/2$. For 1 MHz the $\lambda/2$ antenna will be about 468 ft. The ratio of length in feet, end effect, and frequency holds close enough to set up a general formula to determine the length in feet of any $\lambda/2$ antenna supported at the end by insulators:

$$\lambda/2_{ft} = \frac{468}{f_{MHz}}$$

The length of a 7-MHz antenna is

$$\lambda/2_{ft} = \frac{468}{f_{MHz}} = \frac{468}{7} = 66.9 \text{ ft}$$

If a $\lambda/2$ antenna element is self-supporting at the middle (no end insulators), the end effect is less. A factor of 478 may be used in the formula.

If an antenna wire is close to ground, buildings, trees, etc., the end-to-end capacitance increases and end effect increases, requiring shorter lengths.

When an antenna is a full wave in length, it is composed of two half waves, but there is still only one pair of end effects to consider. The total length of such an antenna is equal to a half wave with end effect plus a half wave without end effect. The factor 492 is used in the formula to compute the half wave without end effect.

A 405-ft antenna is to operate at 1250 kHz. What is its wavelength at this frequency? If 1 m = 3.28 ft, the length of the antenna in meters is 405/3.28, or 123.4 m. The wavelength of 1.25 MHz is 300/1.25, or 240 m. The ratio of 240 to 123.4 m gives the decimal fraction of the wavelength of the antenna, or 123.4/240 = 0.514 wavelength.

20-8 HERTZ AND MARCONI ANTENNAS

Any antenna complete in itself and capable of self-oscillation, such as a half- or full-wavelength wire, is known as a *Hertz* antenna.

When an antenna utilizes the ground (earth) as part of its resonant circuit, it is a *Marconi* antenna. An example of a Marconi antenna is a quarter-wave ($\lambda/4$) vertical antenna, with the ground operating as the missing $\lambda/4$. Most low- and medium-frequency antennas are Marconi types.

Test your understanding; answer these checkup questions.

1. What is the relation in degrees of the electrostatic and electromagnetic fields of an antenna? _____ Is this their relation in space? _____
2. What HF waves are attenuated within a few miles? _____
3. Does wave velocity increase, decrease, or remain the same as it passes from air to ionosphere? _____
4. What wavelength radiations tend to be transmitted entirely between ionosphere and earth? _____
5. What effect do sunspots have on the ionosphere?
6. What is a double-hop signal? _____
7. What are two major causes of fading? _____ _____
8. Where is the skip zone? _____
9. What wave is the same day or night? _____ Why might it appear weaker at night? _____
10. Why do HF communication systems shift frequencies at different times of day? _____
11. Scatter transmission is used at what frequencies? _____
12. Over what areas do ducts often form? _____
13. Is polarization named for the static or magnetic component of the wave? _____
14. What polarization is employed in AM broadcasting? _____ TV and FM broadcasting? _____ Mobile communications? _____ Satellite communications? _____
15. What is the length in feet of an antenna wire for 4 MHz if the antenna is $3\lambda/2$ in length? _____
16. What is the name of antennas that are grounded at one end? _____ That do not use ground? _____

20-9 CURRENT AND VOLTAGE IN A HALF-WAVE ANTENNA

A $\lambda/2$ antenna excited by an RF ac source produces an oscillation of the free electrons in the wire. Since electrons pile up at the ends, the maximum charge (voltage) always occurs at the far ends of the antenna (Fig. 20-4).

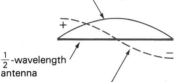

Current amplitude along $\frac{1}{2}$-wavelength antenna wire

$\frac{1}{2}$-wavelength antenna

Voltage amplitude along $\frac{1}{2}$-wavelength antenna wire

FIG. 20-4 V and I distribution on a $\lambda/2$ antenna.

When electrons move from one end of a $\lambda/2$ antenna to the other, the greatest number must move past the midpoint of the wire. Therefore, the center has the maximum current flowing through it. Because the electrons that pile up at the end of the antenna do not move past the end, there is zero current at (through) the far end.

The current distribution in a $\lambda/2$ wire, measured by inserting an ammeter at different points in the antenna, would be very nearly sinusoidal. The voltage at different points would also vary in a sine-wave manner. When the electrons pile up at the far ends, V at the ends is maximum and I is zero. When I at the center is maximum, V at the ends is zero. This is the same 90° V and I phase relation that exists in any other freely oscillating LC circuit.

20-10 RADIATION RESISTANCE

When an antenna is excited into oscillation, it radiates energy. Insofar as space is concerned, the antenna is acting as a source of power. Any source must have an internal resistance or impedance. If the radiated power and the current maximum in an antenna are known, the power formula $P = I^2R$ can be rearranged to $R = P/I^2$. The resistance value computed is known as the *radiation resistance* of the antenna. It is the ratio of radiated power to the square of the antenna center current. Radiation resistance is also the ratio of the voltage at any point on an antenna to the current flowing at that point ($R = V/I$). Since V decreases toward the center of a $\lambda/2$ antenna and I increases, the minimum resistance point will be the center of any $\lambda/2$ antenna.

Theoretically, the radiation resistance of a $\lambda/2$ antenna a quarter-wave high is 73 Ω, but only if constructed of thin wire and away from any reflecting surfaces. As it is lowered toward the ground, the resistance decreases almost linearly to zero. Elevating it to one-third of a wave above ground raises the resistance to about 95 Ω, but the resistance returns to 73 Ω at a half-wave height. At two-thirds of a wave high the resistance drops to about 58 Ω, returning to 73 Ω when three-quarters of a wave high and at every multiple of a quarter-wave higher. The higher above the earth, the less the resistance deviates from the 73-Ω value when not a quarter-wave multiple in height.

Antennas are made of reasonably thick wire, rods, or, in broadcasting, metal towers. The radiation resistance of practical antennas range from about 5 to 600 Ω. The ohmic and skin-effect resistance of the antenna is probably no more than 1 Ω.

If a $\lambda/2$ antenna (some multiple of a quarter-wave high) is cut in two at the center, the two severed points will appear to any transmission line connected to them as a load of about 70 Ω. If the antenna is not cut but two points equally spaced from the center are selected, there will be a resistive value between the two points. If the points are close together, the resistance is low; if far apart, the resistance is high (Fig. 20-5). A $\lambda/2$ antenna fed at its midpoint is known as a *dipole* (see also Sec. 32-12).

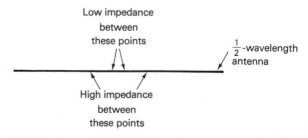

Low impedance between these points

$\frac{1}{2}$-wavelength antenna

High impedance between these points

FIG. 20-5 The farther from the center of a $\lambda/2$ antenna, the higher the impedance points.

Besides the resistive component, an antenna may be reactive: capacitive if it is less than λ/2 in length and inductive if more than λ/2 long.

20-11 TUNING ANTENNAS

To produce optimum operation, an antenna should be tuned to resonance. This tuning is accomplished, basically, by cutting the wire to a λ/2.

If an antenna is electrically shorter than desired, it can be cut apart and a *loading coil* is inserted in series with it (Fig. 20-6). The wire of the coil can

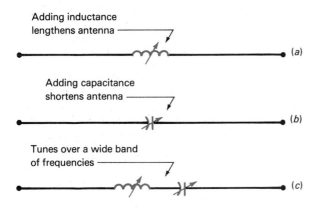

FIG. 20-6 The resonant frequency of an antenna is (*a*) lowered by adding *L* or (*b*) raised by adding *C*. (*c*) A variable *L* and *C* allow the antenna to be tuned over a wide band of frequencies.

be considered as being part of the required length of the antenna but coiled into a confined space. A short antenna may be thought of as having too little inductance (too little X_L) and must be capacitively reactive. To cancel its X_C and attain resonance, X_L (a coil) is needed.

If an antenna is too long, it can be cut apart and a capacitor inserted in series with it. An antenna longer than λ/2 appears inductively reactive. The addition of the proper-value capacitor in series with it cancels the excess X_L, making the antenna resonant, presenting a pure resistance to the feedline.

If both the *L* and *C* inserted in the antenna are variable, it is possible to tune the antenna to resonance over a wide band of frequencies.

Antennas can be used on harmonics of the source ac frequency. A 7-MHz antenna λ/2 long operates as a full-wave 14-MHz antenna, or as a 3λ/2 antenna on 21 MHz, etc.

20-12 BASIC FEEDLINES AND SWR

To produce maximum transference of energy from one circuit to another the impedance of the two circuits must match. An example of impedance matching to produce maximum antenna current and radiation is illustrated in Fig. 20-7.

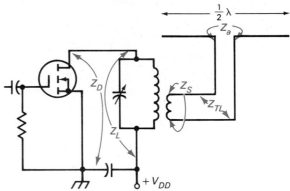

FIG. 20-7 Impedances that must be matched for maximum energy transfer to the antenna from a transmitter.

The impedance of the antenna Z_a (70 Ω) must match the *surge impedance* of the two-wire transmission line Z_{TL}. The impedance of Z_{TL} must match the secondary of the output transformer Z_s, which must couple tight enough to reflect an impedance across the primary Z_L that will match the output impedance of the device, Z_D. A variation of any part in this chain may prevent maximum output.

The two parallel wires are indicated as having an impedance. Any two parallel wires held apart a *constant distance* will have a characteristic, or surge, impedance value. This impedance is a function of the series inductance and shunt capacitance of the parallel wires, the diameter of the wires, and their distance of separation:

$$Z = 276 \log \frac{d}{r}$$

where Z = impedance in Ω
$\quad d$ = center-to-center separation
$\quad r$ = conductor radius (same unit as in d)

Two No. 14 wires (radius, 0.032 in.), held 2 in. apart center to center, form a transmission line with an impedance of 496 Ω. If the wires are thicker, the impedance is lower. If the distance of separation is greater, the impedance increases.

If a 500-Ω line is infinitely long and is connected across a source of ac having 500-Ω impedance, maximum power will be taken from the source and dissipated along the line in the form of heat. No energy will return. If the line is a few yards long and is terminated with a 500-Ω resistor, essentially all the power from the source will be delivered to the resistor. Such a transmission line is an efficient means of transmitting energy from a source to a remote load, provided the transmission line has the proper impedance. A 300-Ω transmission line connected between a 500-Ω source and a 500-Ω load is a mismatch. The load will not draw maximum power from the source.

When the transmission line does not match the load impedance, not all the energy fed down the line flows into the load. Some is reflected back to the source, forming *standing waves* on the line. Every half wave along the line, high-V and low-I points appear. Halfway between these points will be low-V high-I points. The ratio of voltage across the line at the high-V points to that at the low-V points is known as the *voltage standing-wave ratio*, or VSWR. The SWR is also the ratio of the current values at the high and low points on the line. The SWR (VSWR) formulas are

$$\text{VSWR} = \frac{V_{\max}}{V_{\min}} \quad \text{or} \quad \text{SWR} = \frac{I_{\max}}{I_{\min}}$$

When the load impedance matches the line impedance, there will be no standing waves. The current at all points along the line is the same, and SWR is 1:1. The line is said to be *flat*. Changing the length of the line has no effect on its SWR.

If standing waves appear on an antenna transmission line that should be flat, it is necessary either to change the transmission-line impedance until it matches the antenna or to change the antenna impedance until it matches the line. This is important if optimum transmission or reception is desired from an antenna.

The ratio of current (or voltage) delivered to an antenna to that reflected back down the line is the *reflection coefficient*, ρ. It is equivalent to

$$\rho = \frac{\text{SWR} - 1}{\text{SWR} + 1}$$

An impedance mismatch of 70 Ω to 35 Ω produces an SWR of 2:1 and a reflection coefficient of $(2 - 1)/(2 + 1)$, or $1/3$. Since power is proportional to I^2 (or V^2), the power reflected will be the square of ρ, or in this case, $(1/3)^2$, or $1/9$. This means that $8/9$ of the power indicated by meters in the transmission line would actually be delivered to the an-

tenna. The remaining $1/9$ is reflected reactive power (VA, not actually a loss).

Mismatching a transmission line to an antenna results in the line at the transmitter end appearing to have either X_L or X_C, which will detune an *LC* circuit to which it is coupled. A matched antenna system should not detune a final amplifier appreciably when coupled to it.

Open-wire lines held apart by spreaders (long, thin insulators) every few feet can be constructed to have impedances from about 150 to 800 Ω. These are called *balanced lines* because there is always an equal impedance to ground from each wire.

Transmission lines for receivers and low-power transmitters may use two spaced wires held apart by a ribbon of plastic material. An example of this is the flat, 300-Ω twin-lead used in many TV receiving antennas.

When lower-impedance transmission lines are required, coaxial, or concentric, cables can be used. These may consist of a metallic-tube ("hardline") outer conductor, and a copper wire, centered inside the tube with insulating beads, as the inner conductor. A coaxial cable may also consist of a copper wire covered with a pliable plastic insulation, covered in turn with a braided-copper outer conductor. Coaxial cables are made in various impedances, such as 50, 52, 72, 75, 93, and 125 Ω.

Hollow coaxial cables have the advantage of having their working surfaces (the inner surface of the outer conductor and the outer surface of the inner conductor) protected from the weather. This prevents oxidation of the surfaces, which would change the skin resistance as well as the dielectric constant between them and alter the impedance of the line.

The surge impedance of air-dielectric coaxial transmission lines is approximately

$$Z = 138 \log \frac{d_i}{d}$$

where Z = surge impedance in Ω
d_i = inside diameter of hollow tubing
d = diameter of center conductor

SWR in a coaxial transmission line can be measured with an *SWR meter* or *reflectometer*. Such a device may consist of a short length of coaxial line with a short piece of wire parallel to the center conductor inside the coaxial cable, and feeding an SWR-calibrated dc meter through a diode. When SWR is 1:1, the meter does not deflect. As SWR increases, the meter reads higher. The same kind of device, calibrated in watts and with some means of indicating a reference level, acts as an *RF wattmeter*. SWR can also be measured with two directional RF wattmeters forming a reflectometer (Secs. 22-3 and 24-2).

Although it may not be apparent in short, flat lines, when RF is transmitted over long lines, a loss of energy due to skin-effect, radiation, conductor resistance, and dielectric losses occurs. As a result, the value of current flowing into a long, flat transmission line at the transmitter end may be significantly more than that flowing into the antenna at the far end. The higher the frequency, the greater the attenuation.

Standing waves may not be desirable on transmission lines, but they are on resonant antennas as evidenced by high-V points at the ends and low-V points at the center of a dipole.

The efficiency of an antenna system relates almost entirely to the transmission line efficiency because the antenna radiates essentially all of the energy it is fed.

20-13 DIRECTIVITY OF ANTENNAS

A $\lambda/2$ antenna radiates energy in a direction at right angles to the wire itself. A horizontal antenna running north and south radiates maximum energy east and west, up into the sky, and down toward the earth. If the antenna is suspended in free space, well above the ground, the radiation can be represented by vector arrows (Fig. 20-8). If the antenna

FIG. 20-8 Relative radiation from a horizontal $\lambda/2$ antenna in free space (end view of the wire).

is close to ground, the part of the wave striking the ground reflects back upward and outward (Fig. 20-9).

When the antenna is a $\lambda/2$ above ground, the

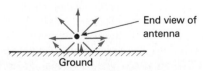

FIG. 20-9 When near ground, energy is reflected, changing the radiation pattern of the antenna.

reflected wave travels to ground, reverses phase, is reflected upward, and reaches the antenna the equivalent of one wave later but 180° out of phase because of the reflection. As a result, the upward radiation is canceled by the reflected wave. When all the vectors of all the angles from such an antenna are combined, the vertical radiation pattern is as shown in Fig. 20-10. The angle of maximum

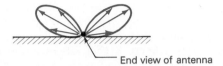

FIG. 20-10 Vertical radiation pattern of a horizontal antenna one $\lambda/2$ above ground, seen from a distance.

radiation is about 30° above the horizon. There is a relatively strong wave being transmitted from a few degrees above the horizon up to more than 50°.

With the antenna only a $\lambda/4$ above ground, the reflected wave returns just in time and phase to add to the vertical radiation. The result is a raising of the angle of maximum radiation (Fig. 20-11).

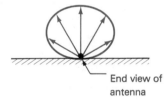

FIG. 20-11 Vertical radiation pattern of a horizontal antenna one $\lambda/4$ above ground, seen from a distance.

With the antenna $3\lambda/4$ above ground, the reflected waves return in time and phase to add to the upward radiation and also to the outward radiation. The radiation pattern is then made up of three lobes (Fig. 20-12). The higher the antenna, the more lobes developed—one lobe for each $\lambda/4$ of height.

When looking down on a horizontal $\lambda/2$ wire

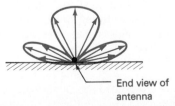

FIG. 20-12 Vertical radiation pattern of a horizontal antenna $3\lambda/4$ of a wave above ground, seen from a distance.

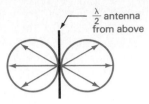

FIG. 20-13 Radiation from a λ/2 horizontal antenna, seen from above (or radiation from a vertical antenna in free space).

from above, the radiation pattern in the horizontal direction forms a figure 8 (Fig. 20-13). Maximum radiation is at right angles to the wire, with no radiation in the direction of the wire. The radiated-energy vectors form a doughnut around the wire if the antenna is suspended in free space. (If the antenna is brought near a reflecting surface, the doughnut shape is altered.)

A vertical λ/2 antenna radiates equally well in all horizontal directions, provided there are no nearby objects to alter its field. The effect of bringing the antenna near ground (making it a λ/4 long) is to raise the angle of maximum radiation (Fig. 20-14),

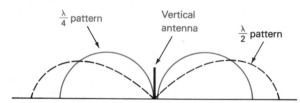

FIG. 20-14 Comparative vertical radiation patterns for vertical λ/2 and λ/4 antennas.

but there is no effect on the circular horizontal radiation pattern (maximum in all directions). A vertical antenna is said to be *omnidirectional* (all-directional) horizontally.

Directly above a vertical antenna there will always be a *cone of silence*. An airplane flying above a vertical transmitting antenna does not hear the station when directly over it. While there is a theoretical cone of silence at the ends of a horizontal λ/2 antenna, the reflected wave from the earth or nearby surfaces closes the cone and some signals will be received off the ends.

▌ Test your understanding; answer these checkup questions.

1. Where are the maximum V points on a λ/2 antenna? _____ Maximum I points? _____
2. Does Fig. 20-4 represent V and I at any one instant? _____
3. If I maximum and P radiated from an antenna are known, what two other facts can be determined? _____
4. What is the theoretical radiation resistance of a horizontal λ/2 dipole a λ/2 high? _____
5. Which reactance would a dipole have if 0.6λ long? _____ If 0.4λ long? _____
6. How is X_C compensated in an antenna? _____ X_L? _____
7. What is the Z_o of a No. 14 wire transmission line if the wires are 4 in. apart? _____
8. Under what condition will a short transmission line have no reflected power? _____ What would the SWR be? _____
9. Is a dipole a center-fed λ/2 antenna only? _____
10. What is the impedance of twin lead? _____
11. Why might more power flow into a transmission line than out of it? _____
12. For best long-distance communication why should a horizontal antenna be λ/2 rather than a λ/4 high? _____
13. Would received signals off the end of a horizontal antenna be horizontally or vertically polarized? _____
14. Would vertical λ/2 or λ/4 antennas have the greater ground-wave range? _____
15. Is the radiation pattern of an antenna the same for transmitting and receiving? _____

20-14 VERTICAL ANTENNAS

The antenna used with most MF (medium-frequency) and LF transmitters is a λ/4 vertical. If a λ/4 vertical wire or mast is erected above a large conducting surface, such as the earth (Fig. 20-15a), the conducting surface will operate as the missing λ/4 (or a great many multiples of quarter waves) to make the antenna resonate as a λ/2 device. The radiation resistance of the λ/4 antenna, between ground and the lower end of the wire, is just half of that of a λ/2 antenna, theoretically 36.5 Ω. The antenna is often *series* fed with 50-Ω coaxial transmission line without excessive SWR, but shunt feed may be more desirable (Sec. 20-24). Usually, λ/4 copper-wire radials will be run out over (or under) the ground every 5° to 15° from the base to make the ground surface more constant electrically and to serve as one terminal for the coaxial feedline. Only if the ground conductivity is very good will a metal rod driven into the earth make a fair ground connection. When a λ/4 whip is installed on top of an automobile, the roof of the car constitutes the *ground plane*. The car and capacitance to the road-

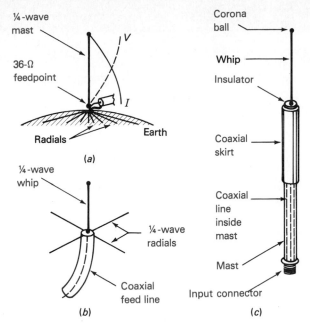

FIG. 20-15 (a) Series-fed λ/4 vertical for LF and MF. (b) VHF λ/4 coaxial-fed ground-plane whip antenna. (c) Coaxial λ/2 vertical antenna.

way act as the other λ/4. When a wire-ground system, 7 to 8 ft above the ground, is used beneath an antenna, it is called a *counterpoise*. Since the ends of a λ/2 antenna have high V and Z (perhaps 2500 Ω) a vertical λ/2 cannot be grounded at the bottom.

In the VHF and UHF range, vertical antennas often employ four or more horizontal λ/4 radials as the ground plane (Fig. 20-15b). If the radials point downward about 45°, the antenna is called a *drooping ground plane*. This has the effect of raising the feed-point resistance to about 50 Ω. If the ground-plane radials are bent straight down and connected together to form a cylinder, sleeve, or skirt (Fig. 20-15c), a λ/2 *coaxial* antenna is formed. These may be seen on tops of buildings. Sometimes a second skirt is added to the support mast a quarter wave below the bottom of the coaxial skirt; this acts as a trap to prevent standing waves from forming on the mast if it is metal.

A small metal or plastic ball at the top of whip antennas is used to prevent *corona discharge* (ionization of the air). Transmitting antennas can develop thousands of volts at their ends. This alternately drives electrons out into the surrounding air molecules and attracts them from the molecules, producing a corona glow and crackling noises, par-

ticularly if the antenna tip is sharp.

A λ/4 vertical, like any other antenna, has a high-V point at the far end, a low-V high-I point a λ/4 down the antenna, and zero radiation off the end of the antenna (upward). Thus, two vertical VHF or UHF antennas, one directly above the other, may have almost no intercoupling.

When the ground-wave field strength of vertical antennas is examined, it is found that if a ¼-wave antenna has 100 mV field strength at 1 mi, a ½-wave vertical will have about 125 mV, a ⅝-wave will have about 140 mV, but a ¾-wave will have only about 80 mV. For maximum ground-wave coverage for mobiles, a ⅝-wave vertical is best. It will be ⅝ wave long (high) with enough added coil at the base to make up ⅛ wave to produce a total of ¾ wave so that it can be grounded directly to the roof of the car. The feed point for a 50-Ω coaxial line is connected a few turns up the coil, to where the SWR is minimum. Like all other single vertical radiators, a ⅝-wave antenna is omnidirectional.

20-15 FULL-WAVE ANTENNAS

When a λ/2 antenna is fed ac, electrons oscillate back and forth along it. If the antenna is a full wavelength, electrons will move from both ends toward the middle during one-half of the excitation cycle (Fig. 20-16a). When the cycle reverses, the electrons travel from the middle toward the ends. Maximum voltage points appear at the middle and at the ends of a full-wave antenna.

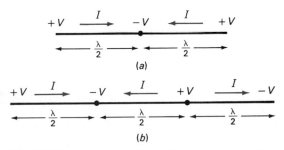

FIG. 20-16 (a) Current flows in opposite directions in two halves of a full-wave antenna at the same instant. (b) V and I in an antenna 3λ/2 long.

A 1-, 1½-, 2-, etc., wave wire will have maximum voltage points every λ/2. Electrons in adjacent half waves are always traveling in opposite directions at any one instant (Fig. 20-16b).

Voltage maximums every λ/2 indicate the development of standing waves on the antenna wire, with maximum current and maximum energy radiation. If the wires are not an exact multiple of a λ/2, lower-amplitude standing waves will be present and the antenna will be a less effective radiator.

Since a full-wave antenna has two equal currents flowing in opposite directions at any one instant, the radiation from one of the half waves exactly equals the radiation from the other. With the polarity of the two being 180° out of phase, the result is zero effective radiation at right angles to the wire. The horizontal radiation pattern of a horizontal full-wave antenna is shown in Fig. 20-17. The angle of maximum radiation is approximately 45° from the line of the wire for lobes 1 and 2.

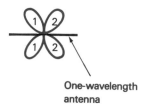

One-wavelength
antenna

FIG. 20-17 The radiation pattern of a full-wave antenna in free space has two doughnut-shaped lobes.

With a 1½-wave antenna, two of the half waves cancel the radiation of each other at right angles to the wire, leaving the third half wave to radiate in this direction. As a result, the major lobes are depressed toward the wire, and the direction of maximum radiation tends to follow the wire, although there is a lobe at right angles to it. The more half waves used, the more the maximum-radiation lobes are formed in the direction of the wire. When about four waves long, an antenna becomes directional *in line with the wire* (Fig. 20-18). If the same antenna is used at different frequencies, it will appear as having a different number of wavelengths and therefore different directional characteristics on each frequency. Note that there is a lobe for each λ/2. A "long wire" antenna is 4λ or more.

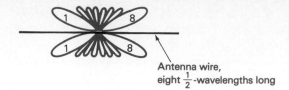

Antenna wire,
eight ½-wavelengths long

FIG. 20-18 The radiation pattern of a 4λ antenna in free space has eight lobes.

20-16 FEEDLINE COUPLING

There are many methods of feeding energy to an antenna. Cutting a λ/2 antenna in two parts and feeding it with a 50- to 72-Ω transmission line is a satisfactory method. This is known as series-type *center-* or *current-feeding* the antenna, since maximum current is present at the feed point.

If the antenna is a full wave in length and is cut in the middle, two points of high impedance are developed, usually having between 2000- and 3000-Ω impedance. Since a two-wire transmission line of more than 700 or 800 Ω is not practical, such an antenna cannot be fed by matching it with a flat transmission line. Instead, a high impedance is developed on a two-wire transmission line by deliberately producing standing waves and thereby high- and low-impedance points on it. Maximum standing waves are developed when the transmission line is some multiple of a λ/4. If the transmission line feeding the center of a full-wave antenna is a multiple of a λ/2 in length, it may be fed as in Fig. 20-19.

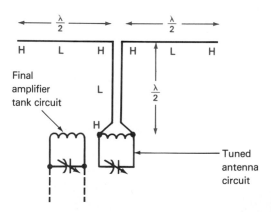

FIG. 20-19 Two λ/2 antennas fed with a λ/2 tuned feeder. (H and L indicate high- and low-impedance or voltage points.)

If cut to a multiple of λ/2 (no end effect), a transmission line repeats its terminal impedance. In Fig. 20-19, by starting at the far end of one of the λ/2 radiators (high-impedance or high-voltage point) and progressing toward the feed point, low impedance appears at the center of each λ/2 wire and high impedance at the feed point. By progressing down the line a λ/2, another point of high impedance is reached. By connecting this high-impedance end of the transmission line across the relatively high impedance of a tuned antenna LC circuit, a satisfactory impedance match is produced and maximum current flows in the tuned transmission line and in the two λ/2 antenna wires.

Starting at the center of the two λ/2 series radiators, but progressing down the transmission line only a λ/4 (or any odd multiple of a λ/4), a point of low impedance is reached. The low-impedance-point terminals can be connected to a few turns of wire (a low-impedance coil) inductively coupled to the tuned LC circuit of a transmitter (Fig. 20-20).

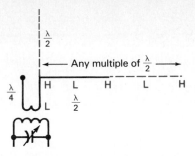

FIG. 20-21 If vertical, a Zepp becomes a "J."

600 Ω) is brought up to the center of a λ/2 dipole. To match the feedline to two points representing 300 or 600 Ω on the antenna, it is necessary to spread out the feeders at the antenna end (Fig. 20-22). The approximate ratio of lengths to match a 600-Ω line to a λ/2 antenna is shown.

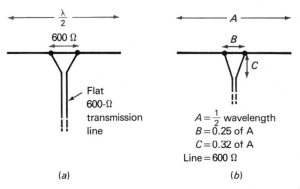

FIG. 20-22 (a) A delta-fed λ/2 antenna and (b) ratio of dimensions to match a 600-Ω line.

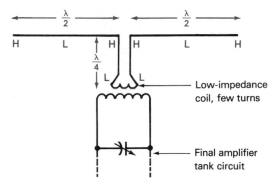

FIG. 20-20 Two λ/2 antenna wires fed with a λ/4 tuned feeder.

The addition of the few turns to the transmission line adds inductive reactance to the circuit. A variable capacitor, not shown, may be added in series with the coil to balance out the X_L.

Any multiple of λ/2 resonant antenna may be end-fed by using a tuned feeder system as above, leaving one end of the feedline unconnected (Fig. 20-21). This *Zepp antenna* (first used on zeppelins) is somewhat unbalanced. A λ/2 Zepp having a λ/4 feedline with the radiating element vertical (shown dashed) is called a "J antenna" and is used often in the VHF and UHF ranges.

Another method used to feed an antenna is the *delta match* (Δ match). A transmission line (300 or

Similar to Δ matching, a coaxial line, 72 Ω, for example, will couple to an antenna by connecting the outer sheath of the cable to the center of the antenna and extending the inner conductor to a point on the antenna equivalent to 72 Ω. This results in a slightly unbalanced *gamma match* (Γ match) (Fig. 20-23a). A balanced system can be produced by using two parallel coaxial lines or a special coaxial cable having two inner conductors parallel to each other. This is termed a *T match* (Fig. 20-23b). Capacitors added in series with the extended inner conductors cancel the X_L of the conductors. These are forms of shunt feed.

A shorted or an open *stub* is sometimes inserted in the middle of a λ/2 antenna to act as an impedance-matching device to a transmission line. Figure 20-24a shows a λ/2 shorted stub connected to a λ/2 antenna. The low impedance of the center of the

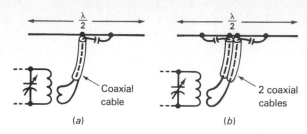

(a)   *(b)*

FIG. 20-23 (*a*) Coaxial transmission line terminated with a gamma match. (*b*) Two coaxial lines forming a T match.

antenna is repeated as a low-impedance point at the shorted end of the λ/2 stub. The whole antenna can be tuned by adjusting the shorting point. By connecting a 600-Ω flat transmission line at two points on the stub at which an impedance of 600 Ω appears, an excellent impedance match between transmission line and antenna system can be produced. A λ/4 open stub (Fig. 20-24*b*) can be utilized in a similar manner. While a stub might be any multiple of quarter waves, the fewer used, the less radiation from the feeder system.

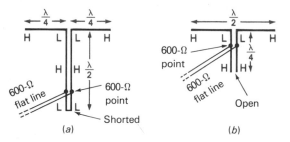

(a)   *(b)*

FIG. 20-24 Two λ/2 dipoles center-fed with (*a*) a λ/2 shorted stub and (*b*) a λ/4 open stub with flat lines coupled to the stub.

An application of stubs is in the elimination of interference to a TV receiver caused by a nearby VHF transmitter. A piece of twin lead is cut to λ/4 at the interfering signal's frequency (using velocity factor, Table 20-1). The insulation is removed from the two wires at one end and they are twisted together to form a shorted stub. The stub is held against the TV receiver twin-lead transmission line and moved back and forth until a point of maximum attenuation is found. The stub is taped to the transmission line and acts as a linear-circuit wave trap for the interfering frequency. If both ends are open but one end is connected to the TV set's antenna terminals, it acts as a wave trap also.

A λ/4 transmission line has a very useful prop-

TABLE 20-1   VELOCITY FACTORS

Dielectric material between wires	Velocity factor
Air-insulated parallel line	0.975
Air-insulated coaxial cable	0.85
Polyethylene parallel line (twin lead)	0.82
Polyethylene coaxial cable	0.66

erty. It can act as an impedance-matching device between high- and low-impedance circuits if it has the proper intermediate impedance (Fig. 20-25). The 70-Ω center impedance of a λ/2 antenna is coupled to a 600-Ω transmission line through a

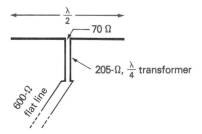

FIG. 20-25 A λ/4 matching transformer.

quarter-wave transformer. The required impedance of the λ/4 section is found by the formula

$$Z_o = \sqrt{Z_1 Z_2} = \sqrt{70(600)} = \sqrt{42,000} = 205 \ \Omega$$

where Z_o = λ/4 line impedance
       Z_1 = impedance connected across one end of the λ/4 line
       Z_2 = impedance across other end

If a λ/4 matching transformer is inserted between two *equal* impedances, it must have the same impedance and the whole line becomes flat with a 1:1 SWR.

Since there are no free ends on a λ/4 matching transformer, there is no end effect and the length of the line is computed by the formula

$$L = \frac{246(\text{vf})}{F}$$

where L = length in ft
       F = frequency in MHz
       vf = *velocity factor* of transmission line (see Table 20-1)

While discussed in terms of two-wire lines, stubs and transmission lines are also possible in coaxial form and in waveguide (Chap. 22) forms.

If transmission lines are constructed of parallel wires with a dielectric material between the conductors, the velocity of the wave traveling down the line is less than in free air. As a result, tuned transmission lines as well as matching transformer lengths must be multiplied by a *velocity factor* to determine the proper working lengths. Table 20-1 lists some velocity factors. Note that the factor does not depend on frequency.

Any solid material dielectric has more loss than air does in a transmission line. Type of dielectric and length of loaded line have no effect on SWR.

A transition from a 300- to a 400-Ω impedance line can be produced by using a tapered 4λ section spaced to have 300 Ω at one end and 400 Ω at the other end (Sec. 20-12). The tapered section is connected between the unequal impedance lines.

A wire 3λ/4 long used as an antenna has high impedance at the far end, but low impedance (about 50λ) at the feed end, which will match the output of many radio devices.

20-17 COLINEAR BEAM ANTENNAS

A shorted λ/4 stub is a λ/2-long wire. Voltages at the open ends of a properly excited stub will be 180° out of phase. If the open ends are connected to two λ/2 horizontal radiators (Fig. 20-26), the array is not a full-wave antenna but *two half waves in phase*, with the currents in both radiating λ/2 elements in the same direction at any given instant.

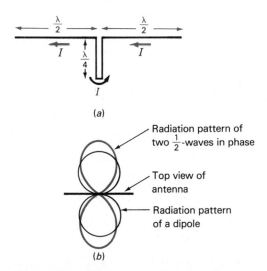

(a)

Radiation pattern of two $\frac{1}{2}$-waves in phase

Top view of antenna

Radiation pattern of a dipole

(b)

FIG. 20-26 (*a*) Colinear array composed of two λ/2 in phase. (*b*) Comparison of its radiation pattern with that of a simple dipole.

This forms a two-element *beam antenna*. It has a maximum radiation in the horizontal direction at right angles to the wire, like a dipole, but the lobes are narrower and longer. A gain of a little more than twice the power, about 4 dB, is produced in the direction of maximum radiation with the same power input to the antenna. No more power is radiated, but energy radiation is more concentrated in a direction at right angles to the antenna wire. It is a double-Zepp antenna.

Two half waves in phase can be fed with a flat transmission line tapped up on the stub or with tuned quarter- or half-wave feeders. The stub can also act as a matching transformer.

Greater energy radiation in a desired direction can be obtained by adding more λ/2 elements, each separated from adjacent elements by a λ/4 shorted stub, by λ/2 open stubs, or by any other means of changing the phase 180° (resonant *LC* circuit). This makes up a *colinear*, or *Franklin*, antenna. It gives a gain whether used horizontally or vertically. When used vertically, it has the advantage of holding the angle of radiation down, toward the horizon, and is omnidirectional.

When colinear elements are connected in phase, the center impedance of the center element is about 200 Ω with five or more elements.

20-18 DRIVEN ARRAYS

The narrowing, or beaming, of the radiation lobe of a multielement array, with its increased energy radiated at right angles to the antenna, is highly effective in producing strong signals at long distances. For reliable point-to-point communication, beam antennas are desirable. For broadcasting, a single vertical antenna will reach listeners in all directions. However, many broadcast stations are located on one side of a population concentration and require an antenna that will beam its signals toward the nearby city and possibly put a minimum signal in some other direction.

Several types of beam antennas are used in broadcasting. Most of them consist of two or three vertical antennas so placed and *driven* that their radiated signals reinforce in certain directions and cancel each other in others.

A simple two-element vertical beam is shown in Fig. 20-27. The two λ/2 vertical radiators are fed by a flat transmission line tapped into a shorted λ/4 stub, with a λ/2 open-wire tuned line between the bases of the two radiators. With the two radiators being fed 180° out of phase, a signal approaching

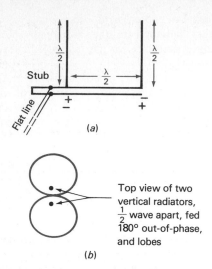

FIG. 20-27 (a) Two driven dipoles fed 180° out of phase. (b) Radiation pattern.

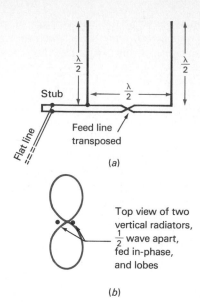

FIG. 20-28 (a) Two driven radiators fed in phase. (b) Radiation pattern.

the reader from one antenna would cancel the signal from the other, resulting in zero radiation at right angles to the plane of the page. The radiated wave from the first antenna expands and follows the wave along the transmission line. The radiated wave from the first antenna arrives in phase with the driven wave in the second antenna, and maximum signal is radiated in a line running through the two radiators (Fig. 20-27b).

When the same two antennas are fed by a *transposed* feedline (Fig. 20-28a), they are being fed in-phase signals. The signal approaching the reader is the sum of the two, or a stronger signal than would be radiated by a single antenna element. By the time the radiated wave from the first antenna reaches the second, it finds that the second antenna is being driven out of phase and zero signal is transmitted in the plane of the page (Fig. 20-28b).

It is also possible to feed λ/4 or other-length radiators similarly. The more radiators fed in phase, or *stacked*, the greater the gain of the system in the maximum signal directions and the less radiation in all other directions.

It is common to speak of the separation of two antennas in degrees. Two antennas separated by a full wavelength are said to be 360° apart. If separated by a λ/2, they are 180°. Two 950-kHz antennas separated by 120° are said to be λ/3 apart. The wavelength of 950 kHz is $\lambda = V/f_{kHz}$ or 300,000/950, or 316 m. One-third of 316 m is 105.3 m, or 345.5 ft of separation.

In broadcast stations two or more antennas may be fed different current values at different phase angles to control directivity of the radiation. Phase differences can be produced by using different-length transmission lines or by using L- or π-type phasing networks. Current variations can be controlled with the coupling networks. Changing either the current amplitude or the phase angle in any of the antenna elements shifts the directional pattern of the array. Such arrays may be tuned by variable coils in the coupling or phasing networks.

How two vertical antennas are fed controls the radiation lobes from them. For example, if separated by λ/4 and fed 45° apart, the signal will be stronger 45° on either side of a line between the two antennas. If fed 90° apart, a *cardioid* (heart-shaped) pattern is produced, with the strongest lobe in the direction of one antenna and a minimum in the direction of the other antenna. If separated 1λ there will be four major lobes developed, which change position as the feeding phase is changed. Thus it is possible to steer lobes by changing radiator spacing and the phase of the RF ac excitation to them.

Test your understanding; answer these checkup questions.

1. How many ground radials should a λ/4 vertical AM broadcast antenna use? _____
2. What is the advantage of drooping ground planes? _____

3. Why are small balls attached to whip antennas?
   _____
4. What is the feed-point Z at the base of a λ/4 vertical? _____ At the center of a λ/2? _____ At the center of a full wave? _____ A λ/4 from the end of a full wave? _____
5. Which radiates N and S, a 3λ/2 or a 4λ/2 horizontal E-W wire? _____
6. What length resonant line would couple an antiresonant circuit to the end of a λ/2 antenna? _____ To the center of the same antenna? _____
7. Why does a delta feedline fan out at the antenna? _____
8. What type of transmission line is used in a gamma match? _____
9. How long is a shorted stub usually? _____
10. What is the Z_o of a λ/4 transformer to match 500 to 50 Ω? _____
11. Does frequency affect the velocity factor of a transmission line? _____
12. Would a radio wave travel faster in an air-insulated or in a solid-dielectric coaxial cable? _____
13. What are beams called if all elements are in line? _____ What is the phase difference between elements? _____ What is used between elements? _____
14. Why do AM broadcast stations usually use vertical antennas? _____
15. What is a transmission line said to be if twisted 180°? _____
16. What is the phase separation of two antennas 3λ/8 apart? _____

20-19 PHASE MONITORS

When there are two or more elements in a commercial antenna array, it is necessary that the current in each tower be maintained within 5% of the licensed value to maintain maximum radiation in a desired direction and minimum radiation in the direction of some other relatively close transmitter's area. If the ratio of currents in a directive array change, the directivity of the transmitted lobe will shift, sending maximum or minimum signals in undesired directions. Broken ground radials can also shift the directivity of an array. To make sure that the radiation pattern remains constant, several field-strength points 1 to 10 miles away are established and the field strengths at these points are checked yearly during the station's proof-of-performance tests. As long as the field strengths remain constant, it is assumed the antenna is operating properly.

RF energy feeding a phased array of two (or more) vertical radiators is transported from the transmitter to a tuning house and its *phasor*, or antenna-tuning circuit, which distributes the RF to the radiators. If the feedlines from the phasor to the antennas are the same length, the two antennas are being fed in phase. If one feedline is a quarter wave longer than the other, the two antennas are fed 90° out of phase. Added L and C in the feedlines can also shift the phase if the feedlines are not exactly the correct length, as can the degree of coupling to the antennas.

Each antenna tower may have an ammeter at its base, but to compute antenna power a single *common-point ammeter* is used at the phasor where the transmission line branches to the towers. The equivalent antenna resistance is also measured at this point. Power is determined by $P = I_a^2 R_a$, using an unmodulated antenna current value. If the current ratios of the two or more elements must be held to 5% and the phase angle to 3°, the antenna is called a *critical phased array*.

The efficiency of an antenna feedline is

$$\text{Efficiency} = \frac{I_a^2 R_a}{P_o} \times 100$$

where I_a is the current at the common point, R_a is the common-point impedance, and P_o is the RF power output of the final amplifier.

To determine the relative phase of the currents in the towers, a *phase monitor* is used. Basically, a phase monitor consists of coaxial lines of equal length bringing in a sampling of the voltages developed in RF pickup coils at the base of the elements (Fig. 20-29). If the two signals fed to RF voltmeters M_1 and M_2 are adjusted to be equal, meter M_3 will read zero when the voltages from the antennas are in phase. The farther the voltages are out of phase, the greater the difference of voltage across M_3 and the greater its deflection. This results in a

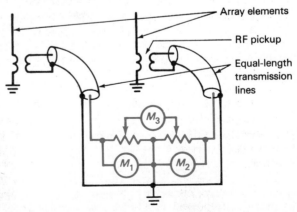

FIG. 20-29 A phase-monitor circuit.

direct indication of the phase difference between the two antennas. These sampling voltages can also indicate if the antenna currents are within the licensed limits. (Transmission lines feeding power to the antenna elements are not shown.)

20-20 PARASITIC ARRAYS

If a driven $\lambda/2$ element is a $\lambda/2$ from another similar *undriven* element (Fig. 20-30), the second has a voltage induced in it by the radiated field from the first and is *parasitically excited*. The current in-

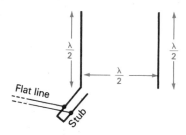

FIG. 20-30 A $\lambda/2$ parasitic element spaced a $\lambda/2$ from a driven element.

duced in the parasitically excited element is 180° out of phase with the wave that produces it (Lenz' law). The parasitic-element current produces a radiated wave that tends to cancel the originating wave, reducing the radiation in the plane of the page, but is in phase with the wave from the first antenna insofar as the reader is concerned. Such an array will radiate at nearly right angles to the plane of the page. Since the reradiated wave cannot be as strong as the driven wave, there is no complete null in any direction.

If the parasitically excited element is moved to within a $\lambda/4$ of the driven element (Fig. 20-31), it intercepts more energy and reradiates a stronger wave of its own. With $\lambda/4$ separation, the radiated wave travels 90° to the parasitic element and reverses phase 180°. Some of it returns 90° to the

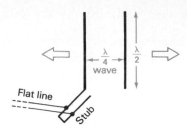

FIG. 20-31 A $\lambda/2$ parasitic element spaced a $\lambda/4$ from a driven element. Maximum radiation indicated by arrows.

driven element, arriving in phase with the next half cycle. Maximum signal is produced toward the left and right (arrows, Fig. 20-31). The parasitic element is acting as a *reflector and director*. The array has a gain of nearly 5 dB in both forward and backward directions. The radiation to the sides (toward the reader) is materially reduced.

A $\lambda/2$ parasitic reflector gives greatest forward gain when spaced about 0.2λ from the driven element. The radiation resistance at the center of the driven element drops to about 40 Ω. The more elements used in parasitic arrays, the *lower* the center impedance of the driven element.

If a parasitic element is a $\lambda/2$ long and is placed within about 0.1λ of the driven element, the induced V and I relationship is such that it acts as a director more than as a reflector and produces nearly 6 dB gain in the forward direction.

If a reflector is tuned for maximum forward gain (5% longer than a $\lambda/2$) and a director is also tuned for maximum forward gain (5% shorter than a $\lambda/2$), a three-element *Yagi* beam results (Fig. 20-32). It can have more than 8-dB gain in the forward direction with a 20-dB *front-to-back* power ratio. For

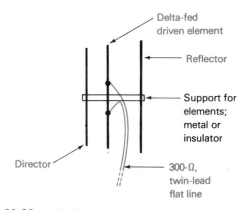

FIG. 20-32 A 3-element Yagi antenna consisting of a driven element, a reflector, and a director.

more forward gain, additional *director* elements are added.

The radiation resistance of a close-spaced three-element Yagi antenna may be about 15 Ω. It can be fed with a λ/4 matching transformer or with the delta-feed system shown. The driven element may be a *folded dipole* to raise its radiation resistance. Figure 20-33 illustrates a dipole and a folded di-

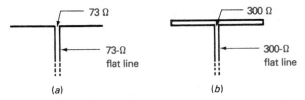

(a)   (b)

FIG. 20-33 (*a*) Center-fed dipole. (*b*) Center-fed folded dipole.

pole. The impedance at the open point of the folded dipole is four times what it would be if the upper wire were not added. With three wires (or if the second wire has twice the surface area of the first), the impedance will be nine times the impedance of the dipole. By formula,

$$Z = 73\eta^2$$

where η = number of wires used
Z = feed-point impedance

When used alone, a folded dipole has a broader bandwidth than a simple dipole. It maintains its center impedance value more nearly constant over a broader spread of frequencies.

Plane-surface-reflector, corner-reflector, and parabolic-reflector antennas are parasitic types. A plane-reflector antenna uses a plane surface about one-wave-square behind the driven element (Fig. 20-34*a*). It has a gain of about 7 dB in the forward direction. Spacing the driven element 0.2λ from the reflector surface produces a center impedance of about 70 Ω.

If the plane surface is bent into a 90° corner, with each side about 2λ long (Fig. 20-34*b*), the antenna is more directive and has about 12 dB of forward gain. Spacing the driven element 0.35λ from the corner produces a center impedance of about 70 Ω.

If the corner-reflector antenna is carefully engineered into a parabolic shape (Sec. 26-5) and the reflecting surface is sufficiently large (Fig. 20-34*c*), a beam width of 0.5° to 3° can be produced, with gains of 30 to 60 dB being possible. Such beams are used in microwave communications. The driven

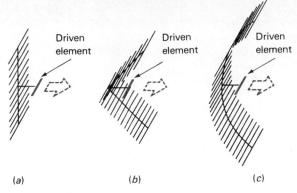

(a)   (b)   (c)

FIG. 20-34 Beam antennas. (*a*) Plane-reflector type, (*b*) 90° corner-reflector type, and (*c*) parabolic-reflector type.

element may be placed at the focal point of a round, parabolic-shaped "dish." For communication with parked satellites, the dish must have both *azimuth* (around horizon) and *elevation* (horizon to zenith) driving motors and controls. Since the gain of such dishes depends on the number of wavelengths of dish diameter, the same reflector will have more gain at higher frequencies. Gains of 40 to over 60 dB are not uncommon. They may have beam widths of a fraction of a degree at high gains. If the satellite drifts, it will be necessary to readjust the azimuth and elevation controls to maintain proper tracking.

The term *beam width* refers to the major lobe of a beam antenna, measured at the 3-dB-down points, or half-power points (where the field strength drops to 0.707 of the received voltage at the peak of the lobe). The angle between the two 3-dB points is the beam width. A λ/2 dipole has a beam width of about 90°.

20-21 LONG-WIRE BEAMS

There are several types of long-wire beam antennas. Since a resonant wire several waves long exhibits directional properties in the line of the wire, it is possible to utilize the lobes of such an antenna to form a *V beam* (Fig. 20-35).

Simplified lobes have been indicated on the two legs of the antenna. Lobes 1 and *A* are in the same directions, as are 4 and *D*, resulting in a maximum radiation to the right and left on the page. Lobes 2 and *B* tend to cancel each other, as do lobes 3 and *C*, resulting in little radiation upward and downward on the page. Such a V beam has high gain in the forward and backward directions. The longer

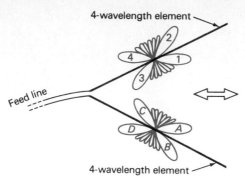

FIG. 20-35 Unterminated V-beam antenna 4λ on a leg.

the legs, in number of wavelengths, the greater the gain. With twice the number of wavelengths there is roughly twice the power, or 3-dB gain, in the direction of maximum radiation. This holds fairly true for the number of elements in most arrays.

The backward radiation of a V beam can be effectively reduced without affecting the forward radiation by terminating the ends of the antenna with noninductive resistors to ground. These resistors must have approximately the impedance of the far end of the antenna (about 800 Ω). When terminated with resistance, the antenna becomes nonresonant, therefore not frequency-selective, and has no standing waves on it! It acts as a properly terminated but improperly *spaced* transmission line, which forces it to radiate energy.

The V beam should not be confused with an *inverted-V* antenna, a λ/2 dipole supported at the center, with both ends dropping toward ground. This antenna radiates horizontally polarized energy at right angles to the plane of the wires and vertically in the plane of the wires.

Another long-wire beam, the *rhombic* (diamond) is composed of four legs (Fig. 20-36) instead of the two of the V beam but operates on the same general theory of addition and cancellation of lobes. It

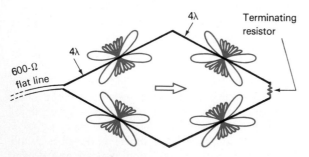

FIG. 20-36 Terminated rhombic-beam antenna 8λ on a leg.

has more gain; it cancels side radiation more; the resistance termination at the far end balances better. A gain of 15 dB or more is possible. A rhombic antenna is not resonant when terminated, as driven or parasitic arrays may be. Single rhombic antennas may operate satisfactorily in one direction over a frequency range from 6 to 30 MHz. The gain may increase at higher frequencies. It is bidirectional when unterminated. Rhombics are used in HF point-to-point communications.

20-22 LOOP ANTENNAS

When speaking of loop antennas, the loop is usually considered as being a 1- to 3-ft vertically wound coil, with a few closely wound turns, and either round or square in shape. The loop coil forms the inductance of a resonant *LC* circuit (Fig. 20-37a). Such a loop antenna is rarely used for

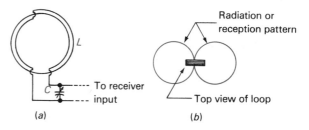

FIG. 20-37 (a) Small loop antenna forms the *L* of a tuned *LC* circuit. (b) Radiation or reception pattern of a loop antenna as seen from above.

transmitting, as its radiating efficiency is low. It finds use as a receiving antenna because of the sharp *nulls* (zero signals) produced.

Radio waves passing across the loop induce a voltage in it. As the loop is rotated 360°, there will be two points where little or no voltage is induced in it and two intermediate points where maximum signal voltages are produced (Fig. 20-37b). When the loop is viewed from above, zero signal is received from a station at the top and bottom of the page and maximum signal from a station at the left or right of the loop. This is the opposite field-strength pattern of a λ/2 dipole. The figure-8 pattern of the loop can be nearly perfect if the loop is balanced electrically (Sec. 25-23).

If the loop antenna is held in a horizontal plane, it becomes omnidirectional, receiving equally well in all directions.

When the sides of a large, single-turn square loop are λ/4 each (Fig. 20-38), the loop acts as two U-shaped λ/2 elements in phase, radiating and re-

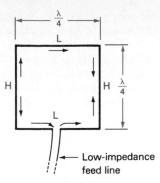

FIG. 20-38 Current flow at a given instant in a center-fed loop having λ/4 sides.

ceiving maximum at right angles to the wires of the loop, but with the vertical currents canceling each other. The loop is shown as center-fed with a low-impedance line. Such a loop is bidirectional but may be made unidirectional by using a parasitic reflector behind it and directors in front of it, making it a form of *cubical quad* antenna. Essentially, it is a folded dipole pulled apart vertically.

20-23 TOP-LOADING

Low-frequency (30–300 kHz) antennas are usually λ/4 Marconi types with ground radials. In the upper MF range (500–3000 kHz) it becomes practical to use λ/4 to 5λ/8 vertical antennas. At the low end of the broadcast band the antennas become quite long, and it is often desirable to *top-load* them. Top-loading may take the form of a metal wheel-like *hat* structure attached to the top of the antenna (Fig. 20-39). This hat increases the length of the antenna

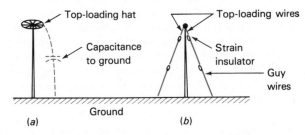

FIG. 20-39 Top-loading a vertical antenna by (*a*) using a top hat and (*b*) using the top of the guy wires.

to the edge of the hat and also increases the capacitance between the top of the antenna and ground. As a result of the increased inductance and capacitance, a short antenna can be made to resonate at a relatively low frequency (see also Sec. 25-6).

Top-loading may also be produced by using the top portion of the antenna guy wires as the top hat, as shown. Antenna insulators are preferably a glazed-surface porcelain. A hard, glossy surface tends to discourage the accumulation of dirt and reduces losses across the insulators. Near salt water, any encrustation may have relatively high conductance and produce leakage as well as corona effect on insulators.

Guy wires are usually cut to lengths that are not harmonically related to the frequency of transmission to prevent pickup of energy and reradiation. The insulators used are always egg or strain types. If an insulator breaks, the two wire loops are still joined and the guy wire does not part. If a common antenna insulator (Fig. 20-1) supporting a horizontal antenna breaks, the antenna comes down.

20-24 COUPLING TO ANTENNAS

There are many methods of coupling an antenna to a transmitter (or receiver). The simplest is by connecting the bottom of the antenna directly to a point on the tank coil (Fig. 20-40). The end of any length

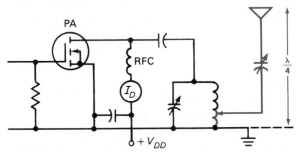

FIG. 20-40 Direct-coupling a λ/4 antenna to an *LC* circuit.

of wire has a resistance and a reactance component to any frequency. If a point can be found on the tank circuit where both antenna and tank R and X values equal or cancel, the antenna will take power. The closer the antenna is connected to the ground end of the coil, the lower the degree of coupling. If the antenna has considerable reactance (not resonant), a capacitor (or inductor) can be added in series with the antenna, as shown, to tune out the reactance. This type of coupling has no means of reducing harmonic radiation.

A π-network coupling circuit is shown in Fig. 20-41. L, C_1, and C_2 act as both a tank-tuning and antenna-coupling network. The base impedance of the antenna must match the impedance across C_2 to

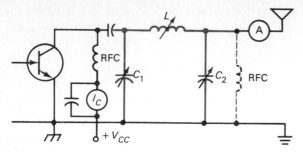

FIG. 20-41 A π-network coupling system to match the collector circuit to the bottom of the antenna.

take maximum power, and the impedance of the collector circuit of the BJT must match the impedance across C_1. Coil L must be the correct inductance to produce resonance in the circuit when C_1 and C_2 are adjusted to their proper values. If C_2 has a large capacitance value, its reactance will be small, resulting in a low voltage-drop across it; the antenna will have little voltage excitation; only a small current will flow into it; and it will take little power. Decreasing C_2 increases the coupling to the antenna. The RFC (or a 10-kΩ resistor) between antenna and ground discharges any static charges picked up by the antenna. It has no effect on tuning.

For better harmonic suppression, a π-network may be followed by an LC L-network.

To tune a π network, start with maximum C_2 value (minimum coupling). Tune C_1 to resonance (minimum I_D). Note the antenna current (I_a). Decrease C_2 somewhat, then redip the I_D. The I_a and I_D values should both rise. Keep decreasing C_2 and retuning C_1 until the I_D minimum is the desired value for the device used. Note the I_a. Repeat this tuning procedure, but use less inductance L. Then try with more inductance. When maximum I_a is obtained with rated I_D, the stage can be considered as tuned to the antenna.

An advantage of a π network is the output capacitor. At the fundamental frequency it helps to tune the circuit, but at the second harmonic it has half the reactance, reducing the feeding of harmonic frequency voltages to the antenna. Higher-order harmonics are bypassed still more.

If the base of a top-loaded $\lambda/4$ vertical is insulated from ground, the impedance to be matched may be about 25 Ω. This can be matched by using two 50-Ω coaxial lines in parallel, connecting the center wires to the base of the antenna and the outer conductors to the central point of the ground-radial system. This is *series feeding*.

It is a good policy to require all antennas that are not directly grounded to have lightning protection in the form of a heavy-duty knife switch to ground the antenna, or to use arcing points.

To match the low impedance at the base of a $\lambda/4$ vertical to a higher-impedance feedline, an L-type impedance-matching network (Fig. 20-42) can be

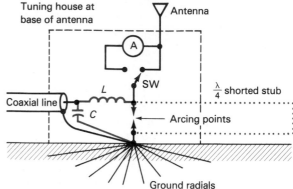

FIG. 20-42 Step-down-ratio L network between line and antenna, including arcing points and a possible $\lambda/4$ shorted stub.

used. For a step-down ratio of impedances the LC configuration must be as shown. This network can present a pure resistance at input and output, but there is always a phase shift in an L network. It can also be designed to match a flat transmission line (pure resistance) to the base of an antenna which is reactive (not exactly a $\lambda/4$).

To match the high impedance at the bottom of an insulated $\lambda/2$ vertical to a lower-impedance transmission line, capacitor C of the L network is moved to the other end of inductor L. This produces the required step-up impedance ratio.

Note the antenna-ammeter. When not in use, the switch is thrown to the right, taking the ammeter out of the circuit to prevent burnout due to lightning striking near or directly on the antenna. The meter may also be removed by a jack and plug arrangement. Between the base of the antenna and ground a lightning gap may be installed. This gap must be just wide enough not to arc with the voltages built up by the transmitter but close enough to arc with the higher voltages of a nearby lightning strike. A $\lambda/4$ shorted stub, shown dotted, connected between the base of an ungrounded antenna and ground presents a very high impedance for RF ac at its resonant frequency, but a low-resistance dc discharge path to ground. It also provides an *even-order* harmonics wave trap to ground.

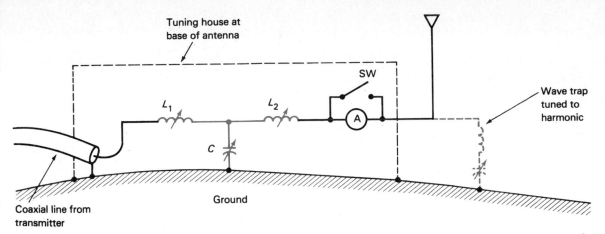

FIG. 20-43 A T-network coupling circuit to match the feed line to the antenna. A wave trap reduces one harmonic.

A π network can produce either step-up or step-down impedance ratios or controlled phase shifts, as can a T-type network (Fig. 20-43). Coil L_1 is tuned for minimum I_p or I_D; coil L_2 is tuned for maximum I_a. A series wave trap from antenna to ground can be used to decrease radiation of any harmonic to which it is tuned.

If the base of a vertical $\lambda/4$ ($3\lambda/4$, $5\lambda/4$, etc.) antenna is connected directly to a ground-radial system, it can be shunt-fed by a half-delta feed system (Fig. 20-44). Whatever the transmission-line imped-

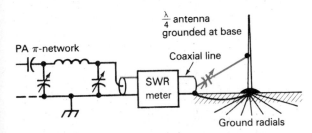

FIG. 20-44 A shunt-fed antenna coupled through an SWR meter to a π-network output circuit.

ance, there is some point on the antenna with the same impedance value. However, the length of the lead up to the desired point on the antenna represents inductance or X_L. To balance out this X_L, a capacitor can be added to the feedline as shown. While the point on the tower can be computed, a series of cut and try tests is usually required to establish the optimum placement and tuning by watching the SWR meter in the line.

Coaxial transmission lines have been shown between transmitter and antenna. They may be of the solid-dielectric or the gas-filled type. Coaxial lines have many advantages: All fields are confined inside the outer conductor, radiation from the line is nil if properly terminated, and the line may be buried if desired. To allow for expansion and contraction, long runs are usually constructed to have a bend or two. To prevent damp air from being drawn into the cable when cooled, air-dielectric lines may be sealed and filled with about 5 psi of nitrogen or dry air. Any loss of pressure indicates a leak. Location of leaks can be determined by watching for bubbles when soapy water is painted on the line or joints. Sections of hollow coaxial line can be butted together, and a sleeve sweated (soldered) over the joint, or the sleeve can be clamped with rings. Screw-type end couplings can be used on solid-dielectric lines. Sharp bends should be avoided with any concentric lines to prevent mechanical pressures and breakage, as well as change of impedance where the line is deformed. Care must be taken that no sharp points are developed inside a cable. An internal arcing point may be indicated by heating at that point under loaded conditions.

Open-line feeders are usually run on top of wooden poles about 8 ft above ground level from transmitter to the tuning house at the base of the antenna. Any excessive lumps of wire or solder form discontinuities and can reflect energy and produce standing waves. Kinks in the wire will also weaken the wire. Flat transmission-line wires must be maintained at the same spacing to prevent change of impedance and resulting reflections. (Not important with tuned lines.)

Automatic tuning of antennas is discussed in Sec. 25-6.

Test your understanding; answer these checkup questions.

1. What is used to determine phase difference between two broadcast antennas? _____
2. Which gives more forward gain, a director or reflector? _____ Which is longer? _____ Which is closer to the driven element? _____ How much do they differ in length from a λ/2? _____
3. How much more feed-point Z does a folded dipole have than a normal dipole? _____
4. Would it be possible to produce 70-Ω twin lead? _____
5. With similar-size parabolic, plane-reflector, and corner-reflector beams, which has greatest gain? _____ Narrowest lobes? _____
6. Why would the wires of a V beam be closer together when used on higher frequencies? _____ How is the beam made unidirectional? _____
7. Why might a rhombic be better than a V beam? _____ _____
8. Are the beams in Fig. 20-34 vertically or horizontally polarized? _____
9. Why is the usual rhombic not frequency-sensitive? _____
10. In what direction is the null of a small-diameter loop in relation to the plane of the loop? _____
11. When is a loop omnidirectional? _____
12. Why are verticals sometimes top-loaded? _____
13. Why is π network superior to direct coupling to an antenna? _____
14. Why might L networks be used between an antenna and transmission line? _____ _____
15. Why are antenna ammeters often shorted out or disconnected from the antenna except when readings are desired? _____

20-25 DETERMINING IMPEDANCE OF AN ANTENNA

Within limits, any length of antenna will take power at any frequency. A 200-ft-high mast can be made to radiate on any frequency in the standard broadcast band. At 560 kHz, it is only about λ/8 high, but at 1600 kHz it is about λ/3. It would be capacitively reactive at the lower frequency and inductively reactive at the higher. What feedline impedance should be used?

To resonate the antenna at 560 kHz, the base connection may be broken and a loading coil inserted to cancel the capacitive reactance of the antenna. A sensitive RF thermocouple ammeter is then added (Fig. 20-45a). A calibrated variable-

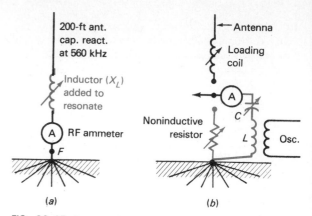

FIG. 20-45 Antenna base resistance may be measured by (a) an impedance bridge or by (b) substitution.

frequency oscillator of several watts can be loosely coupled to the inductor and set accurately to the desired frequency of operation. The loading inductor is varied until the antenna ammeter peaks. The antenna is now resonant. The oscillator is turned off. The antenna circuit is broken at the desired feed-point F, and an RF impedance bridge is inserted and adjusted to a balance. The impedance indicated should be the required feedline impedance. (Being shorter than a λ/4, the base impedance will be quite low, 10 to 20 Ω.)

The ammeter can be inserted anywhere in the antenna when tuning for resonance, but when computing antenna power ($P_o = I_a{}^2R_a$), it must be at the point where the resistance is measured.

Another method of determining antenna resistance is by substitution. As above, the antenna must first be brought to resonance. The circuit between the loading coil and ground is broken, and components are connected as shown in Fig. 20-45b. The oscillator is coupled to the series L and C, and the switch is thrown to the antenna position. The capacitor tunes out the reactance of L and brings the antenna back into resonance. The ammeter reading is noted. The switch is then thrown to the dummy-antenna position, and the noninductive resistor is varied until the ammeter reads the same as when the antenna was connected. The resistor value should equal the antenna resistance. The antenna resistance is usually checked 20 kHz above and below the operating frequency also.

For 1600-kHz operation the antenna above is longer than a λ/4. It is necessary to insert a series capacitor to tune out the inductive reactance of the antenna. A similar method of impedance measurement and substitution can be used.

In all these measurements care must be taken that lead lengths are as short as possible and that stray coupling paths do not exist.

20-26 FIELD STRENGTH

The measure of received strength of a radio signal at a certain point in space is the voltage developed in a 1-m-long wire. The field strength or field intensity is measured in received volts per meter (V/m). Since only a small fraction of a volt will be induced in a remote wire, it is more usual to express the field in millivolts per meter (mV/m) or microvolts per meter (μV/m). (A simple field-strength meter might consist of a metal box with antennas out the top and bottom and a tuned circuit with a rectifier and sensitive meter calibrated in RF millivolts inside.) If a wire 3 m long has 0.001 V induced in it by a certain signal, the field intensity is 0.001/3, or 0.333 mV/m.

The field strength of a received signal is dependent on ground, direct, reflected, and refracted sky waves as well as transmitter power. Field strength will vary at different times of the day unless the path is very short.

The ground wave over seawater decreases almost inversely as the distance from the station, particularly in the low- and medium-frequency ranges. If the field strength at 10 miles is 50 mV/m, at 20 miles, it will be only slightly less than 25 mV/m. As the distance from the station increases to more than 100 miles, the signal decreases more than in a simple inverse proportion. With higher frequencies the signal decreases much more than in a simple inverse proportion until the sky wave begins to be received.

Because power is proportional to both I^2R and V^2/R, twice the current in an antenna represents four times the power. Similarly, twice the voltage represents four times the power. If 1000 W of RF power in an antenna produces 10 mV/m in a remote antenna, 4000 W will produce 20 mV/m. If the field strength doubles at a remote point, it indicates that the power in the transmitting antenna has quadrupled (increased 6 dB).

If the power in an antenna is doubled, the voltage in it (and the field intensity produced by it) is increased by $\sqrt{2}$, or by 1.414. If 1 kW in a transmitting antenna can produce 3 mV/m in a receiving antenna, 2 kW in the same antenna will produce 3(1.414), or 4.24 mV/m.

Field-strength measurements are usually made starting at 1 mile and extending out to at least 20 miles to produce valid average readings.

20-27 FIELD STRENGTH OF HARMONICS

The field strength of harmonic radiations from a transmitter can be measured in volts per meter, but the difference between the fundamental and any harmonics is usually expressed in decibels (Sec. 8-6).

If the fundamental is 147 mV/m and a harmonic is measured as 405 μV/m, the voltage ratio between the two is 147,000/405, or 363:1. In decibels this is

$$\text{dB} = 20 \log \frac{V_1}{V_2} = 20 \log 363 = 51.2 \text{ dB}$$

Because of the different transmission characteristics of different frequencies at the same time of day, a harmonic signal from a transmitter may have a greater field intensity at a distant location than does the fundamental and may interfere with other radio services when the fundamental cannot be heard at all by these other services. Thus, a 5-MHz transmitter may not be heard 2000 miles away during the daytime, but an unattenuated 10-MHz harmonic may be quite readable at that distance.

There are many methods of decreasing harmonic radiation (Sec. 16-16). Figure 20-46 illustrates parallel wave traps connected in an antenna transmission line to attenuate any harmonics to which they are tuned. A series wave trap across the transmission line at any point will effectively decrease radiation of the frequency to which it is tuned.

20-28 EFFECTIVE HEIGHT

The effective height of a transmitting antenna is the height of the antenna's center of radiation above *average terrain*. On a contour chart 8 evenly spaced radials are laid out extending from the antenna site for a distance of 10 miles, beginning toward the north, and radiating outward every 45°. The elevations *above mean sea level* (AMSL) are determined at 0-, 2-, 4-, 6-, 8-, and 10-mile points along each radial. The arithmetic average of these 48 elevations (6 elevations on each of 8 radials) is

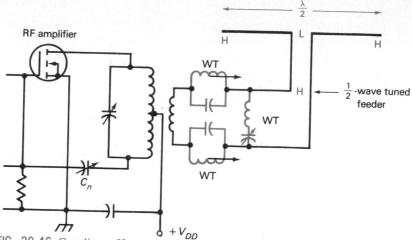

FIG. 20-46 Coupling a Hertz antenna to an RF amplifier. Wave traps are used in the feeders to attenuate harmonic radiation.

calculated. This is the "average terrain" altitude. The "effective height" is the altitude of the center of radiation of the antenna *minus* the computed average terrain altitude.

EXAMPLE: The center of the radiating portion of the antenna is at a 1000-ft altitude. The average terrain calculates to 700 ft. The effective height is 1000 − 700 = 300 ft.

20-29 FIELD GAIN AND ERP

Multielement transmitting antennas may be rated in *field gain* over a dipole (or over an *isotropic radiator*, an imaginary antenna that radiates equally well in all vertical and horizontal directions). If a multielement transmitting antenna produces a 500-mV/m signal in a remote receiving antenna, whereas a simple dipole transmitting antenna will produce only a 250-mV/m signal, the field gain of the multielement antenna is 2 (twice the voltage), or 6 dB.

(The gain of a dipole over an isotropic radiator is 1.64, or 2.15 dB.)

The *effective radiated power*, abbreviated erp, of a transmitter considers the *field gain* of the antenna. As an example, a 60% efficient transmitter has a 1000-W dc input final amplifier. Therefore, the output power to the antenna feedline is 600 W. If the transmission line to the antenna radiating elements is 90% efficient, there must be 540 W of RF being fed to the antenna. Since field gain is expressed in voltage, and power is proportional to *voltage squared*, if the antenna has a field gain of 1.3, the erp is 540 × 1.3², or 912.6 W. (Eirp is the erp when referenced to an isotropic radiator.)

The *power gain* of an antenna can be rated in decibels. As an example, a transmitter has a V_p of 5000 V, an I_p of 3 A, and is 60% efficient (9000-W output). If the antenna has a power gain of 4 (or 36-kW output), but the transmission line has a 3-dB loss (half power), the erp is 18 kW.

If the output of a transmitter is 1000 W and the antenna transmission-line loss is 50 W, with a *power* gain of 4 (or 6 dB) the antenna array has an erp of 4(950), or 3800 W in the direction of its major lobe.

20-30 THE GROUND

The surface of the earth directly below and surrounding an antenna is known as the *ground*. With Marconi antennas the ground is used as one-half of the antenna. With self-resonant Hertz antennas operated high above the earth, the ground may play a less important role. With both types of antennas the ground acts as a reflector of transmitted and received waves and is responsible for a portion of the total transmitted or received signal.

The best ground would be a silver sheet extending several wavelengths in all directions under the antenna. While this is not practical, it is possible to use copper ground radials. Each wire is separated from the next by 5° to 15° of arc and is usually $\lambda/4$ long. [If radials are buried 1 ft (30 cm) in damp ground they may not have to be more than $\lambda/10$ long.] If some of these ground radials are broken off, the resistance of the antenna will change, as will the directivity of the radiation pattern. One of the best practical grounds is a salt marsh.

Without ground radials, the actual or virtual ground may vary from a few inches below the surface of a salt marsh to several feet below the surface of a dry, sandy soil. Because an antenna is erected a $\lambda/4$ above ground physically does not mean that it is operating electrically at that height. Losses result from poor ground areas and connections.

In a boat the ground is either the metal hull or a copper plate attached to the outside of the hull with a heavy ground wire running to the radio equipment. If the wire is too long, the radio equipment will be above ac ground potential, and RF may be induced in the ground lead, which may change the feed-point impedance of the antenna.

20-31 COMPUTING ANTENNA POWER

If an antenna is resonant, it can be considered a resistance. It exhibits neither capacitive nor inductive reactance to the feedline. As a result, the power, voltage, current, and resistance in a tuned antenna can be computed by Ohm's law and power formulas. As examples, if resistance and current at the base of a Marconi antenna are known, the power radiated by the antenna can be found by the formula

$$P_o = I_a^2 R_a$$

The surge impedance of a *flat* transmission line is 500 Ω, and the line has a current of 3 A flowing in it. The power being fed to the antenna is

$$P = I^2R = 3^2 \times 500 = 4500 \text{ W}$$

If the daytime power in a broadcast station antenna is 2000 W and the antenna impedance is 20 Ω, the current is

$$I = \sqrt{\frac{P}{R}} = \sqrt{\frac{2000}{20}} = 10 \text{ A}$$

If the antenna current of this antenna is cut in half for nighttime operation, the transmitter has an antenna power of

$$P = I^2R = 5^2 \times 20 = 500 \text{ W}$$

If the daytime transmission-line current of a 10,000-W transmitter is 12 A, the line impedance is

$$R = \frac{P}{I^2} = \frac{10,000}{144} = 69.4 \ \Omega$$

If it is required to reduce to 5000 W at sunset, the new value of transmission-line current is

$$I = \sqrt{\frac{P}{R}} = \sqrt{\frac{5000}{69.4}} = 8.49 \text{ A}$$

A 72-Ω concentric transmission line is carrying 5000 W. The rms voltage between the inner conductor and the sheath is

$$V = \sqrt{PR} = \sqrt{5000(72)} = \sqrt{360,000} = 600 \text{ V}$$

Since the radiated power of an antenna is directly proportional to the antenna current squared, if the antenna current is doubled, the power is increased 2^2, or 4 times. If the current is increased 2.77 times, the power increases 2.77^2, or 7.67 times.

A transmission line delivers 10,000 W at 4.8 A to an antenna. The line impedance is

$$R = \frac{P}{I^2} = \frac{10,000}{4.8^2} = 434 \ \Omega$$

If the current fed into this transmission line is 5 A, the power fed into it is

$$P = I^2R = 5^2(434) = 10,850 \text{ W}$$

With 10,850 W delivered into the transmission line and 10,000-W output, 850 W must be lost in the line itself.

When antenna resistance and current are known, the *output* power of a transmitter can be determined. When the plate current and voltage are known, the *input* power can be determined. From these two values the efficiency of the amplifier can be found. For example, if the dc input is 1500 V and 0.7 A, the power input to the final amplifier is 1050 W. If the antenna current is 9 A and the antenna resistance is 8.2 Ω, the antenna power is

$$P = I^2R = 9^2(8.2) = 664 \text{ W}$$

The efficiency is determined by output/input, or 664/1050, or 0.632, or 63.2%.

20-32 OMNIDIRECTIONAL ANTENNAS

It has been mentioned that the vertical antenna is omnidirectional in the horizontal plane, with vertical polarization.

VHF and UHF TV and FM emissions are normally horizontally polarized. Transmitting antennas for these services should transmit equally well in all directions and still have horizontal polarization.

A horizontal λ/2 dipole transmits a 2-lobed horizontal pattern with nulls off the ends of the antenna. By bending a dipole into a horizontal loop, it exhibits omnidirectional properties but is not a true omnidirectional antenna.

If two λ/2 dipoles are arranged to form a horizontal X (Fig. 20-47), and the two antennas are fed

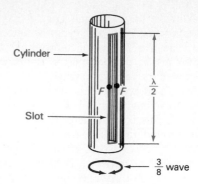

FIG. 20-48 A vertical slotted-cylinder antenna produces a horizontally polarized radiation for VHF or UHF.

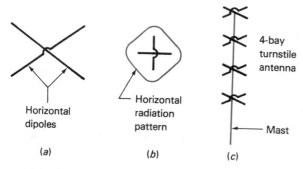

FIG. 20-47 (a) Single-bay turnstile antenna. (b) Horizontal radiation pattern of a turnstile antenna. (c) A 4-bay turnstile (feed lines not shown).

90° out of phase through a λ/4 long transmission line, the resultant horizontally polarized radiation pattern approximates a circle. This forms a basic *turnstile* antenna. For greater gain, turnstile antennas are usually stacked with two or more *bays*, one above the other, as shown. Two dipoles can also be fed 90° out of phase by using two transmission lines, with one a λ/4 longer than the other.

If a basic 2-element turnstile is rotated 90° so that one element is vertical, the other horizontal and they are separated a little, the radiation is said to have *circular polarization* because it will induce signals into receiving antenna wires oriented horizontally, vertically, or in between.

Another omnidirectional antenna is the slotted cylinder. It consists of a metal cylinder with a vertical λ/2 slot cut out of it. The slot is fed with a coaxial transmission line at the points marked *F* in Fig. 20-48. When a cylinder about 3λ/8 in circumference is used, current flows from one side of the slot to the other, around the outside of the cylinder,

producing an omnidirectional pattern. If the circumference is greater than a λ/2, the antenna becomes directional with maximum radiation from the slotted side of the cylinder.

A *discone* (disk-cone) is a vertically polarized, omnidirectional, wideband antenna (Fig. 20-49)

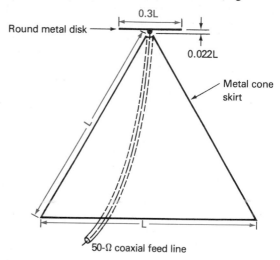

FIG. 20-49 A broad-band discone antenna.

used in the VHF and UHF regions. It has a feedpoint impedance of about 50 Ω, making it simple to excite. The sheet-metal (or wire-mesh, or multiwire vertical rods) cone skirt is about 10% longer than a λ/4 of the lowest frequency to be used. The round, flat metal disk is 30% of the skirt length *L*, the spacing between the disk and cone is about 2.2% of the skirt length, and the diameter of the cone base is equal to the skirt length. The antenna has a radiation pattern about the same as a λ/4 vertical, but has a low SWR over better than a 5 to 1 frequency range.

Other omnidirectional antennas are forms of horizontal loop antennas in which the loop is broken up into three or four dipoles. Each dipole is fed so that the current flow in it is in the same direction as in all dipoles at the same time.

Since most of these omnidirectional antennas are used for VHF and UHF transmissions (FM and TV), which are essentially line-of-sight transmissions, they are usually found on tops of hills or peaks to allow maximum transmission range.

20-33 DANGER OF RF FIELDS

All ac magnetic fields produce some *nonionizing* radiation (*ionizing* radiations are X-ray and higher frequencies), and exposure to them has some effect on human tissues. Frequencies up to about 50 MHz tend to pass through the body, heating it somewhat, but the body cools when out of the field, usually without damage. Above this frequency, damaging effects increase. The upper safe U.S. radiation limit on the surface of the body is considered to be less than 10 mW/cm². Maximum body heating occurs at about 1000 MHz. Radiation above this frequency tends to be reflected from the body surfaces, and heating lessens. It is the *near-field* radiation from an antenna (within $\lambda/2$) which produces maximum radiation absorption. Past this, the radiation decreases inversely as the distance squared. Sterility and cataracts are among the difficulties produced by excessive nonionizing radiation. "Damaging" values of radiation are a function of frequency, proximity, and exposure time and are difficult to determine.

Test your understanding; answer these checkup questions.

1. What are two methods explained to determine the feed-point Z of a vertical antenna? _____ _____
2. In what unit is field intensity measured? _____
3. What formula is used to determine effective height of an antenna? _____
4. How fast does low-frequency field strength attenuate? _____ High-frequency? _____
5. If 4 kW in an antenna produces 50 μV/m in a receiving antenna, what will 16 kW produce? _____ 8 kW? _____ 1 kW? _____
6. Under what condition might a harmonic of a transmission be heard at a distant point when the fundamental cannot be? _____
7. Is field gain measured in voltage or dB? _____
8. What is involved in the erp of an antenna? _____
9. Why do ground radials assure constant feed-point Z? _____
10. What feed requirements must be met to produce a circular radiation pattern when using two crossed dipoles? _____
11. When speaking of antennas, what is a bay? _____
12. In what range of frequencies are most omnidirectional horizontally polarized antennas used? _____ Rhombics? _____ Parabolic reflectors? _____
13. A flat transmission line delivers 5 kW at 10 A. What is its Z? _____ The V across it? _____
14. If the antenna current increases 3.3 times, how much does the radiated power increase? _____ The field strength? _____
15. What is the polarization and advantage of a disconc? _____ _____

ANTENNAS QUESTIONS

1. What length antenna acts like an oscillating *LC* circuit?
2. In relation to a transmitting antenna wire, what are the relative directions of the fields involved?
3. What are the names of the layers of the ionosphere useful for long-distance radio communications? How high is the highest? The lowest?
4. How often does sunspot activity repeat?
5. What can solar winds cause?
6. What are five causes of fading signals?
7. What frequencies have the most similar night and day characteristics? Which tend to skip the furthest?
8. What frequencies are essentially line-of-sight?
9. What causes ducting? Transequatorial scatter?
10. What frequencies are most affected by lightning?
11. What polarization would a vertical antenna have?

Why is polarization important in radio? What signals are affected least by it?
12. Why is a $\lambda/2$ antenna always shorter than that computed by $\lambda = v/f$?
13. What is the length in feet of a $\lambda/2$ antenna cut for 5.5 MHz? What is the length in meters?
14. What is the name of an antenna that does not use ground as part of its length? Which one does?
15. At what point(s) on a $\lambda/2$ antenna is minimum voltage developed? Minimum current?
16. What is the radiation resistance value of a $\lambda/2$ antenna $\lambda/4$ high? Many wavelengths high? When only a few feet high?
17. If a $\lambda/2$ antenna is center-loaded with a coil, what effect does this have on its resonant frequency? If loaded with a series capacitor?

18. Two 2-mm-diameter wires held 6 cm apart center-to-center form what impedance line?
19. If a 210-Ω antenna is coupled to a 70-Ω transmission line, what is developed on the line? What will be the reflection coefficient and the power transmitted down the line?
20. What are the two types of transmission lines used?
21. What is another name for an SWR meter?
22. Which is more efficient for long-distance communication, a horizontal antenna $\lambda/4$ or $\lambda/2$ above ground? Why?
23. What type of antenna is omnidirectional and has a cone of silence?
24. What is the approximate Z value at the base feed point of a vertical $\lambda/4$ antenna? Vertical $\lambda/2$?
25. Why are small metal or plastic balls attached to the top of vertical antennas?
26. How can two vertical antennas be placed to provide minimum intercoupling between them?
27. What is unusual about the currents in the two halves of a full-wave antenna? What is the result?
28. What is the directivity of radiation of an eight-wavelength horizontal wire antenna?
29. What is another name for a "J antenna"?
30. Why are $\lambda/4$ or $\lambda/2$ open-wire lines known as tuned lines? What antenna impedances can they match?
31. If a $\lambda/2$ dipole is fed at the center with a $\lambda/2$ open-wire line, what will be the impedance at the transmitter?
32. What length would an open stub have if connected to the open center of a $\lambda/2$ dipole?
33. Why do gamma matches have series capacitors in the center line lead to the antenna?
34. To couple a 300-Ω twin lead to a 72-Ω dipole, what is the Z of the $\lambda/4$-matching transformer?
35. What would be the length of the line of problem 34 for a frequency of 4 MHz?
36. Describe a Franklin colinear array.
37. Two vertical $\lambda/2$ elements drawn on a page are fed in phase. What amplitude radiated signal is developed in the plane of the page? At right angles to the page? How could the radiations be reversed?
38. What antenna placement and feed produces a cardioid radiation pattern?
39. What is a phasor in a directive antenna array?

40. Draw a diagram of a phase monitor for a 2-element directive array.
41. If a parasitically excited $\lambda/2$ is a $\lambda/4$ from a radiator, in which direction is maximum signal radiated? If a $\lambda/2$ from the radiator?
42. What is a possible gain in dB of a 2-element $\lambda/4$ spaced parasitic array?
43. What is the forward gain in dB of a close-spaced 3-element Yagi, and what is radiation backward compared to that forward? What do these values give?
44. What are the two main features of a folded dipole?
45. What parasitically excited antenna has most gain?
46. At what points on a lobe is beam width measured?
47. What are the two long-wire beams mentioned? How are they made unidirectional? Under what condition are they nonresonant?
48. What is an inverted-V antenna? Is it a beam?
49. What is the main advantage of vertical rotary loop antennas? Of a horizontal loop?
50. Why is top-loading used on vertical antennas?
51. What is the disadvantage of direct coupling a transmitter to an antenna? What is the advantage of using π-network coupling, and how can it be improved?
52. With a π-network coupling system, what component controls the degree of coupling to the antenna?
53. If an antenna has a higher Z value than its feedline, where is the L-network C placed?
54. What is the advantage of using a $\lambda/4$ shorted stub between the base of an antenna and ground?
55. How are antenna ammeters protected from lightning?
56. What are the two methods discussed of measuring the Z of a broadcast antenna?
57. In what units might a field-strength meter be calibrated?
58. How many dB down is a harmonic reading 14 μV/m if the fundamental reads 560 mV/m at 10 miles?
59. Why would it be possible to hear the harmonics but not the fundamental of a distant station?
60. What is the effective height of an antenna?
61. What is an isotropic radiator?
62. If an antenna has a field gain of 5.2 dB over a dipole, what is its gain over an isotropic radiator?
63. How is the erp of a transmitter computed?
64. How is a good ground assured for vertical antennas?
65. Why could $P = V^2/R$ not be used with a transmission line having a high SWR?
66. Why might I_a be reduced at night at a BC station?
67. If a 50-Ω AM broadcast coaxial cable has 50 kW, what is the rms carrier voltage between inner conductor and sheath? Peak voltage at 100% modulation?
68. How must elements of a turnstile antenna be fed?
69. What antenna will have circular polarization?
70. What are the length and diameter of a discone skirt? The diameter of the disk?
71. What frequency produces maximum radiation effects on the body? What damages might result?

ANSWERS TO CHECKUP QUIZ ON PAGE 479

1. (Z bridge, R substitution) **2.** (mV/m or μV/m) **3.** (Radiator altitude minus average terrain altitude) **4.** (Directly proportional to distance) (Greater than simple ratio) **5.** (100 μV/m) (70.7 μV/m) (25 μV/m) **6.** (When fundamental is too weak but harmonic is refracting in) **7.** (In both) **8.** (P_o times field gain) **9.** (Virtual ground constant in any weather) **10.** (Feed 90° apart electrically) **11.** (A section which would be a complete antenna by itself) **12.** (VHF, UHF) (HF, VHF) (UHF, SHF, EHF) **13.** (50 Ω) (500 V) **14.** (10.89 times) (3.3 times) **15.** (Vertical) (Broadband)

21

Two-Way Communications

This chapter discusses various services and equipment involved in two-way communications, either between a base station and mobile units, or between mobiles. Some basic telephone information as it might be applied to microwave links used between control, base, and relay stations is presented. The material can be considered an extension of the basic theory discussed in the chapters on frequency modulation and amplitude modulation. It is fundamental information which service managers of communications equipment depots might want new applicants to understand.

21-1 RADIO SERVICES

Radio services can be categorized as being (1) one-way, or broadcast, such as the Standard AM Broadcast Band stations, the FM Broadcast Band stations, the TV Broadcast stations, time and frequency broadcasts, weather broadcasts, and paging transmissions; (2) two-way communications, such as Citizens' Band (26.96–27.41 MHz, AM or SSB, using shared assigned channels), Amateur (explained in Chap. 32), Aeronautical (Sec. 21-2), and among others, the General Mobile Radio Service (460–470 MHz only, FM, 50-W output or less, for the general public, using shared assigned frequencies).

Besides the General Mobile, there is the *Private Land Mobile Radio Service* stations that fit into the two-way category. These are:

Public Safety, including Police, Fire, Highway Maintenance, Forestry Conservation, and Local Government radio services.

Special Emergency, including Medical, Rescue, Veterinarian, Disaster Relief, School Bus, Beach Patrol, and Paging radio services.

Industrial, including Power Company, Petroleum Company, Forest Products, Motion Picture, Relay Press, Business, Manufacturers, and Telephone Maintenance radio services.

The stations in the *Domestic Public Landline Mobile Radio Service* (DPLMRS) operate similar to the Private Land Mobile Radio Service except that they are connected to common carriers (telephone lines) to allow units to make telephone calls while mobile.

These commercial services are covered in the "FCC Rules and Regulations," Parts 90 and 22.

21-2 AERONAUTICAL COMMUNICATIONS

Radiotelephonic communications for aircraft employ AM and various types of SSB emissions in the HF range in the following bands:

2.86–2.99 MHz
2.182 MHz
3.023 MHz, search and rescue (SAR)
3.281 MHz, lighter-than-air
3.41–3.46 MHz
4.38–4.99 MHz
5.45–5.68 MHz
8.364 MHz, distress, SAR
8.82–8.96 MHz
10.01–11.39 MHz
13.27–13.34 MHz
17.90–17.97 MHz

In the VHF range:

118–136 MHz
121.5 MHz, simplex, emergency and distress
123.1 MHz, SAR
156.8 MHz, distress

In the MF range (CW):

410 kHz, direction finding
457 kHz, working frequency 500 kHz, distress

Information on aviation services can be found in the "FCC Rules and Regulations," Part 87.

21-3 FREQUENCIES, TOLERANCES, AND POWERS

The basic frequency bands involved in two-way mobile-base communications are listed in Table 21-1. The tolerance for transmitters is in percent. To convert to hertz, multiply the tolerance percent by 100 and multiply this by the assigned frequency. For example, for an assigned frequency of 150 MHz, the tolerance of a base station would be $150,000,000 \times 0.000005$, or 750 Hz. The carrier should be as close to 150 MHz as possible but is still within tolerance if less than 750 Hz from 150 MHz when in operation. The tolerance of SSB signals in most services is 50 Hz (10 and 20 Hz for aviation stations) because any greater off-frequency operation produces distorted audio in receivers tuned to the assigned frequency.

Most two-way ground-based voice communications are made with the use of frequency modula-tion. However, special signaling emissions may also be used in the services listed above.

The authorized bandwidth of an FM or AM two-way transmission must contain 99% of the emission. The other 1% might be distributed as shown in Fig. 21-1a. For FM the bandwidth is 20 kHz; for AM, it is 8 kHz. At the band edge, signal power must be down 25 dB, at 250% of the band edge, down 35 dB, and down 80 dB from there out.

The authorized bandwidth of an SSB emission is usually 3500 Hz with the assigned frequency 1400 Hz *higher* than the authorized suppressed carrier frequency (Fig. 21-1b). The carrier should be 40 dB down from the power of the J3E emission itself. At 1.75–5.25 kHz from the assigned frequency, signals must be down 25 dB; at 5.25–8.75 kHz, down 35 dB; and down 60 to 80 dB any further out.

Transmitters in these services should use no more power than necessary for satisfactory operation, and not over: 1.5 kW below 3 MHz, 750 W at 3–25 MHz, 300 W at 25–100 MHz, 350 W at 100–470 MHz, 1000 W erp at 470–512 MHz, and whatever is stated in their FCC authorization from 806 MHz and up.

Police radio service stations may also use CW for interzone or intrazone communications using 2804 and 5195 kHz for calling, and 2808, 2812, 5185, and 5140 kHz for working frequencies.

21-4 BASIC COMMUNICATIONS SYSTEMS

The fundamental two-way communications system has a *base* station with its *control* or *operating*

TABLE 21-1	FREQUENCY TOLERANCE IN TWO-WAY COMMUNICATIONS			
	Fixed and base stations		Mobile stations	
Frequency range, MHz	Over 200-W output, %	200-W output and less, %	Over 2-W output, %	2-W output and less, %
Below 25	0.005	0.01	0.01	0.02
(USB J3E)	50 Hz	50 Hz	50 Hz	50 Hz
25–50	0.002	0.002	0.002	0.005
50–450	0.0005*	0.0005	0.0005	0.005
450–512	0.00025	0.00025	0.0005	0.0005
806–866	0.00015	0.00015	0.00025	0.00026
1427–1435	0.03	0.03	0.03	0.03

Note: 0.0005% = 5 parts per million (ppm); 0.002% = 20 ppm.

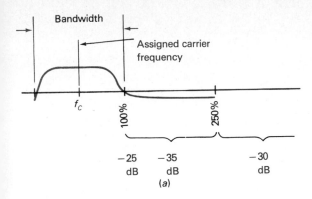

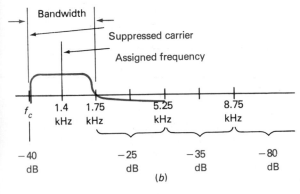

FIG. 21-1 Bandwidths and attenuations for (*a*) FM and (*b*) SSB.

point in a building in a city. One or more mobiles will communicate with the base station by VHF or UHF transmitter-receiver sets. The base station may have its antenna on top of the building in which it is situated. The base transmitter and the control point are usually within 100 ft of each other. The base and mobiles operate on the same frequency and use common antennas which are switched from transmit to receive by a "TR" relay. Reliable range depends on the base station antenna location, height, and directivity and on the power used by base and mobile transmitters. A reliable range might be from 5 to 25 miles.

Further extension of the range requires that the base transmitting and receiving antennas be raised in height. The control point may remain where it is, but a *remote* transmitter, receiver, and antenna system may be mounted on top of the highest building in town, or on top of a nearby hill or mountain peak that commands a view of the area that must be covered. Methods of communicating VF and control signals to the remote station may be by telephone lines, microwave links, or fixed low-power radio relay systems. The mobile range may be extended to over 100 miles.

In some systems one or more dispatch points will be connected to the control point to allow communication with mobiles by different offices located in a city.

Fire, police, and other systems may have so many mobiles from different departments or localities involved that they may require two or more operating frequencies (channels).

Mountain-top *repeaters* require all units to listen on the same frequency but to transmit on another nearby frequency. Signals received from mobiles are coupled directly to the input of an accompanying transmitter and are rebroadcast to other mobiles. This provides much better mobile to mobile communications than is possible between units that may be separated by a few miles in a city where shielding and reflections from buildings may greatly attenuate signals. Repeaters are discussed in Sec. 21-10.

21-5 AN FM COMMUNICATIONS TRANSMITTER

A simplified mobile (or base) station-type FM transmitter is shown in block form in Fig. 21-2. It is one-half of a transmit-receive system. The transmitter feeds its VHF or UHF band signals to a vertical antenna. The crystal oscillator output may be phase-modulated as shown, or direct FM may be developed by feeding AF to a voltage-variable capacitor across the crystal, as discussed in Sec. 19-14. The modulated signal is doubled twice and then tripled before being fed to the driver and power amplifier. The PA output is fed through a low-pass filter to prevent harmonic radiation and then to a *power control* ALC feedback bias system to maintain a constant carrier power output. The FM RF output is fed to the antenna when the push-to-talk (PTT) microphone switch is pressed. The output of such transmitters for mobile operation is between 10 and 100 W. For base stations the output may be as high as 350 W.

Most of the circuits are similar to those discussed in Chap. 19 and previous chapters. When the PTT switch is pressed, the transmit-receive, or TR, relay shifts from the receive to the transmit position. Besides transferring the antenna from the receiver to the transmitter, the power-supply voltage (battery for mobiles) is switched from the receiver to the transmitter circuits, allowing signals to be emitted.

In many services there may be more than one

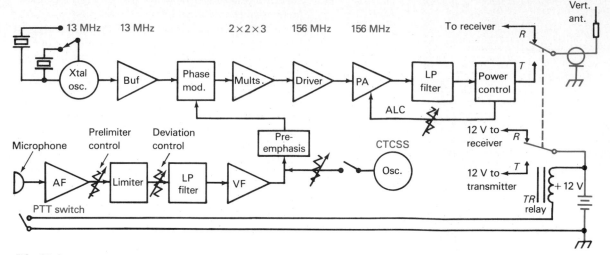

FIG. 21-2 Block diagram of a possible VHF FM mobile (or base) transmitter.

frequency used. If so, the transmitter (and receiver) will have the capability of switching from one crystal to another to select the desired frequency of operation. Since all frequencies will be relatively close together (usually within 100 kHz and within a 3-dB-bandpass limit) no retuning of any of the stages will be required when frequencies are switched.

Many modern transmitters and receivers use frequency synthesizers in conjunction with PLL circuits and may require no frequency multipliers. They generate the assigned frequency for the transmitter directly, as well as the LO frequency for the receiver. The VCO of the PLL circuit can be frequency-modulated to produce the desired modulation.

The tuning of an FM transmitter is similar to that of any other type of transmitter. After proper oscillator frequency has been assured, each stage is tuned for a maximum output signal from the power amplifier. If transmitters have power-control feedback circuits, these must be checked to assure desired power output. If intermediate or final amplifiers are slightly mistuned, the carrier frequency is not affected, but with FM transmitters, the result may be unequal upper or lower sideband output under modulated conditions, as well as an apparent amplitude variation, which must be corrected.

With a 1000-Hz sinusoidal ac fed into the first AF amplifier, the deviation control following the limiter is set so that with just discernible limiting the deviation is at a desired value. In practice, with

a 1000-Hz sinusoidal ac input, deviation is set to about 4.5 kHz. The high peaks of loud voice signals when the dynamic or electret microphone is spoken into will exceed the sine-wave modulation peaks even though the voice peaks are limited, resulting in an approximately 5-kHz deviation. The prelimiter control is set so that with a normal speaking level into a mobile microphone, from a distance of about 1 in., the voice peaks are limited to some degree, producing reasonable modulation density. Stronger voice sounds will limit heavily, producing somewhat more modulation density, but weaker sounds will limit very little, if at all, producing insufficient modulation density to be heard well. Because AF peak limiters or clippers produce AF harmonics, a *postlimiter* 3-kHz low-pass filter is required to prevent transmission of the harmonics. After the filtering the AF may be amplified and then passed through a 6-dB per octave pre-emphasis or integrator circuit (Sec. 19-10).

A direct-FM Pierce oscillator circuit used in communications transmitters is shown in Fig. 21-3. When V_{CC} is keyed on to transmit, the audio input modulates the reverse-biased varactor bias by changing the voltage-drop across R_1, shifting the frequency of the crystal and producing FM. The small variable capacitor across the crystal is used to "warp" (tune) the crystal to the assigned carrier frequency.

Many mobile crystal modules are equipped with a heater and a thermostat that maintain the crystal at a constant temperature, holding the crystal oscillating frequency constant. In others a thermistor

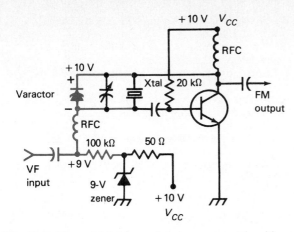

FIG. 21-3 Direct FM by modulating a crystal with a varactor.

(heat-sensitive resistance) senses any warming and develops a change in the varactor bias which compensates for any frequency drift.

When in noisy locations an operator should make sure the microphone face is pointed away from sound sources and may cup the hands around the microphone to shield it from external noises. If the prelimiting voice signal level is too high, external sounds such as wind noise in a moving vehicle will be bothersome to persons receiving such transmissions.

21-6 CTCSS

A circuit used in many communications sets is the *continuous tone coded squelch system* (CTCSS), also known as a *private line* (PL), or *channel guard* system. It is used in selective calling in services having many mobiles but where only one, or one group, of mobiles is to be called and worked. All receivers in such a system must be fitted with CTCSS circuits. If a base station wishes to transmit to one group of mobiles only, it will transmit the required CTCSS tone for that group. With the CTCSS subaudible (67–210 Hz) oscillator on, the CTCSS tone is transmitted as long as the PTT switch of the transmitter is held closed. Assume the CTCSS tone to be 100 Hz. The CTCSS oscillator is made to generate its 100-Hz ac that weakly modulates the carrier (about 10% of the full allowed deviation) at the same time the voice is being transmitted. Because the 100-Hz tone is subaudible, it is not heard by the receiving operator, but it is accepted by the 100-Hz CTCSS filter which unsquelches the receiver, making received voice

signals audible. Other CTCSS mobiles not having the required frequency tone filter module in their receivers will not unsquelch and will remain silenced. If the base station transmits some other CTCSS tone, 86 Hz perhaps, only those mobiles with 86-Hz CTCSS tone detectors in them will unsquelch.

To prevent a CTCSS mobile from transmitting when some other mobile group is operating on the channel, as soon as the PTT microphone is taken "off-hanger" or "off-hook," the receiver goes into normal carrier squelch operation. If someone is using the channel, it will be audible to the person picking up the PTT microphone. If nothing is heard, the operator is clear to transmit on the channel.

A common CTCSS system uses an iron-reed-type oscillator-filter. The fundamental idea can be explained with the simple circuit shown in Fig. 21-4.

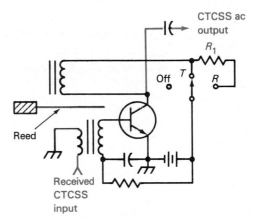

FIG. 21-4 CTCSS reed oscillator-filter circuit.

The iron reed has the length and mass to vibrate at the desired subaudible frequency, such as 100 Hz. The reed forms the coupling between the input and tickler coils of an indeterminate frequency Armstrong-type oscillator. When the switch is set to the transmit position (T), the alternating fields between the coils sets the coupling reed vibrating at its resonant frequency, and the oscillator is forced into this oscillation frequency. The very stable CTCSS output ac is used to modulate the transmitter. When the switch is thrown to the R position, R_1 is added to prevent the circuit from self-oscillating, but it will still amplify. A CTCSS signal being received will be amplified if at the reed frequency, and considerable feedback will be produced. The resulting CTCSS ac is than rectified and used to unsquelch the receiver AF amplifier (Sec. 21-7). A switch in the receiver can disable the CTCSS, rendering the receiver carrier-operated.

Rather than a reed oscillator-filter, an active filter may be used. By incorporating in-phase feedback around an active filter it will oscillate at the frequency of its peak signal passage.

Similar results can be obtained with a *digital private line* (DPL) system. When a base station wants to call a certain mobile, it continuously transmits a subaudible digital coded word as soon as its PTT switch is closed. If the word is the proper one for a mobile, it is recognized by a programmed memory, an unsquelching dc is developed in the mobile receiver, and it unsquelches. The DPL module and a microprocessor in the mobile allow it to transmit back by use of the same DPL code, making the conversation essentially private unless someone else happened to be off-hook.

▌ Test your understanding; answer these checkup questions.

1. Name the three groups of services in the Private Land Mobile Radio Service. _____

2. What does DPLMRS mean? _____
3. What types of emissions do airplanes use? _____
4. Which may have the greater percent tolerance, mobiles or bases? _____
5. How much of the emission must be within the bandwidth of a transmitter? _____
6. How is the assigned frequency related to the suppressed carrier frequency with J3E emissions? _____
7. What service uses HF CW communications? _____
8. Which is likely to have the better coverage, base or repeaters? _____
9. What are the three methods of communicating between base and repeaters? _____ _____
10. Which uses two frequencies, a remote or a repeater station? _____
11. What might be a reliable range mobile-to-mobile? _____
12. What is the main reason for using repeaters? _____
13. Are dispatchers connected to base stations or control points? _____
14. What is an advantage of direct FM over PM transmitters? _____
15. Is deviation controlled before or after the limiter? _____
16. What always follows a VF limiter? _____

17. Where is CTCSS ac inserted in a transmitter? _____
18. Why is retuning not required for multichannel mobiles? _____
19. How close to a mobile microphone should the lips be? _____
20. In Fig. 21-3, why is V_{CC} greater than V_Z? _____
21. What does CTCSS mean? _____ DPL? _____
22. What is the frequency range of CTCSS ac? _____
23. What happens to a CTCSS receiver when it is off-hook? _____

21-7 AN FM RECEIVER WITH CTCSS

Mobile and base receivers are superheterodynes, often using double conversion. A representative FM receiver incorporating a simplified CTCSS circuit is shown in block form in Fig. 21-5. If the communications system uses more than one channel, there will be two or more switchable crystals in the first LO. The LO in this example is multiplied to a frequency 10 MHz from the desired receiving frequency. The 10-MHz IF developed by the heterodyning of the input signal and the LO signal is amplified and fed to a second mixer, where it is translated or down-converted to 450 kHz. It first passes through a 450-kHz 20-kHz-wide bandpass filter. The signal is fed to one or more 450-kHz IF amplifiers and then to two or more IF limiter stages before being detected by either an *LC* or crystal-type discriminator. Modern IF strips may use four monolithic crystal filters coupled together by untuned IF amplifiers and limiter amplifiers, feeding a crystal-type discriminator, ratio detector, or quadrature detector. The audio output from the discriminator or detector is amplified and is then fed to a loudspeaker.

Output from the detector is fed to both a carrier squelch system (Sec. 19-11) that can squelch the AF amplifiers and to a CTCSS circuit. If switched to the CTCSS circuit, as shown, a fixed negative dc voltage through a high resistance is applied to the VF amplifier to squelch it so that no received signals can be heard. Only if the transmitting station adds a CTCSS subaudible tone to its transmission will the CTCSS oscillator-filter pass this frequency, amplify it, and rectify it. The dc so developed is opposite in polarity to the fixed dc squelching voltage, cancels it, unsquelches the VF amplifiers, and allows the receiver to pass any detected voice signals. The transmitting station is now able to call and talk to the selected mobile receiver.

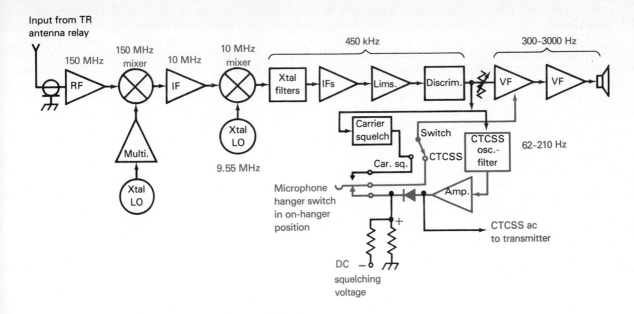

FIG. 21-5 Mobile receiver with CTCSS and carrier squelch.

As soon as the mobile operator takes the microphone off-hanger, the normal carrier squelch circuit is switched in and the receiver operates as if it had no CTCSS circuit. When the microphone is replaced on its hanger, the fixed negative squelch voltage is back in the circuit. If the microphone is left off-hanger, or if the squelch switch is set to the carrier squelch position, CTCSS operation is defeated and the receiver operates with carrier squelch only. On transmit, the CTCSS is keyed on and its output can be fed to the modulation circuits.

Simpler receivers may use a single mixer and 10–20 MHz crystal filters, IFs, limiters, and discriminator. Some receivers may use RF amplifiers, but many may use only narrowband tuned circuits between antenna and mixer. In this case the tuned circuits may be *helical resonators* (Fig. 21-6). The incoming signal excites the first *LC* circuit in its cavity into oscillation when the variable capacitor is tuned to resonate with it. The second cavity is electromagnetically and electrostatically coupled to the first by an open hole or *slot*. The smaller the slot, the looser the coupling and the narrower the bandwidth of the filter. With four to six tuned slot-coupled cavities, a relatively flat bandpass filter of a few hundred kilohertz can be developed. (With a simple *LC* tuned transformer, a passband of over 1 MHz might result.) The narrow bandpass reduces the possibilities of images, as well as intermodulation caused by two local signals mixing in the input

active device. It may also result in decreased noise figure and increased dynamic range (Sec. 18-11). Encircling an inductor with a metal case in this manner acts like a shorted turn around the coil, decreasing its inductance, requiring more turns in the coil, and possibly producing higher *Q* values. The mixer circuit shown is a common one used with BJTs in communications receivers. Helical resonator filters may also be used between an LO or its multipliers and the mixer to prevent oscillator

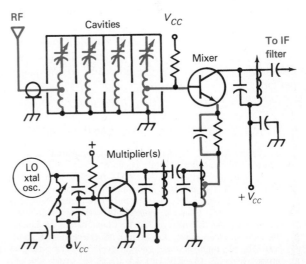

FIG. 21-6 Helical resonators feeding a mixer system.

harmonics from producing unwanted beats with input signals.

A true resonant cavity (Sec. 11-18) may be used when it is necessary that strong near-frequency RF energy be kept out of fixed station receiver or transmitter circuits. Cavities have very high Q values and very narrow passbands. Because they must be essentially a quarter-wavelength long they are quite large for lower VHF band applications. However, they may be found in many pieces of UHF equipment.

When only one antenna is used for simultaneous transmitting and receiving, two or more cavities may be used in a *diplexer* or *duplexer* circuit that prevents the transmitted signal on one frequency from interfering with the received signal on another nearby frequency.

21-8 PAGERS

One application in VHF-UHF communications in which a receiver is the only device used is in *pagers*. These are shirt-pocket-sized, 6- to 10-oz receivers. Some use common carrier squelch circuits, and may be activated by a two-tone signal from the base station which sets off a series of audible beeps (alerting signals) to tell the carrier to report to or call the base station by telephone. Pagers are used in the 30–50, 132–174, 406–420, 450–512, and 929–932 MHz bands. Antennas are inside the plastic receiver cases in most units.

Newer pagers may have microprocessors to control their capabilities. These use subaudible digital calling codes that may allow up to a thousand different pagers to be selectively called, or with different internal plug-in modules (actually ROMs), several different groups of pagers may be selectively called. Pagers may emit audible beeps only or may beep and open the AF amplifier to allow the receiver to pass a voice message to the paged person.

Controls on pagers will be an on-off switch, volume, and sometimes manual squelch setting controls. Some have a beep-voice selection switch, and some have a push-to-hear switch. Many models will operate while in a small charger box that connects the internal battery (through the bottom of the case) to the charger circuit. When in a standby operating condition some pagers alternately turn the V_{CC} on and off at a rapid rate to conserve battery by only having the receiver actually on for a fraction of the time.

Besides one-way pagers, there are also desk-top monitor or *scanner* receivers, capable of receiving from two to hundreds of different programmed frequencies. Some scan the desired frequencies, stopping whenever a carrier is found. Others have a priority frequency, scanning as long as there is nothing on the priority frequency, but switching to it whenever a carrier is detected on that frequency.

21-9 HAND-HELD TRANSMITTER-RECEIVERS

So-called handie-talkie FM transmitter-receiver units are complete portable stations. They have receivers very similar in most features to pagers, plus having miniature transmitters included in the hand-held cases. These units are used in the 30–50, 132–174, 406–512, and 806–854 MHz bands, usually with quarter-wave vertical antennas mounted on the top surface of the case. The quarter-wavelength VHF antennas may be spiral wound on a flexible form ("rubber duckies") to shorten the antennas, or they may use full ¼- or ⅝-wave antennas for added gain.

The transmitter may be a single-channel type, crystal-controlled, with separate receive and transmit crystals. Hand-helds may operate on several different channels by switching to different internal crystals. Usually the transmitting crystal is frequency-modulated and after several stages of multiplication feeds a 1-, 2.5-, or 5-W output final power amplifier. The eyes and face should be kept at least 2 in. (5 cm) from the transmitting antenna at all times, particularly with the higher-power models, to prevent possible nonionization radiation damage to the eyes (cataracts).

Some hand-held equipment use frequency synthesis with PLL circuits to develop both the local oscillator frequencies for the receiving section and the assigned frequency ac for the transmitter.

When operating *simplex*, both transmitter and receiver are on the same frequency. When using repeaters to extend the 1- to 5-mile range of hand-helds, the transmitter will transmit on one frequency, and the receiver will listen on another frequency, from 500 kHz to several megahertz higher or lower, depending on the channel frequencies. With frequency synthesis this is fairly simple to do. Repeaters may increase the range of a hand-held up to 100 miles.

21-10 REMOTE AND REPEATER STATIONS

Remote and repeater stations, mentioned previously, are located on either a high building or on top of a high peak near an area to be serviced by a VHF or UHF communications system. A *remote* or fixed *relay station* is connected to the control point via telephone lines, microwave link, or by a radio relay system. One type of remote will have a fixed transmitter and a receiver, both operating on the same frequency as the mobiles, with a single antenna switched from transmit to receive by a changeover or TR relay. Because of its placement it will have longer receiving and transmitting ranges than would be possible at the lower-altitude control point. Its received signals are relayed down to the control point, and control point audio is relayed up to the remote station transmitter. Either the receiver or the transmitter is operating, but not both at the same time, allowing a single antenna to be used.

A *repeater station* will receive on one frequency but relay its received information to its transmitter, which simultaneously radiates the information on another frequency, 300 kHz–5 MHz higher in the VHF range (600 kHz for amateurs), and 3–12 MHz in the UHF range. For repeater operation, all fixed and mobile equipment must be capable of transmitting and receiving on different frequencies. Mobiles may also be able to switch to simplex (both on the same frequency) for nearby mobile-to-mobile communications.

Repeater systems are highly complex. The system shown in Fig. 21-7 is considerably simplified. It has a 155-MHz receiving antenna and receiver that feeds detected AF to a 150-MHz transmitter having its own antenna. Since antennas radiate minimum energy in line with the elements, with one directly below the other, minimum transmitting energy is picked up by the receiving antenna, preventing overloading and desensitization of the receiver input circuits, or generation of intermodulation products. To further assure minimum transmitter pickup, a 150-MHz *LC* (or cavity) wave trap might be connected across the receiver transmission line as shown. Because of its much higher *Q*, a cavity-type

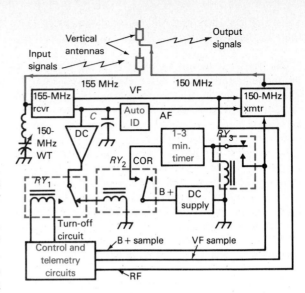

FIG. 21-7 Basics of a repeater system.

wave trap would be more effective. The receiver can then receive weak 155-MHz signals while the transmitter is simultaneously relaying and emitting strong 150-MHz signals. A greater transmit-receive frequency separation (*offset*) may be required for UHF where circuit *Q* values are lower.

The unsquelching dc developed when a carrier is received can charge a time-delay capacitor *C* and be amplified enough to close the normally open (NO) RY_2 carrier-operated relay (COR) through the normally closed (NC) turn-off relay, RY_1. When the COR closes, it starts a timing circuit (set for 1 to 3 min) current to flow, closing RY_3 and applying dc power (B+) to the transmitter. As a result, both dc power and modulating AF are fed to the transmitter. In this way the 150-MHz transmitter retransmits 155-MHz received mobile voice signals to either another mobile or to the base station. If anything holds the transmitter on for an extended period, the timed current stops and locks off. RY_3 opens, B+ is removed from the transmitter, and no signals are rebroadcast until the receiver is squelched and unsquelched again. At the end of a mobile transmission the receiver is squelched but the transmitter continues to transmit until capacitor *C* discharges, usually a fixed period of 1 to 5 s. Such a delay allows mobiles which may fade out for a short period of time to hold the transmitter in operation.

A repeater can run unattended for long periods of time. In case of trouble, however, the control operator, by using remote controls, can activate RY_1 to shut down the transmitter or may call for sample values of B+, RF, and VF, as indicated, to be telemetered (relayed) down to allow evaluation of any difficulty.

The required identification (ID) of the station can be made automatically after each transmission, or every 10, 15, or 30 min, whatever is required of the service. The ID may be developed from either an automatically operated voice tape, or a Morse code tape (F3A), by ROM-controlled AF oscillators, or by a synthesized voice signal. IDs can be keyed on by the squelch voltage when it is developed in the receiver, or when it drops off.

Few modern repeaters use relays. They use biased BJTs as the controlling devices, and many protective circuits are built into the repeater equipment, such as overload, off-frequency, excessive modulation, and insufficient modulation.

21-11 20-dB-QUIETING AND SINAD TESTS

There are two methods of determining and stating the capability of a communications receiver to receive weak signals in noise. With the *20-dB-Quieting* method the receiver is disconnected from the antenna and an *unmodulated* RF signal generator, tuned to the input frequency of the receiver and calibrated in microvolts (μV) or dBm output, is coupled through an impedance matching network (if needed) to the receiver input (Fig. 21-8). A multi-

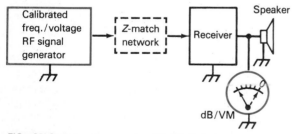

FIG. 21-8 Setup to measure 20-dB-Quieting.

range AF ac voltmeter, calibrated in dB, is connected across the receiver loudspeaker. With the RF signal generator turned off, the receiver gain control is turned up to produce a noise output at a level equal to nearly the rated receiver AF output power (using $P = V^2/R$, where V is the voltmeter reading and R is the impedance of the speaker). The dB meter is switched to a range where the noise develops 0 dB on the meter. As the unmodulated RF output of the signal generator is increased slowly, the noise output of the receiver begins to decrease (remember, all receivers are quieted by a carrier). The RF signal output level is increased until the

receiver noise is reduced 20 dB on the meter (or to one-tenth the starting voltage). The number of microvolts of increased signal from the generator that produces the 20 dB of quieting is the "sensitivity" of the receiver. A good communications receiver should have 20 dB of quieting at 0.2 to perhaps 0.5 μV. If the microvoltage is higher than 0.5, the receiver is not considered to have high sensitivity. In general, it is the front-end circuits which produce noise in receivers. The overall *dynamic range* of the receiver from minimum received signal to maximum AF output should be at least 120 dB.

The *12-dB-SINAD* (Signal Including Noise And Distortion) measurement method requires a special SINAD meter (or distortion analyzer), consisting of a multirange AF voltmeter also calibrated in dB, and a sharp, internal 1000-Hz bandstop filter (Fig. 21-9). The meter is connected across the receiver

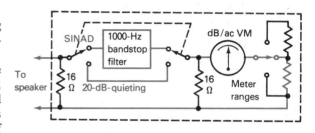

FIG. 21-9 SINAD–20-dB-Quieting meter.

loudspeaker terminals. When a 1000-Hz, 70% FM RF signal is fed to the input of the receiver, the filter in the SINAD meter traps out all of the 1000-Hz modulated signal. The remaining output is receiver noise plus products produced by distortion of the 1000-Hz signal as it passes through the receiver circuits. This is indicated by the meter. With the modulated RF signal generator output at zero, the volume control of the receiver is turned up to produce the rated AF power output in noise, and the SINAD meter is set to read 0 dB at this level. As the signal generator output is increased from zero, the receiver begins quieting. When the meter reads −12 dB signal and set distortion, the output of the signal generator in microvolts will be the 12-dB-SINAD receiver sensitivity. Weak signals at this level will just be readable through the receiver noise. Again, receivers should have sensitivities in the 0.2- to 0.5-μV range. The SINAD meter can also be switched to make 20-dB-Quieting measurements or dB readings, but the SINAD test indicates both noise and the presence of audio or intermodulation distortions.

21-12 DISTORTION ANALYZERS

When audio signals are distorted it is because the fundamental tones fed into an amplifier are developing harmonics and may be beating against other fundamental tones present as the result of nonlinear circuit operation. All spurious signals generated by the fundamentals that are fed into the amplifier are considered to be noise.

Harmonics-type audio distortion can be measured with a *distortion analyzer* (DA) (Fig. 21-10). It is somewhat similar to the SINAD meter. The DA can be used for three purposes: For an ac-dB voltmeter, to measure distortion, and to measure the 12-dB SINAD.

For measurement of distortion, the high-impedance input is "bridged" or connected across the output load of the amplifier to be tested. The mode switch is set to "Level," the meter is switched to 100% distortion, and the notch filter frequency and the input control are set full counterclockwise (off). The amplifier is fed some audio frequency tone, such as 800 Hz, from an AF signal generator which is turned up to produce a desired AF output power to the speaker. The DA input control is adjusted until its meter reads 100% distortion. The mode switch is then set to "Distortion," and the notch filter frequency is tuned and balanced to a minimum (null) reading on the meter. The distortion settings are decreased as the best null is being approached. When the best null is found, the meter will be reading the percent distortion plus noise of the amplifier with the level of input being fed to it. (This served as the original SINAD test before the meter was developed.) Generally, the lower the AF level fed to an amplifier, the less the percent distortion and noise generated.

The ac-dB voltmeter usually has several voltage and decibel ranges rather than the two shown.

▌ Test your understanding; answer these checkup questions.

1. What two methods are used to produce multi-channel mobiles? _____ _____
2. What may be used in receivers to make it unnecessary to tune IF stages? _____
3. What are the three types of FM detectors used in mobiles? _____ _____ _____
4. What type of coupling is used between helical resonators? _____
5. Why are cavities and helical resonators used? _____
6. What is a pager? _____ What outputs does it have? _____ _____
7. What is a receiver called that samples frequencies in some desired order? _____
8. How far should eyes be kept from a 2.5-W handie-talkie antenna? _____
9. Why might a cavity wave trap be better than an *LC* wave trap? _____
10. What is the offset for VHF repeaters? _____ UHF? _____
11. What is an NO *RY*? _____ NC? _____
12. What is a COR in a repeater? _____
13. How long does a repeater transmit after a mobile signal drops out? _____
14. How are repeaters usually identified? _____
15. What kind of system relays remote readings? _____
16. What kind of generator signal is used in 20-dB-Quieting tests? _____ In 12-dB-SINAD tests? _____
17. What ratio of voltage variation is equivalent to 20 dB? _____

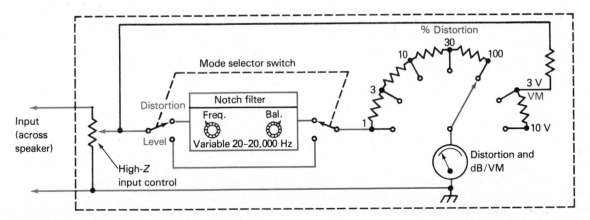

FIG. 21-10 Essentials of a distortion analyzer.

18. Are all receivers quieted when an unmodulated carrier is received? _____
19. What does SINAD mean? _____
20. What causes audio signals to be distorted in a receiver? _____
21. What is the main difference between a DA and a SINAD meter? _____
22. What are the two controls to null a DA? _____ _____

21-13 TRANSMITTER CHECKS

When first installed, when changes are made to it, and at least once a year, every transmitter must be tested. The FCC-required tests are (1) power output or input, (2) modulation deviation or percentage, and (3) frequency. Power output may be tested into the antenna only if the channel is not busy and can be done in a few seconds; otherwise a dummy load is used. An RF wattmeter–SWR meter is added between transmitter and dummy load. When the transmitter is switched on the wattmeter indicates the power output. The output power may also be determined by $P_o = I_a^2 R_a$, where I_a is the value read on a thermocouple antenna ammeter calibrated for VHF or UHF and R is the resistance of the dummy load, or impedance of the antenna at the measured point. When a dummy load is used, the SWR should be nearly 1:1. Into an antenna it may be higher. If the SWR is above 1.3:1, there may be something wrong with the antenna system. Power input is determined by $P_{in} = V_{dc}I_{dc}$ of the final RF amplifier where I_{dc} is the output circuit dc current and V_{dc} is the output circuit dc voltage. Power output may also be determined by $P_{in} \times$ efficiency of the final stage. Transmitters should not be operated at levels greater than 10% above the manufacturers' ratings and preferably at or below these levels.

Modulation tests for an FM transmitter require a calibrated modulation monitor. A sinusoidal 1000-Hz tone (or whistle in an emergency) should produce a maximum of about 4.5 kHz of deviation at the beginning of limiting, and it should be equal in both the positive and negative directions. Voice modulation should then produce 5-kHz deviation. If positive and negative deviations are dissimilar (nonsymmetrical), the cause may be due to mistuning of one or more of the RF stages or to improper functioning of the modulator stage.

For AM transmitters a sinusoidal 1000-Hz input should be able to produce at least 70% modulation as shown on a modulation monitor or oscilloscope. Speech should not be able to drive the negative peaks to zero on the scope. Distortion, audible or noted on a scope, can be checked by observing waveforms as the VF progresses from the microphone to the modulator. A distortion analyzer may be used to check suspected VF distortion. Inability to produce the required modulation levels can often be pinpointed by use of a scope. A spectrum analyzer provides a measurement of the sidebands or the spurious signals generated by modulation.

The peak envelope power, or PEP, of SSB, FM, AM, or CW transmitters can be determined with a calibrated oscilloscope connected across the known antenna transmission line impedance (Sec. 17-36).

Frequency checks are made by special frequency monitors, by using a frequency counter set to a 1-s gating time, but rarely by using a heterodyne frequency meter in the VHF or UHF ranges. The frequency tolerance must be known to determine whether the transmitter is operating within limits. The frequency should be checked when the transmitter is cold, and again after it has been turned on for at least 20 min. Direction of frequency drift is thus established. The transmitter frequency should be set so that when warm it is as close to the assigned frequency as possible without allowing it to be out of tolerance when starting cold. No modulation, voice, tone, CTCSS, etc., should be on the carrier when frequency is being checked, or erroneous readings may result.

Equipment that may be required for these tests includes a dummy antenna, calibrated RF and AF signal generators, deviation monitor, distortion analyzer, RF wattmeter, RF ammeter, AF voltmeter, calibrated oscilloscope, spectrum analyzer, and an adequately regulated power supply for the transmitting equipment.

21-14 RECEIVER CHECKS

Although there are no FCC requirements for receivers, they should be checked periodically. Items to check are sensitivity, squelching level, bandwidth, AF distortion, and AF power output. Manufacturers of radio equipment often supply communications monitors that are used to check out or service their transmitting and receiving equipment. Such a piece of test gear may contain signal sources, deviation or percent modulation indicators, SWR meters, voltmeters, ammeters, dB meters, and RF power meters.

Sensitivity can be determined by the 20-dB-Quieting or the 12-dB-SINAD methods.

Squelching level with no signal input should be approximately midscale of the control knob.

Audio (detected signal) distortion is measured with a distortion analyzer while feeding a 1000-μV, 1000-Hz modulated RF signal to the receiver input. The modulation should be about 60% deviation or percent modulation (3 kHz for 5-kHz authorized systems). Distortion should be less than 5%, although 7% at full rated AF power output may be considered tolerable.

Receiver RF or IF distortion of an undistorted transmitted FM signal may occur if the receiver is not tuned to the center of the signal deviation. It may also be caused by an improper LO crystal frequency, by misalignment of IF or detector tuned circuits, or by too narrow a bandpass for the deviation of the signal being received. For AM received signals, off-frequency, misalignment of IF and detector circuits, too narrow an IF bandpass, or failure of AVC voltage are before-the-detector causes of distortion.

Bandwidth of an FM receiver can be checked by feeding a 1-μV unmodulated RF generator signal into the receiver and tuning the generator across the receiver frequency. The frequencies where the squelch opens and closes again represent the weak signal bandwidth. It should be at least 18 kHz for 5-kHz deviation FM. This test will also indicate whether the receiver passband is centered correctly on the desired channel. If not, the local oscillator crystal may be warped to the proper frequency. With a strong signal input (1000 μV), the bandwidth will be broader but should not produce any off-frequency spurious responses in the receiver as the generator is tuned 100 kHz above and below the received frequency.

With two-way equipment receivers, turning the volume or gain control completely counterclockwise ("off") seldom reduces the signal completely. Full undistorted AF power output should be produced with the gain control advanced about 50%.

An FM *communications monitor* is a single piece of test equipment containing (1) an RF (CW) and FM-modulated RF signal generator developing ac in the 0–10 MHz (AF and IF) range and in the 20–70, 120–170, and 450–500 MHz (HF, VHF, UHF) ranges; (2) an accurately tunable FM receiver for the HF, VHF, and UHF bands; (3) a discriminator meter to indicate how far from a reference frequency the input (or received) signal is; (4) an FM plus or minus deviation indicator calibrated in 0–5 and 0–15 kHz ranges; and (5) an output to an oscilloscope to show modulation. The RF output (CW or 1000-Hz modulated FM) is adjustable over a calibrated voltage range from 0.1 to 1000 μV, which makes it useful in determining receiver sensitivity. It has six or seven frequency selecting switches that operate a PLL synthesizer to produce digital readouts in megahertz, kilohertz, and hertz. By using a decade divider, readouts can be obtained in 0.1-Hz steps on incoming signals to test receiver tone-operated squelch circuits (Sec. 21-7).

21-15 INTERFACING TO TELEPHONE LINES

Technicians must sometimes connect radio equipment to privately or public utility owned telephone lines. All interfacing equipment connected to utility company lines must be registered with the FCC or use an FCC-approved coupler. Technicians should know something about how telephone systems operate.

Figure 21-11 represents a simplified rotary dialing telephone system between a *central office* (CO), or exchange, and a subscriber. Only one pair of wires is used. A CO battery (usually 48 V) provides current to operate the handset carbon microphone and various CO relays (not shown). When a subscriber picks up the handset (microphone M and earphone P), the hanger switch contacts close, as shown. This completes a dc circuit from microphone through the *hybrid coil* and one winding of a CO *repeater* coil (transformer) to battery negative. The other terminal of the microphone connects through the hanger contacts and another repeater coil to ground and battery positive. The microphone is energized, an open operational line in the CO is automatically searched for and selected, and a dial tone ($\pm$400 Hz) is fed back to the subscriber.

If rotary dialing is used, the subscriber line is sequentially opened and closed, twice if a 2 is dialed, eight times if an 8 is dialed, and so forth, returning to a closed condition whenever the dial stops turning. Circuits in the CO sense the dc pulses that are representing numbers and connect the subscriber line to the correct CO (the first three numbers dialed) and then to the called subscriber line (the last four numbers dialed).

With "touch-tone" or DTMF (dual-tone multiple-frequency) dialing, a transistor oscillator in the telephone set develops two different tones for each number pressed on the special touch-tone key pad and feeds these across the line. DTMF telephone dialing uses a pad of 12 buttons, as indicated in

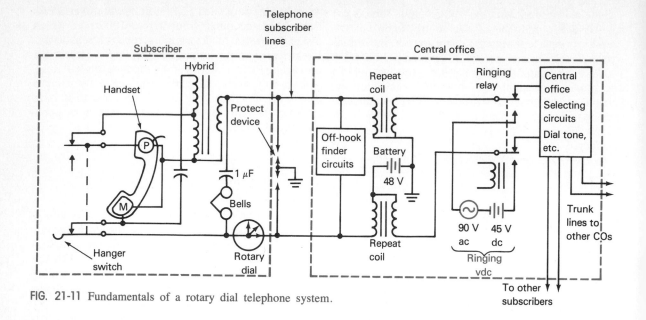

FIG. 21-11 Fundamentals of a rotary dial telephone system.

Fig. 21-12. Some computers and other applications use four extra function keys, totaling 16, shown dashed. Consider each key as being placed over two wires, each carrying the different frequencies shown. For example, if button "2" is depressed, the 1336-Hz wire would contact the 697-Hz wire and both would touch a bottom contact to send these two frequencies down the telephone line. If the keys are pushed, different inductances are actually switched into the oscillator circuit to generate the desired tones. At the CO, or receiving end, the tones may be segregated by tuned filters. The signals are decoded to form either off-on dc pulses that can open or close relays, etc., or to produce binary words (such as 0010 when the "2" button is pressed), that can be made to open or close digital circuits. Decoders may use resonant LC circuits and rectifiers to develop the dc pulses. A single IC with internal active filters and an external 3,579,545-Hz crystal can be used to decode the dual tones and produce 4-bit digital output signals (up to 16 different 4-bit digital data or control words). Since the IC is a CMOS type, the output is a low-current form. To control higher current circuits, such as TTL, the data words may be first fed through a latch to hold them until the following circuits are ready to accept them, when they may then be fed through a CMOS-to-TTL level converter.

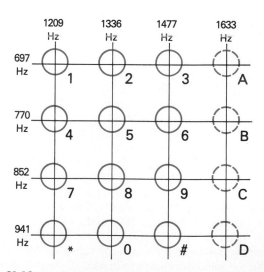

FIG. 21-12 Touch-tone pad. Computer keys are shown dashed.

The hybrid coil in the telephone set is connected so that all the incoming signal ac on the line is fed to the earphone. However, when the microphone is spoken into, the outgoing signal is nearly canceled in the local earphone, but travels unimpeded down the line to the other subscriber. Consider Fig. 21-13a, showing a hybrid circuit. When the micro-

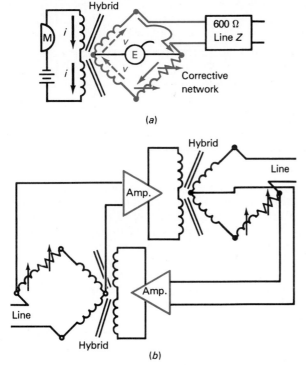

(a)

(b)

FIG. 21-13 Hybrids (a) not completely balanced in handset and (b) balanced in repeater.

phone develops vdc downward (arrows), an emf is induced upward in the two secondaries, downward through the line impedance and through the corrective network. The secondaries form a bridge circuit with the earphone across the points of equal potential, so no microphone signal flows through the earphone. However, complete balance is never used in a handset. If a hybrid were not used, the person speaking would hear his or her voice very loud and might reduce talking volume too much, resulting in weak transmitted signals. (Hybrids are also available in IC form.)

In Fig. 21-13b, two hybrids are used in conjunction with two amplifiers to work as two-way repeater or booster amplifiers for long-mileage telephone lines. In this case the hybrids are completely balanced so that there is no feedback of signals within

the repeater or to the lines. This circuit allows two-way conversation on one "twisted pair" of wires. Normally, four-wire lines are used for two-way communications circuits (two out, two back) to prevent return signals that might occur when unbalance occurs in a hybrid circuit or its twisted pair lines.

To signal that a call is coming in, the CO connects a 20-Hz varying dc ringing voltage across the subscriber's line, ringing the bell (Fig. 21-11) that is connected across the line at all times. The ringing or signaling voltage is applied to the line 1 s on, 4 s off. A simulated ringing signal is fed back on the caller's line to indicate that ringing is taking place.

As an answering handset comes off-hanger, dc flows in the line and a CO relay (not shown) disconnects the ringing circuit. Similarly, dial tone is dropped out at the first break of a rotary dial, or as the first touch-tone signal is fed into the calling lines.

Some of the things a technician should understand are:

1. Subscriber telephone lines have up to 48 V dc across them at all times.
2. Ringing voltages of nearly 90 V dc can wipe out many electronic circuits should they be across a line during ringing times.
3. Only an approved acoustical coupler or *modem* should be used to couple a handset or telephone line to a local device (tape recorder, computer, etc.). A simplified coupler is shown in Fig. 21-14. Telephone earphone (E) signals are picked up by the coupler microphone (M_2) and appear as ac signals between A and B. Local 300–3000 Hz two-tone or VF signals fed to B and to C and to the coupler loudspeaker (L) are picked up by the handset microphone (M_1) and are transmitted along the telephone line. It is usually more reliable to use a data modem to interface a computer to a telephone line. It may

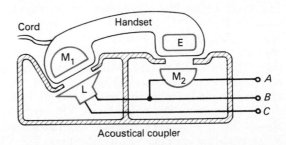

FIG. 21-14 Fundamentals of an acoustical coupler.

use a microcomputer chip with input and output circuit memory banks or have a universal asynchronous receiver-transmitter (UART) IC to change serial ASCII to parallel, or the reverse, for storing and later transmitting information. (It may also use audio filters plus a modulator and a demodulator.)

4. The line level of *speech ac* is measured in volume units (VU) and should not exceed 0 VU as read on a VU meter across a 600-Ω telephone line. (Beware of ringing voltages!)

5. Signals greater than 0 VU may cause enough current and magnetic field around telephone wires to induce "cross-talk" voltages into adjacent lines running in a common cable.

6. The noise level due to cross talk and other noises on a subscriber line should be at least 30 dB down from 0 VU (−30 VU).

7. Subscriber lines are two-wire duplex (two-way) 200–3000 Hz cable pairs. Higher-quality lines use four wires, two for one-direction transmissions and two others for transmissions in the opposite direction.

8. Every subscriber line has some form of lightning or high-voltage air-gap cutout (or a fuse) protective device installed where the lines enter a building. If service is interrupted permanently, this may be the trouble.

21-16 MUX BASEBAND OPERATION

Communications between a control position and a remote or repeater station will include one or more VF signals going each direction plus many control signals. If five different locations use the same repeater site, there should be at least five two-way VF channels plus five sets of two-way control signal paths.

If separate telephone lines to a remote or repeater station are used for receive and transmit information, a four-wire line is desirable. A third VF circuit could be used in a *phantom* circuit, shown in color in Fig. 21-15. If the VF_1 and VF_2 transformers (hybrids) are accurately center-tapped, VF_3 currents, shown in color, will split at the center-taps. The two half-currents will progress along the VF_1 and VF_2 lines, canceling in the center-tapped end transformer's secondary, but feeding through to the VF_3 primary. The result is no VF_3 energy in any of the VF_1 or VF_2 outputs, but unattenuated VF_3 signals between the VF_3 transformers. The VF_3 circuit might be used for order-line-type telephone

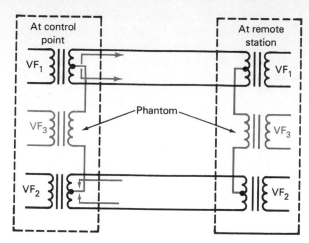

FIG. 21-15 Possible phantom group to place three VF signals on a four-wire line.

communications between control point and remote station or for audible or subaudible telemetry signaling. If a battery were added to the VF_3 circuit, dc relay action might be used to turn equipment on or off remotely.

It is often advantageous to operate a microwave link multiplex system between a control point and a repeater site (as well as between repeater sites). Minimum requirements are a low-power (0.1–20 W) microwave transmitter, receiver, and antenna at the control point, and similar equipment at the repeater site. For each telephone line that might have been required, there will have to be a separate SSB channel to modulate the microwave (950-, 1850-, 2150-, 2450-, 6500-, 12,500-MHz bands and higher) using an FM (AM is also possible) transmitter carrier. The signals that modulate the microwave are all part of what is known as *baseband*. VF is in the range 200–3000 Hz and AF is 20–20,000 Hz, but baseband frequencies are from 0 to perhaps 2.54 MHz or higher. The microwave carrier may be modulated by frequencies from a few hertz to as high as 2.54 MHz (for 600 channels). Only frequencies up to 408 kHz (first seven channel groups) will be included in this basic description of *multiplexed* (multichannel transmissions, or MUX) microwave emissions. (Forms of MUX are also used on some HF-band carriers.)

Part of one baseband frequency spectrum is shown in Fig. 21-16. Between 0 and 4 kHz is the *order-wire* channel, used for telephone conversations between control point personnel and techni-

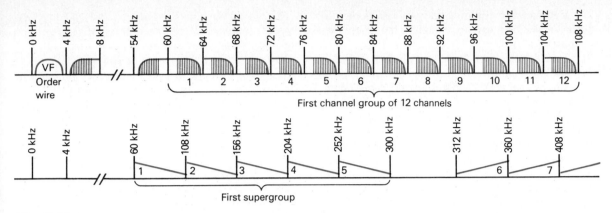

FIG. 21-16 Baseband spectrum through seven channel groups.

cians working at the repeater site. The frequencies between 4 and 60 kHz will be control signals or USB channels based on multiples of 4-kHz carriers and are usually used for ringing, telemetry, and signaling. The 12 channels between 60 and 108 kHz are all 4-kHz LSB channels based on multiple-of-4-kHz carriers. These 12 channels form one *channel group*. The five groups of 12 channels each form one *supergroup*. The second supergroup uses USB, indicated by the reversed slanted lines. All other supergroups up to the normal 10 total use LSB emissions.

A block diagram of one channel of a multiplex system is shown in Fig. 21-17. The control point unit would beam its microwave signals to a similar unit at the repeater site. The hybrid coil makes it possible for one pair of telephone lines to couple two-way (transmit-receive) signals to the microwave radio via channelizing equipment. Incoming VF telephone line signals fed to the balanced modulator of this *modem* (MOdulator-DEModulator) mix with a 5.2-MHz local oscillator and produce the upper and lower sidebands of the 5.2-MHz carrier. The desired SB signals pass through the filter to the mixer, where they are heterodyned against a *channel oscillator*, whose frequency is the channel carrier frequency value above 5.2 MHz. Assume the channel frequency is 100 kHz. Then 5.2 MHz mixed with 5.3 MHz develops a 100-kHz LSB signal and, of course, a 10.5-MHz product. The 3-

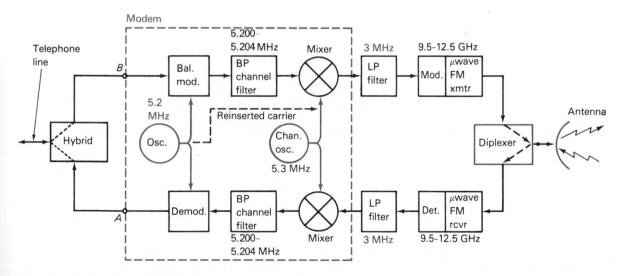

FIG. 21-17 Single-channel MUX system.

MHz low-pass filter allows only the 100-kHz LSB signals to pass through to modulate the microwave transmitter.

If a second location feeds its VF signals into a hybrid at the control point, a second modem, perhaps using a channel 11 crystal, would have a channel oscillator frequency of 5.2 MHz + 105 kHz, or 5.304 MHz. Its LSB signals would also be fed through an LP filter and be added to the microwave modulator stage input circuit. Since the two channel signal frequencies do not overlap, both can modulate the transmitter simultaneously without cross talk.

Returning microwave signals from the repeater site are picked up by the directional microwave antenna at the control point and, by the action of the diplexer or duplexer, are fed to the receiver only. The diplexer, an RF hybrid, also prevents any transmitter signal from passing to the receiver, aided by the fact that transmitting and receiving frequencies are several megahertz apart. Any channel-10 LSB information being received from the repeater-site microwave transmitter is fed through the 3-MHz low-pass filter and is mixed with the channel-10 5.3-MHz oscillator to produce 5.396–5.400 MHz, and 5.204–5.200 MHz SBs. The 5.204–5.200 MHz SBs pass through the bandpass channel filter and mix with the 5.2-MHz oscillator to demodulate as 0–4 kHz VF signals that are fed to the hybrid and to the telephone line.

A similar multiplex system is required at the repeater site. Microwave signals from the control point are received, mixed with the channel oscillator to develop 5.2-MHz SBs, are heterodyned with the 5.2-MHz oscillator and are thereby demodulated as VF signals. These appear at point A and are used to modulate the VHF or UHF repeater transmitter. Signals from mobiles picked up by the repeater receiver are fed as VF signals into point B, developed into SSB signals, filtered, and translated to 100-kHz SBs and modulate the repeater-site microwave transmitter to relay them to the control point.

While the system explained would operate satisfactorily, it is better to reinsert a little carrier, shown dashed, so that it can lock in with the 5.2-MHz oscillator when it is being received. There will also be signaling (ringing) signals, busy indications, fuses, and alarm signals included in a MUX system. Each channel will use a frequency 1 kHz from the channel suppressed carrier as a test tone frequency.

Microwave antennas are directional and may use passive "periscope" or other types of reflectors on high peaks to divert signals to their destinations.

For less elaborate systems a low-power VHF or UHF FM or AM radio communications link can be used to relay voice and telemetering information between a control point and a remote station. If such relay stations use directional antennas, there will be less chance of producing cochannel interference with other systems on the same or adjacent frequencies.

Baseband MUX is used in telephone work using coaxial cables, since twisted pairs will only go up to a few thousand hertz.

Test your understanding; answer these checkup questions.

1. When must transmitters be tested? _____ _____ _____
2. What must be tested in transmitters? _____ _____ _____
3. What are two ways of determining RF power output? _____ _____
4. What should be the SWR into a dummy antenna? _____
5. What are two devices used to measure VHF and UHF transmitter frequencies? _____
6. Under what two conditions should transmitter frequency be measured? _____ _____
7. What are the five items to check with receivers? _____ _____ _____
8. What RF or IF troubles may cause AF distortion? _____ _____
9. What is the function of a hybrid in a telephone handset? _____
10. When are hybrids fully balanced? _____
11. What are the two basic components of an acoustical coupler? _____ _____
12. What is the maximum VF tone that may be fed to a telephone line? _____ Voice signal? _____
13. What is the abbreviation for multiplex? _____
14. What two systems can increase the capabilities of a two-wire telephone line? _____ _____
15. What are the two basic means of telemetering? _____ _____
16. What are the limits of baseband modulating frequencies? _____
17. Where in the baseband is the order-wire channel? _____ Between what frequencies is signaling and telemetering likely to be found? _____
18. What is a channel group? _____ Supergroup? _____

19. Which supergroup does not use LSB? _____
20. What does modem mean? _____ How many oscillators are in one? _____ Mixers? _____ Filters? _____
21. What is the frequency of a modem channel oscillator? _____
22. What is the frequency of the test tone of any MUX channel? _____
23. What would be the bandwidth of an A3 transmitter if modulated by a full baseband MUX signal? _____

21-17 TRUNKED AND CELLULAR SYSTEMS

For many years the telephone company has used a system of searching for idle "trunk lines" from one exchange to another and instantaneously switching calls to these idle lines. This same idea is now applied to 800-MHz-band business communications radio systems.

When several different businesses join together, they may establish perhaps five different repeaters on one antenna tower. The mobiles of the five fleets will normally work on their own channel, although it may be possible to share channels with other fleets. This often results in a long wait for a channel to clear, or stronger signals wiping out weaker ones. If all five channels are in a trunked system, there is very little waiting, and no interference.

One type of *trunked* system uses a separate channel for its automatic control station. All mobiles monitor the control station channel. If a mobile wants to communicate with its dispatcher or one of the other mobiles in its fleet, as soon as it takes its microphone off-hook and presses the PTT switch its transmitter sends out a burst of digital information identifying which fleet it belongs to. The burst is decoded by the control station microprocessor. The control scans all channels to find a clear one. It transmits a burst of digital information to tune all of the mobile receivers and transmitters of that fleet to the free channel and then sends a burst to advise the off-hook mobile to go ahead. This usually takes a fraction of a second. The mobile is now clear to use the free channel to make its call to any station of its fleet. No other fleets will be tuned to this channel so they will be unable to hear the conversations of this fleet. If the channel is unused for a short period of time, the control releases it and it may be used for communications of any other fleet. As soon as the calling mobile hangs up its microphone, its all-clear burst clears the channel and all mobiles return to monitoring the control channel.

If all channels are busy at an off-hook request, the control station signals back with a "busy" light. As soon as a channel clears, the mobile gets a "green light" and then makes its call.

All systems in the 800-MHz band using five or more channels must be trunked. Trunked systems may have 5, 10, 15, or 20 channels and can easily service 80 times the number of mobiles as there are working channels.

Another type of trunked service uses no dedicated control channel station but has a microprocessor and PROMs with plug-in channel modules in each mobile. In effect, each mobile is its own computer, continuously scanning empty channels until its fleet calling digital burst is received on some channel. It then locks onto this channel and is ready to communicate if it is the one that is being called.

A still newer mobile *telephone* communications system is the *cellular* type. Low-altitude repeaters are spaced apart every few miles all over a city. As a mobile moves from one area (cell) to another its signals weaken at one repeater and become stronger at the nearer. Automatically the mobile is shifted to operate at the new stronger-signal frequency of the different repeater. In this way mobile signals can be held at a near maximum value regardless of where the mobile goes in the city. Control of repeaters is accomplished by computer, and the transfer of signals from one cell to the next is so fast that it is not noticeable to the mobile operators using the system. VF signals are transferred intercell on telephone lines.

21-18 TECHNICIAN INFORMATION

Operating the switches and microphone of two-way equipment requires little or no technical skill. When receivers and particularly transmitters are serviced, the technician must understand the FCC rules and regulations and procedures for installing and servicing the equipment. Most commercially obtained equipment will have manufacturer operating explanations and suggestions on installation and servicing. Special test sets are often available from the manufacturers which make servicing much simpler.

Safety is important. All high-voltage leads must be behind metal enclosures, preferably protected by interlocked doors. Voltages above 50 V can cause electrical shocks, which may not be lethal in themselves always, but can cause bodily reactions that can cause falls and traumatic injuries that can be fatal. Voltages as low as 12 V, if contact is made to living tissue, can cause paralysis. RF voltages can cause skin burns, even with 10-W transmitters. Use plastic handled tools. Do not work on live equipment when sleepy or tired. Always use tested safety belts when climbing poles or towers. Use only one hand when servicing operating equipment whenever possible. Do not stand on a wet or grounded floor or lean on grounded objects when servicing live equipment. Remember, the most dangerous tool in number of injuries is the lowly screwdriver!

Access to radio equipment by unauthorized persons should be prevented at all times. Remote stations must be protected with padlocks or other safe locks. Theft of mobile equipment from automobiles is common. Cars carrying mobile equipment must be locked when not being used.

The FCC requires records, called *logs*, be kept of certain radio operations. Operating logs, written or taped, should indicate what communications occurred and are signed for by the control operator. Technical maintenance logs indicate what tests and maintenance were made on transmitters, and they are made out by the person doing the work. This is often a person holding some type of a certificate. Actually, any unlicensed person can make adjustments to transmitting equipment, but only if under the observation of someone who is responsible for and will sign for the technical operation of the system. Logs must be held for a period of at least one year, unless they contain material pertaining to distress traffic or information in litigation, in which case they must be held until release is authorized by the FCC.

Station records must contain statements as to frequency, power, and modulation tests made at the time of installation, every time a transmitter is changed in any way that may affect carrier fre-

quency stability, or output power, and at least once a year thereafter. Mobile equipment may be tested at the bench if similar loading and power sources are used. There is no prescribed form for logs, but they must be orderly and contain the required information and be signed and dated.

Technicians should have on hand, and be familiar with, the "FCC Rules and Regulations" that pertain to the equipment under their care. These might be "Part 90—Private Land Mobile Radio Services," "Part 22—Domestic Private Landline Mobile Radio Services," "Part 94—Private Operational-Fixed Microwave Service," "Part 87—Aviation Services," "Part 68—Connection of Terminal Equipment to the Telephone Network," "Part 13—Commercial Radio Operators," and "Part 17—Construction, Marking and Lighting of Antenna Structures." These are available from the U.S. Government Printing Office, Washington, D.C., 20402.

21-19 STATION LICENSING

Before establishing any radio service it is necessary to submit an application for a *radio station authorization* from the FCC. Forms for this are obtained from either the Washington, D.C. office or any of the field offices of the FCC (Appendix F). After the station has been authorized by a *construction permit* (if required) and before the installation has been completed, a *station license* must be applied for. The engineer-in-charge of the local field office must be notified before on-the-air tests are made. Radio equipment and antennas may be installed, but they cannot be tested on the air until the license has been granted.

While normally an authorization must be obtained before operating a new station, in the case of a mobile unit telephone authorization may be given in emergencies. A mobile may be used as a base station for a period of no more than 10 days in emergencies.

Identification may be made by stations at the end of each completed interchange of information or every 30 min of continuous communication (15 min for stations other than the Public Safety Service) by voice or Morse code at no more than 25 words per minute. Identification can be by the assigned call signs or by special identifiers, provided the FCC is notified of the identifiers. Base stations usually make any required station identifications for its mobile units.

Licensed transmitting equipment is said to have

type approval if the FCC tests its power and frequency characteristics, and *type acceptance* if manufacturers' test data show the equipment to be acceptable. Generally, changes that might affect the power output, frequency stability, or percent modulation of a transmitter must be approved by the FCC. Other types of changes may be made without FCC approval.

21-20 NOISE IN MOTOR VEHICLES

Most communications receivers will satisfactorily amplify and detect a 0.5-μV received signal, whether AM or FM. However, if the *noise* being received has an average value of more than the signal, neither AM nor FM will reproduce the signal satisfactorily. It is important to reduce local noises.

One method of reducing noise generated in land-based electrical equipment is to ground the equipment with as short a ground wire as possible. Another is to bypass the electrical lines to ground with 0.01-μF capacitors, using series RF chokes in lines if necessary. Another aid is to shield the equipment and lines with grounded metallic housings or braid. Still another aid is to connect a 0.1- to 1-μF C in series with a 10-Ω R across making and breaking electric contacts or switches.

In motor vehicles the major source of noise is the ignition system. The popping noise developed by the spark plugs increases in frequency with increased engine speed. It can be decreased by using resistor spark plugs or resistor ignition cables, by making sure all ignition leads are tight, by connecting suppressor resistors in the lead from the distributor to the ignition coil, and by bypassing the primary of the ignition coil to ground. Ignition timing must be correct.

Another source of noise in mobile receivers is a battery-charging generator. When the engine changes speed, the whining noise changes. Bypass the armature terminal of the generator (not the field terminal) with a 0.1- to 1-μF capacitor. It may help to bypass the battery terminal and the armature terminal of the voltage regulator to ground. If alternators cause noise, bypass the ungrounded terminal to ground with a 0.5-μF capacitor.

Gasoline, temperature, and oil gauges can produce clicking noises. The leads to these gauges can be bypassed with a 0.25-μF capacitor at the source and at the dashboard.

An irregular clicking noise that disappears when the brakes are applied is known as *wheel static*.

The front wheels riding on an insulating layer of grease may build up a static charge which sparks across when the voltage increases sufficiently. This can be stopped by using wiping-type front-wheel-static eliminators. Using an antistatic powder on inner tubes will decrease tire static.

If the metal parts of the chassis of a vehicle become loose, a voltage difference may be developed between two parts and noises may occur when the parts work together. Adequate bonding of the major parts of the car, hood, body, chassis, motor cables, rear axles, brake and speedometer cables, rear-axle assembly, doors, and fenders with short flexible-braid conductors is often necessary.

All electrical leads coming through the fire wall may have to be bypassed at the point where they leave the engine compartment.

Test your understanding; answer these checkup questions.

1. In what frequency range are trunked systems used? _____ How many channels may be used in these systems? _____
2. How does a mobile signal its control channel in trunked systems? _____
3. What does a control channel do when signaled? _____
4. How many mobiles might a five-channel trunked system service? _____
5. Who finds clear channels in no-control-channel trunked systems? _____
6. What is the name of the system that transfers mobiles from one repeater to another as they move through a city? _____
7. Could a 12-V battery be dangerous to a technician? _____
8. Normally, for how long must logs be kept for the FCC? _____
9. To license a station, what are the three actions to take? _____ _____ _____
10. In emergencies, may a mobile be used as a base station? _____
11. Who tests for type approval? _____ For type acceptance? _____
12. What usually causes popping noises in mobile receivers? _____
13. What terminal of a dc charging generator should be grounded to reduce radio noise? _____
14. What gauges may have to be bypassed to ground to reduce radio noise in mobiles? _____ _____ _____
15. What is the source of wheel static? _____
16. Why are major parts of an automobile often braid-bonded together in mobiles? _____

1. What are the two basic forms of radio services?
2. How does a DPLMRS differ from other systems?
3. What are the three international distress frequencies?
4. What is the tolerance of most SSB signals?
5. How much of the emission of an FM or AM station must be held within the authorized bandwidth?
6. What is the usual authorized bandwidth for a two-way FM channel? AM? SSB?
7. How may VF and control signals be fed to remote stations?
8. Block diagram a VHF FM mobile transmitter.
9. What does an ALC circuit do?
10. When a PTT switch on a two-way unit is pushed, what does it do?
11. What may result in the output if a multiplier or RF amplifier in an RF transmitter is detuned?
12. What AF is used when tuning a two-way FM transmitter? To what deviation is a transmitter adjusted?
13. What always follows a limiter or clipper stage?
14. Why should mobile microphones always be set for close talking?
15. Draw a diagram of a direct FM circuit using a crystal and a varactor.
16. What is the meaning of CTCSS? What are two other similar meaning terms?
17. What is the frequency range of CTCSS signals?
18. On what does a DPL system depend to unsquelch?
19. Block diagram a CTCSS mobile FM receiver.
20. Draw a diagram of a four-cavity helical resonator filter feeding a VHF receiver front end.
21. Why may a LO be coupled to a mixer through a helical resonator?
22. What may be used to set off a pager receiver?
23. What is the *priority* frequency on a monitor receiver?
24. What is the power output range of handie-talkie units? What types of antennas do they use?
25. When do handie-talkies use different R and T frequencies?
26. How many antennas will a remote or relay station have? A repeater station?
27. On how many frequencies do repeater stations operate? Their mobile stations?
28. What is a COR in a repeater?
29. How and when are repeaters identified?
30. Block diagram the essentials of a repeater system.
31. What are the three units that are required to give a receiver a 20-dB-Quieting test?

32. What minimum dynamic range might a communications receiver have?
33. How would the power input and output be computed to determine dynamic range of a receiver?
34. What does SINAD mean?
35. What two things are used in a 12-dB-SINAD test that are not used in a 20-dB-Quieting test?
36. Draw a diagram of a SINAD–20-dB-Quieting meter.
37. SINAD measures what that 20-dB-Quieting does not?
38. How do SINAD and DA systems differ?
39. Does a DA read both distortion and noise?
40. When must transmitters be checked? What must be checked in them?
41. What maximum P_o should a transmitter produce?
42. What is the maximum deviation of most FM two-way transmitters?
43. How much modulation must two-way AM transmitters be able to develop?
44. When checking frequency of a transmitter at what periods should readings be taken?
45. What equipment should be available to adequately check a piece of two-way FM radio equipment?
46. What are the receiver checks that should be made to two-way radio equipment?
47. To where should the squelch control be adjusted?
48. What receiver distortion levels are tolerable?
49. Why may turning the volume to minimum on two-way sets not reduce the received signal completely?
50. What is included in an FM communications monitor?
51. How is it assured that interfacing equipment connected to utility lines is proper?
52. What power operates a telephone handset?
53. What happens when a "3" is dialed on a telephone?
54. What is DTMF dialing?
55. What does a hybrid coil do in a telephone handset?
56. What are the three separate voltage sources that are applied to a telephone line?
57. What are the basic parts of an acoustic coupler?
58. What is the VF signal level on a telephone line?
59. Diagram a rotary dialing telephone circuit.
60. What are the limits of VF? AF? Baseband?
61. What is the channel width in a MUX system? How many channels in a channel group? A super group?
62. Block diagram a single-channel MUX system.
63. What is an RF hybrid called?
64. What happens in a separate control station frequency trunked system when a mobile picks up its handset and presses PTT?
65. Explain an undedicated control trunked system.
66. How does a cellular communication system work?
67. What two logs must be maintained with two-way systems? How long should they be retained?
68. Before establishing a radio service, what must be done? When may on-the-air tests be made?
69. How do type approval and type acceptance differ?

ANSWERS TO CHECKUP QUIZ ON PAGE 501

1. (800 MHz)(5, 10, 15, 20) **2.** (Digital words) **3.** (Assigns fleet a free channel) **4.** (25 to 50+) **5.** (Mobile on-board computers) **6.** (Cellular) **7.** (Yes) **8.** (1 year) **9.** (Authorization)(Construction)(License) **10.** (Yes) **11.** (FCC)(Manufacturers) **12.** (Ignition, working parts, gauges) **13.** (Armature) **14.** (Gas)(Temperature)(Oil) **15.** (Brakes) **16.** (Reduce working noises)

22 Microwaves

This chapter outlines some of the hardware found in microwave systems, including active devices in use. Some microwave oscillator and amplifier circuits are explained. A multiplex transmitter and receiver system is discussed. Installation procedures and test equipment are outlined. Basic optical fiber information is included. Much of the information is a foundation for marine radar (Chap. 26) and the FCC Element 8 tests.

22-1 MICROWAVE FREQUENCIES

As communication knowledge has increased, the usable spectrum has expanded to higher and higher frequencies. Miniature vacuum tubes reach a practical limit in the range of 1000 MHz (1 GHz), sometimes considered the beginning of the microwave region. Specially designed UHF and lighthouse triodes and newer transistors operate to or above 3 GHz. Specialized microwave vacuum tubes—magnetrons, klystrons, traveling-wave tubes (TWTs), backward-wave oscillators (BWOs), and many solid-state devices now operate well above 40 GHz, some at 100,000 GHz and higher. Above this comes the infrared or heat region, usually considered to be 0.3–4.30 terahertz (THz). Light visible to the human eye ranges from about 4.30 THz (red) to 1 kTHz (violet) (Secs. 1-14, 4-6). In the light region, laser oscillators and LEDs are important communication devices.

Most microwave radio activity ranges from 0.9 to 100 GHz (microwave ovens operate at 2.45 GHz). Technicians of today should be familiar with the basic concepts of equipment operating in this region. Besides being employed in radar systems, point-to-point and earth-satellite-earth radio communications, microwaves are used extensively in research laboratories. Many companies are engaged in building and servicing microwave test equipment and components. The frequencies involved in the VHF through light bands are shown in Fig. 22-1.

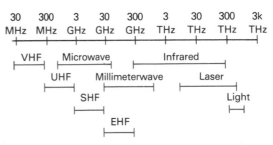

FIG. 22-1 Spectrum showing VHF through light frequencies.

22-2 MICROWAVE TRANSMISSION LINES

Low-frequency ac can be carried effectively by a pair of wires, such as a lamp cord. In the HF spectrum special constant-impedance two-wire transmission lines or coaxial cables are efficient devices for carrying energy. At microwave frequencies a hollow pipe, slightly larger in diameter than a half wavelength, can act as an acceptable confined space down which energy of this or higher frequencies can be propagated. This is similar in some respects to the propagation of radio waves into space from an antenna. Such a hollow metal pipe, called a *waveguide* (WG) may be round or rectangular. Polarization may shift in the round type, however. Most rectangular WG has a height about one-half of the width for the usual voltage and current oscillation mode employed. The width of the guide must be slightly greater than the half wavelength of the ac to be transmitted. A common band of operation is known as the 3-cm, or X band, with lower and upper frequencies of 8.2 and 12.4 GHz, respectively. X-band waveguide is about 0.9 in. wide and 0.4 in. high (3 × 1.5 cm).

Other bands of microwave frequencies have been given labels (Table 22-1)(there is no one standardized letter designation of bands). Frequencies above the operational frequency ($\lambda/2$) of a WG will be propagated, but all lower frequencies will be sharply attenuated. Thus waveguides act as high-pass filters. They have a nominal surge impedance of 50 Ω.

Waveguides may be constructed of brass, copper, or aluminum. Since currents on the walls of waveguides oscillate only on the inner skin, low-loss WG sections will be silver-plated on the inside to reduce skin effect. While currents do flow on the WG walls and voltages are developed between the upper and lower sides, it is not possible to measure them with the usual meters. Instead, the electromagnetic fields developed by the currents are sampled by inserting a pickup probe (antenna) into the WG. The indications obtained can be converted to the desired values.

At RF we think of current as traveling on the surface of the wire and energy as being radiated

Band	Frequency range, GHz	Waveguide size, in.	Waveguide size, cm
975	0.75– 1.12	9.75 × 4.88	24.8 × 12.4
L	1.12– 1.7	6.5 × 3.25	16.5 × 8.26
S	2.6 – 3.95	2.84 × 1.34	7.21 × 3.40
G	3.95– 5.85	1.87 × 0.87	4.75 × 2.21
C	4.9 – 7.05	1.59 × 0.795	4.04 × 2.02
J	5.85– 8.2	1.37 × 0.62	3.48 × 1.57
H	7.05–10.0	1.12 × 0.497	2.84 × 1.26
X	8.2 –12.4	0.9 × 0.4	2.29 × 1.02
M	10.0 –15.0	0.75 × 0.375	1.91 × 0.95
P	12.4 –18.0	0.62 × 0.31	1.57 × 0.79
N	15.0 –22.0	0.51 × 0.255	1.30 × 0.65
K	18.0 –26.5	0.42 × 0.17	1.07 × 0.43
R	26.5 –40.0	0.28 × 0.14	0.71 × 0.36

when the antenna impedance matches the impedance of space. In the microwave case, currents and voltages may be relegated to a secondary role and the radiated electrostatic and magnetic waves, always at right angles to each other, are considered to be carrying the energy from source to load inside the waveguide.

Fields may be set up in WG in several *modes*. The dominant mode is called the TE_{10} (Tee Eee One Oh), meaning transverse electrostatic, less than $\lambda/2$ in width, less than $\lambda/4$ in height. A circular waveguide would be a TE_{11}. The second harmonic of a TE_{10} frequency would see the same WG as a TE_{21} mode.

Waveguides are made into various-length sections. They may be straight (Fig. 22-2), be bent to some desired direction, be twisted to some desired

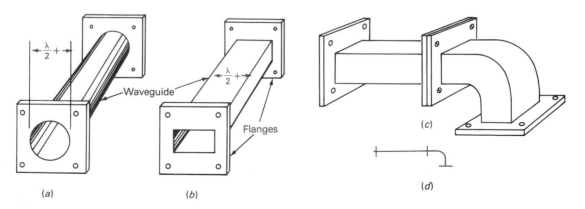

FIG. 22-2 (*a*) Round waveguide. (*b*) Rectangular-waveguide section. (*c*) Straight section coupled to a 90° elbow. (*d*) Section in (*c*) in schematic diagram form.

angle, or even be made flexible. At each end of a waveguide section there is a precisely machined flat metal flange allowing one section to be coupled to another by bolting the flanges together.

Flat flanges may be butted together, but for minimum losses and reflections from the joint, one of the flanges should be a *choke* flange. In a choke flange, part is machined away so that a half-wavelength-long cavity is developed (Fig. 22-3). From

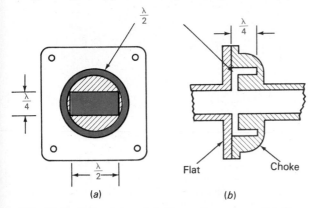

FIG. 22-3 Choke joint. (*a*) Flange-face view. (*b*) Cross section of flat and choke flanges mating.

antenna theory, a half wavelength from a low-impedance point on a stub or transmission line is another low-impedance point. The half-wave cavity in the flange dead-ends (0-Ω impedance) in the flange metal, reflecting 0-Ω impedance back to the spacing between waveguide sections. This represents perfect continuity for the wave between sections and therefore no reflection of energy at the joint. Waveguides are almost always at ground potential.

Coaxial cables also may be used for microwave transmission lines. Solid dielectric coax may have rather high losses at higher microwave frequencies. In addition, coaxial coupling devices present discontinuities and reflections of energy back up the cable. Reflections produce standing waves on the line and prevent full-power transfer from source to load, as well as high- and low-voltage points along the cable, important in high-power applications. If only a few inches long, solid dielectric coaxial cable may be used up to X-band frequencies.

Microstrip is a microwave transmission line consisting of a flat metal base on which is laid an insulator, or dielectric material. A thin metal strip is laid on the dielectric (Fig. 22-4*a*). Strip width and dielectric constant and thickness determine the impedance of this type of transmission line. It is used as a printed-circuit transmission line. Two other planar transmission lines are *stripline* and *coplanar* line, shown in Fig. 22-4*c* and *d*.

High-*Q* microwave filters and tuned circuits can be developed using the *L* and *C* of stripline-type transmission-line techniques.

22-3 SOME WAVEGUIDE DEVICES

As with other transmission lines, terminating a waveguide with a load resistance equal to the surge impedance of the guide produces no reflected energy and maximum transfer of power to the load. If the impedance match is not correct, the reflection may appear to be either capacitive or inductive. To cancel inductive reactance a projection down into the waveguide (Fig. 22-5*a* and *b*) has a capacitive effect. (If it projects down more than a quarter wave, it becomes inductive.) A projection into the

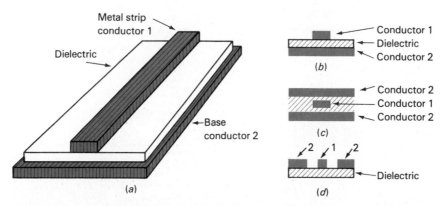

FIG. 22-4 (*a*) Microstrip. (*b*) Microstrip cross-sectional view. (*c*) Stripline cross-sectional view. (*d*) Coplanar transmission line.

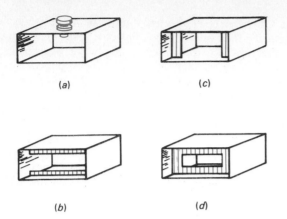

(a)

(c)

(b)

(d)

FIG. 22-5 (a) Adjustable capacitive probe. (b) Capacitive iris. (c) Inductive iris. (d) Resonant or decoupling window.

side of the waveguide (Fig. 22-5c) has an inductive effect. A metal window (Fig. 22-5d) can be made resonant to pass certain frequencies. The optimum placement of a capacitive tuning screw can be determined by replacing a waveguide section with a *slide-screw tuner* section which has a slot down the length of it and is fitted with an adjustable retractable probe that can be moved along the slot. With this device optimum distance for the reactive screw from the reactive load and the best probe depth can be determined by lowest SWR (Sec. 20-12).

One dummy load used in waveguides is a long pyramid of carbonized material with a sharp point to prevent reflections from it (Fig. 22-6a). The

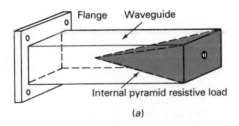

(a)

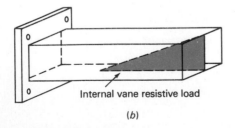

(b)

FIG. 22-6 Low-power waveguide loads. (a) Pyramid type. (b) Resistive vane type.

tapered pyramid absorbs energy, leaving none to be reflected. Another form of low-power resistive load is a tapered resistance-coated strip or vane at the end of the waveguide (Fig. 22-6b). Higher-power dummy loads are air-, oil-, or water-cooled.

When it is desired to lose some fraction of the power flowing along a waveguide, either a flap or a vane *attenuator* may be used (Fig. 22-7). Attenua-

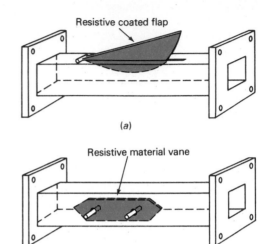

(a)

(b)

FIG. 22-7 Waveguide attenuators. (a) Flap type; maximum attenuation with flap lowered. (b) Vane type; maximum attenuation with vane at center of guide.

tion is often required when making microwave measurements because most such test equipment operates in the milliwatt range, whereas many practical applications operate with several watts of average power, which is enough to burn out the test equipment. These attenuators decrease the signal from 0 dB (no decrease) to more than 30 dB (one-thousandth of the power).

Directional couplers are produced by welding two pieces of waveguide together and opening one or more holes between them (Fig. 22-8). The larger

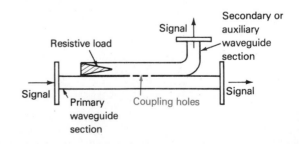

FIG. 22-8 Directional waveguide coupler.

the holes or the more there are of them, the greater the power transfer to the secondary waveguide section. If two holes are a quarter wave apart, the propagation is such that most of the energy induced in the secondary or coupled section is in the forward direction. A dummy load in the backward-direction end absorbs any reflected power, making the coupler a true forward coupler. These devices are rated in decibels. A 3-dB coupler transmits one-half of the power to the secondary section; a 10-dB coupler transmits one-tenth of the power; a 20-dB coupler, one-hundreth of the power. A 1-kW input to a 30-dB coupler couples one-thousandth of the power (1 W) to the secondary section, while 999 W passes through the primary section. Two directional couplers back to back can sample both the energy moving forward and that reflected by the load. From such a *reflectometer*, reflection coefficient and SWR can be found.

One method of determining reflection coefficient and the SWR caused by a mismatched load is illustrated in Fig. 22-9. If the signal generator has a

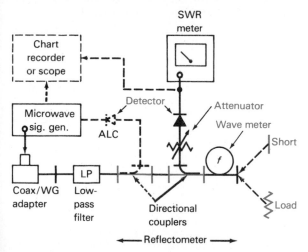

FIG. 22-9 Reflectometer to measure SWR.

coaxial outlet, it is fed to a coax/WG adapter and then to a low-pass filter to remove the possibility of any harmonic energy being read. A forward-direction coupler can be used to produce a feedback voltage to form an automatic leveling control (ALC) if the oscillator is being swept over a band of frequencies. A WG-short is substituted in place of the load, reflecting all transmitted energy back up the waveguide. A reversed direction coupler picks up part of the energy being reflected, which is attenuated a known amount and is displayed as a

voltage on the meter. The load is then connected in place of the short and the attenuator is reduced to zero. The indication on the meter can represent the reflection coefficient, or it can be calibrated directly in SWR. (The calibrations on the meter must match the V/I curve of the diode detector used.) If the load accepts all the energy, no power will be returned and the meter will not read. Reflection coefficient is then infinite and SWR is $(1 + \infty)/(1 - \infty)$, or 1:1. When sweeping a band of frequencies, it is often desirable to chart the SWR along the band. The dashed lines indicate the circuits that would be used.

22-4 COUPLING TO WAVEGUIDES

There are two common methods of coupling energy into (or out of) a WG other than the hole method used in directional couplers. One method (Fig. 22-10a) is similar to link coupling. Energy from a

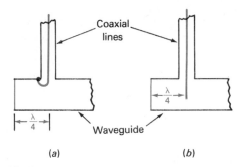

FIG. 22-10 Coupling coaxial line to waveguide. (a) Loop or inductive method. (b) Antenna-probe or electrostatic method.

coaxial line terminates in a single-turn loop connected to a wall of the waveguide. Magnetic fields due to the current in this loop induce voltages in the waveguide space and currents in the walls, allowing energy to radiate down the waveguide. In another method (Fig. 22-10b) the coaxial line terminates in essentially a $\lambda/4$ vertical antenna projecting into the WG space. Energy radiation from this probe is transmitted down the WG. In both cases a reinforcement of the energy transfer occurs if the coupling devices are some odd quarter wavelength from the sealed end of the waveguide. The sealed end then acts as a parasitic reflector. In waveguides containing coupling devices, the near end will often be adjustable or tunable to assure maximum reflection from the end.

To couple from a waveguide to space, which has

an impedance of a few hundred ohms, the waveguide can be flared out in both horizontal and vertical dimensions from its normal 50-Ω impedance to the higher impedance of space. This forms a coupling *horn*, which can be used as a relatively narrow beamwidth transmitting or receiving antenna. Such horns are used to illuminate the parabolic reflecting dishes used in radar, terrestrial microwave antennas, and satellite systems.

22-5 DETECTING DEVICES

To sense the amplitude of the SHF (superhigh-frequency) ac energy in waveguides, either solid-state diodes or bolometers are inserted across a waveguide section. By using a hot-carrier or point-contact diode in a waveguide detector mount across the waveguide (Fig. 22-11a), ac energy is picked

ture coefficient are called *barretters* and are actually only fine resistance wires in a waveguide mount (Fig. 22-11b). Another type of bolometer is the *thermistor*, a small bead of semiconductor substance between two connecting wires (Fig. 22-11c). It has a negative temperature coefficient. Ac energy passing along a waveguide will heat a bolometer element. If the bolometer is used as the fourth arm of a bridge circuit that is balanced when the bolometer is cold, the amount of bridge unbalance indicates the relative value of ac power in the waveguide. The barretter is used for higher-power levels. The thermistor, being quite sensitive, is used for low-power measurements. Bolometers react slowly and are limited to indicating only very low modulating frequencies.

22-6 RESONANT CAVITIES

To produce a resonant *LC* circuit for microwaves, the number of turns of the coil is reduced to one and the capacitance across the circuit is only that between the two ends of the coil wire (Fig. 22-12a). This forms a quarter-wavelength *hairpin* tank. However, the resonant frequency will be somewhat lower than the measured wavelength value. An attempt to reduce the inductance further by paralleling another similar hairpin (Fig. 22-12b) does lower the *L* value but increases the *C* value. The resonant frequency may not change much, but *Q* increases. Note the distribution of current at a

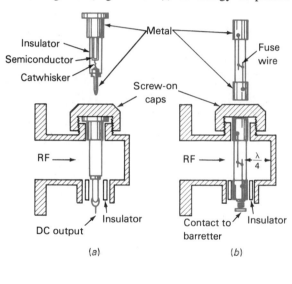

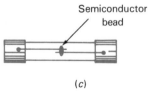

FIG. 22-11 (a)Microwave crystal and placement in a detector mount. (b) Barretter and mount. (c) Thermistor.

up, rectified, and made to operate dc meters or to indicate relative power, or modulation on oscilloscopes. The hot-carrier diodes are preferred because of sensitivity and low noise.

Bolometers are devices that change their resistance when heated. Those with a positive tempera-

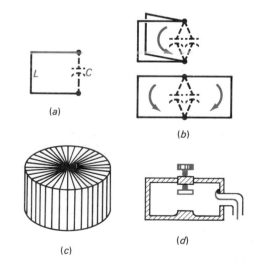

FIG. 22-12 (a) L and C of a λ/4 hairpin. (b) Two paralleled hairpins. (c) Cavity of infinite paralleled hairpins. (d) Capacitive tuning of cavity and low-impedance coupling of a coaxial cable to the cavity.

given instant if the two resonant hairpins are fanned out 180°. They appear as a cross section of a rectangular waveguide.

If many hairpins were added to a central point, a cavity similar to a round tuna can (Fig. 22-12c) would be developed. A tuna can has a radius and height of about 4 cm and should have a natural resonant frequency a little lower than 2 GHz if excited. A larger cavity would resonate at a lower frequency; a smaller, at a higher frequency. The frequency could be varied by a small percent by installing a variable capacitor in the center of the cavity (Fig. 22-12d). This cavity is shown with a coupling loop in it.

If a resonant cavity is made variable by installing a movable plunger in it and is also hole-coupled to a section of waveguide (Fig. 22-13), a *microwave*

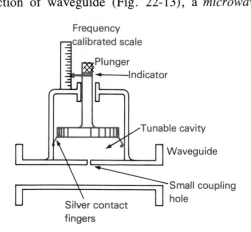

FIG. 22-13 Cavity wavemeter (frequency meter) coupled to a waveguide section.

wavemeter or frequency meter results. Energy in the waveguide passing the hole induces a voltage and current into the cavity. If the cavity is resonant to the frequency, it will absorb some of the energy. This can be noted by a detector coupled to the waveguide system, or to the wavemeter cavity. If the cavity is not resonant, it will not absorb any energy. An indicator on the plunger handle operating along a calibrated scale indicates the frequency of resonance.

▌ Test your understanding; answer these checkup questions.

1. What are the limits of microwaves today? _____
2. What are three transmission lines used for micro-wave frequencies? _____ _____ _____
   Which is most efficient? _____
3. How wide must a WG (waveguide) be? _____
   How high? _____
4. What is the designation of the lowest-frequency microwave band? _____ The highest? _____
5. What is reduced by using choke flanges? _____
6. What is the name of the microwave transmission line that is used with printed circuits? _____
7. Why are attenuators so often used with microwave test equipment? _____
8. Where is a resistive load placed in a directional coupler? _____
9. If the transmission-line $Z = 50 \ \Omega$ and the load is $100 \ \Omega$, what is the value of ρ? _____
   SWR? _____
10. Why are the detectors used in SWR reflectometers? _____
11. What are the three listed methods of coupling into or out of a WG? _____ _____ _____
12. Name two types of bolometers? _____
    _____ Which is a semiconductor? _____
13. Why are bolometers not used to detect VF modulation? _____
14. What would be two advantages of a resonant cavity over a hairpin tank? _____ _____
15. What type of coupling is used between a WG and a WG wavemeter? _____
16. Would a thermistor have a negative or positive coefficient of resistance? _____ Why? _____

22-7 KLYSTRONS

Four important types of VTs used to generate or amplify microwave ac are the *klystron, magnetron, traveling-wave tube* (TWT), and *backward-wave oscillator* (BWO).

There are two basic types of klystrons, reflex (oscillators) and multicavity (amplifiers). The low-power reflex klystrons produce a stream of electrons from a hot cathode that is drawn to a cylindrical cavity with grids on top and bottom (Fig. 22-14). If the cavity is oscillating, at one instant the top grid will be going positive while the bottom grid is going negative. This bunches any electrons traveling through the cavity toward the repeller plate at the top. Since the repeller is negative, it returns the now well-bunched electrons to the cavity in such a phase as to increase the strength of cavity oscillation. If the repeller has the wrong voltage, the bunched electrons will return out of phase and no oscillations can be produced. Various repeller voltages may produce oscillations. Some modes of oscillation will be stronger than others.

The frequency of oscillation can be changed by a few hundred megahertz (in X band) by stretching the grids apart physically by screwing in the tuning

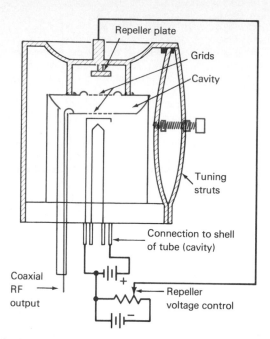

FIG. 22-14 Reflex klystron with coaxial output line.

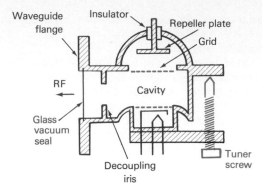

FIG. 22-15 Reflex klystron with waveguide mount.

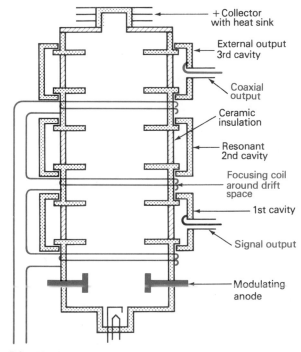

FIG. 22-16 Power-amplifier klystron. Tubes of this type range in height to more than 10 ft (3 m).

strut screw and deforming the cavity at the center. This reduces center-cavity capacitance and raises the frequency. The frequency is also electronically variable by a few megahertz by changing the repeller voltage slightly from the value that produces highest output. Modulating the repeller voltage by 1 or 2 V is a practical and simple method of producing wideband FM in these klystrons.

The original reflex klystrons produced only a few milliwatts of output power and were very inefficient. Newer tubes (Fig. 22-15) can produce much more power. When high power is required, however, 2-, 3-, or 4-cavity klystrons are used. These range in efficiency from about 20 to 50%.

A 3-cavity gridless power klystron is shown in Fig. 22-16. Electrons are emitted from the cathode and are attracted by the collector and modulating anode. However, a magnetic focusing coil around the tube forces the electrons up toward the collec-

tor, past the externally connected cavities. As the electrons stream past the first cavity, input-signal ac oscillations in it begin to bunch them. The bunched electrons move through the first drift space, where they continue to improve their bunching, to the second cavity, which is resonant to the same frequency and produces still greater bunching. The well-bunched electrons pass the third, or output, cavity, giving up most of their energy to this tank circuit. The electrons then pass on to strike the air- or water-cooled anode. By varying the dimensions of the external cavities, the frequency of operation of these tubes can be changed over a wide range by substituting different-size cavities.

Reflex klystrons may be used as the mixer oscillator in microwave superheterodynes or in signal generators. Power-amplifier klystrons can produce 10 to 50 kW of SHF power. They find use in TV transmitters, radar (producing megawatts of peak pulse power), and *tropospheric* beyond-the-horizon *scatter* communications (RF waves striking atmospheric discontinuities refract back to earth a long distance away).

Low-power klystrons are being replaced in many applications with GaAs FET cavity-type oscillators.

22-8 MAGNETRONS

The magnetron was developed to produce high-power microwave pulses for radar (200 MHz to 100 GHz, 1 W to over 1 kW CW, and 10 kW to over 5 MW pulsed). A 25-W plate dissipation magnetron may be visualized as being a cylindrical brass block about 2½ in. in diameter and about 1½ in. thick. A large hole is drilled down the center, and eight smaller holes are drilled between the center and outer edges (Fig. 22-17). Slots interconnect the small holes and the large hole. Both ends are sealed with end plates. The smaller holes form resonant cavities. When the magnetron is operating, electrons take a back-and-forth path along the walls of the cavities (vector arrows in one of the cavities). The magnetron is a vacuum tube.

A cylindrical coated cathode with an internal heater wire is located in the center of the large hole. One heater lead connects to the near side of the heater; the other lead connects to the far side.

A small hook in one of the cavities acts as a pickup loop, extracting microwave energy from this cavity (and thereby all others) when it oscillates. This energy is fed to a short concentric transmission line, which is terminated by a λ/4 radiator protruding into the end of a waveguide.

The magnetron block acts as the anode and is grounded. When the tube is pulsed, the cathode is driven negative by perhaps 10 to 20 kV. This makes the anode relatively positive, and electrons from the hot cathode start moving toward it. However, a strong external horseshoe-shaped permanent magnet, with its north pole at one end of the cathode and its south pole at the other end, produces an intense magnetic field down the central hole. According to the right-hand motor rule (Sec. 3-17), the electrons will be deflected at right angles to the lines of force through which they are passing. This results in an elliptical path for the electrons, as shown, as they progress toward the anode areas.

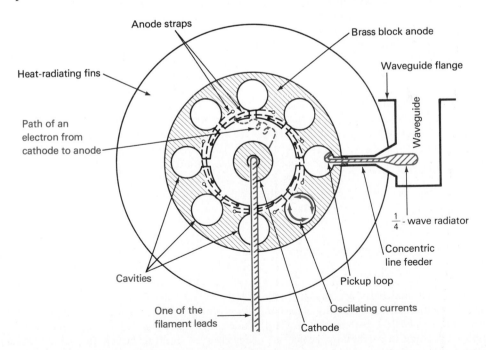

FIG. 22-17 Multicavity magnetron with waveguide output.

The positive potential of the anode accelerates the electrons toward it. This is the same as saying the electrons pick up energy from the difference of potential. As the electrons whirl past the slots between anode areas, they induce voltages between the slot faces which drive currents into oscillation along the surfaces of the cavity walls. In this way the energy of the looping cathode electrons is transferred to oscillating currents in the cavities. All the cavities are of the same size and oscillate at the same frequency. However, adjacent cavities have opposite-direction currents in them at any one time. Strapping every other anode face together (dashed lines) increases the efficiency to about 50% from an unstrapped 35%.

A magnetron may have an average power output of only 20 W, but when it is used in pulsed circuits, it can produce high-peak-power pulses. For example, if a 20-W magnetron is pulsed 1000 times per second and each pulse is only 1 μs long, the total time it is operating is only $^1/_{1000}$ s. Each pulse may therefore have a peak power of 20,000 W. If pulsed with 1-μs pulses only 500 times per second, each pulse could have 40,000 W and not exceed the 20-W average power-output value of the tube.

The frequency of resonance of a cavity can be raised if the volume of the cavity is made smaller by pushing a plug into it. Some magnetrons are made variable in frequency by moving slugs into all the cavities simultaneously. Other magnetrons are electronically tuned by varying the anode voltage on them. For example, one such low-power magnetron can change from 400 to 1200 MHz by changing the anode voltage from 700 to 1900 V.

Magnetrons can be used to produce continuous waves, as in microwave ovens, but since their frequency stability is not particularly good, in communications they are used mostly in radar equipment.

22-9 TWTs AND BWOs

A microwave tube that can tune over a relatively wide band of frequencies electronically is the traveling-wave tube (TWT), or traveling-wave amplifier (TWA). It consists of an electron-focusing gun to deliver a beam of electrons down the center of a spiral (helical) coil to the anode at the far end (Fig. 22-18). The helix, anode, and electron gun are all in a vacuum. The anode and helix are made highly positive in respect to the cathode to pull electrons down the helix center. Outside the evacuated glass tubing are slipped two additional helixes which are 50-Ω-impedance input and output coupling devices.

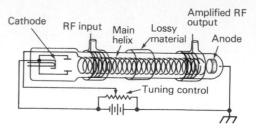

FIG. 22-18 Traveling-wave tube, or TWT.

When a microwave signal is induced into the left end of the main helix, it travels along the surface of this wire at essentially the speed of light. Since the helix is a coil, the actual velocity of propagation down the tube is considerably less than the speed of light. If electrons are fed down the core of the main helix at a slightly higher velocity than the signal wave is traveling on the helix, electrons and wave interact and the electrons inside the main helix become bunched as they progress down the length of the helix. This results in some of the electrons slowing down and losing much of their energy to the induced wave on the main helix. As the energized helix waves pass the output coupling device, they induce energy into it. The gain of a TWT ranges from 30 to 60 dB. Available power outputs range from milliwatts to several hundred watts.

There is no cavity in a TWA or TWT. The only tuning required is to maintain the helix potential at an optimum synchronous value. To prevent the electron beam from being attracted to the positive helix, a strong axial permanent or electromagnetic field is developed down the center of the helix by external magnets (not shown). This magnetic field focuses the electrons and holds them in the center of the main helix area.

To prevent amplified energy from working back up the helix from the output to the input device, which would produce oscillations, a lossy attenuator is wrapped around the glass envelope that surrounds the main helix. Without such an attenuator, and with only one coupling device, the TWT would be essentially the same as a backward-wave oscillator tube. A BWO can operate with relatively high efficiency and is voltage-tunable by varying the helix voltage. For example, one tube can tune over an octave and a half (from 1 to 3 GHz) by varying the helix voltage from 300 to 2000 V. Such a variation of voltage results in an unequal amplitude output over the operating range. It is necessary to use a *leveling* or limiting circuit to maintain equal amplitude output at all frequencies.

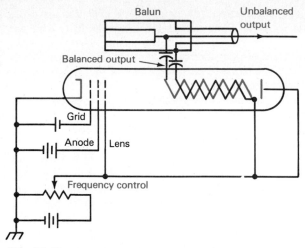

FIG. 22-19 Bifilar-helix backward-wave oscillator, or BWO, with balun.

The BWO in Fig. 22-19 uses a bifilar (two-wire) helix and has a balanced output. To couple to an unbalanced line, such as a coaxial cable, and to convert to some other impedance, if desired, a *balun* (*bal*anced-to-*un*balanced) must be employed. The balun may come with the BWO tube. Basically, a balun is a center-tapped autotransformer (Fig. 22-20*a*). Since a 1:2 turns ratio has a 1:4

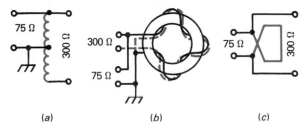

FIG. 22-20 (*a*) Simple coil balun. (*b*) Toroidal coil balun. (*c*) Linear balun.

impedance ratio in a transformer, the balun can change a 75-Ω unbalanced coaxial line to a 300-Ω balanced line. Figure 22-20*b* shows the same circuit wound on a toroidal form. Figure 22-20*c* illustrates a linear balun, in which the 75-Ω input is across one turn of a two-turn coil. A cavity-type balun is shown in Fig. 22-19.

TWTs and BWOs are usually used to amplify or produce continuous waves. The TWT is used extensively as a microwave RF amplifier. The signal it is amplifying is usually modulated with amplitude or frequency modulation, and often with some form of multiplex (Sec. 17-40) to allow a number of chan-

nels of information to be transmitted simultaneously. Since microwave circuits are usually relatively low Q, the transmissions are quite wideband, making multiplex operation fairly simple.

22-10 MICROWAVE SEMICONDUCTORS

Lighthouse-type triodes are limited in frequency by interelectrode capacitances, inductance of leads, and transit time. The same is true of common bipolar transistors, but by improving the geometry and using overlay epitaxial planar types, transistors can be made to amplify or oscillate in the 6- to 10-GHz range. Some microwave FETs can operate up to 40 MHz. Monolithic integrated circuitry is used at lower microwave frequencies.

A highly stable oscillator is produced by loosely coupling a tunnel diode (Sec. 9-3) to a high-Q cavity (Fig. 22-21). By using a short antenna probe

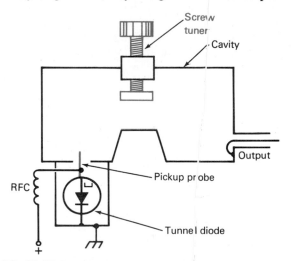

FIG. 22-21 Tunnel-diode cavity oscillator.

protruding into the cavity and by feeding off-center of the cavity, loose coupling is attained. The power output of such oscillators is a few hundred microwatts, but this is sufficient to act as local oscillators for microwave superheterodynes.

One application of a varactor diode is to produce a multiplication of a given ac frequency. It will be remembered that any distortion of a sine wave indicates the presence of harmonics of the sine-wave frequency. The circuit in Fig. 22-22 indicates the LC circuit of an oscillator producing sine-wave ac at frequency f_o. The ac voltage across the LC tank circuit is essentially sinusoidal. This ac is coupled through a high-Q series resonant circuit to the

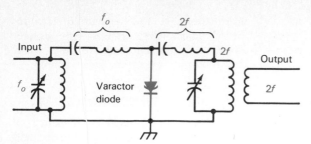

FIG. 22-22 Varactor frequency multiplier.

varactor diode. The ac waveform of the voltage fed to the diode is sinusoidal; but because of the very nonlinear current-voltage relationship in a diode of this type, the voltage-drop across the diode will be very nonlinear (contain considerable harmonics). A second series-resonant circuit, tuned to the second harmonic frequency ($2f$), is used to couple second harmonic energy to the output tank circuit, which is also tuned to $2f$. Any diode can be used in this kind of a circuit, but the rectification by a varactor, because of its nonlinear junction charge storage, is particularly distorted. Efficiency of such a multiplier circuit can be as high as 75%. Had the $2f$ resonant circuit and the $2f$ antiresonant tank been tuned to $3f$, the output circuit would have picked up third harmonic energy; and the output would have been three times the fundamental frequency but at somewhat reduced efficiency. As doublers, varactors can deliver as much as 25 W at 1 GHz and 100 mW at about 35 GHz.

Another diode that can be used in this same type of frequency-multiplier circuit is the *step-recovery* diode. By special doping of its junction, charges store in the junction during forward-bias current-flow time. As the forward-biased pulse starts to fall off, the charges start flowing out of the junction. By the proper doping density, all the charges stored by the high-frequency pulse stop flowing at approximately the same instant. The result is a very fast cessation of current flow some time after the pulse drops off and reverse bias starts. The sharp current break, or step, is rich in harmonics, which is the reason for the step-recovery diode's use. As much as 10 times multiplication can be produced at reasonable efficiencies with step-recovery diodes.

The impact avalanche and transit time diode (IMPATT) is a special microwave diode that utilizes the delay time of attaining an avalanche condition plus transit time to produce the 180° voltage-current condition of negative resistance. As with other microwave diodes, it may use silicon, germanium, or

gallium arsenide. It can oscillate and produce up to 1 W at 10 GHz and about 0.1 W at 100 GHz, but at an efficiency below 10% at the latter frequency.

The hot-carrier, or Schottky, diode is an excellent low-noise detector for use in the higher microwave frequencies.

Microwave PIN diodes can be used in waveguides as switches or modulators. They operate as diodes up to a few megahertz, and then the intrinsic layer produces too much transit time. However, they have a useful high resistance with reverse bias and low resistance with forward bias. A PIN diode mounted across a 50-Ω waveguide and reverse-biased has almost no effect on energy being transmitted down the line. If the junction is forward-biased, the low resistance acts as a wall to the energy, and almost all transmission can be stopped by it. If the bias of the diode is modulated, its resisting effect varies, and an amplitude modulation can be obtained at the output of the waveguide. While one diode cannot switch high powers, several in parallel can be used to pulse several thousand watts of peak power in a waveguide.

22-11 GUNN AND LSA DIODES

These semiconductor devices are called diodes only for the lack of a better term. They do not have junctions but instead depend on other peculiarities of semiconductors to produce negative-resistance effects. They can be called *active-area* or *bulk-effect* devices.

The microwave and millimeter band Gunn diode consists of a thin slice of N-type gallium arsenide between two metal conductors (Fig. 22-23a), assembled in a cylindrical metal and ceramic body. The diode is fitted into a holder inside a cavity and is fed the required dc to make it oscillate. If the voltage across the diode is increased from 300 V/mm to 400 V/mm, at some voltage the electrons from the outer, partly filled energy ring of the atoms of the semiconductor crystal jump across the very narrow forbidden energy gap of gallium arsenide and actually decrease their mobility in the crystal. This produces a negative-resistance effect in the semiconductor. If the voltage is increased further, the current will begin to increase in proportion to the applied dc voltage, and a *VI* curve similar to that of a tunnel diode is produced.

Figure 22-23b represents the placement of a Gunn diode in a cavity that is made tunable over several GHz by making one end wall adjustable.

The other active-area device is the *limited space-*

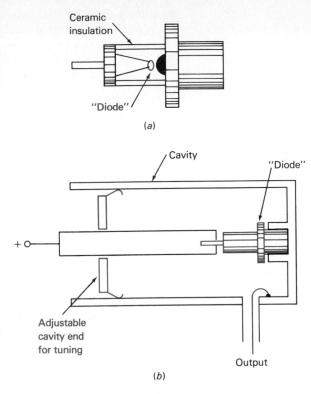

FIG. 22-23 (a) Gunn diode. (b) Gunn-diode oscillator.

charge accumulation (LSA) diode. It is somewhat similar to a Gunn diode except that it is considerably thicker and produces its negative-resistance effect at a dc voltage in the range of 800 V/mm.

Both the Gunn and LSA diodes operate considerably above the 25-GHz region, but the LSA produces several times the power output, in the range of 3 W at 50 GHz. At lower microwave frequencies, peak pulsed powers of 100 kW are possible.

22-12 ISOLATORS AND CIRCULATORS

Ferrites (compounds of oxygen, iron, and several other metals) play an important role in microwaves. They have two important properties. One is *ferromagnetic resonance absorption*. With a magnetic field magnetizing a ferrite rod in a waveguide (Fig. 22-24), energy entering the waveguide at the resonant frequency of the ferrite molecules will be absorbed by the ferrite. Energy entering from the opposite direction will not be affected by the molecular resonance of the ferrite rod. As a result, an on-off waveguide switch can be produced by reversing the magnetism of the electromagnets. A ferrite rod of this type is called an *isolator*. It can

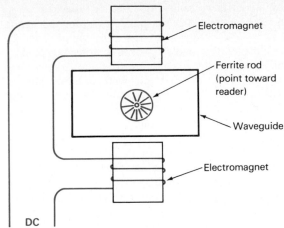

FIG. 22-24 Ferromagnetic resonance absorption switch.

act as a one-way buffer between a microwave oscillator coupled to the end of a waveguide and variations of the loads further down the line. The resonant frequency of the ferrite is controlled by the external-magnetic-field strength.

A second useful property of a ferrite rod is known as *Faraday rotation*. In this case, the frequency of the energy entering the 0° port (Fig. 22-25) must not be near the resonant frequency of

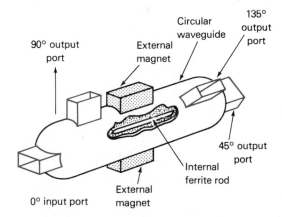

FIG. 22-25 Faraday rotation circulator.

the ferrite molecules. The signal first transfers from rectangular waveguide to the circular waveguide. The energy wave rotates 45° as it flows past the ferrite rod that is in the field of two magnets. The output-port waveguide must therefore be at 45° from the input-port waveguide. If the magnets were removed, no rotation would result and almost no energy would be transmitted out the 45° port because of improper field polarization. If the 45° port has an efficient reflecting terminal attached to it, energy is reflected back, past the ferrite rod again, is rotated 45° more, and now can emerge from the 90° port. If this energy strikes a reflecting load, it will re-enter the circular waveguide and emerge from the 135° port. (If this were reflected and returned to the circular waveguide, it will emerge from the 0° port, either in phase or out of phase, depending on the total travel distance of the wave through the *circulator*.) A purely resistive load, matching the impedance of the waveguide, coupled to the 90° port, produces an isolator. Energy entering the 0° port can emerge only from the 45° port, since energy reflected from any load coupled to the 45° port will be completely dissipated by the resistive load at the 90° port.

Figure 22-26 illustrates one use of a circulator. It allows one parabolic-reflector, focusing-type, high-

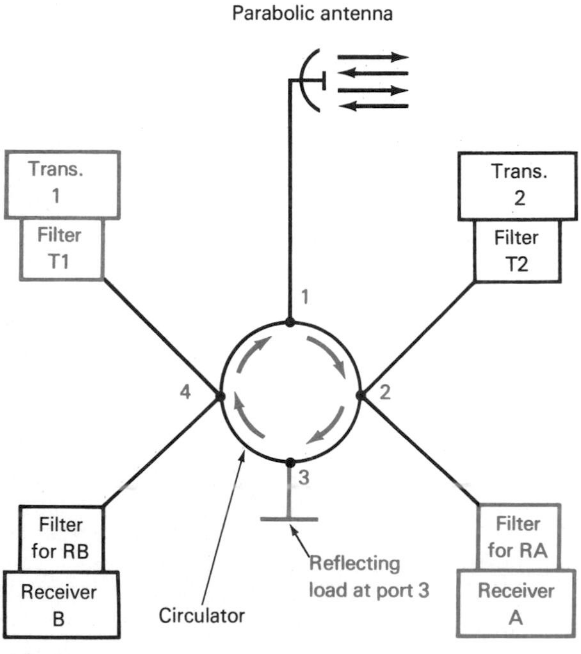

by the RB filter and passes to receiver B. Transmitter 2 energy is reflected from filter RA, from port 3, from the filters of port 4, feeding its signal to the antenna.

22-13 TUNNEL-DIODE AMPLIFIERS

Operation of a tunnel diode, biased to its negative-resistance center point and coupled to a tuned circuit, produces an oscillator at the tuned-circuit frequency. If the input to the tunnel diode can be tuned but the feedback due to the tunnel diode being connected across the tuned circuit cannot reflect energy back to the tuned circuit, the tunnel diode cannot oscillate but will amplify. A circulator can be used (Fig. 22-27) with a tunnel diode to

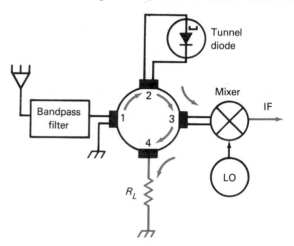

FIG. 22-27 Tunnel-diode amplifier.

produce an RF amplifier ahead of a microwave mixer stage. The bandpass filter feeds the desired-frequency received signals into port 1 and to the tunnel diode in port 2. The amplified signals are fed from port 2 to port 3 and to the mixer input. If any signal is reflected from port 3, it passes to port 4, where it is absorbed by the matched load resistance. There is no way for the tunnel diode to produce feedback of energy, so it cannot oscillate.

Gunn and LSA diodes can be used in similar amplifier circuitry. They have more gain than the tunnel diode, but they have more noise also.

22-14 YIG RESONATORS

A device that can have very high Q at microwave frequencies is a tiny, grown-crystal, highly polished

FIG. 22-26 Circulator to multiplex two transmitters and two receivers to same antenna.

gain (±30-dB), broadband antenna to feed two separate microwave transmitters and receivers at the same time. Each bandpass filter is tuned to its own receiver or transmitter frequency, reflecting all other frequencies. A signal for receiver B is picked up by the antenna, feeds to port 2, is reflected, is reflected again at port 3; but, at port 4, the signal is accepted ball of ferrimagnetic yttrium-iron-garnet (YIG). Such a sphere, smaller than the head of a pin, if exposed to a fixed magnetic field, will rotate until its crystalline domains align with the external field. If the sphere is then anchored and a small loop of wire is wound around it in a direction that allows dc flowing through the loop to produce a magnetic field at right angles to the fixed field, the resultant poles of the YIG sphere shift from the original fixed-field direction. If ac is fed to the loop, the resultant magnetic YIG poles will attempt to *precess*, or rotate around the axis of the fixed field at the frequency of the applied ac. At one particular frequency, determined by the geometry of the YIG sphere and the strength of the fixed field, the precession finds little molecular opposition and the ac sees the YIG sphere as presenting almost no load (high Q) to it.

If the strength of the fixed field is decreased, ease of precession of the poles of the YIG sphere decreases, and the original ac frequency sees the YIG as having a lower Q and being a more lossy device. Now, by lowering the fixed field strength, a value of magnetization that will permit a free resonant precession at a new lower frequency can be found, and the YIG sphere again appears as having a high Q. Conversely, a YIG sphere produces a resonant effect at a higher frequency if the dc magnetic field strength is increased. Thus, the YIG acts as a tank circuit whose resonant frequency is tunable by varying the dc excitation to an electromagnet between whose poles the sphere is mounted.

Figure 22-28a represents a YIG sphere mounted between the north and south poles of an electromagnet. There will be some value of dc excitation which will allow the ac input to see the YIG as a high-Q parallel resonant circuit.

In Fig. 22-28b, the YIG sphere is mounted in a field but is surrounded by two loops at right angles to each other. If the dc excitation produces the proper magnetic field strength, the YIG will allow maximum transformer effect between primary and secondary loops and maximum output at its resonant frequency. At all other input frequencies there will be no magnetic resonant precession of the YIG and there can be no induced emf from primary to secondary because they are at right angles to each other. The device operates as a narrow-bandwidth tunable bandpass filter.

In Fig. 22-28c a YIG sphere forms the tank circuit of a Colpitts-type oscillator circuit. Base and collector are coupled to opposite ends of the single-

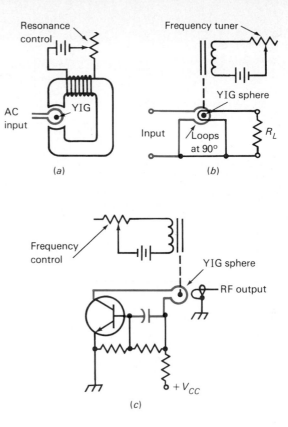

FIG. 22-28 (*a*) Magnetic poles of YIG precess with ac excitation. (*b*) Basic YIG voltage-variable narrow BP filter. (*c*) Basic voltage-variable Colpitts-type YIG-tuned oscillator.

turn inductor. At microwave frequencies, distributed and interelement capacitances make simple Colpitts or ultraudion oscillator circuits possible. The YIG oscillator output frequency is determined entirely by the field strength of the dc-controlled electromagnet. In the SHF range, transistors may be supplanted by tunnel, IMPATT, Gunn, or LSA diodes as the active devices in oscillator circuits. YIG-tuned devices can operate up into the tens-of-gigahertz region.

A YIG can also be used in a frequency multiplier. A 2-GHz RF is fed to a step-recovery diode (Fig. 22-29) which develops a comb of harmonics (4, 6, 8, 10, etc., GHz). The harmonic to which the YIG is made resonant by its dc field excitation is the output frequency. Signal generators with up to 1-mW output can be produced at frequencies higher than 25 GHz. In this case the YIG is operating as a tuned filter.

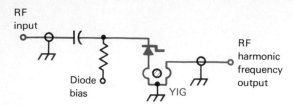

FIG. 22-29 YIG used as a filter in a frequency multiplier (electromagnets not shown).

■ Test your understanding; answer these checkup questions.

1. List five microwave VTs. _____ _____
   _____ _____ _____
2. Which VTs have a repeller? _____ A series of cavities? _____ A helix? _____
   _____ A horseshoe magnet? _____
3. What is the advantage of gridless klystrons?
   _____
4. How are magnetrons tunable? _____
   _____
5. What does TWT mean? _____ BWO? _____
6. Which VT will oscillate at the highest frequency? _____
7. What is used to change impedances and feeds with a BWO? _____
8. What is a microwave use of a varactor? _____
   What diode might be better? _____
9. List four diodes that generate microwave ac.
   _____ _____ _____ _____
10. What does PIN mean? _____ What are PIN diodes used for? _____
11. What does LSA stand for? _____
12. What is wrong with the term "Gunn diode"?
    _____
13. What two properties do ferrites have at microwave frequencies? _____ _____ Which is used as a switch? _____ In circulators? _____
14. What is the advantage of using a circulator with two transmitters and two receivers? _____
15. To what must the fourth port be connected in a tunnel-diode amplifier? _____ The third port in a multiplex circulator? _____
16. What microwave device is said to precess?
    _____
17. What tunes a YIG? _____

22-15 A MICROWAVE MUX TRANSMITTER

There are many bands of frequencies scattered throughout the microwave region that are used for multiplex (MUX) terminal or repeater operations. One band, for example, showing authorized uses and allowable FM bandwidths includes:

1710–1850 MHz, governmental
1850–1990 MHz, operational (5M00FXX, 10M0FXX)*
2110–2130 MHz, common carrier (3M50FXX)
2130–2150 MHz, operational (800HFXX, 1K60FXX)
2150–2160 MHz, common carrier (10M0FXX)
2160–2180 MHz, common carrier (3M50FXX)
2180–2200 MHz, operational (80CHFXX, 1M60FXX)
2200–2290 MHz, governmental

A basic microwave terminal usually consists of a MUX or baseband-modulated FM transmitter and a broadband FM receiver (Sec. 21-16). More complex systems may use two duplicate transmitters and receivers, which can be switched to single standby unit operation in case a fault develops in any section of either system. More complicated systems may use *space diversity* (two antennas with one transmitter and receiver at each location), or *frequency diversity* (two transmitters and receivers with one antenna, but using two separate frequencies). A simple representative 2133 MHz band microwave FM transmitter-receiver terminal will be outlined. (AM may also be used.)

The SSB modulating signals developed by all channels of a multiplex system forms the modulating signal for the microwave transmitter, shown at the top in Fig. 22-30. (Pulse signals may also be used.) An added modulating signal is a low-frequency pilot tone of 20 Hz (or a high-frequency pilot of perhaps 1.5 or 3 MHz), which provides a constant-amplitude modulating signal to be used to check the operation of the system. The amplitude-type baseband modulating signals frequency modulate a self-excited 711-MHz cavity oscillator (Fig. 22-31) having two voltage-variable capacitors (varactors) in it. The FM output developed by the action of one of the varactors is fed through a tunable isolator (acting as a one-way RF filter) to a stripline tuned transistor amplifier. Some of this ac is fed to an automatic frequency control (AFC) detector which compares the pilot signal frequency with the harmonics of a crystal oscillator. If there is a difference between crystal and pilot frequencies, it is detected by a discriminator-like circuit. The dc output of this circuit varies the capacitance of the second varactor in the oscillator cavity, automatically tuning the oscillator until it is back on fre-

_____
*See Appendix E-2.

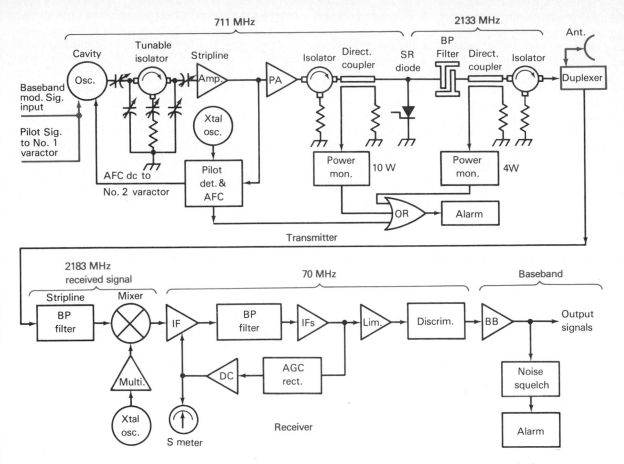

FIG. 22-30 Block diagram of a simplified microwave transmitter and receiver in a two-way terminal.

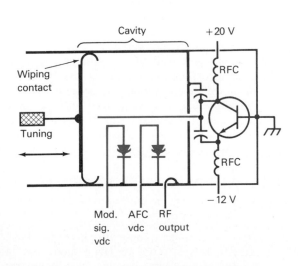

FIG. 22-31 BJT in a microwave cavity oscillator.

quency. The microwave oscillator with AFC has the stability of its low-frequency crystal oscillator.

The 711-MHz FM signal is amplified to about 10-W output and is fed first through a stripline directional coupler that monitors the power at this point, and then to a step-recovery diode that develops many harmonics of the 711-MHz signal. The third harmonic, or 2133 MHz, is fed through a stripline 1700–2300 MHz bandpass filter, through a second directional coupler monitor, to an isolator capable of reducing any *reflected* power at least 20 dB. This helps to reduce the SWR on the antenna transmission line. Stripline circuits are used rather than cavities because they have the necessary lower Q to produce a broader response to pass the 1–10 MHz bandwidth FM signals developed by the baseband modulating frequencies (which range from 20 Hz to possibly 2.54 MHz).

The output of the isolator feeds an antenna

hybrid, diplexer, or duplexer (Sec. 24-2). The duplexer passes the transmitter RF to a parabolic antenna aimed at some other microwave terminal or repeater 5 to 50 mi away. At the same time the duplexer prevents any of the transmitter RF from feeding into the local receiver connected to the same antenna. Use of 90° shifted polarization for transmitting and receiving further isolates the signals.

If the monitored power levels, or the carrier frequency variation, exceeds or drops below predetermined levels, such faults can activate an OR gate to set off either a local light or bell alarm or transmit an alarm signal remotely. One of the baseband channels should be set aside for this use, to turn off the transmitter.

22-16 A MICROWAVE MUX RECEIVER

The usual microwave MUX receiver is a single conversion superheterodyne. It receives signals from a distant microwave transmitter on a frequency displaced perhaps 50–80 MHz from the frequency being used by its own transmitter. As a result of this displacement, plus the action of the duplexer, plus the action of a 5-MHz bandwidth input bandpass filter (tuned to 2183 MHz in this case), plus a 90° antenna polarization difference, there is essentially no interference to the local receiver from the transmitter, although they may be using the same antenna.

The diagram in Fig. 22-30 shows the receiver at the bottom. Signals from the duplexer pass through a stripline bandpass filter selected to pass the desired 2183-MHz received carrier and its sidebands. These are fed to a stripline mixer and are heterodyned against the harmonics of a crystal oscillator to develop 70-MHz IF signals. An *image rejection* bandstop filter tuned 140 MHz above or below 2133 MHz, whichever is required, may also be added *before* the mixer (not shown). The IF signals

are amplified by an untuned amplifier before being passed through a 2-MHz- or wider-bandwidth LCR bandpass filter. The filter bandwidth is determined by how many baseband channels are used to modulate the transmitter. The IF signal is then amplified about 60 dB more. Some of the IF signal is rectified to dc, amplified, and fed back as an automatic gain control (AGC) bias voltage to the first IF stage, to partially counteract signal strength variations due to fading, at the same time providing an S-meter voltage indication of signal strength. The IF signal is limited and fed to an FM discriminator. The output of the discriminator will be the amplitude modulation SSB signals of the original baseband channels at the transmitter input, up to as high as 2.54 MHz if all 600 channels are used.

For lower-frequency microwave bands, the coupling of the RF ac may be by coaxial cable or by waveguide. For the higher-frequency bands, waveguides are usually used if the lines are more than a few inches in length.

22-17 INSTALLING WAVEGUIDES

When waveguides are installed permanently, as in radar, it is important that there be no long horizontal runs which might result in an accumulation of moisture droplets from condensation on the inside walls of the sections. These would attenuate energy transmitted down the line.

Since condensation does occur in exterior waveguides, a small hole is drilled in the elbow of a waveguide section at the lowest point to allow an escape vent for the water. Choke flanges must always be mounted with their half-wave cavities pointed upward so that they will not fill with condensation.

Waveguides must be handled with care. The slightest dent in a wall of a section produces an impedance discontinuity, increasing the SWR and reducing power transfer along the system.

To prevent radiation of undesired signals or interference pickup, waveguides should be firmly fastened to walls and grounded as often as possible.

22-18 MICROWAVE TEST EQUIPMENT

Technicians working in microwaves will be expected to work with:

- *Signal generators.* These use klystron, BWO tube, transistor, YIG, or Gunn diode oscillators (or synthesizers) having CW (continuous-wave), pulse, sine- or square-wave modulated-signal outputs. The generators may be hand-variable or be programmed to sweep a selected band of frequencies. Some generate signals to 40 GHz or more.

- *Calorimetric power meters.* These are oil-cooled bridges with bolometer arms reading directly in watts or in decibels. Their power capabilities range from 10 mW to 10 W.
- *Power meters.* These indicate the resistance change due to microwave energy striking a thermistor mount at the end of a waveguide or coaxial cable, reading directly in watts or in decibels. If the power in the system is greater than can be read directly, accurately calibrated waveguide or coaxial attenuators are used ahead of the thermistor mount.
- *Standing-wave indicators.* These read SWR directly as a carriage and its detector probe are moved along a slotted section of waveguide.
- *Ratio meter.* This device reads the ratio between the two directional couplers of a reflectometer setup, indicating SWR directly, regardless of the frequency being swept by the signal generator.
- *Frequency meters.* These are either the calibrated wavemeter type, heterodyne oscillators beating against the unknown signal and reading out on a frequency counter, or high-speed digital counters.
- *Sampling oscilloscopes.* These can display microwave signals up to about 20 GHz. They may use variable-storage CRTs. A camera may be fastened in front of the CRT to take pictures of waveforms.
- *Spectrum analyzers.* These are swept receivers with a CRT display of the amplitude of received signals versus the frequency band across which the receiver is being tuned. The swept bands may be relatively narrow or broad, but top frequencies are in the 40-GHz range.
- *Function generators.* These are used to modulate signal generators with a variety of waveshapes and pulse widths. PIN diodes across coaxial lines or waveguides produce absorption-type modulation of microwave signals.
- *X-Y recorders.* These are used to plot relatively slow variations of circuit voltages or other parameters by moving a stylus up and down a sheet as the stylus is moved along horizontally. *Strip-chart recorders* have a stylus that is horizontally stationary but moves

up and down a sheet of paper which is being drawn slowly along beneath the stylus.
- *Connectors.* These are used from and to microwave equipment and are of the N, BNC, SMA, GR, APC-7, and banana types. Most of these afford a relatively small impedance discontinuity which prevents reflection of signals in the line.
- *Time domain reflectometry* (TDM). This system utilizes a sampling oscilloscope, sending a very narrow pulse down a line and measuring the time it takes for a return reflected signal from any discontinuities in the line to appear. Time is then converted to distance to locate the discontinuity.

Some of the measurements that are made are:

- Absolute power in a system, using bolometer terminations to a power meter.
- Relative power levels in systems. As a system is tuned or the loading is changed, measurements are taken on a power meter.
- Attenuation produced by a device added in a waveguide system. A reading is taken at the load without the device and then again with the device coupled into the system. The difference between readings is the attenuation produced by the device.
- VSWR measurements. Standing-wave ratio indicates the reflection produced by a mismatched load. Either a detector probe in a slotted section or a reflectometer may be used to produce the voltages that are read.
- Frequency of signals in a system.
- Impedance. Since a matched load produces a 1:1 SWR, analysis of the SWR in a waveguide allows determination of the impedance of a load if the waveguide impedance is known.

22-19 OPTICAL FIBER COMMUNICATIONS

One of the important developments in communications is the use of optical fibers as waveguides. Instead of radiating RF down a microwave waveguide, infrared, visible, or ultraviolet frequency carrier waves are fed into thin, long, clear, round, silica, or halide glass threads. Such optical fibers may have less than 2 dB of attenuation per kilometer. Since the carrier waves are at light and infrared frequencies, each carrier can be modulated by either many thousands of voice frequency channels simultaneously (MUX), by several TV programs at once, or by very fast computer information.

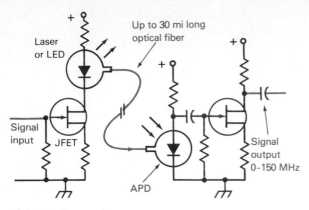

FIG. 22-32 Simple fiber-optic system.

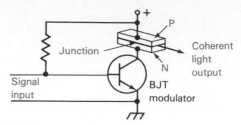

FIG. 22-33 Semiconductor laser circuit to produce modulated coherent light or infrared waves.

Figure 22-32 shows a basic optical fiber circuit in which an analog or digital input signal varies or modulates the I_D of an FET. The result is an LED diode light output that varies in amplitude at the modulation rate. The LED light is focused into one end of a jacketed, or clad, optical fiber. The modulated light travels to the far end of the fiber, where it illuminates the light-sensitive surface of an avalanche photodiode (APD) or phototransistor (Sec. 9-5). The variations of the light intensity causes the APD or transistor current to vary at the modulation rate, which can then be coupled to signal amplifiers. In this case the LED acts as an RF oscillator being amplitude-modulated by a signal. The modulated light signal is radiated along the optical fiber. The photodiode acts as the detector of the modulation. The fiber may be bent in any direction, but the light follows the fiber, and little light is lost.

The incoherent light from an LED requires an optical fiber that is clad with a reflective coating to hold the light waves inside the fiber. With coherent light (all rays similar frequency and polarization), a simple unclad optical fiber might be used. Coherent light can be developed by a *laser* (Lightwave Amplification by Stimulated Emission of Radiation). There are ruby, gas, liquid, chemical, and semiconductor lasers. A basic semiconductor laser transmitter is shown in Fig. 22-33. When current flows through the AlGaAs (aluminum-gallium-arsenide) diode junction, from the P to the N material, weak "light" waves (usually humanly invisible infrared at 800 to 1500 nm wavelengths) are developed in the junction area by electrons driven out of their normal energy levels in the junction molecules, and then falling back into them. The end surfaces of the lasing area are polished and coated with a reflective substance. Waves moving in the lengthwise direction inside the diode are reflected back and forth, picking up more energy with each reflected passage. Waves not reflected back and forth are lost out the sides of the area. When waves pick up sufficient energy they pierce the less densely coated end and radiate outward. Such coherent light is coupled into one end of an optical fiber. The light output can be modulated by varying the current flowing through the diode and the modulating transistor. Diode lasers may be smaller than the head of a pin.

New *photonic* devices combine electronic and optical functions in tiny (0.02 × 0.02 in.) integrated circuit chips. Microscopic 3-mW lasers can develop intense constant-amplitude light carrier waves that can be modulated and monitored by using variable refractive systems and tiny prisms. Several different wavelength carriers can be refracted into the same fiber. Very high modulation rates (100 MHz to 100 GHz, depending on fiber construction and wavelengths used) are possible, increasing the volume of information transmission greatly.

Advantages of optical fiber communications are multiple channels per fiber, while many fibers can be cabled together with no cross talk. With silica optical fiber systems, repeaters (line amplifiers) are required every 20 to 30 miles, roughly similar to coaxial repeater distances in telephone systems. Halide glass fiber runs may be over 100 miles.

22-20 PARAMETRIC AMPLIFIERS

One of the amplifiers with the lowest internal noise is the *parametric* type. The name is derived from the fact that the operating capacitive reactance is a parameter of one of its tuned LC circuits. From the formula $Q = CV$ (Sec. 6-6), the voltage across a charged capacitor is determined by $V = Q/C$, or coulombs/farads. If a charged capacitor has its capacitance decreased, the voltage across it must increase, according to the formula. If energy is used to decrease the capacitance, the increased voltage will also be accompanied by an increase in charge, and both voltage and power will be amplified.

The diagram in Fig. 22-34 illustrates a basic parametric amplifier. The capacitive reactance to be used is the varactor. The input frequency f_i feeds a

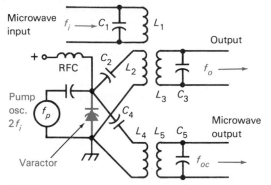

FIG. 22-34 Basic parametric amplifier with both straight through amplification f_o and converted frequency output f_{oc}.

resonant circuit L_1C_1. (In microwave circuits this, and all other LC circuits shown, would actually be cavity tanks.) The input frequency couples through to L_3, providing an output signal f_o which would be the same as the input signal except a little weaker. Some of f_i is also fed across the varactor. If a *pump* oscillator having a frequency of $2f_i$ is also fed to the varactor, its energy will pump up the ac voltages across the varactor and feed a stronger ac to L_3, resulting in a stronger f_o. Since there are no conducting active devices or any resistance in the circuit to produce noises, there is little noise generated by the parametric amplifier. However, if the amplifier is coupled to a mixer, which is usually noisy, it may feed back noise into the amplifier. This can be prevented by using a circulator between the amplifier and any mixer.

If the pump frequency f_p is only slightly higher than f_i, there may be some f_o, but there will also be a down-converted, or idler output from L_5, which can then be heterodyned with a local oscillator to down-convert still further to any desired IF of a microwave superheterodyne. If f_p is somewhat higher, the conversion output circuit L_5 will produce an up-converted output signal. Often the converted signals are used as the output of the amplifier rather than the straight through frequency. The gain of such amplifiers is between perhaps 10 and 30 dB if operated at room temperature. The gain of the amplifier if it and its circulator are immersed in liquid nitrogen will increase several

decibels, and the noise will decrease measurably. Most parametric amplifiers are operated at room temperature unless maximum gain is required.

22-21 SATELLITE COMMUNICATIONS

Satellites having the correct east-west velocity and at an altitude of 22,285 miles (35,800 km) above the equator will have a 24-h orbital period, exactly the same as the rotational period of the earth. They will appear to be stationary above the geographic equator and are said to be *geosynchronous* or *geostationary*. Many such satellites are parked above the equator all around the world, maintaining stations about 4° or 1800 miles apart. Such communications satellites have microwave receiving and transmitting systems, usually in the 4–6, 7–8, 12–14, or 20–30 GHz ranges, although many bands for satellite operation are authorized from 7 MHz to 275 GHz. The "up-link" transmissions are normally at a higher frequency than the "down-link" transmissions (see also Sec. 24-29).

After being launched, the satellites spread out their parabolic antennas, plus two to four solar panels used to absorb energy from the sun to charge the batteries that operate the on-board equipment. *Transponding* (frequency converting, not detecting-modulating *repeaters*) satellites have three major electronic sections: antennas, solar panels, and translating equipment. (They also have radio-controlled mechanical guidance or attitude correction systems.) Radio signals received from earth stations are translated (converted) to a lower microwave frequency, amplified to 4–10 W, and retransmitted back to earth. The *footprint* of the beam from a satellite may be wide enough to cover all of North and South America, or *spot beams* may be narrow enough to illuminate 2 to 20 areas, each only a few hundred miles wide. The usual terrestrial microwave parabolic transmitting antenna has a single horn feed which radiates a single narrow beam of radio energy. If there are three separate horns feeding energy to a satellite's parabolic reflector, three separate spot beams can be transmitted.

A basic satellite system (Fig. 22-35) has an earth transmitting station beaming up-link signals to a satellite which retransmits them down to an earth receiving station. If the path loss earth-to-satellite is −200 dB both up and down, the total path loss is −400 dB. The gains in the system might be something like +50 dB from the earth transmitting antenna, +30 dB from the satellite receiving antenna, +60 dB in the translation equipment with its TWT

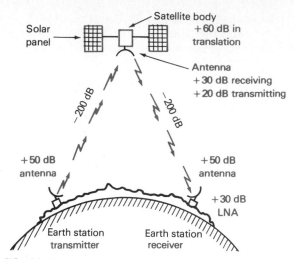

FIG. 22-35 Essentials of a communication satellite system.

power amplifier, $+20$ dB from the satellite transmitting antenna, and $+50$ dB from the earth receiving station antenna, a total of $+210$ dB. Total system loss is therefore -400 and $+210$ dB, or -190 dB. If the power output of the earth station is 1000 W, the earth station received signal is 190 dB down. This is 1000 W with the decimal point moved 19 places to the left, or 1×10^{-16} W. Into a 50-Ω waveguide the received signal would be only $V = \sqrt{RP}$, or the square root of 50×10^{-16}, or 7×10^{-8} V. A signal of 0.07 μV is too little for a receiver. A *low-noise amplifier* (LNA) with a gain of about 30 dB would bring the signal up to over 2 μV. High-frequency BJTs may be satisfactory up to about 5 GHz. Gallium-arsenide FETs have low noise at still higher frequencies. A tunnel-diode amplifier has fairly low noise, but a parametric amplifier may be 6 dB better, and a cooled parametric amplifier (at 20°K, or -253°C) may be 3 dB better still. This simplified discussion disregards sun and space noises, plus losses due to clouds and water vapor in the air. At times all satellites will be eclipsed by the earth, and their solar charging power will be lost for a few hours. At other times satellites will pass between an earth station and the sun. The resulting noise from the sun will drown out all down-link signals for a short time.

One satellite may use several different up- and down-link frequencies. It will also have telemetry radio circuits to actuate the guidance systems aboard and to relay operating parameters to earth. For maintenance of a stabilized attitude, the body of the satellite may be spun at 100 rpm in one direction, while the antenna and solar panels are *despun* at the same rate but in the opposite direction, keeping the antennas always pointed at the same point on earth. While satellite communications may be either analog or digital, many communications are now digital. Digital voice modulation requires about 64 kilobits per second (kb/s), while TV signals require 40 to 90 megabits per second (Mb/s). Both frequency division multiplex and time division multiplex modulation systems are being used.

The *attitude* (position relative to earth's center) of a satellite must be maintained exactly. Radiations from its antennas may be made to have either vertical or horizontal polarization. One communications program may be transmitted with vertical polarization, while a completely different program is broadcast with horizontal polarization from the same reflector. There will be no interference between them at the earth terminal if the earth antennas are properly oriented for vertical or horizontal polarized reception and if the signals are wideband FM.

The delay time to and from a satellite varies with the latitude and longitude of the earth stations, but is approximately 0.25 s, which can be a little disconcerting with two-way voice communications until the parties become used to it.

When satellites die, by using all the gas that operates their attitude jets, or if they lose their power as a result of solar panel failures or malfunction, they begin to drift along above the equator, eventually drifting to either a longitude of 101°W, or to 79°E, where they may all eventually congregate. The life of any new communications satellite should exceed 10 years.

Terrestrial parabolic-reflector antennas used with geosynchronous satellites are essentially fixed and may have 60 dB or more of gain, depending on their size and the frequency used (the higher the frequency, the higher the gain for a given-size antenna). They may automatically adjust or be manually adjustable in azimuth and altitude to maintain maximum signal strength to and from the satellite.

Some satellites (Sec. 25-36) are not parked, but are in a polar or other orbit. Their signals may be transmitted directly down to earth stations. Their earth stations have tracking-type parabolic antennas to keep in communication with the moving satellites for maximum periods. When they move out of radio contact with a desired earth station, they may relay their signals to earth via a geosynchronous satellite.

Although satellites handle voice and digital signals, many relay TV signals, broadcast radio programs, or handle maritime ship-to-shore signals or armed services signals, etc.

| Test your understanding; answer these checkup questions.

1. What does the designation 5000F9 mean? _____
2. How many antennas are used with space diversity? _____ With frequency diversity? _____
3. What kind of signals modulate a MUX transmitter? _____ _____ _____
4. For what is a pilot signal used? _____
5. What are the two uses of varactors in the microwave oscillator shown? _____ _____
6. For what are isolators used? _____
7. For what are directional couplers used? _____
8. What is used as a tripler in the MUX system? _____
9. What might be the output power of a MUX transmitter? _____
10. What kind of antenna is used with microwaves? _____ Could a passive reflecting surface be used to change the direction of radiation and reception? _____
11. What types of receivers are used in microwaves? _____ What is the IF? _____
12. Why are stripline filters rather than cavity types used in MUX equipment? _____
13. What circuit is used for S-meter readings in the MUX system? _____
14. What kind of transmission lines are used in low-frequency microwave systems? _____ High-frequency? _____
15. Why should WG not be installed horizontally? _____
16. How must exterior choke joints be mounted? _____
17. List 12 devices used by microwave technicians. _____
18. List six measurements microwave technicians might make. _____
19. What frequencies can be generated by LEDs? _____ _____
20. List some advantages of optical fibers. _____
21. What device may be used as an optical fiber transmitter? _____ As a detector? _____
22. What type of optical transmitter requires no cladding on the fiber? _____
23. Would it be possible to use X_L in parametric amplifiers? _____
24. What is gained by immersing a parametric amplifier in liquid nitrogen? _____ _____
25. Why does a satellite antenna have more gain receiving than transmitting? _____
26. How are spot beams formed with satellite antennas? _____
27. What is the approximate path loss satellite-to-earth? _____
28. What is the delay time for satellite transmissions from earth transmitter to earth receiver? _____

MICROWAVES QUESTIONS

1. What is the frequency range of microwaves? Infrared? Light?
2. What is a disadvantage when using round rather than rectangular WG?
3. What is the circumference of a 3-cm WG? How does λ/2 relate to this?
4. What is the impedance of most WG?
5. What is always at the end of a WG? Preferably what kind?
6. What determines the impedance of microstrip WG?
7. What is the name of a short WG section having a lengthwise slot in which an adjustable retractable probe may be moved?
8. What is the function of a resistive pyramid or vane at the end of a WG section? What is its function at the end of a directional WG coupler?
9. What is the function of resistive flaps or vanes that can be moved into or out of a WG?
10. Draw a block diagram of a WG system that might be used to measure SWR in a microwave test setup.
11. What are four methods of coupling energy into or out of a WG?
12. List three types of detectors used in WGs. Which has a −TC?
13. Of what might a round cavity be considered to be made?
14. What is the advantage of using cavities in microwaves?
15. What method of coupling is used between a cavity-type wavemeter and a WG section?
16. Name two types of VTs that contain resonant cavities.
17. What type of cavity VT is used as a high-power microwave amplifier? Are the cavities for these tubes internal or external?
18. With what are reflex klystrons being replaced?
19. What type of transmission line is used to extract energy from a magnetron cavity?
20. What is the function of the magnetic field in a magnetron? In a multicavity klystron?

21. What is the phase relationship of cavity currents in adjacent cavities in a magnetron?
22. A magnetron produces 800 0.5-μs square-wave pulses per second, each peaking at 100 kW. What is its P_{avg} output? About how much heat power would its cooling fins have to radiate?
23. If a satellite translator uses a 9-W output amplifier, would it be a TWT or a BWO?
24. What does a TWT have that a BWO does not have? What does this prevent?
25. How is the frequency tuned in a BWO? In a TWT?
26. Draw diagrams of a simple coil balun, a toroidal balun, and a linear balun.
27. What are two diodes that make good frequency multipliers?
28. Draw a diagram of a tunnel-diode cavity oscillator.
29. Draw a diagram of a diode-type frequency multiplier circuit.
30. What diode makes a good low-noise detector at higher microwave frequencies?
31. What does IMPATT mean, and for what are such diodes used?
32. For what are PIN diodes used in microwaves?
33. What microwave "diodes" are not diodes? For what are they used? Which is the more powerful?
34. Name the two properties of microwave ferrites. For what are they used?
35. How many ports does a circulator have? An isolator?
36. Draw a diagram of a circulator used to multiplex two transmitters and two receivers to the same antenna.

37. Draw a diagram of a tunnel-diode microwave amplifier.
38. What does YIG mean? As what does a YIG sphere act?
39. What are the two basic circuits for which YIGs are used?
40. Draw a diagram of a YIG step-recovery diode frequency multiplier.
41. Draw a block diagram of a microwave transmitter-receiver system capable of being modulated by baseband signals and providing detected baseband signals.
42. Draw a diagram of a SHF BJT cavity oscillator capable of FM and AFC.
43. Why is stripline circuitry used sometimes instead of cavities?
44. Besides frequencies used, what is the major difference between a microwave and a VHF FM receiver?
45. Why is outdoor WG rarely installed horizontally?
46. List 10 types of test equipment technicians may use in microwave measurements.
47. List six types of measurements that may be made by microwave technicians.
48. What carrier frequencies are used in optical fiber systems?
49. What is coherent radiation?
50. What does the term "laser" mean?
51. How can different frequency light waves be fed into a single optical fiber?
52. Draw a diagram of a simplified fiber-optic communication system.
53. What is the advantage of a parametric amplifier? What is gained by cooling it?
54. Draw a diagram of a possible parametric amplifier.
55. What is the altitude of a geosynchronous satellite? How long does it take to rotate around the center of the earth?
56. What does a transponding satellite do to received signals?
57. What are the approximate path losses in satellite communications? How are these losses made up?
58. Why do you think spinning a satellite tends to keep its attitude more constant?
59. How can you tell if a telephone conversation is being relayed by satellite?
60. Why are TV programs from satellites always wideband FM types?

ANSWERS TO CHECKUP QUIZ ON PAGE 525

1. (5 MHz FM MUX) 2. (Two)(One) 3. (SSB)(TTY)(Pulse) 4. (Check system)(AFC) 5. (FM)(AFC) 6. (One-way BP filters) 7. (Sampling) 8. (Step-recovery diode) 9. (0.1–20 W) 10. (Parabolic)(Yes) 11. (Single-conversion superheterodyne)(70 MHz) 12. (Broader passband) 13. (AGC) 14. (Coaxial and WG)(WG) 15. (Moisture) 16. (Cavities up) 17. (See Sec. 22-18) 18. (See Sec. 22-18) 19. (Light)(Infrared) 20. (Small)(Broad band)(No cross talk)(Low power)(Low loss) 21. (LED or laser)(Photo diode)(Photo transistor) 22. (Laser) 23. (Yes, inductively exciting a ferrite) 24. (dB)(Less noise) 25. (Up-link is higher frequency) 26. (Multiple feed horns) 27. (200 dB) 28. (0.5 s)

23

Broadcast Stations

This chapter discusses items that should be known by technician-operators expecting to work in any of the various types of broadcast stations. The basic makeup of broadcast stations, components used, operating techniques, log keeping, FM stereo multiplex, and FCC requirements are discussed. It is assumed that the chapters on amplitude modulation and frequency modulation are understood. Recent FCC rules have relieved operators from the requirement of having any type of license to adjust, tune, or operate broadcast transmitters. It is up to management to make sure that all such adjustments are correct and that the transmitter operates within the "FCC Rules and Regulations" (Volume III, Parts 73 and 74).

23-1 BROADCAST STATIONS

A radio broadcast station is one which produces programs and broadcasts them to the general public. The four types are (1) *Standard AM Broadcast*, (2) *FM Broadcast*, (3) *International Broadcast*, and (4) *Television Broadcast* (Chap. 24).

At present, Standard AM Broadcast (BC) stations operate in the 535–1605 kHz *band*. This band is divided into 106 *channels*. The center frequency of each channel, beginning at 540 kHz, is assigned to one or more stations in the country as the carrier or operating frequency. Any AM broadcast transmitter must maintain its frequency with a *tolerance* (Sec. 13-26) of 20 Hz above or below the assigned frequency. Transmitters use an amplitude modulation (AM or A3E) emission with an unmodulated operating carrier power ranging between 100 and 50,000 W, depending on the geographic area the station is expected to serve. While channel-center spacing is every 10 kHz, the permitted bandwidth for channels is 30 kHz to allow modulation with 15-kHz AF. Geographically adjacent stations must be spaced at least three channels apart to provide sideband interference protection.

The service area of a standard AM station is described as *primary* if there is no fading of the signal, *secondary* if there is fading but no objectionable cochannel interference, or *intermittent* if the signal is subject to some interference and fading. The *broadcast day* is considered to have three parts: *experimental period*, midnight to local sunrise; *daytime*, local sunrise to local sunset; and *nighttime*, local sunset to local sunrise. Some stations broadcast daytime only, others nighttime only, and still others for the full broadcast day.

Standard AM BC stations use either single omnidirectional vertical, or multielement vertical antennas in phased arrays to favor listeners or protect other BC stations in certain directions. Generally, antennas are erected on flat lands, preferably those having good ground conditions, such as a salt marsh. To assure constant carrier frequency the crystal ovens of the oscillators are rarely turned off, even if the transmitter is inoperative for many hours.

FM broadcast stations operate in the 88–108 MHz band, and may transmit monaural (180KF3E) or stereo multiplex signals (300KF8E). The FM band is divided into 100 channels, each 200 kHz wide, starting at 88.1 MHz. FM stations are of four basic types: class A, 100 to 3000 W effective radiated power (erp), with up to 300-ft antenna height above average terrain; class B, 5 to 50 kW erp, with 500-ft antenna height; class C, 25 to 100 kW erp, 2000-ft antenna height; and 10-W educational stations. The frequency tolerance of FM BC stations is 2000 Hz above or below the assigned frequency. Minimum mileage between two FM stations and their separation in frequency depends on the class and the power of the stations involved. The greater the power, the further away in miles and in assigned frequency the two stations must be. Generally, FM BC stations are licensed to serve a single community, although sometimes they may be expected to serve two or more nearby communities. Because the stronger of two cochannel FM signals may completely capture any receiver detector, two class A stations may be located as close as 40 miles, two class B stations 75 miles, and two class C stations 100 miles from each other. The closer station is usually the only one heard by a listener in between stations. To receive all sidebands developed by 75 kHz deviation, FM BC receivers require an IF bandpass of about 225 kHz.

Antennas for FM BC stations are usually mounted on top of a high building or on a nearby peak that overlooks the area to be served. Antennas are usually horizontally polarized with several bays (Sec. 20-22). The more elements, the more the radiated lobe is held down toward the horizon and the stronger the signal induced in more distant receiving antennas.

An advantage of broadcast FM over AM is the addition of a 75-μs pre-emphasis circuit to the transmitter audio signal and a 75-μs de-emphasis circuit in the audio section of FM receivers (Sec. 19-10). This greatly improves signal-to-noise ratio (SNR) for the FM system.

There are a few shortwave *international* AM BC stations in the United States that beam their transmissions to the general public in foreign target areas. They operate generally in one or more of seven shortwave bands:

Band A, 5.95–6.20 MHz
Band B, 9.50–9.77 MHz
Band C, 11.70–11.97 MHz
Band D, 15.10–15.45 MHz
Band E, 17.70–17.90 MHz
Band F, 21.45–21.75 MHz
Band G, 25.60–26.10 MHz

International AM BC stations maintain an authorized power output of not less than 50 kW. Because of the variations of transmission paths during different times of the day and year, they shift from one band to another during the day to operate on the frequency that will have the best chance of reaching the desired area. Directional antennas having a minimum power gain of 10 dB in the desired direction are required. Frequency tests, modulation monitors, auxiliary or alternate transmitters, and other technical items are similar to those of standard AM broadcast stations.

23-2 COMPONENTS OF A BROADCAST STATION

A block diagram of a possible broadcast station is shown in Fig. 23-1. The larger dashed section represents a small station with antenna, transmitter, and studios all in one building or area. The transmitter is indicated as being an amplitude-modulated type, in which the final amplifier is modulated. If this were an FM station, the oscillator or buffer would be frequency- or phase-modulated.

Programs may be developed in the local studio and fed live to the transmitter, or they may be taped on cartridges and played later. Programs may also come in from remote feeds such as telephone lines or radio relay systems. The operator may announce and play music from tape cartridges ("carts") or turntables (TT) as well as take care of the transmitter (combination or "combo" operator). News and special announcements may originate from a local announce booth or studio. Except for very small stations, the transmitter is located in some room other than the control room. For technicians assigned to work in this room, there will be a separate frequency monitor, percent of modulation monitor, audible on-the-air monitor, and an RF power output indicator. An *auxiliary* transmitter may be included for emergency use or for EBS transmissions (Sec. 23-12) and must be tested each week. A complete *alternate* transmitter may also be used if the station operates on a 24-h day. There will be an equipment repair room and an area housing auxiliary ac power equipment (diesel or gasoline engine and alternator) in case the local utility ac power is interrupted.

Most stations have a transmitter building with minimal console equipment for use in emergencies, but with more extensive main studios (smaller dashed area in Fig. 23-1) housing most of the program and management equipment and personnel. There will usually be a master audio console used for mixing and switching the various studio programs, announcers, tape cartridges, turntables, and incoming feeds to form a continuous series of programs. If the transmitter is operated unattended,

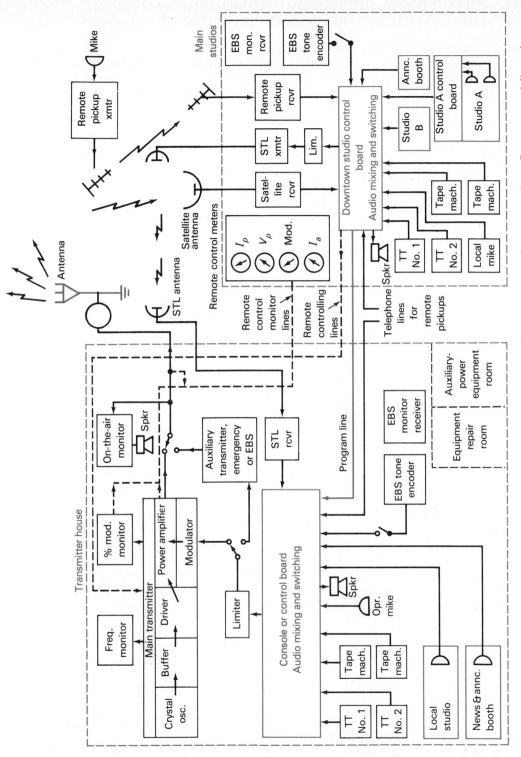

FIG. 23-1 Block diagram of an AM broadcast transmitting station including STL and remote-pickup transmitter. Remote-control lines shown dashed.

there must be available a positive indication at the remote-control point of such transmitter parameters as final amplifier plate voltage and current, modulation percentage, antenna ammeter, etc. These must be checked periodically during the operating hours. In some cases there may be 10 to 20 electrical, temperature, and security parameters being constantly scanned several times a second. These may be relayed to the main operating position and continuously displayed on a monitor CRT. If any parameter exceeds its maximum or minimum tolerance values, that one part of the display may blink and an audible tone alert may also be sounded. Improper functioning is often correctible from the control point by telemetering. If tolerances are exceeded either greatly or slightly for several minutes, the transmitter may shut itself down until a technician can be sent out to the transmitter to correct the difficulty.

Often remote program pickups are made from activities such as athletic or musical events. These may be transmitted to the console via specially balanced, noise-free leased telephone lines or by remote radio transmitters and local receivers. Network programs can come in by leased lines, microwave circuits, or satellite relay.

An operator at the console, or control board, can switch in any of the microphones, tape cartridges, turntables, or remote lines desired. With the *gain controls*, also known as *attenuators, faders,* or *pots* (potentiometers), the amplitude of the program can be varied to keep the highest peaks of modulation above the required 85% and below 100% modulation on negative peaks. This control can be aided by using the automatic level controlling ability of an audio limiter (Sec. 23-7).

Remote-pickup broadcast station systems will consist of a base station located usually at the broadcast station control point. It will communicate with a mobile pickup BC station sent to some remote event, using the minimum power required for satisfactory service, using directional antennas. Frequencies and emissions are:

> 1606–1646 kHz AM
> 25.87–26.47 MHz AM or FM
> 152.87–161.76 MHz AM or FM
> 450.05–455.95 MHz AM or FM

An *aural broadcast STL station* is a fixed station that broadcasts program material from studio to transmitter by radio link. An *aural BC intercity relay station* is used to transmit program material between broadcast stations in different cities. Either of these services may be assigned a 500-kHz channel in the 947–951.5 MHz band, using no more than 200 kHz FM, with directional antennas. Identification of remote pickup, STL and intercity transmitters must be made at the beginning and end of operation and at least every hour. Power output will be only that necessary to provide satisfactory service. Frequency tolerance is 0.005%.

23-3 THE BROADCAST CONSOLE

The nerve center of the broadcast station is its console, control point, or switchboard, a complicated group of audio amplifiers, switches, relays, and gain controls. The operator switches the desired studio, tape deck, turntable, microphone, or remote program through the console into the limiter and modulator stage of the transmitter.

The block diagram in Fig. 23-2 illustrates a simplified console circuit that will demonstrate some of the requirements of broadcast operating. Included are the console, microphone, turntables, tape machines, and loudspeaker for the console operator, a

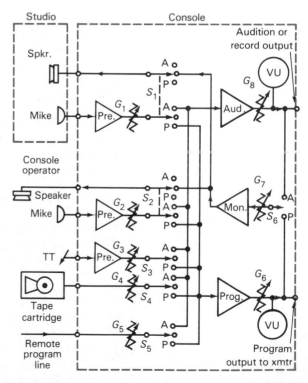

FIG. 23-2 Functional block diagram of a simple broadcast console and studio.

studio with a microphone and loudspeaker, and a remote input line.

For a program originating in the nearby studio, the studio microphone switch S_1 is thrown to the *program*, or P, position. The output of the microphone, through its preamplifier and gain control G_1, is connected to the input of the program amplifier. The output of the program amplifier is fed through the master gain control G_6 to the transmitter (normally through a limiter). At the same time the console operator hears the signal output from the program amplifier via the monitor amplifier, provided the operator's local microphone switch S_2 is in the central or OFF position. The operator also has a visual indication of the signal strength by the VU meter in the output line of the program amplifier. Note that when S_1 is in the P position, the studio speaker is automatically disconnected from the monitor amplifier output, preventing an audio feedback howl from being produced in the studio.

While the studio is on the air (being fed into the program line), the operator can be recording an announcement or program to be used later by switching the microphone, local turntable, tape machine or the remote-line switches to the *audition*, or A, position. This feeds his program into the audition amplifier, the output of which can be fed to a tape recorder. The local monitor speaker and microphone are interconnected by S_2 so that the speaker is disconnected at any time the microphone is in use, preventing a feedback howl.

When the studio program is completed, the operator switches off S_1 and, by throwing the local microphone switch S_2 to P, can make a *station break* (station identification) or other announcement. In many stations there is an announce booth (not shown), in which a staff announcer may make the station break, or it may be from a tape cartridge.

If the next program is coming in from a distant network station, the operator checks the remote-line signal strength a few minutes before the program is scheduled to start by throwing S_5 and S_6 to the A position. This feeds the remote signal through the audition and monitor amplifier and to the local speaker. When it is time for the remote program to start, S_5 is connected to its program position and the program is fed through the program amplifier to the transmitter.

It is usually necessary to provide *talk-back* facilities between the console operator and the people in the studio before they go on the air. This can be done with the audition channel of this console. The operator switches S_2 and S_6 to the audition position to be heard in the studio. When S_2 is switched to the central or OFF position and S_1 is switched to the audition position, the operator can hear anyone speaking in the studio. Talk-back may be used to notify a studio that it is about to be switched on the air. On-the-air may be indicated by switching on a red light in the studio.

When the program is originating at the console, as in a disk-jockey show, the turntable or tape cartridge that is playing the previously announced record is in the program position. The next record may be set up on another turntable (not shown), and its output is fed into the audition channel. By switching the monitor amplifier switch S_6 to the A position, the operator can hear the output of the second turntable. The operator sets the pickup stylus in the outer groove of the record and then spins the turntable by hand until modulation is heard from the record. By backing the turntable past the first sign of modulation the record has been *cued* and it is ready to be played. (Tapes can be manually cued, or they can be automatically cued by special cuing tones recorded on them.) The cued turntable is switched to the program line. A fraction of a second after the motor switch for the turntable has been turned on, the music of the record will begin. In this way there are no long pauses after the announcement of the record as the pickup stylus moves over the first few unmodulated grooves.

The console described here is far simpler than those in actual use, although its basic functions are similar. With this console, the studio is fed the output of the program amplifier, allowing those in the studio to hear what is being broadcast at all times when their studio is not on the air. When the operator is cuing a record, however, the studio is fed the sounds of the cuing rather than the program being broadcast. To prevent this a separate cuing amplifier would be used.

The attenuators G_1 through G_5 may be termed *channel* gain controls. Attenuator G_6 is known as the *program master* gain since it controls any console program output signal. The *audition master* is G_8. The monitor gain is G_7.

The remote program line may be feeding in programs via telephone lines, small portable microwave transmitter-receiver systems, or by hand-held on-the-spot VHF or UHF FM transceivers in communication with a control unit at the station. Network programs may come in by intercity microwave or radio systems or by satellite relays (Sec. 22-21). Eventually satellites on higher microwave frequencies are expected to broadcast directly to small parabolic reflector antennas at the homes of listeners.

Anyone assigned by the station manager may operate any part of a broadcast station. There is no longer a requirement to use licensed operators.

23-4 AUDIO LEVELS

There are three important requirements during a broadcast: (1) Timing. The programs must start and stop when they should. Announcements and station breaks must be squeezed into the time allotted for them. (2) Fidelity. All amplifiers must be operating in such a manner that they do not add distortion to the program. This means frequent maintenance and testing of the console and associated circuits. The input and output impedance of the lines of all equipment must be properly matched and terminated. (3) Maintaining the proper console signal output to modulate the transmitter to 85 to 100%.

Audio levels in broadcasting are expressed in volume units (VU). A VU is equal to a decibel in value. Zero VU of a sinusoidal ac is the equivalent of 1 dBm (1 dB at a standardized power of 1 mW) operating into a 600-Ω audio line. If an ac is increased 10 times in *power*, this represents a 10-dB or 10-VU increase. However, if the ac is voice-frequency (VF) ac from a microphone or amplifier, a zero-VU reading on a meter may be responding to *peak values* perhaps 8 VU higher than indicated by the meter. The exact amount depends on the voice characteristics. Thus, if the telephone company states that no more than zero VU may be used on a line, it expects that the actual peak levels may be in the range of +8 dB over zero VU.

Audio amplifiers are often rated by their dB gain and power output. For example, if an AF amplifier has a gain of 40 dB and the output is 6 W, what input power is required to produce 6 W? Starting with 6 W_o and working down, using only the knowledge that 10 dB equals a tenfold power change, −10 dB represents a power of $^6/_{10}$ or 0.6 W_o. Another −10 dB (or −20 dB) results in 0.06 W_o. Another −10 dB (or −30 dB) computes to 0.006 W_o. Another −10 dB (or −40 dB) indicates the input must be 0.0006 W_{in}. If 0.0006 W_{in} ac is fed into this 40-dB amplifier, it will produce 6-W ac output. What input voltage will this require? Since most studio equipment use 600-Ω impedance lines, from the power formula $P = V^2/R$,

$$V_{in} = \sqrt{PR} = \sqrt{0.0006(600)} = 0.6 \text{ V}_{in}$$

A microphone feeding into a console may have a rating of −85 dBm with a given value of sound striking it. Therefore, the console must be capable of amplifying the signal at least 85 dB to produce a zero-VU-level output. Weaker signals striking the microphone will require more amplifier gain.

A turntable or tape cartridge may have an output signal of −45 dBm and need a two-stage amplifier of 45- to 60-dB gain to bring its output up to the zero VU used to feed the transmitter.

A third input signal, a remote program, may be coming in at a zero-VU level on telephone lines.

To bring up very low levels to a value that can be adjusted with a gain control without introducing excessive amounts of noise, a preamplifier is used. If the mixing (switching and gain controlling) is accomplished at a −10-VU level, the amplifier that follows the mixing must be capable of a gain of at least 10 dB and preferably 20 to 30 dB. The preamplifier must be capable of 50 to 90 dB for the microphone. The preamplifier for the turntable must be capable of a 50-dB gain or more. The remote program must lose 10 dB before being fed into the mixer (Sec. 23-5).

It is desirable to control the volume of a microphone or other input signal in the zero-VU range rather than at the microphone level. All potentiometers introduce some noise as the arm is moved along the resistance. The more amplifiers that follow the gain control, the more this noise is amplified.

Test your understanding; answer these checkup questions.

1. What are the limits of the AM BC band? _____ Of the FM BC band? _____
2. What is the channel width of an AM BC station? _____ FM? _____
3. What is the frequency tolerance of AM BC stations? _____ FM? _____
4. List names and conditions of the three AM BC service areas. _____ _____ _____
5. What are the three parts of the BC day? _____ _____ _____
6. What broadcast stations use vertical antennas on flat ground? _____ Antennas on high peaks? _____
7. Why is FM higher in fidelity than AM? _____
8. What are power and antenna requirements for international BC stations? _____ _____

9. What is a combo operator? _____
10. Would an auxiliary transmitter require the same power output as the main transmitter? _____ Would an alternate transmitter? _____
11. What is the meaning of an STL system? _____
12. Where is telemetering used in BC stations? _____
13. What are three ways programs may be received at BC stations? _____ _____ _____
14. What are other names for gain controls? _____ _____ _____
15. What does an intercity relay station do? _____
16. How often should BC stations be identified? _____
17. What does P mean on a turntable control? _____ A? _____
18. What is a station break? _____ Talk-back system? _____
19. Why is a record or tape cartridge cued? _____
20. What are the three important requirements of a broadcast? _____ _____ _____
21. Does a VU meter accurately indicate voice peaks? _____ A dB meter? _____
22. What VU level is used on telephone lines? _____ When does zero VU equal zero dBm? _____
23. What is the power gain of an amplifier capable of 63-dB gain? _____
24. Why is microphone gain controlled at about zero VU? _____

23-5 ATTENUATOR PADS

It is possible to use the same type of high-gain preamplifiers for all inputs to a console but intentionally lose signal level by inserting *H*- or *T-pad* attenuators between high-level inputs and the preamplifiers. These pads are groups of resistors that act as voltage dividers, at the same time holding the output and input impedances of the circuits being coupled at a constant value, usually 600 Ω.

The T pad in Fig. 23-3 is used with unbalanced lines (one of the lines grounded). If it is desired to drop the voltage to one-half (6 dB) and maintain an impedance match at both transformers, the values

shown can be used. The 600-Ω secondary of the left-hand transformer looks at a 200-Ω resistance in series with two parallel 800-Ω impedances, or 600 Ω. The other transformer sees the same thing. If 1 V is delivered by the input transformer, the ratio of 200:400 Ω in series produces a voltage-drop across the 400-Ω resistance of $2/3$ V. Since the 200-Ω R_2 and the 600-Ω primary are in series across $2/3$ V, the primary sees $600/800$ of $2/3$, or ½ V.

If a loss of 10 dB is desired, R_1 and R_2 may be 300 Ω each and R_3 400 Ω. For a 20-dB loss, R_1 and R_2 may be 500 Ω and R_3 120 Ω. (See an engineering handbook for formulas to develop other losses.)

Most long transmission lines are balanced to ground, as in Fig. 23-4. An H pad is used in this

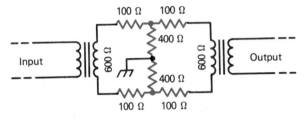

FIG. 23-4 A balanced H-type attenuator pad.

case. The series resistors are half the value of those in the T pad above. The resistance across the line is the same as in the T pad except that it is center-tapped.

Pads are also used to terminate amplifiers feeding long transmission lines to transmitter equipment. Such lines may have differing transmission characteristics for different frequencies. If coupled to a line by a −6- to −10-dB pad, the characteristics of the line and the eventual load will have little effect on the amplifier and less distortion will be produced by the amplifier. Matching impedances in audio equipment is important to reduce distortion.

One application of a T pad is the variable attenuator shown in Fig. 23-5. The arms of all three

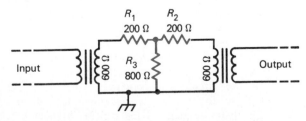

FIG. 23-3 An unbalanced 600-Ω T-pad attenuator.

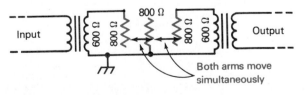

FIG. 23-5 A variable unbalanced T-type attenuator, or gain control.

resistors move at the same time, holding the input and output impedance fairly constant. This attenuator is superior to a simple potentiometer gain control which always changes both input- and output-impedance values as the moving arm is varied. With 800-Ω pots, as shown, at full gain the 600-Ω transformer sees an 800-Ω and a 600-Ω impedance in parallel, which is a little lower than its own impedance. At minimum gain the transformer sees an 800-Ω resistance, which is a little higher than its own impedance. Near midgain, the normal operating point, the impedances match exactly.

When coupling two audio circuits having unequal impedances, either a coupling transformer or an L or U pad (named by configuration) may be used. The L pad shown in Fig. 23-6 matches a 600-Ω

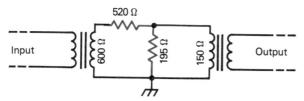

FIG. 23-6 An L-type pad to match a 600-Ω to a 150-Ω circuit.

output to a 150-Ω input. There is no selection of loss in these pads. To match these two impedances only the two resistors shown can be used. The loss is about 18 dB.

For balanced lines, U pads, as in Fig. 23-7, are used. As with H pads, the series-resistance value of the unbalanced L pad is halved and the shunt resistor is center-tapped.

At the receiving end of a transmission line a center-tapped primary transformer may be used. The center tap is grounded. This tends to balance

ANSWERS TO CHECKUP QUIZ ON PAGE 532

1. (535–1605 kHz) (88–108 MHz) **2.** (10 kHz) (200 kHz) **3.** (20 Hz) (2000 Hz) **4.** (Primary; no fade) (Secondary; fading, no interference) (Intermittent; fading and interference) **5.** (Experimental) (Daytime) (Nighttime) **6.** (AM) (FM, also TV) **7.** (Pre- and de-emphasis) **8.** (50 kW) (10-dB gain) **9.** (Announce, play, technician) **10.** (No) (Yes) **11.** (Station-to-transmitter link) **12.** (Remote position to transmitter) **13.** (Telephone lines) (Radio relay) (Satellite) **14.** (Attenuators) (Faders) (Pots) **15.** (Relays programs from one station to another) **16.** (At least once an hour) **17.** (Program position) (Audition) **18.** (Station ID) (Intercom) **19.** (Allow immediate start) **20.** (Timing) (Fidelity) (Signal level) **21.** (No) (No) **22.** (Zero VU max) (Sine-wave ac, 600-Ω lines) **23.** (2 million) **24.** (Less attenuator noise)

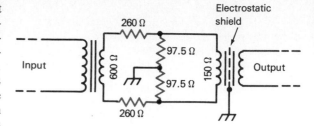

FIG. 23-7 A U-type balanced pad to match dissimilar impedances.

out hum or noises picked up by the lines. The center-tapped shunt resistor of an H or U pad may be a potentiometer to allow a more accurate balance of the lines. A center-tapped transformer is not used in this case. Only one point should be grounded.

To reduce capacitive coupling of noise impulses picked up by transmission lines, the transformer used at the receiving end may have an electrostatic shield between primary and secondary windings. This shield is connected to the core, which is grounded, as shown in Fig. 23-7.

23-6 LINE EQUALIZERS

A short 600-Ω line carrying an audio signal has little attenuation of any of the audio frequencies. If a line exceeds a few hundred meters in length, the inductance of the wires and the capacitance between them do affect the signal, attenuating the high frequencies more than the lows. If the line and transformers have 150-Ω impedances, the high-frequency losses will be less.

The use of a line pad at the output of the amplifier feeding the line helps to flatten the frequency response of the final-amplifier stage, but for high-fidelity transmissions the lines should transmit all audio frequencies equally well. For standard broadcast, this is usually 50 to at least 7500 Hz and for FM and TV, 50–15,000 Hz.

If a line is found to attenuate the higher frequencies, a *line equalizer* can be placed across the receiving end of the line to decrease the amplitude of lower frequencies. This can be accomplished, as shown in Fig. 23-8, by using a parallel-resonant circuit in series with a variable resistance across the line. The circuit is made resonant at some high frequency—5000 Hz, for example. The high impedance of the circuit to frequencies near 5000 Hz allows these frequencies to pass without attenuation. At lower frequencies (as well as higher) it is found that the equalizer presents a lower imped-

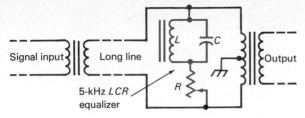

FIG. 23-8 A parallel-resonant type of line equalizer.

ance, and the frequencies are attenuated somewhat. By adjustment of the resistance, it is possible to control the attenuation of the lower frequencies and equalize the frequency characteristics of the line from 50 to 5000 Hz. More than one equalizer may be required. AF equalizers may have 10 or more different frequency controls.

23-7 PEAK-LIMITING AMPLIFIERS

Speech and music produce relatively high peaks of audio voltage at times. If the operator adjusts the gain controls to allow the VU signal level of a musical program to show 75 to 85% modulation, the peaks that occur may produce more than 100% modulation, distortion, and excessive sidebands. In AM BC stations this may actuate overload relays, or may damage tubes and equipment. In voice communication systems, peak limiters, or clippers, are usually used. The distortion that is produced with this type of limiting is too great for broadcast standards, and special higher-fidelity *limiting amplifiers* are used.

The basic idea of a limiting amplifier is shown in Fig. 23-9. Normally a limiting amplifier provides little or no AF gain, but merely limits high audio peaks to values which will not cause overmodula-

tion. The AF input in this circuit can be decreased in value with R_1 to compensate for the natural gain of the MOSFET. The same input signal is being fed to both the MOSFET and the N-channel JFET. Since it is only the peaks of the signal that are to be limited, the peak amplifier bias is adjusted to a class C value by R_2 so that only peak signals over a certain value will cause any I_D to flow. These peak signals are amplified and rectified, and develop negative dc bias voltages across potentiometer R_3, which are then fed to the lower gate of the MOSFET. Negative bias on G_2 decreases the gain of the amplifier, but only during the positive peak voltage times, reducing peak voltage output. The closer to ground the sliding arm of R_3 is, the less peak cancellation that will occur. Thus R_2 adjusts the voltage point where peaks will start to be limited, and R_3 adjusts the extent to which these peaks will be limited. This simple circuit would only limit the positive peaks of the AF signal. To limit both positive and negative peaks a second *P-channel* JFET peak amplifier with a reversed rectifier feeding G_2 would be used.

Through the use of a peak-limiting system in a broadcast station, the gain controls can be raised and the average percent of modulation can be increased, resulting in louder signals in receivers.

23-8 CLASSIFICATIONS OF POWERS

When a broadcast station is licensed for operation, the instrument of authorization will specify a certain *licensed power*, or *authorized operating power*. Operating power is that fed to the antenna by the transmitter. The *tolerance* of the operating power of a station is not more than 5% above nor more than 10% below the authorized power. The operating power may also be known as the *carrier* or *antenna power* and is always the unmodulated value.

The *maximum rated carrier power* is the maximum power at which that model of transmitter can be operated satisfactorily. It is determined primarily by the tubes and plate voltages used in the final amplifier. The maximum rated carrier power capabilities of a transmitter must always be equal to, or exceed, the licensed power. For example, a station licensed for 500 W may emit 500 W from a transmitter that has a capability of 1000 W.

The *plate input power* of a broadcast station transmitter is determined by the product of plate voltage and plate current ($P = V_p I_p$) of the final-amplifier tube or tubes *during a time when no modulation is being applied*.

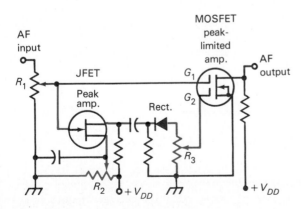

FIG. 23-9 A basic peak-limiter system.

The *operating (output) power* of an AM broadcast station is determined by the *direct method*. This is computed by

$$P_o = I_a^2 R_a$$

where I_a = antenna current with no modulation
R_a = impedance or resistance of antenna where current is measured

EXAMPLE: An antenna has a feed-point resistance of 50 Ω, and the antenna ammeter at the feed point reads 4 A. The power being fed to the antenna is $4^2(50)$, or 800 W.

The *indirect method* may be used to determine the output power of FM broadcast stations and TV aural transmitters. It is employed in standard broadcasting only in an emergency, as when the antenna ammeter becomes defective, or temporarily when authorized installation changes, changes of equipment, or antenna changes are being made. The indirect power is determined at 0% modulation by

$$P_o = V_p I_p F$$

where V_p = plate voltage of final amplifier
I_p = plate current of final amplifier
F = factor between 0.35 and 0.80 (see Table 23-1)

EXAMPLE: A class C final amplifier is plate-modulated. The V_p is 4000 V and the I_p is 1.563 A. The plate power input is 4000(1.563), or 6250 W. With an efficiency factor of 0.8 (from the table) the power output is considered to be 5000 W.

The F factor is determined by the manufacturer, by tests, or by using Table 23-1.

Either plate or grid modulation may be used in AM broadcast stations. Plate may be preferred because it is more efficient and is less critical as far as tuning and antenna matching is concerned.

TABLE 23-1 DETERMINING THE F FACTOR

Factor, (F)	Method of modulation	Carrier power of final amplifier	Class of final amplifier
0.80	Plate	5000 W and over	C
0.70	Plate	250–1000 W	C
0.65	Low level	250 W and over	BC
0.35	Low level	250 W and over	B
0.35	Grid	250 W and over	C

23-9 BROADCAST STATION TESTS

Before a broadcast station is constructed, it is first necessary to obtain a *construction permit* from the FCC. During the construction period, after the transmitter has been installed, it is permissible to test the transmitting equipment for short periods between midnight and local sunrise. These are known as *equipment* tests.

After construction has been completed and until the station license has been issued by the FCC, application can be made to test the station on the air (*service*, or *program*, tests).

The period of the day between midnight and local sunrise is used as the *experimental period*. During this time, broadcast stations may transmit for testing and maintaining equipment, provided that such tests do not interfere with other stations broadcasting on the frequency at that time.

All AM broadcast stations must be capable of transmitting 85 to 95% modulation with a total AF harmonic distortion not to exceed 7.5%. With less than 85% modulation, the total distortion must not exceed 5%. Distortion is measured using modulating frequencies of 50, 100, 400, 1000, 5000, and 7500 Hz, plus any intermediate or higher frequencies found necessary.

When a station is in operation, the percent of modulation must be maintained at as high a level as possible consistent with good quality of transmission but in no case less than 85% on peaks (no more than 100% on negative, or 125% on positive, peaks of modulation with AM transmitters).

On an annual basis all AM broadcast stations must make on all its audio equipment an *equipment performance* test. The response of the transmitter to audio frequencies from 30 to 7500 Hz must be made at 25, 50, 85, and 100% and produced in response-curve form. Harmonic content, carrier amplitude regulation, hum, and spurious radiations must also be measured. These measurements must also be made during the last 4 months preceding filing for station license renewal. An annual *proof-of-performance* test of the antenna radiation pattern of any directional array is also required. Antenna field strengths (patterns) are measured during the daytime when no sky-wave propagation is present. These reports must be filed and kept for a period of 5 years.

All FM broadcast stations of over 10 W must make equipment-performance tests annually. These are AF response measurements of frequencies between 50 and 15,000 Hz; AF harmonic-distortion measurements at 25, 50, and 100% modulation with de-emphasis; output noise of frequency-modulated type between 50 and 15,000 Hz; and output noise of amplitude type between 50 and 15,000 Hz.

No license is required to make any of the station tests. Management must make sure operators are qualified.

The Emergency Broadcast System tests required of broadcast stations are discussed in Sec. 23-19.

23-10 BROADCAST STATION METERS

Ammeters and voltmeters associated with the final radio amplifier stage plate circuit must have an accuracy of at least 2% of the full-scale reading. The full-scale reading of linear and logarithmic (wattmeter) scale meters shall not be greater than five times the minimum indication when the transmitter is in normal operation. Linear meter scales must have at least 40 divisions. Logarithmic scales must have no divisions above $1/5$ scale greater than $1/30$ full scale. Antenna-current (I^2R) meters are thermocouple types and have the same accuracy requirements as the dc meters. Full scale cannot be greater than three times the normal indication, and above $1/3$ scale no division shall be greater than $1/30$ full scale.

A remote-reading antenna ammeter may be used to indicate antenna current if: its thermocouple is installed next to the main ammeter at the base of the antenna; it has the same scale accuracy as the main meter; its calibration is checked at least once a week against the main ammeter; its wiring is shielded.

When directive antennas are used, current ratios in the elements must be held within 5% of the authorized values.

If a required plate-circuit or antenna meter burns out and no substitute conforming to required specifications is available, an appropriate entry must be made in the operating log indicating when the meter was removed from and restored to service. The FCC Engineer-in-Charge of the radio district must be notified when a meter has been found to be defective and again when it has been replaced. It must be repaired or replaced within 60 days. Since *remote* antenna ammeters are not normally required, if such a meter becomes defective, it is necessary only that the ammeter at the base of the antenna be read and logged once daily until the remote meter has been returned to service.

23-11 MONITORING FREQUENCY

Broadcast transmitters have low-temperature-coefficient-type crystals in the oscillator circuit. A second complete standby oscillator stage may be incorporated in the transmitter and may be switched into service if the first should ever drift or fail. Temperature variation inside the temperature-controlled crystal chamber (Sec. 11-14) should not exceed ±1°C.

Carrier frequency is normally being constantly indicated at the transmitter and at the control point by an FCC-approved frequency monitor. If the frequency meter becomes defective, the frequency must be measured at least once a week with some other type of frequency-measuring device. The engineer in charge of the nearest FCC office must be advised when the frequency meter becomes defective and again when it has been corrected and placed back in operation (within 60 days). The carrier frequency meter must be checked at least once a month with an instrument such as a digital counter that compares its frequency indications with the National Bureau of Standards stations WWV, WWVB, or WWVH (Sec. 13-30). With AM BC stations, the carrier frequency is measured at some stage prior to the modulated stage. With FM BC stations, the carrier frequency is measured at the final amplifier stage but during times when there is no modulation being applied.

Frequency-measuring and -monitoring devices may be digital counters or analog-type constant-frequency monitors. The carrier frequency is indicated in the station log daily.

When a constant indication of the frequency of a transmitter is required in an AM broadcast station, an analog-type meter may be used. In standard AM broadcast stations, the transmitter must maintain its assigned frequency within ±20 Hz. This prevents stations on the same frequency but in different parts of the country from producing an audible beat tone in receivers. With a 20-Hz tolerance, the maximum difference of frequency between any two stations is 40 Hz (normally inaudible in receivers). In most cases stations maintain frequency within about 5 Hz, resulting in a maximum beat tone of up to no more than perhaps 10 Hz. This causes the wavering heard on many broadcast-band signals, particularly at night.

Frequency monitors for such stations also use temperature-controlled crystal oscillators. A monitor crystal might be 1 kHz below the assigned fre-

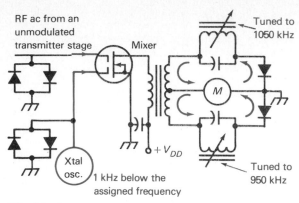

RF ac from an unmodulated transmitter stage

Mixer

Tuned to 1050 kHz

M

+V_{DD}

Xtal osc.

1 kHz below the assigned frequency

Tuned to 950 kHz

FIG. 23-10 Simple AM broadcast station frequency monitor.

quency to develop a beat frequency. Figure 23-10 is a simplified diagram of a carrier frequency monitor circuit. An RF signal from one of the unmodulated stages in the transmitter is limited and fed to one of the gates of the mixer. The output of the monitor crystal oscillating at a frequency 1000 Hz lower than the assigned frequency of the transmitter is limited and fed into the other mixer gate. In the output circuit a difference frequency of 1000 Hz appears. This is fed to a discriminator utilizing two *LC* circuits, one tuned to 1050 Hz and the other to 950 Hz. With the transmitter on frequency the beat tone is 1000 Hz. The two resonant circuits produce equal and opposite rectified dc voltages across the zero-center meter *M*, resulting in zero current through it. The meter reads center scale, which is marked 0 Hz.

If the transmitter shifts up 10 Hz, the beat tone becomes 1010 Hz, the circuit tuned to the higher frequency develops more voltage across it, and a resultant current flows through the meter from *left to right*. The needle moves up to a point on the meter scale which would be marked +10 Hz.

If the transmitter shifts down 5 Hz, the beat tone becomes 995 Hz and more voltage is developed across the circuit tuned to 950 Hz. The resultant current now flows from *right to left* through the meter, moving the indicator to a point which would be marked −5 Hz.

23-12 MONITORING MODULATION

In general, modulation percentages of frequently recurring modulation peaks for both AM and FM should be maintained at 85% or greater but must never exceed 100% of the negative peaks with AM.

Percent of modulation with FM BC stations considers 75-kHz carrier deviation from the assigned carrier frequency to be 100% modulation (37.5 kHz = 50%, etc.). An antenna ammeter in an AM station may increase 6% with 100% sinusoidal modulation, but not at all at an FM station.

Although accurately calibrated analog-meter-type modulation monitors may be used at some standard AM BC station, the indications of modulation given by oscilloscopes are to be preferred. They accurately show the extent of all positive and negative peaks of modulation, as well as certain types of AM distortions, particularly if trapezoidal displays are used. Analog-type modulation monitors will sample some of the modulated RF. They will have (1) a VU-type meter to indicate the modulation peaks, with reversible rectifiers to show positive or negative peaks, (2) a carrier-shift indicator, and (3) a peak-indicating light or alarm that can be set at any value from 50 to 100% negative (or positive) peaks.

A continuously indicating frequency and modulation (deviation) monitor used in some FM BC stations is shown in Fig. 23-11. Its frequency meter face has a zero reading at center scale and frequency calibrations up to at least 2000 Hz on each side of zero in graduations of 100 Hz. The frequency monitor is first calibrated by switching in the 10-MHz-calibration crystal oscillator. This signal is fed through the mixer stage, the 10-MHz IF amplifier, and the limiter to the 10-MHz discriminator. The discriminator is trimmed to read zero on the meter with the calibrating signal. The 30-MHz running crystal is then switched in and fed through a tripler to the mixer. When the transmitter is turned on (assume a 100-MHz transmitter), the 90- and 100-MHz signals mix to produce a 10-MHz resultant. If the transmitter is exactly on frequency, the 10-MHz resultant will read zero on the discriminator meter. If the carrier is off frequency, the meter will read to the left or right of zero, indicating in hertz the carrier-frequency error. The frequency of the calibration crystal can be checked against station WWV or some other primary frequency standard. A different running-crystal frequency is required if the transmitter operates on any frequency other than 100 MHz.

Since the discriminator is an FM detector, its output can (1) be used as an audible monitor signal after passing through a de-emphasis circuit or (2) operate a percent-of-modulation or deviation meter, also after de-emphasis. The percent-of-modulation meter is a VU meter so calibrated that 75-kHz

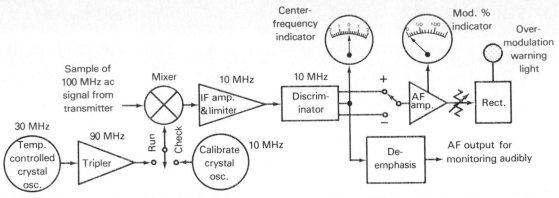

FIG. 23-11 Block diagram of an FM station frequency monitor with modulation indicator and overmodulation warning.

deviation reads 100% modulation. Indications of the deviation meter are directly proportional to the amplitude of the modulating voltage.

If the frequency meter shows the carrier is within tolerance but modulation causes this meter to vary, it indicates that the carrier is deviating in one direction more than in the other. Some possible causes of such nonlinearity are detuned multiplier or amplifier LC tank circuits, detuned antenna, nonlinear modulation of the oscillator, improper neutralization, or an improperly operating AFC system.

If it is noted that a modulation monitor becomes defective, the fact should be logged in the maintenance log and the FCC Engineer-in-Charge of the radio district advised, but the station may continue to operate for a period of 60 days. Log entries are made when the meter is placed back in service.

Broadcast station operators may monitor an outgoing program with a VU meter, although such an analog meter cannot accurately indicate peak values. An LED-type bar-graph indicator (Sec. 18-18) can be used. It may be calibrated in 2-dB steps to read the top 10 to 20 dB of signal amplitudes before overmodulation. The LEDs will respond to the actual peaks that are present. Peaks of over 100% can be made to actuate a different color LED as a warning feature.

| Test your understanding; answer these checkup questions.

1. What are three reasons for using attenuator pads? _____ _____ _____

2. What is the advantage of using variable unbalanced T-type attenuator pads as gain controls? _____

3. If an audio line loses high frequencies what can be added to it to flatten the response? _____

4. To increase loudness of modulation, what may be added to the AF system between console and driver amplifiers? _____

5. How might authorized and antenna powers differ? _____

6. How is plate input power determined? _____ Antenna power? _____ Carrier power? _____

7. What are two ways of determining operating power? _____ _____

8. Who determines the F factor in power computations? _____

9. When do BC stations make equipment tests? _____

10. How often is an equipment performance test required? _____

11. What is a proof-of-performance test? _____ When is it required? _____

12. What is the required accuracy of BC station meters? _____

13. How accurately must antenna currents be maintained in directive arrays? _____

14. If a required antenna ammeter burns out, how soon must it be replaced? _____

15. How often must a carrier frequency meter be checked? _____

16. How often must the carrier frequency be logged? _____

17. Does an AM frequency monitor indicate a carrier or beat frequency? _____

18. Why does an FM antenna ammeter not change with modulation? _____

19. In Fig. 23-11, if the assigned frequency is 88.5 MHz, what is the frequency of the running crystal? _____

20. What might cause an FM frequency monitor to vary with modulation? _____

21. Draw a block diagram of a BC console. Draw diagrams of T- and H-type attenuator pads. A T-type gain control. A U-type pad. A line equalizer. A basic peak limiter. An AM frequency monitor. An FM frequency modulation monitor.

23-13 REMOTE CONTROL

When the transmitter is separated geographically from the control point of the broadcast studios, a remote control system can be used.

Telemetering of remote information may be by wire lines using stepping relays at the transmitter to sample the various parameters. Relative dc values can be used for the telemetering. Special tones can be used on wire lines or by radio relay systems.

Some of the significant rules regarding remote control operation are:

1. Operating and transmitting locations must not be accessible to other than authorized personnel.
2. There must be a positive on-off control of the transmitter at the remote position.
3. Short circuits, open circuits, or other faults in the system must not turn on the transmitter, but must turn off the transmitter.
4. The monitoring equipment at the control point must allow the operator to perform all required functions.
5. All remote-position instruments must be calibrated at least once a week, with the results entered in the maintenance log.
6. Remote position meters must read within 2% of the meters at the transmitter.
7. A modulation level and a negative percent of modulation control must be available at the remote control point. The carrier may be modulated with tones of less than 30 Hz with no more than 6% modulation (a form of multiplex) to signal the control position of the transmitter meter readings. These tones shall be transmitted only on demand of the control point operator.

Any malfunctioning of the remote control system requires direct control of the transmitter by an operator within 1 h.

23-14 AUTOMATIC TRANSMISSION SYSTEM

Some broadcast station transmitters are almost "computerized" when they use an *automatic transmission system* (ATS). A qualified operator must make inspections of the system each calendar month, but at periods not less than 20 days apart.

ATS controlled transmitters must be manually activated at the beginning of each broadcast day. Then, from electronic samples of parameter (meter) values and by the use of electromechanical controllers, the transmitter output power can be computed by either direct or indirect methods and its output power can be automatically adjusted if it deviates from the legal limits. Modulation gain is automatically reduced if modulation peaks of more than 10 excessively high peak bursts occur in any 60-s period. If SCA is used (Sec. 23-26) the SCA modulation must be checked also and must not exceed the authorized limit (with monaural FM, 30%; with FM stereo, 10%).

When ATS is used, transmissions are automatically terminated if the controls fail to correct either an overpower or overmodulation condition in 3 min in situations where the remote turn-off control fails, where the alarm system fails for 3 min, or where the ATS samplings are not being fed to the controlling center. In case the transmitter shuts itself down, it can be operated manually, provided an operator with adequate training is available.

23-15 TAPE EQUIPMENT

Broadcast stations use many magnetic tape recorder–playback cartridges. It is standard practice to record 5- to 60-s announcements or whole programs on tape and hold them for later transmission.

Audio frequency recording tape is made of plastic a few mils thick and ⅛ to ¼ in. wide. One surface is painted with a finely ground magnetic material such as iron oxides or chromium dioxide, which have high degrees of retentivity. The tape is wound onto thin plastic spools.

The basic operation of a 3- to 7-in. open-reel tape recorder is shown in Fig. 23-12a. When the constant-speed rubber drive wheel, or capstan, is forced against the tape and rotatable idler post, the tape is pulled along toward the take-up reel. This reel has a light torque that allows it to reel in the tape that is fed to it. The rewind reel has a slight drag to keep the tape firmly against the heads.

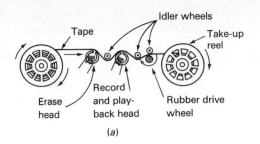

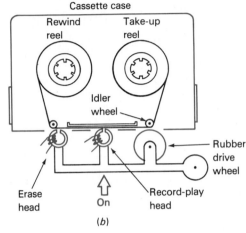

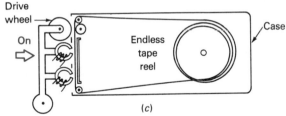

FIG. 23-12 Tape record-play mechanisms: (*a*) open reel, (*b*) cassette, and (*c*) cartridge.

During recording, AF ac fed to the record-head electromagnet produces magnetic flux at the very narrow gap, magnetizing the passing tape. The erase head is fed a supersonic-frequency ac that is used to erase any modulation on the tape before it reaches the record head.

To rewind the tape, drive-wheel pressure is released, drag is placed on the take-up reel, erase and record heads are moved away, and the rewind reel is motor-driven backward at high speed.

During playback, the drive wheel and playback head are engaged and the erase head is disconnected electrically. AF ac induced into the playback coil as the tape passes over the gap is amplified and corresponds to the original AF ac that magnetized the tape during recording.

A tape cassette recorder is somewhat similar in action to the open-reel type, but the smaller reels are enclosed in a plastic case (Fig. 23-12*b*). The erase head, record-play head, and drive wheel are all external to the tape and reels. When the play or record control is pressed, the external heads and drive wheel are forced against the tape through openings in the plastic case. The take-up reel has some torque applied to it through a hole in the case. When the rewind button is pressed, the heads and drive wheel are released and the tape is rewound at high speed onto the rewind reel by motor action through a second reel hole in the cassette case.

A tape cartridge ("cart") is somewhat similar to a cassette mechanism except that it uses an endless tape and only one reel. Tape is fed from the bottom of the reel that lightly holds a relatively few yards of tape. The tape is made to pass by the record-play-drive mechanism and is then laid down on top of the reel, as shown in Fig. 12-12*c*. Unlike the cassette, the drive wheel is in constant rotation as soon as the cartridge is plugged in. Program modulation is produced instantaneously when the engage control is pressed. Cartridges are usually tone-cued. When the advance mechanism is pressed, the tape moves forward until a tone cue is detected, whereupon the tape pressure ceases and the tape is ready to be engaged and played the instant the pressure control is activated. A cartridge machine may hold two or more cartridges. Stations will have several machines.

Stereo tape recorders record simultaneously on the upper and lower halves of the tape. Quadraphonic sound (Sec. 23-26) would require four modulated tracks on the tape.

Generally, the faster the tape moves, the better the fidelity of the sound reproduction. Common tape speeds are 15, 7.5, 3.75, and 1.875 in. per second (see also Sec. 24-9).

The magnetic gap of the record-playback head is filled with a nonmagnetic material, but this area may clog with some of the magnetic material from the tape that passes over it. Such clogging results in a loss of high frequencies, producing distortion during recording as well as playback. Heads should be cleaned with isopropyl alcohol or special cleaning solutions after every few hours of use. If the tape does not press firmly against the gap, high frequencies will also be lost.

23-16 TURNTABLES

Turntables in broadcast stations are usually 12 in. in diameter, made of cast aluminum, and are belt-, rim-, or directly driven by a synchronous motor. The motor and turntable may be spring-suspended to prevent rumble in the pickup.

Pickup heads or cartridges are magnetic or reluctance types with a diamond stylus for long play life.

The speed of turntable rotation may be checked by placing a stroboscopic disk on the turntable and illuminating it with a fluorescent light. The ring of printed squares that appear to stand still indicates the speed at which the turntable is rotating. If the squares do not stop but slide slowly forward, the turntable is rotating too fast. Most turntables will rotate at 33, 45, and 78 rpm.

The frequency response of a pickup head (disk or tape) can be checked with a test record or tape on which are recorded at constant amplitude, frequencies of 50, 100, 400, 1000, 3000, 5000, 10,000 and 15,000 Hz. If the pickup head, its filter, and the amplifiers of the system are operating correctly, the VU meter on the amplifier should read the same amplitude for all frequencies.

23-17 BROADCAST MICROPHONES

These may be of the dynamic, velocity, or condenser types (Secs. 17-26 to 17-28). Many broadcast microphones have transistor amplifiers or tiny AF transformers in them to convert their output impedance to 150, 250, or 600 Ω (or whatever is used in the particular station) so that all microphones can be interchanged without worrying about impedance matching. Generally, the lower the impedance of microphone lines, the less electrical noise they pick up.

When two or more microphones are used in a studio, sometimes the *phase* of their connections may have to be reversed, or their positions changed, because two microphones, with one a half wavelength closer to a sound source, may produce signals that cancel in the amplifier.

23-18 PULSE-WIDTH AMPLITUDE MODULATION

Many AM broadcast transmitters use standard plate or grid modulation (explained in Chap. 17), but other forms of amplitude modulation are possible. One interesting type, a 5-kW transmitter capable of linear modulation to over 10 kHz, uses a series modulator. The modulating current pulses it feeds to the triode RF power amplifier cathode-plate circuit are not varying in amplitude but are varying in *pulse width*. A simplified version of such a modulating system is shown in Fig. 23-13.

The crystal oscillator generates a sinusoidal carrier frequency RF ac which is buffered before being fed into a one-shot multivibrator. For every positive RF ac peak, the multivibrator produces a single rectangular wave pulse of 120° duration (typical class C) at the carrier frequency. These RF pulses are amplified to an output of about 500 W in a high-power transistor RF amplifier. They are then coupled through a broadband toroidal transformer to the grid-cathode circuit of the triode power amplifier. The secondary of this transformer is center-tapped to allow grid neutralization of the PA triode. The PA is grid-leak-biased to class C operation.

Note that the plate of the PA triode is at dc ground potential through the output circuit coils but that the cathode (and grid) are connected to the −12-kV dc potential through the series modulator triode when it conducts. Since the driver is operating near ground potential and the PA cathode-grid circuit is at a high negative potential, the driver transformer must be wound to withstand very high voltages.

The tuned circuits in the PA output to the antenna form a relatively broad (about 50 kHz) bandpass filter at the carrier frequency. This is more than enough to pass the AM carrier and its sidebands.

The program AF ac is fed to one input of a comparator IC. A 70-kHz triangular-wave oscillator signal is fed to the other input. The result is 70-kHz rectangular-wave pulses which reproduce the audio frequency and amplitude signals, but with constant-amplitude 70-kHz pulses that vary only in *width*. For example, the positive half of the AF cycles may produce wider 70-kHz pulses, while the negative half cycles of the AF produce narrower 70-kHz pulses, as shown. The result is 70-kHz pulses being modulated in width according to whatever AF ac is fed to the pulse-developing comparator circuit. The modulated width pulse circuits are near ground potential but must be coupled to the series modulator grid-cathode circuit at the −12-kV potential. The width-modulated pulses are fed to an LED focused on a light-sensitive photodiode through an optic fiber, which adequately insulates the two circuits. The width-modulated light pulses are reconverted with the photodiode to varying-width dc pulses, are amplified, and are fed to the base-emitter circuit of the bipolar transistor. With no positive

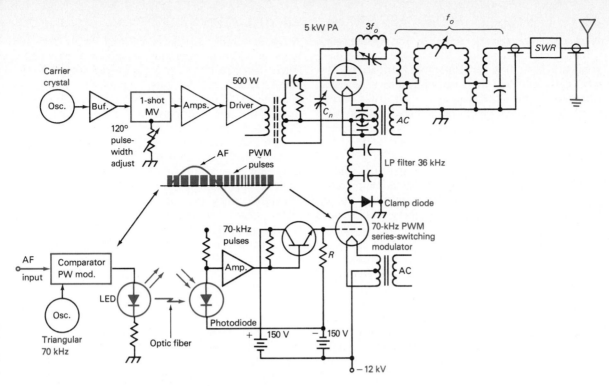

FIG. 23-13 Pulse-width modulation AM broadcast transmitter.

pulse signal driving the base, the -150-V bias supply cuts off the high-μ triode and no plate current can flow to the PA. When the bipolar base is driven positive by a modulating pulse, the transistor becomes a conductor and the $+150$-V bias supply forward-biases the grid of the modulator triode (developing a 300-V drop across resistor R) and passes a dc pulse to the PA.

The PA is constantly being fed carrier-frequency RF ac of class C 120° pulses, but if the series modulator is cut off, there can be no RF output from the amplifier. If the series modulator allows very narrow pulses to flow through to the PA, the output RF will be very low in power. If the modulator allows wider pulses to flow through to the PA, the RF output power is higher. Thus, with no audio input the comparator is biased to put out medium-width pulses, and the modulator feeds medium-width pulses to the PA, producing a medium-power (the carrier 5 kW) RF output. With positive peaks of modulation the pulse widths become wider. For negative peaks of modulation, the pulse widths become narrower. If the pulses are never allowed to go to zero width, there can never be a negative peak of modulation that reaches the 100% value

(zero power output). However, by utilizing still wider pulses than required for 100%, the positive peaks of modulation can rise considerably over 100%, up to perhaps 125%, the maximum allowed. This produces louder signals in receivers than normal 100% modulation would.

The low-pass filter between the series modulator and the PA has a cutoff frequency of about 35 kHz, preventing any of the 70-kHz pulses from producing modulation of the carrier. The clamp diode between modulator plate and ground has no effect while the PA is being driven by a modulator current pulse but discharges the low-pass filter elements during modulator cutoff times.

The LC circuit marked $3f_o$ is tuned to a frequency three times the carrier frequency. This squares off the carrier frequency plate current pulses and raises the efficiency of the PA nearly 10% over normal class C operation (to nearly 90%). The tuned circuits marked f_o reform the output signal into sinusoidal carrier and sideband frequency RF ac.

The SWR-power meter in the antenna indicates whether the antenna impedance matches the output impedance of the PA tuned circuits and tube. It can also indicate the RF power output.

23-19 EMERGENCY BROADCAST SYSTEM

In cases of a national, state, or local emergency, certain broadcast stations (AM, FM, TV), on a voluntary basis form an Emergency Broadcast System (EBS) and will transmit with normal power an emergency action notification (EAN). This consists of a simultaneous two-tone attention signal (853 and 960 Hz for 23 s) followed by an announcement of the details of the emergency and what station listeners should monitor.

At the end of the emergency the EBS stations transmit an emergency action termination (EAT).

The three levels of EBS authorizations are primary stations (national level), primary relay stations (state level), and common program control stations (CPCS—local level).

The priority of EBS traffic is: (1) presidential messages, (2) operational area (local) programming, (3) state programming, and (4) national programming and news.

Notification of an emergency may be received by an EBS station by network facilities, press wires (AP/UPI), or off-the-air monitoring of a primary station or primary relay station.

Standing operating procedures (SOPs) are documents issued by the FCC to EBS stations specifying operational and authentication procedures for the EAN and EBS system. An EBS Plan checklist is posted at each station's duty position. The station operates according to this plan. Stations may broadcast their call letters during an Emergency Action Condition (EAC).

Nonparticipating stations monitor the primary EBS stations using a monitor which gives an audible and/or visual indication when an EAN has been transmitted. All such stations must activate their required EBS 2-tone encoders (except there is no requirement for 10-W or less educational FM stations) when a national-level EAN is received and must transmit the EAN 2-tone signal. After that they announce that there is an emergency, that they are shutting down until an EAT is received, and that listeners should listen to an EBS station in their area.

Tests of the EBS system require a random-time weekly test for each EBS broadcast station, including the 2-tone attention signal and a prepared test script. The news teletype services test their lines to the broadcast stations twice a month. National-level wire communication tests are made at 1- to 3-month intervals. All such transmissions and tests must be entered in either the program or operating log.

23-20 LOGS

A *log* at a radio station is a listing of the date and time of events, programs, equipment parameters (such meter readings are primary concerns of operators), tests, malfunctions, corrections, etc. Logs must be made out by a station employee competent to do so. Each page must be numbered and dated. Time used can be either standard (as EST) or "advanced" time (as EDT). Abbreviations may be used if key letters are shown somewhere on the log. Logs are made out for either the broadcast day (sunrise to midnight), or for 24 h (midnight to midnight). They may be kept manually or by automatic means. Minor corrections to a transmitter should be made *before* logging the meter readings.

Erroneous entries in logs should not be erased. They are lined out, corrected above the error, dated, and initialed by the operator making the error, or if made later, the error should be struck out, corrected, dated, and signed by the corrector.

A *program log* must be signed by the operator when starting on watch and signed off at the end of the watch. It must contain entries of the required transmitter on the air and off the air times and the required on-the-hour station identification (ID). For each program it must contain: program name and title; beginning and ending times; type of program (agriculture, entertainment, news, public affairs, religious, instructional, sports, etc.); source (local, network, etc.); candidate's affiliation (if political); sponsor or who pays for the announcement; who provided the material; duration of commercials (close approximation of time); on whose behalf PSAs (public service announcements) are made; time public notices are announced; statement if material used is taped or recorded; time on and off a network. Entries are to be made for all EBS tests. Manually kept logs may be made at the time of broadcasting or may be made up prior to the broadcast. If automatic devices used for logging malfunction, the logs must be kept manually.

An *operating log* must be signed on and off watch. It may be kept manually or by automatic means. It should contain the following: (1) entries of operating parameters (V_p, I_p, I_a, etc.); (2) the time transmitter is put on and off the air; (3) the time antenna lights are turned on and off (if manually controlled); (4) the time of daily antenna light check and any failures that are noted; (5) the results of power measurements; (6) for directive AM BC antennas, last-stage V_p and I_p readings, and readings of antenna current or common point antenna current, as well as antenna monitor phase indications and sample currents every 3 h or less. (If automatic logging is used, equipment must be checked weekly and logged in the maintenance log.) Some AM BC stations are required to change antenna patterns at sunset to prevent sky-wave interference, requiring log entries at this time. Tests of EBS (unless consistently entered in the program log) are also required. Entries may be made in rough form and transcribed into finished log form.

A *maintenance log* must be made out, as well as signed on and off by a competent technician-operator. It should contain such items as (1) results of auxiliary transmitter tests; (2) frequency measurements and how they are accomplished (monthly, but no more than 40 days between); (3) calibration checks of automatic recording devices, antenna monitors, remote or extension meters; (4) date and time of removal of the modulation monitor or final amplifier and antenna meters; (5) replacement date and time of monitor or meters; (6) checks of weekly antenna current ammeters and remote meters; (7) quarterly antenna checks; (8) records of any experimental operations; (9) any field strength measurements of directional antennas; (10) antenna monitor phase indications; (11) results of required inspections (of antennas, etc.); (12) tests of and repairs to transmitter(s) and the antenna system.

Logs shall be retained by the licensee for 2 years unless they contain information regarding claims, complaints, or disasters, in which case they must be held until notified by the FCC or until the statute of limitations runs out on them.

For broadcast remote pickup base or mobile stations, logs should contain: date, time, and purpose of operation; location of transmitter if mobile or portable; stations with which communications are made; frequency checks made; and pertinent remarks about equipment or transmissions. Logs of communications between mobiles and base stations are kept by the base station. If no base station communications are made, logs are kept by the mobile operator(s).

Test your understanding; answer these checkup questions.

1. How often must remote control instruments be calibrated? _____
2. How can remote readings be telemetered to the studio without using telephone lines? _____
3. How often must ATS systems be inspected? _____

4. With ATS, if overmod or overpower is not corrected in 3 min, what happens? _____
5. What is the advantage of tape cartridges over open reel and cassettes for broadcasting? _____
6. How are tape cartridges usually cued? _____
7. What happens if a magnetic head gap becomes dirty? _____
8. What size turntables are used in BC stations? _____ At what speeds do they rotate? _____
9. What is used to check TT speeds? _____
10. How is TT pickup cartridge response checked? _____
11. Why must BC microphones sometimes have to be phased? _____
12. In the PWM transmitter: What form of ac drives the PA grid? _____ What circuit develops the PWM signals? _____ What is used to transfer PWM signals to the modulator system?
13. How does the RF output of the PWM transmitter differ from normal AM output signals? _____ What is the approximate efficiency of the PA? _____
14. What type of stations are used in an EBS system? _____
15. What is the meaning of EAN? _____ EAT? _____ EAC? _____ SOP? _____
16. How are EBS stations notified of an EAN? _____
17. What do nonparticipating stations do after an EAN? _____
18. What frequencies are used in the EBS attention signal? _____
19. Who makes out logs at BC stations? _____ How are corrections made? _____
20. What are three types of logs used in BC stations? _____ _____ _____
21. How long must logs be kept? _____

23-21 FM STEREO MULTIPLEX

In 1961 the FCC authorized a compatible system of stereophonic FM broadcasting called *stereo multiplex*. It is compatible because it can be received on stereo-type receivers or on standard monaural FM receivers. Most FM broadcast stations now use stereo multiplex.

Stereo is a means of making music or sounds two-dimensional to a listener. It makes sounds that would have approached a listener from the left or right in a concert hall approach from the left or right at home. This effect can be accomplished by having one microphone (L) on the left side of a concert-hall stage and a second microphone (R) on the right side of the stage. Then, by transmitting both L and R signals separately, by detecting them separately, and by feeding them to two separate loudspeakers placed at the left and right, the listener is enabled to hear sounds in much the same way as if at the concert. In the past such a system actually employed two transmitters and two receivers. In FM stereo multiplex this is accomplished by using two microphones, one transmitter with stereo matrixing, one receiver having a stereo matrix, two AF amplifiers, and two loudspeakers.

FM multiplex transmission is the means by which separate L and R signals can be made to frequency-modulate a single radio carrier. FM multiplex reception is the means by which the L and R signals can be separated in the receiving system to feed separate loudspeakers.

If voice-type audio signals having frequencies from 200–3000 Hz are fed to a loudspeaker, they produce intelligible sound to a listener. These voice frequencies can be represented as in Fig. 23-14a. If a 10-kHz steady tone is also fed to the loudspeaker, the listener will hear a high-pitched whistle. This is also indicated.

If the voice frequencies could be made to amplitude-modulate the 10-kHz carrier frequency, the resultant signals would consist of the 10-kHz carrrier plus upper and lower sidebands extending from 7 to 13 kHz (Fig. 23-14b). Although a loudspeaker could reproduce all these frequencies, the listener would hear them not as understandable sounds but as a conglomerate group of high-pitched squeaks—sounds in the range of 7–13 kHz. This is an example of a carrier and sidebands, both in the AF spectrum, but unintelligible because they have not been detected. If they are fed to a 10-kHz receiver and detected, the result would be the 200–3000 Hz signals used to modulate the 10-kHz carrier.

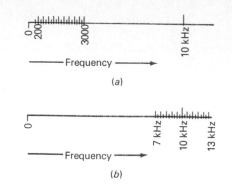

(a)

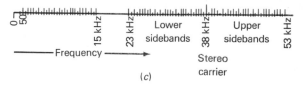

(b)

(c)

FIG. 23-14 (a) VFs and an 10-kHz tone. (b) 10-kHz carrier amplitude modulated by VF. (c) Hi-fi AF and a 38-kHz carrier modulated by hi-fi signals.

By raising the modulated carrier to a supersonic 38 kHz, a listener could hear nothing because this is higher than humans can hear. If this 38-kHz carrier frequency is *amplitude-modulated* by high-fidelity AF (50–15,000 Hz), the sidebands produced would extend 15 kHz above and below 38 kHz, or from 23 to 53 kHz (Fig. 23-14c). None of these supersonic frequencies are audible, either, but they could be detected with a 38-kHz AM detector. This is essentially one of the signals used to modulate an FM stereo multiplex transmitter.

23-22 STEREO MULTIPLEX TRANSMITTERS

A possible system to produce FM stereo multiplex transmissions is shown in Fig. 23-15 in block form. Most circuits have been discussed previously in other applications.

The transmitter is reactance-tube-modulated, but could be phase-modulated. Both L and R microphone signals are fed into a linear AF *adder* circuit and are used to produce a composite L+R main-channel signal (solid color lines) to frequency-modulate the oscillator of the transmitter. This 50–15,000 Hz signal produces modulation that can be detected by standard monaural FM receivers.

Note the polarity indications. The L+R amplifier is being fed the L and R signals in one phase. The L−R amplifier is being fed both L and R signals, but by passing the R signal through a phase inverter

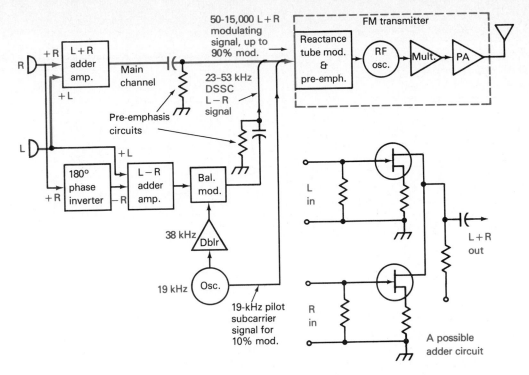

FIG. 23-15 Block diagram of an FM stereo multiplex transmitter system. Right, a simple adder to combine L+R and L−R signals.

it reverses the +R signal to a −R. Thus this adder is fed a +L and a −R producing L−R output. To a listener L+R and L−R signals contain the same frequency components in the same amplitudes and would therefore sound the same to the ear.

The 50–15,000 Hz L−R signal is used to balance-modulate a 38-kHz carrier signal (the second harmonic of the 19-kHz pilot subcarrier in Fig. 23-16). The output from the balanced modulator is a double-sideband suppressed 38-kHz carrier set of supersonic ac signals having frequency components from 23 to 53 kHz. The L−R signal also frequency-modulates the RF oscillator.

The whole multiplex modulating signal for an FM stereo broadcast consists of the monaural 50–15,000 Hz L+R audio ac signals, a supersonic 19-kHz (±2-Hz) ac pilot subcarrier, and the L−R stereophonic ac signals consisting of sidebands from 23 to 53 kHz, but without the 38-kHz carrier, as indicated in Fig. 23-16.

Theoretically there should be no *cross talk* between the monaural and stereophonic channels, but there always is some. By FCC regulations the cross talk must be held at a level of more than 40-dB difference between channels.

Some other stereophonic transmission standards are: The stereophonic subcarrier must be suppressed to a level of less than 1% modulation of the main carrier. Pre-emphasis for the stereophonic subchannel must be identical with that of the main channel. Peak modulation of either channel alone is 45% of the total (pilot subcarrier is 10%). A positive left signal deviates the carrier to a higher frequency.

23-23 STEREO MULTIPLEX RECEIVERS

A possible stereo multiplex receiving system is shown in block form in Fig. 23-17. A normal FM receiver system is employed up to, but not includ-

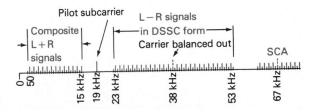

FIG. 23-16 Location of FM stereo multiplex and SCA signals.

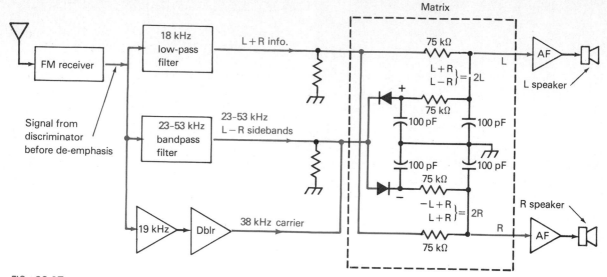

FIG. 23-17 Block diagram of an FM stereo multiplex receiving system, showing the matrix in schematic form.

ing, the de-emphasis circuit (which is added later in the matrix). With the exception of the matrix, all circuitry shown has been explained elsewhere.

The FM receiver discriminator detects from the received FM stereo signal (1) the 50–15,000 Hz composite L+R AF signals, (2) the 19-kHz pilot-subcarrier frequency, and (3) the 23–53 kHz L−R sideband signals.

This complex group of audio and supersonic ac signals is fed to three circuits. One is an 18-kHz low-pass filter. The output of this filter contains only the composite 50–15,000 Hz L+R AF signal. (With monaural transmissions this signal is de-emphasized in the matrix, amplified by both L and R amplifiers, and is fed to both loudspeakers as monaural sound.)

The output of the discriminator is also fed to a 23–53 kHz bandpass filter (or 53-kHz low-pass filter). The output signal of this filter is the DSSC L−R sidebands (to which the 38-kHz carrier signal will be added at the matrix).

The third circuit amplifies only the 19-kHz pilot subcarrier which is fed to a frequency doubler to produce an output of 38 kHz. This 38-kHz carrier, when added to the L−R sidebands, forms a 38-kHz double-SB-with-carrier AM signal.

The matrix has two AM diode detectors that detect the amplitude-modulated 38-kHz L−R signals simultaneously but in opposite polarities. That is, L−R signal information is detected both as a varying dc with positive polarity and as a varying dc with negative polarity.

The positive-polarity L−R signal (now detected into its original 50–15,000 kHz AF form) is added to the composite L+R AF signal in the matrix. Combining the positive L−R and the L+R results in (L+R) + (L−R), or 2L, a strong left output signal from the top terminal of the matrix. Combining the negative L−R and the L+R results in (L+R) + (−L+R), or 2R, a strong right output signal from the bottom terminal of the matrix. Thus, the output of the matrix is the two audio signals that originated in the L and R microphones in the broadcast studio. Each signal is fed to its own amplifier and loudspeaker, forming the stereo reproduction, theoretically with no cross talk between the two channels.

Note the four de-emphasis circuits in the matrix, four 75-kΩ resistors, and four 100-pF capacitors to give a 75-μs de-emphasis—and how both positive and negative L−R detectors mix with the same L+R signal in the de-emphasis circuits.

There is also an AM broadcast band stereo system, called *compatible quadrature amplitude modulation* (C-QUAM). It modulates the same carrier frequency split into two components, each 90° from the other. The R+L audio modulates one carrier. The R−L audio modulates the other. In the receiver an AM-stereo decoder IC has an envelope (diode) detector to detect nonstereo AM signals. This signal is also mixed with the R+L and R−L signals in a matrix to produce R and L output signals, which are fed to two AF amplifiers and speakers to produce the stereo sounds.

23-24 SCA AND QUADRAPHONIC

Besides the multiplexing of stereo information into an FM broadcast transmission, *subsidiary communication authorization* (SCA), can be used to transmit background music, weather, time signals, educational information, and, using digital techniques, can transmit encoded alphanumeric news on FM subcarriers.

Whereas stereophonic multiplex signals consist of AF ac translated up to a subcarrier at 38 kHz with AM sidebands 15 kHz on both sides, SCA information is AF or other ac translated up, using a subcarrier somewhere between 20 and 75 kHz. SCA modulation is always a narrowband 7.5-kHz deviation FM of the subcarrier frequency used. The maximum SCA sideband power must never exceed 30% of the total modulation of a monaural FM carrier. If the main carrier is also being modulated by stereo multiplex, the SCA multiplex must not exceed 10% of the total main-carrier modulation and the SCA information must be in the range of 53–75 kHz, usually with a carrier frequency of 67 kHz as indicated in Fig. 23-16. To produce the SCA signal at the transmitter, an SCA carrier (67 kHz, for example) is modulated, and the resulting carrier and sidebands are fed to the reactance or other modulator without pre-emphasis at the same time that the pre-emphasized 50–15,000 Hz FM program material is fed to the modulator.

An SCA receiver is simpler than a stereo multiplex. The output of an FM broadcast receiver discriminator is fed to two circuits. One is the normal de-emphasis circuit and AF amplifiers for the main-channel program. The other is to a 67-kHz (or whatever SCA carrier frequency is being used) bandpass filter and detector, which extracts the SCA information. This is amplified and fed to a loudspeaker. The listener has the choice of the main-channel program, the SCA program, or both simultaneously.

If a program is picked up with right-front, left-front, right-rear, and left-rear microphones; is broadcast as four signals; is demodulated as four signals; and is fed to four separate speakers in proper relative positions, it represents a *quadraphonic* system. This could be produced with FM, using (1) the main 50–15,000 Hz channel, (2) the stereo channel centered on 38 kHz, (3) a second channel centered on 38 kHz but transmitted in quadrature (at 90°) and detected with a quadrature detector, plus (4) a fourth channel somewhere in the SCA region between 53 and 96 kHz.

23-25 ANTENNA TOWERS

This section deals with lighting and painting of the one or more metal towers used as antennas in broadcast stations. Information pertaining to antenna theory was discussed in Chap. 20.

Antenna towers represent hazards to aircraft. As a result they must be painted with equal-width stripes of aviation orange (TT-P-59) and white (TT-P-102), each stripe approximately one-seventh the height of the tower, but not over 100 ft in width on tall towers. (Paint samples available at Federal Supply Service Center, Washington, D.C.) The top and bottom stripes must be orange. To mark the tower at night (sunset to sunrise), towers up to 150 ft in height must have two steady-burning 116- or 125-W (32.5-candela) lamps in an aviation red light globe at the top of the tower. Control of these lights may be manual, or by a light-sensitive device that will turn on when the north sky illuminance is 35 footcandles (fc) and off when illuminance rises to not less than 58 fc, or by an astronomic dial clock with a time switch.

For towers higher than 150 ft the top beacon light consists of two 620- or 700-W PS-40 Flashing Code Beacon lamps (2000 candelas) with aviation red filters. At half, third, quarter, etc., tower height points (depending on tower heights), flashing 620- to 700-W beacon lights are installed. Midway between flashing and top beacons, or between flashing beacons on taller towers, are installed steady-burning 116- to 125-W red beacon lights. Control of lights is by a device sensitive to north-sky light, described above. If the control device malfunctions, the lights must be turned on manually from sunset to sunrise.

Tower light inspections are required at least once a day, either visually or by observing an automatic indicator. At intervals not to exceed 3 months, inspections of control devices, indicators, and the required light-malfunction alarm system must be made and logged.

If a required tower light fails (except for steady burning intermediate lights), the nearest flight service station or office of the Federal Aviation Administration (FAA) shall be notified within 30 min, advising when correction is expected.

Log entries of tower inspections should include a record of: (1) light on-off times if manually controlled; (2) time of daily checks; (3) date and time of any improper functioning of the light system; (4) date and time of corrections made; (5) when FAA was notified of improper functioning, when FAA was notified of correction of trouble; (6) tower inspections and when repairs were made.

All towers should be cleaned or repainted as often as necessary to maintain good visibility.

All servicing of lights, indicators, control, and alarm systems shall be accomplished as soon as practicable. If the alarm system fails, the lights must burn continuously.

Keys to the antenna house or fence surrounding a tower should be at the transmitter.

A number of spare beacon lights should be on hand in the transmitter house at all times.

23-26 OTHER FCC REQUIREMENTS

Since all radio stations are under the jurisdiction of the FCC, representatives from the commission may request certain information from any station at any reasonable hour. Some of the items that they may wish to examine are the logs, results of equipment-performance tests, or recent antenna and field-intensity measurements.

All broadcast stations are required to identify themselves by transmitting call letters and location at the beginning and end of each broadcast day, as well as on the hour. They may also identify on the 15-, 30-, and 45-min periods of each hour. While it is not necessary to interrupt a continuous program, the identification (ID) announcement should be made within 5 min of the required time.

Any program over 1 min in length that contains recorded portions must be announced as being partly or wholly recorded at either the beginning or at the end of the program.

Whenever a radio station is reimbursed for any program or material being broadcast, it must announce the name of the sponsor.

The station license must be posted in a conspicuous place at the control point of the transmitter.

Normally, broadcast stations may not direct information to any specific person. However, during emergencies in which safety of life and property are involved (hurricanes, floods, earthquakes, etc.), they may address information to dispatch aid. Whenever this is done, the FCC in Washington, D.C. and the Engineer-in-Charge of the radio district must be notified.

No changes may be made to the final amplifier, its tube types, the number of tubes, or the system of modulation without prior approval of the FCC, but changes of audio and RF circuits other than those of the final amplifier and modulator can be made without approval.

Broadcast stations with an operating power of 500 W or less may apply for a *Presunrise Service Authority* to allow them to operate between 6 A.M. local time and local sunrise with antennas that will give them better local coverage, provided that this will not cause interference to other stations.

BROADCAST STATIONS QUESTIONS

1. What is the frequency tolerance of an AM BC station? Of an FM BC station?
2. What is the emission classification of an AM BC station? Of an FM BC station?
3. What is the service area of a primary AM station? A class B FM BC station?
4. What are four reasons why local BC FM may sound better than BC AM?
5. In what band of frequencies will AM BC stations be found?
6. What BC service must use directional antennas?
7. List nine program sources that may feed a signal to a BC transmitter?
8. For what is an auxiliary BC transmitter used? An alternate transmitter?
9. What transmitting information must be fed to an operating position from a remote BC transmitter?
10. What is an STL station? An intercity relay station?
11. What would happen if a studio microphone and

monitor speaker, or console microphone and monitor speaker were both on at the same time?
12. How are tapes cued? Records?
13. What are the three main operating requirements during a broadcast?
14. What should be the maximum meter level of programs on telephone lines?
15. Why is gain controlled at zero VU rather than at the microphone output level?
16. What is the difference between a pad and a gain control?
17. Draw a diagram of a simple single-frequency line equalizer. Does its resonant frequency peak or decrease?
18. What are the results of using a peak-limiter amplifier in an AM BC station?
19. How is operating power of an AM BC station computed? Of an FM BC station?
20. What are the maximum positive and negative peaks

of modulation that an AM BC station may produce? What should the average be?

21. What tests must be made on an annual basis at BC stations?
22. What is the accuracy required of BC station meters?
23. If required meters burn out, what must be done?
24. Within what tolerance must antenna currents be held in directive systems?
25. With what may a constant carrier frequency indication be given at an AM BC station?
26. Draw a diagram of a simple AM BC frequency meter.
27. What may be used to monitor modulation at AM BC stations? At FM BC stations?
28. Draw a block diagram of an FM BC station frequency and modulation monitor system.
29. How does a console operator know that remote transmitter equipment is operating correctly?
30. Within what time period must an operator take over manual control of malfunctioning remote transmitting equipment?
31. What is ATS, and what type of BC stations may use it?
32. What are the three basic types of BC tape machines? Which uses an endless tape?
33. What are the two electrical heads on tape machines called?
34. Why may faster tape travel be more desirable?
35. What is recorded on a frequency-response record or tape?
36. What are the three basic types of microphones used in BC stations, and what impedances are used with them?
37. Draw a semiblock diagram of a pulse-width AM BC station transmitter. Is the waveshape of the 120° pulses important?
38. In the pulse-width transmitter, how are the low-potential AF circuits insulated from the high negative power-supply voltage? What kind of bias is used in the PA?

39. What are the main advantages of the pulse-width transmitter?
40. What does EBS mean? What types of BC stations may use it?
41. What is the EAN signal? How often is it transmitted?
42. What is an EAT signal?
43. What are EBS SOPs?
44. What are the three types of logs at BC stations? Who records them? How long must they be kept?
45. Draw a block diagram of the frequency spectrum involved in a stereo multiplex transmission from 0 to 53 kHz.
46. Draw a block diagram of a possible FM stereo multiplex system.
47. What are two reasons for using 19 kHz as the pilot subcarrier in stereo FM? At what percent modulation is it transmitted?
48. Describe the whole modulating signal used in FM stereo multiplex.
49. Are both L+R and L−R signals pre-emphasized? Which are detected as AM signals?
50. Draw a block diagram of an FM stereo receiving system. Where is the de-emphasis?
51. Will there be any stereo effect if the right and left speakers are stacked one on top of the other?
52. What is C-QUAM? Where would it be used?
53. What is SCA? For what is it used? Where in an FM modulating signal is it found?
54. For aircraft safety, how must antenna towers be outfitted?
55. How often must tower inspections be made? How are tower lights turned on and off?
56. Who is immediately notified if tower lights fail?
57. What items might FCC inspectors want to examine at BC stations?
58. When must BC stations make IDs?
59. What determines whether sponsors' names must be announced?
60. What stages of a transmitter may not be changed without prior approval of the FCC?

24. Television

This chapter first discusses a basic monochrome television transmitting system, and describes some of the specialized circuits and devices that might be found in it, the blanking and synchronizing pulses used, and some modulation requirements. Basic monochrome TV receiving system circuits are briefly developed. The fundamentals of color and some of the additional circuits required in both transmitters and receivers of color telecasts as well as general color TV systems operations are discussed. A complete color TV receiver is explained. It is assumed the reader is familiar with preceding chapters on both AM and FM transmitters and receivers.

24-1 A TV BROADCAST SYSTEM

A television transmitting system involves many of the radio circuits and systems previously discussed. A *monochrome* (black-and-white, or B/W) TV transmitter is actually two separate transmitters coupled to a single antenna (Fig. 24-1). Since transmissions are in the VHF or UHF ranges, the best antenna site is on top of a hill near the area to be served.

The *aural*, or sound, transmitter is essentially the same as an FM broadcast station (Chap. 23), except that 100% modulation is represented by a 25-kHz frequency swing of the carrier instead of 75-kHz deviation. The same 30–15,000 Hz audio capability of the amplifiers is required. The microphone signals are amplified, controlled by an operator, and fed to the FM modulator in the transmitter. The FM transmitter signal (F3E) is monitored and fed to the antenna through a *diplexer*, which prevents any FM signal from being coupled into the visual transmitter.

The *video*, or visual signal, transmitter is amplitude-modulated and receives its modulating voltages from photosensitive camera tubes or video tape machines. These video signals range in frequency from a few hertz to nearly 4.5 MHz. Since no transformer can handle such a wide span of frequencies, *RC*-coupled series, base, gate, or grid modulation is used in the visual transmitter. The video signals from the camera are monitored by a control-room operator, synchronizing pulses are added to them, and they are fed to the visual transmitter, where they amplitude-modulate the carrier. The visual RF signal from the transmitter may be fed to a vestigial sideband filter if the PA stage is grid-modulated, in which some of the below-carrier-frequency sidebands are attenuated. These vestigial sidebands, the carrier, and all the above-carrier-frequency sidebands (type C3F emission) are fed to the antenna through a diplexer, which prevents any of them from entering the FM transmitter.

It is assumed that a TV receiver has been observed with a magnifying glass and that it has been seen that the picture is composed of many illuminated horizontal lines across the face of the picture tube. The illumination of the different parts of each line changes in intensity as the televised scene changes.

It is the task of the TV camera and transmitter to scan, or slice, the scene to be televised into about 495 separate horizontal lines, each with varying intensities, transmitting a stronger signal where the lines are dark and a weaker signal where the lines

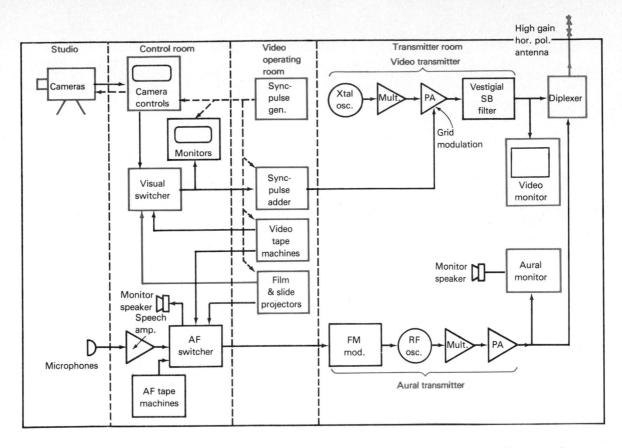

FIG. 24-1 Block diagram of a simple TV transmitting station, including studio, control room, video operating room, and transmitter room. Studio to control room intercom not shown.

are light. This is called *negative-type transmission*. (In positive-type transmission dark scenes produce the weakest output.)

The *synchronizing-pulse generator* generates the necessary pulses to operate cameras, monitors, and projectors. It also adds them to the visual transmitted signal to enable TV receivers to synchronize their sawtooth-waveform oscillators with the sawtooth scanning voltages applied to the camera tube at the transmitter.

The signals radiated from the two transmitters are received by viewers on a superheterodyne TV receiver (Fig. 24-2). The received TV signal is amplified, converted to a 42-MHz IF, and detected. This detected signal contains (1) varying dc components that represent changes in illumination for the lines of the picture (0–4.2 MHz video frequencies); (2) a group of varying dc frequencies centered around 4.5 MHz, which is the frequency-modulated aural information; and (3) horizontal and vertical sync

pulses added to the signal at the transmitter. All this information is amplified by the video amplifier. The frequencies centered around 4.5 MHz are accepted by a tuned 4.5-MHz IF amplifier, detected, amplified, and fed to a loudspeaker. The low-frequency sync pulses are used to synchronize a 60-Hz sawtooth oscillator to sweep the lines down the picture tube at this rate. The high-frequency sync pulses are used to synchronize a 15,750-Hz sawtooth oscillator to sweep lines across the picture tube at this rate. The video information signals are passed to the grid of the picture tube to change the intensity of the illumination as the lines are being written across the face of the tube.

To receive UHF stations, a UHF diode mixer and local oscillator are used. The VHF RF amplifier and mixer (tuned to 42 MHz) may now act as additional IF amplifiers. Functioning of the circuits of complete TV receivers will be discussed later in greater detail.

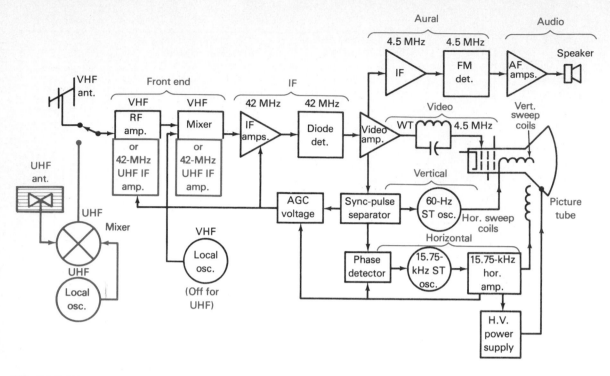

FIG. 24-2 Block diagram of a basic monochrome VHF TV receiver. Color circuits represent a possible arrangement of front-end circuits when receiving UHF TV signals.

24-2 THE TV TRANSMITTING ANTENNA

The transmitted signal from a TV station is essentially 6 MHz wide, requiring a broadband antenna. Both aural and visual TV transmissions are usually horizontally polarized (but may be circularly polarized) (Sec. 20-32) to reduce *ghosts* (reflected signals). A turnstile, slotted-cylinder, or other beam type of antenna can be used. Transmission lines can be balanced, shielded two-wire types, unbalanced coaxial, or, for UHF, waveguide (Sec. 22-2). The standing-wave ratio (SWR) can be determined by SWR bridges or meters, or a pair of directional couplers can be used. A directional coupler samples a small known fraction of the power that is traveling along a transmission line *in one direction only*. The current (or voltage) traveling up toward the antenna shows on one directional coupler. Reflected current shows on the other coupler connected to the line in reverse. The ratio of the two is *reflection coefficient* ρ. Directional couplers used in this manner make up a *reflectometer* (Sec. 22-3). SWR is found by SWR = $(1 + \rho)/(1 - \rho)$.

The diplexer must pass both the visual and the aural signals to the same antenna but prevent either

from coupling to the other transmitter. This is accomplished with a balanced bridge-type circuit (Fig. 24-3). The visual signal is coupled directly to the two center feed points of the antenna through a shielded two-wire transmission line. The shield acts as a center tap of the two wires. Two similar induc-

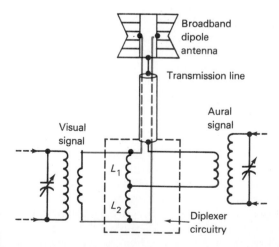

FIG. 24-3 Essentials of a diplexer.

tors, L_1 and L_2 (or two similar capacitors), across the line provide a center tap also. The aural signal is fed between the two center taps. Because of the balance of the system there can be no transmission from transmitter to transmitter, but both transmitters excite the antenna and radiate signals.

24-3 MAGNETIC DEFLECTION AND FOCUSING

In Sec. 13-35 a cathode-ray tube (CRT) that used an electrostatic (positive and negative) means of deflecting the electron beam was discussed. Electrostatic deflection can be used in TV CRTs, but practically all CRT devices in TV systems are magnetically deflected.

A beam of moving electrons, as in a CRT, represents an electric current. The direction of lines of magnetic force around a current is shown by the left-hand rule, in which the thumb points to the current direction and the fingers point in the direction of the lines of force. In Fig. 24-4 an electron beam is moving toward the reader through a magnetic field.

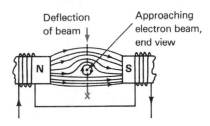

FIG. 24-4 An electron beam moving into a magnetic field is deflected at right angles to the field.

The part of the line of force surrounding the beam at the top has a direction similar to that of the lines of force of the stationary field. Since like lines of force repel, the beam tends to deflect downward. Similarly, the line of force of the beam that is below it is in the opposite direction to the stationary lines. Since unlike lines attract, the beam is attracted downward still more. Note that the electron-beam deflection is at *right angles* to the magnetic field through which it is moving, not toward either magnetic pole. By sending a beam of electrons through the area between two poles of an electromagnet, it is possible to control the amount of deflection of the beam by controlling the strength of the current that produces the magnetic field and the direction of deflection (up or down) by controlling the direction of the current flow through the elec-

tromagnets and therefore the magnetic line direction.

Figure 24-5 illustrates placement of vertical-deflection coils at the neck of a CRT. When current flows through the coils, a horizontal magnetic field

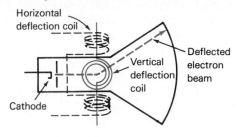

FIG. 24-5 Positions of horizontal- and vertical-deflection coils around the neck of a TV picture tube.

is produced across the neck of the CRT. The amount of vertical deflection will depend on the strength of the magnetic field. The direction of deflection (up or down) will depend on the polarity of the field.

When current flows through the horizontal-deflection coils (dashed), a vertical magnetic field is produced through the tube. The extent of horizontal deflection depends on the strength of the field. Direction of deflection depends on field polarity.

By using both horizontal- and vertical-deflection coils, the beam can be moved to any position on the tube face by applying the proper polarity and strength of current to the two sets of coils. Two horizontal and two vertical coils form a *deflection yoke*.

The electron beam in a CRT may be focused by electrostatic means (Sec. 13-35) or electromagnetically. If a wire carrying a steady current is coiled around the neck of a CRT (Fig. 24-6), the lines of

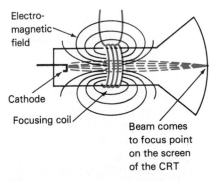

FIG. 24-6 Position of an electromagnetic focusing coil around a CRT.

force of the magnetic field produced are parallel to the axis of the CRT neck. Electrons in the beam tend to converge as they pass through this field of parallel magnetic lines. The current in the coil and therefore the strength of the field determine where the electrons of the beam will come to a point of focus. TV picture tubes usually have magnetic deflection yokes and magnetic focusing coils.

24-4 INTERLACED SCANNING

The current fed to the horizontal-deflection coils has a sawtooth waveform to produce a relatively slow movement of the electron beam across the face of the tube during the writing of a horizontal line, followed by a blanked-out rapid retrace during which nothing is written on the tube face. While it would be simpler to scan from top to bottom of the picture with about 495 sawtooth-produced lines, it is better to *interlace* the scanning lines. Consider the following.

If a motion-picture film is projected with a *frame* (single-picture) frequency of about 15 per second, an illusion of movement is created, but the motions seem jerky and bright areas flicker. Increasing the frame frequency to 24 per second will make motion appear smooth, but a slight flicker can still be noted on more brilliantly lit areas of the picture. Doubling the frame frequency would eliminate the flicker, but too much film would be required. Instead, it is possible to flash each picture on the screen twice by the use of a shutter that opens and closes 48 times per second, and the flicker disappears. Interlaced scanning of the TV picture has a similar effect.

A TV picture has an *aspect ratio* of 4:3; four units wide to three units high. The size of the tube is measured from one corner to the opposite corner, which has no effect on the number of lines received. On larger tubes lines are thicker, longer, and are wider apart.

Satisfactory vertical definition can be produced by breaking a picture into about 495 horizontal lines. (Any number of lines between 482 and 495 may be employed at the discretion of the transmitting station.) By allowing time for 525 lines per picture, an allowance of 30 to 40 lines is included during which the scanning circuits can be brought into proper synchronism to assure that both transmitters and receivers start scanning at the top of the picture at the same time.

To produce satisfactory horizontal picture definition (sufficient dark and light units per line) requires a bandwidth of about 4.2 MHz.

To prevent flicker and still not have to show a complete picture more than 30 times per second, interlaced scanning (Fig. 24-7) is employed. The first time the picture is scanned, only the even-

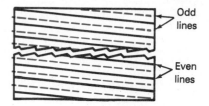

FIG. 24-7 Interlaced scanning. The first and odd numbered field lines are shown broken; the second and even numbered field lines are shown solid.

numbered lines are used. This produces a picture with only 262.5 lines, known as a *field*. The field by itself would be a poorly defined picture vertically. But the picture is immediately scanned again, using the odd lines. This second field is interlaced between the lines of the first. Since the two are presented within $1/30$ s, the eye accepts them as one *frame*, with satisfactory vertical and horizontal definition and showing no flicker on brightly lit areas. In a TV transmission the frame frequency is 30 per second (the field frequency is 60). Accurately timed synchronizing pulses must enable all horizontal lines to start at the correct time to produce straight vertical edges of the picture, as well as to start the half line at the top of the picture properly.

Since each horizontal line is below the preceding one, a gradually changing current must be fed to the vertical-deflection coils to move successive lines continually farther down the CRT. This is known as the *vertical sweep* current. The sweep current must return rapidly to the starting value as soon as a field has been scanned all the way to the bottom. A 60-Hz sawtooth waveform can produce the slow vertical movement downward and the rapid return upward at the end of the cycle.

There are about 24 sweep lines above the TV picture that carry no picture modulation. Lines 17 through 20 may carry test signals indicating modulation reference values, cue, and control signals. Lines 21 through 23 may carry coded information regarding local or network origination, or other identification. (These may be seen when the picture rolls.)

With 525 lines per frame and 30 frames per second, the horizontal sawtooth ac required to sweep the beam horizontally across a B/W camera or picture tube is 525(30), or 15,750 Hz.

24-5 VIDEO CAMERA TUBES

The first TV camera tube was the *iconoscope*. It had an electron gun to provide a thin beam of electrons, a picture screen or plate on which hundreds of thousands of tiny silver islands were laid down. Each island was separated from all others and from the metal backing plate by a thin insulating sheet. The tiny silver capacitors of this *mosaic* would release electrons and become charged when light was focused on them. With a scene focused on the mosaic, the electron-gun beam was scanned across the mosaic. The charged islands would be discharged by the beam through a resistor from plate to ground. The voltage-drop across the resistor formed the video signal for the camera tube. Magnetic deflection coils were used to provide the scanning fields. The iconoscope was relatively large and insensitive.

The next TV camera tube was a somewhat smaller *image orthicon* (IO). It had a photocathode on which the scene was focused. Photoelectrons from the back of the cathode moved to a target plate to produce a charged scene on this plate. The plate was made of a special glass having a controlled conductivity from one of its flat surfaces to the other. Electrons hitting the front of the glass plate charged it positively because it gave off two electrons to a fine positive screen in front of it for each electron hitting it. An electron beam was magnetically scanned across the back of the charged target plate. Those areas having no charge bounced the electrons back down the tube. Lighted areas absorbed the electrons, and none would be left to move back down the tube. Before the electrons were finally collected they were sent through a series of positively charged *dynode* or electron multiplier plates. Impinging electrons liberated two or more electrons from each dynode before they moved to the next more positive plate. In this way the video current was amplified by dynatron action. The IO was still large by modern concepts, needed long warmup times, required heaters when used outdoors in cold weather, and had considerable *latent image*, or "sticking" (bright areas remaining for some time).

The *vidicon* was the next video tube development. It incorporates the basic operating fundamentals of the more refined modern Plumbicon, Vistacon, Saticon, Newvicon, and other tubes.

These are smaller, lighter, less expensive, simpler in design that the IO, and more sensitive without an electron multiplier, and optical lenses cost a fraction of those for larger tubes.

A simplified vidicon-type tube is shown in Fig. 24-8. The light-active area at the front of the tube

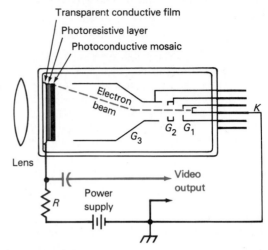

FIG. 24-8 Elements of one of several types of vidicon camera tubes.

consists of three very thin films on the inner surface of the flat end of the tube. The first is a transparent conductive film. Next is a semiconductor photoresistive layer deposited on the conductive film. Lastly, a photoconductive mosaic layer is developed on the semiconductor layer. The middle semiconductor layer has the property of extremely high resistance (50 MΩ) when in the dark but reduces its resistance when photons of light strike it. Thus, areas between the photoconductive mosaic islands and the conductive film act as tiny capacitors whose *dielectric leakage* is directly dependent on the light intensity at these areas.

The cathode K and the three grids develop an electron beam but limit its velocity to a low level, so that as the beam scans, it merely charges the million or so tiny mosaic capacitors. Light on the mosaic discharges the capacitors through the load resistor R. The scanning beam recharges the mosaic capacitors and produces a video signal voltage-drop across R that is proportional to the light intensity present at the spots being scanned.

Focusing can be accomplished electrostatically by controlling the voltages between grids, or magnetically by wrapping a focusing coil around the midsection of the tube. Scanning is produced electromagnetically with horizontal and vertical yoke coils. Centering can be adjusted either by placing permanent magnets around the tube or by dc-biasing the deflection yoke coils.

█ Test your understanding; answer these checkup questions.

1. What is considered 100% modulation for the aural TV transmitter? _____ The video transmitter? _____
2. What device prevents aural RF from entering the video transmitter and vice versa? _____
3. About how many lines make up a TV picture? _____
4. In a TV receiver, what is the video IF? _____ Aural IF? _____ Horizontal sync frequency? _____ Vertical sync frequency? _____
5. What is the width of a TV channel? _____
6. What is the polarization of TV video signals? _____ Aural? _____
7. If a reflectometer shows 50 V radiated and 25 V reflected, what is the SWR? _____
8. What type of deflection is used in an oscilloscope CRT? _____ In TV CRTs? _____
9. What direction deflection is produced by a deflection coil above a CRT? _____ What is the unit that is composed of horizontal- and vertical-deflection coils called? _____
10. What is eliminated by using interlaced scanning? _____
11. What is the aspect ratio of a TV picture? _____
12. What is the B/W field frequency? _____ Frame frequency? _____
13. What was the name of the first TV camera tube? _____ The second? _____ The third? _____
14. Which camera tube has electron multipliers? _____ Which type is used in closed-circuit TV? _____ In portable cameras? _____ In studio cameras? _____

24-6 THE SYNC-PULSE GENERATOR

The timing of the TV transmitted lines is controlled by the synchronizing-pulse (sync-pulse) generator. This complex piece of equipment generates the pulses for monochrome transmissions as shown in Fig. 24-9. The different pulses are:

1. 31,500-Hz *equalizing* pulses having a duration of approximately 2.7 μs each, the leading edges separated by 31.75 μs
2. 15,750-Hz *horizontal sync* pulses, having a duration of about 5.4 μs, the leading edges separated by 63.5 μs
3. 15,750-Hz *horizontal blanking* pulses, having a duration of about 10 μs each
4. 60-Hz *vertical sync* pulses, having a total duration of 190 μs but slotted, or *serrated*, by 4.4-μs spacings every 27.3 μs
5. 60-Hz *vertical blanking* pulses, having a duration of 830 to 1330 μs, depending on how many lines are used in the complete picture
6. Horizontal and vertical *driving* pulses similar to the sync pulses that are transmitted to receivers but slightly out of time to properly synchronize studio cameras and equipment

Note that the *horizontal sync* pulses ride on top of the narrow *horizontal blanking* pulses, except for a few (the number is determined by how many lines

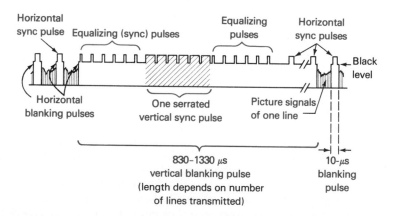

FIG. 24-9 Blanking, synchronizing, and equalizing pulses of a TV transmission.

are used in the picture) that ride on top of the wide *vertical blanking* pulse. The visual modulation is inserted between the horizontal blanking pulses.

The wide vertical blanking pulse includes one serrated vertical pulse with equalizing pulses and slotted horizontal sync pulses.

Basically, the sync generator consists of an oscillator having a frequency of four times the horizontal sweep, or 63,000 Hz. A digital divide-by-2 circuit generates narrow pulses at 31,500 Hz. When shaped properly, these are the equalizing pulses used before and after the vertical sync pulse. These equalizing pulses are also used to serrate, or slot, the vertical sync pulse into six parts. The equalizing pulses are used to produce the horizontal sync pulses at the transmitter by dropping out every other one by a divide-by-2 multivibrator circuit, giving the required monochrome frequency of 15,750 horizontal sync pulses per second.

Serrations keep the horizontal sweep oscillators in receivers in synchronism during the long-duration vertical sync pulse. The vertical sync pulse fed through a low-pass filter triggers the vertical sweep oscillator in receivers. The equalizing and horizontal sync pulses fed through a high-pass filter trigger the horizontal sweep oscillator in receivers. The equalizing pulses allow the receiver horizontal oscillator to hold sync during the periods between the end of the last line of a field and the beginning of the next field line at the top of the picture.

Just before the horizontal sync pulse is produced between lines, a blanking pulse must appear to cut off the electron beam in the camera tube in the transmitter and in the picture tube in the receivers. During the interval of the blanking pulse, the horizontal oscillators return the electron beam to the starting condition for the beginning of the next line. In receivers the sweep current is generated by a local sawtooth oscillator. The sync pulses transmitted as part of the complete TV signal are used to make any necessary correction to this oscillator frequency in order to assure the proper starting time of each line after the blanking pulse drops off. Thus, the equalizing, the horizontal, and the vertical sync pulses are added to the transmitted TV signal only as an aid to receiver oscillator operation; they are not used in picture pickup in the camera.

At the transmitter each piece of equipment has its own local sawtooth sweep oscillators. The *driving pulses* of the sync generator are used to keep these oscillators in proper synchronism. These pulses are similar to the transmitted horizontal and vertical sync pulses, except that they are timed to produce the required time difference for the transmitting equipment.

A sync generator also has a crystal oscillator and dividers to produce an accurate 60 Hz ac in case local power is not available, or when program equipment is operating in the field. With field equipment in operation, the remote sync generator may then develop the pulses for the main transmitter. The sync generator at the main station has an input circuit to allow it to lock in on signals originating from remote sync generators, from network programs, or from field equipment. Digital *frame synchronizers*, which can store two complete frames in memory, allow shifting from one program source to another without losing vertical sync when the programs are operating from different sync sources.

Transmission to the main transmitter from a studio or from a field pickup is usually made by a microwave transmitter using high-gain parabolic directional transmitting antennas to beam the 6–7 or 12–13 GHz band signals.

24-7 MODULATION PERCENTAGES OF THE VISUAL CARRIER

There is never a constant-amplitude carrier in a visual transmission. The only parts of the emission that remain constant are the peaks of the sync pulses and the blanking level (Fig. 24-10). The top

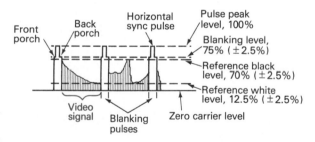

FIG. 24-10 Monochrome horizontal blanking and sync pulses with two lines of picture information.

of any sync pulse is the maximum value of the emitted carrier (the 100% level).

The *blanking*, or *pedestal*, level, on which are found the *front porch* and the *back porch*, is 75% of the peak value (±2.5%).

The *reference black* level, which will produce the darkest black in the transmitted picture, is approximately 70% of the peak level.

The *reference white* level, which will produce the whitest white in the picture, is 12.5% of the peak value. It is undesirable to transmit a whiter white than this, as the power of the emission is then so low that noise at the receiving location may interfere with picture quality. More important, there will be insufficient carrier to beat against the aural FM carrier to create the 4.5-MHz aural IF. This produces a buzzing sound in receivers (often noticeable on weak signals).

The illustration shows two complete lines of video (picture), three blanking pulses with front and back porches, and three horizontal sync pulses on top of the horizontal blanking pedestals. The first line illustrates the line writing of a scene that is black at the left-hand side and white at the right-hand side. The second line is gray at the left-hand side, changes to black at the center, abruptly changes to white, and then to a light gray. Many monitor presentations of video and pulses are shown upside-down from this explanation.

Another modulation scale has the blanking level as 0 and the reference white as 100%. Zero carrier is then 120% and sync pulse peaks are −40%.

24-8 CAMERA CHAINS

A TV camera with an electronic viewfinder mounted on top of it and a camera-control unit with the necessary power supplies make up a *camera chain*. A block diagram of two camera chains and other video equipment necessary for the pickup of a live studio program is shown in Fig. 24-11.

The camera usually has an electronic viewfinder with a 5- or 7-in. picture tube mounted directly above the camera itself and appearing to be part of the camera. The viewfinder operates much the same as a TV-receiver picture tube, except that it receives its picture information directly from the output of the video-frequency amplifiers in the camera and its blanking and driving pulses from the sync generator. The output of the camera is also fed to a nearby control position by low-impedance coaxial lines and to a camera-control unit. This unit has two cathode-ray tubes, one a 7- to 10-in. monitor tube operating like the viewfinder, and the other a 3- to 5-in. oscilloscope that gives a composite dis-

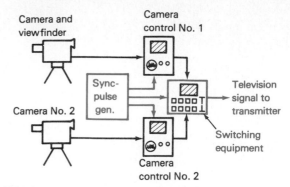

FIG. 24-11 Block diagram of two camera chains, sync-pulse generator, and monitoring equipment.

play of pulses and video continually on its screen. The operator watches the video signals shown on the oscilloscope, controlling them to prevent video modulation peaks from exceeding the reference black or white levels. Once adjusted, automatic gain control circuits can hold signals at the desired levels.

A camera chain might be composed of one or two film projectors and a slide projector, all aimed through semimirrors at a *film multiplexer* and at one camera tube feeding a camera-control unit (Fig. 24-12).

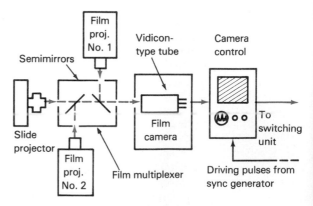

FIG. 24-12 Block diagram of a single film camera with a multiplexer to allow it to pick up pictures from three projectors.

24-9 MOTION-PICTURE PROJECTION FOR TV

The TV frame frequency was selected as 30 per second to allow transmitters and receivers to lock into 60-Hz synchronization to prevent waver of received pictures (which is not a difficulty in modern transistorized equipment). Unfortunately, both 35-

and 16-mm motion pictures have a frame frequency of only 24 per second. To reduce flicker, each film frame must be produced as two or more fields for both motion pictures and TV. A standard 16-mm projector can be used for TV projection by using either an electronically flashed xenon light or a projection lamp with a special shutter to allow the following sequence of operations with a *storage*-type pickup tube such as a vidicon.

1. The film is pulled down into position in the projector. No flashed or shuttered light shines through the film as yet.
2. During the TV vertical blanking-pulse interval, the camera tube scanning beam is blanked out and is returning to the starting point. At this time a flash of light is allowed to shine through the film projecting the picture on the mosaic.
3. The mosaic retains the electrically charged image of the picture.
4. The mosaic is scanned for one field, and the video signals obtained are transmitted.
5. At the end of the field, during the next vertical blanking pulse and scanning retrace, the projector light is again flashed, projecting the same picture on the mosaic.
6. The second (interlaced) field is scanned, and the video signals obtained are transmitted.
7. During the scanning time of the second field the next frame of the film is pulled down into position. During this frame, because of the difference between 24 and 30 frames per second, there is enough time to flash the projector light three times during blanking pulses before the next frame is pulled down.

In this way, the first film frame is scanned twice, producing two fields for the TV picture, but the second frame is scanned as three fields. This allows the correction that is necessary to change 48 fields per second for the film to 60 fields per second for the TV picture.

Accurate synchronization is required between blanking pulses and the operation of the shutter, or the flashing of the xenon light. This necessitates synchronous motors for the shutters, or a multivibrator-keyed light source synchronized by driving pulses from the sync-pulse generator.

24-10 VIDEO TAPE RECORDERS

The original recording and playback of TV programs was by kinescope recording, which consisted simply of photographing the screen of a TV monitor tube with a motion picture camera. Video disk recording can be used, but most recording of TV signals in broadcasting is by magnetic tape recording.

With any magnetic recording (Sec. 23-15), the lower the frequency, the less playback signal there will be. If the signal frequency is doubled, the output signal voltage amplitude will double (6 dB per octave). Below 30 Hz there may be insufficient output signal above the noise level. Above 15,000 Hz the head-gap width may approach the wavelength of the signal and no output will result. Also, the narrower the head gap, the less the depth of magnetization and the lower the output voltage. For 30–15,000 Hz recording, broadcast AF recorders will use 7.5 or 15 in./s tape speeds. With 7.5 in./s and a common-type 0.25-mil (mil = 0.001 in.) recording-playback head gap, 30–15,000 Hz recording is possible. To record TV signals on a *video tape recorder* (VTR) which involves frequencies of 0–4.5 MHz, it is not possible to use an amplitude-modulated recording signal as is used with AF tape recording. Instead, VTRs using 2-in. tapes have the video signals frequency modulate a carrier. The tracks on the VTR tapes are carrying FM, not AM. Such FM signals are laid down with 0.05-mil gap heads in slightly slanted tracks across the tape as the tape travels forward at 15 in./s.

One of the original VTRs, which is still in use, has 2-in.-wide tape, with four recording heads mounted on a round, *quadruplex* headwheel (Fig. 24-13a) that rotates at 240 rps. As the tape is pulled along, each rotating head lays down a single line of video information at a slight slant across the moving tape (Fig. 24-13b). Another *audio* recording head lays down a narrow sound track near one

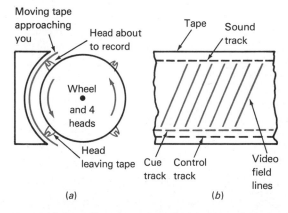

(a) *(b)*

FIG. 24-13 (a) Quadruplex record-play wheel rotating against tape. (b) Tape and placement of recorded tracks.

edge of the tape (this head is situated between the quad-head wheel and the take-up reel). There are other tracks along the edges of the tape, used for cueing the start of commercials or programs, and for control or synchronizing the ac used to drive the capstan during playback. In this way the playback heads are synchronized to trace over the 10-mil-wide recorded video tracks on the tape without slipping into the 5-mil-wide blank guard areas between recorded tracks. Just as the first head completes its magnetic recording of the first video track, the second head is approaching the starting edge of the tape and is ready to lay down the next video track as soon as the horizontal blanking pulse drops off.

The whole video signal from the TV camera, consisting of sync pulses, blanking pulses, and video modulation, is made to "narrowband" frequency-modulate a 7.9-MHz carrier with a modulating index of less than 0.4, thereby producing no more than a single pair of sidebands for any modulating frequency. (The total frequency-modulated bandwidth of pulses and video is about 3 MHz.) The FM signal is fed to the recording heads and modulates the tracks on the tape. On playback, the FM signal on the tracks is picked up by the rotating heads, is detected, and is thereby converted to the original video signals that came from the camera.

The 2-in. tape VTR may play up to 1-h program tapes from large *supply* and *take-up* reels or may play 5- to 60-s commercials from relatively small cassettes (Fig. 24-14). A highly simplified tape cassette containing two spools that hold the tape and with two hinged back covers is shown in Fig. 24-14*a*. When such a cassette, carrying previously recorded tape, is placed into a cassette VTR, the covers are automatically opened (Fig. 24-14*b*), a pair of rollers are mechanically fed inside the tape, and the tape is pushed forward and outward to a position against a capstan, as indicated by the arrows. On playback the quadruplex head is held against the tape as it is pulled by the capstan until the first cueing tone is detected by an AF-type pickup head, whereupon the capstan releases and the tape stops. When the tape-start switch is activated manually or by computer, the capstan is forced against the tape and starts moving, producing playback video and audio signals. Since the blanking and sync pulses might not be accurately reproduced by the VTR heads, these signals may be first stripped and then can be added to the video output by the sync-pulse generator.

The 2-in. tape machines are giving way to the much smaller, less expensive, standardized 1-in. tape machines in studio operations. These are *helical-scan* recorders. One of the most practical of these 1-in. machines has the tape wrapped around a rotating drum having an internal video record-play and a sync head that whirls in a counterclockwise direction in a central slot, while the tape moves with the drum in a clockwise direction at about 9.5 in./s (Fig. 24-15). Each complete rotation of the

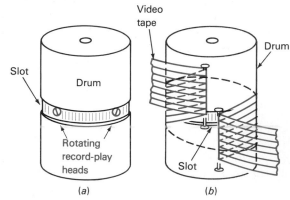

FIG. 24-15 (*a*) Helical-scan drum and placement of rotating heads. (*b*) Tape on drum.

record head lays down a single field line at about a 2° slanted angle along the tape. If the tape is stopped but the head continues to rotate, a still picture would result. If each frame is converted from analog to digital (A/D) and is stored and played twice instead of once, a slow-motion output results. There are four linear tracks along the edges of these tapes. Two tracks can be used for AF (stereo when authorized), the third for cueing or AF, and the fourth for control and synchronizing.

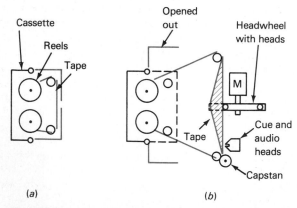

FIG. 24-14 (*a*) A 2-in. tape cassette (*b*) in VTR, with tape pulled into record or playback position.

Not mentioned, but necessary, are erase heads, which precede the rotary record heads and are used when in the "record" condition.

For portable operation, such as electronic news gathering (ENG), small battery-operated cameras containing built-in sync generators are used. They are helical-scan systems with ¾-in. or sometimes ½-in. tapes. Some stations are using ¾-in. tape machines for studio-type recording.

24-11 MONOCHROME TRANSMITTERS

The visual transmitter is required to pass the carrier plus about 4.2 MHz of upper sidebands and about 0.75 MHz of lower sidebands. This is a total bandwidth of about 4.95 MHz. Past these limits the sidebands must be attenuated, at the high-frequency end to prevent visual sidebands from occurring at the aural carrier frequency and at the low-frequency end to prevent sidebands from appearing in the adjacent channel.

A common means of producing the desired transmitted band configuration is to grid-modulate the final amplifier and add a vestigial sideband filter between the output circuit and the antenna diplexer (Fig. 24-1). This filter consists of tuned cavity wave traps with added *LC* circuits to present a constant impedance to the amplifier for all frequencies but still attenuate the sideband signals that must not be fed to the antenna.

Another way to develop the desired bandpass is to modulate a low-level RF amplifier with the video signal. This amplifier must have a flat tuning characteristic over at least 8.4 MHz to produce acceptably the RF carrier and video sidebands 4.2 MHz on each side of the carrier. The linear amplifiers that follow the modulated stage can be overcoupled, stagger-tuned, and wave-trapped to produce the required 4.95-MHz vestigial sideband transmission.

To tune a TV transmitter properly, a *sideband analyzer* is used. This consists of a narrowband receiver that is swept (varactor or reactance-device tuned) across the 6-MHz channel of the transmitter while a complex signal that consists of the carrier and all the video sidebands is being transmitted. The received signal amplitude is displayed on an oscilloscope CRT. The oscilloscope has a horizontal sweep in step with the receiver tuning sweep. The output of the receiver is fed to the vertical-deflection plates of the oscilloscope. The amplitude of the received signal as the receiver sweeps across the channel is displayed on the scope tube. Figure 24-16 illustrates two sideband-analyzer displays,

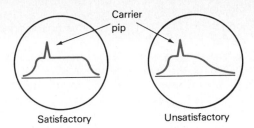

FIG. 24-16 Satisfactory and unsatisfactory presentations on a sideband analyzer of a TV test transmission.

one of a properly tuned amplifier producing both upper and lower sidebands properly and the other of an improperly tuned amplifier in which the upper sidebands are attenuated excessively. This is a form of *spectrum analyzer* (Sec. 13-38).

24-12 MONOCHROME TRANSMISSION REQUIREMENTS

A TV station has one antenna (a turnstile or other horizontally polarized array) common to its two transmitters (FM aural, AM video). These two transmitters always operate with the center frequency of the aural transmitter 4.5 MHz higher than the carrier frequency of the video transmitter (±1 kHz). The whole TV transmission is contained in a 6-MHz-wide channel (Fig. 24-17).

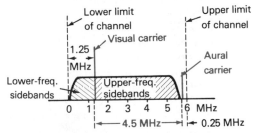

FIG. 24-17 Location of carriers and sidebands in a 6-MHz monochrome TV channel.

The visual carrier is 1.25 MHz above the lower channel limit (±1 kHz). The higher-frequency visual sidebands extend out more than 4 MHz before they are attenuated as the sound carrier frequency is approached. The lower-frequency sidebands exist for only 0.75 MHz before they are attenuated. Since only a vestige of the lower-frequency sidebands is transmitted, the emission is termed *vestigial sideband*. This emission (C3F) requires considerably less bandwidth than a true double sideband TV emission would.

The operating power of the *aural* transmitter is

determined by either the direct or the indirect method (Sec. 17-12) with no modulation applied to the carrier.

The operating power of the *visual* transmitter is determined at the output of the vestigial sideband filter if one is used; otherwise, at the transmitter output terminal. The average power is measured while operating the transmitter into a nonreactive dummy load, having a resistance equal to the transmission-line surge impedance, and while transmitting a black picture. The average power is then determined by direct measurements. When the antenna is coupled to the transmitter, the final-amplifier V_p and I_p values must be the same as with the dummy antenna. Television transmitters are rated in *peak power output*, which is determined by multiplying the average power above by the factor 1.68.

The effective radiated power (erp) of a TV transmitter is the peak power times the antenna field gain (voltage gain over a simple dipole) squared. If a transmitter has a peak power of 10 kW and the antenna has a field gain of 3, the erp is 10,000 × 3², or 90 kW.

Since the transmitted sync pulse peaks always remain at the same amplitude regardless of percent of visual modulation, it is the peaks that are monitored to determine whether the transmitter output is within licensed limits. The peak power must be held within 10% above and 20% below the authorized power, except in emergencies. The *aural* FM transmitter carrier power must be between 10 and 20% of the peak power of the visual transmitter. This produces nearly equally effective visual and aural signals in receivers.

Sometimes, to reduce cochannel interference, one of the TV stations on the channel will have its frequency offset. If the offset is +10 kHz, both aural and video carriers will be 10 kHz higher than the normal assignment would have been.

Frequency allocations for commercial TV bands and channels are shown in Table 24-1. Each channel is 6 MHz wide. The lowest frequency of all channels is given. Only every tenth UHF channel is listed, since these channels are all contiguous. (Note that there is no channel 1.)

▌Test your understanding; answer these checkup questions.

1. What are the three frequencies produced by a B/W sync-pulse generator? _____ _____ _____
2. What are the names of the six signals from a sync-pulse generator? _____ _____ _____ _____ _____ _____
3. Which pulse(s) is (are): Serrated? _____ The longest? _____ The shortest? _____ Of the highest amplitude? _____
4. Which pulses hold the CRT beam cut off? _____ Which impede oscillator starting?
5. What is the waveform of sweep voltages? _____
6. If the pulse peak is 100%, what is the blanking level? _____ Black level? _____ White? _____
7. List the units involved in a camera chain. _____
8. What device allows one camera to accept pictures from three projectors? _____
9. What is the order of scanning four successive motion-picture frames for TV? _____
10. What three methods have been used to record TV pictures? _____ _____ _____
11. What VTR tape widths are used? _____ _____ _____ _____

TABLE 24-1	TV BAND AND CHANNEL FREQUENCY ALLOCATIONS				
Low-VHF band		High-VHF band		UHF band	
Channel designation	Lowest frequency, MHz	Channel designation	Lowest frequency, MHz	Channel designation	Lowest frequency, MHz
2	54	7	174	14	470
3	60	8	180	24	530
4	66	9	186	34	590
	(4 MHz skipped)	10	192	44	650
5	76	11	198	54	710
6	82	12	204	64	770
		13	210	74	830
				83 (highest)	884

12. Why does video recording use FM on the tape? _____
13. Where is audio recorded on video tapes? _____
14. Where would audio erase heads be located? _____
15. With what types of video recording is still or slow motion possible? _____ _____
16. When must vestigial SB filters be used? _____ When not? _____
17. What is used to tune the bandpass of a TV transmitter? _____
18. From the low end of a channel, what is the frequency of the visual carrier? _____ Aural? _____
19. What method of power output measurement is used for the aural transmitter? _____ The visual? _____
20. A 10-kW transmitter feeds an antenna with a field gain of 3.5. What is the erp? _____
21. How much more is the peak power of a TV transmitter than the average power? _____

24-13 LOGS, PERSONNEL, AND SWITCHERS

As with other broadcast stations, a TV station must keep program, operating, and maintenance logs (Sec. 23-20). Daily program logs may be in written, typed, or computer printout forms. They will list station on and off times, program beginning and ending times, commercial times in and out, whether network or local program, and other FCC categories. Computerized logs can also be switched to a TV monitor for viewing at any desired time. Hard copy logs must be kept for 2 years.

Many people are involved in the production of even a simple live TV studio program. On the studio floor there will be a boom-microphone operator (if lapel-type microphones are not used) who keeps a microphone above and in front of the performers, but out of sight of the camera. There will be one or more camera operators and a floor director. In the control room, located either remotely or within sight of the performers, there will be a video console operator who keeps camera signals technically correct and who switches cameras under the direction of a program director. In charge of each program is its producer, who coordinates all activities during that program. There is a one- or two-way intercommunication earphone-microphone or loudspeaker, as warranted, between producer, director, and studio personnel. An audio operator, behind or to the side of the video control position, maintains microphone and other audio signals at proper levels and switches microphones. Many of the gross manual controls of earlier days are now accomplished by automatic gain systems, both video and aural. After switching by operators, only slight touch-up of camera or microphone gains may be necessary. In the motion picture projection room other operators will load and operate the projectors. Many other people are at work in other rooms or studios developing or taping programs for later use. There are also a number of administration, art, sales, remote crew, and other personnel.

Video-audio switchers have been developed that can automatically do such chores as rewind films or tapes, start films or tapes on electronic cue, and even store a complete day of program operations in its internal microprocessor system and memory banks. Automatic switchers are useful during the hour and half-hour periods when many changes of program, commercials, and station breaks are made in a relatively short period of time, which is somewhat difficult to handle manually.

Although maintenance of equipment and its proper logging requires technically trained and skilled people, at this writing there is no longer the requirement that anyone in a TV station hold any type of license. Of course, the station itself must be licensed by the FCC and must operate according to FCC rules and regulations.

24-14 THE TV RECEIVER FRONT END

The *front end* of a superheterodyne receiver includes the antenna input, RF amplifier, local oscillator, and the mixer stage, which must all switch-tune to 81 TV channels.

VHF TV receiving antennas may consist of two Yagi beams, one to pick up low-band and the other for high-band stations. Broadbanding techniques, such as the use of large-diameter antenna elements or the use of a folded-dipole main element, may be employed. V-beam arrays are in use also. In the UHF band, broadband bow-tie, or conical antennas with sheet reflectors, are commonly used because of their ability to accept all frequencies above their fundamental with almost equal response. In weak-signal areas it may be necessary to have a special beam antenna for each frequency to be received. TV receivers may have a 300-Ω antenna input-impedance circuit. This allows use of 300-Ω twin lead as the transmission line, which is relatively inexpensive and simple to install. They may also have a 75-Ω commercial "cable" input.

The RF amplifier in a TV receiver must be capable of producing amplification and have low inherent noise. Any noise generated in this stage develops white dots, termed *snow*, in the picture. Pentodes or BJTs may be used as the RF amplifier, but FETs and triodes have a lower noise figure. A low-noise RF amplifier circuit is the *cascode* (Fig. 24-18a). It has two JFETs in series, the first stage

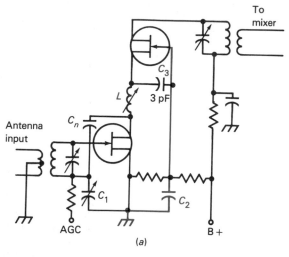

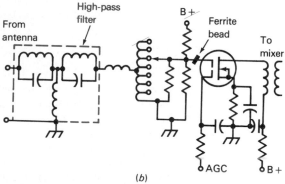

FIG. 24-18 TV receiver RF amplifiers: (a) cascode and (b) MOSFET.

directly coupled to a grounded-gate second stage. C_n is a gate-neutralizing capacitor for the first JFET. C_1 acts as a means of controlling neutralization and bypassing the gate coil to ground.

The amplified signal is fed directly to the source of the second JFET through the small inductance L. The gate of this JFET is bypassed to ground with C_2 and is biased with the voltage divider. Being a grounded-gate amplifier, the second JFET requires no neutralization.

The inductance L is in series with C_3, and with C_2 is made series-resonant to the middle of the upper VHF TV band. Signals near the resonant frequency produce relatively high voltage across C_3 (source-to-gate). This provides added gain at these frequencies, which are normally subject to losses. An RF amplifier also acts as an isolation stage to prevent radiation of local-oscillator signals.

The diagram shows coil and variable capacitor tuning. In TV receivers each channel may have its own separate slug-tuned coil, utilizing distributed capacitance to complete the tuned circuit. When the channel selector in the receiver is changed from one channel to the next, a separate set of coils is connected into the RF amplifier-grid circuit, into the mixer-grid circuit, and into the oscillator circuit. All coils may be separately trimmed to give optimum signal for its channel.

Another method of tuning is the switching of bias voltages across varactor diodes in parallel with inductors to shift the resonant frequency of VHF or UHF LC circuits. Trimming to frequency is accomplished by an AFC circuit.

Figure 24-18b is a MOSFET RF amplifier. After passing through a high-pass filter, the antenna signal is fed to a switch-selected coil for the desired channel and then to one of the MOSFET gates. The ferrite bead on the gate lead stabilizes the circuit. The other gate is fed an AVC (called AGC, for automatic gain control, in TV). Note that B+ may be used instead of V_{CC} or V_{DD} in TV receiver diagrams. The output signal is fed to the mixer.

The mixer stage of the VHF portion of a TV receiver is similar to the mixer stage of any superheterodyne. The local oscillator is usually a separate circuit coupled to the mixer. It is possible to incorporate automatic frequency control (AFC) by adding a 45.75-MHz IF amplifier coupled to a reactance-device circuit or varactor diode across the local oscillator LC circuit (Sec. 19-13) and feeding it the voltage developed at the discriminator. Such a circuit makes a fine-tuning, or trimmer, adjustment unnecessary. BJTs, FETs, VTs and diodes may be used as mixers.

If interference is developed by local HF emissions, a 54-MHz high-pass filter between antenna and antenna input may eliminate it.

A UHF front end usually has no RF amplifier, but may have tuned tank circuits fed by the antenna, a BJT or VT oscillator, and a diode mixer. Figure 24-19 shows a hairpin antenna loop induc-

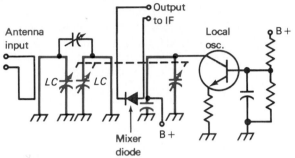

FIG. 24-19 UHF mixer circuit.

tively coupled to a tuned hairpin tank capacitively coupled to a second tuned tank to improve selectivity. The second tank is coupled to a pickup hairpin with a diode mixer in series with it. The BJT oscillator is also coupled to the diode hairpin. The three tuned circuits are ganged. Biased varactors are used instead of the variable tuning capacitors shown here for simplicity.

24-15 THE IF STRIP

The output of the mixer is an intermediate frequency, or IF. TV receivers use a 6-MHz-wide IF, from 41 to 47 MHz. In this band of frequencies the sound carrier is made to fall on 41.25 MHz and the video carrier on 45.75 MHz. The transmitted signal and the desirable bandpass characteristics of a TV-receiver IF section are shown in Fig. 24-20.

Modern TV receivers amplify both the video carrier and sidebands as well as the sound carrier and sidebands with the same IF amplifiers. This is known as an *intercarrier* receiver. In older, so-called *conventional* TV receivers, the sound IF was taken from the mixer stage separately. Slight variations of the local-oscillator frequency due to drift, detuning, etc., can detune the sound noticeably. This does not occur with the intercarrier receiver, because the 4.5-MHz IF is produced by the beating of the video and aural carriers, which are always transmitted 4.5 MHz apart. Synthesized tuning

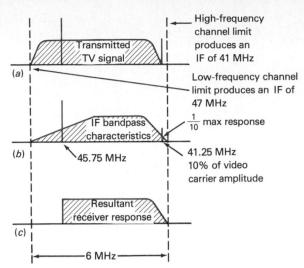

FIG. 24-20 (a) Transmitted carriers and sidebands of a TV signal. (b) Idealized IF response curve in a receiver. (c) Idealized resultant IF receiver response.

would make the conventional system possible again.

The transmitter produces a vestigial lower-sideband emission (Fig. 24-20a). To utilize the lower-frequency sideband, it is desirable to slope the receiver bandpass characteristics, as in Fig. 24-20b. In this way the addition of the upper and lower sidebands near the carrier, which carry the lower video modulating frequencies, produces a sum equal to the maximum of the higher-modulating-frequency sidebands. The result is an idealized response to the transmitted signal, as in Fig. 24-20c. The output from the IF amplifiers contains all the sidebands starting at the carrier out to more than 4 MHz with equal amplitude. Signals displaced farther than this are attenuated.

The slope of the bandpass curve must drop off in such a way that only 10% of the maximum amplification is given to the sound carrier. This prevents the sound signal from forcing its way through to the picture tube and producing dark and light bars across the screen. It also makes the sound-carrier amplitude similar to the whitest white-signal amplitude, allowing the sound and video carriers to beat efficiently to produce a 4.5-MHz sound-signal IF.

Different types of coupling have been used in TV IF stages to produce the required bandpass characteristics, impedance-coupling, single-tuned transformers, double-tuned transformers, autotransformers, and the *bifilar* transformer. A bifilar

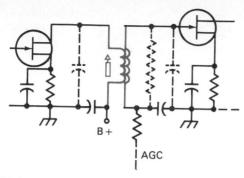

FIG. 24-21 An interstage bifilar transformer circuit.

transformer in Fig. 24-21 has the primary and secondary turns interwound in a form of unity coupling. The lumped input- and output-distributed capacitances (dashed) act as the capacitance of the tuned circuit. A single powdered-iron-core slug broadly tunes the bifilar-transformer circuits. The resistance shown in dashed lines is sometimes used to lower the Q, broaden the tuning, and decrease the possibility of oscillation in the stages. Bifilar-transformer coupling, besides being economical to manufacture, has the advantage of reducing noise impulses on the picture-tube screen that result when capacitive-coupling circuits are employed. A high-amplitude noise pulse charges the coupling capacitor of an impedance-coupled circuit. The discharge of the capacitor through high-resistance coupling resistors represents a relatively long-duration signal that produces white tails on any black noise spots on the screen, making the noise much more noticeable.

The basic IF strip in a TV receiver has two to four amplifier stages, with AGC applied to them. To produce the desired bandpass characteristics the stages are stagger-tuned. The first transformer may be peaked at 43.1 MHz, the second at 45.3 MHz, and the third at 43.1 MHz. This arrangement should result in a bandpass characteristic similar to the curve in Fig. 24-20b. An IC IF strip may use only two external tuned IF transformers.

There are several frequencies other than the desired TV channel signals that may be able to force their way through the front end and into the IF strip to cause visual or audible interference. To prevent this, wave traps are used. To restrict overpowering of the front end by strong local radio signals, a high-pass filter is incorporated in the antenna input circuit (Fig. 24-18b), designed to attenuate all frequencies lower than 54 MHz but pass all higher frequencies without loss.

Besides the internal wave traps in TV IF stages, sometimes wave traps tuned to interfering signal frequencies have to be added across the antenna terminals to reduce lines and flashes on the TV screen or sounds from the loudspeaker. For example, local transmissions from stations in the Public Safety Radio Services operating in the 41–47 MHz or the 150-MHz region may leak through into the IF amplifiers. The antenna wave trap will have to be tuned to the interfering station's frequency. If front-end overloading is present, as from a nearby CB transmitter, the trap will have to be tuned to approximately 27 MHz (Sec. 17-39). If from amateur stations, the trap will have to be tuned to the amateur band in which the station is operating (Sec. 32-5).

Because of the broadness of the IF strip, one source of interference may be the sound carrier of a station operating on the adjacent lower channel. Another source of interference may be the visual carrier of a station operating on the adjacent higher channel. Either series-resonant or parallel-resonant traps tuned to the IF frequencies produced by such interfering signals, namely, 47.25 and 39.75 MHz, may be included in the IF amplifiers. Such traps not only attenuate the undesired response, but also tend to narrow the skirts of the IF bandpass out past the channel limits.

When operating from higher-voltage power supplies, transistor stages may be connected in series across the supply (Fig. 24-22). The input signal feeds a bifilar transformer with two wave traps (47.25 and 39.75 MHz) between it and ground. Note the collector current path through the two BJTs in series and also the series biasing network. If a positive AGC voltage is applied across the first emitter resistor, I_c in both BJTs will decrease, lowering the gain of the stages. The 47-Ω resistors help to prevent oscillation of the relatively low-Q stages. Ferrite beads might be used instead.

Although shown here as discrete BJT, or FET stages, modern TV receivers may have several ICs. One IC may be used for the whole IF section except for the transformers, another for the whole aural section, another for the video amplification and driver, with others for sweep generation and amplification.

24-16 VIDEO CIRCUITS

Signals from the last IF stage are fed to a diode video detector, which rectifies them to a varying dc having a waveform that includes the sync and

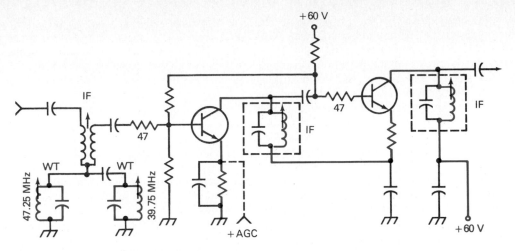

FIG. 24-22 Series-connected two-stage TV IF amplifier with two wave traps.

blanking pulses as well as the video information (Fig. 24-23). Since both the video carrier (45.75 MHz) and the aural carrier (41.25 MHz) are mixed in the diode, their difference frequency, 4.5 MHz, is also present. This 4.5-MHz beat frequency is modulated by the FM produced at the aural transmitter and is also amplitude-modulated by the visual-carrier sidebands, both of which would be amplified by the video amplifier that follows the detector. To prevent the aural signals from producing black sound bars on the picture tube, a 4.5-MHz wave trap can be included in the emitter-follower coupling stage to the video amplifier. If the wave trap is wound with a secondary coupled to

it, 4.5-MHz aural carrier signals for the aural IF stages can be extracted at this point.

Another method of obtaining the 4.5-MHz aural IF is by capacitively coupling from the output of the last IF stage to a separate diode mixer, the 4.5-MHz output of which is fed to the aural IF circuits.

The aural circuits consist of a 4.5-MHz IF amplifier feeding a 4.5-MHz ratio detector, discriminator, or a gated beam FM detector (Sec. 19-8). With either solid-state or VT ratio detectors or discriminators, one or two low-level AF amplifiers are required to drive the AF output amplifier. In many TV receivers all the aural circuitry including the AF power stage may be contained in a single IC.

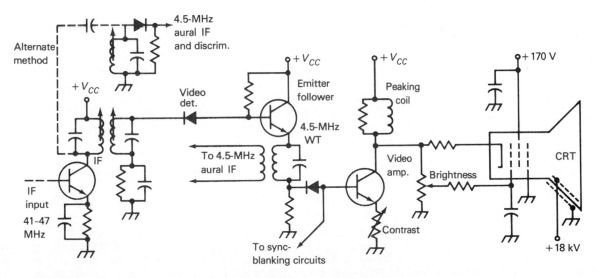

FIG. 24-23 Video detector, amplifiers, and picture tube for monochrome TV.

The video-blanking-sync signals from the second detector diode can be fed to a direct-coupled, wideband (4.2-MHz) video amplifier and then directly to the grid-cathode circuit of the picture tube. These voltages control the intensity of the beam striking the face of the cathode-ray tube. To extend the high-frequency capabilities of the video amplifier past about 2 MHz, a special resonant peaking coil is added in the collector circuit. It forms a parallel-resonant circuit with its distributed capacitance at a frequency of about 4 MHz. This results in a nearly flat amplifier response out to 4.2 MHz.

If the video amplifier is not directly coupled to the picture tube, it is necessary to add a bias that remains just at the blanking level but which will follow any variation of the amplitude of the peak of the sync pulses due to fading. In this way variations of signal voltages of light and dark scenes will not affect the bias to which the video signals are added. Thus, light scenes remain light and dark scenes remain dark. Such a biasing circuit is known as *dc reinsertion*, and may be applied to signals being fed to the picture tube grid.

The amount of CRT cathode-grid bias controls the *brightness* of the picture displayed. The video amplifier signal gain circuit, shown simplified, controls the *contrast* between light and dark signals on the screen. The grids past the control grid on the CRT are the accelerator and focusing electrodes. The Aquadag coating on the inside of the CRT near the screen, sometimes called the *ultor*, is supplied a positive potential of 9 to 18 kV greater than the cathode in monochrome tubes and 22 to 30 kV with color tubes.

Test your understanding; answer these checkup questions.

1. What are the three types of logs kept in TV stations? _____ _____ _____
2. Who is in overall control of a TV production? _____
3. Why is an automatic switcher useful on the hour? _____
4. What are the three TV bands called? _____ _____ _____
5. What circuits make up the front end of a TV set? _____ _____ _____
6. What are the impedances of the input terminals of a TV set? _____ _____
7. Why are MOSFETs best for TV RF amplifiers? _____ Name a popular TV RF amplifier circuit configuration. _____

8. What type of *LC* tanks might be used in UHF circuits? _____
9. On what frequency will the mixer oscillator be operating when receiving channel 2? _____
10. Why might TV IF stages be stagger-tuned? _____
11. What are the two wave-trap frequencies used in a 42-MHz TV IF strip? _____ _____ What wave trap is used in the video amplifier? _____
12. Why may transistor IF stages not use bifilar transformers? _____
13. From what two points might aural IF information be taken in a TV receiver? _____
14. What four signals may a video amplifier amplify? _____ _____ _____ _____
15. What does a peaking coil do in a video amplifier? _____
16. What controls brightness of a CRT picture? _____ Contrast? _____
17. When is dc reinsertion required? _____
18. Why might conventional TV systems be usable today? _____
19. What is used today instead of the discrete component circuits described? _____

24-17 SYNC-PULSE SEPARATION

Thus far, the TV signal has been transmitted, received, converted to an IF, amplified, and detected. The aural signal has been separated, amplified, detected, and fed to a loudspeaker. The video signal, consisting of blanking and sync pulses plus the modulation of each line of picture information, has been amplified and fed to the cathode-grid circuit of the picture tube. The video intensity-modulated CRT electron beam determines how many electrons strike the fluorescent inner coating on the CRT face during horizontal beam traces and thus the amount of light developed on the screen.

The narrow sync pulses on top of the blanking pulses, are also fed to the CRT cathode-grid circuit, but since they represent a "blacker than black" emf they produce no light on the screen.

The video signal is also fed to an active device that biases itself to pass only the sync pulses having amplitudes greater than the blanking-voltage level. Such a circuit, known as a *sync separator*, may employ diodes, transistors, or VTs. In Fig. 24-24, with no signal, the 10-MΩ resistor holds the bias on the FET gate to a fraction of a volt positive, resulting in a relatively high I_D. Incoming video signals charge C_1 enough to produce an average

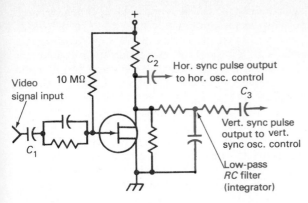

FIG. 24-24 Sync-separator circuit using an FET.

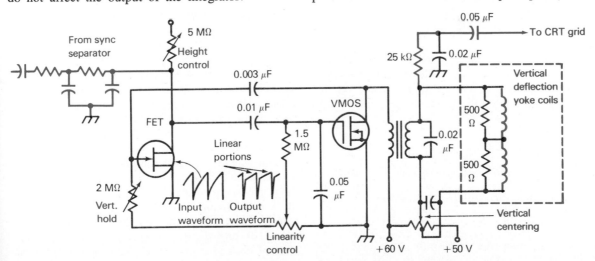

24-18 VERTICAL DEFLECTION

To develop a TV picture on the screen it is necessary to drive the electron beam horizontally across the face of the tube 15,750 times a second and, at the same time, move it down the screen relatively slowly and then back up rapidly 60 times a second. The sawtooth ac that produces the downward (vertical) deflection and the flyback is the simpler of the two deflection systems and will be described first.

A two-device multivibrator-type circuit develops 60-Hz sawtooth ac (Fig. 24-25). Two BJTs or VTs might also be used. This particular circuit consists of an FET coupled to a VMOS power transistor with capacitive feedback to the FET gate. The selection of proper R and C values results in a relatively slow, linear voltage buildup on the VMOS gate, even though the voltage waveform at the gate of the FET may be nonlinear, as indicated. The 0.003-μF capacitor feeds back (in phase) some of the amplified signal, maintaining oscillation of the two-device circuit. The gate-to-gate dc circuit feedback via the three resistors provides a method of controlling linearity of the linear portion of the waveform applied to the VMOS gate. Controlling the drain-load resistance of the FET determines the amplitude of the sawtooth sweep voltage, thereby furnishing a height control for the *raster* (unmodulated lines on the screen).

The 2-MΩ *vertical-hold* rheostat and the 0.003-μF capacitor determine the frequency of oscillation and enable the oscillator to be synchronized by the received vertical sync pulses to produce the correct vertical sweep frequency.

class C bias (zero I_D) for the FET. The only parts of the video signal that can appear above I_D cutoff and produce I_D are the peaks of the horizontal sync pulses, equalizing pulses, and serrated vertical sync pulses. These pulses, which will synchronize the oscillator frequency of the horizontal oscillator at 15,750 Hz, are taken off through a small series capacitor C_2. A high-frequency pulse-passing circuit such as this is known as a *differentiator* circuit.

To synchronize the vertical oscillator at 60 Hz, the sync-separator amplifier output is also passed through a low-pass RC filter, called an *integrator* circuit. The only pulses that can pass through the integrator and C_3 are the low-frequency (60 pulses/s) vertical sync pulses. The narrow serrations on the vertical sync pulses represent high frequencies and do not affect the output of the integrator.

FIG. 24-25 Unbalanced power-multivibrator circuit used to produce sawtooth sweep voltages for vertical deflection.

For vertical centering of the picture on the screen, a dc bias can be introduced into the yoke coils by using a center-tapped potentiometer as shown. Centering can also be accomplished mechanically by positioning of the yoke coils.

If the phase of the synchronizing pulses fed to the oscillator through the *RC* integrator circuit is incorrect, a single amplifier stage may be added to invert the phase.

Note that each coil of the vertical yoke has a 500-Ω damping resistor shunted across it to decrease inductive effects during the rapid-retrace or *flyback* periods. During the flyback time a voltage from the vertical yoke is fed to the CRT grid (or cathode) to blank the screen, preventing any retrace line from the bottom of the screen to the top from showing. Although shown in series, vertical yoke coils may be connected in parallel.

Other semiconductor vertical-sweep circuits may use a multivibrator or a blocking-type oscillator that develops narrow pulses at the sweep frequency. The pulses are amplified by a driver stage, and the required sawtooth waveform is developed by an *RC* circuit at the input of the output amplifier stage. An IC may be used for all this circuitry except for the larger capacitors and the vertical-linearity, hold, and height-control potentiometers.

24-19 HORIZONTAL DEFLECTION

The horizontal-deflection section of a TV receiver has many functions. It generates a 15,750-Hz sawtooth current that sweeps the beam across the screen in synchronism with the sweep of the beam in the camera tube at the transmitter. The same ac that accomplishes the horizontal sweep is also stepped up and rectified to produce the required high Aquadag voltages. A small portion of the horizontal flyback voltage is used to key automatic gain control (AGC) or automatic frequency control (AFC) circuits into operation. Some of the output

voltage may be shaped and added to the vertical yoke to decrease *pincushion* distortion of the raster (pushing out at the center of the top, bottom, and right and left sides of the picture), or it may be shaped and used for convergence correction in color TV receivers. A basic horizontal-deflection circuit is shown in Fig. 24-26. The 15,750-Hz multivibrator oscillator is in the center of the system.

The multivibrator could be synchronized by feeding a positive or negative pulse (from the sync separator) to one of its gate circuits. However, any received static or noise pulses could feed through and upset the synchronization and result in tearing out of portions of the picture. A better system is the one shown, using an FET phase inverter feeding a two-diode phase detector that forms an AFC circuit for the multivibrator. The AFC circuit feeds either a negative or positive biasing voltage to the gate of one of the oscillator transistors. The additional bias forces the oscillator to change its frequency until it agrees with that of the pulses coming from the sync separator.

The 15,750-Hz multivibrator ac appears across the 0.0005-μF capacitor. The basic frequency of oscillation can be varied by adjusting the gate resistor in the second FET, the *horizontal-hold* control. The parallel *LC* tank is resonant to 15,750 Hz and stabilizes oscillations. The frequency of oscillation can be varied slightly by variation of any gate or drain component of this source-coupled multivibrator, however. The sawtooth ac output is fed to a class C biased bipolar power transistor (or SCR) amplifier, which is transformer-coupled to the horizontal-deflection coils. The buildup of current in these coils deflects the electron beam across the face of the picture tube 15,750 times per second.

A flyback pulse of this sawtooth wave is fed back to the AFC circuit to be compared with horizontal sync pulses from the sync-separator via a phase-inverter circuit. Each sync pulse is fed by the phase inverter to one diode as a positive voltage and at the same time as an equal-value negative voltage to the other diode. Thus, both diodes try to charge the 0.01-μF capacitor. Since the charging currents are 180° out of phase and assumably equal, the capacitor receives no charge. The center tap between the two 100-kΩ diode resistors would also be a zero potential insofar as the pulses are concerned. However, if the deflection circuit pulses are ahead of or behind the received sync pulses, the sum of deflection plus sync pulses will result in one diode drawing more current than the other, and the 0.01-μF capacitor will charge either positively or

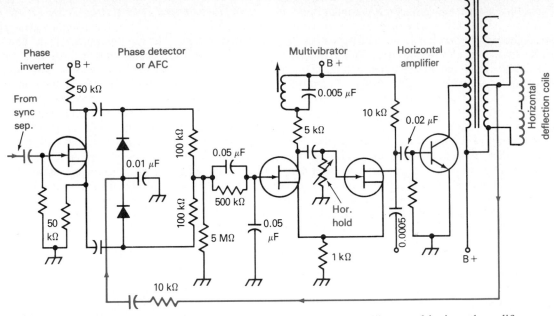

FIG. 24-26 Horizontal-deflection system with AFC, sawtooth multivibrator oscillator, and horizontal amplifier feeding the deflection coils.

negatively. This will result in the center tap of the 100-kΩ resistors having either a positive or a negative potential. Changing the bias of the multivibrator gate will change the time of conduction of one gate of the oscillator circuit and lock the frequency of the oscillation to that of the incoming pulses. The *RC* network between the AFC and multivibrator circuits has a long time constant to prevent rapid changes in oscillation frequency. (Note that this is a phase-locked loop, or PLL circuit.)

24-20 HORIZONTAL OUTPUT CIRCUIT

The horizontal sweep amplifier feeds deflection current to the horizontal-deflection coils, produces the high voltage for the picture tube, and also develops a boosted B+ voltage for the horizontal amplifier and for the first anode of the picture tube (Fig. 24-27).

As the sawtooth ac builds up slowly on the base of the horizontal amplifier, the collector current builds up through the transformer primary and through the damper diode. This linear current change produces a constantly increasing deflection-coil current and magnetic-field strength, resulting in the electron beam's being deflected relatively slowly but evenly across the picture-tube screen. Coil cur-

rent flows in a downward direction during the buildup of the magnetic field.

When the sawtooth ac on the base drops suddenly to a high negative value, I_c in the amplifier drops to zero. The magnetic field around the deflection coils collapses rapidly inducing a high-amplitude *flyback* emf across the transformer secondary.

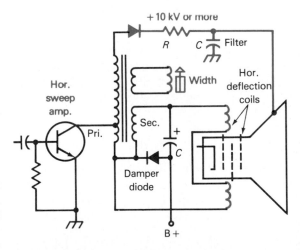

FIG. 24-27 Horizontal amplifier coupling to horizontal deflection coils showing HV supply and width control.

This produces a sharp pulse of high-amplitude flyback current flow upward in the secondary of the transformer and a high induced flyback emf downward in the primary winding, making the top of the primary thousands of volts positive. This is rectified by the high-voltage rectifier, filtered with an RC network, and connected to the CRT Aquadag coating.

During the flyback period, the energy stored in the magnetic field around the deflection coils would tend to produce damped oscillations in the LC circuit consisting of these coils and any distributed capacitance that existed. This would interfere with production of the next sawtooth wave. However, when the collapsing magnetic field induces a current flow downward in the coils, the damper diode passes current, charging the capacitor C. The energy that would have produced oscillations is now contained in the capacitor. The polarity of this charge is such that it is in series with the power-supply voltage. This also results in a source of "boosted" voltage for the horizontal amplifier collector circuit of about twice the power-supply voltage.

The width of the picture can be controlled by shunting a small variable inductance across a few turns of the output transformer, as shown.

The high voltage developed by horizontal flyback and connected to the CRT usually exceeds 10,000 V. When a TV set is serviced, be sure to discharge this circuit to chassis ground after the TV receiver has been turned off before attempting to make any internal adjustments to the set. Care must also be taken not to strike the picture tube, since it can implode violently and throw broken glass and parts many feet.

24-21 AUTOMATIC GAIN CONTROL

AGC in a TV receiver is somewhat similar to AVC in AM receivers. In AM receivers the carrier is rectified, filtered with a long-time-constant RC filter, and fed as a reverse bias to the input circuits of the RF and IF amplifier stages to produce nearly constant average signal at the second detector regardless of the strength of the signal received. AVC is possible in an AM receiver because the *average* voltage of the carrier is constant.

An AVC circuit operating in a similar manner from the second detector of a TV receiver would develop a low bias voltage for light scenes and a high bias voltage for dark scenes, which would not produce the desired pictures. However, if the AGC voltage is developed from the amplitude of only the sync pulses, which do not change when picture amplitudes change, a satisfactory bias voltage will result. The first AGC system employed a sync-separator circuit and a low-pass RC filter. The horizontal and vertical pulses were filtered to a smooth dc and used as the AGC voltage. However, airplanes flying overhead reflected rapidly changing in-phase and out-of-phase signals to the receiver and produced wildly changing AGC amplitudes that sometimes threw the receiver completely out of synchronization. This was overcome by the use of *keyed AGC*.

A basic keyed AGC system is shown in Fig. 24-28. Two sets of pulses are fed to the AGC tube.

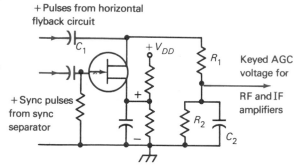

FIG. 24-28 A method of producing keyed AGC voltage.

Positive pulses developed by flyback in the horizontal output transformer are fed to the FET drain. These pulses last for only a few microseconds. Positive horizontal sync pulses from the sync-separator stage are fed to the negatively biased gate. If both gate and drain are keyed positively at the same instant, the FET conducts and electrons flow from source to drain, charging C_1. When the pulses drop off, the electrons on charged C_1 flow down through resistors R_1 and R_2, charging capacitor C_2 negative at the top. This negative voltage is fed to the RF and IF circuits as the AGC voltage. If noise pulses are fed to the gate through the sync separator, they cannot produce I_D unless the drain is also positive. Thus, the keyed AGC discriminates against noise impulses that occur between sync pulses (during the visual modulation). The RC time constant can be relatively short, allowing the AGC voltage to follow reasonably rapid fading, such as that produced by airplanes, making this fading only slightly noticeable. In some cases a dc amplifier is used to increase the AGC effectiveness.

It should be noted that this and all other circuit diagrams shown may not be exactly the same as those found in actual TV receivers. For each simplified basic circuit described here, there are dozens of variations in practical use. For more detailed explanations of TV circuits, refer to TV-receiver texts.

▌Test your understanding; answer these checkup questions.

1. When coupling from a sync separator, what is used for the horizontal sync pulses? _____ Vertical? _____
2. About how many lines would make up a raster? _____ A TV picture? _____
3. What does a vertical-hold control do? _____
4. What is controlled by varying a bias current in yoke coils? _____
5. What is the purpose of a voltage fed from the vertical sweep to the CRT grid? _____ When must it occur? _____
6. What are two ways of obtaining sawtooth ac used in TV sweep circuits? _____ _____
7. Where does pincushion distortion occur? _____
8. What holds the horizontal oscillator on frequency? _____
9. What are four uses of the signal from the horizontal output transformer? _____ _____ _____ _____
10. What does a horizontal-hold control do? _____
11. Carrier amplitude determines AVC voltage. What determines AGC voltage? _____
12. Why is keyed AGC better than simple pulse-voltage AGC? _____
13. What is the result of using a damper diode? _____
14. What are the four types of active devices used in horizontal output stages? _____ _____ _____ _____
15. If solid-state diodes have a maximum inverse voltage rating of only about 1000 V, how could they be used as the high-voltage rectifier in a TV receiver? _____

24-22 COLOR

Radio waves are a form of radiant energy that will travel through space. Light differs from radio waves in frequency or wavelength only. Red light has a wavelength of about 700 nm (700 nanometers, or billionths of a meter; also measured in milli-microns, $m\mu$, and equal to 4.3×10^{14} Hz), green light about 530 nm, and blue light about 450 nm. If the correct proportions of only these three pure (saturated) colors (hues) are added together, any desired color can be produced. For example, a blue, a green, and no red produces a blue-green, called cyan. A green, a red, and no blue results in

yellow. A blue, a red, and no green produces a purple color called magenta. With proper levels of green, blue, and red, all color perception in the eye is activated and white is seen. Thus, to transmit a picture in full color it will only be necessary to scan the picture simultaneously for its blue content, for its green content, and for its red content. When the same percent of each color is projected on a screen at the same time and in the correct places, the picture appears as the original. Bright white areas in the picture will be projected as strong pure green, strong pure red, and strong pure blue colors. Gray areas will be projected as weak green, weak red, and weak blue colors. Black (dark) areas will be projected with no green, no red, and no blue. By proper combinations of hues and intensities, any desired color can be reproduced as well as white, grays, and black.

The color triangle in Fig. 24-29 gives a rough approximation of the distribution of colors it is possible to reproduce by striking picture-tube phosphors with an electron beam. None of the colors can actually be reproduced in a 100% saturated form, but the possible percent is equal to, or better than, that obtained with printing inks.

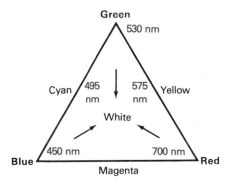

FIG. 24-29 Color triangle with wavelengths of hues.

24-23 TRANSMITTING COLOR SIGNALS

The TV color camera consists of a lens system that focuses the picture to be televised onto and through special *dichroic* glass semimirrors. A dichroic semi-mirror will reflect the color for which it is made and pass all other colors through it. The three reflected color scenes are picked up by three separate camera vidicons (Fig. 24-30). In this way the red, green, and blue color content of a picture is separated into three different signals. The three camera tubes scan their respective color scenes in unison, developing modulated-line information much as a

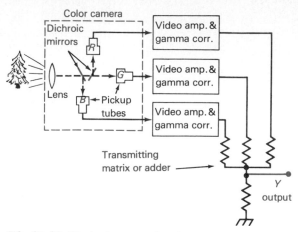

FIG. 24-30 Block diagram of color camera and circuits used to produce Y, or luminance, signals.

monochrome camera tube does. Eventually, these three electrical signals will be made to illuminate red, green, and blue phosphor dots or lines on the face of a color-picture tube and the original scene will be reproduced in color. (Some color cameras also have a third semimirror, not shown, that reflects the scene to a single monochrome vidicon.)

At the transmitter, the outputs of the R, G, and B camera tubes are amplified by a series of video amplifiers and are *gamma*-corrected. Gamma (a measurement of contrast) correction is required because the camera tubes produce a brightness output that does not correspond to the brightness recognition of the human eye. The three gamma-corrected video signals are then fed to a transmitter *matrix*. In the matrix the three color signals are combined into one composite signal with a ratio of 59% green, 30% red, and 11% blue. This is done to produce a signal that will discriminate better against noise and also produce a good rendition of whites, grays, and blacks when the signal is viewed on a monochrome receiver. The output from this part of the matrix is known as the Y, or *luminance*, signal. It contains video frequencies up to 4.2 MHz and would produce a black-and-white picture on any monochrome

ANSWERS TO CHECKUP QUIZ ON PAGE 575

1. (HP filter, or differentiator) (Integrator) **2.** ($\pm$520) (495) **3.** (Vary vertical oscillator frequency) **4.** (Picture centering) **5.** (Assure vertical blank) (During vertical blank pulse) **6.** (Sawtooth oscillator) (Pulses with RC circuits) **7.** (Edges of picture) **8.** (AFC) **9.** (AFC) (Horizontal sweep) (HV) (B+ boost) **10.** (Vary horizontal oscillator frequency) **11.** (Sync pulse amplitude) **12.** (Not subject to impulses between sync pulses) **13.** (Boost B+) **14.** (VT, BJT, MVOS, SCR) **15.** (Many in series)

TV receiver. Thus, the color transmission will be *compatible* (receivable on either a color or a monochrome receiver).

A block diagram of the color section of a TV transmitter is shown in Fig. 24-31. The Y signal is properly proportioned in the luminance matrix and is fed in two directions. One is to the adder, in which luminance, color, sync, and blanking pulses, and a sample of the color subcarrier frequency, called a color *burst*, are added to form the modulating signal for the color TV transmitter.

The positive potential luminance ($+Y$) signal is also fed through a phase inverter, becoming a $-Y$ signal. The blue signal is added to the $-Y$ signal in a $B-Y$ matrix. Similarly, a $R-Y$ signal is developed. These are further added into in-phase *(I)* and quadrature *(Q)*, meaning 90°-out-of-phase, components, each having red, green, and blue voltages with the polarities and relative amplitudes indicated.

The $+Y$ signal controls the brightness of the color picture and also produces the visible signal on monochrome receivers. The I and Q signal modulation is used to carry the color information from the transmitter to color receivers, where it will be added in proper proportions of red, green, and blue to the Y signals and then applied to the three-color TV tube screens. The amplitude of the I and Q signals will determine the saturation (purity) of the colors. The phase developed by the difference in amplitude between the I and Q signals will determine the *hue* (actual color) produced on the receiver screen.

The problem of transmitting the regular black-and-white luminance signal, requiring 4.2 MHz of sidebands, and at the same time the *chrominance* (color) information, without also increasing the bandwidth, was ingeniously solved. Sidebands that are produced by B/W video modulation cluster around frequencies that are harmonics of the line frequency of 15,750. As a result, there are spaces between these clusters that are not used (Fig. 24-32).

It was found that a subcarrier with a frequency of 3,579,545 Hz (3.58 MHz for simplicity) above the video carrier would produce its video sidebands in the spaces between the Y sidebands, as in Fig. 24-32*b*. In determining the desired color subcarrier frequency it was necessary to change the line frequency slightly. A monochrome transmitter uses a line frequency of 15,750 Hz and a field frequency of 60 Hz. A color transmitter uses a line frequency of 15,734.26 Hz and a field frequency of 59.94 Hz. The difference between these frequencies is so

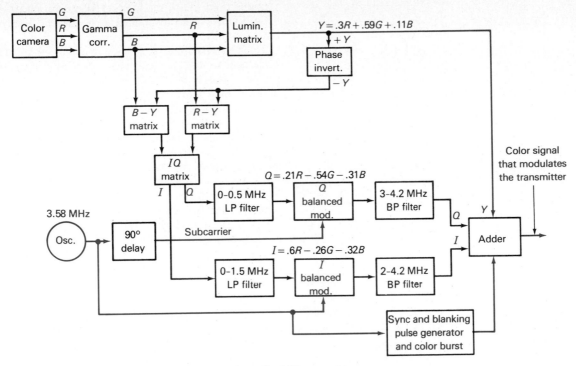

FIG. 24-31 Block diagram of the color system of a TV transmitter.

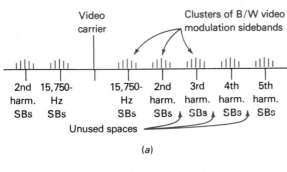

(a)

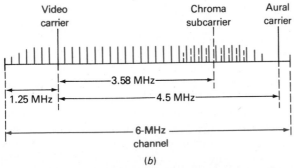

(b)

FIG. 24-32 (a) Monochrome sideband clusters. (b) Interleaved chroma sidebands with luminance sideband clusters.

slight that TV-receiver sweep oscillators will hold synchronism without adjustment of the sync controls.

If the subcarrier frequency is amplitude-modulated by I and Q signals, the sidebands that are produced will fall in the spaces between the sideband clusters produced by the luminance modulation. In this way both the luminance and the chrominance modulation can be transmitted within the same 4.2-MHz bandwidth. The positions of the video carrier and its sidebands, the chrominance subcarrier and its I and Q, sidebands, and the aural-carrier center frequency are shown in Fig. 24-33.

The significant difference between luminance and chrominance bandwidths is shown. It has been found that the fine detail made possible by the highest-frequency sidebands of a TV signal is not necessary in transmission of color. The human eye sees color in large areas and to a certain extent in moderately small areas, but color is not apparent in fine detail. Therefore, the fine detail in a color transmssion can be carried by the Y signal alone, and a narrow chrominance bandwidth may be used. When the chrominance information is inserted where it appears to be high-frequency sidebands to the video carrier, it shows on the monochrome

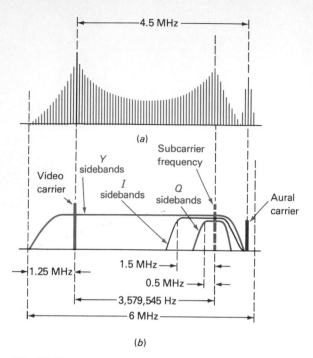

FIG. 24-33 (a) Appearance of color transmission on a spectrum analyzer. (b) Placement of color signal subcarrier and sidebands in a 6-MHz TV channel.

screen as very-small-detail patterns (weak interference signals to the luminance carrier and information). If the interlaced color fields are transmitted properly, a near cancellation of the chrominance signals occurs on the picture-tube screen, making a B/W receiver rendition of color signals very good.

The subcarrier in Fig. 24-33b is shown dashed because it is suppressed and is not transmitted. If it were, it would produce an objectionable pattern on the TV screen.

The I-chrominance signal from the IQ matrix back in Fig. 24-31 is first fed through a 0–1.5 MHz low-pass filter to remove all higher frequencies. (It is still a video-frequency camera signal, not sideband, because it has not yet been used to modulate a carrier.) This filtered signal is fed to a *balanced modulator* that is being fed the 3.58-MHz subcarrier at the same time. The balanced modulator cancels the carrier but leaves the upper and lower sidebands produced by the I signal. These sidebands are ac signals ranging in frequency from 3.58 − 1.5 = 2.08 MHz to 3.58 + 1.5 = 5.08 MHz. They are fed through a 3–4.2 MHz bandpass filter to remove some of the upper sidebands and any

spurious products of modulation. They are added to the Y signal before it modulates the video carrier of the transmitter.

The Q signal is also used to modulate the 3.58-MHz subcarrier in a balanced modulator. It is first passed through a 0–0.5 MHz low-pass filter to remove any higher (finer-detail) frequencies. The carrier that the Q signal will modulate is also 3.58 MHz, but the Q and I carriers are out of phase by 90°. When the I and Q signals are added to the Y signal, the two sets of amplitude-modulated chrominance sidebands are 90° out of phase. Together they form a resultant which acts as phase modulation. The phase angle is determined by the relative amplitude of the two signals. In the receiver the resultant of these out-of-phase signals will produce the hue, or color, of the signal to be displayed, while the amplitudes of the signals will determine the saturation, or depth, of the color.

Two other modulating signals are required in the transmission of color TV signals. One is a short burst of the subcarrier frequency. Eight cycles of this frequency are added to the back porch of the horizontal blanking signal (Fig. 24-34). The color

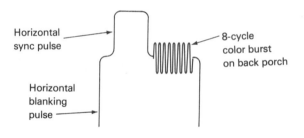

FIG. 24-34 Color burst on back porch of the horizontal blanking pulse.

bursts are used to synchronize a crystal oscillator in the receiver and hold it to the exact frequency and phase of the missing subcarrier. The received sidebands must be mixed with this carrier frequency in the receiver in order that they may be detected properly. Vertical and horizontal synchronizing and blanking pulses are added to the output of the adder circuits before they modulate the transmitter.

It would appear that the green signal has been forgotten. However, since the Y, Q, and I signals all contain components of three colors, in the receiver it is possible to combine the three components and produce a green-chrominance resultant signal even though none is transmitted as such.

24-24 COLOR TV RECEIVERS

Any transmitted color TV signal contains three separate systems of sidebands, all within a band of 4.2 MHz. The Y-signal sidebands extend the full 4.2 MHz. The I and Q 90°-out-of-phase sets of chrominance sidebands occupy the spaces between the Y sideband clusters in the upper 2.5 MHz of the sideband spectrum.

The receiver must develop signals equal to the three original color signals picked up in the camera tubes and reassemble them in their own component colors and intensities in their proper places on the screen of the color-picture tube.

The antenna, front end, IF amplifiers, AGC, detector, video, sweep circuits, and sound systems are very similar to corresponding systems in monochrome receivers except that the color receiver requires more accurate alignment of RF and IF channels to produce a flat response curve. A falling off of response of the higher-frequency sidebands in a monochrome receiver will produce a poorly defined picture. Since all chrominance information is in the higher-frequency end of the IF bandpass, such a loss of response will mean loss of color rendition on the picture tube.

The block diagram in Fig. 24-35 shows the section of the color TV receiver that separates the

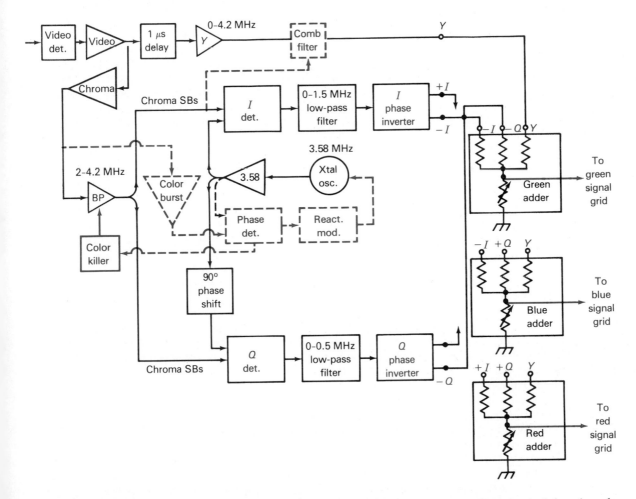

FIG. 24-35 Block diagram of the color signal circuits from the receiver detector to the three signals fed to the color grids in the picture tube.

three color signals. It starts with the composite detected signal from the video (second) detector.

The whole composite signal, including all video frequencies between zero and 4.2 MHz, is passed through a 1-μs delay circuit and is then fed to the luminance or Y amplifier. This delay is required because the chrominance signals passing through narrow-passband LC filters will be delayed about 1 μs.

When a color receiver is receiving a monochrome transmission, the Y signal produces the black-and-white images. The Y signal is fed to the red, green, and blue matrices and their adders. The 0–4.2 MHz Y-signal output is fed with equal amplitude to all three color grids. This results in varying degrees of brightness on the screen (white, light gray, dark gray, and when the Y signal is zero, no illumination, or black).

When a color receiver is receiving a color transmission, the whole composite detected signal (0–4.2 MHz plus the color burst) passes through the *chroma* amplifier and is fed to the 2–4.2 MHz bandpass filter-amplifier. This stage passes on the frequencies of the I and Q sidebands that must beat with the missing subcarriers to produce the color signals for the picture tube.

A 3.58-MHz crystal oscillator is forced into exact synchronization with the color-burst signal through the use of a phase detector and a reactance modulator. If the oscillator is not in exact phase with the color burst, this is sensed by the phase detector, which adds a bias to the reactance device to force the crystal into synchronization. (The accuracy of the color-burst signals transmitted by TV network stations allows the output of any properly operating color TV receiver crystal oscillator to be used as a highly accurate signal source.) This is another application of a phase-locked loop (PLL).

The 3.58-MHz signal is fed to the I detector along with the chroma sidebands. Since this injected carrier has the same frequency and phase as the subcarrier that originally produced the I sidebands in the transmitter's balanced modulator, the output of the I detector or mixer is the same as the I modulating signals. (All Q sidebands were produced by a 90°-out-of-phase subcarrier and will therefore not appear in the output of the I detector.)

The signal from the I detector is now in the form of video frequencies of 0–1.5 MHz. These pass through a 1.5-MHz low-pass filter and to a phase inverter or splitter to produce equal but 180° out-of-phase I signals to feed the three sections of the matrix (+I to +I points, −I to −I points).

The same 3.58-MHz carrier is shifted 90° and is fed to the Q detector, which is receiving the same chroma sidebands as is the I detector. The phase of this carrier allows the Q detector to produce a Q output signal. (Since the Q signals are 90° out of phase with the I signals, no I information can appear in the output of the Q detector.) The Q signal passes through a 0.5-MHz low-pass filter and to a phase inverter. The output of the phase inverter is fed to the three sections of the matrix as indicated.

The Q, I, and Y voltages are mixed in the top section of the matrix in proper proportions to produce a resultant which is equivalent to the green signal picked up by the green camera tube. In the center matrix the proportions of the I, Q, and Y voltages produce the blue signal. In the lower matrix the red signal is developed. The three signals are the same as the three signals produced in the color-camera tubes at the transmitter. It is now necessary to use these three signals to key three scanning beams in such a manner that the proper color proportions will be laid down where they belong on the face of the picture-tube screen. (Although shown as a resistance matrix, diode, transistor, or VT matrices may be used.)

When a TV signal is transmitted in monochrome, no color burst is transmitted. This means that no 3.58-MHz color-burst signals are being fed to the phase detector by the color-burst amplifier. As a result, the color-killer circuit receives no bias from the phase-detector circuit, and a positive flyback pulse from the horizontal yoke produces current flow in the color-killer active device. This current in turn is made to bias the bandpass-filter amplifier past cutoff, so that no signals are fed through it to the I and Q color-signal systems. The only acting signal is the Y signal that produces the B/W pictures.

The chroma sidebands in the upper-frequency sideband range of the luminance signal do cause some interference with the luminance sidebands when receiving either monochrome or color signals. Some receivers add a *comb filter* in the Y circuit (Fig. 24-35) which filters out all frequencies that are multiples of 15,734 Hz (the chroma sideband cluster center frequencies) on both sides of the 3.58-MHz chroma carrier frequency. The clearing of the space between the higher-frequency luminance sideband clusters results in nearly a 25% increase in horizontal resolution of the pictures, as well as reduction of unwanted patterns and cross-color effects in the pictures. Comb filters are also used in video tape recorders and other places at the transmitting station.

The scanning circuits for color are basically the same as for monochrome receivers. The vertical-deflection system consists of a 59.95-Hz oscillator (Fig. 24-36), an amplifier, and the deflection coils

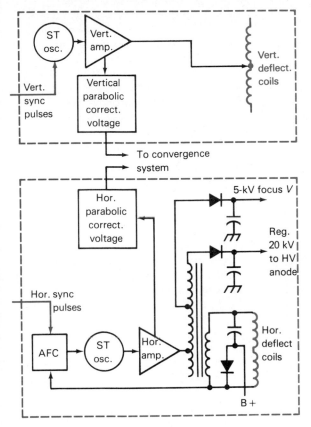

FIG. 24-36 Block diagram of possible vertical- and horizontal-deflection systems for a three-gun color tube.

or yoke. If a three-electron-gun color tube is used (Sec. 24-26), a vertical parabolic voltage is developed in the vertical output and used in the convergence system. Convergence is necessary to make the three separate electron beams converge to a point at the shadow mask just behind the picture screen.

The horizontal-deflection system consists of an AFC circuit, a horizontal oscillator, an amplifier, a damper, a flyback high-voltage rectifier circuit, and a high-voltage regulator. When a three-gun tube is used, a horizontal parabolic convergence voltage, or magnetic field, is developed from the horizontal amplifier stage, and a ±5-kV focusing voltage is required. Note that the *HV* power supply must be regulated.

In large-screen projection TV receivers the three color signals from the matrices are fed to three separate red, green, and blue CRTs. The bright outputs from these tubes are either fed through focusing lenses to the back of a large frosted-glass screen or to mirrors which reflect the three color images to a large front-viewing glass-beaded screen.

Test your understanding; answer these checkup questions.

1. What would be produced by weak blue, weak green, and weak red signals on a color CRT? _____
2. What is the name of a mirror that reflects one color but passes all others? _____ How many such mirrors are required in a color TV camera? _____
3. What is a combining circuit called in TV? _____
4. What is the letter symbol for the luminance signal? _____
5. What is the total $R + G + B$ for a Y signal? _____
6. If a white scene is picked up, what is the Y value? _____ Q value? _____ I value? _____
7. If a black scene is being picked up, what is the Y value? _____ Q? _____ I? _____
8. What determines the purity of received colors? _____ The actual color? _____
9. Which subcarrier leads by 90°, I or Q? _____
10. What is the color horizontal sweep frequency? _____ Vertical sweep? _____ Subcarrier frequency? _____
11. Why can chrominance sidebands have narrower bandwidths than luminance? _____ What bandwidths are used? _____
12. Why are chrominance sidebands used without a carrier? _____
13. Where is a color burst found? _____ What is its frequency? _____ Duration? _____
14. Why is the luminance signal delayed? _____
15. What circuit does a color-killer kill? _____
16. List the color TV receiver systems that are similar to those in a monochrome receiver. _____
17. List the color TV receiver systems that differ from those in a monochrome receiver. _____
18. What are two advantages gained by using comb filters in the Y circuit? _____ _____

24-25 THE SHADOW-MASK TUBE

With three color signals and a suitable deflection system, a color picture can be displayed on a color

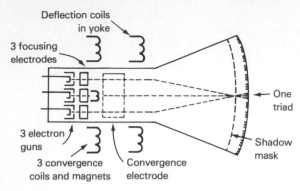

FIG. 24-37 Components of a three-gun color tube.

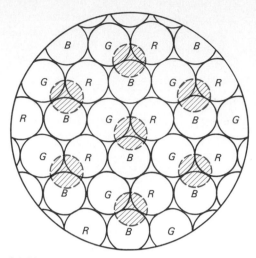

FIG. 24-38 Triads of color dots on the inner face of a three-gun color tube. Shaded areas are holes in the shadow mask behind the color dots.

tube. One tube is the three-gun shadow-mask tricolor type (Fig. 24-37).

The three electron guns when viewed from the front are oriented equidistant and 120° from each other rather than being stacked one on top of the other, as in the illustration. Each projects an electron beam toward the screen end of the tube. Three focusing electrodes are excited with separate potentials (+ 5 kV) to focus their own beams to a tiny spot on the screen. Three small permanent magnets can be spaced around the neck of the tube to aid convergence of the three beams so they all pass through the same tiny holes in the metal *shadow mask* mounted a short distance from the fluorescent screen. The three beams cross in the holes and deconverge, each striking its own phosphor color spot on the screen (Fig. 24-38). Each trio of color-phosphor dots, called a *triad*, is arranged as shown. Each shaded circle represents one hole in the shadow mask behind the dots painted on the inside surface of the tube face.

The metal shadow mask of a medium-size tube may have about 250,000 tiny holes in it. For each

hole there is a green, a red, and a blue phosphor dot on the screen, or a total of 750,000 color dots. When three beams are all passing through one hole, the red-signal beam strikes a red phosphor, the blue-signal beam a blue phosphor, and the green-signal beam a green phosphor. If all beams are of equal intensity, the triad area appears white to the viewer. When only the green-signal and the red-signal electron-gun grids are receiving color signals, only these two phosphors are illuminated and the viewer sees a yellow color. By scanning the picture tube in synchronism with the camera-tube scanning, the original scene can be reproduced in full color on the tricolor picture tube.

24-26 OTHER TV TUBES

Another color CRT in use is the Trinitron. In three-gun shadow-mask tubes there are three guns, three focusing lenses, and a relatively complicated convergence system. The Trinitron has only one gun, requires only one focusing lens, has a vertical-stripe metal *grille* behind the vertically striped phosphers, and has a simpler convergence system. Although it has only one gun, it has three separate cathodes, all in line (top view, Fig. 24-39), to which blue, green, and red signals are fed. The first grid is a single sheet of metal with three holes in it through which electrons from the three cathodes can pass. The second grid is an accelerating anode, pulling electrons out past the three G_1 openings. Once past G_2, the three beams of electrons fall under the

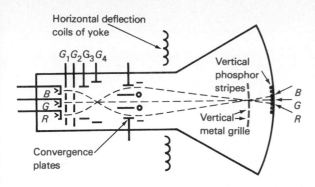

Horizontal deflection coils of yoke

$G_1 G_2 G_3 G_4$

Vertical phosphor stripes

Convergence plates

Vertical metal grille

FIG. 24-39 One-gun Trinitron color TV picture tube, top view.

influence of focusing electrodes G_3 and G_4, acting similar to focusing electrodes in oscilloscope tubes. The three beams are made to converge or focus to a point and then diverge as they continue on. The green beam moves straight through the two zero-charged convergence plates, bending only when passing under the influence of the deflection yokes. The blue and red beams approach the outer negatively charged convergence plates and are repelled back toward each other, meeting the green beam at an opening in the metal grille mounted near the face of the screen, provided the bias on the plates is correct. The blue and red beams deconverge after passing through the grille openings and strike phosphers painted on the inner surface of the screen.

Unlike the shadow-mask tube with its three color dot triads, the Trinitron has its phosphors painted in vertical stripes on the inside of the screen. Mounted near the stripes is the vertical metal grille. If only the green cathode is driven negative (all grids held at a constant negative potential), only the green beam travels through the tube and strikes a green phosphor. If the red and blue cathodes are driven negative, these two colors will be activated and the area will appear magenta in color.

Proper horizontal convergence to produce the correct landing of the beams out at the edges of the screen requires a small parabolic convergence voltage added to the horizontal sweep. This can be aided by proper positioning of the deflection yoke forward or backward. In actual use, the beams are wide enough to strike two adjacent grille openings, producing relatively bright illumination of the color screen. No vertical convergence circuitry is required.

There are a variety of flat-type TV picture screens being developed. One used for small pocket-sized screens is made with back-lighted LCDs (Sec. 12-16). Each *pixel* (picture element or spot) requires a separate MOS transistor to produce the required black, white, and gray gradations of the picture. If there are 56,000 pixels in a black and white picture, 56,000 invisible transistors will have to be laid down on a polycrystalline silicon LCD sheet. For color pictures each pixel must have a red, green, and blue LCD spot grouped together to produce the desired color rendition of the picture. These solid-state screens can be used for pocket-type oscilloscopes and TV sets.

24-27 COMPLETE COLOR TV RECEIVER

As a final recapitulation of TV receivers, the block diagram in Fig. 24-40 represents a simplified but interesting example of one type of color TV circuitry. Note that it uses integrated circuits (ICs) for some subsystems, namely sound, chroma, IF detector, and automatic fine tuning (AFT). Other subsystems and circuits use discrete parts. The picture tube (kinescope) is the only VT used.

A VHF TV signal picked up by the antenna is fed to the RF amplifier. After amplification it mixes with the VHF local oscillator (LO) to develop the 41–47 MHz IF frequencies, which are amplified twice and fed to the diode video detector. The 47.25-MHz trap in the first IF stage reduces any adjacent channel sound carrier. The 41.25-MHz trap in the second IF stage prevents the in-channel sound carrier signal from getting to the detector and video amplifier stages. To produce a keyed AGC, a horizontal blanking pulse from one of the flyback transformer secondaries (labeled BP) allows sampling of the amplitude of the blanking pulse strength at the video detector and develops an AGC bias for the RF and the first IF amplifier.

When receiving UHF signals, they are mixed with the UHF local oscillator to develop an IF in the VHF range. This signal is fed to a switchpoint labeled "UHF" on the VHF tuner switch. The UHF signal then progresses through the receiver as a VHF signal. (UHF TV receivers are thus double-conversion superheterodynes.)

Three signals are taken from the second IF stage: (1) the sound IF signal, which is the result of the mixing of the picture carrier (45.75 MHz in the IF strip) and the sound carrier (41.25 MHz in the IF strip), producing a 4.5-MHz sound IF; (2) the 45.75-MHz visual carrier for an automatic frequency tuning (AFT) subsystem; and (3) the visual

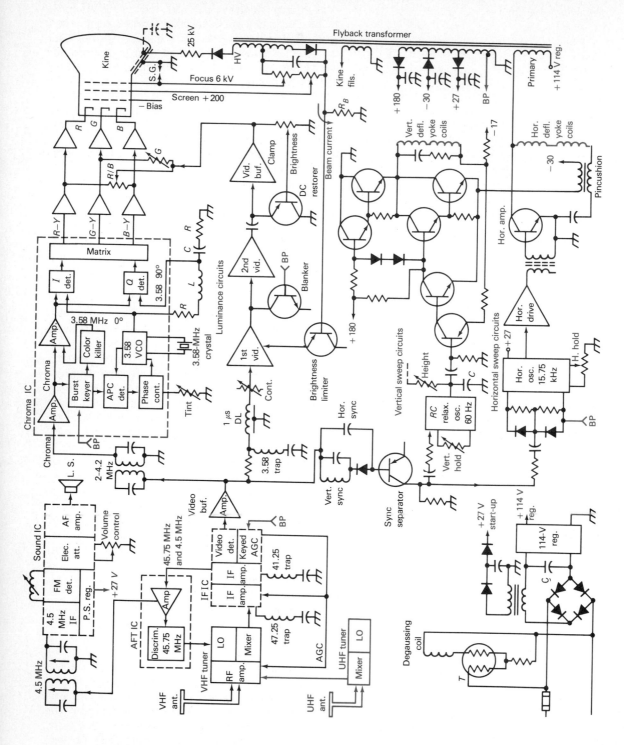

FIG. 24-40 Block diagram of circuits involved in a color TV receiver.

or video sidebands (45.75–41.6 MHz in the IF strip).

By a double-tuned transformer, the 4.5-MHz sound IF signal is coupled through an amplifier to the sound IC. This IC contains a 4.5-MHz IF amplifier, an FM detector, an adjustable electronic attenuator (volume control), and an AF amplifier with power enough to drive the loudspeaker.

The 45.75-MHz visual carrier signal from the IF stage is amplified and fed to a 45.75-MHz discriminator in an AFT subsystem. If the local oscillator drifts, the output voltage from the discriminator feeds a correction, or error, voltage to the voltage-tunable (varactor) LO to bring it back on frequency, keeping the set on frequency and maintaining a constant-color rendition. (Another form of PLL.)

The 41–46 MHz visual IF information is fed to a diode video detector and buffer amplifier. The output of this (0–4.3 MHz video signals) is fed to three subsystems:

(1) The luminance (monochrome) amplifiers, through a 1-μs delay line (DL). A 3.58-MHz trap prevents color burst signals from entering the luminance section. The output of the luminance amplifiers feeds the three picture tube cathodes in correct proportions to produce a brightness only (Y) picture on the screen when monochrome signals are being received. The first two video amplifiers are direct-coupled, but the buffer is capacitively coupled, so dc restoration is required for this stage. Brightness is also controlled in this stage. Picture tube beam current is monitored and fed to the brightness limiter transistor. If the beam current increases, the brightness limiter decreases the gain of the first video amplifier and reduces the beam current, holding a fairly constant overall picture intensity on the screen at all times. The input of the second video amplifier has a blanker transistor in its input, its base being fed by the blanking pulse, BP. This assures adequate blanking of all signals during retrace times.

(2) The video detector output, when the received signal is in color, feeds video signals up to the chroma IC. The video frequencies involved in color, nominally 2–4.3 MHz, are transformer-coupled from the video buffer ahead of the delay line and fed to the first chroma amplifier in the IC. The horizontal blanking or keying pulse, BP, opens the burst keyer to allow it to sample the back porch of the horizontal blanking pulse for a color burst. If a color burst is not present during the horizontal blanking time, the color killer biases off the second chroma amplifier and no color signals are fed to the

picture tube. If there is a color burst, it is fed to the automatic phase control (APC) detector, where it is compared in phase with that of the 3.58-MHz crystal-stabilized voltage-controlled oscillator (VCO). If the VCO output is not exactly in phase with the broadcast color burst, the phase control circuits shift the frequency of the VCO until it is exactly on frequency and in phase. A slight shift of phase can be produced by adjusting the "tint" control on the front of the receiver. The 3.58-MHz output from the VCO is fed directly to the I detector and mixes with the chroma (color) sidebands to produce the I information for the matrix. The VCO output is shifted 90° by an LCR network and is fed to the Q detector which feeds Q information to the matrix. The output of the matrix is $R-Y$, $G-Y$, and $B-Y$ signals. These are amplified and mixed with the luminance signal to produce the proper red, green, and blue signals to feed to the cathodes of the picture tube.

(3) The sync separator transistor accepts both horizontal and vertical sync pulses (when they exceed the black-level bias value). The low-frequency vertical sync pulses pass through a long-time-constant (± 60 Hz) RC circuit to the vertical RC relaxation oscillator. These pulses control the oscillator frequency and thereby the charge-discharge time of capacitor C through the resistance of the height control. The charging of C is roughly sawtooth shaped. This voltage feeds the predrivers, the drivers, and the complementary symmetry-type vertical-deflection amplifiers, which feed the sawtooth wave to the vertical-deflection yoke coils. A small sample of the vertical-deflection yoke current is fed back to the predrivers, which by high gain and operational-amplifier-type operation, makes the sawtooth wave quite linear. The high-frequency horizontal sync pulses pass from the sync separator through a short-time-constant RC-type circuit ($\pm 15,750$ Hz) to an automatic frequency control (AFC) circuit which is keyed on by the blanking pulse, BP, so that only the horizontal sync pulses will control the frequency of the horizontal RC oscillator. The output of this oscillator feeds a driver, which in turn feeds the horizontal output transistor.

The output of the horizontal amplifier does two things. It feeds 15.75-kHz sawtooth current to the horizontal-deflection yoke coils, and also to the primary winding of the flyback transformer. The 15.75-kHz secondary voltages taken from this transformer are (1) an ac blanking pulse, BP; (2) rectified and filtered voltages of +27, −30, and +180 V to operate the transistor circuits of the receiver;

(3) a 25-kV rectified and filtered ultor (Aquadag) dc voltage; (4) a 6-kV rectified and filtered focusing voltage; (5) a dc screen voltage; (6) the beam current sample voltage from across R_B; and (7) an ac filament voltage for the kinescope.

The dc supply for the receiver uses no 60-Hz power transformer. Instead, a full-wave bridge rectifier develops 150 V dc from the 120-V, 60-Hz power line ac, which when regulated operates the horizontal-deflection stage. Because this stage, acting as the primary of the 15.75-kHz transformer, is regulated, all output voltages from the transformer tend to maintain regulated values so that high-voltage regulators are not required. Since the +114-V regulated dc for the horizontal output stage does not operate the horizontal oscillator, there is nothing to amplify when the receiver is first turned on. As a result, a small "start-up" transformer between the bridge rectifier and its filter capacitor utilizes the pulsating dc that exists until the capacitor (C) becomes charged. This pulsating dc acts as a primary voltage for a start-up power supply for the horizontal oscillator circuit. When the "run" voltage supplies come up to operating potential, the start-up supply is no longer needed and, being lower in voltage, no longer feeds current through its series diode to the horizontal oscillator.

The *degaussing* coil around the periphery of the picture tube is fed an ac when the set is first turned on. The two *thermistors* (T) have low resistance when cold. As soon as current flows in the degaussing coil, the thermistors heat and their resistance rises, decreasing degaussing current to zero. The strong ac field produced at turn-on effectively demagnetizes any permanent magnetism that the tube shadow-mask and shield may have picked up, which would produce improper color rendition on the tube screen.

New TV receivers have direct RGB input connections to allow feeding the output signals from a home-type color camera, or from a video tape cassette recorder (VCR), or from a satellite TV receiver's video output, directly to the picture tube. They may also have AF inputs for signals from tape or record players, or from FM or AM receivers, making the TV console a complete sound-video home entertainment center.

24-28 PAY AND CABLE TV

Some TV stations (usually UHF) broadcast *scrambled* or *encoded* "pay" TV signals that cannot be unscrambled without adding a unit that may be rented by the subscriber. A simple method of scrambling is to reduce the amplitude of the blanking levels, the sync pulses, and the color burst to about half of the black level of the picture information being transmitted. Normal TV receivers cannot maintain sync under this condition and a scrambled picture results. The rental unit, which is connected between antenna and TV set, regenerates the required blanking, sync pulses, and color burst, adding them at the proper amplitude between the video lines during horizontal retrace times. The result is a normally blanked and synchronized signal being fed to the receiver. To further defeat the use of the transmitted program by nonsubscribers, the audio signal is transmitted as a double-sideband-suppressed-carrier signal on a subcarrier of twice the horizontal sweep frequency, modulating the aural carrier frequency. Normal receivers cannot hear this audio. The rented unit has a detecting circuit similar to the stereo channel detector of an FM stereo transmission, except for the subcarrier frequency that is used. Of course, there are other scrambling methods, such as inverting the video signals. This is known as *subscription TV*, or *pay TV*, because only those who pay to rent the unscramblers can use the transmitted signals. Pay TV transmissions may have few, if any, commercials during transmission of a motion picture. Part of the day the TV station may operate as a normal broadcast TV station.

Community antenna TV (CATV), or *cable TV*, is used to feed relatively strong, ghostless TV signals to its paying subscribers through a 75-Ω coaxial cable system (Fig. 24-41). Cables may be strung above ground on poles or be run underground. A cable company may also feed satellite TV programs, locally generated TV programs, motion pictures, weather and time information, special sports programs, etc., to its subscribers, in some cases using "cable frequencies."

The frequencies carried by CATV are mostly the normal 12 commercial VHF channels, the low-VHF band (54–88 MHz) and the high-VHF band (174–216 MHZ). All of these channels can be received by any TV set. When there are more than 12 TV programs to be relayed or translated to the VHF TV channels on the cable, a special converter must be rented and be added at the subscriber's TV set. This unit switches these other programs being transmitted on other cable frequencies to one of the 12 VHF TV channels (which also blanks out any other programs from that channel). Cable communication frequencies may be selected from any of the con-

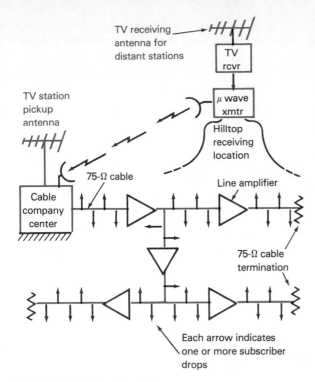

TV receiving antenna for distant stations

TV rcvr

μ wave xmtr

Hilltop receiving location

TV station pickup antenna

Cable company center

75-Ω cable

Line amplifier

75-Ω cable termination

Each arrow indicates one or more subscriber drops

FIG. 24-41 Basic cable TV system.

tiguous 6-MHz channels in the following bands: sub-VHF band, 5.75–47.75 MHz; mid-VHF band, 120–174 MHz; super band, 216–300 MHz; and ultra band, 300–450 MHz.

A cable company may receive a signal off-the-air from a TV station, amplify it with a strip amplifier up to the desired level (48 dBmV), and feed it directly into the cable, or translate it to some other TV channel. Another method is to process the received signal to IF form, modulate a carrier with it, and then feed this modulated signal down the cable. A third method is to demodulate the received signal to its video and audio components and use these to either modulate a carrier, or more likely, to feed the video and audio of a remote receiving location via a microwave relay system aimed at the central cable office where it is used to modulate a carrier oscillator, which then feeds the cable. All 12 channel programs are fed into the same cable but through isolating amplifiers to prevent interaction of the many signals. Subscribers have their TV sets connected to the cable either directly to a 75-Ω input connection on the back of the set, or through a 75- to 300-Ω balun transformer to connect to the 300-Ω contacts. The subscriber drop line is connected to the cable through an *LCR* directional cou-

pler to prevent mismatches at the subscriber termination from affecting the main cable signals. Wherever the signal in a cable has dropped about 22 dB, a line amplifier is added to bring the signal up to the correct level again.

If the coaxial cable outer conductor frays and breaks, or if the couplings oxidize at the point where the subscriber cable is tapped in, or if cables are improperly terminated, an impedance mismatch occurs. This results in development of standing waves not only on the inside of the cable, but on the outside also, forcing the coaxial line to act as an antenna and radiate some or all of the TV signals being carried in the cable. In this way interference can be caused to other radio services in the 7–450 MHz region, and is a serious problem with older cable runs.

For an additional charge, cable companies will also rent unscrambler units to enable subscribers to demodulate relatively new movie or special sports programs that have been scrambled and fed into the coaxial lines using standard TV channels or cable frequencies (which are converted to a receivable channel).

When two-way cable communications are used, they usually involve channels in the sub-VHF band. A 6-MHz-wide band will accommodate a rather sophisticated computer or other digital-type communication system. If more bandwidth is required two or more cable frequency channels can be used. The terminal units in these cases will not be TV sets, of course.

A *master antenna TV* (MATV) system is used in large apartment buildings, etc. One TV antenna system picks up all receivable TV stations. Separate distribution amplifiers then feed all channels to each floor or to each apartment.

24-29 SATELLITE TELEVISION

There are 15 or more North American earth-orbiting geosynchronous satellites (Sec. 22-21), separated by about 4° at this time, that rebroadcast (translate) programs, both TV and audio, back down to earth. An earth receiving station, at a home for example, may consist of a 9- to 14-ft steerable parabolic metallic "dish" that reflects satellite signals up into a small horn-type λ/2 or λ/4 pickup antenna. Received signals are coupled to a 3.7–4.2 GHz broadbanded *low-noise amplifier* (LNA) and then to a *down-converter* to feed a 70-MHz TV IF signal to the remainder of the receiver in the home. The 0–4.5 MHz video signals and the aural signals of a

TV program are both made to *frequency modulate* the earth station transmitter, using about 30-MHz-bandwidth FM centered on one of the twelve 40-MHz-wide channels. The audio signal first frequency modulates a carrier between 5.8 and 8 MHz, usually ebout 6.5 MHz. This 6.5-MHz audio-modulated carrier is added to the 0–4.5 MHz video signal, and both frequency-modulate the transmitter. The TV signal translated by the satellite and received on earth is thus a 30-MHz-bandwidth FM containing both the 0–4.5 MHz video information and the 6.5-MHz frequency-modulated aural carrier. At the receiver the output of the discriminator is an amplitude-type 0–4.5 MHz video signal and a 6.2-MHz FM carrier modulated by the aural information. The video can be fed directly to a video monitor for display. The FM aural information is detected with a 6.5-MHz discriminator and is fed to an audio amplifier system, either mono or stereo. If a home TV receiver is used to display the TV program, the video and detected audio is made to amplitude- and frequency-modulate two oscillators (video and aural, usually for TV channels 2 or 3). The modulated signal is fed to the antenna input of the TV set to be demodulated and displayed when the receiver is tuned to channel 2 or 3.

The signals from the LNA are fed first to a down-converter (mixer) having as its local oscillator a varactor-tuned GaAs FET voltage-tuned oscillator (VTO). The frequency of this oscillator is determined by a dc voltage controlled by a 12-position switch on the receiver panel. In this way the desired channel is selected. Each of the twelve 40-MHz-wide channels can handle both a horizontally and a vertically polarized program signal using the same center frequency, resulting in 24 useful receive channels per satellite. The pickup antenna must be capable of mechanically rotating at least 90° to be properly aligned to either the H or the V signals. The parabolic antenna must be capable of tracking across an arc in the sky from 79°W to 143°W to receive the following presently operating North American satellites:

WESTAR (W2)	79°W	SATCOM (S2)	109°W
SATCOM (S4)	83°W	WESTAR (W5)	123°W
COMSTAR (C3)	87°W	COMSTAR (C4)	127°W
WESTAR (W3)	91°W	SATCOM (S3)	131°W
COMSTAR (C2)	95°W	GALAXY (G1)	135°W
WESTAR (W4)	99°W	SATCOM (S1)	139°W
ANIK (AD)	104°W	SATCOM (S5)	143°W
ANIK (AB)	109°W		

If satellite spacing is reduced to 2° in the future, twice as many "birds" will be receivable, but narrower-beamwidth (larger diameter) receiving dishes may be required.

24-30 AMATEUR TELEVISION

Licensed amateur radio operators may use fast-scan amateur TV (ATV) on the 420-MHz and all higher-frequency amateur bands. Fast-scan is similar to broadcast station TV, using 525 lines, 6-MHz bandwidth, etc. It may be monochrome or color. Video color or monochrome cameras with self-contained sync-pulse generators are available commercially. The output signals from them can be used to modulate a UHF amateur transmitter. The received UHF ATV signals can be fed to a converter that translates them down to TV channel 2 or 3 for viewing with a normal TV set.

Because of its bandwidth of 6 MHz, wide-band TV cannot be used on any of the relatively narrow amateur bands below 420 MHz. By reducing the number of lines from about 490 to 128, and scanning slowly (8.5 s for one complete, noninterlaced frame) the video frequencies can be held within the 3-kHz SSB bandwidth for HF band slow-scan TV (SSTV) operation. The sync-pulse generator must produce (1) a 5-ms horizontal combination sync-blanking pulse, (2) a sawtooth sweep ac at a 15-Hz horizontal rate, (3) a 30-ms vertical sync-blanking pulse, and (4) a sawtooth sweep ac at a 0.125-Hz vertical rate (Fig. 24-42). The pulses are produced by using a 1200-Hz tone. The amplitude-type video camera voltages are converted to varying frequen-

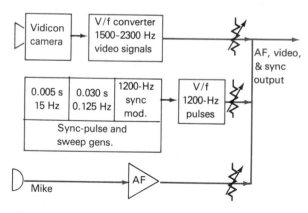

FIG. 24-42 Basic block diagram of a slow-scan video-audio transmitting system.

cies or tones between 1500 and 2300 Hz, with the highest amplitude signals (bright parts of the picture) at 2300 Hz and the lowest amplitude signals (dark parts) at 1500 Hz. This makes the 1200 Hz of the pulse tone ac blacker-than-black and invisible. The video and sync pulses ranging between 1200 and 2300 Hz are added to AF from the microphone and are all fed into the microphone input of a SSB voice transmitter. Voice transmissions are not made during the 8.5-s time of transmission of the still pictures. Note that the camera in SSTV could not be a commercial video camera with fast-scan sync and sweep circuits unless special converters, or digital memories to store the video signals, are used.

At the receiving end, the signal (sounding on a receiver like a machine gun with interspersed AF tones) is tuned in properly while SSB voice modulation is being received. The SSTV signals are fed to a *linear slope filter* (f/V) (Fig. 24-43). The

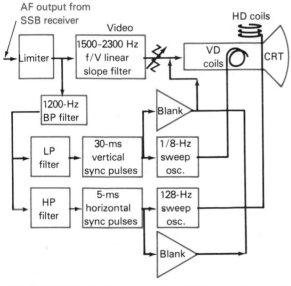

FIG. 24-43 Block diagram of a SSTV monitor (receiving) system.

f/V filter changes 2300-Hz video signals to highest amplitude values (white) and 1500-Hz signals to zero value (black). The output video signals should be the same as the amplitude-modulated video signals that came from the camera. They can be fed to the grid cathode of a long-persistence (P7) phosphor radar-type CRT. Such tubes retain a latent image for a period of about 10 s at reasonably high intensity, fading out completely in about 20 s. Received

SSTV signals may also be stored in a computer memory bank and then repeatedly and rapidly be fed to a P1 phosphor picture tube screen in a standard TV set to provide a nonfading still picture.

The 1200-Hz sync pulses are passed through a bandpass filter. The 30-ms sync pulses, through a low-pass filter, form the blanking pulses and sync the 0.125-Hz vertical sweep oscillator. The sawtooth vertical-deflection (VD) signals are fed to the VD coils of the CRT. The 5-ms sync pulses, through a high-pass filter, form the horizontal blanking pulses and sync the 15-Hz horizontal sawtooth sweep oscillator. The blanking pulses blank the CRT beam during retrace times.

At best, a square 128-*pixel* (horizontal-line picture element) by 128-line picture has a very poor image resolution. Both more modern amateur radio SSTV and commercial telephone-line "freeze-frame" SSTV are more likely to use 17-s pictures with 256 pixels and a 15-Hz line rate. The slower transmitted pictures have very acceptable resolutions.

Some new oscilloscopes use storage-type tubes which can retain a latent image for hours, or days, if desired, and can be used to produce good SSTV pictures.

SSTV can be used on any extra or advanced amateur telephone portions of the 3.5-, 7-, 15-, and 15-m bands, and on all voice frequencies above 28.3 MHz for long-distance communications all over the world. Calling frequencies are 3.845, 7.171, 14.23, 21.34, and 28.68 MHz. Operating procedures are similar to amateur SSB communications. For ATV transmissions, a 2-m transmitter-receiver system is usually used for communicating between the two ATV stations.

Test your understanding; answer these checkup questions.

1. If the blue triad dot is always at the bottom, what position must the blue gun always have? _____
2. In a shadow-mask tube, there are how many cathodes? _____ Guns? _____ First grids? _____
3. In a grille-type tube, there are how many cathodes? _____ Guns? _____ First grids? _____
4. Which picture tubes require horizontal parabolic correction voltage? _____ Vertical? _____

5. What type of convergence is used with three-gun tubes? _____ With one-gun tubes? _____

6. In which tube is one color beam not affected by the convergence field? _____ Which beam is it? _____

7. What might happen if a strong permanent magnet were laid on top of a color TV set? _____

8. In what form are the color phosphors laid down in the three-gun tube? _____ One-gun tube? _____

9. What is the meaning of AFT? _____ APC? _____ VCO? _____ Kinescope? _____

10. Where is a degaussing coil in a TV set? _____

11. Is the TC of a degaussing thermistor positive, negative, or zero? _____

12. In Fig. 24-40, what does "S.G." mean near the kine? _____

13. Is a TV set a single- or double-conversion-type for VHF? _____ UHF? _____

14. What channels are used for pay TV? _____ Cable? _____

15. What does a rented pay TV unit do? _____ How might this be done? _____

16. For normal cable TV what may be connected between the cable and the TV set? _____ When is an electronic unit connected between the cable and the TV set? _____

17. How may long-distance TV signals be relayed to a cable center? _____

18. How might a cable company obtain satellite TV programs? _____

19. What are the sweep rates for amateur SSTV? _____

20. What type of picture does SSTV produce? _____

TELEVISION QUESTIONS

1. Which TV emission is AM? FM?
2. What is used to keep the video and aural signals from intercoupling if both are fed to the same antenna?
3. What is the basic emission classification of the TV video signal? Of the aural signal?
4. Draw a block diagram of a simple TV transmitting station.
5. Why are pulses added to the transmitted TV signal?
6. Draw a block diagram of the circuits used in a TV receiver.
7. What are the only circuits not discussed previously that are found in a TV receiver system?
8. As what may the VHF RF amplifier and mixer be used when receiving UHF stations?
9. Name two possible TV transmitting antennas. What polarization is used?
10. Draw a diagram of a TV transmitting diplexer.
11. What determines direction and distance of deflection in a magnetically deflected electron beam?
12. Would a vertical deflection coil be at the CRT side or top?
13. What is the advantage of using interlaced scanning in TV?
14. What is the aspect ratio of TV pictures?
15. What is the difference between a field and a frame?
16. Why must the horizontal sweep frequency be 15,750 Hz for a B/W frame? What waveform is used for the sweep voltage?
17. What type of camera tube was first used in TV? The second? The present type?
18. In a vidicon, how is deflection produced? Picture centering?
19. Sketch the pulses and picture information as they exist in a TV transmitter.
20. List the six types of pulses in a TV transmitter.
21. Visual modulation is found between what pulses?
22. What rides on top of a vertical blanking pulse?
23. What determines the width of a vertical blanking pulse?
24. Why is the vertical sync pulse serrated?
25. How are remote programs relayed to studios or transmitters?
26. If the pulse-peak level is 100% modulation, what is the reference black level? White level? Blanking level? What are the extremes of the video modulation?
27. Draw a block diagram of two camera chains with associated equipment to produce a TV signal to feed to a transmitter.
28. Draw a block diagram of a three-projector multiplexer to feed pictures to one film camera.
29. How is a 24-picture movie film made to produce 60 fields per second?
30. What are the names of the four tracks that may be recorded on a 2-in. video tape? How many recording heads on the rotating wheel?
31. What type of recorder uses 1-in. tape? What are the five tracks that may be recorded on such tapes?
32. What size tapes are used for portable TV pickup?
33. What is the spectrum analyzer called that is used to tune a TV transmitter?
34. Sketch the location of carriers and SBs in a 6-MHz monochrome TV channel. How far are the visual and aural carriers from the lower-band edge?
35. What is the meaning of vestigial sideband?

36. How is the operating power of the aural transmitter determined? The peak power of the visual transmitter?
37. Name the three TV BC bands. What are the band limits of each?
38. What are the three types of logs that must be kept in TV stations? How may they be made out?
39. Draw diagrams of a JFET cascode TV receiver RF amplifier and a MOSFET RF stage.
40. Draw a diagram of a hairpin-type UHF mixer circuit.
41. Sketch the IF bandpass characteristics that would be used when tuning a TV receiver IF strip.
42. What are the frequencies of the two wave traps used in TV IF strips? What do they trap out?
43. Draw a diagram of a two-stage TV IF strip with series-connected amplifying devices, showing WTs.
44. Draw a diagram of a video detector, amplifiers, and picture tube for a B/W TV set.
45. Why is there a 4.5-MHz WT in a video stage?
46. What three signals are taken from the second detector in a TV receiver?
47. What controls picture brightness? Contrast?
48. What is the function of the sync-separator circuit?
49. Draw a diagram of a vertical-deflection system from sync-pulse input to vertical-deflection coils. What type of circuit is this? What are the four variable controls?
50. What is a raster?
51. What is the frequency of the vertical sweep oscillator in a monochrome set? The horizontal?
52. What are six uses of the horizontal deflection ac?
53. Draw a diagram of a horizontal-deflection system from sync-pulse input to horizontal-deflection coils. What controls are included in the circuit?
54. Draw a diagram of a horizontal sweep amplifier, high-voltage circuit, and yoke coils around a CRT.
55. In Fig. 24-27, what is the function of C?
56. From what part of the TV signal is the AGC voltage developed?
57. What two pulses must be present simultaneously in an AGC circuit to make it a keyed-AGC system?
58. Draw a diagram of a keyed-AGC system.
59. What is the frequency of green light? Blue?

60. What must be projected on a CRT screen to produce a purple color? Yellow? White?
61. What does a dichroic semimirror do?
62. How many vidicon tubes are used in a color camera?
63. Draw a block diagram of a TV camera to transmitter color system. How many resistors would be used in the transmitting matrix?
64. Sketch the interleaved chroma SBs with luminance SB clusters of a color transmission. What are the horizontal and vertical sweep frequencies?
65. Sketch the Y, Q, and I SBs with video and aural carriers in a TV channel.
66. What is the frequency of the ac added to the back porch of the horizontal blanking pulse? How many cycles are used? For what are they used?
67. Draw a block diagram of receiver color signal circuits from the video detector to the three color grids of the CRT.
68. How may spurious signals be cleaned out between Y SB clusters?
69. What are the basic circuits added to color receivers that are not used in monochrome?
70. What are the two common types of color CRTs being used?
71. How can TV video transmissions be scrambled or encoded? Audio? What kind of TV is this called?
72. Why may channel 4 TV programs be received on some other channel when using cable TV?
73. What bands may be received when using cable TV converters?
74. If poor TV signals are received on a cable, what are the possible causes?
75. What are the components that would be attached to a satellite receiving parabolic dish antenna?
76. What type of modulation is used with satellite TV emissions?
77. With SSTV what is the duration of the horizontal sync-blanking pulse, horizontal sweep rate, vertical sync-blanking pulse duration, vertical sweep rate?
78. With SSTV what AF represents a white video signal? A black? What is the total pixel number of a picture?
79. With commercial telephone-line freeze-frame SSTV, what is the pixels/line value? How many pixels per picture?

25

Maritime Radio

This chapter describes radio communication equipment of both telegraphic and telephonic types which may be found aboard seagoing vessels. This includes medium-, high-, and very-high-frequency transmitters and receivers, auto-alarms, teleprinters, direction finders, loran, omega, facsimile, and satellite equipment. (Shipboard radar is discussed in Chap. 26.) The duties of radio operators are outlined briefly. More detailed information will be found in "FCC Rules and Regulations," Part 83.

25-1 RADIO ABOARD SHIP

Small vessels may not carry radio equipment, but ships carrying passengers or those over certain tonnage ratings may be required to have radio equipment aboard for the safety of life at sea. Which ships must carry radio equipment is set by international agreement. Usually, ships of more than 500 gross tons or those carrying more than 12 passengers on international deep-sea voyages are compulsorily radio-equipped.

Rather sophisticated communication equipment may be found aboard larger ships. Smaller vessels, commercial fishing boats, tugs, boats on inland waters, etc., may carry a low-powered radiotelephone transmitter and receiver for communication with other ships, nearby coast stations, or the Coast Guard. Such equipment may operate in the 1619–2850 kHz or 156–162 MHz bands, is fixed-tuned, and requires no radio operator but only a radiotelephone permittee to operate it.

Ships usually have 115 V 60 Hz ac but some may have 115 V dc. As a result, shipboard radio equipment may be manufactured to operate on 115 V ac or dc, or from 12- or 24-V batteries.

As in other radio services which require the sharing of frequencies, the minimum necessary transmitting power should be used at sea for routine communications. For distress communications maximum power output is desirable, except when conservation of batteries is a factor.

25-2 COMPULSORY RADIOTELEGRAPH INSTALLATIONS

When larger ships sailing internationally are required to have radio aboard by the communications act or by the radio safety convention, the minimum *radiotelegraph* requirements are usually (1) a main 405–535 kHz transmitter of at least 160-W A1A and 200-W A2A, (2) a main receiver capable of receiving A1A and A2A on 100–200 and 405–535 kHz bands, (3) a main power supply, usually the electrical system of the ship, and (4) a main antenna with a safety link. In addition, it must have (1) a reserve 405–535 kHz 25-W A2A transmitter, (2) a reserve receiver capable of receiving A1A and A2A on the 100–200 and 405–535 kHz bands, (3) a 6-h minimum reserve power supply, usually batteries, (4) a reserve antenna, (5) an emergency-light system for the operating area, and (6) an automatic radiotelegraph alarm signal keyer. An efficient two-way intercom system is required between the radio room and the ship's bridge. A reliable clock of 5 in. or larger diameter marked off in silence-period segments must be mounted within sight of the radio operator.

If there are an insufficient number of operators to stand 24 h of watch daily, an approved auto-alarm (Sec. 25-16) will be used during the times operators are not on watch.

25-3 COMPULSORY RADIOTELEPHONE INSTALLATIONS

When a ship is required to have radiotelephone because of communications act or safety convention requirements, it must carry *on the bridge* a radiotelephone transmitter capable of at least 50-W PEP A3E or J3E on 2182 kHz (the distress and calling frequency) and 2638 kHz (safety), and at least two other frequencies in the 1605–2850 kHz ship-to-shore and intership band. It must have a device that will transmit the *international radiotelephone alarm signal*, a 30–60 s warbling 2200/1300-Hz 0.25-s series of tones. It must have two receivers, one capable of being switched to any of the assigned transmitter frequencies, and one manually tuned for J3E, R3E, and A3E on any frequency between 1605 and 3500 kHz. This equipment uses a vertical antenna on the bridge. A reserve power source that will operate all the radio equipment plus an emergency-light system for the operating area for a period of 6 h must be provided. A 5-in. minimum-diameter clock must be provided at the operating position, and a spare antenna must be carried.

All ships must carry a VHF bridge-to-bridge transmitter capable of transmitting 1 and 8–25 W of 5-kHz-deviation F3E on 156.3 MHz (safety frequency) and 156.8 MHz (distress and calling), plus other frequencies in the 156–162 MHz band used for ship-to-shore communications in the area in which the ship is to be navigated. It must be capable of reduced power operation of 0.75–1 W and carry a receiver preset to and capable of accurate

and convenient selections of the frequencies of 156.3 and 156.8 MHz, as well as the frequencies of its associated transmitter. The reserve power must be capable of supplying the equipment for 3 h.

25-4 MAIN CW TRANSMITTERS

The main radiotelegraph transmitter aboard a compulsorily equipped ship must be capable of a minimum of 160 W of A1A power output to the main antenna, a minimum of 200 W of A2A output to the main antenna, and break-in operation.

The A2A emission required when transmitting an $\overline{SOS}$ on 500 kHz must be 70 to 100% modulated. The frequency of the modulating tone must be between 300 and 1250 Hz.

The transmitter must be capable of transmitting on the international calling and distress frequency of 500 kHz, on the direction-finding frequency of 410 kHz, and on at least two other working frequencies. The frequency tolerance is 20 Hz. The transmitter must include an antenna ammeter plus a final-amplifier plate voltmeter and milliammeter. It must have some means of reducing the plate input power to 200 W or less for tuning and short-range operation, usually by varying the V_p.

The antenna power is determined by the direct method, $P_o = I_a^2 R_a$ (Sec. 17-12).

A commercial marine radiotelegraph main transmitter in block form is shown in Fig. 25-1.

The RF oscillator stage is either a crystal oscillator or a synthesizer (Sec. 25-12). The operating frequency is selected by changing crystals or adjusting the synthesizer switches. The oscillator drives a broadly tuned (410–535 kHz) 6146B VT RF pentode stage, which in turn drives the parallel 813 pentodes in the power amplifier. This stage is π-network-coupled to the antenna through the keying relay. With the transmitting key up, the relay connects the antenna to the receiver and allows all RF stages to be biased beyond cutoff, preventing any output from the transmitter. When the key is pressed, the relay arms move down, disconnecting the antenna from the receiver and connecting it to the PA stage. At the same time the high bias is grounded and all RF stages start operating. Note that the pretuned powdered-iron PA inductors are ganged to the frequency switch, so that no tuning is required by the operator when switching from one frequency to another. The antenna circuit has a tapped inductance plus a *variometer* (variable inductor) to allow resonating regardless of the length of the antenna used. Variometers tune over a much

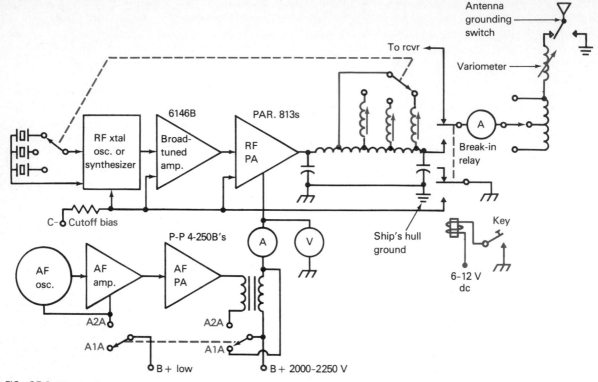

FIG. 25-1 Block diagram of a representative marine main transmitter (from ITT Mackay Marine type 2012).

wider range of frequencies in the MF range than do similar-sized variable capacitors. To transmit A2A (modulated CW, or MCW), the A1A-A2A switch is moved to the A2A position. This turns on a sinusoidal-AF oscillator, amplifier, and power amplifier having two 4-250B power tetrodes in push-pull. The AF output voltage is added in series with the B+ to the RF PA through the modulation transformer, modulating the output stage with a constant AF tone.

25-5 RESERVE TRANSMITTERS

Reserve transmitters must produce four frequencies with at least 25 W of A2A-type RF output in the 410–535 kHz band. A solid-state reserve transmitter is shown in semiblock form in Fig. 25-2. The RF oscillator is a *Butler* circuit having the crystal between the emitters of an amplifier stage and a direct-coupled emitter follower. The frequency of the oscillator, is four times the eventual output frequency. In this way the oscillator can run continuously but cannot be picked up by the local receiver (as it would be if the oscillator operated on

the transmitting or on some subharmonic frequency).

The oscillator drives a dual J-K flip-flop IC (Sec. 12-10) that twice divides the frequency by 2, resulting in a 500-kHz output from a 2-MHz crystal. However, the FF output is square-wave and must be sinusoidally shaped by the following tuned class C RF amplifiers. The final RF power amplifier consists of three parallel transistors feeding a bandpass filter circuit (the parallel *LC* circuits and capacitor to ground) and the antenna tuning circuit. In this case the antenna is the *LC* circuit for the PA stage. Since part of the circuit is a bandpass filter, no harmonics can be transmitted, which would be the case if an amplifier were directly coupled to an antenna circuit. The key activates dc amplifiers that key the FF IC, the driver amplifier, and feeds the antenna relay coil with dc. The modulation is produced by a 500-Hz AF oscillator and amplifier coupled to the collectors of the parallel transistors in the RF PA. The AF stages run continuously. The power supply is a 12-V battery, which must be capable of 6 h of continuous operation with an output of at least 25 W of A2A on both 500 and 410 kHz.

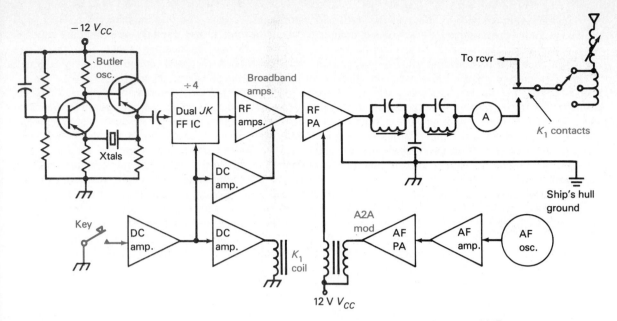

FIG. 25-2 A solid-state reserve transmitter in block form (from ITT Mackay Marine type 2017)

25-6 MARINE ANTENNAS

The main antenna must be as efficient as practicable under the prevailing physical limitations on board ship. It may consist of a single wire hung from the peak of the foremast to the peak of the mainmast, with a lead-in to the radio station (Fig. 25-3*a*). On small vessels it is made as long and kept as high as possible. A *safety link* is incorporated in any such horizontal antenna. It is a planned weak section. A short weak wire and a longer heavy wire are fastened across one another and are inserted in the halyard as illustrated. If the ship is subject to a sudden stress, because of collision, grounding, torpedoing, etc., the weak part of the link breaks as the masts are forced apart and the antenna takes up the slack in the longer, stronger part of the link. This causes the antenna to sag but prevents its falling, and it remains usable.

Unless care is taken in hoisting an antenna, insulators may fracture if they strike a metal object. The wire must not be allowed to kink, since it weakens at such a point and may break under sudden strain.

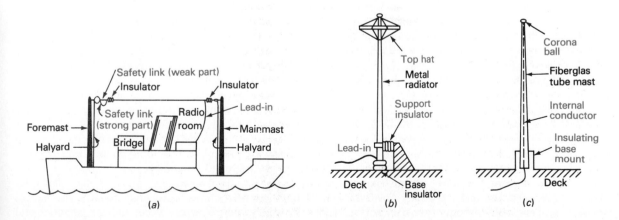

FIG. 25-3 Ship antennas: (*a*) horizontal wire, (*b*) top-loaded vertical, and (*c*) HF vertical.

When the main antenna is used on the 410–515 kHz band, it is operating as a base-loaded quarter-wavelength wire. The far end is the maximum voltage point, and the part leading into the transmitter is the high-current part. When used on high-frequency bands, the far end is always a high-voltage point, with other high-voltage points every half wave along the wire, and high-current points halfway between the voltage points. Ships using high frequencies may use shortwave dipole or vertical antennas for receiving, and the main antenna for transmitting.

Many newer ships use relatively short vertical antennas for both MF and HF communications. A possible MF through HF vertical is shown in Fig. 25-3b. It is a top-loaded vertical pole of stainless-steel tubing, perhaps 40 ft (12.3 m) in length, tapering from about 5 in. at the base to about 2 in. at the top, with a 9-ft-diameter circular spoked 1-in. tubing top hat. These antennas, when used at lower frequencies, have very low feed-point impedance ($\pm 2\ \Omega$) and require impedance loading or transformation to match the relatively higher output impedance of transmitters or input impedance of receivers. Sometimes an unloaded vertical whip is used for HF communications. It may be constructed of tapering Fiberglas with a heavy bronze wire running to a rounded metal top (to discourage corona discharge) (Fig. 25-3c).

All shipboard antennas operating in the MF bands radiate predominantly vertically polarized signals, even if they appear to be horizontal types because of their relatively low height in wavelengths.

Maritime or land-based mobile HF stations may operate on several bands separated widely in frequency. An HF mobile antenna may be a center- (or base-) loaded $\lambda/4$ vertical whip fed by a 50-Ω coaxial transmission line. The normal 37-Ω base impedance of a $\lambda/4$ vertical is reduced to perhaps 8 Ω when a vertical is shortened and an inductor is added at its center to bring it to resonance. For any given narrow band of frequencies and any given whip length, there will be some necessary value of center inductance. In Fig. 25-4a the RF is fed to the 8-Ω base impedance from the 50-Ω coaxial line through a 50- to 8-Ω L-pad impedance matching network. Such shortened antennas are very sharp tuning. The center-loading coil is selected to resonate at the highest frequency of the band being used. To resonate the antenna at any *lower* frequencies in the band an additional base coil can be used. This may be in the form of a multitapped coil, or

preferably a *rotor coil* (contact rides along a coil when the coil turns) capable of producing any required fractional inductance value. The rotor coil can be rotated by hand or be driven by a small permanent-magnet-field dc motor.

If it is required to operate on a second higher-frequency band, there are several possible means of making the same antenna resonate at the second frequency band. One is shown by the dashed lines in Fig. 25-4a. A second whip top section a few inches from the first is used, but it is tapped down the coil so that its total length plus the smaller center coil resonates at the higher band frequency. The tap is set with the rotor coil at zero inductance for the highest end of that band so that the rotor inductor can add inductance and thereby tune the antenna to any frequency in that band. Usually other L-pads will have to be switched in for any second or third bands.

Such antennas can be made *autotunable* (self-tuning) by using a sensing circuit to determine whether the antenna is too long (X_L) or too short (X_C) and moving the rotor coil to a point where X_L and X_C are equal for the transmitting frequency being used. A simplified sensing circuit is shown in Fig. 25-4b. The 1-turn toroid primary is inserted at point X in the antenna. The double diodes may be recognized as forming a discriminator-like circuit. If the antenna is *too long* the V-I phase of the RF ac appears inductively reactive when the antenna is excited by the transmitter. Under this condition the circuit might develop a positive output from U1, shown by the color indicators, turning both Q_2 and Q_3 on, while U2 turns both Q_1 and Q_4 off. The PM motor is now in series with ground through two turned-on (low Z) transistors Q_1 and Q_4, and 12-V positive. The motor is connected so that under this condition it turns the rotor coil in the required direction to reduce the rotor coil inductance. If the antenna is *too short*, U2 would then be fed an input that would produce a positive output from it, driving Q_1 and Q_4 on, while U1 turns Q_2 and Q_3 off. Current now flows through the PM motor in the opposite direction, reversing its rotation and reversing the rotation of the rotor coil, thereby increasing the tuning inductance. When the antenna is tuned to the frequency of the transmitter (SWR is 1:1), or if the transmitter is off, the inputs to U1 and U2 are equal, providing equally positive outputs and biases to Q_1 and Q_3, so that points X and Y are also at equal potentials. Thus, if the transmitter is off, or when the antenna tunes itself to resonance, the PM motor is across zero voltage potential and stops. With the rotor coil turning at about 60 rpm,

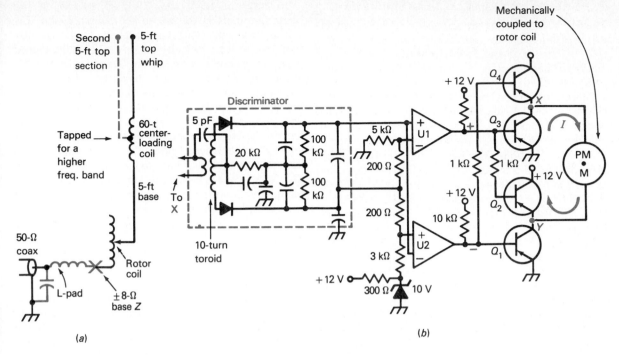

FIG. 25-4 (*a*) HF mobile whip antenna. (*b*) Autotuning sensing circuit.

only a few seconds are required for tuning the antenna after the transmitter starts producing RF output.

| Test your understanding; answer these checkup questions.

1. If a ship has only dc aboard, how could 60-Hz ac be developed for radio equipment? _____
2. What is the minimum power requirement for a main 500-kHz transmitter? _____ A reserve? _____
3. With what part of the ship is the radio station connected by intercom? _____
4. What radiotelephone equipment must all U.S. ships carry? _____
5. What is the required frequency stability of a 500-kHz radio transmitter? _____
6. What is the required modulating frequency for A2A $\overline{SOS}$ transmissions? _____ Modulation %? _____
7. Is shipboard antenna power determined by a direct or by an indirect method? _____
8. In Fig. 25-1, what circuits must be tuned when changing frequency? _____ What type of tuning is used in the PA? _____ Why is break-in used? _____ What are the three required meters? _____ _____ _____

9. What might the 4-250B designation on the power tetrode tubes mean? _____
10. What is a Butler circuit? _____
11. In Fig. 25-2, what does a single J-K FF do? _____ How many circuits must be tuned when the frequency is changed? _____ Why can the oscillator run but not be heard in the receiver? _____
12. For what is a safety link used? _____

25-7 MF VERSUS HF COMMUNICATIONS

During daylight hours the medium frequencies between 405 and 535 kHz provide reliable communication for a 200-W transmitter over a range of 300 to possibly 600 miles under normal conditions. During nighttime when sky waves are not absorbed, the range may extend to more than 2000 miles. However, medium frequencies are subject to high-amplitude static, making long-range communication difficult at times. Ships on long voyages find that frequencies between 4 and 25 MHz provide more satisfactory operation over long distances (Secs. 20-1, 20-2). Most ships engaged in international trade will have high-frequency as well as medium-frequency main transmitters.

25-8 HIGH-FREQUENCY TRANSMITTERS

The HF marine bands are in the 4–25 MHz part of the spectrum and are commonly known as the 4-, 8-, 12-, 16-, 22-, and 25-MHz bands. The first six bands are harmonically related; the 22- and 25-MHz bands are not.

Marine equipment has progressed to rather sophisticated synthesized radiotelegraph, SSB radiotelephone communications with shore-based telephone companies, high-speed *telex* (radioteleprinter), and satellite communications. The transmitter shown in block form in Fig. 25-5 represents a simplified modern marine HF transmitter capable of transmitting A1A (CW), F1A (FSK), A3A (reduced-carrier SSB), A3H (full-carrier SSB), and A3J (suppressed-carrier SSB) signals.

To transmit A1 signals, the CW-SSB and CW-FSK switches are thrown to CW and the desired assigned frequency is selected. Band switches (not shown) determine which CW exciter stages operate as multipliers or amplifiers and to which band the linear amplifier *LC* circuits are tuned. When the key is pressed, the crystal oscillator or synthesizer is keyed on. The transistor amplifiers (or multipliers) amplify (or multiply) the signal that has been generated. Only the last two linear amplifiers are power vacuum tubes. Break-in keying (Sec. 16-3) is always used. The drive to the broadband transistor amplifier in the linear amplifier section is controlled at the output of the last CW exciter multiplier or amplifier. High power (1000 W) and low power (150 W) can be selected by changing the bias on the power amplifiers. If the plates of the tubes dissipate too much heat due to overloading, a light-sensitive cadmium sulfide cell senses it and trips an overload relay (not shown), turning off the linear amplifier power supply.

To transmit F1 teleprinter signals, the output from a separate FSK generator and amplifier (not shown) is fed into the FSK input. The CW-FSK switch is thrown to FSK.

To transmit SSB, the CW-SSB switch on the linear amplifier is thrown to the SSB position. Either push-to-talk (PTT) or voice-operated transmission (VOX) can be selected on the SSB generator panel to key the transmitter on. The microphone AF ac is amplified and fed to balanced modulator 1, producing a double-sideband-without-carrier signal based on 455 kHz. By passing this through a filter only the *upper* sidebands of the modulated carrier remain. These are amplified by linear amplifiers and are fed to balanced modulator 2 (in one of six heterodyne exciters, one for each marine band). The 455-kHz USB signals are mixed in balanced modulator 2 with the proper frequency to develop the desired operating frequency. The USB signals are amplified by the linear-amplifier section and are fed to the antenna.

To prevent overmodulation and also to allow the average microphone level to be raised, an automatic power control (APC) circuit is used. This is shown as a capacitively coupled USB RF output to a diode. The diode is biased so that it will develop dc only when SB peaks exceed a level equal to the maximum desired power. The dc that is developed is fed back as a bias voltage to one of the 455-kHz linear amplifiers, reducing the gain in this stage and thereby the signal amplitude in all following stages. The result of this feedback is to allow no more than the maximum desired peaks but to produce an increase in the average modulation by about 15 dB—a very great increase in volume for receivers.

25-9 RADIOTELEGRAPH FREQUENCIES

Frequency bands used by *ship radiotelegraph* stations for calling, working, and distress traffic are:

- MF or 405–535 kHz band, in kHz
  410 (RDF), 425, 444, 454, 468, 480
  500 (calling and distress)[1]
  512 (calling when 500 occupied by distress)
- HF or 4–25 MHz bands
  Calling, all ships, in MHz
  4.1802–4.187 (in 0.4-kHz steps)
  6.2703–6.280 (in 0.6-kHz steps)
  8.3604–8.374 (in 0.8-kHz steps)
  8.364 safety of life only
  12.5406–12.561 (in 1.2-kHz steps)
  16.7208–16.748 (in 1.6-kHz steps)
  22.228–22.246 (in 2-kHz steps)
  25.071–25.075 (in 2-kHz steps)

---

[1]Silence period 15 to 18 and 45 to 48 min every hour.

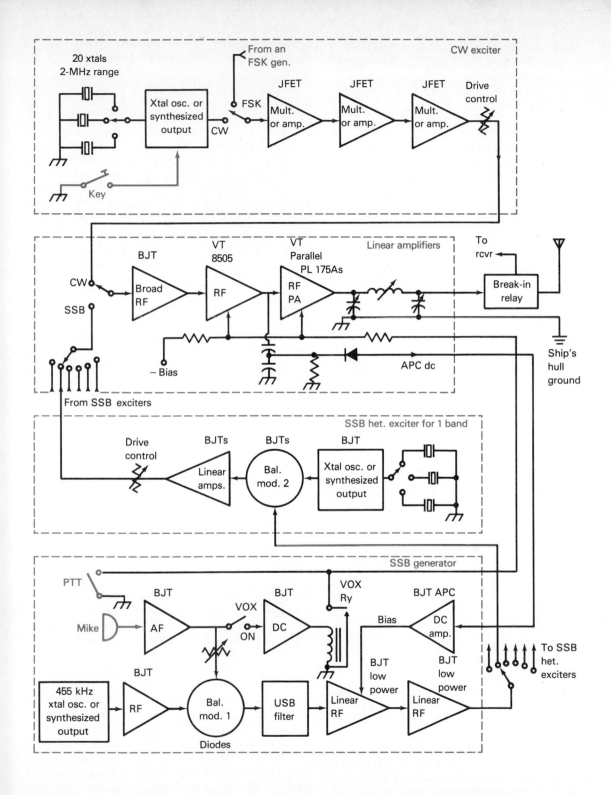

FIG. 25-5 Block diagram of a marine HF transmitter (from ITT Mackay Marine type MRU-27).

Working, high-traffic ships, in MHz
    4.172– 4.176
    6.258– 6.264
    8.342– 8.355
    12.505– 12.532
    16.662– 16.710
    22.189– 22.217
Working, low-traffic ships, in MHz
    4.188– 4.219
    6.282– 6.324
    8.377– 8.359
    12.565– 12.529
    16.754– 16.773
    22.250– 22.261
    25.091– 25.104

Frequency bands in which *coastal stations* communicate with ship radiotelegraph stations are:

- 415–525 kHz band, in kHz
    416–512
- 4–23 MHz band, in MHz
    4.238–4.316
    6.351–6.512.5
    8.502–8.726
    12.745.5–13.119
    16.933.2–17.242.4
    22.407–22.617

25-10 RADIOTELEPHONE FREQUENCIES

The 2–23 MHz or HF bands in which *ship radiotelephone* stations may be assigned for calling, working, and distress traffic are:

- 1.619–2.830 MHz SSB
- **2.182 MHz** International distress[2]
- 4.069–4.434 MHz SSB
- 5.680 MHz SSB
- 6.147–6.512 MHz SSB
- 8.201–8.780 MHz SSB
- 12.358–12.435 MHz SSB
- 16.474–16.593 MHz SSB
- 22.028–22.136 MHz SSB

The 2–23 MHz or HF bands in which *coastal stations* communicate by SSB with ships are:

- 12.379–13.158 MHz
- 16.488–17.283 MHz

The 156–162 MHz or VHF band is divided into 25-kHz channels in which narrowband F3E emissions are used. See Table 25-1.

---

[2]Silence period 3 min on the hour and half hour.

TABLE 25-1	RADIOTELEPHONE CHANNELS		
	Frequency, MHz		
Channel	Ship	Coast	Type of communication
6	156.3	156.3	Intership safety
7	156.35	156.35	Commercial
9	156.45	156.45	Noncommercial
12	156.6	156.6	Port operations
13*	156.65	156.65	Navigational, bridge-to-bridge
15		156.75	Environmental
16	156.8	156.8	Distress, safety, calling
24	157.2	161.8	Public correspondence
68	156.425	156.425	Yachts
WE$_1$		162.55	Weather broadcasts
WE$_2$		162.4	Weather broadcasts

*A continuous listening watch must be maintained without interruption on this channel when vessels are in heavy-traffic areas and on inland waters.

25-11 MARINE MAIN RECEIVERS

The history of radio receivers aboard ships started with *coherer* detectors (encapsulated iron filings which magnetized and adhered to each other when subjected to modulated RF currents and thereby became alternately better and worse conductors). Then came galena or other metal crystal diode detectors. With the vacuum tube came regenerative detectors and TRF receivers. In the 1930s the first superheterodynes appeared aboard ships. They were single-conversion receivers with a BFO for CW detection. To improve image rejection, double- and triple-conversion receivers were developed. Superheterodynes were designed to cover the low frequencies also, allowing them to qualify for main receivers. Today a ship installation may use an all-wave (15 kHz to 30 MHz) synthesized, digital-readout superheterodyne. In addition there must be a 100–535 kHz reserve or main receiver in case the all-wave receiver malfunctions, or to watch 500 kHz while working traffic on HF or distress on 8364 kHz.

A reserve receiver that qualifies as a main receiver covering the two required bands of 100–200 and 400–535 kHz is shown in block form in Fig. 25-6. It is a solid-state regenerative-detector TRF receiver. It has ganged tuning capacitors and band switches (not shown) to cover the required bands. The two protect-circuit series diodes are reverse-

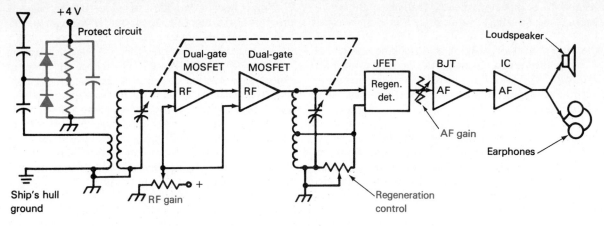

FIG. 25-6 Solid-state 15–560 kHz marine main or reserve TRF receiver (from ITT Mackay Marine type 3018).

biased by 4 V. If an ac signal of more than 4 V is received, the diodes conduct and protect the front-end coils and transistor. The first stage is a tuned dual-gate MOSFET RF amplifier followed by an untuned dual-gate MOSFET amplifier. The RF gain is controlled by varying the bias on the second gates of the MOSFETs. The regenerative detector is a Hartley-type JFET with a potentiometer across the tickler part of the coil as a regeneration control. The detector is followed by a BJT AF preamplifier and an IC AF amplifier feeding earphones or a loudspeaker. The power supply is either a 12-V battery or a rectifier-filter supply. The sensitivity and selectivity may not be quite that of a super-heterodyne, but it is adequate for LF and MF reception with their high noise levels.

> Test your understanding; answer these checkup questions.

1. List the HF CW marine bands. _____ The MF. _____
2. In Fig. 25-4, what would control the CW P_o? _____ The SSB P_o? _____ What does APC stand for? _____ Why would the carrier oscillator not block the transmitting frequency? _____
3. In Fig. 25-4, what does PTT mean? _____ VOX? _____ In what band are the A1A crystals? _____ The SSB carrier oscillator? _____ The SSB channel crystals? _____ What has to be changed in the SSB generator when shifting from band to band? _____ What does closing the PTT or VOX switches do? _____

4. What is the result of using APC? _____
5. On which side of the calling frequencies are the low-traffic working bands? _____ High-traffic bands? _____
6. When is 512 kHz used? _____ For what is 410 kHz used? _____
7. Are coastal stations found above or below ship working frequencies? _____
8. Are ship HF radiotelephone bands in the same general bands as ship radiotelegraph? _____
9. What VHF channel is used for intership safety? _____ For distress, safety, and calling? _____ For bridge-to-bridge communication? _____ On which channel must a continuous watch be maintained? _____
10. What are the two radiotelegraph distress frequencies? _____
11. In Fig. 25-5, what basic receiver circuit is this? _____ What frequency bands must it cover? _____ What does it cover? _____ Why are MOSFETs particularly useful in RF amplifiers? _____ What is the maximum rms RF voltage that will be fed to the receiver circuits? _____

25-12 MARINE SUPERHETERODYNES

The IF frequencies of the original marine superheterodynes were usually in the 455-kHz MF region to produce a narrow IF passband. Unfortunately, signals near the first IF frequency tend to force their way through to the mixer and interfere with other signals being received. Double conversion, with the first IF in the 2–8 MHz region, reduces this effect for MF signals and also results in better image signal rejection. But with two oscillators in the receiver, unexpected beats, called spurious responses or "birdies," appear at various places on the dial. A difficulty in older superheterodynes was the requirement of ganging and tracking one or two RF amplifier tuned circuits with the mixer and local oscillator. A 15-kHz to 30-MHz "third generation" superheterodyne circuit overcomes most of these objections (Fig. 25-7).

If a received signal is too strong because of proximity, it is attenuated by a resistive attenuator and is fed to the RF-tuned circuit and amplifier. After being amplified and fed through a 30-MHz low-pass filter, signals mix with frequencies produced by overtone crystals ranging in frequency from 39 to 67 MHz in 2-MHz steps. A 1-MHz signal would mix with the 39-MHz crystal to produce a 38-MHz difference or IF frequency. A 29-MHz signal would mix with the 67-MHz crystal to produce a 38-MHz IF frequency also. From this it can be deduced that, to receive signals from 15 kHz to 30 MHz in 2-MHz bands, the first IF must be from 37 to 39 MHz. For this reason the first IF strip consists of a bandpass amplifier and a 2-MHz-wide bandpass filter, passing all 37–39 MHz signals from the mixer. Note that the front-end oscillator frequencies are well above any received signal frequencies, and except for a trimmer control to peak the RF signal, there are no circuits that must be gang-tuned.

The received-signal frequency-determining circuit, a variable-frequency oscillator (VFO) tuning the 2-MHz frequency band of 3–5 MHz, is mixed with a 47.94-MHz crystal oscillator to form a 42.94–44.94 MHz injection signal. This signal and the IF_1 signal beat in the second mixer to form 5.94-MHz second-IF signals. These are amplified and fed to a third mixer, where they are heterodyned with a 5.485-MHz crystal to form a 455-kHz third IF frequency. Either bandpass filter, 400 Hz or 2.8 kHz, can be selected for CW or SSB signals.

With no filter the bandpass is wide enough for AM emissions. An amplifier feeds the IF_3 signal to outside auxiliary equipment (teleprinter, scope, etc.).

The detector is a diode for AM signals, with a variable BFO for CW or SSB reception. One of two crystals can also be used to produce the proper beat frequency for USB or LSB reception. A shunt-type noise clipper can be switched on between the detector and the first AF amplifier. AVC is fed to the RF amplifier and the IF_3 amplifier. A sensitive dc meter reads the AVC voltage as an indication of RF signal strength, or the AF voltage can be rectified to indicate the AF signal strength.

The advantages of this receiver are all-wave reception, excellent sensitivity and selectivity, practically no image signals or birdies, and only one tuning circuit plus a trimmer to peak signals. Points of difficulty might be possible drifting of the VFO and the variable BFO.

A fourth-generation receiver is shown in block form in Fig. 25-8. The received signal passes the protect circuit, is attenuated as necessary, is then trimmed to a peak, and is fed through a 30-MHz low-pass filter to the RF amplifier(s). The signal is then mixed with a *frequency synthesizer* acting as a first local oscillator.

A frequency synthesizer puts together different frequencies produced by a stable source by multiplying, dividing, and/or mixing the source frequency to produce desired output frequencies. This particular synthesizer has a temperature-controlled 8-MHz crystal oscillator as a standard which shifts less than a hertz once it reaches operating temperature. The desired frequency is dialed on the digital frequency selector (12.3456 MHz shown). This operates six binary-coded decimal circuits that cause the major and minor phase-lock loop (PLL) circuits to synthesize the desired frequency and lock on it. Since this frequency synthesizer is being used not as a calibrated signal generator, but as a local oscillator for a superheterodyne, the frequency output is offset enough to beat against the incoming signal to produce a first IF frequency in the 92–122 MHz range.

After conversion into the 92–122 MHz IF, the received signal is mixed in a second mixer with an 84-MHz crystal oscillator from the frequency synthesizer that also is feeding into the major PLL circuit. Should the 84-MHz crystal drift a few hertz high, this will increase the major loop output frequency a like amount so that there is no effective change in the 92–122 MHz signal. Thus, the 84-MHz oscillator does not have to be temperature-

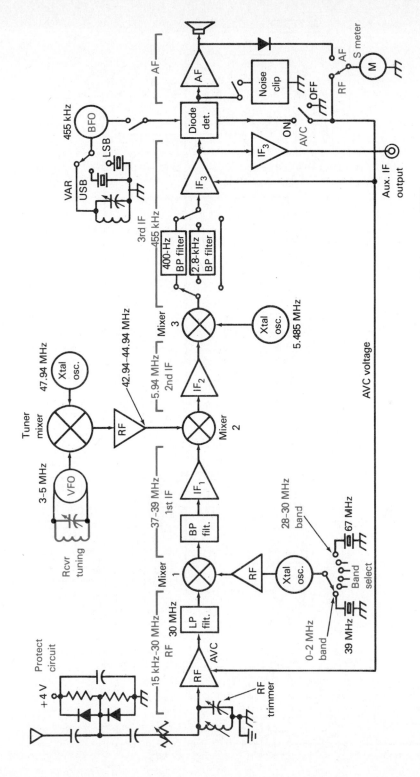

FIG. 25-7 Triple conversion marine third generation superheterodyne (from ITT Mackay Marine type 3010).

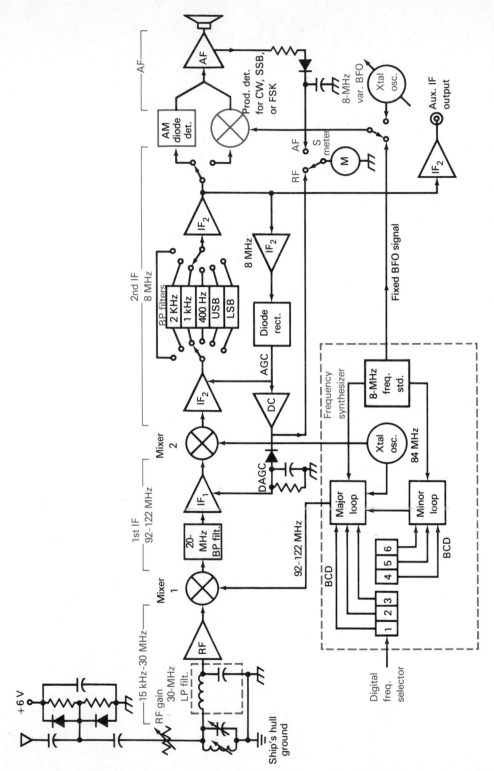

FIG. 25-8 Simplified modern solid-state fourth generation superheterodyne using digital frequency synthesis for the local oscillators (from ITT Mackay Marine type 3020).

controlled. The output of the second mixer is an 8-MHz IF, which is chosen so that the 8-MHz standard can be used as a BFO. The desired bandpass is selected from five different filters plus a wide-band-pass circuit. After further amplification the 8-MHz signal is fed to an emitter-follower amplifier for external equipment and to either a diode detector to detect AM signals or a product detector for SSB, CW, or FSK detection. The product detector acts as a mixer, and is fed either a fixed BFO signal from the 8-MHz standard or the output from an 8-MHz crystal oscillator which can be tuned 1 kHz.

The frequency synthesizer can be set to within 100 Hz of any desired frequency. If a transmitter radiates a 12.3456-MHz signal, the receiver frequency selector is set to that frequency and the product detector will produce a zero-beat output. This is desirable for SSB. However, to receive a CW station on this frequency, if the operator prefers to listen to code at a 700-Hz pitch, the frequency select is set to 12.3456, and the variable BFO is tuned 700 Hz higher (or lower) than 12.3456 MHz.

To provide an AVC or AGC voltage for AM signals, the 8-MHz IF is amplified, rectified, filtered, and fed to the first IF amplifiers as a delayed AGC, and to the second IF strip amplifiers as an undelayed AGC. This particular receiver uses solid-state devices only, for minimum power drain and fault-free operation.

25-13 BRIDGE-TO-BRIDGE RADIOTELEPHONE

In 1973 it was made mandatory for all U.S.-flag power-driven vessels of over 300 gross tons, or of 100 gross tons and carrying one or more passengers, or any towing vessel 26 ft or over in length operating in navigable waters of the United States to have an 8- to 25-W 1K6ØF3E transmitter and accompanying receiver operating on channel 13 in the 156–162 MHz band. The transmitter must also be capable of operating at a power level of 0.75 to 1 W output. Such type-accepted equipment must be mounted on or operated from the navigational bridge. The transmitters have a pre-emphasis curve of 6 dB per octave from 300 to 3000 Hz. The receivers must have a similar de-emphasis response curve. Only the master or the person in charge of navigating the vessel is to operate the equipment. A continuous watch must be maintained on the *simplex* (transmit and receive on the same frequency) channel 13 (156.65 MHz) as long as the vessel is

within 100 miles of U.S. shores. Foreign vessels may have portable equipment brought aboard by pilots. Because a continuous watch must be maintained, a bridge-to-bridge radiotelephone system must have two receivers if it is desired to communicate on other channels regarding port operations, commercial traffic, public correspondence, or international distress, or to listen to weather broadcasts on 162.4 or 162.55 MHz.

An example of the sophistication that may be found in some equipment is the ITT Mackay Marine type 222 VHF radiotelephone system. This is an all-channel solid-state transmitter, two receivers, a synthesizer, a scanning system, and a ringer. It features automatic shift from simplex operation (used on 156.3 and 156.375–156.875 MHz) to duplex operation (used on all other channels, with the receiving frequency always 4.6 MHz higher than the transmitting frequency) at any time that the receiving channel is switched. The system has one transmitter but two separate receivers feeding a single audio circuit. One receiver uses the transmitting antenna, and the other uses its own separate antenna. The synthesizer supplies both the fundamental frequency for the transmitter and the local oscillator signal for the receivers. The system may also incorporate a channel-scanner circuit that continually samples up to six channels to see if there are any calls coming in on them. A coded tone transmission from a calling station can activate a ringer circuit on the called vessel only.

The bridge-to-bridge systems provide instant intercommunication, as well as afford assured response to distress traffic. VHF channels now carry much of the traffic that was previously carried by the 1600–2850 kHz radiotelephones, since port authorities, telephone company facilities, etc., are available on VHF. For long-haul voice communications the 4–24 MHz bands are used.

25-14 SURVIVAL RADIO EQUIPMENT

Ships must carry *portable*, floatable, survival-craft radio transmitter-receiver units capable of 3-W output on 500, 2182, and 8364 kHz. The pretuned transmitter must be capable of sending groups of 3 or more SOS signals automatically. On 500 kHz the automatic transmission consists of the radiotelegraph auto-alarm signal (Sec. 25-15) followed by SOSs. On 8364 kHz the SOSs are followed by a 30-s dash. Frequency stability on 500 and 2182 kHz is ±50 Hz, and on 8364 it is ±420 Hz. The unit must contain earphones, erectable wire or collapsible rod antenna, an antenna resonating control, a telegraph key, a hand-crank type generator power supply, a receiver that covers 492–508 kHz untuned, 2179–2185 kHz untuned, and 8320–8745 kHz, tunable, plus an attached ground wire with a weight to be dropped into the water. The unit must not exceed 60 lb.

A portable survival transceiver is shown in block form in Fig. 25-9. The frequency selection switch is in the MF (500-kHz) position. With the key up, the relay coil is energized and the antenna is connected to the MF tuned circuit of the receiver, which feeds 500-kHz signals to the mixer or amplifier stage. Since the local oscillator is not connected, the stage acts as a 500-kHz amplifier, as does the next stage. This feeds 500-kHz signals to the diode detector, which beats against the BFO signal to produce the audio frequency to be amplified and fed to the earphones. The receiver operates as a TRF on 500 kHz.

When the key is pressed, the relay coil is de-energized and the TR relay arms move up, shifting antenna and dc from receiver to transmitter. A 500-kHz signal is now generated by the MF crystal oscillator, is amplified, and is fed to the antenna. A 550-Hz AF oscillator modulates the PA to produce an A2A emission. The variometer in the antenna circuit tunes the MF power amplifier to resonance for maximum power output.

Power-supply voltages are generated by hand-cranking a dc generator through rotation-increasing gears. If the M–A switch is moved from manual to automatic, a geared-down code wheel is rotated, which makes and breaks electric contacts as a wiper moves over projections along the wheel edge to produce automatic auto-alarm signals, $\overline{SOS}$, and/or a long dash to allow nearby vessels to take direction-finder bearings on the lifeboat.

When the master switch is moved to HF, the local oscillator in the receiver section is turned on and the antenna is connected to an 8364-kHz tuned circuit which is coupled to the mixer input. The LO

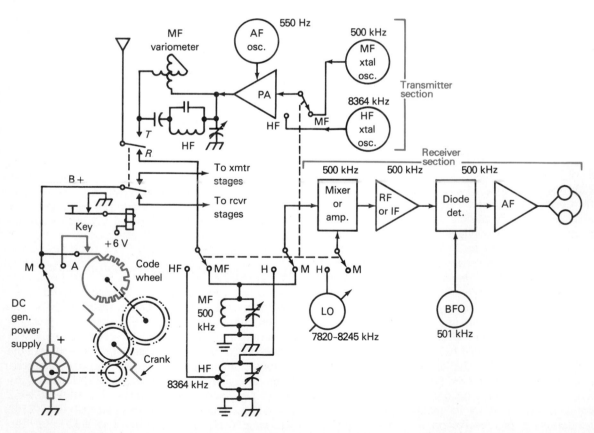

FIG. 25-9 Block diagram of a portable 500–8364-kHz survival transceiver (from ITT Mackay Marine type 401).

is tunable from 8320 to 8745 kHz, allowing the operator to locate answering signals if not on 8364 kHz. On HF the receiver operates as a super-heterodyne.

Note that the MF and HF transmitting antenna tuning circuits are in parallel. For HF ac the variometer has so much inductance it looks like an RFC and has little effect on circuit operation. For MF ac the HF-tuned circuit and its series capacitor appear as a small capacitor across the variometer and have little effect on circuit operation. The variable capacitor tunes the antenna to 8364 kHz.

Nonportable lifeboat equipment is permanently installed aboard at least one cabin-type motor lifeboat in a housing large enough to hold the equipment and the operator. The A2A transmitter must be capable of at least 70% modulation and not less than 30-W output on 500 kHz, and 40-W output on 2182 and 8364 kHz. It is powered by batteries (6 or 12 V) of sufficient capacity to operate the equipment for at least 6 h. It must be capable of manual operation on either frequency and have the same automatic transmissions as the portable types. It must have an antenna at least 20 ft high, an artificial antenna for testing, and earphones.

Ships that carry paying passengers must carry an Emergency Position Indicating Radio Beacon (EPIRB) transmitter. It is mounted on an open part of the bridge. If the ship sinks, the EPIRB floats free, turns upside down, and rotates an internal mercury switch, activating it. It then transmits 700-Hz decreasing-tone-modulated signals on 121.5 and 243 MHz alternately. These signals are used for search and rescue (SAR) operation. All EPIRBs must be FCC-licensed. The frequency 121.5 MHz is also the aircraft distress frequency.

Test your understanding; answer these checkup questions.

1. In Fig. 25-6, why does this receiver have high sensitivity? _____ Selectivity? _____ Image rejection? _____ What would be the image frequency when tuned to 30 MHz? _____ 2 MHz? _____ How many band selector crystals are needed? _____
2. In Fig. 25-7, over what band of frequencies must the frequency synthesizer generate frequencies to be able to receive signals from nearly zero frequency to 20 MHz? _____
3. Why might Fig. 25-7 be better for A3J than Fig. 25-6? _____
4. Why might DAGC be better at the front end of a receiver than AGC? _____

5. What is the purpose of the 20-MHz BP filter in Fig. 25-7? _____ The 400-Hz filter? _____ The 2-kHz filter? _____
6. How can a crystal oscillator be made variable? _____
7. If the dial in Fig. 25-7 is adjusted to 07.5412, to what frequency should the receiver then be tuned? _____ Within how many hertz of this value will it be? _____
8. What is the only LC circuit in Fig. 25-7 that must be band-switched? _____ Why? _____
9. What does 2K80J3E mean? _____ 6K00A3E? _____ 200HA1A? _____ 600HF1B? _____
10. What is meant by simplex? _____ Duplex? _____
11. Why is a synthesizer used with VHF radiotelephone sets? _____
12. What is the P_o of a portable survival transmitter on 500 kHz? _____ On 8364 kHz? _____ Of a nonportable lifeboat transmitter on 500 kHz? _____ On 8364 kHz? _____
13. In Fig. 25-9, what CW letters are shown on the code wheel? _____ Can the BFO be turned off? _____ Why is the dc generator geared up? _____

25-15 AUTO-ALARM KEYERS

When no radio operator is standing watch on a compulsorily equipped ship at sea, a device known as an *auto-alarm* (AA) must be in operation. By international agreement, prior to sending an $\overline{SOS}$ on 500 kHz, all ships will transmit (if time allows) an AA signal consisting of alternate 4-s dashes and 1-s spaces for a period of 1 min. Whenever possible, the AA signal will be followed by a 2-min silence before any $\overline{SOS}$ message is transmitted.

An AA receiver registers dash-space transmissions, rejecting them if they are not timed correctly but accepting them if they are reasonably close to the required lengths. After registering three or four successive, properly made dashes and spaces, the AA rings a bell in the radio station, on the navigation bridge, and in the radio operator's living quarters. Note that the AA does not stand watch for an $\overline{SOS}$ but for the AA signal that should precede any $\overline{SOS}$.

While the auto-alarm signal may be transmitted manually by an operator watching the sweep second hand of a clock, an alarm may be transmitted by automatic means. The automatc AA keyer in Fig. 25-10 is operated from an emergency battery. When the switch is turned on, a reed-type 60-Hz vibrator is set into vibration. When the reed is in the UP

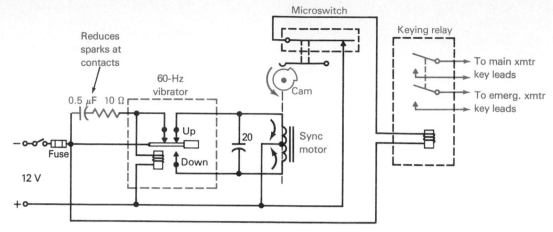

FIG. 25-10 An auto-alarm signal keyer used to transmit an AA signal automatically (RCA AR8651).

position, current flows downward through the motor; when it is in the DOWN position, current flows upward through the motor. Thus, the motor is fed ac.

The synchronous motor rotates at 3600 rpm, which is reduced by gears to a speed of 12 rpm (5 s per revolution) to rotate a circular cam. One-fifth of the outer rim of the cam is raised, as indicated. When the raised portion strikes the *microswitch* (light-pressure switch), the keying-relay circuit is opened and the relay contacts open. For 4 s the keying-relay circuit remains closed, followed by a 1-s open period. One of the two sets of contacts on the keying relay is permanently connected across the key leads of the main transmitter, and the other across the key leads of the reserve transmitter.

An electronic auto-alarm keyer is shown in Fig. 25-11. Transistors Q_1 and Q_2 form an asymmetrical astable multivibrator. Since the ratio of resistances in the two circuits is about 5:1, one transistor conducts four times longer than the other. The dc amplifier Q_3 coupled to the collector of Q_2, operates a sensitive relay, closing it for 4 s and opening it for 1 s as the two transistors conduct alternately. The relay keys the transmitter(s).

25-16 AUTO-ALARMS

Auto-alarm receiving equipment must allow for imperfectly made dash spaces, as well as interfering signals adding to the received dash length. Consequently, they must accept dashes from about 3.5 to 6 s long and spaces from about 0.1 to 1.5 s.

The auto-alarm receiver is turned on by throwing the main antenna switch to the AA position. (This also disables all transmitters.) The receiver must detect any A2A emission that is modulated by fre-

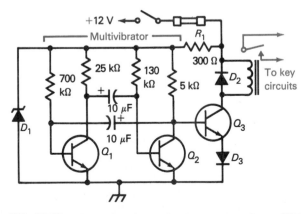

FIG. 25-11 Automatic electronic AA keyer (from ITT Mackay Marine type 5103).

quencies between 300 and 1350 Hz, must have 100-μV or better sensitivity, and must not overload with less than 1 V of RF input. The receiver is fixed-tuned to 500 kHz (± 8 kHz bandwidth) and is either a superheterodyne or a TRF type. An oscillator that generates approximately 100 μV on 500 kHz must be incorporated in the input of the receiver to be used for testing.

After three or four properly received AA dashes and spaces, audible alarm bells are sounded in the radio room, bridge, and radio operator's quarters. The only way that the bells can be silenced is by pushing a reset switch button on the AA panel in the radio room. After silencing the bells, the radio operator listens on the main receiver for distress traffic on 500 kHz. If none is heard after a period of several minutes, it is assumed that the alarm was false. This is noted in the radio log and the AA is placed back in service.

Audible warnings will be sounded and a fault light will glow on the panel if the AA receiver circuitry malfunctions (relay falls out due to decreased current, as from a burned-out filament) or if the line voltage varies excessively. If the bells do not stop when the reset button is pressed, there is a fault in the equipment and the AA must be turned off and serviced.

Whenever a constant dash of 3.5 s or longer is received the counting circuits activate a red light on the AA panel and on the bridge. When the dash exceeds 6 s, the light goes out. A blinking 4-s light indicates proper-length dashes being received.

Prior to signing off watch, the radio operator turns on the AA, tests it, and makes an entry as to this fact in the radio log. While the ship is at sea, the AA and all other emergency equipment, batteries, generators, fuel, etc. must be checked daily. Prior to leaving port they must also be tested. To test the AA, the operator pushes the bridge-bell-off switch with one hand and the test signal button with the other, observing the dash-present light to indicate proper operation of the equipment. The bells may be allowed to ring for a short time to test them.

Several methods have been used to time received AA signals—the charging of *RC* circuits operating stepping relays which ring the bells after receiving four properly made dashes and spaces, dc motors operating electromagnetic drums that are engaged to slowly rotating axles which when rotated to the timing, received dashes and spaces gating on tone generators with counter circuits to count the number of cycles of ac produced to determine the dash and space times, etc. The receiver and timing circuits normally operate from the ship's main power, but the AA bells operate from storage batteries.

Early AA equipment had manual sensitivity controls. If static levels increased, a false alarm resulted. New models use slow-rise and fast-fall AGC voltages which improve operation and make manual adjustments unnecessary.

One modern AA system (ITT Mackay Marine type 5003) is a solid-state TRF receiver with a 16-kHz-bandpass 500-kHz filter feeding five tuned RF amplifiers. A diode detector and a 1350-Hz low-pass AF filter develop a slow-rise and fast-fall AGC voltage for the RF amplifiers. Incoming dashes, converted to dc by the diode detector, key a 2-Hz pulse oscillator into operation. A counter consisting of a series of flip-flop stages begins to react to the pulses. After the seventh pulse (3.5 s), the short-dash circuit drops out. After the twelfth pulse (6 s), the long-dash circuit stops the oscillator and the logic circuits wait for another beginning dash to start. If the dash is between 3.5 and 6 s long, it is accepted in a dash-storage flip-flop. If the space is proper, the next dash is counted; and if it is correct, it is stored in dash-storage also. When four proper dashes are stored, the alarm circuits are activated, and the bells ring.

Should the power supply fail, the line fuse burn out, or the line voltage be interrupted or drop below 65 V, a relay coil across part of the power supply receives insufficient voltage excitation. Its arm releases, connecting the alarm bells across the storage battery. If the power-line voltage rises too much, the bells also ring.

25-17 THE MAIN-ANTENNA SWITCH

Besides a main antenna, ships have an auxiliary antenna. The antenna switch has several positions:

- *Ground*. The main antenna is switched to ground when the radio watch is secured or when the ship is in an electrical storm.
- *Main transmitter*. In this position, the main antenna is connected to the main receiver and to the main transmitter through a break-in relay.
- *Emergency transmitter*. The main antenna is connected to the emergency receiver and to the emergency transmitter through a break-in relay.
- *AA*. The main antenna is connected to the auto-alarm receiver, and power is connected to

the AA equipment. All transmitters are disabled by interconnecting circuits.

- *DF (direction finder)*. The main antenna is open-circuited or an auxiliary antenna is connected to the main receiver, depending on the antenna conditions when the DF was calibrated. The transmitters are disabled.
- *AA-DF*. (May be labeled AA only.) Main or auxiliary antenna is connected to the auto-alarm, as when the DF was calibrated. Power is applied to the AA equipment.
- *HF (high-frequency) transmitter*. Main antenna is connected to the HF transmitter. (It may be connected to the HF receiver and to the HF transmitter through a break-in relay if a separate antenna is not used on the HF receiver.) The auxiliary antenna may be connected to the main receiver when the main antenna is used on the HF transmitter.

25-18 TELEPRINTERS

A piece of equipment used aboard ships is a *teleprinter* or *telex*. It has a keyboard similar to a typewriter. When a letter key is pressed, the machine develops a series of open-closed circuit conditions that form a code for that particular letter or function (punctuation, spacing, line feed, etc.). The on-off Baudot code (Sec. 12-20) can be used to key a transmitter, normally as a 170-Hz FSK (F2B) emission. At the receiving end the open-closed (*space* and *mark*) conditions are received and converted to off-on signals. These are mechanically set up by the receiving machine as a particular letter, actuating the proper type-bar to print the letter on paper.

When the timing of the mark-space code is done mechanically, the code generators and printers must run at the same speed. For this reason teleprinter internal parts are moved by synchronous constant-speed motors. Teleprinters may also operate from microprocessor-based systems using electronic timing circuits.

Each letter consists of seven parts: a *start* signal (always a space), followed by five mark- or space-coded periods, and finally a *stop* signal. For 60-word-per-minute (wpm) machines the mark-space code segments are formed at a rate of 22 ms (22 milliseconds) each, with a 31-ms stop signal. For 100-wpm machines the code segments are at a rate of 13.5 ms, and the stop is 19 ms.

A teleprinter, also known as a Teletype (trade name) machine, may have a page printer, CRT

monitor and printer, or a tape printer. In the latter case, the letters are printed on narrow paper tape. In the other cases after about 65 letters, it is necessary to receive *carriage return* (CR) and *line feed* (LF) function signals to make the printer return to the left-hand edge of the page or CRT, and also drop one line down. These functions do not cause a printout, so that they are not indicated on tape copy.

To transmit, the mark-space signals generated by an operating teleprinter machine can be processed into the standard 170-Hz frequency shift at the desired transmitting frequency. This is then fed to amplifier stages of a transmitter (FSK, Fig. 25-4), resulting in an F2B emission. Converting on-off to FSK might be by biased varactors or biased diodes across oscillator circuits (Sec. 16-12).

To receive, either a separate TTY (TeleTYpe) receiver is used or the IF output from a station receiver (Figs. 25-6 and 25-7) may be used. The received signal can be converted to off-on dc currents in several ways. The original method used a *polar relay* having separate coils on both sides of its magnetized armature (Fig. 25-12), along with

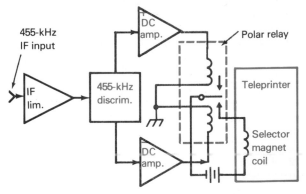

FIG. 25-12 Rudimentary polar-relay IF-type RTTY terminal unit, or TU.

separate permanent biasing magnets on each side. If the magnets are properly positioned, the armature is magnetically biased halfway between the coils and the contacts on both sides of the armature are open. A small current in either coil will cause the contacts to close according to the direction of the current. In a 455-kHz IF receiver, a 455-kHz limiter might feed a 455-kHz discriminator which would develop + and − signals from received FSK signals. These voltages can be amplified by dc amplifiers and actuate the field coils of a polar relay. The contacts of the polar relay key a mark current into the selector magnet coil in the teleprinter and no current for the space part of each letter. The machine mechanically selects the letter from the coded pulse input and causes the proper type-bar to strike the paper. Instead of activating the polar relay, the amplified + and − signals can act as on-off bias for amplifying devices in which the TTY selector magnet coil forms the output circuit load.

Another form of TU operates from two-tone audio signals. If an F1 signal is being received on a CW detector, the mark and space signals will produce two separate tones. For example, if the mark signal is tuned in to produce a 2100-Hz tone, the space signal will produce 2100 + 170 Hz, or a 2270-Hz tone. (Space-high, mark-low is standard for marine teleprinters.) In Fig. 25-13 after passing

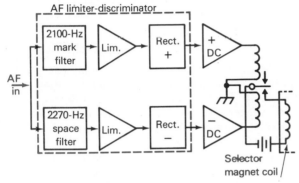

FIG. 25-13 Simple polar-relay AF-type RTTY TU.

through tone filters (2100 and 2270 Hz), the two signals are limited to prevent fading from affecting the operation of the system. Then both tones are rectified, one to a positive voltage and the other to a negative. Both voltages are amplified and can be made to either actuate a polar relay or key an electronic circuit to feed on-off dc pulses to the selector magnet coil in the teleprinter. To prevent the machine from chattering constantly when no signal is present, some form of timed bias must be used to switch the TU to a mark condition if the last received signal was a space. The first space-frequency signal starts the machine's timing circuits.

A block diagram of a telex station is shown in Fig. 25-14. Some systems transmit two-tone audio frequency shift (AFSK) instead of FSK. The key is used to send a CW or FSK identification.

This description has been simplified. A wide variety of sophisticated transmitting and receiving systems are in actual use in telex stations.

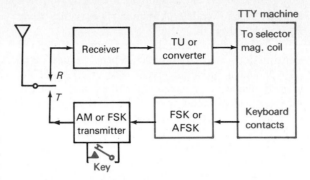

FIG. 25-14 Basic components of a teleprinter station.

25-19 SITOR

An essentially error-free teleprinter system called SImplex Teleprinter On Radio (SITOR or "simplex TOR") is used for ship-to-shore communications. A teleprinter can be used to punch a message onto a tape in Baudot, the 5-element code. The tape is fed into a SITOR converter, which changes the 5-element Baudot to a special 7-element code. The two extra elements are used as *parity* checks to reduce the possibility of accepting mutilated letters. The 7-element code is then transmitted by the sending ("master") station. The receiving ("slave") station converts the 7-element code back into Baudot to allow the slave-station teleprinter to print what has been sent, *provided* the parity checks indicate that the received letters are correct. Because letters may be mutilated during reception, transmissions are always made in groups of three letters (or characters, or characters and spaces) followed by an idle time. If a 3-letter group is correctly received, it is printed and the slave station transmits a single-letter control signal (CS) back to the master station (within 0.02 s) telling it to continue with the next 3-letter group. However, if any of the three received letters is mutilated (indicated by incorrect parity checks), the slave station transmits a different CS letter, one which requests the master station to repeat the last 3-letter group. A group may be repeated up to 32 times, or until an unmutilated 3-letter group is received and printed. Only when 7-element coded letters are correctly received will they be converted to Baudot at the slave station and fed to its teleprinter. Should any of the returning slave CS letters be mutilated in reception at the master station, the master sends a special 3-letter group to the slave which will make it send back a CS requesting a repeat of the previous group. (Amateurs use a very similar system called AMateur Teleprinting Over Radio, or AMTOR.)

There are two generally used types of SITOR transmissions, an automatic repeat request type (ARQ) and a broadcast type. The ARQ type, described above, is a selective calling and synchronization system using specially coded call signs for each ship or station. Only a given ship or station's SITOR equipment will be able to synchronize and decode transmissions addressed to its call sign. A broadcast-type transmission, such as press, weather, etc., can be received by any SITOR receiver. In this case, no return CSs are transmitted. To reduce errors, broadcast transmissions send each three-letter group twice. If a letter is mutilated the first time it is sent, in all probability it will be received correctly the second time the group is transmitted. If not, a blank will be left on the teleprinter page in place of the mutilated letter, rather than printing an incorrect character.

SITOR transmissions take place in the HF bands, with the ship and shore stations on any one band being separated by about 1 MHz. Under good conditions transmission speeds of over 60 words per minute are possible. Charges are $3 a minute with a minimum of 3 minutes to U.S. coast stations, with somewhat higher rates to foreign countries.

▌ Test your understanding; answer these checkup questions.

1. What makes up an AA signal? _____ When should the SOS be sent? _____ Why?

2. In Fig. 25-10, what would the 20-μF capacitor do? _____ The 0.5-μF C and 10-Ω R? _____ If the cam has the microswitch closed when the keyer is turned off, might this leave a transmitter on? _____

3. In Fig. 25-11, what is the function of D_1? _____ D_2? _____ D_3? _____ R_1? _____ Is the 12 V positive or negative? _____

4. What tolerance do AA receivers have for their dash and dot lengths? _____

5. When is an AA used? _____ How is an AA turned on? _____ What is its required bandwidth? _____

6. Where are audible AA warnings sounded? _____ Where are they shut off? _____ What does a red light indicate? _____

7. What might cause constant audible alarms from an AA? _____ Intermittent alarms? _____

8. What powers an AA set? _____ Its bell system? _____

9. What circuit in a modern AA system makes manual setting of the RF gain control unnecessary? _____ What are its general characteristics? _____

10. List seven positions of a radio-room antenna switch. _____ _____ _____ _____ _____ _____ _____ Would all ships have all of these? _____ Why? _____

11. What are three possible uses of RTTY aboard ships? _____ _____ _____

12. Are all coded letters and functions of a teleprinter of equal time length? _____

13. What is fed to a teleprinter to make it print? _____ What comes from it when transmitting? _____ Does it also print when transmitting? _____

14. What are the two types of discriminators used in RTTY TUs? _____ _____

15. What does SITOR mean? _____

25-20 SHIPBOARD RADIO OPERATORS

Freighters and tankers usually employ one radio operator who stands 8-h watch daily, using the auto-alarm for the other 16 h. The operator is a ship's officer and eats with the other ship's officers. A Radiotelegraph First Class (T1) license or a Second Class (T2) license with an endorsement of 6 months' satisfactory service is required. The minimum age for a T1 license is 21 years.

Passenger vessels usually employ three watch-standing operators. The chief operator usually has passenger privileges. On larger ships there may be a chief and three watch-standing operators. When radar is used aboard the ship, one or more of the operators must have radar endorsements (FCC Element 8, Chaps. 22 and 26) on their licenses to allow them to service radar equipment. On many ships radio operators are the only technically trained personnel and may service all of the electronic equipment on board.

Both T1 and T2 licenses authorize the holder to operate any *radiotelephone* equipment aboard the vessel.

The radio operator is aboard ship to enable the ship to communicate in case of emergency, to receive distress messages from other ships, to transmit weather reports, to handle messages pertaining to shipping business, and to send and receive routine messages for passengers or crew.

The main duty of the operator is to keep an efficient watch by earphones or loudspeaker on the international calling and distress frequency of 500 kHz. The operator must log any signals heard on this frequency during the two 3-min silence periods from 15 to 18 min and from 45 to 48 min after each hour and also log at least one other call heard on this frequency every 15 min. If no signals are heard, a log entry to this effect is made. All log entries are in 24-h Greenwich mean time (GMT), which is also called universal coordinated time (UTC), or Zulu (Z) time.

The radio-room clock is an 8-day windup type marked with 12 h, plus additional markings up to 24 h (Fig. 25-15), with a sweep second hand. It has

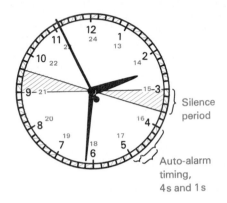

FIG. 25-15 Shipboard radio-room 24-h clock face.

the two 3-min silence periods marked off in red, and around the minute or second scale it has twelve 4-s red arcs separated by 1-s spaces to aid in sending an AA signal. The clock is compared daily with a standard time transmission from such stations as WWV or WWVH.

The operator must abide by all federal as well as international telecommunication laws, rules, and regulations, as applicable. An operator who observes a ship station flagrantly violating international radio regulations may make a report of the violation on a form supplied by the radio company servicing the ship and submit it to the FCC, Washington, D.C.

On single-operator ships the operator stands an aggregate of 8 h daily, using the auto-alarm for the off hours. Watch hours are determined by the geographical area in which the ship is operating. Ship stations whose service is not continuous should not close before finishing all operations resulting from a distress call or an urgency or safety signal or before exchanging, insofar as possible, all traffic for the ship to or from any coastal station within its range. The operator normally advises the nearest coastal station the hours of closing and reopening the sta-

tion, unless the ship is in foreign waters, in which case such transmissions are prohibited.

Radio operators should know that New York City and San Francisco are the principal Atlantic and Pacific Coast ports or shipping terminals. They are also the major centers of telecommunications with Europe or the Pacific. The greatest traffic is handled out of New York, both shipping and radio. The greatest number of telecommunication channels are between New York and Europe.

25-21 POSITIONS AND TIME

The earth, being a nearly round sphere rather than having a flat plane surface, presents some difficulties to a navigator who must sail his vessel from one point on the sphere to another. To help him, the earth is divided into the Northern Hemisphere, with the North Pole at the "top," and a Southern Hemisphere with the South Pole at the "bottom." An imaginary line running from the North to the South Pole is known as a *line of longitude*, or a *meridian*. The sphere is divided into 360°, with a starting point of 0° being the imaginary longitudinal line running through Greenwich (near London), England. Progressing westward from 0°, the meridians are termed *west*. New York is located on a meridian about 75° west of Greenwich, the Panama Canal about 80° W, Los Angeles 120° W, and Honolulu about 157° W. Moving east of the 0° meridian, Cairo is 30° E, Bombay 73° E, Shanghai and Manila 120° E, and Tokyo about 142° E. At the 180th, the east and west meridians meet in the *international date line*. Here, halfway around the world from Greenwich, the day begins. When it is noon tomorrow in Greenwich, it is midnight on the date line and today is considered to be starting. (The date line runs through no populated areas, being mostly over the ocean.) When a ship passes over this line from west to east, it skips a day, possibly going from 6 P.M. Wednesday to 6 P.M. Thursday. When traveling in the opposite direction, it will have two identical days, perhaps two Fridays.

Since there are 360° of longitude and there are 24 h in a complete rotation of the earth, $1/24$ of the total rotation, or 15°, represents how far the sun travels in 1 h. Therefore, each 15° of longitude represents 1 h that the traveler must turn his clock back if going westward, or forward if going eastward. A ship keeps its clocks on *local*, or sun, time. The local time of a ship traveling east or west may change a few minutes per day near the equator to more than an hour in high-latitude regions where the lines of longitude are closer together.

Since Shanghai is 120°, or 8 h, east of Greenwich, when it is 1700 GMT (12 h plus 5 h, or 5 P.M.) Monday in London, it is 17 h plus 8 h, or 25 o'clock, or 1 A.M., in Shanghai on the next day, Tuesday. Since New York is 75° W, or 5 h behind GMT, when it is 1 A.M. Tuesday in Shanghai, it is 12 noon Monday in New York (and 9 A.M. in San Francisco).

The *equator* is an imaginary line girdling the world in the middle. From the equator to either pole is one-fourth of a circle, or 90°. Imaginary lines parallel to the equator are known as lines of *latitude*, or *parallels*. Since lines of latitude do not converge, the distance between 5° of latitude anywhere in the world is the same. A degree is considered to be composed of 60 equal parts called minutes. A minute of latitude is equal to 1 nautical mile (approximately 1⅛ land miles). Latitude lines above or below the equator are said to be so many degrees north or south. The position of a point on the earth, Colón, Panama, for example, can be indicated accurately by its longitude and latitude. Thus, "Colón is at 80° W and 9° N" specifies only one place in the world where it can be found. The island of Guam, located in the Pacific Ocean, is at approximately 10° N and 146° E; Sydney, Australia is at 33° S and 151° E; and so on.

All radio-room clocks are kept on GMT. One-operator ships usually have the operator standing watch according to GMT for the area or region of the world in which the ship is traveling. Three-operator ships usually have the operators standing watch by local time to prevent confusion at mealtimes, but the log and message times are kept in GMT.

ANSWERS TO CHECKUP QUIZ ON PAGE 612

1. (12 dashes, 11 spaces) (2 min after AA signal) (Give alerted operators time to get to receiver) **2.** (Shape ac) (Reduce contact sparking) (No) **3.** (Hold V_{CC} constant) (Damp out inductive V of coil) (Bias Q_3 to allow cutoff) (V drop so D_1 can operate) (+ for NPNs) **4.** (3.5–6 s, 0.01–1.5 s) **5.** (When operator is off watch) (Throw antenna switch) (16 kHz) **6.** (Bridge, radio room, operator quarters) (At AA) (Dash over 3.5 s long) **7.** (AA signal combinations of static and signals, faults, low line V) (Line-V variations) **8.** (Ship's line) (Battery) **9.** (AGC V) (Slow rise, rapid fall) **10.** (Ground, main transmitter, emergency transmitter, AA, DF, AA-DF, HF transmitter) (No) (Some have no HF) **11.** (Weather, news, messages) **12.** (Yes) **13.** (Coded pulses) (Open-closed circuit) (Yes) **14.** (IF) (AF) **15.** (Simplex teleprinter on radio)

The three *radio regions* of the world are (1) Europe, Russia, and Africa; (2) North, Central, and South America; and (3) Australia and Asia.

▌Test your understanding; answer these checkup questions.

1. How much shipboard experience is required for a First Class license? _____ To be sole operator on a ship? _____
2. Who must have a radar-endorsed license? _____
3. What is the minimum number of log entries in 1 h? _____
4. What is meant by GMT? _____ By UTC? _____ Are they the same? _____
5. In Fig. 25-15, what time(s) is (are) shown? _____
6. During what time(s) during any hour is routine traffic and calling not allowed on 500 kHz? _____ Why? _____
7. How are watch hours determined on one-operator ships? _____ On three-operator ships? _____
8. Is the equator a line of latitude or longitude? _____
9. Where does 0° longitude start? _____ What is another name for 180° longitude? _____
10. How many hours difference between San Francisco and Greenwich? _____ Which is ahead? _____
11. According to what time: Is a radio log kept? _____ Do operators eat? _____
12. What are the three radio regions of the world? _____ _____ _____

25-22 LOOP ANTENNA CHARACTERISTICS

A *radio direction finder* (RDF or DF) is on the bridge of every ship. It is used to determine the direction of a transmitted signal approaching the ship. It gives good direction indications for distances up to 50 mi, and fair at 100 mi. At times it is used for ships in distress that are several hundred miles away. It is normally operated by bridge personnel but may be serviced at sea by radio operators. The older rotating DF loop antenna is constructed in either circular or square form. Figure 25-16 illustrates the basic properties of a loop antenna.

A vertically polarized signal, such as most LF and MF signals, striking the loop wires induces currents in the vertical sides S_1 and S_2 (Fig. 25-16a). These currents are equal in amplitude and polarity if the transmitter is an equal distance from the two sides (as when the radio waves are traveling toward the loop from the position of the reader or striking

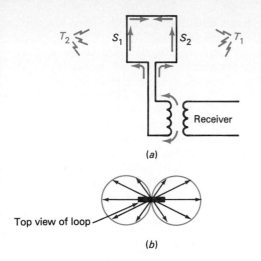

(a)

(b)

Top view of loop

FIG. 25-16 (a) Basic balanced RDF loop circuit. (b) Reception pattern of the loop.

it from behind the page). Two equal induced currents flowing upward, as shown, would cancel at the top of the loop and in the receiver, leaving a zero resultant, or *null*. On the opposite half cycle, cancellation also occurs. Rotating the loop 180° places the wires in a similar relative position, and another null is received.

With a transmitter at T_1 or T_2, the currents induced in S_1 and S_2 will not be of equal amplitude during the major portion of any cycle because they are different distances from the transmitter, and a current equal to the difference between the two induced currents will flow in the receiver. While this *difference current* may be weak, multiturn loops with a diameter of about 3 ft produce satisfactory pickup for sensitive receivers.

Looking down on the loop in Fig. 25-16b signals approaching at a 90° angle from the plane of the loop will be received as nulls. Signals approaching in the direction of the plane of the loop will be maximums. Signals from other directions will be intermediate in amplitude. The loop is said to have a *bilateral*, or *bidirectional* (two-way), *figure-8* response when rotated 360°, giving a good null in two directions 180° apart and two maximums 180° apart (Fig. 25-16b).

Note that the horizontal portions of the loop are considered to be picking up no signal at all from approaching radio waves. Any signal induced in the top portion should be canceled by that induced in the bottom portion as long as the wave is traveling parallel to the earth, regardless of how the loop is turned. If a signal is approaching the loop from above the horizon, however, a horizontal difference current will be induced in these portions of the loop. This appears as a residual signal, shifting the null or making a complete null impossible, or both. This occurs with sky-wave signals but is nonexistent for ground-wave signals.

Practical loop antennas are 10- to 15-turn coils of 3-ft diameter encased in a hollow brass or aluminum case. The shield is broken only at the top, at which point an insulating segment is inserted (Fig. 25-17). Without the insulating segment no signal

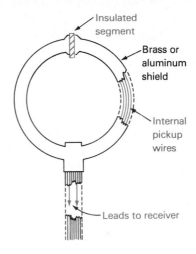

FIG. 25-17 A shielded RDF loop antenna.

would be able to penetrate to the pickup wires inside. The insulation allows oscillating currents to be induced in the metal shield, inducing voltages into the internal pickup wires. The loop is rotated by hand. Its shaft terminates below deck at the operating position and is coupled mechanically to a wheel.

A loop antenna that is properly balanced electrically produces a perfect figure-8 reception pattern. Small differences in wire lengths, or in distributed capacitance between the two sides and ground (shield), will unbalance a loop. Unbalance results in one reception lobe being greater than the other, a lack of complete nulls, and a shifting of the nulls from their normal 180° positions. Figure 25-18a illustrates S_1 as being shorter electrically and picking up less signal than S_2, resulting in noncancellation. A possible reception pattern is shown in Fig. 25-18b. In the unbalanced condition neither minimum is suitable for determining direction. The lack of a deep, sharp null on stations within 50 mi indicates loop unbalance and erroneous bearings.

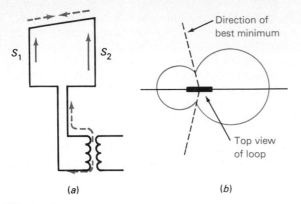

S_1 S_2

Direction of best minimum

Top view of loop

(a)   (b)

FIG. 25-18 (a) Unblanaced loop results in a difference current in the receiver. (b) Reception pattern of an unbalanced loop.

25-23 BALANCING A LOOP

By proper manufacturing, a loop with little electrical unbalance can be constructed. If in a clear space or in the central portion of an airplane (where nearby portions of the body on both sides are sym-

ANSWERS TO CHECKUP QUIZ ON PAGE 614

1. (1 year) (6 months) **2.** (One operator if radar aboard) **3.** (Four) **4.** (Greenwich mean time) (Universal coordinated time) (Yes) **5.** (0231:56 or 1431:56) **6.** (Silence periods) (All listen for SOSs) **7.** (By geographic position) (By local ship or sun time) **8.** (Latitude) **9.** (Greenwich) (Date line) **10.** (8 h) (Greenwich) **11.** (GMT, or UTC) (Local ship's) **12.** (1 = Europe-Russia-Africa) (2 = Americas) (3 = Australia-Asia)

metrical) errors tend to balance out and no electrical balancing may be required.

On a ship there may be no place where such an electrically symmetrical condition is available, so some form of electrical balance must be used. The loop in Fig. 25-19a uses a differential variable capacitor to balance both sides of the loop to ground. Rotating this capacitor increases the capacitance to ground from the left side as the capacitance to the right side is decreased, and vice versa. This can make up for electrical unbalance due to greater capacitance to ground from one side of the loop than from the other. The center of the loop coil is connected to the shield (ground) to aid in balancing.

The second loop (Fig. 25-19b) uses a differential capacitor and a short vertical antenna. The reception pattern of the vertical antenna is equal in all horizontal directions. Signals from this *sense antenna* can be fed by the differential capacitor to the side of the loop that is not picking up sufficient signal because of improper electrical balance.

The third loop (Fig. 25-19c) employs a rotatable coil to induce the required amplitude and phase-signal voltage into one side of the loop to produce balance.

It is necessary to balance the loop on each bearing taken. A balance adjustment may not hold for more than a few degrees of loop rotation.

The usual procedure for balancing a loop is to tune in the signal, rotate the loop for the best minimum obtainable, and then rotate the balancing control to the weakest response possible. If the null

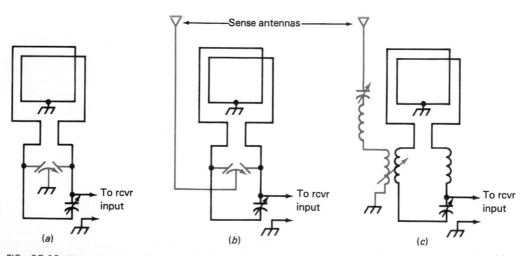

Sense antennas

To rcvr input

To rcvr input

To rcvr input

(a)   (b)   (c)

FIG. 25-19 Three means of balancing a loop: (a) with a differential capacitor to ground, (b) with a differential capacitor to sense antenna, and (c) inductive balancing with a sense antenna.

is not a zero signal, the loop must be rotated to a still greater minimum and the balancing control adjusted until a sharp null results. The sharp null should indicate an accurate bearing, or direction to the station being received.

25-24 UNIDIRECTIONAL BEARINGS

A loop alone gives bidirectional bearings. If the loop is intentionally unbalanced, a distorted figure-8 pattern results. If the distortion is sufficient, the amplitudes of the two maximums are noticeably different. The stronger maximum is used to indicate the direction of the station and is known as a *unidirectional*, or *unilateral*, bearing. Since it is a maximum-signal indication, it is very broad. It is possible to sense the direction of the signal first (take a unidirectional bearing) and then take an accurate bidirectional null bearing on the station.

The indicator on a DF is directly coupled to the rotator wheel and has two pointers (Fig. 25-20).

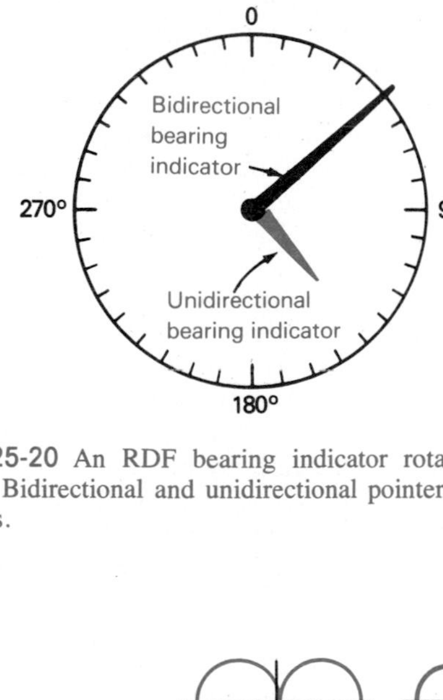

FIG. 25-20 An RDF bearing indicator rotates with the loop. Bidirectional and unidirectional pointers are at right angles.

One is long and points out the bearing of the null on a compass ring marked in degrees. The other is the unidirectional indicator, is short, and points at right angles to the bearing indicator.

The unbalancing of the loop is accomplished by pressing a sense switch that couples the sense antenna to one side of the loop. The sense antenna is usually 20 to 30 ft long, is erected as close to the loop as practical, and must be vertical.

The amount of sense-antenna signal affects the shape of the loop reception pattern (Fig. 25-21a). A small value of sense signal (Fig. 25-21b) decreases one maximum a little and increases the other. With still more sense signal (Fig. 25-21c), one of the maximums can be canceled entirely, resulting in a *cardioid*, or heart-shaped, reception pattern. Still greater sense signal (Fig. 25-21d) produces a pattern with no minimums.

While the cardioid would seem to be a good bearing indicator, its null is frequency-sensitive and is not sharply enough defined to make it practical.

25-25 DF ERRORS

There are several factors which may cause significant changes in DF bearings:

COASTLINE REFRACTION, *or land effect*. Radio waves crossing from land to water at any angle other than a right angle will refract, or bend, toward the coastline. The error may not be appreciable if the transmitting station is within a few hundred yards of the coast but may be significant if the transmitter is inland several miles.

NONOPPOSITE MINIMUMS, *or antenna effect*. This is a form of loop unbalance that is due to stray signal pickup through earphone cords or power lines, or by improper shielding or grounding of the loop. To lessen this effect, earphones are bypassed and low-pass filters are installed in any power lines

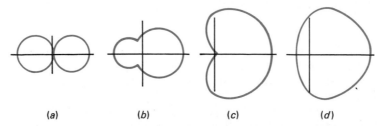

FIG. 25-21 Reception pattern with four values of sense signal. (*a*) Balanced loop. (*b*) Small value of sense signal. (*c*) A cardioid with more sense signal. (*d*) With excessive sense signal.

feeding the DF. The shield constructed around the loop coil materially decreases the effect.

RERADIATION. Signals striking the loop may also strike nearby metal objects, inducing currents in them that produce reradiated signals. In this case, the loop receives the same signal from two directions, at two different amplitudes, and displaced in phase. The loop responds to the vector sum of the two signals, which represents a deviation error when compared with the true direction of the approaching signal. Such signals also require an adjustment of the balancing control to feed in an opposite-phase signal to the loop circuit to attain a sharp incorrect direction null. Bearings taken within ½ mile of large bridges, ships, or other metal objects may be subject to considerable error because of reradiation.

GREAT-CIRCLE ERROR. Since both Tokyo and San Francisco are at approximately 37° N, it would be assumed by looking at the usual map that the shortest route between the two would be directly east or west along the 37th parallel. However, the world is a sphere, and the shortest track between the two cities takes a route northward up to almost the 50th parallel. Thus, for all normally used maps or charts (except great-circle charts), radio waves, which are received via the most direct route, appear to travel a curved course rather than a straight line. The only time they conform to the usual maps is when they are traveling either north or south or along the equator. If the transmission distance is only a few miles, great-circle curvature can be ignored, but for distances in excess of 100 mi it may have to be considered in DF navigation.

POLARIZATION ERRORS. All ground-wave signals striking a loop induce currents in the vertical portions, with any induced currents in the horizontal portions balancing out. However, signals approaching a loop from above induce nonbalancing currents in the horizontal portions that add vectorially with the vertical signals and result in loss of null or null deviation. With frequencies used in marine DF work (280–535 kHz) the ground wave is so much stronger than the sky wave for 50 mi or more that any downward-approaching sky waves may be too weak to be significant during the daytime. During nighttime hours the sky wave refracts from the ionosphere and is receivable for long distances. When the sky-wave amplitude approaches the ground-wave amplitude, the null of a loop antenna will no longer be reliable. This is known as *night effect*. Reception is unreliable for navigation with sky-wave signals alone. The varia-

tion of the ionosphere at night produces a fading sky-wave condition that may make the null appear to vary back and forth many degrees in a short period of time. During the periods of ½ h before and ½ h after sunrise and sunset the ionosphere is varying so wildly that bearings at even 50 mi may be affected.

25-26 CALIBRATING A DF

A shipboard DF will normally be quite accurate for signals approaching from ahead, from astern, or from either beam. Between these points it may deviate several degrees. These are known as *quadrantal* errors. By running the vessel in a circle within sight of a radio transmitting antenna, it is possible to take simultaneous visual and radio bearings every 5° or 10° as the ship turns. Comparison of the two bearings indicates how far the DF bearings are in error. This is known as *calibrating* the DF. By building into the DF indicator system a mechanical means of producing opposite errors, it is possible to produce a DF indicator with no appreciable error. Adding the mechanical opposite errors is known as *compensating* the DF.

Another form of compensation is sometimes used. The masts, deck, smokestack, plus guy and other wires aboard ship, form a circuit that may introduce considerable quadrantal error in a DF. This error may be reduced by closing the mast-deck-stack circuit with a wire connected between the top of the mast and the top of the stack. (Sometimes it increases the error.)

Because the DF is sensitive to signals reflected from nearby objects, it is important that the calibration be carried on when the ship is in its sea-going condition insofar as halyards, guy wires, masts, and booms are concerned. If the main radio antenna is installed over the loop, the calibration of the DF may shift if the resonant frequency of the radio antenna is changed. To prevent this, the DF is interlocked with the main-antenna switch in the radio station. When the bridge officers turn on the DF, a red light shows in the radio room. The operator throws his main-antenna switch to the DF position. This disconnects his main antenna, closes a circuit that allows the DF to start operating on the bridge, and connects his watch-standing receiver to an auxiliary antenna that has no effect on the DF. (The antenna switch is always in this position when the DF is calibrated.)

All U.S. ships of 1600 gross tons or more to be

navigated in the open sea must carry a radio direction finder.

Marine RDFs must be capable of operating on 500 kHz to take bearings on ships in distress, on the international DF frequency of 410 kHz when requesting DF bearings of other stations, and on the frequencies used by the marine radio-beacon stations operating between 285 and 320 kHz. If the DF is calibrated at a frequency of 410 kHz, the farther from this frequency the greater the quadrantal error it may have.

▌Test your understanding; answer these checkup questions.

1. Why are signal nulls used in RDF work? _____ How many are there in one loop rotation?

_____

2. Seen from above, what horizontal reception pattern does a small vertical loop have? _____
3. Under what conditions do the horizontal portions of a loop pick up difference currents? _____ What effect does this have? _____
4. Why must there be a break in loop shielding?

_____

5. What are the results of an improperly balanced loop? _____ _____
6. List three methods of balancing a loop?

_____ _____ _____

7. For what are unidirectional bearings used? _____ How can a loop be made to give them? _____
8. Why might it be better to take unidirectional bearings before taking bidirectional ones? _____ Are they always required? _____
9. What kind of an antenna is a sense antenna? _____ Where should it be located? _____
10. What happens to a radio wave path as it moves outward across a coastline at less than 90°?

_____

11. What is the result of antenna effect? _____ Of reradiation of signals? _____
12. What path is always taken by radio waves? _____ When would bearings on such waves plot correctly on charts? _____
13. Does a ground wave fade? _____ Does it vary in strength from day to night? _____
14. What causes night effect? _____ At what times of the day are DF bearings least accurate?

_____

15. At what angles are quadrantal errors maximum? _____ How are they corrected?

_____ _____

16. Why must the main switch in the radio room be thrown to DF or AA-DF before the DF can be used? _____

25-27 THE GONIOMETER RDF

A more modern nonrotating, double-loop DF is called a *goniometer*. Its shielded loops can be 100 ft from the DF equipment and be coupled to it by a transmission line (Fig. 25-22). The two separate

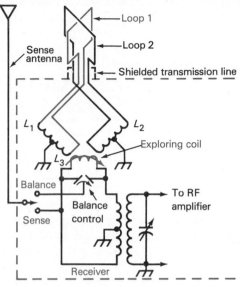

FIG. 25-22 Goniometer loops coupled to a receiver.

fixed loops are at right angles to each other, on top of a mast, and centered above the navigation bridge. Such a centered position eliminates most of the quadrantal errors to which lower rotating-loop installations are subject. Each loop has its own transmission line that terminates in a fixed coil in the receiver. The two coils are installed at right angles to each other, as are the loops. A rotary *exploring* coil, L_3, can be adjusted to pick up signals from either L_1 or L_2 or from both.

If loop 1 is receiving a null from a station, then loop 2, at right angles, will be receiving a maximum. If the exploring coil is rotated for maximum coupling to L_1, with no signal in loop 1, no signal is picked up by the exploring coil. Since L_3 is 90° from coil L_2, it is completely decoupled and can pick up no signal from this coil either. Therefore, no signal is heard in the receiver. When the exploring coil is rotated in either direction, the coupling between it and L_2 increases and signals are received.

If the transmitted wave approaches at an angle of 45° from both loops, the exploring coil will receive equal-amplitude and opposite-phase signals from L_1 and L_2 when it is midway between them. There is zero signal output when the exploring coil is at 45° from L_1 and L_2. Thus, the exploring coil is pointing in the direction of the approaching signal whenever a null is being received. An indicator arrow is attached to the shaft of the exploring coil that sweeps around a 360° compass scale.

A goniometer DF is balanced, sensed, and operated by methods similar to those used with rotating-loop DFs.

25-28 DF RECEIVERS

The receiver used in conjunction with the DF antenna is either an AM-CW superheterodyne or a TRF type. Since an aural null is the indication of direction, no AVC circuit is used. An RF gain control but no AF control is provided.

Some DF receivers use a form of visual indicator of the amplitude of the received signal.

On old ships having only 110 V dc for power, a storage battery may be used for the filament supply with a dynamotor operating from the storage battery as the plate-voltage supply. On ac-powered ships VT filaments operate from ac and the plate supply is a rectifier-filter circuit. Transistorized DFs may use self-contained batteries with a battery charger.

25-29 DF MAINTENANCE

Besides the usual electronic maintenance of transistorized or VT equipment, a DF of the rotating-loop or -coil type requires periodic checks. The rotating loop or coil mechanism requires greasing or oiling once or twice a year, and the silver-plated slip rings and brushes that connect the rotating loop or coil to the receiver require cleaning with a dry cloth from time to time. The insulating segment between the tubular sections at the top of the loops must not be painted, or loss of signals will result.

25-30 AN AUTOMATIC DIRECTION FINDER

The newest DF is a solid-state automatic direction finder (ADF). Within 2 s of the time it is turned on and the frequency of a desired station is selected, the indicator swings to the correct bearing of the station, or an unambiguous digital readout is shown. This is accomplished by the use of a servomotor rotating a goniometer exploring coil until a minimum signal is reached. At this point the error signal developed by the received signal drops to a minimum and the indicator stops on the correct bearing.

A block diagram of an ADF is shown in Fig. 25-23. Assume that a bearing is desired on a 400-kHz station. The 100-Hz-per-step frequency synthesizer local oscillator is set by thumbwheel dials to a reading of 400.0 kHz. The frequency synthesizer develops a local oscillator frequency of 8 MHz (the IF) + 400 kHz, or 8.4 MHz for the dialed 400-kHz signal. This also electronically switches in the 300–550 kHz bandpass filter. Note that all stages up to the mixer are broad-tuned types. Among many signals picked up by the loop antenna is the desired 400-kHz signal. This is amplified (along with all others) and is fed to a ring-type balanced modulator. A modulating frequency of 75 Hz is also fed to this modulator. The result is a 75-Hz-modulated 400-kHz carrier signal, but with the carrier canceled. The two 75-Hz sidebands, 400,075 and 399,925 Hz, are fed to the adder circuit. If the exploring coil is picking up a strong signal from the station, strong 75-Hz sidebands of 400 kHz are developed. If the goniometer coil happens to be on the correct bearing (nulled), no signal or 75-Hz sidebands will be sent to the adder.

The same 400-kHz signal is also picked up by the sense antenna. This signal is always present, although it is shifted in phase by 90° before being amplified and fed to the adder in order to prevent 180° ambiguity of bearing indications.

The same 75 Hz ac fed to the balanced modulator is amplified and excites one of the windings of a two-phase 75-Hz ac servomotor coupled to the goniometer coil. Only when 75 Hz ac is fed to the second phase winding will the servomotor turn. When the goniometer coil is rotated to zero signal by the servo, the servo is fed no receiver 75 Hz ac, and the servo and indicator stop. The zero-signal bearing is indicated on the compass ring.

The 400-kHz sense antenna signal may be considered as a carrier that is fed through the whole RF system. It picks up the 75-Hz sidebands as modula-

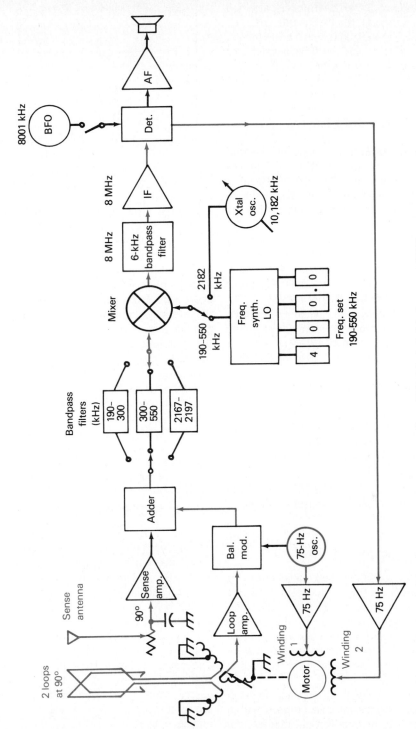

FIG. 25-23 Simplified block diagram of an ADF system (from ITT Mackay Marine type 4005).

tion in the adder. The local oscillator converts this carrier and sidebands to 8 MHz, which is then fed through a 6-kHz bandpass filter. By this filter all other signals in the 300–550 kHz band are rejected. Only the 400-kHz carrier and its 75-Hz sidebands pass through the 6-kHz filter and are amplified. The 8-MHz IF signal is rectified by a diode detector. The 75-Hz output component is amplified and fed to the second-phase or control winding of the servomotor, producing motor and goniometer coil rotation.

If the signal being received is AM, its modulation is also detected and fed through the AF amplifier to a loudspeaker or earphones. If the signal is CW an 8.001-MHz crystal oscillator can be turned on to produce a 1-kHz beat tone which can be heard in the loudspeaker or earphones.

The ADF can be used as a simple receiver by disabling the loop circuits. The sense antenna and its amplifiers feed signals to the mixer. Any station to which the synthesizer frequency set is dialed can then be heard on the loudspeaker.

If manual control is required (when noises confuse the automatic circuits), after determining the approximate bearing automatically, one phase of the 75 Hz ac can be reduced to zero by disabling the sense antenna. The goniometer coil can then be rotated by hand to an exact audible null. The human ear can identify a null through noise better than an automatic circuit can.

This DF will also take bearings on the radiotelephone calling and distress frequency of 2182 kHz by switching from the frequency synthesizer to a 10,182-kHz variable crystal oscillator for the local oscillator. The oscillator can be tuned ±15 kHz to allow trimming in case the transmitting station is not exactly on frequency.

25-31 ADCOCK ANTENNAS

The loop-type direction finders are satisfactory for shipboard work because the frequencies used are low and a strong ground wave exists for a considerable distance. At higher frequencies the groundwave range decreases and a loop becomes inaccurate at distances over 10 to 30 mi because of loss of ground wave. Only sky-wave signals are received at distances.

For shortwave use or for long-distance low-frequency use, an *Adcock* antenna can be used with a DF receiver. It consists of two vertical dipoles separated by perhaps 10 to 20 ft. From the centers of the two verticals the signal is coupled to a two-wire

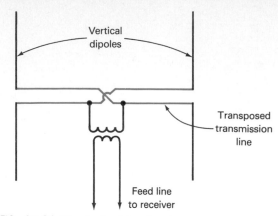

FIG. 25-24 The basic Adcock antenna.

horizontal transmission line that is coupled in turn to the receiver (Fig. 25-24). The transmission line balances out any signals striking it, leaving the vertical portions of the antenna as the only signal-accepting elements. Waves approaching the antenna, either in a direction parallel to the earth or downward from the ionosphere, induce currents in the vertical elements. The sky and ground waves can be out of phase, but accurate bearings can be taken as long as the sky wave has not been changed in travel direction by the ionosphere. Because of their large size, Adcock direction finders are found at land stations only.

Rather than rotate such a large antenna, direction finders may be constructed by using two pairs of Adcock verticals installed at right angles and fed to a set of fixed coils similar to a goniometer DF. The exploring coil is the only part that rotates.

Adcock antennas are used in VHF marine DFs.

25-32 DETERMINING POSITION BY RDF

A vessel can determine its position by taking a bearing on two shore transmitters. For example, a vessel takes a bearing on two shore stations S_1 and S_2 whose positions are shown on a chart (Fig. 25-25). The bearing of S_1 happens to be due east, or 90° (90° from true north). The bearing of S_2 happens to be due south, or 180° from true north. On the chart, a pencil line is drawn through the position of S_1 at an angle of 90° (east and west). The angle is taken from the *compass rose* on the chart and transferred to the position of S_1 by a device known as *parallel rules*. A line is also drawn through S_2 at an angle of 180°. The position indicated by the crossing of the two lines of position is known as a *fix*.

Lighthouses and light vessels along coastlines

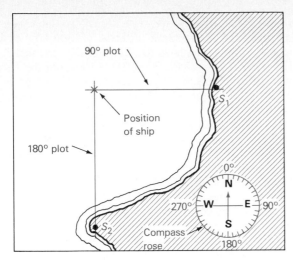

FIG. 25-25 A fix of a ship's position by RDF bearings on two shore stations.

may have radio-beacon transmitters to allow navigators to take DF bearings on them. During clear weather many of them do not operate, but during foggy weather they will be in operation, allowing navigators to check their positions. These beacon stations transmit an identifying signal, listed in the publication "Radio Aids to Navigation," aboard all ships, such as dash-dash-dot, or perhaps dot-dot-dash-dot, to enable the navigator to identify the station on which he is taking a bearing.

While a bearing on a single station gives only a *line of position*, it is possible to cruise a few minutes and then take a second bearing on the same station. These two bearings through the position of the beacon station are plotted on a chart. Since the ship's course and speed and the elapsed time are known, the position of the ship at the time of both bearings can be plotted.

The position of a ship can be determined by two RDF stations ashore taking bearings on radio transmission from the ship. Plotting the two bearings on a chart, as in Fig. 25-25, indicates the position of the ship. Such a service was previously available to ships through the United States Coast Guard but has been discontinued. Some foreign countries still operate RDF stations, however.

▍Test your understanding; answer these checkup questions.

1. Where is the rotating loop in a goniometer-type RDF system? _____ Must it be balanced? _____

2. Why might goniometer DFs have less quadrantal error than rotating loops? _____
3. Name four methods by which aural nulls may be detected by the operator. _____
_____ _____ _____
4. What are three important points regarding rotating-loop RDF system maintenance? _____
_____ _____
5. How is it that none of the 120–550 kHz ADF circuits are tuned to the desired station? _____ What is the only hand-tuned circuit in the ADF? _____
6. What forms the error signal that actuates the ADF servomotor? _____ What stops the servomotor? _____
7. When would the 8.001-MHz ADF oscillator be used? _____ Is it required during ADF operation? _____
8. What is the local oscillator frequency to receive 500 kHz? _____ To receive 2182 kHz? _____
9. Which ADF antenna(s) is (are) used for manual operation? _____ For simple receiver operation? _____ For ADF operation? _____
10. Does the ADF adder ever receive any loop signals? _____
11. What is heard by earphones in the ADF? _____ Why is an AF signal desirable? _____
12. Why are Adcock DFs better than loops? _____
13. What is the point where two bearings cross on a chart called? _____ How are bearings laid out on a chart? _____ _____
14. If two bearings are determined from a radio beacon or station over a period of a few minutes, what else must be known to determine the ship's position? _____

25-33 LORAN A

The RDF has been used as an aid to navigation since the early 1920s. During World War II a new aid, called *loran* (LOng RAnge Navigation), was developed. Whereas RDF has maximum reliable range of about 100 mi, loran A, in the 1900-kHz range, could develop a usable line of position (LOP) for about 700 mi during the day, and 1400 mi during the night. Although loran A is no longer used by the United States, the newer loran C on 100 kHz is providing good coastwise navigational fixes.

A loran unit aboard a ship (or airplane) consists of an antenna, a receiver that includes time-between-pulses determining circuits with a numerical indicator of the time differential, and a CRT on which the pulses are shown.

A representative loran A transmitting system consists of a *master* double-pulsed transmitter and two slave single-pulsed transmitters, separated by 100 miles or more. The master transmits its 40-μs pulses at a high (H = 33³/₉ pps), low (L = 25 pps), or slow (S = 20 pps) pulse recurrent rate. The first slave station receives the master's first pulse, delays a short period of time, and then transmits its 40-μs pulse. The master pulses again. This pulse is received by the second slave, which delays and then transmits its 40-μs pulse. Thus, each master pulse has a delayed echo pulse from a slave station.

Consider the following simplified explanation. A loran receiver picks up both the master pulse and the desired slave pulse and displays them both on a single sweep on its CRT. If the pulse recurrence rate (33³/₉, for example) is also the CRT sweep rate, the pulses will stand still on the screen (Fig. 25-26). The spacing between pulses would repre-

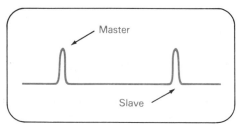

FIG. 25-26 Master and slave signals shown on a single CRT sweep.

sent a reception time differential. This time differential in microseconds indicates that the receiving ship must be on one hyperbolic line of position (LOP) that crosses the base line between the master and that slave (Fig. 25-27). The ship must have loran A charts showing correct hyperbolic lines, each labeled with a master-slave basic pulse recurrence rate and a difference time in microseconds, as

shown. The master-slave-1 hyperbolic lines are marked 2H4 plus a time difference (3300, for example). The 2 indicates the frequency of channel 2 (1 = 1950, 2 = 1850, 3 = 1900 kHz), and the H4 indicates the specific high rate number 4, which is 33⁷/₉ pps. All hyperbolic lines marked 2H3 are used with the master and slave number 2. The H3 indicates the third specific high rate of 33⁶/₉ pps. If the CRT sweep is set at the H4 rate, only the master pulse and the number 1 slave pulse will stand still on the screen, no matter how many other pulses are being received. If the sweep is set to the H3 rate, only the master and the number 2 slave pulses (both pulsing at the H3 rate) will stand still.

Controls on the loran A receiver can be set to select the channel frequency and the pulse recurrence rate for any desired master and slave. Gain controls allow adjustment of the two pulses to equal amplitudes. The slave pulse can then be moved to the left by operating a time-adjust control until the slave is superimposed on the master pulse. The amount that the timing circuit must be changed to make the left edges of the master and slave pulses exactly coincide is marked off in microseconds. The number shown by the time readout control indicates on which hyperbolic line of position the ship is (2H4-3300, for example).

If the master-slave frequency and pulse rate are changed to 2H3, the master and slave-2 pulses now are the only pulses that stand still on the CRT. When they are brought into coincidence, the time-adjust readout might be 1600. Therefore, the ship must be at the latitude and longitude where the 2H4-3300-μs line and the 2H3-1600-μs line cross, or at point X in the illustration.

If one of the pulses blinks, it means that there is trouble at that station and the LOP time is questionable.

25-34 LORAN C

To increase the ground-wave range, loran C was developed. It operates on a lower frequency, exactly 100 kHz. While the basic theory of loran A and C is similar, there are significant differences. The loran C usable ground wave is about 1500 miles day or night. Instead of single pulses, both master and slave loran C stations transmit 8-pulse groups (about 40 RF cycles in each pulse). Each pulse increases to a peak value and then decreases to zero again in about 400 μs. The master transmits a 9th pulse to identify it; if blinking or missing, it indicates trouble.

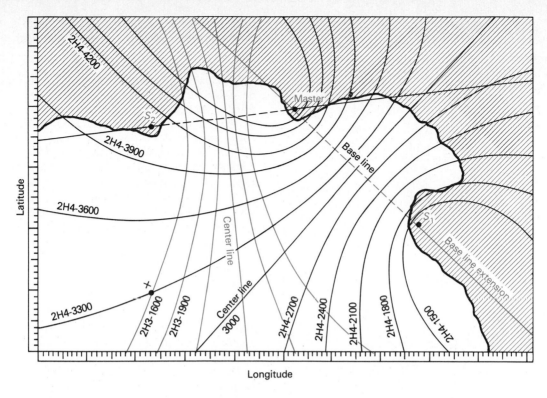

FIG. 25-27 A loran chart using one master to double-pulse two slave stations and showing station identification and time differences on the curves.

Instead of the three basic pulse rates of loran A (H, L, S), loran C uses four basic group repetitive intervals, which are either expressed in letter-number groups in older equipment, or in microseconds of group separation. The specific repetitive intervals are shown in Table 25-2. These can be dialed up by controls on the loran C receiver. Whatever master and slave stations the specific rate applies to will stand still on a two-pedestal presentation if shown on a CRT (Fig. 25-28a). The master pulses are shown already on the higher pedestal. The W-slave pulses have been moved onto the lower pedestal. The X-slave station pulses are also synchronized and are standing still, but are not moved to a pedestal until their master-slave LOP is wanted. The next operation moves the W-slave pulses directly under the master pulses (Fig. 25-28b) by changing the internal timing circuits and electronically magnifying both sets of pulses. The next switching operation magnifies the first pulses of both master and slave still more and superimposes them. If they are not the same amplitude, the M- or

TABLE 25-2	CORRELATION OF OLD AND NEW LORAN C LOP DESIGNATIONS			
	SS1 = 9990	SH1 = 5990	SL1 = 7990	SC1 = 4990
	SS2 = 9980	SH2 = 5980	SL2 = 7980	SC2 = 4980
	SS3 = 9970	SH3 = 5970	SL3 = 7970	SC3 = 4970
	SS4 = 9960	SH4 = 5960	SL4 = 7960	SC4 = 4960
	SS5 = 9950	SH5 = 5950	SL5 = 7950	SC5 = 4950
	SS6 = 9940	SH6 = 5940	SL6 = 7940	SC6 = 4940
	SS7 = 9930	SH7 = 5930	SL7 = 7930	SC7 = 4930
	SS8 = 9920	SH8 = 5920	SL8 = 7920	SC8 = 4920
	SS9 = 9910	SH9 = 5910	SL9 = 7910	SC9 = 4910

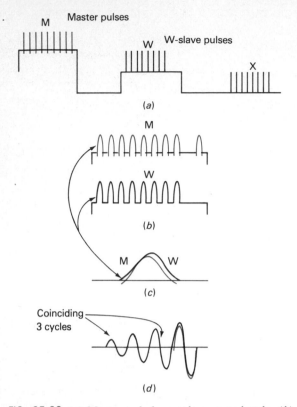

Master pulses

M

W-slave pulses

W

X

(a)

M

W

(b)

M   W

(c)

Coinciding
3 cycles

(d)

FIG. 25-28 (a) Master and slave pulses on pedestals. (b) Pedestals magnified with slave moved below master. (c) Magnified 1st pulses brought to rising-edge coincidence. (d) Still greater magnification of 1st pulse for first 3-cycle coincidence.

S-GAIN controls are trimmed until they are of equal height on the CRT. If the rising edge of the pulses do not coincide, a FINE timing control is adjusted until they do. The final switching adjustment of the time control is the CM (cycle matching), by which the first three cycles of the still further expanded first pulses are brought into coincidence by operating the FINE timing control. The timing number shown when the proper CM adjustment has been made is the time representation of the distance between master and slave as far as the ship is concerned. This number indicates a loran-C chart hyperbolic line of position. Where the master and slave-W line and the master and slave-X lines cross is the position, or fix, of the ship. Because of the cycle-matching accuracy of loran C, the position of the ship may be more precisely determined than by loran A. A loran C LOP might be identified as SS2-W-35000 (or 9980-W-35000) indicating basic rate SS, specific rate 2, slave W, and 35,000-μs

time differential. Note that loran C has a five-digit timing readout whereas loran A has only a four-digit readout, indicating a tenfold more accurate LOP.

By using sky waves with loran C the range can be extended to about 3000 mi. Sky-wave pulses usually show a variation in pulse outline that shifts as the ionosphere varies. As long as the leading edges of sky-wave pulses are made to coincide, LOPs will be reasonably accurate. Sky-wave corrections may have to be added as explained on both loran A and C charts.

As with all equipment using a CRT, high voltages are involved. Turn the equipment off before opening enclosures and exercise care while servicing.

Modern loran C sets may have no CRT. They automatically synchronize the manually selected master-slave pair pulses and compute the LOPs. With the internal microprocessor they can compute the point of LOP crossings and show the latitude and longitude of the crossing with two alphanumeric LCD or LED displays, such as N42°24.51' and W135°46.23'.

The length of a resonant wire antenna for loran C at 100 kHz would be 750 m long, about half a mile! Instead of a resonant linear wire, an *active antenna* is used. An active antenna consists of a short wire coupled to a semiconductor amplifier. A 1-m wire appears as an impedance of about 150,000 Ω to a frequency of 100 kHz. If the receiver has an input impedance of 50 Ω, an extreme mismatch would occur and little signal would be developed in the receiver. By feeding a 1-m wire to a high-impedance gate circuit of an FET, and by using a low-Z source-follower circuit as the output, input and output impedances are matched and signal strength increases. The amplification of the FET further increases the signal. If tuned circuits are used in the amplifier input or output, it becomes a narrowband rather than a wideband antenna. Broadband systems are subject to undesirable intermodulation products. Active antenna systems are used in many LF and VLF *receiver* systems. Transmitters, of course, must use relatively long loaded linear antennas.

There are other systems of position determination, such as Raydist and Mini-Ranger. They transmit an RF pulse and measure the time it requires to return from a known-position *transponder* (receiver-transmitter) on shore to develop an equidistance arc on a chart. Where two such arcs cross from two transponders is the position of the transmitting ship.

Test your understanding; answer these checkup questions.

1. What does loran mean? _____ What is the loran A nighttime range? _____ Daytime?

2. What is the minimum number of loran stations required for a navigational line of position? _____ For a fix? _____

3. What is the distance between loran A masters and slaves? _____ For loran C? _____

4. Besides receiving equipment, what is required to use loran navigation? _____

5. What shape does a loran line of position have? _____

6. What letters indicate the three basic pulse rates? _____ How many loran A specific pulse rates are there? _____

7. In respect to master pulses, when do slaves transmit? _____ When are masters double-pulsed? _____ When do they blink? _____

8. What does 1S5-1600 indicate? _____

9. What parts of pulses must coincide? _____

10. Are ground or sky waves preferable? _____ Which extend farther? _____ Which do not change? _____

11. If a pulse appears twice on the CRT, what would it indicate? _____ If three times? _____ If the pulse is deformed? _____

12. What is the frequency of loran C? _____ What are two advantages of loran C over A? _____ _____

13. What letters indicate the basic rates of loran C? _____ What would SL2-W-26000 mean? _____

14. How many slaves does a loran C master have? _____

25-35 OMEGA RADIO NAVIGATION

The lowest-frequency application of VLF radio transmissions is the Omega navigation system operating on 10.2 kHz. With eight transmitters dispersed around the world, a complete global-type navigation is possible. Each Omega station transmits eight dashes approximately 1 s long in 10 s on 10.2, 11.05, 11.33, and 13.6 kHz. The first station transmits a 0.9-s 10.2-kHz dash. After 0.2 s wait, the second station transmits a 1.0-s 10.2-kHz dash while the first station is transmitting on one of the other frequencies. After another 0.2-s pause the third station transmits a 1.1-s 10.2-kHz dash, etc., while all other stations are transmitting dashes on other frequencies. In this way only one of the eight stations is transmitting the 10.2-kHz signal at any one time. The various other frequencies help iden-

tify the stations. With transmissions at 10.2 kHz the usable ground-wave range is thousands of miles day or night.

Omega navigation relies on the fact that 10.2 kHz has a wavelength of 16 nautical miles. By having a 10.2-kHz oscillator in the receiver and comparing this with received dashes from Omega transmitters, each time the ship moves 8 mi nearer to a transmitter the phase of signals being compared changes 180°. By keeping track of how many phase changes occur over a period of hours or days, the receiver will be able to indicate how much closer (or farther away) the ship is to the transmitter. By comparing phases on two Omega stations, a line of position can be determined at any time, provided that the proper position was fed into the unit at the beginning of the voyage and provided that the unit operates continuously.

The unit must be turned on at least 30 min before sailing. After 15 min of warm-up the circuits are tested to see that the internal oscillator is synchronized with received dashes, so that the receiver circuits shift from station A to station B signals as those stations transmit their 10.2-kHz signals. The proper hyperbolic LOP of constant phase difference for two receivable stations (A and B, for example) according to the ship's present location is determined. This LOP information is fed into the equipment computer in numerical form. The hyperbolic LOP for the present ship's location, according to two other stations (C and D, or perhaps B and C), is found and is fed into the receiver. As long as the ship stays at this position, the readout of the LOP of either pair that was fed into the receiver will show on the indicator. It will be some value, such as 756.42. If the readout device is switched to the other station pair, the readout might be 321.87, for example. However, as soon as the ship sails, phase changes begin to occur for all stations (all eight are received and information regarding them is continually stored in memory). After the ship has moved a few miles, the LOP of 756.42 may change to a readout of 758.01. By finding on a chart the LOP for the new number, the navigator can tell on what new LOP the ship is. To get a fix it is necessary to switch the readout to the other pair of stations being used. Where the second LOP crosses the first on the chart is the present position of the ship. If a ship moves out of the usable area of one station, it will be necessary to shift to some other more readable pair. Modern units automatically compute latitude and longitude and display these digitally. Accuracy is better than 2 nautical miles (nm).

25-36 SHIPBOARD SATELLITE USES

Among the uses of satellites aboard ship, INMARSAT employs geostationary satellites for voice, data, facsimile, and teleprinter communications, uplinking to satellites on 1.64 GHz, which then transmit a down-link to a fixed earth station on 4 GHz (Sec. 22-21). The return up-link circuit from earth station to satellite on 6 GHz is translated to 1.54 GHz and is down-linked at this frequency to the ship. Communications are always ship-satellite-earth, never directly ship-satellite-ship in the merchant marine. Voice modulation may use time division multiplex, while teleprinters often use binary phase-shift keying. The antenna system must automatically lock in on the satellite regardless of the *roll* (side-to-side motion), *pitch* (bow up-and-down motion), and *yaw* (heading change). Marine INMARSAT (originally MARISAT) satellite transmissions are nonpolarized, using circular polarization both receiving and transmitting with parabolic reflector antennas having over 35 dB of gain. Although it might be thought that satellite communications would never be affected by fading, signals entering the ionosphere tend to break up into two components which may take different paths and produce phase cancellations at the receiving point. This can cause dropouts of teleprinter signals, for example.

Another use of satellites aboard ship is *satellite navigation* (Sat-Nav), provided by the Doppler-type U.S. Navy Navigation Satellite System (NNSS), also called *Transit*. The 8 to 18 Sat-Nav OSCAR or NOVA satellites in use are in a circular polar orbit about 550 nautical miles (nm = 6080 ft) in altitude. Their orbital period is 107 minutes at a speed of 14,600 *knots* (nm/h). Each satellite broadcasts on 399.968 MHz (and also 149.988 MHz), using 1.5 to 5 W or more of RF output power to its omnidirectional antenna. The receiving antenna aboard ship is an omnidirectional quarter-wave vertical with eight or more radials forming a ground plane.

The antenna is mounted in an area with clear visibility to the horizon in all directions. Coupled to the antenna base is an amplifier to provide about 30 dB of gain for received signals which are then fed to the microprocessor-based receiving-computing unit below deck.

The exact position of each satellite in longitude, latitude, and altitude is known at all times and is updated every few hours by ground stations. Updating information is broadcast by phase modulation every 2 min by the satellite. The satellite transmitting frequency is checked and corrected continually. Every even minute the satellite also transmits a time marker.

When the satellite is directly overhead (also directly east or west), the received frequency will be exactly 399.968 MHz. As a satellite approaches, the radio waves from it to the ship are compressed and are received as a higher frequency. When moving away, each RF wave is stretched out and the received frequency is lower at the ship (Doppler effect). Doppler frequency change is most rapid as the satellite passes overhead, as indicated by the frequency curve in Fig. 25-29. If the satellite passes to the east or west, the doppler change is less abrupt. When the satellite is on the horizon east or west there is very little Doppler.

One of the requirements of the NNSS receiver-computer unit is to count the number of cycles between its own 2-min time markers. It counts the cycles of a beat frequency of the received Doppler-shifted signals against an accurate 400-MHz onboard oscillator (represented by the shaded areas in Fig. 25-29). After acquisition of the satellite signal the beat tone continually decreases, giving lower total beat signal cycle counts at the end of each 2-min period aboard ship. If satellite position, speed, time, altitude, and Doppler change are known, there is only one hyperbolic line on the surface of the earth where this information is a constant, giving an exact line-of-position. In the next 2-min interval a new beat frequency cycle total is counted, and a new LOP is available for the unit to compare with the first. Where the two LOPs cross is the ship position, provided it has not moved in the interval. Since it usually has moved, its course from its gyrocompass and speed from its speed log must be continuously fed into the Transit receiver-computer unit to allow updated corrections to be made. The Transit equipment is continuously producing an accurate dead reckoning position, updated and corrected every 2 min that a satellite signal is being received. When the satellite signal drops out, the

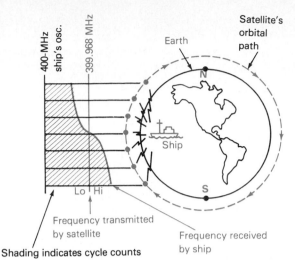

FIG. 25-29 Polar-orbiting satellite and navigation information for one pass.

Labels on figure:
- 400-MHz ship's osc.
- 399.968 MHz
- Earth
- Satellite's orbital path
- N
- S
- Ship
- Lo Hi
- Frequency transmitted by satellite
- Frequency received by ship
- Shading indicates cycle counts

unit continues to update the dead reckoning position until another satellite is acquired.

When a ship starts on a voyage, the Transit equipment must be fed the correct GMT time, latitude, and longitude of the ship; antenna height; and if not already in its memory, the *geoidal height* (distance from the center of the earth if it were actually round). Position readouts may be given by printed copy, by CRT readout, or by LED figures, or combinations of these.

As each satellite passes over one of the four ground stations, its frequency, orbit, and speed are measured. Corrections are sent to other ground stations which inject this information into the satellite receiver when it passes within range. This allows the satellite to broadcast correctional data for continual updating of the ship receiver-computer units.

The internal computer can also be used to determine correct course, distance, and time to a desired destination. It can also compute true speed, as well as drift due to currents or wind.

Daytime accuracy of Sat-Nav is 600 m and nighttime, about 100 m. However, if both Sat-Nav and Omega are combined in one system, accuracy is under 100 m at any time, at any place on earth.

The Navstar global positioning system (GPS), developed for aircraft, provides location in three dimensions, but in the future may be used aboard ships also.

In the near future ship-satellite-earth distress alerts may be made on public correspondence channels of INMARSAT satellites. Subsequent distress traffic will be by satellite to ships near enough to help, or by present terrestrial systems on MF (500 kHz CW), HF (2182 kHz SSB, 8364 kHz CW), and VHF (156.8 MHz FM). Satellites will use radiotelephone and radioteleprinter communications for distress and add data type transmissions for general correspondence. Digital selective ship calling systems should be implemented within a few years.

25-37 FACSIMILE TRANSMISSIONS

A radio-operated device being used to provide weather maps or page printouts is a facsimile (fax) recorder. With it, weather charts or maps transmitted by HF shore-based stations and received from geostationary applied technology satellites (ATS) can be received aboard ships at sea. Only HF SSB and 135-MHz satellite fax emissions will be discussed.

A facsimile (A3C or F3C emission) picture is roughly comparable to a single frame of a TV picture, except that dark and light portions of each line are laid down across a sheet of special paper instead of across a CRT. The dark portions are developed electrochemically by applying a dc through the paper as the line is being scanned. Instead of the one-thirtieth of a second required for a TV frame, it may take 10 min to produce a fax picture or map.

A basic flat-copy scanner used to develop the video signals that are transmitted is shown in simplified block diagram form in Fig. 25-30. In this system it will be assumed that a standard 120-rpm line scanning rate and 96 lines per inch (LPI) are used. The receiving equipment can be automatically started by radio with the transmission of a 5-s 300-Hz tone and stopped with a 5-s 450-Hz tone.

The picture information is obtained through the use of a hollow, rotating black-coated 19-in.-long glass drum having a single-turn helical scratch on it from one end to the other. Only through the thin, scratched line can light penetrate into the drum. Through a lens system an image of the map as it moves along is projected up onto a flat metal surface with a very narrow slit across it, as shown in the illustration. Any light that passes through the slit falls on the rotating helix. As it turns once, at 120 rpm, the helix allows light-dark variations to pass through the rotating scratch on its surface. These are the video signals for one horizontal line of the map. With mirrors and lenses, the light impulses passing through the helix opening are re-

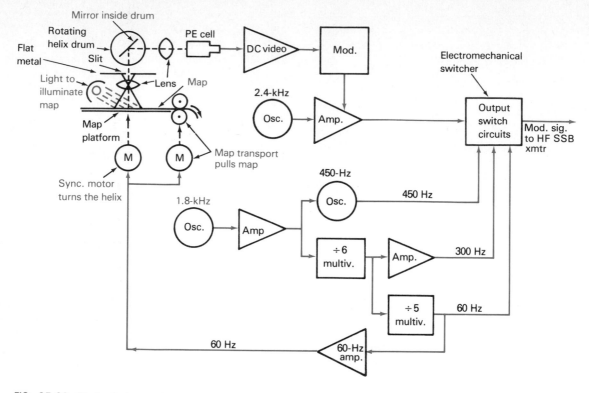

FIG. 25-30 Block diagram of basic facsimile scanner (from Alden model 9165KJL/AN-GXT-2).

flected down the center of the drum to a photoelectric (PE) cell and are amplified. These video signals are used to modulate a 2.4-kHz carrier, with signals from black areas producing maximum carrier strength and signals from lighter areas developing less carrier strength. During picture transmission times, the video-modulated 2.4-kHz carrier can be fed as AF signals to the AF input of a HF SSB transmitter. This is the signal that is picked up by fax receivers.

A 1.8-kHz crystal (or tuning fork) oscillator can act as a frequency standard. Its stabilized output is amplified and used to hold a divide-by-6 multivibrator in synchronism to produce the ac for a 300-Hz start signal. It also locks in a 450-Hz phase-shift oscillator to produce the ac for the stop signal. The 300 Hz ac synchronizes a divide-by-5 multivibrator to produce an exact 60-Hz ac output. The 60 Hz ac is amplified and operates the synchronous helix motor as well as the map transport rollers that pull the map slowly under the video-pickup PE-cell system.

An electromechanical switcher acts as a timing device through cams and contacts to transmit first the 5-s 300-Hz start signal followed by 25 s of *phasing signal* (95% of each line in black, then 5% of white), then a 1-s burst of 60 Hz, and then the map signals start. At the end-of-map scanning the stop signal is switched in for 5 s and the equipment can then shut down.

There are various transport lines-per-inch speeds (LPI), helix rotation speeds (rpm), and carrier frequencies other than the standards mentioned. As an example, if the starting signal is 852 Hz instead of 300 Hz, the helix motor will be switched to rotate at 240 rpm instead of 120.

The video-modulated 2.4-kHz carrier modulates a HF SSB transmitter. This results in a form of FSK, with black signals producing lower frequency sidebands than white signals. (This is reversed with some fax transmissions.) As a result, a receiver tuned to the transmitter carrier demodulates the video as tones that range from about 2300 Hz (white) down to about 1500 Hz (black), which sounds to a listener like a warbling tone. (Slow-scan amateur TV uses somewhat similar values: black = 1500 Hz, white = 2300 Hz, 128 lines/picture, 66.7 ms per line including 5 ms for horizontal sync, 30 ms for veritical sync, totaling 8.53 s per picture.)

25-38 FACSIMILE RECEPTION

The facsimile equipment aboard ships usually has its own HF receiver for SSB map transmissions from shore stations and possibly a 135.6-MHz down-link FM receiver to pick up the transmissions being relayed from geostationary satellites. The satellite transmissions are narrowband FM (10 kHz) and are relatively noise-free. The receiver feeds its signal to a facsimile recorder, such as the Alden[1] model 519/4A shown in simplified block form in Fig. 25-31.

To reproduce maps or pictures, a moist, electrosensitive paper is pulled slowly between a 120-rpm rotating drum and an essentially stationary

---

[1]Alden Research Center, Westborough, MA 01581.

ruler-straight steel electrode blade that extends under and across the width of the 19-in.-wide moving paper. On the nonconductive revolving drum is mounted a single spiral or helical turn of wire. The whole rotating assembly is called a *helix*, with the wire turn functioning as one electrode and the steel blade as the other. When paper is between the two electrodes and 60 V is applied, the paper is electrochemically activated between the recording blade and the helical wire, marking the paper at this point with a dark spot. The darkness of the marking is directly proportional to the voltage between electrodes.

The rotation of the helix results in a left-to-right traveling point of contact between helix wire and blade. If a constant 60 V is applied between helix and blade, a dark line is developed across the paper. If the voltage is modulated, the horizontal line will consist of darker and lighter spots. As the right end of the helix wire passes beneath the upper electrode, the left end of the single-turn helix is back at the left side of the paper again, ready to start the next line. Since the paper is constantly being pulled ahead slowly, one line is never laid down over the preceding one.

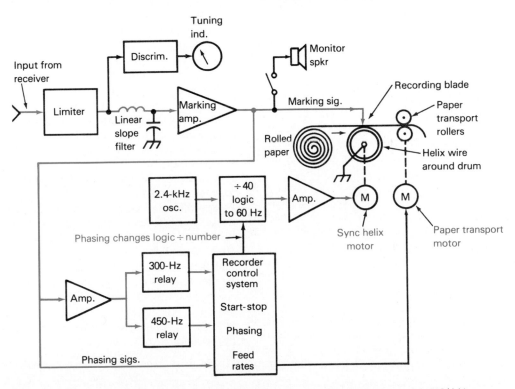

FIG. 25-31 Simplified block diagram of a facsimile recorder (from Alden model 519/4A).

The detected 1500–2300 Hz signals from the receiver are first limited, then are passed through a linear-slope filter that passes 1500 Hz 10 times stronger than 2300 Hz, and are amplified. The received signal has three functions: It is the marking voltage that marks the paper. It can be fed to a speaker for aural monitoring. It feeds control signals to the recorder control system. Through an amplifier it can actuate either 300- or 450-Hz resonant relays to produce start and stop signals for the recorder control. The 25-s phasing signals that follow a start signal can increase the divide-ratio of a logic circuit that normally divides the 2400-Hz crystal or tuning fork oscillator by 40 to produce 60 Hz ac. The 60 Hz ac drives the synchronous helix motor in exact synchronism with the scanner helix at the transmitter. The phasing signals should reduce the 60-Hz frequency and slow the helix to move the left margin of the map to the left side of the recording paper. If for some reason they do not, a manual control can be actuated a few times to reduce the 60-Hz frequency and move the margin of the map toward the left edge of the paper.

The 1-s 60-Hz synchronizing signal that follows the phasing signals is not used in the system being discussed, but it can produce map alignment and start-signaling for other systems.

When the map is completed, the 450-Hz stop signal is transmitted, shutting down the transmitter and, if on automatic mode, the recorder circuitry also.

Inasmuch as the received map video signals may be considered as either FM or FSK, with signals varying from about 2300 to 1500 Hz, the signals, after limiting, can be fed to a 1900-Hz discriminator and to a zero-center tuning indicator meter. The signals are tuned in correctly when the meter needle wavers least during map reception.

To prevent rapid wearing due to the helix wire wiping against it through the paper, the blade is made in the form of an endless steel loop moving along at about ⅛ in. per minute. It should last 3 to 6 months with relatively constant use.

The recorder paper feed speed must be set to the same rate as that of the map feed rate of the scanner. Rolls of 19-in.-wide electrosensitive paper are packaged in 170-ft lengths.

Transmissions from parked satellites are relayed signals fed up to the satellites from ground stations. These are cloud-cover pictures picked up from weather satellites overlaid on charts of that portion of the earth's surface.

25-39 SMALL VESSEL INSTALLATIONS

Small coastal, bay, and river boats may have radio and other electronic equipment aboard. Some of the radio units used are transistorized two-way HF (2–16 MHz) SSB transceivers, AM-SSB CB equipment, and VHF FM sets (Sec. 25-13). There is no CW aboard unless possibly some licensed amateur equipment (Chap. 32).

HF antennas are usually 10- to 18-ft vertical whips, often Fiberglas rods with internal wires and base or center loading coils. Wire antennas for HF may be installed from the bow up to the peak of a mast, sometimes continuing down to the stern (insulated back stay for sailboats) and are end-fed. VHF antennas may be quarter-wave verticals, or coaxial half waves mast-top-mounted, and fed by coaxial cable. Antennas must be ruggedly fastened, and be as clear of metal rigging as possible. Metals must be selected for the damp environment. Only brass, copper, or stainless-steel hardware can be used. Iron or steel rapidly rusts and deteriorates.

For HF equipment a good, low-resistance ground is a necessity. Wood, Fiberglas, and even metal hulled vessels require several square feet of unpainted copper ground sheeting mounted on the outer hull surface, well below the water line, and near the radio equipment. One of the brass mounting bolts holding the copper sheeting is brought through the hull and is used as the ground connection. In some cases a $12 \times 5 \times \frac{1}{2}$ in. or larger bar of sintered, porous bronze may be bolted to the hull, with one mounting bolt used as the ground connection. (Sailboats ground to the lead keel.)

Over a period of time any external metal plates, propellors, etc., are eaten away by *electrolysis*, caused by RF or other currents from ship to water. Copper parts can be edged with zinc strips. The zinc will then be the metal that is eaten away. The zinc strips must be replaced periodically. Furthermore, different types of metals anywhere on the vessel must never be bolted together, or electrolysis will be set up. All stainless-steel metal wires and parts should be bonded together, then all copper, and then all brass. From each, a heavy lead should be run to the central ground connection. All metal masts and rigging must be grounded.

Test your understanding; answer these checkup questions.

1. On what frequency are Omega signals measured? _____ How many dashes are sent in 10 s?

_____ Omega receivers keep track of how many stations simultaneously? _____

2. What is measured in the Omega system? _____
3. If the Omega value 546.72 is shown, what does this mean? _____
4. What Sat-comm frequencies are used aboard ships? _____
5. What is meant by yaw? _____
6. Do down-link signals fade? _____
7. NNSS satellites have what orbits? _____ Frequency? _____ Timing intervals? _____
8. When is Doppler effect greatest? _____
9. What is counted in Sat-Nav? _____ What is developed by this? _____
10. Besides lat. and long. what can a Transit unit indicate? _____
11. Where is a fax scanner found? _____ What are the standard scanning and LPI rates? _____
12. What tone is used to start fax? _____ To stop it? _____
13. What tones carry map information? _____
14. What is 1.8 kHz used for in a scanner? _____ Could it be used as the video-modulated carrier? _____

15. To what is the output of a scanner fed? _____
16. Basically, what is a scanner helix? _____ A recorder helix? _____
17. What does a scanner switcher switch in? _____ _____ _____ _____ _____
18. With what type transmitter is fax transmitted on HF? _____ On VHF? _____
19. The helix wire forms one end of a circuit; what forms the other? _____ What emf is required across them to form a dark spot? _____
20. In a recorder, what is used to make 1500 Hz ac stronger than 2300 Hz? _____ To generate 60 Hz ac? _____
21. If the tuning meter reads full scale, what would this indicate? _____ Midscale? _____ Low-scale? _____
22. Why is the blade an endless loop? _____
23. What should bring the left margin of a map to the left side of the paper in a recorder? _____
24. What is the makeup of the phasing signal? _____
25. How do wooden hulled vessels ground equipment? _____
26. What causes electrolysis of metals? _____

MARITIME RADIO QUESTIONS

1. When is maximum power output required of a maritime transmitter?
2. What ships require radio operators aboard?
3. What radiotelephone equipment should all ships carry?
4. Why should all radiotelegraph ships have A2A emission transmitters? What minimum power?
5. Draw a block diagram of a marine main CW transmitter.
6. Draw a block diagram of a marine reserve transistor CW transmitter.
7. What polarization do all 500-kHz or MF marine antennas have? Why?
8. What must be added between a 50-Ω coaxial line and the base of a center-loaded vertical whip? What is used to tune such an antenna to resonance?
9. For what is the sensing circuit of an antenna autotuning circuit looking?
10. Draw a diagram of an autotuning sensing circuit.
11. Why are HF transmitters sometimes more desirable than MF on ships?
12. Draw a block diagram of a general-purpose HF marine transmitter capable of CW, SSB, and FSK.
13. How many marine MF bands are there? HF calling bands? High-traffic HF working bands? Low-traffic HF working bands?

14. What are the four distress frequencies for ships?
15. Why might a TRF receiver be used as a reserve marine receiver?
16. What is the basic difference between second and third generation superheterodynes? What range of frequencies might the third generation have?
17. Draw a block diagram of a third generation marine superheterodyne.
18. How does a fourth generation superheterodyne differ from a third?
19. What is the frequency and power of bridge-to-bridge communication equipment? Who operates it? What emission does it use?
20. What is the power and operating frequency of portable survival-craft radio equipment? What type of power supply is used? What can it transmit?
21. What is unusual with the hand keying circuit of the survival radio equipment shown?
22. Lifeboat emergency radio equipment has what power output emission capability and operating frequencies?
23. What does EPIRB mean? What signals are transmitted, and on what frequencies?
24. What is the AA signal? When and where is it sent?
25. What are the two methods discussed of developing automatic AA signals?

26. What dash lengths and space durations does an AA receiver accept?
27. What does a properly made and received AA signal do?
28. How is an AA alarm silenced?
29. How often must AA equipment be tested at sea?
30. What may cause AA bells to ring besides AA signals?
31. What are the seven possible main antenna switch positions?
32. What is the basic function of a TTY receiving TU? Of the transmitting converter?
33. What does SITOR mean? What is it? How does it work?
34. What are the two possible types of SITOR transmissions?
35. What is the age requirement for T1 licenses?
36. How many silence periods are there in one hour, and when do they occur?
37. What are two other methods of expressing "GMT"?
38. Which lines converge, latitude or longitude?
39. Signals approaching at right angles to the plane of a loop produce what output signal? What is the shape of the reception pattern of a loop?
40. What produces residual signal in a loop?
41. What must be done to a loop antenna to produce a true null? How may this be done?
42. How is a unidirectional bearing obtained with a loop?
43. What are 5 DF errors?

44. In which direction from the bow are quadrantal errors found?
45. In what frequency band do most marine loops operate?
46. What is the name of the stationary two-loop RDF system?
47. What is different about an RDF TRF or superhet from the usual AM/CW receiver?
48. What type of loop system does an ADF use?
49. Draw a block diagram of a synthesized-frequency ADF. What frequency bands does it cover?
50. Why is an Adcock system better for HF and VHF DF work?
51. How may the position of a vessel be plotted on a nautical chart?
52. What does loran mean? On what frequency is modern loran? The original loran?
53. What are the two or more stations of a loran system called?
54. What shape LOP is developed in loran systems?
55. What is the approximate range of MF DFs? Of loran A? Of loran C?
56. What is the theory of measurements of loran systems?
57. Besides navigation for what can loran C be used? What kind of antenna is used with loran C receivers?
58. What is the name of the navigation system that requires comparison of 1-s dashes? On what frequencies does it operate?
59. What type of satellite is used with INMARSAT communications? What emissions are relayed by it?
60. What type of satellites are used with Sat-Nav? What does Sat-Nav measure to determine LOPs?
61. What navigational system provides three-dimensional fixes?
62. What is usually transmitted by facsimile? How long may one fax transmission take? What is a common line rate?
63. What type of transmission is usually used with fax on HF? What bandwidth emission is usually used?
64. What produces the dark spots on fax receiving paper?
65. What types of metals are used aboard vessels?
66. What causes underwater metals to be eaten away?
67. How is a good ground made on a metal hull vessel? On a wooden hull? What should be added to it?
68. What type of HF antennas are usually used on small vessels?

ANSWERS TO CHECKUP QUIZ ON PAGE 632

1. (10.2 kHz) (8) (8) **2.** (Phase and phase reversals) **3.** (546.72 phase reversals) **4.** (1.54 and 1.64 GHz) **5.** (Turning to a side) **6.** (Yes) **7.** (Circular) (399.965 MHz) (2 min) **8.** (Overhead) **9.** (Beat cycles) (LOP) **10.** (Course, distance, destination time, speed, drift) **11.** (At origin of fax signals) (120) (96) **12.** (300 Hz) (450 Hz) **13.** (1500–2300 Hz) **14.** (Develop 300, 450, 60 Hz) (Yes) **15.** (Transmitter) **16.** (Opaque-sided drum with 1-turn scratch) (Insulated drum with 1-turn wire) **17.** (300 Hz, phasing signal, 450 Hz, 60 Hz, map signals) **18.** (SSB) (FM) **19.** (Blade) (60 V) **20.** (Linear-slope filter) (2.4-kHz oscillator and logic dividers) **21.** (1500 signal or mistuned) (1900 or average) (2300 signal or mistuned) **22.** (Wears slower) **23.** (Phasing signals or manual switch) **24.** (25 s of solid black lines 95% across paper) **25.** (External ground plates) **26.** (RF or other currents to water)

26

Radar

This chapter describes the general purpose of a radar set and then discusses possible basic radar transmitter, receiver, and antenna circuits. The component circuits of a complete PPI transmit-receive-display system are explained. Nonmarine, small boat, and collision radar are discussed. This chapter and the chapter on microwaves form the basis of the FCC radar endorsement.

26-1 PRINCIPLES OF RADAR

Radar is an electronic system of determining the RAdio Direction And Range of anything that will reflect microwave or millimeter radio waves. It is one application of the microwave theory presented in Chap. 22.

The use of echoes as an aid to navigation is not new. When running in fog near a rugged shoreline, ships may sound a short blast on their whistles, fire a shot, or strike a bell. The time between the origination of the sound and the returning echo indicates how far the ship is from the cliffs. If sound travels 1100 ft/s and an echo is heard after 2 s, it indicates a total sound-travel distance, outward and return, of 2200 ft. The ship must be half this distance, or 1100 ft, from shore. The direction from which the echo approaches indicates the *bearing* of the shore.

Today, ships transmit a short pulse (PØN emission) of radio energy and receive the echo produced when the wave is reflected from an object. By using an antenna with a narrow lobe, the *direction* of the reflecting target can be accurately determined. By measuring the time between pulse transmission and reception of the reflected signals, the *range* of the target is determined.

Besides being employed as a marine navigational aid, radar is used by aircraft, by airfields, by the armed services as a means of locating enemy targets and aiming guns, and by police.

The indicator in a ship radar set is a cathode-ray tube (CRT), calibrated in nautical miles (nm = 6080 ft), statute miles (5280 ft), or in kilometers.

Radio waves travel 162,000 nm/s (186,000 statute mi/s, or 300,000 km/s). Thus, in 1 μs they travel 0.162 mi. To travel 1 nautical mile a radio signal requires 1/0.162, or 6.17 μs. A *radar mile* is considered to be 12.3 μs, since the wave must travel for this period of time, to and from a target, 1 mile away. If a target is 10 mi away, it takes 123 μs from the time of transmission before the echo signal returns and is displayed on the CRT.

The frequencies generally used for marine radar are in the SHF (superhigh-frequency) part of the radio spectrum, either in the 3–3.25 GHz band (10-cm, or S band) or in the 9.32–9.5 GHz band (3-cm, or X band).

For the 10-cm band, a half-wave antenna is only 5 cm long. A parabolic reflector 2 m wide can form a radiated beam with a horizontal width of 2° to 3°. In the 3-cm band, a dipole is only 1½ cm in length, and a 2-m reflector can form a 1° to 2° beam. At these frequencies the effective range is about 30% more than line of sight.

Marine radar sets are made to operate with a full-scale range of ¼ to 100 mi and a minimum range of less than 50 yards from the antenna.

The number of pulses transmitted per second, the *pulse repetition rate* (PRR), varies between about 800 and 2000 per second. The lower PRR is used for longer ranges. Longer pulses (0.5 μs) represent greater power output from the transmitter to produce stronger echo signals from distant objects. The pulse width is about 0.15 μs (or less) for short-range indications.

26-2 A BASIC RADAR SYSTEM

Before the operation of a more complex radar system is considered, the functioning of a simplified one (Fig. 26-1) will be explained.

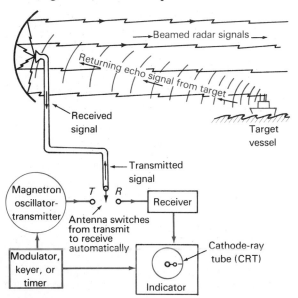

FIG. 26-1 Block diagram of the elements of a simple radar system.

The heart of the system is the modulator (keyer, or timer). In this unit a 0.2-μs dc pulse is formed and fed to a magnetron oscillator, resulting in the transmission of a 0.2-μs pulse of SHF radio energy. The timer pulse is also fed to the indicator, starting a dot moving horizontally, tracing from the center of the CRT face to one edge. If the maximum range of the radar set is to be 10 mi, the time for the barely visible dot to move to the edge of the scope will be 10 × 12.3, or 123 μs. The timer pulse is also fed to the grid (or cathode) of the

CRT, producing a bright spot at the center of the tube at the start of each trace.

After each RF pulse has been transmitted, the antenna automatically switches to the receiver and waits for the return of the echo signal. If a target is 5 mi away and in the beam of the antenna, after 61.5 μs a weak echo signal is received by the antenna, fed to the receiver, amplified, and fed to the grid of the indicator scope. This signal increases the voltage developing the moving dot wherever it happens to be on the trace line, and a bright spot appears. The distance from the center spot to the echo spot is an indication of the range of the target. In this case the target *blip* will appear halfway across the 10-mi trace. The bearing of the target is indicated by the direction the antenna is pointing when it picks up the strongest echo signal.

As soon as the barely visible dot has traveled for 123 μs, the indicator tube is desensitized and the dot returns to the center without producing any visible indication.

With a PRR of 800, every 1250 μs a new target blip is registered at the same point on the trace. If the target is approaching, the distance between the center spot ("main bang") and the blip on the trace shortens. In this way a constant check can be maintained on the range of the target.

The parabolic antenna reflector has a radiated beam of about 2° horizontal width for good *bearing resolution*, and a vertical beam height of 15° to 20° to allow for ship roll. Bearing resolution is the ability to separate adjacent targets the same distance away. *Range resolution* is the ability to distinguish two or more targets in the same direction but at different distances.

If antenna rotation were controlled manually, it would be difficult to keep the target in the 2° beam should the target or the radar ship be moving. It is necessary to improve this basic radar system to make it a practical aid to navigation.

If the antenna rotates horizontally at a constant speed, 10 times per minute, it will make one rotation in 6 s. With a PRR of 1000, it will fire 6000 pulses per rotation, or 17 pulses per degree. If horizontal-deflection coils are rotated physically around the neck of the CRT in exact synchronism with the rotation of the antenna, targets can be shown on the indicator face in exact relation to their range and bearing. This is known as a plan-position-indication (PPI) type of presentation.

Radar CRTs use electromagnetic deflection and vary in diameter from 7 to 16 in. They differ from television tubes in screen *persistence*. The radar-tube faces are coated with a little fluorescent and considerable phosphorescent material. The phosphors retain a latent image for 10 s or more, longer than is required to make one antenna rotation. As a result, a PPI presentation forms a constant, fairly well illuminated plan, map, or chart of the targets in all horizontal directions from the ship. A block diagram of the component parts of the system is shown in Fig. 26-2. The motor that rotates the antenna is shown mechanically coupled to the deflection coils around the neck of the indicator CRT.

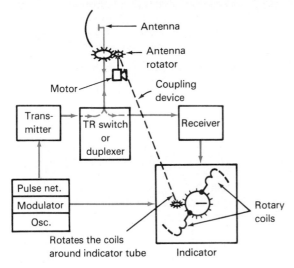

FIG. 26-2 Block diagram of the elements of a PPI navigational radar system.

26-3 A RADAR TRANSMITTER

A step-by-step explanation of the important circuits in a marine radar system will be given. Most of the circuits in the receiver and indicator have been discussed in earlier chapters. New circuits will be explained. The circuitry has been simplified as much as possible to give a general understanding. It is impossible to cover all the various types and models of modern radar sets.

This radar transmitter consists of a blocking oscillator, a modulator, a pulse-forming circuit, and a magnetron oscillator (Fig. 26-3). The blocking-oscillator circuit depends on the C_g and R_g values to determine how many times per second it will operate. The circuit shown is an Armstrong oscillator, but any oscillator circuit will block if high values of grid-leak R and C are used. As the circuit starts to oscillate, plate current (I_p) flows for an instant, inducing a charge in C_g well beyond I_p cutoff.

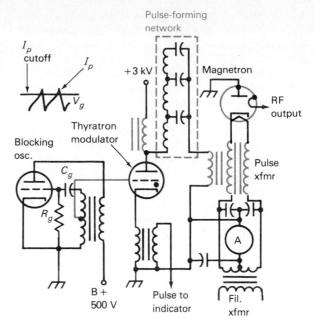

FIG. 26-3 A basic radar transmitter system.

During the I_p-off time, C_g discharges slowly through R_g and the bias voltage decreases until finally I_p can begin to flow again. Circuit regeneration develops a rapid rise in I_p. These short-duration I_p pulses induce high-amplitude pulses in the grid winding of the transformer. The number of pulses of current per second in this stage determines the PRR of the whole radar system.

The grid of the hydrogen thyratron (or an SCR) modulator is triggered by the positive pulse, and the tube ionizes. (Hydrogen ionizes and deionizes very rapidly.) Ionization discharges the capacitors of the pulse-forming network that have charged to 3000V through the charging reactor during the nonoperating period of the oscillator. The discharge produces a narrow *square-wave* pulse of current through the primary of the pulse transformer, inducing a high voltage in the two secondaries. These two voltages are of equal value and are in phase, raising the cathode of the magnetron (Sec. 22-8) to a negative potential of several thousand volts above the magnetron plate without changing the potential between the two filament terminals. As a safety measure, the plate cavities of the magnetron with their metal cooling fins are grounded and the high-voltage pulses are fed to the cathode. The *average* value of magnetron I_p, perhaps 2 to 5 mA, is indicated by the meter between ground and filament transformer. The square-wave pulse of current from the pulse transformer excites electrons into oscillation in the cavities of the magnetron. These oscillations are coupled to the antenna and form the RF output burst.

A pulse-forming network, as shown, may also be called an *artificial transmission line*. The line can produce a square-wave output pulse when triggered with a short burst of energy, provided line impedance is matched to load impedance. Pulse length will be a function of the values of L and C used in the pulse-forming network (pulse length = $2\sqrt{LC}$; L in henrys; C in farads). An artificial transmission line can be called a *delay line* because a pulse voltage fed across the input end will appear a few microseconds later across the output end (delay time = $\sqrt{LC}$). If an electric impulse travels 300 m in 1 μs, a 300-m transmission line will delay the voltage 1 μs. An artificial line with similar values of series L and shunt C will delay a pulse the same amount.

The transformer in the cathode of the thyratron pulser tube has a primary of two or three turns, which induces pulse voltages into the secondary. These pulses are fed to the indicator system to trigger the circuit that starts the trace moving across the CRT screen. Earlier radar systems employed a rotating spark gap to develop the narrow high-voltage pulses.

Test your understanding; answer these checkup questions.

1. What does "radar" mean? _____
2. In what distance units are marine radars calibrated? _____
3. What time unit is the equivalent of a radar mile? _____
4. In what frequency bands does marine radar operate? _____ _____
5. What does PRR mean? _____ What rates are used? _____
6. Why must short-range radar have narrow pulse widths? _____
7. To what two circuits would a radar timer feed signals? _____ _____
8. What is a target display on a CRT called? _____
9. Why must a radar antenna be pointed directly at a target? _____
10. What is the main bang on a PPI radar set? _____
11. What is the ability to separate adjacent equidistant targets called? _____

12. What is the approximate rotational rate of a radar antenna? _____ About how many pulses are transmitted per degree? _____
13. In what two ways do radar CRTs differ from TV CRTs? _____ _____
14. How would a "sector scan" differ from a PPI presentation? _____
15. What circuits make up the radar transmitter? _____ _____ _____
16. What type of oscillator determines the PRR? _____ What other circuit might be used? _____
17. Why are hydrogen thyratrons used in radar? _____
18. Why is a magnetron anode grounded? _____
19. What are two other names for a pulse-forming network? _____ _____
20. What determined the PRR in old-time radar sets? _____

26-4 AVERAGE POWER AND DUTY CYCLE

At the marine radar frequencies, 3 and 9 GHz, the most practical transmitting oscillator tube is the magnetron. Voltage pulses fed to it may range from 5 to more than 20 kV, with output pulses of perhaps 15 kW (megawatts for military radar). What power rating must such tubes have?

If a radar transmitter has a PRR of 900 pps, each pulse has a duration of 0.5 μs, and it has a peak pulse power of 15 kW, the total emission duration must be 900×0.0000005, or 0.00045 s. The transmitter is on 0.00045 s each second. It has a *duty cycle* of 0.00045. The average power output of the transmitter is the peak power times the fraction of a second it operates, or $15,000 \times 0.00045 =$ 6.75 W. The tube is transmitting pulses with peaks of 15 kW but is transmitting an average power of only 6.75 W. If the magnetron is 50% efficient, the average dc power input is only 13.5 W. From this, average power output is

$$P_{av} = P_{peak} \times PRR \times \text{pulse width}$$

$$P_{av} = P_{peak} \times \text{duty cycle}$$

Two formulas for duty cycle are

$$\text{Duty cycle} = \frac{P_{av}}{P_{peak}} = PRR \times \text{pulse width}$$

The duty cycle is also the ratio of the pulse width to the time between the beginning of two pulses (called the *pulse repetition time*, or PRT). With a PRR of 900, the PRT is $^1/_{900}$ s, or 0.00111 s. For the transmitter above, the duty cycle must be only 0.0000005/0.00111, or 0.00045.

What would be the peak power and duty cycle of a radar transmitter with a pulse width of 1 μs, a PRR of 900, and an average power of 18 W? By solving for P_{peak} from the P_{av} formula,

$$P_{peak} = \frac{P_{av}}{\text{PRR} \times \text{pulse width}}$$

$$= \frac{18}{900(0.000001)} = 20,000 \text{ W}$$

$$\text{Duty cycle} = \frac{P_{av}}{P_{peak}} = \frac{18}{20,000} = 0.0009$$

26-5 THE ANTENNA SYSTEM

A radar antenna system may consist of a waveguide from the magnetron to the antenna and a *duplexer* near the magnetron (Fig. 26-4). The end of the conductor from the magnetron acts as a quarter-

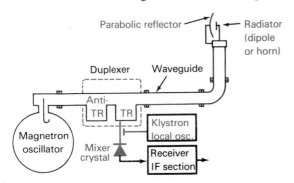

FIG. 26-4 Block diagram of a radar magnetron, duplexer and components coupled to it, waveguide, and antenna.

wavelength antenna, transmitting RF energy along the waveguide. At the far end, the walls of the waveguide may be expanded to form a horn, and the usually vertically polarized RF energy is radiated from the horn into a parabolic reflector. In some cases the waveguide terminates in a tiny dipole antenna. Any coaxial-cable runs are kept as short as possible because the losses are far greater than for waveguide lines.

Radio waves of 3 or 9 GHz propagate in much the same way as light rays. They can be focused into a narrow beam with a metal parabolic reflecting surface by placing the radiator at the focal point (Fig. 26-5). By shaping the reflector the emitted wave can be formed into the desired 2° horizontal beam width and 15° vertical beam height. Radiator and reflector rotate at about 10 rpm, in synchronism with the sweep coils rotating around the CRT.

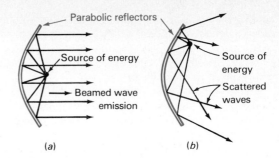

FIG. 26-5 Parabolic reflectors. (*a*) With source at the focal point, where a narrow beam is formed. (*b*) Waves scatter when the source is not at the focal point.

The horn (or dipole) radiator is usually covered with polystyrene or some other plastic to protect it from weather. The plastic must not be painted. An excessive amount of soot or dirt on either the polystyrene cover or the active surfaces of the reflector may decrease transmission and reception. Radiated energy is considerable with some radar equipment. Anyone in the beam of a high-powered radar antenna for a few minutes may die within a few days. Radar beams have been known to ignite a shipment of photograph flashbulbs and to start fires.

Radar antennas are of two basic types. One is the parabolic reflector, described above, and the other is the slotted waveguide. Round parabolic reflectors are used for millimeter radar. Shipboard microwave parabolic reflectors are made in more of an oval shape, with the wide dimension horizontal. All of the reflecting surface is at an angle from the RF radiator to reflect parallel radio waves to form a horizontally narrow beam lobe.

A slotted waveguide antenna consists of a piece of waveguide 20 or more wavelengths long, with slots machined in one side of the waveguide wall (Fig. 26-6). Through these slots phased RF energy radiates. With proper dimensioning of the slots, the resulting radiation can be formed into a narrow

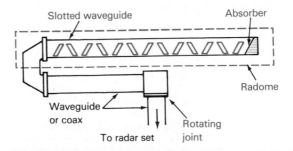

FIG. 26-6 Slotted-waveguide radar antenna.

beam of any desired polarization. A slotted guide radar antenna resembles a "T," with the top horizontal part rotating. The rotating part is encased in a plastic protective enclosure, called a *radome*, which must be kept clean.

Radar signals at 3 GHz may not be greatly affected by rain storms, but at 9 GHz signal losses of 20 dB or more may be noted. A very heavy rain storm shows as a haze on the CRT.

Placement of the radar antenna is important; it must be as high and as much in the clear as possible. The higher it is, the farther it can "see" targets. If the radiated beam is above masts, booms, and stacks, these objects will not shadow, or reflect signals and produce false blips on the CRT.

26-6 THE DUPLEXER

The radar transmitter and receiver use the same antenna system. Some means must be provided to prevent the powerful transmitter signal from feeding directly into the receiver and burning out the input circuit. This is accomplished by using a resonant transmit-receive (TR) cavity with a special gas-filled spark-gap TR tube in it (Fig. 26-7). The cavity is coupled to an opening in the antenna waveguide. Signals in the waveguide excite the tuned cavity into oscillation. A coupling loop in the cavity feeds these signals to the crystal-diode mixer. The klystron (or Gunn diode) local oscillator is also coupled to the mixer circuit.

When the transmitter emits a pulse from the magnetron, high-powered waves induce enough voltage across the TR cavity to ionize the TR tube, and it conducts. This effectively shorts out the cavity, detunes it, and prevents high-amplitude signals from forming in it. At the conclusion of the transmitted pulse, the TR tube deionizes and the cavity can receive any returning echo signals. To make the TR tube more sensitive, a dc *keep-alive* voltage is

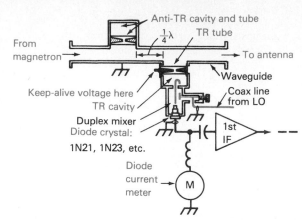

FIG. 26-7 Details of the ATR, TR crystal mixer, and waveguide.

applied across it at all times. This voltage is not quite high enough to support ionization, but a small increase in voltage across the gap will ionize it. When the TR tube ages, it requires a greater signal voltage to produce ionization, it does not protect the crystal diode in the mixer circuit of the receiver, and the diode burns out. When a diode must be replaced, the TR tube is also replaced.

The distance from the magnetron feed point to the TR cavity ("box") is very critical. A less critical means of coupling the TR box to the waveguide is to use a second tuned cavity, placing it exactly one quarter wavelength from the TR box opening. This second cavity is called an *anti-TR* (ATR) *box*. It has a TR tube in it also, but since it does not have to protect any circuits, it has no keep-alive voltage. The ATR ionizes during each transmitted RF pulse, presenting a low impedance across its opening in the waveguide. This allows the pulse moving up the waveguide to pass unattenuated. When the pulse ceases, the ATR tube deionizes and produces a high-impedance point (mismatch) in the waveguide. Now echo signals coming down the waveguide can no longer pass to the magnetron but are reflected at the ATR opening and dispose of their energy in the TR cavity, which is the input circuit of the receiver. The TR and anti-TR cavities, the λ/4 of waveguide between them, and the mixer section form the *duplexer*.

26-7 THE RADAR RECEIVER

Radar receivers are superheterodynes, with an IF of at least 30 MHz. This requires a local oscillator operating at a frequency 30 MHz or more from the

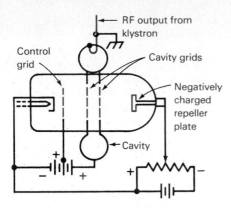

FIG. 26-8 SHF reflex klystron oscillator circuit.

transmitting frequency (3 or 9 GHz). A reflex klystron is often the LO (Fig. 26-8).

A radar receiver consists of a crystal-diode mixer stage, a klystron (or Gunn diode) local oscillator, six or more 30-MHz IF amplifiers, a diode second detector, and video amplifiers capable of amplifying signals up to several megahertz (Fig. 26-9). Such a

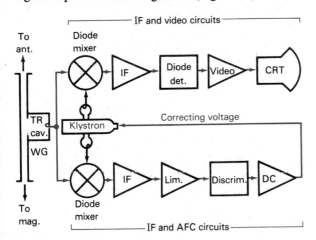

FIG. 26-9 Block diagram of a radar receiver with AFC.

wide range is required because a pulse of 0.5-μs duration represents a frequency of 1 MHz. The square waveshape of the pulse requires many harmonics of 1 MHz to reproduce the square waveshape. A bandwidth of 10 MHz can be attained with a 30-MHz or higher IF strip.

The automatic-frequency control (AFC) system keeps the klystron circuit oscillating on the correct frequency. The output of a second crystal mixer, similar to that of the receiver mixer, feeds an IF amplifier, limiters, and a 30-MHz discriminator cir-

cuit. When the transmitter and local oscillator are separated by the proper IF difference, zero discriminator output voltage is delivered to the klystron repeller. Should either the transmitter or the klystron drift in frequency, a dc output voltage is produced by the discriminator, is amplified, and is made to vary the repeller voltage of the klystron, forcing it to oscillate at a frequency exactly 30 MHz from the transmitter frequency. When a receiver mixer crystal must be replaced, the AFC crystal should be changed at the same time.

The IF stages of a radar receiver are basically similar to other broadband superheterodynes. However, to prevent strong returning echoes from nearby sea waves, known as *sea return,* from being shown on the CRT, the receiver is desensitized immediately after the transmitted pulse. The IF stages must then return to maximum sensitivity in 10 to 15 μs. Sea-return elimination may be accomplished by using a thyratron, SCR, or other keying device, in a *sensitivity-time-control* (STC) circuit (Fig. 26-10).

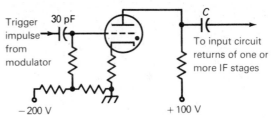

FIG. 26-10 Thyratron STC circuit to develop a decaying bias for the IF amplifiers of a radar receiver.

When a positive trigger impulse from the modulator is applied to the thyratron grid, it overcomes the bias. The tube ionizes, conducts heavily, and discharges capacitor C. This drives electrons from the right-hand plate of C to the input circuits of the IF stages, reverse biasing and desensitizing them. At the completion of the trigger pulse, the thyratron deionizes and capacitor C begins charging, reducing the reverse bias on the input circuits to normal in a few microseconds. Thus, as the transmitted pulse is produced, the IF stages are highly biased and insensitive. As time progresses, the bias falls off and in 10 to 15 μs normal sensitivity returns. STC is controlled by a panel knob.

The signal from the IF stages is detected by a diode and fed to the video stages, where the video ac is *limited* to reduce *blooming* (excessive diameter blips on the CRT screen).

The output impulses of the video amplifiers are

fed to either the grid or the cathode of the CRT, producing the visible echo signals on the screen. A manual gain control in the video stages acts as the brilliance control for the scope presentation. A manual gain control in the IF amplifiers acts as the *sensitivity* control which is adjusted to produce a just-visible trace on areas of the CRT screen where no targets are being displayed.

1. With a PRR of 800 and a 1-μs pulse width, if the peak power is 100 kW, what is the average power? _____ The duty cycle? _____
2. What is the PRT with a PRR of 800? _____ What does the answer mean? _____
3. What is used to carry RF energy from magnetron to antenna in a radar set? _____
4. What are two methods of illuminating a parabolic reflector with RF? _____ _____
5. Would a radar reflector be parabolic-shaped vertically or horizontally? _____ Why? _____ _____
6. In what way are radar emissions similar to microwave ovens? _____
7. What method of coupling is used between the klystron LO and the diode mixer in Fig. 26-7? _____
8. To what tube(s) is a dc keep-alive voltage applied? _____
9. If a mixer diode burns out, what are replaced? _____
10. What is the cavity between magnetron and mixer cavity called? _____
11. Radar receivers use what IFs? _____ Why? _____ What kind of LO? _____ What solid-state types might be used? _____
12. Do ATR tubes aid transmitting or receiving? _____
13. To what is the AFC voltage applied in a klystron? _____
14. What is the control that desensitizes a radar receiver for 10 to 15 μs called? _____
15. How is blooming prevented? _____ How is brilliance controlled? _____ Sensitivity? _____

26-8 CIRCUITS OF THE INDICATOR

The radar indicator consists of a CRT and the circuits necessary to produce a sawtooth sweep current for the deflection coils that rotate around the CRT neck, plus range-marker pulses. At first the CRT

grid will be considered biased so that no electrons can strike the face of the tube.

The originating pulse for the deflection system comes from the thyratron shown in Fig. 26-3. The pulse transformer signal, in negative phase, is fed to a *one-shot* multivibrator *gating* circuit (Fig. 26-11). Q_1 conducts heavily because of the positive

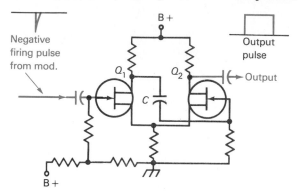

FIG. 26-11 One-shot multivibrator oscillator to develop a square-wave intensifier pulse.

bias on it. The resulting source bias of this stage holds Q_2 in nonconduction. A negative pulse from the thyratron drives Q_1 into nonconduction for an instant charging C, and Q_2 conducts for a period of time. When C discharges far enough, Q_1 starts conducting again and Q_2 cuts off once more. For each pulse from the modulator a negative-going square-wave *intensifying* pulse is produced from Q_2. This pulse is reversed in phase through an amplifier and fed to the grid of the CRT through an adder circuit, overcoming the bias on the grid and allowing a few electrons to strike the face of the tube.

The leading edge of the intensifying pulse triggers a sawtooth circuit into one cycle of oscillation. The sawtooth current is amplified and fed through slip rings and brushes to the deflection coils that rotate around the CRT neck (Fig. 26-12). With the intensifying pulse operating, deflection coils rotating, and the sawtooth current deflecting the beam, a weak glow appears all around the face of the tube as the coils rotate.

When echo signals are received and fed in a positive phase to the CRT grid through the adder circuit, they produce much brighter spots than are produced by the intensifying pulse. These target signals blips are readily visible.

The negative intensifying pulse is also fed to a *range-marker* circuit (Fig. 26-13). Q_1 has no bias and conducts heavily, producing a strong stationary

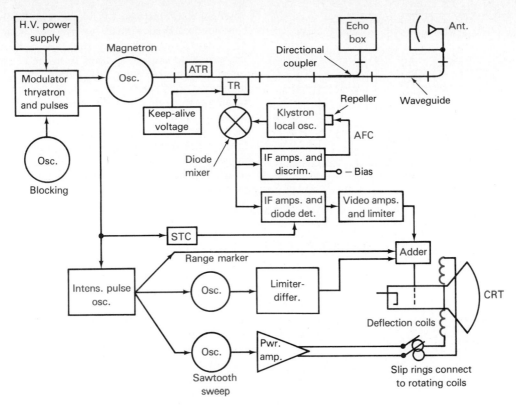

FIG. 26-12 Block diagram of a complete PPI marine radar system.

magnetic field around the coil L. As the negative intensifying pulse suddenly biases Q_1 into nonconduction, the magnetic field collapses and the high-Q LC circuit is driven into damped oscillations. This ac is fed to amplifier Q_2. The amplified output voltage is fed through resistor R_1 to a zener diode, which acts essentially as a short circuit to the negative half of the ac and limits the positive excursion to its zener-voltage rating. The result is half-wave-rectified, constant-amplitude square-wave pulses. These are differentiated (coupled through a short-time-constant R_2C_1 circuit), which develops a narrow positive pip on the rise of the square wave and a narrow negative pip (dashed) on the fall of the square wave. The negative pip is shorted to ground by diode D_2, leaving only the positive pip to be fed to the adder circuit and to the CRT grid. (A Schmitt trigger can also be used to generate narrow pulses.)

An LC oscillation frequency of 80.7 kHz produces one complete cycle every 12.3 μs, the time equivalent to 1 radar mile. If these multiples of 80.7-kHz pips are added to each outward moving sweep on the CRT, they will produce concentric rings of illumination, each separated from the next by the equivalent of 1 mi. By counting range markers, the range of any observed target blip can be accurately estimated. A gain control on the output of the range-marker circuit controls the intensity of the rings. On longer-range presentations, the markers may be generated every 3, 5, or 10 mi by lowering the LC circuit oscillation frequency.

A separate variable frequency oscillator can provide a *variable range marker* (VRM) if calibrated in nautical miles instead of in frequency. It allows an accurate range readout of any target seen on the CRT. An accurate bearing of a target can be determined by moving a rotary *cursor* line over the target and main bang. Where the cursor falls on the 360° scale around the CRT is the target bearing.

The CRT used in radar has 10-s persistence instead of the few hundredths of a second of TV tubes. A TV receiver employs both vertical- and horizontal-deflection coils, but in a radar indicator there is only one set of coils. These are gear-driven around the neck of the tube 10 times per minute, in synchronism with the rotation of the antenna.

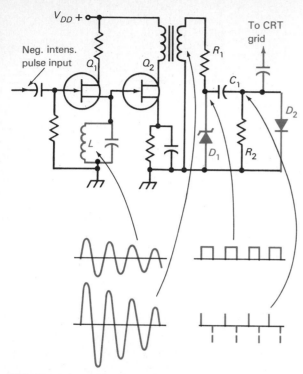

FIG. 26-13 Possible circuit to develop range-marker signals and the waveforms developed.

With the exception of the synchronizing system between antenna and sweep coils on the CRT, a complete radar system has been outlined.

Summarizing (Fig. 26-12), the sequence began with the blocking oscillator generating a series of pulses at approximately 1000 pps. These pulses were shaped and used to fire a magnetron, as well as to trigger the indicator circuits. The magnetron emitted a strong RF pulse into the slowly rotating antenna reflector. The TR and ATR tubes protected the receiver and allowed only received echo signals to enter the mixer cavity. The received pulse signals were amplified, detected, and fed to the CRT. At the same time, the trigger pulse started an inten-

sifying pulse that was fed to the grid of the CRT, enabling received signals to produce target blips. The intensifying pulse also started a sawtooth wave that produced the sweep trace from the center of the screen out to the edge, 1000 times per second as the sweep coils rotate. This results in a presentation of all radar targets in their relative position around the ship. The bearing of targets is determined by their angle from the top of the screen; range is how far they are from the center. To determine distance, range markers can be turned on. To reduce sea-return echoes, the STC circuit can be adjusted to the lowest value that does not produce blurred bright areas near the center of the screen.

26-9 ANTENNA SYNCHRONIZATION

If the antenna were mounted on or just above the indicator, it would be possible to use a single vertical drive shaft to rotate the antenna and the deflection coils in perfect synchronization. In larger vessels this is not feasible. Some sort of *synchro system* is needed. A basic form consists of a *selsyn* generator and a selsyn motor. The two units are similar; each has a rotor and three stator windings interconnected (Fig. 26-14).

The connections of the stator windings make this appear to be a three-phase system, but there is only a single-phase 115-V 60-Hz power line ac fed to both motor and generator rotors. The magnetic fields from these rotors induce voltages in the stators. As long as the two rotors are resting at the same relative angle between similar field coils, the voltages induced in the stator coils will equal each other and a condition of balance occurs. If the motor is held in position and the generator rotor is moved clockwise by hand, the voltages induced in the two sets of stators will no longer be similar. This results in a magnetic pulling, counterclockwise by the generator rotor and clockwise by the motor. When the generator is turned, the motor will respond to the proportionate magnetic changes produced in its stator fields and will follow the angular rotation of the generator rotor.

Coupling the rotating radar antenna to a selsyn generator and the selsyn motor to the rotating mechanism that drives the deflection coils around the neck of the CRT provides a possible means of synchronizing antenna and deflection-coil rotations. However, in operation, rotation of the selsyn motor must always lag that of the generator by a few degrees. The lag angle may change with variations

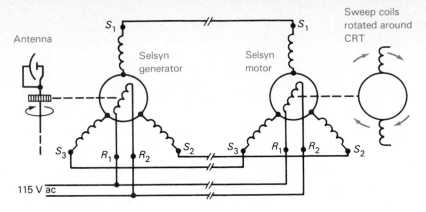

FIG. 26-14 A basic selsyn system. Dashed lines represent mechanical coupling.

of friction. The possibility of change of lag angle and an inherent lack of sensitivity of selsyns make other forms of synchros more desirable.

The system in Fig. 26-15 uses the emf induced in the unexcited selsyn-type *control transformer* (CT) rotor winding as a correcting voltage. If the two rotors are in the same angular position, no voltage will be induced in the CT rotor. As the rotors are varied in angular placement, the CT-rotor-induced voltage will change, but the rotor itself does not try to turn. However, any 60-Hz ac induced in the CT is first shifted in phase 90°, is amplified, and is fed to one winding of a two-phase ac motor (Sec. 27-21). The power-circuit ac is fed to the other winding of this motor. With both phases applied, the motor rotates, turning the CT rotor and the deflection coils of the CRT. If the antenna tends to rotate faster than the deflection coils, a greater voltage is induced in the CT rotor. This correction voltage increases the speed of the two-phase motor, and the deflection coils speed up. The CT rotor must always lag the generator somewhat, but in this system the amplifiers reduce the lag or variation in lag to a small value, resulting in satisfactory synchronization for the radar system. The amplifier and driving motor are known as a *servomechanism*. The rotation of the CRT coils is completely dependent on the rotation of the antenna. If the antenna motor stops, the selsyn generator is no longer rotated by the antenna rotation, and the correcting voltage is no longer developed. Without both phases the two-phase motor and the CRT coils stop.

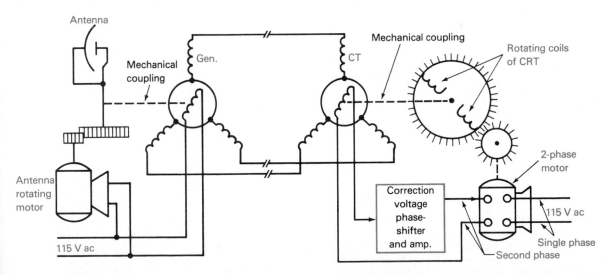

FIG. 26-15 Servomechanism to synchronize rotation of the radar antenna and the deflection coils around the CRT.

26-10 HEADING FLASH

As the radar antenna turns toward the bow of the ship, it will trip a microswitch, which feeds a short positive pulse of voltage to the grid of the CRT. This results in a trace being made from the center of the screen to the edge. Such a *heading flash* indicates, on the CRT, the direction the ship is taking on the chartlike presentation of the PPI screen. The circuit can be turned on or off by a control on the indicator panel. This is known as "Course-Up" mode. If a gyrocompass is installed in the indicator, the display can be constantly rotated so that the top of the screen is always true north ("North-Up" mode). The display is thus oriented as a normal chart would be.

26-11 ECHO BOX

When at sea (no targets) or whenever it is desired to test the overall sensitivity and operation of the radar set, an *echo box* is used. This is a high-Q cavity resonant to the transmitter frequency coupled into the waveguide with a directional coupler (Fig. 26-12). Each transmitted pulse shock-excites the cavity into oscillation, and it produces a damped-wave output. The coupling to the echo box is adjusted to ring for about 12 μs. As long as it is active, it produces a tapering-off signal that illuminates the screen, outward from the center, for a distance equal to about a mile. If tubes or crystals become weakened or the system is not operating properly, the distance indicated is less, or little echo-box signal will be seen.

Since an echo box will blot out all targets in a 1-mi radius, it is necessary to decouple it by some mechanical means or to detune it far enough so that it will not ring. One method uses a plunger that tunes the box through resonance as it is pushed down, producing a flash on the radial traces during the time the box is resonant.

26-12 OPERATING THE RADAR SET

The master of the vessel or any person designated by him may operate the radar set. Such a person may replace fuses or receiver-type tubes in the set, although this duty usually falls to one of the radio operators. However, whenever the equipment requires maintenance other than fuses and tubes, only persons holding radio licenses with radar endorsements or persons working under direct supervision of such a license holder may make adjustments to, service, or install radar equipment.

Each radar installation must have an installation and maintenance record, kept at the radar station. This record will include the date and location of installation and the name and license number of the person installing it. All subsequent maintenance, tubes, fuses, oiling, interference reports, tuning, etc., must be noted, with date and action taken, and signed by the person responsible. The station licensee, usually through the master, is jointly responsible with the operator concerned for the faithful and accurate making of such entries.

It is required that at least one set of instructions for the use and operation of the particular type of radar being used, as well as the FCC publication "Part 83—Stations on Shipboard in the Maritime Services," be on board the vessel.

A radar transmitter is one of the few RF emissions that require no specific identifying emissions or call letters, although it must be FCC-licensed.

26-13 RADAR INTERFERENCE

Interfering signals received on a radar set are due to other radar transmitters that are operating in the same area. Such interference takes the form of curved dotted lines across the screen.

The radar transmitter can produce interference to radio receiving or electronic devices in its vicinity if they are not properly shielded and grounded. Improper bonding or grounding of radar equipment, connecting cables and waveguides, or inadequate bypassing of the input power lines may result in interference.

Interference with a radio receiver by radar is characterized by a harsh tone having a frequency of the PRR, about 1000 Hz. The noise may increase and decrease as the radar antenna rotates or be steady if originating at the radar set itself. If grounding and bypassing power lines and other circuits do not help, it may be necessary to change the position of the receiving antenna. Rotation of the radio-direction-finder loop or coil may indicate nulls on interference produced by a radar set.

Any motors or generators in the radar set having slip rings, or armatures and brushes may cause noise in receivers that may be picked up on all frequencies used at sea (100 kHz to 160 MHz).

On a loran screen, radar interference appears as either vertical *grass* or *spikes*. These spikes may drift in one direction or the other and may seem to synchronize for short periods of time.

On an auto-alarm receiver, radar interference will sound the same as on any other receiver when earphones are used. It presents a constant signal which may activate and hold the first dash counter circuit. The red light on the bridge and in the radio station will glow, indicating trouble.

Intercommunication, motion-picture, or public address systems on the vessel may also pick up radar impulses if they are not properly shielded and grounded or if they have poor connections at some points in them.

26-14 BASIC RADAR MAINTENANCE

Although cabinet enclosures may be protected by interlocks that remove high voltages when opened, some VT circuits with up to 200 or 300 V may not be interlocked. Interlocks should never be jumpered or shorted to operate the high-voltage systems with the enclosure doors open.

In most cases, faulty operation of a radar set is the result of weak tubes, faulty TR tubes or diodes, open fuses, or malfunctioning printed circuit boards. Tubes may be checked with a tube tester, or similar tubes may be substituted, one by one. When a TR tube weakens, the mixer diode fails and its current drops. When mixer diodes are replaced, always replace the TR tube. Diodes are sensitive to high static voltages. Under certain conditions, an operator may attain a static charge. When a cartridge-like diode (about the size of a .22-caliber shell) is pushed into its socket, the operator may discharge through it, burning it out. To prevent this, the operator should always touch the mixer cavity with one hand while inserting the diode with the other hand. When a diode is handed from one person to another, it should be kept in its metal-foil capsule.

The magnetron current should be checked periodically. No I_p may indicate an open filament or no modulator pulses. If the current is abnormally high, it may mean a gassy magnetron or a high PRR. The permanent magnet used in conjunction with the magnetron is quite strong. There is danger, when a magnetron is being removed or installed, that steel tools may be grabbed by the magnet and cause damage to the tube. The filament leads and the output circuit have long glass seals that may be fractured by mechanical jarring.

Remember that the filament leads to the magnetron have several thousand volts on them when the set is in operation.

If the permanent magnet weakens, the magnetron current will increase, the output power will lessen, and the frequency of operation may change so that the AFC will not hold the receiver in tune.

If the AFC circuit, the adder, the intensifying pulse circuit, or the magnetron is functioning improperly, bright pie sections may appear on the screen.

Most equipment will have a series of test-point jacks into which a test meter can be inserted to test the operation of the various circuits. It will be necessary to check the instruction booklets that accompany the equipment to determine what the readings should be.

If the sensitivity control is turned to maximum but little or no grass or signals appear and the diode current is low, the diode may be removed and checked by measuring its resistance in both directions with a sensitive ohmmeter adjusted to read "R times 1000." If the front-to-back ratio is less than about 100:1, the crystal should be replaced. No crystal current may also indicate a defective klystron. Turn off the equipment before changing this tube, which may have several hundred volts on the shell of its cavity. When klystrons or TR tubes are replaced, it is usually necessary to readjust the tuning screws in the associated cavities to bring the set up to optimum performance.

A CRT is dangerous to service because of high voltages applied to it, plus the possibility of implosion. Heavy gloves and a face mask should be worn when changing such tubes.

Motors, generators, and synchros should be checked every 200 to 300 h of operation. They should be cleaned, and any brushes should be checked and replaced if necessary. Oil or grease should be applied where necessary.

Before a ship leaves the dock, the radar set should be turned on and tested. At this time it should be dusted thoroughly and observed closely. If any signs of overheating of any component or improper functioning of mechanical parts are noted, corrective steps should be taken immediately.

When it is determined that a PC board is inoperative and a replacement is not available, sometimes the faulty part on it can be located by looking for overheated parts. Often an ohmmeter can be used to locate shorted or open transistors, resistors, or capacitors. Voltage readings may be made in some cases to indicate where the problem is.

26-15 NONMARINE RADAR

Although there are many different types of radar (gunfire control, landing approach, distance measuring, etc.), only two others not used in the merchant marine will be mentioned.

One type of radar used near airports for surveillance of all flying aircraft in the area is basically similar to the PPI system discussed. To prevent the radar from displaying trees, buildings, bridges, and hills in its range, it can use a discriminator as the second detector in its receiver. All echo signals returning at exactly the same frequency as transmitted produce zero output from the discriminator. However, an approaching target will compress any radio waves striking it. The returned wave will be slightly higher in frequency. Any target moving away returns waves that have a longer wavelength and a lower frequency. This is Doppler effect. The discriminator might give a positive signal for all approaching targets and a negative one for departing targets. Thus, on Doppler radar, moving targets produce blips, but all stationary targets produce no signal on the screen. If a series of chart signals of the area are fed electronically to the CRT the watcher can see the position of targets in relation to ground points. If any aircraft is carrying a *transponder,* when the radar signal strikes an antenna on the aircraft, this signal is detected and keys a small transmitter. It returns a short burst of RF to the radar receiver which is displayed as an identifying blip out past the target indication on the CRT. By switching to an amplitude detector, the CRT will display both moving and stationary targets.

Some automobile speed-check radar sets may operate on a different principle. They may use a low-power CW emission from a directive antenna. An adjacent antenna is used to pick up the transmitted signal and the returning echo signal. The faster a target is approaching the greater the phase shift between transmitted and received signals. A phase-shift detector produces a direct readout in miles per hour on a meter and can also print out a speed indication on a paper tape. It operates in the X or K band (10.5–10.55 or 24.1–24.2 GHz).

26-16 SMALL BOAT RADAR

Large ships may have stand-up-type consoles that enclose the PPI radar transmitter, the receiver RF, IF and video circuits, the CRT, and the power supply all in one cabinet. The antenna unit may be mounted on a mast, may be over 100 ft away, and is fed with waveguide. Small vessels may have the 10 to 24 rpm rotating antenna, transmitter, and receiver RF, IF and detector circuits, all mounted as one unit 10 to 30 ft above the deck. Target signals in video form, and antenna rotary information are sent down to the indicator unit below by coaxial and electrical cables. With semiconductors, radar set size has been reduced greatly. The indicator may be no larger than a portable TV set. It may be mounted in the cabin of small boats, or on the bridge of larger vessels, possibly with repeater indicators in the chart room or on the flying bridge.

Small boat radar is usually 9-GHz equipment to allow small rotating slotted-waveguide antennas to be used. Usable ranges will be ¼ to 80 mi or more, depending on antenna height, power, and pulse width (0.15-μs short range, 0.5-μs long range). Distance to the horizon for a given antenna height is approximately

$$D = \sqrt{2h}$$

where D = distance in miles

$\quad\quad h$ = height in feet

The distance to a 30-ft-high target vessel from a 20-ft-high radar antenna is thus $\sqrt{2(30)}$, or 7.74 mi, plus $\sqrt{2(20)}$, or 6.32 + 7.74, or 14.1 mi.

Small boats, buoys, channel markers, etc., may have small metallic *corner reflectors* mounted on their highest points to provide strong radar echos to increase their visibility on radar CRTs.

26-17 COLLISION AVOIDANCE RADAR

A collision avoidance system (CAS) can be added to an already installed PPI radar set, or it may be a single complete unit. It requires a radar set, input from a satellite navigation receiver, input from the ship's gyrocompass, and speed input from the ship's engine revolutions or from a *speed log.* The speed of a ship can be determined by a Doppler log. If a sonic radiating device developing a frequency such as 30 kHz is mounted below the water line, and a detector is mounted somewhat forward of it, when the ship is dead in the water the detector picks up 30-kHz signals. When the ship is moving forward, the water movement compresses the supersonic waves and they appear to the detector to have a higher frequency. The Doppler differential between 30 kHz and the received signal frequency can be converted to a speed-log indication, which can be fed to the bridge and to the CAS.

All of the required data is fed to a digital computer in the CAS. By internal hardware and software, heading flashes, bearing cursor flashes, selected target brackets, possible area of danger (PAD) ellipses, plus target courses and speeds can be computed. A few scans after a target is bracketed the computer determines the target's course, its PAD ellipse, and the closest point of approach (CPA). Desired targets are bracketed on the CRT by manual operation of a *joystick* control that can move bracket lines around any target at any place on the CRT, thereby giving the computer the range of these targets. If targets are computed to be on a collision course with the ship and are approaching within what may be a dangerous range, both an audible and a flashing visual warning on the CRT are given.

The theory of a joystick deflecting system can be explained with Fig. 26-16. In Fig 26-16*a* a poten-

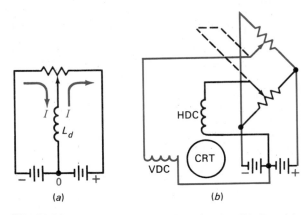

FIG. 26-16 (*a*) A basic deflection circuit. (*b*) Joystick system.

tiometer is across a center-tapped dc source. A magnetic deflection coil L_d is between the dc center tap and the pot. When the pot is centered, the voltage-drops across the coil are equal and opposite, resulting in zero coil current and therefore no magnetic deflection. In Fig. 26-16*b* a vertical- and a horizontal-deflection coil are both operating from their own pots, and mechanically 90° from each other. If pot 1 is moved, the deflection will be vertical, above or below the screen center. If pot 2 is moved, deflection will be horizontal, right or left. A joystick mechanical control lever that can move either or both pots simultaneously (dashed lines in Fig. 26-16*b*) can position a beam anywhere on the face of the CRT.

The CRT presentation may be North-Up from

gyrocompass information, or Course-Up, in which case the direction the ship is heading is indicated by a dashed cursor from the CRT center to the top of the round screen. If the ship changes course, the dashed cursor can be made to rotate to show the variation in heading from the original course.

Test your understanding; answer these checkup questions.

1. How does a one-shot multivibrator differ from the common multivibrator? _____
2. How much intensifying pulse is used? _____
3. If echo signals are developed as negative pulses, to what part of the CRT would they be fed? _____
4. What starts the range-marker circuit ringing? _____
5. Why is a selsyn pair not too successful for radar antenna synchronization? _____
6. What is fed to the rotor of a selsyn motor? _____ Generator? _____
7. What is fed to a control transformer rotor? _____ To a synchro-system coil-turning motor? _____
8. List six components of the servo system described. _____ _____ _____ _____ _____ _____
9. What is used to indicate the bow of the ship on the CRT? _____
10. When is an echo box used? _____
11. What qualification must a person have before he is eligible to make repairs to a radar set? _____
12. List seven types of electronic equipment aboard ship that may be interfered with by radar. _____ _____ _____ _____ _____ _____
13. What device can be used to test diodes? _____ What ratio should be obtained? _____
14. What is the advantage of using Doppler radar? _____ Why is it not used on ships? _____
15. How could Doppler radar give speed indications? _____ What are other applications of Doppler radar? _____
16. What is included in the antenna unit of a small boat radar? _____ In what frequency band does it operate? _____ Does it give North-Up or Course-Up indications? _____
17. What is the horizon distance from 10 ft above the water? _____
18. What is the control that can move a mark or bracket to any place on the CRT screen? _____ Of what does it consist? _____

1. What does "radar" mean?
2. What is a "radar mile"?
3. What might be the minimum and maximum ranges of marine radar?
4. What is meant by PRR? What PRR is used in marine radar?
5. What is required for good bearing resolution? Range resolution?
6. What kind of an antenna is used with a PPI radar?
7. In what two ways do radar CRTs differ from oscilloscope CRTs?
8. What is a common antenna rotation speed for a radar set?
9. What are the three parts of a marine radar transmitter?
10. Draw a diagram of a simple marine radar transmitter.
11. What are two other names for a pulse-forming network?
12. What is the P_{pk} and duty cycle if pulse width is 0.4 μs, PRR is 1100, and P_{av} is 15 W?
13. What is included in the whole antenna system of a radar set?
14. What happens if the radiator of a radar antenna is not at the focal point of its parabolic reflector?
15. Besides a parabolic reflector, what other type of antenna is used for marine radar?
16. If a radar antenna is encased, what is the protective enclosure called?
17. What is a radar duplexer?
18. Which TR tube requires a keep-alive voltage?
19. If a mixer diode burns out, what else should be replaced?
20. Why is an AFC circuit important in a radar set?
21. What devices may be used as the LO of a radar receiver?
22. Draw a block diagram of a radar receiver, TR cavity to CRT.
23. What does STC mean? What does it do? Why is it needed?
24. What produces the slight glow on the radar CRT when no targets are being received?
25. What are the circuits of the radar indicator?
26. Draw a block diagram of a PPI radar system.
27. What is a VRM? Why is it used?
28. What is the basic motor-generator synchro system called?
29. Draw a diagram of a practical system of rotating the radar antenna and the CRT coils in sync. How is the sync held close?
30. What are the two possible heading flash modes?
31. To what is an echo box coupled? What is its use? How does it function?
32. What is required of the person making technical adjustments to a radar set?
33. What would curved dotted lines across a radar CRT indicate?
34. What does radar interference on a receiver sound like?
35. What are some things that might have to be serviced on a radar set?
36. What type of radar uses a discriminator as its 2nd detector? Where are these used?
37. What radar may measure phase difference between transmitted and echo signals?
38. What radar equipment is usually mounted on top of the mast of a small vessel? In the cabin?
39. What is the distance at which a small boat with a 15-ft-high radar antenna can pick up another small boat having a 15-ft mast?
40. What should all small boats have on the top of their masts if at sea? Why?
41. What is a CAS? What does it require?
42. What theory is used to determine ship speed through water?
43. How does a CAS determine the PAD ellipses on its radar CRT?
44. What is used to move a bracket around a target on a CRT?

ANSWERS TO CHECKUP QUIZ ON PAGE 649

1. (Not free-running; produces output only when triggered) **2.** (Just enough to produce light flicker on screen) **3.** (Cathode) **4.** (Intensifying pulse) **5.** (Variable lag angle) **6.** (Power-line ac) **7.** (Same) **7.** (Nothing; emf is induced into it) (Power ac shifted 90°) **8.** (Generator) (CT) (Phase shifter) (Amplifier) (2-ϕ motor) (Coil assembly) **9.** (Heading flash) **10.** (Testing only) **11.** (Radar endorsement) **12.** (Commercial receivers) (Loran) (RDF) (Auto-alarm) (Public address system) (Motion pictures) (Intercoms) **13.** (Ohmmeter) (Better than 10:1) **14.** (Shows only moving targets) (Navigators must see everything) **15.** (Calibrate discriminator output) (Ground-speed indicator) (Missile velocity) (Water log) **16.** (Antenna, transmitter, receiver minus video) (9 GHz) (Course-Up) **17.** (4.47 mi) **18.** (Joystick) (Two or four pots)

27

Motors and Generators

This chapter describes the significant points regarding ac generators, dc generators, dc motors, and ac motors. Servicing and maintenance suggestions are included at the end of the chapter.

27-1 ELECTRIC MACHINES

A motor converts energy of one form into mechanical rotational or twisting power, called *torque*. Examples of motors are gasoline or diesel engines, which convert the expansion of gas by heating into torque, a steam engine, which converts the expansion of hot steam into torque, and an electric motor, which converts electricity into twisting effort by the interaction of magnetic fields.

A generator converts mechanical rotational power into electric energy and may be called a *prime source* of emf. The two basic forms are the dc generator and the ac generator, or *alternator*. All generators require a *prime mover* (motor) of some type to produce the rotational effort by which a conductor can be made to cut through magnetic lines of force and produce an emf. The simplest electric machine is the alternator.

27-2 ALTERNATORS

An alternator consists basically of (1) a magnetic field, (2) one or more rotating conductors, and (3) a mechanical means of making a continuous connection to the rotating conductors (Fig. 27-1).

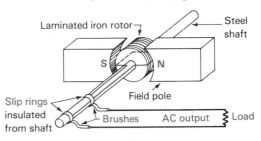

FIG. 27-1 Parts of a simple rotating-armature alternator.

The magnets, called the *field poles*, produce lines of force. The laminated soft-iron *rotor*, also called the *armature*, presents a highly permeable path, resulting in fewer leakage lines and a greater magnetic-field strength in the gaps between the field-pole faces and the rotor. One end of the rotor conductor terminates at a brass *slip ring* on the insulated shaft to which the rotor is fixed. The other end of the conductor terminates at a second slip ring. Held against the slip rings by spring tension are two brushes made of copper, brass, or carbon.

If the shaft is rotated by a motor, the iron rotor and the conductor on it turn. The conductor has an emf induced in it as it passes a field pole. As a wire rotates down past the north pole, the emf induced will be in one direction. As the conductor continues on upward across the face of the south pole, the same lines of force are being cut but in the opposite direction, resulting in an opposite-polarity induced emf. The continuous rotation of the conductor produces an alternating emf at the slip rings and brushes.

To produce a greater emf, the rotor may be wound with a many-turn coil. The emf induced in each turn adds to that of all the other turns, and a higher-amplitude ac emf is produced. The number of magnetic lines of force produced by a permanent magnet is rather limited. A much more intense field can be produced by making the field poles of iron and wrapping them with several hundred turns of wire. When the field coils are excited by dc they form strong electromagnets. The greater the current flowing through the field-pole coils the greater the number of lines of force produced and the higher the output voltage of the rotor coil. Thus, a practical means of controlling the output emf of an alternator is to vary the dc excitation to the field coils. Another means of varying the output voltage is to vary rotation speed (varies ac frequency).

A diagram of an *externally excited field* alternator is shown in Fig. 27-2. The interlocked circles rep-

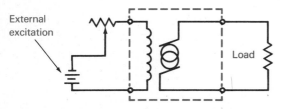

FIG. 27-2 Alternator with an externally excited field.

resent slip rings with brushes pressing against them. The coil represents the two field windings. The rheostat varies the excitation current to the field and controls the output voltage. The prime mover is not indicated.

If a *rotating field coil* is excited with dc (Fig. 27-3), the rotor part becomes a strong electromagnet. As the rotor moves at the position shown, a strong magnetic field is induced in the *stator* poles. The magnetic-field path is completed through the frame of the alternator. As the rotor moves to a point 90° from that shown, the stator poles are left without any magnetic field. Continued rotation of the rotor alternately produces and stops magnetic flux in the stator poles. The expanding and contracting alternating magnetic fields in the

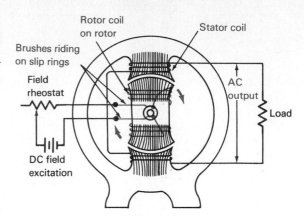

FIG. 27-3 Components of a rotating field alternator.

stator poles induce an ac emf into the turns wound around them, which forms the output emf of the machine.

The rotor is usually made into 4-pole form, with opposite poles in parallel. Only one pair of slip rings is required. The stationary part of the machine also has four poles, with all stationary armature windings in series for maximum voltage output.

Heavy-duty alternators are usually three-phase (Sec. 10-28) types, with rotating fields, dc-excited through slip rings and brushes. Instead of having stator pickup-coil pairs 180° apart as in Fig. 27-3, there are three pairs of coils spaced at 120°. As the rotating field turns, it induces ac into each pair of coils, developing voltage or current peaks 120° apart. The coils may be connected either star or delta. The alternator shown in Fig. 27-4 represents a low-power star-connected alternator used in auto-

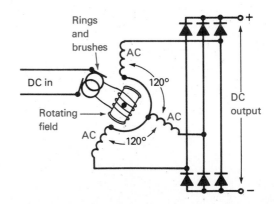

FIG. 27-4 Star-connected 3-φ alternator with a bridge rectifier used in automobiles to charge batteries.

mobile battery charging systems, with a 6-diode bridge rectifier circuit. The dc output is connected across the battery through cutout relays, and is controlled in strength by varying the rotating field current. When the battery is charged and its voltage is high, field coil current is held low. If the battery is discharged and its voltage is low, the field coil current is made greater to increase the alternator output voltage and current, which charges the battery.

A third type of ac generator is the *inductor* alternator (Fig. 27-5). With the toothed soft-iron rotor

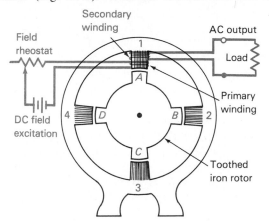

FIG. 27-5 Simplified inductor alternator.

in the position shown, the dc excitation on the stator-pole piece produces a magnetic flux that takes the path from pole 1 into tooth *A*, out of teeth *B* and *D*, and back to pole 1. In this position a maximum number of lines of force are in the iron pole piece. When the toothed rotor moves 45°, the magnetic path between pole 1 and the rotor teeth lengthens, greatly lessening the number of lines of force in the field pole. As the rotor continues to rotate, the alternately stronger and weaker magnetism in the pole induces an ac into the secondary winding.

Inductor alternators have primary and secondary windings on all pole pieces, of course. The windings are series-connected to add to the effect of the simple single pole illustrated.

In practical applications, inductor alternators have many pairs of pole pieces, with a tooth on the rotor for each pole piece. They are used when 500-Hz or higher-frequency ac is required.

Iron field poles of all motors and generators, both ac and dc, are laminated (Sec. 5-17).

27-3 ALTERNATOR VOLTAGE OUTPUT

An alternator operates as an inductance. With a resistive load, the output voltage may be somewhat less than the no-load output, but if the load is inductive, its inductive reactance adds to the reactance of the alternator and the output voltage sags considerably. If the load is capacitive, however, the capacitive reactance counteracts the inductive reactance inherent in the alternator, and a voltage output in excess of the no-load value may result.

27-4 ALTERNATOR FIELD EXCITATION

All alternators require dc field excitation. This can be from a battery, rectified ac, or, in some cases, a dc generator attached to the same prime mover that rotates the alternator. As soon as the prime mover rotates, the generator (exciter) produces dc, which excites the alternator field coils, allowing ac to be generated.

27-5 PARALLELING ALTERNATORS

When it is necessary to connect a second alternator across a line already being fed by one 120-V ac alternator, either to aid the first in carrying the load or to allow the first to be removed from the line without interruption of service, the second machine must be synchronized and cut in at an instant when the *voltage* and *phase* of the two are equal.

The voltage of the second alternator can be controlled by the field rheostat and its voltage phase compared with the existing line voltage. There are several methods by which the phase can be compared. One is by using a 240-V electric light connected as in Fig. 27-6.

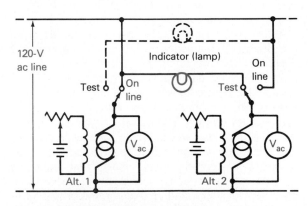

FIG. 27-6 Circuit to determine when two alternators are in phase and may be paralleled.

Alternator 1 is supplying the line. Alternator 2 is to be coupled to the line also. Alternator 2 is first brought to 120-V output. With the switch as shown, the voltage across the lamp will be something between 240 V (alternators 180° out of phase) and zero (machines in phase). The speed of the prime mover of alternator 2 must be adjusted until the light varies in intensity slowly. This indicates a slow shift of phase between the two machines. The switch is thrown to the line position while the lamp is dark (indicating that the two ac voltages are in phase). An ac voltmeter, or an oscilloscope-type device might also be used as the indicator. Division of the load between the alternators is adjusted by the torque delivered by the respective prime movers.

27-6 DC GENERATORS

The basic dc generator is similar to the basic alternator except for the substitution of a *commutator* for the slip rings (Fig. 27-7). When the shaft is rotated (by some type of motor), the rotor and

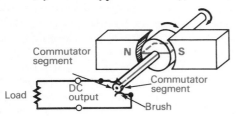

FIG. 27-7 Elements of a simple dc generator.

conductor turn and ac is induced in the armature coil. With the conductors rotating, and in the position shown, a maximum emf will be produced between the brushes.

When the armature rotates to a point 90° from that shown, the conductors will be moving with the magnetic lines of force, not across them, and no voltage will be induced in the conductors. At this instant, the commutator segments have rotated, and each brush is making contact with both segments. The brushes are now shorting the segments together but at a time when zero voltage is being induced in the armature. If the armature windings and the commutator segments are not properly oriented, the brushes may short the segments when there is an emf between them, and sparking at the brushes will occur. Such sparking is one of the difficulties experienced with dc motors and generators, prarticularly in radio applications.

As the armature rotates another 90°, the emf induced in the conductors is again at a maximum and again in a direction out of the top segment and brush and into the bottom brush and segment.

With continual rotation of the armature, the ac emf induced in the armature conductors is made to enter and leave the brushes in the same direction at all times. The action of commutator and coil results in a full-wave-rectified output (Fig. 27-8a). This is a form of mechanical rectification of the armature ac.

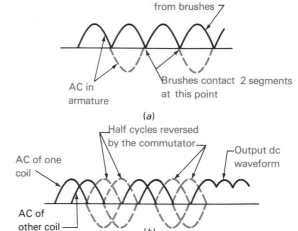

FIG. 27-8 (a) Pulses of a 2-pole, 2-segment commutator dc generator. (b) Pulses of a 2-pole, 2-coil, 4-segment commutator dc generator.

When a 4-segment commutator with two armature coils wound at right angles to each other is used, the dc produced does not drop to zero at any time (Fig. 27-8b). The more pairs of commutator segments and armature coils, the less variation in the output dc waveform.

Test your understanding; answer these checkup questions.

1. What is an electrical prime mover called? _____
2. What is an alternator? _____
3. List the four basic components of an alternator.
   _____ _____ _____ _____
4. In Fig. 27-1, if the conductor is moving down past the north pole, what is the current direction in the load? _____
5. What are two methods of increasing the output voltage of an alternator? _____ _____
   Why is only one used? _____

6. Why would a rotating-field alternator be better than a rotating-armature type? _____

7. What type of alternator operates on a transformer principle? _____

8. What type of alternator is used for high-frequency ac generation? _____

9. Why are field poles always laminated? _____

10. What kind of load tends to increase the output voltage of an alternator? _____

11. List three ways of exciting alternator fields. _____ _____ _____

12. In Fig. 27-6, should alternator 2 be switched "on line" when the lamp is dark or bright? _____

13. If alternator 1 ac is applied to the horizontal plates of an oscilloscope and alternator 2 ac is applied to the vertical plates, when should alternator 2 be switched on line? _____

14. List the four components of a dc generator? _____ _____ _____ _____

15. Is ac or dc developed in the armature of a dc generator? _____

27-7 THE EXTERNALLY EXCITED DC GENERATOR

As with the alternator, the magnetic field of a dc generator is normally supplied by an electromagnet rather than by a permanent magnet, except for small units. The diagrammatic symbol of a dc generator with a separately excited field is shown in Fig. 27-9. The external excitation is from a battery

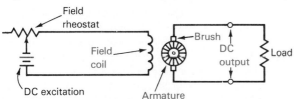

FIG. 27-9 Externally excited dc generator circuit.

and is controlled by a *field rheostat*. The brushes and the multisegment commutator of the armature are indicated in symbol form.

Voltage output depends on speed of armature rotation and magnetic-field strength. The greater the field-coil current, the greater the output voltage of the generator, up to the point of field-pole saturation. Increasing field-coil current past saturation will not increase the output voltage.

The output voltage is also subject to an internal voltage-drop because of resistance in the windings of the armature. When the output current increases, the internal voltage-drop increases. As a result, the

separately excited generator has a drooping voltage output with increase of load.

Note: If the field coils are replaced by *permanent magnets*, the output polarity of such a p-m generator will be determined by the direction of rotation of the armature, or polarity of field magnets.

27-8 THE SERIES DC GENERATOR

When the field coils have a few turns of heavy wire and are connected in series with the armature (Fig. 27-10), a *series generator* results. With no load, it has almost no voltage output. With the load discon-

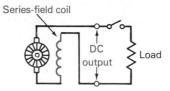

FIG. 27-10 Series-field dc generator circuit.

nected, no current flows through the field coils, even with armature rotation. The rotating conductors cut only the residual lines of force of the field poles, producing very low output voltage.

When a load is connected across the output, current begins to flow in the circuit. The field poles build up a stronger magnetism, and the voltage output of the generator increases. The heavier the load, the greater the field current and the higher the output voltage of a series generator up to the point where saturation of the field poles occurs. Because of its varying voltage output with a varying load, a series generator finds few applications.

27-9 THE SHUNT DC GENERATOR

A practical dc generator connects the field coils across the armature (Fig. 27-11) to form a self-excited *shunt generator*. The shunt field coils have many turns of fine wire to produce a maximum number of ampere-turns of magnetic force with as little drain on the output current as possible.

The rheostat in the field of the shunt generator

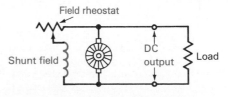

FIG. 27-11 Shunt-field dc generator circuit.

controls the current flow through the field coils. This determines the magnetic-field strength and controls the voltage output. Increased field current produces increased voltage output.

A shunt generator has a sagging voltage characteristic under load due to the voltage-drop in the armature when current flows through it, which decreases the voltage excitation to the shunt field.

27-10 THE COMPOUND DC GENERATOR

In some applications the load varies over a relatively wide range. To maintain a constant-voltage output from the source, neither the series generator with its increasing output voltage under load nor the shunt generator with its decreasing output voltage under load is suitable. But it is possible to use both series and shunt windings on each field pole (Fig. 27-12). By selecting the proper proportions of each

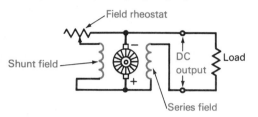

FIG. 27-12 Compound-field dc generator circuit.

winding, a *flat-compounded* generator can be produced. If the series field coils have too many turns, an *overcompounded* machine (rising voltage with increase of load) results. If the series field coils have too few turns, an *undercompounded* machine results. If the power lines carrying the current to a distant load are long, a slightly overcompounded generator may be desirable to compensate for voltage-drop across the resistance in the line. The voltage output of a compound generator is controlled by the shunt-field excitation current.

If the series- and shunt-field effects are additive, the machine is said to be *cumulatively* compounded. If the field effects oppose, a *differentially* com-

pounded machine results, which has a sharply decreasing output voltage under load, making it unsuitable in normal applications.

27-11 OUTPUT VOLTAGE REGULATION

It is possible to electronically voltage regulate a shunt dc generator. Consider the basic circuit shown in Fig. 27-13. The field coil plus the output

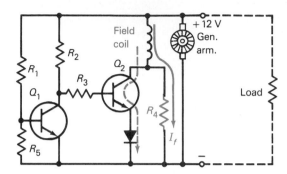

FIG. 27-13 A possible output voltage regulator for a shunt generator.

voltage control resistor R_4 are shunted across the armature. The desired unloaded output might be 12 V. With no load connected, R_1 through R_5 are selected to produce cutoff of Q_2 so that it feeds none of its I_C to the field coil (which would increase the output voltage). The voltage-drop across R_4 might be 3 V.

When a load is connected, the voltage output of the generator decreases, or sags, somewhat. This is felt by the base of Q_1 as a lessened positive potential. As a result, the Q_1 collector current decreases, there is less voltage-drop across R_2, and the base of Q_2 is fed an increased forward (+) bias. The Q_2 collector current increases, resulting in more field current I_f, an increase in magnetic-field strength, and a rise in the output voltage of the generator.

27-12 COMMUTATING POLES, OR INTERPOLES

In the description of a simple dc generator it was stated that the brushes move from one segment of the commutator to the next during a time when no voltage difference is being produced between them to prevent sparking at the brushes.

For any given armature-rotation speed and output current, the brushes can be adjusted to a *neutral point* where there is negligible sparking. The neutral point is determined by the interaction of the magnetic field of the field poles and the magnetic field of the armature. If the output current (or rotation speed) of the generator changes, the magnetic field of the armature shifts and the neutral position changes. To reduce sparking, the brushes must be rotated forward or backward around the commutator to the neutral point. With a changing load a physical shifting of the brushes is not feasible.

With no load, the mechanical neutral (position of the brushes) and the magnetic neutral are exactly halfway between field poles. With a load, the magnetic neutral shifts a few degrees in the direction of armature rotation in a generator (backward in a motor). An electromagnet *commutating pole*, or *interpole*, can be placed between field poles at the mechanical neutral (Fig. 27-14). Its winding is

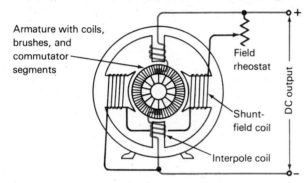

Armature with coils, brushes, and commutator segments
Field rheostat
DC output
Shunt-field coil
Interpole coil

FIG. 27-14 Interpole connections on a shunt-wound dc generator (or motor).

connected in series with the armature. When the armature (output) current increases, the interpole current increases, producing an opposite field to that of the armature. This cancels the effect of the armature field and holds the magnetic neutral at the mechanical-neutral point. The brushes can now be set at the mechanical neutral, and minimum sparking will occur at all load values. Interpoles are used in almost all larger dc generators and motors.

27-13 BRUSH SPARKING

In radio, any nearby electric sparking can produce interference (RFI) to received signals. Sparking at a commutator or slip ring pits it and wears out the brushes at an abnormal rate. It is necessary to lessen all sparking as much as possible. Causes of sparking at the brushes are worn or dirty commutators, worn or improperly fitting brushes, overloading the machine, an open armature coil, a short-circuited interpole coil, or improper neutral positioning.

To prevent interference to local reception, bypass

capacitors can be connected between brushes, or from each brush of a generator or motor to ground. A low-pass filter composed of an RF choke in series with the ungrounded line and bypass capacitors from each side of the choke to ground may be used. A similar low-pass filter may be used in the field-coil lines. This also prevents RF ac from a transmitter from entering a generator or motor and possibly breaking down insulation in the machine.

27-14 DC MOTORS

A dc motor is essentially the same machine as a dc generator. A series dc generator will operate as a series dc motor. A shunt generator will operate as a shunt motor, and a compound generator will operate as a compound motor.

27-15 SERIES DC MOTORS

Whereas series-wound generators are rarely used, many series-wound motors are in use.

When current is fed to a series motor (Fig. 27-15), the low resistance of the field and armature

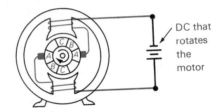

DC that rotates the motor

FIG. 27-15 Series motor. As the armature rotates, the commutator segments slide under the brushes.

windings allows a high current to flow through them and strong magnetic fields to form around them. The armature and the stator fields oppose each other and produce a pushing effort on the armature conductors, forcing them to turn away from the stator field poles. The twisting effort experienced by the armature is called *torque*.

In the figure, the brushes are in contact with the two commutator segments marked *A*. When current flows, the armature coils (not shown) connected to these segments are pushed away from the field poles, perhaps in a clockwise direction, and segments marked *B* are moved into position under the brushes. Now current flows through the *B* segments and their coil, producing further clockwise torque, moving the *C* segments under the brushes, and so on. Continuous rotation of the armature results.

If the current direction through the motor is reversed, the polarities of the field poles *and* the armature conductors are both reversed, resulting in a pushing effort between them in the same direction as before. Because of this, a series motor will rotate when excited by an alternating current, its direction of rotation being the same for both halves of each cycle. The lower the frequency of the ac, the greater the torque that can be produced. If the frequency is high, the inductive reactance of the field and armature coils limits the current that can flow and limits the torque of the motor.

When dc is applied to the series motor, the current flowing has nothing to oppose it except the resistance in the circuit. This results in high starting current and torque.

As the armature starts to rotate, the movement of the armature conductors through the magnetic field of the field poles induces a counter emf in the armature wires. If the counter emf could rise to the value of the source emf, the motor could no longer increase in speed. However, the counter emf can never equal the source emf in a *series* motor. The motor may accelerate until it flies apart by centrifugal force *if operated without a load*.

When starting a series motor, a rheostat may be included in series with it (Fig. 27-16). This adds

torque makes larger series-type motors useful for such heavy-duty jobs as starting electrical buses. The poor speed regulation of a series motor makes it undesirable for use in motor-generator sets.

The direction of rotation of a series motor can be reversed by reversing the current through either the armature or the field coil but not through both. Figure 27-17 illustrates a double-pole double-throw

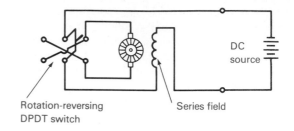

FIG. 27-17 Polarity-reversing switch in armature circuit to reverse motor rotation.

switch used as a rotation-reversing switch. *Note*: If the field coils are replaced with *permanent magnets*, the direction of motor rotation will be determined by the polarity of the dc fed to the armature.

27-16 SHUNT DC MOTORS

The shunt-type motor has fairly good speed regulation and is used in many applications.

A shunt motor with a 3-terminal starter is shown in Fig. 27-18. To start the motor the field rheostat

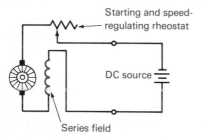

FIG. 27-16 Series dc motor circuit with a rheostat for speed adjustment and starting.

resistance to the circuit and reduces the starting current. As the motor picks up speed, counter emf is developed and the rheostat resistance can be lessened. The rheostat can adjust the speed of the motor to a certain extent. Small motors may require no starting resistors, but larger motors always do.

Small series motors are used in fans, blowers, electric drills, or any application in which a high speed of rotation is required but in which good speed regulation is not important. The high starting

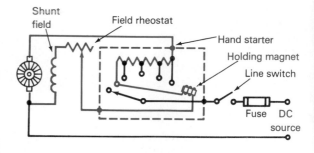

FIG. 27-18 A 3-terminal (no field release) manual starter connected to a shunt motor.

is first adjusted to zero resistance (low running speed) to produce maximum counter emf in the armature. The line switch is then closed, and the starter-rheostat arm is moved to the second contact. Current begins to flow through the armature, field, starter resistor, and the holding magnet. Without the starting resistance the current through the armature would be excessive. The magnetic fields of the armature and the shunt fields repel each other, and the armature starts to turn. As the armature conductors move, a counter emf is induced in them. As the speed increases, the counter emf increases and the current through the armature begins to decrease. The starter-control arm is moved from contact to contact slowly to allow the counter emf at each point to build up as the resistance of the starter is being decreased.

If the starter rheostat is advanced too rapidly, counter emf will not develop fast enough, current through the machine may be excessive, and the starter rheostat, armature, or fuses in the line may burn out. If the starter rheostat is moved too slowly, it may become overheated, particularly if there is any load on the motor. In general, a motor should be started with as little load on it as possible.

When the iron rheostat arm is brought up to the full ON position, the holding magnet holds it in the running position against the tension of a spring on the arm.

When the line switch is opened and the motor is disconnected from the source, the motor becomes a dc generator, with the same polarity as it had when it was operating as a motor. As a generator, it supplies current to the holding magnet until it slows down to the point where its emf will not produce enough current to energize the magnet coil. The rheostat arm then snaps back to the OFF position by the action of the arm spring.

A shunt motor has less starting torque than a series motor but builds up a counter emf equal to the source voltage minus the voltage losses in the armature. This limits the speed to a definite value. To increase the speed of a shunt motor, the generation of the counter emf in the armature must be reduced. This can be done by *decreasing* the field-coil current. When the resistance in series with the shunt field is increased, the field current is less and the counter emf decreases. With less counter emf, more current flows through the armature, which picks up speed until it can again produce a counter emf almost equal to the source voltage.

If a shunt motor running without load has its shunt field opened, the counter emf will fall to nearly zero and the armature will continually accelerate until the machine flies apart. If the motor is under load, loss of shunt-field current results in stopping of the motor, burnout of the armature or

fuse, or opening of any overload relay in the line. With the starter shown, an open field circuit de-energizes the holding magnet and the starter arm snaps to the OFF position.

In large motors, commutating poles (not shown) are used to assure a constant neutral position for the brushes.

The shunt motor is usually reversed by reversing the current through the armature (and interpoles, if any). The inductive reactance in the shunt winding limits current through this field so much that a shunt motor will not usually run on ac.

27-17 COMPOUND DC MOTORS

The compound dc motor has a series and a shunt field and, usually, interpoles. A simple schematic diagram with a motor starter is shown in Fig. 27-19. The rheostat in series with the shunt field

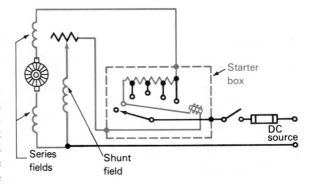

FIG. 27-19 A 3-terminal manual starter connected to a compound motor.

controls the counter emf generated in the armature and therefore the speed of the machine. Maximum resistance results in maximum speed. Starting and reversing direction are similar to those of a shunt motor.

It was mentioned that the differentially compounded machine produced a generator with a badly sagging voltage output under load. However, a differentially compounded motor may have good speed regulation.

27-18 ELECTRONIC SPEED REGULATORS

There are a variety of regulating circuits to maintain a constant dc motor speed. The simple circuit shown in Fig. 27-20 illustrates one approach to regulating the speed of a shunt dc motor. Assume the motor is running at the desired speed with little

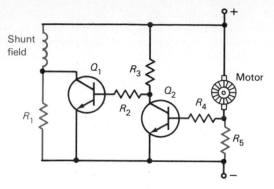

FIG. 27-20 A possible speed regulator for a shunt-wound dc motor.

or no load coupled to it. The field coil and R_1 are in series across the armature, and the small resistance R_5 is in series with the armature. With no load, the armature current and the voltage-drop across R_5 will be low. Resistances R_2, R_3, and R_4 are selected to produce a nearly saturated Q_1 I_C value through the field coil. When the load is connected, the speed of the armature decreases and the armature current increases. The voltage-drop across R_5 and the forward bias on Q_2 increase. The Q_2 I_C value through R_3 increases, lowering the positive bias on Q_1, and the Q_1 I_C value decreases. With the lessening of current through the field coil, the speed of the motor increases.

Small dc motors use permanent-magnet fields. An increase in current through the armature by an increased voltage causes the motor speed to increase. There are many circuits that will regulate the speed of such motors. One uses the ac generated by a mechanical or optical tachometer attached to the motor shaft and compares this with the frequency of a constant-frequency oscillator (Fig. 27-21). The difference frequency is used to control

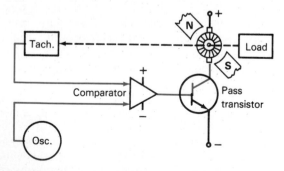

FIG. 27-21 A possible speed control for a p-m motor.

the bias on a pass transistor in series with the motor and the power source. If the motor tends to go slower the forward bias increases, the transistor conductance increases, and the motor speeds up.

| Test your understanding; answer these checkup questions.

1. What is the advantage of multisegment over two-segment commutators? _____
2. In Fig. 27-9, what makes the machine rotate? _____
3. What are two methods of increasing output V of a dc generator? _____ _____ Are both practical? _____
4. What is the main disadvantage of a separately excited dc generator? _____ A series generator? _____ A shunt generator? _____
5. In Fig. 27-11, is maximum output produced with maximum or minimum rheostat resistance? _____
6. Which has the larger field-coil wire, a series or a shunt generator? _____
7. What are the three types of compound dc generators? _____ _____ _____ Which would be best to feed a local load? _____ A distant load? _____
8. What do commutating poles reduce? _____
9. What value bypass capacitors do you think might be best to reduce radio noise from a dc generator? _____
10. In Fig. 27-15, how many coils would be wound on the armature? _____ Why would this not be a good dc generator? _____
11. Why might a series motor operate on dc or ac? _____
12. Why might a shunt motor not operate on ac? _____
13. If the motor in Fig. 27-18 is running unloaded, will the starter arm snap off as soon as the line switch is opened? _____ Why? _____
14. Will an unloaded series motor reach a reasonable running speed? _____ A shunt motor? _____ A compound motor? _____
15. Basically, what determines the running speed of a series motor? _____ Shunt motor? _____ Compound motor? _____

27-19 AC MOTORS

In a few cases, as where dc is the primary source of available power or where a wide range of speed control is desired, dc motors may be employed. However, most modern motors operate from an ac source. While there are many different types of ac motors, only three basic types will be discussed, the *universal*, the *synchronous*, and the *squirrel-cage* motors.

27-20 UNIVERSAL MOTORS

The series-type dc motor will rotate when dc or low-frequency ac is applied to it. Such a *universal motor* is used in fans, blowers, food mixers, portable electric drills, and other applications in which a high speed under light load or a slow speed with high torque is required.

One of the difficulties with universal motors is the radio interference or noise caused by commutator sparking. This noise may be reduced by bypassing the two brushes to the frame of the motor and grounding the frame.

27-21 SYNCHRONOUS MOTORS

A single-phase alternator works as a motor under certain circumstances. If the field is excited by dc and ac is fed to the slip rings and rotor coil, the machine will *not* start rotating. While the rotor-coil field may be alternating magnetically, during one half cycle it will try to move in one direction but during the other half cycle it will try to move in the opposite direction. The net result is no movement. The machine will only heat and may burn out.

The rotor of a 2-pole alternator must make one complete rotation to produce one cycle of ac. It must rotate 60 times per second, or at 3600 rpm, to produce 60 Hz ac. If such an alternator can be rotated at 3600 rpm by some outside mechanical device, such as a dc motor, and the armature is then excited with a 60 Hz ac, it will continue to rotate as a synchronous motor at 3600 rpm. As long as the load is not too heavy, a synchronous motor will run at its synchronous speed and at this speed only. If the load becomes too great, the motor will slow down, lose synchronism, and come to a halt.

The synchronous speed of any synchronous motor in rpm can be determined by the formula

$$\text{Sync speed} = \frac{\text{Hz} \times 60}{\text{pairs of poles}}$$

An example of a synchronous motor is an electric clock. As long as the ac is maintained at the correct *frequency*, the clock keeps correct time. The voltage amplitude is not important. Turntables and tape decks also use synchronous or electronically regulated motors.

27-22 SQUIRREL-CAGE MOTORS

Most motors that operate from single-phase ac have a squirrel-cage rotor. A simple form is shown in Fig. 27-22. A practical squirrel-cage rotor is considerably more massive and has a laminated iron core.

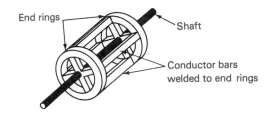

FIG. 27-22 Basic rotor of a squirrel-cage ac motor.

The conductors running the length of the squirrel cage are copper or aluminum and are welded to the metal end rings. Each conductor forms a shorted turn with the conductor on the opposite side of the cage. When this cage is between two electromagnetic field poles that are being magnetized by an ac, an alternating emf is induced in the shorted turns, a heavy current flows in them, and a strong counter field is produced which bucks the field that produced the current (Lenz' law). Although the rotor may buck the field of the stationary poles, there is no reason for it to move in either one direction or the other, and so it remains stationary. It is similar to the synchronous motor, in that it is not self-starting either. What is needed is a *rotating field* rather than simply an alternating one.

How the field is made to have a rotary effect names the type of squirrel-cage motor. A *split-phase* motor uses an additional pair of field poles that are fed out-of-phase currents, allowing the two sets of poles to develop maximum current and magnetic fields at slightly different times. The out-of-phase windings on the out-of-phase field poles could be fed by two-phase ac and produce a rotating magnetic field, but for single-phase operation the second phase is usually developed by connecting a capacitor (or resistor if the extra poles are highly inductive) in series with the out-of-phase winding (Fig. 27-23). This shifts the phase about 20° and produces a maximum magnetic field in the phased winding that leads the magnetic field in the main winding. The effectively moving maximum strength of the magnetic field passing from one pole to the next one attracts the squirrel-cage rotor with

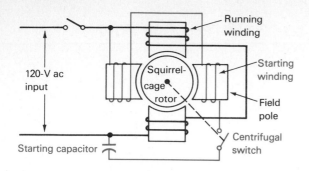

FIG. 27-23 Capacitor-start squirrel-cage motor.

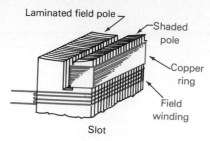

FIG. 27-24 Details of the end of a field pole of a shaded-pole ac motor.

its induced currents and fields, rotating it. This makes the motor self-starting. The split-phase winding can be left in the circuit, or can be cut out by a centrifugal switch that disconnects it when the motor reaches a running speed. Once the motor starts rotating, it operates better without the split-phase winding. Not being a synchronous motor, it does not have to maintain a synchronous speed. The rotor of a split-phase induction motor always slips behind what would be the synchronous speed. If the synchronous speed were 1800 rpm, the squirrel-cage rotor with a certain load might rotate at 1750 rpm. The heavier the load on the motor, the more the rotor slips. If the rotor is constructed with two flat sides, it *will* maintain a synchronous speed. Under optimum operating conditions a split-phase motor with the phased or starting poles disconnected may operate at approximately 75% efficiency.

Another method of producing a rotary field in a motor is to *shade* the field poles. This is accomplished by slotting the field poles and connecting a copper ring around one part of the slotted pole (Fig. 27-24). As an alternation is increasing in amplitude in the field coil, the magnetic field expands and induces an emf and current in the copper ring. This produces a magnetic field around the ring that

bucks the magnetism in the part of the pole surrounded by the ring. Maximum magnetic field is developed at this time in the unshaded part of the pole, and minimum in the shaded part. As the field-current cycle reaches a maximum, the magnetic field no longer moves, the copper ring has no current induced in it, and the ring has no effect. Maximum magnetic field is now developed across the whole pole face. As the alternation decreases in amplitude, the field collapses and induces an emf and a current in the ring in the opposite direction, which produces a maximum field in the shaded part of the pole face. Thus the maximum magnetic field moves from the unshaded over to the shaded part of the field pole as the cycle progresses. This maximum-field movement produces the necessary rotating field in the motor to make the squirrel-cage rotor self-starting. The efficiency of shaded-pole induction motors is not high, ranging from 30 to 50%.

The main advantage of all squirrel-cage motors, particularly in radio applications, is the lack of commutator and brushes, or slip rings and brushes, resulting in interference-free operation.

27-23 POLYPHASE MOTORS

When 3-phase ac is available, a 3-ϕ motor is usually desirable. Each of the three phases is fed to separate field coils wound around field poles in the motor. Since each phase is 120° from all other phases, current maximums appear in one field pole, then 120° later in the next pole, 120° later in the third pole, and so on. This action produces a true revolving field. All such motors are self-starting, are quite efficient, are usually squirrel-cage rotor types, and operate at a nearly constant speed, although they have some slip.

At the instant of starting, a high current flows.

Either series resistances are used in two of the three lines feeding the motor to limit starting current, or the line voltage is reduced until the machine approaches running speed.

When rotating at normal speed, the induced currents and fields in the rotor occur at such a time and phase that they do not cancel the field-coil inductance. With inductance in the field coils, their reactance limits the line current to a low value. When a load is applied, the rotor begins to fall behind the rotating fields. This increases the current induced in the rotor, which in turn reduces the inductance (and X_L) of the field windings, demanding more line current. The result is greater torque and horsepower from the motor as it tries to maintain a nearly synchronous speed.

A 2-pole 3-ϕ motor (two poles per phase) operating on 60 Hz ac will rotate at a speed of 3600 rpm; a 4-pole motor will rotate at 1800 rpm.

27-24 MOTOR-GENERATORS

When it is desired to convert one form of electricity to another, a motor and a generator can be coupled together. The motor may be either ac or dc, and the generator may be either ac or dc. For example, 1500 V dc may be required, but the power lines carry 120 V ac. A 120-V ac motor coupled to a 1500-V dc generator will provide the necessary high-voltage dc. A diagram of such a motor-generator set is shown in Fig. 27-25.

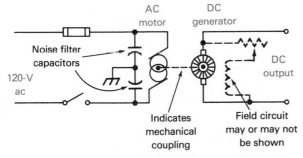

FIG. 27-25 An ac motor and dc generator system diagram.

When motor-generator sets are dc-to-dc, as low-voltage to high-voltage, the driving motor is either shunt or compound-wound for good speed regulation. When starting up the system, it should be operated at no load until up to running speed. As with any shunt-type motor, the field resistor is at minimum resistance at the start and is increased as the motor approaches the desired operating speed. To vary the output voltage, the field rheostat of the generator is used as the control.

When the motor-generator set is ac-to-ac, the conversion of energy may be either from 1-ϕ to 3-ϕ, or vice versa, or from one frequency to another. (Transformers convert voltages of the same frequency.)

A motor-generator set is quite heavy and bulky and makes an audible noise. It is rather costly to purchase, and maintenance is required. Being partly mechanical, it requires lubrication. Commutator or slip rings need cleaning. Brushes must be replaced periodically. Motor-generator sets have fairly good voltage regulation, although the output voltage may be somewhat limited because of difficulty of insulating the armature and commutator components in dc machines. Sparking at brushes may cause radio interference. In comparison, a VT or solid-state rectifier is lighter and less expensive, has almost unlimited voltage possibilities, operates without noise or vibration, is more efficient, produces no radio interference and may be operated in any position.

27-25 DYNAMOTORS

A *dynamotor* is a form of a dc motor-generator but with a common field coil for both motor and generator. The motor normally operates from a 6- to 28-V battery. The output dc voltage ranges from about 225 to 1000 V.

A dynamotor is lighter and smaller than a motor-generator because of the common field winding. One end of the common armature core carries the commutator segments for the motor. The other end carries the commutator segments for the generator. The two sets of armature coils are separate but are wound into the same slots in the armature. Figure 27-26 shows a schematic and pictorial diagram of a dynamotor.

The output voltage of a dynamotor cannot be changed efficiently. If the primary source voltage is changed, the output will change, but best efficiency and voltage regulation are produced with a specific input voltage. The overall efficiency ranges from 50 to 60%, somewhat less than a motor-generator. The voltage regulation of most dynamotors is fair but does not compare with that of a properly designed motor-generator. The dynamotor's high output power for its weight is produced by its high speed-of-rotation characteristic.

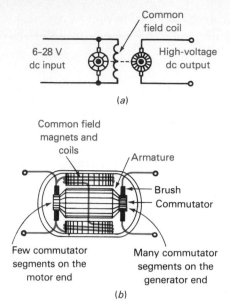

Common
field coil

6-28 V
dc input

High-voltage
dc output

(a)

Common field
magnets and
coils

Armature

Brush

Commutator

Few commutator
segments on the
motor end

Many commutator
segments on the
generator end

(b)

FIG. 27-26 *(a)* Schematic diagram of a dynamotor. *(b)* Component placement in the device.

In general, the dc-to-dc application of dynamotors has been superseded by use of the transistor power oscillator, whose output ac is stepped up, rectified, filtered, and often regulated electronically (Sec. 10-27).

When the output winding is terminated with slip rings and brushes, such a dc-to-ac machine is known as a *rotary converter*.

27-26 RATING GENERATORS AND MOTORS

DC generators are rated in *watts* of output. When an alternator feeds power to a circuit, it is not known what power factor will be involved. If the load is completely resistive (*V* and *I* in phase), the power output will be equal to $V \times I$ and will be the true power. Should the load be completely reactive (*V* and *I* out of phase by 90°), the current drawn from the machine may be the maximum that the wires of the alternator can stand without overheating but the true power delivered to and used by the load is zero ($VI \cos \theta = V \times I \times 0$) watts. Current is flowing and producing a field, but the field collapses and returns the energy to the circuit. The wires are carrying current, but no power is being lost. In such a case, an alternator normally thought of as capable of 2-kW output when used on electric lights alone will be delivering no true power to a reactive load

but may still be running at a maximum current value. For this reason, alternators are always rated in apparent power (voltamperes, or VA). Large alternators are rated in kVA.

It is customary to rate motors in output horsepower: 746 W = 1 hp. A 3-hp dc motor delivers the equivalent of 2238 W of turning power. However, the machine is not 100% efficient and therefore must be drawing more than this number of watts from the line. If the machine is 85% efficient, the 2238 W must be 85% of the input power, or

$$0.85(P_i) = 2238 \text{ W}$$

$$P_i = \frac{2238}{0.85} = 2633 \text{ W}$$

From this, the formula for input power is

$$P_i = \frac{P_o}{\%}$$

where P_i = input power in W
P_o = output power in W
% = percent efficiency (as a decimal)

If the motor above is operating from a 110-V dc line, from $P = VI$, the line current would be

$$I = \frac{P}{V} = \frac{2633}{110} = 23.9 \text{ A}$$

Alternating-current motors are also rated in horsepower, but besides their efficiency the power factor must be considered.

A 7-hp ac motor operating at full load, with a power factor of 0.8 and 95% efficiency, will have what power input? Power factor can be considered a type of ac efficiency rating. The power output developed is equivalent to 7(746), or 5222 W. The input power is 5222/0.8(0.95), or 6871 W. If the motor is operating from a 120-V line, the line current is $I = P/V$, or 6871/120, or 57.3 A.

27-27 MAINTENANCE OF MOTORS AND GENERATORS

It is usually an assignment of an operator at a radio installation to see that any motors and generators are in proper operating condition at all times. The following are some suggestions regarding such maintenance.

Oil on rubber insulation causes the insulation to soften and weaken. Care must be exercised when oiling motors and generators that no oil is allowed to drip onto such wires.

The bearings of rotary machines must be oiled

periodically to prevent overheating and freezing. If a bearing overheats because of lack of oil, it should be flushed while still running, with light oil until cool.

Sparking at the brushes indicates trouble. The brushes may be worn and be making poor contact. The slip rings or commutator may be dirty. An armature coil may be shorted or open. The machine may be running with too heavy a load. The interpoles may be inoperative or incorrectly wired.

If slip rings or commutators become dirty, they can be cleaned with a piece of heavy canvas and a liquid solvent, or they may be sanded smooth with a very fine grade of sandpaper. Emery paper and steel wool should not be used to smooth a commutator. Metal particles from them may lodge between segments. The mica insulation between commutator segments may wear more slowly than the copper or brass segments, leaving the brushes to ride on insulated ridges instead of on the segments. The mica insulation should be *undercut* below the level of the surface of the commutator segments.

Motors and generators should be cleaned and dusted regularly and given a close visual examination. The length of the brushes should be checked. If worn, replace with brushes of the correct size and type. In an emergency, a larger brush may be sanded down to fit the machine. The spring tension holding the brush against the commutator should produce adequate pressure.

When the commutator segments show signs of excessive wear, the armature may have to be taken out of the machine and the commutator turned down on a lathe. The mica insulation must be undercut before turning the commutator. After turning a commutator, remove all metal particles that may be deposited on the mica insulation between segments. If an armature coil becomes shorted, the armature will heat and the brushes will spark.

If a motor or motor-generator switch is closed but the machine does not start, check the line voltage, fuses, and overload relay. If the overload relay will not hold in when the motor is switched on, the bearings of the machine may be frozen, the armature shaft may be locked, the armature may have a shorted turn, or the field-coil circuit may be open or burned out. If the fuses or overload relays are not open and the line voltage is normal, a brush may not be making contact with the commutator or slip ring because of dirt under the brush or insufficiently undercut segments. There may be a poor electrical connection leading into the motor.

27-28 MAGNETOHYDRODYNAMICS

One important development in generating electricity directly from heat is magnetohydrodynamics (MHD). If a hot, conductive liquid metal, or a *plasma* (hot ionized gas), is pumped through an insulating-material pipe in the direction of the colored arrows shown in Fig. 27-27, and past a strong mag-

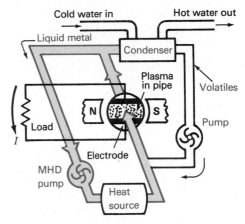

FIG. 27-27 Basic MHD dc generating system.

netic field, an emf will be developed between any two metal electrodes inside the pipe. According to the left-hand generator rule, electrons will be pushed from the bottom electrode to the top electrode. If a load is connected across the plates, a current will flow. A dc potential of 3 to 5 V at a relatively high current can be generated in such a mechanically simple system. The electrical output is directly proportional to the magnetic field strength and the velocity of the conductive metallic plasma.

The heat source that heats the liquid metal can be solar, geothermal, or industrial waste. The closed liquid metal system can use an MHD pump, the reverse of the MHD generator, to keep the liquid metal moving. The easily liquefied metal might be mercury, sodium, potassium, indium, tin, or gallium. This is mixed with a volatile fluid such as hexane, toluene, or pentane to produce the conductive metallic steamlike plasma that is pumped past the magnets. Once past the magnets, the volatile parts are separated from the liquid metal, are pumped through a condenser, and are then fed back to mix with heated liquid metal again. Cold water pumped through the condenser produces a usable hot water output.

DC generated can be changed to ac by a converter and stepped up or down.

Test your understanding; answer these checkup questions.

1. What are the three basic ac motor types? _____ _____ _____
2. What is a universal motor? _____
3. If an alternator is to be used as a synchronous ac motor, what must be fed to it and under what conditions? _____ _____ _____
4. What is the synchronous speed of a 2-pole alternator being fed 50 Hz ac? _____
5. Will an electric clock keep accurate time if the line voltage drops 20%? _____ Rises 20%? _____
6. How is a squirrel-cage motor made synchronous? _____ What is it said to have if not synchronous? _____
7. What are the two types of squirrel-cage motors discussed? _____ _____

8. Why do squirrel-cage motors produce no radio noise? _____
9. How many field poles does a 2-pole 1-ϕ motor have? _____ A 2-pole 3-ϕ motor? _____
10. Are 3-ϕ motors normally synchronous? _____
11. In what units are large alternators rated? _____ DC generators? _____
12. Does a 1-hp dc motor use more, less, or exactly 746 W? _____ Why? _____
13. A 2-hp ac motor when operating has a pf of 0.9 and is 80% efficient. What is the power input? _____ Power output? _____
14. Why is emery cloth not used to clean electric machines? _____
15. What is it called when the mica insulation between commutator segments is reduced in height? _____
16. What does MHD mean? _____ What type of current does it generate? _____

MOTORS AND GENERATORS QUESTIONS

1. What is an alternator? A generator?
2. What are the three basic parts of a simple alternator?
3. Name the sliding connections of an alternator?
4. What is an externally excited alternator? How is its output voltage controlled?
5. What is excited in a rotating-field alternator?
6. Diagram an automobile alternator charger.
7. Where are the primary and secondary windings on an inductor alternator? For what are they usually used?
8. What output effect may occur if a capacitive load is connected across an alternator?
9. What type of voltage is used to excite the fields of alternators?
10. Diagram a circuit by which two alternators can be paralleled. When is the second alternator connected?
11. Name the parts of a dc generator.
12. What is the advantage of using many coils and commutator segments in a dc armature?
13. If a battery is used to produce its field magnetism, what type of dc generator is the machine called?
14. What determines the polarity of the output of an externally excited dc generator? Of a p-m generator?
15. Diagram a series-field dc generator connected to a load. What is the main disadvantage of this machine? What relative size wire is used on its field?
16. Draw a diagram of a shunt dc generator with a means of varying its output voltage. What happens to its voltage under load? What relative size wire is used on the shunt field?
17. Draw a diagram of a compound generator with a means of varying its output voltage. What is the advantage of this machine?
18. Draw a diagram of a possible electronically regulated dc shunt generator.
19. What can be added to a dc generator to prevent brush sparking under varying loads?
20. What may be added to a motor to reduce RFI?
21. Does reversing the dc applied to a series dc motor reverse its rotation? To a shunt dc motor?
22. What will a series dc motor do if operated without a load?
23. Diagram a series dc motor with a rotation-reversing switch. What is being reversed in this circuit?
24. What determines rotation direction of p-m motors?
25. What is important to remember when using a resistive starter on a dc motor?
26. What limits the unloaded speed of a shunt dc motor?
27. What excitation change speeds up a shunt motor?
28. What is the advantage of a compound motor?
29. Draw a diagram of a possible electronic speed regulator for a small dc shunt motor.
30. Draw a semiblock diagram of a possible speed-regulating circuit for a small p-m dc motor.
31. Which dc motor will operate on ac? Its name?
32. What is the advantage of a synchronous motor?
33. Name two types of squirrel-cage 1-ϕ ac motors? Are they synchronous? Which is more efficient?
34. Why is a polyphase ac motor always self starting?
35. What rpm is produced by a 6-pole, 3-ϕ ac motor?
36. What are the uses of a motor-generator set?
37. What is a dynamotor? A rotary converter?
38. In what electrical power unit is a dc motor or generator rated? An ac motor or generator?
39. How many watts in 1 hp?
40. A 3-hp 115-V ac motor is 85% efficient at 0.92 pf. What is the line power and current?
41. M-G bearings overheat. What should be done?
42. What is important about the insulating material between commutator segments on rotary machines?
43. What is MHD? What does an MHD unit do?

28

Batteries

This chapter outlines some of the many types of primary and secondary cells used in batteries and explains some of the charging methods employed with secondary cells.

28-1 PRIMARY VERSUS SECONDARY CELLS

An electric *battery* is a combination of two or more *cells*. A cell is either not rechargeable and called a *primary* cell, or is rechargeable and called a *secondary* cell. Cells and batteries store energy in such a way that they can produce electric energy when required to do so.

When a dc voltage is required, a cell or battery may be used as the power supply. In VT radios, batteries used to heat a tube filament are called *A batteries*; those used to supply plate potential, *B batteries* (V_{bb}, B+); and those that supply grid bias voltage, *C batteries* (V_{cc}, C−). Either primary or secondary cells may be used as A, B, or C batteries. Sometimes the terms B+, meaning battery, may be used with transistors.

28-2 PRIMARY CELLS

Almost any two dissimilar metals immersed in a dilute acid or alkaline solution (Fig. 28-1) will produce a difference of potential between them. The chemical action that takes place in such a cell pro-

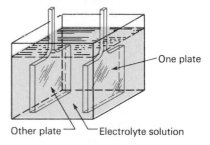

FIG. 28-1 An electric cell consists of two dissimilar plates and an acidic or alkaline electrolyte.

duces a negative charge on one plate and a positive charge on the other. The solution into which the plates are immersed is known as an *electrolyte*. Such a cell may produce a few tenths of a volt to nearly 3 V, depending upon the metals and chemicals used. Unless the materials employed are selected properly, a continuous chemical action takes place and the cell wears itself out in a short time.

When carbon is the positive electrode or pole, zinc is the negative electrode, and a solution of sal ammoniac and zinc chloride is the electrolyte, a 1.5-V cell known as a *Leclanché* cell is produced. This cell will slow or reduce its chemical action when not in use so much that it may have a *shelf life* (retain its charge) of 1 to 3 years.

When the Leclanché cell is under load, the chemical action produces hydrogen atoms at the positive electrode. This forms a nonconductive hydrogen-gas sheath around the surface of the carbon; current flow through the cell is hindered; and the cell is said to be *polarized*.

The hydrogen gas can be made to combine with oxygen to form water by adding manganese dioxide, which is rich in oxygen, to the electrolyte. In this way, the cell is depolarized and a relatively heavier current can flow through it. This is the basic theory of operation of the common *dry cell*, which is actually quite damp inside.

Dry cells are made in three basic forms. The oldest form uses a cylindrical zinc can as the negative pole (Fig. 28-2a). A mixture of ground carbon, manganese dioxide, sal ammoniac, and zinc chloride forms a thick black paste electrolyte nearly

filling the can. A carbon-rod positive pole is held in the center of the electrolyte. A layer of sealing wax seals the top of the cell. A small air space is left between the wax and the electrolyte. The cell may be covered with cardboard or other insulating covering.

When several cylindrical cells of this type are packaged to form a battery, the space between cylinders is wasted. To reduce this waste, the cells may be made in layers as in Fig. 28-2b.

Another form is the *inside-out* dry cell. It uses an outer cylinder with a coating of carbon on its inner surface. A connection from the carbon coating is brought to the top of the cell as the positive connection. The same electrolyte is employed. The negative plate consists of a sheet-zinc structure extending up into the electrolyte (Fig. 28-2c). The zinc is connected to the metal base of the cell to form the negative pole.

As the chemical action of any dry cell continues, the zinc is eaten away by chemical interaction with the electrolyte. This produces holes in zinc cans and the electrolyte oozes out, or the sides of the cell may swell. The inside-out cell has the carbon, which is not affected by the chemical reaction, on the outside and the chemically active zinc inside. Thus, the cell does not puncture and leak. Other "leakproof" cells use a steel outer protective jacket.

The condition of charge of a dry cell is determined by measuring the voltage across it, preferably with a load on the cell. A new cell has a voltage of 1.55 V. As a cell ages, the voltage across it decreases. When the voltage without load drops 15%, to about 1.32 V, the cell is usually considered as being near the end of its chemical life. Under a heavy load the cell voltage drops more. A cell under load may not be considered as being at the end of its chemical life until it drops below 0.9 V. When a primary cell reaches the end of its chemical life, it is discarded. Recharging a zinc-carbon cell by forcing a current backward through it may depolarize it somewhat but does not actually recharge it by returning the chemicals to their original form. (This cell must not be confused with the newer completely sealed nickel-cadmium *rechargeable dry cell*.)

Batteries in receivers may produce noise when they polarize because of small variations in current through the cell.

The shelf life of a dry cell is increased by storing the cell in a dry, cool place, usually near the floor. If a cell is allowed to heat, internal chemical reactions increase and the cell life decreases. Although

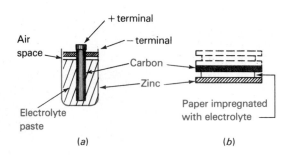

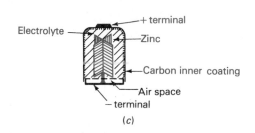

FIG. 28-2 Types of dry-cell construction: (*a*) zinc can, (*b*) layer, and (*c*) inside-out.

dry cells are not damaged if exposed to low temperatures, they will not produce normal current if too cold (chemical action slowed).

A large No. 6 dry cell (2⅝ by 6½ in.) has a short-circuit current of approximately 32 A. The largest size of flashlight cell, type D, has a short-circuit current of about 6.5 A. The small penlite size produces about 4.5 A. (Short-circuit testing is not recommended.)

28-3 LEAD-ACID BATTERIES

One secondary cell in general use has a lead dioxide (peroxide) (PbO_2) positive plate, a pure, spongy lead (Pb) negative plate, and a dilute sulfuric acid (H_2SO_4) electrolyte (Fig. 28-3). A group of these

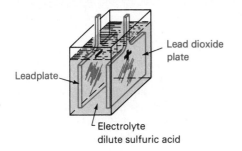

FIG. 28-3 A lead-acid cell has a lead negative plate and a lead dioxide positive plate with a sulfuric acid electrolyte.

cells in series forms a *lead-acid* battery. These batteries are used in automobiles and as power supplies for mobile and emergency equipment.

In practice, several positive and negative plates are interleaved (Fig. 28-4). Since a greater negative than positive plate area is required, there is always one more negative plate.

To prevent the positive and negative plates from

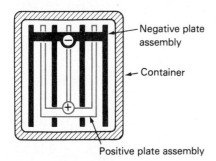

FIG. 28-4 How lead-acid cell plates are interleaved in a container (top view). Separators not shown.

touching each other, thin wooden, glass, rubber, or plastic, or matted spun-glass separator sheets (not shown) are inserted between adjacent plates. The solid separators are vertically grooved to allow gas bubbles that form on the plates during the charging process to float to the surface and escape. The glass separators hold gases between the plates.

The plates and electrolyte are enclosed in a hard-rubber, plastic, or glass container. A screw-type vent cap in the top cover of each cell allows access to the electrolyte, which must be ¼ to ½ in. above the plates at all times.

When a load is connected across a charged lead-acid cell (2.1 V), the chemical action that takes place internally as the cell discharges can be represented by

$$
\begin{array}{lll}
(+\text{plate}) & PbO_2 \rightarrow PbSO_4 \\
(\text{Electrolyte}) & H_2SO_4 \rightarrow H_2O \\
(-\text{plate}) & Pb \rightarrow PbSO_4
\end{array}
$$

During the time the cell is discharging, the sulfuric acid is combining chemically with both the lead dioxide and the pure lead plates, changing them to lead sulfate. When both plates have been changed to lead sulfate, they are no longer different materials, no potential difference is produced between them, and the battery is completely dead. The sulfuric acid, having combined with the plates, leaves only water as the electrolyte.

A lead-acid cell should never be allowed to become completely discharged. During discharge, lead sulfate forms as crystals on the plates, causing them to swell, which will bend, or buckle, the plates, possibly splitting solid separators or even the battery case itself.

To recharge a lead-acid cell, a source of dc emf greater than that of the cell must be used. This forces a current backward through the cell, reversing the chemical action of discharging, above. The sulfate in both plates is chemically broken down, reappearing as H_2SO_4 in the electrolyte, leaving the positive plate PbO_2, the negative plate Pb, and the cell is recharged.

Lead-acid cells must be recharged as soon as possible after being discharged. The lead sulfate first forms small, soft crystals on the plates. If allowed to age, the crystals grow and turn hard. Soft crystals respond to recharging and easily change back into the sulfuric acid, lead, and lead dioxide. Hard crystals require a long, slow recharging to complete the chemical reversal. *Sulfated* batteries that are allowed to remain in a semicharged or discharged state for long periods may never completely recharge or may become completely useless.

Jarring a discharged solid-separator battery knocks crystals off the plates; they fall to the bottom of the cell and are lost to the plates. An old battery will have a thick layer of this sludge in the pockets built into the case beneath the plates. If sludge builds up and covers the bottom edges of any two positive and negative plates, the cell becomes shorted and useless. A cell also becomes shorted if sludge can form a conducting path through a cracked separator.

A battery becomes warm when charged or discharged rapidly. This heat is caused by the internal chemical action and the I^2R loss when current flows through the lead-antimony (or lead-calcium) grids that hold the active materials of the plates. Some battery manufacturers add silver and nickel to the grids to decrease their resistance and the I^2R loss. A battery should not be discharged or recharged at a rate that will heat it over 110°F (42°C). Except for its heating, a lead-acid battery can be discharged at as high a rate as the load demands.

28-4 SPECIFIC GRAVITY

A voltage reading of a lead-acid cell under load gives a fair indication of state of charge, but measurement of the *specific gravity* of the electrolyte is better.

The specific gravity (sp gr) of a substance is a comparison of its weight with the weight of water. Water has 1.000 sp gr. Pure sulfuric acid has 1.835 sp gr. It weighs 1.835 times as much as water per unit volume. The electrolyte solution in a lead-acid cell ranges from 1.210 to 1.300 sp gr for new, fully charged batteries. This is about 20% acid and 80% water. The higher the specific gravity, the less internal resistance of the cell and the higher the possible output current. To start an automobile motor, 75 A at 12 V may be needed. An automobile battery has a fully charged specific gravity of about 1.280. However, the higher the specific gravity, the more active the chemical action and the faster the battery will discharge itself.

In stationary services, as standby or emergency power supplies, or in a service where a relatively low-current drain over a long period of time is required, lead-acid batteries may use 1.210 to 1.220 sp gr. This gives a longer life but a higher internal resistance and less current capability.

The voltage of a lead-acid cell varies with the specific gravity. For example,

Sp gr	No-load voltage
1.300	2.2
1.280	2.1
1.220	2.05

In most cases, a battery should be recharged before its specific gravity drops 100 points. A 1.280 sp gr battery should be recharged when it reaches a 1.180 sp gr.

When the specific gravity is 1.280, the freezing temperature of the electrolyte is −90°F. With a 1.180 sp gr, the freezing point is −6°F. With a 1.100 sp gr, the electrolyte will freeze at about 18°F. If a cell does freeze, it may split its case and be ruined. A charging battery will not freeze because of the internal heat developed by the charging process. While cold prevents the chemical action from occurring rapidly, it does not necessarily damage a battery.

28-5 HYDROMETERS

Specific gravity is measured with a *hydrometer* of the syringe type (Fig. 28-5), having a compressible rubber bulb at the top, a glass barrel, and a rubber

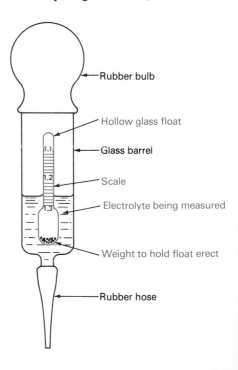

FIG. 28-5 Hydrometer measures specific gravity of an electrolyte.

hose at the bottom. A weighted, calibrated, hollow glass float is inside the barrel.

To measure the specific gravity, the cell cap is unscrewed and the hydrometer hose is dipped into the electrolyte above the plates. The bulb is compressed *slowly* and then released, drawing some of the electrolyte into the barrel. The heavier the electrolyte, the higher the float rides. The sp gr is indicated by the highest calibration that is visible below the surface of the electrolyte (1.260 is shown).

Many hydrometers have a thermometer in the base of the barrel to indicate the temperature of the electrolyte being measured. Since heating the electrolyte causes it to expand, warm electrolyte is lighter per unit volume. Therefore, the warmer the solution (above 70°) the greater the correction to add to the sp gr reading to express what it would be at the reference temperature of 70°F ($\Delta 3$°F = $\Delta 0.001$ sp gr). If its temperature is lower than 70°, the solution is more dense and the correction must be subtracted from the hydrometer indication. The corrections are marked on the *thermometer* scale.

When hydrometer readings are taken, care must be exercised to prevent drops of the electrolyte from spilling on the top of the cells or contacting clothing or hands. The sulfuric acid is very corrosive and will eat holes in fabrics and burn human skin. When an acid solution is spilled on flesh or clothing, immediate flushing with water is recommended. The acid can then be counteracted by applying a weak base or alkali solution such as dilute ammonia or baking soda.

Batteries that are frequently tested for specific gravity may lose electrolyte (a drop or two) each time the battery is measured. When the fully charged specific-gravity reading drops 20 or 30 points, it may be advisable to add sulfuric acid to the electrolyte. CAUTION: When mixing electrolyte, always pour the acid into the water. Never pour water into concentrated acid, or the heat produced by ionization will cause a corrosive steam and acid explosion.

28-6 WATER FOR BATTERIES

The cap on a storage cell has a hole in it to relieve internal gas that develops, but the hole is small enough to prevent dirt from falling into the electrolyte. There is little evaporation of water from a battery unless it overheats. Whenever a battery is charging, the chemical action produces hydrogen gas on one plate surface and oxygen gas on the other. These gases bubble to the surface and escape through the vent hole in the cap. Thus, water (H_2O) is lost to the cell when the gases leave.

The water that escapes must be replaced to maintain the proper electrolyte level. *Any impurity* in the added water will combine chemically with the sulfuric acid or the plates and form a stable compound that will not enter into the charge or discharge action of the battery. The presence of impurities on a cell plate, either developed there during manufacture or added later, produces *local action* at that point. The part of the plate involved in the local action is effectively lost, and the cell decreases its capacity to some extent. To be safe, *only distilled water* should be added to a cell!

If the active materials are carefully balanced and matted spun-glass separators are used, a cell can be produced that exhausts no gas and therefore requires no water replacement.

28-7 CAPACITY OF A BATTERY

The *capacity* of a battery is rated in ampere-hours (Ah). The number of ampere-hours produced in an 8-h period to bring a lead-acid cell down to 1.75 V is one standard of measurement used. If the battery is forced to discharge faster than this, the number of ampere-hours produced will be somewhat less. If discharged slower, the ampere-hours obtained will be greater because of lessened internal heat.

The capacity of a storage battery determines how long it will operate at a given discharge rate. An 80-Ah battery must be recharged after 8 h of an average 10-A discharge. When fully charged, it may be capable of 1000-A discharge for a short period without damage. If electrolyte temperature exceeds 110°F, the discharge rate should be reduced.

The capacity required of a storage battery to operate an emergency radiotelegraph transmitter for 6 h under a continuous transmitter load of 70% of the telegraph-key-down demand of 40 A plus a continuous emergency-light load of 1.5 A can be computed:

$$\begin{aligned} \text{Capacity} &= \text{amperes} \times \text{hours} \\ &= (0.7 \times 40 \times 6) + (1.5 \times 6) \\ &= 168 + 9 = 177 \text{ Ah} \end{aligned}$$

It is important that lifeboat or emergency batteries of any kind be kept fully charged. If under load an emergency transmitter appears to have unusually low battery voltage, corroded battery terminals may be causing resistance in the battery line, producing a voltage-drop, or the batteries may not be fully charged. This could be disastrous.

When a lead-acid cell is constructed, pink lead peroxide is molded into the grids of the positive plates. The battery is then *formed* by *cycling*, which is an alternate charging and discharging until the pink oxide has been changed to brown lead dioxide. With each cycle the ampere-hour output of the battery increases. After a user purchases the battery, its capacity may continue to increase for a few discharge-charge cycles. With sufficient cycling, all the lead oxide is finally converted to lead dioxide and the battery reaches its maximum capacity. *Dry-charge* batteries are manufactured with brown lead dioxide in the plates. To activate such cells it is necessary only to add electrolyte.

▌ Test your understanding; answer these checkup questions.

1. Why would transistor circuits not use A batteries? _____ C batteries? _____ Do they use B batteries? _____
2. Why does touching a fork to a tooth filling sometimes produce an electric shock? _____
3. What are the basic constituents of a Leclanché cell? _____ _____ _____
4. What forms around the + electrode that hinders current in a Leclanché cell? _____ What is this effect called? _____
5. Why does an inside-out cell not leak? _____
6. What is the fully charged voltage of a dry cell? _____ At what voltage is the cell discharged? _____
7. What is the polarity of the outer plate of a lead-acid cell? _____ Why? _____
8. Why do lead-acid cell caps always have vent holes? _____
9. What is the constituent of a charged lead-acid cell's + plate? _____ − plate? _____ Electrolyte? _____
10. What is the constituent of a discharged lead-acid cell + plate? _____ − plate? _____ Electrolyte? _____
11. What causes plates to buckle? _____
12. Why should cells be recharged as soon as possible? _____
13. What might be two causes of shorting in a lead-acid cell? _____ _____
14. What is affected if a lead-acid cell is overcharged? _____
15. If a lead-acid cell reads 1.220 sp gr, what should be done? _____

16. Does specific gravity read high or low on a cold day? _____ If the battery is charging? _____
17. What will counteract spilled battery acid? _____
18. What kind of water must be used in batteries? _____ What does this prevent? _____
19. In what unit is the capacity of a battery expressed? _____

28-8 CHARGING BATTERIES

When a current is forced backward through a lead-acid battery, the chemical action of discharge is reversed and the battery recharges.

A *constant-voltage* charging circuit is shown in Fig. 28-6. Note the positive-to-positive and nega-

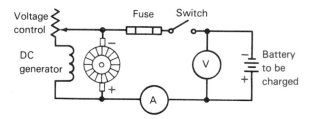

FIG. 28-6 A constant-voltage battery-charging circuit.

tive-to-negative connection of the battery and charging source. The dc generator has an output voltage 5 to 25% greater than the full-charge voltage of the battery, depending on the rate of charge desired. With a discharged battery, the differential between battery and charging voltages is great and relatively high current flows. As the battery charges, the differential between battery and charging voltages is smaller and less charging current is produced.

EXAMPLE: A three-cell, 6.3-V discharged storage battery has an open-circuit voltage of 1.8 V per cell and an internal resistance of 0.1 Ω in each of the cells. The potential needed to produce an initial charging rate of 10 A for this 5.4-V 0.3-Ω internal-resistance battery will be the voltage in excess of 5.4 that will produce 10 A through the 0.3 Ω, or

$$V = IR = 10(0.3) = 3 \text{ V}$$

Since

$$V_{bat} = 5.4 \text{ V}$$

$$V_{chg} = 3 + 5.4 = 8.4 \text{ V}$$

A *constant-current* charging circuit is shown in Fig. 28-7. In this circuit the charging source has a potential several times that of the battery. The charging resistor limits the value of current that will flow. With low resistance the charging current will be high. With high resistance the charging current

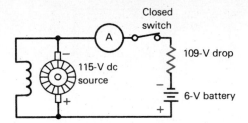

FIG. 28-7 A constant-current battery-charging circuit.

will be low. Since the difference of battery voltage between full charge and discharge is relatively small in comparison with the charging voltage, the charging current will decrease very little as the battery nears full charge.

EXAMPLE: If a 3-A charging rate is required, the charging resistor value is computed by Ohm's law. If the battery has 6 V and the source 115 V, the charging resistance must produce 115 − 6 V, or a 109-V drop across it when 3 A flows:

$$R = \frac{V}{I} = \frac{109}{3} = 36.3 \ \Omega$$

The minimum power rating of the charging resistor will be

$$P = VI = 109(3) = 327 \ W$$

A 500-W resistor would probably be used to allow a margin of safety. Note the excessive heat loss with this type of charging.

In an emergency, a 100-W 120-V lamp in series with a 6- or 12-V battery across a 120-V dc line will produce an approximate 1-A charging rate. (Use a high-voltage semiconductor diode in series with the lamp if the source is 120 V ac.)

If the polarity of a charger is reversed the battery will discharge very rapidly.

Some electronic battery-charging circuits (Fig. 28-8) use a multitapped transformer and a silicon diode half-wave rectifier. When the voltage differ-

ential between the ac voltage of the transformer secondary exceeds the voltage of the battery, current flows through the circuit, charging the battery. Diode rectifiers employed as battery chargers cannot discharge the battery if the output voltage falls below the battery voltage, since current cannot flow backward through a diode. (A dc generator used to charge a battery will operate as a motor if its output voltage drops below the battery voltage and will discharge the battery.)

Quick-charge 12-V battery chargers may use a starting current of 20 to 30 A. These chargers employ high-current dc generators or low-resistance solid-state (or gaseous tube) rectifiers in a full-wave center-tap or bridge-rectifier circuit.

Different values of charging current are used with batteries. A *trickle charge* is usually about one-hundredth of the ampere-hour rating. Thus, a 120-Ah battery may use a 1.2-A trickle charge. A *low charge* may be about one-tenth of the ampere-hour rating of the battery, while a *high charge* may be as much as one-half of the ampere-hour rating.

As charging progresses, the charging rate must be tapered off to prevent excessive heat and electrolysis (loss of water due to electrolytic action). When the battery becomes fully charged, there is no further chemical action between the plates and electrolyte but the water of the electrolyte continues to be driven off in the form of hydrogen and oxygen gas. A constant watch should be maintained to determine the amount of *gassing* produced. If no hydrometer is available, an indication of probable full charge is given by rapid gassing with a low charging rate. A trickle charge of several hours is usually beneficial at the end of a rapid charge.

Frequently it is possible to operate a receiver or a transmitter from a storage battery and charge the battery at the same time. In this case, the battery acts as a voltage-regulating device. The energy to operate the equipment is coming from the charging source, but if the charging source drops off, the battery feeds energy to the equipment. This is essentially what happens in mobile radio equipment. Unshielded leads from a battery to a transmitter should be kept as short as possible to prevent pickup of RF ac. Bypassing to ground at battery and transmitter reduces such pickup.

When constant operation is required at sea using batteries, it is possible to use one battery and charge another at the same time. When one becomes discharged, the other is switched in and the first is placed on charge (Fig. 28-9). If it does not add noise to the system, a single battery can be

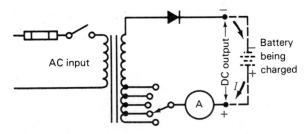

FIG. 28-8 Solid-state battery-charging circuit.

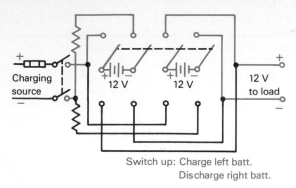

FIG. 28-9 Circuit to charge one battery while the other is being discharged.

Switch up: Charge left batt.
Discharge right batt.

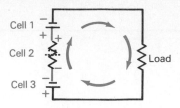

FIG. 28-10 Cell polarities in a discharging circuit with one dead cell.

"floated" across a charger that charges at a little more than the load current drain value.

If the polarity of a charging source is reversed, the battery will discharge and the voltage across the battery will read less than normal. The voltage across a charging battery will always be higher than its fully charged voltage.

With a constant-voltage charger, incorrect polarity across the battery will produce an extremely high current and any overload relay in the circuit will refuse to stay closed or a fuse will burn out.

If one of the cells of a battery becomes faulty, shorts out, and discharges itself, the cell then acts as a series resistance. When the battery is under a load, the voltage-drop across this cell will be found to be the reverse of its original polarity. Referring to Fig. 28-10, the dead cell is represented by a resistor. If it is remembered that current flows through a resistor from its negative terminal to its positive, the polarity will be as shown. A battery can produce very little output current with one dead cell.

A test for a dead cell is the gas produced in it by electrolysis when the battery is under a heavy load. With cell caps off, if a copper bar is held across the terminals of a 6-V storage battery that is known to be operating improperly and one of the cells gasses profusely, that cell is faulty.

There are several methods of determining the polarity of a charging line. When a voltmeter reads correctly across a line, the terminal of the meter marked positive will be connected to the positive line. When wires from the positive and negative terminals of a source of dc are held in a glass of salt water, the wire having the greater number of bubbles developed around it by electrolysis is the negative terminal. If two copper wires are stuck into a raw potato, a blue color will develop around the positively charged wire.

When batteries must supply power to remote equipment on top of mountains, etc., a solar-cell charger (Fig. 28-11) may be used to keep unat-

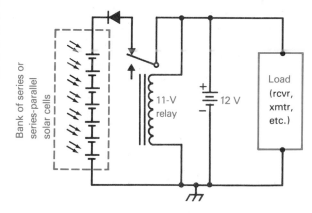

FIG. 28-11 Battery charger using solar cells.

tended batteries charged. With no sunlight the solar cells produce no voltage, but because of the polarity of the series diode, the solar cells in this case are disconnected from the battery. When activated by the infrared or visible frequency energy from the sun, the solar cells pick up emf. When this emf value plus the voltage-drop value across the diode exceeds the voltage of the battery (12 V), charging current starts to flow through the diode and battery. If the charging battery's voltage rises too high (due to overcharge), the sensitive 13-V relay pulls down and disconnects the solar cells. When the battery voltage decreases due to normal discharge, the relay arm is released and the solar-cell charger is reconnected across the battery. The relay arm spring tension setting determines at what voltage the relay arm moves back up.

28-9 MAINTAINING LEAD-ACID CELLS

There are several important points regarding the maintenance of lead-acid batteries:

1. Keep flames and sparks away from a charging or recently charged battery. The mixture of H and O given off during charge is explosive.
2. Be careful when using a hydrometer. Avoid spilling acid.
3. Keep the tops of the cells clean and free from moisture to prevent leakage across the surface of the cell top.
4. Keep the terminals of cells coated with petroleum jelly to prevent corrosion.
5. Keep cell tops on, or slightly loose, while charging to prevent electrolyte droplets from spraying out of the cell.
6. Maintain the proper electrolyte level.
7. Use only distilled water to replace lost water.
8. Take a hydrometer reading weekly.
9. Test-operate the battery at least once a month.
10. Trickle-charge unused batteries at least one full day a month.
11. Always bring batteries up to full charge after a discharge.
12. Use only chemically pure sulfuric acid (diluted) if necessary to add new electrolyte.
13. Provide adequate ventilation while charging.
14. If a battery box is used, clean the inside surfaces once a year to remove sulfuric acid droplets that may form during charging.
15. Do not overcharge a battery, as this causes the grids to deteriorate.
16. Batteries may be stored several months if first fully charged and then kept refrigerated (not frozen), or for several weeks if kept cool.
17. If necessary to store for a year or more, bat-

teries should be fully charged, the electrolyte removed, and the cells flushed with clear water and then filled with distilled water.
18. Plates and wooden separators must never be allowed to dry.
19. When removing caps, do not turn them over or place on an unclean surface.

28-10 EDISON CELLS

A lighter, more rugged secondary cell is the *Edison*, or *nickel-iron-alkaline*-type cell. When fully charged, it has a positive plate of nickel and nickel hydrate in small perforated nickel-plated steel tubes. It has a negative plate with iron as the active material in small pockets in a nickel-plated steel plate. The electrolyte is potassium hydroxide, a little lithium hydrate, and water.

While discharging, the electrolyte transfers oxygen, chemically, from the nickel-oxide positive plate to the iron negative plate, producing iron oxide (rust). When an Edison cell is recharged, the iron oxide is reduced to iron and the oxygen appears in the positive plate as a higher oxide of nickel.

The electrolyte has about 1.220 sp gr but does not change during charge and discharge, making a hydrometer useless. The voltage of the cell under load is the best indication of state of charge. An ampere-hour meter may also be used in the circuit to indicate state of charge.

The Edison cell has a fully charged voltage of approximately 1.4 V with no load. When a load is applied, the voltage decreases to about 1.3 V. As discharge progresses, the voltage drops off more or less linearly to about 1.1 V. When it reaches 1.0 or 0.9 V, the cell should be recharged. It can be completely discharged without damaging it. In fact, it can be recharged to the opposite polarity and will return to normal operation when charged properly. It may be stored for any length of time in a completely discharged condition without damage.

A lead-acid cell has a corrosive acid electrolyte. The Edison cell has a corrosive alkaline electrolyte, and with higher internal resistance for the same capacity, produces less current under load.

While charging, Edison cells produce H and O gases, and the electrolyte requires replenishing with distilled water. The electrolyte must never drop below the level of the plates. If the battery is trickle-charged, it will not reach full charge. If the electrolyte becomes contaminated, the capacity will be lowered. If the battery heats, the capacity decreases, but the current output increases.

To prevent electrolyte absorption of carbon dioxide from the air, the cells have one-way capped cell vents that stop outside air from entering.

28-11 OTHER CHEMICAL CELLS

Many other types of cells are available. Most of these have been developed in an attempt to improve on the older Leclanché cell.

Manganese-alkaline-zinc primary cells have a manganese-dioxide positive electrode, a zinc negative electrode, and a potassium hydroxide electrolyte. The nominal output is 1.5 V. These cells have low internal resistance, relatively high current capabilities, and a long shelf life, and they operate well at low temperatures. They find use in such applications as photoflash or small-motor power sources.

Mercury (or Ruben) primary cells have a cathode of mercuric oxide, an anode of zinc, and an electrolyte of potassium hydroxide and zincate. They are highly efficient and have a high capacity-volume ratio, a low internal resistance, a constant voltage under load, a long shelf life, and good high-temperature characteristics. They are made in 1.35- and 1.40-V no-load voltage ratings, operating at approximately 1.25 to 1.31 V under load.

Silver-oxide—alkaline-zinc primary cells have a silver-oxide cathode, a zinc anode, and a potassium (or sodium) hydroxide electrolyte. They have a flat 1.55-V characteristic under light load, are useful as a reference voltage source, and have a low internal resistance. They are used in hearing aids and, with the sodium-hydroxide electrolyte, in watches that operate continuously for a year or more.

Chargeable alkaline secondary cells are hermetically sealed units with manganese dioxide cathodes, zinc anodes, and potassium hydroxide as the electrolyte. They are relatively inexpensive and can be recharged many times. They have a nominal 1.5 V unloaded, and they may be overcharged without damage but should not be discharged below 0.9 V. The charging rate is about 2½ times the ampere-hours of discharge. They are called rechargeable dry cells.

Nickel-cadmium (NiCad) secondary cells are usually *sintered* (finely ground and then formed) plate devices with a nickel hydroxide cathode, a cadmium anode, and a potassium hydroxide electrolyte. They have an operating voltage of about 1.25 V and a low internal resistance, holding their voltage relatively constant until nearly discharged. They should be charged at 1.4 times the ampere-hours of discharge. They also are called *rechargeable dry cells*. These cells suffer from *memory effect*. If not allowed to discharge completely, they soon lose their full capacity. To prevent this, such cells must often be deeply discharged and then immediately recharged completely. This is important for batteries used in emergency communications, as in hand-held transceivers.

Polymer secondary cells are being developed that use thin films (polyacetylene or polyparaphenylene plastics) as positive and negative plates. The dielectric is a dopant that allows the polymers to become conductive, one plate accepting negative ions and the other positive ions when the cell is charged. When discharged the ions return to the electrolyte. Such cells may be about 15 times as light as lead-acid cells for the same power output. Voltage output may be 2.5 to 3.7 V.

28-12 NONCHEMICAL CELLS

There are a number of cells that do not operate on the principle of chemical reactions and are therefore neither primary nor secondary types. When light strikes the junction between a strip of selenium (a semiconductor) and iron (Fig. 28-12), an emf of

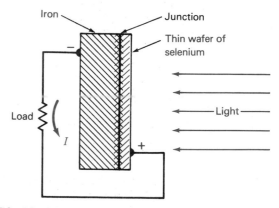

FIG. 28-12 Selenium or photovoltaic cell. The selenium is so thin that light energy penetrates to the junction.

about 0.4 V is developed between the two strips. This is the basis on which *photovoltaic*, or *selenium, cells* operate. When photons (light energy) strike the selenium, electron-hole (− and +) pairs are developed at the junction, resulting in a drift of electrons across to the iron. A single selenium cell may be useful as a photographic exposure meter. Many cells in series form a solar battery to deliver a small amount of power as long as they are illuminated. Their efficiency is only about 1%.

A practical *solar battery* is made of silicon cells in series. Each cell consists of two wafer-thin silicon strips, with the face of each infused with an opposite-polarity impurity. The strips are pressed together, forming a PN junction. When illuminated by sunlight, these junctions produce an emf, as do selenium cells, but they have an efficiency of nearly 10%. A square meter of cells make up a battery that can produce more than 50 W of power. Efficiency is expected to approach 20% in the future. Gallium arsenide solar cells can produce 1 V at 10 A per cell with an efficiency of nearly 20%.

Atomic, or *nuclear*, *cells* are somewhat similar to solar cells. A PN junction is bombarded by beta particles (electrons) from radioactive material painted on one side of the semiconductor wafers. Strontium-90 gives off an almost constant stream of high-velocity beta particles. In 20 years the intensity of the beta emission decreases by half. These cells produce about 0.3 V.

As previously explained, current flowing through an electrolyte chemically dissociates the hydrogen and oxygen of the water in the electrolyte. Hydrogen gas is given off at the negative electrode and oxygen at the positive electrode. This is called electrolysis. The reverse chemical process is also possible. That is, H and O gases can be made to combine, producing a current of electrons and water. This is the basis of operation of a *fuel cell* (Fig. 28-13), in which hydrogen gas under pressure is forced into the left chamber, which has a porous conductive electrode partitioning off the right side of its area. On the other side of this negative-pole electrode is a potassium hydroxide electrolyte. Oxygen gas under pressure is forced into the right chamber, which also has a porous conductive electrode wall between the oxygen and the electrolyte.

The potassium hydroxide electrolyte (KOH + H_2O) breaks up into positive potassium ions, K^+ (lacking an electron), and negative hydroxyl ions, OH^- (having one excess electron). These ions filter into the porous electrodes.

Because of the catalysts embedded in the electrodes, wherever the hydrogen gas contacts the surface of its electrode, it combines with the negative hydroxyl ion to form water, H_2O, plus one free electron (e^-).

Where the oxygen gas contacts the surface of its electrode, it combines with water and electrons returning from the load to form hydroxyl ions, OH^-, which travel into the electrolyte and over to the

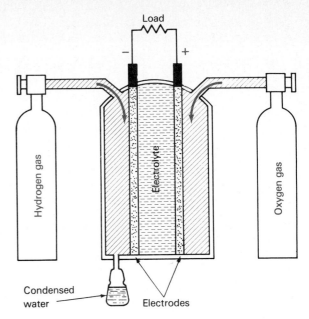

FIG. 28-13 Fuel cell uses hydrogen and oxygen as fuel.

negative (hydrogen) electrode. This completes the electrical system. Current continues to flow as long as hydrogen and oxygen are fed as fuel to the cell. A residue of pure condensed water is produced in the hydrogen system. The catalysts in the electrodes are usually platinum and palladium. Increasing the temperature of these cells increases their current output. At about 250°C, cells must be maintained under pressure but will produce about six times as much current as they do at room temperature. Other fuels, such as alcohol, gasoline, methane, kerosene, ammonia, and hydrazine, are also used but must be operated at high temperatures.

These cells can produce about 1000 A/m² of electrode area (100 A/ft²) at about 0.85 V, with about 75% efficiency. At one-third this load value the output of the cell is about 1 V.

Test your understanding; answer these checkup questions.

1. Which has the higher voltage, a constant-voltage or a constant-current charger? _____
2. How much current produces a trickle charge? _____ Low charge? _____ High charge? _____
3. What does rapid gassing of a charging cell indicate? _____ Of a cell in a heavily loaded battery? _____
4. What is wrong if the voltage-drop across a cell under load reads reversed? _____

5. What is another name for an Edison cell?
   _____

6. What is used to determine the state of charge of an Edison cell? _____

7. What is the fully charged voltage of an Edison cell? _____ Discharged voltage? _____

8. What can be used to counteract spilled electrolyte from an Edison cell? _____

9. Why do Edison cell caps have one-way vent holes?
   _____

10. What cell is used in wristwatches? _____

11. Name two "rechargeable dry cells." _____
    _____

12. Which two cells mentioned could be classified as solar cells? _____ _____

13. What are cells that depend on beta-particle bombardment called? _____

14. What two fuels are used in the fuel cell that was explained?_____ _____ What is the advantage of using these fuels? _____

15. List 19 points on maintaining lead-acid cells.
    _____

16. Does Fig. 28-9 show a constant-current or -voltage charger? _____

BATTERIES QUESTIONS

1. What is a primary cell? A secondary cell?
2. What is the composition of a Leclanché cell?
3. What are the three possible forms of dry cells or batteries?
4. What is meant by polarization of a dry cell?
5. Why do cells not produce full current if too cold?
6. What is the chemical composition of a charged lead-acid cell? Of a discharged cell? Its full-charge voltage?
7. What may happen to a lead-acid cell if not recharged?
8. What is the meaning of specific gravity? What is the specific gravity of a fully charged lead-acid cell? Of a fully discharged one?
9. What is the advantage of using a lower sp gr electrolyte in a lead-acid cell? Disadvantage?
10. What device is used to measure sp gr? What are its four parts?
11. How should spilled H_2SO_4 be treated?
12. Why should only distilled water be used in batteries?
13. In what unit is the capacity of a battery rated?
14. Does a battery have more capacity if discharged rapidly or slowly, or does rate have any bearing on capacity?
15. What is a dry-charge battery?
16. Which requires the greater voltage, a constant-current or a constant-voltage charger?

17. Draw a diagram of a battery charger capable of producing several different rates of charge.
18. What does rapid gassing indicate if a battery is being recharged?
19. Draw a diagram of a circuit to charge one battery while another is being discharged. What is another possible way that a constantly charged battery can be made available?
20. How can it be determined which cell of a battery is dead?
21. Draw a diagram of a solar-cell battery charger circuit.
22. List 19 precautions regarding recharging of lead-acid cells.
23. What is another name for an Edison cell? What is its fully charged voltage value?
24. What are manganese-alkaline-zinc cells noted for?
25. What is another name for a mercury cell? For what is it noted?
26. For what is a silver-oxide–alkaline-zinc cell noted?
27. For what are rechargeable alkaline dry cells noted?
28. For what is a nickel-cadmium cell known?
29. For what are selenium cells noted?
30. For what are atomic or nuclear cells noted?
31. How are fuel cells recharged? How can their output current be increased?

29

Radio Rules and Regulations

This chapter generalizes on some of the basic radio rules and regulations of the FCC as they pertain to commercial radio stations and operators. It outlines what commercial radio operator licenses (certificates) may be required for some types of radio operating. The material forms the basis for FCC license test Elements 1 and 2 required for Radiotelegraph licenses and for some of the questions in the commercial General Radiotelephone license. Amateur radio rules and regulations that may differ from these are discussed in Chap. 32.

29-1 RADIO RULES AND REGULATIONS

To prevent intolerable interference, radio stations throughout the world operate according to agreements set up at international communication meetings. In 1934, the United States combined laws of many previous acts and agreements into the Communications Act of 1934. The Communications Act set up general laws to be followed in the United States that would coincide with communication agreements with other countries, such as the International Telecommunications Union (ITU). To execute and enforce the provisions of this act, the Federal Communications Commission was constituted. The FCC has developed a series of rules and regulations for different types of communications services. The first seven volumes contain information of general radio communication interest and are outlined in Appendix F. Each part is subject to change as the art of radio progresses in its particular field. It may be found that a rule regarding broadcast station operation last year may not exist or may be changed this year. It is important that owners and operators of radio systems keep themselves informed about the rules and regulations referring to their particular service.

Anyone who willfully does anything prohibited by the Communications Act or knowingly omits to do anything required by the act is subject, upon conviction, to a fine of not more than $10,000 or imprisonment for a period of 1 year on the first offense and 2 years on the second offense.

Anyone willfully violating any *rule, regulation, restriction,* or *condition* set up by the FCC by authority of the Communications Act or by international treaty to which the United States is a party is subject, upon conviction, to a fine of not more than $500 for each and every day during which such offense occurs.

One of the many functions of the FCC is the issuance of operator and station licenses to those qualifying for them. This requires the administration of operator examinations. The FCC also has authority to and does make inspections of licensed United States radio stations of all types whenever necessary to assure operation in accordance with FCC rules and regulations.

Messages transmitted by radio are subject to secrecy provisions of law. No persons receiving such messages, except amateur, distress, broadcast, and messages preceded by CQ—"attention all stations"—may divulge their content to anyone but the legal addressee, his agent, or attorney, nor may they use information so gained to their advantage.

29-2 OPERATOR-LICENSE ELEMENTS

All commercial radio operator license or certificate examinations given by the FCC will be composed of one or more of the following "elements":

- *Element 1: Basic Law*. Twenty questions on provisions of laws, treaties, and regulations with which every radio operator should be familiar (covered in this chapter).
- *Element 2: Basic Operating Practice*. Twenty questions on marine radio operating procedures and practices in communicating by means of radiotelephone stations (covered in this chapter and Chap. 30).
- *Element 3: General Radiotelephone*. One hundred questions on technical, legal, and other matters applicable to the operation of radiotelephone stations other than broadcast (most chapters other than those on BC and TV).
- *Element 5: Radiotelegraph Operating Practice*. Fifty questions on operating procedures and practices for radiotelegraph stations (Chap. 31).
- *Element 6: Advanced Radiotelegraph*. One hundred questions on technical, legal, and other matters applicable to the operation and practices of all radiotelegraph stations, radio navigational aids, message traffic routing, accounting, etc. (most chapters except those on BC and TV).
- *Element 8: Ship Radar Techniques*. Fifty questions on specialized theory and practice for the installation, servicing, and maintenance of ship radar equipment (Chaps. 22 and 26).

License test questions are multiple-choice types.

29-3 LICENSES AND PERMITS

All United States commercial and amateur radio stations must be licensed by the FCC. Some must be operated by FCC-licensed operators or permit-

tees. (Exclusions: armed-service and government stations and some special communications handie-talkie equipment of 100 mW or less.) For amateur requirements, see Chap. 32.

Any device that radiates RF energy which can be detected and used for more than a few hundred feet must usually be either licensed or type-accepted by the FCC before it can be operated.

The various classes of commercial radio *operator* licenses (certificates) or permits for which the FCC gives tests are:

1. General Radiotelephone Operator license (Element 3, which includes a few Element 1- and 2-type questions).
2. First Class Radiotelegraph Operator certificate (Elements 1, 2, 5, 6, plus code tests of 20 groups per minute and 25 wpm plain language).
3. Second Class Radiotelegraph Operator certificate (Elements 1, 2, 5, 6, plus code tests of 16 groups per minute and 20 wpm).
4. Third Class Radiotelegraph Operator certificate (Elements 1, 2, 5, plus code tests of 16 groups per minute and 20 wpm).
5. Marine Radio Operator permit (Elements 1 and 2).
6. Restricted Radiotelephone Operator permit (ability to send and receive messages in English, be familiar with FCC laws, and be able to keep a log).

The following license endorsements may be added to FCC licenses or certificates:

1. Ship Radar endorsement (on 1st or 2nd Radiotelegraph, or on General Radiotelephone).
2. Six Months Service endorsement (on 1st or 2nd Radiotelegraph to allow licensee to be sole operator on ships).
3. Restrictive endorsements (physical handicaps, language, etc.).

Radiotelegraph licenses are required to operate and service internationally operating merchant marine type radiotelegraph or radiotelephone equipment. General Radiotelephone licenses are required to service Maritime, Aviation, and International Public Fixed Radio Services. A Marine Radio Operator permit is required for persons using maritime radiotelephone equipment, such as captains of fishing vessels, etc. Restricted Radiotelephone Operator permits are required to operate radiotelephone communications transmitting systems ashore.

No operator license is required to operate any standard AM, FM, or TV broadcast stations, any mobile radio services, or any cable TV services. It is up to the management of such services to make sure that their equipment is operated according to FCC rules and regulations, and can demonstrate this if requested to by an employee of the FCC.

All licenses and permits require an ability to transmit and receive spoken messages in English and are issued to citizens of the United States, or to anyone legally eligible for employment in the United States.

To obtain a commercial license it is necessary to take a test at one of the field engineering offices of the FCC, located in many of the larger cities of the country. An application form 756 must be filled out before the license examination will be given. Each office sets specific days in February, May, August, and November for license examinations. This information should be obtained by phone or mail before appearing at the nearest office for an examination (Appendix F).

If when taking any commercial license examination an element is failed (less than 75%), no test containing that element may be taken for 60 days. For Amateur licenses, 30 days.

Radiotelegraph licenses are issued at present for a period of 5 years. To renew a license, a renewal application may be obtained by mail and filled out. The original license and the renewal application may then be presented to the nearest office in person or by mail, within the last year of the license term. During the period when the license and the renewal application are in the mail or at the FCC office, the operator may continue operating by posting an exact, signed copy of his renewal application in lieu of the original license or permit. Operators who do not renew their licenses before expiration may apply for renewal up to 5 years after the license has expired, but during that time they have no operating authority until the renewed license is issued.

An operator's license normally must be posted at the place where the operator is on duty. If working at two different stations, the operator must post their license or permit at one station and a *verified statement* (FCC Form 759) at the other.

A licensee or permit holder may apply for a *verification card* (FCC Form 758-F). This card may be carried on the person of the operator in lieu of the original license or permit when operating any station at which posting of an operator license is not required.

If a license or permit is lost, mutilated, or destroyed, the FCC should be notified and a duplicate requested. If the license is lost, the application for duplicate should state that reasonable search has been made and that, if the original is found, either the original or duplicate will be returned for cancellation. A signed copy of the application for duplicate license must be used in lieu of the license until the duplicate is issued.

When a licensee qualifies for a higher-grade FCC license (or permit), the lower-grade license must be turned in to the FCC to be marked "canceled." The canceled license is then returned to the licensee.

29-4 SUSPENDED LICENSES

The FCC has authority to suspend licenses upon proof of violations of acts, treaties, or FCC rules; failure to carry out a lawful order of the master or person lawfully in charge of a ship or aircraft; willfully damaging or permitting radio apparatus to be damaged; transmitting superfluous radio communications or signals, obscene language, false or deceptive signals, or call signs not assigned by proper authority; willfully or maliciously interfering with any radio communications; obtaining or helping anyone to obtain an operator license by fraudulent means.

Such a suspension becomes effective 15 days after the licensee receives the notice of suspension. In this 15-day period the licensee may make application for a hearing on the suspension order. The suspension will be held in abeyance until the conclusion of the hearing, and the licensee may continue to operate until that time at least.

29-5 NOTICES OF VIOLATIONS

The FCC maintains monitoring stations. Listening operators check all receivable signals. If a radio station violates any provision of the Communications Act or FCC rules and regulations, it will be served with a *notice of violation* calling the facts to the station's attention and requesting a written statement concerning the matter within 10 days (unless another period is specified). The answer must be addressed to the office of the FCC originating the notice and be a full explanation of what occurred, what steps have been taken to prevent future violations or what new apparatus has been or will be installed, and the name and license number (if any) of the operator. If an answer cannot be sent or an acknowledgment made within the required period by reason of illness or other unavoidable circumstances, acknowledgment and answer must be made at the earliest practical time, along with an explanation of the delay.

29-6 WHO MAY OPERATE TRANSMITTERS

In general, any system that transmits International Morse code requires an operator holding a Radiotelegraph license to both operate or service it. Radiotelephone transmitters used for any international communications require a General Radiotelephone license to service the equipment, although aboard ship (and for some other radiotelephone applications) a Radiotelegraph license may be used to operate, maintain and service radiotelephone transmitters. International broadcast stations require General Radiotelephone license operators to operate, maintain and service the transmitters, but anyone can use the microphones. Aviation communications transmitters require General Radiotelephone licenses to maintain and service them.

29-7 CLASSIFICATIONS OF RADIO COMMUNICATIONS

Normal transmissions from radio stations may be classified as *routine*. Transmissions made during times of emergency have a higher priority: (1) *distress*, (2) *urgency*, (3) *safety*.

A mobile station in *distress* is in need of immediate assistance. By radiotelegraphy the distress signal is the well known $\overline{\text{SOS}}$. By radiotelephone the distress signal is the word MAYDAY (from the French *m'aider*), usually transmitted three or more times to attract attention. A distress message should contain the name of the station, particulars of its position, nature of the distress, kind of assistance desired, and any other information which might facilitate aid. The mobile station in distress is responsible for the control of distress-message traffic. However, if the station in distress is not itself in a position to transmit the distress message or if the master of a station observing the one in distress believes that further help is necessary, then the observing station can send a distress message. Distress messages are not subject to the secrecy provisions of law, as are most other radio communications. Any false or fraudulent signals of distress are prohibited by law and punishable by a fine of up to $10,000 and up to 1 year in prison.

Radio messages with an *urgency* classification refer to a situation that requires immediate attention and might conceivably become distress in nature. By radiotelegraphy the urgency signal is the transmission of the three letters XXX. By radiotelephone the urgency signal is the spoken word PAN repeated three or more times.

Radio communications with a *safety* classification refer to meteorological information, storms, hurricanes, etc., or to other navigational warnings. By radiotelegraphy the safety signal is TTT. By radiotelephone the safety signal is the word SECURITY (from the French *sécurité*).

The 1959 Geneva Treaty designates any emission which endangers the functioning of a radio navigation or other safety service, or degrades, obstructs or interrupts a radio communication service as *harmful interference* and subject to legal action.

29-8 RADIO LOGS

Most communication systems are required to keep logs of their transmissions. Log entries normally show the station, date, time, operator on duty, station with which the communication was carried on, and an indication of what communications occurred or what traffic was handled.

Logs are made out by those legally competent to do so. If an error is made in a log, it may not be erased. The error should be struck out, the correct entry made above the error, and then the correction initialed and dated by the operator who made the original entry. Logs are usually required to be kept for at least 1 year. If they contain information pertaining to distress traffic, they should be kept for at least 3 years. If they contain information pertaining to an FCC investigation they must be kept until authority to destroy them is received in writing.

Logs are also discussed in Chaps. 19, 23, 24, 30, 31, and 32.

Test your understanding; answer these checkup questions.

1. Is the Communications Act of 1934 a national or international set of laws? _____
2. What is the penalty for violating a rule, regulation, restriction, or condition set up by the FCC? _____ Of willfully doing anything prohibited by the Communications Act? _____
3. What four communications are not subject to secrecy provisions? _____ _____ _____ _____
4. How long should logs be held? _____

5. What are the four FCC operator licenses?

_____ _____ _____ _____
The permit? _____
6. What is the lowest passing grade for any FCC element? _____
7. Which FCC elements are endorsements for licenses? _____
8. If an FCC commercial license test is failed, how long must the applicant wait before retaking? _____
9. Is it possible to hold two grades of radiotelegraph licenses simultaneously? _____ A telephone and telegraph simultaneously? _____

10. If an operator is served with a suspension, when is the suspension effective? _____
11. Can a First Class telegraph license holder tune or operate an amateur station? _____
12. What is the priority order of transmissions of messages? _____ _____ _____
13. What are the telegraph indicators of the three highest priority types of messages? _____ _____ _____ The telephone? _____ _____ _____
14. Under what condition may incorrect entries in a radio log be erased? _____
15. What is an FCC form 756? _____

RADIO RULES AND REGULATIONS QUESTIONS

1. What is the organization that determines and polices national radio policies? The international organization?
2. What is the penalty for anyone willfully failing to act according to the Communications Act?
3. What is the penalty for willfully violating FCC or ITU rules, regulations, restrictions, or conditions?
4. What transmissions are not subject to secrecy provisions?
5. Name the six FCC test elements.
6. Name the four FCC operator licenses and indicate what elements are in the test.
7. What are three endorsements that may be added to operator licenses?
8. In general, what radio equipment requires FCC licensed operators?
9. Who is eligible to obtain FCC operator licenses or permits?
10. What are the age restrictions on FCC operator licenses?

11. In what months are commercial license tests given at FCC offices?
12. What is the passing score for any FCC license element test?
13. For what term of time are commercial licenses issued?
14. What must be done if a license is lost or mutilated?
15. List the grounds for suspension of operator licenses.
16. When does a suspension go into effect? A notice of violation?
17. What does a Radiotelegraph First Class license entitle the licensee to operate aboard ship?
18. List the four classifications of radio communications and their identifiers, if any.
19. What is harmful interference?
20. What is usually shown on radio logs?
21. Who makes out logs?
22. How long are logs normally kept? If they contain distress traffic? If they contain information regarding an FCC investigation?

30

Radio-telephone Communication

This chapter presents some basic radiotelephone operating practices and procedures. The use of microphones and operational words, and the makeup of informal and formal voice messages, are outlined. This chapter should answer questions asked in the FCC Radiotelegraph Element 2 tests and some questions in the Radiotelephone Element 3 tests.

30-1 RADIOTELEPHONE OPERATION

Radiotelephone communications take place from ship to ship, ship to shore, land station to land station, mobile to mobile, mobile to land station, aircraft to ground, and so on.

Police, fire, and taxi services represent an informal type of radio communication. Mobiles talk to their base station or to other mobile units of the same system. The base station may be on one frequency, the mobiles on another, base and mobiles may be on the same frequency, or each station may have two or more frequencies on which it may operate. The base station has an assigned call, such as KMA539, but the mobile units may be known only as "car 216" or some such designation. Logs of communications are kept at the base station only.

Many commercial vessels, fishing boats, and pleasure craft use the 2000–3000 kHz or the 156–162 MHz VHF bands for radiotelephonic communications ship-to-ship or ship-to-shore. These services may use a more formal marine procedure, each station maintaining watch on the calling frequency, 2182 kHz or 156.8 MHz.

Every licensed station is assigned a *call* to distinguish it from all others. The call may consist of letters only or a combination of letters and numbers. When signing a call, the operator should do so clearly and distinctly.

Since there are innumerable mobile and fixed stations communicating every day, it is important that some logical procedure be followed to reduce confusion and interference. Some good operating procedures are outlined in this chapter.

30-2 MESSAGE PRIORITIES

Urgent messages take preference over routine traffic. Message priorities are

1. Distress calls, messages, and traffic
2. Communications preceded by the urgency signal
3. Communications preceded by the safety signal
4. Communications about radio direction finding

ANSWERS TO CHECKUP QUIZ ON PAGE 682

1. (National) 2. (Up to a $500 fine)(Up to a $10,000 fine and 2 yr imprisonment)3. (Amateur, distress, CQ, BC) 4. (1 year) 5. (1st, 2nd, and 3rd Telegraph, General phone)(Marine) 6. (75%) 7. (8)8. (60 days) 9. (No)(Yes) 10. (15 days if not appealed)11. (Only if person holds proper amateur license) 12. (Distress, urgent, safety, routine) 13. (SOS, XXX, TTT)(MAYDAY, PAN, SECURITY) 14. (None) 15. (Request to take an exam)

684

5. Messages relative to the navigation of aircraft
6. Messages relative to navigation, movements and needs of ships, and official weather messages
7. Government messages for which priority has been claimed
8. Service messages relating to previously transmitted messages or to the operation of the communication service
9. All other routine communications

30-3 DISTRESS

A distress call has absolute priority. Stations hearing it must immediately cease any transmissions capable of interfering with it and must listen on the frequency used for the distress call. A *distress call* sent by radiotelephony consists of

1. The *distress signal* MAYDAY spoken three times
2. The words THIS IS, followed by the identification of the mobile station in distress, the whole being repeated three times

The *distress message* must follow the distress call as soon as possible and should contain

1. The distress call
2. The name of vessel, aircraft, or vehicle in distress
3. Position, nature of distress, and kind of assistance desired
4. Any other information which might aid rescue

The position is given in latitude and longitude or by bearing in degrees from true north and distance in miles from a known geographical point.

An operator hearing a distress call must listen on the distress frequency for the distress message and should acknowledge receipt of the distress message if the station in distress is in the vicinity. An acknowledgment by radiotelephone might be (1) call of the station in distress; (2) the words THIS IS, followed by the call of the acknowledging station three times; (3) the words ROGER YOUR MAYDAY MESSAGE; (4) the word OUT.

Every mobile station that acknowledges receipt of a distress message must, on the order of the person responsible for the ship, aircraft, etc., transmit as soon as possible (1) its name, (2) its present position, and (3) the speed at which it is proceeding to the station in distress.

A station must ensure that it will not interfere with emissions of other stations better situated to render assistance.

Any operator in the mobile service who has knowledge of distress traffic must follow such traffic, even if taking no part in it. For the duration of the distress traffic no station must use the distress frequency for other types of calls or traffic. If a transmitting operator is told that he or she is interfering with distress traffic, the operator must cease transmitting immediately and listen for distress signals.

An operator too far away to assist in a distress must take all possible steps to attract the attention of stations which might be in a position to render assistance.

Stations not directly involved in distress traffic may continue normal service on frequencies that will not interfere with the distress traffic after the distress traffic is well established.

Any land station receiving a distress message must, without delay, advise authorities who might participate in rescue operations. The land station must maintain silence on the distress frequency unless involved in the distress traffic. If it appears that the distress call and message have not been acknowledged, all steps should be taken to attract the attention of stations in position to render assistance.

If a radio watch is required on a distress frequency, the receiver must be tuned to this frequency as soon as traffic has been completed on other frequencies.

When distress traffic is ended, an announcement must be made on the distress frequency, such as (1) the words MAYDAY ALL STATIONS, three times; (2) the words THIS IS, followed by the call letters of the station transmitting the message; (3) the name of the station in distress; (4) the words DISTRESS TRAFFIC ENDED. OUT.

30-4 URGENCY SIGNALS

The *urgency signal* by radiotelephone is PAN, spoken three times, followed by the words THIS IS and the call of the transmitting station. It indicates an urgent message (one which concerns the safety of a ship or person but is not quite of distress priority).

Mobile stations hearing the urgency signal must continue to listen for at least 3 min. If no urgency message is heard, they may then resume their normal service. Operations under urgency conditions are similar to distress traffic.

30-5 SAFETY MESSAGES

The *safety signal* by radiotelephone is SECURITY, spoken three times, followed by the words THIS IS and the call of the transmitting station three times. It indicates a storm warning, danger to navigation, or other navigational-aid message is to follow.

All stations hearing the safety signal must continue to listen on the frequency until they are satisfied that the message is of no importance to them. They must make no transmissions likely to interfere with the message.

A *safety communication* is not necessarily the same as a message following the safety signal. A safety communication pertains to any distress, urgency, or safety messages which, if delayed in transmission or reception, might adversely affect the safety of life or property. Stations handling paid radio messages cannot charge for forwarding any safety-communication messages.

30-6 INTELLIGIBILITY

Communication by radiotelephone may be hampered by static, fading, interference by other stations, noise in the receiving room, noise picked up by the transmitting microphone, unusual voice accents, colloquialisms, improper enunciation or pronounciation of words, and by rapid speaking. To improve intelligibility a transmitting operator should speak slowly and clearly, using well-known words and phrases and simple language. Unusual or important words should be repeated or spelled out if it is known that the receiving operator is experiencing any difficulty in reception.

Speaking too far from the microphone may result in weak, hard-to-hear signals. Shouting into the microphone produces a distorted output signal that may be difficult to understand. Most communication microphones are constructed for close talking but in a normal tone of voice. If there is considerable local noise, it may help intelligibility to cup the hands around the microphone and speak directly into it in a moderate voice. Directing the front of the microphone away from noise sources helps.

Distortion of the voice is produced by fading signals and by improper functioning of the transmitting circuits. In the latter case, the fault must be found by a licensed operator or serviceman. In many cases, a distorted transmission is readable if the operator speaks slowly and distinctly.

30-7 PHONETIC ALPHABET

When words are spelled out in a radiotelephone communication, confusion can result because many letters sound similar unless clearly heard. For ex-

ample, the letters B, C, D, E, G, P, T, V, and Z all have the same *ee* ending sound. The word *get*, when spelled out, might be copied as b-e-t, p-e-t, b-e-d, etc. To prevent this confusion each letter of the alphabet may be represented by a well-known word. Thus, *golf* for G, *echo* for E, and *tango* for T might be used. When the word *get* is spelled out, using this phonetic alphabet it is spoken "golf echo tango." The receiving operator writes down the first letter of each word and receives "get."

There have been many phonetic alphabets in the past, using names, cities, and other words. Unfortunately, non-English-speaking people mispronounced English words so confusion still resulted. An international phonetic alphabet using words familiar in most languages is shown in Table 30-1.

30-8 OPERATIONAL WORDS

In the course of radiotelephone operation many special words or phrases have a definite meaning and are desirable to use, such as:

Words	Meaning
Roger	I received your message.
Over	I have completed transmitting and await your reply.
Go ahead	Same as *over*.
Out	I have completed my communication and do not expect to transmit again.
Clear	I have no further traffic. (Sometimes used in place of *out*.)
Stand by	Wait for another call or further instructions.
Break	I am changing from one part of the message to another [address to text . . .]. (Also used to request the receiving operator to indicate if he has received the portion of the message transmitted thus far.)

| TABLE 30-1 | INTERNATIONAL PHONETIC ALPHABET |||

A	Alpha	J	Juliet	S	Sierra
B	Bravo	K	Kilo	T	Tango
C	Charlie	L	Lima	U	Uniform
D	Delta	M	Mike	V	Victor
E	Echo	N	November	W	Whiskey
F	Foxtrot	O	Oscar	X	X-ray
G	Golf	P	Papa	Y	Yankee
H	Hotel	Q	Quebec	Z	Zulu
I	India	R	Romeo		

Words	Meaning
Words twice	Transmit each word or phrase twice, or, I will transmit each word or phrase twice.
Read back	Read the message back to me.
I spell	I will spell [usually phonetically] the word I just said.
Say again all after	Repeat all words transmitted after . . . [give last correctly received word].
Say again all before	Repeat all words transmitted before . . . [give first correctly received word].
Say again . . . to . . .	Repeat all words transmitted from . . . [word before missing portion] to . . . [word after missing portion].

30-9 CALLING AND WORKING

Station KBBB wishes to communicate with station KAAA. KBBB transmits on a frequency known to be monitored by KAAA, saying,

KAAA KAAA KAAA   THIS IS   KBBB KBBB KBBB   OVER

The called station answers, saying,

KBBB   THIS IS   KAAA   OVER

Note that only on the first call is it desirable to call and sign three times. The fewer calls the better.

A formal type of message from KBBB to KAAA might be transmitted as follows:

KAAA   THIS IS   KBBB   MESSAGE NUMBER ONE FROM NEW YORK MAY ONE NINETHIRTY A.M. BREAK MACPHERSON UNIT TWENTYTHREE   BREAK   ADVISE EXPECTED ARRIVAL NEW YORK   BREAK SIGNED WILLIAMS   BREAK   (or END OF MESSAGE)   THIS IS   KBBB   OVER

The operator at KAAA should copy the message in a form similar to:

KBBB 1 NEW YORK MAY 1 1985 0930 ⟍
MACPHERSON ⟩   Preamble
   UNIT 23 ⟨ Address
ADVISE EXPECTED ARRIVAL NEWYORK ⟋   Text
     Signature → WILLIAMS
     Service   → 0950   LC

Note the terminology for the different parts of the message. The service indicates the time the operator (initials LC) received the message.

If a "check" count is used, the total number of words in the address, text, and signature are usually counted. In the message above, the preamble would read ". . number one check eight from . ." etc. (The check in amateur messages includes only the words in the text.)

The receiving operator acknowledges the message and also signs out:

KBBB   THIS IS   KAAA   ROGER YOUR MESSAGE NUMBER ONE   OUT

After contact has been established, continuous two-way communication is usually desirable without identification on each transmission until termination of the contact (unless of 15-min duration or more).

A mobile station calling a particular station by radiotelephone must not continue for a period of more than 30 s in each instance. If the called station is not heard to reply, it should not be called again until after an interval of 1 min (emergencies excepted). A coast station must not call for more than 1 min and must wait 3 min between calls.

If it is desired to call a vessel within sight when its identity is not known, an operator may call on 156.8 MHz, saying, in effect, "Calling the green, two-masted yacht passing Point Conception (3 times), this is WMBD WMBD WMBD. Over." If the yacht hears, it should answer on the same frequency.

In heavy radio traffic, the duration of communications between two stations should not exceed 5 min (excluding distress or emergency communications). Calls and messages should be spaced to allow other stations to communicate.

30-10 RADIOTELEPHONE STATION IDENTIFICATION

Both mobile and fixed stations should identify with their FCC assigned call letters in English.

If a mobile station has no assigned call letters, it must use its full name, transmitted in English, as "Yacht Cleopatra." Coastal stations may identify themselves by location, as "Washington marine operator," if this has been approved.

Mobile or coastal stations should identify at the beginning and end of communications as well as at the beginning and end of tests. Mobiles must transmit their call letters at intervals not exceeding 15 min whenever transmission is sustained for a long period. Stations in the Public Safety Radio Service (fire, police) transmit identifying calls at the end of each exchange of transmissions, or at 30-min intervals, as the licensee prefers.

30-11 GOOD OPERATING PRACTICES

Before making a call, or testing, an operator should monitor the transmitting frequency to ensure that interference will not be caused to communications already in progress. An operator should never transmit on an unmonitored frequency for the same reason.

To allow maximum use of a frequency, all communications should be as brief as possible and no unnecessary calls and transmissions should be made. Transmitters and receivers should be capable of changing frequency rapidly and should be in constant readiness to make or answer a call.

With routine-type messages, if receiving and transmitting conditions are poor (static, fading, interference) it may be better to wait for improved conditions rather than tie up a frequency with slow-moving nonemergency traffic.

A radiotelephone transmitter should be tested at least once a day to assure proper operation of the equipment. If no other communications are made during a day, a suitable test may consist of turning on the transmitter briefly, saying, "This is [call letters] testing." If all meters indicate normal values, it is assumed the transmitter is operating properly. However, an operator should not press the push-to-talk switch on a transmitter except when intending to speak into the microphone. Radiation from AM or FM transmitters may cause interference even if voice signals are not being transmitted.

An operator must always operate a station according to the provisions of the station license except during distress or emergency.

When two or more groups of stations are sharing the use of one frequency or channel, it is good practice to leave an interval between transmissions in case one of the other sharing stations desires to transmit emergency traffic.

It must be remembered that radiotelephone transmissions may be received by many unauthorized persons and are not confidential. It is sometimes necessary to choose the phrasing of messages carefully to attain a desired secrecy of meaning.

Anyone may use the microphone of a marine transmitter. However, if a licensed operator is responsible for transmissions made by a station, if anyone uses obscene language, it is that operator's duty to stop the transmission immediately.

Operators must prevent unauthorized use of transmitting equipment. Transmitters in mobile stations in a public place must not be left unattended. The transmitter should be turned off when an operator is away from a car.

The license of the operator of a radiotelephone station should be posted in plain view in the operating room. The 1959 Geneva Treaty states that this license will be available to authorities of any country in which the ship or vehicle is operating.

In all commercial communication services, a log must be kept of all transmissions made, tests or maintenance performed, messages handled, and distress, urgency, and safety messages heard. Each sheet is to be serially numbered and signed by the operator or person authorized to do so. Logs for international voyages are kept in GMT or universal coordinated time (UTC); others may be in local time or local 24-h time (1:30 P.M. being 1330 h).

30-12 CALLING AND WORKING FREQUENCIES

In the Maritime Mobile Radio Service *calling* and *working* frequencies are used. Original contacts between two stations are made on a calling frequency, such as the international radiotelephone distress and general calling frequency of 2182 kHz. This frequency is reserved for short calls and answers or for distress, urgency, or safety communications. However, it may be used for short tests of equipment or to broadcast short lists of stations for which a coastal station has traffic.

Normally, as soon as contact between two stations has been accomplished on the calling frequency, operations are shifted to a correspondence channel (or "working frequency") on which all routine-type messages must be transmitted. Communications in the 2000–3000 kHz range must be safety traffic, never social or personal messages.

To contact a coast station (KSA) for routine traffic, the operator (of WINB) would first check the working frequency of KSA and then call on a calling frequency, such as 2182 kHz

KSA KSA KSA   THIS IS   WINB WINB
WINB   ANSWER ON . . . . . (Frequency)
OVER.

As soon as KSA has acknowledged the call, WINB shifts to the frequency designated and completes the communication on that frequency.

The frequency 156.8 MHz is the short-range international radiotelephone frequency for calling, safety, intership, and harbor-control purposes for the maritime mobiles in the 156–162 MHz band.

The United States Coast Guard monitors the maritime calling frequencies (2182 kHz and 156.8 MHz for radiotelephone, 500 and 8364 kHz for radiotelegraph). The call NCU means "Calling all Coast Guard stations." For example, "NCU NCU NCU THIS IS WINB WINB WINB OVER" on 2182 kHz, is a call to any Coast Guard station to answer on that frequency. For distress traffic only, the Coast Guard can be called on the government frequency, 2670 kHz.

Ship stations licensed for radiotelephone operation only in the 1600–3500 kHz band must maintain an efficient watch on 2182 kHz when not in operation on another frequency.

Use of the calling frequencies (2182 kHz and 156.8 MHz) should be as brief as possible, in no case over 3-min duration for one exchange of communication (emergency traffic excepted).

30-13 ANTENNA-TOWER LIGHTS

Radio transmitting or receiving towers on land high enough above surrounding terrain to be considered dangerous to aerial navigation, are required to carry one or more warning lights on them. These lights may burn steadily or be rotating beacons. For further information, see Sec. 23-25 and "FCC Rules and Regulations," Volume I, Part 17.

30-14 COAST STATIONS

Coast stations must maintain an accurate log during their hours of service of the following items: "on duty" and "off duty" entry of the operator on watch; the call signs of all stations worked and the time in UTC (except on inland waters, where a local standard time may be used); any interruptions of the watch, including reasons and when the watch was resumed; all distress, urgency, and safety messages intercepted or transmitted copied in full into the log; all tests made; a daily comparison of the required station clock or clocks with standard time signals; measurements of transmitter frequency; service or maintenance work performed on the transmitter; and entries on antenna-tower lights. Log sheets are numbered in sequence, dated, carry the call sign of the station, and are signed by the operator(s) on duty. (*Signature* indicates a minimum of first initial and last name, written, not printed.) Similar logs ere also kept by compulsorily radio-equipped ships.

A coast station communicates with ship or aircraft stations, and with other coast stations only to facilitate communications with ship stations. Except for safety communications and short calls on calling frequencies, a coast station operates on its assigned working frequency as much as possible.

Most voice communications are of a private type (Private Land Mobile Radio Service, etc.). When paid messages are handled for the general public, as ship-to-shore or shore-to-ship, the stations are said to be *open to public service*.

▌ Test your understanding; answer these checkup questions.

1. List the three top priority messages in order. _____ _____ _____ What is the lowest-priority message? _____
2. When should a radiotelephone distress message be sent? _____
3. What should a distress message contain? _____
4. What should an operator do if he hears a distress message in the area? _____ If out of the area? _____
5. How would KAAA conclude distress traffic? _____
6. Which of these indicates a safety communication to follow: PAN, SECURITY, MAYDAY? _____
7. For best communication, how close should an operator's mouth be from the microphone? _____
8. How would the name "Dean" be spelled phonetically? _____
9. In formal messages, what is the preamble? _____ Text? _____ Address? _____ Service? _____ Signature? _____
10. Basically, when should stations identify? _____ _____ _____
11. When and how often should a radiotelephone transmitter be tested? _____ _____
12. Where should operator licenses be posted? _____
13. After communication is established on a calling frequency, to what frequency should the stations move? _____
14. What are the two international radiotelephone calling frequencies? _____ _____
15. What is a coast station? _____
16. Where is complete information regarding antenna lighting found? _____
17. Where might an operator listen for a coast station for routine-type traffic? _____ For emergency traffic? _____

1. Who keeps the log in services using an informal type of communication?
2. What are the two radiotelephone calling frequencies?
3. List the nine message-type priorities in commercial communications.
4. What is included in a radiotelephone distress call?
5. What is included in a radiotelephone distress message?
6. What are two ways of specifying a ship's position?
7. How does a receiving station acknowledge a radiotelephone distress call?
8. After acknowledging a distress call, what should nearby ships transmit?
9. How is a distress incident terminated on the air?
10. What is the urgency signal by radiotelephone? What does it indicete?
11. What is the safety signal by radiotelephone? What does it signify?
12. What is a safety communication?

13. List six items that help to improve intelligibility of radiotelephone transmissions.
14. How would the word "Xingu" be transmitted phonetically?
15. What word or phrase means: "I received your message." "I have no further traffic." "Repeat all words after. . . ."?
16. Name the five parts of a formal radiotelephone message.
17. What is the "check" of a message? What does it include?
18. After communications have been established how often do the stations identify themselves?
19. How do mobile or fixed stations usually identify themselves?
20. Why is it bad practice to transmit on an unmonitored frequency?
21. How often should a radiotelephone transmitter be tested?
22. Pressing the microphone switch ON may produce no interference on the channel when using what type of emission?
23. When may a station not be operated in accordance with the provisions of the station license?
24. Why should a period of about 1 s be left after AM or FM transmissions between two stations?
25. What should be done if someone uses obscene language over a radio transmitter?
26. Why should a transmitter be turned off when an operator leaves a vehicle for any time?
27. What time is used in logs?
28. What kind of traffic is worked on calling frequencies? On working frequencies?
29. List nine items that a required log might contain.
30. What does it mean if a station is open to public service?

ANSWERS TO CHECKUP QUIZ ON PAGE 689

1. (Distress, urgent, safety)(Routine) **2.** (Immediately after distress signal) **3.** (Signal, call, name of craft, position, nature of distress, assistance required, etc.) **4.** (Acknowledge)(Listen, help if needed) **5.** (MAYDAY all stations this is KAAA SS Blank distress traffic ended. Out) **6.** (All) **7.** (Within an inch usually) **8.** (Delta-echo-alpha-November) **9.** (Origin, data, time, number)(Body of message)(Where going)(Received time and operator)(Sender's name) **10.** (Calling, conclusion, every 15 min) **11.** (When channel clear)(Daily) **12.** (In view of operating position) **13.** (Working) **14.** (2182 kHz and 156.8 MHz) **15.** (Communicates with ships and aircraft) **16.** ("FCC Rules and Regulations," Vol. I, Part 17) **17.** (Coast working frequency)(International calling frequency)

31

Radio-telegraph Communication-

This chapter discusses the fundamentals of commercial maritime and, to some extent, amateur radiotelegraph operating, including the Morse code, calling and answering procedures, emergency calls, and radio messages. Some basic keying devices are explained. The material should answer many of the FCC Elements 5 and 6 Radiotelegraph questions.

31-1 FUNDAMENTALS OF OPERATING

There is a basic similarity of radiotelephone and radiotelegraph regulations. The list of message priorities is the same for both (Sec. 30-2). However, radiotelegraph stations must be operated by licensed persons, and operator licenses should be posted at the operating position.

When operators exercise good common sense in communicating, they will usually be conforming with regulations. Such regulations have been developed to enable the greatest number of stations to handle the most traffic, in the least time possible, with the least confusion. Long calls, failure to listen on a frequency before transmitting, transmitting on a frequency already in use, employing either more or less transmitting power than necessary (the minimum necessary should always be used), sending faster than the receiving operator can copy, or sending faster than an operator can legibly telegraph, using improperly functioning equipment, and using abbreviated procedures with stations not understanding them are common operating faults that can delay communications.

31-2 RADIOTELEGRAPH LICENSES

There are three types of radiotelegraph licenses: First Class, Second Class, and Third Class (Chap. 29). All require passing FCC Elements 1, 2, and 5, plus code tests. Holders of the First and Second Class Radiotelegraph licenses must pass Element 6, which authorizes them to make technical adjustments to radiotelegraph and radiotelephone transmitters.

With radiotelegraphy there is no time when an unlicensed person may transmit with a telegraph key, with the exception of a dire emergency such as in a lifeboat with no operator aboard and adrift at sea. Anyone may operate radiotelephone transmitters, if authorized by the person in charge or if under the supervision of a licensed operator.

31-3 THE MORSE CODE

The telegraphic code used for commercial and amateur radiotelegraphic communication is the International, or Continental, Morse code, consisting of dots and dashes (Table 31-1). (American Morse code letters consist of dots, dashes, and spaces.) In International Morse code a *dot* is made by pressing the telegraph key down and allowing it to spring back up immediately. The length of a dot is the basic time unit. A *dash* is made by pressing the key

TABLE 31-1

THE INTERNATIONAL, OR CONTINENTAL, MORSE CODE

A · –	N – ·	1 · – – – –
B – · · ·	O – – –	2 · · – – –
C – · – ·	P · – – ·	3 · · · – –
D – · ·	Q – – · –	4 · · · · –
E ·	R · – ·	5 · · · · ·
F · · – ·	S · · ·	6 – · · · ·
G – – ·	T –	7 – – · · ·
H · · · ·	U · · –	8 – – – · ·
I · ·	V · · · –	9 – – – – ·
J · – – –	W · – –	0 – – – – –
K – · –	X – · · –	
L · – · ·	Y – · – –	
M – –	Z – – · ·	

. (period)	· – · – · –
, (comma)	– – · · – –
? (question mark)($\overline{\text{IMI}}$)	· · – – · ·
/ (fraction bar)	– · · – ·
: (colon)	– – – · · ·
; (semicolon)	– · – · – ·
((parenthesis)	– · – – ·
) (parenthesis)	– · – – · –
' (apostrophe)	· – – – – ·
- (hyphen or dash)	– · · · · –
$ (dollar sign)	· · · – · · –
" (quotation marks)	· – · · – ·
or,	· – · · – ·
Error sign (8 dots)	· · · · · · · ·
Separation indicator also known as $\overline{\text{BT}}$	– · · · –
End of transmission of a message ($\overline{\text{AR}}$)	· – · – ·
Invitation to transmit	– · –
Wait ($\overline{\text{AS}}$)	· – · · ·
End of work ($\overline{\text{SK}}$)	· · · – · –
Starting signal	– · – · –

down and holding it for a period of three basic time units. The spacing between two dots or between a dot and dash in the same letter is equal to one time unit. The spacing between two letters in one word is equal to three units. The spacing between two words is equal to seven units.

A number and a fraction are transmitted as number, hyphen, and fraction. For example, 45½ is transmitted: 45-½. (Sometimes · – · · – is used instead of the hyphen.)

Four dashes (– – – –) may be used as the two letters *ch*, or to indicate end-of-paragraph.

Two question marks are sometimes used instead of the eight-dot error signal shown.

There are several other codes used in communica-

tions. American Morse code is used on telegraph lines (*landlines*). Baudot code is used in teleprinter communications. ASCII code is used for communicating between computers. Some languages, such as Japanese and Russian, have their own radiotelegraph codes.

31-4 FREQUENCIES USED

Radiotelegraph-equipped ships on the high seas maintain either a constant or a specified number of hours watch on the international calling and distress frequency of 500 kHz (600 m). All coastal radiotelegraph stations maintain a constant watch on 500 kHz. Most of these stations also keep watch on one or more HF (high-frequency, 4-, 6-, 8-, 12-, 16-, 22-, and 25-MHz) bands (Sec. 25-9). An old maritime band between 90 and 160 kHz, had a calling frequency on 143 kHz.

The calling frequency 8364 kHz is for aircraft, lifeboats, and other survival craft for communication with stations of the Maritime Mobile Service. This frequency is monitored by the United States Coast Guard and Navy and, being the center of the 8-MHz marine band, is constantly scanned by coastal stations listening for calls by ships.

31-5 CALLING BY RADIOTELEGRAPH

An original call is made by radiotelegraph by transmitting the call sign of the station wanted three times, followed by the letters DE ("from"), the call sign of the station three times, and the letter K ("go ahead"). For example,

KAAA KAAA KAAA DE KBBB KBBB KBBB K

(On 500 kHz, when stations are close geographically and are known to each other they may call only once and sign once.)

In the bands between 4 and 25 MHz, when the conditions of establishing contact are difficult, the call signs may be transmitted more than three times but not more than eight times.

Calls are made on a frequency that is known to be monitored by the station to be called. If within a few hundred miles, the frequency of 500 kHz is used for ship-to-ship or ship-to-shore calls. If farther away, high-frequency ship-to-shore calling frequencies are used.

Except for distress traffic, all stations handle traffic on working frequencies, not on calling frequencies. A ship station calling a coastal station listens for the coastal station on its assigned working frequency and notifies the coast station to listen for the ship on one of its assigned working frequencies. Thus, "KFS DE KDMW QSW 463" means "KFS from KDMW, I am going to send on 463 kHz. Listen for me there." If KFS hears the call, it acknowledges on its working frequency and no other transmission may be made on the calling frequency.

In the 405-535 kHz band, a coastal station will shift to 500 kHz to call a ship, or to call CQ ("attention all stations") and advise that it is going to transmit a list of traffic on hand on its working frequency. HF coastal stations always remain on their working frequencies, sending traffic lists often.

If a coast station has traffic for one or two ships, it may transmit their call signs on the calling frequency. If for three or more ships, the coast station will send a traffic list on its working frequency. It will first send CQ on the calling frequency and the Q signal (Appendix G) to listen on its working frequency. For example,

CQ CQ CQ DE KPH KPH KPH TFC QSS 426

This indicates a traffic list will be transmitted by KPH on 426 kHz.

The letters CP followed by two or more call signs on a calling frequency are a general call to those stations only and indicate that no reply is expected from them. This call precedes a general broadcast or information to a special group of receiving stations, such as press, on a working frequency.

31-6 ANSWERING BY RADIOTELEGRAPH

If KBBB calls KAAA on a calling frequency, KAAA should answer on the same frequency, advising KBBB which working frequency to use. For example,

KAAA DE KBBB TFC K
KBBB DE KAAA QSY 428 K
DE KBBB R (Or DE KBBB QSL)

The last transmission indicates that KBBB understands (R = Roger, I agree with your statement).

If a station hears another calling but cannot make out the call sign, it should transmit the Q signal meaning "By whom am I being called?" as

QRZ? DE KAAA K

If a station hears another calling but is busy, if possible it should answer the calling station and transmit $\overline{AS}$ ($\cdot - \cdot \cdot \cdot$) or the Q signal QRX, followed by a number indicating how many minutes to wait.

31-7 TUNING AND TESTING

When it becomes necessary to test or tune a transmitter on the air, a time should be chosen when the frequency to be used is idle. The tuning should be accomplished as rapidly as possible. At the conclusion a radiotelegraph station transmits a series of Vs as a test signal, followed by DE and its call sign.

31-8 STATION IDENTIFICATION

Besides the normal station identification that occurs during calling and answering, radiotelegraph stations usually transmit a station identification by their call signs in Morse code at the completion of each transmission, at the conclusion of an exchange of transmissions, or every 15 min (10 min for amateurs) if a transmission is sustained for a long period.

31-9 SILENCE PERIODS

All ships and coastal stations operating on 500 kHz must listen, but not transmit on, 500 kHz during the 15- to 18- and 45- to 48-min periods of every hour of operation. These are known as *silence periods*. Ships in distress should repeat their distress messages at no more than 16 wpm (words per minute) during these periods to assure reception.

During silence periods routine transmissions are forbidden in the band 485 to 515 kHz.

31-10 DISTRESS

One of the main reasons why a ship carries radio equipment is to enable it to signal its condition and position in case of emergencies or distress.

The *distress signal* when sent by radiotelegraph is $\cdot \cdot \cdot - - - \cdot \cdot \cdot$ transmitted three times. This is referred to as an SOS but is actually a single character ($\overline{SOS}$) selected for its easy identification. It is sent only on the authority of the master or person responsible for the ship, aircraft, or other vehicle and only when the distressed station is in grave and imminent danger and requests immediate assistance.

In a distress situation, if time allows, the auto-alarm signal should be transmitted for 1 min on 500 kHz. After a 2-min interval, to allow operators summoned by the auto-alarm signal to assume a watch, the *distress call* is transmitted. This consists of the distress signal 3 times, DE, and the call sign of the station 3 times. The distress call is followed by the *distress message* transmitted at no more than 16 wpm. The *distress message* consists of (1) the distress signal; (2) the name of the ship, aircraft, or vehicle in distress; (3) particulars of position, nature of the distress, kind of assistance desired, speed and course if underway; and (4) any other information which might facilitate rescue.

After the distress message, the mobile station should transmit two 10-s dashes, followed by its call sign, to permit direction-finding stations to determine its position.

The distress message should be repeated on 500 MHz during the next silence period.

All subsequent distress messages must contain the $\overline{SOS}$ indicator in their preamble.

Distress signals, calls, and messages should be transmitted by A2A (modulated-code) if possible.

Stations that cannot use 500 kHz use their normal calling frequency for distress transmissions.

The station in distress controls distress traffic unless it delegates control to another station.

An operator hearing a distress call must cease transmitting and listen on the frequency of the distress call. If the station in distress is without doubt in the vicinity, after the completion of the distress message, the operator must acknowledge receipt of the message by transmitting (1) the call of the distressed station (three times), (2) DE followed by the call of the receipting station (three times), and (3) R R R $\overline{SOS}$.

Stations receiving a distress message not in their vicinity must allow a short interval of time before acknowledging receipt to permit closer stations to acknowledge.

When distress traffic is well established, stations not involved and in no position to help may continue normal service on other frequencies, provided they do not interfere with any distress traffic.

When all distress traffic has ceased or when silence is no longer necessary, a station which has controlled the distress traffic must send a message to terminate the distress condition. Such a message will take the form (1) $\overline{SOS}$ CQ CQ CQ DE followed by the call of the station sending, (2) time of the message, (3) name and call of distressed station, and (4) QUM $\overline{SK}$.

31-11 THE URGENCY SIGNAL

In radiotelegraphy, the *urgency signal* is XXX sent 3 times, usually on 500 kHz. It indicates that the calling station has a very urgent message to transmit concerning the safety of a ship, aircraft, or other vehicle or of some person on board or within sight. It must be authorized by the person responsible for the transmitting ship.

An urgency message from a ship may be addressed to a specific station. The urgency signal, with the approval of the responsible authority, may be transmitted from a coast station and addressed to all ships.

Mobile stations hearing the urgency signal must continue to listen for at least 3 min. If no urgency message has been heard by then, they may resume their normal service.

An auto-alarm signal may be transmitted before an urgent cyclone warning by a coastal station authorized to do so by its government. A period of 2 min should elapse between transmission of the AA signal and the cyclone-warning message.

31-12 THE SAFETY SIGNAL

In radiotelegraphy, the *safety signal* is TTT (3 times) followed by the station call (3 times). It indicates that the station is about to transmit a message concerning the safety of navigation or important meteorological warnings and is sent on the distress frequency. The safety signal is usually transmitted during the last minute of a silence period. The safety message is then transmitted at the conclusion of the silence period.

Operators hearing a safety signal must continue to listen to the message until they are satisfied that it is of no importance to them. They may then resume normal service on frequencies that will not interfere with the safety transmission.

31-13 RADIOTELEGRAPH LOGS

All ship stations authorized to use telegraphy must maintain an accurate radiotelegraph log. The first page is a *title page*. At the completion of the voyage the following information is placed on it: the name of the ship, the call letters, the period of time covered by the log, the number of pages, a statement whether any distress message entries are contained in it and on what pages, the operator's signature and mailing address, and his license number, class, and date of issuance.

Each page is numbered serially for the voyage and contains the name of the ship, call letters, and name of the operator on watch. The entry "on watch" is made at the beginning of a watch, followed by the operator's signature. The entry "off watch" is made when an operator is being relieved or terminating a watch, followed by the signature.

The log is kept in GMT or UTC (EST for ships in the Great Lakes) and must contain all calls or tests transmitted by the ship, stations contacted, and serial numbers of messages handled, stating times and frequencies. A positive entry with respect to reception on 500 kHz should be made at least once in each 15 min. Entries stating whether or not the international silence period was observed shall be made twice per hour, noting any signals heard during these periods. All distress, urgency, and safety signals heard must be entered, with complete text of distress messages if possible. Any harmful interference noted should be logged. Once a day the position of the ship and a comparison of the radio station clock with standard time, including errors noted, must be entered. Times of arrival and departure from ports are logged. Failures of equipment and corrections taken should be noted. Results of emergency-equipment tests, the battery specific gravity when placed on and taken off charge, and the quantity of fuel for emergency generators must be entered. On cargo vessels, the time the auto-alarm was placed in service and when out of service, as well as the setting of the sensitivity control (if any), results of tests, alarms, and false alarms must be logged. (See also Sec. 29-8.)

Test your understanding; answer these checkup questions.

1. What words are counted in commercial radio messages? _____ In amateur messages? _____
2. When only might an unlicensed person transmit radiotelegraph code? _____
3. What is the basic unit of Morse code? _____ How long is a dash? _____ The space between letters? _____ Between words? _____
4. Must a calling station always call at least 3 times and sign 3 times? _____
5. What are two marine CW calling frequencies? _____ _____
6. What does CQ mean? _____ QSW? _____ QSY? _____ QRZ? _____
7. On HF bands, why might a ship call KPH KPH KPH DE WAAA WAAA WAAA KPH K? _____

8. If a coast station tells a ship on 500 kHz "up," what does it mean? _____
9. What is the test signal for CW? _____
10. A ship is in distress at 0952 UTC. When should the operator send the SOS message? _____ At what speed? _____ What should be sent after the message? _____
11. What does QUM mean? _____ XXX? _____ TTT? _____ SK?
12. Besides prefacing an SOS message, when may an AA signal be sent? _____
13. What is considered a minimum signature on a log sheet? _____
14. How often must log entries be made while at sea? _____
15. How long must logs be kept if they contain distress messages? _____

31-14 COMMERCIAL RADIOTELEGRAPH MESSAGES

Radiotelegraph messages from ship to ship or from ship to shore have a preamble, address, text, signature and servicing (Chap. 30). A word is 10 letters or less. Any word with 11 letters is counted and charged as 2 words. A 21-letter word is counted as 3 words. An example of a paid ship-to-ship message, as sent from KBBB to KAAA, might be

P 1 KBBB CK 13/16 SS ARROW 12 2145Z BT

FRED MANGELSDORF
SS SPEAR BT

MAKING ARRANGEMENTS MEET YOU HILO TWENTYFIRST BT

STEVE AND FREDA AR
1 KAAA ES 2213 12

In the **preamble**, P 1 KBBB indicates a paid message NR 1 from the station of origin, KBBB, the SS *Arrow*. The check (CK) is the number of words paid for in the address, text, and signature. There are 13 words but they are charged for as 16. The 12 indicates the day of the month. The 2145Z indicates the filing time 2145 UTC. The serial numbering of messages begins at midnight (0000 h) UTC and continues to the next midnight (2400 h) UTC. (Amateurs may use monthly or yearly numbering.)

Stations on inland waters may use local time. The date and time may be sent as a date-time group, 122145Z. The same preamble might also be transmitted in one of the following forms:

P 1 13/1 SS ARROW/KBBB DATE 2145 UTC

SS ARROW/KBBB NR 1 CK 13/16 122145Z

NR 1 KBBB 13/16 P SS ARROW 2145 GMT 12

The **address** includes the name of the person to whom the message is being sent and sufficient address to deliver it.

The **text** is the body of the message.

The **signature**, if any, is the person sending the message. If no signature is to be sent, the words *no sig* are transmitted in place of the signature. The sign $\overline{AR}$ indicates "end of message." $\overline{BT}$ indicates a break in the message.

The **servicing** on the message to Fred Mangelsdorf is printed on a line beneath the signature by the transmitting operator. It indicates the message was sent as NR 1 to KAAA by operator Emery Simpson at 2213 on the twelfth. Somewhere on every message the full date (month, day, and *year*) must be shown. If it is not in the preamble, it must be in the servicing.

The received message should be copied in a form similar to

P 1 KBBB CK 13/16 SS ARROW
                    FEBRUARY 12 1985 2145 UTC

FRED MANGELSDORF
SS SPEAR

MAKING ARRANGEMENTS MEET YOU HILO TWENTYFIRST

                    STEVE AND FREDA
                    12 2213 BW

Note that the operating signals $\overline{BT}$ and $\overline{AR}$ are not shown on the received message. Also, the receiving operator has typed in the complete date in the preamble *while the message is being sent*!

A ship-to-shore message is addressed to some destination on land. It is transmitted to a coastal station and has a form similar to that of the ship-to-ship message.

A message filed at a telegraph station ashore for a ship is relayed by telegraph to a coastal station and transmitted to the ship as soon as it can be reached. An example of a shore-to-ship message as received by a ship operator might be as follows:

P 1 RENO NEV CK 14/15 FEB 12 1985
                                   8:45 AM

SASSER
SS ARROW
SANFRANCISCORADIOKPH

MEET   YOU ON ARRIVAL EUREKA FRIDAY AFTERNOON WITH RONADLO

                    JANE
                    2301 12 BW

Note that the name of the coastal station transmitting the message is counted (2 words). If the call letters of the coastal station are all that is transmitted, the receiving operator fills in the complete name of the station, and it would count as only 1 word (CK 14). Also note that the receiving operator double-spaces after the fifth word *(meet)* and drops to the next line after the tenth word (and after every multiple of 10 words in long messages) to facilitate counting check while copying.

The punctuation in the filing time may be transmitted as a period, although the letter R also may be used. It may be copied as a period or colon.

When a message contains words that appear to the receiving (or transmitting) operator as possibly improper, they should be confirmed. For example, the last word in the text of the shore-to-ship message appears suspicious. At the completion of the transmission of the message the receiving (or transmitting) operator might send CFM RONADLO K, meaning "Confirm the spelling of the word Ronadlo." If the questioned spelling is correct, the transmitting operator responds with R, or preferably C, meaning "yes." If the receiving operator confirms a word incorrectly, the transmitting operator must correct him, as N RONALDO (if this happens to be the actual spelling). The N indicates "no."

31-15 COUNTING WORDS IN MESSAGES

In any language, any word, combination of letters and numbers ("code," or "cipher" words for security), or name in the address, text, or signature up to 10 letters long counts as only one word. If it has 11 or more letters, it is charged for as two words (21 letters is three words).

When transmitted, punctuation marks such as period, comma, colon, question mark, parentheses, and quotation marks are counted as separate words. Parentheses and quotation marks require two transmitted characters for each single word count.

31-16 MESSAGE CHARGES

Each word in the address, text, and signature of a radiogram is charged for. (Domestic or landline telegraph messages charge for the text only.) Between United States ships and shore stations the rates are in dollars and cents. When a United States ship has traffic with a foreign station, the rates (other than its own ship charges) are in gold francs, with some number of gold francs equaling $1 (subject to change).

For a regular, full-rate type radiogram (P) the charges per word are 8¢ to the transmitting ship, 8¢ (about 0.04 gf) to any receiving ship, 17½¢ to the coastal station (varying rates in gold francs to foreign-country coastal stations), and 7¾¢ landline charge to any place in the United States. Thus, a ship-to-ship message between United States ships costs 16¢ per word, and a ship-to-shore message costs 33¼¢ per word (subject to change).

31-17 TYPES OF MESSAGES

- CODH, or DHCO. *Company deadhead* messages refer to traffic, licenses, repairs, supplies, etc., exchanged by radio between ships and offices of the same radio service. There are no charges.
- DH, or PDH. These complimentary franked messages may require payment of only the landline charges, the ship and coastal charges being franked (transmitted free). The sender must possess a frank card.
- DH MEDICO. *Deadhead medical* messages are free (deadhead) messages pertaining to medical or surgical advice for persons aboard

a ship and are normally addressed to MARINE HOSPITAL in a city served by the coastal station receiving them. (May be sent to U.S.C.G. coastal stations.)

- DH OPR. *Operator deadhead* messages are personal messages of radio officers and may be forwarded through the coast stations of the same radio service subject only to landline charges.
- GOVT. *Government radiograms* by accredited officials of the United States government on official government business are charged full rates.
- HYDRO. *Hydrographic* messages report menaces to navigation and are addressed to HYDRO WASHINGTON. There are no charges insofar as the ship operator is concerned. (May be sent to U.S.C.G. coastal stations.)
- MSG. *Master's* messages are from masters of vessels and pertain to ship's business. Only coastal and landline charges are collected.
- OBS. *Observer* messages are meteorological reports from ships at sea to the U.S. Weather Bureau addressed OBSERVER in the city where the report is to be sent. There are no charges insofar as the ship operator is concerned.
- PC. *Acknowledged delivery of message.* Coastal station notifies sender via telegraph of time and date message was delivered to the ship destination. J1PC means advise sender if undelivered after day 1. J12hrPC means after 12 h, etc.
- PRESSE, or PX. *Press* messages transmitted by authorized members of the press when addressed to newspapers, magazines, or broadcast stations are charged for at a rate 50% of the standard radiogram rate. The word *presse* is also transmitted as the first word of the address and is charged for. The minimum charge is for 14 words.
- AMVER. Automated Mutual-assistance VEssel Rescue system messages used for downed aircraft, medico, man overboard, fire, overdue, precautionary, and sinking emergencies. Sent to U.S.C.G. coast stations. Participating ships' courses and sailing data stored in computer in N.Y. to allow dispatch of closest vessel to any point of emergency. Addressed to "AMVER NEWYORK" or other coastal city. No charges. Check count not required.
- ST. *Paid service* messages are made out in the more formal preamble, address, text, and signa-

ture form. This type of service adds words to or corrects a message because of errors made by the originator of the message. The sender must pay standard rates for only the words in the text that are required to make the correction.

- SVC, or A. *Service* messages refer to previous messages, the operation of the communication system, or the nondelivery of messages. Since they refer to previous traffic, they have priority over other routine messages. They carry no charges, and may or may not carry a serial number. An example of a service message might be:

SVC 3 (or A 3)
KAAA

RE OUR NR 1 TO JONES SF 12TH
CORRECT FIRST WD TXT
READ MAILING

KBBB 13

This message advises the operator of KAAA to change the N 1 message of KBBB of the 12th of this month to make the first word of the text read "mailing" instead of as originally transmitted. The signature of the SVC is the call letters of the sending ship and date of the service message.

- TR. *Position reports* may be requested by land stations or mobile stations. The message indicates approximate position and next place of call when furnished by the master of the ship or vehicle.

31-18 TRANSMITTING SPEED

The average speed of radiotelegraphic communications is probably about 20 wpm. This is a comfortable telegraphic hand-key speed. Sending 25 to 30 wpm with a hand key is possible. However, with a semiautomatic key, called a *bug*, or an electronic keyer, good operators may work easily at speeds of 35 to 40 wpm under fair to good conditions. The same operators may not be able to average 15 wpm when static, fading, or interference is bad. The limiting factor is always how well the receiving operator can read the transmitted signal. Transmitting operators must gauge their sendings to that speed which the receiving operator can copy

without breaking too often. When receiving conditions are poor, an operator should not hesitate to send QRS or QSZ in order to attain accuracy, the most important factor in communications.

In distress, urgency, and safety traffic, the speed should not exceed 16 wpm to assure a maximum number of operators' copying correctly.

31-19 TRANSMITTING MORSE CODE

There are four types of mechanisms with which Morse code is transmitted by hand. The first is the *telegraph key* (Fig. 31-1). The operator places the

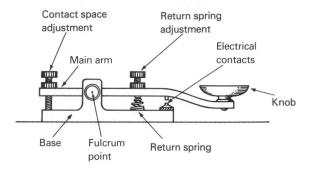

FIG. 31-1 Telegraph straight, or hand key. Electrical connections are made by pushing the top contact down onto the lower contact.

first finger on the top of the knob and the thumb at the side of the knob and, with a downward pressure of the first finger, closes the contacts. A quick make and break produces a dot. A dash is produced by holding the contacts closed for three times as long as for a dot. For speeds over about 18 wpm, two fingers are used on the top of the knob. The contacts are adjusted for a maximum opening of about $1/16$ in. The spring is adjusted to the desired, comfortable upward pressure. For higher speeds a slightly lighter spring tension, and less contact spacing, with two fingers on the knob, may be better.

The second mechanism is a semiautomatic key, called a *bug* (Fig. 31-2). When the knob is pressed to the left with the first finger, the contacts close, as with a telegraph key. When the paddle is pressed to the right with the tip of the thumb, the vibrating arm makes a series of dots. The speed of the dots is controlled by the placement of the weights on the vibrating arm. Dots are made automatically, but the operator makes each dash separately. The dot contact moves about $1/8$ in. and the dash contact about $1/32$ in. before making contact.

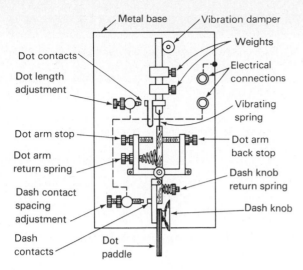

Dot contacts
Metal base
Vibration damper
Weights
Electrical connections
Dot length adjustment
Vibrating spring
Dot arm stop
Dot arm back stop
Dot arm return spring
Dash knob return spring
Dash contact spacing adjustment
Dash knob
Dash contacts
Dot paddle

FIG. 31-2 Semiautomatic telegraph key, or bug.

An *electronic keyer* not only makes the dots automatically by an internal multivibrator, but also makes a series of dashes when the dash knob is pressed. It can produce very precise sending.

An *electronic keyboard* is like a typewriter, except when the letter A is pressed the output circuit will key a transmitter with a dot and dash. Other electronic devices can receive Morse code over the air, convert it to dc pulses, and through a computer, display the received letters on a CRT or on some form of printer.

When an error in sending has been made, the transmitting operator should immediately send a series of eight dots, stop, and start at the beginning of the word in which the error was made. If the error occurs on the first letter of a word, after the error signal is sent the operator should repeat the word before the word in which the error was made.

31-20 CALL LETTERS

The nationality of a station is indicated by the first or first two letters (or number and letter) of its call sign. All calls beginning with K, W, N, and AA to AL are those of United States stations.

G and 2A calls are assigned to Great Britain, F calls to France and its colonies and protectorates, R and U calls to the Union of Soviet Socialist Republics, V calls to British colonies, 4U to United Nations, and so on.

In general, 3-letter calls indicate coastal stations in the maritime service or land stations in other

services. Ship station calls may be 4-letter or 4- or 5-number/letter calls. Aeronautical mobile stations are usually 5-letter calls. Radiotelephone land or coast stations are assigned 3-letter or 5- to 7-letter/number calls, as KMA539. Mobile radiotelephones are also assigned letter/number calls.

31-21 OPERATING SIGNALS

To facilitate communications between operators speaking the same or different languages, a group of Q signals which have the same meanings internationally have been developed (Appendix G). Radiotelegraph abbreviations or signals that are in more or less common use by operators are listed in Appendix H.

■ Test your understanding; answer these checkup questions.

1. What items make up a preamble of a ship-to-shore message? _____ Of a shore-to-ship message? _____
2. What time is used in ship messages? _____ Land messages? _____
3. What is included in the servicing of a transmitted message? _____ Received message? _____
4. Why should a receiving operator double-space after each fifth word? _____
5. What punctuation in the text of a paid message is not counted? _____
6. What three kinds of charges are made for paid ship-to-shore messages? _____ _____ _____
7. What is another designation for a SVC message? _____ Paid service? _____ Franked message? _____ Press? _____ Company deadhead? _____ A Master's message? _____
8. What would J2PC mean on a message? _____ J24hrPC? _____
9. Besides an operator's own abilities, list what limits how fast that operator should transmit a message. _____
10. If an error is made in the first letter of a word, how is it corrected? _____ If in the fourth letter? _____
11. What type of stations might have calls such as WSL? _____ KH4267? _____ KDOZ? _____ KNL123? _____ WNGPL? _____
12. What is meant by the operating signal K? _____ C? _____ AA? _____ AB? _____ CFM? _____ TU? _____ WA? _____ ETA? _____
(See Appendix H)

1. How are rules and regulations developed?
2. What type of communications always require the operator to have a license?
3. What is the basic unit of timing in Morse code? How many in a dash? In the spacing between dots or dashes? In the spacing between words? Between letters?
4. List four codes used in communicating in the United States.
5. What are the two marine CW calling and distress frequencies?
6. Give an example of the basic CW calling procedure between two MF ship stations.
7. How does a land station advise three or more ships it has traffic for them?
8. What are two CW signals that tell the other station to wait?
9. On what frequency does a called station usually answer?
10. Explain a tune-up procedure for a CW station.
11. How often must working CW stations identify?
12. What are the silence periods that must be maintained by stations working in the 485–515 kHz band? What does an operator do about them?
13. What is the distress signal for CW? How is a distress call made?
14. How is a distress message sent?
15. Why are two 10-s dashes sent after an $\overline{\text{SOS}}$ signal?
16. What type of emission is used for CW distress traffic?
17. How is 500 kHz cleared when a distress incident is over?
18. What is the CW urgency signal? What should an operator do if one is heard?
19. What is the CW safety signal? When is it normally transmitted?
20. What should a radiotelegraph log title page contain at the completion of a voyage?
21. What log entries are made by CW operators?
22. What information is in the preamble of a CW message?
23. The service on a transmitted message reads: 3 WBBB RS 1401 7. What does this mean?
24. When does an operator double space after a word in a message?
25. What does an "R" signify in a series of numbers?
26. What two punctuation marks are sent twice but are counted once.
27. What is the breakdown for the U.S. money charges between two U.S. ships per word in a radio message?
28. What is the breakdown for U.S. ships per word to any place in the U.S. through a U.S. coastal station?
29. List 14 types of messages that may be sent to or from ships.
30. What is a comfortable hand-key sending speed? Bug speed?
31. What are the four types of keying devices that may be used to send CW?
32. What should an operator who makes an error on the first letter of a word do?
33. What indicates nationality in a radio call sign?

ANSWERS TO CHECKUP QUIZ ON PAGE 699

1. (Type of message indicator, number, ship name, ship call, date, time)(Type of message, number, originating city, date, time) **2.** (UTC or GMT)(Local of sender) **3.** (Number, call of station sent to, operator initials, date) (Time, date, initials) **4.** (To facilitate count of total check) **5.** (Fraction-to-follow) **6.** (Ship, coastal, landline) **7.** (A)(ST)(DH or PDH)(PX)(DHCO)(MSG) **8.** (Advise if not delivered in 2 days)(If not delivered in 1 day) **9.** (Receiving conditions, receiving operator ability, type keyer used) **10.** (Error sign, repeat last correctly sent word)(Error sign, repeat from first letter of the word) **11.** (Land)(Mobile-phone)(Ship or land)(Base phone)(Aeronautical) **12.** (Go ahead)(Yes)(All after)(All before)(Confirm)(Thank you)(Word after)(Estimated time of arrival)

32

Amateur Radio

This chapter outlines the amateur operating phase of radio and the various grades of licenses that are available. Some of the more important rules and regulations of the Amateur Radio Service are included. The required electric, electronic, and radio theory that may be required to pass written amateur license tests are distributed throughout much of this book. (See note on licenses and typical questions in Preface.)

32-1 THE AMATEUR RADIO SERVICE

Since the beginning of the twentieth century many persons have been interested in experimenting with their own radio transmitting and receiving equipment without any pecuniary interest. When such persons obtain FCC licenses, they are known as amateurs solely because they accept no money for their on-the-air operations and communications. Many of the foremost radio and electronic engineers and technicians operate as amateurs during times when they are not occupied professionally. (Amateur should not be confused with citizens' band operations; CB operations require no license, call signs, or tests of any kind and are limited to 4-W AM or 12-W PEP SSB on frequencies in the 27-MHz region. Communications are not allowed between amateur and CB stations.)

Rules governing the Amateur Radio Service are specified in Part 97 of the "FCC Rules and Regulations," which may be purchased from the Superintendent of Documents, Government Printing Office, Washington, D.C. Every amateur should have a copy of either these rules or a copy of "The FCC Rule Book," which may be purchased from the American Radio Relay League, Newington, Connecticut, 06111, USA, and which may be in an easier to read form as well as being more up to date.

Amateurs are subject to essentially the same penalties for violating FCC rules as commercial operators are (Sec. 29-1).

FCC rules and regulations are designed to provide an Amateur Radio Service with fundamental purposes to (1) provide emergency or disaster communications if necessary, (2) contribute to the advancement of the radio art, (3) improve skills in the communication and technical phases of radio, (4) provide a reservoir of trained operators and technicians, and (5) enhance international goodwill.

The beginning of amateur radio operations dates back to the turn of the century. In 1914 the American Radio Relay League (ARRL) was formed, and since then it has acted as one of the strongest proponents of the radio amateur. Its magazine, *QST*, carries information regarding amateur activities, experimental circuits, and other items of interest to amateurs. Among other activities, the ARRL HQ station W1AW transmits code practice daily (see "Schedules" in *QST*). As the FCC adds new subjects for amateur licenses, they may appear in *QST*. Some other periodicals dedicated to amateur information are *Ham Radio, 73, CQ,* and *Worldradio*.

32-2 LICENSES AND TESTS

There are two amateur licenses, one is the *operator* license, and the other, available only after passing an operator license test, is the *station* license with its accompanying call sign. Usually, both licenses are applied for at the same time.

There are five grades of operator licenses at this time. From the simplest they are the Novice, Technician, General, Advanced, and Extra Classes. These are discussed separately later.

All license tests are given by volunteer amateur operators using questions made up from subjects outlined by the FCC. Contact local amateurs, amateur radio clubs, or the closest FCC office for further information on obtaining an amateur radio license (Appendix F).

Amateur license examinations consist of two or more numbered test *elements*. These are:

1. *a*. Code test, 5 words per minute (wpm).
   b. Code test, 13 wpm.
   c. Code test, 20 wpm.
2. Test on basic laws, rules, and regulations for beginners, plus elementary radio theory.
3. Test on general amateur practice and regulations involving radio operation, apparatus, treaties, statutes, and rules affecting amateur stations and operators.
4. *a*. Test on intermediate-level radio theory and operation as applicable to modern amateur radio techniques, including but not limited to radiotelephony and radiotelegraphy.
   b. Test on advanced radio theory and operation as applicable to modern amateur techniques, including radiotelephony, radiotelegraphy, and transmissions of energy for measurements and observations applied to propagation, for the radio control of remote objects, and for similar experimental purposes.

Amateur license tests are given by a volunteer examiner (VE) team and are corrected by the VE, who notifies the applicant of passing or failing, forwards test papers to a volunteer examiner coordinator (VEC) body, which relays them to the FCC. About five weeks after the test, station and operating licenses and call letters will be sent to the applicant. When *upgrading* a license, the necessary element test(s) are given by a VE, and if passed a "Certificate of Successful Completion of Examination" is issued to the applicant, along with a special test-session *identifier code* (two letters selected from WA-WZ, KA-KZ, NA-NZ, or AA-AL). The code appears on the certificate and must be used as a suffix to the amateur's call on the air, as W6ECU/WA. When notified by the FCC of successful upgrading, the code is no longer used for on-air communications.

License terms are for 10 years. Renewals should be applied for between 90 and 30 days before the termination of the license period. Should renewal application not be made at the proper time, a 5-year grace period is allowed for operator licenses and a 2-year period for station licenses. During this period no transmissions may be made by the operator, but renewal may be applied for without having to take a reexamination. After the 5-year grace period, a reexamination test for the operator license is required.

32-3 STATION LICENSES

Station licenses with call signs will be issued only to licensed amateur radio operators, to bona fide radio clubs, and to an appropriate military recreation person. Special-event licenses and calls may be issued for a period of 30 days to applicants who are Advanced or Extra Class license holders.

Station licenses must specify one geographical land location as the control point of the station. If the location is moved or if the name of the operator or organization is changed, the FCC, Gettysburg, PA 17325, should be notified of the new address.

32-4 CALL SIGNS

Amateur station call signs are made up of one or two letters, a number, and one to three letters, specified as 1×2 ("1 by 2"), 1×3, 2×1, 2×2, and 2×3, with $\times$ being the area number in the call sign. The call sign district or area numbers of the 48 contiguous states and the District of Columbia are:

1 for MI, CT, MA, NH, RI, VT
2 for NJ, NY
3 for DE, MD, PA, DC
4 for AL, FL, GA, KY, NC, SC, TN, VA
5 for AR, LA, MS, NM, OK, TX
6 for CA
7 for AZ, ID, MT, NV, OR, UT, WA, WY
8 for MI, OH WV
9 for IL, IN, WI
Ø for CO, IA, KS, MN, MO, NE, ND, SD

The letters which may be used to prefix United States amateur station calls are: AA to AL, K, KA to KZ, N, W, WA to WZ. As an example, the author's call sign is a 1×3, W6BNB. Examples of 2d district call-sign prefixes would be W2, K2, WA2, N2, AA2, etc.

The call-sign prefix for Hawaii and the Pacific Islands is KH6. Alaska uses KL7, and Puerto Rico is KP4.

Extra Class operators may, if they wish apply for 1×2 or 2×1 calls, which are considered premium calls by some amateurs.

32-5 THE AMATEUR BANDS

Licensed amateurs may operate on a number of possible bands. Some of the bands are broken up into segments. Extra Class licensees can operate on any amateur frequency. Advanced Class licensees are restricted from operating on a few frequencies. General Class licensees are still further limited. Technician Class licensees can operate on any Novice frequencies but also have full amateur privileges on all frequencies above 50 MHz. Novice and Technician Class amateurs have only narrow segments on four of the HF bands on which they may operate, using CW emissions only. The amateur bands, as of this writing, are shown in Table 32-1, which indicates frequencies, emissions allowed, and the classes of licenses required to operate on the specified bands of frequencies.

32-6 EMISSION REQUIREMENTS

For most amateur emissions, the maximum RF power output is limited to 1500-W peak-envelope-power (PEP). One exception is the 200-W PEP for *all amateurs* operating in any Technician-Novice section of any HF band. Also, until 1990, A3E emission maximums will remain 1000-W dc power *input* (about 2000-W PEP) to the final RF amplifier, at which time they will decrease to 1500-W PEP. PEP can be measured directly with a through-line peak-reading RF wattmeter or by calculation of the power from $P_{pep} = V^2/R$, where V is the peak voltage across a feeder by oscilloscope and R is the impedance of the feedline.

Parts of the 160-meter band may be lost to the AM broadcast band in a few years. See "FCC Rules and Regulations," Part 97, or "The FCC Book." VHF and UHF repeater stations are limited in power depending on frequency and antenna altitude, and can be determined by FCC rules and regulations. The 30-m band is not a true amateur band yet. The maximum power that may be used there is 250-W dc input to the final amplifier.

At all times, amateur stations should use the minimum output power to carry out communications. If 25 W produces good readable signals at the receiving end, there is no reason to use higher power, as it may needlessly interfere with other amateurs on the band. Below 144 MHz all transmitting equipment must use adequately filtered power supplies to prevent hum modulation of any kind on any radiated signals. Spurious radiations above and below the carrier must be reduced or eliminated in accordance with good engineering practice. Such radiations include key clicks from CW emissions, overmodulation products of A3E, J3E, and F3E emissions, and parasitic oscillations in transmitter circuits.

There are three digital codes, Baudot, AMTOR, and ASCII, that may be used on HF amateur bands. They usually are produced in F1B form. When using such FSK emissions the shift must not exceed 1000 Hz below 50 MHz. Below 28 MHz 300 bauds is allowed. On the 28-MHz band 1200 bauds is permitted. In the 50- and 144-MHz bands 19.6 kilobauds is allowed, with 56 kilobauds on all higher bands. These are in addition to the International Morse code which may be used on any amateur frequency.

32-7 AMATEUR STATIONS

An amateur station may consist of only a radiotelegraph transmitter with a low-pass filter at its output, a key, a receiver with earphones or loudspeaker, an antenna, a transmit-receive switch, and possibly an SWR meter and an antenna tuner, as shown in block form in Fig. 32-1. Some separate means of

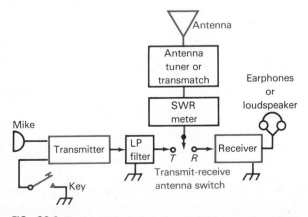

FIG. 32-1 Basic components of a simple amateur station.

TABLE 32-1 FREQUENCIES, EMISSIONS, AND CLASSES

Amateur frequencies		Emissions allowed (see Appendix E)	Who may use (Extra, Advanced, General, Technician, or Novice Classes)
	(160-m band)		
kHz			
1800–2000		A1A-A3E	E-A-G
	(80-m band)		
3500–3525		A1A-F1B	E
3525–3700		A1A-F1B	E-A-G
3700–3750		A1A-F1B*	E-A-G-T-N
3750–3775		A1A-A3E-J3E-A3C C3F-F3E-F3C-F3F	E
3775–3800		Same	E-A
3850–4000		Same	E-A-G
	(40-m band)		
7000–7025		A1A-F1B	E
7025–7100		A1A-F1B	E-A-G
7100–7150		A1A-F1B*	E-A-G-T-N
7150–7225		A1A-A3E-J3E-A3C C3F-F3E-F3C-F3F	E-A
7225–7300		Same	E-A-G
	(30-m band)		
MHz			
10.100–10.109		A1A-F1B	E-A-G
10.115–10.150		A1A-F1B	E-A-G
	(20-m band)		
14.000–14.025		A1A-F1B	E
14.025–14.150		A1A-F1B	E-A-G
14.150–14.175		A1A-A3E-J3E-A3C C3F-F3E-F3C-F3F	E
14.175–14.225		Same	E-A
14.225–14.350		Same	E-A-G
	(16-m band)		
18.068–18.168		Not available at this writing	
	(15-m band)		
21.000–21.025		A1A-F1B	E
21.025–21.100		A1A-F1B	E-A-G
21.100–21.200		A1A-F1B*	E-A-G-T-N
21.200–21.225		A1A-A3E-J3E-A3C C3F-F3E-F3C-F3F	E
21.225–21.300		Same	E-A
21.300–21.450		Same	E-A-G

TABLE 32-1 (Continued)

Amateur frequencies	Emissions allowed (see Appendix E)	Who may use (Extra, Advanced, General, Technician, or Novice Classes)
(12-m band)		
24.890–24.990	Not available at this writing	
(10-m band)		
28.0–28.1	A1A-F1B	E-A-G
28.1–28.2	A1A-F1B*	E-A-G-T-N
28.2–28.5	A1A-F1B	E-A-G
28.5–29.7	A1A-A3E-J3E-A3C C3F-F3E-F3C-F3F	E-A-G
(6-m band)		
50.0–50.1	A1A	E-A-G-T
50.1–50.4	A1A-A2A-A3E-J3E-A3C C3F-F1B-F2A-F2D-F3E F3C-F3F (FØ above 51 MHz)	E-A-G-T
(2-m band)		
144.0–144.1	A1A	E-A-G-T
144.1–144.8	NØN-A1A-A2A-A2D-A3E J3E-A3C-C3F-A3F-F1B F2A-F2D-F3E-F3C-F3F	E-A-G-T
(1.3-m band)		
220–225	Same as 144.1–144.8	E-A-G-T
(70-cm band)		
420–450	Same	E-A-G-T
GHZ		
1.24–1.30	Same	E-A-G-T
2.30–2.45	Same	E-A-G-T
3.30–3.50	Same + P1B	E-A-G-T
5.65–5.925	Same as 3.3–3.5 GHz	E-A-G-T
10.0–10.5	Same as 144.1–144.8	E-A-G-T
24.0–24.5	Same as 3.3–3.5 GHz	E-A-G-T
48.0–50.0	Same	E-A-G-T
71.0–76.0	Same	E-A-G-T
165–170	Same	E-A-G-T
240–250	Same	E-A-G-T
300 and above	Same	E-A-G-T

*F1B not available to T or N, and *all classes* limited to 200-W PEP output.

checking the transmitter frequency should be available. If transmitter and receiver are built into a single cabinet, the system is called a *transceiver*.

Actual powers used for amateur communications range from a fraction of a watt rms to 750-W rms output, which is 1500-W PEP. Transceivers usually have a maximum output of 50- to 200-W rms.

More advanced stations may include equipment to send and receive SSB, FM, AM, radioteleprinter, HF slow-scan (3-kHz bandwidth) still-picture TV transmission (SSTV), VHF or UHF slow- or fast-scan TV, facsimile, etc. Many stations use separate power amplifier stages between a transceiver and the antenna to produce the maximum power output values.

Some amateurs prefer to communicate by CW, others by telephone, by radioteleprinter (RTTY), by computer, or by TV. Handling messages or "traffic" is an important and helpful part of amateur radio, particularly during times of emergency. Talking long distances (DX) to other amateurs, or communicating via space amateur radio stations (Orbiting Satellites Carrying Amateur Radio, or OSCAR), such as using a 2-m up-link and a 10-m down-link, or bouncing signals off the moon (EME or moonbounce), are all interesting phases of amateur radio. Radioteleprinters may be used on any frequencies where F1B (FSK) or F3B (AFSK) is allowed.

Amateur antennas may be single wires, dipoles, verticals, or fixed or rotatable beams. Antennas may be coupled to transmitters or receivers by transmission lines, possibly using an impedance-changing "balun" (balanced-to-unbalanced wideband transformer), or by a *transmatch* to match transmitter to antenna feedline impedances (Sec. 32-12).

Antennas must conform to local building codes. Also, if they are more than 200 ft high, the FAA and FCC must be notified and lighting may be required. Antenna towers must not be more than 1 ft in height for every 100 ft from a listed airfield having any runway of 3200 ft or more length (1 ft per 50 ft from some other airfields) and 1 ft per 25 ft from listed heliports. Refer to Part 17 of the "FCC Rules and Regulations" regarding construction, marking, and lighting of unusually high antenna towers. The usual amateur antenna not exceeding by 20 ft any nearby buildings or trees requires no special height considerations.

Repeaters (Sec. 21-10) are receiver-transmitter systems, used both by amateurs and commercially. A 2-m mobile may have only a 10-mi range on flat ground. If a receiver is installed on a nearby hilltop and its output operates a transmitter on top of the hill, but on an adjacent frequency (usually 600 kHz above or below the receiver's frequency), mobiles can transmit on the repeater's receiving frequency, listen on the repeater's transmitting frequency, and communicate up to 100 mi or more quite satisfactorily. When a signal is received by the repeater's receiver, it operates its *carrier-operated relay*, which turns on the transmitter, allowing the received modulation to modulate the repeater transmitter. *Local-control* repeaters are usually at an amateur's home location. The amateur has direct control of the repeater's operation. *Remote control* uses a telephone line or radio link to tone-operate or control the repeater transmitter from a distance. *Automatic-control* repeaters operate unattended, but have automatic shutdown circuits in them in case the transmitter malfunctions. A control-link transmitter, usually on another band, sends controlling signals to the repeater. Control operators must shut down the transmitter if anything improper is being transmitted. A diagram of a complete repeater system is known as a system network diagram.

Repeater frequencies are 29.5–29.7 MHz, 52–54 MHz, 144.5–145.5 and 146–148 MHz, 220.5–225 MHz, 420–431 and 433–435 MHz, and all higher amateur frequencies. Control-link transmitters may only operate on frequencies above 220.5 MHz, and must use frequencies other than the repeater's receiving frequency, except for an emergency transmitter shutdown signal, which may operate into the receiver's input frequency.

Beacon stations, used to test ionospheric propagation, may be automatically controlled.

The mean power of any spurious emissions from amateur equipment operating below 30 MHz must be at least 40 dB below the output power of the transmitter, without exceeding 50 mW (30 dB down for 5-W or less transmissions). For 30–225 MHz equipment spurious radiations must be 60 dB or more down (25-W output or less, 40 dB down).

32-8 OPERATING PROVISIONS

Any amateur station must have a licensed amateur operator at the control position (except repeaters). If a licensed operator other than the licensee operates an amateur station, both operators share responsibility for all operations. A station may only be operated as permitted by the operator privileges of the class of license held by the operator at the control point. A nonamateur *third party* may communicate over an amateur radiotelephone station provided a licensed control operator continuously monitors and supervises communications.

If the control operator has privileges higher than those of the licensee and is operating in a higher privilege portion of a band, the licensee's call sign followed by a slant (/) plus the control operator's call sign must be used as identification (such as "WA4xxx/W4xx").

Whenever operating transmitting equipment, an operator must have an operator license (or a photocopy) on hand and available for inspection. If lost or destroyed, a duplicate license may be applied for.

An amateur station must identify (ID) itself at least once every 10 min during a communication ("QSO") and when completing the QSO. The only time the other station's call need be transmitted is during the call-up that leads to the QSO. However, it is considered common courtesy to transmit the other station's call sometimes during a "rag-chew" QSO. A repeater station must ID by voice (English) or CW (not over 20 wpm) every 10 min of its operation. An ID is usually made using the type emission being used in the QSO. Voice (if in a voice part of the band) or CW may be used with any type of emission but is required for facsimile or SSTV transmissions.

Radio communications for remotely controlling other amateur radio stations, or automatically relaying radio signals of other amateur stations in a system of stations, or to intercommunicate with other amateur stations in a system of stations is known as *auxiliary operation*.

A remotely controlled transmitter or an auxiliary link station is identified by the call sign of the controlling station unless automatic emissions contain the proper call sign. Remote equipment must be available to the licensee only and be tagged with an address so the FCC can contact the licensee. Remote-control apparatus used between a remote-control station and a control point is called a control link.

Except for tune-up or beacon transmissions on frequencies below 50 MHz, unmodulated carriers transmitting no intelligence are prohibited.

Amateur operation is *fixed* if the station is at the licensed address, *mobile* if in an automobile, aircraft, boat, etc., and *portable* if at some fixed location other than the licensed address.

If operating portable or mobile out-of-district, an amateur should sign his or her call, followed by the words "portable," or "mobile," and the number of the district in which operations are occurring [when using CW, with a slant (/) and the number of the district].

Amateurs are prohibited from: broadcasting; handling some forms of third-party traffic; transmitting music, codes, ciphers, or obscene, profane, or indecent language; producing false, unidentified, or intentionally interfering signals; damaging any licensed radio apparatus; obtaining a license fraudulently; handling traffic contrary to federal, state, or local laws; rebroadcasting anything including other amateur signals (except repeaters); making one-way transmissions other than for certain code practice transmissions, net operations, and emergency communications. Amateurs should always use courteous operating procedures on the air. No unidentified transmissions shall be made. Except in repeater systems, automatic retransmission of amateur signals are prohibited.

Amateurs report each other's radiotelegraph signals by using an RST system, where R stands for readability (1 = Unreadable; 5 = Perfectly readable), S stands for strength (1 = Extremely weak; 9 = 54 dB above S-1, using 6 dB per S unit), and T stands for tone (1 = Extremely poor; 9 = Very pure). For example, RST 579 indicates a perfectly readable, reasonably strong, and very pure tone signal.

Radiotelephone reports are given using only the RS portion of the RST system. For example, "4 and 9" indicates somewhat unreadable for some reason, but quite strong.

What band to use for communications depends on which one will provide the desired distance at that time of day and year. Usually daytime frequencies are 7–29 MHz, while nighttime frequencies are 1.8–14 MHz. Local communications may use any band. The frequency selected should not interfere with any communications in progress and should not produce any sidebands out of the amateur bands. Amateurs listen for a "general call" (CQ), but hearing none may call CQ themselves. CQs should be answered on or close to the frequency of the other station. CQ calls should be made at the desired communication speed.

A *log* is a written notation of transmissions made by a station. There are no longer requirements for amateurs to keep daily logs. Many do, however, to enable them to refer back to previous communications with other stations. Some keep logs in looseleaf form, some on computers, and some on file cards. Information usually includes date, time, frequency band, signal strengths, names, powers used, traffic handled, items of interest discussed, etc. FCC-required logs need only contain (1) call sign and signature of station licensee, (2) starting and ending (if any) dates of station operation, (3) starting and ending dates of any portable operation, and (4) time and date and signature of control operator when other than station licensee.

Calling and working procedures, Q signals (Appendix G), and the phonetic alphabet (Table 30-1) in the amateur service generally follow those of commercial services outlined in Chaps. 30 and 31. If an answering station is on or near the frequency of a calling station, a short call (W6ECU DE W6BNB W6BNB) is all that may be necessary. When calling a station that is removed somewhat in frequency, the called station's call letters should be repeated more than once. When calling CQ, a satisfactory procedure is to send CQ five times, then DE (from) and the call sign of the transmitting station two or three times, repeating CQs and call signs for a period of perhaps half a minute.

When operating aboard ships or aircraft, the amateur installation and operation must be approved by the captain. It must be separate from all other radio equipment, must be wired in accordance with safe practices, must not interfere with any other radio equipment aboard, and must not in any way constitute a hazard to life or property.

During emergencies special frequencies may be set aside by the FCC for emergency use only. Nonemergency operations must not be held on these frequencies during these times.

When operating a control transmitter for model craft (1 W maximum), a Transmitter Identification Card (FCC Form 452-C) or a plate indicating the owner's name and address should be attached to the transmitter.

32-9 NOTICES OF VIOLATIONS

If an amateur station causes interference to local radio broadcast reception on receivers of good engineering design, the FCC may require that silent periods be observed from 8 to 10:30 P.M. local time daily, plus 10:30 to 1:30 P.M. on Sundays, on frequencies that cause the interference. Steps to minimize such interference should be taken.

Upon receipt of any Notice of Violation from the FCC, the amateur must reply within 10 days or the license may be revoked. Explanations must be made to explain why the violation occurred and steps taken to prevent a recurrence. If the amateur is sick or away from the licensed address, reply

must be made as soon as possible, explaining also the delay in answering.

32-10 INTERNATIONAL OPERATIONS

Communications with amateurs in foreign countries are forbidden if the countries so notify (none at this writing). Communications must be in plain language and be of a technical and personal character only. Third-party traffic with foreign amateur stations is forbidden unless bilateral agreements exist between the countries. From time to time, amateur radio magazines run lists of countries that permit third-party traffic at that time. Communications may be made in any of the common world languages. At the conclusion of foreign QSOs the foreign station call sign must also be transmitted.

A bilateral agreement exists between Canada (and some other countries) and the United States that allows visiting amateurs to use radio transmitting equipment in the other country. However, in most cases, guest amateurs must register and receive a permit from any FCC field office before operating equipment in this country. Guest amateurs using radiotelegraph must identify their station by call letter followed by a slant (/) sign and the prefix and call-area number of the host country in which operation is occurring (such as W6BNB/VE7). For radiotelephone the call sign is spoken followed by the words "portable," or "mobile," as appropriate, and the call-area number of the host country. Present operating location must be mentioned frequently. Host-country frequency, power, and emission limitations must be observed by guest amateurs.

International "region 2" includes North and South America and Greenland. "Region 1" includes Africa and Northern Europe. "Region 3" includes Southern Asia and Australian areas.

32-11 RACES

Under normal circumstances amateur radio stations may communicate only with other amateur stations. During emergencies amateurs who are enrolled in the *Radio Amateur Civil Emergency Service* (RACES) may communicate with other RACES stations, and may also be authorized to communicate with certain governmental or military radio stations to handle emergency traffic on amateur frequencies. During times of civil defense operations the RACES is expected to be called upon to provide whatever communications their volunteer amateur operators and their equipment may be able to provide. Any amateur must be previously registered with a RACES organization before being allowed to communicate with any RACES station during an emergency condition. Any traffic relating to safety of life or property is considered *emergency communication*.

Messages handled by RACES stations may concern public safety, national defense, safety of life, protection of property, maintenance of law and order, alleviation of human suffering, and combating armed attack or sabotage.

RACES on-the-air training and tests may not exceed a total time of one hour per week during non-emergency times. All messages which are transmitted in connection with drills or tests shall be clearly identified as "drill" or "test" messages in the body of the message.

All authorized frequencies and emissions of the Amateur Radio Service are available to the RACES on a shared basis. However, during a War Emergency Powers declared emergency, the only RACES frequencies authorized are, in megahertz,

1.800–1.825	1.975–2.000		
3.500–3.550	3.930–3.980	3.984–4.000	3.997*
7.079–7.125	7.245–7.255		
14.047–14.053	14.220–14.230	14.231–14.350	
21.047–21.053	21.228–21.267		
28.550–28.750	29.237–29.273	29.450–29.650	
50.350–50.750	53.350–53.750		53.300*
144.50–145.71	146.00–148.00		
220–225	420–450		
1240–1300	2390–2450		

32-12 SPECIAL AMATEUR EQUIPMENT

Essentially all of the commercial radio subjects discussed throughout this book might be applicable to some form of amateur radio. A specialized piece of amateur equipment is an antenna tuner. Amateurs operate on bands of frequencies and may want to peak their antenna for optimum operation on any given frequency in any one of several bands. This can be accomplished by using an antenna-tuner, also called a *transmatch*.

In Fig. 32-2a the common 50-Ω RF output circuit from a transmitter is coupled to a link coupling coil that is coupled to the LC_1C_2 resonating circuit. The high-impedance end of a balanced tuned feedline, from a Zepp antenna, for example, can be coupled across the LCC circuit through the series capacitors. If the transmission lines are slightly

---

*For contacting military units.

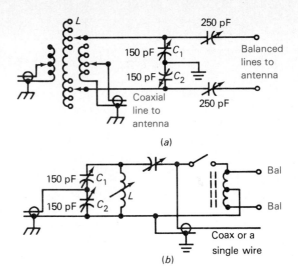

FIG. 32-2 Amateur station antenna tuners.

long, they can be electrically shortened by reducing these capacitances. If the feedline has 150- to 600-Ω impedance, it may be tapped directly across the proper number of turns of L. If a low-impedance coaxial feedline is used, it can be link-coupled to the resonant tuned circuit to reduce harmonic radiation. If the tap positions on L are changed, the tuned circuit can be made to resonate on other bands.

In Fig. 32-2b another type of tuner is shown. With the switch open the output can be adjusted to match a low-impedance unbalanced coaxial line, or it can be adjusted to match the impedance of almost any length of single-wire antenna on several bands. With the switch closed the 1:1 balun transformer produces a balanced output for balanced tuned or untuned feedlines.

Multiband antennas (sometimes used aboard ship) might be considered amateur radio-type antennas. There are several of these. One is a multidipole (Fig. 32-3a) fed at the center with a 70-Ω coaxial cable. The top $\lambda/2$ is 134-ft long for 3.5 MHz. The mid-dipole is about 66-ft long for 7 MHz. The shortest dipole is about 33-ft long for 14 MHz. Any frequency fed to the antenna will pick its resonant length $\lambda/2$ dipole and excite it into oscillation. This is actually a 4-band antenna because 21 MHz, the third harmonic of 7 MHz, makes the 7-MHz dipole a $3\lambda/2$ antenna being fed at the center of its mid-$\lambda/2$ section, and it resonates well. Multiband verticals use two or more adjacent vertical $\lambda/4$ rods connected together near the base.

The antenna shown in Fig. 32-3b is a 3-band

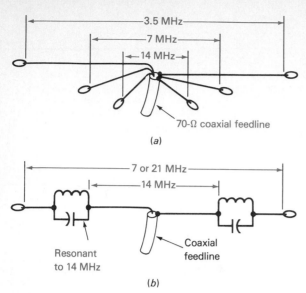

FIG. 32-3 (*a*) Multiband dipole antenna. (*b*) Trapped 3-band dipole.

trapped dipole. The tuned *LC* circuits are resonant to 14 MHz and appear as an RFC to this frequency. They appear as lumped inductors to 7-MHz RF ac, acting as loading coils for this frequency, shortening the overall length of the 7-MHz antenna materially. The 21-MHz band sees the 7-MHz antenna as a 3λ/2 center-fed antenna. Multiband trapped λ/4 vertical antennas are used also.

32-13 NOVICE CLASS LICENSES

The test for the Novice, or beginners, Class license consists of FCC amateur test Elements 1*a* and 2 (Sec. 32-2).

An applicant must find a local General, Advanced, or Extra Class license holder that is willing to give a Novice test, perhaps located through a local amateur radio club. An FCC application Form 610 must be obtained from an FCC field office (although some amateur volunteers may have a supply). It must be filled out and given to the volunteer examiner. The volunteer examiner must then make up a 5-wpm Morse code sending and receiving test, Element 1*a*, and give it to the applicant. A code test consists of the demonstration by the applicant to the volunteer examiner that at least 1 min of clean sending out of 5 min can be produced with any type of hand-operated key provided by the applicant (other than a keyboard). The receiving test may be either a 1-min correct copy out of 5 min of 5-wpm plain-language letters and numbers

transmitted code, or a written test on a 5-min-long transmitted message. If this is passed, the volunteer must then make up a 20-question test from a list of Element 2 question-items provided by the FCC for the Novice test. If the applicant passes at least 15 of the 20 questions on the test, the volunteer certifies on Form 610 that the applicant has been examined and has passed the required tests. Form 610 is sent to FCC, Gettysburg, Pennsylvania 17325. The license and call sign will be made out by the FCC and returned to the applicant in a few weeks.

If either the code or the theory test is failed, the applicant must wait 30 days before taking another Novice test.

Novices may communicate with any other licensed amateur but the Novice may not transmit outside the Novice bands.

32-14 TECHNICIAN CLASS LICENSES

The test for the Technician Class license consists of Elements 1*a*, 2, and 3 (Sec. 32-2). Credit will be given for both Elements 1*a* and 2 if the applicant holds a valid Novice Class license.

The applicant must locate a volunteer examination coordinator (VEC), either by checking with local amateurs, amateur radio clubs, or the FCC may know the closest VEC to the applicant. The VEC will assign a volunteer examiner to the applicant, who will make up a 5-wpm code test, Element 1*a*, and written tests on Elements 2 and 3 from FCC provided test materials. If these tests are passed successfully, the volunteer will so certify on the Form 610 and send it to the FCC, Gettysburg, PA 17325, as with a Novice test. If a valid Novice license is held, anyone passing the Technician class examination can immediately use the Novice call on Technician bands.

32-15 GENERAL CLASS LICENSES

A test for a General Class license consists of Elements 1*b*, 2, and 3 (Sec. 32-2). It is the simplest all-band amateur license. It differs from a Technician Class license test in code speed only, requiring a 13-wpm test.

As with other amateur licenses, the test will be given by a volunteer examiner. Credit will be given for Element 2 if a valid Novice Class license is held, or Elements 2 and 3 if a valid Technician Class license is held. Applicants successfully passing a General Class license examination may immediately use their Novice or Technician call on the General Class frequencies.

32-16 ADVANCED CLASS LICENSES

The Advanced Class license test consists of Elements 1*b*, 2, 3, and 4*a*. Credit is given for elements included in any other valid amateur license held. The procedure is the same as with all other amateur licenses. The only new requirement is Element 4*a* (Sec. 32-2).

32-17 EXTRA CLASS LICENSES

The Extra Class license test consists of Elements 1*c*, 2, 3, 4*a*, and 4*b* (Sec. 32-2). Credit is given for elements included in any other valid amateur license held. The Extra Class license is the highest class amateur license and offers full amateur privileges to its holder. It differs from the Advanced Class license in code speed (20 wpm instead of 13 wpm) and in an additional more advanced theory examination. The Extra Class tests are given in a manner similar to all other amateur licenses.

AMATEUR RADIO QUESTIONS

1. Where can complete rules governing the Amateur Radio Service be found?
2. Where can over-the-air code practice be heard?
3. What are the five grades of amateur licenses?
4. What is the lowest code speed test given for amateur licenses?
5. What is the term of an amateur license? The grace period?
6. When an applicant takes an amateur license test, does it provide a station or operator license, or both?
7. How many call-sign districts are there in the contiguous United States, Alaska, Hawaii, and Puerto Rico?
8. What call-sign prefixes are used by U.S. amateurs?
9. What is meant by a 2×3 call?
10. List the HF amateur bands available at this time.
11. What class licensee can use all frequencies of all amateur bands?
12. What is the maximum power for most amateur frequencies? What sections of which bands are limited to 200-W PEP?
13. How much power should be used for radio communications?
14. What are the three digital codes used by amateurs?
15. Draw a block diagram of a simple amateur station.
16. List eight forms of communications used in amateur radio.
17. What is a transmatch?
18. Where is information found about antenna lighting and painting?
19. How are repeaters remotely controlled?
20. If an Extra Class amateur transmits at a Novice's station, what frequencies and emissions can be used? How is the station identified?
21. How often must all amateur stations be identified?
22. How must FAX or SSTV be identified?
23. What type of amateur transmissions are prohibited?
24. For what type emission are RST reports given? How would a pure-tone, strong, easily read signal be indicated?
25. What is the general call to any one interested in a QSO? With what emissions is it used?
26. Is it necessary to log amateur communications?
27. When first tuning up on a frequency should a short or long CQ be given?
28. Within what period must a written reply be made to any Notice of Violation?
29. In general, is 3rd party traffic allowed with foreign countries?
30. Is it necessary to notify the Canadian government if a visiting U.S. amateur is going to operate in Canada?
31. In what international region is the United States? Australia? France?
32. What does RACES mean? Who may operate in it during an emergency condition?
33. Draw a diagram of an antenna tuner that can be used to feed either a balanced feeder or a coaxial line.
34. Draw diagrams of a 4-band multidipole antenna and a trapped 3-band antenna.
35. What elements must be passed to be issued a Novice Class license?
36. A General Class licensed amateur must pass what test element(s) to obtain an Advanced Class license? An Extra Class license?

Appendixes

GREEK ALPHABET

Upper case	Lower case	Name	English equivalent	Upper case	Lower case	Name	English equivalent
A	α	Alpha	a	N	ν	Nu	n
B	β	Beta	b	Ξ	ξ	Xi	x
Γ	γ	Gamma	g	O	o	Omicron	ŏ
Δ	δ	Delta	d	Π	π	Pi	p
E	ϵ	Epsilon	ĕ	P	ρ	Rho	r
Z	ζ	Zeta	z	Σ	σ, ς	Sigma	s
H	η	Eta	ē	T	τ	Tau	t
Θ	θ	Theta	th	Υ	υ	Upsilon	u
I	ι	Iota	i	Φ	ϕ, φ	Phi	ph, f
K	κ	Kappa	k	X	χ	Chi	ch
Λ	λ	Lambda	l	Ψ	ψ	Psi	ps
M	μ	Mu	m	Ω	ω	Omega	ō

APPENDIX B

STANDARD COMPONENT VALUES

Carbon resistors, 0.1- to 3-W types, are available with ±5% tolerance in all of the values listed below. They are also available in ±10% tolerance, but only in the values shown in bold figures. Capacitators, in picofarads (pF), are also generally available in the 10% values shown.

Ohms	Ohms	Ohms	Ohms	Ohms	Ohms	Ohms	Ohms	Ohms	Ohms
2.7	13	**68**	360	**1,800**	9,100	**47,000**	0.24 M	**1.2 M**	6.2 M
3.0	**15**	75	**390**	2,000	**10,000**	51,000	**0.27 M**	1.3 M	**6.8 M**
3.3	16	**82**	430	**2,200**	11,000	**56,000**	0.30 M	**1.5 M**	7.5 M
3.6	**18**	91	**470**	2,400	**12,000**	62,000	**0.33 M**	1.6 M	**8.2 M**
3.9	20	**100**	510	**2,700**	13,000	**68,000**	0.36 M	**1.8 M**	9.1 M
4.3	**22**	110	**560**	3,000	**15,000**	75,000	**0.39 M**	2.0 M	**10.0 M**
4.7	24	**120**	620	**3,300**	16,000	**82,000**	0.43 M	**2.2 M**	11.0 M
5.1	**27**	130	**680**	3,600	**18,000**	91,000	**0.47 M**	2.4 M	**12.0 M**
5.6	30	**150**	750	**3,900**	20,000	**0.1 M**	0.51 M	**2.7 M**	13.0 M
6.2	**33**	160	**820**	4,300	**22,000**	0.11 M	**0.56 M**	3.0 M	**15.0 M**
6.8	36	**180**	910	**4,700**	24,000	**0.12 M**	0.62 M	**3.3 M**	16.0 M
7.5	**39**	200	**1,000**	5,100	**27,000**	0.13 M	**0.68 M**	3.6 M	**18.0 M**
8.2	43	**220**	1,100	**5,600**	30,000	**0.15 M**	0.75 M	**3.9 M**	20.0 M
9.1	**47**	240	**1,200**	6,200	**33,000**	0.16 M	**0.82 M**	4.3 M	**22.0 M**
10	51	**270**	1,300	**6,800**	36,000	**0.18 M**	0.91 M	**4.7 M**	
11	**56**	300	**1,500**	7,500	**39,000**	0.20 M	**1.0 M**	5.1 M	
12	62	**330**	1,600	**8,200**	43,000	**0.22 M**	1.1 M	**5.6 M**	

MILITARY PRECISION STANDARD VALUES

Carbon or wire-wound precision (1%) resistors are available from ⅛ to 10 W in the values indicated below. Only one decade is shown. For other decade values, multiply by 0.1, 10, 100, 1,000, etc. Preferred values are shown in bold figures.

1.00	**1.47**	**2.15**	**3.16**	**4.64**	**6.81**
1.02	1.50	2.21	3.24	4.75	6.98
1.05	1.54	2.26	3.32	4.87	7.15
1.07	1.58	2.32	3.40	4.99	7.32
1.10	**1.62**	**2.37**	**3.48**	**5.11**	**7.50**
1.13	1.65	2.43	3.57	5.23	7.68
1.15	1.69	2.49	3.65	5.36	7.87
1.18	1.74	2.55	3.74	5.49	8.06
1.21	**1.78**	**2.61**	**3.83**	**5.62**	**8.25**
1.24	1.82	2.67	3.92	5.76	8.45
1.27	1.87	2.74	4.02	5.90	8.66
1.30	1.91	2.80	4.12	6.04	8.87
1.33	**1.96**	**2.87**	**4.22**	**6.19**	**9.09**
1.37	2.00	2.94	4.32	6.34	9.31
1.40	2.05	3.01	4.42	6.49	9.53
1.43	2.10	3.09	4.53	6.65	9.76

TABLE OF NATURAL TRIGONOMETRIC FUNCTIONS

Angle,°	sin	tan	cot	cos	
0.0	.00000	.00000	∞	1.00000	90.0
.1	.00175	.00175	572.96	1.00000	.9
.2	.00349	.00349	286.48	0.99999	.8
.3	.00524	.00524	190.98	.99999	.7
.4	.00698	.00698	143.24	.99998	.6
.5	.00873	.00873	114.59	.99996	.5
.6	.01047	.01047	95.489	.99995	.4
.7	.01222	.01222	81.847	.99993	.3
.8	.01396	.01396	71.615	.99990	.2
.9	.01571	.01571	63.657	.99988	.1
1.0	.01745	.01746	57.290	.99985	89.0
.1	.01920	.01920	52.081	.99982	.9
.2	.02094	.02095	47.740	.99978	.8
.3	.02269	.02269	44.066	.99974	.7
.4	.02443	.02444	40.917	.99970	.6
.5	.02618	.02619	38.188	.99966	.5
.6	.02792	.02793	35.801	.99961	.4
.7	.02967	.02968	33.694	.99956	.3
.8	.03141	.03143	31.821	.99951	.2
.9	.03316	.03317	30.145	.99945	.1
2.0	.03490	.03492	28.636	.99939	88.0
.1	.03664	.03667	27.271	.99933	.9
.2	.03839	.03842	26.031	.99926	.8
.3	.04013	.04016	24.898	.99919	.7
.4	.04188	.04191	23.859	.99912	.6
.5	.04362	.04366	22.904	.99905	.5
.6	.04536	.04541	22.022	.99897	.4
.7	.04711	.04716	21.205	.99889	.3
.8	.04885	.04891	20.446	.99881	.2
.9	.05059	.05066	19.740	.99872	.1
3.0	.05234	.05241	19.081	.99863	87.0
.1	.05408	.05416	18.464	.99854	.9
.2	.05582	.05591	17.886	.99844	.8
.3	.05756	.05766	17.343	.99834	.7
.4	.05931	.05941	16.832	.99824	.6
.5	.06105	.06116	16.350	.99813	.5
.6	.06279	.06291	15.895	.99803	.4
.7	.06453	.06467	15.464	.99792	.3
.8	.06627	.06642	15.056	.99780	.2
.9	.06802	.06817	14.669	.99768	.1
4.0	.06976	.06993	14.301	.99756	86.0
.1	.07150	.07168	13.951	.99744	.9
.2	.07324	.07344	13.617	.99731	.8
.3	.07498	.07519	13.300	.99719	.7
.4	.07672	.07695	12.996	.99705	.6
.5	.07846	.07870	12.706	.99692	.5
.6	.08020	.08046	12.429	.99678	.4
.7	.08194	.08221	12.163	.99664	.3
.8	.08368	.08397	11.909	.99649	.2
.9	.08542	.08573	11.664	.99635	.1
5.0	.08716	.08749	11.430	.99619	85.0
.1	.08889	.08925	11.205	.99604	.9
.2	.09063	.09101	10.988	.99588	.8
.3	.09237	.09277	10.780	.99572	.7
.4	.09411	.09453	10.579	.99556	.6
.5	.09585	.09629	10.385	.99540	.5
.6	.09758	.09805	10.199	.99523	.4
.7	.09932	.09981	10.019	.99506	.3
.8	.10106	.10158	9.8448	.99488	.2
.9	.10279	.10334	9.6768	.99470	.1
6.0	.10453	.10510	9.5144	.99452	84.0
	cos	cot	tan	sin	Angle,°

Angle,°	sin	tan	cot	cos	
6.0	.10453	.10510	9.5144	.99452	84.0
.1	.10626	.10687	9.3572	.99434	.9
.2	.10800	.10863	9.2052	.99415	.8
.3	.10973	.11040	9.0579	.99396	.7
.4	.11147	.11217	8.9152	.99377	.6
.5	.11320	.11394	8.7769	.99357	.5
.6	.11494	.11570	8.6427	.99337	.4
.7	.11667	.11747	8.5126	.99317	.3
.8	.11840	.11924	8.3863	.99297	.2
.9	.12014	.12101	8.2636	.99276	.1
7.0	.12187	.12278	8.1443	.99255	83.0
.1	.12360	.12456	8.0285	.99233	.9
.2	.12533	.12633	7.9158	.99211	.8
.3	.12706	.12810	7.8062	.99189	.7
.4	.12880	.12988	7.6996	.99167	.6
.5	.13053	.13165	7.5958	.99144	.5
.6	.13226	.13343	7.4947	.99122	.4
.7	.13399	.13521	7.3962	.99098	.3
.8	.13572	.13698	7.3002	.99075	.2
.9	.13744	.13876	7.2066	.99051	.1
8.0	.13917	.14054	7.1154	.99027	82.0
.1	.14090	.14232	7.0264	.99002	.9
.2	.14263	.14410	6.9395	.98978	.8
.3	.14436	.14588	6.8548	.98953	.7
.4	.14608	.14767	6.7720	.98927	.6
.5	.14781	.14945	6.6912	.98902	.5
.6	.14954	.15124	6.6122	.98876	.4
.7	.15126	.15302	6.5350	.98849	.3
.8	.15299	.15481	6.4596	.98823	.2
.9	.15471	.15660	6.3859	.98796	.1
9.0	.15643	.15838	6.3138	.98769	81.0
.1	.15816	.16017	6.2432	.98741	.9
.2	.15988	.16196	6.1742	.98714	.8
.3	.16160	.16376	6.1066	.98686	.7
.4	.16333	.16555	6.0405	.98657	.6
.5	.16505	.16734	5.9758	.98629	.5
.6	.16677	.16914	5.9124	.98600	.4
.7	.16849	.17093	5.8502	.98570	.3
.8	.17021	.17273	5.7894	.98541	.2
.9	.17193	.17453	5.7297	.98511	.1
10.0	.17365	.17633	5.6713	.98481	80.0
.1	.17537	.17813	5.6140	.98450	.9
.2	.17708	.17993	5.5578	.98420	.8
.3	.17880	.18173	5.5026	.98389	.7
.4	.18052	.18353	5.4486	.98357	.6
.5	.18224	.18534	5.3955	.98325	.5
.6	.18395	.18714	5.3435	.98294	.4
.7	.18567	.18895	5.2924	.98261	.3
.8	.18738	.19076	5.2422	.98229	.2
.9	.18910	.19257	5.1929	.98196	.1
11.0	.19081	.19438	5.1446	.98163	79.0
.1	.19252	.19619	5.0970	.98129	.9
.2	.19423	.19801	5.0504	.98096	.8
.3	.19595	.19982	5.0045	.98061	.7
.4	.19766	.20164	4.9594	.98027	.6
.5	.19937	.20345	4.9152	.97992	.5
.6	.20108	.20527	4.8716	.97958	.4
.7	.20279	.20709	4.8288	.97922	.3
.8	.20450	.20891	4.7867	.97887	.2
.9	.20620	.21073	4.7453	.97851	.1
12.0	.20791	.21256	4.7046	.97815	78.0
	cos	cot	tan	sin	Angle,°

Angle, °	sin	tan	cot	cos	
12.0	.20791	.21256	4.7046	.97815	**78.0**
.1	.20962	.21438	4.6646	.97778	.9
.2	.21132	.21621	4.6252	.97742	.8
.3	.21303	.21804	4.5864	.97705	.7
.4	.21474	.21986	4.5483	.97667	.6
.5	.21644	.22169	4.5107	.97630	.5
.6	.21814	.22353	4.4737	.97592	.4
.7	.21985	.22536	4.4373	.97553	.3
.8	.22155	.22719	4.4015	.97515	.2
.9	.22325	.22903	4.3662	.97476	.1
13.0	.22495	.23087	4.3315	.97437	**77.0**
.1	.22665	.23271	4.2972	.97398	.9
.2	.22835	.23455	4.2635	.97358	.8
.3	.23005	.23639	4.2303	.97318	.7
.4	.23175	.23823	4.1976	.97278	.6
.5	.23345	.24008	4.1653	.97237	.5
.6	.23514	.24193	4.1335	.97196	.4
.7	.23684	.24377	4.1022	.97155	.3
.8	.23853	.24562	4.0713	.97113	.2
.9	.24023	.24747	4.0408	.97072	.1
14.0	.24192	.24933	4.0108	.97030	**76.0**
.1	.24362	.25118	3.9812	.96987	.9
.2	.24531	.25304	3.9520	.96945	.8
.3	.24700	.25490	3.9232	.96902	.7
.4	.24869	.25676	3.8947	.96858	.6
.5	.25038	.25862	3.8667	.96815	.5
.6	.25207	.26048	3.8391	.96771	.4
.7	.25376	.26235	3.8118	.96727	.3
.8	.25545	.26421	3.7848	.96682	.2
.9	.25713	.26608	3.7583	.96638	.1
15.0	.25882	.26795	3.7321	.96593	**75.0**
.1	.26050	.26982	3.7062	.96547	.9
.2	.26219	.27169	3.6806	.96502	.8
.3	.26387	.27357	3.6554	.96456	.7
.4	.26556	.27545	3.6305	.96410	.6
.5	.26724	.27732	3.6059	.96363	.5
.6	.26892	.27921	3.5816	.96316	.4
.7	.27060	.28109	3.5576	.96269	.3
.8	.27228	.28297	3.5339	.96222	.2
.9	.27396	.28486	3.5105	.96174	.1
16.0	.27564	.28675	3.4874	.96126	**74.0**
.1	.27731	.28864	3.4646	.96078	.9
.2	.27899	.29053	3.4420	.96029	.8
.3	.28067	.29242	3.4197	.95981	.7
.4	.28234	.29432	3.3977	.95931	.6
.5	.28402	.29621	3.3759	.95882	.5
.6	.28569	.29811	3.3544	.95832	.4
.7	.28736	.30001	3.3332	.95782	.3
.8	.28903	.30192	3.3122	.95732	.2
.9	.29070	.30382	3.2914	.95681	.1
17.0	.29237	.30573	3.2709	.95630	**73.0**
.1	.29404	.30764	3.2506	.95579	.9
.2	.29571	.30955	3.2305	.95528	.8
.3	.29737	.31147	3.2106	.95476	.7
.4	.29904	.31338	3.1910	.95424	.6
.5	.30071	.31530	3.1716	.95372	.5
.6	.30237	.31722	3.1524	.95319	.4
.7	.30403	.31914	3.1334	.95266	.3
.8	.30570	.32106	3.1146	.95213	.2
.9	.30736	.32299	3.0961	.95159	.1
18.0	.30902	.32492	3.0777	.95106	**72.0**
	cos	cot	tan	sin	Angle, °

Angle, °	sin	tan	cot	cos	
18.0	.30902	.32492	3.0777	.95106	**72.0**
.1	.31068	.32685	3.0595	.95052	.9
.2	.31233	.32878	3.0415	.94997	.8
.3	.31399	.33072	3.0237	.94943	.7
.4	.31565	.33266	3.0061	.94888	.6
.5	.31730	.33460	2.9887	.94832	.5
.6	.31896	.33654	2.9714	.94777	.4
.7	.32061	.33848	2.9544	.94721	.3
.8	.32227	.34043	2.9375	.94665	.2
.9	.32392	.34238	2.9208	.94609	.1
19.0	.32557	.34433	2.9042	.94552	**71.0**
.1	.32722	.34628	2.8878	.94495	.9
.2	.32887	.34824	2.8716	.94438	.8
.3	.33051	.35020	2.8556	.94380	.7
.4	.33216	.35216	2.8397	.94322	.6
.5	.33381	.35412	2.8239	.94264	.5
.6	.33545	.35608	2.8083	.94206	.4
.7	.33710	.35805	2.7929	.94147	.3
.8	.33874	.36002	2.7776	.94088	.2
.9	.34038	.36199	2.7625	.94029	.1
20.0	.34202	.36397	2.7475	.93969	**70.0**
.1	.34366	.36595	2.7326	.93909	.9
.2	.34530	.36793	2.7179	.93849	.8
.3	.34694	.36991	2.7034	.93789	.7
.4	.34857	.37190	2.6889	.93728	.6
.5	.35021	.37388	2.6746	.93667	.5
.6	.35184	.37588	2.6605	.93606	.4
.7	.35347	.37787	2.6464	.93544	.3
.8	.35511	.37986	2.6325	.93483	.2
.9	.35674	.38186	2.6187	.93420	.1
21.0	.35837	.38386	2.6051	.93358	**69.0**
.1	.36000	.38587	2.5916	.93295	.9
.2	.36162	.38787	2.5782	.93232	.8
.3	.36325	.38988	2.5649	.93169	.7
.4	.36488	.39190	2.5517	.93106	.6
.5	.36650	.39391	2.5386	.93042	.5
.6	.36812	.39593	2.5257	.92978	.4
.7	.36975	.39795	2.5129	.92913	.3
.8	.37137	.39997	2.5002	.92849	.2
.9	.37299	.40200	2.4876	.92784	.1
22.0	.37461	.40403	2.4751	.92718	**68.0**
.1	.37622	.40606	2.4627	.92653	.9
.2	.37784	.40809	2.4504	.92587	.8
.3	.37946	.41013	2.4383	.92521	.7
.4	.38107	.41217	2.4262	.92455	.6
.5	.38268	.41421	2.4142	.92388	.5
.6	.38430	.41626	2.4023	.92321	.4
.7	.38591	.41831	2.3906	.92254	.3
.8	.38752	.42036	2.3789	.92186	.2
.9	.38912	.42242	2.3673	.92119	.1
23.0	.39073	.42447	2.3559	.92050	**67.0**
.1	.39234	.42654	2.3445	.91982	.9
.2	.39394	.42860	2.3332	.91914	.8
.3	.39555	.43067	2.3220	.91845	.7
.4	.39715	.43274	2.3109	.91775	.6
.5	.39875	.43481	2.2998	.91706	.5
.6	.40035	.43689	2.2889	.91636	.4
.7	.40195	.43897	2.2781	.91566	.3
.8	.40355	.44105	2.2673	.91496	.2
.9	.40514	.44314	2.2566	.91425	.1
24.0	.40674	.44523	2.2460	.91355	**66.0**
	cos	cot	tan	sin	Angle, °

TABLE OF NATURAL
TRIGONOMETRIC FUNCTIONS (continued)

Angle, °	sin	tan	cot	cos	
24.0	.40674	.44523	2.2460	.91355	66.0
.1	.40833	.44732	2.2355	.91283	.9
.2	.40992	.44942	2.2251	.91212	.8
.3	.41151	.45152	2.2148	.91140	.7
.4	.41310	.45362	2.2045	.91068	.6
.5	.41469	.45573	2.1943	.90996	.5
.6	.41628	.45784	2.1842	.90924	.4
.7	.41787	.45995	2.1742	.90851	.3
.8	.41945	.46206	2.1642	.90778	.2
.9	.42104	.46418	2.1543	.90704	.1
25.0	.42262	.46631	2.1445	.90631	65.0
.1	.42420	.46843	2.1348	.90557	.9
.2	.42578	.47056	2.1251	.90483	.8
.3	.42736	.47270	2.1155	.90408	.7
.4	.42894	.47483	2.1060	.90334	.6
.5	.43051	.47698	2.0965	.90259	.5
.6	.43209	.47912	2.0872	.90183	.4
.7	.43366	.48127	2.0778	.90108	.3
.8	.43523	.48342	2.0686	.90032	.2
.9	.43680	.48557	2.0594	.89956	.1
26.0	.43837	.48773	2.0503	.89879	64.0
.1	.43994	.48989	2.0413	.89803	.9
.2	.44151	.49206	2.0323	.89726	.8
.3	.44307	.49423	2.0233	.89649	.7
.4	.44464	.49640	2.0145	.89571	.6
.5	.44620	.49858	2.0057	.89493	.5
.6	.44776	.50076	1.9970	.89415	.4
.7	.44932	.50295	1.9883	.89337	.3
.8	.45088	.50514	1.9797	.89259	.2
.9	.45243	.50733	1.9711	.89180	.1
27.0	.45399	.50953	1.9626	.89101	63.0
.1	.45554	.51173	1.9542	.89021	.9
.2	.45710	.51393	1.9458	.88942	.8
.3	.45865	.51614	1.9375	.88862	.7
.4	.46020	.51835	1.9292	.88782	.6
.5	.46175	.52057	1.9210	.88701	.5
.6	.46330	.52279	1.9128	.88620	.4
.7	.46484	.52501	1.9047	.88539	.3
.8	.46639	.52724	1.8967	.88458	.2
.9	.46793	.52947	1.8887	.88377	.1
28.0	.46947	.53171	1.8807	.88295	62.0
.1	.47101	.53395	1.8728	.88213	.9
.2	.47255	.53620	1.8650	.88130	.8
.3	.47409	.53844	1.8572	.88048	.7
.4	.47562	.54070	1.8495	.87965	.6
.5	.47716	.54296	1.8418	.87882	.5
.6	.47869	.54522	1.8341	.87798	.4
.7	.48022	.54748	1.8265	.87715	.3
.8	.48175	.54975	1.8190	.87631	.2
.9	.48328	.55203	1.8115	.87546	.1
29.0	.48481	.55431	1.8040	.87462	61.0
.1	.48634	.55659	1.7966	.87377	.9
.2	.48786	.55888	1.7893	.87292	.8
.3	.48938	.56117	1.7820	.87207	.7
.4	.49090	.56347	1.7747	.87121	.6
.5	.49242	.56577	1.7675	.87036	.5
.6	.49394	.56808	1.7603	.86949	.4
.7	.49546	.57039	1.7532	.86863	.3
.8	.49697	.57271	1.7461	.86777	.2
.9	.49849	.57503	1.7391	.86690	.1
30.0	.50000	.57735	1.7321	.86603	60.0
	cos	cot	tan	sin	Angle, °

Angle, °	sin	tan	cot	cos	
30.0	.50000	.57735	1.7321	.86603	60.0
.1	.50151	.57968	1.7251	.86515	.9
.2	.50302	.58201	1.7182	.86427	.8
.3	.50453	.58435	1.7113	.86340	.7
.4	.50603	.58670	1.7045	.86251	.6
.5	.50754	.58905	1.6977	.86163	.5
.6	.50904	.59140	1.6909	.86074	.4
.7	.51054	.59376	1.6842	.85985	.3
.8	.51204	.59612	1.6775	.85896	.2
.9	.51354	.59849	1.6709	.85806	.1
31.0	.51504	.60086	1.6643	.85717	59.0
.1	.51653	.60324	1.6577	.85627	.9
.2	.51803	.60562	1.6512	.85536	.8
.3	.51952	.60801	1.6447	.85446	.7
.4	.52101	.61040	1.6383	.85355	.6
.5	.52250	.61280	1.6319	.85264	.5
.6	.52399	.61520	1.6255	.85173	.4
.7	.52547	.61761	1.6191	.85081	.3
.8	.52696	.62003	1.6128	.84989	.2
.9	.52844	.62245	1.6066	.84897	.1
32.0	.52992	.62487	1.6003	.84805	58.0
.1	.53140	.62730	1.5941	.84712	.9
.2	.53288	.62973	1.5880	.84619	.8
.3	.53435	.63217	1.5818	.84526	.7
.4	.53583	.63462	1.5757	.84433	.6
.5	.53730	.63707	1.5697	.84339	.5
.6	.53877	.63953	1.5637	.84245	.4
.7	.54024	.64199	1.5577	.84151	.3
.8	.54171	.64446	1.5517	.84057	.2
.9	.54317	.64693	1.5458	.83962	.1
33.0	.54464	.64941	1.5399	.83867	57.0
.1	.54610	.65189	1.5340	.83772	.9
.2	.54756	.65438	1.5282	.83676	.8
.3	.54902	.65688	1.5224	.83581	.7
.4	.55048	.65938	1.5166	.83485	.6
.5	.55194	.66189	1.5108	.83389	.5
.6	.55339	.66440	1.5051	.83292	.4
.7	.55484	.66692	1.4994	.83195	.3
.8	.55630	.66944	1.4938	.83098	.2
.9	.55775	.67197	1.4882	.83001	.1
34.0	.55919	.67451	1.4826	.82904	56.0
.1	.56064	.67705	1.4770	.82806	.9
.2	.56208	.67960	1.4715	.82708	.8
.3	.56353	.68215	1.4659	.82610	.7
.4	.56497	.68471	1.4605	.82511	.6
.5	.56641	.68728	1.4550	.82413	.5
.6	.56784	.68985	1.4496	.82314	.4
.7	.56928	.69243	1.4442	.82214	.3
.8	.57071	.69502	1.4388	.82115	.2
.9	.57215	.69761	1.4335	.82015	.1
35.0	.57358	.70021	1.4281	.81915	55.0
.1	.57501	.70281	1.4229	.81815	.9
.2	.57643	.70542	1.4176	.81714	.8
.3	.57786	.70804	1.4124	.81614	.7
.4	.57928	.71066	1.4071	.81513	.6
.5	.58070	.71329	1.4019	.81412	.5
.6	.58212	.71593	1.3968	.81310	.4
.7	.58354	.71857	1.3916	.81208	.3
.8	.58496	.72122	1.3865	.81106	.2
.9	.58637	.72388	1.3814	.81004	.1
36.0	.58779	.72654	1.3764	.80902	54.0
	cos	cot	tan	sin	Angle, °

TABLE OF NATURAL
TRIGONOMETRIC FUNCTIONS (continued)

Angle, °	sin	tan	cot	cos		Angle, °	sin	tan	cot	cos	
36.0	.58779	.72654	1.3764	.80902	**54.0**	**40.5**	.64945	.85408	1.1708	.76041	**49.5**
.1	.58920	.72921	1.3713	.80799	.9	.6	.65077	.85710	1.1667	.75927	.4
.2	.59061	.73189	1.3663	.80696	.8	.7	.65210	.86014	1.1626	.75813	.3
.3	.59201	.73457	1.3613	.80593	.7	.8	.65342	.86318	1.1585	.75700	.2
.4	.59342	.73726	1.3564	.80489	.6	.9	.65474	.86623	1.1544	.75585	.1
.5	.59482	.73996	1.3514	.80386	.5	**41.0**	.65606	.86929	1.1504	.75471	**49.0**
.6	.59622	.74267	1.3465	.80282	.4	.1	.65738	.87236	1.1463	.75356	.9
.7	.59763	.74538	1.3416	.80178	.3	.2	.65869	.87543	1.1423	.75241	.8
.8	.59902	.74810	1.3367	.80073	.2	.3	.66000	.87852	1.1383	.75126	.7
.9	.60042	.75082	1.3319	.79968	.1	.4	.66131	.88162	1.1343	.75011	.6
37.0	.60182	.75355	1.3270	.79864	**53.0**	.5	.66262	.88473	1.1303	.74896	.5
.1	.60321	.75629	1.3222	.79758	.9	.6	.66393	.88784	1.1263	.74780	.4
.2	.60460	.75904	1.3175	.79653	.8	.7	.66523	.89097	1.1224	.74664	.3
.3	.60599	.76180	1.3127	.79547	.7	.8	.66653	.89410	1.1184	.74548	.2
.4	.60738	.76456	1.3079	.79441	.6	.9	.66783	.89725	1.1145	.74431	.1
.5	.60876	.76733	1.3032	.79335	.5	**42.0**	.66913	.90040	1.1106	.74314	**48.0**
.6	.61015	.77010	1.2985	.79229	.4	.1	.67043	.90357	1.1067	.74198	.9
.7	.61153	.77289	1.2938	.79122	.3	.2	.67172	.90674	1.1028	.74080	.8
.8	.61291	.77568	1.2892	.79016	.2	.3	.67301	.90993	1.0990	.73963	.7
.9	.61429	.77848	1.2846	.78908	.1	.4	.67430	.91313	1.0951	.73846	.6
38.0	.61566	.78129	1.2799	.78801	**52.0**	.5	.67559	.91633	1.0913	.73728	.5
.1	.61704	.78410	1.2753	.78694	.9	.6	.67688	.91955	1.0875	.73610	.4
.2	.61841	.78692	1.2708	.78586	.8	.7	.67816	.92277	1.0837	.73491	.3
.3	.61978	.78975	1.2662	.78478	.7	.8	.67944	.92601	1.0799	.73373	.2
.4	.62115	.79259	1.2617	.78369	.6	.9	.68072	.92926	1.0761	.73254	.1
.5	.62251	.79544	1.2572	.78261	.5	**43.0**	.68200	.93252	1.0724	.73135	**47.0**
.6	.62388	.79829	1.2527	.78152	.4	.1	.68327	.93578	1.0686	.73016	.9
.7	.62524	.80115	1.2482	.78043	.3	.2	.68455	.93906	1.0649	.72897	.8
.8	.62660	.80402	1.2437	.77934	.2	.3	.68582	.94235	1.0612	.72777	.7
.9	.62796	.80690	1.2393	.77824	.1	.4	.68709	.94565	1.0575	.72657	.6
39.0	.62932	.80978	1.2349	.77715	**51.0**	.5	.68835	.94896	1.0538	.72537	.5
.1	.63068	.81268	1.2305	.77605	.9	.6	.68962	.95229	1.0501	.72417	.4
.2	.63203	.81558	1.2261	.77494	.8	.7	.69088	.95562	1.0464	.72297	.3
.3	.63338	.81849	1.2218	.77384	.7	.8	.69214	.95897	1.0428	.72176	.2
.4	.63473	.82141	1.2174	.77273	.6	.9	.69340	.96232	1.0392	.72055	.1
.5	.63608	.82434	1.2131	.77162	.5	**44.0**	.69466	.96569	1.0355	.71934	**46.0**
.6	.63742	.82727	1.2088	.77051	.4	.1	.69591	.96907	1.0319	.71813	.9
.7	.63877	.83022	1.2045	.76940	.3	.2	.69717	.97246	1.0283	.71691	.8
.8	.64011	.ε 317	1.2002	.76828	.2	.3	.69842	.97586	1.0247	.71569	.7
.9	.64145	.83613	1.1960	.76717	.1	.4	.69966	.97927	1.0212	.71447	.6
40.0	.64279	.83910	1.1918	.76604	**50.0**	.5	.70091	.98270	1.0176	.71325	.5
.1	.64412	.84208	1.1875	.76492	.9	.6	.70215	.98613	1.0141	.71203	.4
.2	.64546	.84507	1.1833	.76380	.8	.7	.70339	.98958	1.0105	.71080	.3
.3	.64679	.84806	1.1792	.76267	.7	.8	.70463	.99304	1.0070	.70957	.2
.4	.64812	.85107	1.1750	.76154	.6	.9	.70587	.99652	1.0035	.70834	.1
40.5	.64945	.85408	1.1708	.76041	**49.5**	**45.0**	.70711	1.00000	1.0000	.70711	**45.0**

	cos	cot	tan	sin	Angle, °		cos	cot	tan	sin	Angle, °

TABLE OF LOGARITHMS
(FOUR-PLACE MANTISSAS)

No.	0	1	2	3	4	5	6	7	8	9
10	0000	0043	0086	0128	0170	0212	0253	0294	0334	0374
11	0414	0453	0492	0531	0569	0607	0645	0682	0719	0755
12	0792	0828	0864	0899	0934	0969	1004	1038	1072	1106
13	1139	1173	1206	1239	1271	1303	1335	1367	1399	1430
14	1461	1492	1523	1553	1584	1614	1644	1673	1703	1732
15	1761	1790	1818	1847	1875	1903	1931	1959	1987	2014
16	2041	2068	2095	2122	2148	2175	2201	2227	2253	2279
17	2304	2330	2355	2380	2405	2430	2455	2480	2504	2529
18	2553	2577	2601	2625	2648	2672	2695	2718	2742	2765
19	2788	2810	2833	2856	2878	2900	2923	2945	2967	2989
20	3010	3032	3054	3075	3096	3118	3139	3160	3181	3201
21	3222	3243	3263	3284	3304	3324	3345	3365	3385	3404
22	3424	3444	3464	3483	3502	3522	3541	3560	3579	3598
23	3617	3636	3655	3674	3692	3711	3729	3747	3766	3784
24	3802	3820	3838	3856	3874	3892	3909	3927	3945	3962
25	3979	3997	4014	4031	4048	4065	4082	4099	4116	4133
26	4150	4166	4183	4200	4216	4232	4249	4265	4281	4298
27	4314	4330	4346	4362	4378	4393	4409	4425	4440	4456
28	4472	4487	4502	4518	4533	4548	4564	4579	4594	4609
29	4624	4639	4654	4669	4683	4698	4713	4728	4742	4757
30	4771	4786	4800	4814	4829	4843	4857	4871	4886	4900
31	4914	4928	4942	4955	4969	4983	4997	5011	5024	5038
32	5051	5065	5079	5092	5105	5119	5132	5145	5159	5172
33	5185	5198	5211	5224	5237	5250	5263	5276	5289	5302
34	5315	5328	5340	5353	5366	5378	5391	5403	5416	5428
35	5441	5453	5465	5478	5490	5502	5514	5527	5539	5551
36	5563	5575	5587	5599	5611	5623	5635	5647	5658	5670
37	5682	5694	5705	5717	5729	5740	5752	5763	5775	5786
38	5798	5809	5821	5832	5843	5855	5866	5877	5888	5899
39	5911	5922	5933	5944	5955	5966	5977	5988	5999	6010
40	6021	6031	6042	6053	6064	6075	6085	6096	6107	6117
41	6128	6138	6149	6160	6170	6180	6191	6201	6212	6222
42	6232	6243	6253	6263	6274	6284	6294	6304	6314	6325
43	6335	6345	6355	6365	6375	6385	6395	6405	6415	6425
44	6435	6444	6454	6464	6474	6484	6493	6503	6513	6522
45	6532	6542	6551	6561	6571	6580	6590	6599	6609	6618
46	6628	6637	6646	6656	6665	6675	6684	6693	6702	6712
47	6721	6730	6739	6749	6758	6767	6776	6785	6794	6803
48	6812	6821	6830	6839	6848	6857	6866	6875	6884	6893
49	6902	6911	6920	6928	6937	6946	6955	6964	6972	6981
50	6990	6998	7007	7016	7024	7033	7042	7050	7059	7067
51	7076	7084	7093	7101	7110	7118	7126	7135	7143	7152
52	7160	7168	7177	7185	7193	7202	7210	7218	7226	7235
53	7243	7251	7259	7267	7275	7284	7292	7300	7308	7316
54	7324	7332	7340	7348	7356	7364	7372	7380	7388	7396
No.	0	1	2	3	4	5	6	7	8	9

No.	0	1	2	3	4	5	6	7	8	9
55	7404	7412	7419	7427	7435	7443	7451	7459	7466	7474
56	7482	7490	7497	7505	7513	7520	7528	7536	7543	7551
57	7559	7566	7574	7582	7589	7597	7604	7612	7619	7627
58	7634	7642	7649	7657	7664	7672	7679	7686	7694	7701
59	7709	7716	7723	7731	7738	7745	7752	7760	7767	7774
60	7782	7789	7796	7803	7810	7818	7825	7832	7839	7846
61	7853	7860	7868	7875	7882	7889	7896	7903	7910	7917
62	7924	7931	7938	7945	7952	7959	7966	7973	7980	7987
63	7993	8000	8007	8014	8021	8028	8035	8041	8048	8055
64	8062	8069	8075	8082	8089	8096	8102	8109	8116	8122
65	8129	8136	8142	8149	8156	8162	8169	8176	8182	8189
66	8195	8202	8209	8215	8222	8228	8235	8241	8248	8254
67	8261	8267	8274	8280	8287	8293	8299	8306	8312	8319
68	8325	8331	8338	8344	8351	8357	8363	8370	8376	8382
69	8388	8395	8401	8407	8414	8420	8426	8432	8439	8445
70	8451	8457	8463	8470	8476	8482	8488	8494	8500	8506
71	8513	8519	8525	8531	8537	8543	8549	8555	8561	8567
72	8573	8579	8585	8591	8597	8603	8609	8615	8621	8627
73	8633	8639	8645	8651	8657	8663	8669	8675	8681	8686
74	8692	8698	8704	8710	8716	8722	8727	8733	8739	8745
75	8751	8756	8762	8768	8774	8779	8785	8791	8797	8802
76	8808	8814	8820	8825	8831	8837	8842	8848	8854	8859
77	8865	8871	8876	8882	8887	8893	8899	8904	8910	8915
78	8921	8927	8932	8938	8943	8949	8954	8960	8965	8971
79	8976	8982	8987	8993	8998	9004	9009	9015	9020	9025
80	9031	9036	9042	9047	9053	9058	9063	9069	9074	9079
81	9085	9090	9096	9101	9106	9112	9117	9122	9128	9133
82	9138	9143	9149	9154	9159	9165	9170	9175	9180	9186
83	9191	9196	9201	9206	9212	9217	9222	9227	9232	9238
84	9243	9248	9253	9258	9263	9269	9274	9279	9284	9289
85	9294	9299	9304	9309	9315	9320	9325	9330	9335	9340
86	9345	9350	9355	9360	9365	9370	9375	9380	9385	9390
87	9395	9400	9405	9410	9415	9420	9425	9430	9435	9440
88	9445	9450	9455	9460	9465	9469	9474	9479	9484	9489
89	9494	9499	9504	9509	9513	9518	9523	9528	9533	9538
90	9542	9547	9552	9557	9562	9566	9571	9576	9581	9586
91	9590	9595	9600	9605	9609	9614	9619	9624	9628	9633
92	9638	9643	9647	9652	9657	9661	9666	9671	9675	9680
93	9685	9689	9694	9699	9703	9708	9713	9717	9722	9727
94	9731	9736	9741	9745	9750	9754	9759	9763	9768	9773
95	9777	9782	9786	9791	9795	9800	9805	9809	9814	9818
96	9823	9827	9832	9836	9841	9845	9850	9854	9859	9863
97	9868	9872	9877	9881	9886	9890	9894	9899	9903	9908
98	9912	9917	9921	9926	9930	9934	9939	9943	9948	9952
99	9956	9961	9965	9969	9974	9978	9983	9987	9991	9996
No.	0	1	2	3	4	5	6	7	8	9

PREVIOUS EMISSION CLASSIFICATIONS
(WARC 1959)

Amplitude-Modulated

A0 Carrier; no modulation or information
A1 Telegraphy; on-off; no other modulation
A2 Telegraphy; on-off; with amplitude-modulated tone
A3 Telephony; carrier with double SB
A3A Telephony; reduced carrier with SSB
A3J Telephony; suppressed carrier with SSB sideband
A3H Telephony; full carrier with SSB
A3B Telephony with two independent SBs
A3Y Digital voice modulation
A4 Facsimile (slow-scan TV)
A5C Television with vestigial sideband
A9B Telephony or telegraphy with independent sidebands
A9Y Nonvoice digital modulation

Frequency- or Phase-Modulated

F1 Telegraphy; frequency-shift-keyed
F2 Telegraphy; frequency-modulated tone
F3 Telephony; frequency- or phase-modulated
F3Y Digital voice modulation
F9Y Nonvoice digital modulation
F4 Facsimile

F5 Television
F6 Telegraphy; four-frequency diplex

Pulse-Modulated

P0 Radar (pulsed carrier without information)
P1D Telegraphy; on-off keying of pulsed carrier
P2D Telegraphy; pulsed-carrier tone-modulated
P2E Telegraphy; pulse-width tone-modulated
P2F Telegraphy; phase or position tone-modulated
P3D Telephony; amplitude-modulated pulses
P3E Telephony; pulse-width-modulated
P3F Telephony; pulses phase- or position-modulated

NOTE: If number precedes emission designation, it indicates allowable bandwidth in kHz. Thus, 0.1A1 is slow-speed radiotelegraph on-off emission; 2.1A2 is slow-speed, 1000-Hz tone-modulated telegraph; 6A3 is 3000-Hz voice-modulated double-sideband; 3A3J is suppressed-carrier single-sideband 3000-Hz voice modulation; 10A3, sound broadcasting of 5000-Hz music and speech; 5750A5C is broadcast TV video.

PRESENT EMISSION CLASSIFICATIONS
(WARC 1979)

The Appendix E-1 classifications were used by the FCC from 1960 to 1980 and will still be found in many publications. The following are some abbreviated new WARC 1979 international emission classifications.

A complete classification consists of three parts:

(1) a 3-number-1-letter group, indicating bandwidth; (2) a letter-number-letter group, being the basic emission classification; and (3) a 2-letter group, giving additional description of the emission. The basic group will probably be the one most often seen.

---

1. *Bandwidth:* H = Hz; K = kHz; M = MHz; G = GHz. The placement of these letters between numbers indicates the decimal point. Examples: 1H00 = 1 Hz; 25K5 = 25.5 kHz; 7M35 = 7.35 MHz; 10G3 = 10.3 GHz.

2. Basic emission:

First letter (type of mod)	Number (signal nature)	Second letter (info type)
A = Dbl SB + carr	0 = No mod	A = Teleg, aural
B = 2 indep SBs	1 = Keyed carr, digi	B = Teleg, machine
C = Vestigial SB	2 = Mod tone, digi	C = Facsimile
F = Freq mod	3 = Analog, voice/music	D = Telemetry data
G = Phase mod	7 = Multi-chan, digi	E = Telephony, BC
H = 1 SB + carr	8 = Multi-chan, analog	F = TV video
J = SSSC	9 = Chans with analog and digital	W = Combination of above
K = Amp mod, puls		
L = Width mod, puls		
M = Position mod, puls		
N = No mod		
P = Unmod puls		
R = 1 SB + rdcd carr		

ABBREVIATIONS: Mod = modulation; Dbl = double; Digi = digital; Carr = carrier; Rdcd = reduced; SSSC = single-sideband suppressed carrier; Puls = pulse; Chan = channel; Teleg = telegraphy; BC = broadcast quality

3. Added descriptions:

First letter	Second letter
A = Code, different duration elements	A = 2-condition code
B = Code, same duration elements	F = FDM
G = BC, monophonic	N = None
H = BC, stereo	T = TDM
J = Commercial quality	W = Combination of above
K = Frequency inversion	
M = Monochrome	
N = Color	

ABBREVIATIONS: FDM = frequency division multiplex; TDM = time division multiplex

Examples of emission classifications:

1. Morse teleg, 25 wpm = A1A or 100HA1AAN
2. AM Morse, 1-kHz tone mod, keyed = A2A or 2K10A2AAN
3. AM telephony, voice = A3E or 6K00A3EJN
4. AM BC speech and music = A3E or 10K0A3EGN
5. AM telephony, SSSC voice = J3E or 2K70J3EJN
6. AM TV video with FM sound = C3F−F3E or 6M25C3F−75K0F3EGN
7. FM telegraphy, 100 baud, 170-Hz shift FSK = F1B or 304HF1BBN
8. FM voice, 3-kHz AF, 5-kHz deviation = F3E or 16K0F3EJN
9. FM BC, 15-kHz AF, 75-kHz deviation = F3E or 180KF3EGN
10. FM stereo BC with telephony mux = F8E or 300KF8EHF
11. Radar, unmodulated pulses = P0N or 3M00P0NAN
12. Television, video and aural = 6M00C3F−75K0F3EEG

APPENDIX F

FCC RULES AND REGULATIONS AND FIELD OFFICES

The "FCC Rules and Regulations" are sold in volumes by the Superintendent of Documents, Government Printing Office, Washington, DC 20401. The most useful volumes are:

Volume I
    Part 0: Commission Organization
    Part 1: Practice and Procedure
    Part 13: Commercial Radio Operators
    Part 17: Construction, Marking, and Lighting of Antenna Structures
    Part 19: Employee Responsibilities and Conduct
Volume II
    Part 2: Frequency Allocations and Radio Treaty Matters; General Rules and Regulations
    Part 5: Experimental Radio Services (Other than Broadcast)
    Part 15: Radio Frequency Devices
    Part 18: Industrial, Scientific, and Medical Equipment
Volume III
    Part 73: Radio Broadcast Services
    Part 74: Experimental, Auxiliary, and Special Broadcast Services

    Parts 76, 78: Cable Television
Volume IV
    Part 81: Stations on Land in the Maritime Services and Alaska—Public Fixed Stations
    Part 83: Stations on Shipboard in the Maritime Services
Volume V
    Part 87: Aviation Services
    Part 90: Private Land Mobile Radio Services
    Part 94: Private Operational-Fixed Microwave Service
Volume VII
    Part 21: Domestic Public Radio Services (Other than Maritime Mobile)
    Part 23: International Fixed Public Radiocommunication Services
    Part 25: Satellite Communications

NOTE: Parts 95 (Citizens Radio Service), 97 (Amateur Radio Service), and 99 (Disaster Communications Service) are in pamphlet form and are for sale at the Government Printing Office.

The following are the mailing addresses for FCC field offices:

ALASKA, Anchorage
1011 East Tudor Road, Room 240
PO Box 102955
Anchorage, Alaska 99510
Phone: Area Code 907-563-3899

CALIFORNIA, La Mesa
  (San Diego)
7840 El Cajon Blvd.
Suite 405
La Mesa, California 92041
Phone: Area Code 619-293-5478

CALIFORNIA, Long Beach
3711 Long Beach Blvd.
Suite 501
Long Beach, California 90807
Phone: Area Code 213-426-4451

CALIFORNIA, San Francisco
423 Customhouse
555 Battery Street
San Francisco, California 94111
Phone: Area Code 415-556-7701

COLORADO, Denver
12477 West Cedar Drive
Denver, Colorado 80228
Phone: Area Code 303-234-6977

FLORIDA, Miami
Koger Building—Suite 203
8675 NW 53rd St.
Miami, Florida 33166
Phone: Area Code 305-350-5542

FLORIDA, Tampa
Interstate Building—Suite 601
1211 N. Westshore Boulevard
Tampa, Florida 33607
Phone: Area Code 813-228-2872

GEORGIA, Atlanta
Room 440, Massell Building
1365 Peachtree Street N.E.
Atlanta, Georgia 30309
Phone: Area Code 404-881-3084

HAWAII, Honolulu
7304 Prince Kuhio Federal
  Building
300 Ala Moana Blvd.
P.O. Box 50023
Honolulu, Hawaii 96850
Phone: Area Code 808-546-5640

ILLINOIS, Chicago
3940 Federal Building
230 South Dearborn Street
Chicago, Illinois 60604
Phone: Area Code 312-353-0195

LOUISIANA, New Orleans
1009 F. Edward Herbert Federal
  Building
600 South Street
New Orleans, Louisiana 70130
Phone: Area Code 504-589-2095

MARYLAND, Baltimore
George M. Fallon Federal
   Building
Room 1017, 31 Hopkins Plaza
Baltimore, Maryland 21201
Phone: Area Code 301-962-2728

MASSACHUSETTS, Boston
1600 Customhouse
165 State Street
Boston, Massachusetts 02109
Phone: Area Code 617-223-6609

MICHIGAN, Detroit
1054 Federal Building & U.S.
   Courthouse
231 W. Lafayette Street
Detroit, Michigan 48226
Phone: Area Code 313-226-6078

MINNESOTA, St. Paul
691 Federal Building
316 N. Robert Street
St. Paul, Minnesota 55101
Phone: Area Code 612-725-7810

MISSOURI, Kansas City
Brywood Office Tower,
   Room 320
8800 East 63rd Street
Kansas City, Missouri 64133
Phone: Area Code 816-926-5111

NEW YORK, Buffalo
1307 Federal Building
111 W. Huron Street at Delaware
   Avenue
Buffalo, New York 14202
Phone: Area Code 716-846-4511

NEW YORK, New York
201 Varick Street
New York, New York 10014
Phone: Area Code 212-620-3437

OREGON, Portland
1782 Federal Office Building
1220 S.W. 3rd Ave.
Portland, Oregon 97204
Phone: Area Code 503-221-4114

PENNSYLVANIA, Philadelphia
One Oxford Valley Office Bldg.
2300 East Lincoln Highway,
   Room 404
Langhorne, Pennsylvania 19047
Phone: Area Code 215-752-1324

PUERTO RICO, Hato Rey
   (San Juan)
Federal Building & Courthouse,
   Room 747
Avenida Carlos Chardon
Hato Rey, Puerto Rico 00918
Phone: Area Code 809-753-4567

TEXAS, Dallas
Earle Cabell Federal Bldg.
Room 13E7, 1100 Commerce
   Street
Dallas, Texas 75242
Phone: Area Code 214-767-0761

TEXAS, Houston
5636 Federal Building
515 Rusk Avenue
Houston, Texas 77002
Phone: Area Code 713-229-2748

VIRGINIA, Norfolk
Military Circle
870 North Military Highway
Norfolk, Virginia 23502
Phone: Area Code 804-441-6472

WASHINGTON, Seattle
3256 Federal Building
915 Second Ave.
Seattle, Washington 98174
Phone: Area Code 206-442-7653

NOTE: License tests are given the second week of February, May, August, and November. Check with nearest office to determine when they give tests and make an appointment 30 days beforehand.

Q SIGNALS

When a *Q* signal is terminated with a question mark, it is asking the question being answered by the *Q*-signal statements shown.

Signal	Meaning
QRA	The name of my station is . . .
QRB	The approximate distance between our stations is . . . miles.
QRC	The accounts for charges of my station are settled by . . .
QRD	I am bound for . . . from . . . .
QRE	ETA at . . . is . . . hours.
QRF	I am returning to . . . .
QRG	Your exact frequency (or, that of . . .) is . . . kHz (or MHz).
QRH	Your frequency varies.
QRI	The tone of your transmission is: (1) Good. (2) Variable. (3) Bad.
QRK	Readability is: (1) Unreadable. (2) Readable now and then. (3) Readable with difficulty. (4) Readable. (5) Perfectly readable.
QRL	I am busy (or, busy with . . .). Please do not interfere.
QRM	I am being interfered with.
QRN	I am troubled by static.
QRO	Increase power.
QRP	Decrease power.
QRQ	Send faster (. . . wpm).
QRR	I am ready for automatic operation. Send at . . . wpm.
QRS	Send more slowly.
QRT	Stop sending.
QRU	I have nothing for you.
QRV	I am ready.
QRW	Please inform . . . that I am calling him on . . . kHz.
QRX	I will call you again at . . . hours.
QRY	Your turn is number . . . .
QRZ	You are being called by . . . .
QSA	Your signals are: (1) Scarcely perceptible. (2) Weak. (3) Fairly good. (4) Good. (5) Very good.
QSB	Your signals are fading.
QSC	I am a cargo vessel.
QSD	Your keying is defective.
QSG	Send . . . messages at a time.

Signal	Meaning
QSJ	The charge to be collected per word to . . . including my internal charge is . . . francs.
QSK	I can hear you between my signals.
QSL	I am acknowledging receipt.
QSM	Repeat the last telegram which you sent me [or, telegram(s) number(s) . . .].
QSN	I did hear you (or, on . . . kHz).
QSO	I can communicate with . . . direct or by relay through . . . .
QSP	I will relay to . . . free of charge.
QSQ	I have a doctor on board or, . . . (name of person) is on board.
QSS	I will send on . . . kHz and listen on . . . kHz.
QSU	Send or reply on this frequency (with emissions of class . . .).
QSV	Send a series of V's on this frequency (or, . . . kHz).
QSW	I am going to send on this frequency (or, . . . kHz) (with emissions of class . . .).
QSX	I am listening to . . . (call sign) on . . . kHz.
QSY	Change to transmission on another frequency (or, on . . . kHz).
QSZ	Send each word or group twice (or, . . . times).
QTA	Cancel telegram number . . . as if it had not been sent.
QTB	I do not agree, I will repeat the first letter of each word.
QTC	I have . . . telegrams for you (or, for . . .).
QTE	Your TRUE bearing from me is . . . degrees (at . . . hours) (or, Your TRUE bearing from . . . was . . . degrees at . . . hours; or, The TRUE bearing of . . . from . . . was . . . degrees at . . . hours).

Signal	Meaning	Signal	Meaning
QTF	The position of your station according to the bearings taken by the direction-finding stations which I control was . . . latitude . . . longitude.	QUG	I am forced to land immediately [or, I shall be forced to land at . . . (position or place)].
QTG	I am going to send two dashes of 10 s each followed by my call sign (repeated . . . times).	QUH	The present barometric pressure at sea level is . . . (units).
QTH	My position is . . . latitude . . . longitude (or, according to any other indication).	QUI	My navigation lights are working.
		QUJ	The TRUE course for you to steer toward me (or, . . .) with no wind is . . . degrees at . . . hours.
QTI	My TRUE track is . . . degrees.		
QTJ	My speed is . . . knots.	QUK	The sea at . . . (place) is . . . .
QTK	The speed of my aircraft is . . . knots.	QUL	The swell at . . . (place) is . . . .
QTL	My TRUE heading is . . . degrees.	QUM	The distress traffic is ended.
QTN	I departed from . . . (place) at . . . hours.	QUN	My position, TRUE COURSE, and speed are . . . .
QTO	I have left dock (or, I am airborne).	QUO	Please search for . . . [(1) aircraft, (2) ship, (3) survival craft] in the vicinity of . . . latitude . . . longitude (or according to any other indication).
QTP	I am going to enter dock (or, I am going to alight).		
QTQ	I am going to communicate with your station by means of the International code of signals.	QUP	My position is indicated by . . . [(1) searchlight, (2) black-smoke trail, (3) pyrotechnic lights].
QTR	The correct time is . . . hours.	QUQ	Please train your searchlight on a cloud, occulting if possible, and if my aircraft is seen or heard, deflect the beam upwind and on the water (or land) to facilitate my landing.
QTS	I will send my call sign for . . . minute(s) now (or, at . . . hours) (on . . . kHz) so that my frequency may be measured.		
QTU	My station is open from . . . to . . . hours.	QUR	Survivors . . . [(1) are in possession of survival equipment dropped by . . . , (2) have been picked up by rescue vessel, (3) have been reached by ground rescue party].
QTV	Stand guard for me on the frequency of . . . kHz (from . . . to . . . hours).		
QTX	I will keep my station open for further communication with you until further notice (or, until . . . hours).	QUS	Have sighted . . . [(1) survivors in water, (2) survivors on rafts, (3) wreckage] in position . . . latitude . . . longitude (or according to any other indication).
QUA	Here is news of . . . (call).		
QUB	Here is the information requested on . . . visibility . . . cloud height . . . wind . . . at		
		QUT	Position of incident is marked (by . . .).
QUC	The number (or other indication) of the last message I received from you [or from . . . (call sign)] is	QUU	Home ship or aircraft [(1) . . . (call sign) to your position by transmitting your call sign and long dashes on . . . kHz, (2) . . . (call sign) by transmitting on . . . kHz] courses to steer to reach you.
QUD	I have received the urgency signal sent by . . . (call sign).		
QUF	I have received the distress signal sent by . . . (call sign).		

RADIOTELEGRAPH OPERATING SIGNALS

Signal	Meaning	Signal	Meaning
AA (?AA)	All after . . .	K	Invitation to transmit
AB (?AB)	All before . . .	MN or MIN	Minute(s)
ABV	Repeat figures in abbreviated form	N	No
ADS	Address	NIL	I have nothing for you
AR	End of message	NW	Now
AS	Wait	OK	We agree
BK	Used to interrupt a transmission in progress	PBL	Preamble
BN	All between . . . and . . .	R	Received
BQ	A reply to an RQ	REF	Reference to . . .
C	Yes	RPT	Repeat
CFM	Confirm	RQ	Indication of a request
CL	I am closing my station	SIG	Signature
COL	Collate, or I collate	SK	End of transmission
CP	Call to two or more stations	SYS	See your service message
CQ	General call to all stations	TFC	Traffic
CS?	Call sign?	TU	Thank you
DE	From	TXT	Text
ER or HR	Here	VA or SK	End of work
ETA	Estimated time of arrival	W or WD	Word
ITP	The punctuation counts	WA	Word after . . .
JM	Make a series of dashes if I may transmit, dots if not to transmit	WB	Word before . . .
		73	Best regards

Phillips Code

In the late 1800s telegraph operators started using a method of shorthand which was developed by Walter P. Phillips for telegraphic purposes as well as for newspaper and court reporting. There were over 5000 abbreviations used. Some of them are still used in communication circuits. A few of these words are listed to illustrate formation of abbreviations. The telegrapher sends the Phillips abbreviation and the receiving operator types out the full word or words.

Ab	About	Alw	Always	C	See	Deg	Degree
Abb	Abbreviate	Amt	Amount	Cd	Could	Dg	Doing
Abbd	Abbreviated	Ao	At once	Cf	Chief	Dld	Delivered
Abbg	Abbreviating	Ar	Answer	Cfm	Confirm	Dols	Dollars
Abbn	Abbreviation	Ax	Ask	Chg	Charge	Ea	Each
Abd	Aboard	B	Be	Cld	Called	Enh	Enough
Abs	Absent	Bc	Because	Clr	Clear	Eqm	Equipment
Abv	Above	Bd	Board	Cm	Come	Es	And
Ads	Address	Bf	Before	Ctd	Connected	Eu	Europe
Af	After	Bk	Break	Cy	Copy	F	Of the
Agn	Again	Btn	Between	D	In the	Fm	From
Ak	Acknowledge	Bun	Bulletin	Dd	Did	Fo	For

Fri	Friday	Mda	Monday	Rd	Read	Tuy	Tuesday
Fw	Follow	Mfg	Manufacturing	Rpy	Reply	Tx	This is
G	From the	Mk	Make	Rt	Are the	Ty	They
Gd	Good	Mng	Morning	Ru	Are you	U	You
Gg	Going	Mo	Month	Ry	Railway	Uk	Understand
Gn	Gone	Mvg	Moving	Sdy	Sunday	Ur	Your
Gv	Give	N	Not	Sig	Signature	V	Of which
H	Has	Ni	Night	Snd	Send	Vy	Very
Hr	Here	Numd	Numbered	Std	Standard	W	With
Hv	Have	Nw	Now	Stn	Station	Wb	Will be
Hb	Have been	O	Of	Suy	Saturday	Wda	Wednesday
Iw	It was	Opr	Operator	Svc	Service	Wi	Will
J	By which	Oth	Other	T	The	Wl	Well
Jr	Junior	P	Per	Tbl	Trouble	Wn	When
K	Out of the	Pby	Probably	Tdy	Today	Wo	Who
Kmn	Communication	Pc	Percent	Tg	Thing	Wrd	Word
Kp	Keep	Pd	Paid	Thd	Thursday	Wt	What
Kppg	Cooperating	Pfd	Preferred	Tm	Them	Wtv	Whatever
Lg	Long	Pkj	Package	Tn	Then	X	In which
Lic	License	Pls	Please	Tr	There	Xj	Explain
Ltr	Letter	Q	On the	Ts	This	Y	Year
Lvg	Leaving	Qsn	Question	Tse	These	Ya	Yesterday
M	More	R	Are	Tt	That	Z	From which
Md	Made	Rcd	Received				

Answers to Even-Numbered End-of-Chapter Questions

CHAPTER 1

2. Molecule. (1-1)
4. Same. (1-2)
6. Yes. Proton 1800 times heavier. (1-2)
8. Hot. (1-3)
10. Free-electron movement. (1-4)
12. Source, load, connecting wires, control device. (1-6)
14. Ampere. (1-7)
16. 5.3×10^5. (1-7)
18. Volt. (1-8)
20. Chemical, electromagnetic, thermal, piezoelectric, magnetostriction, static, photoelectric, MHD. (1-8)
22. Sparking or corona. (1-9)
24. Corona. (1-9)
26. Lightning, neon lights, fluorescent lights. (1-10)
28. Decreases in strength over time. (1-11)
30. Material, length, area, temperature. (1-12)
32. 100. (1-12)
34. Zero TC. (1-12)
36. 68 MΩ, 10%. (1-13)
38. Meter, liter, gram. (1-14)
40. 3.5×10^6, 2×10^6. (1-14)
42. Confined area accumulates heat. (1-15)
44. Yes. (1-17)
46. Desoldering tool, copper-wire braid. (1-17)

CHAPTER 2

2. 160 Ω. (2-3)
4. 280 mA. (2-3)
6. 58.5 W. (2-7)
8. Watt-second. (2-4)
10. $1.62. (2-4)
12. $P = VI$, I^2R, V^2/R. (2-5)
14. One-half W, preferably 1 W. (2-5)
16. R, I, P. (2-6)
18. Low-current fuse. (2-7)
20. Slow-blow, chemical. (2-7)
22. Series. (2-10)
24. Kirchhoff's voltage law. (2-10)
26. Top. (2-11)
28. When load current is small. (2-11)
30. 0.144 A, 10.22 V. (2-11)
32. 0.00208 S. (2-12)
34. Kirchhoff's current law. (2-13)
36. $R_1 = 2.27$ A, 45.4 V. $R_2 = 1.365$ A, 54.6 V. $R_3 = 0.91$ A, 54.6 V. (2-15)
38. Maximum power transfer. (2-16)
40. 22.86 V, 0 A, 2 A, 2 A. (2-18)

CHAPTER 3

2. Always at right angles. (3-2)
4. Center of core. (3-2)
6. No. (3-2)

8. South. (3-2)
10. Collapse into wire. (3-2)
12. Maxwell. (3-3)
14. Magnetomotive force, $\mathscr{F} = NI$. (3-4)
16. Accepts more lines of force. (3-5)
18. Permeability, μ (3-5)
20. Permeance, $\mathscr{P} = 1/\mathscr{R}$. (3-5)
22. Crystals of fully magnetized material. (3-6)
24. Saturated. (3-6)
26. Magnetize slightly. Slight reverse magnetism. (3-7)
28. Opposition of domains to demagnetize. (3-8)
30. Good permanent magnet. (3-8)
32. Insulators. (3-9)
34. Demagnetizes it. (3-10)
36. English, cgs, mksa. (3-12)
38. Accepts any nearby lines of force. (3-14)
40. S end of compass points to N pole. (3-16)
42. Motion, field, induced emf. (3-18)
44. Field strength, conductor speed, number of conductors. (3-18)
46. NO or SPST. (3-20)
48. *NI*. (3-20)
50. Fast, small. (3-20)

CHAPTER 4

2. Alternator. (4-2)
4. Square, sawtooth, triangular, sinusoidal. (4-2)
6. 360°. 180°. (4-3)
8. 0.86 max. 0.5 max. Max. (4-3)
10. 90°. 180°. 270°. 360°. (4-3)
12. 10 V_{dc}. (4-4)
14. 10.6 V_{rms}. (4-4)
16. 155.5 V_{pk}. 311 V_{p-p}. (4-4)
18. 100 W. (4-4)
20. 0.636 V_{pk}. 0.9 V_{rms}. (4-5)
22. 4560 kHz. (4-6)
24. 262 Hz. (4-6)
26. 10 kHz to 300 GHz. (4-6)
28. Degrees lagging or leading. (4-7)
30. 4 MHz. 25.6 MHz. 25 kHz. (4-8)
32. 2.5 μs. (4-8)

CHAPTER 5

2. Counter emf. (5-1)
4. In the source emf direction (5-2)
6. Yes. Yes. (5-2)
8. Proportional to turns squared. (5-3)
10. Adjustable ferrite core in coil. (5-3)
12. 5 T_c, or 0.00625 s. (5-4)
14. Current variations. (5-6)
16. 10^8 lines. (5-7)
18. 2.83 H. (5-7)

20. High coefficient of coupling. (5-8)

22. $L_t = L_1 + L_2 + 2M$. (5-9)

24. Reduces it. Increases it. (5-11)

26. 5 V. (5-12)

28. 29.6 Ω. 18.8 MΩ. (5-12)

30. 9.55 A. (5-12)

32. No current or field change. (5-13)

34. None. (5-15)

36. Tertiary. (5-16)

38. Laminate core. (5-17)

40. Hysteresis. (5-18)

42. Turns ratio. Also coupling coefficient for air-core transformers. (5-21)

44. 1.23 A. (5-24)

46. Core losses greater, primary X_L too high limits I_p. (5-26)

CHAPTER 6

2. Plate, dielectric, plate, connecting wires. (6-1)

4. Exponential. (6-1)

6. 4.7 ms. 23.5 ms. (6-1)

8. 0.844 μF. (6-2)

10. Anything over 15 V. Anything over 30 V. (6-4)

12. 1.44 Ws. No. Skin resistance too high to pass enough current. (6-5)

14. Small, high capacitance. Used on dc only. (6-7)

16. Adjustable. Variable. (6-8)

18. Vacuum, air, mica, ceramic, paper, plastic, electrolytic. (6-9)

20. Yes. Yes. (6-11)

22. Capacitive reactance. (6-13)

24. 4.55×10^{-11} F, or 45.5 pF. (6-13)

26. No. No resistance so no power loss. (6-13)

28. Half of one. Half of one. (6-10, 6-12)

30. Inversely. Directly. (6-15)

32. I leads V by 90°. Not at all. (6-16)

34. Working voltage, capacitance, dielectric, size, cost, temperature, variable/fixed, temperature coefficient. (6-17)

36. Celsius. (6-18)

CHAPTER 7

2. $Z^2 = R^2 + X^2$. $Z = \sqrt{R^2 + X^2}$. (7-2)

4. 2.5 A. 60 V. 80 V. (7-2)

6. 29.15 V. 30.9° lagging. (7-4)

8. VA. (7-5)

10. 0.89105 or 89.1%. 26.99° lagging. (7-6)

12. 386.4 W. (7-6)

14. Add opposite reactance to circuit. (7-6)

16. 25.24 Ω. 1.189 A. 16.65 V. 24.97 V. 19.79 W 35.67 VA. 0.55467 pf. 56.3° leading. (7-7)

18. Capacitive. More X_C volts. (7-8)

20. 80.6 Ω. (7-10)

22. 85.7-ΩX_C. (7-12)

24. $I_R = I_{X_L} - I_{X_C}$. (7-12)

26. 6.4 A. 15.63 Ω. 0.78125 pf. 38.6° lagging. (7-13)

28. 0.3333 S. 0.025 S. 0.02 S. 0.00833 S. 0.02166 S. 46.15 Ω. (7-15)

30. $Z = 40.46 \,\underline{/-50°}\,$ Ω. (7-17)

CHAPTER 8

2. 45,015 Hz. (8-1)

4. Frequency increases 1.414 times. (8-1)

6. Series LC. Parallel LC. (8-2, -3)

8. 180°. Same current or ∅°. (8-2)

10. Allow resonating at different frequencies. (8-2)

12. 180°. Same voltage, or ∅°. (8-3)

14. 0 Ω. ∞ Ω. (8-3)

16. 514 Ω. (8-3)

18. Oscillating, or flywheel. (8-3)

20. 10. 100. (8-4)

22. Add R in series or in parallel with either leg. (8-4)

24. Mantissa. Characteristic. (8-5)

26. 17.96 dB. (8-5)

28. 1 mW. (8-5)

30. 37.5 dB. (8-5)

32. BW inversely proportional to Q. (8-6)

34. Loose. (8-7)

36. Overcoupling, or stagger tuning of pri and sec. (8-7)

38. Antiresonant. Resonant. (8-8)

40. Narrow-bandstop, or wave trap. (8-8)

42. Add sections, or use m-derived filters. Increases insertion losses, pop-up past ∞-attenuation point. (8-8)

44. Center of shunt element grounded. (8-8)

46. L, T, π, H, U. (8-8)

CHAPTER 9

2. Power output. Size, weight, efficiency, cost. (9-1)

4. Semiconductors. (9-2)

6. Intrinsic, or I. (9-2)

8. Electron fills it and hole appears where electron was. (9-2)

10. Silicon. Withstands higher voltage and heat. (9-2)

12. Minority carriers. (9-3)

14. Zener. (9-3)

16. Negative resistance part of its curve. Heavy doping. (9-3)

18. Light-emitting diode. Solid-state lamp. (9-4)

20. Limit current through it. (9-4)

22. PIN is instantaneous. (9-4)

24. Base. Emitter, collector. (9-5)

26. Yes, if base is forward-biased, but not optimally. (9-5)

28. Light energy. (9-5)

30. $A_i = 80$. (9-6)

32. − to E to B to +. − to E to C to +. (9-6)

34. See Fig. 9-14. (9-6)

36. Where gain drops −3 dB from gain at 1 kHz. (9-6)

38. Current gain and output load voltage. (9-7)

40. Common-base. Current gain. No. Yes. (9-9)

42. See Fig. 9-19. (9-9)

44. See Fig. 9-20. (9-10)

46. JFETs. Depletion MOSFETs. Enhancement MOSFETs. (9-11)

48. $\mu = dV_{DS}/dV_{GS}$. (9-11)

50. $g_m = di_D/dV_{GS}$. (9-11)

52. See Fig. 9-22. (9-11)

54. Insulation layer between gate and channel. (9-12)

56. Depletion. Enhancement. (9-12)

58. Yes, but not optimally. (9-12)

60. Complementary N- and P-channel MOSFETs in a balanced device. (9-12)

62. Pulsating, oscillating, firing SCRs. (9-13)

64. Silicon-controlled rectifier. Forward bias on gate. Reduce V_{AA} to essentially zero. (9-14)

66. Has turn-off gate. (9-14)

68. Avalanches at 25–35 V to provide delayed conduction. (9-15)

70. 100,000 internal units. (9-17)

72. N- and P-type areas. Filament or cathode, plate or anode. (9-18)

74. See Fig. 9-36. (9-19)

76. $\mu = dV_p/dV_g$. (9-19)

78. Melt anode. (9-19)

80. 1.125 W. (9-22)

82. Beam-power tetrodes. (9-24)

84. See Fig. 9-41. (9-23)

86. CE, CB, CC. CS, CG, CD. Common-cathode, grounded-grid, cathode-follower. (9-25)

88. See Figs. 9-47 and 9-48. (9-25)

CHAPTER 10

2. Transformer, rectifier, and filter system. (10-1)

4. Average current is half of full-wave. Harder to filter. (10-2)

6. Full-wave. (10-3)

8. Ripple. (10-4)

10. 56.5 V. 28.3 V. 28.3 V. (10-4)

12. Round off forward corner. Elongate rear pulse base. (10-6)

14. Mercury-vapor rectifiers. (10-8)

16. Too much ripple. (10-9)

18. Choke saturates, improves voltage regulation. Swinging. (10-11)

20. See Fig. 10-14. (10-14)

22. Safety discharge and better voltage regulation. No, 20 V is not dangerous, and low-V supplies usually regulated. (10-16)

24. Reduce series R in circuit. Use semiconductor rectifiers. Use some L filter. Use swinging choke. Sufficient capacitance. Correct bleeder value. Full- rather than half-wave. (10-17)

26. See Fig. 10-16. (10-19)

28. See Fig. 10-17. (10-21)

30. 5 to 15 mH. (10-21)

32. Shunt regulator. (10-22)

34. Load yes; transformer no. (10-22)

36. See Fig. 10-22. (10-23)

38. Regulation corrects ripple variations also. (10-23)

40. See Fig. 10-23b. (10-24)

42. Loss of voltage regulation and high voltage to load. (10-24)

44. $R_s = 155.5\ \Omega$, one-half W. R_b, 250 Ω, one-fourth W. (10-25)

46. Vibrator type. (10-27)

48. 120 Hz. 360 Hz. (10-28)

50. 120°. (10-28)

52. 1:1. 1:1. 1:0.578. 1:1.73. (10-28)

54. Open-Δ. (10-28)

56. See Fig. 10-36. (10-28)

58. See Fig. 10-40. (10-29)

60. See Fig. 10-43. (10-29)

62. Visual examination. Ohmmeter checks of systems. Ohmmeter checks of components. (10-30)

64. Panel meter readings. (10-30)

66. Fuses, capacitors, resistors, rectifiers, active devices. (10-30)

CHAPTER 11

2. Flywheel. (11-2)

4. Amplification of some of the ac and in-phase feedback. (11-3)

6. To match the low-Z of BJT circuits. (11-4)

8. Degeneration. Regeneration. (11-4)

10. Parasitic oscillations. (11-5)

12. See Fig. 11-5. (11-4)

14. TBTC, TGTD, TGTP. (11-6, 11-7)

16. Appears inductive but is actually also capacitive. (11-8)

18. LC capacitance center-tapped instead of the inductance. (11-9)

20. Gate is insulated, no gate current. (11-10)

22. Fig. 11-14. (11-11)

24. Material, thickness, slab dimensions, angle of cut, pressure on plates, type of circuit used, temperature. (11-12)

26. Thickness, shear, longitudinally, flexurally, or torsionally. (11-12)

28. Zero TC. (11-13)

30. Bimetallic element, thermometer, thermocouple. (11-14)

32. Colpitts. (11-16)

34. Overtone. (11-16)

36. Alternating current. (11-16)

38. Hairpin tanks, coaxial tanks, resonant cavities. (11-17)

40. Negative resistance. (11-18)

42. R, C, and bias voltage. (11-20)

44. Sinusoidal. (11-21)

46. Receiver, bias voltage, output circuit current, RF indicator, oscilloscope, neon lamp, lead pencil, electronic counter. (11-23)

48. Colpitts crystal. No. (11-25)

50. VCO, reference oscillator, phase detector, LP filter, dc amplifier. (11-27)

52. To smooth any ripple of dc from the phase detector. (11-27)

54. Add extra crystal and switch point to Fig. 11-42. (11-27)

56. Only one forward-biased diode completes the crystal circuit to ground. (11-28)

58. PLL synthesizer. (11-29)

CHAPTER 12

2. 0–2 V, 4–6 V. 0–4 V, 9–15 V. (12-1)

4. 4. 4. (12-2)

6. 0110, 6. 0001. 1, 1001, 9. (12-3)

8. 2. 3. (12-3)

10. ECL. CMOS. CMOS. (12-4)

12. High. (12-5)

14. Low. (12-7)

16. Low. (12-8)

18. Low. (12-9)

20. See Figs. 12-4 through 12-9.

22. Ignores clocking pulses. (12-10)

24. See Fig. 12-9. (12-10)

26. Digital word storage device. (12-11)

28. High output when an enabling and clock pulse arrive. (12-11)

30. Bus. (12-11)

32. Programmable read-only memory, or PROM. (12-12)

34. Loses data if power is shut off. (12-12)

36. 250 kHz. Square-wave. (12-14)

38. See Fig. 12-17*b*. (12-14)

40. Polarized filter, invisible bar section, nematic fluid, invisible bar section, polarized filter, reflector plate. (12-16)

42. Digital less affected by noise pulses. (12-17)

44. 7 wires for 7 different binary digits. (12-17)

46. +, either · or ×, $\oplus$

48. See Fig. 12-23. (12-19)

50. 16. 6 × 6 or 3 × 12. (12-20)

52. To transmit data on wires or by radio. To convert received serial to parallel for a parallel printer. (12-21)

54. See Fig. 12-26. (12-21)

56. TTL. Uses 5-V power supply. Crystal. (12-22)

58. 64 k. Enables registers, memories, etc. Loads data. (12-22)

60. ROM. (12-23)

62. By 5 × 7 (or more) dot groups per character. By lines of dots for graphics. (12-23)

64. Directly related. (12-24)

66. Convert parallel ASCII to serial Baudot. Convert parallel ASCII to serial Morse. Convert to serial and slow down. (12-25)

68. Most amplify input signals. (12-26)

70. Changes sine-wave ac to square-wave pulses. (12-26)

CHAPTER 13

2. Portables. (13-1)

4. Horseshoe magnet, round iron core piece, rotating coil mechanism, calibrated scale, case. (13-2)

6. Make meter read same regardless of physical position. (13-2)

8. Galvanometer. (13-2)

10. Shunt takes part of circuit current. (13-4)

12. 15.79 Ω. (13-5)

14. Full-scale I. Ω/V. (13-6)

16. Laboratories. (13-8)

18. 20 mV. 10,000 − 200 = 9800 Ω. 1 MΩ. (13-9)

20. Prevent rheostat variation from burning out meter. (13-11)

22. See Fig. 13-8. (13-11)

24. Voltmeter, ohmmeter, milliammeter. (13-12)

26. See Fig. 13-10. 9 MΩ, 900 kΩ, 90 kΩ, 10 kΩ. 10 MΩ/V. 10 kΩ/V. (13-13)

28. Make scale linear. (13-13)

30. 73.8 V peak. 52.17 V rms. (13-15)

32. Add C across rectifier output. (13-16)

34. 1 mW. (13-17)

36. Type A. Type B. (13-17)

38. Electromagnet instead of permanent magnet. (13-19)

40. AC voltmeter. AC ammeter. AC or dc wattmeter. (13-19, -21)

42. Field coils and moving coil. Field coils. Moving coil. (13-21)

44. Electric energy in kWh. At every power line entrance box. (13-22)

46. 6500 Ω. (13-24)

48. 0.00001, or 0.001%. (13-25)

50. Vibrating reed. (13-26)

52. I_G or I_g. Resonant frequency of LC circuits. (13-28)

54. Oscillator that is checked against an NBS or other primary frequency standard. (13-29)

56. PLL frequency synthesizer. Use atomic clock for its reference oscillator. (13-29)

58. Oscillator isolated from antenna, crystal checkpoint oscillator, tone-modulated output, louder signals. (13-30)

60. 1-s AND gate. Oscillator accuracy. (13-31)

62. Loosely couple a link-coupling loop to LC circuit. Touch counter probe to output of oscillator circuit. (13-31)

64. Cycles of an internal oscillator or on-off RC pulsing circuit. (13-33)

66. Between focusing and accelerating anodes. Grid bias voltage. (13-34)

68. HDPs. VDPs. (13-34)

70. Instantaneous. (13-34)

72. Grid or cathode of CRT. (13-36)

74. Storage. Sampling. (13-36)

76. AC waveform. (13-37)

78. Test amplifiers or receivers. (13-38)

CHAPTER 14

2. First, or input. (14-2)
4. See Fig. 14-3. (14-4)
6. When trying to develop maximum output V or P. (14-4)
8. 25 μF minimum. (14-4)
10. Twice as much primary current. Devices appear to have half the impedance of one.
12. Reduce even-order harmonics, higher efficiency, more power output. (14-6)
14. Split-load or paraphase. See Fig. 14-6. (14-7)
16. 120 Ω. (14-8)
18. P-m or dynamic, electrodynamic, electrostatic. (14-9)
20. All types. All but resistance. (14-10)
22. Z of choke high but R low, allowing more I flow to produce high counter plus induced emfs. (14-10)
24. 2.89:1 or 3:1. Secondary. (14-11)
26. See Fig. 14-11. (14-12)
28. Amplitude or harmonic, frequency, phase, intermodulation. (14-13)
30. Inverse, degenerative, negative. (14-14)
32. 9.01 times. (14-14)
34. See Figs. 14-16 and 14-17. (14-15)
36. $C_m = C_{GS} + (A)C_{GD}$. $C_m = C_{BE} + (A)C_{BC}$. (14-17)
38. See Fig. 14-20. (14-18)
40. See Fig. 14-23. (14-19)
42. See Fig. 14-25. (14-19)
44. 10^6 or more. Z_f/Z_i. (14-20)
46. 60 mV. (14-20)
48. Nearly zero ohms. (14-20)
50. See Fig. 14-29. (14-21)
52. Three-pole bandpass filter. (14-21)
54. Increase output power several times. (14-23)
56. 120 Hz. 60 Hz. (14-25)
58. Grounding mike cable at both ends. (14-25)

CHAPTER 15

2. 500 kHz to 300 MHz. (15-1)
4. See Figs. 15-1, 15-2. (15-2)
6. Common-drain or source-follower, common- or grounded-gate. Common-collector or emitter-follower, common- or grounded-base. Common-plate or cathode-follower, common- or grounded-grid. (15-3)
8. See Fig. 15-6a. (15-3)
10. Holdover from old days. (15-4)
12. $+1$ V. $+1.33$ V. $+1.58$ V. $+2$ to $+6.67$ V. (15-5)
14. Class C. (15-6)
16. Antiresonant. (15-7)
18. Maximum RF output for minimum dc input power. (15-8)
20. Class C MOSFETs and VMOSs. (15-9)
22. Minimize even-order harmonics. (15-11)

24. Shunt. (15-12)
26. Negative. (15-14)
28. Double variable capacitor with two sets of rotor plates on a common shaft. (15-15)
30. Zero. (15-17)
32. Single-ended. Push-pull. (15-23)
34. Push-pull. Push-push. (15-23)
36. 16.9 μH. 0.169 μH. (15-24)
38. RF bypass C's to ground. (15-25)
40. 8. (15-26)
42. See Fig. 15-33. (15-28)
44. The push-push I_D will increase as drive to it peaks, resulting in decreased TP_7 voltage. Its own I_D dips at resonance. If the output load is tuned to resonance the TP_7 voltage will decrease. (15-28)
46. 1070 pF, 0.357 μH, 5720 pF. (15-29)
48. Cathode hot, RF drive zero, I_p zero. (15-30)

CHAPTER 16

2. A1A. A1A. (16-1)
4. Transmit and receive on same frequency. T and R on 2 frequencies. (16-3)
6. Master oscillator power amplifier. Better frequency stability with higher power output. (16-4)
8. Class C. All classes, A, AB, B, or C possible. (16-4)
10. Antenna current. (16-4)
12. Abrupt frequency variation. Drift. (16-6)
14. Emitter, base, collector. Source, gate, drain. (16-8)
16. Odd order. Even order. (16-9)
18. Unfiltered corners of dots and dashes. (16-9)
20. See Fig. 16-10. (16-10)
22. Needs a separate bias supply. (16-11)
24. See Fig. 16-14. Reverses frequency shift. (16-12)
26. By keying the divide-by circuitry. (16-12)
28. Prevent interaction of RF fields, which might change oscillator frequency, radiate harmonics, burn out low-power devices. (16-13)
30. Use bias supply, cathode-R biasing, clamp tube. (16-14)
32. Drifted 9 Hz. (16-15)
34. Inductive coupling, high-Q input circuit, do not overdrive, classes A or AB, mica bypasses, VHF wavetrap or parasitic choke or ferrite beads at plate, Q of no less than 12 for tank, Faraday shield, loose antenna coupling, low-pass filter to antenna, ground one wire of any link loop, shield stages, use push-pull, use π-network coupling. (16-16)
36. Construct a second shield can around the first. (16-17)
38. See Fig. 16-21. (16-19)
40. Switch in fixed bandpass filters for each of the bands to be used in the mixer and all amplifiers. Gang-tuning all stages is possible but difficult to align. (16-20)
42. No signal on the operating frequency. (16-20)

44. 32 dB. 160 W. (16-20)

46. Determine lowest level stage that appears to have improper voltage or current readings. Shut off transmitter and look for the trouble. (16-22)

48. 50 V. 100 V. (16-23)

CHAPTER 17

2. Nothing. (17-2)

4. 262 Hz. Different harmonics produced by them. (17-3)

6. Young people. (17-3)

8. Telephone companies. (17-4)

10. Biased PIN diodes across waveguides or AF or RF circuits. (17-5)

12. See Fig. 17-6. (17-6)

14. Capable of modulating very high frequencies; linear. (17-6)

16. Increase and decrease with modulation. Flywheel effect of output LC tank. (17-7)

18. AF ac. (17-8)

20. % mod = $(V_{max} - V_{min})/(V_{max} + V_{min}) \times 100$. (17-9)

22. Louder received signals but very little distortion. (17-9)

24. See Fig. 17-12. (17-10)

26. Simultaneously modulate the screen grid. (17-11)

28. Mixing or heterodyning. Any nonlinear circuit. (17-13)

30. 250 W in 1 SB. 180 W in 1 SB. (17-13)

32. 6 kHz. 10 kHz. 30 kHz. (17-14)

34. See Fig. 17-17. AF ac. (17-15)

36. High modulating frequencies not limited by a transformer. Series modulation. (17-6, 17-16)

38. Stage ahead of final PA is modulated. Final PA is plate- (collector-, drain-) modulated. (17-17)

40. Oscillator, buffer, driver, modulated amplifier. (17-19)

42. See Fig. 17-26. To provide a constant nonreactive load for modulated stage. (17-19)

44. 35%. 60%. (17-21)

46. Negative carrier shift. (17-22)

48. From 25 to 35%. (17-24)

50. P-m loudspeaker. (17-26)

52. Condenser. Such high-Z a circuit cannot be carried any distance in low-Z coaxial cable without losing high frequencies. (17-28)

54. Part semiconductor, part VT. Yes, most new high-power transmitters. (17-29)

56. See Fig. 17-35. (17-29)

58. Power saved, bandwidth halved, distortion fading not a problem. SSB more complex, and is difficult to tune in. (17-31)

60. AF and RF. USB and LSB but no carrier. (17-32)

62. See Fig. 17-38. (17-32)

64. Diodes, BJTs, FETs, VTs. (17-33)

66. See Fig. 17-40. (17-33)

68. See Fig. 17-41. (17-34)

70. A or AB, although B can be used. (17-36)

72. Broad horizontal band. Pattern of two sine waves 180° out of phase. "Christmas trees." (17-36)

74. See Fig. 17-46. (17-36)

76. Voice-operated transmission. Frequency-division multiplex. Time-division multiplex. (17-37)

78. See Fig. 17-49. (17-37)

80. AM and SSB. (17-39)

82. See Fig. 17-50. (17-39)

84. See Fig. 17-52. Barrier voltage of diode and setting of R_2. (17-40)

86. See Fig. 17-53. Diode barrier voltage and bias applied to it. (17-40)

CHAPTER 18

2. Rectify down to nearly zero volts rather than to 0.6 or 0.3 V. (18-2)

4. Tap down tuned LC circuit, use loose input coupling. (18-3)

6. Essentially linear. Almost square-law. Linear. (18-5)

8. Grid-leak. Autodyne or self-heterodyne. (18-7)

10. Changing RC values in the circuit. (18-8)

12. Sensitivity, selectivity, prevents oscillations of detector from being radiated, simpler, fewer spurious signals. (18-9)

14. Mixer stage, local oscillator, IF stage(s). (18-10)

16. Yes, if its harmonics do not fall on any frequencies to be received. (18-10)

18. Mixer stage. (18-10)

20. See Fig. 18-14. (18-10)

22. NF = $(S_o/N_o)/(S_i/N_i)$. (18-11)

24. See Fig. 18-15. AVC voltage applied. (18-11)

26. See Fig. 18-16. (18-12)

28. Armstrong. (18-12)

30. Larger. (18-12)

32. See Fig. 18-18b. See Fig. 18-18a substituting a JFET for the pentode, deleting screen and suppressor circuits. (18-13)

34. Vary supply V, bias V, use variable-biased PIN diodes across input circuits. (18-13)

36. See Fig. 18-20. (18-14)

38. Half-wave, full-wave, voltage-doubling rectifiers, power detector. (18-15)

40. See Fig. 18-21. Add diode in series with AVC line to C_1. (18-15)

42. AVC not functional until carrier attains a certain amplitude. (18-15)

44. AVC uses rectified IF signals; AGC usually uses rectified AF signals. (18-16)

46. See Fig. 18-22. Zero-beat signal or throw SW_1 and SW_2. (18-16)

48. See Figs. 18-24, 18-25. (18-18)

50. See Fig. 18-28. To set clipping level. (18-19)

52. Series and parallel. Parallel. (18-21)

54. See Fig. 18-30. (18-23)

56. Different inductors are switched into tuned circuits. (18-25)

58. See Fig. 18-33. Keying in second transmitter mixer, use 500-Hz bandpass crystal filter, use variable or crystal BFO 700 Hz off IF center frequency. (18-27)

60. See Fig. 18-34. Feed a 1400-kHz sig gen into input, turn dial to 1400. Adjust osc. tracking C to sig max. Turn dial and sig gen to 600 kHz and adjust LO trimming C to max sig. Repeat 1400 and 600 kHz checks until dial markings track with input sigs, then adjust input tuning to max sig output. (18-28)

CHAPTER 19

2. FM broadcasting, TV audio, satellite TV video and audio. Two-way communications, amateur radio. (19-2)

4. MI = f_{dev}/f_{af}. 5. (19-3)

6. 100 W. (19-3)

8. Voice or music modulation. (19-3)

10. Slope detection. (19-4)

12. It is sensitive to amplitude variations as well as frequency. (19-5)

14. See Fig. 19-5*b*. IF. (19-6)

16. See Fig. 19-7. (19-8)

18. Transmitter. Receiver. Less noise in output of receiver. R and C in AF circuits. (19-10)

20. Yes. No, there is no carrier transmitted. No, AFC takes time to change the frequency and would develop chirps. (19-9)

22. See Fig. 19-12. IF ac, dc reverse bias. (19-12)

24. Unmodulated IF signal to output of limiter. Tune for maximum V_{dc} across one of the output resistors, then tune for zero V_{dc} across both resistors. (19-12)

26. See Fig. 19-13. Called direct but is a form of phase. (19-12)

28. No. Receiver discriminator output is the only dc needed to operate the reactance modulator stage. (19-13)

30. AF amplitude and frequency. AF amplitude. (19-15)

32. See Fig. 19-17. (19-16)

34. See Fig. 19-18. Remove harmonics above 3 kHz generated by clipping AF cycles, and make the AF output inversely proportional to frequency. (19-16)

CHAPTER 20

2. Static parallel to wire, magnetic around the wire, always 90° from each other. (20-1)

4. Short term, 27 days, long term, 11 years. (20-2)

6. Ionospheric variation, two sky waves going in and out of phase, ground and sky waves in and out of phase, reflections from airplanes, one or both transmitters (or receivers) moving. (20-3)

8. 100 MHz and up. (20-4)

10. VLF through HF. (20-5)

12. End effects. (20-7)

14. Hertz. Marconi. (20-8)

16. 73 Ω. 73 Ω. Approaches 0 Ω. (20-10)

18. 491 Ω. (20-12)

20. Open-wire. Coaxial cable. (20-12)

22. λ/2, because no energy is wasted in a vertical direction. (20-13)

24. 36.5 Ω. Perhaps 2500 Ω. (20-14)

26. One directly over the other. (20-14)

28. In line of wire. (20-15)

30. Resonant lengths. Almost any impedance. (20-16)

32. λ/4, 3λ/4, 5λ/4, etc. (20-16)

34. 147-Ω open-wire line. (20-16)

36. λ/2 elements separated by a λ/4 stub. (20-17)

38. 45° apart and fed at 90° (by two unequal length lines). (20-18)

40. See Fig. 20-29. (20-19)

42. 5 dB. (20-20)

44. Four times the feed-point Z and bandwidth of a dipole. (20-20)

46. Half-power or 3-dB from maximum points. (20-20)

48. Drooping dipole. No. (20-21)

50. Allows shorter antenna tower to be used. (20-23)

52. Output C. (20-24)

54. Lightning or static protection. (20-24)

56. Impedance bridge when antenna is resonant. Substitution method. (20-25)

58. 92 dB down. (20-27)

60. Altitude of center of radiator minus average terrain altitude. (20-28)

62. 7.32 dB. (20-29)

64. Copper radials buried in ground every 5 to 10°. (20-30)

66. Longer path at night, signal might interfere with distant stations on same frequency. (20-31)

68. 90° out of phase. (20-32)

70. 10% greater than λ/4 for lowest operating frequency. 30% of skirt length. (20-32)

CHAPTER 21

2. Mobiles can be connected to telephone lines. (21-1)

4. 50 Hz. Usually synthesized, sometimes crystal. (21-3)

6. 20 kHz. 8 kHz. 3.5 kHz. (21-3)

8. See Fig. 21-2. (21-5)

10. Activates relay type system to change antenna and power supply from receiver to transmitter. (21-5)

12. 1 kHz. 4.5 kHz. (21-5)

14. Prevent wind and other noises from overriding voice sounds. (21-5)

16. Continuous Tone Coded Squelch System. Private line, channel guard. (21-6)

18. Specific digital word being received. (21-6)

20. See Fig. 21-6. (21-7)

22. Two-tone signals or digital calling codes from the base station. (21-8)

24. 1 to 5 W. Shortened spiral-wound λ/4, full λ/4 or 5λ/8 verticals. (21-9)

26. One. Two. (21-10)

28. Carrier-operated relay. (21-10)

30. See Fig. 21-7. (21-10)

32. 120 dB. (21-11)

34. Signal Including Noise And Distortion. (21-11)

36. See Fig. 21-9. (21-11)

38. DA has variable frequency bandstop or notch filter and a balance control. (21-12)

40. When installed, when changes made to it, once a year. Power output or input, modulation %, frequency. (21-13)

42. 5 kHz. (21-13)

44. When cold and after at least 20 min warmup. (21-13)

46. Sensitivity, squelching level, bandwidth, AF distortion, AF power output. (21-14)

48. 5% or less half medium volume, 7% at maximum. (21-14)

50. Calibrated RF signal generator, tunable FM receiver, discriminator meter, deviation indicator, scope output. (21-14)

52. CO battery. (21-15)

54. Dual-Tone Multiple-Frequency pad dialing. (21-15)

56. 48-V battery, 400-Hz dial tone, 90-V 20-Hz vdc ringing voltage. (21-15)

58. Ø VU. (21-15)

60. 200–3000 Hz, 20–20,000 Hz, 0–2.54 MHz for 600 channels. (21-16)

62. See Fig. 21-17. (21-16)

64. Its digital burst is decoded. Control scans for free channel, sends digital information to all fleet mobiles to tune to free channel, and activates a green light, or "busy" light if no free channels. (21-17)

66. Many low-power repeaters in one city. As mobile travels, the repeater receiving strongest signal shifts mobile's frequency to its channel. (21-17)

68. Obtain FCC radio station authorization, construction permit, station license. When license has been granted. (21-18)

CHAPTER 22

2. Polarization may rotate. (22-2)

4. 50 Ω. (22-2)

6. Strip width, dielectric constant and thickness. (22-2)

8. Dummy load. Absorb reflected energy. (22-3)

10. See Fig. 22-9. (22-3)

12. Diodes, barretters, thermistors. Thermistor. (22-5)

14. High Q. (22-6)

16. Reflex klystrons, magnetrons. (22-7, 22-8)

18. GaAs FET cavity oscillators. (22-7)

20. Make electron stream deviate or loop. Hold electron stream in straight line to anode. (22-8, 22-7)

22. 40 W. About 40 W if 50% efficient, plus filament heat. (22-8)

24. Losser ring around its helical coil. Oscillations. (22-9)

26. See Fig. 22-20. (22-9)

28. See Fig. 22-21. (22-10)

30. Schottky. (22-10)

32. Modulators or gates. (22-10)

34. Ferromagnetic resonance absorption and Faraday rotation. On-off action in isolators, shift wave polarization 45° in circulators. (22-12)

36. See Fig. 22-26. (22-12)

38. Yttrium-iron-garnet. V-tunable SHF resonant circuit. (22-14)

40. See Fig. 22-29. (22-14)

42. See Fig. 22-31. (22-15)

44. Stripline input circuits. (22-16)

46. Signal generators, power meters, SWR indicators, ratio meters, frequency meters, sampling oscilloscopes, spectrum analyzers, function generators, X-Y recorders, TDM systems. (22-18)

48. Infrared, visible, ultraviolet. (22-19)

50. Lightwave Amplification by Stimulated Emission of Radiation. (22-19)

52. See Fig. 22-32. (22-19)

54. See Fig. 22-34. (22-20)

56. Antenna signals translated to lower frequency, amplified, radiated back to ground. (22-21)

58. Gyroscopic action tends to hold it steady. (22-21)

60. One program on vertical and one on horizontal polarization on same frequency possible because of FM capture effect. (22-21)

CHAPTER 23

2. 30KØA3EGN or 1ØKØA3EGN. 18ØKF3EGN. (23-1, Appendix E-2)

4. Usually higher fidelity AF. Pre-emphasis and de-emphasis circuits used. More noise-free. FM capture effect. (23-1)

6. International BC. (23-1)

8. Emergency or EBS. Used when main transmitter is inoperative for any reason. (23-2)

10. Studio Transmitter Link relaying programs to a transmitter. Relays programs between BC stations in different cities. (23-2)

12. Manually, or by recorded tones or digital information on them. Manually using a monitor system. (23-3)

14. Ø VU. (23-4)

16. Gain control is a variable pad. (23-5)

18. Louder listener signals, higher % modulation. Makes riding gain easier. (23-7)

20. 125% positive, no more than 100% negative. At least 85%. (23-9)

22. Linear meters 2%, logarithmic scales 5%. (23-10)
24. 5%. (23-10)
26. See Fig. 23-10. (23-11)
28. See Fig. 23-11. (23-12)
30. Within 1 h. (23-13)
32. Open reel, cassette, cartridge. Cartridge. (23-15)
34. Higher fidelity. (23-15)
36. Dynamic, velocity, condenser. 150, 250, or 600 Ω. (23-17)
38. Fiber-optic circuit. Grid-leak. (23-18)
40. Emergency Broadcast System. AM, FM, TV. (23-19)
42. Emergency Action Termination announcement. (23-19)
44. Program, operating, maintenance. By anyone competent to do so. 2 years. (23-20)
46. See Fig. 23-15. (23-14)
48. 50–15,000 Hz, 19 kHz, 23–53 kHz SBs. (23-22)
50. See Fig. 23-17. In the matrix. (23-23)
52. Compatible QUadrature Amplitude Modulation. AM BC band. (23-23)
54. Painted with white and orange stripes, with red lights at tops and flashing lights on sides of tall towers. (23-25)
56. FAA. (23-25)
58. On the hour and half-hour. (23-26)
60. PA and modulator. (23-26)

CHAPTER 24

2. Diplexer. (24-1)
4. See Fig. 24-1. (24-1)
6. See Fig. 24-2. (24-1)
8. First IF stages. (24-1)
10. See Fig. 24-3. (24-2)
12. Side. (24-3)
14. 4:3. (24-4)
16. 525 lines times 30 frames. Sawtooth. (24-4)
18. Electromagnetic yoke coils. Magnetically or by dc-biasing the yoke coils. (24-5)
20. Equalizing, horizontal sync, horizontal blanking, vertical sync, vertical blanking, driving. (24-6)
22. Equalizing, serrated vert., and hor. sync pulses. (24-6)
24. Allow vert. sync pulses to hold sync. (24-6)
26. 70%. 12.5%. 75%. 12.5%–70%. (24-7)
28. See Fig. 24-12. (24-8)
30. Video, sound, cueing, control. 4. (24-10)
32. ¾ or ½ inch. (24-10)
34. See Fig. 24-17. 1.25 and 5.75 MHz. (24-11)
36. Direct or indirect methods. 1.68 times RF power into dummy load while transmitting a black signal. (24-12)
38. Program, operating, maintenance. Written, typed, computerized. (24-13)
40. See Fig. 24-19. (24-14)
42. 47.25 and 39.75 MHz. Lower channel sound carrier, higher channel visual carrier. (24-12)
44. See Fig. 24-23. (24-16)

46. Video, aural IF, sync and blanking signals. (24-16)
48. To separate hor. from vert. sync pulses and couple them to the hor. and vert. oscillators. (24-17)
50. Unmodulated lines showing on the screen. (24-18)
52. Hor., stepped up HV for CRT, AGC circuit, AFC circuit, pincushion correction, convergence in color TV. (24-19)
54. See Fig. 24-27. (24-20)
56. Sync pulses only. (24-21)
58. See Fig. 24-28. (24-21)
60. Red and green. Green and red. Equal values of red, green, and blue. (24-22)
62. 3. (24-23)
64. See Fig. 24-32b. 15,734.26 Hz. 59.94 Hz. (24-23)
66. 3.58 MHz. 8. Sync the 3.58-MHz oscillator. (24-23)
68. Add comb filter in Y circuit. (24-24)
70. Three-gun shadow-mask, and one-gun Trinitron. (24-25, 24-26)
72. Cable companies may translate received TV signals to any channel they wish on their cables. (24-28)
74. Poor cable connections, bad cable amplifiers, poor-quality cable modulation systems. (24-28)
76. FM for both video and aural signals. (24-29)
78. 2300 Hz. 1500 Hz. 16,384 pixels. (24-30)

CHAPTER 25

2. Those carrying 12 or more passengers on international voyages. (25-1)
4. Required for $\overline{SOS}$ transmissions. 200-W output. (25-4)
6. See Fig. 25-2. (25-4)
8. L-pad. Rotor inductor and SWR meter. (25-6)
10. See Fig. 25-4. (25-6)
12. See Fig. 25-5. (25-8)
14. 500 kHz, 2182 kHz, 8364 kHz, 156.8 MHz. (25-9, 25-10)
16. IF is above the highest received frequency. 10 kHz to 30 MHz. (25-12)
18. Uses a synthesizer for all oscillators except possibly the BFO. (25-12)
20. 3 W, 500 kHz, 8364 kHz. Hand-cranked generator. CW, AA signal, $\overline{SOS}$. (25-14)
22. 30-W A2A on 500 kHz, 40-W A2A on 2182 and 8364 kHz. (25-14)
24. 1 min of 4-s on, 1-s off A2A transmission on 500 kHz. (24-15)
26. 3.5–6 s, 0.1–1.5 s. (25-16)
28. Button pushed on AA cabinet. (25-16)
30. Power line variations, open fuse. (25-16)
32. Convert two-tone signals to positive and negative voltages. Convert keyboard coded contacts to FSK signals. (25-18)
34. ARQ, broadcast to several stations. (25-19)

36. 2, at 15–18 and 45–48 m after the hour. (25-20)

38. Longitude. (25-21)

40. Sky waves or reflected signals. (25-22)

42. Intentionally unbalance the loop. (25-23)

44. 45°, 135°, 225°, 315°. (25-26)

46. Goniometer. Rotates an exploring coil. (25-27)

48. Goniometer. (25-30)

50. Transmission lines pick up no residual horizontal signal components. (25-31)

52. LOng RAnge Navigation. Loran C, 100 kHz. Loran A, 1900-kHz range. (25-33)

54. Hyperbolic. (25-33)

56. Superimpose master and slave for time-difference readout. (25-33, -34)

58. Omega. 10.2–13.6 kHz. (25-35)

60. Circular polar orbiting. Doppler effect frequencies. (25-36)

62. Pictures, maps. 10 min. 96 lines/inch. (25-37)

64. Burning due to electric current flow between helix wire and blade. (25-38)

66. Electrolysis. (25-39)

68. Verticals, center- or base-loaded. (25-39)

CHAPTER 26

2. 12.3 μs, the time taken for a signal to reach a target 1 mile away and return. (26-1)

4. Pulse Repetition Rate. 800 to 2000 pulses/s. (26-1)

6. Rotating parabolic or slotted waveguide. (26-2)

8. 10 rpm. (26-2)

10. See Fig. 26-3. (26-3)

12. 34 kW, 0.00044. (26-4)

14. Wide beamwidth, loss of sensitivity. (26-5)

16. Radome. (26-5)

18. Mixer section TR tube. (26-6)

20. Magnetron is not frequency stable. (26-7)

22. See Fig. 26-9. (26-7)

24. Intensifying pulse. (26-8)

26. See Fig. 26-12. (26-8)

28. Selsyn. (26-9)

30. Course-Up, North-Up. (26-10)

32. Radar endorsement on any commercial FCC license. (26-12)

34. Harsh 1000-Hz signal. (26-13)

36. Doppler. Airport traffic surveillance. (26-15)

38. Antenna, transmitter, receiver RF, IF and detector. Indicator and its circuits, power supply. (26-16)

40. Corner reflectors. So other vessels are sure to see them on their radar sets. (26-16)

42. Doppler log. (26-17)

44. Joystick. (26-17)

CHAPTER 27

2. Magnetic field, armature with rotating conductors, electromechanical sliding connections. (27-2)

4. Field coil is fed dc. DC excitation varied. (27-2)

6. See Fig. 27-4. (27-2)

8. AC output voltage may rise. (27-3)

10. See Fig. 27-6. When indicator shows zero voltage difference. (27-5)

12. Smoother dc developed. (27-6)

14. Field excitation polarity and rotor direction. Polarity of magnets and rotor direction. (27-7)

16. See Fig. 27-11. Sags somewhat. Small. (27-9)

18. See Fig. 27-13. (27-11)

20. Add bypass capacitors from dc leads to ground, or use LP *LC* filters in series with the load. (27-13)

22. Continually accelerate to destruction. (27-15)

24. Current direction through motor. (27-15)

26. The counter emf developed in the armature. (27-16)

28. Better speed regulation. (27-17)

30. See Fig. 27-21. (27-18)

32. It operates at the same speed at all times. (27-21)

34. Yes. Phase rotates around the field poles. (27-23)

36. Change dc to some other *V* dc, or dc to ac, or ac to dc, or ac to some other ac voltage or frequency. (27-24)

38. Watts. Volt-amperes, or VA. (27-26)

40. 2862 W, 24.89 A. (27-26)

42. Must not be shorted across, and must be undercut. (27-27)

CHAPTER 28

2. Carbon + cathode, zinc − anode, sal ammoniac-zinc chloride electrolyte. 1.55 V. (28-2)

4. Formation of a gas in the cell. (28-2)

6. PbO_2 + plate, Pb − plate, H_2SO_4 electrolyte. Pb + plate, Pb − plate, water electrolyte. 2.1 V. (28-3)

8. Weight of liquid compared to weight of water. 1280 usually. 1180 usually. (28-5)

10. Hydrometer. Rubber bulb, glass barrel, calibrated glass float, rubber hose (sometimes with thermometer in it). (28-5)

12. To prevent local action or chemical contamination of plate surfaces and electrolyte. (28-6)

14. Slowly. (28-7)

16. Constant-current. (28-8)

18. Battery recharged, giving off H_2 and O_2. (28-8)

20. Sp gr very low, or *V* across cell is low or reversed under load. (28-8)

22. See Sec. 28-9.

24. 1.5 V., low internal *R*, high *V*, long shelf life. (28-11)

26. 1.55 V reference *V* source, low internal *R*. (28-11)

28. 1.25 V, rechargeable, low internal *R*, have a memory effect. (28-11)

30. 0.3 V, 20-year half-life. (28-11)

CHAPTER 29

2. Up to $10,000 and 1 year for first offense, 2 years for second. (29-1)
4. Amateur, distress, broadcast, CQ messages. (29-1)
6. General Radiotelephone, 3; Radiotelegraph 1st, 1, 2, 5, 6 plus code test; Radiotelegraph 2nd, same but slower code test; Radiotelegraph 3rd, 1, 2, 5 plus slower code test. (29-3)
8. If used in international communications, Aviation, Merchant Marine, International BC. (29-3)
10. Must be 21 years to be 1st class Radiotelegraph licensee. (29-3)
12. 75% for each element. (29-3)
14. Notify FCC and make out application for duplicate. Use copy of application as license in interim.
16. 15 days or until conclusion of a hearing. 10 days or until conclusion of a hearing. (29-4, -5)
18. Distress, $\overline{SOS}$ or MAYDAY; Urgency, XXX or PAN; Safety, TTT or SECURITY; Routine. (29-7)
20. Station, date, time, operator on duty, with whom communicated, what type of communications occurred. (29-8)
22. 1 year. 3 years. Until authorized in writing to destroy. (29-8)

CHAPTER 30

2. 2882 kHz and 156.8 MHz. (30-1)
4. MAYDAY (3); THIS IS; Call (3). (30-3)
6. Latitude and longitude, distance and direction from a known point. (30-3)
8. Its name, present position, speed toward ship in distress. (30-3)
10. PAN. Concerns safety of person or ship. (30-4)
12. Pertains to any distress, urgency, or safety type traffic. (30-5)
14. X-RAY INDIA NOVEMBER GOLF UNIFORM. (30-7)
16. Preamble, address, text, signature, service. (30-9)
18. Usually only at the end of traffic or every 15 min while in communication. (30-9)
20. May interfere with communications on that frequency. (30-11)
22. SSB. (30-11)
24. Allow break-in time for any emergency traffic. (30-11)
26. Prevent unauthorized use. (30-11)
28. Calling, distress, possibly urgent or safety. All other traffic. (30-12)
30. The station will handle paid messages for the general public.

CHAPTER 31

2. Radiotelegraph. (31-2)
4. International Morse, American Morse, Baudot, ASCII. (31-3)

6. KAAA KAAA KAAA DE WBBB WBBB WBBB K. (31-5) __
8. QRX and $\overline{AS}$. (31-5)
10. Wait for clear channel, tune up as rapidly as possible, send V V V, sign call. (31-7)
12. 15–18 and 45–48 min after every hour. Logs any signals heard. (31-9)
14. $\overline{SOS}$ DE call; position; nature of distress; what assistance required; course and speed if underway; other details. (31-10)
16. Modulated CW, or A2A. (31-10)
18. Monitor frequency for at least 3 min for message before resuming normal service. If message is received act on it similar to distress traffic. (31-11)
20. Ship name, call letters, dates in log, number of pages, statement regarding distress traffic in it, operator signature, home address, license number and date of issuance. (31-13)
22. Message number, type of message, check, station of origin, date, time. (31-14)
24. After every multiple of five words. (31-14)
26. Parentheses, quotation marks. (31-15)
28. 8¢ + 17.5¢ + 7.75¢ = 33.25¢. (31-16)
30. 18–20 wpm. 20–40 wpm. (31-18)
32. Make error sign, repeat last correctly sent word and continue. (31-19)

CHAPTER 32

2. ARRL's W1AW station. (32-1)
4. 5 wpm. 20 wpm. (32-2)
6. Usually both. (32-3)
8. AA-AL, K, KA-KZ, N, W, WA-WZ, KH, KL, KP. (32-4)
10. 160-, 80-, 40-, 30-, 20-, 15-, 10-meter bands. (32-5)
12. 1500-W PEP. Technician-Novice sections of 80-, 40-, 15-, and 10-meter bands. (32-5)
14. Baudot, AMTOR, ASCII. (32-6)
16. CW, SSB, FM, AM, RTTY, SSTV, FSTV, FAX. (32-7)
18. "FCC Rules and Regulations," Part 17. (32-7)
20. Extra class frequencies and emissions. Novice call, slant, extra call, if on a non-Novice frequency. (32-8)
22. Voice or CW. (32-8)
24. CW. 599. (32-8)
26. No. (32-8)
28. 10 days. (32-9)
30. No. (32-10)
32. Radio Amateur Civil Emergency Service. Only those previously enrolled in it. (32-11)
34. See Fig. 32-3. (32-12)
36. Element 4a. Elements 4a and 4b. (32-16, 32-17)

Index

SUPPLEMENT
FOR
ELECTRONIC
COMMUNICATION
Fifth Edition

**EXAMINATION-TYPE
QUESTIONS FOR**

- Commercial General Radiotelephone License
  Commercial Radiotelegraph Licenses
  FCC- and Industry-Sponsored Certification
  Amateur Radio License Information

- Amateur Radio License Elements

Preface for Supplement

This supplement contains two separate groups of questions. The first group may be used when preparing to pass FCC (Federal Communications Commission) commercial grade license examinations, both radiotelephone and radiotelegraph, as well as to aid in study for Amateur Radio license examinations. These questions also include information believed by the FCC to be important for operators or technicians to understand when working in broadcast stations and when servicing or installing two-way radio equipment (services no longer requiring FCC licensing). Each question is followed by numbers indicating the chapter and section in *Electronic Communication*, Fifth Edition, where the subject is discussed.

The second section contains questions on subjects necessary to pass the tests for Novice, Technician, General, Advanced, and Extra class Amateur Radio licenses. These questions are also keyed to both chapter and section where the subjects are discussed in *Electronic Communication*, Fifth Edition.

The meanings of the bold letters that follow each question in the first group of questions are:

A Advanced class Amateur license question
B Broadcast station information question
E Extra class Amateur license question
G General and Technician classes Amateur licenses question
N Novice class Amateur license question
P Radiotelephone license for international equipment question
R Radar endorsement question
T Radiotelegraph commercial license question
V Television station information question

If studying for the General Radiotelephone license, for example, readers should scan questions suffixed with a **P**, and be sure they know the answers to those questions.

The FCC is at present giving tests for the Commercial General Radiotelephone license, required for technicians who work on radio transmitting and receiving equipment involved in international communications. Although no license is required any longer to service or install locally used radio equipment, it is still necessary for everyone training for such operations to have the requisite background and to be familiar with the answers to the **P** questions. Those in training for broadcast technician work should know the answers to the **B** questions. The FCC is also testing for several grades of Radiotelegraph licenses used aboard ships plying international waters; those interested should review the **T** questions. (See Chap. 29 in *Electronic Communication*, Fifth Edition for additional license information.)

The FCC issues the licenses for Amateur Radio operators and stations after the applicants have satisfactorily passed license tests given by authorized volunteer examiners selected from groups of qualified Amateur Radio operators.

In all cases it is necessary to make prior appointments to take tests given by both the FCC and the volunteer Amateur examiners. Information about examination schedules can be determined by contacting the nearest FCC field office (Appendix F in *Electronic Communication*, Fifth Edition). Preferably, applicants for Amateur Radio licenses can obtain testing information from local Amateur Radio clubs in their area.

The many questions in this supplement may also form a well of test material for instructors making up tests on the various chapter subjects included.

Note the use of the symbol letter E to signify electromotive force, or voltage, in FCC test questions rather than the more modern usage of the symbol letter V.

QUESTIONS

CURRENT, VOLTAGE, AND RESISTANCE

1. What produces an electrostatic field? (1-2) **P**
2. What is stored in an electrostatic field? (1-2) **P**
3. What is an electron? A proton? (1-2) **T**
4. What makes a substance an insulator? (1-3, 1-7) **N**
5. With respect to electrons, what makes a substance a conductor? (1-3, 1-7) **BPN**
6. Explain the relationship between the atom, electrons, and current. (1-3, 1-4, 1-7) **BP**
7. In what manner does the resistance of a copper conductor vary with variations in temperature? **T**
8. What other expression describes electron flow? (1-7) **PT**
9. Define the term "coulomb." (1-7) **PT**
10. What is an ampere? (1-7) **PN**
11. If a coulomb passes a point in 0.25 s, what is the current value? (1-7) **PT**
12. What produces an electrostatic field? (1-8) **BP**
13. What other expression describes a difference of potential, or electromotive force? (1-8) **PN**
14. What is the unit of measurement of emf? (1-8) **PTN**
15. If an emf exists between two points with no flow of current, what kind of field is present? (1-8) **T**
16. List eight means by which an emf may be produced. (1-8) **T**
17. What should be looked for if a corona discharge occurs? (1-9) **PTA**
18. What is an ion? (1-10) **T**
19. What is meant by direct current? (1-11) **PN**
20. What is meant by electrical resistance? (1-12) **BPTN**
21. What is the result if resistance is added to an electric circuit? (1-12) **PN**
22. What is an ohm? (1-12) **PTN**
23. What is the relationship between wire size and resistance of the wire? (1-12, 1-15) **BP**
24. List some of the best insulator and conductor materials. (1-12) **PG**
25. What factors affect the resistance of a conductor? (1-12) **T**
26. If the diameter of a wire of given length is dou-

bled, how will the resistance be affected? (1-12) **T**
27. How many leads does a potentiometer have? A rheostat? (1-12) **T**
28. Name five conducting materials in order of descending conductivity. (1-12) **T**
29. What type of resistor would normally be expected to have the smallest tolerance? (1-12) **P**
30. Describe a carbon resistor. (1-12) **PN**
31. Describe a wirewound resistor. (1-12) **PN**
32. What symbols are used with resistive components? (1-12) **PG**
33. What is meant by standardized values of resistors? (1-13) **P**
34. What color-coded number is represented by yellow-gray-white? By blue-red-brown? (1-12) **PT**
35. What is indicated by a resistor that is color-coded yellow-violet-orange-silver? (1-13) **PT**
36. What is meant by the tolerance of a resistor? (1-13) **PT**
37. What is meant by the metric prefixes: mega; kilo; centi; milli; micro; pico; nano; micromicro? (1-14) **PTN**
38. Make the following transformations: 7 kΩ to Ω; 14 kV to V; 25 mA to A; 75 nA to μA. (1-14) **P**
39. How are AWG size numbers related to current-carrying capacity? (1-15) **PA**
40. Why is rosin used as a soldering flux in electronic construction work? (1-16) **T**
41. List at least two essentials for a good soldered connection. (1-16) **PA**
42. List three precautions which should be taken in soldering electrical connections to assure a permanent junction. (1-16) **PA**
43. Explain how etched-wiring printed circuits are used with respect to determining wiring breaks, excessive heating, and removal or installation of components. (1-17) **PG**

DIRECT-CURRENT CIRCUITS

1. State the three forms of Ohm's law. (2-2) **PTG**

2. Give examples of the use of Ohm's law. (2-3, 2-11, 2-15) **PTG**

3. If the voltage in a circuit is doubled and the resistance is increased four times, what will be the final current? (2-3) **T**

4. What is the unit of electric power? (2-4) **BPTN**

5. What is the formula to determine power when the known factors are: voltage and resistance; current and resistance; voltage and current? (2-4) **BPTN**

6. If the resistance is tripled across a constant emf, how is the power dissipation affected? (2-4) **T**

7. What is the unit of electric energy? (2-4) **BPTN**

8. How many watts in 1 horsepower (hp)? (2-4) **PT**

9. What one factor differs between electric power and energy? (2-4) **BPTN**

10. What is the formula for joules? (2-4) **P**

11. What should be the minimum power dissipation rating of a 10,000-Ω resistor to be connected across a potential of 100 V? (2-4) **BPT**

12. What is the heat dissipation, in watts, of a 40-Ω resistor with 0.25 A passing through it? (2-4) **BPT**

13. If the power input to a radio receiver is 75 W, how many kWh does it consume in 24 h? (2-4) **BPT**

14. How much would it cost to operate three 100-Ω 120-V lamps in parallel for 24 h if energy costs 7¢ per kWh? (2-4) **BPT**

15. A power company charges 6¢ per kWh for electricity. How much would it cost to operate two parallel 120-V lamps, each with 100-Ω resistance, for 12 h? (2-4) **BPT**

16. What is the formula to determine current if the known factors are: resistance and power; power and voltage? (2-5) **BPT**

17. What is the formula to determine voltage if the known factors are: power and current; resistance and power? (2-5) **BPTN**

18. What is the formula to determine resistance if the known factors are: voltage and power; power and current? (2-5) **BPT**

19. What is the maximum rated current-carrying capacity of a 1000-Ω 100-W resistor? (2-5) **T**

20. What is meant by a short circuit? (2-7) **PN**

21. What is meant by an open circuit? (2-7) **PN**

22. What is the symbol of a fuse? (2-7) **BPTN**

23. What are the functions of fuses? (2-7) **BPTN**

24. What are slow-blow and chemical fuses? (2-7) **BPTN**

25. Name the meter that measures: emf; current; power; energy; resistance. (2-8) **TG**

26. Name two meters that have three terminals on them. (2-8) **T**

27. Diagram how the total current in three branches of a parallel circuit can be measured by one ammeter. (2-8, 2-13) **PT**

28. Diagram how a voltmeter and ammeter can be connected to measure power in a dc circuit. (2-8) **PTG**

29. Describe and explain the use of the: wattmeter; voltmeter; ammeter; ohmmeter; watthourmeter. (2-8) **PG**

30. What is an oscilloscope and how is it sometimes used? (2-8) **PG**

31. What type of circuit is used in voltage dividers? (2-10) **PG**

32. How are components connected in a series circuit? (2-10, 2-11) **P**

33. What is Kirchoff's voltage law? (2-10, 2-17) **P**

34. Diagram how to connect batteries in series. (2-10) **BPTN**

35. What is the sum of all the voltage-drops around a simple dc circuit, including the source? (2-10) **PT**

36. Two 5-W 100-Ω resistors are in series. What is the total power-dissipation capability? (2-11) **PTG**

37. A circuit has a flow of 5 A. The internal source resistance is 1 Ω. The external resistance is 80 Ω. What is the terminal voltage of the source? (2-11) **PT**

38. Two resistors are in series. The current through them is 100 mA. R_1 is 50 Ω, and R_2 has a voltage-drop of 60 V across it. What is the total impressed emf? (2-11) **PT**

39. A 6-V battery has an internal resistance of 1 Ω. What current will flow when a 5-W 6-V lamp is connected across it? (2-11) **PTG**

40. A vacuum tube has a filament rated at 0.3 A and 6 V. If it is to operate from a 12.6-V battery, what is the necessary series resistor value? (2-11) **T**

41. What is electrical conductance? (2-12) **PT**

42. What is the unit of conductance? (2-12) **PT**

43. What is the conductance of a circuit if 1 A flows when 10 V is applied across it? (2-12) **PT**

44. How are resistors connected if they are said to be "in parallel"? (2-13) **PTN**

45. What type of circuit might be called a current divider? (2-13) **PTG**

46. What is Kirchhoff's current law? (2-13) **P**

47. What is the total resistance of a parallel circuit having 10- and 15-Ω branches? (2-13) **PT**

48. Two 10-W 100-Ω resistors are in parallel. What is the total power dissipation capability? (2-13) **PT**
49. If resistors of 6, 8, and 12 Ω are connected in parallel, what is the total resistance? (2-13) **PT**
50. A 1-kΩ 50-W resistor, a 2-kΩ 20-W resistor, and a 500-Ω 10-W resistor are connected in parallel. What maximum current through this parallel combination will not exceed the wattage rating of any of the resistors? (2-13) **T**
51. Diagram two cells: in series across a load; in parallel across a load. (2-14) **BPTG**
52. What method of connection should be used to obtain the maximum: no-load voltage from cells; current from similar cells? (2-14) **PT**
53. Draw a diagram of a 12-V battery with three resistors of 10, 120, and 300 Ω, respectively, in a pi-network. What is the: current through each resistor; voltage-drop across each resistor; power dissipated by each resistor; total current; total power dissipation? (2-15) **BP**
54. Draw diagrams showing how to connect three resistors of equal value so that the total R will be: three times the R of one unit; one-third the R of one unit; two-thirds the R of one unit; 1½ times the R of one unit. (2-15) **T**
55. Resistors of 20 and 25 Ω are in parallel. In series with them is an 11-Ω R. In parallel with this combination is a 20-Ω R. The total current is 5 A. What is the current in the 11-Ω R? In the 25-Ω R? (2-15) **T**
56. What is meant by impedance matching? (2-16) **BPG**
57. Why is impedance matching between a source and a load important? (2-16) **BPG**
58. A 500-Ω impedance source will develop maximum power into a load having how much resistance (impedance)? (2-16) **BPTG**
59. Can impedance matching always be attained and is it always desirable? (2-16) **BP**
60. State Thevenin's theorem. (2-18) **P**
61. Explain how it is possible to replace a voltage source and resistive voltage divider with an equivalent circuit consisting of a voltage source and one resistor. (2-18) **PA**
62. What is Norton's theorem? (2-18) **P**

MAGNETISM

1. Name at least five pieces of equipment which make use of electromagnetism. (3-1) **T**
2. What determines the direction of the lines of force of an electromagnet? (3-2) **T**
3. What factors influence the direction of magnetic lines of force produced by an electromagnet? (3-2) **BP**
4. How can the direction of flow of dc in a conductor be determined? (3-2, 3-16) **BT**
5. What are two units by which magnetic flux can be measured? (3-3) **P**
6. What are two units by which magnetic-flux density can be measured? (3-3) **P**
7. What is meant by ampere-turns, and what is its symbol? (3-4) **T**
8. What are two units by which mmf can be measured? (3-4) **T**
9. What produces the energy that is stored in electromagnetic fields? (3-4) **P**
10. Define permeability. (3-5) **PT**
11. What is the permeability of air? (3-5) **P**
12. What does the ratio B/H indicate? (3-5) **P**
13. Define the term "reluctance." (3-5) **BP**
14. What is the reciprocal of reluctance? (3-5) **P**
15. Explain the theory of molecular or domain alignment as it affects magnetic properties of materials. (3-6) **BP**
16. What is residual magnetism? (3-7) **PT**
17. In what types of materials does residual magnetism not occur? (3-7) **PT**
18. What is meant by the hysteresis of a material? (3-8) **PT**
19. What is the name of the force which cancels all residual magnetism? (3-8) **P**
20. What loss causes the core of a relay or transformer to heat when ac is fed to their coils? (3-8) **P**
21. Do ferrite cores have high or low retentivity? (3-9) **P**
22. Are ferrite cores conductors, insulators, or semiconductors? (3-9) **PA**
23. What is the symbol for a weber, and to what is 1 weber equal? (3-12) **P**
24. What is the symbol for a tesla, and to what is 1 tesla equal? (3-12) **P**
25. What effect does one magnetic north pole have on: another north pole; a south pole? (3-13) **P**
26. Will there be any external field if an electromagnet is completely encased in an iron box? (3-15) **P**
27. Why should a compass be kept away from strong alternating currents? (3-16) **T**

28. What are two examples of electrons being deflected by magnetic fields? (3-17) **T**
29. State Lenz's law. (3-18) **P**
30. Which factors determine the amplitude of the emf induced in a conductor which is cutting magnetic lines of force? (3-18) **BP**
31. If a nickel rod is magnetized, will it expand, contract, or remain the same length? What effect does this explain? (3-19) **PT**
32. What materials are used for relay contacts and why? (3-20) **T**
33. Explain the method of cleaning relay contacts and why it is necessary that the original contact shape be maintained. (3-20) **BP**
34. What is meant by self-wiping contacts as used in connection with relays? (3-20) **T**
35. Explain the operations of the: break-contact relay; make-contact relay. (3-20) **BP**
36. What are two advantages of reed relays? (3-20) **PT**

ALTERNATING CURRENT

1. What is meant by an alternating current? (4-2) **PN**
2. What is meant by: sine wave; square wave; triangular or sawtooth wave? (4-2) **BPA**
3. Draw two cycles of a sine wave on an amplitude vs time graph. (4-3) **BP**
4. How many degrees in 1 cycle? (4-3) **BP**
5. How many degrees in an alternation? (4-3) **BPTG**
6. What is meant by the root-mean-square value of a sine wave? (4-4) **PG**
7. What is the relation between the effective value of an ac current and the heating value of the current? (4-4) **TG**
8. Give a practical application of rms voltage. (4-4) **BPG**
9. For a sine wave, what is the ratio of: peak to effective; peak to average? (4-4, 4-5) **T**
10. What are the relationships between rms, effective, peak, and peak-to-peak values of sinusoidal ac? (4-4) **BPG**
11. Define an average current value in an ac circuit. In what practical situation might it be used? (4-5) **BP**
12. What is meant by an audio frequency? (4-6) **PN**
13. What is meant by a radio frequency? (4-6) **PN**
14. What are the meanings of VLF, LF, MF, HF, VHF, UHF, SHF, EHF? (4-6) **PG**
15. What is the frequency range of: VLF, LF, MF, HF, VHF, UHF, SHF, EHF? (4-6) **BPG**
16. What is meant by: hertz; kilohertz; megahertz; gigahertz? (4-6) **PN**
17. Name four materials which are good insulators at radio frequencies, and four which are not good insulators at radio frequencies but are satisfactory for power frequencies. (4-6) **BP**
18. What is meant by ac phase and why is it important? (4-7) **BP**
19. List the fundamental and the next five harmonics of 800 kHz. (4-8) **BP**
20. How long would it take for a 5-MHz vector to rotate: 45°; 90°; 280°? (4-8) **BP**
21. What is the relationship between frequency and wavelength? (4-8) **PN**
22. What is the relationship between frequency and period? (4-8) **P**
23. What is meant by a harmonic? (4-8) **BPN**
24. What is meant by an octave? (4-8) **BP**

INDUCTANCE AND TRANSFORMERS

1. What is a circuit said to have if it opposes any change in circuit current? (5-1) **B**
2. What component has the property of electrical inductance? (5-1, 5-2) **TG**
3. What is the unit of measurement of inductance? (5-1, 5-2) **TG**
4. Explain the exponential charge and discharge of *RL* circuits. (5-1, 5-4) **E**
5. Describe the appearance, characteristics, applications, and symbols of various types of inductors. (5-1, 5-6) **PG**
6. State a definition of inductance. (5-2) **P**
7. Define: henry; picohenry; microhenry; millihenry. (5-2, 1-14) **PG**

8. What is the effect of adding an iron core to an air-core inductor? (5-3) **T**
9. What is the relationship between the number of turns and the inductance of a coil? (5-3) **BPT**
10. What is the relationship between inductance of a coil and the permeability of its core? (5-3) **BP**
11. What is meant by the time constant of an LR circuit? (5-4) **PE**
12. How do the values of R and L in an RL network affect its time constant? (5-4) **BP**
13. Explain energy storage in electromagnetic fields. (5-5) **PTA**
14. What is the formula for the energy stored in the magnetic field surrounding an inductor? (5-5) **B**
15. Explain choke coils, ferrite beads, and toroid coils. (5-6) **PG**
16. What is the formula used to compute mutual inductance? (5-7) **P**
17. What is meant by unity coupling? (5-8) **T**
18. If the mutual inductance between two coils is 0.03 H and the coils have inductances of 40 and 90 mH, what is the coefficient of coupling? (5-8) **B**
19. What is the total inductance of two coils in series and with no mutual coupling? (5-9) **TG**
20. If two coils of equal inductance are in series with unity coefficient of coupling, with fields in phase, what is the total inductance? (5-9) **PG**
21. What effect does mutual inductance have on the total inductance of two coils connected in series? (5-7, 5-9) **B**
22. What is the formula for total inductance if inductors are connected in series; in parallel? (5-9, 5-10) **PG**
23. What is the total inductance of two coils in parallel and having no mutual inductance? (5-10) **T**
24. What is the effect if a turn is shorted in an inductor? (5-11) **T**
25. What is the total reactance of two inductors in series with zero mutual inductance? (5-12) **B**
26. What is the reactance of a 4-H choke at a frequency of: 500 Hz; 5 kHz? (5-12) **PA**
27. What is the reactance of a 10-mH coil at 2 MHz? (5-12) **PA**
28. What is the formula for inductive reactance? (5-12) **PG**
29. A series inductor acting along in an ac circuit has what properties? (5-12) **T**
30. Explain inductive reactance. (5-12, 5-13) **BPG**
31. What may cause an ac current to lag its voltage? (5-13) **B**
32. What E-I phase relationship is caused by inductance? (5-13) **BP**
33. Explain the relationship of turns ratio, voltage ratio, and current ratio of transformers. (5-15, 5-21, 5-23) **PG**
34. Describe the appearance, symbols, and operation of power, AF, and RF transformers. (5-15 thru 5-26) **P**
35. What prevents high-current flow in the primary of an unloaded power transformer? (5-15) **BP**
36. Explain how self-inductance and mutual inductance produce transformer action. (5-15, 5-26) **BP**
37. Why are laminated-iron cores used in audio and power transformers? (5-17) **B**
38. What factors determine the core losses in a transformer? (5-17, 5-18) **T**
39. What are eddy currents and how are they reduced? (5-17) **BPT**
40. What is hysteresis in a transformer? (5-18) **T**
41. What circuit constants determine the copper losses of a transformer? (5-19) **T**
42. List four ways that power is lost in an iron-core transformer. (5-17 through 5-20) **BP**
43. How is power lost in an air-core transformer? (5-19, 5-20) **BP**
44. What factors determine the no-load voltage ratio of a power transformer? (5-21) **BPT**
45. What is the E_s of a transformer with 300 T_p, 40 T_s, and an E_p of 100 V? (5-21) **T**
46. What is the relationship between the turns ratio and the primary-secondary voltage ratio of a power transformer? (5-21) **BPT**
47. In an iron-core transformer, what is the relationship between turns ratio and primary-secondary current ratio? (5-23) **BPT**
48. If a power transformer with a 1:4 step-up ratio is placed under load, what will be the approximate ratio of primary to secondary currents? (5-23) **BPG**
49. What factors determine the efficiency of a power transformer? (5-24) **T**
50. If a transformer has an E_p of 4800 V, an E_s of 240 V, and an efficiency of 90% at 20 A output, what is the primary current value? (5-24) **BT**

51. If part of the secondary winding of a power transformer were shorted, what would be the immediate effect? (5-26) **T**
52. What would happen if a 500-Hz power transformer were connected across a 60-Hz source of the same voltage? (5-26) **T**
53. What would happen if a 120-V 60-Hz power transformer were connected across a 120-V 300-Hz source? (5-26) **T**
54. What would be the effect if dc were applied to the primary of an ac power transformer? (5-26) **T**

CAPACITANCE

1. What is the basic unit of capacitance? (6-1) **PTG**
2. What component has capacitance? (6-1) **PG**
3. What symbols are used for various types of capacitors? (6-1, 6-8) **PG**
4. How many picofarads are in: 1 μF; 1 nF? (6-1) **PG**
5. What is meant by the time constant of an RC circuit? (6-1) **PTE**
6. Explain how the values of resistance and capacitance in an RC network affect its time constant. (6-1) **BPE**
7. A capacitor is charged to 10 V and is then connected in parallel with a resistor. What will be the voltage across it after 1 RC time constant? (6-1) **BPE**
8. What is the relationship between storage of charge and capacitance of a single capacitor? Of two capacitors in series? (6-1, 6-6, 6-15) **BPA**
9. Explain the exponential charge and discharge of an RC circuit. (6-1) **PE**
10. How does the capacitance of a capacitor vary with area of plates, spacing between plates, and dielectric material? (6-2) **BP**
11. What is the capacitance of a three-plate capacitor with 2-in.^{2} plates separated by 0.003 in. of insulation having a dielectric constant of 7? (6-2) **BT**
12. If the specific inductive capacity of a capacitor's dielectric is changed from 1 to 4, what is the capacitance change? (6-2) **T**

13. What factors determine a capacitor's breakdown-voltage rating? (6-4) **T**
14. Where is the charge in a capacitor stored? (6-5) **T**
15. What is the formula for quantity of charge in a capacitor? (6-6) **T**
16. What factors determine the charge stored in a capacitor? (6-6) **BP**
17. Two 0.5-pF capacitors are given. One is charged to 600 V and disconnected from the charging source. The charged capacitor is then connected in parallel with the uncharged capacitor. What voltage will now appear across the two capacitors? (6-6) **T**
18. What is the meaning of electrolyte? (6-7) **P**
19. What is the principle of operation of electrolytic capacitors? (6-7) **BPTG**
20. What is the desirable feature of electrolytic capacitors? (6-7) **BPTG**
21. What polarity precaution should be observed when connecting electrolytic capacitors in a circuit? (6-7) **BPT**
22. Describe the following types of capacitors: disc, electrolytic, mica, ceramic, polystyrene, tantalum, air variable, trimmer. (6-8, 6-9) **PG**
23. Which types of dielectrics are used in capacitors having rotor and stator plates? (6-8) **T**
24. What types of capacitors operate best in higher frequency circuits? (6-9) **PG**
25. What is the formula used to determine capacitive reactance? (6-13) **BPTG**
26. What is the capacitance of a capacitor if its X_c at 4 MHz is 20 Ω? (6-13) **PTG**
27. What is the reactance of a capacitor at 1200 kHz if its reactance is 100 Ω at 300 kHz? (6-13) **PTG**
28. What is the total capacitance of three similar capacitors: in series; in parallel? (6-14, 6-15) **BPTG**
29. If capacitors of 3, 4, and 6 pF are connected in parallel, what is the total capacitance? (6-14) **PTG**
30. What is the total reactance when two capacitors with equal capacitances are connected in series? (6-15) **PTG**
31. What formula is used to determine the total capacitance of three or more dissimilar capacitors connected in series? (6-15) **PTG**
32. If capacitors of 5, 8, and 10 μF are connected in series, what is the total capacitance? (6-15) **PTG**

33. The voltage-drop across an individual capacitor of a group of capacitors in series across a potential is proportional to what factors? (6-15) **T**
34. How many 300-V 4-μF capacitors would be necessary to obtain a combination rated at 1200 V and 2 μF? (6-15) **T**
35. Why is a high-value fixed resistor shunted across each series capacitor when connected across high-voltage circuits? (6-15) **T**
36. What may cause an ac current to lead its voltage? (6-16) **B**
37. Show by a simple graph what is meant when it is said that the current in a circuit leads the voltage. What would cause this? (6-16) **T**
38. What is the value, tolerance, and voltage rating of an EIA mica capacitor whose colors are (left to right): first row, white, brown, green; second row, black, gold, orange? (6-18) **T**

ALTERNATING-CURRENT CIRCUITS

1. What are the properties of the following when acting alone in an ac circuit: an inductor; a capacitor; a resistor? (7-1) **PE**
2. State the three Ohm's law formulas for ac circuits. (7-2) **PTG**
3. In what unit is impedance measured? (7-2) **BPTG**
4. What are some formulas to compute impedance? (7-2, 7-7, 7-8, 7-13) **PG**
5. What is the impedance of a choke coil if its resistance is 8 Ω and 0.4 A flows through it when 220 V at 50 Hz is applied across it? (7-2) **TE**
6. How does an inductor affect the voltage-current phase relationship of a circuit? (7-3) **BPA**
7. In a circuit having a 200-Ω X_L and a 200-Ω R, what will be the phase angle? (7-3) **PA**
8. How is phase angle between E and I determined if R and X are known? (7-3, 7-7) **BPA**
9. Explain how to determine the sum of two vector quantities which have the same reference point, but the directions of which are: 180° out of phase; in phase; 90° out of phase? How does this pertain to electric currents or voltages? (7-4) **BPE**

10. The product of E and I in an RC ac circuit indicates what kind of power? (7-5) **PG**
11. What is the meaning of power factor? (7-6) **BP**
12. What factors must be known in order to determine the power factor of an ac circuit? (7-6) **BP**
13. How is power factor determined if the phase angle is known? (7-6) **BPA**
14. What is the meaning of the terms: leading power factor; lagging power factor? (7-6) **TA**
15. Why is the phase angle in a circuit important? (7-6) **T**
16. How can low-power factor in an electric power circuit be corrected? (7-6, 7-7) **PA**
17. If a 220-V 60-Hz line delivers 200 W at 90% pf to a load, what is the phase angle, and how much current flows? (7-6) **TE**
18. How is the impedance of a series RLC group calculated? (7-8) **BPTA**
19. Draw a vector diagram of the voltages in a series circuit having X_L, X_C, and R, with X_L predominating. (7-8) **BPE**
20. What is the impedance of a series ac circuit if R = 20 Ω, X_L = 30 Ω, and X_C = 60 Ω? (7-8) **BPTA**
21. If 10 A flows in a series circuit composed of 5-Ω R, 25-Ω X_C, and 12-Ω X_L, what is the: voltage-drop across each component; total source voltage? (7-8) **B**
22. What is the Z value at 1 kHz of a series circuit having X_C = 9900 Ω, R = 12,500 Ω, and L = 1.57 H? (7-8) **BP**
23. A 110 V ac is across a series circuit of 25-Ω X_C, 10-Ω X_L, and 15-Ω R. What is the phase angle? (7-8) **PA**
24. If 2 A flows in an ac series circuit made up of a 12-Ω R, a 20-Ω X_L, and a 40-Ω X_C, what is the source voltage? (7-8) **PE**
25. A series circuit contains a 7-Ω R, a 10-Ω X_L, and X_C is unknown. What value X_C will produce a 15-Ω Z? (7-9) **B**
26. What is the relationship between: resistance and conductance; impedance and admittance; susceptance and reactance? (7-12) **BP**
27. A 7-Ω R is in series with a parallel combination of a 20-Ω X_C and a 10-Ω X_L. What is the impedance of the circuit? Is it capacitive or inductive? (7-12) **BE**
28. If 115 V ac is applied across a 30-Ω R, a 20-Ω X_L, and a 10-Ω X_C all in parallel, what is the: source

current value; impedance? (7-15) **BPE**

29. How can the phase angle of a parallel *RL* circuit be determined? (7-13) **BPE**
30. How is the impedance of a parallel *RLC* group calculated? (7-15) **BPE**
31. If the source frequency were doubled in the circuit shown in Fig. 7-24, what would be the total impedance? (7-16) **BP**
32. Diagram a 100-V 500-Hz source, and a 1-μF capacitor in series with the source, followed by a T-network having a 2-mH *L*, a 100-Ω *R*, a 1-mH *L*, and a 200-Ω load. What are the: currents through each component; voltage-drop across each component; total real and apparent power values? (7-17) **BPE**

RESONANCE AND *LC* FILTERS

1. What is meant by series resonance? (8-1) **BPTA**
2. What is the formula for the resonant frequency of an *L* and *C*: in series; in parallel? (8-1) **BPT**
3. A parallel *LC* circuit is resonant at 25 MHz. If the values of both *L* and *C* are doubled what will be the new resonant frequency? (8-1) **BP**
4. What changes in circuit capacitance will double the resonant frequency of an *LC* circuit? (8-1) **BPT**
5. If a 2-MHz *LC* circuit has its *L* and *C* values both halved, what will be the resultant resonant frequency? (8-1) **T**
6. A 200 μH *L* and a 200 pF *C* are in parallel. What is the resonant frequency? (8-1) **TA**
7. What value of capacitance must be shunted across a coil having 50 μH in order that the circuit will resonate at: 4 MHz; 3.5 MHz? (8-1) **PA**
8. What is the value of total reactance in a series-resonant circuit at resonance? (8-2) **TA**
9. A series-resonant circuit has 10-Ω *R* and reactances of 100 Ω. What is its impedance at: resonance; half the frequency; twice the frequency? (8-2) **T**
10. What is the impedance of a series circuit at reso-

nance if it is composed of pure reactances? (8-2) **BP**
11. A series *LC* circuit with 50-Ω X_L, 50-Ω X_C, and 5-Ω *R* is across 100 V ac. What is the voltage-drop across the coil? The capacitor? (8-2) **PA**
12. A 4-H *L* is in parallel with a 2-μF *C*. If there is no *R* in either leg, what is the *Z* value at resonance? (8-3) **PT**
13. What effect will a small amount of *R* in the *C* branch of a parallel-resonant circuit have on the circuit *Z*? (8-3) **P**
14. What is the value of reactance across the terminals of the capacitor of a 4-MHz parallel-resonant circuit at the resonant frequency if there is zero *R* in both legs of the circuit? (8-3) **P**
15. Under what condition will the voltage-drop across a parallel-tuned circuit be a maximum? (8-3) **PA**
16. In what manner is energy stored in parallel *LC* circuits? (8-3) **P**
17. Give four formulas by which the *Z* of a parallel resonant circuit can be calculated. (8-3) **PE**
18. If an antiresonant circuit has reactances of 3500 Ω and a series impedance of 20 Ω, what is its impedance? (8-3) **PE**
19. What is skin effect, and how does it affect conductor resistance at higher radio frequencies? (8-4) **BPA**
20. What is meant by the *Q* of a circuit? (8-4) **BPA**
21. How may the *Q* of a parallel-resonant circuit be increased or decreased? (8-4) **PA**
22. How does resistance affect circuit *Q*? (8-4) **BPA**
23. What effect does a loading resistance have on the *Q* of a tuned *RF* circuit? (8-4) **PA**
24. What is the *Q* of a resonant *LC* circuit having 1000-Ω reactors and an effective load resistance of 5 kΩ? (8-4) **PA**
25. What is a decibel? (8-5) **BPTG**
26. What is the formula for determining the decibel loss or gain of power in a circuit? (8-5) **BPG**
27. How many dB gain does an amplifier have if 2-mW input produces 4-W output? (8-5) **PG**
28. What is the dB gain if power is increased: 2 times; 10 times; 200 times; 1000 times; 1,000,000 times? (8-5) **PG**
29. What is the formula for determining the dB loss or gain of voltage in a circuit? (8-5) **BPG**
30. An amplifier has a voltage gain of 20 dB and 15-V output. What is the voltage input? (8-5) **BP**

31. What is the dB gain of an amplifier with 0.02 V across a 100-Ω input if it produces 3-V output across a 200-Ω load? (8-5) **PG**
32. What is meant by bandwidth of *LCR* circuits? (8-6) **PG**
33. What are four methods of expressing bandwidth? (8-6) **PG**
34. How is bandwidth affected by circuit *Q*? (8-6) **BP**
35. How is bandwidth determined if resonant frequency and *Q* are known? (8-6) **PA**
36. What effect does a loading resistance have on the bandwidth of a tuned *RF* circuit? (8-6) **PA**
37. For what are filters usually used? (8-8) **BP**
38. Why is a wave trap used in receivers? (8-8) **P**
39. Draw a diagram of a wave trap in an antenna circuit to attenuate an interfering signal. (8-8) **P**
40. What are some uses of low-pass filters? (8-8) **T**
41. Draw diagrams and explain the characteristics of these filters: bandstop; high-pass; low-pass; band-pass. (8-8) **BPG**
42. Describe the characteristics of the: constant-*k* filter; *m*-derived filter. (8-8) **BPA**
43. Describe the characteristics of a: pi-section filter; T-type filter; L-type filter. (8-8) **BPA**
44. Draw a diagram of a composite filter having low-pass, constant-k, and m-derived sections. (8-8) **BPA**

ACTIVE DEVICES

1. What are advantages and disadvantages of transistors and vacuum tubes? (9-1) **B**
2. Explain doping of germanium and silicon. (9-2) **BPA**
3. What are the differences between P-type and N-type semiconductors with respect to current and resistance? (9-2) **BPA**
4. What are some differences between silicon and germanium diodes? (9-3) **PTG**
5. What is the difference between forward- and reverse-biasing of diodes? (9-3) **BPA**
6. Explain the operation of the following diodes: junction; zener; varactor; tunnel; hot-carrier; PIN. (9-3, 9-4) **PA**
7. How does a light-emitting diode (LED) function? (9-4) **PA**
8. Why does an LED use a series resistor in normal operation? (9-4) **BPA**
9. Why are LEDs now used in place of pilot lamps? (9-4) **BPA**
10. Describe some devices which operate on photoconductive effects. (9-4) **PE**
11. In transistors what is the function of the: emitter; base; collector? (9-5) **PA**
12. What is the difference between forward and reverse biasing of transistors? (9-5) **BPA**
13. What is meant by NPN and PNP bipolar junction transistors? (9-5) **PA**
14. Describe the physical structure of two types of transistors and explain their operation as amplifiers. (9-5, 9-11) **PTA**
15. What is a junction-tetrode transistor and how does it differ from earlier transistors in operating frequency? (9-5) **BP**
16. Explain the functioning of an NPN transistor as an amplifier. (9-5) **BPTA**
17. Diagram and explain an NPN common-emitter amplifier with fixed bias. (9-6, 9-7) **BPTA**
18. Diagram and explain a common-emitter amplifier with self-bias. (9-6) **BPA**
19. Explain a common-emitter class A transistor amplifier in terms of its bias network, signal gain, input and output impedances, voltages, currents, and power. (9-6, 9-7) **PA**
20. Describe the following transistors: power type BJT; JFET; MOSFET; unijunction. (9-6 to 9-13) **PA**
21. How do silicon and germanium transistors differ? (9-6) **BPA**
22. What is the gain factor of a transistor? (9-6) **B**
23. Explain the cutoff frequency of a common-emitter transistor amplifier. (9-6) **P**
24. What is meant by: V_{CE}; V_{EB}; I_C; V_{CBO}? (9-6, 9-8) **PA**
25. Why is stabilization of a transistor amplifier usually necessary? (9-6) **BP**
26. What is meant by transistor dissipation? (9-6) **PA**
27. What is the function of a heat sink? (9-6) **PA**
28. How might a thermistor be used for the stabilization of a transistor amplifier? (9-6) **P**
29. What is meant by ambient temperature? (9-6) **P**

30. How is the alpha cutoff frequency related to the collector-base circuit? (9-6) **BP**

31. Draw the symbols of common BJT, FET, and MOSFET transistors. (9-6, 9-11, 9-12, 9-13) **PA**

32. Explain the differences between common-emitter, common-base, and common-collector circuits. (9-6, 9-9, 9-10) **BP**

33. What is the significance of handbook-listed transistor characteristics? (9-8) **BPA**

34. Why do high-power transistors require heat sinks? (9-8) **BP**

35. What are some precautions when soldering transistors and repairing printed circuits? (9-8) **B**

36. Explain a common-collector class A transistor amplifier in terms of its bias network, signal gain, and input and output impedances. (9-10) **PA**

37. What is meant by alpha and alpha cutoff frequency of a transistor? (9-9) **BP**

38. Describe gains and impedances of common-base and common-collector amplifiers. (9-9, 9-10) **PA**

39. Explain the types, characteristics, applications, and symbols of field-effect transistors such as the: N-channel FET; P-channel FET; depletion MOSFET; enhancement MOSFET. (9-11, 9-12) **PE**

40. Describe common-source, common-gate and common-drain amplifiers and their impedances. (9-11) **PE**

41. Describe the operation of an SCR and a triac. (9-14, 9-15) **PA**

42. What components might be incorporated in an IC? (9-17) **P**

43. What is meant by space charge? (9-18) **BPT**

44. What is the composition of filaments, heaters, and cathodes in VTs? (9-18) **T**

45. What is thermionic electron emission? (9-18) **BPT**

46. When an ac filament supply is used, why is a filament center-tap usually provided for the plate (and grid) return circuits? (9-18) **T**

47. Why is it important to maintain transmitting-tube filaments at recommended voltages? (9-18) **BPTA**

48. What are advantages of indirectly heated cathodes? (9-18) **BPT**

49. What is meant by a soft tube? (9-18) **T**

50. What does a blue haze between cathode and anode of a vacuum tube indicate? (9-18) **BP**

51. Describe the appearances, applications, and symbols of the various basic vacuum tubes. (9-18, 9-19, 9-23, 9-24) **N**

52. Describe the function of the following in VTs: filament; cathode; heater; control grid; screen grid; suppressor grid; plate. (9-18 to 9-24) **BPT**

53. Describe operation and characteristics of: vacuum diodes; vacuum triodes; vacuum tetrodes; beam-power tetrodes; pentodes; remote vs sharp-cutoff tubes. (9-18 to 9-24) **BP**

54. Describe the physical structure of a triode VT. (9-19) **TG**

55. What is the primary purpose of the control grid of a triode? (9-19) **TG**

56. Explain the operation of a triode as an amplifier. (9-19) **TG**

57. What is meant by the load on a tube? (9-19) **TG**

58. What is an A battery, B battery, and C battery? (9-19) **T**

59. What is meant by amplification factor or μ? (9-19, 9-20) **BPT**

60. What is meant by the voltage gain of an amplifier stage, and how can gain be achieved? (9-19) **BP**

61. What occurs in the grid circuit when the grid is driven positive? (9-19) **BPG**

62. What is meant by plate saturation? (9-19) **BPT**

63. What is meant by maximum plate dissipation? (9-19) **T**

64. Draw an $E_g I_p$ curve for a triode and explain the relation of I_p to E_g. (9-20) **BP**

65. Draw an $E_p I_p$ graph of a typical triode for three different bias voltages. (9-20) **BP**

66. What factors determine grid bias? (9-20) **BPT**

67. What is meant by plate current cutoff bias? (9-20) **BPG**

68. In what ways are transistors and VTs comparable? (9-20) **BPG**

69. What is the most important factor when choosing a VT for voltage amplification? (9-21) **T**

70. What is another term meaning mutual conductance and what do they both mean? (9-21) **PT**

71. What is the plate resistance of a VT? (9-21) **BP**

72. What load condition is required for maximum possible power output from any power source? (9-21) **PTG**

73. What is secondary emission in a VT? (9-22) **BPT**

74. What is the purpose of a screen grid? (9-23) **TG**

75. How do tetrodes and triodes compare as to plate current and interelectrode capacitance? (9-23) **BP**

76. What are the characteristics of tetrodes? (9-23) **TG**
77. Why is a transformer rarely employed in the plate circuit of a tetrode? (9-23) **T**
78. Describe beam-power tubes. (9-23) **TG**
79. What does a suppressor grid do? (9-24) **PT**
80. Diagram and explain a grounded-cathode pentode class A AF amplifier with battery biasing. (9-24) **BP**
81. Why should the cathode of an indirectly heated VT be maintained at nearly the same potential as the heater? (9-25) **T**
82. Draw a diagram and explain the following circuits: grounded-grid; cathode follower. (9-25) **P**
83. What are some common tube ratings found in manufacturers' manuals? (9-26) **BP**
84. In a VT, what are grid-plate and output capacitances? (9-26) **BP**
85. What is a pentagrid converter? A duo-triode? (9-26) **BP**
86. What are some indications when VTs are defective? (9-28) **T**
87. What are possible causes of low I_p in VTs? (9-28) **BP**
88. What may cause VT-plate overheating? (9-28) **TG**
89. What conditions might shorten VT life? (9-28) **BPG**
90. Describe the uses of acorn and lighthouse tubes. (9-28) **BP**
91. Why are special tubes required at UHF and above? (9-28) **BPG**
92. Describe the characteristics of gaseous rectifiers. (9-29) **P**
93. What are two advantages of mercury vapor diodes? (9-29) **P**
94. Explain battery-charger rectifier tubes. (9-29) **T**
95. Describe the functioning and use of a thyratron. (9-29) **P**

POWER SUPPLIES

1. Describe the types, characteristics, applications, and symbols of power-supply type diodes. (10-2, 10-3, 10-13, 10-22) **BPG**
2. What are the merits and limitations of high-vacuum and mercury-vapor diodes in power supplies? (10-2, 10-3, 10-13) **BPT**
3. What is meant by the peak-inverse-voltage rating of a diode? (10-2, 10-14) **BPT**
4. Diagram and explain a bridge-rectifier circuit. (10-3, 10-23, 10-29) **BPG**
5. List some advantages of a full-wave rectifier as compared to a half-wave rectifier. (10-3, 10-14) **TG**
6. What are the merits and limitations of silicon and germanium diodes in power supplies? (10-3) **BPG**
7. What advantages does a bridge-rectifier circuit have over a conventional center-tap full-wave rectifier? (10-3) **BPG**
8. How can two dc voltages be obtained from one rectifier circuit? (10-4, 10-25) **BPT**
9. What is the function of a power-supply filter? (10-5, 10-6) **TG**
10. What are the approximate values of power-supply filter inductors and capacitors? (10-5, 10-7) **PT**
11. Diagram and explain voltages of a half-wave rectifier power supply with a capacitive-input pi-section filter. (10-7) **BP**
12. What are the characteristics of capacitor-input compared to choke-input filter systems? (10-7, 10-8) **BPT**
13. How is a minimum acceptable voltage rating for power-supply filter capacitors determined? (10-7) **BP**
14. Draw diagrams and explain common power-supply circuits. (10-7, 10-9, 10-19, 10-24) **BPTG**
15. Diagram and explain a power supply with an *RC* filter. (10-9) **BPG**
16. Diagram and explain a bridge-rectifier power supply. (10-9) **PG**
17. What effect does a large value of dc flow have on a filter choke? (10-10) **BP**
18. Why is low resistance desirable in a filter choke? (10-10) **T**
19. What is the effect of loose laminations in a filter choke? (10-10) **T**
20. What is the purpose of an air gap in the core of a filter choke? (10-10) **T**
21. What is a swinging choke and where is it used? (10-11) **BPT**
22. What action makes possible the high-conduction

currents of a hot-cathode mercury-vapor rectifier? (10-13) **T**

23. What is the voltage-drop across a mercury-vapor rectifier when carrying current? (10-13) **B**
24. What is arc-back or flashback in a tube? (10-13) **B**
25. Why should the cathode in a mercury-vapor rectifier tube reach normal operating temperature before the plate voltage is applied? (10-13) **T**
26. What type of rectifier tubes must work into inductive input filters? (10-13) **B**
27. Diagram and explain voltages of a mercury-vapor-diode full-wave rectifier power supply with choke-input filter. (10-13) **BP**
28. Why are small resistors placed in series with each plate lead of mercury-vapor rectifiers when in parallel? (10-13) **T**
29. How is peak-inverse-voltage computed for full-wave and half-wave rectifier circuits? (10-14) **BP**
30. With rectifiers having a peak-inverse-voltage rating of 12,000 V, what is the maximum allowable total secondary voltage of a center-tapped transformer to be used in a full-wave rectifier circuit? (10-14) **T**
31. How may a series circuit of silicon diodes be protected against unequal peak-reverse voltages? (10-14) **BP**
32. List advantages and disadvantages of high-vacuum vs. mercury-vapor rectifiers. (10-14) **T**
33. What is the ratio of the frequencies in the output and input circuits of a 1-phase full-wave rectifier? (10-15) **T**
34. What is the ripple frequency of a 3-ϕ 60-Hz: full-wave rectifier; half-wave rectifier? (10-15, 10-29) **P**
35. How could varying the value of the bleeder in a power supply affect the ripple voltage? (10-16) **BP**
36. Why are bleeder resistors used in high-voltage power supplies? Would carbon or wire-wound types be preferable? (10-16) **BPT**
37. What is meant by power-supply voltage regulation, and what effect does the load have on its percentage? (10-17) **BPT**
38. How is power-supply regulation percentage calculated? (10-17) **BP**
39. What is the percent regulation of a power supply with a no-load voltage of 120 V and a full-load voltage of 115 V? (10-17) **BP**
40. If a power supply has a regulation of 10% when the output voltage at full load is 250 V, what is the output voltage at no load? (10-17) **BPT**
41. A power supply has 150-V output at no load and 15% regulation. What is the output voltage at full load? (10-17) **B**
42. What causes poor power-supply regulation? (10-17) **BPT**
43. Diagram a full-wave center-tap rectifier and filter-power supply for supplying plate voltage to a VT receiver. (10-19) **PT**
44. What are approximate values of power-supply filter inductance encountered in practice? (10-19) **T**
45. How can power factor in an electric power circuit be corrected? (10-20) **T**
46. Why is a capacitor sometimes placed in series with the primary of a power transformer? (10-20) **T**
47. Diagram a voltage-doubler power supply using two half-wave rectifiers. (10-21) **T**
48. Diagram and explain voltages of a silicon-diode voltage-doubler power-supply circuit with a resistive load. (10-21) **BP**
49. Draw a diagram of a zener-diode voltage-regulated power-supply labeling input and output voltages. (10-22) **BPA**
50. How is the value of the resistor in series with a zener diode (or VR tube) computed to form a voltage-regulating circuit? (10-22) **PA**
51. Diagram and explain voltage regulators, both discrete and integrated circuit types. (10-22, 10-23, 10-24) **PA**
52. Diagram and explain a cold-cathode VR tube connected as a voltage regulator. (10-22) **BPTA**
53. Diagram and explain a voltage regulator with pass transistor and a zener diode to produce a given output voltage. (10-24) **BPA**
54. Diagram a circuit to protect against damage to the load in the event of a pass-transistor short circuit. (10-24) **BP**
55. Explain three-terminal IC regulators. (10-24) **BP**
56. If a 10-V IC regulator has a 500-Ω resistor across it and a 500-Ω resistor between its ground terminal and actual ground, what is the ground to +10-V regulated terminal voltage value? (10-24) **BP**
57. A power supply is to furnish 500 V at 50 mA to one circuit and 400 V at 20 mA to another. The bleeder current is 10 mA. What value resistor

should be between the 500- and 400-V taps of the voltage divider? (10-25) **P**

58. What are selenium and copper oxide rectifiers and what are some of their uses? (10-26) **BPT**

59. Diagram and explain a switching circuit that can develop high voltages for mobile radio equipment. (10-27) **P**

60. Diagram and explain a nonsynchronous vibrator power supply with bridge rectifiers and a capacitive-input pi-section filter. (10-27) **PT**

61. Diagram and explain an inverter. (10-27) **P**

62. In what circuits in a radio station are 3-ϕ circuits sometimes employed? (10-28) **BT**

63. What are the advantages of 3-ϕ power over 1-ϕ for broadcast use? (10-28) **BT**

64. What system of connections for a 3-ϕ 3-transformer bank will produce maximum secondary voltage? (10-28) **BT**

65. Show by diagrams the delta and the wye methods of connecting 3-ϕ transformers. (10-28) **BT**

66. Diagram a 3-ϕ transformer with delta primary and Y secondary. (10-28) **BT**

67. Diagram how three power transformers can be connected for 3-ϕ operation using star primaries and delta secondaries. (10-28) **BT**

68. Show how only two transformers can be connected for full operation on a 3-ϕ circuit. (10-28) **BT**

69. Three 20:1-ratio 1-ϕ transformers are connected across 220-V 3-ϕ lines with primaries in delta. If the secondaries are in Y, what is the output voltage? (10-28) **BT**

70. What would happen if the pass transistor of a transistorized regulator short-circuited? (10-30) **BP**

71. What does a blue haze in the space between filament and plate of a high-vacuum rectifier indicate? (10-30) **BP**

72. If the plates of a power-supply rectifier tube suddenly becomes red-hot, what might be causes and remedies? (10-30) **BP**

73. Explain what circuit components are likely to be faulty when changes in power-supply performance are observed. (10-30) **P**

74. How might it be determined what component is at fault when improper circuit performance is noted? (10-30) **P**

75. Why should the metallic case of a high-voltage transformer be grounded? (10-30) **T**

76. What are some applications of vacuum tubes in power supplies? (10-2, 10-3, 10-19) **BPN**

OSCILLATORS

1. What is flywheel effect in a tank circuit and how does it relate to *RF* oscillators? (11-2) **BPTA**

2. What is meant by shock excitation of a circuit? (11-2) **T**

3. Describe how an active device can produce sine wave oscillations in a circuit. (11-3) **TA**

4. Diagram and explain the operation of an Armstrong oscillator. (11-4) **BPTA**

5. Explain how grid-leak bias is developed in an oscillator. (11-5) **P**

6. What is the direction of electron flow in the grid circuit of a vacuum tube? (11-5) **PA**

7. Diagram and explain by what means feedback is obtained in a TPTG oscillator. (11-6) **BPA**

8. Diagram and explain series and shunt feed in oscillators. (11-6, 11-7) **BPT**

9. What is the purpose of an RF choke? (11-7) **PTG**

10. Diagram and explain a tuned-gate or tuned-grid Armstrong oscillator. (11-7) **PTA**

11. What is the meaning of VFO? (11-8) **TA**

12. Diagram and explain the operation of a Hartley oscillator. (11-8) **BPTA**

13. Diagram a Hartley oscillator with a shunt-fed plate or collector circuit. (11-8) **TA**

14. Diagram and explain the operation of a Colpitts oscillator. (11-9) **BPTA**

15. Diagram a Colpitts oscillator with shunt-fed plate circuit. (11-9) **T**

16. How do Colpitts and Hartley oscillators differ? (11-9) **PA**

17. Why is a high ratio of C to L employed in the grid circuit of some oscillators? (11-9, 11-24) **TA**

18. What is the meaning of PTO? (11-9) **TA**

19. Diagram and explain the operation of an electron-coupled oscillator. (11-11) **BPTA**

20. What will be the effect of applying a dc potential to the opposite plane surfaces of a quartz crystal? (11-12) **PTA**

21. Diagram the equivalent circuit of a quartz crystal. (11-12) **BA**
22. Diagram and explain a crystal oscillator. (11-12, 11-16) **BPA**
23. What are the advantages of crystal over *LC* oscillators? (11-12) **BPTA**
24. What determines the fundamental frequency of oscillation of a quartz crystal? (11-12) **BPA**
25. How may the oscillating frequency of a crystal be adjusted? (11-12) **BPA**
26. What may result if a high degree of coupling exists between plate and grid of a crystal oscillator? (11-12) **BPT**
27. Why might a separate source of plate power be desirable for the crystal oscillator of a transmitter? (11-12) **BPTA**
28. Explain quartz crystals in terms of their appearance, applications, and symbol. (11-12, 11-13) **PTN**
29. Explain some methods of determining if oscillation exists in a crystal oscillator circuit. (11-12, 11-23) **PA**
30. What crystalline substance is widely used in crystal oscillators? (11-12) **PTA**
31. Diagram and explain a crystal controlled triode or FET oscillator. (11-12, 11-16) **PTA**
32. Why is an additional feedback capacitor sometimes necessary in a crystal oscillator? (11-12) **PT**
33. For maximum stability, where should the tuned circuit of a crystal oscillator be tuned? (11-12) **PA**
34. What does low-temperature-coefficient crystal mean? (11-13) **T**
35. What does positive temperature coefficient of a crystal mean? (11-13) **T**
36. What does negative temperature coefficient of a crystal mean? (11-13) **T**
37. A 500-kHz crystal, calibrated at 40°C and having a TC of -25 ppm/°C, will oscillate at what frequency when its temperature is 60°C? (11-13) **P**
38. A transmitter is on 4 MHz, using a 1-MHz crystal with a TC of -5 Hz/°C/MHz. If the crystal temperature increases 7°, what is the change in the output frequency? (11-13) **T**
39. Why is the temperature of a quartz crystal usually maintained constant? (11-14) **PT**
40. Diagram and explain a mercury-thermometer crystal heater control circuit. (11-14) **BP**
41. Diagram and explain a thermocouple type of crystal heater control circuit. (11-14) **B**

42. What is the maximum allowable temperature variation when using X- or Y-cut crystals in broadcasting? (11-14) **B**
43. Why are crystal heaters often left on all night even when the station is not on the air? (11-14) **B**
44. What cleaning agents may be used to clean the surface of a quartz crystal? (11-15) **PT**
45. Diagram and explain the operation of a Pierce oscillator. (11-16) **BPT**
46. Diagram and explain a crystal oscillator using a pentode tube. (11-16) **PT**
47. What are characteristics and uses of overtone and third-mode crystal oscillators? (11-16) **BPA**
48. How do fundamental cuts differ from overtone cuts in quartz crystals? (11-16) **BP**
49. Explain the principles of operation of a cavity resonator. (11-17) **PE**
50. What is the wavelength of a 6-MHz wave? (11-17) **PTG**
51. What type of oscillator depends on secondary emission from the anode for its operation? (11-18) **T**
52. Diagram and explain a dynatron oscillator. (11-18) **T**
53. Diagram an AF oscillator using an iron-core choke. (11-19) **PA**
54. Diagram and explain two ways of obtaining sawtooth waves. (11-20, 11-21) **BA**
55. Diagram and explain the operation of multivibrators. (11-21) **BPTA**
56. How do multivibrators differ from oscillators such as the Hartley? (11-21) **TA**
57. What determines the operating frequency of a multivibrator? (11-21) **TA**
58. Define parasitic oscillation. (11-22) **BPTA**
59. What are the effects of parasitic oscillations? (11-22) **BPTA**
60. How are parasitic oscillations detected and prevented? (11-22) **BPA**
61. For what are ferrite beads used? (11-22) **PA**
62. Name some devices used to indicate oscillation. (11-23) **PA**
63. Explain some factors involved in the stability of a crystal or *LC*-tank oscillator. (11-24) **BPA**
64. What are some factors which may cause frequency drift in oscillators? (11-24) **PA**
65. What may be the effects of shielding *RF* inductances? (11-24) **PA**

66. Describe two methods of synthesizing ac. (11-26, 11-27) **PE**
67. What is the meaning of VCO? (11-27) **BPE**
68. Describe how one crystal in a synthesizer can generate many different frequencies. (11-27) **BPE**
69. Block diagram and explain a PLL synthesizer having a frequency divider, a frequency-phase comparator, a low-pass filter, a reference oscillator, and a VCO. (11-27) **BPE**
70. Diagram a crystal oscillator in which two or more frequencies may be selected by diode switching. (11-28) **BPA**
71. What are some applications of vacuum tubes in oscillators? (11-6, 11-8, 11-16) **PN**
72. Describe the four basic types of signal generators. (11-29) **PG**
73. What waveforms do signal generators produce? (11-29) **PG**

DIGITAL FUNDAMENTALS

1. What are the high and low voltage levels for TTL and CMOS digital devices? (12-1) **PE**
2. What is the binary numbering system and why is it used in electronic circuits? (12-2) **P**
3. All forms of electronic computations require what single form of mathematics? (12-3) **PE**
4. Draw an AND gate logic symbol. (12-4) **BPE**
5. Construct an AND gate truth table. (12-4) **BPE**
6. Draw an OR gate logic symbol. (12-5) **BPE**
7. Construct an OR gate truth table. (12-5) **BPE**
8. Draw an inverter logic symbol. (12-6) **BPE**
9. Construct an inverter truth table. (12-6) **BPE**
10. Draw a NAND gate logic symbol. (12-7) **BPE**
11. Construct a NAND gate truth table. (12-7) **BPE**
12. Draw a NOR gate logic symbol. (12-8) **BPE**
13. Draw an XOR gate logic symbol and truth table. (12-9) **PE**
14. Draw an XNOR gate logic symbol and truth table. (12-9) **PE**
15. Draw a logic diagram, the symbol, and explain an RS-type flip-flop. (12-10) **PE**
16. Draw a logic diagram, the symbol, and explain a D-type flip-flop. (12-10) **PE**
17. What type of FF can be used to enable or latch data from a bus to a register? (12-10) **PE**
18. Which FF is the most versatile? (12-10) **PE**
19. Diagram and explain a 4-bit register. (12-11) **PE**
20. What are the meanings of ROM, RAM, PROM, and EPROM? (12-12) **PE**
21. What does a RAM have that a ROM does not? (12-12) **PE**
22. What logic device is used to address 16 memories? (12-12) **P**
23. Explain volatile and nonvolatile memories. (12-12) **P**
24. What do you know about 3000-series ICs? 7400-series? (12-1 through 12-12) **PE**
25. What is the binary coded decimal for 8362? (12-13) **PE**
26. What is the minimum FFs required for a modulus 10 counter or divider? (12-14) **PE**
27. Explain LED digital readouts. (12-15) **BP**
28. Explain LCD digital readouts. (12-16) **PE**
29. In what form is the output given from a modulus 10 counter? (12-14) **PE**
30. Explain analog vs digital signals. (12-17) **PE**
31. Diagram and describe an A/D conversion system. (12-17) **P**
32. Diagram and describe a D/A conversion system. (12-18) **P**
33. Express De Morgan's laws in Boolean logic. (12-19) **PE**
34. Draw and explain a half adder. (12-19) **PE**
35. Draw and explain a full adder. (12-19) **PE**
36. List at least 12 circuits that might be found in a microprocessor chip. (12-20) **P**
37. In what systems is ASCII used? (12-20) **P**
38. How is the letter "M" expressed in binary ASCII? (12-20) **PE**
39. What systems use Baudot code? (12-20) **P**
40. How is the letter "M" expressed in binary Baudot? (12-20) **PE**
41. Why is it important that digital logic circuitry be shielded from RF sources and receivers? (12-20) **BP**
42. What is meant by data rate? (12-21) **P**
43. What is meant by baud rate? (12-21) **P**

44. What are the three common pulse-type codes? (12-24) **TE**
45. Discuss the use of gates in some circuits. (12-22) **P**
46. Explain the operation of a logic probe. (12-22) **PE**

MEASURING DEVICES

1. What are some limitations of analog and digital meters? (13-1) **PA**
2. What types of meters are used as panel meters? (13-1, 13-34) **P**
3. Explain D'Arsonval meters in terms of their appearance, applications, and symbols. (13-2) **PN**
4. Sketch the construction of a D'Arsonval-type meter and label the parts. (13-2) **BPT**
5. What is a shunt on an ammeter? (13-4) **T**
6. If two ammeters are in series, how is the total current determined? (13-4) **T**
7. If two ammeters are in parallel, how is the total current determined? (13-4) **T**
8. A 0-1-mA meter with an internal resistance of 30 Ω is shunted with a 3-Ω resistor. When the meter reads 0.5 mA, what is the line current? (13-5) **T**
9. What voltmeter absorbs no power from the circuit under test?
10. What does 20,000-Ω/V mean? (13-9) **BP**
11. To convert a 0-1-mA meter to read 150 V full scale, what value multiplier should be used? (13-9) **BP**
12. How may a dc milliammeter be used in an emergency to indicate voltage? (13-9) **T**
13. What is a multiplier in a voltmeter? (13-9) **T**
14. If two voltmeters are in series across a circuit, how can the total voltage of the circuit be determined? (13-9) **T**
15. What are two important factors about voltmeters when measuring voltages in high resistance circuits? (13-10) **P**
16. Diagram and explain a simple ohmmeter. (13-11) **BPTG**

17. What four types of meters make up a multimeter? (13-12) **PG**
18. When using a multimeter in an unknown circuit, what current range should always be used first? What voltage range? (13-12) **P**
19. What are some applications of VTs in radio? (13-13, 13-28) **TN**
20. Diagram and explain a simple VTVM. (13-13) **BP**
21. Diagram and explain a simple EVM. (13-13) **P**
22. Does the scale of an ac ammeter usually indicate peak, effective, or average values? (13-14) **T**
23. By what factor must the voltage of an ac circuit as indicated on the scale of an ac voltmeter be multiplied to obtain the peak value? (13-14) **T**
24. A dc voltmeter is used to measure effective ac voltages by the use of a bridge rectifier. By what factor must the meter readings be multiplied to give rms readings? (13-15) **T**
25. Why are copper-oxide meter rectifiers not suitable for measuring RF? (13-15) **T**
26. What meter is suitable for measuring peak ac voltages? (13-16) **B**
27. VU meters are normally placed across AF lines of what characteristic impedance? (13-17) **B**
28. Draw a diagram of a circuit used to desensitize a VU meter in order to produce a lower reading. (12-17) **BP**
29. Where might a T-pad attenuator be used? (13-17) **B**
30. To what power is the standard "0 VU" equal? (13-17) **BP**
31. What is meant by dBm? (13-17) **P**
32. What is a thermocouple? (13-18) **T**
33. What meter can measure RF currents? (13-18) **BA**
34. Describe a thermocouple ammeter. (13-18) **T**
35. How may the range of a thermocouple ammeter be increased? (13-19) **T**
36. Describe a dynamometer-type alternating-current meter. (13-19) **T**
37. Describe a repulsion-type meter. (13-20) **T**
38. What type of angular scale deflection does a repulsion ammeter have? (13-20) **T**
39. The product of ac voltmeter and ammeter readings in an ac circuit provides what value? (13-21) **T**
40. Describe the construction, operation, and char-

acteristics of a wattmeter. (13-21)   **BPTG**

41. Describe the meter that measures energy. (13-22)   **T**

42. Describe an ampere-hour meter and explain its use. (13-23)   **T**

43. Explain the operation of a Wheatstone bridge. (13-24)   **BP**

44. If the known values of a bridge are 50 and 100 Ω, and if adjusting the third value to 50 Ω produces balance, what is the unknown resistance? (13-24)   **BP**

45. What is the meaning of frequency tolerance? (13-25)   **T**

46. A ship is assigned a frequency of 8 MHz and a tolerance of 20 Hz. If the oscillator operates at one-eighth the output frequency, what is the maximum permitted deviation of the oscillator in hertz which will not exceed the tolerance? (13-25)   **T**

47. How accurate should a frequency meter be to measure to within ±0.0005% tolerance? (13-25)   **BPABPA**

48. If the frequency of a 150-MHz transmitter is measured with a frequency meter with an accuracy of ±0.00005%, what are the highest and lowest frequencies that could be certified to be within a tolerance of 0.0005%? (13-25)   **BP**

49. Draw a simple schematic diagram of an absorption-type wavemeter. (13-27)   **BPT**

50. Explain the operation and possible applications of an absorption wavemeter. (13-27)   **BPT**

51. What are advantages and disadvantages of using absorption-type wavemeters in comparison to other types of frequency meters? (13-27)   **T**

52. What precautions should be used when an absorption-type frequency meter is used to measure the output of a self-excited oscillator? (13-27)   **T**

53. If a wavemeter having an error proportional to the frequency is accurate to 200 Hz when set at 1000 kHz, what is its error when set at 1800 kHz? (13-27)   **T**

54. Explain how a wavemeter can be used to measure frequency. (13-27)   **TA**

55. Diagram, explain and give possible applications of a grid-dip meter. (13-28)   **BPA**

56. What is the solid-state version of a grid-dip meter called? (13-28)   **BPA**

57. How can a grid-dip meter be used to measure frequency? (13-28)   **PA**

58. How can the resonant frequency of a parallel *LC* circuit be measured? (13-28)   **P**

59. What is the primary standard of frequency for RF measurements for all licensed radio stations? (13-29)   **TA**

60. With measuring equipment available, is it possible to measure a frequency of 10 MHz to within 1 Hz? (13-29)   **T**

61. What device may develop a frequency of 10 MHz from an oscillator operating on 10 kHz? (13-29)   **BPA**

62. Block diagram and explain a secondary frequency standard. (13-29)   **BP**

63. How can a secondary frequency standard be calibrated against a WWV signal? (13-29)   **PTA**

64. How can the operating frequency of a transmitter be determined by the use of a secondary frequency standard? (13-29)   **TA**

65. What precautions should be taken before using a heterodyne-type frequency meter? (13-30)   **T**

66. How can the frequency of a transmitter be determined with a heterodyne meter using earphones? (13-30)   **BPA**

67. What is the meaning of zero beat as used in conjunction with frequency-measuring equipment? (13-30)   **BPTA**

68. For a heterodyne zero-beat circuit, what is the best ratio of signal emf to heterodyne oscillator emf? (13-30)   **PA**

69. Block diagram and explain a frequency meter with: antenna; variable oscillator; mixer detector; AF amplifier with earphones; crystal oscillator; AF modulator. (13-30)   **BPT**

70. When should calibration check points be used with a heterodyne frequency meter? (13-30)   **BP**

71. How is the crystal in a heterodyne frequency meter checked against WWV? (13-30)   **BP**

72. A heterodyne meter is calibrated 20–40 MHz and has usable harmonics up to 640 MHz. How can it be used to measure in the 150-MHz range? (13-30)   **T**

73. If the tuning dial of a heterodyne meter indicates a signal between 2-dial frequency marks in the calibration book, how could the fre-

quency value be interpolated? (13-30) **BP**

74. Explain how to read the vernier on a frequency meter having a vernier scale. (13-30) **T**

75. An absorption wavemeter indicates 500 kHz. A heterodyne frequency meter dial reads 375. The frequency-meter calibration book indicates dial readings of 371.5, 376, for frequencies of 499.6, 499.8 kHz, respectively. What is the frequency of the transmitter? (13-30) **T**

76. How can a heterodyne frequency meter be used as an RF signal generator? (13-30) **BP**

77. Under what conditions would the AF modulator of a heterodyne frequency meter be used? (13-30) **BP**

78. When should a heterodyne frequency meter be recalibrated? (13-30) **BP**

79. Block diagram and describe the operation of a frequency counter. (13-31) **PA**

80. Describe how horizontal and vertical deflection takes place in an oscilloscope. (13-34) **PG**

81. What are some uses of oscilloscopes? (13-34 to 13-36) **PG**

82. Describe the operation of a simple oscilloscope, including waveforms involved and how trace intensity is varied. (13-34, 13-35) **BPG**

83. For what is a spectrun analyzer used? (13-37) **BPE**

84. What does a logic probe indicate? (13-38) **PE**

85. How may transistors be tested? (13-39) **P**

86. How may vacuum tubes be tested? (13-39) **P**

AUDIO-FREQUENCY AMPLIFIERS

1. Define the audio frequency bands used in communications. (14-1) **PN**

2. What is meant by high-fidelity audio? (14-1) **B**

3. What causes noise in transistors and electrical conductors and resistors, and shot-effect noise in tubes? (14-2) **BP**

4. Why is correct bias important in an AF amplifier? (14-3, 14-4) **T**

5. What is the gain of a 25-μ device having an R_d of 50 kΩ and an R_L of 75 kΩ? (14-3) **BT**

6. Why is noise often produced when an AF signal is distorted? (14-3) **BP**

7. Diagram and describe the characteristics of a stage operating as a class A amplifier. (14-3, 14-4) **TA**

8. Under what conditions does amplifier stage gain approach the μ value of the active device used? (14-3) **BP**

9. Does dc input current normally flow in a class A amplifier employing one JFET or VT? (14-3) **TA**

10. What is the maximum rms AF voltage which can be applied to the input of a class A amplifier biased to -10 V? (14-3) **TA**

11. What is the difference between battery and self-bias? (14-3, 14-4) **BP**

12. In an amplifier, what is the relationship between distortion, class of operation, portion of the curve in use, and amplitude of the input signal? (14-3) **BP**

13. List causes of distortion in an AF amplifier. (14-3, 14-4, 14-5, 14-7, 14-13) **T**

14. What determines the class of operation of amplifiers? (14-4) **BPA**

15. What factors determine the bias for amplifying devices? (14-4) **BP**

16. What is the effect of incorrect bias in a class A AF amplifier? (14-4) **TA**

17. Explain how to determine the self-bias resistance for an AF amplifier. (14-4) **BPT**

18. A grounded-cathode VT has a μ of 30, E_p of 100 V, and I_p of 20 mA. What are the self-bias R and C values for 100 Hz as the lowest frequency? (14-4) **BP**

19. Describe the characteristics of classes A, AB, and B AF amplifiers. (14-4, 14-5, 14-6) **BA**

20. What is the difference between stage gain and amplifier gain, and how is each determined? (14-4) **B**

21. Diagram and explain the operation of two tubes in parallel in a class A amplifier. (14-4) **BPA**

22. Diagram and explain two devices in a push-pull amplifier. (14-4, 14-5) **BPA**

23. In an amplifier, when the input voltage varies, how does an output circuit ammeter vary if operation is in classes A, AB, and B? (14-4, 14-6) **BP**

24. In a class A AF amplifier, what is the main advantage of push-pull as compared to parallel operation? (14-5) **TA**

25. Diagram and explain a class B push-pull AF amplifier using transistors. (14-5, 14-22) **TA**

26. Why is a push-pull AF amplifier preferable to a single active device stage? (14-5) **T**

27. How may even-order harmonic energy be reduced in the output of an AF amplifier? (14-5) **T**

28. Diagram methods of using single-ended stages to drive a push-pull output stage. (14-5, 14-7) **BP**

29. Draw an input-voltage vs output-current curve of a triode device and indicate the operating points for classes A, AB, B, and C. (14-5) **TA**

30. Operation over which portion of a triode device curve produces the least distortion? (14-5) **BPA**

31. What is the advantage of class A AF operation as compared to other classes? (14-6) **TA**

32. Why is it necessary to use two active devices in a class B AF amplifier? (14-6) **TA**

33. During what portion of the excitation voltage cycle does output current flow in a class B device? (14-6) **TA**

34. What are the approximate efficiencies of classes A, AB, and B AF amplifiers? (14-6) **TA**

35. Why does a class B AF amplifier normally require more driving power than a class A stage? (14-6) **BPA**

36. Why is class C not used in AF amplifiers? (14-6) **TA**

37. Regarding AF amplifiers, what is the meaning of class A_1, A_2, AB_1, AB_2, B_1, B_2? (14-6) **PA**

38. What is the advantage of transformer over resistance coupling in AF amplifiers? (14-7) **T**

39. Diagram and explain a split-load phase inverter. (14-7) **T**

40. What would be the result of saturation of an output transformer core? (14-7) **BP**

41. Describe the operation of the two electrical-to-sound transducers. (14-8, 14-9) **P**

42. Why might headphones used in radio communications have high-impedance windings? (14-8) **T**

43. Why should polarity be observed when connecting p-m earphones directly in the output circuit of an AF amplifier? (14-8) **T**

44. What are two ways low-impedance earphones might be connected in the output of a VT amplifier? (14-8) **BPT**

45. Why are permanent magnets used in earphones? (14-8) **T**

46. Diagram and explain an application of a cathode-follower triode AF amplifier. (14-8) **BP**

47. What is the voltage gain of a cathode-, source-, and emitter-follower amplifier? (14-8) **PT**

48. How does increasing the capacitance of an RC-coupled AF amplifier affect the signal? (14-10) **BPA**

49. What is the plate voltage of an RC-coupled stage with a supply voltage of 260 V, I_p of 1 mA, and R_L of 100 kΩ? (14-10) **T**

50. Diagram and explain impedance coupling. (14-10, 14-12) **BPT**

51. Explain matching impedances by using a transformer. (14-11) **BPTG**

52. What factors should be considered when feeding a known impedance speaker with a transformer in an AF stage? (14-11) **BP**

53. What ratio transformer is required to match a 600-Ω line to an 8-Ω speaker? (14-11) **P**

54. What is the turns ratio of a transformer needed to match a 3.2-Ω speaker to a 5000-Ω output circuit? (14-11) **BP**

55. How can a low-impedance speaker be coupled to an AF amplifier without using an output transformer? (14-12) **B**

56. Describe four forms of distortion that may be present in an AF amplifier. (14-14) **B**

57. What is the purpose of deliberately introduced degenerative feedback in AF amplifiers? (14-14) **B**

58. Diagram and explain negative voltage and current feedback. (14-14) **BP**

59. What would be the effect on the output if the cathode-resistor bypass capacitor were removed? (14-14) **BP**

60. What should be the output-input feedback phase at a nominal midfrequency of an AF amplifier? (14-14) **B**

61. What is the formula used to determine amplifier gain if negative feedback is involved? (14-14) **B**

62. Diagram a triode-type AF amplifier inductively coupled to a loudspeaker. (14-14) **BP**

63. Diagram RC coupling between two active devices in an AF amplifier. (14-14, 14-15, 14-22, 14-24) **T**

64. Diagram a common-source or cathode AF amplifier with self-bias. (14-14, 14-25) **BP**

65. How is gain or volume controlled in an AF amplifier system? (14-15) **BPT**
66. Diagram and explain two commonly used tone-control circuits. (14-16) **BP**
67. What is the result of Miller effect in AF amplifiers? (14-17) **B**
68. Diagram a method of direct-coupling between stages of an AF amplifier. (14-18, 14-19) **T**
69. Diagram and explain a Darlington configuration. (14-19) **BP**
70. Diagram and explain an NPN transistor directly connected to a PNP transistor. (14-19) **PE**
71. Diagram and explain a differential amplifier. (14-19) **B**
72. Diagram and explain a cascode amplifier. (14-19) **B**
73. Describe the operation of an operational amplifier IC. (14-20) **BPE**
74. What are the internal circuits of an op amp? (14-20) **BPE**
75. What would be the ratio of input-to-output feedback resistances to produce a gain of 200 in an op amp? (14-20) **BPE**
76. What is the frequency response of an op amp with and without feedback? (14-20) **BPE**
77. What are some uses of op amps? (14-20) **BPE**
78. What is the output voltage of an op amp having an input of 0.2 V with 1000 Ω and a feedback resistor of 3000 Ω? (14-20) **BPE**
79. What is meant by high-pass, low-pass, and band-pass active filter circuits? (14-21) **PG**
80. Explain the operation of an active AF band-pass IC op amp. (14-21) **PE**
81. What is the purpose of decoupling networks in the output circuits of a multistage AF amplifier? (14-22) **BPE**
82. List three means used to prevent interaction between stages of a multistage AF amplifier. (14-22) **T**
83. How is a thermistor used in stabilization of a transistor AF amplifier? (14-23) **PE**
84. What are some applications of VTs in radio? (14-24) **TN**
85. Diagram and explain a push-pull tetrode class B AF amplifier transformer coupled to a loudspeaker. (14-24, 14-5) **BP**
86. Diagram and explain an amplifier stage's desired frequency response by proper selection of bypass and coupling capacitors. (14-25) **PE**

87. What would be the effect of a leaking coupling capacitor in an *RC* coupled amplifier? (14-25) **PTG**
88. How are interstage connecting leads susceptible to other circuit currents? (14-25) **BP**
89. What circuit components might be affected by electromagnetic fields? (14-25) **BP**
90. What are the causes and corrections of hum and self-oscillation in AF amplifiers? (14-25, 14-36) **BPN**
91. Diagram a resistance load in the plate circuit of a VT and indicate the direction of I_p. (14-25) **T**
92. What is the purpose of a center-tap connection on a filament transformer? (14-25) **BT**
93. Diagram a pentode AF amplifier with *RC* coupling to the next stage. (14-25) **BP**
94. Is the bias normally positive or negative in a VT AF amplifier? (14-25) **T**
95. Diagram a triode AF amplifier inductively coupled to a loudspeaker. (14-25) **T**
96. What are some results if components fail in an AF amplifier? (14-26) **P**
97. How might a faulty component be determined from improper circuit performance? (14-26) **P**
98. What may be used to locate trouble in an AF amplifier? (14-26) **P**
99. How might an oscilloscope be used to locate trouble in an AF amplifier? (14-26) **PG**

RADIO-FREQUENCY AMPLIFIERS

1. What are the limits of radio frequencies? (15-1) **PN**
2. What are the bandwidths that RF stages may be required to amplify? (15-1) **PG**
3. How is circuit Q involved in RF amplifiers? (15-1) **PG**
4. Why must some RF amplifiers be neutralized? (15-2, 15-14) **TG**
5. Why should input wiring be separated as far as possible from output wiring in RF amplifiers? (15-2) **PG**

6. Why are bypass capacitors used across the self-bias resistors of an RF amplifier? (15-2) **TG**
7. Describe the operation and characteristics of classes A, AB, B, and C amplifiers. (15-2, 15-5) **BTA**
8. Diagram and explain a grounded-emitter RF amplifier. (15-2, 15-28) **BPG**
9. What type of tube is generally employed in RF voltage amplifiers? (15-4) **BP**
10. Does a pentode usually require neutralization when used as an RF amplifier? (15-4) **TG**
11. What are the biasing differences between RF voltage and power amplifiers? (15-5) **BPG**
12. How is the power input to a pentode RF amplifier determined with an ammeter in the cathode circuit? (15-5) **BPG**
13. What are some applications of VTs in radio? (15-5, 15-23) **TN**
14. Define a class C amplifier. (15-6) **TG**
15. Diagram and explain a class C RF power amplifier with battery bias. (15-6) **B**
16. Does grid current flow in a class C VT RF amplifier? (15-6) **T**
17. Why do some tubes have three pins or prongs connected to the cathode? (15-6) **PTG**
18. During what approximate portion of the excitation voltage cycle is output circuit current present in class C operation? (15-6) **T**
19. A triode has an E_p of 1250 V, I_p of 150 mA, and μ of 25. What grid-bias value produces class C operation? (15-6) **T**
20. Why is the efficiency of a class C amplifier higher than classes B or A? (15-6) **BPG**
21. What are RF chokes and for what are they used? (15-6) **BPG**
22. When tuning the output-tank circuit of an RF amplifier, what output-current value indicates resonance? (15-7) **TG**
23. In a class C RF amplifier, as the output circuit is tuned through resonance, what effect would be observed on the input bias current? (15-7) **T**
24. What load condition is required for maximum power output from an RF amplifier? (15-8) **PTG**
25. In a class C RF amplifier, what ratio of load to plate impedance gives greatest efficiency? (15-8) **PG**
26. Why is grid-leak type biasing not practical in AF amplifiers? (15-9) **BPG**
27. What might be caused by a loss of drive in an RF power amplifier? (15-9) **BPG**
28. What is the function of a grid-leak resistor in a class C amplifier? (15-9) **TG**
29. Diagram and explain grid-leak bias in an RF amplifier. (15-9) **BP**
30. Why is a cathode resistor sometimes used in class C RF amplifiers? (15-9) **BPG**
31. In an RF amplifier, if I_g is 10 mA, R_g is 5 kΩ, what is the bias voltage? (15-9) **TG**
32. Diagram and explain an RF power amplifier with two tetrode tubes in parallel. (15-10) **BPG**
33. Diagram and explain an RF power amplifier with two tetrode tubes in push-pull. (15-11) **BPG**
34. Diagram and explain a series-fed plate circuit in an RF amplifier. (15-12) **BP**
35. Diagram and explain a shunt-fed plate circuit in an RF amplifier. (15-12) **BP**
36. In a shunt-fed plate circuit of a VT RF amplifier, what would be the result of a short circuit in the plate RFC? An open circuit? (15-12) **T**
37. How can a lack of neutralization cause spurious emissions from a transmitter? (15-12) **BPG**
38. In a capacitor-coupled RF amplifier, what symptoms would result if a coupling capacitor shorted? Opened? (15-13) **BPG**
39. Diagram inductive coupling between RF amplifier stages. (15-13) **T**
40. Diagram and explain capacitive or impedance coupling between RF amplifier stages. (15-13) **T**
41. Diagram and explain link coupling between two RF amplifier stages. (15-13) **PT**
42. What is the advantage of link coupling? (15-13) **PT**
43. What are advantages of tetrodes or pentodes over triodes as an RF amplifier? (15-14) **BPTG**
44. Why is an RF amplifier neutralized? (15-14) **BPG**
45. How can it be determined whether an RF power amplifier is self-oscillating? (15-14) **BPG**
46. When is neutralization of a triode not required? (15-14, 15-23) **PG**
47. Diagram and explain a plate-neutralized triode RF amplifier. (15-15) **BT**
48. Describe how an RF amplifier stage can be neutralized and the precautions to be taken. (15-15 through 15-22) **T**
49. Diagram and explain a grid-neutralized triode RF amplifier. (15-16) **BP**

50. Why is it necessary to remove the power-supply voltage from a stage being neutralized? (15-17) **BP**
51. Name three devices used as indicators when neutralizing an RF amplifier. (15-17) **T**
52. Explain a step-by-step procedure for neutralizing an RF amplifier stage. (15-17, 15-18) **BPT**
53. What may indicate when an RF amplifier stage is not properly neutralized? (15-17, 15-18) **PTG**
54. What is the function of a doubler amplifier? (15-23) **TG**
55. What factors are important in the operation of a frequency doubler? (15-23) **PTG**
56. What class amplifier is used in frequency doublers? (15-23) **BPG**
57. What circuits are used to produce frequency multiplication? (15-23) **BPG**
58. Diagram and explain a harmonic generator. (15-23) **BPTG**
59. What is the relationship between the fundamental frequency, harmonics, and octaves? (15-23) **PG**
60. A circuit has two tubes with grids in push-pull and plates in parallel. What relation will hold between input and output frequencies? (15-23) **PG**
61. Which harmonics are developed by push-push stages? (15-23) **BP**
62. Diagram and explain a push-push stage. (15-21) **BPG**
63. Which harmonics are produced by push-pull stages? (15-23) **BP**
64. What is the crystal frequency of a transmitter having three doubler stages and an output of 16 MHz? (15-23) **PG**
65. How may coils and capacitors be designed to resonate at a given frequency? (15-24) **PA**
66. If the period of one complete cycle of a radio wave is 0.000002 s, what are the wavelength and frequency? (15-24) **TA**
67. A doubler has an input of 2 MHz and a plate inductance of 40 μH. What capacitance is needed for resonance? (15-24) **PA**
68. When an ac filament is used, why is a filament center-tap usually used for grid and plate return circuits? (15-25) **T**
69. What currents will be indicated by a milliammeter connected between the center-tap of a filament transformer of a tetrode and negative high voltage (ground)? (15-25) **TG**

70. What are the advantages and disadvantages of grounded-grid type amplifiers? (15-26) **BG**
71. Diagram and explain a grounded-grid RF amplifier with an *LC* plate circuit load. (15-26) **BP**
72. What type of triode RF power amplifier requires no neutralization? (15-26) **BPG**
73. Why are grounded-grid amplifiers useful at VHF? (15-26) **BP**
74. List some circuit factors of an RF amplifier that should be considered at VHF which would not be of concern at VLF. (15-27) **BP**
75. Diagram and explain a grounded-base RF amplifier. (15-28) **BPG**
76. Diagram and explain a grounded-emitter RF doubler. (15-28) **BPG**
77. What components in a transistorized RF amplifier might short-out without stopping circuit operation? (15-28) **P**
78. What are some possible results if components fail in an RF amplifier? (15-28, 15-29) **PG**
79. How might the component at fault be determined from improper circuit performance? (15-28, 15-29) **P**
80. Describe the use of a pi-type impedance matching network. (15-28) **TG**
81. What components are most likely to break down in electronic equipment? (15-29) **BP**
82. What may cause an RF power amplifier tube to have excessive plate current? Insufficient plate current? (15-29) **TG**
83. In a series-fed plate circuit of an RF amplifier, what would be the effect of a short circuit of the plate-supply bypass capacitor? (15-29) **T**
84. What are some possible causes of insufficient plate current in a pentode RF amplifier? (15-29) **BP**

BASIC TRANSMITTERS

1. What are the lowest radio frequencies useful in radio communication? (16-1) **T**
2. What is meant by a carrier? (16-1) **BPG**

3. Define a type A1 emission. (16-1, 16-3)   **TN**
4. Define emission types NØN, A1A, A2A, B, and F1A. (16-1, 16-2, 16-3, 16-11, 16-12)   **T**
5. How are sidebands involved in radiotelegraph emissions? (16-1, 16-9)   **TG**
6. What is chirp, what causes it, and how can it be reduced? (16-2)   **TN**
7. How might the keying of a simple emergency transmitter be accomplished? (16-2, 16-3, 16-4)   **T**
8. What is the effect of excessive coupling between the output of a simple oscillator and an antenna? (16-2)   **TG**
9. What would be the effect of a swinging antenna on the output of a self-excited oscillator transmitter? (16-2)   **TG**
10. A transmitter uses full-wave rectified 500 Hz ac with no filtering as the plate supply. How is the emission classified? (16-2)   **T**
11. What is meant by a self-rectified circuit radiotelegraph transmitter? (16-2)   **T**
12. What is meant by full break-in operation at a radiotelegraph station, and how is it accomplished? (16-3)   **TG**
13. Diagram a simple radiotelegraph transmitter keyed by a keying relay. (16-3)   **T**
14. How are RF emission, signal, and information related? (16-3)   **TG**
15. What is the primary function of the power-amplifier stage of a radiotelegraph transmitter? (16-4)   **PTA**
16. Block diagram a simple radiotelegraph transmitter. (16-4)   **TN**
17. Diagram a method of coupling the RF output of the final PA stage of a transmitter to an antenna. (16-4, 16-13, 16-16)   **T**
18. Why must final amplifiers be neutralized in some transmitters? (16-4)   **BTG**
19. What are advantages and disadvantages of self-excited-oscillator and MOPA transmitters? (16-4)   **TG**
20. What are causes and corrections of transmitter spurious emissions other than harmonics? (16-4, 16-9, 16-16)   **BPG**
21. What class of amplifier should be employed for best plate efficiency in the final amplifier stage of a radiotelegraph transmitter? (16-4)   **T**
22. What causes a backwave and how can it be corrected? (16-4)   **TN**
23. What produces a superimposed hum on a carrier and how might it be overcome? (16-4)   **PTN**
24. How may dc voltmeters and ammeters in transmitters be protected against damage due to stray RF? (16-4)   **TG**
25. Explain basic transmitter tune-up procedures. (16-5, 16-7, 16-13, 16-20)   **BTN**
26. What precautions should be observed in tuning a transmitter to avoid damage to components? (16-5, 16-7)   **TA**
27. What is the result of Miller effect in RF amplifiers? (16-5)   **PTA**
28. Should the antenna of an MOPA be adjusted to resonance before the plate-tank circuit of the final stage? Why? (16-5)   **TA**
29. What effect on the I_p of the final amplifier will be observed as the antenna circuit is brought into resonance? (16-5, 16-13)   **TA**
30. Describe how to obtain maximum power output from a final RF amplifier into a tunable antenna with adjustable coupling to the plate circuit, using an I_p meter and a tube manual. (16-5)   **TA**
31. What is meant by the frequency tolerance of a transmitter? (16-6)   **P**
32. For what may phase-locked loops be used? (16-6)   **PE**
33. Why is a dummy antenna used in testing a transmitter? By what other name is it known? (16-7)   **BPTG**
34. How is a dummy antenna used and constructed? (16-7)   **PG**
35. How are power input and output of transmitters measured? (16-7)   **PG**
36. List points in a transmitter where keying may be accomplished. (16-8, 16-10, 16-11, 16-12, 16-19, 16-20)   **TG**
37. What are key clicks, their causes, and cures? (16-9)   **N**
38. Describe a means of reducing sparking at the key contacts of a telegraph transmitter. (16-9)   **TN**
39. Diagram a key-click filter suitable for use when a transmitter is keyed in the negative high-voltage circuit. (16-9)   **TN**
40. What frequency sine waves combine to form a square wave? (16-9)   **BPA**
41. How does transmitted code speed (words per minute) effect the bandwidth of the emission? (16-9)   **TG**
42. What is meant by data rate and baud rate? (16-9)   **P**

43. How does information rate compare with bandwidth? (16-9) **PE**
44. What is meant by a blocked grid? (16-11) **BPT**
45. Diagram how a radiotelegraph transmitter may be keyed by the grid-blocking method. (16-11) **T**
46. In a radiotelegraph transmitter with grid-bias keying, if the key is up but there is RF output, what could be the trouble? (16-11) **T**
47. What are the characteristics of the emission of a radiotelegraphy transmitter which uses a chopper to obtain A2 type emission? (16-11) **T**
48. What is meant by frequency-shift keying and how is it accomplished? (16-12) **PTG**
49. What is the purpose of a buffer-amplifier stage in a transmitter? Block diagram a three-stage transmitter. (16-13) **BPTN**
50. Diagram and explain capacitive coupling between an oscillator and buffer amplifier. (16-13, 16-19) **BP**
51. Describe the order in which the circuits should be adjusted in placing a three-stage transmitter in operation. (16-13) **TG**
52. What is the purpose of shielding between RF amplifier stages? (16-13, 16-16) **PA**
53. Diagram and explain a pi-type impedance matching network. (16-13) **BA**
54. What is meant by a fundamental frequency? A harmonic? A multiplier? (16-15) **PG**
55. Block diagram a radiotelegraph transmitter with the oscillator on 1 MHz and the PA on 4 MHz. (16-15) **TG**
56. What is the crystal frequency of a transmitter having three doubler stages and an output of 14,060 kHz? (16-15) **TG**
57. What is meant by harmonic radiation? (16-16) **TN**
58. Why is it important that transmitter harmonics be attenuated? (16-16) **BPTN**
59. What are methods of attenuating harmonics in transmitters? (16-16) **BPTN**
60. What is another name for a Faraday screen? (16-16) **T**
61. Diagram and explain link coupling with a low-pass filter between a final amplifier and an antenna. (16-16) **BP**
62. What are indications of parasitic oscillations in transmitters? (16-16) **BPG**
63. Describe how to measure harmonic attenuation of a transmitter. (16-16) **BP**

64. What materials can be used to shield magnetic fields? Electrostatic fields? (16-17) **T**
65. Why should all exposed metal parts of a transmitter be grounded? (16-17) **T**
66. Explain the function of interlocks. (16-17) **T**
67. Diagram a crystal oscillator coupled to an amplifier. (16-19) **T**
68. How is frequency translation (mixing) used in transmitters? (16-20) **PG**
69. How is the power output of a VT radiotelegraph transmitter ordinarily adjusted? (16-20) **T**
70. What are some indications of a defective tube? (16-21) **T**
71. What would indicate subnormal filament emission of a tube in a transmitter? (16-21) **BTG**
72. How might the component at fault be determined from improper circuit performance? (16-21) **P**
73. An MOPA has been operating normally. Suddenly the antenna ammeter reads zero, but plate and grid meters indicate normally. What could be the cause? (16-21) **T**
74. If the I_p of the final RF amplifier suddenly increases and radiation decreases, but the antenna is in good order, what are possible causes? (16-21) **T**
75. How are dangerous voltages in equipment made inaccessible to accidental contact? (16-21) **BG**
76. Explain the operation of the following relays: time-delay; overload; recycling. (16-21) **BP**
77. What are some results of component failure in a transmitter? (16-21, 16-22) **P**
78. What are some emergency repairs that may be made to faulty transmitter components? (16-22) **T**
79. What electrical shocks may be dangerous? (16-23) **BPTN**

AMPLITUDE MODULATION AND SSB

1. What is a radio carrier? (17-1) **TG**

2. What is meant by amplitude modulation? (17-1) **PG**

3. What is an A3 emission? (17-1) **PG**

4. What causes sound and how is it transmitted in air? (17-3) **B**

5. What form of energy is contained in a sound

6. What characteristic of a sound wave determines its pitch? (17-3) **B**

7. What frequencies are considered to be in the AF range? (17-3) **BPG**

8. Diagram and explain a single-button carbon microphone circuit. (17-4) **BT**

9. What are advantages of carbon button microphones? (17-4) **T**

10. What may cause packing of granules in a single-button microphone? (17-4) **T**

11. Diagram a microphone connected to an AF amplifier. (17-6) **T**

12. What might cause FM in an AM transmitter? (17-6) **BP**

13. What is meant by dynamic instability? (17-6) **B**

14. What is the purpose of a speech amplifier? (17-6) **T**

15. Diagram and explain how AM modulators operate. (17-6, 17-8, 17-10, 17-11, 17-15, 17-16) **PG**

16. Explain amplitude modulation methods. (17-6 through 17-11) **BA**

17. What class of amplifiers are used in modulated stages? (17-6, 17-10, 17-16, 17-19, 17-21) **P**

18. What is a modulated envelope? (17-7) **PG**

19. What is meant by the peak amplitude of a modulated signal? (17-7, 17-8, 17-10) **BG**

20. Explain amplitude modulation transmissions. (17-8) **BG**

21. What is the ratio of unmodulated carrier power to instantaneous peak power at 100% modulation? (17-8, 17-10) **BE**

22. What is meant by percentage of modulation and how may it be measured? (17-9, 17-18) **P**

23. Why is a high percentage of modulation desirable? (17-9) **B**

24. What is the relationship between modulation percentage and output carrier envelope waveform? (17-9) **BP**

25. What is the relationship between the power in the sidebands and intelligibility of the received signal? (17-9, 17-13) **BP**

26. What undesirable effects result from over-modulation? (17-9) **BPG**

27. What produces splatter in an AM transmission? (17-9, 17-40) **BG**

28. A 50% efficient modulated RF amplifier has 100 W output if the modulator is 66% efficient. What modulator input power is required to produce 100% sinusoidal modulation? (17-9) **B**

29. What are the results of using an AF peak limiter on the output signal of an A3 transmitter? (17-9, 17-10, 17-40) **BE**

30. What is the relationship between the AF power output of the modulator and the dc plate-circuit input of the modulated amplifier under 100% sinusoidal plate modulation? How does this differ when voice modulation is employed? (17-10) **BPT**

31. How is the load impedance of a modulator determined if it is modulating the plate circuit of a class C RF stage? (17-10) **B**

32. What percent increase in average RF power is obtained at 100% sinusoidal modulation? (17-10) **B**

33. A class C amplifier with E_p of 1000 V and I_p of 200 mA is modulated by a class A amplifier with 10-kΩ plate impedance. What is the proper turns ratio for the modulation transformer? (17-10) **B**

34. What are the advantages and disadvantages of class B modulators? (17-10) **B**

35. Diagram both a class B modulator high-level modulating a triode RF amplifier and a beam power RF amplifier. (17-11, 17-12) **B**

36. A high-level modulated transmitter operates with E_p of 6.7 kV and I_p of 2.3 A. If modulated 100% by a sine-wave AF signal, what is the modulator power output? (17-10) **B**

37. Diagram and explain a high-level modulated tetrode. (17-11) **BP**

38. How is the operating power of an AM transmitter determined using antenna resistance and current? (17-12) **BP**

39. Describe how transmitters can be aligned or tuned and checked for power and modulation. (17-12, 17-17, 17-29, 17-36) **P**

40. What contains the intelligence in AC emissions? (17-13) **B**

41. If a 2.5-MHz radio wave is amplitude-modula-

ted by a 1-kHz sine-wave tone, what frequencies are emitted? (17-13) **T**

42. What causes sidebands in amplitude modulation? (17-13) **BPG**

43. What formula determines sideband power in AM emissions? (17-13) **B**

44. In AM, what is the relationship of sideband power, carrier power, and modulation percentage? (17-13) **B**

45. During 100% sinusoidal AM, what percent of the average output power is in sidebands? (17-13) **B**

46. What are sidebands? (17-14, 17-31) **BPG**

47. Why does exceeding 100% negative peak modulation in AM cause excessive bandwidth? (17-14) **BP**

48. What is the bandwidth of an emission? (17-14, 17-31) **BPG**

49. What is the bandwidth of an A3 or A2 emission having a 600-Hz modulating frequency and a 500-Hz carrier frequency? (17-14) **TG**

50. Draw a diagram of a Heising modulation system capable of 100% modulation. (17-15) **T**

51. What is the purpose of the plate choke in Heising modulation? (17-15) **T**

52. Why is a parallel *RC* network used in the dc plate supply of the modulated RF amplifier using Heising modulation? (17-15) **T**

53. Diagram a grid-bias modulation circuit for an RF amplifier stage. (17-16) **B**

54. Explain bias adjustment of a grid-modulated stage. (17-16) **T**

55. Is grid current present in a conventional grid-bias modulated stage? (17-16) **T**

56. Is efficiency of a grid-bias modulated stage at maximum at zero or 100% modulation? (17-16) **T**

57. Compare characteristics of plate and grid modulation. (17-16) **T**

58. Define low- and high-level modulation. (17-17) **BT**

59. How can A3 percent modulation be measured? (17-18) **BP**

60. Sketch oscilloscope envelope display for 0, 50, 100, and over 100% modulation. (17-18) **B**

61. For what can oscilloscopes be used? (17-18, 17-36) **PA**

62. Describe how to connect and use an oscilloscope to monitor the output of an A3 transmitter. (17-18, 17-36) **BG**

63. Sketch a trapezoidal pattern showing 50% modulation without distortion. (17-18) **B**

64. What can a spectrum analyzer show with modulated signals? (17-18) **BPE**

65. How does a linear RF amplifier differ from other amplifiers? (17-19) **BP**

66. Linear amplifiers following a modulated stage operate in what class(es)? (17-19) **T**

67. Why are linear RF amplifiers not normally biased to class A? (17-19) **B**

68. If a class B linear were excited to saturation with no modulation, what would be the effect when modulated? (17-19) **T**

69. Doubling the excitation on a class B linear gives what increase in RF power output? (17-19) **B**

70. Diagram and explain a low-level plate modulated pentode with a push-pull linear amplifier. (17-19) **BP**

71. Draw a block diagram showing the stages in a complete AM transmitter. (17-19) **BPG**

72. What may be the cause of a decrease in plate and antenna current during modulation if a class B linear is used? (17-19) **B**

73. What is a Doherty linear amplifier, for what is it used, and what is its efficiency? (17-21) **B**

74. What is carrier shift? What is another name for it? (17-22, 17-23) **BP**

75. Diagram and explain a circuit that can measure carrier shift (17-22) **B**

76. What may cause positive carrier shift in a linear RF amplifier? (17-22, 17-23) **T**

77. What might cause a dip in antenna current when AM is applied to a modulated amplifier? (17-22) **BP**

78. What form of distortion is present if the average carrier power or antenna current of an AM transmission decreases during modulation? (17-22) **BP**

79. What are the effects of overexcitation of a class B RF amplifier grid circuit? (17-23) **B**

80. What may cause positive or negative carrier shift during modulation? (17-23) **T**

81. What may cause asymmetrical modulation of an A3 transmitter? (17-23) **B**

82. What are effects of insufficient RF excitation to a high-level modulated RF amplifier? (17-23) **T**

83. If a transmitter is adjusted for maximum power output for radiotelegraph operation, why must the plate voltage be reduced if amplitude-modulated? (17-24) **T**

84. If a 7-A carrier is modulated 100% by a sinusoidal tone, how much does the antenna current increase? (17-25) **BT**

85. List four ways an emf may be generated by sound waves. (17-26, 17-28) **T**

86. Sketch and explain the operation of a crystal microphone. (17-27) **BP**

87. What precautions should be observed when using and storing crystal microphones? (17-27) **T**

88. What types of microphones have a high-impedance output? (17-27, 17-28) **BT**

89. What are advantages and disadvantages of dynamic and velocity microphones? (17-27) **BP**

90. Explain unidirectional, bidirectional, and omnidirectional microphones. (17-28) **B**

91. What is meant by phasing of microphones, and when is it necessary? (17-28) **B**

92. Diagram and explain an A3 transmitter having a microphone, preamplifier, speech amplifier, class B modulator, crystal oscillator, buffer amplifier, class C modulated amplifier, and an antenna. (17-29) **BP**

93. Explain steps to tune an AM transmitter. (17-29) **BP**

94. If a speech amplifier is overexcited, what would be the effect on the output signal of an A3 transmitter? (17-29) **T**

95. Does the I_p fluctuate or remain steady in a class B modulator? In a class C modulated amplifier? (17-29) **T**

96. What would be the effect of a shorted turn in a modulation transformer? (17-29) **BPT**

97. Diagram and explain a pi-type impedance matching network coupled to a bipolar transistor. (17-30) **PA**

98. What are the different types of double sideband modulations? (17-31) **PG**

99. Explain selective fading. (17-31) **TA**

100. What is SSB? (17-31) **PG**

101. Explain the principles of SSB emissions. (17-31) **BPG**

102. Explain full, reduced, and suppressed carrier emissions. (17-31) **P**

103. How does A3J bandwidth and required power compare with A3 power and sidebands? (17-31) **BPG**

104. What is meant by A3, A3A, and A3J? (17-31) **BPG**

105. With 100% sinusoidal modulation, what percent of the output power is in the sidebands with A3? With A3J? (17-31) **P**

106. How is frequency translation related to mixing? (17-32) **PG**

107. What is the result of frequency mixing? (17-32) **P**

108. How is frequency translation accomplished? (17-32) **P**

109. In what type of transmitters are balanced modulators used? (17-32, 17-35) **PA**

110. Block diagram a simple SSB transmitter. (17-32, 17-35) **PG**

111. Block diagram a filter-type SSB transmitter with a 20-kHz oscillator and output at 6 MHz. (17-32) **BPG**

112. Diagram and explain function and purpose of balanced modulators. (17-33) **PA**

113. Explain crystal lattice SSB filters. (17-34) **PA**

114. What are two-tone tests? (17-36) **TG**

115. Explain multiplexing briefly. (17-37) **P**

116. What is meant by FDM and TDM? (17-37) **P**

117. Explain VOX operating circuits. (17-37) **PTG**

118. List causes and effects of, and steps that can reduce transmitter intermodulation. (17-38) **PA**

119. Two base station transceivers are installed at the same location. One operates on 151.2 MHz and the other on 151.1 MHz. On what nearby frequencies will a receiver in the area receive intermod interference? (17-38) **BPA**

120. Explain CB transmitter service and maintenance. (17-39) **P**

121. What tests and adjustments may be made to CB units? (17-39) **P**

122. Who may modify CB transmitters? (17-39) **P**

123. What is a CB performance tester? (17-39) **P**

124. How can improper grounding affect the harmonic radiation from a CB or other transmitter? (17-39) **P**

125. In which TV channel(s) would a harmonic of a 27.005-MHz CB transmission fall? (17-39) **BP**

126. Why may low-pass filters be connected at the output of CB transmitters? (17-39) **P**

127. Explain why a 27-MHz CB transmission may be heard on an audio device, and how this can be eliminated. (17-39) **P**
128. Explain AF and RF speech processing and companders. (17-40) **PG**
129. What is ALC? (17-40) **PG**

AMPLITUDE-MODULATION RECEIVERS

1. Explain in a general way how AM signals are received. (18-2, 18-9, 18-10) **BG**
2. Explain the purpose and function of detectors. (18-2, 18-4, 18-6, 18-8) **BPTA**
3. Diagram and explain a diode detector coupled to an AF amplifier. (18-2) **PTA**
4. What substances were used as crystals in crystal detectors? (18-3) **T**
5. Diagram and explain a crystal-detector receiver. (18-3) **TA**
6. Explain what receiver sensitivity is and in what unit it is measured. (18-3, 18-8) **BPE**
7. Diagram and explain a power detector. (18-4) **TA**
8. What operating conditions determine that a device is being used as a power detector? (18-4) **TA**
9. List and explain the characteristics of a square-law detector. (18-5) **T**
10. Diagram and explain a grid-leak detector. (18-6) **TA**
11. What effect does a signal have on the I_p of a grid-leak detector? (18-6) **T**
12. What is the advantage of a diode over a grid-leak detector? (18-6) **T**
13. How do grid-leak and power detectors compare in terms of sensitivity and selectivity? (18-4, 18-6) **T**
14. Diagram and explain the operation of a regenerative detector. (18-7) **TA**
15. What controls determine the selectivity of a 3-circuit detector? (18-7) **TA**

16. Why is an RF filter used in the output of a detector? (18-7) **T**
17. Why should a regenerative detector not be directly coupled to an antenna? (18-7) **TG**
18. How is a regenerative detector adjusted to receive radiotelegraph signals through interference? (18-7, 18-26) **T**
19. Do oscillators operating on adjacent frequencies tend to synchronize or drift apart? (18-7) **BPTG**
20. What is meant by locking two oscillators together? (18-7) **BPT**
21. What might be the cause of low sensitivity of a 3-circuit regenerative receiver? (18-7) **T**
22. Describe the operation of mixing type circuits. (18-7, 18-12, 18-15) **PTG**
23. Name the circuits in a receiver in which frequencies are mixed. (18-7, 18-12, 18-15) **P**
24. Describe the principle of operation of a super-regenerative detector. (18-8) **T**
25. Why is an RF amplifier used in a regenerative receiver? (18-9) **TG**
26. What type of radiotelephone receiver requires no oscillator? (18-9) **P**
27. Why are stages of a receiver often electrostatically shielded from each other? (18-9, 18-23) **BPT**
28. Block diagram an AM receiver. (18-9, 18-24) **BPTG**
29. Block diagram a radiotelegraph receiver. (18-9) **PTN**
30. Draw a diagram of a TRF receiver. (18-9) **T**
31. Block diagram and explain a single-conversion superheterodyne AM receiver. (18-10, 18-12) **BPT**
32. Block diagram and explain a superheterodyne capable of receiving CW signals. (18-10) **PT**
33. A superheterodyne is receiving a 1000-kHz signal. The mixer is on 1500 kHz. What is the IF? (18-10) **BPTG**
34. A superheterodyne is tuned to 2738 kHz. The IF is 475 kHz. What is the 2nd detector frequency? The LO? (18-10, 18-12, 18-14) **BPT**
35. Explain the operation of a synchronous detector. (18-10, 18-15) **PA**
36. In what receiver circuit does translation occur? (18-10) **PG**
37. What is meant by zero-beating a received signal? (18-10) **TN**

38. What is the purpose of a tuned RF amplifier ahead of the mixer in a receiver? (18-11)   **TG**

39. What is the "dynamic range" of a receiver? (18-11)   **PE**

40. Explain signal-to-noise ratio. (18-11)   **PE**

41. What is meant by the noise figure and sensitivity of a receiver? (18-11)   **PE**

42. Explain the purpose and function of mixers. (18-12, 18-14, 18-15)   **PG**

43. How many spurious signals can be created in a receiver? (18-12, 18-23, 18-29)   **BP**

44. What is the difference between active and passive mixers? (18-12)   **PG**

45. What is the advantage of a double-balanced mixer? (18-12)   **PG**

46. What causes intermodulation interference? (18-12)   **PA**

47. What can produce cross modulation? (18-12) **PA**

48. What factors are involved in receiver desensitizing? (18-12)   **PA**

49. Why do some superheterodynes use a crystal in the first detector? (18-12)   **T**

50. Explain the function and purpose of an IF amplifier. (18-13)   **PA**

51. What is the advantage of using special iron cores in IF transformers? (18-13)   **TA**

52. What is meant by IF bandwidth? (18-13) **BPTA**

53. Diagram and explain the second detector and AVC circuitry of an AM receiver. How could the circuit be modified to give delayed AVC? (18-14)   **BP**

54. How is the AVC switch set to receive: CW; SSB; AM? (18-14, 18-26)   **TG**

55. A product detector would most likely be used to demodulate what type of emissions? (18-14, 18-28)   **BPG**

56. Describe the operation of an AGC circuit. (18-14, 18-15)   **P**

57. What is the purpose of an oscillator operating on a frequency near the IF in a receiver? (18-15)   **TG**

58. Diagram and explain a BFO circuit. (18-15)   **BPTA**

59. Diagram and explain a squelch circuit. (18-16) **P**

60. What is an S meter and how does it operate? (18-18)   **TN**

61. What type of modulation is contained in static or lightning radio waves? (18-18)   **BP**

62. What is the advantage of a variable or tuned AF amplifier in a radiotelegraph receiver? (18-19)   **TG**

63. What is meant by bandwidth of a receiver and what determines it? (18-20)   **PG**

64. Why are bandpass filters used in a receiver? (18-19, 18-20)   **BP**

65. How are crystal filters used in receivers? (18-20)   **BPTG**

66. What are monolithic crystal filters and where are they used? (18-20)   **P**

67. If broadcast signals interfered with 500-kHz signals aboard ship, how could this be eliminated? (18-21)   **T**

68. Diagram a wave trap in an antenna circuit. (18-21)   **T**

69. Why may a station sometimes be heard at two places on the dial? (18-22)   **T**

70. What is meant by image response, and in what kind of a receiver is it developed? (18-22, 18-24)   **T**

71. Why should a 400-kHz IF receiver have an RF amplifier? (18-22)   **T**

72. Explain the relationship of signal, oscillator, and image frequencies in a superhet. (18-22)   **BPT**

73. A mixer is tuned to 2000 kHz and the local oscillator is on 1300 kHz. What frequency would cause an image? (18-22)   **BPT**

74. What might cause a 120-MHz aircraft transmission to be received on an FM broadcast-band receiver with a 10.7-MHz IF? (18-22)   **BPT**

75. How does double conversion affect images? (18-22, 18-24)   **BPT**

76. What is the advantage of using a high IF? (18-22)   **BPTG**

77. Diagram and explain a superheterodyne with an AVC circuit. (18-23)   **T**

78. Diagram and explain a superheterodyne with AVC. (18-23)   **T**

79. Draw a diagram of a complete SSB receiver. (18-24)   **PG**

80. Why is a superheterodyne not used near its first IF? (18-24)   **T**

81. What is meant by double conversion in a receiver? (18-24)   **PG**

82. Explain the principles involved in detecting FSK. (18-24)   **PG**

83. How is a receiver adjusted to receive CW? AM? (18-25)   **TG**

84. Explain how SSB emissions are detected. (18-25) **BPT**
85. What adjustment should be made if a receiver blocks on the reception of strong signals? (18-25) **T**
86. How is a receiver adjusted for best response to weak A1 signals? Strong A1 signals? (18-25) **T**
87. Why are unused portions of inductances in receivers sometimes shorted? (18-25) **T**
88. After long periods of listening to CW signals of constant tone, what adjustments can the operator make to relieve hearing fatigue? (18-25) **TG**
89. What is diversity reception and how is it accomplished? (18-26) **BP**
90. Explain how to align an AM receiver using a: speaker; oscilloscope; VTVM. (18-28) **BP**
91. How can an oscilloscope be used when checking a receiver? (18-28, 18-29) **PG**
92. How might signal generators be used with receivers? (18-28, 18-29) **PG**
93. What meter is suitable for measuring receiver AVC voltage? (18-28) **PG**
94. Explain how to test components in a receiver. (18-29) **P**
95. List some precautions to be observed when soldering transistors and repairing printed circuits. (18-29) **BG**
96. How might AVC voltage help in troubleshooting? (18-29) **BP**
97. What may cause noisy operation of a receiver? (18-29) **TG**
98. How can faulty components be determined from improper circuit performance? (18-29) **P**
99. What is a signal tracer and how is it used? (18-29) **PTG**
100. If the active device in the only RF stage of a receiver burned out and no spares were available, how could temporary repairs be made to permit operation? (18-30) **T**

FREQUENCY MODULATION

1. What radio frequencies are most affected by lightning or static waves? (19-1) **BPG**
2. What type of modulation is largely contained in static and lightning radio waves? (19-1) **T**
3. What kind of radio receivers to not respond to static interference? (19-1) **T**
4. What are four fields of communications that use FM? (19-2) **BPG**
5. What is meant by the deviation of an FM signal? (19-3) **BPG**
6. What determines the deviation ratio of an FM emission? (19-3) **BPG**
7. What factors determine the modulation index of an FM transmission? (19-3) **BPG**
8. To what is deviation proportional in a PM or FM transmitter? (19-3, 19-15) **BPG**
9. When a carrier is frequency-modulated, what is developed on both sides of the carrier? (19-3) **BPG**
10. In FM, what determines bandwidth? (19-3) **BP**
11. In FM, what produces the sidebands? (19-3) **BPG**
12. In FM, what is the relationship between number of sidebands and amplitude of the modulating voltage? (19-3) **BP**
13. In FM, what is the relationship between number of sidebands and bandwidth? (19-3) **BP**
14. In FM, what is the relationship between number of sidebands and modulating frequency? (19-3) **BP**
15. In FM, what is the relationship between spacing of sidebands and modulating frequency? (19-3) **BP**
16. In FM, what is the relationship between percent modulation and percent deviation? (19-3) **BP**
17. In FM, what is the relationship between percent modulation and number of sidebands? (19-3) **BPG**
18. How do wide-band and narrow-band FM systems compare with respect to frequency deviation and bandwidth? (19-3) **BPG**
19. What is meant by emission types F1A? F1B? F3E? (19-3) **BPG**
20. What is meant by emission types F3C? F3F? (19-3) **BP**
21. What would be the emission designation of a voice communication FM transmitter with 20-kHz bandwidth authorization? (19-3) **BP**
22. The transmission-line current of an FM station is 8.5 A without modulation. What should it be with 100% modulation? (19-3) **B**
23. Explain a phase-locked-loop FM detector. (19-4) **BP**

24. What is the result of overmodulation of an FM emission? (19-5, 19-13) **PG**
25. Explain how FSK may be detected. (19-5) **PT**
26. Diagram and explain a Foster-Seeley discriminator. (19-6) **T**
27. Why is a limiter-discriminator sensitive to frequency changes rather than amplitude changes? (19-6, 19-9) **BP**
28. Diagram and explain a ratio detector. (19-7) **BP**
29. Explain the operation of a quadrature detector. (19-8) **BP**
30. In an FM receiver, what stage is between the discriminator and the last IF amplifier? (19-9) **BP**
31. What is different about the last two IF stages in an FM receiver? (19-9) **P**
32. Why are pre-emphasis and de-emphasis used? (19-10) **BP**
33. Diagram and explain a de-emphasis circuit. (19-10, Fig. 19-12) **BP**
34. Diagram and explain a pre-emphasis circuit in a phase-modulated FM transmitter. (19-10, Fig. 19-18) **BP**
35. Block diagram and explain the operation of a broadcast FM receiver. (19-11) **PT**
36. Block diagram and label blocks of VHF communications superheterodyne. (19-11) **BP**
37. Diagram and explain a limiter, discriminator, and differential squelch in an FM receiver. (19-11) **BP**
38. What is meant by capture effect in relation to FM receivers? (19-11) **PA**
39. How might signal generators be used with receivers? (19-12) **PG**
40. Describe a step-by-step procedure for aligning an FM superheterodyne. (19-12) **BP**
41. How can receiver limiter current be used to determine if front-end overload is occurring? (19-12) **BP**
42. Diagram and explain a reactance modulator. (19-13) **BPA**
43. What is meant by direct FM? Indirect FM? (19-13, 19-15) **PG**
44. Do the harmonics of an FM transmission contain usable modulation? (19-13) **BPA**
45. Block diagram and explain the operation of an FM transmitter. (19-13, 19-15) **PTA**
46. If an FM transmitter has two doublers and one tripler, what is the carrier swing when the oscillator deviates 2 kHz? (19-13) **BP**

47. If a PM transmitter has two triplers and one doubler with an output of 154.31 MHz, what is the crystal frequency? (19-13) **BP**
48. How is good frequency stability of a reactance modulator achieved? (19-13) **BP**
49. Explain how an AFC circuit operates in an FM transmitter. (19-13) **ABP**
50. How might a varactor be used to generate direct FM of a crystal oscillator? (19-14) **BPG**
51. Describe the operation of the Armstrong FM system of obtaining phase modulation. (19-15) **B**
52. What is the difference between frequency and phase modulations? (19-15) **BP**
53. What occurs to a waveform if it is phase modulated? (19-15) **BP**
54. Diagram and explain the circuits involved in a phase modulator system. (19-16) **BPG**
55. Diagram and explain an audio peak limiter used in an FM transmitter. (19-16) **BP**
56. What are some of the instruments used in making FM transmitter measurements? (19-16) **P**

ANTENNAS

1. What are the two fields that radiate from any antenna and how do they differ. (20-1) **BP**
2. What is the lowest radio frequency useful in radio communications? (20-1) **T**
3. How can line-of-sight propagation be used in radio? (20-1) **PG**
4. What is meant by radio-path horizon? (20-1) **PA**
5. How can knife-edge refraction be used in radio? (20-1) **BP**
6. Explain the four ionospheric layers and their effects. (20-2) **TG**
7. How does velocity of signal propagation differ in different mediums? (20-2) **BP**
8. Explain skywaves, ground waves, and their part in skip distance. (20-2) **PN**
9. What effect do angle of radiation, density of ionosphere, and frequency have on the length of the skip zone? (20-2, 20-4) **BPG**

10. How are operating frequency and ground-wave coverage related? (20-2) **BP**
11. What causes radio wave absorption? (20-2) **PG**
12. What is MUF and its limits? (20-2) **PG**
13. What advantages may be expected from the use of HF in radio communications? (20-2, 20-4) **T**
14. How are sunspots and sky-wave coverage related? (20-2) **BPG**
15. What makes auroral communications possible? (20-2) **BPA**
16. What range of frequencies is involved in sporadic-*E* ionization clouds? (20-2) **TA**
17. What role do meteors play in radio communications? (20-2) **TE**
18. Explain the meaning of SID. (20-2) **TG**
19. What results when a sky wave meets a ground wave? (20-3) **BPA**
20. What are the frequency ranges, distance characteristics, and allocations of the VLF, LF, MF, VHF, UHF, SHF, and EHF bands? (20-4) **BP**
21. What radio frequencies are useful for continuous long-distance communications? (20-4) **T**
22. How do regular daily propagation characteristics vary? (20-4) **TG**
23. Explain ducting and tropospheric bending? (20-4) **PG**
24. What frequencies have substantially straight-line propagation characteristics and are unaffected by the ionosphere? (20-4) **T**
25. What is meant by scatter communications? (20-4) **PG**
26. Explain transequatorial propagation. (20-4) **E**
27. What is the chief cause of atmospheric noise? (20-5) **P**
28. How can radio equipment be protected from lightning? (20-5) **BPTN**
29. What is meant by polarization of a radio wave and how does it affect the reception of radio waves? (20-6) **BPTA**
30. What are the three forms of polarization of antennas? (20-6) **PG**
31. What is the velocity of a radio wave in space? (20-7) **BP**
32. How are frequency and wavelength related? (20-7) **PN**
33. What is the wavelength of a 5-MHz ac in meters and in centimeters? (20-7) **BPG**
34. What is the working resonant length of a 4 MHz dipole in feet? (20-7) **PN**
35. How do physical and electrical lengths of an antenna differ? (20-7) **BPTG**
36. A 1400-kHz vertical antenna is 380 feet high. What is its physical height in wavelengths? (20-7) **B**
37. What is the difference between Hertz and Marconi antennas? (20-8) **T**
38. What is meant by a voltage node? Loop? (20-9) **PA**
39. Explain *E* and *I* relationships in $\lambda/2$ and $\lambda/4$ antennas. (20-9, 20-14) **BP**
40. Diagram how *E* and *I* vary along a $\lambda/2$ antenna. (20-9) **T**
41. What effect do the values of *E* and *I* at any point on an antenna have on the *Z* at that point? (20-10) **BP**
42. What is meant by radiation resistance? (20-10) **BPG**
43. Explain the characteristic and feedpoint impedances of an antenna. (20-10, 20-14, 20-24) **PG**
44. What are the feedpoint impedances of a $\lambda/2$ dipole and a $\lambda/4$ vertical? (20-10, 20-14, 20-24) **TG**
45. What would be the effect on the resonant frequency if the length of a Hertz antenna were reduced? (20-11) **T**
46. How can operating on a lower frequency than the resonant frequency of an antenna be accomplished? (20-11) **TG**
47. How is the resonant frequency affected if a capacitor is added in series with an antenna? With a loading coil? (20-11) **BPT**
48. If a loading coil of a resonant antenna is removed, is the antenna then capacitive or inductive? (20-11) **BP**
49. For what are parallel-conductor lines used? (20-12) **PTN**
50. What is meant by the characteristic, or surge, impedance of a transmission line? (20-12, 20-16) **BPG**
51. How is the impedance of a 2-wire nonresonant transmission line computed? (20-12) **BPT**
52. If the conductors in a two-wire RF transmission line are replaced by larger wires, how is the surge impedance affected? (20-12) **BPT**
53. What is meant by balanced and unbalanced feed lines? (20-12) **BG**
54. How do frequency and line length affect transmission-line attenuation? (20-12) **BPG**

55. What determines the basic efficiency of an antenna system? (20-12) **BA**

56. Why is impedance matching of a transmission line to the antenna important? (20-12) **BPTG**

57. What is meant by standing waves and SWR of a transmission line? (20-12) **BPN**

58. Does minimum or maximum VSWR indicate best matching of antenna to transmission line? (20-12) **BPN**

59. Which would be possibly acceptable, an SWR of 10:1 or 3:1? (20-12) **BPN**

60. What are possible causes of unacceptable SWR readings? (20-12) **TN**

61. How can SWR be minimized? (20-12) **BP**

62. What does a reflectometer indicate? (20-12) **PG**

63. How can RF power to an antenna be measured? (20-12, 20-25) **PG**

64. Explain the use of a directional wattmeter in measuring VSWR and transmitter power. (20-12) **BP**

65. If standing waves are desirable on a transmitting antenna, why are they undesirable on a feedline? (20-12) **BPG**

66. If the feedling Z matches the antenna feed-point Z, what effect will doubling the line length have on SWR? (20-12) **B**

67. For what are coaxial cables used? (20-12) **BPN**

68. What should be considered in choosing a solid-dielectric cable over a gas-pressurized cable for use as a feedline? (20-12, 20-24) **BP**

69. What is hardline? (20-12) **PN**

70. What are the properties of $\lambda/4$ and $\lambda/2$ sections of feedline when shorted, matched, and open? (20-12, 20-16) **PE**

71. How are radiation patterns, directivity, and major lobes of antennas related? (20-13, 20-15, 20-17, 20-20, 20-21) **PG**

72. How should a horizontal dipole be oriented to transmit N and S? (20-13) **TG**

73. What antenna has omnidirectional characteristics? (20-13) **BT**

74. Describe the directional characteristics of a horizontal Hertz, vertical Hertz, and vertical Marconi. (20-13, 20-22) **BPT**

75. Draw the radiation pattern of a $\lambda/2$ horizontal antenna $3\lambda/4$ above ground. (20-13) **T**

76. How are the dimensions of a $\lambda/4$ vertical computed? (20-14) **PN**

77. What is the feedpoint Z of a $\lambda/4$ ground-plane antenna? (20-14) **BP**

78. Discuss series- and shunt-feeding of $\lambda/4$ antennas with respect to Z matching. (20-14, 20-24) **BP**

79. What types of antennas are used in mobile work? (20-14) **PA**

80. What constitutes the ground plane if a whip antenna is mounted on the metal roof of an automobile? If mounted near the rear bumper? (20-14) **BP**

81. Sketch a coaxial whip antenna and identify the whip, insulator, skirt, trap, support mast, coaxial line, and input connector. (20-14) **BP**

82. What is corona discharge from an antenna? (20-14) **BP**

83. Explain directivity and physical characteristics of $5\lambda/8$ verticals. (20-14) **BP**

84. How can two vertical UHF antennas be oriented to produce maximum isolation between them? (20-14) **PA**

85. Discuss the use of loading devices at the base, center, and top of vertical antennas. (20-23, 25-6) **PA**

86. How is the electrical length of a feedline computed? (20-16) **BTA**

87. What is the velocity factor of a transmission line and why is it important? (20-16) **BPA**

88. Explain delta, gamma, and stub feedline matching. (20-16) **TE**

89. What is meant by stub tuning? (20-16) **BP**

90. Discuss how open and shorted stubs may attenuate interfering signals at a receiver. (20-16) **BP**

91. Explain the properties of a $\lambda/4$ section of line. (20-16) **BP**

92. What Z value should a $\lambda/4$ matching line have to match a 500-Ω feeder to a 70-Ω antenna? (20-16) **BT**

93. Discuss directivity and physical characteristics of colinear vertical dipoles. (20-17) **BP**

94. How are antenna gain and beamwidth related? (20-17, 20-18, 20-20) **PA**

95. What is meant by a driven element of an array? (20-18) **BA**

96. How are radiation patterns of phased verticals affected by their spacing? (20-18) **PE**

97. What is the direction of maximum radiation

from two verticals spaced 180° with equal currents in phase? (20-18) **B**

98. What might cause a change in the directional-antenna patterns of an AM station? (20-18, 20-29) **B**

99. If two towers of a 1100-kHz array are separated by 120°, what is the tower separation in feet? (20-18) **B**

100. Directional antenna-tower-monitor sample current ratios must be within which percentages? (20-19) **B**

101. What factors can cause a change in the directional pattern of an AM broadcast station? (20-19) **B**

102. How does a directional antenna array at an AM broadcast station reduce radiation in some directions and increase it in others? (20-19) **B**

103. What is meant by the terms "critical" phased array, "phasor," and "common-point resistance"? (20-19) **B**

104. What are proof-of-performance tests and when are they made? (20-19) **B**

105. Explain the use of a broadcast station antenna monitor. (20-19) **B**

106. A final RF stage with an E_p of 5000 V and an I_p of 1.25 A feeds a four-tower directional array. If the common-point current is 10 A and the common-point resistance is 45 Ω, what is the transmitter efficiency? (20-19) **B**

107. Explain parasitic element operation in antennas. (20-20) **BPA**

108. Discuss construction and characteristics of a Yagi antenna. (20-20) **PG**

109. Discuss directivity and characteristics of some parasitic arrays. (20-20) **BPG**

110. What is a corner-reflector antenna? (20-20) **BP**

111. Explain folded, or multiple-wire, dipoles. (20-20) **PA**

112. What is the significance of the bandwidth of an antenna? (20-20) **P**

113. What are the gain, beamwidth, and tracking requirements for space communications antennas? (20-20) **PE**

114. Discuss the directivity and physical characteristics of V-beam antennas. Of rhombic antennas. (20-21) **BPTE**

115. Discuss vertical loop antennas. (20-22) **BP**

116. What are the characteristics of horizontal loops? (20-22) **BP**

117. What is a quad antenna and what are its characteristics? (20-22) **G**

118. Why are some antennas top-loaded? (20-23) **B**

119. What material is best suited for use as an antenna strain insulator? (20-23) **T**

120. Why are insulators placed in antenna guy wires? (20-23) **BP**

121. What is the effect on a transmitter if antenna insulation becomes dirty or salt-encrusted? (20-23) **T**

122. Diagram and explain the coupling of an RF amplifier to a λ/4 Marconi antenna other than by link or transmission line. (20-24) **BP**

123. How is the degree of coupling varied between a PA and an antenna when using a pi-network? (20-24) **TA**

124. How are L networks used in antenna coupling? (20-24) **BA**

125. Diagram an RF amplifier and transmission line to shunt-feed a λ/4 antenna. (20-24) **BP**

126. Discuss solid-dielectric and gas-filled coaxial cables for antennas. (20-24) **B**

127. Why is dry air or nitrogen often used in coax lines? (20-24) **B**

128. Why are antenna ammeters removed from the circuit when not being used? (20-24) **B**

129. How are antenna ammeters protected? (20-24) **B**

130. Explain how to adjust a T-network with two coils and a capacitor to properly couple a transmitter to its antenna. (20-24) **B**

131. Diagram a coupling method for an antenna, including Z matching and harmonic attenuation. (20-24) **B**

132. Describe installation of transmission lines, including Z, bends, kinks, cutting, and connections. (20-24) **B**

133. If R and I at the base of an antenna are known, how is the radiated power determined? (20-25, 20-30) **T**

134. What is the relationship between antenna current and radiated power of an antenna? (20-25, 20-26) **T**

135. What are some methods of measuring power output and input of transmitters? (20-25, 20-28, 20-30) **BPTG**

136. Describe how to tune an antenna by the substitution method and by the RF bridge method. (20-25) **B**

137. What is meant by field strength? (20-26) **BP**

138. What is a field strength-meter and in what units is it calibrated? (20-26) **BPG**

139. At what distances are field strength measurements made? (20-26, 20-27) **P**

140. How are transmitted power and field intensity at a receiving point related? (20-26) **BP**

141. If antenna current is halved, what is the percent change in field intensity at a receiving point? (20-26) **T**

142. A 500-kHz transmitter produces a field strength of 100 μV/m at 100 mi. What is the theoretical strength at 200 mi? (20-26) **T**

143. If a transmitter increases power from 150 W to 300 W, what would be the percent change in field intensity? (20-26) **T**

144. If field intensity doubles, what increase takes place in the transmitter antenna current? (20-26) **B**

145. How are decibels used in expressing harmonic field strength? (20-27) **P**

146. A transmitter has a field strength of 100 mV/m and a second harmonic of 200 μV/m at a distant point. What is the attenuation in dB? (20-27) **P**

147. How can harmonic radiation from a transmitter cause interference at distances where the fundamental signal cannot be heard? (20-27) **T**

148. A field intensity of 25 mV/m develops 2 V in a certain antenna. What is its effective height? (20-28) **B**

149. How is height above average terrain calculated? (20-28) **PB**

150. What is an isotropic radiator and for what is it used? (20-29) **BPE**

151. What is the power gain of an antenna? (20-29) **BP**

152. What is meant by erp? (20-29) **BPA**

153. How is erp determined, given system gains and losses? (20-29) **BPA**

154. What is the erp if the output of a transmitter is 50 kW, the coax line loss is 300 W, and antenna power gain is 3 dB? (20-29) **B**

155. A transmitter has a final RF stage E_p of 5000 V, I_p of 3 A, and 60% efficiency. If the antenna has a power gain of 4 and the transmission line has a 3 dB loss, what is the erp? (20-29) **B**

156. What is meant by the ground system? (20-30) **T**

157. What is the importance of the ground radials of a standard broadcast antenna? (20-30) **B**

158. What is likely to result if a large number of antenna radials become broken? (20-30) **B**

159. In a 2-MHz marine installation, why does the ground system of the boat affect the antenna impedance? (20-30) **BP**

160. If the antenna current of a station is 8 A for 5 kW, what is the current value for 1 kW? (20-31) **B**

161. If the daytime input power to an antenna having a resistance of 20 Ω is 5 kW, what is the nighttime input power if it is half the daytime power? (20-31) **B**

162. The P_{in} to a 50-Ω concentric line is 5 kW. What is the I value? (20-31) **B**

163. The power input to a 72-Ω coaxial line is 10 kW. What is the rms voltage between inner conductor and sheath? (20-31) **P**

164. Explain the operation of a turnstile TV transmitting antenna. (20-32) **B**

165. What is meant by a circularly polarized antenna? (20-32) **P**

166. Are radio-frequency waves harmful to humans? (20-34) **BPTG**

TWO-WAY COMMUNICATIONS

1. Explain DPLMRS-type mobile systems. (21-1, 21-4) **P**

2. What are some transmitter carrier frequency tolerances for VHF and UHF mobile services? (21-3) **P**

3. What are the bandwidth and frequency deviations of VHF and UHF mobile services? (21-3) **P**

4. What types of installations are used in the VHF and UHF mobile ranges? (21-4, 21-5, 21-7) **P**

5. What is a control point in mobile services? (21-4) **PG**

6. How is operation of two-way radio systems affected by antenna heights, antenna gains, terrains, urban environment, and frequency used? (21-4) **BP**

7. Draw and label a block diagram of a press-to-talk voice-modulated PM transmitter with a crystal multiplication of 12. (21-5) **BP**

8. What types of microphones are used in communications transmitters? (21-5) **PG**

9. Explain how to tune and align an FM VHF mobile transmitter. (21-5) **P**

10. What might be the effect on the transmitted frequency if a tripler stage is tuned slightly high in frequency? (21-5) **BP**

11. Explain carrier- and tone-actuated squelch. (21-6, 21-7) **PA**

12. What is CTCSS, how is it used, and how does it affect frequency deviation? (21-6) **BP**

13. What is meant by selective calling? (21-6) **P**

14. How is subaudible tone squelch defeated to assure channel monitoring before transmitting? (21-6) **P**

15. What range of frequencies is used for subaudible tones? (21-6) **PE**

16. For what purpose are helical resonators used? (21-7) **P**

17. What may be the effects of shielding applied to RF inductances? (21-7) **PA**

18. Explain mobile relay stations (repeaters) and their timers. (21-10) **PG**

19. What can cause the desensitizing of a receiver? (21-10) **PA**

20. Why is vertical separation better than an equal distance of horizontal separation to reduce receiver desensitization in land-mobile antenna operation? (21-10) **BP**

21. Discuss the merits of RF cavities as opposed to *LC* filters for filtering between transmitter output and receiver input of a repeater. (21-10) **BP**

22. What are the general requirements for transmitting the identification announcements for stations in the mobile service? (21-10) **BP**

23. How is receiver sensitivity measured? (21-11) **PE**

24. Explain how to make SINAD measurements. (21-11) **P**

25. What transmitter measurements are required by the FCC for stations in the land mobile services? (21-13) **BP**

26. How are transmitter frequency, power, and modulation checks made? (21-13) **P**

27. What equipment is necessary for installation and maintenance of an FM mobile transmitter? (21-13) **T**

28. Would whistling or speaking produce the greater deviation of a Safety Service FM transmitter? (21-13) **BP**

29. Under what usual conditions of maintenance and/or repair should a transmitter be retuned? (21-13) **BP**

30. What might cause nonsymmetrical deviation of FM signals? (21-13) **BP**

31. What can be used to measure spurious signal output? (21-13) **P**

32. What problem is encountered when measuring the frequency of an FM carrier containing a CTCSS tone? (21-13) **BP**

33. What is a communications monitor? (21-14) **P**

34. Describe microwave installations used in VHF and UHF communication services. (21-16) **P**

35. For what are cavity-type duplexers used? (21-16) **P**

36. Explain RF and remote wireline control links. (21-16) **P**

37. Discuss the uses of directional antennas in eliminating land mobile cochannel interference problems. (21-16) **BP**

38. What are some examples of mixing or translation? (21-16) **AP**

39. Discuss baseband modulation of a microwave transmitter. (21-16) **BP**

40. What are trunked communication systems? (21-17) **P**

41. Who may adjust transmitting apparatus? (21-18) **BP**

42. In general, what are an operator's responsibilities in the land mobile radio service? (21-18) **BP**

43. What kind of service and maintenance records must be kept by a radio technician or operator? (21-18) **BP**

44. What are the FCC requirements regarding the records required to be kept by stations in the land mobile service? (21-18) **BP**

45. How long should radio station logs be kept? (21-18) **BP**

46. When an operator makes FCC required measurements, what information should be written into

the log? (21-18) **BP**

47. In what form should radio station records be kept? (21-18) **BP**

48. How can unauthorized access to radio equipment be prevented? (21-18) **PN**

49. Explain communication equipment installation safety precautions. (21-18) **P**

50. What legal requirements must be met before installing and operating a radio station on the air? (21-19) **BP**

51. What radio station authorizations are required? (21-19) **BP**

52. If the license for a radio system has been applied for but not yet been granted, may a licensed installer make short tests on the air with the transmitter? (21-19) **BP**

53. What is the difference between type approval and type acceptance? (21-19) **BP**

54. What types of changes in authorized stations must be approved by the FCC and what types do not require FCC approval? (21-19) **BP**

55. Discuss the various causes and means of suppressing vehicle noises. (21-20) **PE**

MICROWAVES

1. What is a lighthouse tube and for what frequency range was it designed? (22-1, 9-27) **PR**

2. Describe the construction and purpose of a waveguide. (22-2) **PR**

3. Why are rectangular waveguides used in preference to round ones? (22-2) **R**

4. Why are waveguides used in preference to coaxial lines for the transmission of microwave energy in most SHF radar installations? (22-2) **PR**

5. Discuss with respect to waveguides: relation between frequency and size; modes of operation; coupling of energy into the waveguide; general principles of operation. (22-2, 22-4) **BPR**

6. What are waveguides, and in what type of radio circuits do they find application? (22-2, 22-16) **BPT**

7. Why are choke joints often used in preference to flange joints to join sections of waveguides? (22-2) **R**

8. Draw and explain a longitudinal section of a waveguide choke joint, indicating the $\lambda/4$ slot dimensions. (22-2) **R**

9. What is stripline, and for what is it used? (22-2) **P**

10. Discuss the use and operation of the reflectometer. (22-3) **PG**

11. Discuss methods of coupling to and from waveguides. (22-3, 22-4) **BP**

12. What are cavity resonators and in what type of radio circuits do they find application? (22-6, 22-7, 22-10, 22-11, 22-15) **RT**

13. Describe the physical structure of a klystron tube. (22-7) **BPR**

14. Explain the principle of operation of a klystron. (22-7) **BPR**

15. What determines the operating frequency of a klystron oscillator? (22-7, 22-8) **RT**

16. In what part of the spectrum are magnetron tubes used as oscillators? (22-8) **BP**

17. What determines the operating frequency of a magnetron? (22-8) **RT**

18. Draw a cross-sectional diagram of the magnetron and show the anode, cathode, and direction of electron flow under the influence of a magnetic field. (22-8) **R**

19. Describe the physical structure of the multicavity magnetron and explain how it operates. (22-8) **BPR**

20. Diagram and explain the construction and principles of operation of a traveling-wave tube. (22-9) **BP**

21. What is a microwave circulator and how is it used? (22-12) **BP**

22. What is a microwave isolator and how is it used? (22-12) **BP**

23. Diagram and describe operation of a varactor-diode multiplier. (22-10, 22-15) **BP**

24. What is the function of a Gunn diode?? (22-10) **BP**

25. Discuss some methods of modulation of the microwave baseband which allow a number of channels to be transmitted. (22-15) **BP**

26. What precautions should be taken in the installation and maintenance of a waveguide to ensure proper operation? (22-17) **BPR**

27. Why are long exterior horizontal waveguide runs not desirable? (22-17) **BPR**
28. Why should the interior of a waveguide be clean, smooth, and dry? (22-17) **R**
29. Why is a small hole sometimes drilled on the underside of a waveguide elbow near the point where the waveguide enters a radar transmitter? (22-17) **R**
30. What precautions should be taken when installing vertical sections of waveguides with choke-coupling flanges to prevent moisture from entering the waveguide? (22-17) **BPR**
31. How are optical fibers used in communications? (22-19) **BP**
32. Describe a simple laser transmitter. (22-19) **BP**
33. What are satellite communications? (22-20) **BP**
34. Discuss parametric amplifiers used in microwaves. (22-21) **BP**

BROADCAST STATIONS

1. What is the meaning of a broadcast band? channel? (23-1) **B**
2. What is the frequency tolerance of a standard AM broadcast station? (23-1) **B**
3. An AM BC station is licensed for 1260 kHz. What are the minimum and maximum frequencies its carrier may operate on, according to FCC rules? (23-1) **B**
4. What type of antenna site is best for an AM BC station? (23-1) **B**
5. What is the frequency tolerance of FM BC stations? (23-1) **B**
6. What type of antenna site is best for an FM BC station? (23-1) **B**
7. Explain why high-gain antennas are used at FM BC stations. (23-1) **B**
8. What is the meaning of erp? (23-1) **BP**
9. How wide a frequency band must the IF amplifiers of an FM BC receiver pass? (23-1) **B**
10. What is the frequency tolerance of an international BC station? (23-1) **B**

11. Why are crystal ovens often left on all night even though the station is not on the air? (23-1) **BP**
12. What do the following terms mean as related to broadcast stations: daytime; nighttime; broadcast day; primary service area; experimental period. (23-1) **B**
13. Block diagram and explain briefly a complete broadcast station. (23-2) **B**
14. What is an STL system? (23-2) **B**
15. What is the basic difference between STL and intercity relay stations? (23-2) **B**
16. What is the frequency tolerance of STL and intercity relay stations? (23-2) **B**
17. What type of antenna must be used with STL and intercity relay stations? (23-2) **B**
18. What is the uppermost power limit for remote pickup BC stations? STL stations? Intercity relay stations? (23-2) **B**
19. What is an auxiliary broadcast transmitter and how frequently must it be tested? (23-2) **B**
20. What are preamplifiers and where are they used in BC stations? (23-3, 23-4) **BP**
21. Who may operate the audio console, turntables, and tape machines in a BC station? (23-3) **B**
22. How are VU and dBm related? (23-4) **BP**
23. A preamplifier with a −30-dBm output is connected through a mixer with 20-dB loss to a line amplifier. What must be the dB gain of the line amplifier to feed a +6 dBm to a 600-Ω line? (23-4) **B**
24. If an amplifier has an overall gain of 50 dB and the output is 1 W, what is the input power? (23-4) **BP**
25. Why should impedance be matched in speech-input equipment? (23-5) **BP**
26. What are the purposes of H- and T-pad attenuators? (23-5) **BP**
27. Why are electrostatic shields used between windings in BC line transformers? (23-5) **BP**
28. Why are grounded-center-tap transformers frequently used to terminate program wire lines? (23-5) **BP**
29. What are line equalizers, why are they used, and where in the line are they placed? (23-6) **BP**
30. Why are limiting amplifiers used in BC stations rather than peak-clippers? (23-7) **B**
31. Where are limiting amplifiers used in BC stations? (23-7) **B**

32. What is meant by authorized power? (23-8) **BP**
33. What methods are used to determine operating power at FM BC stations? (23-8) **B**
34. Explain the direct method of calculating operating power of BC stations and give an example. (23-8) **BP**
35. Explain the indirect method of calculating operating power of BC stations and give an example. (23-8) **BP**
36. How is operating power determined if the antenna ammeter at a BC station becomes defective? (23-8) **BP**
37. What are the permissible power tolerances of a broadcast station? (23-8) **B**
38. Once a BC station has received a construction permit, are there additional requirements before commencing equipment tests? (23-9) **B**
39. During what time period may a BC station transmit signals for testing and maintenance purposes? (23-9) **B**
40. May a BC station licensed "daytime only" operate during the experimental period without specific authorization? (23-9) **B**
41. What percent modulation capability is required of a BC station? (23-9) **B**
42. When a BC station is operated at 85% modulation, what is the maximum permissible combined audio harmonic output? (23-9) **B**
43. What AF range must an FM BC station be capable of transmitting? (23-9) **B**
44. What is the time of day to measure the field strength of a directional AM BC station at the nighttime pattern's monitoring points? (23-9) **B**
45. What equipment performance measurements must be made at all BC stations on an annual basis? (23-9) **B**
46. During what period preceding the date of filing for a renewal of the station license should equipment performance measurements be made? (23-9) **B**
47. How does a proof-of-performance test differ from the annual equipment performance measurements? (23-9) **B**
48. What must be included in the annual equipment performance measurements of BC stations? (23-9) **B**
49. What is the required accuracy of the plate ammeter and voltmeter of the last stage of a BC station? (23-10) **B**
50. What is the required accuracy of the transmission line ammeter in a BC station? (23-10) **B**
51. If the plate ammeter in the last stage of a BC transmitter burns out, what should be done? (23-10) **B**
52. Why do AM BC station frequency monitors receive their energy from an unmodulated stage of the transmitter? (23-11) **B**
53. What should an operator do if the frequency or modulation monitor becomes defective? (23-11, 23-12) **B**
54. What is the FCC's requirement for maintenance of percent of modulation for AM stations? (23-12) **BP**
55. What is the required frequency range of the indicating device on the frequency monitor of an FM BC station? (23-12) **B**
56. Explain the operation of an FM frequency-modulation monitor. (23-12) **B**
57. What meter in an FM BC station is a deviation meter? (23-12) **B**
58. What is the maximum permissible percent of modulation for FM BC stations? (23-12) **B**
59. What is the frequency swing if an FM BC transmitter is modulated 90% by a 1-kHz sine-wave tone? (23-12) **B**
60. If an FM transmitter without modulation was within frequency tolerance, but with modulation is out of tolerance, what might be some possible causes? (23-12) **BP**
61. If the transmission-line meter in an FM BC station reads 105% of the licensed value under 100% modulation by a 1-kHz tone, what will it read with no modulation? (23-12) **B**
62. What are FCC requirements for remote control of BC stations? (23-13) **B**
63. What should an operator do if remote-control devices malfunction? (23-13) **BP**
64. What two factors can cause loss of high frequencies in tape recordings? (23-15) **B**
65. How does dirt on the playback head of a tape recorder affect the audio output? How are such heads cleaned? (23-15) **B**
66. What are turntable "wow" and "rumble" and how are they prevented? (23-16) **B**
67. What type of playback stylus is used in BC station turntables and why? (23-16) **B**
68. Explain the use of a stroboscope disk in checking turntable speed. (23-16) **B**

69. How is the frequency response of a pickup unit of turntables and tape recorders tested? (23-16) **B**

70. Define the following: EBS; Emergency Action Notification; Emergency Action Termination; Emergency Broadcast System plan. (23-19) **B**

71. What provisions must all AM, FM, and TV BC stations make for receiving EANs and EATs? (23-19) **B**

72. Describe the EAN attention signal. (23-19) **B**

73. What type of station ID shall be given during an EAC? (23-19) **B**

74. What notice shall be given by a BC station when operating during a local emergency? (23-19) **B**

75. What is the EBS checklist and where should it be located? (23-19) **B**

76. How often and at what times must EBS tests be made? (23-19) **B**

77. When is suspension of transmissions required? (23-19) **B**

78. How many times and when must the station's operating log be signed by an operator who goes on duty at 10 A.M. and off duty at 6 P.M.? (23-20) **B**

79. Should minor corrections to a transmitter be made before or after logging the meter readings? (23-20) **BP**

80. How and by whom may station logs be corrected? (23-20) **BP**

81. When may key letters or abbreviations be used in a log? (23-20) **BP**

82. What should be included in the maintenance log of a BC station? (23-20) **B**

83. What entries shall be made in a BC station's operating log? (23-20) **B**

84. How long must a BC station keep its operating logs? (23-20) **B**

85. What records of operation must be maintained for each licensed remote pickup BC station? (23-20) **B**

86. What is meant by fidelity of an audio amplifier and why is it important in a BC station? (23-21) **B**

87. How wide is the FM stereophonic subchannel? (23-21) **B**

88. Block diagram and explain a complete stereo multiplex FM BC station. (23-22) **B**

89. Explain the following stereo multiplex terms: cross talk; left stereo channel; main channel; pilot subcarrier; stereo separation; stereo subcarrier; stereo subchannel. (23-21, 23-22) **B**

90. What is the frequency and tolerance of the pilot subcarrier in an FM stereo broadcast? (23-22) **B**

91. With what signals is the main carrier modulated in an FM stereo transmission? (23-22) **B**

92. What is SCA and what are some possible uses of it? (23-24) **B**

93. What are the transmission standards of SCA multiplex operations? (23-24) **B**

94. Under what two general conditions must antenna structures be painted and lighted? (23-25) **BP**

95. What colors must antenna structures be painted and where can paint samples be obtained? (23-25) **B**

96. If a station's tower lights are not continually monitored by an alarm device, how often should they be visually checked? (23-25) **BP**

97. If required tower lights go out at night and cannot be turned on again, who must the operator notify? (23-25) **BP**

98. What items regarding operation of tower lighting should be entered in the maintenance log? (23-25) **BP**

99. How soon after a defect in tower lights is noted should the defect be corrected? (23-25) **BP**

100. If a flashing beacon at an intermediate level on an antenna tower burns out, what should be done? (23-25) **BP**

101. What action should be taken if the tower lights at a station malfunction and cannot be immediately repaired? (23-25) **BPG**

102. If tower lights are required and are controlled by a light-sensitive device but the device malfunctions, when should the tower lights be on? (23-25) **BP**

103. Light-sensitive devices used to control tower lights should face which direction? (23-25) **BP**

104. How often should control and alarm circuits associated with antenna-tower lights be checked for proper orientation? (23-25) **BP**

105. What inspections and records must be kept of tower lights and associated equipment? (23-25) **BP**

106. How often should antenna towers be painted? (23-25) **BP**

107. Is it necessary to have available replacement lamps for antenna tower lights? (23-25) **BP**

108. Who keeps the keys to the fence which surrounds the antenna base at a BC station and where are they kept? (23-25) **BP**

109. What information must be made available to authorized FCC employees? (23-26) **BP**

110. At what place must a BC station license be posted? (23-26) **BP**

111. When and how often must BC stations identify themselves and with what information? (23-26) **B**

112. What changes to a BC transmitter require FCC approval and what changes do not? (23-26) **BP**

113. Under what conditions may a standard BC station use its facilities for communication directly with individuals or other stations? (23-26) **B**

114. What hours may be included in a Presunrise Service Authority? (23-26) **B**

TELEVISION

1. What are the two emission designations of a TV transmission? (24-1) **VPE**

2. What kind of antenna site is technically best for a TV station? (24-1) **V**

3. Does the video transmitter at a TV BC station employ FM or AM? Which does the aural transmitter employ? (24-1) **V**

4. Why is grid modulation more desirable than plate in TV video transmitter power amplifiers? (24-1) **V**

5. Block diagram and explain a complete TV broadcast station. (24-1) **V**

6. What AF range is required of the aural transmitter of a TV BC station? (24-1) **V**

7. Where in TV transmitters are sawtooth waves employed and for what reasons? (24-1, 24-4, 24-6) **V**

8. Define the following: monochrome transmission; negative transmission; aural transmitter; video transmitter; frequency swing; vestigial sideband transmission. (24-1) **V**

9. What is meant by 100% modulation of the aural transmission of a TV transmitter? (24-1) **V**

10. If the aural transmitter of a TV station is modulated 80% by a 5-kHz sine wave, what is the frequency swing of the aural carrier? (24-1) **V**

11. What emission type is known as A5? (24-1) **VA**

12. What polarization is generally used for the aural portion of TV signals? The visual portion? (24-2) **V**

13. Explain a turnstile antenna. (24-2) **V**

14. Why is a diplexer a necessary part of most TV transmitters? (24-2) **VP**

15. Diagram a typical bridge-type diplexer. (24-2) **VP**

16. What is a reflectometer or directional coupler and how is it used in TV transmission systems? (24-2) **VG**

17. Explain the use of a directional wattmeter in measuring VSWR and transmitter power. (24-2) **VP**

18. How is the deflection direction of an electron beam determined when it is moving through a magnetic field? (24-3) **V**

19. Define and explain: aspect ratio; field; field frequency; frame; frame frequency; interlaced scanning; scanning line. (24-4) **V**

20. What is a mosaic plate in a TV camera tube? (24-5) **V**

21. What is an image orthicon camera tube? (24-5) **V**

22. Explain the operation of vidicon-type camera tubes. (24-5) **VE**

23. Why did vidicons supercede image-orthicons? (24-5) **V**

24. What is the purpose of synchronizing pulses in a TV broadcast signal? (24-6) **V**

25. Sketch the equalizing, blanking, and synchronizing pulses of a TV transmission and indicate where the line writing appears. (24-6) **V**

26. What is meant by Reference black level? Reference white level? Blanking level? (24-7) **V**

27. What percent of the sync peak is the blanking level? (24-7) **V**

28. What is a monitor picture tube at a TV BC station? (24-8) **V**

29. What is the frame frequency of motion pictures and of TV? (24-9) **V**

30. What is the gain per octave of any tape recording? (24-10) **V**

31. Do VTRs use FM or AM recorded tracks? (24-10) **V**

32. How many recording heads on a 2-inch tape recorder? On a helical-scan recorder? (24-10) **V**

33. Describe the tracks recorded on 2-inch tapes and on helical-scan 1-inch tapes. (24-10) **V**

34. Describe how to couple a VHF visual transmitter to its load circuit. (24-11) **V**

35. Define or explain: standard TV signal; aural center frequency; peak power; TV transmission standards. (24-12) **V**

36. Explain vestigial sideband transmissions. (24-12) **V**

37. For the 82-88 MHz channel, what is the video carrier frequency? Aural? (24-12) **V**

38. How is operating power determined for a TV visual transmitter? Aural? (24-12) **V**

39. Within what limits is the operating power of a TV aural or visual transmitter required to be maintained? (24-12) **V**

40. Under what condition should the meters of a TV visual transmitter be read to determine operating power? (24-12) **V**

41. What are the frequency tolerances for TV transmissions? (24-12) **V**

42. What is meant by TV BC band? Channel? (24-12) **V**

43. What is meant by a TV frequency offset and why is it done? (24-12) **V**

44. A 794-800 MHz TV channel station has a +10-kHz offset. What is the aural carrier frequency? (24-12) **V**

45. What items must be included in a TV station's operating and maintenance logs? (24-13) **V**

46. What are the operator requirements for TV stations? (24-13) **V**

47. Explain some common circuits used in the front end of a TV receiver. (24-14) **V**

48. Sketch the amplitude characteristics of an idealized monochrome TV transmission. (24-15) **V**

49. What happens if the front end of a TV receiver is overloaded by a nearby 27-MHz CB transmitter? (24-15) **VP**

50. Why is interference a problem and what are some solutions when a 40-MHz-band transmitter is operated near a TV receiver? (24-15) **VP**

51. Describe the output signals from the 2nd detector of a TV receiver. (24-16) **V**

52. Which sync pulse separator circuit uses a differentiator circuit? An integrator circuit? Why? (24-17) **V**

53. Explain the input, contents, and output of a vertical deflection circuit. (24-18) **V**

54. Explain the input, contents, and output of a horizontal deflection circuit. (24-19) **V**

55. Describe what the various outputs from the horizontal deflection transformer are used for. (24-20) **V**

56. What are AGC amplifiers and why are they used? (24-21) **V**

57. Draw a block diagram of the color portion of a color-TV BC transmitter and explain basic operation of the circuits. (24-22) **V**

58. Describe the composition of the chrominance subcarrier and sidebands used in TV transmissions. (24-22) **V**

59. What is the purpose of the color burst? (24-23) **V**

60. Where on the synchronizing waveform is the color burst? (24-23) **V**

61. What is the frequency of the chrominance subcarrier in relation to the video carrier? (24-23) **V**

62. Sketch the idealized color-TV transmission in one channel. Label aural and visual carriers and color subcarrier. (24-23) **V**

63. What is meant by the luminance portion of a TV emission? (24-23) **V**

64. Explain how video signals from a video detector can feed either monochrome or color signals to the RBG resistor matrix. (24-24) **V**

65. How do scanning circuits for color receivers differ from monochrome? (24-24) **V**

66. What is a comb filter and for what can it be used? (24-24) **V**

67. Explain the operation of a shadow-mask color tube. (24-25) **V**

68. Explain the operation of a metal grille color tube. (24-26) **V**

69. Where in a color TV receiver would an AFT circuit be found? A degaussing coil? A 3.48-MHz crystal? An APC detector? (24-27) **V**

70. What type of TV reception always requires an unscrambler? What type sometimes does? (24-28) **V**

71. What TV channels are used for pay TV? For cable TV? (24-28) **VP**
72. What may happen if a cable TV cable oxidizes at a connection? (24-28) **VP**
73. Describe a basic slow-scan transmitting and receiving system. (24-30) **PA**
74. In what range of frequencies do TV satellites operate? (24-29) **V**
75. Is the IF of satellite TV signals higher or lower than the broadcast frequency? (24-29) **V**
76. What type of modulation is used for the video portion of satellite TV broadcasts? (24-29) **V**

MARINE RADIO

1. What amount of power should be employed for routine marine radio communications? (25-1) **T**
2. What is indicated if a voltmeter between the negative side of a ship's dc line and ground reads the full line voltage? (25-1) **T**
3. With what type of emission and on what frequency should a radiotelegraph transmitter transmit a distress call? (25-2, 25-4) **T**
4. For how long should the emergency power supply of a ship station be capable of energizing the emergency radiotelegraph installation? (25-2, 25-5) **T**
5. Between what points on a ship is a reliable intercom system required? (25-2) **T**
6. Where are compulsory radiotelephone installations located? What emission do they use? What frequencies? (25-3) **T**
7. What is the formula to determine the power output of the main CW transmitter? (25-4) **T**
8. How is the power output of a marine radiotelegraph transmitter ordinarily adjusted? (25-4) **T**
9. Why might marine transmitters employ variometers rather than variable capacitors as the tuning elements? (25-4) **T**
10. Why are iron-compound cylinders used in the inductances of certain marine radiotelegraph transmitters? (25-4) **T**
11. In what band of frequencies does a reserve transmitter operate? What emission is used? What power output? (25-5) **T**

12. Sketch a shipboard antenna for 500 kHz showing the supporting insulators, safety link, and lead-in. (25-6) **T**
13. At what point(s) on a shipboard antenna system will the maximum potential be found? (25-6) **T**
14. What care should be taken in hoisting an antenna to avoid damage to the antenna wire and insulators? (25-6) **T**
15. What is the approximate feed-point impedance of a top-loaded MF vertical antenna? (25-6) **T**
16. What makes long range MF communications difficult at night? (25-7) **T**
17. What does an APC circuit do in a HF ship antenna? (25-8) **T**
18. What are the radiotelegraph bands in the MF region? In the HF region? (25-9) **T**
19. What are the frequency bands for marine radiotelephone calling and working? (25-10) **PT**
20. What bands of frequencies must a marine main receiver cover? (25-11) **T**
21. If broadcast signals interfered with reception of signals on 500 kHz how might they be reduced or eliminated? (25-12) **T**
22. What does a "4th generation" superheterodyne receiver have that is unique to it? (25-12) **T**
23. In what frequency band do bridge-to-bridge radiotelephones operate? With what powers? (25-13) **P**
24. List the frequencies used in aircraft and EPIRB distress and search-and-rescue operations. (25-14) **P**
25. What is the purpose of an auto-alarm signal keying device on a ship? (25-15) **T**
26. Describe the dashes and spaces which compose the international AA signal. (25-16) **T**
27. How is the ship's transmitter prevented from operating when the AA receiver is in use? (25-16, 25-17) **T**
28. What factors determine the setting of the sensitivity control of older AA receivers? (25-16) **T**
29. To what band of freqencies is an AA receiver tuned? (25-16) **T**
30. What signal causes an approved AA receiver to ring the warning bell? (25-16) **T**
31. With an AA receiver, what factors cause: the bell to sound; the warning light to operate; the bells to ring intermittently? (25-16) **T**
32. When the AA bell rings, what should the operator do? (25-16) **T**

33. Where is the control button located which silences the AA warning bells? (25-16) **T**
34. If the release button stops the AA bell ringing, what could be the cause? (25-16) **T**
35. If the release button does not stop the AA bell, what could be the causes? (25-16) **T**
36. If a VT heater burns out in an AA, what causes the warning bells to ring? (25-16) **T**
37. While in port, how frequently should the emergency equipment be tested? (25-16) **T**
38. When a vessel is at sea, how frequently must the AA be tested? The emergency battery? (25-16) **T**
39. How frequently must emergency generator fuel tanks be checked at sea? (25-16) **T**
40. What is the purpose of an auxiliary antenna installed on a vessel having a DF? (25-17) **T**
41. What types of emissions are used with teleprinters? (25-18) **TG**
42. Explain regarding teleprinters: AF shift keying; mark; space; shift. (25-18) **TG**
43. Why do SITOR communications have few errors? (25-19) **T**
44. How frequently should ship radiotelegraph log entries be made at sea? (25-20) **T**
45. What action may an operator take if it is observed that a ship station is flagrantly violating radio regulations and causing harmful interference? (25-20) **T**
46. How frequently must the radio room clock be adjusted and compared with standard time? (25-20) **T**
47. Why must a radio room clock aboard ship be required to have a sweep second hand? (25-20) **T**
48. Under what conditions may a ship station close if its service is not required to be continuous? (25-20) **T**
49. When might a ship station not be required to notify its coast station of its closing and reopening? (25-20) **T**
50. What age requirement must a person meet before being issued a radiotelegraph operator license? (25-20) **T**
51. What experience under a radiotelegraph license is required before one's acting as chief or sole operator is permitted on a cargo ship? (25-20) **T**
52. What is the principal United States port at which navigation lines terminate on the Pacific Coast? Atlantic Coast? (25-20) **T**
53. To what continent do the greatest number of telecommunication channels from the United States extend? (25-20) **T**
54. What city is the major telecommunication center of the United States? (25-20) **T**
55. What is the GMT time and the day of the week in Shanghai when it is Wednesday noon in New York City? (25-21) **T**
56. What is the approximate latitude and longitude of Colón, Panama? (25-21) **T**
57. In what ocean is Guam located? (25-21) **T**
58. What is indicated by the bearing obtained by the use of a bilateral RDF? (25-22) **TE**
59. What is the directional reception pattern of a loop antenna? (25-22) **TE**
60. Why are RDF loops shielded? (25-22) **T**
61. Describe the construction and operation of a shielded RDF loop. (25-22) **TE**
62. Discuss direction-finding techniques and methods for locating a source of radio signals. (25-22, 25-23, 25-24, 25-32) **PTE**
63. How many maximum signals will be received when a unilateral RDF loop is rotated 360°? (25-23) **TE**
64. What is the function of the balancing capacitor in a DF system? (25-29) · **TE**
65. What is the principal function of a vertical antenna associated with a bilateral RDF? A unilateral RDF? (25-23, 25-24) **TE**
66. What is indicated by the bearing obtained by the use of a unilateral RDF? (25-24) **TE**
67. How is a unilateral effect obtained with a DF? (25-24) **TE**
68. What figure represents the reception pattern of a properly adjusted unilateral RDF? (25-24) **TE**
69. What factors may affect the accuracy of a DF that has been properly installed, calibrated, and compensated? (25-25, 25-26) **T**
70. In what frequency-band do United States marine radio-beacon stations operate? (25-26) **T**
71. What is the purpose of an auxiliary antenna installed on a ship having a DF? (25-26) **T**
72. What is a compensator that is used with RDFs? What is its purpose? (25-26) **T**
73. What type of DF systems use two nonrotatable loops? (25-27) **TE**
74. Besides batteries, what maintenance may be required of RDFs? (25-29) **T**
75. Explain how an ADF operates. (25-30) **T**
76. Explain the use of Adcock antennas for use on

HF bands. (25-31)   **TE**

77. Sketch how a fix on a ship station can be obtained by taking DF bearings. (25-32)   **TE**
78. During daytime, for approximately what maximum distance from loran A and C stations are lines of position accurate? (25-33, 25-34)   **T**
79. What is the purpose of blinking in a loran system and how can it be recognized at the receiver? (25-33, 25-34)   **T**
80. What are the pulse characteristics and the operating frequency used in loran A and C systems? (25-33, 25-34)   **T**
81. How are loran A stations identified by an operator? loran C? (25-33, 25-34)   **T**
82. What is the relationship of loran master and slave stations? (25-33, 25-34)   **T**
83. Sketch relative positions of master-slave pairs of a loran system and indicate LOPs of each pair. (25-33)   **T**
84. To obtain a fix of a ship's position how many loran stations are required? (25-33)   **T**
85. What precautions should be observed when servicing loran receivers? (25-34)   **T**
86. What is measured with omega navigation to determine position location? (25-35)   **T**
87. How are ship-to-ship communications made via satellite? (25-36)   **T**
88. What is the frequency of transmission of NNSS satellites? In what orbit are NNSS satellites? How is a LOP obtained? (25-36)   **T**
89. What is meant by Doppler effect in relation to satellite navigation? (25-36)   **T**
90. Discuss the operation of facsimile transmission and reception. (25-37, 25-38)   **TA**
91. What types of VHF marine equipment are used in smaller vessels and what are the requirements for installation? (25-39)   **P**

RADAR

1. In what part of the RF spectrum do marine radar systems operate? (26-1)   **RT**

2. Explain briefly the principle of operation of a radar system. (26-1, 26-2)   **R**
3. What is the emission designation for radar transmissions? (26-1)   **BPT**
4. Within what frequency bands do ship radar transmitters operate? (26-1)   **R**
5. How many nautical miles does a radar pulse travel in 1 μs? (26-1)   **RT**
6. What is the distance to a target if it takes 246 μs for a radar pulse to travel to and from the target and be displayed? (26-1)   **R**
7. In what radio circuits do magnetron oscillators find application? (26-2, 26-3)   **RT**
8. At what approximate speed does the antenna of a marine radar rotate? (26-2, 26-16)   **RT**
9. Draw a block diagram of a radar system and label the antenna, duplexer, transmitter, receiver, modulator/timer, and indicator. (26-2)   **RT**
10. Explain the principle of operation of the cathode ray PPI tubes used in radar. (26-2)   **R**
11. What is meant by bearing resolution? Range resolution? (26-2)   **R**
12. What circuit in a radar set determines PRR? (26-3)   **R**
13. What circuit elements determine the operating frequency of a self-blocking oscillator? (26-3) **R**
14. What is the purpose of an artificial transmission line? (26-3)   **R**
15. Draw a simple diagram of an artificial transmission line showing inductance, capacitance, source of power, load, and electronic switch. (26-3)   **R**
16. Why is the anode of a radar magnetron normally maintained at ground potential? (26-3)   **R**
17. What was the purpose of the rotary spark gap that was used in older radar sets? (26-3)   **R**
18. What is the average plate-power input to a radar transmitter if the peak pulse is 25 kW, the pulse length is 1 μs, and the PRR is 1000 pps? (26-4)   **RT**
19. What is the peak power of a radar pulse if the pulse width is 1 μs, PRR is 800, and P_{avg} is 40 W? What is the duty cycle? (26-4)   **R**
20. Describe how a radar beam is formed by a parabolic reflector. (26-5)   **R**
21. How are waveguides terminated at the radar antenna reflector? (26-5)   **R**
22. Why is a microwave antenna dish usually fed with waveguide instead of coaxial cable? (26-5)   **BP**

23. Why does heavy rain or snow sometimes disrupt a microwave signal? (26-5)   **BPT**

24. What is the best location of the radar antenna assembly aboard ship? (26-5)   **R**

25. What effect might soot or dirt on the antenna reflector have on the operation of a ship radar? (26-5)   **R**

26. Is there any danger in testing or operating radar equipment aboard ship when explosive or inflammable cargo is being handled? (26-5)   **R**

27. Draw a simple block diagram of a radar duplexer system and label waveguide, TR box, ATR box, receiver, and transmitter. (26-5)   **R**

28. Describe briefly the construction and operation of radar TR and ATR boxes. What is the purpose of a keep-alive voltage? (26-6)   **R**

29. What adjustment is made to a radar set by the operator to reduce sea return? (26-7)   **RT**

30. What is the purpose of a klystron in a radar set? (26-7)   **R**

31. Draw a simple block diagram of a radar receiver and label the signal crystal, local oscillator, AFC, IF amplifier, and discriminator. (26-7) **R**

32. What is a nominal IF commonly found in radar receivers? (26-7)   **R**

33. What type of detector is used in radar receivers? (26-7)   **R**

34. What is a purpose of a discriminator in a radar set? (26-7)   **R**

35. What is sea return on a radarscope? (26-7)   **R**

36. Explain the purpose of the STC circuit in a radar set. (26-7)   **R**

37. For what purpose is a one-shot multivibrator used in a radar set? (26-8)   **R**

38. Describe how heading flash and range-marker circles are produced on a radar PPI scope. (26-8, 26-10)   **R**

39. Diagram how a synchrogenerator located in a radar antenna assembly is connected to a synchromotor located in the indicator to drive the deflection coils. Label leads and voltages. (26-9)   **R**

40. What is an echo box, its purpose, principle of operation, and indications given by it when the set is operating correctly and incorrectly? (26-11)   **R**

41. Who may operate a ship radar station? (26-12) **R**

42. Under what conditions may a person who does not hold a radio license operate a ship radar station? (26-12)   **R**

43. What entries are required in the installation and maintenance record of a ship radar station? (26-12)   **R**

44. Who has the responsibility for making and who may make entries in the installation and maintenance record of a ship radar station? (26-12) **R**

45. May fuses and receiving-type tubes be replaced in ship radar equipment by a person whose operator license does not have a ship radar endorsement? (26-12)   **R**

46. What are the FCC license requirements for the operator who is responsible for the installation, servicing, and maintenance of ship radar equipment? (26-12)   **R**

47. Describe how various types of interference from a radar installation sound on a radio communications receiver. (26-13)   **R**

48. Why is radar interference to a radio receiver a steady tone? (26-13)   **R**

49. Why is it important that all units of a radar installation be thoroughly bonded to the ship's electrical ground? (26-13)   **R**

50. On what frequencies may radar interfere with communications receivers on ships? (26-13) **R**

51. In checking an RDF for radar interference, why is it a good policy to check for interference while the DF loop is being rotated? (26-13)   **R**

52. Is there likelihood of radar interference to radio receivers if long connecting lines are used in the radar set? (26-13)   **R**

53. What steps might be taken to eliminate steady-tone interference to radio receivers or loran receivers displays? (26-13)   **R**

54. How are the various types of radar interference recognized in: AA equipment; RDF equipment; loran equipment? (26-13)   **R**

55. Name types of equipment aboard ship that might suffer interference from a radar set. (26-13)   **R**

56. What precautions should a service or maintenance operator observe when replacing a CRT in a radar set? (26-13)   **R**

57. What precautions should a radar serviceman take when making repairs or adjustments to a radar set to prevent injury? (26-14)   **R**

58. What may cause bright flashing pie sections to appear on a radar PPI scope? (26-14)   **R**

59. What tests may a radar serviceman make to determine whether the radar mixer crystal is defective? (26-14)   **R**

60. In a radar set what indicates: a defective magne-

tron; a weak magnet in the magnetron; a defective crystal? (26-14) **R**

61. What care should be taken when handling silicon crystal rectifier cartridges for replacement in radar receivers? (26-14) **R**

62. What precautions should a radar serviceman take when working with or handling a magnetron to prevent weakening or damage to the magnetron? (26-14) **R**

63. How is Doppler effect used in police radar systems? (26-15) **P**

64. What type of antennas are used with small-boat radar sets? (26-16) **RT**

65. What can be added to small boats to produce better radar target signals as to their location? (26-16) **T**

66. Assuming a radar set is available, what are the three input data required in a collision-avoidance system? (26-17) **T**

MOTORS AND GENERATORS

1. What is a prime mover? A prime source? (27-1) **BPT**

2. How may the output voltage of a separately excited ac generator, at constant output frequency, be varied? (27-2) **BPT**

3. To what part of a rotating-field alternator is dc applied? (27-2) **PT**

4. What type of load may cause a rising voltage alternator output under load? (27-3) **PT**

5. How might large alternators obtain their field excitations? (27-4) **BPT**

6. What conditions must exist before two ac alternators can be operated in parallel? (27-5) **BT**

7. What is the purpose of a commutator on a dc generator? On a dc motor? (27-6, 27-15) **BPT**

8. Why are the field poles and armatures of motors and generators laminated? (27-6, 27-15) **T**

9. What determines the maximum output voltage of an externally excited dc generator? (27-7) **BT**

10. What are the principles of operation and characteristics of a series-wound dc generator? (27-8) **T**

11. Diagram and explain the principle of operation and characteristics of a dc shunt generator. (27-9, 27-12) **T**

12. Diagram and explain the principles of operation and characteristics of a dc compound generator. (27-10) **T**

13. What is the purpose of interpoles or commutating poles in a dc motor? (27-12) **T**

14. List some causes of excessive sparking at the brushes of a dc motor or generator. (27-13) **BPT**

15. If the RF chokes in the power leads of a motor-generator were shorted, what problems would result? (27-13) **BP**

16. How may RF interference caused by sparking at motor or generator brushes be minimized? (27-13) **BPT**

17. Why are the brushes of a high voltage dc machine often bypassed to ground? (27-13) **BPT**

18. Why might it be dangerous to operate a series dc motor without a load? (27-15) **T**

19. Describe the operation and characteristics of a series-wound dc motor. (27-15) **PT**

20. What determines the speed of a dc series motor? (27-15) **BPT**

21. Diagram three types of dc motors, including their starting devices. (27-15, 27-17) **T**

22. Describe the operation and characteristics of a shunt-wound dc motor. (27-16) **T**

23. If the field of a shunt-wound dc motor were opened while the machine is running under no load, what would be the probable results? (27-16) **BPT**

24. What is counter-emf in a dc motor? (27-16) **T**

25. What are the characteristics and the principle of operation of a compound-wound dc motor? How is its speed regulated? (27-19) **T**

26. Name the three basic types of ac motors. (27-19) **T**

27. Why will series dc generators operate as ac motors but shunt-types will not? (27-20) **P**

28. What determines the speed of a synchronous motor? (27-21) **BPT**

29. What determines the speed of an induction motor? (27-22) **BPT**

30. What makes a shaded-pole ac motor self-starting? (27-22) **P**

31. Why are polyphase ac motors self-starting? (27-23) **BP**
32. What are comparative advantages and disadvantages of motor-generator and transformer-rectifier power supplies? (27-24) **BPT**
33. Describe the function and operation of a dynamotor. (27-25) **BPT**
34. What may cause a motor-generator bearing to overheat? (27-27) **BPT**
35. In what units are dc motors and generators rated? Why? AC motors and generators? Why? (27-26) **BPT**
36. What may cause sparking at brushes? (27-27) **PT**
37. What materials may be used to clean the commutator of a motor or generator? (27-27) **BPT**
38. Explain the functioning of a simple magnetohydrodynamic dc generating system. (27-28) **P**

BATTERIES

1. How does a primary cell differ from a secondary? (28-1) **BPT**
2. What form of energy is stored in a lead-acid cell? (28-1) **T**
3. What precautions should be observed when storing dry-cell batteries? (28-2) **PT**
4. What is an electrolyte? (28-2) **PT**
5. Name two types of radio equipment in which an electrolyte is used. (28-2) **T**
6. What is polarization of a primary cell and how may its effect be counteracted? (28-2) **T**
7. How may a dry cell be tested to determine its condition of charge? (28-2) **PT**
8. What material is used in the electrodes of a common dry cell? (28-2) **T**
9. Sketch the construction of a lead-acid cell. (28-3) **PT**
10. What is the chemical composition of the negative plate of a lead-acid cell? The positive plate? The electrolyte? (28-3) **BPT**
11. What is the fully charged voltage of a lead-acid cell? (28-3) **PT**
12. What is the cause of heat developed in a storage cell during charge or discharge? (28-3) **PT**
13. What may cause sulfation of a lead-acid cell? (28-3) **BPT**
14. What are the effects of sulfation? (28-3) **BPT**
15. What may cause the plates of a lead-acid cell to buckle? (28-3) **PT**
16. What is the specific gravity of an electrolyte? How is it used to determine state of charge of a lead-acid cell? (28-3, 28-4, 28-5) **BPT**
17. What will result if a lead-acid cell is discharged at an excessively high current rate? (28-3) **BP**
18. How do the characteristics of nickel cadmium and lead-acid cells differ? (28-3, 28-11) **BP**
19. What is meant by specific gravity? (28-4) **PT**
20. An emergency lead-acid cell has a sp gr of 1.12. What should be done? (28-4) **PT**
21. How can it be determined if a lead-acid cell is fully charged? (28-4, 28-5) **BPT**
22. What is the effect of low temperature on the operation of a lead-acid battery? (28-4) **PT**
23. Why is low internal resistance desirable in a storage cell? (28-4) **PT**
24. What precautions should be taken when lead-acid cells are subjected to low temperatures? (28-5) **PT**
25. What may be used to neutralize a storage-cell acid electrolyte? (28-5) **PT**
26. What is the effect of local action in a lead-acid cell and how may it be decreased? (28-6) **PT**
27. Why should care be taken in the selection of water to be added to a storage cell? (28-6) **PT**
28. What should be done if the electrolyte in a lead-acid cell becomes low? (28-6, 28-7) **PT**
29. What is the only type of water to be added to a lead-acid cell? (28-6) **PT**
30. In what unit is the capacity of a battery rated? (28-7) **BPT**
31. What battery capacity is required to operate a 50-W emergency telegraph transmitter for 6 h, assuming a transmitter load of 70% of the key-down demand of 30 A, if the emergency light load is 0.7 A? (28-7) **T**
32. How may the polarity of a charging source be determined? (28-8) **T**
33. If emergency batteries are placed on charge but the overload circuit breakers will not stay closed, what is the trouble? (28-8) **T**
34. What is indicated if the voltage polarity of a cell

in a charging battery is found to be reversed? (28-8) **T**

35. Without a hydrometer, how can the condition of charge of a storage battery be determined? (28-8) **T**

36. Why does the charging rate to a storage cell being charged from a fixed-voltage source decrease as charging progresses? (28-8) **T**

37. If the charging current through a storage battery is maintained at the normal rate but its polarity is reversed, what will result? (28-8) **BPT**

38. Diagram a battery-charging circuit using a silicon diode. A tube-type diode. (28-8) **PT**

39. Why should unshielded leads from a battery to a transmitter be kept short? (28-8) **PT**

40. In a dc powered vessel, if it were impossible to keep the receiver-storage battery charged and at the same time maintain the required watch, what remedy may be found? (28-8) **T**

41. Diagram a circuit which can be used to charge one battery while a second is under load using a 115-V dc line. (28-8) **T**

42. If a storage battery shows 12.4 V on open circuit and 12.2 V when the charging switch is closed, what does this tell you? (28-8) **T**

43. What value resistor should be connected in series with a 12-V battery to charge it at a 3-A rate from a 110-V dc source? At a 1-A rate? (28-8) **T**

44. A discharged storage battery of six cells has an open-circuit voltage of 1.9 V/cell and an internal resistance of 9.2 Ω/cell. What potential is required to produce an initial charging rate of 10 A? (28-8) **T**

45. Discuss the difference in charging methods between nickel-cadmium and lead-acid cells. (28-8, 28-11) **BP**

46. Describe the care which should be given a group of storage cells to maintain them in good operating condition. (28-9) **BPT**

47. What steps may be taken to prevent corrosion of lead-acid cell terminals? (28-9) **BP**

48. Why should adequate ventilation be provided in the room housing a large number of storage cells? (28-9) **T**

49. Why should the tops of storage batteries be kept clean and free from moisture? (28-9) **T**

50. What are the differences between Edison and lead-acid batteries? (28-10) **T**

51. What is the approximate fully charged voltage of an Edison cell? (28-10) **T**

52. What is the chemical composition of the negative plate of an Edison cell? Positive plate? Electrolyte? (28-10) **T**

53. How is the condition of charge of an Edison cell determined? (28-10) **T**

54. Describe three causes of a decrease in capacity of an Edison cell. (28-10) **T**

55. Why should an Edison battery not be charged at less than the current specified by the manufacturer? (28-10) **T**

56. Describe the voltage, capacity, care, and recharging of a nicad battery. An alkaline cell battery. (28-11) **PG**

RADIO RULES AND REGULATIONS

Element 1 (Commercial)

1. What government agency inspects radio stations in the United States? (29-1) **BPTG**

2. When may an operator divulge the contents of an intercepted message? (29-1, 29-7) **BPTG**

3. Under what conditions may messages be rebroadcast? (29-1) **BPTG**

4. What are the penalties provided for violating (a) a provision of the Communications Act of 1934 and (b) a rule of the FCC? (29-1) **BPTG**

5. Where and how are FCC operator certificates (licenses) and permits obtained? (29-3) **BPTG**

6. Who may apply for an FCC operator certificate? (29-3) **BPTG**

7. If a license or permit is lost, what action must be taken by the operator? (29-3) **BPTG**

8. What is the term in years for radio operator certificates? (29-3) **BPTG**

9. When may an operator certificate be renewed? (29-3) **BPTG**

10. When a licensee qualifies for a higher grade of

FCC license or permit, what happens to the lesser-grade license? (29-3) **BPTG**
11. What messages and signals may not be transmitted? (29-4) **BPTG**
12. May an operator deliberately interfere with any radio communication or signal? (29-4, 29-7) **BPTG**
13. If licensees receive a notice of suspension of their licenses, what must they do? (29-4) **BPTG**
14. What are the grounds for suspension of operator certificates? (29-4) **BPTG**
15. If licensees are notified that they have violated an FCC rule or a provision of the Communications Act of 1934, what must they do? (29-5) **BPTG**
16. What type of communication has top priority in the mobile service? (29-7) **BPT**
17. What is meant by *harmful interference?* (29-7) **BPTG**
18. Who keeps the station logs? (29-8) **BPTG**
19. Who corrects station logs? (29-8) **BPT**
20. How may errors in the station logs be corrected? (29-8) **BPT**

RADIOTELEPHONE COMMUNICATION

Element 2 (Commercial)

1. Why is a station call sign transmitted? (30-1) **PTG**
2. In the case of a mobile radio station in distress, what station is responsible for the control of distress-message traffic? (30-3) **PT**
3. What information must be contained in distress messages? (30-3) **PT**
4. What procedure should be followed by a radio operator in sending a distress message? (30-3) **PT**
5. What is a good choice of words to be used in sending a distress message? (30-3) **PT**
6. What do *distress, safety,* and *urgency* signals indicate? (30-3) **PT**

7. What are the international distress, safety, and urgency signals? (30-5) **PT**
8. What actions should be taken by a radio operator who hears (a) a distress message, and (b) a safety message? (30-3, 30-5) **PT**
9. A radio operator is required to stand watch on an international distress frequency. When may the operator cease to listen? (30-3) **PT**
10. When must a coast station not charge for messages it is requested to handle? (30-5) **PT**
11. How should a microphone be treated when used in a noisy location? (30-6) **PTG**
12. What may happen to the received signal if an operator shouts into a microphone? (30-6) **PTG**
13. Why should an operator use well-known words and phrases? (30-6) **PTG**
14. What is meant by a phonetic alphabet? (30-7) **PTG**
15. What are the meanings of *clear, out, over, roger, words twice, repeat,* and *break?* (30-8) **PTG**
16. In regions of heavy traffic, why should an interval be left between radiotelephone calls? (30-9) **PT**
17. How long may a radio operator in the mobile service continue attempting to make contact with a station which does not answer? (30-9) **PT**
18. How often should a called station be called? (30-10) **PT**
19. When may an operator use a station without regard to certain provisions of the station license? (30-11) **PTG**
20. What should an operator do if profanity is being used on the station? (30-11) **PTG**
21. Who bears the responsibility if an operator permits an unlicensed person to speak at a station? (30-11) **PTG**
22. What precautions should be observed in testing a station on the air? (30-11) **PTG**
23. What should an operator do when a transmitter is left unattended? (30-11) **PTG**
24. Why should radio transmitters be off when signals are not being transmitted? (30-11) **PTG**
25. What is the proper way to send a test message? Why? How often? (30-11) **PTG**
26. In the mobile service, why should radiotelephone messages be as brief as possible? (30-11) **PT**
27. Why should a radio operator listen before transmitting on a shared channel? (30-11) **PTG**

28. How does a licensed operator of a station exhibit authority to operate a station? (30-11) **PTG**
29. Does the 1959 Geneva Treaty give other countries authority to inspect United States vessels? (30-11) **PT**
30. What is the importance of the frequencies 2182 kHz and 156.8 MHz? (30-12) **PT**
31. What are the requirements for keeping watch on 2182 kHz? (30-12) **PT**
32. What is the difference between *calling* and *working* frequencies? (30-12) **PT**
33. Under what circumstances may a coast station make contact with a land station by radio? (30-14) **PT**
34. What is meant by a *station open to public service*? (30-14) **PT**

RADIOTELEGRAPH COMMUNICATION

Element 5 (Commercial)

1. What is meant by "a station open to public correspondence"? (31-1) **BPT**
2. Where should the operator on duty at a manually operated radiotelegraph station post his or her license? (31-1) **T**
3. Is the holder of a Radiotelegraph Third Class operator permit authorized to make technical adjustments to radiotelegraph or radiotelephone transmitters? (31-2) **T**
4. List three classes of stations which may not be operated by the holder of a Third Class Radiotelegraph Certificate. (31-2) **T**
5. How should numbers involving a fraction be transmitted in order to avoid confusion? (31-3) **TG**
6. If a radiotelegraph operator makes an error in transmitting message text, how is this error indicated? (31-3, 31-19) **TG**
7. What radiotelegraph signal is used as a call to all stations? (31-5) **TN**
8. Describe a procedure of radiotelegraph transmission in which one station calls another. Give an example. (31-6) **TG**
9. Describe a procedure of radiotelegraph transmission in which one station answers the call of another. Give an example. (31-6) **TG**
10. If, upon being called by another station, a called station is busy with other traffic, what should the operator of the called station do? (31-6) **TG**
11. When testing a radiotelegraph transmitter, what signals are generally transmitted? (31-7) **TG**
12. What are the requirements for station identification at radiotelegraph stations in the Public Safety Radio Service? (31-8) **T**
13. Radiotelegraph code distress, urgency, or safety messages must not be sent in excess of what speed? (31-9, 31-18) **T**
14. What are the radiotelegraph distress, urgency, and safety signals? (31-10, 31-12) **T**
15. What is meant by the preamble of a radiotelegraph message, and what information is in it? (31-14) **TG**
16. In addition to a *preamble*, what parts does a radiotelegraph message contain? (31-14) **TG**
17. What is meant by a service prefix or indicator in a radiotelegraph message? (31-14) **TG**
18. What is the meaning of *word count*, or *check*, in a radiotelegraph message? (31-14) **TG**
19. At what time does the serial numbering of radiotelegraph messages begin? Does the numbering period vary in other services? (31-14) **TG**
20. Why might code words be confirmed or collated? (31-14) **T**
21. Why are code or cipher groups used in radiotelegraph messages? (31-15) **T**
22. What are purposes of service messages? (31-17) **TG**
23. Should the speed of radiotelegraph transmissions be in accordance with the desire of the transmitting or receiving operator? (30-18, 31-1) **TN**
24. How should a telegraph key be adjusted for good operation? Is it the same for slow and high speeds? (31-19) **TG**
25. How is an automatic key, or bug, properly adjusted? (31-19) **TG**
26. Why are Q signals or other procedure signals used in radiotelegraph communications? (31-21) **TG**

27. What is meant by the following: QRA, QRM, QRN, QRT, QRZ, QSA, QSV, QUM, QRL? (Appendix G) **TG**
28. If receiving conditions are bad and you desire the transmitting operator to send each word twice, what operating signal would you use? (Appendix G) **TG**
29. If the signal strength of a signal is reported on a 1 to 5 scale, what number indicates a very strong signal? (Appendix G, QSA) **TG**
30. What is meant by the operating signals: R, AS, IMI, C, BT, K, AR, DE? (Appendix H) **TN**

Element 6

31. How much power should be employed for routine communications? (31-1) **TG**
32. What are the priorities of the various types of radio communications? (31-2) **T**
33. Under what circumstances might a station be operated by an unlicensed person? (31-2) **TG**
34. In the transmission of the International Morse code, what are the relative time lengths of dashes, dots, and spaces? (31-3) **TG**
35. You intercept CQ WSV TFC QSY 408 AS. What does this message mean? (31-5) **T**
36. If called by another station and the called station is unable to proceed with accepting traffic, what should the operator of the called station do? (31-6) **TG**
37. When must the international silence periods be observed? (31-9) **T**
38. At what times are routine transmissions in the 485–515 kHz band forbidden? (31-9) **T**
39. Describe how to send a distress call. (31-10) **T**
40. What station should be in control of distress traffic? (31-10) **T**
41. What are the international radiotelegraph distress frequencies for the mobile service? (31-10) **T**
42. With what type of emission should distress calls be made? (31-10) **T**
43. What transmission should precede the sending of a distress call? (31-10) **T**
44. What space of time should elapse between the transmission of the auto-alarm signal and the distress call? (31-10) **T**
45. Upon hearing an SOS, what should an operator do? (31-10) **T**
46. After a distress call has been transmitted, every distress-traffic message should contain what symbol in the preamble? (31-10) **T**
47. Under what circumstances is a station not required to listen to distress traffic? (31-10) **T**
48. During what period is a distress message repeated following the initial transmission? (31-10) **T**
49. How long must mobile stations listen after they have heard an urgency signal? (31-11) **T**
50. What interval of time must elapse between the end of the auto-alarm signal and an urgent cyclone warning? (31-11) **T**
51. Under what circumstances and by whom may the international auto-alarm signal be transmitted to announce an urgent cyclone warning? (31-11) **T**
52. During what periods must the safety signal be transmitted? (31-12) **T**
53. Upon hearing a safety signal, what should the operator at the receiving station do? (31-12) **T**
54. Under what circumstances must log entries regarding observance of the international silence period be made? (31-13) **T**
55. What time system is used in making log entries with respect to the observance of the international silence period? (31-14) **T**
56. For how long must a station log containing entries incident to a disaster be retained? (31-13) **T**
57. Explain cable count and standard service abbreviations, and show the difference between cable count and domestic word count. (31-14, 31-16) **T**
58. Construct a plain-language radiotelegram and indicate what portions constitute the preamble, address, text, and signature. (31-14) **TG**
59. Explain the meaning and use of the following indicators: RP, PC, TR, MSG, CDE, OPS, PDH, CODH. (31-17) **T**
60. If a distress call sign is composed of five letters, what type of craft transmitted the signal? (31-20) **T**

Amateur Radio License Elements

AMATEUR RADIO

The following group of Amateur Radio license questions are only representative types, but they cover the subjects that may be asked in Amateur license tests. Actual tests are continually being compiled from the areas of information covered by the sample questions. The 97.xx references at the end of questions indicate sections of the "FCC Rules and Regulations," Part 97. Part 97 is available through the Superintendent of Documents, Washington, DC, 20402, and is also in the ARRL's "FCC Rule Book." As in previous questions, the 1-1 through 32-17 references indicate the chapters and sections in *Electronic Communication*, Fifth Edition, where the subjects are discussed.

The term "Element 1" in Amateur Radio licensing tests refers to Morse Code tests. Specifically, Element 1*a* is a 5 words-per-minute (wpm) code test, sending and receiving. Element 1*b* is a 13 wpm test. Element 1*c* is a 20 wpm test.

The Novice grade license requires passing the code Element 1*a* test plus the theory test known as Element 2, selected from the subjects shown in the Element 2 questions below.

The Technician grade license requires passing Elements 1*a*, 2, and 3.

The General grade license requires passing Elements 1*b*, 2, and 3.

The Advanced grade license requires passing Elements 1*b*, 2, 3, and 4*a*.

The Extra grade license requires passing Elements 1*c*, 2, 3, 4*a*, and 4*b*.

(See Chap. 32, *Electronic Communication*, Fifth Edition, for further information on Amateur licensing.)

QUESTIONS

Element 2 (For Novice and All Amateur Licenses)

1. What is the Amateur Radio Service? (97.3a) (32-1)
2. What is an Amateur Radio operator? (97.3c) (32-1)
3. What is an Amateur Radio station? (97.3e) (32-1)
4. What are Amateur radiocommunications? (97.3b) (32-1)
5. What are Amateur operator and station licenses? (97.3d) (32-2)
6. Who is the control operator at an Amateur station? (97.3o) (32-8)
7. What is meant by "third party" and third-party traffic? (97.3v) (32-8)
8. What are the Novice bands, frequencies, and emissions? (97.7e) (32-5)
9. What does the term "frequency band" mean? (97.7e) (32-5)
10. What emission may be used by Novices? (97.7e) (32-5)
11. What are the five basic purposes of Amateur radio? (97.1) (32-1)
12. What are the rules regarding unidentified communications? (97.123) (32-8)
13. What are the rules regarding intentional interference? (97.125) (32-8)
14. What are the rules regarding paid communications? (97.112) (32-1)
15. What are the rules regarding false signals? (97.121) (32-8)
16. Explain United States Amateur radio station call signs. (32-4)
17. With whom may Novice operators communicate on amateur bands? (32-13)
18. What are the logging requirements? (97.103, 97.105) (32-8)
19. What are the rules regarding station identifications? (97.84) (32-8)
20. What maximum power may be used by Novices? (97.67) (32-6)
21. What is the procedure to respond to an official notice of violation? (97.137) (29-5)
22. What are the requirements of control operators? (97.79) (32-8)
23. Explain the RST signal reporting system. (32-8)
24. What telegraphy speed should be used? (31-18, 32-8)
25. What is meant by zero-beating a received signal? (18-10)
26. Explain how a transmitter might be tuned. (16-5, 16-13, 21-5)
27. What do the following telegraphy abbreviations mean? CQ, DE, K, SK, R, AR, 73, QRS, QRZ, QTH, QSL, QRM, QRN? (32-8, Appendixes G, H)
28. Explain sky waves; skip zones; ground waves. (20-1, 20-2)
29. What measures should be taken to prevent use of radio station equipment by unauthorized persons? (21-18)
30. How might an antenna system be protected from lightning? (20-5, 20-24)
31. What is a ground system for a radio station? (20-14)
32. What are some antenna installation safety procedures other than using safety belts and watching for overhead wiring? (20-24)
33. What is meant by a harmonic radiation? (4-8)
34. What are possible causes and cures if Amateur transmissions interfere with local receivers or other equipment? (16-16, 16-17)
35. What is meant by SWR? What are acceptable values? (20-12)
36. What might cause undesirable SWR readings? (20-12)
37. What is meant by voltage? (1-8)
38. How do alternating and direct currents differ? (1-11, 4-2)
39. How do conductors and insulators differ? (1-3, 1-7)
40. How do open circuits and short circuits differ? (2-7)
41. In what unit is power measured? (2-4) Energy? (2-4)
42. How are frequency and wavelength related? (4-8, 15-24)
43. What frequencies are considered radio frequencies? (4-6, 15-1)
44. What are considered to be audio frequencies? (4-6, 14-1)
45. What is the unit of electromotive force? (1-8) Current? (1-7) Power? (2-4) Frequency? (4-6)
46. Define mega, kilo, centi, milli, micro, pico. (1-14)

47. What are the appearance, uses, and symbols of: Quartz crystals? (11-13, 11-14, 11-15) Meters? (13-2, 13-4, 13-11) Vacuum tubes? (9-18, 9-19, 9-24, 9-25, 9-29) Fuses? (2-7)

48. Explain the function of the stages of a simple block diagram of a telegraph transmitter. (16-13, 16-20)

49. Explain the function of the stages of a block diagram of a simple CW receiver. (18-9, 18-10, 18-24)

50. Explain the function of the various components of a Novice station, including transmitter, receiver, antenna switching, antenna feedline, antenna, and telegraph key. (32-7)

51. What is the meaning of type A1 or A1A emissions? (16-1, Appendix E)

52. What is a backwave and how is it cured? (16-4)

53. What causes key clicks and how are they cured? (16-9)

54. What causes a chirp in a CW transmitter and how is it cured? (16-2)

55. What causes hum and how is it cured? (14-25)

56. What causes harmonics and how are they cured? (14-13, 16-16)

57. What are spurious emissions and how are they cured? (16-16)

58. How is the length of a half-wave dipole computed? (20-7) A quarter-wave vertical? (20-14)

59. Explain the use of coaxial cables and parallel conductor feedlines. (20-13, 20-16)

Element 3 (Technical/General and Higher Classes)

1. What is the control point of an Amateur station? (97.3p) (32-8)

2. What are emergency communications? (97.32, 97.107) (32-11)

3. What are Amateur transmitter power limitations? (97.67) (32-6)

4. What are the rules regarding identifying transmissions? (97.79, 97.84) (32-8)

5. How may third parties participate in Amateur radio? (97.79) (32-8)

6. What are the rules on domestic and international third party traffic? (97.114) (32-8)

7. What are the only permissible one-way transmissions? (97.91) (32-8)

8. List the frequency bands, emissions, and powers available to Technician class licensees. (97.7) (32-5)

9. List the frequency bands, emissions, and powers available to General class licensees. (97.7) (32-5)

10. What types of emissions are allowed on what portions of Amateur bands? (97.61) (32-5)

11. What determines what frequencies an Amateur will use? (97.63) (32-8)

12. What are the rules regarding radio-controlled model crafts? (97.65) (32-8)

13. On what bands and frequencies are radioteleprinters (F1B) used? (97.69) (32-8) (Table 32-1)

14. List seven prohibited practices on the Amateur bands. (97.113, 97.115, 97.117, 97.119) (32-8)

15. Explain calling and answering procedures for: radiotelegraphy (31-6), radiotelephony (30-9, 30-10), radioteleprinting (32-8).

16. Explain the use of repeaters. (21-10, 32-7)

17. What is VOX operation? (17-37, 18-27)

18. What is meant by full break-in telegraphy? (16-3)

19. Should operating courtesy always be used on Amateur bands? (Yes, 32-8)

20. Besides selecting the direction of desired communications, what is the advantage of using rotating-beam antennas? (20-20)

21. What are the restrictions on international communications? (32-10)

22. What are the limitations on emergency preparedness drills? (32-11)

23. Explain the ionospheric layers D, E, F1, and F2. (20-2)

24. What is meant by radio wave absorption? (20-1, 20-2)

25. What is meant by MUF? (20-2) SID? (20-2)

26. Explain the regular daily variations that can be expected with the various Amateur bands. (20-2, 20-4, 32-5)

27. What are scatter communications? (20-4)

28. What effect do sunspot cycles have on wave propagation? (20-2)

29. What is a line-of-sight communication? (20-1)

30. What is meant by tropospheric bending and ducting? (20-4)

31. What are some safety precautions regarding household and other electrical wiring? (16-23)
32. How should dangerous voltages in equipment be made inaccessible to accidental contact? (16-23, 17-29, 21-18)
33. What is a two-tone test? (17-36)
34. Explain how RF amplifiers can be neutralized. (15-2, 15-15 to 15-22)
35. How are dc input and RF output power measured? (16-7, 17-12, 17-29, 17-36)
36. What is an oscilloscope and how might it be used? (13-34 to 13-36, 14-26, 16-19, 17-18, 17-36)
37. Explain multimeters. (13-12)
38. Explain signal generators. (11-29)
39. Explain signal tracers. (13-38)
40. Besides oxidized connections, what may cause audio rectification and distortion in consumer electronic products? (14-13, 16-16)
41. What is another term meaning reflectometer and where is it used? (20-12, 22-3, 24-2)
42. Explain speech processors, both AF and RF. (17-40)
43. An electronic T-R switching system accomplishes the same function as what magnetic device? (21-5, 25-4)
44. Explain an antenna tuner or matching network. (32-12)
45. How can an oscilloscope be used to monitor signal output? (13-34 to 13-36, 14-26, 16-19, 17-18, 17-36, 22-18)
46. What is a dummy antenna and when should it be used? (15-8, 16-7, 22-3)
47. What is a field strength meter? (20-26)
48. What is an S meter? (18-18)
49. Explain dc, ac, and RF wattmeters. (2-8, 7-5, 20-12)
50. What is impedance and how is it measured? (7-2, 7-7, 7-8, 7-15)
51. Explain resistance and how it is measured. (1-12, 13-11, 13-24)
52. What are the two types of reactance and how are they computed? (5-12, 6-13)
53. What is inductance and what component has it? (5-1, 5-2)
54. What is capacitance and what component has it? (6-1)
55. Why is impedance-matching important? (2-16, 14-11, 15-8)
56. Explain the electrical units of ohm (1-12), microfarad, picofarad (6-1), henry, millihenry, micro-

henry (1-14, 5-2), and decibel (8-5).
57. Express Ohm's law and explain some of its uses. (2-2, 2-3)
58. What are current and voltage dividers? (2-10, 2-13, 10-25)
59. How is electrical power calculated? (2-4, 2-5, 2-6, 7-5, 7-6)
60. How are transformer turns, voltage, current, and impedance ratios related? (5-15, 5-21 to 5-23, 14-11)
61. Explain the results when resistors are in series or in parallel. (2-9 to 2-11, 2-13)
62. Explain capacitors in series and in parallel. (6-14, 6-15)
63. Explain inductors in series and in parallel. (5-9, 5-10)
64. What is meant by the rms value of a sine-wave ac? (4-4)
65. Explain the physical appearance, types, characteristics, applications, and symbols for resistors (1-2), capacitors (6-1, 6-2, 6-7 to 6-9), inductors (5-3, 5-6), and transformers (5-15, 5-16, 5-21, 5-22, 5-26).
66. Explain the types of diodes used in power supplies. (9-3, 10-2)
67. Explain the operation of ac-to-dc power supplies. (10-2, 10-3, 10-4, 10-7)
68. What are high-pass, low-pass, and bandpass filters, their makeup and uses? (8-8, 17-34, 18-14)
69. Block diagram and explain briefly the stages in transmitters and receivers of the AM (17-19, 17-29, 18-9, 18-10), SSB (17-24, 17-32, 18-24), and FM (19-10, 19-13, 21-5, 21-7) types.
70. Explain what is meant by A0, A3, F1, F2, F3 emissions. (Appendix E)
71. What is a signal and the information it can carry? (16-3)
72. Explain double sideband amplitude modulation. (17-6 to 17-10)
73. Explain single sideband amplitude modulation. (17-31 to 17-33)
74. Explain frequency modulation. (19-1, 19-3)
75. Explain phase modulation. (19-15)
76. What is meant by a carrier? (16-1, 17-1)
77. What are the relationships between sidebands and bandwidth? (17-13, 17-14)
78. In regard to modulation, what is meant by envelope (17-7), deviation (19-3), overmodulation (17-9), and splatter (17-9)?
79. Explain how radio-frequency ac's can be mixed

(17-13, 18-12), translated (17-32, 18-10), and multiplied (15-23, 16-15).

80. Explain radioteleprinting using FSK and AFSK, and the meaning of mark, space, and shift. (16-12, 25-18)
81. What is a Yagi form of antenna and of what is it composed? (20-20)
82. What are the feedpoint impedances of half-wave dipoles (20-10), and quarter-wave vertical antennas? (20-14)
83. What is meant by radiation patterns, directivity, and major lobes of common antennas? (20-13, 20-15, 20-18, 20-21)
84. What is a quad antenna and how is it constructed? (20-22)
85. How are the physical dimensions of common antennas computed? (20-7)
86. What is meant by the characteristic impedance of an antenna? (20-10)
87. What are the advantages and disadvantages of standing waves on antenna systems? (20-12)
88. What is the significance of the standing wave ratio? (20-12)
89. What is meant by balanced and unbalanced feed lines? (20-12, 20-16)
90. What is the disadvantage of an antenna-feedline mismatch? (20-12)
91. What does feedline attenuation or loss mean? (20-12, 20-31)

Element 4a (Advanced and Extra Classes)

1. List the frequency bands, emissions, and powers available to Advanced class amateurs. (97.7, 97.61)(32-5)
2. What are the rules regarding retransmission of signals by Amateur stations? (97.3, 97.113, 97.126)(32-8)
3. How are repeaters used in amateur radio? (97.3, 97.85, 97.61)(32-7)
4. What is an auxiliary amateur radio station? (97.3, 97.86, 97.61)(32-8)
5. What are the rules for remote control of amateur stations? (97.3, 97.88)(32-8)

6. What are the rules for automatic control of amateur stations? (97.3)(32-8)
7. What is a control link? (97.3)(32-8)
8. What is a network diagram of a repeater system? (97.3)(32-7)
9. What are the proper methods of identifying all types of amateur emissions? (97.84)(32-8)
10. What are the logging requirements for mobile and fixed stations? (97.103)(32-8)
11. What are the height limitations for amateur antennas? (97.45)(32-7)
12. How is the height above average terrain calculated for an Amateur antenna? (97.45, 97.67)(20-28)
13. Who must be notified when high antennas are constructed? (FAA, 23-25)
14. What is facsimile transmission and how is it identified? (25-37, 25-38, CW, phone)
15. What are the basics of slow-scan TV transmissions? (24-30)
16. What is the result of sporadic-E layer activity? (20-2)
17. What causes selective fading? (20-3)
18. What effect does the aurora borealis have on radio waves? (20-2)
19. What is meant by the radio-path horizon? (20-1)
20. What devices are used to measure frequencies and how are they used? (13-26, 13-30, 13-31, 13-32)
21. What are grid-dip meters or solid-state dip meters and how are they used? (13-28)
22. What are some performance limitations such as accuracy, frequency response, and stability of oscilloscopes? (13-34 to 13-36) Of frequency counters? (13-31, 13-32)
23. What are some causes of intermodulation interference in receivers? (17-38, 18-11, 18-12)
24. What causes the desensitizing of receivers? (18-12)
25. What is meant by capture effect in FM receivers? (19-11)
26. What is the meaning of reactive power? (7-5)
27. How do series and parallel resonant circuits differ in response? (8-2, 8-3)
28. What is meant by skin effect? (8-4, 11-17)
29. How is energy stored in electromagnetic and electrostatic fields? (5-2, 6-1)
30. Given component values, how are resonant frequency, bandwidth, and Q of RLC circuits computed? (8-1, 8-4, 8-6)

31. Given resistance and reactance, how is the phase angle determined? (7-3)
32. If the phase angle is known, how is power factor determined? (7-6)
33. Given system gains and losses, how is effective radiated power determined? (20-29)
34. What is Thevenin's theorem and how might it be used? (2-18)
35. Explain the following diodes: junction (9-3), varactor (9-4), tunnel (9-3), hot-carrier (22-5), point contact (22-5), PIN (9-4)
36. Discuss the following transistors: NPN, PNP, junction, unijunction, power, germanium, silicon. (9-1, 9-5, 9-8, 9-13)
37. What are silicon-controlled rectifiers and triacs? (9-14, 9-15)
38. How does a light-emitting diode operate? (9-4) A neon lamp? (10-22)
39. How are crystals used in SSB filters? (17-34)
40. Explain discrete and integrated voltage regulator circuits. (10-22 to 10-24)
41. What are the characteristics of classes A, AB, B, and C amplifiers? (14-4, 14-6, 15-5)
42. What are pi, L, and pi-L impedance-matching circuits? (20-24)
43. Describe the characteristics of the following filters: constant-*k, m*-derived, low-pass, high-pass, band-stop, notch, pi-section, T-section, L-section. (8-8, 20-24)
44. Describe the various common oscillators and their stabilities. (11-3, 11-6, 11-8 to 11-12, 11-16)
45. Explain the use and operation of AM modulators. (17-6, 17-10)
46. Explain the use and operation of FM modulators. (19-13 to 19-15)
47. Explain the use and operation of balanced modulators. (17-32, 17-33)
48. Describe the operation and tuning of final RF amplifiers. (15-29, 16-5)
49. Describe the operation of AM detectors. (18-2 to 18-8, 18-15, 18-16)
50. Describe the operation of mixers and heterodyne detectors. (18-10, 18-12, 18-16)
51. Describe the operation of RF and IF amplifiers. (18-11, 18-13, 18-14)
52. Describe a common-emitter class A transistor amplifier, its biasing, signal gain, input and output impedances. (9-6)
53. Describe a common-collector class A transistor amplifier, its biasing, signal gain, input and output impedances. (9-10)
54. Explain how to construct a voltage regulator with a pass transistor and zener diode to produce a given output voltage. (10-24)
55. How might coils and capacitors be selected to resonate at a given frequency? (15-24)
56. Define emission types A4, A5, F4, F5. (Appendix E)
57. Describe the different types of modulation methods used in Amateur radio. (AM, 17-8, 17-9, 17-10) (SSB, 17-31, 17-32) (FM, 19-3, 19-13, 19-14) (PM, 19-15, 19-16)
58. What is meant by deviation ratio and modulation index? (19-3)
59. What is meant by electromagnetic radiation? (20-1) Wave polarization? (20-6)
60. How do sine, square, and sawtooth waveforms differ? (4-2, 16-9)
61. What is the root-mean-square value applied to ac? (4-4)
62. Explain peak envelope power relative to the average value. (4-4, 17-36)
63. What is meant by a signal-to-noise ratio? (18-11)
64. What is the relationship of antenna gain and beamwidth? (20-20)
65. What is a trapped antenna? (32-12)
66. Explain the operation of parasitic and driven elements in antennas. (20-20)
67. What is antenna radiation resistance? (20-10)
68. Explain multiwire folded dipoles. (20-20)
69. What is the velocity factor used with feedlines? (20-16)
70. What are the requirements of electrical lengths of feedlines? (20-16)
71. Explain the meaning of voltage and current nodes on antennas. (20-9)
72. What types of antennas are used for mobile operation? (20-14)
73. Explain base-, center-, and top-loading of vertical antennas. (20-23, 25-6)

Element 4b (Extra Class Licenses)

1. What are the frequency bands and limitations for Extra class amateurs? (97.61, 97.95)(32-5, 32-7)

2. In what way are amateurs involved in space communications? (97.3)(32-7)
3. What are the requirements of purity of emissions? (97.73)(32-7)
4. What are the requirements for Amateur operation aboard ships or aircraft? (97.101)(32-8)
5. Explain the meaning and operations of RACES. (97, subpart F)(32-11)
6. With what other points may amateurs communicate? (97.89)(32-11)
7. Explain the use of satellites carrying Amateur Radio equipment. (32-7)
8. Explain fast-scan Amateur television. (24-30)
9. What is meant by moonbounce and EME? (32-7)
10. How can communications be affected using meteor bursts? (20-2)
11. What are transequatorial scatter communications? (20-4)
12. What is a spectrum analyzer and what does it display? (13-37)
13. What is a logic probe and what does it indicate? (13-38)
14. Explain what may be done to vehicles if they produce radio noises. (21-20)
15. Explain methods for location of the source of radio signals by direction-finding techniques. (25-22 to 25-32)
16. What is photoconductive effect? (9-4)
17. What are exponential charge and discharge rate curves? (5-1, 6-1)
18. Explain time constants for *RL* and *RC* circuits. (5-4, 6-1)
19. Explain the use of Smith Charts. (Not covered in this text.)
20. How can impedance be determined in *RLC* networks? (7-2, 7-7, 7-8, 7-12)
21. Explain real and imaginary numbers and the use of complex numbers. (7-17)
22. Explain enhancement, depletion, MOS, CMOS, N-channel and P-channel field-effect transistors. (9-11, 9-12)
23. What are operational amplifiers? (14-20)
24. Explain phase-locked loop integrated circuits. (11-27)
25. Discuss some of the 7400-series TTL digital integrated circuits. (12-4 to 12-12)
26. Discuss some of the 4000-series CMOS ICs. (12-6 to 12-8, 12-10)

27. Explain the operation of a cathode ray tube. (13-34, 24-3) A vidicon. (24-5)
28. Explain the following logic circuits: flip-flop (12-10), multivibrator (11-21, 12-10), AND (12-4), OR (12-5), NAND (12-7), NOR (12-8).
29. Explain digital frequency divider circuits as used in counters and for crystal markers. (12-14, 13-31)
30. Discuss active audio filters using integrated op-amps. (14-21).
31. Explain receiver-noise figure and sensitivity. (18-11)
32. What is and what determines receiver selectivity? (18-3, 18-9, 18-10)
33. What is the dynamic range of a receiver? (18-11, 21-11)
34. What determines the voltage gain and frequency response of IC op-amps? (14-20)
35. Explain the input impedance and operation of an FET common-source amplifier. (9-11, 9-12)
36. Of what might a receiver preselector be composed? (18-24)
37. How might an AF stage have a desired frequency response by proper selection of bypass and coupling capacitors? (14-25)
38. What are the various types of pulse modulation? (12-18, 17-37)
39. What are digital signals? (12-24)
40. Explain the fundamentals of narrow-band voice modulation. (17-41)
41. What is the relationship of pulse or information rate versus bandwidth? (17-37, 26-7)
42. What might be desirable gain, beamwidth, and tracking capabilities for antennas used for space radio communications? (20-20)
43. What is an isotropic radiator and when is it used? (20-29)
44. Explain the resultant patterns of phased vertical antennas in which their spacing in wavelengths may be changed. (20-18)
45. What are rhombic antennas? Their advantages and disadvantages? (20-21)
46. Explain how a feedline can be coupled to antennas by delta, gamma, and stub feeds. (20-16)
47. Discuss properties of $1/8$, $1/4$, $3/8$, and $1/2$ wavelength sections of feedlines. (20-16)
48. What are the properties of shorted and open feedlines? (20-16)